HANDBOOK OF
LESS-COMMON
NANOSTRUCTURES

HANDBOOK OF LESS-COMMON NANOSTRUCTURES

Boris I. Kharisov • Oxana Vasilievna Kharissova
Ubaldo Ortiz-Méndez

CRC Press
Taylor & Francis Group
Boca Raton London New York

CRC Press is an imprint of the
Taylor & Francis Group, an **informa** business

CRC Press
Taylor & Francis Group
6000 Broken Sound Parkway NW, Suite 300
Boca Raton, FL 33487-2742

First issued in paperback 2019

ISBN-13: 978-1-4398-5343-6 (hbk)
ISBN-13: 978-0-367-38163-9 (pbk)

Library of Congress Cataloging-in-Publication Data

Kharisov, Boris I.
    Handbook of less-common nanostructures / Boris I. Kharisov, Oxana Vasilievna Kharissova, Ubaldo Ortiz-Méndez.
        p. cm.
    Includes bibliographical references and indexes.
    ISBN 978-1-4398-5343-6 (hardback)
    1. Nanostructured materials--Handbooks, manuals, etc. I. Kharissova, Oxana Vasilievna. II. Ortiz Méndez, Ubaldo. III. Title.

TA418.9.N35K485 2012
620.1'15--dc23

2011045047

This book is dedicated to the memory of the outstanding Russian chemist, Professor Alexander D. Garnovskii (1932–2010).

Жизнь коротка, наука - вечна! / Life is short, science is eternal. / ¡La vida es corta, la ciencia es eterna!

Contents

PART I   Introduction to Nanostructures

PART II   Less-Common Nanostructures

PART III Selected Intriguing Topics in Nanotechnology

PART IV   Nanometals and Nanoalloys

Abbreviations

AA	L-ascorbic acid
ALD	atomic layer deposition
AN	aniline
apab	4-amino-3-(pyridin-4-ylmethyleneamino)benzoate
APG	alkyl polyglucoside
APTES	3-aminopropyltriethyloxysilane
BC	bacterial cellulose
bdc	1,4-benzenedicarboxylate
BDD	boron-doped diamond
BDDNF	boron-doped diamond nanorod forest
BEA	beta-zeolite
BIBA	2-bromoisobutyric acid
Bibp	4,4'-bis(1-imidazolyl)biphenyl

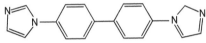

[bmim][Cl]	(1-n-butyl-3-methylimidazolium chloride)
BPEIS	{(biphenyl ester)3-(isoprene)9-(styrene)9}
bpy, bipy	bipyridine
BTC, 1,3,5-BTC	1,3,5-benzene tricarboxylate
CFCO	catalyst-free combust-oxidized process
Chitosan	

[C16Mim]Cl	1-hexadecyl-3-methylimidazolium chloride
CNFs	carbon nanofoams
CNTs	carbon nanotubes
cod	1,5-cyclooctadiene
cot	1,3,5-cyclooctatriene
CTAB, CTABr	cetyltrimethylammonium bromide
CTAH	cetyltrimethylammonium halides
CVD	chemical vapor deposition
DBM	dibenzoylmethane
DEC	1-[3-(dimethylamino)propyl]-3-ethylcarbodiimide
DEG	diethylene glycol
DETA	diethylenetriamine
DMA	dimethylaniline
DoTAC	dodecyltrimethylammonium chloride
DTPA	diethylene triamine pentaacetic acid
ECD	electrochemical deposition
ECR	electron-cyclotron resonance

EDA	ethylenediamine
EDS	energy dispersive spectra
EFLAL	laser ablation in liquid
enH$_2$	ethylenediammonium
FA	formic acid
FCA	ferrocenyl-2-crotonic acid
Fcc	face-centered cubic structure
GNP	gold nanoparticles
GO	graphene oxide
GOx	glucose oxidase
Hap	hydroxyapatites
H$_2$ATIBDC	5-amino-2,4,6-triiodoisophthalic acid
H$_2$bbim	2,2′-bibenzimidazole
H$_2$BDC	1,3-benzenedicarboxylic acid
HbimbdcH$_2$	1-(2-benzimidazolyl)-3,5-benzenedicarboxylic acid
1,3,5-H$_3$BTC	1,3,5-benzentricarboxylic acid
HCNTs	helical carbon nanotubes
HCS	hard carbon spherule
H$_2$-DBED	*N,N″*-dibenzylethylenediamine
H-DDP	2,6-diphenylphenol
H-DMBS	*t*-butyldimethylsilanol
H-DMP	2,6-dimethylphenol
HETPHEN	heteroleptic bisphenanthroline metal complexes
HFCVD	hot filament chemical vapor deposition
Hhpzc	3-hydroxy-pyrazine-2-carboxylic acid
HMT	hexamine, hexamethylenetetramine
HNTs	halloysite nanotubes
H-O(*t*-Bu)	*t*-Bu alcohol
H-PS	benzenethiol
H$_2$salten	bis(3-salicylideneaminopropyl)amine
H$_3$sglu	N-(2-hydroxybenzyl)-L-glutamic acid
H$_3$Sser	N-(2-Hydroxybenzyl)-L-serine
Hthd	2,2,6,6-tetramethyl-3,5-heptanedione
H-TPS	triphenylsilanol
H-TPST	triphenylsilanethiol
IBA	isobutyric acid
ILs	ionic liquids
ITO	indium tin oxide
LA	laser ablation
LVCC	laser vaporization controlled condensation
MA	methacrylic acid
MCM-41	"mobile crystalline material" ("mobil composition of matter") on SiO$_2$ basis
MCNPs	metal coordination nano-polymers
MH	magnesium hydroxide
MIBA	2-methylisobutyric acid
MMA	Me methacrylate
MNPs	magnetic nanoparticles
MOEP	metal octaethylporphyrin
6-MPH	6-mercaptopurine
MPTES	mercaptopropyltriethoxysilane
MWCNTs	multiwalled carbon nanotubes

MWI	microwave irradiation
MW-ST	microwave-assisted solvothermal technique
Nafion	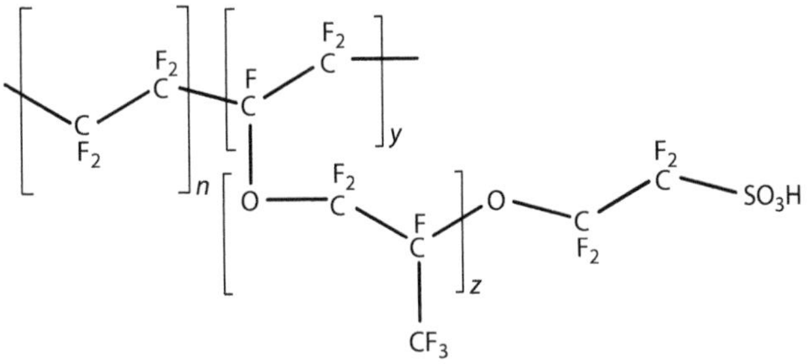

NB	nitrobenzene
*n*BA	*n*-butylacrylate
NHS	*N*-hydroxysuccinimide
NOLF	nano-onion-like fullerenes
NSL	nanosphere lithography
ODP	octadecyl dihydrogen phosphate
OLF	onion-like fullerenes
ONeP	neo-pentoxide
o-PD	*o*-phenylenediamine
OPPOSS	octaphenyl polyhedral oligomeric silsesquioxane
OPV	oligo-(phenylene vinylene)
ORR	oxygen reduction reaction
O/W	oil-in-water
PAA	polyacrylic acid or polyacrylic acid sodium salt or poly(amic acid)
PAAc	polyacrylic acid
PACVD	plasma-assisted chemical vapor deposition
PAM	polyacrylamide
P(AMPS-co-MMA)	poly(2-acrylamido-2-methylpropanesulfonic acid Me methacrylate)
PANI	polyaniline
PB	Prussian blue
pbbt	1,1′-(1,3-propylene)-*bis*-1H-benzotriazole
p-BDC	1,4-benzenedicarboxylic acid
PC	polycarbonate
p-CAN	*p*-chloroaniline
p-CNB	*p*-chloronitrobenzene
pda	3-(3-pyridyl)acrylate acid
2-PDD	*N*-phenyl-1,2-phenylenediamine
4-PDD	*N*-phenyl-1,4-phenylenediamine
PDLLA	random polylactide, poly(D,L-lactide)
PDMS	poly(dimethylsiloxane)
PDVB	polydivinylbenzene
PEDOT	poly(3,4-ethylenedioxythiophene)
PEG	polyethylene glycol
PEGMA	poly(ethylene glycol) methacrylate
PEO	poly(ethylene oxide)
PET	poly(ethylene terephthalate)
PFSEA	perfluorosebacic acid
P3HT	poly(3-hexylthiophene)
PI	polyimide

PLLA	poly(-lactide)
*p*MA	*p*-mercaptoaniline
PMEO$_2$MA	poly(2-(2-methoxyethoxy)ethyl methacrylate)
PMMA	polymethylmetacrylate
POEM	poly(oxyethylene) methacrylate
POM	polyoxymethylene
p-PFS	poly(nBA)/pentafluorostyrene
PPy	polypyrrole
PS	polystyrene
PS-b-PAA	polystyrene-b-poly(acrylic acid)
PS-b-PEO	poly(styrene-block-ethylene oxide)
PU	polyurethane
PVA	poly(vinyl alcohol)
PVA-mod-PEDOT	poly(3,4-ethylenedioxythiophene)
PVD	physical vapor deposition
PVdF-NFM	polyvinylidene fluoride nanofibrous membranes
PVP	poly(vinyl pyrrolidone)
pyr	pyrazine
PZS	poly(cyclotriphosphazene-co-4,4′-sulfonyldiphenol)
QD	quantum dot
QDSC	quantum dot–sensitized solar cell
RhB	rhodamine B
SCCNT	stacked-cup carbon nanotubes
SDSn	sodium dodecylsulfonate
SERS	surface-enhanced Raman spectroscopy
SNTs	silica nanotubes
Span-60	nonionic surfactant sorbitan monostearate
SQUIDs	superconducting quantum interference devices
SSQZ	phenylsilsesquiazanes
TAA	thioacetamide
TBAOH	tetrabutylammonium hydroxide (C$_4$H$_9$)$_4$NOH
TEA	triethanolamine
TEM	transmission electron microscopy
TEOS	tetraethylorthosilicate
6-ThioGH	6-thioguanine
TMA	trimethylaluminium
TMAH	tetramethylammonium hydroxide
TOP	trioctylphosphine
TOPO	tri-*n*-octylphosphine oxide
TPG	tetrapropylgermane
TPT	*tetrakis*(4-pyridyl)thiophene
TPyP	*tetrakis*(4-pyridyl)porphine
TPyTa	*tris*(4-pyridyl)-1,3,5-triazine
TTABr	tetradecyltrimethylammonium bromide
US	ultrasound/ultrasonic
VONTs	vanadium oxide nanotubes
XRD	x-ray diffraction
XPS	x-ray photoelectron spectroscopy
ZSM-5	"zeolite sieve of molecular porosity," an aluminosilicate zeolite Na$_n$Al$_n$Si$_{96-n}$· 16H$_2$O ($0 < n < 27$)

Preface

When the area of nanotechnology began to develop intensively as an independent field in the frontiers of physics, chemistry, materials chemistry and physics, medicine, biology, and other disciplines two decades ago, terms such as "nanoparticle," "nanopowder," "nanotube," "nanoplate," and other terms related to shape rapidly became very common. For instance, a simple search using SciFinder results in hundreds of thousands of articles with the keywords "nanoparticle" and "nanotube." At the same time, during recent years, researchers have reported a large number of the nanostructure types mentioned earlier and the discovery of more rare species, such as "nanodumbbells," "nanoflowers," "nanorices," "nanolines," "nanotowers," "nanoshuttles," "nanobowlings," "nanowheels," "nanofans," "nanopencils," "nanotrees," "nanoarrows," "nanonails," "nanobottles," and "nanovolcanoes," among many others. The naming of a discovered rare nanoform is commonly left to the imagination of researchers. Since any novel nanoform/nanostructure could, theoretically, have useful, unexpected, and unpredictable applications (e.g., graphene, discovered not long ago[1]), each new achievement, reproducible or not, is welcome due to the importance of nanotechnology in current and future applications. Without a good understanding of the reasons for shape formation, approaches to the synthesis of nanostructures can be hard to carry out.

According to the available literature, there is no universal generalization of rare and common nanostructures. Several existing classifications are related to the dimensionality of the nanostructure itself and its components[2,3] (for instance, 0D clusters and particles, 1D nanotubes and nanowires, 2D nanoplates and layers, 3D core-shell nanoparticles, and self-assembled massive, intermediate dimensional nanostructures as fractals or dendrimers) or based on the triad symmetry group-shell composition-structural formula of the shell (here nanostructures are divided into branches, classes, and subclasses determined by the symmetry group of a shell and the sets of the quantum numbers of a structure).[4] In this book, we present a nonformal classification that is not directly related to dimensionality and the chemical composition of nanostructure-forming compounds or composites but is based mainly on the less-common nanostructures. The classification is on page xxiv.

This book is focused on the examination of less-common nanostructures (i.e., published mainly in the range of $1 \div 100$ reports) that correspond to the shapes mentioned earlier. Such structures, as will be shown later, possess unusual shapes and high surface area, which make them very useful for catalytic, medical, electronic, and many other applications. At the same time, the main classic nanostructures are briefly described to provide a comparison with rare nanoforms.

We will be pleased to receive suggestions, corrections, and critical notes in order to improve future editions of this handbook. Also, information about any novel nanostructural type that is not described in this book will be always welcome.

Classic Carbon-Based Nanostructures	Conventional Non-Carbon Nanostructures
Carbon nanotubes, fullerenes, nanodiamonds, graphene, and graphane*	Simple and core-shell nanoparticles, non-carbon nanotubes, nanometals, nanowires, nanorings, nanobelts, nanopowders, nanocrystals, nanoclusters, nanofibers, and nanodots/quantum dots

Relatively Rare Nanostructures

Simple linear 1D, 2D, and 3D nanostructures
Nanolines, nanopencils, nanodumbbells, nanopins, nanoshuttles, nanopeapods, nanochains, nanowicks, nanobars, and nanopillars

Various prolonged 3D nanostructures
Nanobricks, nanocones, nanoarrows, nanospears, nanospikes, nanonails, nanobowlings, nanobones, nanobottles, nanotowers, nanoarmors, and nanopins

Circle- and ball-type nanostructures
Nanowheels, nanoballs, nanoeggs, nanograins, nanorices, and nanospheres

Nanocage-type structures
Nanocages, nanoboxes, nanocubes, and nanocapsules

"Nanovegetation" world
Nanotrees, nanopines, nanopalms, nanobushes, nanograsses, nanoacorns, nanokelps, nanomushrooms, nanoflowers, nanobouquets, nanoforests, nanocorns, nanoleaves, nanobroccoli, nanomulberry, nanocactus, nanospines, nanosheaves, nanoonions, and nanodewdrops

"Nanoanimal" world
Nanourchins, nanoworms, nanolarvae, and nanosquamae

"Nanohome" objects
Nanobrushes, nanobrooms, nanocombs, nanocarpets, nanofans, nanowebs, nanospoons, nanoforks, nanobowls, nanotroughs, nanocups, nanospindles, and nanofuns

"Nanotechnical" structures and devices
Nanosaws, nanosprings (nanocoils/nanospirals), nanoairplanes, nanopropellers, nanowindmills, nanoboats, nanobridges, nanocars, nanobatteries, nanotweezers, nanomeshes, nanofoams, nanobalances, nanojunctions, nanopaper, nanorobots, nanothermometers, E-nose, E-tongue, E-eye, and Nano Electromechanical Systems (NEMS)

Nanostructures classified as polyhedra
Nanotriangles, nanotetrahedra, nanosquares, nanorectangles, nanopyramids, nanooctahedra, nanoicosahedra, nanododecahedra, nanocubes, nanoprisms, and nanocuboctahedra

Other rare nanostructures
Nano–New York, nanopaper, nanovolcanoes, nanosponges, nanostars, nanoglasses, and nanodrugs

* Readers also will see some other (more rare) carbon nanostructures throughout the text.

REFERENCES

1. Geim, A. K.; Novoselov, K. S. The rise of graphene. *Nature Materials*, 2007, *6* (3), 183–191.
2. Pokropivny, V. V.; Skorokhod, V. V. Classification of nanostructures by dimensionality and concept of surface forms engineering in nanomaterial science. *Materials Science and Engineering: C*, 2007, *27* (5–8), 990–993.
3. Pokropivny, V. V.; Skorokhod, V. V. New dimensionality classifications of nanostructures. *Physica E: Low-Dimensional Systems and Nanostructures*, 2008, *40* (7), 2521–2525.
4. Kustov, E. F.; Nefedov, V. I. Theory and classification system of nanostructures. *Doklady Physical Chemistry*, 2007, *414* (2), 150–154.

Acknowledgments

We would like to acknowledge Professors Dr. Sci. Sergei S. Berdonosov, Igor V. Melikhov, and Marina A. Yurovskaya (all from Moscow State University, Russia) and the late Professors Dr. Sci. Yurii E. Alexeeev and Alexander D. Garnovski (both from Southern Federal University, Rostov-na-Donu, Russia) for providing useful comments during the preparation of this book.

We are very grateful to many professors and researchers worldwide for giving us permission to reproduce images from their publications.

Among many other researchers, we would like to mention the following specialists: Professors Nazario Martin (the University of Complutense, Spain), Guowei Yang (Zhongshan University, China), Yury Gogotsi (Drexel University, United States), Masaki Ozawa and Anke Krueger (both from Institut fur Organische Chemie der Julius-Maximilians-Universitat, Germany), Kian Ping Loh (National University of Singapore, Singapore), Naomi J. Halas (Rice University, United States), Kuo Chu Hwang (National Tsing Hua University, Taiwan), Sergey K. Gordeev (Central Research Institute of Materials, Russia), Guozhen Shen (the University of Southern California, United States), Evgenii A. Gudilin (Moscow State University, Russia), Shaojun Dong (Changchun Institute of Applied Chemistry, China), Tsun-Kong Sham (the University of Western Ontario, Canada), Baodan Liu (Nanoscale Materials Center, Japan), Santanu Karan (Indian Association for the Cultivation of Science, India), Dmitri Golberg (National Institute for Materials Science, Japan), Chen-Sheng Yeh (National Cheng Kung University, Taiwan), Xiaogang Peng (the University of Arkansas, United States), Ting-Ting Kang (the University of Fukui, Japan), Tetsu Yonezawa and Ryuichi Arakawa (both from the University of Tokyo, Japan), and Baomei Wen and John J. Boland (both from Trinity College, Ireland).

We would also like to thank the American Chemical Society, the American Physical Society, Elsevier Science, the American Institute of Physics, Springer, Wiley InterScience, Moscow State University Press, IOP Science, the Royal Society of Chemistry, the Chemistry Society of Japan, Taylor & Francis, Hindawi Publishing Corporation, and the *Journal of Ceramic Processing Research* for permission to reproduce images from their publications.

Authors

Boris I. Kharisov, Dr Hab, is currently a professor and researcher at the Universidad Autónoma de Nuevo León (UANL), Monterrey, Mexico. He took part in liquidating contaminated material in the aftermath of the Chernobyl accident, working in the contaminated zone in 1987. He received his MS in radiochemistry in 1986 and his PhD in inorganic chemistry in 1993 from Moscow State University, Russia, and his Dr Hab in physical chemistry in 2006 from Rostov State University, Russia. He is a member of the Mexican Academy of Science, National System of Researchers (Level II), and Materials Research Society. He is the coauthor of 4 books and 117 articles; he also has 2 patents registered in his name. Kharisov is the coeditor of three invited special issues of international journals, *Polyhedron* (1 issue) and *Journal of Coordination Chemistry* (2 issues). He is also a member of the editorial boards of four journals. He specializes in coordination and inorganic chemistry, phthalocyanines, ultrasound, and nanotechnology.

Oxana Vasilievna Kharissova, PhD, is currently a professor and researcher at the UANL, Monterrey, Mexico. She received her MS in crystallography in 1994 from Moscow State University, Russia, and her PhD in 2001 in material science from the UANL, Mexico. She is a member of National System of Researchers (Level I) and the Materials Research Society. She is the coauthor of 1 book and 57 articles; she also has 2 patents registered in her name. Kharissova specializes in nanotechnology (carbon nanotubes, nanometals, and fullerenes) and crystallography.

Ubaldo Ortiz-Méndez, PhD, is currently the academic provost for the UANL, Monterrey, Mexico. He received his BSc in physics in 1981 from the UANL and his PhD in materials engineering in 1984 from l'Institut National des Sciences Appliquées (INSA), Lyon, France. He received the UANL Research Award in 1996, 2000, and 2001 for his work in research and publications as well as the TECNOS Award from the State Government of Nuevo Leon in 1994 and 2000 for several of his research works. For his academic achievement, the Education Board of the French Republic gave him the proclamation of Knight of the Order of Academic Palms in 2009. He teaches at the UANL, where he has served for over 20 years.

Part I

Introduction to Nanostructures

1 Methods for Obtaining Nanoparticles and Other Nanostructures

1.1 GENERAL REMARKS ON NANOPARTICLE FABRICATION

Techniques applied for obtaining nanostructures are very distinct and can be classified as *gas phase, liquid phase, aerosol phase, and solid phase* in terms of the phase of medium for preparation or *physical, physicochemical, chemical, and biological* in terms of the types of processes used (Tables 1.1 and 1.2). Currently, the most common are physical and physicochemical processes, where various equipment should be used (*i.e.,* laser, sputtering, microwave, ultrasonic waves, etc.), as well as "wet-chemical" methods. Biological techniques are not so widespread, but they intensively develop. The major part of these procedures are discussed in all the chapters.* The selection or choice of a necessary experimental method depends on a series of factors, for instance, availability of equipment, type and shape of formed nanoparticles, possibility of synthesis of necessary products, and cost of production (e.g., carbon nanotubes are generally fabricated by CVD) among others.

Certain techniques are more appropriate for nanostructures of a definite dimensionality. In this context, strategies to obtain 0D–3D nanostructures are always under consideration; thus, strategies for achieving 1D growth, contributing to an anisotropic growth of crystals, have been emphasized.[1] In another report,[2] the route of the *monolayer colloidal crystal template* was noted as a promising, alternative process for the synthesis of micro- or nanostructures with differently designed morphologies, in particular 2D-ordered arrays (e.g., nanoparticle arrays, pore arrays, nanoring arrays, nanobowl arrays, hollow sphere arrays, etc.; additionally, even 1D nanostructures of ordered nanorod, nanopillar, nanowire arrays, etc., could be produced).

In this chapter, we discuss both classic and exotic physical and physicochemical methods, applied by the majority of researchers and nanotechnologists worldwide.

In case of *vapor-phase synthesis* of nanoparticles,[3] the following methods are presented in more detail in Table 1.2.

1.2 EXAMPLES OF SEVERAL IMPORTANT METHODS FOR THE SYNTHESIS OF NANOOBJECTS

1.2.1 VAPOR AND PLASMA-BASED TECHNIQUES

Flame synthesis. As an example, single-crystalline ZnO nanowires were grown directly on zinc-plated steel substrates at high rates (microns/minute) using a flame-synthesis method, with no catalysts.[16] Tests were performed using an axisymmetric inverse jet diffusion flame (Figure 1.1), where air is ejected from a center jet, and nitrogen-diluted methane is ejected from an outer co-flow annulus. The growth of the nanostructures was found to be very sensitive to local gas-phase chemical species concentrations (O_2, H_2, CO_2, H_2O) and temperature (800–1500 K).

---

* See Chapter 22, dedicated to the preparation of nanostructured elemental metals, and Chapter 14, devoted to the use of ultrasound for nanoparticle formation.

TABLE 1.1

Main Methods for Nanoparticle Preparations

Method	Description
Gas-phase physical preparation	Requires large-volume and high-temperature process. High equipment cost. Excludes additional sintering process. High-purity product, environmentally friendly. No effective stabilizers and less controllable.
Electrically heated generators	Electrical heating for evaporation of bulk materials in tungsten heater into low-pressure inert gas (He, Ne, Xe). Transported by convection and thermophoresis to cool environment. Subsequent nucleation and growth. Suitable for substances having a large vapor pressure at intermediate temperatures up to about 1700°C.
Laser processes	Use of (pulsed) laser instead of electrical heating. Energy efficiency improved but expensive energy cost. Laser: excimer laser (193, 248, 308 nm), Nd:YAG laser (532 nm), ruby laser, and CO_2 laser. Short pulses: 10–50 ns.
Arc (DC) plasma	Spark (arc) discharge. High current spark across the electrodes produced by breakdown of flowing inert gas vaporizes a small amount of the electrode metals.
Wire electrical explosion	Overheating (resistive or Joule heating) of a metal wire through which a strong pulse current flows, followed by explosion of metal vapor so evaporated.
Sputtering	Plasmas (high-temperature plasma [electrons and ions: high temperature] e.g., arc discharge, AC(RF) plasma; low-temperature plasma [electrons: high T but ions: low T] e.g., sputtering, microwave plasma). Glow discharge to produce energetic particles (ions). Momentum transfer to a target, resulting in the ejection of surface atoms or molecules to produce the sputtering species.
Gas-phase chemical preparation	Monomers are chemically produced from low-boiling but highly reactive precursors.
Hot-wall tubular reactor	Tubular reactor with heating provided through hot wall. Temperature: directly controllable. Use of low-boiling precursors such as organometallics or metal carbonyls, followed by its decomposition to yield a condensable material. Flexibility in producing a wide range of materials.
Laser pyrolysis/ photothermal synthesis	Spot heating and rapid cooling. Highly localized and efficient. IR (CO_2) laser: absorbed by precursors or by an inert photosensitizer.
Thermal plasma synthesis	Thermal plasma produced by arc discharge, microwaves, laser or high-energy particle beams, or electroless radio frequency (RF) discharge. Precursors: solid, liquid, gas.
Flame synthesis	Requires fuels for combustion as heat supply. Evaporation and chemical reaction of precursors in a flame produced by fuel combustion. Difficult to control the process. Flame temperature: 1000°C–2400°C. Residence time in flame: 10–100 ms.
Liquid-phase preparation	*Physical*: solvent removal, crystallization/not adequate for nanoparticle formation. *Chemical*: liquid involving reaction. *Advantages*: Highly developed, high controllability, produces very sophisticated products. *Disadvantages*: more art than technology, requires lots of chemical species and process steps, for example, aging, filtration, washing, drying, sintering (thermal treatment).

TABLE 1.1 (continued)
Main Methods for Nanoparticle Preparations

Method	Description
Sol-gel processing	*Precursors*: metal alkoxides, $M(OR)_n$, in organic solvent; metal salts (chloride, oxychloride, nitrate, etc.) in aqueous solution. *Basic mechanism*: hydrolysis, polycondensation. *Characteristics of sol-gel processes*: low processing temperature, molecular-level homogeneity, useful in making complex metal oxides, temperature-sensitive organic-inorganic hybrid materials, and thermodynamically unfavorable or metastable materials.
Confined growth	Growth in structured materials (in polymer matrix, layered materials, porous materials, carbon nanotubes) or in microemulsion.
Droplet-to-particle conversion (aerosol-phase preparation)	Conventional spray pyrolysis, assisted spray pyrolysis (salt- and polymer-assisted). *Salt-assisted* (solution of precursor + water-soluble stable salt; salts should melt but not decompose in processing temperatures; nanoparticles form and grow dispersed in salt, crystallinity increased due to enhanced solubility and mass transfer of precursors in the melt). *Polymer-assisted* (solution: metal salt + polymer + water; polymers used: PVP, PVA, dextran).

The *arc discharge*. It can be used in gaseous as well as in liquid phase; it is applied, in particular, for the synthesis of carbon nanotubes in a Krätschmer reactor (Figure 1.2).[17] The authors emphasized that achieving stable discharge plasma was found to be the main factor in generating an environment favorable to nanotube growth. This is not so easy, as the anode is consumed and therefore has to be tracked toward the cathode continuously. Furthermore, the stability of an electric arc is limited due to its moving nature on the cathode and anode surface. Additionally, after a few minutes, the resulting uneven consumption of the anode and build-up of material on the cathode side cause a further instability in the DC arc. The rate of synthesis of a lab Krätschmer reactor can surpass 100 mg/min. This is also the most widely used one and is based on a relatively simple basic setup. As noted in Ref. [18] among plasma-based techniques for carbon nanostructure synthesis, the arc discharge technique is probably the most practical from both scientific and technological standpoints. In fact, it has a number of advantages in comparison to other techniques, such as fewer defects and high flexibility of carbon nanostructures produced. Arc-grown nanotubes demonstrated the lowest emission capability degradation compared with those produced by other techniques.

As an example of *arc discharge in a liquid phase*, Cu nanoparticles with a mean diameter of 10–15 nm were prepared and self-assembled *via* discharge of bulk copper rods in a cetyltrimethyl-ammonium bromide (CTAB)/ascorbic acid solution.[19] Two copper rod electrodes with 3 mm diameter were placed on two side-necks of the flask, in which one end of the copper rod was submerged into 40 mL liquid and the gap between the ends of the two rods was about 2 mm. When an alternating electrical source (with a voltage of about 1.5 kV, discharge current of 10 A, pulsing frequency of 50 Hz, and input power of ca. 10 kW) was supplied to the electrodes, pulsed arc discharge was observed at the gap in the liquid phase. Ascorbic acid was used as a protective agent to prevent the nascent Cu nanoparticles from oxidation in the solution; otherwise, spindlelike nanostructures with a central lateral dimension of 30–50 nm and a length of up to 100 nm dominate the products (Figure 1.3). The as-prepared nanostructures were a mixture of Cu_2O and CuO rather than the expected pure Cu species. Such a low-temperature and nonvacuum method, exhibiting the characteristics of both physical and chemical processes, provides a versatile choice for economical preparation and assembly of various metal nanostructures.

Other *plasma-based techniques*. Synthesis of nanostructured metals (pure metallic nanoparticles of metals, nanostructured Mg, and Li-based hydrides) and compounds (nanowires of amorphous silicon oxynitrides [a-SiO$_x$N$_y$]) was carried out by *hydrogen plasma metal reaction* (HPMR, Figure 1.4) and *RF cold plasma* (Figure 1.5), respectively.[20] It was established that plasma is an effective method to control reactivation and to synthesize nanostructured materials such as nanoparticles, nanowires,

TABLE 1.2

Generalization of Vapor-Phase Techniques in Nanoparticle Synthesis

Method	Brief Description	Examples of Prepared Nanoparticles
Methods using solid precursors		
Inert gas condensation	Heating a solid to evaporate it into a background gas, then mixing the vapor with a cold gas to reduce the temperature. Including a reactive gas, such as oxygen, in the cold gas stream, oxides or other compounds of the evaporated material can be prepared.	PbS/Ag,[4] Si/In, Ge/In, Al/In, and Al/Pb[5]
Pulsed laser ablation	Use of a pulsed laser to vaporize a plume of material that is tightly confined, both spatially and temporally. This method can generally only produce small amounts of nanoparticles. However, laser ablation can vaporize materials that cannot readily be evaporated.	Si[6]
Spark discharge generation	Vaporization of metals charging electrodes made of the metal to be vaporized in the presence of an inert background gas until the breakdown voltage is reached. The arc (spark) formed across the electrodes then vaporizes a small amount of metal. This produces very small amounts of nanoparticles but does so relatively reproducibly.	Ni[7]
Ion sputtering	Vaporizing a solid is *via* sputtering with a beam of inert gas ions. This process must be carried out at relatively low pressures ($\sim$1 mTorr), which makes further processing of the nanoparticles in aerosol form difficult.	Metal nanoparticles[8]
Methods using liquid or vapor precursors		
Chemical vapor synthesis	Vapor-phase precursors are brought into a hot-wall reactor under conditions that favor nucleation of particles in the vapor phase rather than deposition of a film on the wall. It is called *chemical vapor synthesis* or chemical vapor condensation in analogy to the chemical vapor deposition (CVD) processes used to deposit thin solid films on surfaces.	W[9]
Spray pyrolysis	Use a nebulizer to directly inject very small droplets of precursor solution. This has been called *spray pyrolysis*, *aerosol decomposition synthesis*, *droplet-to-particle conversion*, etc. Reaction often takes place in solution in the droplets, followed by solvent evaporation.	Cu[10]
Laser pyrolysis/ photothermal synthesis	Heating the precursors to induce reaction and homogeneous nucleation by absorption of laser energy. This allows highly localized heating and rapid cooling, since only the gas (or a portion of the gas) is heated, and its heat capacity is small. Heating is generally done using an infrared (CO_2) laser, whose energy is either absorbed by one of the precursors or by an inert photosensitizer such as sulfur hexafluoride.	Si[11]
Thermal plasma synthesis	Injection of the precursors into a thermal plasma. This generally decomposes them fully into atoms, which can then react or condense to form particles when cooled by mixing with cool gas or expansion through a nozzle.	SiC and TiC[12]
Flame synthesis	Particle synthesis within a flame, so that the heat needed is produced *in situ* by the combustion reactions. This is by far the most commercially successful approach to nanoparticle synthesis— producing millions of metric tons per year of carbon black and metal oxides. However, the coupling of the particle production to the flame chemistry makes this a complex process that is rather difficult to control.	TiO$_2$[13]

TABLE 1.2 (continued)
Generalization of Vapor-Phase Techniques in Nanoparticle Synthesis

Method	Brief Description	Examples of Prepared Nanoparticles
Flame spray pyrolysis	Instead of injecting vapor precursors into the flame, it is possible to directly spray liquid precursor into it. This process, generally called *flame spray pyrolysis*, allows use of precursors that do not have sufficiently high vapor pressure to be delivered as a vapor.	Si[14]
Low-temperature reactive synthesis	For particular materials, it is possible for vapor-phase precursors to react directly without external addition of heat and without significant production of heat. ZnSe nanoparticles were produced this way from dimethylzinc-trimethylamine and hydrogen selenide by mixing them in a counter-flow jet reactor at r.t. Apparently, the heat of reaction was sufficient to allow crystallization of the particles without substantially increasing the gas temperature.	ZnSe[15]

Source: Data from multiple data references.

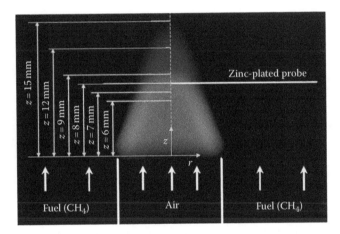

FIGURE 1.1 Methane inverse co-flow jet diffusion flame (synthesis of ZnO nanostructures). Blue chemiluminescence (CH*) marks the reaction zone, and orange emission corresponds to soot intermediates. (Reproduced from *Chem. Phys. Lett.*, 449, Xu, F. et al., Flame synthesis of zinc oxide nanowires, 175–181, Copyright 2007, with permission from Elsevier.)

nanotubes and nanopyramids. In addition, by using different types of plasma at different conditions, it is possible to control nanoparticles' size, distribution, and shape.

Variations of *laser-based techniques*. Metal nanoparticles can be prepared by a novel technique (Figure 1.6) that consists of the laser ablation of a solid target immersed in a water solution of a metal salt.[21] Silicon was chosen as the most adequate target to synthesize silver and gold nanoparticles from a water solution of either $AgNO_3$ or $HAuCl_4$. The laser used for ablation is a third-harmonic ($\lambda = 355$ nm) Q-Switch Nd:YAG ultraviolet (UV) laser, giving pulses <40 ns wide at a repetition rate of 5 KHz. The power density per pulse used in the experiments was around 40 J/cm². In a related report,[22] a unique technique, the facile electrical-field-assisted laser ablation in liquid (EFLAL) without any catalyst or organic additives (Figure 1.7), to controllably fabricate the mass production of GeO_2 micro- and nanoparticles with various shapes (Figure 1.8) was reported.

Atomic layer deposition (ALD). Atomic layer deposition of complex nanostructures was extensively described in an excellent review,[23] where the thin film growth cycle for a binary compound (TiO_2)

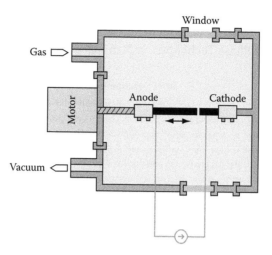

FIGURE 1.2 Schematic of a standard DC Krätschmer reactor (arc discharge). (Reproduced with permission from Springer Science+Business Media: Arc discharge and laser ablation synthesis of singlewalled carbon nanotubes, 2006, 1–18, Hornbostel, B. et al. In *Carbon Nanotubes*, Popov, V.N. and Lambin, P. (Eds.), Springer, Berlin, Germany.)

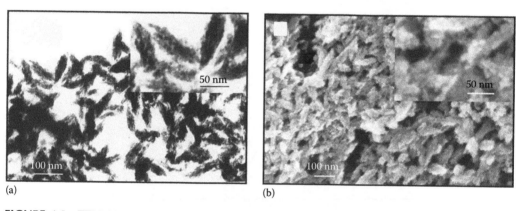

(a)   (b)

FIGURE 1.3 TEM (a) and SEM (b) images of the spindle-like nanostructures produced from Cu arc-discharge in deionized water. (Reproduced from *J. Solid State Chem.*, 177, Xie, S.-Y. et al., Preparation and self-assembly of copper nanoparticles via discharge of copper rod electrodes in a surfactant solution: A combination of physical and chemical processes, 3743–3747, Copyright 2004, with permission from the Elsevier.)

from gaseous precursors ($TiCl_4$ and H_2O) was presented as an example (Figure 1.9). This method is now a mature technique that has demonstrated its unique ability for processing ultrathin overlayers on complex-shaped substrates. ALD is already the method of choice in the semiconductor industry for depositing only a few nanometer-thick highly insulating oxide films. This technology has also become an attractive coating technique for the development of nanostructured materials with optimized chemical or physical properties driven by fundamental research. In addition to the booming activities in the deposition of high-*k* dielectrics for the semiconductor industry, the application of ALD in combination with nanostructured materials for conventional optics, solar cells, fuel cells, nanoparticle coatings, and jewelry is under discussion and in some cases has already reached the commercial stage.

The applications of *ion beams* in different energy regimes (equipment is shown in Figure 1.10) to create or modify nanostructures were discussed[24] on the basis of a large number of experiments. Atom beams of a few keV can be used to synthesize nanocomposites or to create ripples at the surface. Implantation and ion beam mixing with a few hundred keV ions are useful for obtaining

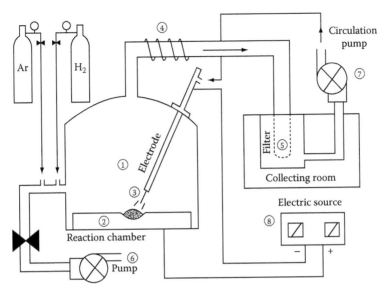

FIGURE 1.4 Scheme of HPMR plasma equipment. (Reproduced from Lai, W. et al., Synthesis of nanostructured materials by hot and cold plasma, http://www.ispc-conference.org/ispcproc/papers/82.pdf, accessed January 21, 2011.)

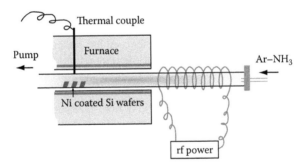

FIGURE 1.5 Scheme of the cold RF plasma. (Reproduced from Lai, W. et al., Synthesis of nanostructured materials by hot and cold plasma, http://www.ispc-conference.org/ispcproc/papers/82.pdf, accessed January 21, 2011.)

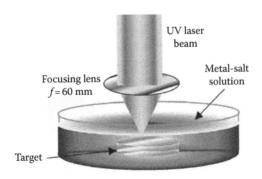

FIGURE 1.6 Schematic diagram of the experimental setup for nanoparticle production. (Reproduced from *Superlattices Microstruct.*, 43(5–6), Jimenez, E. et al., A novel method of nanocrystal fabrication based on laser ablation in liquid environment, 487–493, Copyright 2008, with permission from the Elsevier.)

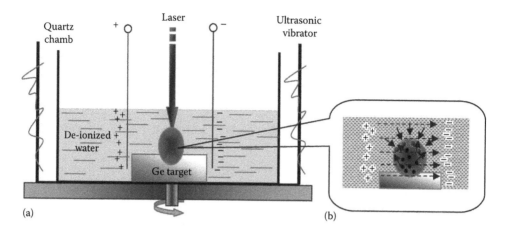

FIGURE 1.7 Illustration of the electrical-field-assisted pulsed-laser ablation in liquid (a) and a detailed depiction of the reactivity field instance (b). (Reproduced with permission from Liu, P. et al., Controllable fabrication and cathodoluminescence performance of high-index facets GeO$_2$ micro- and nanocubes and spindles upon electrical-field-assisted laser ablation in liquid, *J. Phys. Chem. C*, 112, 13450–13456. Copyright 2008 American Chemical Society.)

buried nanoparticles. Ion irradiation is a particularly suitable means to create or modify nanostructures, since ions with low energies (in the range where the energy loss is dominantly *via* elastic collisions) induce a collision cascade on a shallow thickness of the material and high-energy ions perturb the structure of the target in narrow channels (with diameters of the order of 10 nm) up to a few microns in depth, mainly through electronic energy transfer processes. In addition, the nanopatterning of thin films can be achieved by a focused ion beam. The induced transformations are often unusual, because of the short duration of the interaction, typically of 10^{-13} s for the cascade and of 10^{-12} and 10^{-10} s for the relaxation of the lattice and electron gas, respectively.

Various types of *chemical vapor deposition* (CVD) methods are common. Thus, well-aligned and randomly grown multiwall nanotubes (MWNTs) were fabricated by the *radio-field-induced self-bias hot-filament CVD* method (Figure 1.11).[25] It was illustrated that the growth of MWNTs can be modeled either as "tip growth" or "base growth" depending on the size of the catalytic particles involved. A high yield of CNT growth was obtained when oxidized metal alloys were used. It was suggested that the larger surface area due to oxidization caused a proliferation in the nucleation sites for CNT growth. A relative method, *plasma-enhanced CVD*, was used, in particular, for obtaining carbon nanostructures.[26]

Crystalline nanometer-sized Cu$_2$O and CuO particle formation was studied by *vapor thermal decomposition* of copper(II) acetylacetonate in a vertical laminar flow reactor at ambient pressure.[27] The introduction of oxygen into the system was found to increase the decomposition rate and removed impurities from particles. The size of primary particles produced varied from 10 to 200 nm. Particle crystallinity was found to depend on both the oxygen concentration and the furnace temperature. A model (Figure 1.12) that took into account the detailed chemical reaction mechanisms during particle formation was proposed. Gas-phase synthesis can also lead to nanostructured particulate films.[28]

1.2.2 ELECTROCHEMICAL METHODS

Among variations of these methods, the most common techniques are *electrospinning* (used mainly for nanofiber fabrication) and *electrodeposition*. Thus, a method for electrospinning was applied to create conducting polymeric composite nanofibers using a poly(dimethyl siloxane) (PDMS)-based microfluidic device (Figure 1.13).[29] In addition, nanofibers of poly(vinylpyrrolidone) (Figure 1.14) and its composite with polypyrrole were prepared using one-step and two-step

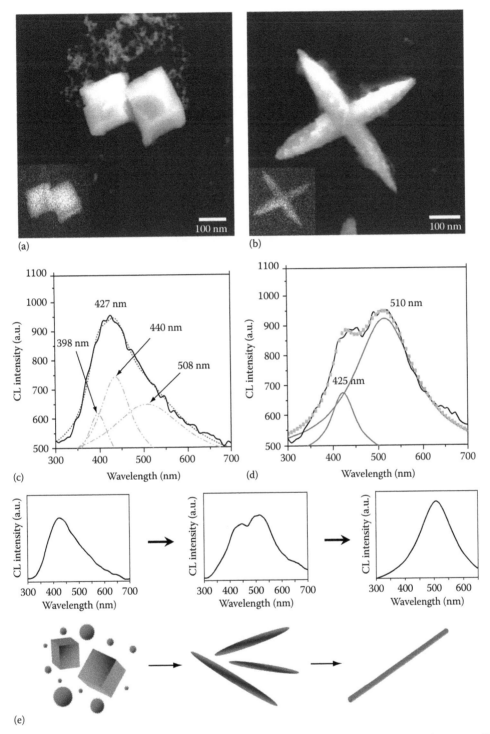

FIGURE 1.8 (a) SEM image of two single cubes; the inset is the corresponding cathodoluminescence (CL) image. (b) SEM image of two spindles and the inset of the corresponding CL image. The corresponding CL spectrum of cubes (c) and spindles (d). (e) An illustration of the luminescence shift of GeO_2 nanostructures with various shapes. (Reproduced with permission from Liu, P. et al., Controllable fabrication and CL performance of high-index facets GeO_2 micro- and nanocubes and spindles upon electrical-field-assisted laser ablation in liquid, *J. Phys. Chem. C*, 112, 13450–13456. Copyright 2008 American Chemical Society.)

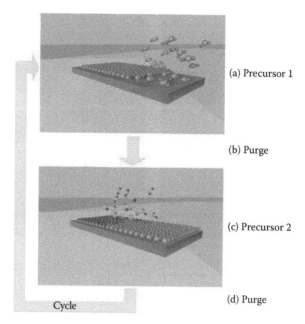

FIGURE 1.9 Schematic of an ALD process. One ALD cycle consists of four separate steps. In step (a) the substrate is exposed to precursor molecules (precursor 1), which adsorb ideally as a monolayer on the surface. In step (b), the excess of precursor 1 in the gas phase is removed by inert gas purging. In step (c), the substrate is exposed to precursor 2, which reacts with the adsorbed precursor 1 to form a layer of the desired material. In step (d), the excess of precursor 2 and the reaction by-products are removed by purging. This cycle is repeated (arrow) until the desired thickness of the deposit is obtained. (Knez, M. et al.: Synthesis and surface engineering of complex nanostructures by atomic layer deposition. *Adv. Mater.* 2007. 19. 3425–3438. Copyright Wiley-VCH Verlag GmbH & Co. KGaA. Reproduced with permission.)

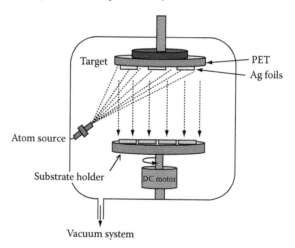

FIGURE 1.10 A schematic sketch of the atom beam sputtering setup. (Reproduced with permission from Avasthi, D.K. and Pivin, J.C., *Current Sci.*, 98(6), 780–792. Copyright 2010 Indian Academy of Science.)

microfluidic electrospinning. It was demonstrated that the morphology and dimension of the nanofibers can be modified by adjusting the polymer concentration, surface tension, salt, strength of the potential, and feed rate. An example of the second electrochemical method is preparation of monodispersed single-crystalline gold nanocubes of highly uniform size and lengths of the edges of about 30 nm (Figures 1.15 and 1.16).[30] The growth solution was prepared from two cationic surfactant solutions as micelle templates with added acetone solvent.

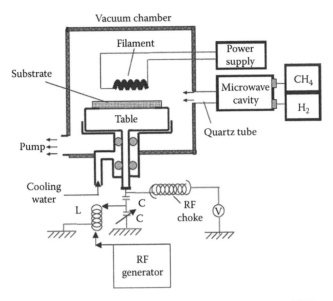

FIGURE 1.11 A schematic diagram of the MW/RF enhanced hot-filament PECVD system. (Reproduced with permission from Chen, S.Y. et al., *J. Phys. D: Appl. Phys.*, 37, 273. Copyright 2004 Institute of Physics Science.)

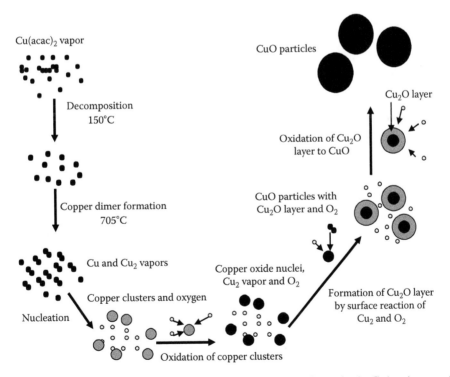

FIGURE 1.12 The schematic presentation of copper (II) oxide particle formation by $Cu(acac)_2$ vapor decomposition in the presence of oxygen. (Reproduced from Nasibulin, A.G. et al., *Aerosol Sci. Technol.*, 36, 899. Copyright 2002 Taylor & Francis.)

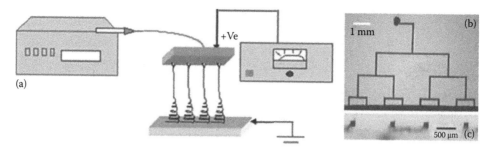

FIGURE 1.13 (a) Schematic diagram of microfluidic electrospinning using a branching microchannel architecture to simultaneously spin multiple fibers directly from the channel outlets. (b) Close-up of top side of the poly(dimethyl-siloxane) (PDMS) device with channels filled with food coloring to enhance contrast. (c) Side-on cross-section of the device, showing the channel outlets. (Srivastava, Y. et al.: Multijet electrospinning of conducting nanofibers from microfluidic manifolds. *J. Appl. Polym. Sci.* 2007, 106, 3171–3178. Copyright Wiley-VCH Verlag GmbH & Co. KGaA. Reproduced with permission.)

1.2.3 Microwave, Ultrasonic, and UV-Irradiation Techniques

Microwave irradiation (MWI). As a "nonconventional reaction condition,"[31] it has been applied in various areas of chemistry and technology to produce or destroy diverse materials and chemical compounds as well as to accelerate chemical processes. The advantages of its use are as follows[32]:

1. Rapid heating is frequently achieved.
2. Energy is accumulated within a material without surface limits.
3. Economy of energy is achieved due to the absence of a necessity to heat environment.
4. Electromagnetic heating does not produce pollution.
5. There is no direct contact between the energy source and the material.
6. Suitability of heating and possibility of automation.
7. Enhanced yields, substantial elimination of reaction solvents, and facilitation of purification relative to conventional synthesis techniques.
8. This method is appropriate for green chemistry and energy-saving processes.

Substances or materials have different capacity to be heated by MWI, which depends on the nature of a substance and its temperature. Generally, chemical reactions are accelerated in microwave fields, as well as those by ultrasonic treatment, although the nature of these two techniques is completely different. Microwave heating (MWH) is widely used to prepare various refractory inorganic compounds and materials (double oxides, nitrides, carbides, semiconductors, glasses, ceramics, etc.)[33] as well as in organic processes[34,35]: pyrolysis, esterification, and condensation reactions. Recent excellent reviews have described different aspects of microwave-assisted synthesis of various types of compounds and materials, in particular organic[36–38] and organometallic[39] compounds, polymers, and applications in analytical chemistry[40] among others. A host of nanostructures have also been prepared by this route. For instance, nanostructures composed of the mixed valent $La_{0.325}Pr_{0.300}Ca_{0.375}MnO_3$ complex oxide were prepared by MWH of precursors (metal nitrates).[41]

A typical reactor used for organic or organometallic syntheses[42] is presented in Figure 1.17, which can be easily implemented using a domestic microwave oven. Due to some problems occurring during microwave treatment, for example, related to the use of volatile liquids (they need an external cooling system *via* copper ports), original solutions to these problems are frequently reported in the literature. More modern laboratory MW reactors[43] are shown in Figure 1.18.

Some years ago, an alternative method for performing microwave-assisted organic reactions, termed *enhanced microwave synthesis* (EMS), was examined in an excellent review.[44] By externally

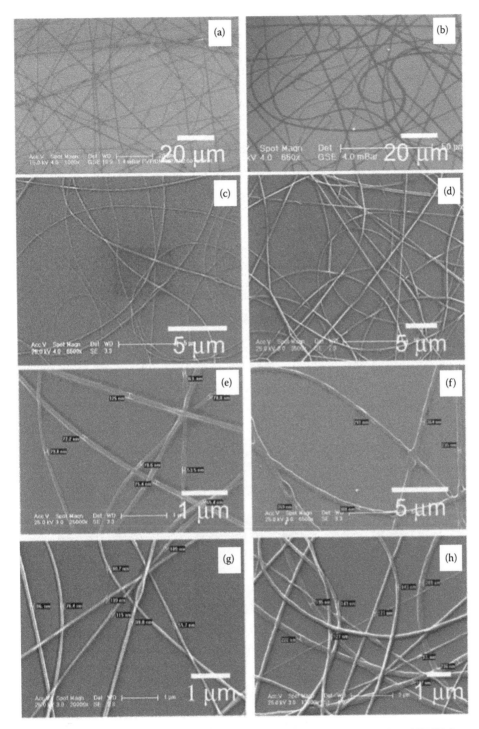

FIGURE 1.14 Variation in the morphology and dimension of PVP nanofibers in ethanol/DMF electrospun under different processing conditions: (a) 2% PVP (w/v), (b) 6% PVP (w/v), (c) 4% PVP (w/v) in ethanol/DMF with 10% (w/v) FeCl$_3$, (d) 4% PVP (w/v) (no FeCl$_3$), (e) 4% PVP (w/v) formed at a feed rate of 0.007 mL/min, (f) 4% PVP (w/v) formed at a feed rate of 0.1 mL/min, (g) 4% PVP (w/v) formed at 10 kV, (h) 4% PVP (w/v) formed at 20 kV. (Srivastava, Y. et al.: Multijet electrospinning of conducting nanofibers from microfluidic manifolds. *J. Appl. Polym. Sci.* 2007, 106, 3171–3178. Copyright Wiley-VCH Verlag GmbH & Co. KGaA. Reproduced with permission.)

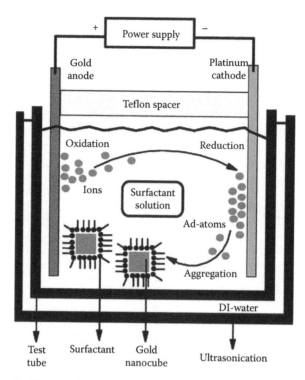

FIGURE 1.15 Schematic diagram of the electrochemical apparatus for the synthesis of gold nanocubes. (Reproduced from *Mater. Lett.*, 60, Huang, C.-J. et al., Electrochemical synthesis of gold nanocubes, 1896–1900, Copyright 2006, with permission from the Elsevier.)

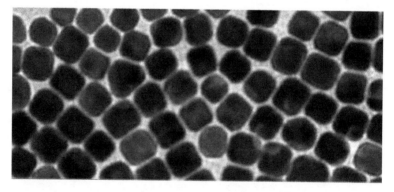

FIGURE 1.16 Gold nanocubes. (Reproduced from *Mater. Lett.*, 60, Huang, C.-J. et al., Electrochemical synthesis of gold nanocubes, 1896–1900, Copyright 2006, with permission from the Elsevier.)

cooling the reaction vessel with compressed air, while simultaneously administering MWI, more energy can be directly applied to the reaction mixture. In *conventional microwave synthesis* (CMS), the initial microwave power is high, increasing the bulk temperature (TB) to the desired set point very quickly. However, on reaching this temperature, the microwave power decreases or shuts off completely to maintain the desired bulk temperature without exceeding it. When MWI is off, classical thermal chemistry takes over, resulting in loss of the full advantage of microwave-accelerated synthesis. With CMS, MWI is predominantly used to reach TB faster. Microwave enhancement of chemical reactions will take place only during application of microwave energy. This source of energy will directly activate the molecules in a chemical reaction. EMS ensures that a high, constant level of microwave energy is applied.

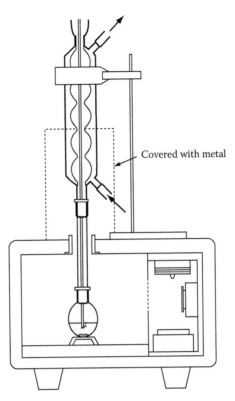

Covered with metal

FIGURE 1.17 Typical MW-reactor for organic and organometallic synthesis. (Reproduced with permission from Matsumura-Inoue, T. et al., *Chem. Lett.*, 23, 2443. Copyright 1994 Chemistry Society of Japan.)

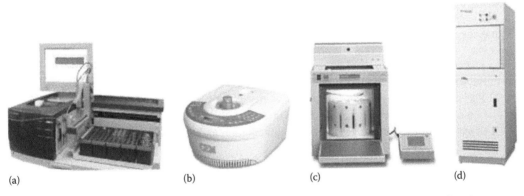

(a) (b) (c) (d)

FIGURE 1.18 Microwave reactors for chemical syntheses. (a) Emrys Liberator (Biotage, Sweden, www.biotage.com); (b) CEM Discover BenchMate (CEM, Matthews, NC, www.cem.com) Copyright CEM Corporation; (c) Milestone Ethos TouchControl (Milestone, Italy, www.milestonesci.com); (d) Lambda MicroCure2100 BatchSystem (Lambda, Plano, TX, www.microcure.com). (Wiesbrock, F. et al.: Microwave-assisted polymer synthesis: State-of-the-art and future perspectives. *Macromol. Rapid Commun.* 2004. 25. 1739–1764. Copyright Wiley-VCH Verlag GmbH & Co. KGaA. Reproduced with permission.)

Ultrasonic treatment. It is also used for nanoparticle synthesis, although not so frequently. Ultrasound (US), as it is well known, is a part of the sound spectra with a frequency of ~16 kHz, which is out of the normal range of human hearing. The effects produced by US are derived from the creation, expansion, and destruction of small bubbles, which appear under US irradiation of a liquid phase. After creation, these bubbles grow 2–10 times the initial diameter of a few microns and undergo radial vibration. The circulating liquid cools these unstable cavities, so, after a few cycles of rarefaction and compression, they collapse violently in 10^{-5}–10^{-7} s. This phenomenon, named *cavitation,* produces high temperatures and pressures in the liquid. The temperature of cavitation varies from 1,000 to 10,000 K, more frequently in the range of 4500–5500 K. It should be noted that acoustic irradiation is a mechanical energy (not quantum), which is transformed to thermal energy. Contrary to photochemical processes, this energy is not absorbed by molecules. Due to the extensive range of cavitation frequencies, many reactions are not well reproducible. Therefore, each publication related to the use of US generally contains a detailed description of equipment (dimensions, frequency used, intensity of US, etc.). Sonochemical reactions are usually marked)))), in accordance with internationally accepted usage. For successful application of US, the influence of various factors can be summarized as follows:

1. *Frequency.* Increase in frequency leads to decrease in the production and intensity of cavitation in liquids. This fact can be explained as follows: at high frequencies, the time necessary for a bubble to appear as a result of cavitation and to have sufficient size to affect the liquid phase is too low.
2. *Solvent.* Cavitation produces considerably minor effects in viscous liquids or those with higher surface pressures.
3. *Temperature.* Increase in temperature allows the cavitation to be performed at lower acoustic intensities. This is a consequence of increasing vapor pressure of the solvent with increasing temperature.
4. *Application of gases.* In the case of application of gases (poor or very soluble), the intensity of cavitation decreases due to the formation of a large number of additional nuclei in the system.
5. *External pressure.* Increase in external pressure leads to increase in the intensity of destruction of cavitation bubbles, that is, the effects of US in this case are more rapid and violent in comparison with normal pressure.
6. *Intensity.* In general, the increase in intensity of US intensifies the effects produced.

With respect to metal surfaces, the US action can be described as follows:

1. The acoustic flow is a movement of the liquid, induced by the sound wave (a conversion of the sound to kinetic energy) and is not a cavitation effect.
2. The formation of asymmetric cavities on the metal surface is a direct result of destruction of the low-life bubbles near the surface. As a result of the cavitation, the deformation of surface takes place, together with fragmentation and decrease in the size of appearing particles.

The use of US for the fabrication of nanostructures is described here.* Among other elements and compounds, nanostructured metals, carbides, nitrides, oxides, and sulfides (Figure 1.19) can be ultrasonically obtained.[45] In the case of using *ultraviolet irradiation,* the fabrication of Au particles with triangular, pentagonal, and hexagonal shapes from $HAuCl_4 \cdot 3H_2O$ as a precursor under UV irradiation has been reported (Figure 1.20).[46]

1.2.4 HIGH-PRESSURE METHODS

Hydro- and solvothermal methods. These are widely applied to synthesize nanostructures, both classic (nanowires, nanobelts) and less-common nanoforms. As an example, the solvothermal method (based on thermal decomposition of organometallic compounds in organic solvents and

---

* See Chapter 14 for more information.

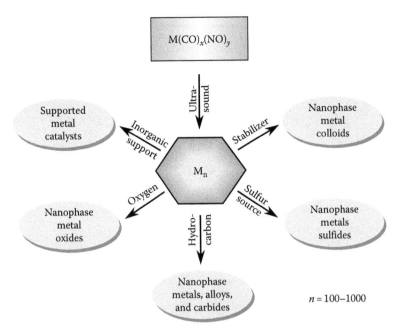

FIGURE 1.19 Sonochemical synthesis of nanostructured materials. (Reproduced with permission from Abedini, R. and Mousavi, S.M., Preparation and enhancing of materials using ultrasound technique: Polymers, catalysts and nanostructure particles, *Pet. Coal*, 52(2), 81, 2010.)

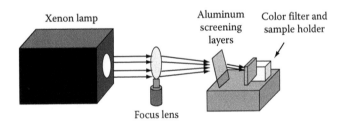

FIGURE 1.20 Diagram of the UV-experimental setup. (Reproduced with permission from Hsu, H.-Y. et al., *NNIN REU Res. Accomplishments*, 68, 2004.)

applied for the synthesis of various types of nanosized metal oxides with large surface area, high crystallinity, and high thermal stability) was employed to synthesize ZnO nanostructure from zinc acetate dihydrate as the zinc source at 80°C for 24 h in autoclave (Reaction 1.1).[47] An interesting correlation between the aspect ratio of the ZnO products and physical properties of the solvent was observed. In addition, the different environments during ZnO preparation led to the different morphology with width and length in the range of 8.9–69 nm and 68–108 nm, respectively, and 0.17–0.93 in terms of the aspect ratio:

$$Zn(CH_3COO)_2 + 2ROH \rightarrow ZnO + 2CH_3COR + H_2O \tag{1.1}$$

Ionothermal synthesis. It is one of the most promising methods with the use of ionic liquids (ILs), was developed into a versatile and advantageous synthesis technique.[48] ILs have unique properties such as negligible vapor pressure and thermal stability and also tunable properties and designable structures. Compared with traditional hydro/solvothermal synthesis, ionothermal synthesis has two advantageous features: the ionic reaction medium can uniquely influence the course of chemical reactions; ILs can act as an "all-in-one" solvent or template for the synthesis of inorganic

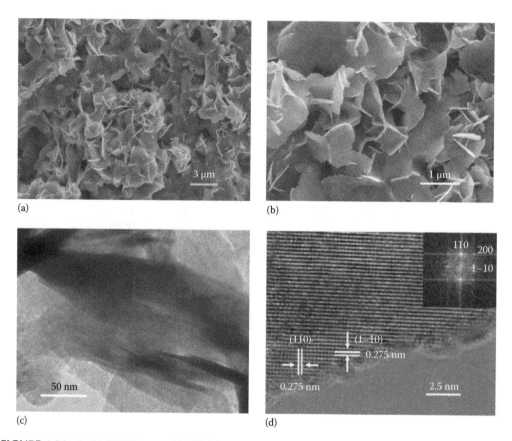

FIGURE 1.21 (a, b) SEM images; (c) TEM image; and (d) HR-TEM image of BiOCl nanoplates obtained at 180°C for 24 h. Inset: its corresponding fast Fourier transform (FFT) image. (Reproduced with permission from Ma, J. et al., Ionothermal synthesis of BiOCl nanostructures via a long-chain ionic liquid precursor route, *Cryst. Growth Des.*, 10(6), 2522–2527. Copyright 2010 American Chemical Society.)

materials, which enables the synthesis of inorganic materials with novel and improved properties. It is expected that this ionothermal synthesis using long-chain IL could be used to fabricate other polar nanomaterials with novel morphologies and improved properties in the fields of optics, electrochemistry, and catalysis. The BiOCl nanoplates synthesized by this route (Figure 1.21) were found to have a high adsorption capacity and high adsorption efficiency compared with some other nanomaterials in the neutral media. In a typical synthesis, 1.0 mmol of $Bi(NO_3)_3 \cdot 5H_2O$ and 10 mmol of IL 1-hexadecyl-3-methylimidazolium chloride ($[C1_6Mim]Cl$) were mixed together and then transferred into a 20 mL Teflon-lined stainless autoclave. The autoclave was heated at 180°C (or alternatively 120°C and 200°C) for 12 h and then naturally cooled to room temperature. The resulting precipitates were collected and washed with ethanol and deionized water thoroughly and dried at 50°C in air.

1.2.5   USE OF MICROFLUIDIC CHIPS

The synthesis of 1D titanium oxide nanostructures was accelerated by performing the reaction in a *microfluidic environment* as opposed to a classical batch process.[49] The use of continuous-flow, microfluidic reactors as environments in which to perform synthetic processes confers several advantages over conventional "macroscale" techniques. In simple terms, the high surface area-to-volume ratios and reduced diffusional dimensions that characterize microfluidic systems allow reactions to be performed in a rapid and controllable manner. Local variations in reaction conditions

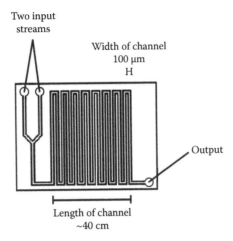

Two input
streams

Width of channel
100 μm
H

Output

Length of channel
~40 cm

FIGURE 1.22 Schematic diagram of the microfluidic chip used for the synthesis of TiO$_2$ nanostructures. (Reproduced with permission from Cottam, B.F. et al., Accelerated synthesis of titanium oxide nanostructures using microfluidic chips, *Lab. Chip*, 7, 167–169. Copyright 2007 The Royal Society of Chemistry.)

(such as concentration and temperature) are minimized, and, therefore, control of both nucleation and particle growth can be used to improve monodispersity and efficiency in nanoparticle synthesis. To synthesize TiO$_2$, solutions of precursors (titanium tetraisopropoxide and trimethylamine N-oxide dehydrate) were concurrently pumped into a two-input/one-output y-shaped microfluidic chip, a schematic of which is shown in Figure 1.22 (channel dimensions: 60 μm depth × 100 μm width × 40 cm length, corresponding to a volume of ~17.0 μL). The chip was placed in an oven, set at 90°C; the use of an oven rather than a more conventional hot plate provided improved thermal stability and reproducibility. The flow regime in the chip was plug flow, with ~100 mm plugs of each solution alternating along the length of the microchannel.

1.2.6 SYNTHESIS IN REVERSED MICELLES

Recent advances in nanoparticle synthesis with reversed micelles were reviewed in an excellent article.[50] In particular, reactions in water-in-oil microemulsions and water/supercritical (sc) fluid microemulsions were discussed. The last method promises to be a highly useful route for controlled nanoparticle synthesis due to the added control variables afforded by tuneability of the solvent quality (density) through pressure and temperature. The technique, known as *RESOLV* (*r*apid *e*xpansion of a *s*upercritical *s*olution into a *l*iquid *s*olvent) (Figure 1.23), where a stable microemulsion of silver cations in sc-CO$_2$ is expanded through a nozzle into a solvent containing reducing agent, was used to produce silver nanoparticles with good size control.[51]

1.2.7 HOT-PLATE METHOD

A surprisingly simple method to synthesize metal oxide (α-Fe$_2$O$_3$ [Figure 1.24], Co$_3$O$_4$, and ZnO) nanostructures by directly heating metal-foil- or metal-film-coated substrates in air using a hotplate (called the *hotplate method*) has been reported.[52] The appropriate metal films were first deposited directly onto the substrates by RF magnetron sputtering, and the coated substrates were subsequently heated on a hotplate in air. Typical film thickness ranged from 300 to 900 nm. Successful attempts have been achieved on a wide variety of substrates, such as a plain silicon wafer, a glass slide, quartz, a silica microsphere, atomic force microscopy (AFM) tips, and electrochemically etched W tips. This technique may provide an alternative method or open a new approach to the simple fabrication of metal oxide nanostructures at low temperatures, especially for those metals with high melting points.

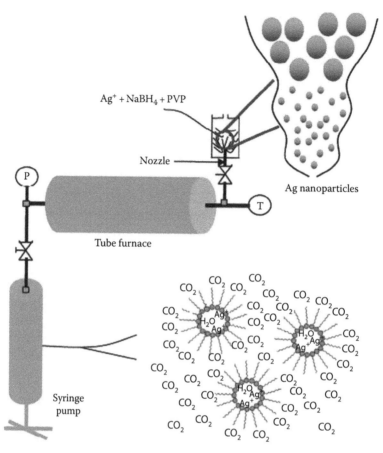

FIGURE 1.23 Experimental setup for RESOLV. (Reproduced from *J. Supercrit. Fluids*, 34, Meziani, M.J. et al., Nanoparticle formation in rapid expansion of water-in-supercritical carbon dioxide microemulsion into liquid solution, 91–97, Copyright 2005, with permission from the Elsevier.)

1.2.8 OTHER CHEMICAL ROUTES

The "wet-chemistry" techniques are widely applied in the preparation of nanostructures due to no necessity to use high-cost equipment, limiting their use in laboratories worldwide. The precursors for inorganic nanostructures and nanocomposites can be elemental metals, metal oxides, salts, complexes, etc. Thus, AuPt alloy nanoparticles were fabricated from an organometallic complex precursor.[53] In general, bimetallic nanostructures can be formed as *alloys* (if the reduction rate of their precursors is comparable) or *core-shell nanoparticles* (in the opposite case) (Figure 1.25).[54]

The use of support for the formation of nanoparticles is also a classic route: for instance, synthesis of nanostructures on the surface of solid was carried out by the *molecular layering method*.[55] Another interesting example is a poly(methyl methacrylate) (PMMA)-mediated *nanotransfer printing technique* (Figure 1.26), which aims at creating arbitrary purpose-directed nanostructures with various nanoscale building blocks.[56] Its strategy is based on the utilization of the PMMA film as a macroscopic mediator for handling tiny nanoscale building blocks. This nanoscale bricklaying technique paves the way to generate purpose-directed nanostructures with homo- or heterogeneous building blocks, which facilitates exploring their fundamental properties and building novel devices.

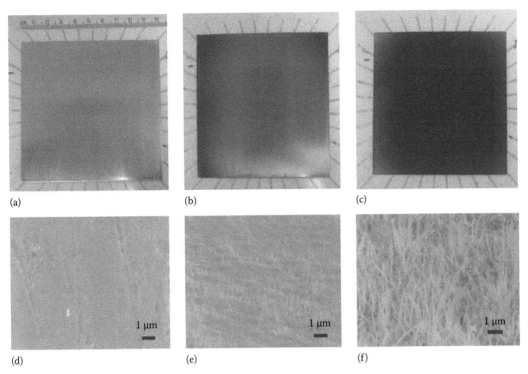

FIGURE 1.24 Fabrication of α-Fe$_2$O$_3$ nanoflakes on Fe foil. Optical images of the Fe foil (a) before heating, (b, c) after heating at 300°C for 10 min and 24 h, respectively (d through f) Corresponding scanning electron microscopy (SEM) images of the foil surfaces are shown in (a through c). The size of the nanoflakes shows a dramatic increase after heating for a long duration. The growth is rapid and over a large area (10 cm × 10 cm). (Yu, T. et al., Substrate-friendly synthesis of metal oxide nanostructures using a hotplate. *Small.* 2006. 2(1). 80–84. Copyright Wiley-VCH Verlag GmbH & Co. KGaA. Reproduced with permission.)

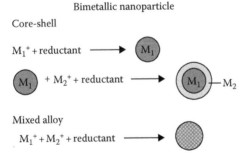

FIGURE 1.25 Possibility of a mixed alloy or core-shell nanostructure formation. (From Toshima, N. and Yonezawa, T., Bimetallic nanoparticles—Novel materials for chemical and physical applications, *New J. Chem.*, 1179, 22, 1998. Reproduced from http://libattery.ustc.edu.cn/chinese/ppt/%283-2%29%20Metal%20 synthesis.pdf.)

For nanocrystal/nanoparticle formation, simple laboratory equipment is used. Thus, the CdCuS nanocrystals were synthesized by chemical route method in a Kibbs apparatus (Figure 1.27) from an aqueous solution of cadmium chloride (CdCl$_2$) and copper chloride (CuCl$_2$) as precursors and H$_2$S gas, yielding a brownish solution. The product was then dispersed in a PMMA matrix.[57] The formed nanocomposite polymer films were irradiated by a swift heavy ion (SHI) (100 MeV, Si^{+7} ions beam) at different fluences of 1×10^{10} and 1×10^{12} ions/cm^2.

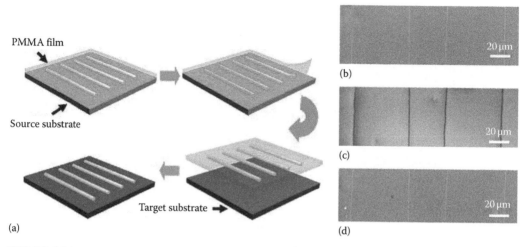

FIGURE 1.26 (a) Illustration of the procedures of PMMA-mediated nanotransfer printing technique. Spin coating a PMMA film on source substrate, peeling off the film from substrate, attaching the film to target substrate, and finally removing the mediator. SEM images of SWNT array (b) on SiO_2/Si substrate before transfer, (c) embedded in PMMA film, and (d) transferred to another SiO_2/Si substrate. (Reproduced with permission from Jiao, L. et al., Creation of nanostructures with poly(methyl methacrylate)-mediated nanotransfer printing, *J. Am. Chem. Soc.*, 130, 12612–12613. Copyright 2008 American Chemical Society.)

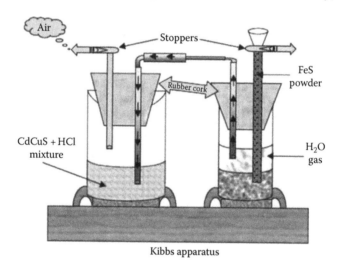

FIGURE 1.27 The schematic diagram of Kibbs apparatus. (Reproduced from Agrawal, S. et al., Swift heavy ion irradiation effect on Cu-doped CdS nanocrystals embedded in PMMA, *Bull. Mater. Sci.*, 32(6), 569–573. Copyright 2009 Indian Academy of Science.)

1.2.9 Biochemical and Self-Assembly Methods

In analyzing a series of methods available to synthesize nanoparticles, challenges still exist such as synthesis under mild conditions and controlling of shape and size of the nanoparticles. The use of proteins as *biotemplates* appears to be a promising route to synthesize nanoparticles. Certain proteins such as ferritin[58] (Figure 1.28), ferritin-like-protein (FLP), chaperonin, and cowpea chlorotic mottle virus (CCMV)[59] have cavities in the center. The protein cavity can be used as a template for the growth of nanoparticles. Thus, it is possible to prepare nanoparticles with uniform size and shape. Recently, iron vanadate, phosphate, molybdate, and arsenate nanoparticles were synthesized using ferritin templates.[60] It has even been reported that hollow nanoparticles can be synthesized.[61]

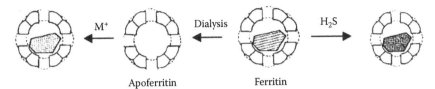

Apoferritin   Ferritin

FIGURE 1.28 Scheme of nanosynthesis using a biotemplate. The iron core of ferritin can be sulfurated to form FeS nanoparticles. The core can be removed by dialysis and incubated with other metal ions to form nanoparticles of different compositions. (Reproduced from Macmillan Publishers Ltd., *Nature*, Meldrum, F.C. et al., Synthesis of inorganic nonophase materials in supramolecular protein cages, 349, 684–687, copyright 1991.)

Elemental metal nanostructures can also be produced biologically.* Thus, biosynthesis of silver nanoparticles using *Penicillium fungi* was reported.[62] Formation of colloidal silver particles can be easily followed by changes in UV–Vis absorption. Their formation proceeds *via* an extracellular mechanism. The most important feature of *Penicillium fungi* is the fact they are widespread in the waste biomass from pharmaceutical industry. Such a cheap source of material gives an opportunity for cost-effective preparation of various silver-based nanostructures. In a related report,[63] the bacterial strain *Escherichia coli* was used for the biosynthesis of silver nanoparticles.

Various aspects of nanoparticle formation from a variety of materials such as *proteins, polysaccharides, and synthetic polymers* were reviewed in Ref. [64]. Nanoparticles can be prepared most frequently by three methods: (1) *dispersion of preformed polymers* (a common technique used to prepare biodegradable nanoparticles from poly(lactic acid) (PLA); poly(D,L-glycolide), PLG; poly(D,L-lactide-co-glycolide) (PLGA), and poly(cyanoacrylate) (PCA)); (2) *solvent evaporation* (after the formation of a stable emulsion of polymer and drug solution in an organic solvent, the organic solvent is evaporated either by reducing the pressure or by continuous stirring); and (3) *ionic gelation or coacervation of hydrophilic polymers* (here, the positively charged amino group of chitosan interacts with the negatively charged tripolyphosphate to form coacervates with a size in the range of nanometers; coacervates are formed as a result of electrostatic interaction between two aqueous phases, whereas, ionic gelation involves the material undergoing transition from liquid to gel). *Spontaneous emulsification or solvent diffusion* method is a modified version of the solvent evaporation method, where the water-miscible solvent along with a small amount of the water-immiscible organic solvent is used as an oil phase. Due to the spontaneous diffusion of solvents, an interfacial turbulence is created between the two phases, leading to the formation of small particles. As the concentration of water-miscible solvent increases, a decrease in the size of particle can be achieved. However, other methods such as supercritical fluid technology (requiring specially designed equipment and more expensive, although this is an alternative to prepare biodegradable micro- and nanoparticles because supercritical fluids are environmentally safe) and particle replication in nonwetting templates have also been described in the literature for the production of polymer nanoparticles.

Methods based on self-assembly. Active microelectronic arrays and DNA-modified components may allow scientists and engineers to direct self-assembly of 2D and 3D molecular electronic circuits and devices within the defined perimeters of larger silicon or semiconductor structures[65] (Figure 1.29). The authors noted that electronically directed DNA self-assembly technology could encompass a broad area of potential applications from nearer term heterogeneous integration processes for photonic and microelectronic device fabrication to the longer term nanofabrication of true molecular electronic circuits and devices. These and relative methods can result in unusual and very useful devices. Thus, Figure 1.30 depicts a single molecule bridging the gap between two metallic contacts, forming the smallest and ultimate limit of an electronic device. The molecule is designed

---

* See Chapter 22.

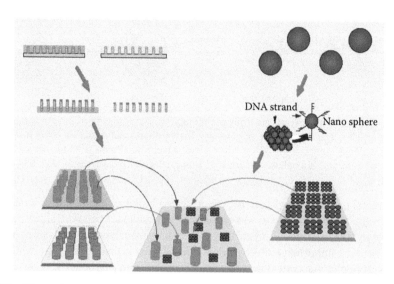

FIGURE 1.29 Directed nanofabrication on a chip (Nanogen, Inc., San Diego, CA). (Reproduced from Tirrell, M. et al., Chapter 4: Synthesis, assembly, and processing of nanostructures. Nanotechnology Research Directions: IWGN Workshop Report, 1999, National Science and Technology Council, Washington, DC, http://www.wtec.org/loyola/nano/IWGN.Research.Directions/IWGN_rd.pdf. With permission from the World Technology Evaluation Center, Inc.)

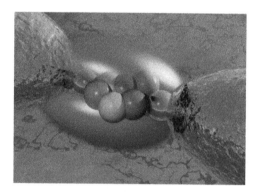

FIGURE 1.30 A single molecule bridging the gap between two metallic contacts, forming the smallest and ultimate limit of an electronic device. (Reproduced from Tirrell, M. et al., Chapter 4: Synthesis, assembly, and processing of nanostructures. Nanotechnology Research Directions: IWGN Workshop Report, 1999, National Science and Technology Council, Washington, DC, http://www.wtec.org/loyola/nano/IWGN. Research.Directions/IWGN_rd.pdf. With permission from the World Technology Evaluation Center, Inc. ©1999 Mark Reed.)

with end groups (dull gold spheres) of sulfur atoms, which automatically assembly onto the gold wire contacts. The blue fuzz above and below the atoms represents the electron clouds, through which the current actually flows.

1.3 "GREEN" ASPECTS OF NANOPARTICLE SYNTHESIS

Environmentally friendly synthetic methodologies. These have gradually been implemented as viable techniques in the synthesis of a range of nanostructures. In particular, the application of green-chemistry principles to the synthesis of complex metal oxide and fluoride nanostructures was discussed, paying attention to advances in the use of the molten-salt synthetic methods, hydrothermal protocols, and template-directed techniques as environmentally sound, socially responsible, and cost-effective methodologies that allow us to generate nanomaterials without the need to sacrifice sample quality, purity, and crystallinity, while allowing control over size, shape, and morphology.[66] The authors underlined the main features of the green methods:

1. Use of cost-effective, *nontoxic precursors*, if at all possible. Minimization of the use of carcinogenic reagents and solvents (if possible, utilization of aqueous solvents). No experiments carried out with either pyrophoric, flammable, or unstable precursors.
2. Use of *relatively few* numbers of reagents.
3. *Minimization* of reaction steps—reduces waste, reagent use, and power consumption.
4. Development of reactions with *little if any by-products* (if possible, high-yield processes with an absence of volatile and toxic byproducts).
5. Room-temperature (or *low-temperature*) synthesis under ambient conditions, if at all possible.
6. Efficiency of *scale-up*.

As discussed, the techniques for the synthesis of nanostructures are very distinct. Some of them require sophisticated equipment with a cost of hundreds of thousands of dollars or more (for instance, the CVD equipment, applied to produce carbon nanotubes or nanodiamonds). On the contrary, wet-chemical methods are available in any laboratory worldwide. The size and shape of formed nanostructures obviously depend on reaction conditions as in standard chemical reactions: concentration of precursors, temperature, pressure, process duration, mixing rate, etc. Both classic and less-common nanostructures, described throughout the book, can be obtained, frequently together with each other. Applying certain strategies, it is possible (of course, not in all cases) to get purpose-directly nanostructures of desirable dimensionality, shape, length, width, etc. A considerable number of reported nanostructures have been obtained accidentally. Other relevant information on applied synthesis methods can be seen in each chapter.

REFERENCES

1. Xia, Y.; Yang, P.; Sun, Y.; Wu, Y.; Mayers, B.; Gates, B.; Yin, Y.; Kim, F.; Yan, H. One-dimensional nanostructures: Synthesis, characterization, and applications. *Advanced Materials*, 2003, *15* (5), 353–389.
2. Li, Y.; Cai, W.; Duan, G. Ordered micro/nanostructured arrays based on the monolayer colloidal crystals. *Chemistry of Materials*, 2008, *20* (3), 615–624.
3. Swihart, M. T. Vapor-phase synthesis of nanoparticles. *Current Opinion in Colloid and Interface Science*, 2003, *8*, 127–133.
4. Maisels, A.; Kruis, F. E.; Fissan, H.; Rellinghaus, B.; Zahres, H. Synthesis of tailored composite nanoparticles in the gas phase. *Applied Physics Letters*, 2000, *77*, 4431–4433.
5. Ohno, T. Morphology of composite nanoparticles of immiscible binary systems prepared by gas-evaporation technique and subsequent vapor condensation. *Journal of Nanoparticle Research*, 2002, *4*, 255–260.
6. Makimura, T.; Mizuta, T.; Murakami, K. Laser ablation synthesis of hydrogenated silicon nanoparticles with green photoluminescence in the gas phase. *Japanese Journal of Applied Physics*, 2002, *41*, L144–L146.
7. Weber, A. P.; Seipenbusch, M.; Kasper, G. Application of aerosol techniques to study the catalytic formation of methane on gasborne nickel nanoparticles. *Journal of Physical Chemistry A*, 2001, *105*, 8958–8963.
8. Urban, F. K. III; Hosseini-Tehrani, A.; Griffiths, P.; Khabari, A.; Kim, Y.-W.; Petrov, I. Nanophase films deposited from a highrate, nanoparticle beam. *Journal of Vacuum Science and Technology B*, 2002, *20*, 995–999.
9. Magnusson, M. H.; Deppert, K.; Malm, J.-O. Single-crystalline tungsten nanoparticles produced by thermal decomposition of tungsten hexacarbonyl. *Journal of Materials Research*, 2000, *15*, 1564–1569.
10. Kim, J. H.; Germer, T. A.; Mulholland, G. W.; Ehrman, S. H. Size-monodisperse metal nanoparticles via hydrogen-free spray pyrolysis. *Advanced Materials*, 2002, *14*, 518–521.
11. Ledoux, G.; Amans, D.; Gong, J.; Huisken, F.; Cichos, F.; Martin, J. Nanostructured films composed of silicon nanocrystals. *Materials Science and Engineering C*, 2002, *19*, 215–218.

12. Heberlein, J.; Postel, O.; Girshick, S. L.; McMurry, P.; Gerberich, W.; Iordanoglou, D.; Di Fonzo, F. et al. Thermal plasma deposition of nanophase hard coatings. *Surface and Coatings Technology*, 2001, *142–144*, 265–271.

13. Wegner, K.; Stark, W. J.; Pratsinis, S. E. Flame-nozzle synthesis of nanoparticles with closely controlled size, morphology, and crystallinity. *Materials Letters*, 2002, *55*, 318–321.

14. Madler, L.; Kammler, H. K.; Mueller, R.; Pratsinis, S. E. Controlled synthesis of nanostructured particles by flame spray pyrolysis. *Journal of Aerosol Science*, 2002, *33*, 369–389.

15. Sarigiannis, D.; Peck, J. D.; Kioseoglou, G.; Petrou, A.; Mountziaris, T. J. Characterization of vapor-phase-grown ZnSe nanoparticles. *Applied Physics Letters*, 2002, *80*, 4024–4026.

16. Xu, F.; Liu, X.; Tse, S. D.; Cosandey, F.; Kear, B. H. Flame synthesis of zinc oxide nanowires. *Chemical Physics Letters*, 2007, *449*, 175–181.

17. Hornbostel, B.; Haluska, M.; Cech, J.; Dettlaff, U.; Roth, S. Arc discharge and laser ablation synthesis of singlewalled carbon nanotubes. In *Carbon Nanotubes*. Popov, V. N.; Lambin, P. (Eds.). Springer, Berlin, Germany, 2006, pp. 1–18.

18. Keidar, M.; Shashurin, A.; Volotskova, O.; Raitses, Y.; Beilis, I. I. Mechanism of carbon nanostructure synthesis in arc plasma. *Physics of Plasmas*, 2010, *17*, 057101.

19. Xie, S.-Y.; Ma, Z.-J.; Wang, C.-F.; Lin, S.-C.; Jiang, Z.-Y.; Huang, R.-B.; Zheng, L.-S. Preparation and self-assembly of copper nanoparticles via discharge of copper rod electrodes in a surfactant solution: A combination of physical and chemical processes. *Journal of Solid State Chemistry*, 2004, *177*, 3743–3747.

20. Lai, W.; Zheng, J.; Yang, R.; Xie, L.; Li, X. Synthesis of nanostructured materials by hot and cold plasma. http://www.ispc-conference.org/ispcproc/papers/82.pdf. Accessed January 21, 2011.

21. Jimenez, E.; Abderrafi, K.; Martınez-Pastor, J.; Abargues, R.; Valdes, J. L.; Ibanez, R. A novel method of nanocrystal fabrication based on laser ablation in liquid environment. *Superlattices and Microstructures*, 2008, *43* (5–6), 487–493.

22. Liu, P.; Wang, C. X.; Chen, X. Y.; Yang, G. W. Controllable fabrication and cathodoluminescence performance of high-index facets GEO_2 micro- and nanocubes and spindles upon electrical-field-assisted laser ablation in liquid. *Journal of Physical Chemistry C*, 2008, *112*, 13450–13456.

23. Knez, M.; Nielsch, K.; Niinistö, L. Synthesis and surface engineering of complex nanostructures by atomic layer deposition. *Advanced Materials*, 2007, *19*, 3425–3438.

24. Avasthi, D. K.; Pivin, J. C. Ion beam for synthesis and modification of nanostructures. *Current Science*, 2010, *98* (6), 780–792.

25. Chen, S. Y.; Miao, H. Y.; Lue; J. T.; Ouyang, M. S. Fabrication and field emission property studies of multiwall carbon nanotubes. *Journal of Physics D: Applied Physics*, 2004, *37*, 273–279.

26. Jasek, O.; Synek, P.; Zajıckova, L.; Elias, M.; Kudrle, V. Synthesis of carbon nanostructures by plasma enhanced chemical vapour deposition at atmospheric pressure. *Journal of Electrical Engineering*, 2010, *61* (5), 311–313.

27. Nasibulin, A. G.; Richard, O.; Kauppinen, E. I.; Brown, D. P.; Jokiniemi, J. K.; Altman, A. S. Nanoparticle synthesis by copper (II) acetylacetonate vapor decomposition in the presence of oxygen. *Aerosol Science and Technology*, 2002, *36*, 899–911.

28. Wegner, K.; Barborini, E.; Piseri, P.; Milani, P. Gas-phase synthesis of nanostructured particulate films. *KONA*, 2006, *24*, 54–69.

29. Srivastava, Y.; Marquez, M.; Thorsen, T. Multijet electrospinning of conducting nanofibers from microfluidic manifolds. *Journal of Applied Polymer Science*, 2007, *106*, 3171–3178.

30. Huang, C.-J.; Wang, Y.-H.; Chiu, P.-H.; Shih, M.-C.; Meen, T.-H. Electrochemical synthesis of gold nanocubes. *Materials Letters*, 2006, *60*, 1896–1900.

31. Giguere, R. A. In: *Organic Synthesis: Theory and Application*. Hudlicky, T. (Ed.). Vol. 1, JAI Press Inc., London, U.K., 1989, pp. 103–172.

32. Roussy, G.; Pearce, J. A. *Foundations and Industrial Applications of Microwave and Radio Frequency Fields*. John Wiley & Sons, Chichester, U.K., 1995.

33. Ahluwulia, V. K. *Alternate Energy Processes in Chemical Synthesis: Microwave, Ultrasonic, and Photo Activation*. Alpha Science Int'l Ltd, Abingdon, U.K., 2007, 280 pp.

34. Oliver Kappe, C.; Dallinger, D.; Murphree, S. *Practical Microwave Synthesis for Organic Chemists: Strategies, Instruments, and Protocols*. Wiley-VCH, Weinheim, Germany, 2009, 310 pp.

35. Leadbeater, N. E. *Microwave Heating as a Tool for Sustainable Chemistry*. CRC Press, London, U.K., 2010, 282 pp.

36. Martínez-Palou, R. Ionic liquid and microwave-assisted organic synthesis: A "green" and synergic couple. *Journal of the Mexican Chemical Society*, 2007, *51* (4), 252–264.

37. Oliver Kappe, C. Controlled microwave heating in modern organic synthesis. *Angewandte Chemie International Edition*, 2004, *43*, 6250–6284.

38. Besson, T.; Thiery, V.; Dubac, J. Microwave-assisted reactions on graphite. In *Microwaves in Organic Synthesis*, 2nd edn. Andre, L. (Ed.). Wiley-VCH, Weinheim, Germany, 2006, Vol. 1, pp. 416–455. http://onlinelibrary.wiley.com/book/10.1002/9783527619559

39. Shi, S.; Hwang, J.-Y. Microwave-assisted wet chemical synthesis: Advantages, significance, and steps to industrialization. *Journal of Minerals and Materials Characterization and Engineering*, 2003, *2* (2), 101–110.

40. Kubrakova, I. V. Effect of microwave radiation on physicochemical processes in solutions and heterogeneous systems: Applications in analytical chemistry. *Journal of Analytical Chemistry*, 2000, *55* (12), 1113–1122. Translated from *Zhurnal Analiticheskoi Khimii*, 2000, *55* (12), 1239–1249.

41. Leyva, A. G.; Stoliar, P.; Rosenbusch, M.; Lorenzo, V.; Levy, P.; Albonetti, C.; Cavallini, M. et al. Microwave assisted synthesis of manganese mixed oxide nanostructures using plastic templates. *Journal of Solid State Chemistry*, 2004, *177*, 3949–3953.

42. Matsumura-Inoue, T.; Tanabe, M.; Minami, T.; Ohashi, T. A remarkably rapid synthesis of ruthenium(II) polypyridine complexes by microwave irradiation. *Chemistry Letters*, 1994, *23*, 2443–2446.

43. Wiesbrock, F.; Hoogenboom, R.; Schubert, U. S. Microwave-assisted polymer synthesis: State-of-the-art and future perspectives. *Macromolecular Rapid Communications*, 2004, *25*, 1739–1764.

44. Hayes, B. L. Recent advances in microwave-assisted synthesis. *AldrichChimica Acta*, 2004, *37* (2), 66–76.

45. Abedini, R.; Mousavi, S. M. Preparation and enhancing of materials using ultrasound technique: Polymers, catalysts and nanostructure particles. *Petroleum and Coal*, 2010, *52* (2), 81–98.

46. Hsu, H.-Y.; El-Sayed, M.; Eustis, S. Photochemical synthesis of gold nanoparticles with interesting shapes. *NNIN REU Research Accomplishments*, 2004, 68–69. http://www.nnin.org/doc/2004NNINreuHsu.pdf

47. Yiamsawas, D.; Boonpavanitchakul, K.; Kangwansupamonkon, W. Preparation of ZnO nanostructures by solvothermal method. *Journal of Microscopy Society of Thailand*, 2009, *23* (1), 75–78.

48. Ma, J.; Liu, X.; Lian, J.; Duan, X.; Zheng, W. Ionothermal synthesis of BiOCl nanostructures via a long-chain ionic liquid precursor route. *Crystal Growth and Design*, 2010, *10* (6), 2522–2527.

49. Cottam, B. F.; Krishnadasan, S.; deMello, A. J.; deMello, J. C.; Shaffer, M. S. P. Accelerated synthesis of titanium oxide nanostructures using microfluidic chips. *Lab on a Chip*, 2007, *7*, 167–169.

50. Eastoe, J.; Hollamby, M. J.; Hudson, L. Recent advances in nanoparticle synthesis with reversed micelles. *Advances in Colloid and Interface Science*, 2006, *128–130*, 5–15.

51. Meziani, M. J.; Pathak, P.; Beacham, F.; Allard, L. F.; Sun, Y.-P. Nanoparticle formation in rapid expansion of water-in-supercritical carbon dioxide microemulsion into liquid solution. *Journal of Supercritical Fluids*, 2005, *34*, 91–97.

52. Yu, T.; Zhu, Y.; Xu, X.; Yeong, K.-S.; Shen, Z.; Chen, P.; Lim, C.-T.; Thong, J. T.-L.; Sow, C.-H. Substrate-friendly synthesis of metal oxide nanostructures using a hotplate. *Small*, 2006, *2* (1), 80–84.

53. Xu, J.; Zhao, T.; Liang, Z.; Zhu, L. Facile preparation of AuPt alloy nanoparticles from organometallic complex precursor. *Chemistry of Materials*, 2008, *20*, 1688–1690.

54. Toshima, N.; Yonezawa, T. Bimetallic nanoparticles—Novel materials for chemical and physical applications. *New Journal of Chemistry*, 1998, *22*, 1179–1201.

55. Malkov, A. A.; Malygin, A. A.; Lee, C.-T. Temperature factor in the synthesis of nanostructures on the surface of solid by the molecular layering method. *Journal of Industrial and Engineering Chemistry*, 2006, *12* (5), 739–744.

56. Jiao, L.; Fan, B.; Xian, X.; Wu, Z.; Zhang, J.; Liu, Z. Creation of nanostructures with poly(methyl methacrylate)-mediated nanotransfer printing. *Journal of the American Chemical Society*, 2008, *130*, 12612–12613.

57. Agrawal, S.; Srivastava, S.; Kumar, S.; Sharma, S. S.; Tripathi, B.; Singh, M.; Vijay, Y. K. Swift heavy ion irradiation effect on Cu-doped CdS nanocrystals embedded in PMMA. *Bulletin of Materials Science*, 2009, *32* (6), 569–573.

58. Meldrum, F. C.; Wade, V. J.; Nimmo, D. L.; Heywood, B. R.; Mann, S. Synthesis of inorganic nonophase materials in supramolecular protein cages. *Nature*, 1991, *349*, 684–687.

59. Granier, T.; Gallois, B.; Dautant, A.; Langlois d'Estaintot, B.; Precigoux, G. Comparison of the structures of the cubic and tetragonal forms of horse-spleen apoferritin. *Acta Crystallographica Section D: Biological Crystallography*, 1997, *53*, 580–587.

60. Polanams, J; Ray, A. D.; Watt, R. K. Nanophase iron phosphate, iron arsenate, iron vanadate and iron molybdate minerals synthesized within the protein cage of ferritin. *Inorganic Chemistry*, 2005, *44*, 3203–3209.

61. Kim, J.; Choi, S. H.; Lillehei, P. T.; Chu, S.; King, G. C.; Watt, G. D. Cobalt oxide hollow nanoparticles derived by bio-templating. *Chemical Communications*, 2005, 4101–4103.

62. Sadowski, Z.; Maliszewska, I. H.; Grochowalska, B.; Polowczyk, I.; Koźlecki, T. Synthesis of silver nanoparticles using microorganisms. *Materials Science-Poland*, 2008, *26* (2), 419–424.
63. Natarajan, K.; Selvaraj, S.; Murty, V. R. Microbial production of silver nanoparticles. *Digest Journal of Nanomaterials and Biostructures*, 2010, *5* (1), 135–140.
64. Mohanraj, V. J.; Y Chen, Y. Nanoparticles—A review. *Tropical Journal of Pharmaceutical Research*, 2006, *5* (1), 561–573.
65. Tirrell, M.; Requicha, A.; Hagnauer, G. Chapter 4: Synthesis, assembly, and processing of nanostructures. Nanotechnology Research Directions: IWGN Workshop Report, 1999, National Science and Technology Council, Washington, DC, http://www.wtec.org/loyola/nano/IWGN.Research.Directions/chapter04.pdf.
66. Mao, Y.; Park, T.-J.; Zhang, F.; Zhou, H.; Wong, S. S. Environmentally friendly methodologies of nanostructure synthesis. *Small*, 2007, *3* (7), 1122–1139.

2 Brief Description of Some Classic Nanostructures

2.1 CARBON-BASED NANOSTRUCTURES

2.1.1 Carbon Nanotubes

It is little known[1] that as far back as in 1952 Radushkevich and Lukianovich published clear images (Figure 2.1) of 50 nm diameter tubes made of carbon in the *Russian Journal of Physical Chemistry.*[2] Later results, obtained by Oberlin et al.,[3] clearly showed hollow carbon fibers with nanometer-scale diameters using a vapor-growth technique. Additionally, the authors showed a TEM image of a nanotube consisting of a single wall of graphene. Later, Dresselhaus[4] referred to this image as a single-walled nanotube (SWNT). Results of chemical and structural characterization of carbon nanoparticles produced by a thermocatalytical disproportionation of carbon monoxide were reported in Ref. [5]. Since the formal discovery of carbon nanotubes (CNTs) in 1991,[6] a number of device applications, such as full-color displays, field-effect transistors, and molecular computers, have been envisioned.[7,8] These applications are highly dependent on the electronic properties of the CNTs, which can be tuned by their helicity, diameter, and defects presence. In recent years, novel strategies have been devoted to modify physical properties of the CNTs by surface modification with organic, inorganic, and biological species.[9–14] During the last 20 years, the number of reports on carbon nanotubes, one of the hottest topics in nanotechnology, corresponds to hundreds of thousands.

Structural studies of carbon nanotubes continue being important due to a variety of applications for these materials. It is known that their geometry (Figure 2.2) and a major part of their properties depend on diameter and chiral angle (θ) (Figure 2.3), also known as helicity. These two parameters are completely determined by two Hamada indexes[15] (n, m). There are two types of nanotubes (Figure 2.4):

- SWNTs
- Multiwalled nanotubes (MWNTs)

A carbon nanotube may be viewed as a graphite sheet that is rolled up into a nanometer-scale tubular from (i.e., SWNT) or with additional graphene tubes that form around the core of an SWNT (i.e., MWNT).[17] Since the graphene sheet can be rolled up with varying degrees of twist along its length, carbon nanotubes can have a variety of chiral structures. Depending on their diameter and the helicity of the arrangement of graphitic rings in the walls, carbon nanotubes have been demonstrated to possess unusual electronic, photonic, magnetic, thermal, and mechanical properties.[18] Over the years, basic research and applications exploration have been greatly accelerated by the synthesis techniques of high-quality nanotube materials. Synthesis methods[19] for the formation of carbon nanotubes include arc discharge,[20] laser ablation,[21] microwave heating (Figure 2.5),[22] gas-phase catalytic growth from carbon monoxide,[23] and chemical vapor deposition from hydrocarbons,[24] among other techniques. Silicon crystals, quartz glass, porous silicon dioxide, and aluminum oxide were used as substrates for the growth of carbon nanotubes. Carbon nanotubes collected from these substrates were used to make carbon nanotube composites,[25] for gas[26] and electrochemical energy[27] storage, and as templates for making nanotubes of other materials,[28] among many other uses, the most intriguing of which, in our opinion, was the creation of the smallest incandescent lamp

FIGURE 2.1 First carbon nanotubes, discovered in 1952.

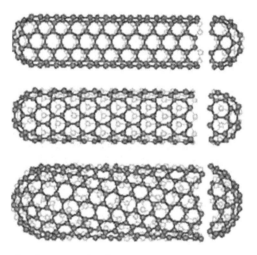

FIGURE 2.2 Classification of carbon nanotubes from (top to bottom): chair, zigzag, and helicoidal or chiral.

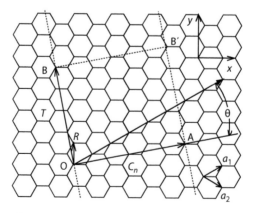

FIGURE 2.3 Unrolled nanotube.

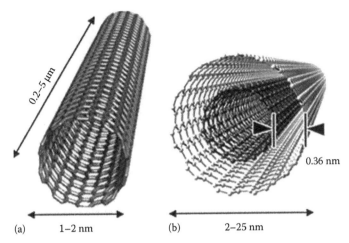

FIGURE 2.4 Conceptual diagram of single-walled carbon nanotube (SWCNT) (a) and multiwalled carbon nanotube (MWCNT) (b) delivery systems showing typical dimensions of length, width, and separation distance between graphene layers in MWCNTs. (Reproduced from *Physica B*, 323(1–4), Iijima, S., Carbon nanotubes: Past, present, and future, 1–5, Copyright 2002, with permission from Elsevier.)

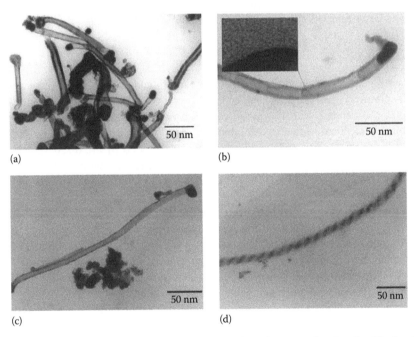

FIGURE 2.5 (a) TEM images of Fe-filled CNTs grown by microwave heating for 20 min, (b) straight nanotube with bamboo compartments grown by microwave heating for 20 min, (c) straight nanotube grown from microwave heating method for 30 min, (d) helical nanotubes grown at the nanorope (30 min heating). (Reproduced from Kharissova, O.V. et al., *Mech. Adv. Mater. Struct.*, 16(1), 63. Copyright 2009 Taylor & Francis.)

(1.4 × 13 nm).[29] A completed device (Figure 2.6) consisted of an arc-discharge grown, multiwalled carbon nanotube on a 12 nm thick Si_3N_4 membrane that spans a hole in a 2 mm × 2 mm × 0.2 mm silicon chip. The nanotube is contacted with 80 nm of Au over 50 nm of Pd using electron beam (E-beam) lithography. When the current passes through the CNT, it is heated and starts to emit photons, whose light can be seen directly by unaided eyes. The researchers constructed an unusual equipment to study a "transition state" between thermodynamical (determining laws for enormous

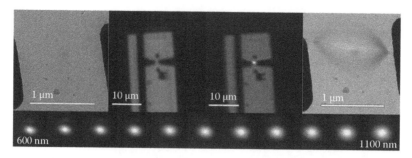

FIGURE 2.6 The carbon nanotube lamp. The four upper images from left to right are "before" (TEM), "backlight only" (optical), "backlight with nanotube" (optical), and "after" (TEM). The 4 µm × 4 µm lower images are 4.7 V bias "nanotube only," taken with bandpass filters centered at wavelengths 600–1100 nm (50 nm increments). The lamp is visible to the unaided eye. Higher resolution TEM images (not shown) indicate that the 1.4 µm long nanotube has 11 walls and an outer diameter of 13 nm. (Reproduced with permission from Fan, Y. et al., Probing Planck's law with incandescent light emission from a single carbon nanotube, *Phys. Rev. Lett.*, 102, 187402, 2009. Copyright 2009 by the American Physical Society.)

number of particles) and quantum-mechanical (operating separated elemental particles) systems. According to its size, the lamp made of one nanotube does not yet correspond to a thermodynamical system, but at the same time it is not a quantum-mechanical object. Currently, there are no relevant laws in physics describing the behavior of such objects.

2.1.2 FULLERENES

The fullerenes[30–32] represent an allotropic form of carbon, additionally to graphite and diamond, having high-symmetry geometry and exceptional properties. The smallest (without taking into account unstable fullerenes with the number of carbon atoms less than 60) fullerene C_{60} possesses a geometry identical to that of a football. It was discovered[33] by Harold Kroto, Richard E. Smalley, and Robert F. Curl in 1985. Fullerenes can be prepared by arc discharge using graphite electrodes in a noble gas atmosphere or irradiating a graphite surface with laser. Reaction products (C_{60}, C_{70}, C_{76}, C_{78}, C_{82}, C_{84}, etc., Figure 2.7) can be extracted with organic solvents and separated chromatographically.

The structure of a buckminsterfullerene C_{60} (Figure 2.8) is a truncated icosahedron made of 20 hexagons and 12 pentagons, with a carbon atom at the vertices of each polygon and a bond along each polygon edge. The van der Waals diameter of a C_{60} molecule is about 1 nm. The nucleus to nucleus diameter of a C_{60} molecule is about 0.71 nm. The C_{60} molecule has two bond lengths. The 6:6 ring bonds (between two hexagons) can be considered "double bonds" and are shorter than the 6:5 bonds (between a hexagon and a pentagon). Its average bond length is 1.4 Å.

The fullerenes are soluble in certain organic solvents and insoluble in polar solvents or those containing hydrogen bonds. Their chemistry is very rich[34–39]: they can add hydrogen, halogens, oxygen, metals, and radicals, can be polymerized, can form complexes with transition metals, or can take

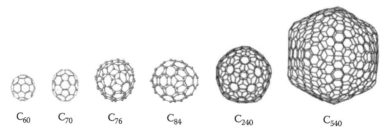

C_{60} C_{70} C_{76} C_{84} C_{240} C_{540}

FIGURE 2.7 Some fullerenes.

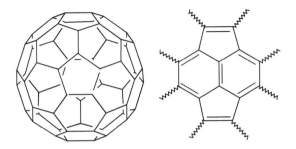

FIGURE 2.8 C$_{60}$ structure.

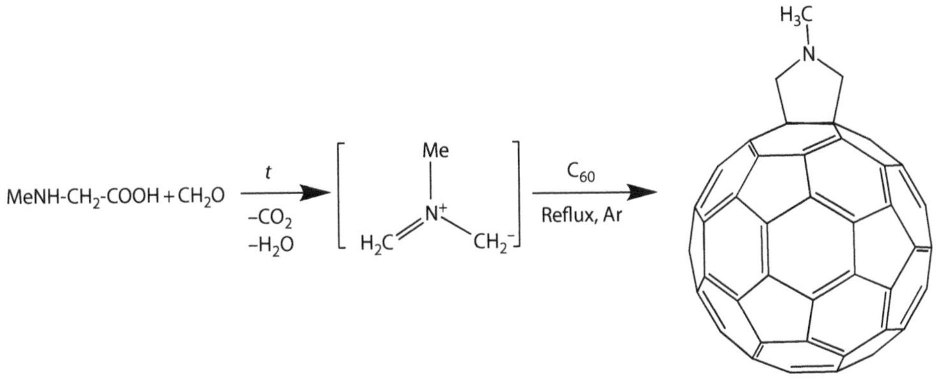

MeNH-CH$_2$-COOH + CH$_2$O $\xrightarrow[\substack{-CO_2 \\ -H_2O}]{t}$ $\left[\substack{Me \\ | \\ H_2C= \overset{+}{N} \diagdown CH_2^-} \right]$ $\xrightarrow[\text{Reflux, Ar}]{C_{60}}$

FIGURE 2.9 Preparation of *N*-methylfulleropyrrolidine.

part in electron-transfer reactions. As an example, the pyrrolidino[3′,4′:1,2][60] fullerenes, more common as fulleropyrrolidines[40,41] (Figures 2.9 and 2.10), have many applications in medicine, solar cells, materials chemistry,[42] etc. (see also Chapter 19).

2.1.3 NANODIAMONDS*

Nanodiamond (ND, Figure 2.11) is a new member of nanocarbons, which consist of nanosized tetrahedral networks. The history of the discovery of ND is unique, since ND synthesis was accidentally discovered in the USSR three times over 19 years starting from 1963 by shock compression of nondiamond carbon modifications in blast chambers.[43] The term *nanodiamond* is currently used broadly for a variety of diamond-based materials at the nanoscale ranging from single diamond clusters to bulk nanocrystalline films.[44] It is generally accepted that *nanocrystalline diamond* (NCD) consists of facets less than 100 nm in size, whereas a second term *ultrananocrystalline diamond* (UNCD) has been coined to describe material with grain sizes less than 10 nm. These differences in morphology originate in the growth process.[45]

Nanometer-sized diamond has been found in meteorites, proto-planetary nebulae, and interstellar dusts, as well as in residues of detonation and in diamond films. It is known that primitive chondritic meteorites contain up to approximately 1500 ppm of nanometer-sized diamonds, containing isotopically anomalous noble gases, nitrogen, hydrogen, and other elements. These isotopic anomalies indicate that meteoritic NDs probably formed outside our solar system prior to the Sun's formation (they are thus presolar grains).[46] It appears that some interstellar emission bands in the 3–4, 7–10, and approximately 21 µm spectral regions could originate from NDs.[47]

Diamond thin films have outstanding optical, electrical, mechanical, and thermal properties, which make these attractive for applications in a variety of current and future systems. In particular,

---

* See Chapter 18 for more information.

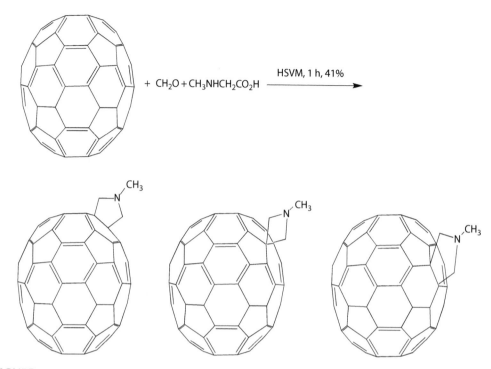

FIGURE 2.10   Fulleropyrrolidines de C_{70}.

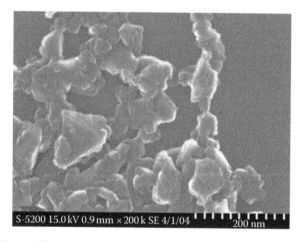

FIGURE 2.11   Nanodiamond image.

the wide band gap and optical transparency of diamond thin films make them an ideal semiconductor for applications in current and future electronics. Various aspects of the NDs have been recently emphasized in a series of reviews[48–54] and books,[55,56] in particular such topics as ND preparation by chemical vapor deposition (CVD), NCD film formation from hydrogen-deficient and hydrogen-rich plasma, nanocomposite films, mechanical behavior of NCD films, field emission characteristics,[57] diamond nanowires and their biofunctionalization of nucleic acid molecules,[58] or use of NCD to study noncovalent interaction.[59]

FIGURE 2.12 Graphene.

2.1.4 GRAPHENE* AND GRAPHANE

A new allotropic form of carbon, 2D graphene (Figure 2.12) (nanographenes are also known as polycyclic aromatic hydrocarbons [PAHs]), discovered in 2004, is a rapidly rising star on the horizon of materials science and condensed-matter physics. This strictly 2D material exhibits exceptionally high crystal and electronic quality, and, despite its short history, has already revealed a cornucopia of new physics and potential applications, which are briefly discussed here. Graphene represents a conceptually new class of materials that are only one atom thick (it is just one layer of carbon atoms,[60] a similar structure to graphite, but is a single isolated sheet of carbon), and, on this basis, offers new inroads into low-dimensional physics, which has never ceased to surprise and continues to provide a fertile ground for applications. Strictly 2D crystals, such as planar graphene, have for a long time been considered as a thermodynamically unstable form with respect to the formation of curved structures such as fullerenes or nanotubes. Geim and Novoselov described[61] that small size (≪1 mm) and strong interatomic bonds ensure that thermal fluctuations cannot lead to the generation of dislocations or other crystal defects even at elevated temperature. The extracted 2D crystals become intrinsically stable by gentle crumpling in the third dimension; such 3D warping (observed on a lateral scale of ≈10 nm) leads to a gain in elastic energy but suppresses thermal vibrations (anomalously large in 2D), which above a certain temperature can minimize the total free energy.

Despite the fact that graphene was discovered recently, its structure, properties, and applications have already been generalized in a series of monographs[62–65] and reviews[66–69] (among which we note an excellent work of Mullen on graphenes as potential material for electronics[70]). Various patents are dedicated to obtaining graphenes.[71–73] Among experimental works, we note a number of investigations of Novoselov[74–78] who precisely discovered this material. At present, the graphene area (together with carbon nanotubes) is one of the hottest topics in physics and nanotechnology.[79] In particular, the recently fabricated fluorographene is close to Teflon, and it is inert and stable up to 400°C.[80]

In addition, it was calculated that it could be possible to obtain an excellent dielectric, adding a hydrogen atom to each carbon atom in the graphene cell (Figure 2.13). An electric impulse was applied to graphene in hydrogen flow.[81] The formed composite named graphane can also serve for prolonged hydrogen storage.

---

* See Chapter 17 for more information.

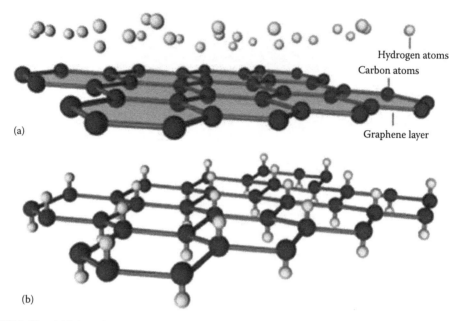

(a)

Hydrogen atoms
Carbon atoms

Graphene layer

(b)

FIGURE 2.13 Addition of hydrogen atoms to graphene (a), resulting in graphane (b).

2.2 CONVENTIONAL NONCARBON NANOSTRUCTURES

2.2.1 SIMPLE AND CORE-SHELL* NANOPARTICLES

The term "nanoparticle" or "nanosize particle" entered scientific language 20 years ago. However, the nanosize criterion is still the object of many discussions for researchers. According to the JUPAC convention, the maximum size limit of nanoparticles corresponds to 100 nm, although this magnitude is very conditional and formal. Two main types of nanoparticles are distinguished: (1) the *nanoclusters* or *nanocrystals* and (2) the *nanoparticles* themselves. The first type includes particles with ordered structure (frequently centrosymmetrical) and size 1–5 nm containing up to 1000 atoms. The second type corresponds to nanoparticles with a size of 5–100 nm consisting of 10^3–10^8 atoms. Several nanostructures (e.g., nanowires) can exceed these limits in their length, but their properties in a definite direction remain nanocrystalline. Additionally, nanoparticles can be divided into zero- (0D) (clusters, fullerenes, quantum dots), mono- (1D) (nanowires, whiskers, nanotubes, nanobelts), bi (2D) (thin films with a thickness of up to hundreds of nanometers, *Langmuir-Blodgett* films, self-assembled monolayers, nanosheets), and tridimensional (3D) (nanoparticles themselves, core-shell nanoparticles, self-assembled nanoobjects), respectively (Figures 2.14 through 2.20).

Bimetallic nanoparticles are of great interest because of the modification of properties observed not only due to size effects but also as a result of the combination of different metals,[89,90] either as an alloy or as a core-shell structure,[91] modifying the catalytic, electronic, and optical properties of the monometallic nanoparticles.[92,93] One of these characteristics is the optical properties due to the resonance of surface plasmons with visible light at well-defined frequencies.[94] The specific resonance frequency depends on a number of parameters, such as nanoparticle composition,[95,96] morphology,[97,98] concentration,[99] solvent refractive index,[100] surface charge,[101] and temperature.[102] These effects are observed only in the visible range for a few metals, among which gold and silver have been studied most often, mainly because of their chemical stability.

---

* See also Chapter 24.

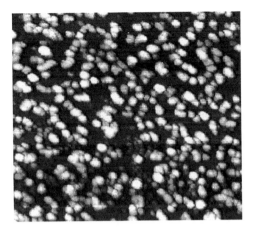

FIGURE 2.14 Ge nanocluster. (Reproduced with permission from Li, A.P. et al., *Phys. Rev. B*, 69, 245310, 2004. Copyright 2004 by the American Physical Society.)

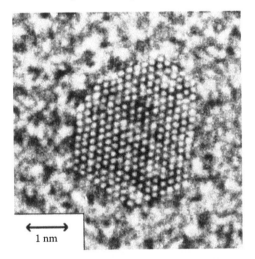

FIGURE 2.15 Cobalt cluster. (Reproduced from Jamet, M. et al., *Phys. Rev. Lett.*, 86(20), 4676–4679, 2001. Copyright 2001 by the American Physical Society.)

FIGURE 2.16 Cu_2O nanowhickers. (Reproduced from *Mater. Lett.*, 62, Qu, Y. et al., Synthesis of Cu_2O nano-whiskers by a novel wet-chemical route, 886–888, Copyright 2008, with permission from Elsevier.)

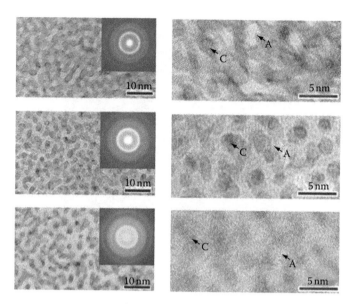

FIGURE 2.17 TEM images of $Co_xPt_yAl_kO_l$ thin films. A and C indicate the amorphous phase and crystalline phase, respectively. (From Park, H.S. et al., *IEEE Trans. Magn.*, 41(10), 3724, © 2005 IEEE; from Jamet, M. et al., Magnetic anisotropy of a single cobalt nanoclusters, *Phys. Rev. Lett.*, 86(20), 4676–4679, © 2001 IEEE.)

FIGURE 2.18 CdS/CdSe nanosheets. (Reproduced with permission from Kim, Y.L. et al., *Nanotechnology*, 20, 095605, 7 pp. Copyright 2009 IOP Science.)

When more than one metal atom type is present in the composition of the nanoparticles, the actual distribution of the metal atoms within the particle is of prime importance for correct interpretation of the optical properties.

The case of gold and silver (Figure 2.21) is particularly difficult to control, because they can form perfectly miscible alloys over the whole composition range, and their crystalline lattice constants are extremely similar (4.078 Å for Au; 4.086 Å for Ag), so there is still a debate as to whether or not pure core-shell geometries can be obtained. For this purpose, many colloidal methods of synthesis have been approached to obtain bimetallic nanoparticles,[103–105] such as homogeneous reduction in aqueous solutions[106] or phase transfer reactions,[107] with sodium citrate, hydrazine, NaBH$_4$, and ethylene glycol (EG) as reducing agents, each of them yielding products with different physicochemical

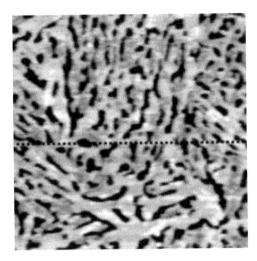

FIGURE 2.19 *Langmuir-Blodgett* films of cadmium arachidate. (Reproduced from *Chem. Phys. Lett.*, 405, Kundu, S. et al., Manipulating headgroups in Langmuir–Blodgett films through subphase pH variation, 282–287, Copyright 2005, with permission from Elsevier.)

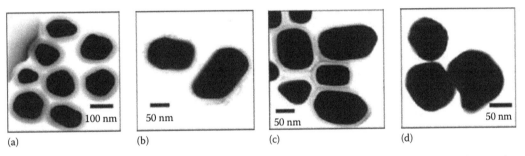

(a) (b) (c) (d)

FIGURE 2.20 TEM images of core-shell Ag@SiO$_2$ nanoparticles. Panels (a) through (d) show the samples with different thickness of the SiO$_2$ coating at 35, 15, 11, and 2 nm (±1 nm), respectively. The diameter of the Ag is 130 ± 10 nm for all the samples. (Reproduced with permission from Aslan, K. et al., Fluorescent core-shell Ag@SiO$_2$ nanocomposites for metal-enhanced fluorescence and single nanoparticle sensing platforms, *J. Am. Chem. Soc.*, 129, 1524–1525. Copyright 2007 American Chemical Society.)

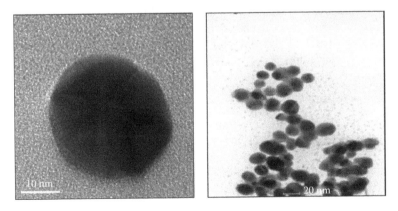

FIGURE 2.21 Bright field transmission electron micrographs of silver-coated gold (Au@Ag) nanoparticles obtained through silver reduction in the presence of 17 nm Au nanoparticles.

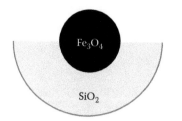 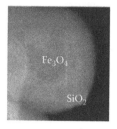

FIGURE 2.22 Scheme of typical core-shell nanoparticles.

and structural characteristics.[108] Among these, the polyol method has been reported to produce the core-shell type of nanoparticles as the final product,[109,110] easily changing composition and surface modifiers. This technique does not require an additional reducing agent because the solvent by itself reduces the metallic species.[111]

Obviously, core-shell nanoparticles can be formed not only by metals but also by oxides (Figure 2.22) and a series of other compounds.

2.2.2 Nanometals*

The area of nanometals as a part of the topic "nanomaterials"[112] is nowadays being rapidly developed. During the last four decades, great efforts have been dedicated to the preparation and use of activated zero-valent, more exactly, elemental metals in organic and organometallic reactions. Unfortunately, without stabilization the formed activated metals fuse together, losing their special shape and properties. A major part of them is not sufficiently active due to a series of factors: high particle size (up to 359 mesh) and oxide and other films on their surface. However, these factors can be partially or completely eliminated, if metals are subjected to diverse activation techniques discussed here. The necessity for metal activation is caused by a host of their applications in a series of chemical processes, related to catalysis, organic and organometallic synthesis, etc. In comparison with classic methods using metal salts or carbonyls, application of metals in the elemental form frequently leads to unique compounds, which cannot be obtained by traditional routes.

During the last two decades, the use of zero-valent activated and nonactivated metals as precursors in organic or organometallic processes has been described in detail in some monographs[113-118] and reviews.[119-121] Activated elemental metals in nanoforms can be divided into active metal *nanopowders*,[122] *nanowires*,[123,124] *nanolines*,[125] *nanodumbbells*,[126] *nanoclusters*,[127-129] *nanosheets*,[130] *nanorods*,[131-133] *nanoalloy*,[134,135] *nanobelts*,[136] nanotubes[137] (Figure 2.23), and *nanofilms*,[138] which have a host of applications, in particular as catalysts,[139-142] magnetic materials,[143] nanocomposites,[144,145] chemical sensors,[146-148] degradation of toxic chemicals,[149] or even as possible carriers of isotopes for medical applications.[150]

2.2.3 Gallery of Other Conventional Nanostructures

Other very common nanostructures are formed by a variety of inorganic compounds (metals, oxides, salts, etc.), as well polymers; some representative examples are shown in Figures 2.24 through 2.33. We note that the strategies for achieving 1D growth, contributing to an anisotropic growth of crystals and resulting nanowires (Figures 2.24 and 2.25), nanobelts (Figure 2.28), nanorods (Figure 2.32), and nanotubes, are extensively revised in an excellent review.[151]

---

* See also Chapters 22 and 24 on nanometals and core-shell metal nanoparticles for more detail.

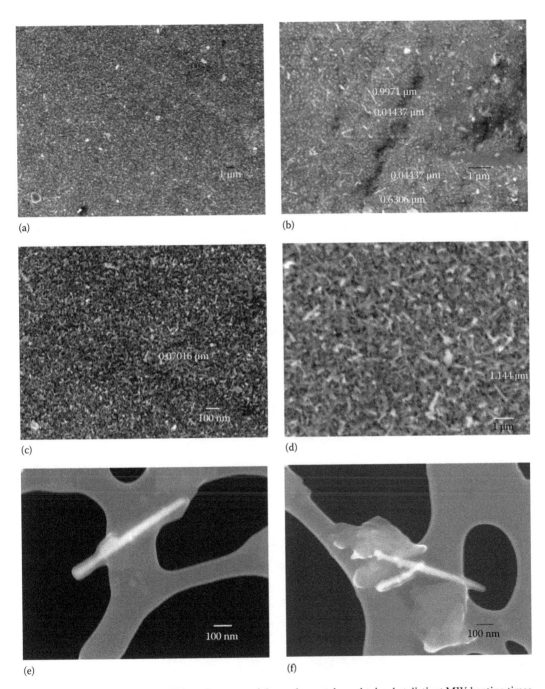

(a)　　　　　　　　　　　　　　(b)

(c)　　　　　　　　　　　　　　(d)

(e)　　　　　　　　　　　　　　(f)

FIGURE 2.23 SEM images of bismuth nanoparticles and nanotubes, obtained at distinct MW-heating times in vacuum. (a) 5 min (×4000), (b) 10 min (×10,000), (c) 10 min (×10,000), (d) 15 min (×80,000), (e) 10 min (×80,000), and (f) 15 min (×70,000). (Reproduced from *Mater. Chem. Phys.*, 121, Kharissova, O.V. et al., A comparison of bismuth nanoforms obtained in vacuum and air by microwave heating of bismuth powder, 489–496. Copyright 2010, with permission from Elsevier.)

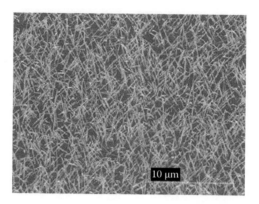

FIGURE 2.24 InSb nanowires grown in a simple closed quartz tube. (Reproduced from *J. Cryst. Growth*, 304, Park, H.D. et al., Growth of high quality, epitaxial InSb nanowires, 399–401. Copyright 2007, with permission from Elsevier.)

FIGURE 2.25 Mn₂O₃ nanowires, prepared by thermal decomposition of MnCO₃. (Reproduced with permission from Wang, H.-z. et al., *Chem. Res. Chin. Univ.*, 26(1), 5. Copyright 2010.)

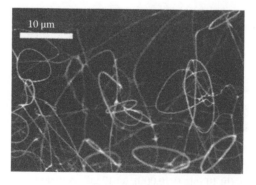

FIGURE 2.26 ZnSe nanorings, synthesized by thermal evaporation of ZnSe powders. (With permission from Leung, Y.P. et al., Synthesis of wurtzite ZnSe nanorings by thermal evaporation. *Appl. Phys. Lett.*, 88, 183110. Copyright 2006, American Institute of Physics.)

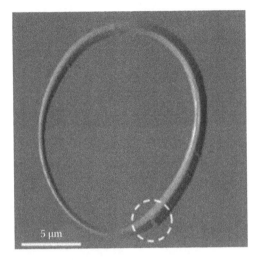

FIGURE 2.27 ZnO nanorings, obtained by physical evaporation of ZnO powder. (Reproduced with permission from Hughes, W.L. and Wang, Z.L., Controlled synthesis and manipulation of ZnO nanorings and nanobows, *Appl. Phys. Lett.*, 86, 043106. Copyright 2005, American Institute of Physics.)

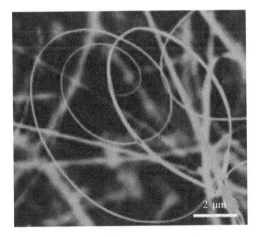

FIGURE 2.28 Spiral of a ZnO nanobelt with increased thickness along the length, synthesized by a solid-state thermal sublimation process. (Wang, Z.L., Self-assembled nanoarchitectures of polar nanobelts/nanowires, *J. Mater. Chem.*, 15, 1021, 2005. Reproduced with the permission of The Royal Society of Chemistry.)

├───┤ 1 mcm   ├───┤ 300 nm

├─┤ 300 nm   ├─┤ 200 nm

FIGURE 2.29 ZnO:CeO$_2$ nanopowders, obtained from Zn(CH$_3$COO)$_2 \cdot$2H$_2$O and cerium nitrate with further calcinations. (Reproduced from *Appl. Surf. Sci.*, 255, Fonseca de Lima, J. et al., ZnO:CeO$_2$-based nanopowders with low catalytic activity as UV absorbers, 9006–9009, Copyright 2009, with permission from Elsevier.)

FIGURE 2.30 Poly(ethylene imine)/poly(vinyl pyrrolidone) nanofibers, obtained *via* electrospinning. (Reproduced from *Polymer*, 50, Patel, P.A. et al., Rapid synthesis of polymer-silica hybrid nanofibers by biomimetic mineralization, 1214–1222. Copyright 2009, with permission from Elsevier.)

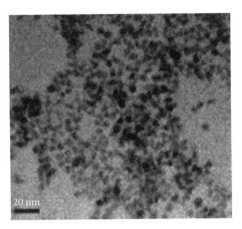

FIGURE 2.31 TEM images of CdTeSe quantum dots, prepared using chemical aerosol flow synthesis. (Reproduced with permission from Bang, J.H. et al., Quantum dots from chemical aerosol flow synthesis: Preparation, characterization, and cellular imaging, *Chem. Mater.*, 20, 4033–4038. Copyright 2008 American Chemical Society.)

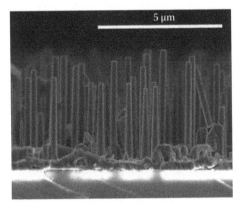

FIGURE 2.32 Zn-doped InN nanorods, fabricated by MOCVD method. (Reproduced with permission from Song, H. et al., Well-aligned Zn-doped InN nanorods grown by metal-organic chemical vapor deposition and the dopant distribution, *Cryst. Growth Des.*, 9(7), 3292–3295. Copyright 2009 American Chemical Society.)

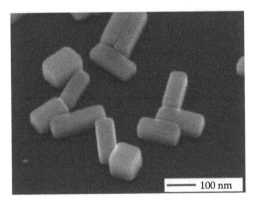

FIGURE 2.33 Ag nanobars,* obtained by polyol synthesis. (Reproduced with permission from Wiley, B.J. et al., Synthesis and optical properties of silver nanobars and nanorice, *Nano Lett.*, 7(4), 1032–1036. Copyright 2007 American Chemical Society.)

* See more details in Section 3.8.

REFERENCES

1. Monthioux, M.; Kuznetsov, V. L. Who should be given the credit for the discovery of carbon nanotubes? *Carbon*, 2006, *44*, 1621–1625.
2. Radushkevich, L. V.; Lukianovich, V. M. About carbon structure, formed by thermal decomposition of carbon monoxide at iron contact. *Zhurn. Fiz. Khim.*, 1952, *XXVI* (1), 88–95. http://carbon.phys.msu.ru/publications/1952-radushkevich-lukyanovich.pdf (dead link).
3. Oberlin, A.; Endo, M.; Koyama, T. Filamentous growth of carbon through benzene decomposition. *Journal of Crystal Growth*, 1976, *32* (3), 335–349.
4. Dresselhaus, M. Structure, properties and applications of carbon nanotubes. Nanostructures, Seminar Series at MIT, 2003. http://web.mit.edu/tinytech/Nanostructures/Spring2003/MDresselhaus/MDresselhaus.htm
5. *Izv. Akad, Nauk SSSR, Metals*, 1982, (3), 12–17. Cited in http://en.wikipedia.org/wiki/Carbon nanotube
6. Iijima, S. Helical microtubes of graphitic carbon. *Nature (London)*, 1991, *354*, 56–58.
7. Rueckes, T.; Kim, K.; Joselevich, E.; Tseng, G. Y.; Cheung, C. L.; Lieber, C. M. Carbon nanotube-based nonvolatile random access memory for molecular computing. *Science*, 2000, *289*, 94–97.
8. Saito, S. Carbon nanotubes for next-generation electronics devices. *Science*, 1997, *278*, 77–78.
9. Yang, W.; Thordarson, P.; Gooding, J. J.; Ringer, S. P.; Braet, F. Carbon nanotubes for biological and biomedical applications. *Nanotechnology*, 2007, *18* (41), 412001.
10. Wei, H. F.; Hsiue, G. H.; Liu, C.Y. Surface modification of multi-walled carbon nanotubes by a sol–gel reaction to increase their compatibility with PMMA resin. *Composites Science and Technology*, 2007, *67* (6), 1018–1026.
11. Rivas, G. A.; Rubianes, M. D.; Rodriguez, M. C.; Ferreyra, N. F.; Luque, G. L.; Pedano, M. L.; Miscoria, S.A.; Parrado, C. Carbon nanotubes for electrochemical biosensing. *Talanta*, 2007, *74* (3), 291–307.
12. Tran, M. Q.; Tridech, C.; Alfrey, A.; Bismarck, A.; Shaffer, M. S. P. Thermal oxidative cutting of multi-walled carbon nanotubes. *Carbon*, 2007, *45* (12), 2341–2350.
13. Daniel, S.; Rao, T. P.; Rao, K. S.; Rani, S. U.; Naidu, G. R. K.; Lee, H. Y.; Kawai, T. A. Review of DNA functionalized/grafted carbon nanotubes and their characterization. *Sensors and Actuators: B. Chemical*, 2007, *122* (2), 672–682.
14. Shanmugharaj, A. M.; Bae, J. H.; Lee, K. Y.; Noh, W. H.; Lee, S. H.; Ryu, S. H. Physical and chemical characteristics of multiwalled carbon nanotubes functionalized with aminosilane and its influence on the properties of natural rubber composites. *Composites Science and Technology*, 2007, *67* (9), 1813–1822.
15. Hamada, N.; Sawada, S.; Oshiyama, A. New one-dimensional conductors—Graphitic microtubules. *Physical Review Letters*, 1992, *68* (10), 1579–1581.
16. Iijima, S. Carbon nanotubes: past, present, and future. *Physica B: Condensed Matter*, 2002, *323* (1–4), 1–5.
17. Harris, P. J. F. *Carbon Nanotubes and Related Structures*. Cambridge University Press, Cambridge, U.K., 2001.
18. Saito, R.; Dresslehaus, G.; Dresselhaus, M. S. (Eds.). *Physical Properties of Carbon Nanotubes*. Imperial College Press, London, U.K., 1998.
19. http://en.wikipedia.org/wiki/Carbon_nanotube.
20. Journet, C.; Maser, W. K.; Bernier, P.; Loiseau, A.; de la Chapelle, M. L.; Lefrant, S.; Lee, R.; Fischer, J. E. Large-scale production of single-walled carbon nanotubes by the electric-arc technique. *Nature*, 1997, *388*, 756–758.
21. Rinzler, A. G.; Liu, J.; Dai, H.; Nicolaev, P.; Huffman, C. B.; Rodríguez-Macias, F. J.; Boul, P. J. et al. Large scale purification of single-wall carbon nanotubes: Process, product, and characterization. *Applied Physics A*, 1998, *67*, 29–37.
22. Kharissova, O. V.; Garza Castañón, M.; Hernández Pinero, J. L.; Ortiz Méndez, U.; Kharisov, B. I. Fast production method of Fe-filled carbon nanotubes. *Mechanics of Advanced Materials and Structures*, 2009, *16* (1), 63–68.
23. Nicolaev, P.; Bronikowski, M. J.; Bradley, R. K.; Fohmund, F.; Colbert, D. T.; Smith, K. A.; Smalley, R. E. Gas-phase catalytic growth of single-walled carbon nanotubes from carbon monoxide. *Chemical Physics Letters*, 1999, *313*, 91–97.
24. Li, W. Z.; Xie, S. S.; Qian, L. X.; Chang, B. H.; Zou, B. S.; Zhou, W. Y.; Zhao, R. A.; Wang, G. Large-scale synthesis of aligned carbon nanotubes. *Science*, 1996, *274*, 1701–1703.
25. Qian, D.; Dickey, E. C.; Andrews, R.; Rantell, T. Load transfer and deformation mechanisms in carbon nanotube-polystyrene composites. *Applied Physics Letters*, 2000, *76*, 2868–2870.

26. Chen, P.; Wu, X.; Lin, J.; Tan, K. L. High H_2 uptake by alkali-doped carbon nanotubes under ambient pressure and moderate temperatures. *Science*, 1999, *285*, 91–93.
27. Britto, P. J.; Santhanam, K. S. V.; Rubio, A.; Alonso; J. A.; Ajayan, P. M. Improved charge transfer at carbon nanotube electrodes. *Advanced Materials*, 1999, *11*, 154–57.
28. Han, W.; Redlich, P.; Ernst, F.; Ruehle, M. Synthesizing boron nitride nanotubes filled with SiC nanowires by using carbon nanotubes as templates. *Applied Physics Letters*, 1999, *75* (13), 1875–1877.
29. Fan, Y.; Singer, S. B.; Bergstrom, R.; Regan, B. C. Probing Planck's law with incandescent light emission from a single carbon nanotube. *Physical Review Letters*, 2009, *102*, 187402.
30. http://en.wikipedia.org/wiki/Fullerene.
31. Fowler, P. W.; Manolopoulos, D. E. *An Atlas of Fullerenes*. Dover Publications, Mineola, NY, 2007, 416 pp.
32. Rietmeijer, F. J. M. *Natural Fullerenes and Related Structures of Elemental Carbon*. Springer, Berlin, Germany, 2006, 295 pp.
33. Kroto, H. W.; Heath, J. R.; O'Brien, S. C.; Curl, R. F.; Smalley, R. F. C_{60}-buckminsterfullerene. *Nature*, 1985, *318* (6042), 162–163.
34. Hirsch, A. (Ed.). *Fullerenes and Related Structures*. Springer-Verlag, Berlin, Germany, 1999, 246 pp.
35. Hirsch, A.; Brettreich, M.; Wudl, F. *Fullerenes: Chemistry and Reactions*. Wiley-VCH, Weinheim, Germany, 2005, 440 pp.
36. Sidorov, L. N.; Yurovskaya, M. A.; Borschevskii, A. Y.; Trushkov, I. V.; Ioffe, I. N. *Fullerenes*. Examen, Moscow, Russia, 2005, 688 pp.
37. Langa, F.; Nierengarten, J.-F. (Eds.). *Fullerenes: Principles and Applications*. Royal Society of Chemistry, Cambridge, U.K., 2007, 300 pp.
38. Abrasonis, G.; Amer, M. S.; Blanco, R.; Chen, Z. *Fullerene Research Advances*. Kramer, C. N. (Ed.). Nova Science Pub Inc., New York, 2007, 305 pp.
39. Prassides, K. *Fullerene-Based Materials: Structures and Properties*. Springer, Berlin, Germany, 2004, 294 pp.
40. Maggini, M.; Menna, E. Addition of azomethyne ylides: Fulleropyrrolidines. In: *Fullerenes: From Synthesis to Optoelectronic Properties*. Series: Developments in Fullerene Science, Vol. 4. Guldi, D. M.; Martin, N. (Eds.). Springer, Berlin, Germany, 2003, 447 p.
41. Kharisov, B. I.; Kharissova, O. V.; Jimenez Gomez, M.; Ortiz Mendez, U. Recent advances in the synthesis, characterization, and applications of fulleropyrrolidines. *Industrial and Engineering Chemistry Research*, 2009, *48* (2), 545–571.
42. Margadonna, S. (Ed.). *Fullerene-Related Materials*. Springer, Berlin, Germany, 2008, 700 pp.
43. Danilenko, V. V. On the history of the discovery of nanodiamond synthesis. *Physics of the Solid State*, 2004, *46* (4), 595–599.
44. Shenderova, O. A.; Zhirnov, V. V.; Brenner, D. W. Carbon nanostructures. *Critical Reviews in Solid State and Materials Sciences*, 2002, *27* (3–4), 227–356.
45. Williams, O. A.; Nesladek, M.; Daenen, M.; Michaelson, S.; Hoffman, A.; Osawa, E.; Haenen, K.; Jackman, R. B. Growth, electronic properties and applications of nanodiamond. *Diamond and Related Materials*, 2008, *17* (7–10), 1080–1088.
46. Huss, G. R. Meteoritic nanodiamonds: Messengers from the stars. *Elements*, 2005, *1* (2), 97–100.
47. Jones, A. P.; d'Hendecourt, L. B. Interstellar nanodiamonds. *Astronomical Society of the Pacific Conference Series*, 2004, *309*, 589–601.
48. Khan, Z. H.; Husain, M. Nanodiamond: Synthesis, transport property, field emission and applications. *Material Science Research India*, 2006, *3* (1a), 1–22.
49. Tanaka, A. Tribology of carbon composites. *Toraiborojisuto*, 2009, *54* (1), 16–21.
50. Hu, X.; Li, M.; Sun, Z.; Wang, Q.; Fan, D.; Chen, L. Research status and prospection of synthetic nanodiamonds. *Guanli Gongchengban*, 2009, *31* (2), 301–304, 317.
51. Osawa, E. Nano carbon materials. Fullerenes and diamond. *Seramikkusu*, 2004, *39* (11), 892–909.
52. Quiroz Alfaro, M. A.; Martinez Huitle, U. A.; Martinez Huitle, C. A. Nanodiamonds. *Ingenierias*, 2006, *IX* (33), 37–43.
53. Freitas Jr., R. A. A simple tool for positional diamond mechanosynthesis, and its method of manufacture. http://www.MolecularAssembler.com/Papers/DMSToolbuildProvPat.htm, 2003–2004.
54. Freitas Jr., R.A. How to make a nanodiamond. http://www.kurzweilai.net/articles/art0632.html?printable=1, 2006.
55. Ho, D. *Nanodiamonds: Applications in Biology and Nanoscale Medicine*. Springer, Berlin, Germany, 2009, 380 pp.

56. Shenderova, O. A.; Haber, D. M. *Ultrananocrystalline Diamond: Synthesis, Properties, and Applications.* William Andrew, New York, 2007, 620 pp.
57. Tjong, S. C. Properties of chemical vapor deposited nanocrystalline diamond and nanodiamond/amorphous carbon composite films. *Nanocomposite Thin Films and Coatings,* 2007, 167–206.
58. Yang, N.; Uetsuka, H.; Williams, O. A.; Osawa, E.; Tokuda, N.; Nebel, C. E. Vertically aligned diamond nanowires: Fabrication, characterization, and application for DNA sensing. *Physica Status Solidi A: Applications and Materials Science,* 2009, *206* (9), 2048–2056.
59. Kong, X.-L. Nanodiamonds used as a platform for studying noncovalent interaction by MALDI-MS. *Chinese Journal of Chemistry,* 2008, *26* (10), 1811–1815.
60. Katsnelson, M. I.; Novoselov, K. S. Graphene: New bridge between condensed matter physics and quantum electrodynamics. *Solid State Communications,* 2007, *143* (1), 3–13.
61. Geim, A. K.; Novoselov, K. S. The rise of graphene. *Nature Materials,* 2007, *6* (3), 183–191.
62. Gogotsi, Y. (Ed.). *Carbon Nanomaterials.* CRC Press, Boca Raton, FL, 2006, 344 pp.
63. Jorio, A.; Dresselhaus, M. S.; Dresselhaus, G.; Gogotsi, Y. (Eds.). *Carbon Nanotubes: Advanced Topics in the Synthesis, Structure, Properties and Applications,* 1st edn., Springer, Berlin, Germany, 2008, 744 pp.
64. Haug, R. (Ed.). *Advances in Solid State Physics,* 1st edn., Springer, Berlin, Germany, Vol. 47, 2008, 363 pp.
65. Bharat Bhushan, B. (Ed.). *Springer Handbook of Nanotechnology,* 1st edn., Springer, Germany, 2008, 363 pp.
66. Van Noorden, R. Moving towards a graphene world. *Nature,* 2006, *442,* 228–229.
67. Müller, S.; Müllen, K. Expanding benzene to giant graphenes: Towards molecular devices. *Philosophical Transactions of the Royal Society A: Mathematical, Physical and Engineering Sciences,* 2007, *365* (1855), 1453–1472.
68. Ando, T. Exotic electronic and transport properties of graphene. *Physica E: Low-Dimensional Systems and Nanostructures,* 2007, *40* (2), 213–227.
69. Aida, T.; Fukushima, T. Soft materials with graphitic nanostructures. *Philosophical Transactions of the Royal Society A: Mathematical, Physical and Engineering Sciences,* 2007, *365* (1855), 1539–1552.
70. Wu, J.; Pisula, W.; Mullen, K. Graphenes as potential material for electronics. *Chemical Reviews,* 2007, *107,* 718–747.
71. Gruner, G.; Hu, L.; Hecht, D. Graphene film as transparent and electrically conducting material. U.S. Patent 20070284557, 2007. http://www.freepatentsonline.com/20070284557.html
72. Prud'homme, R. K.; Ozbas, B.; Aksay, I. A.; Register, R. A.; Adamson, D. H. Functional Graphene—Rubber nanocomposites. U.S. Patent Application Filed—Invention # 07-2323-1, 2006.
73. Jang, B. Z. Highly conductive nano-scaled graphene plate nanocomposites and products. U.S. Patent 20070158618, 2007. http://www.freshpatents.com/Bor-Z-Jang-cndirb.php
74. Abanin, D. A.; Novoselov, K. S.; Zeitler, U.; Lee, P. A.; Geim, A. K.; Levitov, L. S. Dissipative quantum hall effect in graphene near the Dirac point. *Physical Review Letters,* 2007, *98* (19), 196806.
75. Schedin, F.; Geim, A. K.; Morozov, S. V.; Hill, E. W.; Blake, P.; Katsnelson, M. I.; Novoselov, K. S. Detection of individual gas molecules adsorbed on graphene. *Nature Materials,* 2007, *6* (9), 652–655.
76. Meyer, J. C.; Geim, A. K.; Katsnelson, M. I.; Novoselov, K. S.; Obergfell, D.; Roth, S.; Girit, C.; Zettl, A. On the roughness of single- and bi-layer graphene membranes. *Solid State Communications,* 2007, *143* (1), 101–109.
77. Novoselov, K. S.; Jiang, Z.; Zhang, Y.; Morozov, S. V.; Stormer, H. L.; Zeitler, U.; Maan, J. C.; Boebinger, G. S.; Kim, P.; Geim, A. K. Room-temperature quantum Hall effect in graphene. *Science,* 2007, *315* (5817), 1379.
78. Pisana, S.; Lazzeri, M.; Casiraghi, C.; Novoselov, K. S.; Geim, A. K.; Ferrari, A. C.; Mauri, F. Breakdown of the adiabatic Born-Oppenheimer approximation in graphene. *Nature Materials,* 2007, *6* (3), 198–201.
79. van den Brink, J. Graphene: From strength to strength. *Nature Nanotechnology,* 2007, *2* (4), 199–201.
80. Nair, R. R.; Ren, W.; Jalil, R.; Riaz, I.; Kravets, V. G.; Britnell, L.; Blake, P. et al. Fluorographene: A two-dimensional counterpart of teflon. *Small,* 2010, *6* (24), 2877–2884.
81. Elias, D. C.; Nair, R. R.; Mohiuddin, T. M. G.; Morozov, S. V.; Blake, P.; Halsall, M. P.; Ferrari, A. C. et al. Control of graphene's properties by reversible hydrogenation: Evidence for graphane. *Science,* 2009, *323,* 610–613.
82. Li, A. P.; Flack, F.; Lagally, M. G.; Chisholm, M. F.; Yoo, K.; Zhang, Z.; Weitering, H. H.; Wendelken, J. F. Photoluminescence and local structure of Ge nanoclusters on Si without a wetting layer. *Physical Review B: Condensed Matter,* 2004, *69,* 245310.
83. Jamet, M.; Wernsdorfer, W.; Thirion, C.; Mailly, D.; Dupuis, V.; Mélinon, P.; Pérez, A. Magnetic anisotropy of a single cobalt nanocluster. *Physical Review Letters,* 2001, *86* (20), 4676–4679.

84. Qu, Y.; Li, X.; Chen, G.; Zhang, H.; Chen, Y. Synthesis of Cu_2O nano-whiskers by a novel wet-chemical route. *Materials Letters*, 2008, *62*, 886–888.

85. Park, H. S.; Shindo, D.; Mitani, S.; Takanashi, K. Magnetic microstructures of nano-granular CoPt–Al–O thin films studied by Lorentz microscopy and electron holography. *IEEE Transactions on Magnetics*, 2005, *41* (10), 3724–3726.

86. Kim, Y. L.; Jung, J. H.; Kim, K. H.; Yoon, H. S.; Song, M. S.; Bae, S. H.; Kim, Y. The growth and optical properties of CdSSe nanosheets. *Nanotechnology*, 2009, *20*, 095605, 7 pp.

87. Kundu, S.; Datta, A.; Hazra, S. Manipulating headgroups in Langmuir–Blodgett films through subphase pH variation. *Chemical Physics Letters*, 2005, *405*, 282–287.

88. Aslan, K.; Wu, M.; Lakowicz, J. R.; Geddes, C. D. Fluorescent core-shell $Ag@SiO_2$ nanocomposites for metal-enhanced fluorescence and single nanoparticle sensing platforms. *Journal of the American Chemical Society*, 2007, *129*, 1524–1525.

89. Thomas, J. M.; Raja, R.; Johnson, B. F. G.; Hermans, S.; Jones, M. D.; Khimyak, T. Bimetallic catalysts and their relevance to the hydrogen economy. *Industrial and Engineering Chemistry Research*, 2003, *42*, 1563–1570.

90. Bronstein, L. M.; Chernyshov, D. M.; Volkov, I. O.; Ezernitskaya, M. G.; Valetsky, P. M.; Matveeva, V. G.; Sulman, E. M. Structure and properties of bimetallic colloids formed in polystyrene-block-Poly-4-vinylpyridine micelles: Catalytic behavior in selective hydrogenation of dehydrolinabol. *Journal of Catalysis*, 2000, *196*, 302–314.

91. Kharisov, B. I.; Kharissova, O. V.; José Yacamán, M.; Ortiz Mendez, U. State of the art of the bi- and trimetallic nanoparticles on the basis of gold and iron. *Recent Patents on Nanotechnology*, 2009, *3* (2), 81–98.

92. Mulvaney, P.; Giersig, M.; Henglein, A. Surface chemistry of colloidal gold: Deposition of lead and accompanying optical effects. *Journal of Physical Chemistry*, 1992, *96*, 10419–10424.

93. Chushak; Y. G.; Bartell, L. S. Freezing of Ni–Al bimetallic nanoclusters in computer simulations. *Journal of Physical Chemistry B*, 2003, *107*, 3747–37451.

94. Mulvaney, P. Surface plasmon spectroscopy of nanosized metal particles. *Langmuir*, 1996, *12*, 788–800.

95. Link, S.; Wang, Z. L.; El-Sayed, M. A. Alloy formation of gold-silver nanoparticles and the dependence of the plasmon absorption on their composition. *Journal of Physical Chemistry B*, 1999, *103*, 3529–3533.

96. Rodriguez-Gonzalez, B.; Sanchez-Iglesias, A.; Giersig, M.; Liz-Marzan, L. M. Au/Ag bimetallic nanoparticles. Formation, silica coating and selective etching. *Faraday Discussions*, 2004, *125*, 133–144.

97. Liz-Marzan, L. M. Nanometals: Formation and color. *Materials Today*, 2004, *7*, 26–31.

98. Cao, E.; Schatz, G. C.; Hupp, J. T. Synthesis and optical properties of anisotropic metal nanoparticles *Journal of Fluorescence*, 2004, *14* (4), 331–341.

99. Liz-Marzan, L. M.; Mulvaney, P. The assembly of coated nanocrystals. *Journal of Physical Chemistry B*, 2003, *107* (30), 7312–7326.

100. Mulvaney, P.; Underwood, S. Effect of the solution refractive index on the color of gold colloids. *Langmuir*, 1994, *10*, 3427–3430.

101. Ung, T.; Dunstan, D.; Giersig, M.; Mulvaney, P. Spectroelectrochemistry of colloidal silver. *Langmuir*, 1997, *13* (6), 1773–1782.

102. Liz-Marzan, L. M.; Mulvaney, P. $Au@SiO_2$ colloids: Temperature effect on the surface plasmon absorption. *New Journal of Chemistry*, 1998, *22*, 1285–1288.

103. Mallin, M. P.; Murphy, C. J. Solution-phase synthesis of sub-10 nm Au–Ag alloy nanoparticles. *Nano Letters*, 2002, *2* (11), 1235–1237.

104. Mulvaney, P.; Giersig, M.; Henglein, A. Electrochemistry of multilayer colloids: Preparation and absorption spectrum of gold-coated silver particles. *Journal of Physical Chemistry*, 1993, *97*, 7061–7064.

105. Rivas, L.; Sanchez-Cortes, S.; García-Ramos, J. V.; Morcillo, G. Mixed silver/gold colloids: A study of their formation, morphology, and surface-enhanced Raman activity. *Langmuir*, 2000, *16* (25), 9722–9228.

106. Srnova-Sloufova, I.; Lednicky, F.; Gemperle, A.; Gemperlova, J. Core–shell (Ag)Au bimetallic nanoparticles: Analysis of transmission electron microscopy images. *Langmuir*, 2000, *16* (25), 9928–9935.

107. Schierhorn, M.; Liz-Marzan, L. M. Synthesis of bimetallic colloids with tailored intermetallic separation. *Nano Letters*, 2002, *2*, 13–16.

108. Sun, Y.; Xia, Y. Multiple-walled nanotubes made of metals. *Advanced Materials*, 2004, *16*, 264–268.

109. Sun, Y.; Wiley, B.; Li, Z.-Y.; Xia, Y. Synthesis and optical properties of nanorattles and multiple-walled nanoshells/nanotubes made of metal alloys. *Journal of the American Chemical Society*, 2004, *126*, 9399–9406.

110. Srnova-Sloufova, I.; Vlckova, B.; Bastl, Z.; Hasslett, T. L. Bimetallic (Ag)Au nanoparticles prepared by the seed growth method: Two-dimensional assembling, characterization by energy dispersive x-ray analysis, x-ray photoelectron spectroscopy, and surface enhanced Raman spectroscopy, and proposed mechanism of growth. *Langmuir*, 2004, *20* (8), 3407–3415.

111. Oldenburg, S. J.; Jackson, J. B.; Westcott, S. L.; Halas, N. J. Infrared extinction properties of gold nanoshells. *Applied Physics Letters*, 1999, *75*, 2897.

112. Mackerle, J. Nanomaterials, nanomechanics and finite elements: A bibliography (1994–2004). *Modelling and Simulation in Materials Science and Engineering*, 2005, *13* (1), 123–158.

113. Fürstner A. (Ed.). *Active Metals. Preparation, Characterization, Applications*. VCH, Weinheim, Germany, 1996, 465 pp.

114. Cintas, P. *Activated Metals in Organic Synthesis*. CRC Press, Boca Raton, FL, 1993, 250 pp.

115. Garnovskii, A. D.; Kharisov, B. I. (Eds.). *Direct Synthesis of Coordination and Organometallic Compounds*. Elsevier Science, Amsterdam, the Netherlands, 1999, 244 pp.

116. Garnovskii, A. D.; Kharisov, B. I. (Eds.). *Synthetic Coordination and Organometallic Chemistry*. Marcel Dekker, New York, 2003, 513 pp.

117. Fedlheim, D. L.; Foss, C. A. Metal nanoparticles: Synthesis characterization and application. Marcel Dekker, New York, 2002; 360 pp.

118. Klabunde, K. J. *Nanoscale Materials in Chemistry*. VCH, Weinheim, Germany, 2001, 304 pp.

119. Fürstner, A. Chemistry of and with highly reactive metals. *Angewandte Chemie International Edition English*, 1993, *32* (2), 164–189. Published on-line in 2003.

120. Garnovskii, A. D.; Kharisov, B. I.; Gójon-Zorrilla, G.; Garnovskii, D. A. Direct synthesis of coordination compounds starting from zero-valent metals and organic ligands. *Russian Chemical Reviews*, 1995, *64* (3), 201–221.

121. Rao, C. N. R.; Kulkarni, G. U.; Govindaraj, A.; Satishkumar, B. C.; Thomas, P. J. Metal nanoparticles, nanowires, and carbon nanotubes. *Pure and Applied Chemistry*, 2000, *72* (1–2), 21–33.

122. Park, K.; Kim, H. J.; Suh, Y. J. Preparation of tantalum nanopowders through hydrogen reduction of TaCl$_5$ vapor. *Powder Technology*, 2007, *172* (3), 144–148.

123. Rao, C. N. R.; Deepak, F. L.; Gundiah, G.; Govindaraj, A. Inorganic nanowires. *Progress in Solid State Chemistry*, 2003, *31* (1), 5–147.

124. Chin, W. S.; Liu, C. Method of preparing nanowire(s) and product(s) obtained therefrom. U.S. Patent 20070221917, 2007.

125. Owen, J. H. G.; Miki, K.; Bowler, D. R. Self-assembled nanowires on semiconductor surfaces. *Journal of Materials Science*, 2006, *41* (14), 4568–4603.

126. Huang, C.-J.; Chiu, P.-H.; Wang, Y.-H.; Chen, W.-R.; Meen, T.-H.; Yang, C.-F. Preparation and characterization of gold nanodumbbells. *Nanotechnology*, 2006, *17* (21), 5355–5362.

127. Aiken, J. D.; Finke, R. G. A review of modern transition-metal nanoclusters: their synthesis, characterization, and applications in catalysis. *Journal of Molecular Catalysis A: Chemical*, 1999, *145* (1), 1–44.

128. Finney, E. E.; Finke, R. G. Nanocluster nucleation and growth kinetic and mechanistic studies: A review emphasizing transition-metal nanoclusters. *Journal of Colloid and Interface Science*, 2008, *317* (2), 351–374.

129. Bansmann, J.; Baker, S. H.; Binns, C.; Blackman, J. A.; Bucher, J.-P.; Dorantes-Dávila, J.; Dupuis, V. et al. Magnetic and structural properties of isolated and assembled clusters. *Surface Science Reports*, 2005, *56* (6), 189–275.

130. He, Y.; Wu, X.; Lu, G.; Shi, G. A facile route to silver nanosheets. *Materials Chemistry and Physics*, 2006, *98* (1), 178–182.

131. Kuchibhatla, S. V. N. T.; Karakoti, A. S.; Bera, D.; Seal, S. One dimensional nanostructured materials. *Progress in Materials Science*, 2007, *52* (5), 699–913.

132. Perez-Juste, J.; Pastoriza-Santos, I.; Liz-Marzan, L. M.; Mulvaney, P. Gold nanorods: Synthesis, characterization and applications. *Coordination Chemistry Reviews*, 2005, *249* (17), 1870–1901.

133. Wang, P. I.; Zhao, Y. P.; Wang, G. C.; Lu, T. M. Novel growth mechanism of single crystalline Cu nanorods by electron beam irradiation. *Nanotechnology*, 2004, *15* (1), 218–222.

134. Rodríguez-López, J. L.; Montejano-Carrizales, J. M.; Pal, U.; Sánchez-Ramírez, J. F.; Troiani, H. E.; García, D.; Miki-Yoshida, M.; Yacamán, J. M. Surface reconstruction and decahedral structure of bimetallic nanoparticles. *Physical Review Letters*, 2004, *92* (19), 196102.

135. Kim, W.-B.; Park, J.-S.; Suh, C.-Y.; Kil, D.-S.; Lee, J.-C. Method for manufacturing alloy nanopowders. U.S. Patent 20070209477A1, 2007.

136. Chen, Y; Gong, R.; Zhang, W.; Xu, X.; Fan, Y.; Liu, W. Synthesis of single-crystalline bismuth nanobelts and nanosheets. *Materials Letters*, 2005, *59* (8), 909–911.

137. Kharissova, O. V.; Osorio, M.; Kharisov, B. I.; José Yacamán, M.; Ortiz Méndez, U. A comparison of bismuth nanoforms obtained in vacuum and air by microwave heating of bismuth powder. *Materials Chemistry and Physics*, 2010, *121*, 489–496.

138. Zhang, X.; Xie, H.; Fujii, M.; Takahashi, K.; Ikuta, T.; Ago, H.; Abe, H.; Shimizu, T. Experimental study on thermal characteristics of suspended platinum nanofilm sensors. *International Journal of Heat and Mass Transfer*, 2006, *49* (21), 3879–3883.

139. Yamada, Y. M. A.; Uozumi, Y. Development of a convoluted polymeric nanopalladium catalyst: α-alkylation of ketones and ring-opening alkylation of cyclic 1,3-diketones with primary alcohols. *Tetrahedron*, 2007, *63* (35), 8492–8498.

140. Niu, Y.; Crooks, R. M. Dendrimer-encapsulated metal nanoparticles and their applications to catalysis. *Comptes Rendus Chimie*, 2003, *6* (8), 1049–1059.

141. Liu, H.; Song, C.; Zhang, L.; Zhang, J.; Wang, H.; Wilkinson, D. P. A review of anode catalysis in the direct methanol fuel cell. *Journal of Power Sources*, 2006, *155* (2), 95–110.

142. Karen, B.; Brian, H.; Christopher, L.; Jhierry, L. Palladium alloy catalysts for fuel cell cathodes. WO07042841A1, 2007.

143. Yurkov, G.; Baranov, D. A.; Dotsenko, I. P.; Gubin, S. P. New magnetic materials based on cobalt and iron-containing nanoparticles. *Composites Part B: Engineering*, 2006, *37* (6), 413–417.

144. Armelao, L.; Barreca, D.; Bottaro, G.; Gasparotto, A; Gross, S.; Maragno, C.; Tondello, E. Recent trends on nanocomposites based on Cu, Ag and Au clusters: A closer look. *Coordination Chemistry Reviews*, 2006, *250* (11), 1294–1314.

145. Atanassova, P.; Bhatia, R.; Sun, Y.; Hampden-Smith, M. J.; Brewster, J.; Napolitano, P. Alloy catalyst compositions and processes for manufacturing. U.S. Patent 20070160899A, 2007.

146. Huang, X. J.; Choi, Y. K. Chemical sensors based on nanostructured materials. *Sensors and Actuators: B. Chemical*, 2007, *122* (2), 659–671.

147. Atashbar, M. Z.; Singamaneni, S. Room temperature gas sensor based on metallic nanowires. *Sensors and Actuators: B. Chemical*, 2005, *111*, 13–21.

148. Brust, M.; Kiely, C. J. Some recent advances in nanostructure preparation from gold and silver particles: A short topical review. *Colloids and Surfaces A: Physicochemical and Engineering Aspects*, 2002, *202* (2), 175–186.

149. McDowall, L. *Degradation of Toxic Chemicals by Zero-Valent Metal Nanoparticles—A Literature Review.* Human Protection and Performance Division, DSTO Defence Science and Technology Organisation, Victoria, Australia, 2005.

150. Kucka, J.; Hruby, M.; Konak, C.; Kozempel, J.; Lebeda, O. Astatination of nanoparticles containing silver as possible carriers of [211]At. *Applied Radiation and Isotopes*, 2006, *64* (2), 201–206.

151. Xia, Y.; Yang, P.; Sun, Y.; Wu, Y.; Mayers, B.; Gates, B.; Yin, Y.; Kim, F.; Yan, H. One-dimensional nanostructures: synthesis, characterization, and applications. *Advanced Materials*, 2003, *15* (5), 353–389.

152. Park, H. D.; Prokes, S. M.; Twigg, M. E.; Ding, Y.; Zhong Lin Wang. Growth of high quality, epitaxial InSb nanowires. *Journal of Crystal Growth*, 2007, *304*, 399–401.

153. Wang, H.-z; Zhao, H.-l.; Liu, B.; Zhang, X.-t.; Du, Z.-l.; Yang, W.-s. Facile preparation of Mn_2O_3 nanowires by thermal decomposition of $MnCO_3$. *Chemical Research in Chinese Universities*, 2010, *26* (1), 5–7.

154. Leung, Y. P.; Choy, W. C. H.; Markov, I.; Pang, G. K. H.; Ong, H. C.; Yuk, T. I. Synthesis of wurtzite ZnSe nanorings by thermal evaporation. *Applied Physics Letters*, 2006, *88*, 183110.

155. Hughes, W. L.; Wang, Z. L. Controlled synthesis and manipulation of ZnO nanorings and nanobows. *Applied Physics Letters*, 2005, *86*, 043106.

156. Wang, Z. L. Self-assembled nanoarchitectures of polar nanobelts/nanowires. *Journal of Materials Chemistry*, 2005, *15*, 1021–1024.

157. Fonseca de Lima, J.; Martins, R. F.; Neri, C. R.; Serra, O. A. $ZnO:CeO_2$-based nanopowders with low catalytic activity as UV absorbers. *Applied Surface Science*, 2009, *255*, 9006–9009.

158. Patel, P. A.; Eckart, J.; Advincula, M. C.; Goldberg, A. J.; Mather, P. T. Rapid synthesis of polymer-silica hybrid nanofibers by biomimetic mineralization. *Polymer*, 2009, *50*, 1214–1222.

159. Bang, J. H.; Suh, W. H.; Suslick, K. S. Quantum dots from chemical aerosol flow synthesis: Preparation, characterization, and cellular imaging. *Chemistry of Materials*, 2008, *20*, 4033–4038.

160. Song, H.; Yang, A.; Zhang, R.; Guo, Y.; Wei, H.; Zheng, G.; Yang, S.; Liu, X.; Zhu, Q.; Wang, Z. Well-aligned Zn-doped InN nanorods grown by metal-organic chemical vapor deposition and the dopant distribution. *Crystals Growth and Design*, 2009, *9* (7), 3292–3295.

161. Wiley, B. J.; Chen, Y.; McLellan, J. M.; Xiong, Y.; Li, Z.-J.; Ginger, D.; Xia, Y. Synthesis and optical properties of silver nanobars and nanorice. *Nano Letters*, 2007, *7* (4), 1032–1036.

Part II

Less-Common Nanostructures

3 Simple, Linear 1D, 2D, and 3D Nanostructures

The following closely related nanostructures can be classified as simple, linear 1D, 2D, and 3D nanostructures: nanolines, nanopencils, nanodumbbells, nanopins, nanoshuttles, nanopeapods, nanobricks, nanochains, nanowicks, nanobars, and nanopillars. Some of them are very rare, such as, for instance, nanopencils or nanoshuttles.

3.1 NANOLINES

Nanolines, which are sometimes confused with nanowires, are mainly represented by elemental metals (e.g., Au[1–3]), although several examples of metal oxides,[4] binary salts, as well as organic-containing moieties have also been reported. Among metals, bismuth nanolines (Figure 3.1) are the most frequent and are systematically studied by Owen,[5,6] Bowler,[7] and other researchers.[8,9] The number of reports on the nanolines of this metal exceeds that on other elements and compounds, so Bi nanolines are described in detail in the Chapter 26. Among other metals, Fe nanolines (Figure 3.2) were deposited with a pitch of 186 nm, a full width at half maximum (FWHM) of 50 nm, and a height of up to 6 nm, using direct-write atom lithography.[10] Alternatively, iron nanolines were fabricated by depositing an atomic beam of iron through a far-off resonant laser standing wave onto a glass-ceramic substrate.[11] The resulting nanolines exhibited a period of 186 nm, a height above the background of 8 nm, and a full width at half maximum of 95 nm. These nanostructures covered a surface area of $\sim 1.6 \times 0.4\,mm^2$, corresponding to ~ 8600 iron lines with a length of $\sim 400\,\mu m$. Additionally, Fe nanolines[12] in the carbon nanotubes were also reported. The morphological evolution of Cu nanolines induced by a focused-ion-beam at normal bombardment was investigated by *in situ* SEM.[13] A periodic array of particles was observed when the width of the lines reached a certain value. When the line width was below this value, an unstable mode whose wave vector was found to be parallel to the line axis developed and a chain of periodic particles formed. When the width was above this critical value, the sputtering etching only led to the decrease in width. As applications of such nanolines, it has been mentioned[14] that copper nanolines fabricated by the damascene process are commonly used as interconnects in advanced electronic devices.

Chromium nanolines were fabricated by depositing an atomic beam of ^{52}Cr through an off-resonant laser standing wave with a wavelength of 425.55 nm on a silicon substrate.[16] The resulting nanolines exhibited a period of 215 ± 3 nm with a height of 1 nm. Also, the Yb growth on a vicinal Si(100) surface was studied.[17] The switchover of the 2D layer orientation, similar to the case of the Bi nanolines on Si(100), was observed depending on the growth procedure. In addition, the structure and morphology of the Yb silicide phase were found to depend critically on the growth conditions, and the ability to grow very long, unidirectional Yb silicide nanowires was demonstrated. Additionally, the reaction of the rare earth metal Ho with the Ge(001) surface at 440°C was studied by scanning tunneling microscopy (STM).[18] The self-assembly of ultrafine nanolines growing along substrate <110> directions was observed. The presence of nanoscale trenches associated with well-ordered lines of missing dimer defects was noted. Possible applications of these nanostructures could be interconnects or templating in nanoscale devices; in addition, they may provide insight into the nucleation mechanism of coarser nanowires. In addition, Ru[19] (together with its nanowires) and In[20] nanolines are known. Metal alloys were also reported in nanoline form; thus, high orderly CoPt

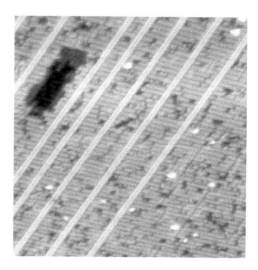

FIGURE 3.1 Bi lines grown on Si(001). Image size 47 × 47 nm. (Reproduced from *Appl. Surf. Sci.*, 244, Miwa, R.H. et al., The geometry of Bi nanolines on Si(001), 157–160, Copyright 2005, with permission from Elsevier.)

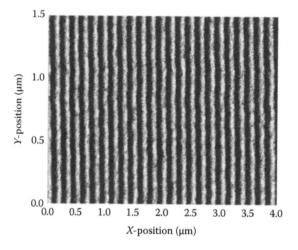

FIGURE 3.2 Example of a 4 × 1.5 μm AFM scan of the Fe nanolines. The height (*gray*) scale ranges over 6 nm. (Reproduced with permission from Smeets, B. et al., *Appl. Phys. B*, 1, 2009, arXiv.org, e-Print Archive, Physics, arXiv:0908.2733v1 [physics.atom-ph]. Cornell University Library, http://aps.arxiv.org/PS_cache/arxiv/pdf/0908/0908.2733v1.pdf, doi 10.1007/s00340-009-3867-3.)

nanoline arrays with a diameter of 50 nm were fabricated by electrodeposition in the anodic aluminum oxide template.[21] It was shown the CoPt alloy nanoline arrays are fcc structures. The results indicate that nanowires had obvious anisotropy with their easy axis parallel to nanoline arrays. The coercivity of materials can be greatly improved by changing the CoPt alloy microstructure during heat treatment. Other alloys in nanoline forms (Ag/Cu,[22] Cu/Ni/Cu,[23] Ni/Fe, and metallic Co[24]) were examined too.

In the case of silicon and its oxide, silicon nanolines[25,26] (SiNLs, Figure 3.3), with a line width ranging from 24 to 90 nm, having smooth and vertical sidewalls and the aspect ratio (height/line width) from 7 to 18 were fabricated[27–29] by an anisotropic wet etching process. The early stages (submonolayer) of silicon deposition onto the anisotropic Ag(110) surface led to the formation of flat-lying, individual Si nanostripes, with a high aspect ratio, all oriented

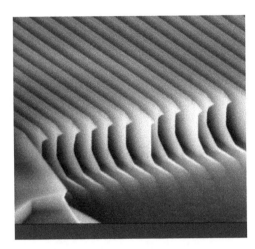

FIGURE 3.3 SEM images of parallel silicon nanolines, with 74 nm line width and 510 nm height; the pitch distance is 180 nm. (Reproduced from Kang, M.K. et al., *J. Nanomater.*, Article ID 132728, 11 pp. Copyright 2008 Hindawi Publishing Corporation.)

along the Ag[-100] direction.[30] This Si grating was subsequently used as a template for selective 1D growth of 1D Co nanolines. Two types of multistack nanolines (MNLs), Si-substrate (Si)/ siliconoxynitride (SiON)/amorphous Si (a-Si)/SiO$_2$ and Si/SiO$_2$/polycrystalline Si (poly-Si)/SiO$_2$, were used to measure the collapse force and to investigate their collapse behavior by an AFM.[31] The Si/SiON/a-Si/SiO$_2$ MNL showed a larger length of fragment in the collapse patterns at a smaller collapse force; meanwhile, the Si/SiO$_2$/poly-Si/SiO$_2$ MNL demonstrated a smaller length of fragment at a higher applied collapse force. The different collapse behaviors were attributed to the magnitude of adhesion forces at the stack material interfaces and the mechanical strength of MNLs. Multiple instability states, for example, grouped collapse, single collapse, wavy, and grouped wavy states, were observed in hydrogen silsesquioxane (HSQ, its monomer formula is H$_8$Si$_8$O$_{12}$) nanolines (Figure 3.4) defined by electron beam (E-beam) lithography (EBL).[32] It was shown that the critical aspect ratio of the HSQ lines dramatically increased when the line pitch reduced to sub-100 nm, which was in contrast to theoretical models for capillary forces and swelling strain. Stable high aspect ratio HSQ nanolines over metal pads were used to make working Si nanowire transistors on Si on insulator substrates.

A few metal oxides, sulfides, silicides, and nitrides[33] were described as nanolines, for example, those of ZnO[34] on the Cu-Zn alloy. Thus, the soft E-beam lithography (soft-EBL) fabricated poly-crystalline ZnO nanolines showed reproducible response to ppm-level H$_2$ and NO$_2$ even at r.t., due to the intrinsic Joule heating effect in such nanodevices.[35] Joule heating was found to increase the nanoline temperature to around 72°C, which enhanced the oxidation–reduction reaction at the ZnO surface and allowed a faster photoresponse than the thin film. Additionally, TiO$_2$ nanolines (diameter 30–40 nm and length 300–400 nm) were prepared from TiCl$_4$ as a precursor.[36] Netted sphere-like CdS nanostructures consisting of the dendritic nanolines (Figure 3.5) were prepared in ethylenediamine using L-cysteine and cadmium nitrate tetrahydrate as precursors by a solvothermal process at 150°C for 8 h.[37] It was noted that the reaction time, the molar ratio of precursors, solvent, and temperature can be used as the additional means to control the size and morphology. A top-down method for fabricating nickel monosilicide (NiSi) nanolines with smooth sidewalls and line widths down to 15 nm was offered.[38] It was revealed that the r.t. electrical resistivity of the NiSi nanolines remained constant as the line widths were reduced to 23 nm. The resistivity at cryogenic temperatures was found to increase with decreasing line width. The possibility of designing Fe silicide wires along Bi nanolines on the hydrogenated Si(001) surface was studied by spin-density functional calculations.[39]

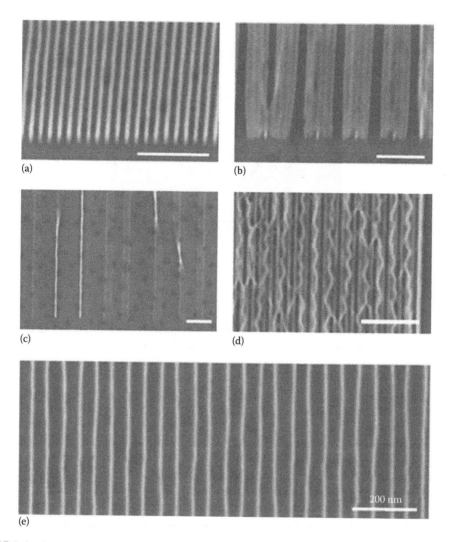

FIGURE 3.4 SEM 45° tilted view of HSQ lines in four stability states: (a) stable, (b) clustering or grouped collapsed lines, (c) completely collapsed lines forming free ribbons on the surface, (d) wavy lines with a spatial wavelength of 150–200 nm, and (e) 12 nm wide lines of 50 nm pitch and aspect ratio of 11. Scale bar in images (a) through (d) is 500 nm. (Reproduced with permission from Regonda, S. et al., *J. Vac. Sci. Technol., B*, 26(6), 2247. Copyright 2008 American Vacuum Society.)

Oxygen-containing salts are represented by ferrates and titanates. Thus, multiferroic $BiFeO_3$ was fabricated by the soft-EBL technique.[40] Its nanolines patterned on $SrTiO_3$ substrate exhibited a bamboo-like structure, which can serve as an excellent model system for investigating the contribution of grain boundaries to the leakage current. $SrTiO_3$ nanoline structures ($SrTiO_3$(001) surface) (Figure 3.6) were used as a support to grow palladium nanocrystals and annealed in ultrahigh vacuum at 620°C.[41] Among other inorganic nanolines, boron nitride[42], GaN,[43] potassium titanate[44] and its nonstoichiometric analogue $K_2Ti_8O_{15}$,[45] Bi_2Te_3,[46] $F_xC_yO_z$ (x: y: z = 1–8: 1–18: 0–1),[47] etc., have been fabricated and characterized.

A few coordination and organic compounds, as well as polymers (such as poly(Me methacrylate) [PMMA][48]) and even DNA, are known as nanolines. Thus, the r.t. growth and ordering of (porphyrinato)nickel(II) (or nickel (II) porphine, NiP) molecules on a Ag(111) surface were investigated, finding NiP molecules forming a second layer self-assemble into well-ordered and uniformly separated nanolines at r.t.[49] These nanolines (Figure 3.7) consisted of hexagonally ordered NiP molecules and

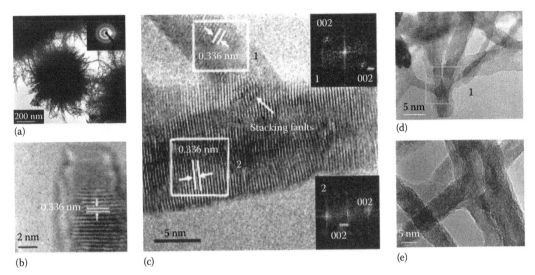

FIGURE 3.5 (a) TEM image of an individual urchin-like CdS crystal. The inset is the electron diffraction (ED) pattern of a whole urchin-like CdS crystal in (a); (b) high-resolution TEM image of a nanoneedle taken from an urchin-like CdS crystal in (a); (c) high-resolution TEM image of the dendritic line of a netted sphere-like CdS nanostructure. Insets 1 and 2 are FFT patterns of the high-resolution TEM images of the marked parts 1 and 2 in (c), respectively; (d) TEM image of the aggregating line tips of a netted sphere-like CdS nanostructure; (e) TEM image of the intersectant lines of a netted sphere-like CdS nanostructure. (Reproduced with permission from Zhao, P. and Huang, K. Preparation and characterization of netted sphere-like CdS nanostructures, *Cryst. Growth Des.*, 8(2), 717–722. Copyright 2008 American Chemical Society.)

were found to be 1–4 molecules wide, depending on the molecular coverage. Polyterthiophene nanostructures consisting of periodic nanolines were prepared using the precursor polymer approach in conjunction with nanoimprint lithography.[50] It was established that nanoimprinted lines of conductive polyterthiophene exhibited high electrochromic contrast at 437 nm. In addition, optical diffraction measurements from both parallel arrays of gold nanowires and also nanoparticle-decorated DNA nanolines were analyzed to characterize the sensitivity of these surfaces to changes in local interfacial refractive index and the generation of surface plasmon polaritons.[51,52] DNA nanoline patterns with the same dimensions were created on streptavidin-coated polymer surfaces by transferring biotinylated ssDNA from master patterns of gold nanowires.

3.2 NANOPENCILS

Nanopencils, a close relative to widespread nanorods or nanopillars and frequently grown together with them, are mainly represented by ZnO[53–56] (this is one of the most famous compounds in nanotechnology, which forms a wide variety of nanostructures). Thus, a single-step solvent-, catalyst-, and template-free synthesis process to prepare luminescent pencils of ZnO either in micro- or in nanosize diameters from a single precursor was demonstrated.[57] The technique, which consisted of the thermolysis of a $Zn(OAc)_2 \cdot 2H_2O$ precursor in a closed stainless steel reactor at 700°C under autogenic pressure (6.5 MPa), yielded carbon sphere–decorated ZnO micropencils (Figure 3.8), which were found to have r.t. luminescence (PL) with well-defined emission peaks at the green, yellow, orange, and red regions of the visible spectra while suppressing the blue region.

Substrates, used for the growth of ZnO nanopencils and related nanostructures, are mainly silicon, alumina, and sapphire. Thus, well-aligned ZnO nanorod and nanopencil arrays were synthesized in a high density on a ZnO/Si substrate by a low-temperature aqueous solution technique using $Zn(NO_3)_2 \cdot 6H_2O$ and hexamethylenetetramine $(C_6H_{12}N_4)$, in which the hexamethylenetetramine acted as a pH buffer to regulate the pH value of the solution and the slow supply of OH⁻ ions

FIGURE 3.6 (a) STM image of nanoline-structured $SrTiO_3(001)$, $157 \times 154\,nm^2$. (b) Triple-row nanolines order into a (9×2) surface reconstruction $(14 \times 10\,nm^2)$. (c) Surface with triple-row nanolines, double-row nanolines, and nanodots $(21 \times 17\,nm^2)$. (d) Pd nanocrystals on $SrTiO_3(001)$ $(80 \times 80\,nm^2)$. (e) 3D representation of a Pd nanocrystal on double-row and triple-row nanolines (V_s) +0.8 V, I_t) 0.3 nA). (Reproduced with permission from Silly, F. and Castell, M.R., Encapsulated Pd nanocrystals supported by nanoline-structured $SrTiO_3(001)$, *J. Phys. Chem. B*, 109, 12316–12319. Copyright 2005 American Chemical Society.)

FIGURE 3.7 NiP nanolines. (Reproduced with permission from Krasnikov, S.A. et al., *Nanotechnology*, 20(13), 135301/1. Copyright 2009 IOP Science.)

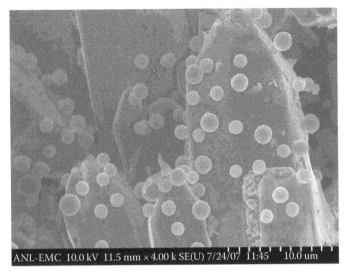

ANL-EMC 10.0 kV 11.5 mm × 4.00 k SE(U) 7/24/07 11:45 10.0 um

FIGURE 3.8 ZnO pencils decorated with carbon spheres. (Reproduced with permission from Pol, V.G. et al., Facile synthesis of novel photoluminescent ZnO micro- and nanopencils, *Langmuir*, 24(23), 13640–13645. Copyright 2008 American Chemical Society.)

(Reactions 3.1 through 3.4).[58] It was revealed that the as-synthesized nanorods and nanopencils were single crystalline, with a hexagonal phase, and with growth along the [0001] direction. Possible growth mechanism for the formation of ZnO nanorod and nanopencil arrays is shown in Figure 3.9. In a related work,[59] ZnO nanopencils were synthesized on a silicon wafer without catalysts at a low temperature of 550°C through a simple two-step pressure-controlled thermal evaporation. It was

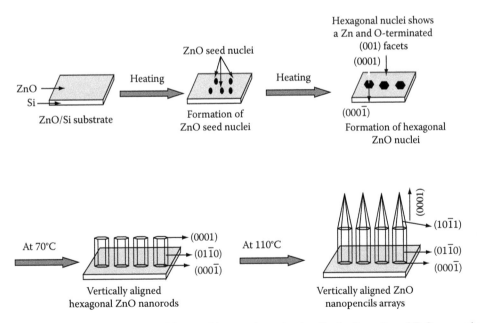

FIGURE 3.9 Schematic illustration of the possible growth mechanism for the formation of ZnO nanorod and nanopencil arrays grown at different reaction temperatures. (Reproduced with permission from Ahsanulhaq, Q. et al., *Nanotechnology*, 18(11), 115603/1. Copyright 2007 IOP Science.)

shown that the nanopencils were single crystals growing along the [0001] direction and the pen tips subtend a small angle with multiple surface perturbations:

$$(CH_2)_6 N_4 + 6H_2O \rightarrow 6HCHO + 4NH_3 \tag{3.1}$$

$$NH_3 + H_2O \leftrightarrow NH_4^+ + OH^- \tag{3.2}$$

$$Zn^{2+} + 2OH^- \rightarrow Zn(OH)_2 \tag{3.3}$$

$$Zn(OH)_2 \rightarrow ZnO + H_2O \tag{3.4}$$

Also, vertically aligned ZnO nanonails and nanopencils (Figure 3.10) were synthesized[60] on a silicon substrate using a modified thermal-evaporation process, without using a catalyst or pre-deposited buffer layers. In addition, ZnO pencil-like nanorods and a variety of other vertically aligned ZnO nanostructures, such as nanorods and nanowalls, were grown on sapphire substrates coated with Au nanocolloidal solutions by atmosphere-pressure CVD using Zn powder and H_2O as source materials.[61] Shapes of the nanorods grown on the sapphire substrates coated with the diluted

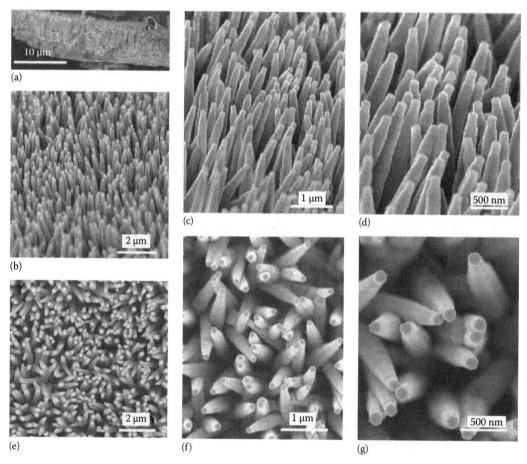

FIGURE 3.10 SEM images of vertically aligned ZnO nanopencils synthesized at 700°C. (a) Side-view image; (b through d) tilt-view images; (e through g) top-view images. (From Shen, G. et al., Characterization and field-emission properties of vertically aligned ZnO nanonails and nanopencils fabricated by a modified thermal-evaporation process. *Adv. Funct. Mater.* 2006. 16. 410–416. Copyright Wiley-VCH Verlag GmbH & Co. KGaA. Reproduced with permission.)

solution of enhanced metalorganic decomposition for NiO film depended strongly on the degree of dilution of the solution. In the case of use of alumina as support, ZnO nanopencils were grown on R-plane Al_2O_3 by MOCVD technique; meanwhile, vertical aligned nanowires were observed on A- and C-plane alumina.[62] In general, this ZnO nanoform was found to serve as an efficient field emitter,[59] attributed to the sharp tip and surface perturbations on the nanopencils.

Other oxides are considerably lesser presented. Thus, SnO_2 nanopencils (together with nanoleaves) were synthesized on single silicon substrates using Au-Ag alloying catalyst-assisted carbothermal evaporation.[63] In comparison with SnO_2 nanopencils, the new peak at 456 nm in the measured PL spectra of SnO_2 nanoleaves was observed, implying that more luminescence centers exist in these nanostructures. The authors concluded that the diffusion behavior of surface atoms under different ambient conditions and the different supersaturating degree of alloying droplets would be responsible for the formation of different morphologies of SnO_2 nanostructures. Also, tungsten oxide nanopencil array as field emission material[64] and In_2O_3[65] (Figure 3.11) are known. In addition, a nanopencil of silicon oxide on a carbon nanotube (Figure 3.12) was fabricated[66] by conformal deposition of silicon oxide on a carbon nanotube and subsequent "sharpening" to expose its tip. Its application as a wear-tolerant probe for ultrahigh density data storage was proposed.

Organic nanopencils are very rare. For example, styrene-Me methacrylate diblock copolymer [P(S-b-MMA)] nanopencils were prepared using anodized aluminum membranes as templates.[67] TEM images of the symmetrical (50:50) block copolymer showed concentric structures with the

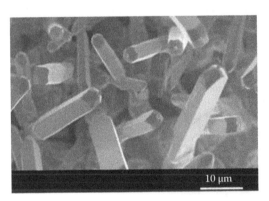
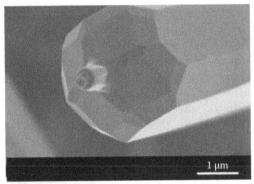

FIGURE 3.11 In_2O_3 nanopencils. (Reproduced from Qurashi, A. et al., *Arab. J. Sci. Eng.*, 35(1C), 125–145, 2010.)

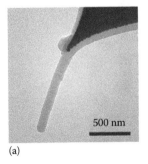
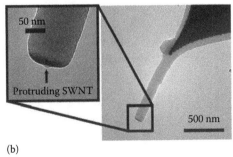

(a) (b)

FIGURE 3.12 TEM images of the nanopencil. (a) Nanopencil before SWNT electrode exposure initial length: 980 nm. (b) Nanopencil after electrode exposure length: 870 nm. The inset shows the clean SWNT electrode protruding from the SiO_x coating. (Reproduced with permission from Tayebi, N. et al., Nanopencil as a wear-tolerant probe for ultrahigh density data storage. *Appl. Phys. Lett.*, 93(10), 103112/1–103112/3 Copyright 2008, American Institute of Physics.)

PMMA layer located outermost. TEM images of an asymmetrical P(S-b-MMA) block copolymer (70:30) in the octyltriethoxysilane-modified membranes showed that PMMA islands were hexagonally packed in the middle.

3.3 NANODUMBBELLS

Nanodumbbells, closely related to nanoshuttles and nanorods, are known mainly for metals (mainly Au[68,69]), alloys, metal oxides and several metal-containing compounds, and for DNA.[70] Thus, gold nanodumbbells were fabricated *via* a simple electrochemical method using a micelle template formed by two surfactants in the presence of surfactant cetyltrimethylammonium bromide, co-surfactant tetradecyltrimethylammonium bromide, and acetone solvent. The shape of Au nanoparticles can be modified to form dumbbell structure by the addition of acetone solvent during the electrolysis.[71–73] The modulation of localized surface plasmons in gold nanodumbbells (Figure 3.13) through stepwise silver coating, along with a detailed discussion regarding the experimental parameters affecting the final core-shell morphology, was described in Ref. [74]. In addition, the rod-like tobacco mosaic virus and citrate-covered gold nanoparticles of 6 nm diameter self-assemble into a metal-virus nanodumbbell.[75] Through the reaction of Te nanowires with sodium tetrachloroaurate in the presence of hexadecyltrimethylammonium bromide (CTAB) over reaction times of 10, 20, and 60 min, gold-tellurium nanodumbbells (Figure 3.14) were obtained, among other nanoforms, including closely related nanopeapods (Figure 3.15)[76] (see Section 3.5).

In addition to gold, Co and Co-Ni nanoparticles were synthesized by reducing mixtures of cobalt and nickel acetates in an NaOH solution in 1,2-propanediol (Reaction 3.5).[77] One of a series of products, $Co_{50}Ni_{50}$ nanodumbbells were formed, consisting of a central column richer in Co capped with two terminal platelets richer in Ni. The shape of the dumbbells (Figure 3.16) was found to depend strongly on the basicity; long dumbbells were obtained for the lowest NaOH concentration, and short dumbbells and diabolos for the highest. In addition, the generation of reactive oxygen species (ROS) at the surface of differently sized $CoPt_3$ spherical nanocrystals and $CoPt_3$/Au nanodumbbells was systematically studied.[78] Additionally, synthesis of colloidal nanostructures combining magnetic (FePt) with a narrow-gap semiconductor (PbS and PbSe) as core-shells or nanodumbbells and study

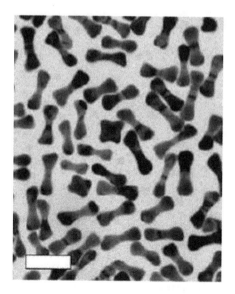

FIGURE 3.13 Gold nanodumbbells. Scale bar 100 nm. (Reproduced with permission from Fernanda Cardinal, M. et al., Modulation of localized surface plasmons and SERS response in gold dumbbells through silver coating, *J. Phys. Chem. C*, 114(23), 10417–10423. Copyright 2010 American Chemical Society.)

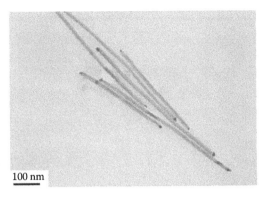

FIGURE 3.14 Au-Te nanodumbbells. (Reproduced with permission from Lin, Z.-H. and Chang, H.-T., Preparation of gold-tellurium hybrid nanomaterials for surface-enhanced Raman spectroscopy, *Langmuir*, 24(2), 365–367. Copyright 2008 American Chemical Society.)

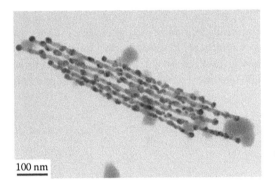

FIGURE 3.15 Au-Te nanopeapods. (Reproduced with permission from Lin, Z.-H. and Chang, H.-T., Preparation of gold-tellurium hybrid nanomaterials for surface-enhanced Raman spectroscopy, *Langmuir*, 24(2), 365–367. Copyright 2008 American Chemical Society.)

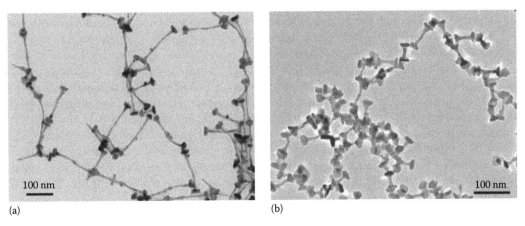

(a)   (b)

FIGURE 3.16 (a) Long and (b) short nanodumbbells of $Co_{50}Ni_{50}$. (Reproduced with permission from Ung, D. et al., Growth of magnetic nanowires and nanodumbbells in liquid polyol, *Chem. Mater.*, 19(8), 2084–2094. Copyright 2007 American Chemical Society.)

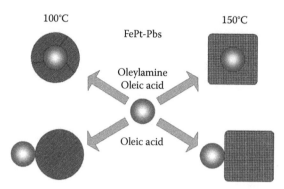

FIGURE 3.17 Morphology-directing synthesis conditions for FePt-PbS nanostructures. (Reproduced with permission from Lee, J.-S. et al., "Magnet-in-the-semiconductor" FePt-PbS and FePt-PbSe nanostructures: Magnetic properties, charge transport, and magnetoresistance, *J. Am. Chem. Soc.*, 132(18), 6382–6391. Copyright 2010 American Chemical Society.)

of their optical, magnetic, electrical, and magnetotransport properties were carried out.[79] Figure 3.17 provides an illustration of the parameters controlling the FePt-PbS nanostructure morphology. The arrays of magnet-in-the-semiconductor nanostructures showed semiconductor-type transport properties with magnetoresistance typical for magnetic tunnel junctions, thus combining the advantages of both functional components. Multicomponent colloidal nanostructures can be used as the building blocks for design of multifunctional materials for electronics and optoelectronics.

$$(Co^{II})_{alkoxide} + (Co^{II}, Ni^{II})_{hydroxyacetate} \underset{OH^-}{\overset{}{\rightleftharpoons}} Co^{II}_{solution} + Ni^{II}_{solution} \underset{Growth}{\overset{Nucleation}{\longrightarrow}} CoNi^0$$

$$\uparrow 1,2\text{-propanediol}$$

$$Co(CH_3CO_2)_2 + Ni(CH_3CO_2)_2 \tag{3.5}$$

Oxide nanodumbbells can be an intermediate phase depending on reaction conditions. Thus, an inverse microemulsion method was developed not only for synthesizing low-cost TiO_2 nanocrystals but also for making these nanocrystals self-assemble into various nanoparticles at 85°C. By variation of the volume ratios of oil to water in reverse microemulsions, the morphologies of obtained samples turned from nanoclusters to nanospherules, then grew into nanodumbbells, and became nanorods at last, as directly observed at TEM.[80] Site-selective assembly of magnetic nanoparticles onto the ends or ends and sides of gold nanorods with different aspect ratios to create multifunctional nanorods decorated with varying numbers of magnetic particles led to hybrid nanoparticles, designated as Fe_3O_4-Aurod-Fe_3O_4 nanodumbbells and Fe_3O_4-Aurod necklace like.[81] These hybrid nanomaterials can be used for multiplex detection and separation because of their tunable magnetic and plasmonic functionality. Nanodumbbells have also been reported for ZnO,[82] CdSe,[83] and Au-CdSe (Figure 3.18).[84] In the case of the last composite, light photocatalysis using highly controlled hybrid gold-tipped CdSe nanorods (nanodumbbells) was discussed.[85] Under visible light irradiation, charge separation takes place between the semiconductor and metal parts of the hybrid particles (Figure 3.19). The charge-separated state was then utilized for direct photoreduction of a model acceptor molecule, methylene blue, or alternatively, retained for later use to perform the reduction reaction in the dark.

At last, three rigid-rod conjugated organic–inorganic hybrid nanodumbbells, $(Bu_4N)_4$ $[O_{18}Mo_6N(C_6H_4)NMo_6O_{18}]$, $(Bu_4N)_4[O_{18}Mo_6N(C_{12}H_8)NMo_6O_{18}]$, and $(Bu_4N)_4[O_{18}Mo_6N(C_{14}H_{12})$ $NMo_6O_{18}]$, bearing terminal polyoxometalate cages, with different lengths of ~2–3 nm and different substituents on the rod, were synthesized (Reaction 3.6) *via* 1-pot reaction of octamolybdate and

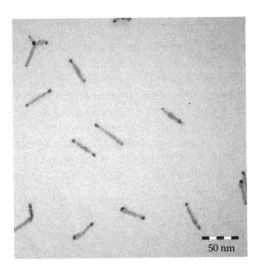

FIGURE 3.18 Nanodumbbells (Au-CdSe). (Reproduced with permission from Costi, R. et al., Electrostatic force microscopy study of single Au-CdSe hybrid nanodumbbells: Evidence for light-induced charge separation, *Nano Lett.*, 9(5), 2031–2039. Copyright 2009 American Chemical Society.)

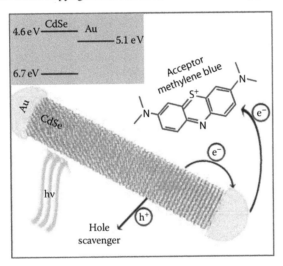

FIGURE 3.19 Scheme of a light-induced charge separation mechanism in a nanodumbbell in which the photo-generated electron-hole pair separates so that the electron resides at the gold tip and the hole at the CdSe nanorod. The scheme also depicts the transfer of the hole to the scavenger and the reduction of the MB molecule upon electron transfer from the gold tip. The inset shows the energy band alignment between CdSe (4 nm dots) and Au. (Reproduced with permission from Costi, R. et al., Visible Light-induced charge retention and photocatalysis with hybrid CdSe-Au nanodumbbells, *Nano Lett.*, 8(2), 637–641. Copyright 2008 American Chemical Society.)

the appropriate aromatic diamine dihydrochloride with N,N-dicyclohexylcarbodiimide (DCC) as a dehydrating agent.[87]

$$3(Bu_4N)_4(Mo_8O_{26}) + 2ClH_3N-Ar-NH_3Cl + 6DCC \xrightarrow{CH_3CN}$$

$$\longrightarrow 2(Bu_4N)_4[O_{18}Mo_6 \equiv N-Ar-N \equiv Mo_6O_{18}] + 6DCU + 4Bu_4NCl \quad (6)$$

$$(DCU = N,N'\text{-dicyclohexylurea})$$

$$Ar = \text{—⟨◯⟩—} , \text{—⟨◯⟩⟨◯⟩—} , \text{—⟨◯⟩⟨◯⟩—} \quad (3.6)$$

3.4 NANOSHUTTLES

Nanoshuttles are mainly reported for oxides, sulfides, and more rarely for nonmetals, elemental metals, and their other salts. Thus, Au/Ag nanoshuttles (Figure 3.20) with sharp tips at both ends were synthesized in glycine solution by chemically depositing silver on gold nanorods.[88] Simple amino acids (serine and lysine) with different functionalities and structural conformation were used as good crystal growth modifiers for controlled growth of tellurium crystals by a hydrothermal approach from sodium telluride as a precursor.[89] This way, shuttle-like scrolled Te nanotubes (Figure 3.21) with two sharp and flexible tails and nanowires with very sharp tips can be selectively synthesized. Their proposed growth mechanism is shown in Figure 3.22. It was demonstrated that the choice of an amino acid with a suitable combination of functional groups

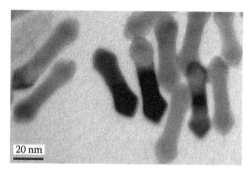

FIGURE 3.20 Nanoshuttle (Au/Ag). (Reproduced with permission from Li, M. et al., *Opt. Express*, 16(18), 14288–14293. Copyright 2008 Optical Society of America.)

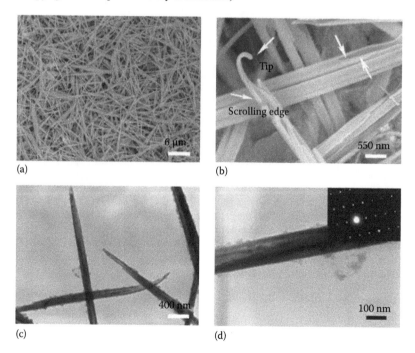

FIGURE 3.21 SEM and TEM images of the nanotubes: (a) a general view of shuttle-like scrolled nanotubes of tellurium, (b) a high magnification image (white arrows indicate the flexible tips and the scrolling edges), (c) a typical TEM image of the scrolled nanotubes, and (d) a typical nanotube and the electron diffraction pattern taken on it along the [110] axis. (Reproduced with permission from He, Z. et al., Amino acids controlled growth of shuttle-like scrolled tellurium nanotubes and nanowires with sharp tips, *Chem. Mater.*, 17(11), 2785–2788. Copyright 2005 American Chemical Society.)

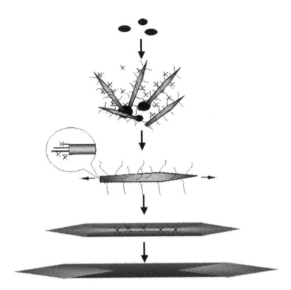

FIGURE 3.22 Proposed growth mechanism for the formation of shuttle-like scrolled nanotubes using serine as additive. (Reproduced with permission from He, Z. et al., Amino acids controlled growth of shuttle-like scrolled tellurium nanotubes and nanowires with sharp tips, *Chem. Mater.*, 17(11), 2785–2788. Copyright 2005 American Chemical Society.)

such as $-NH_2$, $-COOH$, and $-OH$ can induce the formation of different tellurium nanostructures by a facile hydrothermal approach.

CuO shuttle-like and flower-like nanocrystals were synthesized through a one-step, low-temperature solution-phase method in the presence of a cation surfactant, hexadecyl trimethyl ammonium bromide.[90] Alternatively, shuttle-like copper oxide, prepared[91] by a hydrothermal decomposition process, was then immobilized on the surface of a glassy carbon electrode modified with a film of poly(thionine). The modified electrode displayed excellent amperometric response to Hg(II), with a linear range from 40 nM to 5.0 mM and a detection limit of 8.5 nM at a signal-to-noise ratio of 3. The sensor exhibited high selectivity and reproducibility and was applied to the determination of Hg(II) in water samples. A nanocomposite membrane, comprising of nanosized shuttle-shaped cerium oxide (CeO_2), single-walled carbon nanotubes (SWNTs), and hydrophobic room temperature ionic liquid 1-butyl-3-methylimidazolium hexafluorophosphate, was developed on the glassy carbon electrode for electrochemical sensing of the immobilization and hybridization of DNA.[92] Additionally, a solvothermal method for growing ZnO nanostructures with different morphologies, including nanoshuttles (Figure 3.23),[93] is based on the following Reaction 3.7, combining two different coordination agents and adjusting the ratio of solvents and the reaction time:

$$Zn^{2+} + CH_3(CH_2)_{15}N(Br)(CH_3)_3/C_6H_{12}N_4 \rightarrow Zn^{2+}\text{-amino complex} \rightarrow Zn(OH)_2 \rightarrow ZnO \quad (3.7)$$

Anatase TiO_2 nanoshuttles (Figure 3.24) were prepared[94] *via* a hydrothermal method under alkaline conditions by employing titanate nanowires as the self-sacrificing precursors. It was established that a radical structural rearrangement took place from titanate wires to anatase TiO_2 shuttles during the hydrothermal reaction on the basis of a dissolution-recrystallization process. The shape and phase transformation process were found to be dependent on the hydrothermal reaction time. In addition, the delicate fishbone-like composite structure of titanate nanowire/anatase nanoshuttles may find practical applications in photocatalysts and separation. Also, ZnS nanospheres and nanoshuttles were obtained[95] in a simple reverse micelle system containing cyclohexane as oil phase, aqueous solution of Na_2S and $Zn(OAc)_2$ as reactant, *n*-pentanol as co-surfactant and surfactant (peregals). It was established that the obtained ZnS nanospheres and nanoshuttles were of hexagonal primitive

FIGURE 3.23 ZnO nanoshuttle. (Reproduced with permission from Wen, B. et al., Controllable growth of ZnO nanostructures by a simple solvothermal process, 112(1), 106–111. Copyright 2008 American Chemical Society.)

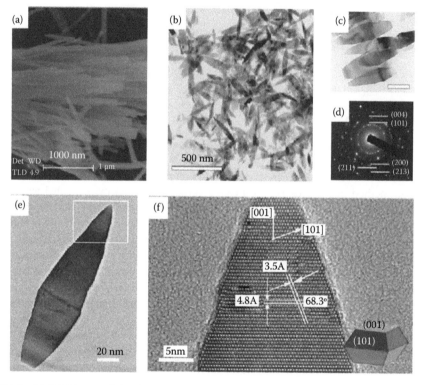

FIGURE 3.24 (a) SEM image of the titanate nanowires precursor. (b–d) TEM and ED results of the anatase TiO$_2$ nanoshuttles. (e, f) TEM and corresponding HRTEM image for a single shuttle. Inset of (f) denotes the relationship of (001) and (101) facets. (Reproduced with permission from Wang, H. et al., Synthesis of anatase TiO$_2$ nanoshuttles by self-sacrificing of titanate nanowires, *Inorg. Chem.*, 48(20), 9732–9736. Copyright 2009 American Chemical Society.)

phase with calculated cell parameters $a = 0.3823$ nm and $c = 56.2$ nm. The nanoshuttles were about 60 nm in diameter with a length of about 110 nm. Uniform shuttle-like Sb$_2$S$_3$ nanorod bundles were synthesized *via* a polyvinylpyrrolidone (PVP)-assisted solvothermal approach under alkaline condition, using antimony chloride (SbCl$_3$) and thiourea (CH$_4$N$_2$S) as the starting materials in ethanol.[96] It was shown that the shuttle-like Sb$_2$S$_3$ bundles were composed of nanorods with a size distribution of 20–40 nm and growing along *c*-axis. In addition, a luminescent nanomaterial, monoclinic wolframite-type HgWO$_4$ nanoshuttles (length 200–300 nm, diameter 100–150 nm), was prepared through ultrasonic method.[97] Other nanoshuttles have been reported for TiO$_2$,[98] derivatized SiO$_2$,[99] SnO$_2$,[100] CdS,[101] and CaCO$_3$.[102] Biomedical uses of nanoshuttles were reviewed in Ref. [103].

3.5 NANOPEAPODS

"Peapod" is the descriptive term applied to a supramolecular hybrid assembly in which nanotubes are filled with a 1D chain of molecules.[104] The archetypal peapod is that in which C_{60} molecules[105] are contained within carbon nanotubes (Figure 3.25).[106] Nanopeapods are mainly examined for carbon nanotubes and fullerenes (or their endohedral metal complexes), although several examples of other inorganic compounds have been reported. For CNTs and fullerenes encapsulated in them, comprehensive investigations have been carried out, including, for example, studies of charge transfer,[107] binding fullerene nanotube,[108] atomic structure[109] and electronic properties,[110] influence of electromechanical effects,[111] fundamental mechanical principles and conventional applied mathematical modeling,[112] structural evolutions,[113] electrical transport,[114] density functional study,[115] among many others. Thus, the dynamics of C_{60} fullerene molecules inside SWNTs was studied using inelastic neutron scattering.[116] C_{60} vibrations and their sensitivity to temperature were identified. Additionally, a clear signature of rotational diffusion of the C_{60} was evidenced, which persisted at lower temperature than in 3D bulk C_{60}. A fascinating structural transformation occurring inside SWNTs was found to be the fullerene coalescence, which is responsible for forming stable zeppelin-like carbon molecules.[117] Sequences of fullerene coalescence induced by electron irradiation on pristine nanotube peapods were revealed. It was indicated that the merging of fullerenes resulted in stable but corrugated tubules (5–7 Å in diameter) confined within SWNTs. The process occurred *via* the polymerization of C_{60} molecules followed by surface reconstruction, which can be triggered either by the formation of vacancies (created under electron irradiation) or by surface-energy minimization activated by thermal annealing. The rotation of fullerene chains in SWNT peapods was studied using low-voltage high resolution TEM.[118] Anisotropic fullerene chain structures (i.e., C_{300}) were formed *in situ* in carbon nanopeapods *via* E-beam-induced coalescence of individual fullerenes (i.e., C_{60}). It was established that the large asymmetric C_{300} fullerene structure exhibited translational motion inside the SWNT and unique corkscrew-like rotation motion. The axial stability of single-walled carbon nanopeapods was studied based on an elastic continuum shell model.[119]

In the case of metal endo-fullerenes, after heating nanopeapods of the SWNT encapsulating the $La_2@C_{80}$, the atomic wires of La atoms inside CNTs were obtained.[120] The La atoms formed dimers and were linearly arranged inside the CNTs. The valence state of the La in atomic wires was the same as that in the pristine $La_2@C_{80}$ peapods. In addition, a direct evidence of the dynamic behavior of the confined atoms in metallofullerenes was given[121] by observing individual atoms using the high-resolution TEM. Rapid movement of Gd atoms in $Gd_2@C_{92}$ peapod was identified. By comparison with simulation, the amplitude of the motion is roughly quantified as 0.2 nm at r.t. and was reduced to almost half at 100 K. In a related work,[122] $M@C_{82}$ (M = Gd, Dy) and their composites with SWCNTs (Figure 3.26) were studied. Cesium-doped C_{60} nanopeapods were prepared from SWNTs and characterized by electron-energy-loss spectroscopy.[123] The ordered Cs_2C_{60} structure

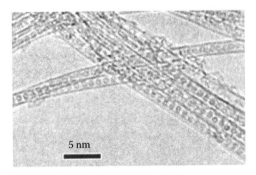

5 nm

FIGURE 3.25 Typical HRTEM image of C_{60} nanopeapods. (Reproduced with permission from Okazaki, T., et al., Optical band gap modification of single-walled carbon nanotubes by encapsulated fullerenes, *J. Am. Chem. Soc.*, 130, 4122–4128. Copyright 2008 American Chemical Society.)

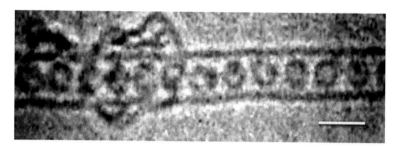

FIGURE 3.26 A HRTEM image of (Gd@C$_{82}$)@SWNT. The scale bar is approximately 2 nm. (Reproduced from Kitaura, R. et al., Magnetism of the endohedral metallofullerenes M@C82 (M = Gd, Dy) and the corresponding nanoscale peapods: Synchrotron soft x-ray magnetic circular dichroism and density-functional theory calculations, *Phys. Rev. B*, 76, 172409. Copyright 2007 by the American Physical Society.)

was observed inside the carbon nanotubes as well as cesium ions on the exterior of the carbon nanotubes. Single-atom migration of Tb, Gd, or Ce metal from cage to cage through an intentionally induced atomic path within a carbon nanopeapod was demonstrated.[124] Additionally, carbon nanotube peapods can be applied to a nanoelectroemitter when encapsulated endo-fullerenes are electroemitted from the carbon nanotube under applied external electric fields.

Fullerenes can be encapsulated by not only CNTs but also in other nanotubes. Thus, *ab initio* simulations of the electronic properties of a chain of C$_{60}$ molecules encapsulated in a boron-nitride nanotube,[125] so-called BN nanopeapod, were carried out.[126,127] It was demonstrated that this structure can be effectively doped by depositing potassium atoms on the external wall of the BN nanotube. The resulting material became a true metallic 1D crystal. In a related report,[128] dedicated to B-N fullerenes, the atomic structures, energies of formation, electronic structures, and thermal stabilities (in the temperature range $T = 0$–$3000\,K$) of boron-nitrogen nanopeapods B$_{12}$N$_{12}$@N-NT were simulated using the self-consistent density functional tight-binding method. The B$_{12}$N$_{12}$@N-NT nanopeapods are regular linear ensembles of B$_{12}$N$_{12}$ boron-nitrogen fullerenes encapsulated in boron-nitrogen nanotubes (BN-NT), such as the (14, 0) nonchiral zigzag BN nanotubes, the (8, 8) nonchiral armchair BN nanotubes, and the (12, 4) chiral BN nanotubes.

In addition to carbon and boron-nitrogen nanopeapods, noble metal composites are known. Thus, inorganic nanopeapod-faceted Au nanoparticles inside MgO nanowires were obtained by Au self-assembling into a nanoparticle chain during the vapor-liquid-solid growth of the MgO nanowires for which gold also served as the catalyst.[129] It was shown that such Au@MgO nanopeapods formed not only under metalorganic CVD conditions but also under our conventional vapor transport deposition condition. The growth and optical properties of a noble nanocomposite Au$_2$Si@SiONW, consisting of a self-organized chain of Au silicide nanocrystals pea-podded in an Si oxide nanowire, were studied.[130] Tellurium nanowires, prepared from TeO$_2$ and hydrazine as reducing agent, allowed reacting with sodium tetrachloroaurate in the presence of CTAB over reaction times of 10, 20, and 60 min, resulting in gold-tellurium nanodumbbells, gold-tellurium nanopeapods, and gold pearl-necklace nanomaterials, respectively.[77] Silica-coated silver nanowires can be chemically treated to produce a peapod architecture in which silver peas were embedded in silica pods (Reaction 3.8).[131] This architecture is potentially useful for chemical sensing, plasmonic, or catalytic applications. In addition, Pt@CoAl$_2$O$_4$ nanopeapods (Figures 3.27 and 3.28) consisting of well-defined PtNP chains embedded in CoAl$_2$O$_4$ were synthesized by the pulsed electrodeposition of Co/Pt multilayered nanowires into an anodic Al oxide (AAO) membrane and a subsequent high-temperature solid-state reaction between Co/Pt nanowires and AAO.[132] The pulse durations for Pt and Co depositions control the size and separation of PtNPs. Among other peapod-like nanostructures, 1D nanohybrids made of Fe@CaS nanopeapods homogeneously coated with protecting boron nitride sheaths were synthesized through a solid-liquid-solid reaction at ~1500°C.[133] Their formation mechanism is shown in Figure 3.29. The structures had diameters of 150–180 nm and lengths in the range of tens

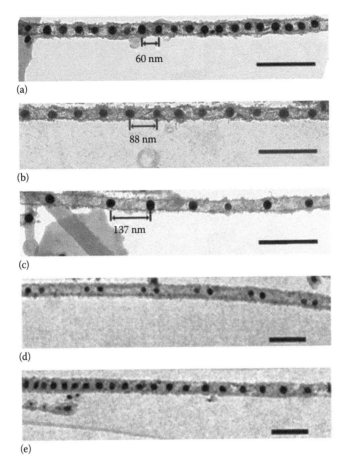

(a)

(b)

(c)

(d)

(e)

FIGURE 3.27 (a through c) TEM micrographs of Pt@CoAl$_2$O$_4$ inorganic nanopeapods with regularly variable "pea" separations; (d, e) TEM micrographs of Pt@CoAl$_2$O$_4$ inorganic nanopeapods with periodically modulated separations. Scale bars = 200 nm. (Liu, L. et al., Tailor-made inorganic nanopeapods: Structural design of linear noble metal nanoparticle chains. *Angew. Chem. Int. Ed.* 2008, 47(37), 7004–7008. Copyright Wiley-VCH Verlag GmbH & Co. KGaA. Reproduced with permission.)

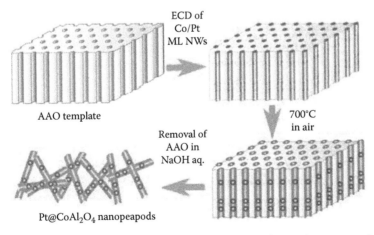

FIGURE 3.28 Schematic illustration of the fabrication of Pt@CoAl$_2$O$_4$ inorganic nanopeapods. (Liu, L. et al., Tailor-made inorganic nanopeapods: Structural design of linear noble metal nanoparticle chains. *Angew. Chem. Int. Ed.* 2008, 47(37), 7004–7008. Copyright Wiley-VCH Verlag GmbH & Co. KGaA. Reproduced with permission.)

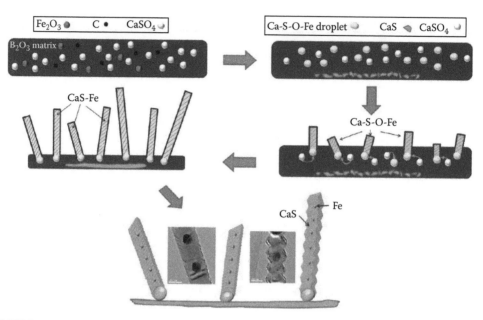

FIGURE 3.29 Schematic illustration of the nanopeapod growth mechanism. (Reproduced with permission from Lin, J. et al., Ferromagnetic Fe@CaS nanopeapods with protecting BN tubular sheaths, *J. Phys. Chem. C*, 113(33), 14818–14822. Copyright 2009 American Chemical Society.)

of micrometers. The nanopeapods were found to be ferromagnets at r.t. with a coercive force of 64 Oe. Also, *in situ* observations of the shape transformation of $Sn-SnO_2$ coaxial nanocables induced by E-beam irradiation were performed in a transmission electron microscope. $Sn-SnO_2$ coaxial nanocables, induced by E-beam irradiation, spontaneously transformed into nanopeapods with Sn peas and SnO_2 pods through the melting and condensing mechanism of Sn in SnO_2 nanoshells.[134] Nanopeapods then transformed into SnO_2 nanotubes covered by Sn islands through the nanojet mechanism. Nanostructures consisting of SnO_2 trunks and Sn branches finally formed through size-confinement basal growth.

$$R = (CH_3)_2Si(OCH_3)_2$$

$$\boxed{Ag} \xrightarrow{HS(CH_2)_2Si(OCH_3)_3} \boxed{Ag} \quad (8)$$

$$\downarrow Na_2Si_3O_7$$

Silica coating process

$$(3.8)$$

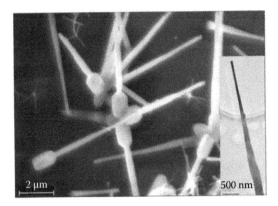

FIGURE 3.30 ZnO nanopins. (Reproduced with permission from Zhang, Y. et al., *Front. Phys. China*, 1, 72, 2006.)

3.6 NANOPINS

This rare nanostructural type is mainly described for zinc oxide (Figure 3.30).[135] Thus, various morphologies of ZnO nanostructures were obtained through a method incorporating electrochemical corrosion with three modes: liquid membrane and above and below the H_2O line in partial immersion.[136] The evolution of ZnO nanostructures such as nanorods, nanowires, nanopins, and nanodendrites was observed. In a related work,[137] well-aligned ZnO nanorods and nanopins were synthesized on an Si substrate using a one-step simple thermal evaporation of a mixture of Zn and $Zn(OAc)_2$ powder under controlled conditions. Two different growth mechanisms were offered to interpret the formation of ZnO nanostructures. For ZnO nanorods, better crystallinity and optical quality were established than the ZnO nanopins (this ZnO nanostructure is used as a field emitter[138,139]). Several interesting nonstandard structures (including nanopins, see Figure 3.31) on TiO_2 basis were observed during the solution synthesis of large single crystals of octahedron-like anatase TiO_2 from $TiCl_3$ as precursor.[140] Systematic control of the crystal growth of rutile nanopins on the microanatase octahedron single crystal resulted in a nanomicro chestnut-like TiO_2 structure. The important point in the crystal growth was the control of the chemical reaction in the solution. Slow oxidation of Ti(III) → Ti(IV) and decomposition of sodium dodecyl sulfate (SDS) into SO_4^{2-} (Reactions 3.9 through 3.11) resulted in the maintenance of a low degree of supersaturation, which was essential to achieve the conditions necessary for crystal growth (the sulfate species was

FIGURE 3.31 Rutile nanopins. (Reproduced with permission from Hosono, E. et al., One-step synthesis of nano-micro chestnut TiO_2 with rutile nanopins on the microanatase octahedron. *ACS Nano*, 1(4), 273–278. Copyright 2007 American Chemical Society.)

formed by the hydrolysis of SDS and adsorbed onto the TiO_2). In addition, semiconducting oxide In_2O_3 nanostructures (nanowires, nanobouquets, and nanopins) were prepared by reduction of In_2O_3 powder at 650°C through vapor-phase transport process.[141] Additionally, nanopine oxide structures can serve as precursors to synthesize other nanostructures; thus, Fe_3N nanodendrites directly by reduction-nitriding of α-Fe_2O_3 nanopine dendrites in a mixed stream of H_2-NH_3.

$$C_{12}H_{25}OSO_3^-Na^+ \rightarrow C_{12}H_{25}OH + HSO_4^-Na^+ \tag{3.9}$$

$$HSO_4^- + H_2O \rightarrow H_3O^+ + SO_4^{2-} \tag{3.10}$$

$$TiOH^{2+} + O_2 \rightarrow Ti(IV) \text{ oxo species with } SO_4^{2-} + O^{2-} \rightarrow \text{anatase } TiO_2 \text{ with sulfate species} \tag{3.11}$$

Synthesis of Fe-Al nanopins were fabricated using arc discharge.[142] These nanopins grew according to a vapor-liquid-solid mechanism and were found to be composed of a spherical base of ~20–100 nm and a needle-like tip of about several hundred nanometers. The spherical base was mainly composed of α-Fe and FeAl core coated with a thin Al_2O_3 layer, while the needle-like part contains only Al and O and corresponded to Al_2O_3. Also, high-density single-crystalline orthorhombic TiSi nanorods and nanopins (Figure 3.32) were prepared on a Ti_5Si_3 layer by CVD, using SiH_4 and $TiCl_4$ as the precursors (Reactions 3.12 and 3.13).[143] The maximum density of TiSi nanopins was obtained at the deposition temperature of 730°C.[144] The nanopins were 0.7–1 μm long in total, with the quadrate tip about 200 nm long and 50 nm × 50 nm in area. Both nanostructures grew along the [110] direction of the orthorhombic TiSi crystal from the bottom, with the tip pushed upward, and the growth process can be defined as a self-induced growth.

$$5TiCl_4(g) + 5SiH_4(g) \rightarrow Ti_5Si_3(s) + 2SiCl_4(g) + 12HCl(g) + 4H_2(g) \tag{3.12}$$

$$2TiCl_4(g) + 3SiH_4(g) \rightarrow 2TiSi(s) + SiCl_4(g) + 4HCl(g) + 4H_2(g) \tag{3.13}$$

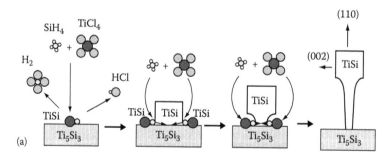

(a)

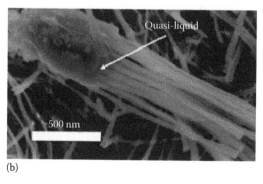

(b)

FIGURE 3.32 TiSi nanopins: (a) formation mechanism and (b) TiSi growth on quasi-liquid alloy. (Reproduced with permission from Du, J. et al., Growth mechanism of TiSi nanopins on Ti5Si3 by atmospheric pressure chemical vapor deposition. *J. Phys. Chem. C*, 111(29), 10814–10817. Copyright 2007 American Chemical Society.)

3.7 NANOCHAINS

Nanochains, which can be artificially considered as distorted nanodumbbells or nanoshuttles, connected to each other, are widespread nanostructures.[145] Chain-like nanostructures are represented by a variety of inorganic compounds, in particular by elements (free or doped, metals, alloys, or nonmetals). Thus, titanium nanochains were fabricated using crown ether as a template exploiting a simple sliding technique.[146] The nanochains of Ti were formed with an average diameter of the sphere nodes around 80 nm separated by organic molecules with a space of 1–5 nm. These metallic nanochains are potentially applicable for making electronic and optical devices. Uniform and aligned nickel nanochains with interlaced nanodisc structure were synthesized using a template-free magnetic-field-assisted method at r.t.[147] These nanochains were found to be composed of nano-discs with diameters of ~90 nm and thicknesses of ~10 nm. It was established that Ni nanochains had a coercivity of about 300 Oe and effective anisotropy of about five times more than the bulk value. Ni@Ni$_3$C nanochains, Ni@NiSG nanochains with a spin glass (SG) surface layer, and Ni@NiNM nanochains with a nonmagnetic (NM) surface layer were also discovered.[148] 1D germanium nanostructures (nanochains, nanowires, microsphere-string nanowires, and microrods) with different morphologies, sizes, and microstructures were synthesized in a small-diameter quartz cavity by evaporation of Ge powders.[149] The nanochains and nanowires grew along [111] direction, whereas the microspheres and microrods were polycrystalline. In addition, Cu nanoparticle chains (reported also in Ref. [150]) encapsulated in Al$_2$O$_3$ nanotubes were generated (Figures 3.33 and 3.34) in a controlled manner by reduction of CuO nanowires embedded in Al$_2$O$_3$ at a sufficiently high temperature.[151] The particle diameters and chain lengths corresponding to the inner diameters and lengths of the tubes, respectively, were controlled by the size of the CuO nanowire templates. Magnetic FeNi$_3$ nanochains were synthesized by reducing iron(III) acetylacetonate and nickel(II) acetylacetonate with hydrazine in ethylene glycol solution without any template under microwave irradiation.[152] The proposed mechanism of their formation is shown in Figure 3.35. The size of the aligned nano-spheres in these chains could be adjusted from 150 to 550 nm by increasing the amounts of the precursors; their length was found to be about several tens of micrometers. The as-prepared nanochains

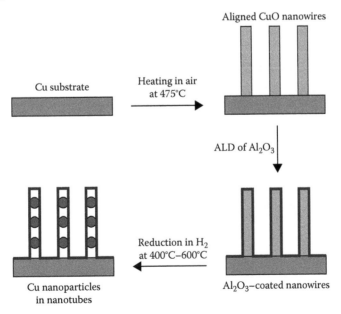

FIGURE 3.33 A schematic diagram of the preparation process of Cu nanoparticle chains confined in Al$_2$O$_3$ nanotubes. (Reproduced with permission from Qin, Y. et al., Rayleigh-instability-induced metal nanoparticle chains encapsulated in nanotubes produced by atomic layer deposition, *Nano Lett.*, 8(1), 114–118. Copyright 2008 American Chemical Society.)

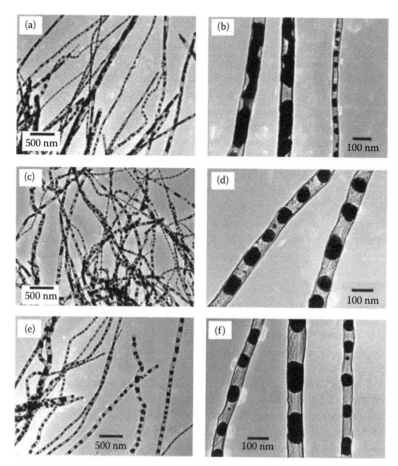

FIGURE 3.34 TEM images of Cu nanoparticle chains prepared by reduction of CuO nanowires with a 5 nm Al_2O_3 shell in H_2 for 1 h at various temperatures: (a, b) sample prepared at 400°C; (c, d) sample prepared at 500°C; (e, f) sample prepared at 600°C. Panels (b), (d), and (f) are TEM images corresponding to panels (a), (c), and (e), respectively, at higher magnification. Note that the white shadows near some tubes result from defocused features due to different heights. (Reproduced with permission from Qin, Y. et al., Rayleigh-instability-induced metal nanoparticle chains encapsulated in nanotubes produced by atomic layer deposition, *Nano Lett.*, 8(1), 114–118. Copyright 2008 American Chemical Society.)

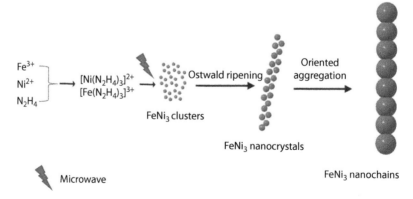

FIGURE 3.35 Illustration of a proposed mechanism for the formation of $FeNi_3$ nanochains. (Reproduced from Jia, J. et al., Magnetic nanochains of $FeNi_3$ Prepared by a template-free microwave-hydrothermal method, *ACS Appl. Mater. Interfaces*, 2(9), 2579–2584. Copyright 2010 American Chemical Society.)

are nontoxic to zebrafish larvae. These FeNi$_3$ alloyed nanochains show enhanced coercivity and saturation magnetization. Additionally, Mn[153] and Co[154] nanochains have been also prepared.

Noble metal nanochains are widespread. Thus, micrometer-length platinum nanochains (PtNCs) constructed with spherical Pt nanoparticles (PtNPs) of about 5 nm were synthesized by the reduction of a Pt-EDTA chelate complex with sodium borohydride (NaBH$_4$).[155] Comparative studies of electrocatalysts on their basis demonstrated that the PtNCs catalyst had a higher electrochemical active surface area (ECSA, almost 2 times) and better catalytic activity (almost 10 times) in comparison with the PtNPs catalyst. Gold nanochains were prepared through the reduction of HAuCl$_4$ by NaBH$_4$ in 1-butyl-3-methylimidazolium tetrafluoroborate ionic liquid.[156] This ionic liquid not only acted as a reaction medium but also as a modifier in the preparation of nanogold. In addition, a one-step polyol process was used to prepare linear assemblies of Pd nanoparticles (highly linearly ordered Pd NP chains with ~8.5 nm in diameter and the interparticle distance ~1.4 nm).[157] Ag,[158] NiAg,[159,160] TePt,[161] and PtCo[162] nanochains are also known.

Chain-like nanostructures on carbon basis are quite distinct: connected carbon nanocapsules, carbon nanotubes containing C$_{60}$, or other species inside. Thus, an aligned array of chain-like carbon nanocapsules was prepared by a simple CVD process.[163] It was revealed that the carbon nanochains were filled with Fe$_3$C particles other than Fe; also, the nanochain was formed by the converging of conjoint multinanocapsules. Also, the adsorption of C$_{60}$ on a bilayer film of pentacene on Ag(111) was studied with STM and spectroscopy.[164] At low coverage, C$_{60}$ molecules form extended linear structures due to the templating effect of the pentacene bilayer. The C$_{60}$ molecules in the chains adsorbed at bridge sites between two neighboring pentacene molecules. Other elemental chain-like nanostructures are represented by B-doped Si nanochains (an excellent nanostructure to build single-charge tunneling transistors).[165]

Oxide nanochains are known, for example, those of silica (Figure 3.36).[166] Metal oxide nanotubes can embed inorganic species (e.g., elemental metals) forming nanochains. Thus, a flexible assembly method for producing linear metal nanoparticle chains embedded in nanotubes by annealing metal nanowires confined in nanotubes based on the Rayleigh instability was demonstrated.[167] The technique consisted of two alternative routes (Figure 3.37). As an example, Au nanochains embedded in TiO$_2$ nanotubes are shown in Figure 3.38. Particle spacing, diameter, shape of the nanochains, tube diameter, and shell thickness can be tuned by atomic layer deposition (ALD) significantly.

Hybrid In$_2$O$_3$/Ag nanochains (Figure 3.39) and nanoparticles were synthesized (Reactions 3.14 through 3.16) by the solvothermal reaction of indium nitrate and silver nitrate (AgNO$_3$) with or without PVP in an ethylene glycol solution at 230°C for 20 h followed by annealing at 450°C for 5 h.[168]

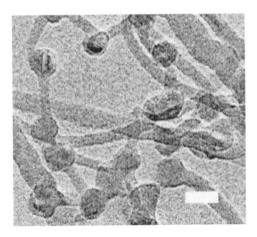

FIGURE 3.36 Nanochain of SiO$_x$. (Reproduced with permission from Barsotti, R.J., Jr. et al., Imaging, structural, and chemical analysis of silicon nanowires, *Appl. Phys. Lett.*, 81(15), 2866–2868. Copyright 2002, American Institute of Physics.)

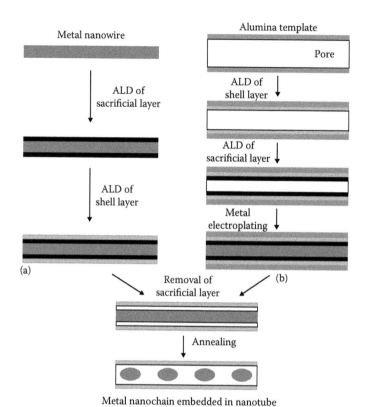

FIGURE 3.37 Schematic of two fabrication approaches for nanochains embedded in nanotubes. (a) ALD of sacrificial and shell layers on metal nanowires leading to unaligned nanochains after removing the sacrificial layers and subsequent annealing. (b) ALD of shell and sacrificial layers in the pores of alumina templates followed by metal electrodeposition leading to well-aligned nanochains after removing the sacrificial layers and templates and subsequent annealing. (Reproduced with permission from Qin, Y. et al., General assembly method for linear metal nanoparticle chains embedded in nanotubes, *Nano Lett.*, 8(10), 3221–3225. Copyright 2008 American Chemical Society.)

A surfactant-induced formation mechanism was proposed to account for their growth behavior. Ag-coupled N-doped TiO$_2$ (Ag-N-TiO$_2$) nanochains were synthesized by a combination of sol-gel and hydrothermal methods.[169,170] These nanostructures were observed to strongly absorb visible light over wide wavelength range (440–800 nm) as well as UV light, exhibiting visible-light photocatalytic activity of the highest quantum efficiency (46%) ever reported. It was demonstrated that the HPA-incorporated Ag-N-TiO$_2$ nanochains are very useful for solar cells in which electron transfer takes place similar to the Z-scheme mechanism of the plant photosynthetic system with good light harvesting efficiency over the entire visible spectrum. Additionally, hierarchically nanostructured ZnO flowers composed of bundled nanochains were synthesized by a facile wet chemical method.[171] Studies of electrochemical performance of these ZnO products by their co-immobilization onto the surface of glassy carbon electrode showed both high sensitivity and selectivity for the direct detection of dopamine in the presence of L-ascorbic acid without any visible interference.

$$HOCH_2CH_2OH \leftrightarrow CH_3CHO + H_2O \tag{3.14}$$

$$CH_3CHO + AgNO_3 + H_2O \rightarrow Ag + HNO_3 + CH_3COOH \tag{3.15}$$

$$2In(NO_3)_3 \rightarrow In_2O_3 + 6HNO_3 \tag{3.16}$$

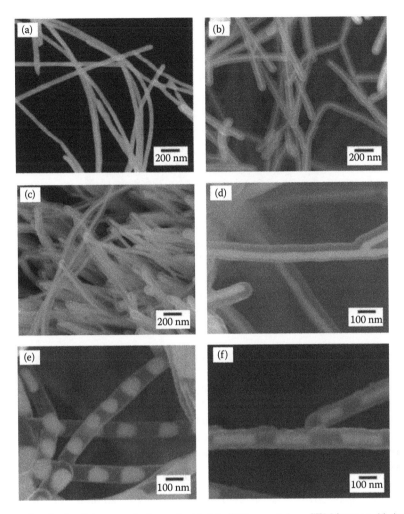

FIGURE 3.38 Synthesis of Au nanochains embedded in TiO_2 nanotubes. SEM images: (a) Au nanowires; (b) Au nanowires coated first by Al_2O_3 (20 nm) and then TiO_2 (10 nm); (c, d) confined Au nanowires with surrounding free volume after removing Al_2O_3; (e) Au nanochains produced by annealing Au nanowires confined in 10 nm thick TiO_2 shells after removing the 20 nm thick Al_2O_3; (f) Au nanochains produced by annealing Au nanowires confined in 20 nm thick TiO_2 shells after removing the 5 nm thick Al_2O_3. Annealing was performed at 550°C for 5 h in air. (Reproduced with permission from Qin, Y. et al., General assembly method for linear metal nanoparticle chains embedded in nanotubes, *Nano Lett.*, 8(10), 3221–3225. Copyright 2008 American Chemical Society.)

Among other metal oxide nanostructures, highly uniform coalescent moniliforme-shape α-Fe_2O_3 straight nanochains (Figure 3.40) built up of single crystal building blocks were synthesized *via* a facile one-pot solution-phase route at 180°C using $FeCl_3$ as precursor.[172] Their formation mechanism is shown in Figure 3.41. An exceptionally high TM (Morin transition temperature) was observed. Additionally to oxides, some oxygen-containing metal salts are known, for instance $MgAl_2O_4$.[173]

Except oxides, binary inorganic nanochains of borides and carbides and their composites have been fabricated. Thus, the CoB-silica nanochain hydrogen storage composite was prepared by *in situ* reduction of cobalt salt on the surface of amine-modified silica nanospheres.[174] It was demonstrated that this composite possessed amorphous nanochains structure by a series of nanospheres connecting in 1D. The material as electroactive negative electrodes showed high reversible

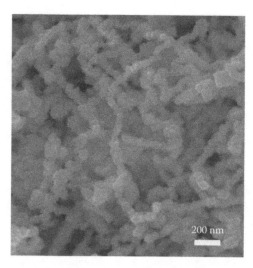

FIGURE 3.39 In$_2$O$_3$/Ag nanochains. (Reproduced with permission from Du, J. et al., A controlled method to synthesize hybrid In$_2$O$_3$/Ag nanochains and nanoparticles: Surface-enhanced Raman scattering, *J. Phys. Chem. C*, 113(23), 9998–10004. Copyright 2009 American Chemical Society.)

discharge capacity (about 500 mAh/g in the first cycle) and good cycling stability. A unique nanochain-structured mesoporous tungsten carbide (m-NCTC) was synthesized through a simple combined hydrothermal reaction–post-heat-treatment approach.[175] When loaded with Pt, the nanostructure (Pt/m-NCTC), as a catalyst, demonstrated high unit mass electroactivity and high resistance to CO poisoning for methanol oxidation, and it was much superior to Pt/C, one of the known excellent electrocatalysts. The authors emphasized that its high reaction activity and strong poison resistivity were very likely due to the unique mesoporous nanochain structure and high specific surface area (113 m^2/g). SiC[176] and SiC/SiO$_2$ were also reported. Thus, 1D nanochains and SiC/SiO$_2$ 2D *X*-junction and *Y*-junction nanochains (Figure 3.42) were synthesized under microwave heating (Reactions 3.17 through 3.20).[177] The SiC/SiO$_2$ nanochains consisted of 3C-SiC strings with diameters of 20–80 nm and periodic SiO$_2$ beads with diameters of 100–400 nm. Both SiC strings and SiO$_2$ beads produced significant photoluminescence, and the presence of SiO$_2$ beads enhanced the emissions from SiC strings.

$$C(s) + SiO_2(s) \rightarrow SiO(g) + CO(g) \tag{3.17}$$

$$SiO_2 + Si(s) \rightarrow 2SiO(g) \tag{3.18}$$

$$C(s) + CO_2(g) \rightarrow 2CO(g) \tag{3.19}$$

$$SiO(g) + 3CO \rightarrow SiC(g) + 2CO_2(g) \tag{3.20}$$

Sulfides and phosphides are also common as nanochains. Thus, highly regular clew-like Co$_4$S$_3$ nanostructure intertwisted from ultralong nanochains that were self assembled by irregular nanoparticles were synthesized by a simple and mild chemical solution route.[178] The presence of a magnetic anisotropy potential barrier was revealed even at $T > 350$ K, which causes the blocking temperature, T_B, to exceed 350 K. A method for self-assembly of CdS nanochains was proposed by direct deposition of CdS on an unfixed DNA template with 2-aminoethanethiol as a capping agent (Figure 3.43).[179] Additionally, ZnS and Zn$_x$Cd$_{1-x}$S,[180] TlGaTe$_2$[181] and CdTe[182] were also reported.

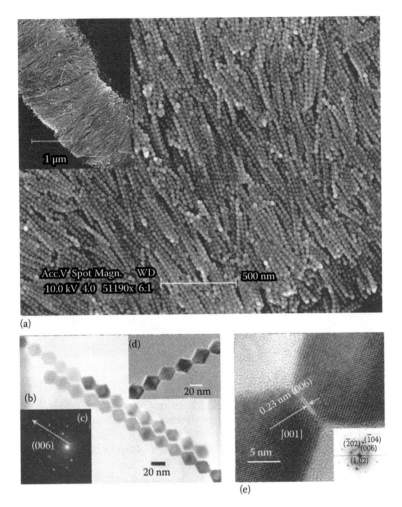

(a)

(b)

(c)

(d)

(e)

FIGURE 3.40 (a) SEM images of the moniliforme-shaped α-Fe$_2$O$_3$ nanostructures. (b) TEM image of two nanochains abutting each other side by side. The corresponding electron diffraction pattern is shown in part (c). (d) TEM image of the truncated hexagonal bipyramid particle units. (e) Lattice-resolved HRTEM image taken from the contact area of two contiguous particles and the associated FFT pattern. (Reproduced with permission from Meng, L.-R. et al., Uniform α-Fe$_2$O$_3$ nanocrystal moniliforme-shape straight chains, *Cryst. Growth Des.*, 10(2), 479–482. Copyright 2010 American Chemical Society.)

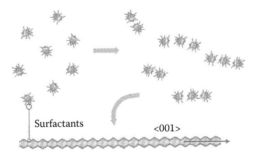

FIGURE 3.41 Illustration of the formation mechanism of the α-Fe$_2$O$_3$ nanostructures. (Reproduced with permission from Meng, L.-R. et al., Uniform α-Fe$_2$O$_3$ nanocrystal moniliforme-shape straight chains, *Cryst. Growth Des.*, 10(2), 479–482. Copyright 2010 American Chemical Society.)

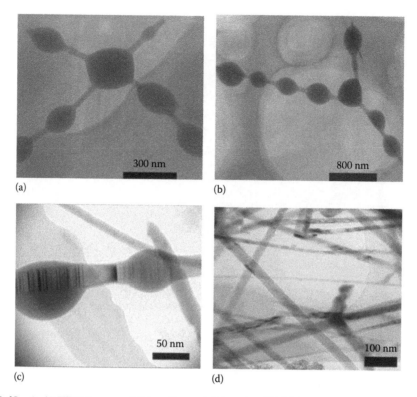

(a)   (b)

(c)   (d)

FIGURE 3.42 (a, b) TEM images of *X*-junction and *Y*-junction SiC/SiO$_2$ nanochain; (c) TEM image of a single nanochain; (d) TEM image of the nanochains after being washed with diluted HF solution for 6 h, showing only the nanowires. (Reproduced with permission from Wei, G. et al., Synthesis and properties of SiC/SiO$_2$ nanochain heterojunctions by microwave method, *Cryst. Growth Des.*, 9(3), 1431–1435. Copyright 2009 American Chemical Society.)

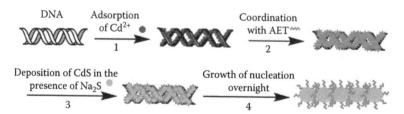

FIGURE 3.43 Schematic depiction of the CdS nanochains assembly on nonfixed DNA scaffold. (Reproduced with permission from Ge, C. et al., Luminescent cadmium sulfide nanochains templated on unfixed deoxyribonucleic acid and their fractal alignment by droplet dewetting, *J. Phys. Chem. C*, 112(29), 10602–10608. Copyright 2008 American Chemical Society.)

The diameter-modulated single crystalline gallium phosphide (GaP) nanochains were synthesized by a facile method within a confined reaction zone.[183] By varying the Ga concentration in the reaction zone, the size of knots of GaP nanochains can be manipulated. Their potential applications may be in light sources, laser or light emitting display devices. Also, GaP/GaO$_x$ core-shell nanowires and nanochains (Figure 3.44) were synthesized in a large quantity by thermal evaporation of mixtures of GaP and Ga powders at high temperature.[184] Both products were composed of GaP nanowires and GaO$_x$ amorphous shells. A distinct response between the core-shell nanowire and nanochain devices was observed under the illumination of a UV light, which might be attributed to the nature of the nanowires and nanochains.

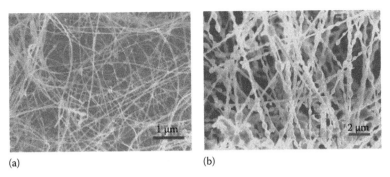

(a)   (b)

FIGURE 3.44 SEM images showing the typical morphologies of the as-synthesized GaP (a) nanowires and (b) nanochains, respectively. (Reproduced with permission from Zeng, Z.M., GaP/GaOx core-shell nanowires and nanochains and their transport properties, *J. Phys. Chem. C*, 112(47), 18588–18591. Copyright 2008 American Chemical Society.)

Organic nanochains are rare; for example, high-quality necklace-like polyaniline nanochains assembled by elliptical nanoparticles were synthesized in chitosan aqueous solution by a facile dispersion polymerization method.[185] These nanochains coated by a layer of chitosan are typical doped polyaniline in its emeraldine salt form, which is easy to form stable polyaniline dispersion in water.

3.8 NANOBARS

Nanobars, closely related to nanorods, are known both for metals and inorganic compounds, for example, such noble metals as silver,[186,187] (Figure 3.45) gold, palladium and their composites. Thus, the synthesis of a single crystalline rectangular silver bar using polyacrylamide (PAM) and $AgNO_3$ by a hydrothermal process was reported.[188] At relatively low temperatures (above 380 K) and high pressure, amide group of PAM is hydrolyzed with the liberation of ammonia (NH_3), which produces a reducing atmosphere, resulting in nucleation sites producing the assembly of silver nanocrystals along the chain. Growth mechanism for Pd nanobars synthesized by reducing Na_2PdCl_4 with L-ascorbic acid in an aqueous solution in the presence of bromide ions as a capping agent was studied.[189] It was revealed that the growth at early stages of the synthesis was dominated by particle coalescence, followed by shape focusing *via* recrystallization and further growth *via* atomic addition. Upon thermal annealing, the nanobars evolved into a more thermodynamically

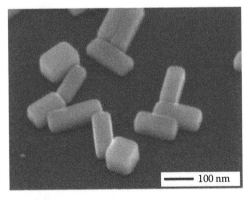

FIGURE 3.45 Silver nanobars. (Reproduced with permission from Wiley, B.J. et al., Synthesis and optical properties of silver nanobars and nanorice, *Nano Lett.*, 7(4), 1032–1036. Copyright 2007 American Chemical Society.)

favored shape with enhanced truncation at the corners. In a related work,[190] palladium nanocubes and nanobars with a mean size of ~23.8 nm were readily synthesized with H_2PdCl_4 as a precursor and tetraethylene glycol as both a solvent and a reducing agent in the presence of PVP and CTAB in 80 s under microwave irradiation. The formation of $PdBr_4{}^{2-}$ due to the coordination replacement of the ligand Cl^- ions in $PdCl_4{}^{2-}$ ions by Br^- ions in the presence of bromide was responsible for the synthesis of Pd nanocubes and nanobars. The authors noted that a milder reducing power, a higher viscosity, and a stronger affinity of TEG were beneficial to the larger sizes of Pd nanocubes and nanobars. The mechanisms responsible for the formation of Pd nanobars and nanorods, as well as the morphological changes in an aging process, are shown in Figure 3.46.[191] Au@Pd core-shell nanobars (Figure 3.47)[192] and Ni nanobar/lead zirconate titanate film[193] are also known.

Nonmetals are represented by carbon nanostructures and selenium. Thus, the effect of magnetism and perpendicular external electric field strengths on the energy gap of length confined bilayer graphene nanoribbons (or nanobars) was studied[195] as a function of ribbon width and length using a first principles density functional electronic structure method and a semilocal

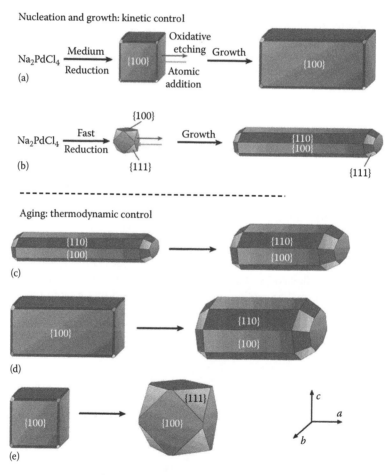

FIGURE 3.46   Schematic illustration of the mechanisms responsible for the formation of nanobars and nanorods, as well as the morphological changes in an aging process: (a) nucleation and formation of nanobars at a medium reduction rate; (b) nucleation and formation of nanorods at a fast reduction rate; (c) decrease in aspect ratio for nanorods; (d) evolution of nanobars into nanorods; and (e) evolution of nanobars with an aspect ratio of ~1 (i.e., nanocubes) into cuboctahedrons. (Reproduced with permission from Xiong, Y. et al., Synthesis and mechanistic study of palladium nanobars and nanorods, *J. Am. Chem. Soc.*, 129(12), 3665–3675. Copyright 2007 American Chemical Society.)

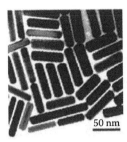
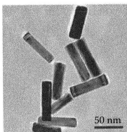
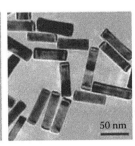
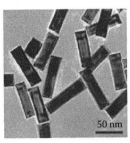

FIGURE 3.47 Au@Pd core-shell nanobars. (Reproduced with permission from Zhang, K. et al., Enhanced optical responses of Au@Pd core/shell nanobars, *Langmuir*, 25(2), 1162–1168. Copyright 2009 American Chemical Society.)

exchange-correlation approximation. It was found that the gaps decreased with the applied electric fields due to the large intrinsic gap of the nanobar. Magnetism between the layers played a major role in enhancing the gap values resulting from the geometrical confinement, hinting at an interplay of magnetism and geometrical confinement in finite-size bilayer graphene. Bovine serum albumin (BSA) was used as a shape-directing agent to synthesize crystalline Se nanobars (Figure 3.48) and amorphous nanospheres in aqueous phase at a relatively low temperature of 85°C.[196] Na_2SeO_3 was used as the Se source to achieve nanoselenium following hydrazine reduction (Reaction 3.21). It was established that well-defined multifacet nanobars were produced when the amount of Na_2SeO_3 was at least six times greater than that of BSA (on the basis of per residue), while amorphous spheres were formed with nearly a 1:1 ratio. It was shown that the shape-directing ability of unfolded BSA helped to achieve the formation of crystalline nanobars, while its soft template effect directed the nanosphere formation. Alternatively, selenium nanobars and nanorods were prepared[197] in a biomineralization process through biomembrane bi-templates of rush at r.t. according to the mechanism shown in Figure 3.49. The offered method was mild, convenient, and was "green" because of the absence of surfactant and organic solvent:

$$N_2H_4(aq) + SeO_3^{2-}(aq) \rightarrow Se(s) + N_2(g) + 2OH^-(aq) + H_2O \qquad (3.21)$$

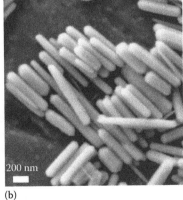

(a)   (b)

FIGURE 3.48 (a) Low-resolution FESEM image of several bundles of Se nanobars. (b) Magnified image of several multifaceted Se nanobars. (Reproduced with permission from Kaur, G. et al., Biomineralization of fine selenium crystalline rods and amorphous spheres, *J. Phys. Chem. C*, 113(31), 13670–13676. Copyright 2009 American Chemical Society.)

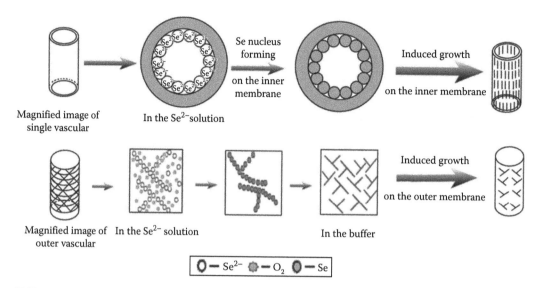

FIGURE 3.49 Schematic illustration of the possible mechanism of formation of 1D Se nanomaterials. (From Li, L. et al.: Simultaneously inducing synthesis of semiconductor selenium multi-armed nanorods and nanobars through bio-membrane bi-templates. *Sci. China Ser. B Chem.* 2004, 47(6), 507–511, http://www.springerlink.com/content/g88q504t424gj244/fulltext.pdf. Copyright Wiley-VCH Verlag GmbH & Co. KGaA. Reproduced with permission.)

A series of metal oxides obtained as nanobars are known, mainly ZnO.[198] Thus, a ZnO nanobar array, fabricated[199] from zinc acetate y hexamethylenetetramine, had controllable diameter (20–60 nm), good optical performance and hydrophobic performance, and 0 contact angle with water of 105°–115° and can be used in gas sensor material, sensor, optical detector field, and self-cleaning device and solid lubricant field.

Ta_2O_5 nanobars anchored on micron-sized carbon spheres were prepared by the thermal decomposition of pentaethoxy tantalum, $Ta(OEt)_5$.[200] Bamboo-like α-Fe_2O_3 nanobars were prepared from iron nitrate as the source of Fe, NH_3 as precipitant, and poly-glycol as dispersant with further calcination at 450°C for 2 h.[201] The products were shown to be composed of Fe_2O_3 nanoparticles with sizes at 50–100 nm and bamboo-like Fe_2O_3 nanobars ~10 nm in diameter and exhibited typical magnetic hysteresis loops of ferromagnetism materials. ZrO_2 nanostructures were prepared[202] by a hydrothermal route, resulting in the formation of nanobars and hexagonal-shaped nanodiscs at different preparation conditions. The structural analysis confirmed that the as-prepared ZrO_2 product was of pure monoclinic phase (m-ZrO_2) with a crystallite size of ~25 nm, consisting of monodispersed nanoparticles of uniform composition, high purity, and crystallinity. In addition, TiO_2[203] nanobars are known.

Bi- and trielemental alloys/composites have also been reported. Thus, magnetostrictive nanobars based on Fe-B alloy[204] were synthesized using a template-based electrochemical deposition method.[205] These Fe-B nanobars (Figure 3.50) were found to be amorphous, possessing a good thermal stability, showing no significant compositional variation along the length direction, and covered by an oxidation layer of a typical thickness of ~10 nm, related to the passivation of nanobars in air. Si/Ge/Si[206] nanoscale bars were also fabricated.

A belt-like nanostructure of iron silicide, called by authors as nanobars, was synthesized on a silicon (001) substrate by microwave plasma method.[207] Also, large-scale production of GaMnN nanobars, by ammoniating Ga_2O_3 films doped with Mn under a flowing ammonia atmosphere at 1000°C (Reactions 3.22 through 3.24), was offered.[208] The Mn-doped GaN sword-like nanobars (Figure 3.51) were observed as a single-crystal hexagonal structure, containing Mn up to 5.43 atom %

FIGURE 3.50   SEM pictures of Fe-B nanobars. (Reproduced from Cheng, Z.-Y. et al., *Adv. Sci. Technol.*, 54(*Smart Materials & Micro/Nanosystems*), 19. Copyright 2008 Scientific.Net.)

FIGURE 3.51   Typical SEM images of Mn-doped GaN nanobars at different magnifications. (Reproduced from Xue, C.-S. et al., *Chin. Phys. Lett.*, 27(3), 038102/1. Copyright 2010 IOP Science.)

and having a thickness of about 100 nm and a width of 200–400 nm. The GaN nanobars showed two emission bands with a well-defined PL peak at 388 and 409 nm, respectively. In addition, two sized $In_{0.25}Ga_{0.75}As$ nanobar arrays, grown on a GaAs substrate by MOCVD[209] and lead sulfide[210] were reported:

$$2NH_3(g) \rightarrow N_2 + 3H_2(g) \tag{3.22}$$

$$Ga_2O_3(s) + 2H_2(g) \rightarrow Ga_2O(g) + 2H_2O(g) \tag{3.23}$$

$$Ga_2O(g) + 2NH_3(g) \rightarrow 2GaN(s) + 2H_2(g) + H_2O(g) \tag{3.24}$$

Oxygen-containing salts are represented by a few examples. Thus, different nanostructures of bismuth subcarbonate $(BiO)_2CO_3$, one of the commonly used antibacterial agents against *Helicobacter pylori*, such as cube-like nanoparticles, nanobars, and nanoplates, were fabricated from bismuth nitrate *via* a simple solvothermal method.[211] It was found that the solvents and precursors had an influence on these morphologies. Additionally, $Zn_xCd_{1-x}O$ ($0 < x < 0.15$) nanobars were formed (Reactions 3.25 and 3.26) in a solution of $ZnCl_2 + CdCl_2 + KCl$ in the presence of tartaric acid at 75°C.[212] Organic matter is practically unknown in nanobar form, except phthalocyanines.[213] Nanobar principal applications, mainly for drug discovery, were reviewed in Ref. [214]:

$$O_2 + 2H_2O + 4e^- \rightarrow 4OH^- \tag{3.25}$$

$$(1-x)Zn^{2+} + xCd^{2+} + 2OH^- \rightarrow (1-x)Zn(OH)_2 \cdot xCd(OH)_2 \rightarrow Zn_{1-x}Cd_xO + H_2O \tag{3.26}$$

3.9 NANOWICKS

A nanowick can be defined as a strip full of dense arrays of nanotubes for liquid delivery.[215] In case of carbon nanotubes, a miniature wicking element is capable for liquid delivery and potential microfluidic chemical analysis devices. The delivery function of nanowicks enables novel fluid transport devices to run without any power input, moving parts, or external pump. The intrinsically nanofibrous structure of nanowicks provides a sieving matrix for molecular separations and a high surface-to-volume ratio porous bed to carry catalysts or reactive agents. Examples of diode-type nanowicks are shown in Figure 3.52. Additionally, dense arrays of bismuth single-walled nanotubes (Figure 3.53), obtained by hydrothermal synthesis from bismuth powder as a precursor at 220°C,

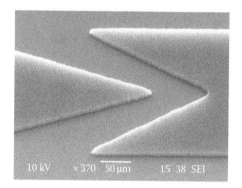

FIGURE 3.52 SEM images of a typical diode-like wick. (Zhou, J. and Huang, X., Parametric verification of one-way lithographic wicks, *Lab Chip*, 9, 1667–1669, 2009. Reproduced by permission from The Royal Society of Chemistry.)

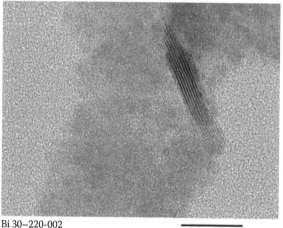

Bi 30–220-002
Print mag: 1960000x @ 7.0 in

20 nm

FIGURE 3.53 Bismuth nanowick.

could also be attributed to a nanowick structure.[216] A series of useful applications of CNTs nanowicks were emphasized in a NASA report.[217]

Also, a manipulation stage was used to electrically contact individual nanotube bundles coated with metal nanoparticles for *in situ* studies in a TEM.[218] When electric current was passed through a bundle, unusual mass transport was observed along that bundle. Nanocrystals melted and disappeared from a given section, with a correlated growth of similar nanoparticles further along the bundle. The authors noted that this unusual phenomenon, termed *nanowicking*, may provide a method for controlled nanoscale mass transport.

3.10 NANOPILLARS

Pillar-like objects are widespread nanostructures possessing a series of useful applications in device development, so in this section we limit their discussion by several representative examples. Thus, elemental metal/alloy nanopillars have been reported for a series of transition metals. For example, the growth of arrays of pagoda-topped tetragonal Cu nanopillars (Figure 3.54), (length 1–6 μm; width 150 ± 25 nm) with (100) side faces on Au/glass was achieved by a simple electrochemical reduction of $CuCl_2$(aq) by Al(s) in aqueous dodecyltrimethylammonium chloride.[220] The proposed mechanism of their formation is shown in Figure 3.55. It was established that the Cu nanopillars can emit electrons under relatively low electric field strength; this could be employed for interesting nanodevice applications in the future. Additionally, a nanoscale pillar consisting of a Co/Cu/Co layered structure (Figure 3.56) was fabricated by means of EBL to study perpendicular transport

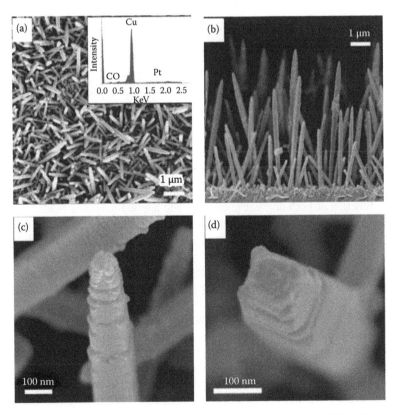

FIGURE 3.54 SEM images of Cu nanopillars on Au/glass: (a) top view (inset: EDS); (b) side view; (c) side view; and (d) top view of a pagoda-shaped tip. (Reproduced with permission from Chang, I.-C., Growth of pagoda-topped tetragonal copper nanopillar arrays, *ACS Appl. Mater. Interfaces*, 1(7), 1375–1378. Copyright 2009 American Chemical Society.)

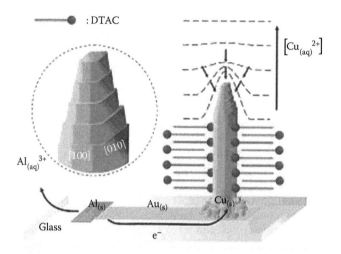

FIGURE 3.55 Proposed growth mechanism of pagoda-topped tetragonal Cu nanopillars on a Au/glass electrode. (Reproduced with permission from Chang, I.-C., Growth of pagoda-topped tetragonal copper nanopillar arrays, *ACS Appl. Mater. Interfaces*, 1(7), 1375–1378. Copyright 2009 American Chemical Society.)

FIGURE 3.56 (a) Schematic illustration of the Co/Cu/Co pillar (cross section); (b) SEM image of the pillar (top view). (Reproduced with permission from Yanga, T. et al., Spin-injection-induced intermediate state in a Co nanopillar, *J. Appl. Phys.*, 97, 064304. Copyright 2005, American Institute of Physics.)

properties as a function of both dc electric currents and applied magnetic fields.[221] The nanopillar exhibited sharp transitions in magnetoresistance associated with magnetization reversal between antiparallel and parallel configurations of the two Co layers. Among a series of other metallic composites, NiFe-silicon[222] and GaSb[223] are known. Nonmetallic nanopillars are represented by carbon and silicon. Thus, the study of tubulization process of amorphous carbon nanopillars, observed *in situ* by TEM, revealed that amorphous carbon nanopillars were transformed into graphitic tubules by annealing at 650°C–900°C in the presence of iron nanoparticles.[224] The tubulization mechanism was found to be a solid–(quasiliquid)–solid mechanism where the carbon-phase transformation is a kind of liquid-phase graphitization of amorphous carbon catalyzed by liquefied metal-carbon alloy nanoparticles. Electron transport in silicon nanopillars was investigated, with a view to developing vertical electron emission, electroluminescent, and photoluminescent devices.[225] These arrays of nanopillars were fabricated in highly doped single crystal silicon and polysilicon materials. The pillar height was 100 nm in single crystal silicon material and 40 nm in polysilicon material, and the diameter of the pillars was 30 nm.

The most important "nanotechnological" oxides—ZnO and TiO₂—have also been fabricated as nanopillars. Thus, high-quality ZnO nanopillars with good uniformity were fabricated using ALD[226] on the (0001) sapphire. The ZnO nanopillars had high crystallinity with the [0001]

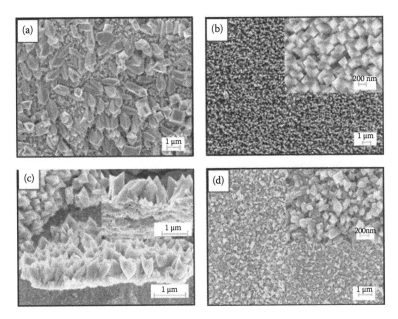

FIGURE 3.57 FESEM images of vertically oriented TiO_2 nanopillar arrays, (a), (b), and (d) images of the sample synthesized at 200°C with 12 h, with the TMAOH concentration of 0.5, 1, 1.5 M, respectively. (c) Cross-sectional FESEM image of the TiO_2 nanopillar arrays (TMAOH concentration is 1 M; inset shows the bottom parts of the nanopillar arrays). (Reproduced from *Appl. Surf. Sci.*, 256, Dong, X. et al., Oriented single crystalline TiO_2 nano-pillar arrays directly grown on titanium substrate in tetramethylammonium hydroxide solution, 2532–2538, Copyright 2010, with permission from Elsevier.)

orientation and exhibited a significant UV luminescence at r.t. Oriented single crystalline titanium dioxide (TiO_2) nanopillar arrays (Figure 3.57) (single crystalline anatase) were directly synthesized on the Ti plate in tetramethylammonium hydroxide (TMAOH) solution by one-pot hydrothermal method.[227] It was shown that the TiO_2 nanopillar with a tetrahydral bipyramidal tip grew vertically on the titanium substrate. The special morphology of the TiO_2 nanopillar arrays was caused by the selective absorption of the tetramethylammonium (TMA) through hydrogen bonds on the lattice planes parallel to (001) of anatase TiO_2. The schematic of growth of a TiO_2 nanopillar crystal is shown in Figure 3.58.

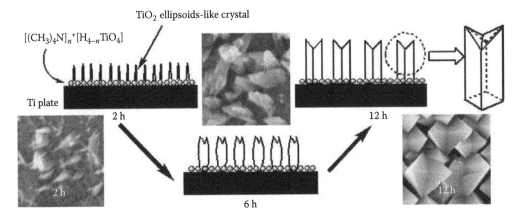

FIGURE 3.58 Schematic growth of TiO_2 nanopillar crystal. (Reproduced from *Appl. Surf. Sci.*, 256, Dong, X. et al., Oriented single crystalline TiO_2 nano-pillar arrays directly grown on titanium substrate in tetramethylammonium hydroxide solution, 2532–2538, Copyright 2010, with permission from Elsevier.)

The growth of InP/InAs/InP core-shell nanopillars was carried out by metalorganic vapor-phase epitaxy (MOVPE) using indium droplets to catalyze deposition.[228] Indium arsenide quantum wells were deposited on the nanopillars at 420°C and a V/III ratio of 120 and then capped with a thin layer of InP. Photoluminescence spectra at 77 K yielded a single intense band at 1750 nm (0.7 eV) with a full width at half maximum of 350 nm. In a related work,[229] the morphology, structure, and chemical properties of $In_xGa_{1-x}N$ nanopillars directly grown on Si (111) substrates, by molecular beam epitaxy, were investigated. Initial nanostructures merged through subgrain boundaries to form final nanopillars. A very low InN mole fraction was revealed near the interface with the substrate, owing to high desorption rates from the elevated growth temperature and gradually higher in incorporation rates near the tips of the nanopillars.

In addition to inorganic pillar-like nanostructures, a few organic nanopillars have been reported, in particular those of poly(3-hexylthiophene).[230] Nanopillars are generally applied in solar cells[231] (in particular CdS/CdTe photovoltaics[232]) and nanodevices, for instance, as a transmission-type electron multiplier on the basis of silicon nanopillars.[233]

REFERENCES

1. Ishizaki, T.; Saito, N.; Ishida, K.; Takai, O. Fabrication of Au nanolines by direct metal-drawing technique using scanning probe microscope. *Hyomen Gijutsu*, 2004, *55* (12), 964–965.
2. Olliges, S.; Gruber, P. A.; Orso, S.; Auzelyte, V.; Ekinci, Y.; Solak, H. H.; Spolenak, R. In situ observation of cracks in gold nano-interconnects on flexible substrates. *Scripta Materialia*, 2007, Volume Date 2008, *58* (3), 175–178.
3. Bieri, N. R.; Chung, J.; Poulikakos, D.; Grigoropoulos, C. P. Manufacturing of nanoscale thickness gold lines by laser curing of a discretely deposited nanoparticle suspension. *Superlattices and Microstructures*, 2004, *35* (3–6), 437–444.
4. Tatte, T.; Talviste, R.; Paalo, M.; Vorobjov, A.; Part, M.; Kiisk, V.; Saal, K.; Lohmus, A.; Kink, I. Preparation and applications of transition metal oxide nanofibres and nanolines. *NSTI Nanotech, Nanotechnology Conference and Trade Show, Technical Proceedings*, Boston, MA, June 1–5, 2008, Vol. 3, pp. 109–111.
5. Owen, J. H. G.; Bianco, F.; Koester, S. A.; Mazur, D.; Bowler, D. R.; Renner, Ch. One-dimensional Si-in-Si(001) template for single-atom wire growth. *Applied Physics Letters*, 2010, *97*, 093102, 3 pp.
6. Javorsky, J.; Owen, J. H. G.; Setvin, M.; Miki, K. Electronic structure of Bi lines on clean and H-passivated Si(100). *Journal of Physics: Condensed Matter*, 2010, *22* (17), 175006/1–175006/5.
7. Rodríguez-Prieto, A.; Bowler, D. R. Atomic-scale nanowires on Si(001): Cu on Bi. *Physical Review B: Condensed Matter and Materials Physics*, 2010, *82* (4), 041414/1–041414/4.
8. Alzahrani, A. Z.; Srivastava, G. P. Self-assembled Bi nanolines on the InAs(100) surface: A theoretical study. *International Journal of Nano and Biomaterials*, 2009, *2* (1/2/3/4/5), 155–163.
9. Srivastava, G. P.; Miwa, R. H. Nanostructure formation aided by self-organized Bi nanolines on Si(001). *Applied Surface Science*, 2008, *254* (24), 8075–8082.
10. Smeets, B.; van der Straten, P.; Meijer, T.; Fabrie, C. G. C. H. M.; van Leeuwen, K. A. H. Atom lithography without laser cooling. *Applied Physics B, Laser and Optics*, 2009, 1–7, arXiv.org, e-Print Archive, Physics, arXiv:0908.2733v1 [physics.atom-ph]. Cornell University Library, http://aps.arxiv.org/PS_cache/arxiv/pdf/0908/0908.2733v1.pdf. doi 10.1007/s00340-009-3867-3.
11. Myszkiewicz, G.; Hohlfeld, J.; Toonen, A. J.; Van Etteger, A. F.; Shklyarevskii, O. I.; Meerts, W. L.; Rasing, T. H.; Jurdik, E. Laser manipulation of iron for nanofabrication. *Applied Physics Letters*, 2004, *85* (17), 3842–3844.
12. Wang, W., Wei, J., Wang, K., Lu, R., Kang, F., Zhang, X., Wu, D. Process for filling in-situ associated iron nanotube in thin-wall carbon nanotube. Chinese Patent CN 1868868 A 20061129, 2006, 11 pp. Application: CN 2006-10012164.
13. Wei, Q.; Li, W.; Sun, K.; Lian, J.; Wang, L. Morphological instability of Cu nanolines induced by Ga+-ion bombardment. In situ scanning electron microscopy and theoretical model. *Journal of Applied Physics*, 2008, *103* (7, Pt. 1), 074306/1–074306/9.
14. Huber, T. E.; Trottman, P.; Halpern, J. B. Influence of quantum-mechanical boundary roughness resistance on copper nanolines. *Condensed Matter*, 2009, 1–14, arXiv.org, e-Print Archive, arXiv:0906.3747v1 [cond-mat.mes-hall]. Cornell University Library, http://aps.arxiv.org/ftp/arxiv/papers/0906/0906.3747.pdf.

15. Miwa, R. H.; MacLeod, J. M.; Srivastava, G. P.; McLean, A. B. The geometry of Bi nanolines on Si(001). *Applied Surface Science*, 2005, *244*, 157–160.

16. Zhang, W.-T.; Li, T.-B. Atom lithography using chromium(52) atomic beam for fabricating nanostructures. *Chinese Physics Letters*, 2006, *23* (11), 2952–2955.

17. Peraelae, R. E.; Kuzmin, M.; Laukkanen, P.; Ahola-Tuomi, M.; Punkkinen, M. P. J.; Vaeyrynen, I. J. Ytterbium on vicinal Si(100). Growth and properties of the 2D wetting layer and the Yb silicide phase. *Surface Science*, 2009, *603* (1), 102–108.

18. Bonet, C.; Tear, S. P. Self-assembly of ultrafine nanolines upon Ho reaction with the Ge(001) surface. *Applied Physics Letters*, 2006, *89* (20), 203119/1–203119/3.

19. Xu, J., Lu, F.; Liu, J. Process for preparation of metal Ru nanowire using as catalyst for water/organic two-phase reaction system. *Chinese Patent* CN 1903427 A 20070131, 2007, 6 pp. Application: CN 2005–10012266.

20. Miwa, R. H.; Srivastava, G. P. In nanolines and nanoclusters on self-assembled Bi-lines. *Surface Science*, 2006, *600* (18), 4048–4051.

21. Su, Y.; Tang, J. Preparation of CoPt alloy nanoline arrays and the influence of heat treatment on its magnetic properties. *Jinshu Rechuli*, 2006, *31* (9), 10–13.

22. Miyoshi, K.; Fujikawa, S.; Kunitake, T. Fabrication of nanoline arrays of noble metals by electroless plating and selective etching process. *Colloids and Surfaces, A: Physicochemical and Engineering Aspects*, 2008, *321* (1–3), 238–243.

23. Lyons, E. S.; O'Handley, R. C.; Ross, C. A. Effect of nano-patterning on anisotropy of Cu/Ni/Cu nanolines. *Journal of Applied Physics*, 2006, *99* (8, Pt. 3), 08R105/1–08R105/3.

24. Uhlig, W. C.; Shi, J. Superparamagnetic transitions in ultrathin film NiFe nanolines. *Journal of Applied Physics*, 2004, *95* (11, Pt. 2), 7031–7033.

25. Li, B.; Zhao, Q.; Huang, H.; Luo, Z.; Kang, M. K.; Im, J.-H.; Allen, R. A.; Cresswell, M. W.; Huang, R.; Ho, P.S. Indentation of single-crystal silicon nanolines: Buckling and contact friction at nanoscales. *Journal of Applied Physics*, 2009, *105* (7, Pt. 1), 073510/1–073510/7.

26. McLean, A. B.; Hill, I. G.; Lipton-Duffin, J. A.; MacLeod, J. M.; Miwa, R. H.; Srivastava, G. P. Nanolines on silicon surfaces. *International Journal of Nanotechnology*, 2008, *5* (9/10/11/12), 1018–1057.

27. Huang, H.; Li, B.; Zhao, Q.; Luo, Z.; Im, J.; Kang, M. K.; Allen, R. A.; Cresswell, M. W.; Huang, R.; Ho, P. S. Nanoindentation of Si nanostructures: buckling and friction at nanoscales. *AIP Conference Proceedings (Stress-Induced Phenomena in Metallization)*, Austin, TX, 2009, Vol. 1143, pp. 204–212.

28. Kang, M. K.; Li, B.; Ho, P. S.; Huang, R. Buckling of single-crystal silicon nanolines under indentation. *Journal of Nanomaterials*, 2008, Article ID 132728, 11 pp.

29. Li, B.; Kang, M. K.; Lu, K.; Huang, R.; Ho, P. S.; Allen, R. A.; Cresswell, M. W. Fabrication and Characterization of patterned single-crystal silicon nanolines. *Nano Letters*, 2008, *8* (1), 92–98.

30. Sahaf, H.; Dettoni, F.; Leandri, C.; Moyen, E.; Masson, L.; Hanbuecken, M. Self-assembled Si nanostripe grating at the molecular scale as a template for 1D growth. *Surface and Interface Analysis*, 2010, *42* (6–7), 687–691.

31. Kim, T.-G.; Wostyn, K.; Mertens, P. W.; Busnaina, A. A.; Park, J.-G. Collapse behavior and forces of multistack nanolines. *Nanotechnology*, 2010, *21* (1), 015708/1–015708/7.

32. Regonda, S.; Aryal, M.; Hu, W. Stability of HSQ nanolines defined by e-beam lithography for Si nanowire field effect transistors. *Journal of Vacuum Science and Technology, B: Microelectronics and Nanometer Structures—Processing, Measurement, and Phenomena*, 2008, *26* (6), 2247–2251.

33. Zhang, X.; Wen, G.; Huang, X.; Zhong, B.; Bai, H.; Xing, S. Preparation of silicon nitride nanolines and nanobelts. Patent CN 101224876, 2008, 6 pp.

34. Huo, K.; Hu, Z.; Hu, Y.; Ma, Y.; Chen, Y. Method for growing zinc oxide one-dimensional nanomaterial on zinc alloy material directly. Chinese Patent CN 1868892 A 20061129, 2006, 21 pp. Application: CN 2006-10040324.

35. Fan, S.-W.; Srivastava, A. K.; Dravid, V. P. Nanopatterned polycrystalline ZnO for room temperature gas sensing. *Sensors and Actuators, B: Chemical*, 2010, *B144* (1), 159–163.

36. Zhao, X.; Xiang, L.; Yin, J.; Zhang, T. Method for preparing montmorillonite electrorheological fluid material with titania containing surface nano-lines. Patent CN 101538504, 2009, 12 pp.

37. Zhao, P.; Huang, K. Preparation and characterization of netted sphere-like CdS nanostructures. *Crystal Growth and Design*, 2008, *8* (2), 717–722.

38. Li, B.; Luo, Z.; Shi, L.; Zhou, J. P.; Rabenberg, L.; Ho, P. S.; Allen, R. A.; Cresswell, M. W. Controlled formation and resistivity scaling of nickel silicide nanolines. *Nanotechnology*, 2009, *20* (8), 085304/1–085304/7.

39. Miwa, R. H.; Orellana, W.; Srivastava, G. P. Iron silicide wires patterned by Bi nanolines on the H/Si(001) surface. Spin density functional calculations. *Physical Review B: Condensed Matter and Materials Physics*, 2008, *78* (11), 115310/1–115310/7.

40. Sun, T.; Pan, Z.; Dravid, V. P.; Wang, Z.; Yu, M.-F.; Wang, J. Nanopatterning of multiferroic BiFeO$_3$ using "soft" electron beam lithography. *Applied Physics Letters*, 2006, *89* (16), 163117/1–163117/3.

41. Silly, F.; Castell, M. R. Encapsulated Pd nanocrystals supported by nanoline-structured SrTiO$_3$(001). *Journal of Physical Chemistry B*, 2005, *109*, 12316–12319.

42. Liu, C.; Tang, D.; Cheng, H. Preparation of quasi-one-dimensional boron nitride nanostructure. Chinese Patent CN 101062765, 2007, 12 pp. Application: CN 2006–10046469 20060429.

43. Xu, B.; Li, C.; Liang, J.; Zhai, L.; Hao, H.; Liu, G.; Wang, F.; Yang, D.; Ma, S. Preparation process for gallium nitride nanolines. Chinese Patent CN 1803627 A 20060719, 2006, 20 pp. Application: CN 2005-10048111.

44. Zhang, J.; Wu, Z.; Dang, H. Method for preparing potassium titanate nano-lines. Chinese Patent CN 1796288 A 20060705, 2006, 8 pp. Application: CN 2004-10102439.

45. Xia, W.; Wang, D.; Zhang, S.; Wan, S.; Meng, X.; Wang, H. A new compound—Nanoline K$_2$Ti$_8$O$_{15}$. *Huaxue Wuli Xuebao*, 2003, *16* (1), 1–2.

46. Zhao, X.; Ji, X.; Zhang, Y. Bi$_2$Te$_3$ base compound nanoline and its preparing method. Chinese Patent CN 1554574 A 20041215, 2004, Application: CN 2003-10122821.

47. Zhang, J. Fluorocarbon nanotube or nanoline and its direct rapid assembly method. Chinese Patent CN 1546535 A 20041117, 2004, Application: CN 2003-10109263 20031210.

48. Johnson, W. L.; Kim, S. A.; Geiss, R.; Flannery, C. M.; Soles, C. L.; Wang, C.; Stafford, C. M.; Wu, W.-L.; Torres, J. M.; Vogt, B. D.; Heyliger, P. R. Elastic constants and dimensions of imprinted polymeric nanolines determined from Brillouin light scattering. *Nanotechnology*, 2010, *21* (7), 075703/1–075703/8.

49. Krasnikov, S. A.; Beggan, J. P.; Sergeeva, N. N.; Senge, M. O.; Cafolla, A. A. Ni(II) porphine nanolines grown on a Ag(111) surface at room temperature. *Nanotechnology*, 2009, *20* (13), 135301/1–135301/6.

50. Choi, J.; Kumar, A.; Sotzing, G. A. Nanopatterned electrochromic conjugated poly(terthiophene)s via thermal nanoimprint lithography of precursor polymer. *Journal of Macromolecular Science, Part A: Pure and Applied Chemistry*, 2007, *44* (12), 1305–1309.

51. Corn, R. M.; Halpern, A. R.; Kim, D.; Chen, Y. Fabrication of plasmon-coupled gratings with nanowires, nanolines and nanoparticles for bioaffinity sensing. *Abstracts of Papers, 240th ACS National Meeting*, Boston, MA, August 22–26, 2010, ANYL-47.

52. Chen, Y.; Kung, S.-C.; Taggart, D. K.; Halpern, A. R.; Penner, R. M.; Corn, R. M. Fabricating nanoscale DNA patterns with gold nanowires. *Analytical Chemistry*, 2010, *82* (8), 3365–3370.

53. Ahn, C. H.; Mohanta, S. K.; Lee, N. E.; Cho, H. K. Enhanced exciton-phonon interactions in photoluminescence of ZnO nanopencils. *Applied Physics Letters*, 2009, *94* (26), 261904/1–261904/3.

54. Xiao, J.; Wu, Y.; Zhang, W.; Bai, X.; Yu, L.; Li, S.; Zhang, G. Enhanced field emission from ZnO nanopencils by using pyramidal Si(100) substrates. *Applied Surface Science*, 2008, *254* (17), 5426–5430.

55. Liu, J.; Huang, X.; Li, Y.; Ji, X.; Li, Z.; He, X.; Sun, F. Vertically aligned one-dimensional ZnO nanostructures on bulk alloy substrates: direct solution synthesis, photoluminescence, and field emission. *Journal of Physical Chemistry*, 2007, *111* (13), 4990–4997.

56. Wang, R. C.; Liu, C. P.; Huang, J. L. ZnO hexagonal arrays of nanowires grown on nanorods. *Applied Physics Letters*, 2005, *86*, 251104.

57. Pol, V. G.; Calderon-Moreno, J. M.; Thiyagarajan, P. Facile synthesis of novel photoluminescent ZnO micro- and nanopencils. *Langmuir*, 2008, *24* (23), 13640–13645.

58. Ahsanulhaq, Q.; Umar, A.; Hahn, Y. B. Growth of aligned ZnO nanorods and nanopencils on ZnO/Si in aqueous solution: Growth mechanism and structural and optical properties. *Nanotechnology*, 2007, *18* (11), 115603/1–115603/7.

59. Wang, R. C.; Liu, C. P.; Huang, J. L.; Chen, S.-J.; Tseng, Y.-K.; Kung, S.-C. ZnO nanopencils: Efficient field emitters. *Applied Physics Letters*, 2005, *87* (1), 013110/1–013110/3.

60. Shen, G.; Bando, Y.; Liu, B.; Golberg, D.; Lee, C.-J. Characterization and field-emission properties of vertically aligned ZnO nanonails and nanopencils fabricated by a modified thermal-evaporation process. *Advanced Functional Materials*, 2006, *16*, 410–416.

61. Terasako, T.; Saito, D.; Taira, K.; Nishinaka, A.; Yamaguchi, T.; Shirakata, S. Possibility of shape control of ZnO nanostructures grown by atmospheric-pressure CVD utilizing catalytic materials. *e-Journal of Surface Science and Nanotechnology*, 2009, *7*, 78–83.

62. Kar, J. P.; Das, S. N.; Choi, J. H.; Lee, T. I.; Myoung, J. M. Study of the morphological evolution of ZnO nanostructures on various sapphire substrates. *Applied Surface Science*, 2010, *256* (16), 4995–4999.

63. Wang, B.; Ouyang, G.; Li, I. L.; Xu, P. Growth and photoluminescence of SnO$_2$ nanoleaves and nanopencils synthesized by Au-Ag alloying catalyst assisted carbothermal evaporation. *Nano Science and Nano Technology: An Indian Journal*, 2009, *3* (2), 21–27.

64. Xu, N.; Li, Z.; Deng, S.; Chen, J. Method for growing tungsten oxide nanopencil array as field emission material by multistage heating. Patent CN 101353816, 2009, 15 pp.

65. Qurashi, A.; Faisal Irfan, M.; Wakas Alam, M. In$_2$O$_3$ nanostructures and their chemical and biosensor applications. *The Arabian Journal for Science and Engineering*, 2010, *35* (1C), 125–145.

66. Tayebi, N.; Narui, Y.; Chen, R. J.; Collier, C. P.; Giapis, K. P.; Zhang, Y. Nanopencil as a wear-tolerant probe for ultrahigh density data storage. *Applied Physics Letters*, 2008, *93* (10), 103112/1–103112/3.

67. Kim, T.; Moon, S. I.; Xiang, H.; Shin, K.; Russell, T. P.; McCarthy, T. J. Block copolymer nano-pencils prepared using anodized aluminum membranes. *PMSE Preprints*, 2004, *91*, 751.

68. Huang, C.-J.; Chiu, P.-H.; Wang, Y.-H.; Chen, W.-R.; Meen, T.-H.; Yang, C.-F. Preparation and characterization of gold nanodumbbells. *Nanotechnology*, 2006, *17* (21), 5355–5362.

69. Landon, P. B.; Gilleland, C. L.; Synowczynski, J.; Hirsch, S. G.; Glosser, R. Synthesis of gold nano-wire and nano-dumbbell shaped colloids and AuC$_{60}$ nano-clusters. *Journal of Materials Science: Materials in Electronics*, 2007, *18* (Suppl. 1), S415–S418.

70. Rahaie, M.; Ghai, R.; Babic, B.; Dimitrov, K. Synthesis and characterization of DNA-based micro- and nanodumbbell structures. *Journal of Bionanoscience*, 2009, *3* (2), 73–79.

71. Chiu, P.-H.; Huang, C.-J.; Yang, C.-F.; Meen, T.-H.; Wang, Y.-H. Structural and morphological properties of gold nanodumbbells. *Journal of the Electrochemical Society*, 2009, *156* (2), E35–E39.

72. Huang, C. J.; Chiu, P. H.; Wang, Y. H.; Chen, K. L. Fabrication of gold nanocubes by the electrochemical method. *Advances in Materials Research*, 2008, *55–57 (Smart Materials)*, 661–664.

73. Huang, C.-J.; Chiu, P.-H.; Wang, Y.-H.; Yang, C.-F. Synthesis of the gold nanodumbbells by electrochemical method. *Journal of Colloid and Interface Science*, 2006, *303* (2), 430–436.

74. Fernanda Cardinal, M.; Rodriguez-Gonzalez, B.; Alvarez-Puebla, R. A.; Perez-Juste, J.; Liz-Marzan, L. M. Modulation of localized surface plasmons and SERS response in gold dumbbells through silver coating. *Journal of Physical Chemistry C*, 2010, *114* (23), 10417–10423.

75. Balci, S.; Noda, K.; Bittner, A. M.; Kadri, A.; Wege, C.; Jeske, H.; Kern, K. Self-assembly of metal-virus nanodumbbells. *Angewandte Chemie International Edition*, 2007, *46* (17), 3149–3151.

76. Lin, Z.-H.; Chang, H.-T. Preparation of gold-tellurium hybrid nanomaterials for surface-enhanced Raman spectroscopy. *Langmuir*, 2008, *24* (2), 365–367.

77. Ung, D.; Soumare, Y.; Chakroune, N.; Viau, G.; Vaulay, M.-J.; Richard, V.; Fievet, F. Growth of magnetic nanowires and nanodumbells in liquid polyol. *Chemistry of Materials*, 2007, *19* (8), 2084–2094.

78. Krylova, G.; Dimitrijevic, N. M.; Talapin, D. V.; Guest, J. R.; Borchert, H.; Lobo, A.; Rajh, T.; Shevchenko, E. V. Probing the surface of transition-metal nanocrystals by chemiluminescence. *Journal of the American Chemical Society*, 2010, *132* (26), 9102–9110.

79. Lee, J.-S.; Bodnarchuk, M. I.; Shevchenko, E. V.; Talapin, D. V. "Magnet-in-the-semiconductor" FePt-PbS and FePt-PbSe nanostructures: Magnetic properties, charge transport, and magnetoresistance. *Journal of the American Chemical Society*, 2010, *132* (18), 6382–6391.

80. Li, X.; Li, T.; Wu, C.; Zhang, Z. Self-assembly of rutile (α-TiO$_2$) nanoclusters into nanorods in microemulsions at low temperature and their photocatalytic performance. *Journal of Nanoparticle Research,* 2007, *9* (6), 1081–1086.

81. Wang, C.; Irudayaraj, J. Multifunctional magnetic-optical nanoparticle probes for simultaneous detection, separation, and thermal ablation of multiple pathogens. *Small*, 2010, *6* (2), 283–289.

82. Zhang, Z.; Liu, S.; Chow, S.; Han, M.-Y. Modulation of the morphology of ZnO nanostructures *via* aminolytic reaction: From nanorods to nanosquamas. *Langmuir*, 2006, *22* (14), 6335–6340.

83. Salant, A.; Amitay-Sadovsky, E.; Banin, U. Directed self-assembly of gold-tipped CdSe nanorods. *Journal of the American Chemical Society*, 2006, *128* (31), 10006–10007.

84. Shaviv, E.; Banin, U. Synergistic effects on second harmonic generation of hybrid CdSe-Au nanoparticles. *ACS Nano*, 2010, *4* (3), 1529–1538.

85. Costi, R.; Saunders, A.; Elmalem, E.; Salant, A.; Banin, U. Visible light-induced charge retention and photocatalysis with hybrid CdSe-Au nanodumbbells. *Nano Letters*, 2008, *8* (2), 637–641.

86. Costi, R.; Cohen, G.; Salant, A.; Rabani, E.; Banin, U. Electrostatic force microscopy study of single Au-CdSe hybrid nanodumbbells: Evidence for light-induced charge separation. *Nano Letters*, 2009, *9* (5), 2031–2039.

87. Zhu, Y.; Wang, L.; Hao, J.; Xiao, Z.; Wei, Y.; Wang, Y. Synthetic, structural, spectroscopic, electrochemical studies and self-assembly of nanoscale polyoxometalate-organic hybrid molecular dumbbells. *Crystal Growth and Design*, 2009, *9* (8), 3509–3518.

88. Li, M.; Zhang, Z. S.; Zhang, X.; Li, K. Y.; Yu, X. F. Optical properties of Au/Ag core/shell nanoshuttles. *Optics Express*, 2008, *16* (18), 14288–14293.

89. He, Z.; Yu, S.-H.; Zhu, J. Amino acids controlled growth of shuttle-like scrolled tellurium nanotubes and nanowires with sharp tips. *Chemistry of Materials*, 2005, *17* (11), 2785–2788.

90. Wang, J.; He, S.; Li, Z.; Jing, X.; Zhang, M.; Jiang, Z. Self-assembled CuO nanoarchitectures and their catalytic activity in the thermal decomposition of ammonium perchlorate. *Colloid and Polymer Science*, 2009, *287* (7), 853–858.

91. Yin, Z.; Wu, J.; Yang, Z. A sensitive mercury (II) sensor based on CuO nanoshuttles/poly(thionine) modified glassy carbon electrode. *Microchimica Acta*, 2010, *170* (3–4), 307–312.

92. Zhang, W.; Yang, T.; Zhuang, X.; Guo, Z.; Jiao, K. An ionic liquid supported CeO_2 nanoshuttles-carbon nanotubes composite as a platform for impedance DNA hybridization sensing. *Biosensors and Bioelectronics*, 2009, *24* (8), 2417–2422.

93. Wen, B.; Huang, Y.; Boland, J. J. Controllable growth of ZnO nanostructures by a simple solvothermal process. *Journal of Physical Chemistry C*, 2008, *112* (1), 106–111.

94. Wang, H.; Shao, W.; Gu, F.; Zhang, L.; Lu, M.; Li, C. Synthesis of anatase TiO_2 nanoshuttles by self-sacrificing of titanate nanowires. *Inorganic Chemistry*, 2009, *48* (20), 9732–9736.

95. Cheng, L.-Y.; Chen, Y.; Wu, Q.-S. Morphology exchange and optical properties of monodispersed ZnS nanospheres and nanoshuttles. *Huaxue Xuebao*, 2007, *65* (17), 1851–1854.

96. Zhang, L.; Chen, L.; Wan, H.; Zhoul, H.; Chen, J. Preparation of shuttle-like Sb_2S_3 nanorod-bundles via a solvothermal approach under alkaline condition. *Crystal Research and Technology*, 2010, *45* (2), 178–182.

97. Jia, R.; Wu, Q.; Ding, Y. Ultrasonic preparation and optical properties of $HgWO_4$ nanoshuttles. *Nano*, 2007, *2* (1), 15–19.

98. Wang, J.; Han, X.; Zhang, W.; He, Z.; Wang, C.; Cai, R.; Liu, Z. Controlled growth of monocrystalline rutile nanoshuttles in anatase TiO_2 particles under mild conditions. *Crystal Engineering Communications*, 2009, *11* (4), 564–566.

99. Wang, L.-S.; Wu, L.-C.; Lu, S.-Y.; Chang, L.-L.; Teng, I.-T.; Yang, C.-M.; Ho, J. A. Biofunctionalized phospholipid-capped mesoporous silica nanoshuttles for targeted drug delivery: Improved water suspensibility and decreased nonspecific protein binding. *ACS Nano*, 2010, *4* (8), 4371–4379.

100. Chen, M.-m.; Zheng, T.-l.; Wang, Z.-l.; Li, J.-j.; Gu, C.-z.; Yejing, Y. X. Field emission properties of tin oxide nanoshuttle arrays. *Yejing Yu Xianshi*, 2007, *22* (6), 672–676.

101. Yang, X.-H.; Wu, Q.-S.; Li, L.; Ding, Y.-P.; Zhang, G.-X. Controlled synthesis of the semiconductor CdS quasi-nanospheres, nanoshuttles, nanowires and nanotubes by the reverse micelle systems with different surfactants. *Colloids and Surfaces, A: Physicochemical and Engineering Aspects*, 2005, *264* (1–3), 172–178.

102. Sun, J.-H.; Ding, Y.-H.; Shang, T.-M. Calcium carbonate nanoparticles prepared by lamellar liquid crystal as template and shape evolution during growth. *Wuji Huaxue Xuebao*, 2007, *23* (1), 75–80.

103. Vogel, V.; Hess, H. Nanoshuttles: Harnessing motor proteins to transport cargo in synthetic environments. *Lecture Notes in Physics*, 2007, *711 (Controlled Nanoscale Motion)*, 367–383.

104. Luzzi, D. E. Synthesis, structure, and properties of fullerene and nonfullerene nanopeapods. *Abstracts of Papers, 225th ACS National Meeting*, New Orleans, LA, March 23–27, 2003, COLL-370.

105. Okazaki, T.; Shinohara, H. Nano-peapods encapsulating fullerenes. Rotkin, S. V.; Subramoney, S. (Eds.). *Applied Physics of Carbon Nanotubes*, 2005, 133–150.

106. Okazaki, T.; Okubo, S.; Nakanishi, T.; Joung, S.-K.; Saito, T.; Otani, M.; Okada, S.; Bandow, S.; Iijima, S. Optical band gap modification of single-walled carbon nanotubes by encapsulated fullerenes. *Journal of the American Chemical Society*, 2008, *130*, 4122–4128.

107. Cho, Y.; Han, S.; Kim, G.; Lee, H.; Ihm, J. Orbital hybridization and charge transfer in carbon nanopeapods. *Physical Review Letters*, 2003, *90* (10), 106402/1–106402/4.

108. Yumura, T.; Kertesz, M.; Iijima, S. Local modifications of single-wall carbon nanotubes induced by bond formation with encapsulated fullerenes. *Journal of Physical Chemistry B*, 2007, *111* (5), 1099–1109.

109. Kitaura, R.; Shinohara, H. Carbon-nanotube-based hybrid materials. Nanopeapods. *Chemistry—An Asian Journal*, 2006, *1* (5), 646–655.

110. Enyashin, A. N.; Ivanovskii, A. L. Atomic structure and electronic properties of nanopeapods: Isomers of endohedral dititanofullerenes $Ti_2@C_{80}$ in carbon nanotubes. *Zhurnal Neorganicheskoi Khimii*, 2006, *51* (9), 1576–1585.

111. Krive, I. V.; Ferone, R.; Shekhter, R. I.; Jonson, M.; Utko, P.; Nygaard, J. The influence of electromechanical effects on resonant electron tunneling through small carbon nano-peapods. *New Journal of Physics*, 2008, *10*, 043043.

112. Baowan, D.; Thamwattana, N.; Hill, J. M. Encapsulation of C_{60} fullerenes into single-walled carbon nanotubes: Fundamental mechanical principles and conventional applied mathematical modeling. *Physical Review B: Condensed Matter and Materials Physics*, 2007, *76* (15), 155411/1–155411/8.

113. Gloter, A.; Suenaga, K.; Kataura, H.; Fujii, R.; Kodama, T.; Nishikawa, H.; Ikemoto, I.; Kikuchi, K.; Suzuki, S.; Achiba, Y.; Iijima, S. Structural evolutions of carbon nano-peapods under electron microscopic observation. *Chemical Physics Letters*, 2004, *390* (4–6), 462–466.

114. Pati, R.; Senapati, L.; Ajayan, P. M.; Nayak, S. K. Theoretical study of electrical transport in a fullerene-doped semiconducting carbon nanotubes. *Journal of Applied Physics*, 2004, *95* (2), 694–697.

115. Liu, Y.; Jones, R. O.; Zhao, X.; Ando, Y. Carbon species confined inside carbon nanotubes: A density functional study. *Physical Review B: Condensed Matter and Materials Physics*, 2003, *68* (12), 125413/1–125413/7.

116. Rols, S.; Cambedouzou, J.; Chorro, M.; Schober, H.; Agafonov, V.; Launois, P.; Davydov, V.; Rakhmanina, A. V.; Kataura, H.; Sauvajol, J.-L. How confinement affects the dynamics of C_{60} in carbon nanopeapods. *Physical Review Letters*, 2008, *101* (6), 065507/1–065507/4.

117. Hernandez, E.; Meunier, V.; Smith, B. W.; Rurali, R.; Terrones, H.; Buongiorno Nardelli, M.; Terrones, M.; Luzzi, D. E.; Charlier, J.-C. Fullerene coalescence in nanopeapods: A path to novel tubular carbon. *Nano Letters*, 2003, *3* (8), 1037–1042.

118. Warner, J. H.; Ito, Y.; Zaka, M.; Ge, L.; Akachi, T.; Okimoto, H.; Porfyrakis, K.; Watt, A. A. R.; Shinohara, H.; Briggs, G. A. D. Rotating fullerene chains in carbon nanopeapods. *Nano Letters*, 2008, *8* (8), 2328–2335.

119. Sohi, A. N.; Naghdabadi, R. Stability of single-walled carbon nanopeapods under combined axial compressive load and external pressure. *Physica E: Low-Dimensional Systems and Nanostructures*, 2009, *41* (3), 513–517.

120. Guan, L.; Suenaga, K.; Okubo, S.; Okazaki, T.; Iijima, S. Metallic wires of lanthanum atoms inside carbon nanotubes. *Journal of the American Chemical Society*, 2008, *130* (7), 2162–2163.

121. Suenaga, K.; Taniguchi, R.; Shimada, T.; Okazaki, T.; Shinohara, H.; Iijima, S. Evidence for the intramolecular motion of Gd atoms in a $Gd_2@C_{92}$ nanopeapod. *Nano Letters*, 2003, *3* (10), 1395–1398.

122. Kitaura, R.; Okimoto, H.; Shinohara, H. Magnetism of the endohedral metallofullerenes $M@C_{82}$ (M = Gd, Dy) and the corresponding nanoscale peapods: Synchrotron soft x-ray magnetic circular dichroism and density-functional theory calculations. *Physical Review B*, 2007, *76*, 172409.

123. Sato, Y.; Suenaga, K.; Bandow, S.; Iijima, S. Site-dependent migration behavior of individual cesium ions inside and outside C_{60} fullerene nanopeapods. *Small*, 2008, *4* (8), 1080–1083.

124. Urita, K.; Sato, Y.; Suenaga, K.; Gloter, A.; Hashimoto, A.; Ishida, M.; Shimada, T.; Shinohara, H.; Iijima, S. Defect-induced atomic migration in carbon nanopeapod: Tracking the single-atom dynamic behavior. *Nano Letters*, 2004, *4* (12), 2451–2454.

125. Trave, A.; Ribeiro, F. J.; Louie, S. G.; Cohen, M. L. Energetics and structural characterization of C_{60} polymerization in BN and carbon nanopeapods. *Physical Review B*, 2004, *70*, 205418.

126. Timoshevskii, V.; Cote, M. Doping of C_{60}-induced electronic states in BN nanopeapods: Ab initio simulations. *Physical Review B: Condensed Matter and Materials Physics*, 2009, *80* (23), 235418/1–235418/5.

127. Li, X.; Yang, W.; Liu, B. Fullerene coalescence into metallic heterostructures in boron nitride nanotubes: A molecular dynamics study. *Nano Letters*, 2007, *7* (12), 3709–3715.

128. Enyashin, A. N.; Ivanovskii, A. L. Atomic and electronic structures and thermal stability of boron-nitrogen nanopeapods: $B_{12}N_{12}$ fullerenes in BN nanotubes. *Physics of the Solid State*, 2008, *50* (2), 390–396.

129. Zhou, W. W.; Sun, L.; Yu, T.; Zhang, J. X.; Gong, H.; Fan, H. J. The morphology of Au@MgO nanopeapods. *Nanotechnology*, 2009, *20* (45), 455603/1–455603/6.

130. Wu, J. S.; Dhara, S.; Wu, C. T.; Chen, K. H.; Chen, Y. F.; Chen, L. C. Growth and optical properties of self-organized Au_2Si nanospheres pea-podded in a silicon oxide nanowire. *Advanced Materials*, 2002, *14* (24), 1847–1850.

131. Hunyadi, S. E.; Murphy, C. J. Tunable one-dimensional silver/silica nanopeapod architectures. *Journal of Physical Chemistry B*, 2006, *110* (14), 7226–7231.

132. Liu, L.; Lee, W.; Scholz, R.; Pippel, E.; Goesele, U. Tailor-made inorganic nanopeapods: Structural design of linear noble metal nanoparticle chains. *Angewandte Chemie, International Edition*, 2008, *47* (37), 7004–7008.

133. Lin, J.; Huang, Y.; Tang, C.; Bando, Y.; Shi, Y.; Takayama-Muromachi, E.; Golberg, D. Ferromagnetic Fe@CaS nanopeapods with protecting BN tubular sheaths. *Journal of Physical Chemistry C*, 2009, *113* (33), 14818–14822.

134. Wang, B.; Yang, Y. H.; Yang, G. W. Electron-beam irradiation induced shape transformation of $Sn-SnO_2$ nanocables. *Nanotechnology*, 2006, *17* (24), 5916–5921.

135. Zhang, Y.; Huang, Y.-h.; He, J.; Dai, Y.; Zhang, X.-m.; Liu, J.; Liao, Q.-L. Quasi one-dimensional ZnO nanostructures fabricated without catalyst at lower temperature. *Frontiers of Physics in China*, 2006, *1*, 72–84.

136. Zhong, K.; Xia, J.; Li, H. H.; Liang, C. L.; Liu, P.; Tong, Y. X. Morphology evolution of one-dimensional-based ZnO nanostructures synthesized via electrochemical corrosion. *Journal of Physical Chemistry C*, 2009, *113* (35), 15514–15523.

137. Yin, S.; Chen, Y.; Su, Y.; Jia, C.; Zhou, Q.; Li, S.; Xin, M.; Kong, W.; Zhang, X.; Lu, Y. Controllable synthesis and photoluminescence properties of ZnO nanorod and nanopin arrays. *Journal of Nanoscience and Nanotechnology*, 2008, *8* (2), 993–996.

138. Wei, L.; Zhang, X.; Lou, C.; Zhu, Z. An improved planar triode with ZnO nanopin field emitters. *IEEE Electron Device Letters*, 2007, *28* (8), 688–690.

139. Wei, L.; Zhang, X.; Zuoya, Z. Application of ZnO nanopins as field emitters in a field-emission-display device. *Journal of Vacuum Science Technology, B: Microelectronics and Nanometer Structures— Processing, Measurement, and Phenomena*, 2007, *25* (2), 608–610.

140. Hosono, E.; Fujihara, S.; Imai, H.; Honma, I.; Ichihara, M.; Zhou, H. One-step synthesis of nano-micro chestnut TiO_2 with rutile nanopins on the microanatase octahedron. *ACS Nano*, 2007, *1* (4), 273–278.

141. Ji, X. H.; Zhai, J. W. Synthesis of In_2O_3 nanowires, nanobouquets and nanopins. *Second IEEE International Nanoelectronics Conference, INEC 2008*, Shanghai, China, 2008, pp. 375–376.

142. Zhang, W. S.; Brueck, E.; Li, W. F.; Si, P. Z.; Geng, D. Y.; Zhang, Z. D. Synthesis, characterization and magnetic properties of Fe-Al nanopins. *Physica B: Condensed Matter*, 2005, *370* (1–4), 131–136.

143. Du, J.; Du, P.; Hao, P.; Huang, Y.; Ren, Z.; Weng, W.; Han, G.; Zhao, G. Self-induced preparation of TiSi nanopins by chemical vapor deposition. *Nanotechnology*, 2007, *18* (34), 345605/1–345605/4.

144. Du, J.; Du, P.; Hao, P.; Huang, Y.; Ren, Z.; Han, G.; Weng, W.; Zhao, G. Growth mechanism of TiSi nanopins on Ti_5Si_3 by atmospheric pressure chemical vapor deposition. *Journal of Physical Chemistry C*, 2007, *111* (29), 10814–10817.

145. Durrani, Z. A. K.; Rafiq, M. A. Electronic transport in silicon nanocrystals and nanochains. *Microelectronic Engineering*, 2009, *86* (4–6), 456–466.

146. Pendleton, A.; Kundu, S.; Liang, H. Controlled synthesis of titanium nanochains using a template. *Journal of Nanoparticle Research*, 2009, *11* (2), 505–510.

147. Li, P.; Cui, Y.; Behan, G.; Zhang, H.; Wang, R. Room temperature synthesis and one-dimensional self-assembly of interlaced Ni nanodiscs under magnetic field. *Journal of Physics D: Applied Physics*, 2010, *43* (27), 275002/1–275002/6.

148. Chen, W.; Zhou, W.; He, L.; Chen, C.; Guo, L. Surface magnetic states of Ni nanochains modified by using different organic surfactants. *Journal of Physics: Condensed Matter*, 2010, *22* (12), 126003/1–126003/5.

149. Wei, D.; Chen, Q. The temperature dependence of 1D germanium nanostructures grown in a small-diameter quartz tube cavity by vapor deposition. *Journal of Crystal Growth*, 2010, *312* (16–17), 2315–2319.

150. Chepok, A. O. Passage of an electromagnetic signal through copper nanochains. *Trudy Odesskogo Politekhnicheskogo Universiteta*, 2009, (1), 143–147.

151. Qin, Y.; Lee, S.-M.; Pan, A.; Goesele, U.; Knez, M. Rayleigh-instability-induced metal nanoparticle chains encapsulated in nanotubes produced by atomic layer deposition. *Nano Letters*, 2008, *8* (1), 114–118.

152. Jia, J.; Yu, J. C.; Wang, Y.-X. J.; Chan, K. M. Magnetic nanochains of $FeNi_3$ prepared by a template-free microwave-hydrothermal method. *ACS Applied Materials and Interfaces*, 2010, *2* (9), 2579–2584.

153. Rudenko, A. N.; Mazurenko, V. V.; Anisimov, V. I.; Lichtenstein, A. I. Weak ferromagnetism in Mn nanochains on the CuN surface. *Physical Review B: Condensed Matter and Materials Physics*, 2009, *79* (14), 144418/1–144418/9.

154. Shi, X.-L.; Cao, M.-S.; Yuan, J.; Fang, X.-Y. Dual nonlinear dielectric resonance and nesting microwave absorption peaks of hollow cobalt nanochains composites with negative permeability. *Applied Physics Letters*, 2009, *95* (16), 163108/1–163108/3.

155. Chen, X.; Xie, J.; Hu, J.; Feng, X.; Li, A. EDTA-directed self-assembly and enhanced catalytic properties of sphere-constructed platinum nanochains. *Journal of Physics D: Applied Physics*, 2010, *43* (11), 115403/1–115403/6.

156. Li, Z.; Gu, A.; Zhou, Q. Preparation of gold nanochains in 1-butyl-3-methylimidazolium tetrafluoroborate ionic liquids. *Xiyou Jinshu Cailiao Yu Gongcheng*, 2009, *38* (8), 1454–1457.

157. Feng, C.; Guo, L.; Shen, Z.; Gong, J.; Li, X.-Y.; Liu, C.; Yang, S. Synthesis of short palladium nanoparticle chains and their application in catalysis. *Solid State Sciences*, 2008, *10* (10), 1327–1332.

158. Leroux, F.; Gysemans, M.; Bals, S.; Batenburg, K. J.; Snauwaert, J.; Verbiest, T.; Van Haesendonck, C.; Van Tendeloo, G. Three-dimensional characterization of helical silver nanochains mediated by protein assemblies. *Advanced Materials*, 2010, *22* (19), 2193–2197.

159. Min, Y. L.; Zhang, K.; Chen, Y. C.; Zhao, Y.; Zhang, Y. G. Simple method to synthesize nickel-gold composition nanostructure and their applications. *Materials Chemistry and Physics*, 2010, *119* (1–2), 11–14.

160. Min, Y. L.; Zhang, K.; Chen, Y. C.; Zhang, Y. G.; Pang, M. L. Novel 1D Ni-Ag composition nanostructure: Synthesis, properties and antioxygenation improvement. *Colloids and Surfaces, A: Physicochemical and Engineering Aspects*, 2010, *353* (1), 92–96.

161. Zhang, B.; Hou, W.; Ye, X.; Fu, S.; Xie, Y. Response to comment on 1D tellurium nanostructures: Photothermally assisted morphology-controlled synthesis and applications in preparing functional nanoscale materials. *Advanced Functional Materials*, 2009, *19* (20), 3193–3194.

162. Che, X.; Yuan, R.; Chai, Y.; Li, J.; Song, Z.; Li, W. Amperometric glucose biosensor based on Prussian blue-multiwall carbon nanotubes composite and hollow PtCo nanochains. *Electrochimica Acta*, 2010, *55* (19), 5420–5427.

163. Su, J.; Gao, Y.; Che, R. Synthesis and microstructure of Fe_3C encapsulated inside chain-like carbon nanocapsules. *Materials Letters*, 2010, *64* (6), 680–683.

164. Dougherty, D. B.; Jin, W.; Cullen, W. G.; Dutton, G.; Reutt-Robey, J. E.; Robey, S. W. Local transport gap in C_{60} nanochains on a pentacene template. *Physical Review B: Condensed Matter and Materials Physics*, 2008, *77* (7), 073414/1–073414/4.

165. Ma, D. D. D.; Chan, K. S.; Chen, D. M.; Lee, S. T. Single-charge tunneling in uncoupled boron-doped silicon nanochains. *Chemical Physics Letters*, 2010, *484* (4–6), 258–260.

166. Barsotti, R. J. Jr.; Fischer, J. E.; Lee, C. H.; Mahmood, J.; Adu, C. K. W.; Eklund, P. C. Imaging, structural, and chemical analysis of silicon nanowires. *Applied Physics Letters*, 2002, *81* (15), 2866–2868.

167. Qin, Y.; Liu, L.; Yang, R.; Gosele, U.; Knez, M. General assembly method for linear metal nanoparticle chains embedded in nanotubes. *Nano Letters*, 2008, *8* (10), 3221–3225.

168. Du, J.; Huang, L.; Chen, Z.; Kang, D. J. A controlled method to synthesize hybrid In_2O_3/Ag nanochains and nanoparticles: Surface-enhanced Raman scattering. *Journal of Physical Chemistry C*, 2009, *113* (23), 9998–10004.

169. Yoon, M.; Lee, H.; Lee, J. Highly visible light-sensitive TiO_2 photocatalysts toward bioinspired solar cells. *Abstracts of Papers, 240th ACS National Meeting*, Boston, MA, August 22–26, 2010, PHYS-655.

170. Yoon, M.-J. Surface modifications and optoelectronic characterization of TiO_2-nanoparticles: Design of new photo-electronic materials. *Journal of the Chinese Chemical Society*, 2009, *56* (3), 449–454.

171. Xia, C.; Wang, N.; Wang, L.; Guo, L. Synthesis of nanochain-assembled ZnO flowers and their application to dopamine sensing. *Sensors and Actuators, B: Chemical*, 2010, *B147* (2), 629–634.

172. Meng, L.-R.; Chen, W.-M.; Chen, C.-P.; Zhou, H.-P; Peng, Q.; Li, Y.-D. Uniform α-Fe_2O_3 nanocrystal moniliforme-shape straight chains. *Crystal Growth and Design*, 2010, *10* (2), 479–482.

173. Zhang, Y.; Li, R.; Zhou, X.; Cai, M.; Sun, X. Self-organizing growth of $MgAl_2O_4$ based heterostructural nanochains. *Journal of Physical Chemistry C*, 2008, *112* (27), 10038–10042.

174. Han, Y.; Wang, Y.; Wang, Y.; Jiao, L.; Yuan, H. Characterization of CoB-silica nanochains hydrogen storage composite prepared by in-situ reduction. *International Journal of Hydrogen Energy*, 2010, *35* (15), 8177–8181.

175. Wang, Y.; Song, S.; Shen, P. K.; Guo, C.; Li, C. M. Nanochain-structured mesoporous tungsten carbide and its superior electrocatalysis. *Journal of Materials Chemistry*, 2009, *19* (34), 6149–6153.

176. Zhao, H.; Shi, L.; Li, Z.; Tang, C. Silicon carbide nanowires synthesized with phenolic resin and silicon powders. *Physica E: Low-Dimensional Systems and Nanostructures*, 2009, *41* (4), 753–756.

177. Wei, G.; Qin, W.; Zheng, K.; Zhang, D.; Sun, J.; Lin, J.; Kim, R.; Wang, G.; Zhu, P.; Wang, L. Synthesis and properties of SiC/SiO_2 nanochain heterojunctions by microwave method. *Crystal Growth and Design*, 2009, *9* (3), 1431–1435.

178. Zhang, Y.; Guo, L.; Liu, K.; He, L.; Chen, J. Synthesis of uniform clew-like cobalt sulfide nanochains by mild solution chemical route and their magnetic property. *Xiyou Jinshu Cailiao Yu Gongcheng*, 2009, *38* (Suppl. 2), 1003–1006.

179. Ge, C.; Xu, M.; Fang, J.; Lei, J.; Ju, H. Luminescent cadmium sulfide nanochains templated on unfixed deoxyribonucleic acid and their fractal alignment by droplet dewetting. *Journal of Physical Chemistry C*, 2008, *112* (29), 10602–10608.

180. Xiao, F.; Liu, H.-G.; Wang, C.-W.; Lee, Y.-I.; Xue, Q.; Chen, X.; Hao, J.; Jiang, J. ZnS, Zn_xCd_{1-x}S and CdS nanoparticles at the air/water interface. *Nanotechnology*, 2007, *18* (43), 435603/1–435603/9.

181. Sardarli, R. M.; Samedov, O. A.; Abdullayev, A. P.; Huseynov, E. K.; Salmanov, F. T.; Safarova, G. R. Specific features of conductivity of γ-irradiated $TlGaTe_2$ crystals with nanochain structure. *Semiconductors*, 2010, *44* (5), 585–589.

182. Niu, H.; Zhang, L.; Gao, M.; Chen, Y. Amphiphilic ABC triblock copolymer-assisted synthesis of core/shell structured CdTe nanowires. *Langmuir*, 2005, *21*, 4205–4210.

183. Fu, L.-T.; Chen, Z.-G.; Zou, J.; Cong, H.-T.; Lu, G.-Q. Fabrication and visible emission of single-crystal diameter-modulated gallium phosphide nanochains. *Journal of Applied Physics*, 2010, *107* (12), 124321/1–124321/5.

184. Zeng, Z. M.; Li, Y.; Chen, J. J.; Zhou, W. L. GaP/GaOx core-shell nanowires and nanochains and their transport properties. *Journal of Physical Chemistry C*, 2008, *112* (47), 18588–18591.

185. Li, Y.; Zhang, C.; Li, G.; Peng, H.; Chen, K. Self-assembled necklace-like polyaniline nanochains from elliptical nanoparticles. *Synthetic Metals*, 2010, *160* (11–12), 1204–1209.

186. Zeng, J.; Zheng, Y.; Rycenga, M.; Tao, J.; Li, Z.-Y.; Zhang, Q.; Zhu, Y.; Xia, Y. Controlling the shapes of silver nanocrystals with different capping agents. *Journal of the American Chemical Society*, 2010, *132* (25), 8552–8553.

187. Chau, Y.-F.; Lin, Y.-J.; Tsai, D. P. Enhanced surface plasmon resonance based on the silver nanoshells connected by the nanobars. *Optics Express*, 2010, *18* (4), 3510–3518.

188. Mondal, B.; Majumdar, D.; Saha, S. K. Synthesis of single crystalline micron-sized rectangular silver bar. *Journal of Materials Research*, 2010, *25* (2), 383–390.

189. Lim, B.-K.; Kobayashi, H.; Camargo, P. H. C.; Allard, L. F.; Liu, J.-Y.; Xia, Y.-N. New insights into the growth mechanism and surface structure of palladium nanocrystals. *Nano Research*, 2010, *3* (3), 180–188.

190. Yu, Y.; Zhao, Y.; Huang, T.; Liu, H. Microwave-assisted synthesis of palladium nanocubes and nanobars. *Materials Research Bulletin*, 2010, *45* (2), 159–164.

191. Xiong, Y.; Cai, H.; Wiley, B. J.; Wang, J.; Kim, M. J.; Xia, Y. Synthesis and mechanistic study of palladium nanobars and nanorods. *Journal of American Chemical Society*, 2007, *129* (12), 3665–3675.

192. Zhang, K.; Xiang, Y.; Wu, X.; Feng, L.; He, W.; Liu, J.; Zhou, W.; Xie, S. Enhanced optical responses of Au@Pd core/shell nanobars. *Langmuir*, 2009, *25* (2), 1162–1168.

193. Chung, T.-K.; Wong, K.; Keller, S.; Wang, K.-L.; Carman, G. P. Electrical control of magnetic remanent states in a magnetoelectric layered nanostructure. *Journal of Applied Physics*, 2009, *106* (10), 103914/1–103914/5.

194. Wiley, B. J.; Chen, Y.; McLellan, J.; Xiong, Y.; Li, Z.-Y.; Ginger, D.; Xia, Y. Synthesis and optical properties of silver nanobars and nanorice. *Nano Letters*, 2007, *7* (4), 1032–1036.

195. Sahu, B.; Min, H.; Banerjee, S. K. Effects of magnetism and electric field on the energy gap of bilayer graphene nanobars. *Condensed Matter*, 2009, 1–6, arXiv.org, e-Print Archive, arXiv:0910.2719v1 [cond-mat.mtrl-sci], Cornell University Library.

196. Kaur, G.; Iqbal, M.; Bakshi, M. S. Biomineralization of fine selenium crystalline rods and amorphous spheres. *Journal of Physical Chemistry C*, 2009, *113* (31), 13670–13676.

197. Li, L.; Wu, Q.; Ding, Y.; Li, P. Simultaneously inducing synthesis of semiconductor selenium multi-armed nanorods and nanobars through bio-membrane bi-templates. *Science in China Series B: Chemistry*, 2004, *47* (6), 507–511. http://www.springerlink.com/content/g88q504t424gj244/fulltext.pdf.

198. Wang, S.; Han, G.; Ding, H.; Chen, X. Method for preparing zinc oxide nano bar from zinc quantum dot. Chinese Patent CN 101249980 A 20080827, 2008, 7 pp. Application: CN 2008–10035110.

199. Wu, X.; Chen, H.; Gong, L. Method for large-scale synthesis of ZnO nanobar array. Patent CN 101798106, 2010, 10 pp.

200. George, P. P., Gedanken, A. Ta₂O₅ nanobars and their composites: Synthesis and characterization. *Journal of Nanoscience and Nanotechnology*, 2008, *8* (11), 5801–5806.

201. Chen, C.-S; Liu, Y.; Huang, B. Y.; Lie, T.; Chen, X.-H. Preparation, characterization and property of bamboo-like α-Fe₂O₃ nanobars. *Guangpuxue Yu Guangpu Fenxi*, 2009, *29* (10), 2871–2874.

202. Kumari, L.; Du, G. H.; Li, W. Z.; Vennila, R. S.; Saxena, S. K.; Wang, D. Z. Synthesis, microstructure and optical characterization of zirconium oxide nanostructures. *Ceramics International*, 2009, *35* (6), 2401–2408.

203. Sanz, R.; Jaafar, M.; Hernandez-Velez, M.; Asenjo, A.; Vazquez, M.; Jensen, J. Patterning of rutile TiO₂ surface by ion beam lithography through full-solid masks. *Nanotechnology*, 2010, *21* (23), 235301/1–235301/6.

204. Cheng, Z.-Y.; Li, S. Q.; Zhang, K. W.; Fu, L. L.; Chin, B. A. Novel magnetostrictive microcantilever and magnetostrictive nanobars for high performance biological detection. *Advances in Science and Technology*, 2008, *54* (*Smart Materials & Micro/Nanosystems*), 19–28.

205. Li, S.; Fu, L.; Wang, C.; Lea, S.; Arey, B.; Engelhard, M.; Cheng, Z.-Y. Characterization of microstructure and composition of Fe-B nanobars as biosensor platform. *MRS Symposium Proceedings (Nanoscale Magnets—Synthesis, Self-Assembly, Properties and Applications)*, Warrendale, PA, 2007, Volume Date 2006, Vol. 962E, Paper # 0962-P09-14.

206. Park, Y.; Atkulga, H. M.; Grama, A.; Strachan, A. Strain relaxation in Si/Ge/Si nanoscale bars from molecular dynamics simulations. *Journal of Applied Physics*, 2009, *106* (3), 034304/1–034304/6.

207. Xu, B.-X.; Zhang, Y.; Zhu, H.-S.; Shen, D.-Z.; Wu, J. L.; Xue, Z. Q.; Wu, Q. D. Self-assembling iron silicide nanobars and structure on silicon wafer by microwave plasma method. *Proceedings of SPIE-The International Society for Optical Engineering (Micro- and Nanotechnology: Materials, Processes, Packaging, and Systems II)*, Sydney, Australia, 2005, Vol. 5650, pp. 365–372.

208. Xue, C.-S.; Liu, W.-J.; Shi, F.; Zhuang, H.-Z.; Guo, Y.-F.; Cao, Y.-P.; Sun, H.-B. Fabrication of Mn-doped GaN nanobars. *Chinese Physics Letters*, 2010, *27* (3), 038102/1–038102/4.

209. Wang, B.; Chua, S.-J. MOCVD behaviors of two-sized InGaAs ordered nano-bar arrays grown selectively on a GaAs substrate. *MRS Symposium Proceedings (Low-Dimensional Materials–Synthesis, Assembly, Property Scaling, and Modeling)*, 2007, Vol. 1017E, Paper # 1017-DD11-12.

210. Bakshi, M. S.; Kaur, G.; Possmayer, F.; Petersen, N. O. Shape-controlled synthesis of poly(styrene sulfonate) and poly(vinyl pyrolidone) capped lead sulfide nanocubes, bars, and threads. *Journal of Physical Chemistry C*, 2008, *112* (13), 4948–4953.

211. Cheng, G.; Yang, H.; Rong, K.; Lu, Z.; Yu, X.; Chen, R. Shape-controlled solvothermal synthesis of bismuth subcarbonate nanomaterials. *Journal of Solid State Chemistry*, 2010, *183* (8), 1878–1883.

212. Li, G.-R.; Bu, Q.; Zheng, F.-L.; Su, C.-Y.; Tong, Y.-X. Electrochemically controllable growth and tunable optical properties of $Zn_{1-x}Cd_xO$ alloy nanostructures. *Crystal Growth and Design*, 2009, *9* (3), 1538–1545.

213. Wang, M.; Yang, Y.-L.; Deng, K.; Wang, C. Phthalocyanine nanobars obtained by using alkyloxy substitution effects. *Chemical Physics Letters*, 2007, *439* (1–3), 76–80.

214. Ozkan, M. Quantum dots and other nanoparticles: What can they offer to drug discovery? *Drug Discovery Today*, 2004, *9* (24), 1065–1071.

215. Zhou, J. Nanowicking: Multi-scale flow interaction with nanofabric structures. PhD thesis, California Institute of Technology, Pasadena, CA, 2005, 129 pp.

216. Kharissova, O. V. Unpublished results.

217. Nanowicks: Fiber geometries could be tailored for pumping, filtering, mixing, separating, and other effects. *NASA Tech Briefs*, NASA's Jet Propulsion Laboratory, Pasadena, CA, October 2007, pp. 19–20. http://ntrs.nasa.gov/archive/nasa/casi.ntrs.nasa.gov/20100011126_2010012638.pdf.

218. Regan, B. C.; Aloni, S.; Huard, B.; Fennimore, A.; Ritchie, R. O.; Zettl, A. Nanowicks: Nanotubes as tracks for mass transfer. *AIP Conference Proceedings (Molecular Nanostructures)*, 2003, Vol. 685, pp. 612–615.

219. Zhou, J.; Huang, X. Parametric verification of one-way lithographic wicks. *Lab Chip*, 2009, *9*, 1667–1669.

220. Chang, I.-C.; Huang, T.-K.; Lin, H.-K.; Tzeng, Y.-F.; Peng, C.-W.; Pan, F.-M.; Lee, C.-Y.; Chiu, H.-T. Growth of pagoda-topped tetragonal copper nanopillar arrays. *ACS Applied Materials and Interfaces*, 2009, *1* (7), 1375–1378.

221. Yanga, T.; Kimura, T.; Otani, Y. Spin-injection-induced intermediate state in a Co nanopillar. *Journal of Applied Physics*, 2005, *97*, 064304.

222. Vasic, R.; Brooks, J. S.; Jobiliong, E.; Aravamudhan, S.; Luongo, K.; Bhansali, S. Dielectric relaxation in nanopillar NiFe-silicon structures in high magnetic fields. *Current Appl. Phys.*, 2006, *7*(1), pp. 34–38. arXiv:cond-mat/0508432v1, http://arxiv.org/abs/cond-mat/0508432.

223. Nerbo, I. S.; Le Roy, S.; Foldyna, M.; Kildemo, M.; Sondergard, E. Characterization of inclined GaSb nanopillars by Mueller matrix ellipsometry. *Journal of Applied Physics*, 2010, *108* (1), 014307/1–014307/8.

224. Ichihashi, T.; Fujita, J.-i.; Ishida, M.; Ochiai, Y. *In situ* observation of carbon-nanopillar tubulization caused by liquidlike iron particles. *Physical Review Letters*, 2004, *92* (21), 215702, 4 pp.

225. Rafiq, M. A.; Mizuta, H.; Uno, S.; Durrani, Z. A. K. Fabrication of vertical nanopillar devices. *Microelectronic Engineering*, 2007, *84*, 1515–1518.

226. Wu, M.-K.; Chen, M.-J.; Tsai, F.-Y.; Yang, J.-R.; Shiojiri, M. Fabrication of ZnO nanopillars by atomic layer deposition. *Materials Transactions*, 2010, *51* (2), 253–255.

227. Dong, X.; Tao, J.; Li, Y.; Zhu, H. Oriented single crystalline TiO_2 nano-pillar arrays directly grown on titanium substrate in tetramethylammonium hydroxide solution. *Applied Surface Science*, 2010, *256*, 2532–2538.

228. Evoen, V.; Gao, L.; Pozuelo, M.; Chowdhury, S.; Tatebeyashi, J.; Liang, B.; Kodambaka, S.; Huffaker, D. L.; Hicks, R. F. InP/InAs core-shell nanopillars on InP(111)B. *AIChE Annual Meeting, Conference Proceedings*, Nashville, TN, November 8–13, 2009, evoen1/1-evoen1/12.

229. Kehagias, Th.; Kerasiotis, I.; Vajpeyi, A. P.; Hausler, I.; Neumann, W.; Georgakilas, A.; Dimitrakopulos, G. P.; Komnninou, Ph. Electron microscopy of InGaN nanopillars spontaneously grown on Si(111) substrates. *Physica Status Solidi C: Current Topics in Solid State Physics*, 2010, *7* (5), 1305–1308.

230. Santos, A.; Formentin, P.; Pallares, J.; Ferre-Borrull, J.; Marsal, L. F. Fabrication and characterization of high-density arrays of P3HT nanopillars on ITO/glass substrates. *Solar Energy Materials and Solar Cells*, 2010, *94* (7), 1247–1253.
231. Fan, Z.; Ruebusch, D. J.; Rathore, A. A.; Kapadia, R.; Ergen, O.; Leu, P. W.; Javey, A. Challenges and prospects of nanopillar-based solar cells. *Nano Research*, 2009, *2*, 829–843.
232. Fan, Z.; Razavi, H.; Do, J.-w.; Moriwaki, A.; Ergen, O.; Chueh, Y.-L.; Leu, P. W. et al. Three-dimensional nanopillar-array photovoltaics on low-cost and flexible substrates. *Nature Materials*, 2009, *8*, 648–653.
233. Qin, H.; Kim, H.-S.; Blick, R. H. Nanopillar arrays on semiconductor membranes as electron emission amplifiers. *Nanotechnology*, 2008, *19*, 095504, 5 pp.

4 Various Prolonged 3D Nanostructures

4.1 NANOARROWS

Nanoarrows, an extremely rare nanostructural type, are fabricated by various methods, such as template-assisted nanoarea selective growth on GaN substrates such as InGaN[1,2] (Figure 4.1), metalloorganic CVD (ZnCdSe alloy three-bladed nanoarrowheads, Figure 4.2),[3] or use of oxidizing and reducing ambients (In_2O_3).[4] The reducing reagent, ethanol, made the growth environment indium rich, resulting in the growth of indium-filled In_2O_3 tubular nanoarrow structures. In_2O_3 nanoarrows (obtained also by CVD) and other nanoforms have important potential applications as functional blocks in nanodevices.[5] ZnO[6] nanoarrows were also obtained, among a large series of other structural types.

A simple method such as drying can also lead to unexpected results. Thus, drying CdS nanocrystals with cubic, triangular, and hexagonal geometries, synthesized using simple wet-chemistry techniques, leads to their self-assembly to form complex structures such as linear rods, nanoarrows, and dimers.[7]

4.2 NANOBONES

Bone-like nanostructures are also extremely rare and were discovered for inorganic compounds (for instance ZnO[8]), which were found to form various nanoforms. Thus, uniform bone-like MgO nanocrystals were prepared via a solvothermal process using commercial Mg powders as the starting material in the absence of any catalyst or surfactant followed by subsequent calcinations of the preformed precursor $Mg(OH)_x(EtO)_y$.[9] It was established that the product consisted of a large quantity of bone-like nanocrystals with lengths of 120–200 nm (Figure 4.3). The widths of these nanocrystals at both ends were in the range of 20–50 nm, which were 3–20 nm wider than those of the middle parts. On the basis of these nanostructures, the authors expect that they hold promise for the design of optical devices owing to their strong PL emission centered at 410 nm and also that they may have significant scientific and technological applications as building blocks for many other functional devices due to the new nanostructures.

Two simple methods (Figure 4.4), the thermal evaporation method and the solution method, were developed to synthesize a variety of SiC nanoarchitectures.[10] SiC nanowires, nanopyramids, and nanobones (Figure 4.5) were obtained by the first technique, while nanokelps, nanoflowers, and nanocombs were achieved by the second route. In the case of nanobones, the synthesis conditions are as follows: U = MWCNTs, D = Si + Zn + SiO_2, T = 1550°C, 3 h. These SiC nanoarchitectures may have potential applications in nanoelectronics, nanooptics, nanocomposites, catalysts, and other areas of nanoscience and nanotechnology.

4.3 NANOBOTTLES

Nanobottle structures are known mainly for oxides, for example, for ZnO and SiO_2, and sulfides. Thus, nanosized hollow SiO_2 spheres with holes in the wall (denoted as SiO_2 nanobottles) were prepared by the assembly of functional polymer nanospheres with tetraethoxysilane through hydrothermal methods, coupled with the removal of the core by programmed calcination.[11] ZnO nanobottles and nanorods were synthesized with a facile solvothermal treatment of an ethylene glycol solution of

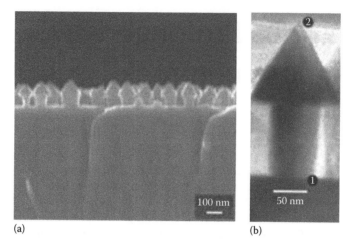

(a)   (b)

FIGURE 4.1   (a) SEM image of InGaN nanorods with pyramidal tips formed on GaN using nanoscale selective area growth. (b) Nanoarrow-like TEM cross-sectional image of the InGaN nanorod. (Reproduced from Wang, Y. et al., High-density arrays of InGaN nanorings, nanodots, and nanoarrows fabricated by a template-assisted approach, *J. Phys. Chem. B*, 110, 11081–11087. Copyright 2006 American Chemical Society.)

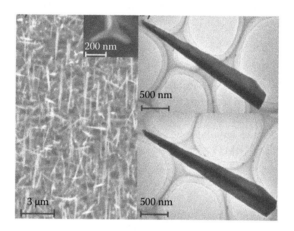

FIGURE 4.2   ZnCdSe alloy three-bladed nanoarrowheads. (Reproduced from Liu, Z. et al., Quadra-twin model for the growth of nanotetrapods and related nanostructures, *J. Phys. Chem. C*, 112(24), 8912–8916. Copyright 2008 American Chemical Society.)

$Zn(NO_3)_2 \cdot 6H_2O$, and NaAc in the presence of polyethylene glycol-10,000.[12] These nanostructures had hexagonal wurtzite structures with high crystallinity and a stronger absorption at 200–380 nm wavelengths. Potential use was found to be related to their photocatalytic activity against methyl orange as the pollutant object. In a related work,[13] well-aligned nanotip-decorated ZnO nanobottles grown on ITO-coated glass substrates were synthesized by the thermal evaporation technique with the upside down arrangement of the substrate compared with the common method. It was found that the nanobottles decorated with nanotips showed a much better field emission property. In addition, Cu-In sulfide heterostructured nanocrystals were fabricated from the thermal decomposition of a mixture of copper oleate and indium oleate complexes in dodecanethiol.[14] By varying the reaction temperature and time, Cu-In sulfide nanocrystals with acorn, bottle (40 × 110 nm) (Figure 4.6), and larva shapes can be prepared.

A few examples of elemental nanobottles, having very useful possible applications, have been discovered. Thus, carbon nanobottle with guest molecules and C_{60} was fabricated.[15] It was revealed that C_{60} filled at the extremities of CNTs can act as caps to seal them. Releasing the incorporated

(a)

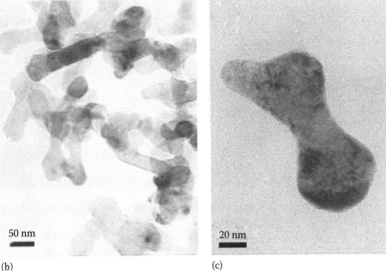

(b)   (c)

FIGURE 4.3 (a) FE-SEM image of the MgO product; (b) TEM image of the product; (c) TEM image of one typical bone-like MgO nanocrystal. (Reproduced with permission from Springer Science+Business Media: *J. Nanopart. Res.*, Large-scale synthesis of single-crystalline MgO with bone-like nanostructures, 8(6), 2006, 881–888, Niu, H. et al.)

molecules, CNT nanobottles could be applied in compound synthesis, drug delivery, and even in materials storage. In addition, a nanocapsule, which combines the advantages of a high-pressure vessel and adsorbents, was developed for the storage of a large methane mass content and for safekeeping.[16] It was a system of combined nanotubes forming bottle-like pores, the entrance to which was closed by a positively charged endohedral complex (K@C_{60}) with the help of an electric field. In r.t., the nanocapsule can retain the amount of methane adsorbed under charging conditions. These nanocapsules can retain ~17.5 wt.% of methane at an internal pressure of 10 MPa and a temperature of 300 K. Also, with surface plasmon polaritons manipulation of 2D arrays of subwavelength bottle-shaped cavities on gold surface, by tuning the geometry of such a nanobottle (Figure 4.7), it was found to be possible to control the resonant frequencies and near-field patterns of different surface plasmon resonances.[17] The plasmonic band structures were not sensitive to the sizes and depths of the nanobottles but depended strongly on the polarization. These nanobottle arrays can be useful for making plasmonic devices.

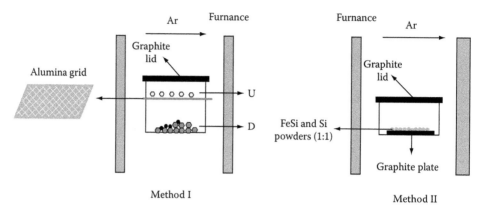

Method I   Method II

FIGURE 4.4 Schematic diagram of equipment for the experimental setup. (Reproduced with permission from Springer Science+Business Media: *Appl. Phys. A*, Thermal evaporation and solution strategies to novel nanoarchitectures of silicon carbide, 88(4), 2007, 679–685, Wu, R.B. et al.)

FIGURE 4.5 Low (left) and high (right) magnification FESEM images of SiC nanobones. (Reproduced with permission from Springer Science+Business Media: *Appl. Phys. A*, Thermal evaporation and solution strategies to novel nanoarchitectures of silicon carbide, 88(4), 2007, 679–685, Wu, R.B. et al.)

FIGURE 4.6 Cu-In sulfide nanobottles. (Reproduced with permission from Choi, S.-H. et al., One-pot synthesis of copper-indium sulfide nanocrystal heterostructures with acorn, bottle, and larva shapes, *J. Am. Chem. Soc.*, 128, 2520–2521. Copyright 2006 American Chemical Society.)

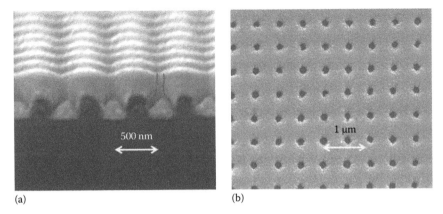

(a) (b)

FIGURE 4.7 (a) Cross section of the nanobottle array with an aperture of 160 nm; the line outlines the bottle shape. (b) Plane view SEM image of the nanobottle array. (Reproduced with permission of the Optical Society of America Iu, H. et al., *Opt. Express*, 16(14), 10294–10302, 2008.)

In addition, nanometer-sized, glass nanobottles were made by using polymer nanostructures as template, to capture and contain biomolecules in their free state on surfaces.[18] The polymer nanostructures were formed by film symmetry breakup during microwave irradiation under liquid environment.

4.4 NANOBRICKS

Brick-like objects are relatively rare nanostructures in the nanoworld. These and other nanoforms useful in nanolandscaping were examined in Ref. [19]. According to a few reports on this nanoform, the palladium nanobricks (Figure 4.8), synthesized using the simple reduction of K_2PdCl_6 using ascorbic acid in the presence of an aqueous-phase mixture of CTAB and hexamine (HMT) at r.t., could find extensive use in catalysis, surface-enhanced Raman scattering, and magnetic and optoelectronics because of their unique shapes and surface structure.[20] The authors noted that the shape of the Pd nanoparticles can be finely controlled by the simple manipulation of the concentration ratio between the CTAB and HMT in the reaction. Carbon nanocubes and nanobricks were

FIGURE 4.8 FESEM image of Pd nanoparticles prepared using cetyltrimethylammonium bromide (CTAB) with a concentration of 75 mM. (Reproduced with permission from Umar, A.A. and Oyama, M., Synthesis of palladium nanobricks with atomic-step defects, *Cryst. Growth Des.*, 8(6), 1808–1811. Copyright 2008 American Chemical Society.)

prepared by pyrolyzing rice powder at 600°C under N_2 atmosphere.[21] Solid-state electronic spectrum showed several bands in the UV and visible region, and excitation at 336 and 474 nm generated photoluminescence response in the UV and visible region.

A few oxides have been reported in nanobrick form. Thus, structural and photoluminescence properties of stacking Ta_2O_5 nanobrick arrays, synthesized by hot filament metal-oxide vapor deposition, were studied.[22] It was shown that these nanobricks were arranged in a large-area array, on average ~20.7 nm wide. Their photoluminescence spectra showed very strong green-light emissions, which emerged from the trap levels of the oxygen vacancies within the Ta_2O_5 bandgap. In addition, various ZnO nanostructures doped with In and Ga cationic substituents were grown by the vapor-phase transport process.[23] During the growth, Zn/ZnO_x was adsorbed on the surface of Ag nanograins and self-catalyzed to form ZnO nanoparticles. Hexagonal-faced nanobricks and nanorods were grown by increasing the ZnO vapor concentration. However, nanodisks rather than nanobricks were grown when In_2O_3 was doped. The unique morphology of TiO_2 nanorods grown in a two-step thermal evaporation process under a controlled atmosphere was observed.[24] The nanorods were found to be composed of nanobricks. In addition, different morphologies of copper oxide nanostructures such as nanofibers, nanospindles, nanocones, and nanobricks were synthesized by microwave-assisted method.[25] It was established that the intensity and the time of microwave irradiation, especially the intensity, played critical roles in the formation of the products.

Nanobricks of metal salts can be formed as *intermediate* nanostructures in formation processes of structures possessing other shapes. Thus, for instance, olivary Sb_2S_3 microcrystallines were synthesized via a hydrothermal process at 180°C for 24 h using HCl, $SbCl_3$, and Na_2S as starting materials.[26] It was revealed that the irregular-shaped nanobricks (being intermediate products) self-assembled into hollow olivary Sb_2S_3 microstructures (Figure 4.9). Another example related to the formation (at 900°C) of multitwinned Zn_2TiO_4 nanowires based on a solid–solid reaction of ZnO nanowires with a conformal shell of TiO_2, which was deposited by atomic layer deposition, was studied.[27] The mechanism was found to be as follows (Figure 4.10). First, spinel Zn_2TiO_4 nanobricks formed on the surface of the ZnO nanowires. Subsequently, the unconsumed ZnO core was desorbed or evaporated through the gaps of the bricks, very much like a chemical desorption of surface atoms of ZnO. Lastly, the remaining loosely connected bricks tended to attach to each

FIGURE 4.9   A schematic illustration of the formation of olivary Sb_2S_3. (Reproduced with permission from Han, Q. et al., Template-free route to Sb2S3 crystals with hollow olivary architectures, *Cryst. Growth Des.*, 8(2), 395–398. Copyright 2008 American Chemical Society.)

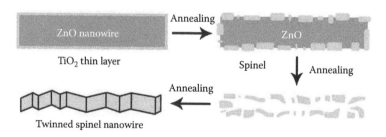

FIGURE 4.10   Schematic diagram of the growth process for multitwinned Zn_2TiO_4 nanowires from ZnO/TiO_2 core/shell nanowires by annealing at high temperatures. (Reproduced with permission from Yang Y. et al., Multitwinned spinel nanowires by assembly of nanobricks via oriented attachment: A case study of Zn_2TiO_4, *ACS Nano*, 3(3), 555–562. Copyright 2009 American Chemical Society.)

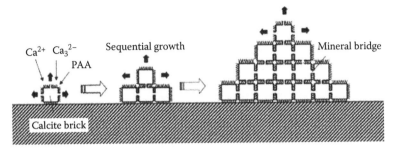

FIGURE 4.11 Schematic illustration of the sequential growth of calcite bricks covered with PAA. Nanoscale calcite bricks are formed by competitive growth of the crystal with adsorption of PAA. An oriented architecture is achieved by the sequential growth through mineral bridges. (Reproduced with permission from Miura, T. et al., Emergence of acute morphologies consisting of ISO-oriented calcite nanobricks in a binary poly(acrylic acid) system, *Cryst. Growth Des.*, 6(2), 612–615. Copyright 2006 American Chemical Society.)

other, coalesce, and finally evolve into the multitwinned spinel nanowires. The oriented assembly of the individual bricks was strongly dependent on annealing conditions. This mechanism differed dramatically from those proposed for twinned nanowires grown in the presence of metal catalysts.

Acute spines and cones of calcite similar to biominerals in mollusks were grown[28] in an aqueous solution system containing low- and high-molecular-weight poly(acrylic acid) (PAA). Precipitation was induced by the dissolution of CO_2 into a 20 mM $CaCl_2$ aqueous solution containing two kinds of PAA molecules (Mw: 2,000 [PAA2k] and 250,000 [PAA250k]. The tapering morphologies consisting of polymer-mediated calcite bricks with a preferred crystallographic orientation were produced by the self-organized crystal growth in a diffusion field. Growth scheme for these calcite bricks is shown in Figure 4.11. w-ZrAlN[29] nanobricks are also known.

4.5 NANOBOWLINGS AND NANONAILS

Almost all publications on nanonails (Figure 4.12) or closely related nanobowlings (Figure 4.13) are devoted to ZnO,[6] with such applications as sensors for the detection of hydrazine[30] and glycose[31,32] or potential applications for UV laser, UV emitters, field emission displays, piezoelectric nanogenerator,[33,34] etc. ZnO nanonails were synthesized, in particular, by nanoparticle-assisted pulsed-laser ablation deposition,[35] thermal vapor transport and condensation method (leading even to nanonail nanoflowers [Figure 4.14])[36] at a low temperature without a metal catalyst using pure Zn powders as raw material and O_2/Ar powders as source gas[37] (zinc powder evaporation is an efficient way of synthesizing a wide range of high-quality ZnO nanostructures (Figure 4.15) at relatively low

FIGURE 4.12 Nanonail (ZnO). (Reproduced with permission from Shen, G. et al., Growth of self-organized hierarchical ZnO nanoarchitectures by a simple In/In$_2$S$_3$ controlled thermal evaporation process, *J. Phys. Chem. B*, 109(21), 10779–10785. Copyright 2005 American Chemical Society.)

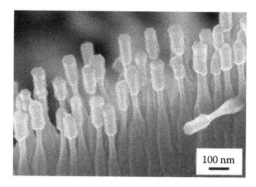

FIGURE 4.13 Nanobowling (ZnO). (Reproduced with permission from Shen, G. et al., Growth of self-organized hierarchical ZnO nanoarchitectures by a simple In/In$_2$S$_3$ controlled thermal evaporation process, *J. Phys. Chem. B*, 109(21), 10779–10785. Copyright 2005 American Chemical Society.)

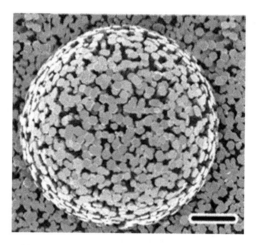

FIGURE 4.14 Medium magnification top view of nanonail flower. Scale bar = 5 μm. (Reproduced with permission from Lao, J.Y. et al., ZnO nanobridges and nanonails, *Nano Lett.*, 3(2), 235–238. Copyright 2003 American Chemical Society.)

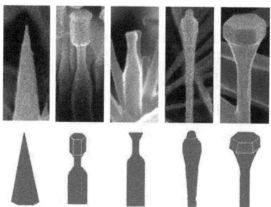

Nanocone   Nanobowling Nanobottle   Nanoarrow   Nanonail

FIGURE 4.15 Schematic illustration of the construction units for different ZnO nanostructures. (Reproduced from Shen, G. et al., Growth of self-organized hierarchical ZnO nanoarchitectures by a simple In/In$_2$S$_3$ controlled thermal evaporation process, *J. Phys. Chem. B*, 109(21), 10779–10785. Copyright 2005 American Chemical Society.)

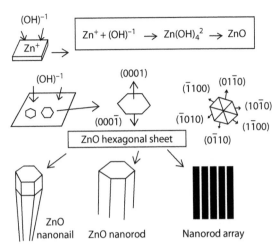

FIGURE 4.16 Schematic view of the growth mechanism of the ZnO nanoforms. (Reproduced with permission from Kar, S. et al., Simple solvothermal route to synthesize ZnO nanosheets, nanonails, and well-aligned nanorod arrays. *J. Phys. Chem. B*, 110(36), 17848–17853. Copyright 2006 American Chemical Society.)

temperature[38]), or the solvothermal approach using ethanol as the solvent (leading to various nanoforms, whose formation mechanism is shown in Figure 4.16).[39] InN[40] is a practically unique example for reported non-ZnO nanonails.

4.6 NANOCONES*

Cone-like species are relatively not rare nanostructures, and they are widely represented in the available literature by a variety of distinct compounds, such as ZnO[41] (Figure 4.17), carbon,[42] silicon,[43] boron nitride,[44] γ-MnO$_2$,[45] silicon carbide,[46] and much more. In this section, we discuss some

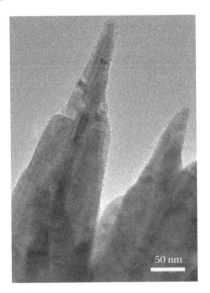

FIGURE 4.17 Nanocone (ZnO). (Reproduced from Shen, G. et al., Growth of self-organized hierarchical ZnO nanoarchitectures by a simple In/In$_2$S$_3$ controlled thermal evaporation process, *J. Phys. Chem. B*, 109(21), 10779–10785. Copyright 2005 American Chemical Society.)

* Sometimes referred to as nanohorns.

representative examples of nanocones. Elemental cone-like nanostructures have been fabricated for a series of metals, in particular Ni,[47] Cu,[25] Nb, and Hf.[48] Ge nanocones were grown by an Au-catalyzed process in a CVD chamber using germane (GeH_4) as the source gas.[49] The growth temperature was varied for different segments of the growth, and a corresponding variation of taper (angle of cone sidewalls relative to the axial direction) in the segments was observed. This dependence on temperature of the growth morphology was attributed to the relative influence of competing growth mechanisms at different temperature regimes: catalyzed unidirectional growth at lower temperatures approaching the Au-Ge eutectic point, and increasingly isotropic and facet-bounded epitaxial growth at higher temperatures (400°C).

Single crystalline boron nanocones were prepared by a simple CVD method using thermal evaporation of B/B_2O_3 powder precursors in an Ar/H_2 gas mixture at the synthesis temperature of 1000°C–1200°C and with Fe_3O_4 nanoparticles as catalyst.[50] The length of the boron nanocone was about 5 μm, and the diameter of the nanocone tip was about 50 nm. Boron nanocones with good field emission properties are promising candidates for applications in optical emitting devices and flat panel displays. A number of reports were devoted to little-explored carbon nanocones (as promising practical phononic devices): conical structures made predominantly from carbon and that have at least one dimension on the order of 1 μm or smaller. Nanocones have height and base diameter on the same order of magnitude; this distinguishes them from tipped nanowires, which are much longer than their diameter. Carbon nanocone has a high asymmetric geometry and is characterized by the cone angle[51] (the largest cone angle observed experimentally and theoretically was 113° [Figure 4.18]). In addition, pentagonal carbon nanocones were found to be constructed from a graphene sheet by removing a 60° wedge and joining the edges, producing a cone with a single pentagonal defect at the apex.[52]

Large-scale production of conical carbon nanostructures is possible through pyrolysis of hydrocarbons in a plasma torch process.[53] The resulting carbon cones occurred in five distinctly different forms, and disk-shaped particles are produced as well. The carbon nanocones were found to exhibit several interesting structural features: instead of having a uniform cross section, the walls consist of a relatively thin inner graphite-like layer with a noncrystalline envelope, where the amount of the latter can be modified significantly by annealing. In addition to applications of carbon nanocone phononic devices, mentioned earlier, the carbon nanomaterial (BET surface area of 15–40 m^2/g), containing 70 wt.% nanodisks and 30 wt.% nanocones, was used as a filter material for a gas mixture containing NO_2 and ozone, which was permeable to NO_2 and impermeable to ozone.[54]

Si-containing nanocones are also common. Thus, a new type of surface-enhanced Raman Spectroscopy (SERS) substrates, based on metal (Au or Ag) coated Si nanocones fabricated by a

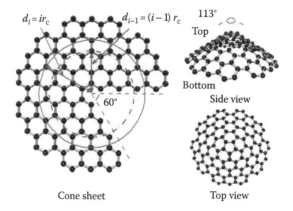

FIGURE 4.18 Schematic picture of the carbon nanocone. d is the distance from the atom to the center point in the cone sheet. (Reproduced with permission from Yang, N. et al., Carbon nanocone: A promising thermal rectifier, *Appl. Phys. Lett.*, 93, 243111. Copyright 2008 American Chemical Society.)

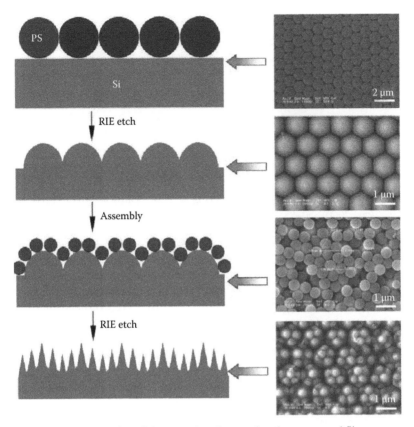

FIGURE 4.19 Schematic illustration of the procedure for creating the corrugated Si nanocone arrays on an Si substrate using polystyrene spheres. (Reproduced with permission from Springer Science+Business Media: *Nano Res.*, Biomimetic corrugated silicon nanocone arrays for self-cleaning antireflection coatings, 3, 2010, 520–527, Wang, Y. et al.)

Bosch etching process, was reported.[55] A duplication of the Si nanocones by a cross-linked polymer using 3D nanoimprint lithography was also carried out, demonstrating better enhancement factors for both 633 nm excitation and 785 nm excitation with analytical enhancement factors of over 1011. Corrugated silicon nanocone arrays were fabricated (Figure 4.19) on a silicon wafer by two polystyrene(PS)-sphere-monolayer-masked etching steps to create high-performance antireflective coatings.[56] The fluorinated corrugated Si nanocone array surface exhibited superhydrophobic properties with a water contact angle of 164°. In addition, a well-aligned growth of a CNT at the tip of SiO_2 nanocone using the CVD method was described,[57] showing that the Fe particle at the tip of a nanocone worked as the catalyst for CNT growth. Initially, a number of self-organized SiO_2 nanocones were grown via thermal annealing of $MnCl_2$ on an Si substrate in the presence of H_2; the CNT grew from the tip of the nanocone where Fe particles accumulated after the reduction of $FeCl_3$ at 950°C. It was revealed that the alignment of the nanotube at the tip of the SiO_2 nanocone can be controlled by orientation of the nanocone in the reaction tube. In addition, nanocone-presenting SiGe antireflection layers were fabricated using only ultrahigh-vacuum CVD.[58] *In situ* thermal annealing was adopted to cause SiGe clustering, yielding a characteristic nanocone array on the SiGe surface. It was established that the SiGe nanocones had uniform height and distribution.

The mixed-composition (Si–N–C–O) nanocones (Figure 4.20) were produced by gas phase mixing and deposition of plasma-sputtered silicon, nitrogen, carbon, and oxygen species on a central backbone nucleated by the Fe–Pt catalyst particle.[59] The conical shape of the nanocones was established to be formed by gas-phase deposition of amorphous Si–N–C–O on a core nucleated by the

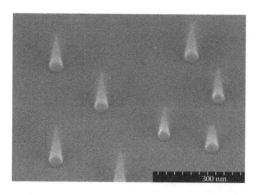

FIGURE 4.20 FESEM image of nanocones on a silicon substrate viewed at 45° tilt. (Reproduced from Cui, H. et al., *J. Mater. Res.*, 20(4), 850–855. Copyright 2005 Materials Research Society.)

catalyst particle during its growth. However, the nanocone had an amorphous internal structure due to the damaging impact of ion bombardment and silicon deposition during growth. A large turn-on electric field was found to be required for the nanocones to emit electrons.

Oxides and mixed oxo (hydroxo)halides were also discovered in nanocone forms. Thus, a nanocone ZnO thin film (Figure 4.21), prepared by electron beam evaporation on an Si(100) substrate, had a hexagonal wurtzite structure and was preferentially oriented along the c-axis perpendicular to the substrate surface.[60] The surface of the film was covered with dense and uniform nanocone ZnO grains, whose average height was about 50 nm. The formation of the nanocone structure mainly results from annealing treatment. Studies of the observed aging effect of the photoluminescent behavior of this nanocone ZnO thin film showed that the green emission was related to oxygen vacancy defects. Highly ordered α-Al_2O_3 microcones, readily generated by a facile thermal oxidation method,[61] were found to be composed of a large amount of nano-cones, indicating a complex form and remarkable repetitive growth patterns of the superstructure microcones. The photoluminescence spectrum revealed a strong red luminescence at 693 nm and three small-shoulder emission bands at 677, 706, and 712 nm, respectively. MnOOH and MnO_2 polymorphs were synthesized using a reaction between Mn^{2+} and MnO_4^- in water at low tempera-ture (95°C).[62] Depending on the acidity of the aging medium, γ-MnO_2 compounds were shown to be obtained as nanorods (final pH = 2.0) or hollow nanocones (3.6 ≤ final pH ≤ 4.5). The external faces of the cones originated from the heterogeneous oriented attachment of α-MnOOH nanorods on early cones. In addition, SnO_2 nanocrystals with various morphologies, including nanocones, were synthesized via a hydrothermal method assisted by the surfactant CTAB.[63] Also, CdClOH subnanocone crystals were synthesized on a large scale by a facile solution-based method using polymers (polyacrylamide [PAM]) as crystal growth modifiers.[64] It was revealed that the PAM played a key role in the formation of CdClOH subnanocones. Additionally, the as-prepared CdClOH sub-nanocones could be further transformed into CdS hollow subnano-cones by an anion-exchange reaction.

AlN nanocones have been a subject of discussion of researchers due to their absorbing proper-ties. Thus, a series of aluminum nitride (AlN) nanostructures, including nanocones, were synthe-sized by a variation of calcination conditions of an Al(OH) (succinate) complex, which contained a very small amount of iron as a catalyst, under a mixed gas flow of nitrogen and CO (1 vol.%).[65] The capability (Figure 4.22) of such AlN-based nanocones (as well as its other nanostructures, such as nanocages, nanotubes, and nanowires) to store hydrogen was found;[66,67] each Al atom was capable of binding one H_2 molecule in quasi-molecular form, leading to 4.7 wt.% hydrogen, irrespective of the topology of the nanostructures. These materials do not suffer from the clustering problem that often plagues metal-coated carbon nanostructures and are ideal for applications under ambient thermodynamic conditions.

(a)

(b)

FIGURE 4.21 2D (a) and 3D (b) surface morphology of the prepared ZnO thin film. (Reproduced from *Appl. Surf. Sci.*, 255, Xu, L. et al., Preparation of nanocone ZnO thin film and its aging effect of photoluminescence, 5957–5960, Copyright 2009, with permission from Elsevier.)

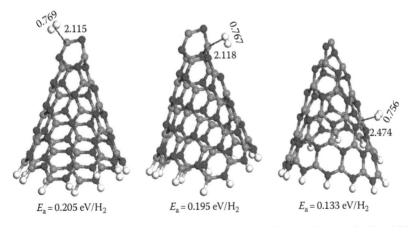

$E_a = 0.205$ eV/H$_2$ $E_a = 0.195$ eV/H$_2$ $E_a = 0.133$ eV/H$_2$

FIGURE 4.22 Adsorption of a single hydrogen molecule on different Al sites in the AlN nanocone. (Reproduced with permission from Wang, Q. et al., Potential of AlN nanostructures as hydrogen storage materials, *ACS Nano*, 3(3), 621–626. Copyright 2009 American Chemical Society.)

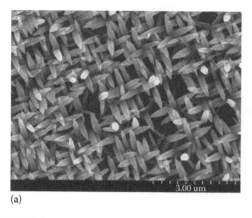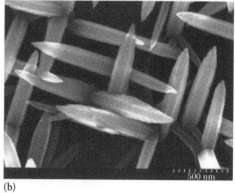

(a)   (b)

FIGURE 4.23   SEM images of as-deposited ZnO on the Si(001) substrate. (a) Low- (bar marker = 3 μm), and (b) high-magnification (bar marker = 500 nm) images. (Reproduced with permission from Mu, G. et al., Tilted epitaxial ZnO nanospears on Si(001) by chemical bath deposition, *Chem. Mater.*, 21(17), 3960–3964. Copyright 2009 American Chemical Society.)

4.7   NANOSPEARS

Nanospears (i.e., tapered nanorods) are very rare. Thus, nearly monodispersive nickel 3D nanostructures, exhibiting a catalytic effect, were synthesized using an improved hydrothermal route with poly-(N-vinyl-pyrrolidone) as capping reagent.[68] The cubic nickel crystals formed possessed flower-like 3D nanostructures with nanospearheads growing radially from the core and a uniform diameter of about 1.5 μm, the root in 120 ± 20 nm, the tip in 155 nm, the length in 400 ± 50 nm, and the thickness in 102 nm. Spear-like In-O-N nanostructures were synthesized on an Si substrate via a simple thermal evaporation process by ammoniating the In-containing reactants at 1100°C.[69] All the In-O-N nanostructures contain a long, thin pole with a diameter ranging from 20 to 100 nm and a huge spear-like tip at the end. In addition, epitaxial ZnO nanospears (Figure 4.23) can be deposited onto degenerate p-type Si(001) from an alkaline supersaturated solution of Zn(II) at 70°C using chemical bath deposition (Reactions 4.1 through 4.3).[70] The ZnO grew with a nanospear morphology, with spears that are 100–200 nm in diameter and about 1 micrometer in length. The nanospears grew along the ZnO[0001] direction (c-axis), but they are tilted with respect to the normal of the substrate surface. The ZnO/Si heterojunction leads the way to the integration of a large bandgap oxide semiconductor with traditional semiconductor devices. It may also produce efficient photoelectrochemical and photovoltaic solar cells:

$$Si + 4H_2O \rightarrow H_2SiO_4{}^{2-} + 2H_2 + 2H^+ \tag{4.1}$$

$$Zn(OH)_4{}^{2-} + 2H^+ \rightarrow ZnO + 3H_2O \tag{4.2}$$

$$Zn(OH)_4{}^{2-} + Si + H_2O \rightarrow ZnO + H_2SiO_4{}^{2-} + 2H_2 \tag{4.3}$$

4.8   NANOSPIKES

Nanospikes (frequently a part of nanoflowers or related nanostructures) are known for a few inorganic phases or compounds, such as Au[71] and TiO$_2$.[72] This type of nanostructures is obtained, in particular, by lithography[73] or ion beam irradiation (Si, Ge, and GaSb)[74] and by more simple chemical techniques. Thus, a simple method, based on colloid seed-engaged or seed-mediated replacement or deposition reactions 4.4 through 4.6, was offered to prepare shell-type Ag-Au bimetallic colloids with hollow interiors.[75] It was demonstrated that heat treatment can lead to the evolution

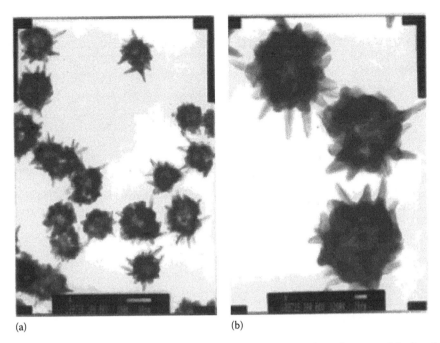

(a)   (b)

FIGURE 4.24   (a) Typical TEM image of as-prepared shell-type Ag-Au bimetal nanoparticles bearing nano-spikes; bar = 100 nm. (b) An enlarged TEM image showing the well-defined nanospikes and shell-type feature; bar = 20 nm. (Reproduced with permission from Jin, Y. and Dong, S., One-pot synthesis and characterization of novel silver-gold bimetallic nanostructures with hollow interiors and bearing nanospikes, *J. Phys. Chem. B*, 107(47), 12902–12905. Copyright 2003 American Chemical Society.)

of the shell-type Ag-Au bimetal colloids to spontaneously generate novel shell-type nanostructures bearing nanospikes (bimetallic nanostructures with hollow interiors and bearing nanospikes, Figure 4.24). The strong absorption of the as-prepared Ag/Au bimetallic nanoshells (and even bearing nanospikes) in the near-infrared region (800–1200 nm, the transparent window of tissues) makes these nanomaterials ideal candidates for photothermally triggered drug release in tissues. The formation of dense arrays of noble and other metal nanospikes also occurred under laser ablation of bulk targets (Ag, Au, Ta, Ti) immersed in liquids such as H_2O or EtOH.[76,77] Similarly, exposing a germanium surface to femtosecond laser pulses in an environment of SF_6 led to nearly regular arrays of nanospikes atop conical microstructures:[78]

$$(\text{in NH}_2\text{OH}) \; \text{Ag}_{core} + \text{Au}^{3+} \rightarrow \text{Ag}_{core}\text{Au}_{shell} \tag{4.4}$$

$$3\text{Ag}_{(s)} + \text{AuCl}_4^- \rightarrow \text{Au}_{(s)} + 3\text{Ag}_{(aq)}^+ + 4\text{Cl}_{(aq)}^- \tag{4.5}$$

$$(\text{in NH}_2\text{OH}) \; (\text{Ag}_{core}\text{Au}_{shell}) + \text{Ag}^+ \rightarrow (\text{Ag}_{core}\text{Au}_{shell})\text{Ag} \tag{4.6}$$

Direct electrodeposition technique was employed to fabricate ZnO nanospikes (Figure 4.25) and nanopillars on indium-tin oxide glass substrates at 70°C without using any template, catalyst, or seed layer (Reactions 4.7 through 4.10).[79,80] Both ZnO nanospikes and nanopillars exhibited highly crystalline ZnO wurtzite structure with a preferred (0001) plane orientation. The ZnO nanospikes with tapered tips of 20–50 nm diameter provide a favorable geometry to facilitate excellent field-emission performance. In addition, magnetic Fe/ZnO composites with the shape of a sea urchin were hydrothermally synthesized at 80°C from metallic iron and $Zn(NO_3)_2 \cdot 6H_2O$ as precursors without the use of a surfactant; they contained the Fe nanoparticles, encapsulated inside a shell of self-assembled ZnO nanospikes (Figure 4.26).[81] Because of the encapsulation of the Fe nanoparticles,

FIGURE 4.25 ZnO nanospikes. (Reproduced from Pradhan, D. and Leung, K.T., Controlled growth of two-dimensional and one-dimensional ZnO nanostructures on indium tin oxide coated glass by direct electrode-position, *Langmuir*, 24, 9707–9716. Copyright 2008 American Chemical Society.)

(a)

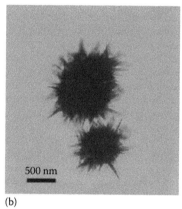

(b)

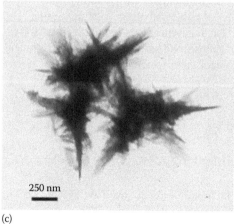

(c)

FIGURE 4.26 TEM images of Fe/ZnO composites collected after (a) 3 h, (b) 10 h, and (c) 24 h of hydrothermal treatment at 80°C. (Reproduced from Yang, Z.X. et al., Novel photoluminescence properties of magnetic Fe/ZnO composites: Self-assembled ZnO nanospikes on Fe nanoparticles fabricated by hydrothermal method, *J. Phys. Chem. C*, 113(51), 21269–21273. Copyright 2009 American Chemical Society.)

the Fe/ZnO composites are stable in air and high in magnetization. Among other applications for nanospikes, SnO_2 thin-film CO gas sensors on nanospiked polyurethane (PU) polymer surfaces, showing sensitive responses to the CO gas at r.t. because of the sharp structures of the nanospikes, were fabricated.[82] At the same time, the sensors of SnO_2 thin film coated on smooth surfaces showed no response to the CO gas at r.t.:

$$Zn(NO_3)_2(aq) \leftrightarrow Zn^{2+} + 2NO_3^- \tag{4.7}$$

$$NO_3^-(aq) + H_2O(l) + 2e \leftrightarrow NO_2^-(aq) + 2OH^- \tag{4.8}$$

$$Zn^{2+}(aq) + 2OH^-(aq) \rightarrow Zn(OH)_2(s) \tag{4.9}$$

$$Zn(OH)_2(s) + 2OH^- \rightarrow ZnO(s) + H_2O \tag{4.10}$$

REFERENCES

1. Wang, Y.; Zang, K.; Chua, S.; Sander, M. S.; Tripathy, S.; Fonstad, C. G. High-density arrays of InGaN nanorings, nanodots, and nanoarrows fabricated by a template-assisted approach. *Journal of Physical Chemistry B*, 2006, *110*, 11081–11087.
2. Zang, K. Y.; Wang, Y. D.; Chua, S. J.; Wang, L. S.; Tripathy, S.; Thompson, C. V. Nanoheteroepitaxial lateral overgrowth of GaN on nanoporous Si(111). *Applied Physics Letters*, 2006, *88*, 141925-1–141925-3, 2006.
3. Liu, Z.; Zhang, X.; Hark, S. K. Quadra-twin model for growth of nanotetrapods and related nanostructures. *Journal of Physical Chemistry C*, 2008, *112* (24), 8912–8916.
4. Kumar, M.; Singh V. N.; Mehta B. R.; Singh, J. P. Tunable synthesis of indium oxide octahedra, nanowires and tubular nanoarrow structures under oxidizing and reducing ambients. *Nanotechnology*, 2009, *20* (23), 235608.
5. Yan, Y.; Zhang, Y.; Zeng, H.; Zhang, J.; Cao, X.; Zhang, L. Tunable synthesis of In_2O_3 nanowires, nanoarrows and nanorods. *Nanotechnology*, 2007, *18* (17), 175601/1–175601/6.
6. Shen, G.; Bando, Y.; Lee, C.-J. Growth of self-organized hierarchical ZnO nanoarchitectures by a simple In/In_2S_3 controlled thermal evaporation process. *Journal of Physical Chemistry B*, 2005, *109* (21), 10779–10785.
7. Warner, J. H.; Tilley, R. D. Synthesis and self-assembly of triangular and hexagonal CdS nanocrystals. *Advanced Materials*, 2005, *17* (24), 2997–3001.
8. Seo, H. W.; Wang, D.; Tzeng, Y.; Sathitsuksanoh, N.; Tin, C. C.; Bozack, M. J.; Williams, J. R.; Park, M. Growth and characterization of ZnO nanonail. *MRS Proceedings (Progress in Compound Semiconductor Materials IV—Electronic and Optoelectronic Applications)*, Boston, MA, 2005, Volume Date 2004, Vol. 829, pp. 157–162.
9. Niu, H.; Yang, Q.; Tang, K.; Xie, Y. Large-scale synthesis of single-crystalline MgO with bone-like nanostructures. *Journal of Nanoparticle Research*, 2006, *8* (6), 881–888.
10. Wu, R. B.; Yang, G. Y.; Pan, Y.; Wu, L. L.; Chen, J. J.; Gao, M. X.; Zhai, R.; Lin, J. Thermal evaporation and solution strategies to novel nanoarchitectures of silicon carbide. *Applied Physics A: Materials Science and Processing*, 2007, *88* (4), 679–685.
11. Zhang, G.; Yu, Y.; Chen, X.; Han, Y.; Di, Y.; Yang, B.; Xiao, F.; Shen, J. Silica nanobottles templated from functional polymer spheres. *Journal of Colloid and Interface Science*, 2003, *263* (2), 467–472.
12. Yang, Q.; Liu, Q.; Chen, Q.; He, M.; Huang, H.g.; Chen, S.; Zhu, E. Photocatalytic degradation of methyl orange using ZnO nanobottles and nanorods. *Huagong Xinxing Cailiao*, 2009, *37* (5), 78–81.
13. Cheng, C.-L.; Chao, S.-H.; Chen, Y.-F. Enhancement of field emission in nanotip-decorated ZnO nanobottles. *Journal of Crystal Growth*, 2009, *311* (19), 4381–4384.
14. Choi, S.-H.; Kim, E.-G.; Hyeon, T. One-pot synthesis of copper-indium sulfide nanocrystal heterostructures with acorn, bottle, and larva shapes. *Journal of American Chemical Society*, 2006, *128*, 2520–2521.
15. Ren, Y.; Pastorin, G. Incorporation of hexamethylmelamine inside capped carbon nanotubes. *Advanced Materials*, 2008, *20* (11), 2031–2036.

16. Vakhrushev, A. V.; Suyetin, M. V. Methane storage in bottle-like nanocapsules. *Nanotechnology*, 2009, *20*, 125602.
17. Iu, H.; Li, J.; Ong, H. C.; Wan Jones, T. K. Surface plasmon resonance in two-dimensional nanobottle arrays. *Optics Express*, 2008, *16* (14), 10294–10302.
18. Dong, O.; Yeung, K. L. Immobilization of biomolecules in nanobottles. In *NSTI Nanotech 2007, Nanotechnology Conference and Trade Show*. Laudon, M.; Romanowicz, B. (Eds.). Santa Clara, CA, May 20–24, 2007, Vol. 1, pp. 294–296.
19. Krupenkin, T. Nanograss, nanobricks, nanonails, and other things useful in your nanolandscaping. *Abstracts of Papers, 237th ACS National Meeting*, Salt Lake City, UT, March 22–26, 2009, POLY-333.
20. Umar, A. A.; Oyama, M. Synthesis of palladium nanobricks with atomic-step defects. *Crystal Growth and Design*, 2008, *8* (6), 1808–1811.
21. Sonkar, S. K.; Saxena, M.; Saha, M.; Sarkar, S. Carbon nanocubes and nanobricks from pyrolysis of rice. *Journal of Nanoscience and Nanotechnology*, 2010, *10* (6), 4064–4067.
22. Ma, Y.-R.; Lin, J.-H.; Ho, W.-D.; Devan, R. S. Strong room-temperature photoluminescence of β-crystalline Ta_2O_5 nanobrick arrays. In *Nanotech Conference and Expo 2010: An Interdisciplinary Integrative Forum on Nanotechnology, Biotechnology and Microtechnology*. Laudon, M.; Romanowicz, B. (Eds.). June 21–24, 2010, Anaheim, CA, Vol. 1, pp. 45–48.
23. Yang, S.-H.; Hong, S.-Y.; Tsai, C.-H. Growth mechanisms and characteristics of ZnO nanostructures doped with In and Ga. *Japanese Journal of Applied Physics*, 2010, *49* (6, Pt. 2), 06GJ06/1–06GJ06/6.
24. Wu, J.-M.; Shih, H. C.; Wu, W.-T. Growth of TiO_2 nanorods by two-step thermal evaporation. *Journal of Vacuum Science and Technology, B: Microelectronics and Nanometer Structures—Processing, Measurement, and Phenomena*, 2005, *23* (5), 2122–2126.
25. Yan, S.; Shen, K.; Zhang, Y.; Zhang, Y.; Xiao, Z. Synthesis of copper oxide nanostructures with controllable morphology by microwave-assisted method. *Journal of Nanoscience and Nanotechnology*, 2009, *9* (8), 4886–4891.
26. Han, Q.; Lu, J.; Yang, X.; Lu, L.; Wang, X. A Template-free route to Sb_2S_3 crystals with hollow olivary architectures. *Crystal Growth and Design*, 2008, *8* (2), 395–398.
27. Yang Y.; Scholz, R.; Fan, H. J.; Hesse, D.; Gosele, U.; Zacharias, M. Multitwinned spinel nanowires by assembly of nanobricks *via* oriented attachment: A case study of Zn_2TiO_4. *ACS Nano*, 2009, *3* (3), 555–562.
28. Miura, T.; Kotachi, A.; Oaki, Y.; Imai, H. Emergence of acute morphologies consisting of ISO-oriented calcite nanobricks in a binary poly(acrylic acid) system. *Crystal Growth and Design*, 2006, *6* (2), 612–615.
29. Rogstroem, L.; Johnson, L. J. S.; Johansson, M. P.; Ahlgren, M.; Hultman, L.; Oden, M. Age hardening in arc-evaporated ZrAlN thin films. *Scripta Materialia*, 2010, *62* (10), 739–741.
30. Hahn, Y.-B.; Umar, A. Chemical sensor based on zinc oxide nanostructures for detection of hydrazine. U.S. Patent Application Publication US 2009178925, 2009, 17 pp. A1 20090716 US 2008-37912 20080226.
31. Umar, A.; Kim, S. H.; Kim, J. H.; Park, Y. K.; Hahn, Y. B. Fabrication of electrochemical bio-sensors for the detection of glucose and hydrazine using ZnO nanonails grown by the thermal evaporation process. In NSTI Nanotech 2007, *Nanotechnology Conference and Trade Show*. Laudon, M., Romanowicz, B. (Eds.). Santa Clara, CA, May 20–24, 2007, Vol. 1, pp. 332–335.
32. Umar, A.; Rahman, M. M.; Kim, S. H.; Hahn, Y. B. ZnO nanonails: Synthesis and their application as glucose biosensor. *Journal of Nanoscience and Nanotechnology*, 2008, *8* (6), 3216–3221.
33. Dev, A.; Chaudhuri, S.; Dev, B. N. ZnO 1-D nanostructures: Low temperature synthesis and characterizations. *Bulletin of Materials Science*, 2008, *31* (3), 551–559.
34. Umar, A.; Hahn, Y.-B. Ultraviolet-emitting ZnO nanostructures on steel alloy substrates: Growth and properties. *Crystal Growth and Design*, 2008, *8* (8), 2741–2747.
35. Guo, R.; Nishimura, J.; Higashihata, M.; Nakamura, D.; Okada, T. Substrate effects on ZnO nanostructure growth via nanoparticle-assisted pulsed-laser deposition. *Applied Surface Science*, 2008, *254* (10), 3100–3104.
36. Lao, J. Y.; Huang, J. Y.; Wang, D. Z.; Ren, Z. F. ZnO nanobridges and nanonails. *Nano Letters*, 2003, *3* (2), 235–238.
37. Song, X.; Zhang, Y.; Zheng, J.; Li, X. Low-temperature synthesis of ZnO nanonails. *Journal of Physics and Chemistry of Solids*, 2007, *68* (9), 1681–1684.
38. Zhang, Y.; He, J.; Huang, Y.; Gu, Y.; Ji, Z.; Zhou, C. Zinc powder evaporation: An efficient way of synthesizing a wide range of high-quality ZnO nanostructures at lower temperature. *MRS Proceedings (Micro- and Nanosystems—Materials and Devices)*, 2005, Vol. 872, pp. 513–518.
39. Kar, S.; Dev, A.; Chaudhuri, S. Simple solvothermal route to synthesize ZnO nanosheets, nanonails, and well-aligned nanorod arrays. *Journal of Physical Chemistry B*, 2006, *110* (36), 17848–17853.

40. Kuo, S.-Y.; Chen, W.-C.; Kei, C. C.; Hsiao, C. N. Fabrication of nanostructured indium nitride by PA-MOMBE. *Semiconductor Science and Technology*, 2008, *23* (5), 055013/1–055013/5.

41. Ma, J. H.; Yang, H. Q.; Song, Y. Z.; Li, L.; Xie, X. L.; Liu, R. N.; Wang, L. F. Controlled growth and photoluminescence of highly oriented arrays of ZnO nanocones with different diameters. *Science in China, Series E: Technological Sciences*, 2009, *52* (5), 1264–1272.

42. Yang, N.; Zhang, G.; Li, B. Carbon nanocone: A promising thermal rectifier. *Condensed Matter*, 2009, 1–17, arXiv.org, e-Print Archive, arXiv:0906.1049v1 [cond-mat.mes-hall].

43. Li, Q.-T.; Li, Z.-G.; Xie, Q.-L.; Gong, J.-L.; Zhu, D.-Z. Controlled evolution of silicon nanocone arrays induced by Ar⁺ sputtering at room temperature. *Chinese Physics Letters*, 2009, *26* (5), 056102/1–056102/4.

44. Zou, Y. S.; Chong, Y. M.; Ji, A. L.; Yang, Y.; Ye, Q.; He, B.; Zhang, W. J.; Bello, I.; Lee, S. T. The fabrication of cubic boron nitride nanocone and nanopillar arrays via reactive ion etching. *Nanotechnology*, 2009, *20* (15), 155305/1–155305/5.

45. Portehault, D.; Cassaignon, S.; Baudrin, E.; Jolivet, J.-P. Twinning driven growth of manganese oxide hollow cones through self-assembly of nanorods in water. *Crystal Growth and Design*, 2009, *9* (6), 2562–2565.

46. Liu, Z.; Ci, L.; Srot, V.; Jin-Phillipp, N. Y.; Van Aken, P. A.; Ruehle, M.; Yang, J. C. Crystalline silicon carbide nanocones and heterostructures induced by released iron nanoparticles. *Applied Physics Letters*, 2008, *93* (23), 233113/1–233113/3.

47. Nagaura, T.; Wada, K.; Inoue, S. Hexagonally ordered Ni nanocone array; controlling the aspect ratio. *Materials Transactions*, 2010, *51* (7), 1237–1241.

48. Morant, C.; Marquez, F.; Campo, T.; Sanz, J. M.; Elizalde, E. Niobium and hafnium grown on porous membranes. *Thin Solid Films*, 2010, *518* (23), 6799–6803.

49. Cho, H. S.; Kamins, T. I. *In situ* control of Au-catalyzed chemical vapor deposited (CVD) Ge nanocone morphology by growth temperature variation. *Journal of Crystal Growth*, 2010, *312* (16–17), 2494–2497.

50. Wang, X. J.; Tian, J. F.; Bao, L. H.; Shen, C. M.; Gao, H. J. Single crystalline boron nanocones. *MRS Proceedings (Nanotubes, Nanowires, Nanobelts and Nanocoils)*, Warrendale, PA, 2009, Vol. 1142, Paper #: 1142-JJ10-28.

51. Yang, N.; Zhang, G.; Li, B. Carbon nanocone: A promising thermal rectifier. *Applied Physics Letters*, 2008, *93*, 243111.

52. Khalifeh, M. H.; Yousefi-Azari, H.; Ashrafi, A. R. A method for computing the Wiener index of one-pentagonal carbon nanocones. *Current Nanoscience*, 2010, *6* (2), 155–157.

53. Naess, S. N.; Elgsaeter, A.; Helgesen, G.; Knudsen, K. D. Carbon nanocones: Wall structure and morphology. *Science and Technology of Advanced Materials*, 2009, *10* (6).

54. Pauly, A.; Dubois, M.; Guerin, K.; Hamwi, A.; Brunet, J.; Varenne, C.; Lauron, B. Use of carbon nanomaterials as a filtration material impermeable to ozone. WO 2010000956, 2010, 20 pp.

55. Wu, W.; Hu, M.; Ou, F. S.; Williams, R. S.; Li, Z. Rational engineering of highly sensitive SERS substrate based on nanocone structures. *Proceedings of SPIE (Advanced Environmental, Chemical, and Biological Sensing Technologies VII)*, Orlando, FL, 2010, Vol. 7673, pp. 76730O/1–76730O/6.

56. Wang, Y.; Lu, N.; Xu, H.; Shi, G.; Xu, M.; Lin, X.; Li, H. et al. Biomimetic corrugated silicon nanocone arrays for self-cleaning antireflection coatings. *Nano Research*, 2010, *3*, 520–527.

57. Kumar, R.; Choi, C.; Hwang, I.-C.; Thakur, Nh. Carbon nanotube growth at the tip of SiO₂ nanocone. *Materials Science and Engineering, C: Materials for Biological Applications*, 2009, *29* (8), 2384–2387.

58. Chang, Y.-M.; Dai, C.-L.; Cheng, T.-C.; Hsu, C.-W. Nanocone SiGe antireflective thin films fabricated by ultrahigh-vacuum chemical vapor deposition with in situ annealing. *Thin Solid Films*, 2010, *518* (14), 3782–3785.

59. Cui, H.; Yang, X.; Meyer, H. M.; Baylor, L. R.; Simpson, M. L.; Gardner, W. L.; Lowndes, D. H.; An, L.; Liu, J. Growth and properties of Si–N–C–O nanocones and graphitic nanofibers synthesized using three-nanometer diameter iron/platinum nanoparticle-catalyst. *Journal of Materials Research*, 2005, *20* (4), 850–855.

60. Xu, L.; Shi, L.; Li, X. Preparation of nanocone ZnO thin film and its aging effect of photoluminescence. *Applied Surface Science*, 2009, *255*, 5957–5960.

61. Li, P. G.; Lei, M.; Tang, W. H. Raman and photoluminescence properties of α-Al₂O₃ microcones with hierarchical and repetitive superstructure. *Materials Letters*, 2010, *64* (2), 161–163.

62. Portehault, D.; Cassaignon, S.; Baudrin, E.; Jolivet, J.-P. Selective heterogeneous oriented attachment of manganese oxide nanorods in water: Toward 3D nanoarchitectures. *Journal of Materials Chemistry*, 2009, *19* (42), 7947–7954.

63. Jia, T.; Wang, W.; Fu, Z.; Huang, F.; Wang, H. Controlled growth of SnO_2 micro/nanocrystals with different morphologies via a solution method. *Advanced Materials Research*, 2009, *66 (Advanced Synthesis and Processing Technology for Materials)*, 155–158.

64. Xiong, Y.; Peng, Y.; Liu, Z. Controlled synthesis of CdClOH sub-nanocones by low-temperature solution process and their transformation into CdS hollow sub-nanocones. *Chinese Journal of Chemistry*, 2009, *27* (11), 2178–2182.

65. Jung, W.-S. Morphologically controlled growth of aluminum nitride nanostructures by the carbothermal reduction and nitridation method. *Bulletin of the Korean Chemical Society*, 2009, *30* (7), 1563–1566.

66. Wang, Q.; Sun, Q.; Jena, P.; Kawazoe, Y. Potential of AlN nanostructures as hydrogen storage materials. *ACS Nano*, 2009, *3* (3), 621–626.

67. Wang, Q.; Sun, Q.; Jena, P. Hydrogen storage in AlN-based nanostructures. *Preprints of Symposia—American Chemical Society, Division of Fuel Chemistry*, 2009, *54* (2), 751.

68. Li, Q.; Du, W. Large-scale synthesis and catalytic properties of nearly monodispersive nickel 3D nanostructures. *Xiyou Jinshu Cailiao Yu Gongcheng*, 2009, *38* (12), 2080–2084.

69. Liu, B.; Chen, X.; Yao, J. Structure analysis and optical study of In-O-N nanospears. *Nanotechnology*, 2007, *18* (19), 195604/1–195604/4.

70. Mu, G.; Gudavarthy, R. V.; Kulp, E. A.; Switzer, J. A. Tilted epitaxial ZnO nanospears on Si(001) by chemical bath deposition. *Chemistry of Materials*, 2009, *21* (17), 3960–3964.

71. Plowman, B.; Ippolito, S. J.; Bansal, V.; Sabri, Y. M.; O'Mullane, A. P.; Bhargava, S. K. Gold nanospikes formed through a simple electrochemical route with high electrocatalytic and surface enhanced Raman scattering activity. *Chemical Communications*, 2009, (33), 5039–5041.

72. Song, Y.-Y.; Lynch, R.; Kim, D.; Roy, P.; Schmuki, P. TiO_2 nanotubes: Efficient suppression of top etching during anodic growth. *Electrochemical and Solid-State Letters*, 2009, *12* (7), C17–C20.

73. Rose, J.; Baugh, D. Nanorings, nanopillars and nanospikes on Si(111) by modified nanosphere lithography: Fabrication and application. *MRS Proceedings (Group-IV Semiconductor Nanostructures)*, 2005, Vol. 832, pp. 299–309.

74. Cuenat, A.; Aziz, M. J. Spontaneous pattern formation from focused and unfocused ion beam irradiation. *MRS Proceedings (Self-Assembly Processes in Materials)*, 2002, Vol. 707, pp. 113–118.

75. Jin, Y.; Dong, S. One-pot synthesis and characterization of novel silver-gold bimetallic nanostructures with hollow interiors and bearing nanospikes. *Journal of Physical Chemistry B*, 2003, *107* (47), 12902–12905.

76. Truong, S. L.; Levi, G.; Bozon-Verduraz, F.; Petrovskaya, A. V.; Simakin, A. V.; Shafeev, G. A. Generation of nanospikes *via* laser ablation of metals in liquid environment and their activity in surface-enhanced Raman scattering of organic molecules. *Applied Surface Science*, 2007, *254* (4), 1236–1239.

77. Truong, S. L.; Levi, G.; Bozon-Verduraz, F.; Petrovskaya, A. V.; Simakin, A. V.; Shafeev, G. A. Generation of Ag nanospikes via laser ablation in liquid environment and their activity in SERS of organic molecules. *Applied Physics A: Materials Science and Processing*, 2007, *89* (2), 373–376.

78. Nayak, B. K.; Gupta, M. C.; Kolasinski, K. W. Spontaneous formation of nanospiked microstructures in germanium by femtosecond laser irradiation. *Nanotechnology*, 2007, *18* (19), 195302/1–195302/4.

79. Pradhan, D.; Kumar, M.; Ando, Y.; Leung, K. T. Fabrication of ZnO nanospikes and nanopillars on ITO glass by templateless seed-layer-free electrodeposition and their field-emission properties. *ACS Applied Materials and Interfaces*, 2009, *1* (4), 789–796.

80. Pradhan, D.; Leung, K. T. Controlled growth of two-dimensional and one-dimensional ZnO nanostructures on indium tin oxide coated glass by direct electrodeposition. *Langmuir*, 2008, *24*, 9707–9716.

81. Yang, Z. X.; Zhong, W.; Au, C. T.; Du, X.; Song, H. A.; Qi, X. S.; Ye, X. J.; Xu, M. H.; Du, Y. W. Novel photoluminescence properties of magnetic Fe/ZnO composites: Self-assembled ZnO nanospikes on Fe nanoparticles fabricated by hydrothermal method. *Journal of Physical Chemistry C*, 2009, *113* (51), 21269–21273.

82. Huo, H.-B.; Yan, F.-D.; Wang, C.; Ren, H.-Z.; Shen, M.-Y. Low-cost self-cleaning room temperature SnO_2 thin film gas sensor on polymer nanostructures. *Proceedings of SPIE (Pt. 2, Sensors and Smart Structures Technologies for Civil, Mechanical, and Aerospace Systems 2010)*, San Diego, CA, 2010, Vol. 7647, pp. 76474R/1–76474R/8.

5 Circle and Ball-Type Nanostructures

5.1 NANOWHEELS

Wheel-like nanostructures as nanoparticle forms are very rare. In addition to the description of a nanowheel as a geometric form of nanoparticles, this term is also used to discuss molecular parts in organic or coordination compounds,[1,2] which can be used, in particular, in the creation of nanocars.[3] A few examples of nanowheels are as follows. Various platinum nanostructures with size and shape control (globular nanodendrites, flat dendritic nanosheets, foam-like nanospheres, porous nanocages, nanowheels [Figures 5.1 and 5.2], nanowire networks, and hollow nanospheres) were synthesized by using a series of templates with/without porphyrin photocatalysts.[4] The porphyrin molecules rapidly reduced the platinum complex into a tunable number of initial seed particles under white light irradiation. This provided a nearly equal growth time for each seed, leading to final products with uniform and predictable sizes. Hierarchical ZnO nanoarchitectures, such as microtrepangs, microbelts, nanoflowers, nanocombs, nanowheels (Figure 5.3), and nanofans assembled by ZnO nanocones, nanobowling pins, nanobottles, nanoarrows, and nanonails, have had their growth controlled by the thermal evaporation of Zn and a mixture of In and In_2S_3.[5] These novel hierarchical ZnO nanoarchitectures may be attractive building blocks for creating optical or other nanodevices. A flow system was employed to study the steps underlying the assembly of molybdenum oxide wheel of 3.6 nm diameter.[6] Crystallization of an intermediate structure in which a central $\{Mo_{36}\}$ cluster appears to template the assembly of the surrounding $\{Mo_{150}\}$ wheel was observed (Figure 5.4). In addition, various ZnSe nanostructures (nanowires, nanorings, nanowheels, and tricrystal nanobelts) were fabricated by thermal evaporation method.[7,8]

In addition, a very rare example of a large polynuclear lanthanide complex nanowheel (Figure 5.5) with a very symmetric ring structure leading to a system containing europium ions in two different coordination environments was described.[1]

5.2 NANOEGGS

This nanostructural type is very rare and is developed mainly by the research group of *N. Halas* from Rice University, TX, for the creation of plasmonic nanoparticles. At present, a large variety of structures, including nanoeggs, have been synthesized and characterized, whose plasmon resonances may be varied over the entire visible to mid-infrared (IR) part of the electromagnetic spectrum (see Figure 5.6).[10] Thus, an individual Au nanoshell was controllably reshaped (Figure 5.7) into a reduced-symmetry nanoegg, then a semishell or nanocup by an electron-beam (E-beam)-induced ablation method, transforming its plasmonic properties.[11] The splitting of plasmon modes and the onset of electroinductive plasmons upon controlled, incremental opening of the outer metallic layer of the nanoparticle were observed.

The near- and far-field optical properties of three variants of a core-shell nanoparticles, nanoshells, nanoeggs[12] and nanocups, were examined.[13] Nanoshells, consisting of a spherical silica core coated with a thin gold shell, converted to nanoeggs (Figure 5.8) by offsetting the core within the shell. The absorption and scattering spectra of a nanoegg revealed the emergence of multipolar peaks strongly red shifted relative to those of nanoshells and larger near-field enhancements. These and other observations may lead to new opportunities to tailor near- and far-field properties

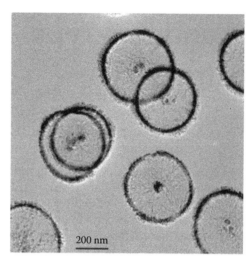

FIGURE 5.1 Platinum nanowheel. (Reproduced with permission from Song, Y. et al., Synthesis of platinum nanowheels using a bicellar template, *J. Am. Chem. Soc.*, 130, 12602–12603. Copyright 2008 American Chemical Society.)

FIGURE 5.2 SEM image of platinum nanowheels with a small percentage of by-products without defined morphology. Reaction conditions: 10 mM Pt(II), 0.5 mM of CTAB and FC7, and 75 mM AA in 20 mL of water at 25°C. (Reproduced with permission from Song, Y. et al., Size- and shape-controlled platinum nanostructures. *Abstracts of Papers, 238th ACS National Meeting*, Washington, DC, August 16–20, 2009, INOR-377. Copyright 2009 American Chemical Society.)

of plasmonic nanoparticles for specific applications, such as high-performance surface-enhanced spectroscopy, bioimaging, and nanoparticle-based therapeutics.

Other nanoeggs are represented by gold-nanoshell magnetic nanoeggs with surface-immobilized vancomycin (Van-Fe$_3$O$_4$@Au), in which the temperature of a suspension rose from 23°C to about 55°C over 3 min under illumination by near-IR (NIR) light.[15] The cell growth of nosocomial pathogenic bacteria, including antibiotic-resistant strains, was targeted by the Van-Fe$_3$O$_4$@Au nanoeggs and can be effectively inhibited. Additionally, a drug delivery system (DDS) to enhance inorganic-coated (with CaCO$_3$) all-*trans* retinoic acid therapy (atRA) for external treatments of photodamaged skin was developed.[16,17] Nanoparticles of atRA with an egg-like structure in nanoscale (nanoegg) were prepared using boundary-organized reaction droplets. Both irritate symptoms and the physicochemical instability of atRA were improved by a complete encapsulation of the atRA molecules. The nanoegg particles significantly increased the permeability of atRA to the stratum corneum.

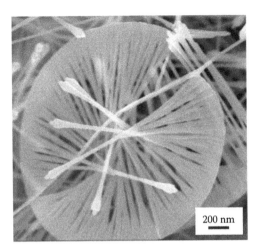

FIGURE 5.3   ZnO nanowheel. (Reproduced with permission from Shen, G. et al., Growth of self-organized hierarchical ZnO nanoarchitectures by a simple In/In$_2$S$_3$ controlled thermal evaporation process, *J. Phys. Chem. B*, 109(21), 10779–10785. Copyright 2005 American Chemical Society.)

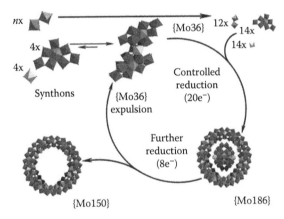

FIGURE 5.4   Conceptual representation of the MB assembly showing the building block "synthons" (which are assigned on the basis of structural considerations) that form the template complex. The {Mo$_{36}$}, {Mo$_{150}$}, and {Mo$_{36}$ ⊂ Mo$_{150}$} complexes have each been isolated separately. Continuous flow-reaction conditions, along with a finely tuned reducing environment, are required to trap the template complex. (Reproduced from Miras, H. N. et al., *Science*, 327(5961), 72–74. Copyright 2010. With permission.)

5.3   NANOBALLS

Nanoballs are closely related to nanospheres, but this first term is considerably lesser used in comparison with the last. Elemental nanoballs are represented by carbon and metals.[18] Thus, a swirled fluidized bed chemical vapor deposition (SFCVD) reactor was manufactured and optimized to produce carbon nanostructures (in the absence of a catalyst at temperatures higher than 1000°C) on a continuous basis using *in situ* formation of floating catalyst particles by thermal decomposition of ferrocene.[19] Ni, Ti, Ag, and Au metallic nanoballs (Figure 5.9) were produced in a gas/liquid mixed dual-phase system during plasma electrolysis from the cathode mother materials with a certain size controllability without contamination from electrolytes.[20]

Sub-micron-size bismuth balls with diameters between 100 and 2000 nm were formed on an anodic aluminum oxide (AAO) template using thermal expansion and thermal oil reflow processes.[21] Because of their round shape, the contact angle of the Bi nanoballs on the AAO surface was large after hot oil reflow, making it easy to remove them using ultrasound and to separate different sizes

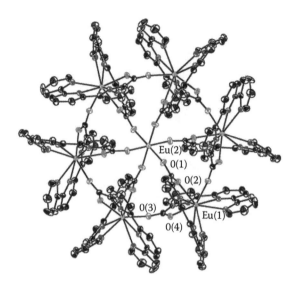

FIGURE 5.5 Crystal structure of the cation [Eu ⊂ (EuL$_2$)$_6$]$^{9+}$ (2,2′:6′,2″-terpyridine-6-carboxylic acid [HL]). (Reproduced with permission from Bretonniere, Y. et al., Cation-controlled self-assembly of a hexameric europium wheel, *J. Am. Chem. Soc.*, 124(31), 9012–9013. Copyright 2002 American Chemical Society.)

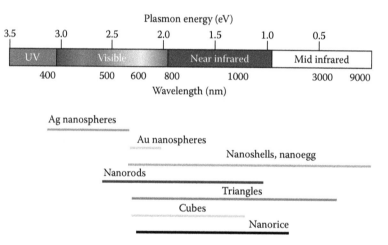

FIGURE 5.6 Nanoparticle resonances. A range of plasmon resonances for a variety of particle morphologies. (Reproduced with permission from Macmillan Publishers Ltd. *Nat. Photonics*, Lal, S. et al., Nano-optics from sensing to waveguiding, 1 (November), 641–648. Copyright 2007.)

of balls using centrifugal force. Porous Pt nanoballs were fabricated using soft templates made by a quaternary system as nanoreactors that form giant hexagonal liquid crystals.[22] Both confinement effect and slow reduction of the Pt salt were found to be necessary to obtain the porous nanoball. Also, 1D Ni/Ni$_3$C core-shell nanoball (Figure 5.10) chains with an average diameter of around 30 nm were synthesized by means of a mild chemical solution method using a soft template of trioctylphosphine oxide (TOPO).[23] It was revealed that the uniform Ni nanochains were capped with Ni$_3$C thin shells by about 1–4 nm in thickness and each Ni core consists of polygrains. Among other metals, nanoballs were described for Pd.[24]

Metal oxide nanoballs are common. Thus, tin oxide nanoplates (average size of 9.36 nm) and nanoballs (up to 4.51 nm) were fabricated using a cationic surfactant of cetyltrimethylammonium bromide (CTABr) as an organic supramolecular template and tin(IV) chloride as an inorganic precursor *via* the hydrothermal and conventional heating methods.[25] Urea, which decomposes to

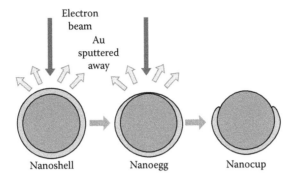

FIGURE 5.7 E-beam-induced ablation of Au nanoshells. Schematic illustrating nanoshell ablation process resulting in the transformation of a nanoshell to nanoegg to nanocup. (Reproduced with permission from Lassiter, J. B. et al., Reshaping the plasmonic properties of an individual nanoparticle, *Nano Lett.*, 9(12), 4326–4332. Copyright 2009 American Chemical Society.)

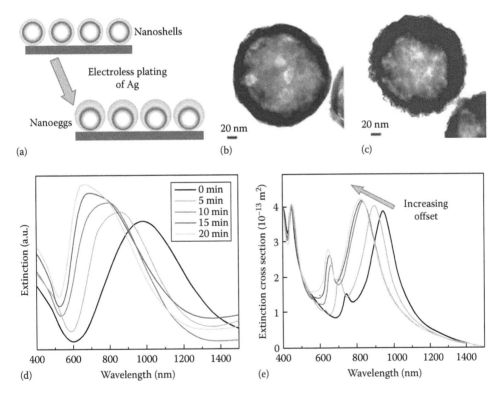

FIGURE 5.8 Synthesis of nanoshells with a nonconcentric core. (a) Schematic of nanoegg fabrication. Monolayers of silica–Au core-shell nanoparticles were first immobilized on PVP-functionalized glass slides. The nanoshells used in this set of experiments are 94 ± 9 nm in core radius and 9 ± 1 nm in shell thickness. By controlling the reaction time, the offset extent can be controlled. Longer reaction time results in the formation of nanoeggs with larger offset cores. (a,b) TEM images of a nanoshell (b) and nanoegg (c). (d) Experimentally measured evolution of extinction spectra of oriented monolayer nanoegg film during metallization with unpolarized optical excitation at normal incidence. (e) Calculated normal incidence extinction spectra (FDTD) of reduced symmetry nanoparticles as a function of increasing offset. In this set of calculations, the nanoeggs have a silica core of 94 nm in radius. The thinnest part of the shell is 9 nm, and the thickest part of the shell is 9 nm for concentric nanoshell and varied to be 15, 21, 27, and 33 nm for nanoeggs. (Reproduced with permission from Wang, H. et al., Symmetry breaking in individual plasmonic nanoparticles, *PNAS*, 103(29), 10856–10860. Copyright 2006 National Academy of the United States of America.)

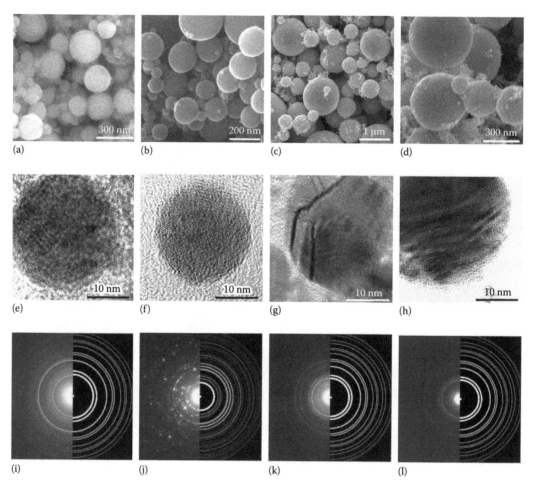

FIGURE 5.9 Typical micrographs of Ni, Ti, Ag, and Au nanoballs. (a through d) SEM images of (a) Ni, (b) Ti, (c) Ag, and (d) Au electrode surfaces. (e through h) TEM images of (e) Ni, (f) Ti, (g) Ag, and (h) Au spherical nanoparticles. (i through l) Electron diffraction patterns of (i) Ni, (j) Ti, (k) Ag, and (l) Au. The right side of each diffraction pattern is a numerical simulation assuming powder samples. (Reproduced with permission from Toriyabe, Y. et al., Controlled formation of metallic nanoballs during plasma electrolysis, *Appl. Phys. Lett.*, 91(4), 041501/1–041501/3. Copyright 2007 American Institute of Physics.)

ammonium and hydroxide ions during hydrolysis, was used as the source of slow homogeneous precipitation of Sn^{4+} with OH^- to control the particle size. It was shown that the size of the SnO_2 nanoparticles decreased with increasing reaction time using the conventional heating method, while no significant change was observed with the hydrothermal method. Pure ZnO nanorods, nanoneedles, nanoparticles, and nanoballs were synthesized on fused quartz substrates upon irradiation of a droplet of methanolic zinc acetate dihydrate solution by an IR continuous wave CO_2 laser for a few seconds.[26] Their photoluminescence spectra were dominated by an intense UV emission around 390 nm. Doped ZnO composites are represented by Fe-doped ZnO nanoballs $Zn_{1-x}Fe_xO$ ($x = 0.1$), prepared by the microemulsion route.[27] The observed absorption and photoluminescence properties of the product were both relevant to a Fe-doped situation. Mn-doped ZnO nanoballs are also known.[28] In addition, the structure and electronic properties of Al, B, Ga, and In oxide fullerene shape nanoballs were calculated at DFT-B3LYP and HF levels of theory.[29]

A series of metal salt nanoballs with distinct properties and applications have been reported. Thus, superparamagnetic fluorescent nanocomposites based on doped $LaPO_4$ and Fe_3O_4 nanoparticles, accessible through a facile one-pot method, adopted a Koosh ball structure with both

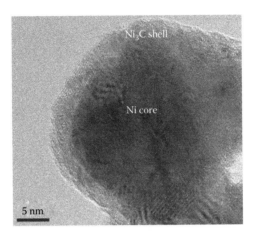

FIGURE 5.10 High-magnification TEM image of a single nanoball from the top of a chain, showing the core-shell structure. (Reproduced with permission from Zhou, W. et al., Ni/Ni₃C core–shell nanochains and its magnetic properties: One-step synthesis at low temperature, *Nano Lett.*, 8(4), 1147–1152. Copyright 2008 American Chemical Society.)

fluorescent and superparamagnetic properties for individual components maintained in the final nanostructure.[30] POM/MoS$_2$ nanoballs composite, prepared by adding MoS$_2$ nanoballs synthesized from Na$_2$MoO$_4$ and CH$_3$CSNH$_2$ into polyoxymethylene (POM),[31] was used as the polymeric layer in the three-layer self-lubrication materials, showing that the POM with MoS$_2$ nanoballs presented better tribological properties than that with micro-MoS$_2$. The reason for the stable self-lubrication properties of POM/MoS$_2$ nanoballs composite was ascribed to the forming-destroying of debris clusters in a long-time sliding process. In addition, MoS$_3$ nanoballs were also described.[32] Also, monodispersed CdS nanoballs were synthesized by γ-irradiation of CdCl$_2$, Na$_2$S$_2$O$_3$, and polyvinyl-pyrrolidone in aqueous solution at room temperature.[33] With these monodispersed CdS nanoballs, CdS-SiO$_2$ core-shell nanostructures were prepared by hydrolysis of tetraethylorthosilicate without adding a coupling agent. These final core-shell nanoballs are useful as luminescence detecting material for biological systems. ZnS nanoballs were similarly obtained.[34] In addition, glass nanoballs can be used for hydrogen storage materials.[35]

For the nanoball area, we note an elevated number (in comparison with other nanostructures) of examined coordination compounds. Thus, a one-pot, direct synthesis of photosensitive phthalocyanine (Pc) nanoballs composed of ZnPc cores was carried out without using any preorganized structure, emulsifier, or template.[36] The structural stability of self-assembled Cu(II) hydroxylated nanoballs [Cu^{2+}(5-OH-bdc)$_2$L$_2$]$_{12}$ (Figure 5.11) [(5-OH-bdc)$_2$ = 5-hydroxybenzene-1,2-dicarboxylate; L = di-Me sulfoxide, methanol, or water ligand] was investigated.[37,38] The nanoball was found to be moderately stable in solution with minimal perturbations up to 30°C and up to 3.5 kbar and stable over a pH range 5–10. Among a series of other examples, we emphasize [(Tp4-pyCuI(MeCN))$_8$M$^{II}_6$(X)$_{10}$(MeCN)$_2$] X$_2$ · xMeCN (Tp4-py = tris[3-(4-pyridyl)pyrazol-1-yl]hydroborate, M = Cu, Zn, Mn, Fe, Cd; X = ClO$_4$, BF$_4$).[39] Biologically important nanoballs are represented by DNA.[40]

5.4 NANOSPHERES

The term *nanosphere* is closely related to nanoball and it is not really rare; about 10,000 reports on nanotechnology using sphere-like nanoparticles are known, including organic, bioorganic, and polymeric nanospheres having a series of various applications, from drug delivery to hydrogen storage; here we present only the most recent and representative inorganic examples. Thus, carbon nanospheres were synthesized with 4% yield through the chemical reactions of calcium carbide and oxalic acid at 65°C–250°C without using catalysts.[41] A nonenzymic sensor for hydrogen peroxide

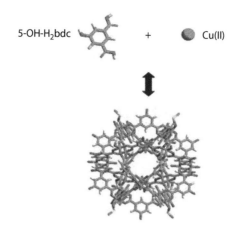

5-OH-H$_2$bdc   +   Cu(II)

FIGURE 5.11 Schematic diagram illustrating the assembly of a discrete OH nanoball from Cu^{2+} ions and the 5-OH-H$_2$bdc ligand. The OH nanoball also contains DMSO solvent molecules coordinated axially to the Cu^{2+} ion in this structure. (Reproduced with permission from Larsen, R.W. et al., Spectroscopic characterization of hydroxylated nanoballs in methanol, *Inorg. Chem.*, 46(15), 5904–5910. Copyright 2007 American Chemical Society.)

was fabricated by dispersing platinum hollow nanospheres onto polypyrrole (PPy) nanowires to form a PPy-Pt hollow sphere nanocomposite on a glassy carbon electrode.[42] It was revealed that the electrode had a large electroactive surface area and small resistance to electron transfer, exhibiting good stability and excellent repeatability. Foam-like platinum nanospheres[43] are shown in Figure 5.12. Raspberry hollow Pd nanosphere (HPN)-decorated carbon nanotube (CNT) was developed for electro-oxidation of methanol, ethanol, and formic acid in alkaline media.[44] The electrocatalyst was fabricated simply by attaching HPNs onto the surface of CNT that had been functionalized by polymer wrapping. It is found that this novel hybrid electrocatalyst exhibited excellent electrocatalytic properties and can be further applied in fuel cells, catalysts, and sensors. Another metallic palladium nanosphere catalyst with multi-shell-layer structure is known.[45] Ag/polypyrrole (PPy) core-shell nanospheres (a shell thickness of 10–12 nm and a core diameter of 20–40 nm) were fabricated through the redox reaction between pyrrole monomer and silver nitrate in the presence of polyvinylpyrrolidone (PVP) and by using the Ag colloidal nanoparticles acting as the seedings.[46] Additionally, a surface-enhanced Raman spectroscopy (SERS) active substrate of 60–80 nm diameter was developed through the assembly of Ag nanocrystals (AgNCs) into Ag nanospheres (AgNSs) and can be used in the detection (Figure 5.13) of pathogenic bacteria (*Escherichia coli* O157, *Salmonella typhimurium*, and *Staphylococcus aureus*): it could detect cells as few as 10 colony-forming units/mL (CFU/mL).[47] Other metals have also been described as nanospheres: Bi,[48] Au,[49] Nb and Hf,[50] etc.

Among metal oxide nanospheres, ZnO[51] and its composites are common. Thus, uniform ZnO/ZnFe$_2$O$_4$ fluorescent magnetic composite hollow nanospheres can be fabricated using carbonaceous nanospheres as templates, which were synthesized *via* a hydrothermal approach from FeCl$_3$·6H$_2$O and ZnCl$_2$ as precursors and further calcinations (Reaction 5.1 and Figure 5.14).[52] These composited nanoarchitectures showed fluorescent and magnetic properties useful for electronics, magnetism, optics, and catalysis compared with their single counterparts and may find potential applications in many fields, such as catalyst supports, catalysis, drug delivery, chemical/biological separation, sensing, etc.

$$Zn(OH)_2 + 2Fe(OH)_3 \rightarrow ZnFe_2O_4 + 4H_2O \qquad (5.1)$$

Titania (TiO$_2$) nanoparticles, surface-modified *via* atom transfer radical polymerization with hydrophilic poly(oxyethylene) methacrylate (POEM) (which can coordinate to the TiO$_2$ precursor, titanium(IV) isopropoxide) and further subjected to a sol-gel process and calcination at 450°C, formed TiO$_2$ nanospheres with hierarchical pores.[53] Uniform carbon-coated MoO$_2$ nanospheres assembled from small primary nanocrystals have been synthesized by a one-pot hydrothermal method

FIGURE 5.12 (a) TEM image. (b) HAADF scanning TEM image (Reproduced with permission from Song, Y. et al., Foamlike nanostructures created from dendritic platinum sheets on liposomes, *Chem. Mater.*, 18, 2335–2346. Copyright 2006 American Chemical Society.) (c) SEM image of foam-like platinum nanospheres.

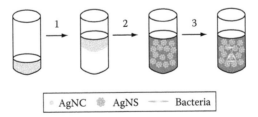

FIGURE 5.13 Schematic of the formation of SERS active AgNSs substrate for pathogen detection: (1) SDS aqueous solution, (2) sonication for 60 min and evaporation of cyclohexane at 70°C, and (3) pathogenic bacteria detection. (From Wang, Y. et al., Silver nanosphere SERS probes for sensitive identification of pathogens, *J. Phys. Chem. C*, 114(39), 16122–16128. Copyright 2010 American Chemical Society.)

followed by thermal annealing.[54] These core-shell MoO_2@Carbon nanospheres exhibited significantly improved electrochemical performance for high-rate reversible lithium storage. Mesoporous silica spheres were prepared in the preformed emulsion-templated porous polyacrylamide (PAM),[55] further removing the PAM by calcination. The hierarchical structures of porous nanospheres within macroporous structures or on nanofibers are of potential interest to researchers in nanomaterials, porous polymers, supported catalysis, and controlled delivery. Cobalt cluster-silica nanospheres (15–30 nm) were synthesized using a $Co(NH_3)_6Cl_3$ template method in a polyoxyethylene-nonylphenyl

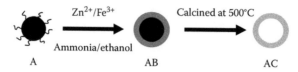

FIGURE 5.14 Schematic illustration of the formation of magnetic-fluorescent $ZnO/ZnFe_2O_4$ hollow nanospheres. A, carbon nanospheres; B, amorphous $Zn(OH)_2$ and $Fe(OH)_3$; C, $ZnO/ZnFe_2O_4$ hollow nanospheres. (Reproduced with permission from Qian, H.-S. et al., $ZnO/ZnFe_2O_4$ magnetic fluorescent bifunctional hollow nanospheres: Synthesis, characterization, and their optical/magnetic properties, *J. Phys. Chem. C*, 114, 17455–17459. Copyright 2010 American Chemical Society.)

ether/cyclohexane reversed micelle system followed by *in situ* reduction in aqueous $NaBH_4/NH_3BH_3$ solutions.[56] These cobalt-silica nanospheres are found to have a high catalytic activity for the hydrolysis of ammonia borane that generated a stoichiometric amount of hydrogen and can be efficiently cycled and reused 10 times without any significant loss of the catalytic activity. In addition, silane-modified magnetite nanoparticles (nanospheres with average diameter 25 nm) were synthesized by chemical coprecipitation and subsequent surface modification with 3-aminopropyltriethyloxysilane (APTES) and mercaptopropyltriethoxysilane (MPTES).[57] Lanthanide oxide sphere-like nanoparticles are represented by CeO_2[58] and Y_2O_3 nanospheres; the last compounds were prepared[59] by a two-step process involving cathodic deposition of yttrium hydroxide thin films in a chloride bath followed by thermal conversion of hydroxide into crystalline Y_2O_3 in air.

Several metal salt nanospheres are known for nanosized tin telluride compounds (nanoalloys), prepared by a chemical reduction process (quasi-spherical morphology, 40–50 nm) and hydrothermal (nearly nanospheres, 30–40 nm) methods. It was indicated that $N_2H_4 \cdot H_2O$ played a crucial role in the formation of nanosized rode-like SnTe compounds. Bi_2WO_6 hollow nanospheres were prepared in a sacrificial template approach by converting the reaction medium from aqueous solution to a mixed solvent of water and ethanol.[60] Also, calcium phosphate (CaP)/block copolymer hybrid porous nanospheres were synthesized by a simple solution method using $CaCl_2$ and $(NH_4)_2HPO_4$ in the presence of a block copolymer at room temperature.[61] The as-prepared CaP/PLLA-mPEG hybrid porous nanospheres were explored as drug carriers. These CaP/block copolymer hybrid porous nanospheres exhibit a great potential for application in drug delivery. In addition, $NaYF_4$:Yb, Er micro-/nanocrystals with different sizes and morphologies, such as nanospheres, short flexural nanorods, and half opened microtubes, were synthesized in reverse microemulsion under a solvothermal condition using the quaternary reverse microemulsion system, CTAB/1-butanol/cyclohexane/aqueous solution.[62] It was revealed that the morphology of the product can be tailored by varying the reaction time.

5.5 NANOGRAINS

The term *nanograin* cannot be attributed to a definitive/independent nanostructural type, and it very rarely appears in articles' titles, despite about 1000 available reports examining nanograins. Nanograins usually figure as a part of nanocomposites, films, layers, alloys, etc., describing mainly their internal composition and particle size as well as nucleation processes where grains first appear and grow. Several elemental metals, for instance, Al,[63] alloys, and supported metals have been studied in nanograin form. Thus, attempts at generating nanograins through uniaxial single compression were made by deforming copper samples at 298 and 77 K.[64] At a reduced temperature of 77 K, nanograins were generated in dynamically deformed samples, whereas quasi-statically deformed samples showed forest dislocations and twins. C-Pd films, obtained by a physical vapor deposition (PVD) process, were composed of fullerenes, amorphous carbon, and palladium nanograins and can be applied as active layers in gas sensor applications.[65] $Fe_{61}Co_{21}Nb_3B_{15}$ alloy, prepared[66] from a powder mixture by a mechanical alloying process, revealed the formation, after 48 h of milling, of a highly disordered amorphous structure where nanometer-sized iron borides were embedded. A mechanical recrystallization process gave rise to the formation of α-Fe and α-FeCo nanograins on further milling. In addition,

AgCl cores coated with Ag nanograins were found to have high photocatalytic activity under visible light for degradation of Methylene blue.[67] Bimetallic composite nanoparticles of PtM (M = Au, Cu or Ni) supported on γ-Fe_2O_3 were synthesized by a radiolytic method employing a 4.8 MeV E-beam to reduce aqueous ions and formed metallic nanograins of 3 nm diameter stabilized on the support of γ-Fe_2O_3 particles of 13 nm diameter.[68] A significant Pt-M coordination was noted. All the PtM systems exhibited catalytic activities of CO oxidation higher than those of monolithic Pt on γ-Fe_2O_3. Pd-C films (known as very promising materials for hydrogen sensors and storage),[69] composed of carbon and palladium nanograins and obtained in a two-step method, exhibited multiphase structure containing fullerene nanograins, amorphous carbon, and palladium nanocrystals.

Among metal oxides, zinc oxide[70] and its composites[71] have been examined in nanograin form. Thus, the study of a CdO-ZnO nanocomposite, synthesized by a sol-gel pyrolysis method based on polymeric network of polyvinyl alcohol (PVA), showed that this nanocomposite at optimum conditions had excellent linear nanoclusters created from nanograins.[72] Each nanograin was made of a CdO core that was completely covered by ZnO layers. The nanocomposite was used as a sensing agent of CO gas. Fe_2O_3 is also a common oxide, studied in nanograin form. Thus, the hematite crystal shapes had a sequential variation from nanoplates to nanograins[73] with an increase in aspect ratios (Figure 5.15).

(a)   (b)

(c)   (d)

FIGURE 5.15 SEM images of hematite products grown from the solvent of ethanol with an addition of (a) 0.3 mL, (b) 0.7 mL, (c) 1.2 mL, and (d) 2.5 mL of distilled water. The volume of ethanol is kept at 10.0 mL, and the use of sodium acetate is 0.8 g. (Reproduced with permission from Chen, L. et al., Continuous shape- and spectroscopy-tuning of hematite nanocrystals, *Inorg. Chem.*, 49(18), 8411–8420. Copyright 2010 American Chemical Society.)

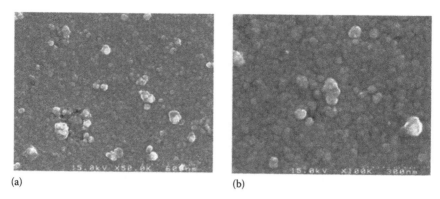

(a)   (b)

FIGURE 5.16 Scanning electron micrographs of electrodeposited ruthenium oxide electrode of magnifications (a) 50KX and (b) 100KX. (From Gujar, T.P. et al., *Int. J. Electrochem. Sci.*, 2, 666. Copyright 2007. Reproduced with permission.)

In a related study,[74] the visible light photocatalytic activity of α-Fe_2O_3 nanograin chains coated by anatase TiO_2 nanolayer, as a photocatalyst thin film for inactivation of *E. coli* bacteria, was investigated for the solutions containing 10^6 colony-forming units/mL of the bacteria, without and with H_2O_2 (60 μM). Thin films of the α-Fe_2O_3 nanograins with the grain size of 40–280 nm were grown on glass substrates by postannealing of the thermal evaporated Fe_3O_4 thin films at 400°C in air. In addition, the crystalline structure of ruthenium oxide (RuO_2) was electrochemically deposited onto tin-doped indium oxide (ITO) electrode and used as electrodes to form a supercapacitor in a 0.5 M H_2SO_4 electrolyte.[75] The structural and surface morphological observations revealed the formation of nanograins (Figure 5.16) of RuO_2 of tetragonal crystal structure. Barium titanate ($BaTiO_3$) films were prepared on TiN- and Ti-coated Si as well as bulk-Ti substrates by a hydrothermal method at <100°C.[76] Both titanium nitride and titanium films prepared by dc sputtering exhibited nanograins, while bulk titanium possessed micrograins.

An ultrasensitive nanostructured sensor that can detect 50 ppt of NH_3 gas in air consisted of nanograins of a *p*-type conductive polymer, polyaniline (PANI), enchased on an electrospun *n*-type semiconductive TiO_2 fiber surface (Figure 5.17).[77] The resistance of the *p-n* heterojunctions combining with the bulk resistance of PANI nanograins can function as electric current switches when NH_3 gas was absorbed by PANI nanoparticles. This sensor was 1000 times more sensitive than the best PANI sensor reported in the literature. Additionally, poly[aniline(AN)-co-5-sulfo-2-anisidine(SA)]

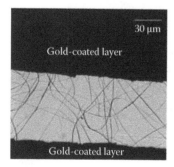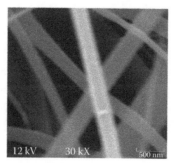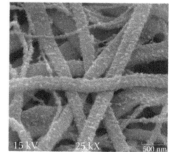

FIGURE 5.17 An optical microscope image of the sensor (left) and high-magnification scanning electron microscopy images of TiO_2 microfibers (middle) and TiO_2 microfibers enchased with PANI nanograins (right). (Reproduced with permission from Gong, J. et al., Ultrasensitive NH_3 gas sensor from polyaniline nanograin enchased TiO_2 fibers, *J. Phys. Chem. C*, 114(21), 9970–9974. Copyright 2010 American Chemical Society.)

nanograins with rough and porous structure demonstrate ultrastrong adsorption and highly efficient recovery of silver ions.[78] These nanograins have great application potential in the noble metals industry, resource reuse, wastewater treatment, and functional hybrid nanocomposites.

5.6 NANORICE

Nanorice[79] species have been reported for Ag,[80] Au,[81,82] superparamagnetic Fe_xO_y,[83] and polymers,[84] among other compounds. Thus, the formation process[85] of hematite-Au core-shell nanorice particles is shown in Figure 5.18. The observed electric current-assisted assembling

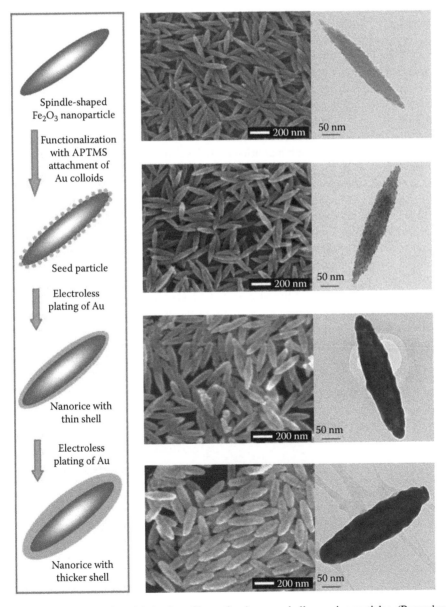

FIGURE 5.18 Schematics of the fabrication of hematite-Au core-shell nanorice particles. (Reproduced with permission from Wang, H. et al., Nanorice: A hybrid plasmonic nanostructure, *Nano Lett.*, 6(4), 827–832, http://pubs.acs.org/doi/pdf/10.1021/nl060209w. Copyright 2006 American Chemical Society.)

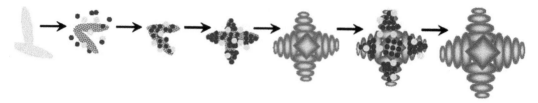

FIGURE 5.19 Schematic representation of mesostar formation mechanism. (Reproduced with permission from Bardhan, R. et al., Au nanorice assemble electrolytically into mesostars, *ACS Nano*, 3(2), 266–272. Copyright 2009 American Chemical Society.)

(Figure 5.19) of Au nanorice particles, consisting of prolate hematite cores and a thin Au shell, to star-shaped mesostructures was considered as the first reported observation of nanoparticle assembly into larger ordered structures under the influence of an electrochemical process (H_2O electrolysis).[86] Rapid and ultrasensitive immunoassays were developed by using biofunctional superparamagnetic Fe_3O_4/ZnO/Au nanorices as Raman probes.[87] The labeled proteins can be separated rapidly and purified with a common permanent magnet. Manipulation of the nanorice probes using an external magnetic field can enhance the assay sensitivity by several orders of magnitude and reduce the detection time from 1 h to 3 min, showing significant value for potential applications in biomedicine, food safety, and environmental defense. Among other applications of nanorices, magnetic immunonanorice particles can be used for the capture and detection of *E. coli* bacteria.[88]

Optical excitation of high-order surface plasmon resonance modes in individual Ag nanorice particles (Figure 5.20), obtained from $AgNO_3$ as a precursor, PVP as surfactant, and poly(ethylene glycol) 600 (PEG 600) as reducing agent, was studied using dark-field scattering spectroscopy.[89] High-order surface plasmon resonance modes in individual Ag nanorice particles were observed. The multipolar plasmon resonances in the visible to NIR range make nanorice particles suitable as substrates in surface-enhanced spectroscopy applications. In addition, various shapes and sizes of colloidal Co ferromagnetic nanoparticles, in particular nanorice species, were fabricated by the polyol process using Co(acac)₃, 1,2-hexadecanediol, oleylamine, and oleic acid within octylether. The rice-shaped Co nanoparticles with a size of 30 nm and aspect ratio of 1.8 can be fabricated into a self-assembled form without the assistance of surfactants.[90]

Among other compounds, 700 nm length YF_3:Yb^{3+}/Er^{3+} nanorice materials (Figure 5.21) were obtained.[91] These nanoparticles are good candidates for biolabeling applications, especially for deep *in vivo* imaging due to the deep penetration of NIR radiation into tissue. A periodic mesoporous organosilica with nanorice morphology (surface area of 753 m²/g, length of nanorice particles

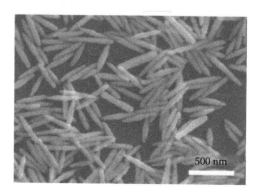

500 nm

FIGURE 5.20 SEM image of Ag nanorice. (Reproduced with permission from Wei, H. et al., Multipolar plasmon resonances in individual Ag nanorice, *ACS Nano*, 4(5), 2649–2654. Copyright 2010 American Chemical Society.)

FIGURE 5.21 Luminescent $YF_3:Yb^{3+}/Er^{3+}$ nanorice. (Reproduced with kind permission from Springer Science+Business Media: *Nano Res.*, One-pot synthesis and strong near-infrared upconversion luminescence of poly(acrylic acid)-functionalized YF3:Yb^{3+}/Er^{3+} nanocrystals, 3(5), 2010, 317–325, Wang, L. et al.)

ca. 600 nm and width of ca. 200 nm.), composed of D and T sites in the ratio 1:2, was synthesized by a template assisted sol-gel method using a chain-type precursor.[92] In addition, good crystalline nanorices composed of helical PANI molecules and their 2D shoulder to shoulder self-organized hexagonal microplates were obtained by a self-assembling process.[93]

REFERENCES

1. Bretonniere, Y.; Mazzanti, M.; Pecaut, J.; Olmstead, M. M. Cation-controlled self-assembly of a hexameric europium wheel. *Journal of the American Chemical Society*, 2002, *124* (31), 9012–9013.
2. Gorbachev, M. Yu.; Dobrova, B. N.; Dimoglo, A. S. The relative stability of nano-"chromic wheels" $[Cr_n(OCH_3)_{2n}(CH_3COO)_n]$ depending on the number of their coordination units. *Polyhedron*, 2009, *28* (6), 1169–1173.
3. Morin, J.-F.; Sasaki, T.; Shirai, Y.; Guerrero, J. M.; Tour, J. M. Synthetic routes toward carborane-wheeled nanocars. *Journal of Organic Chemistry*, 2007, *72* (25), 9481–9490.
4. Song, Y.; Garcia, R. M.; Dorin, R. M.; Wang, H.; Qiu, Y.; Medforth, C. J.; Shelnutt, J. A. Size- and shape-controlled platinum nanostructures. *Abstracts of Papers, 238th ACS National Meeting*, Washington, DC, August 16–20, 2009, INOR-377.
5. Shen, G.; Bando, Y.; Lee, C.-J. Growth of self-organized hierarchical ZnO nanoarchitectures by a simple In/In$_2$S$_3$ controlled thermal evaporation process. *Journal of Physical Chemistry B*, 2005, *109* (21), 10779–10785.
6. Miras, H. N.; Cooper, G. J. T.; Long, D.-L.; Boegge, H.; Mueller, A.; Streb, C.; Cronin, L. Unveiling the transient template in the self-assembly of a molecular oxide nanowheel. *Science*, 2010, *327* (5961), 72–74.
7. Jin, L.; Wang, J.-b.; Jia, S.-f.; Xu, Z.-l.; Yan, X.; Deng, L.-Z.; Cai, Y.; Lu, P.; Leung, Y. P.; Choy, W. C. H. Fabrication and characterization of ZnSe nanostructures and related interfaces. *Dianzi Xianwei Xuebao*, 2009, *28* (4), 325–334.
8. Jin, L.; Choy, W. C. H.; Leung, Y. P.; Yuk, T. I.; Ong, H. C.; Wang, J.-b. Synthesis and analysis of abnormal wurtzite ZnSe nanowheels. *Journal of Applied Physics*, 2007, *102* (4), 044302/1–044302/6.
9. Song, Y.; Dorin, R. M.; Garcia, R. M.; Jiang, Y. B.; Wang, H.; Li, P.; Qiu, Y.; Van Swol, F.; Miller, J. E.; Shelnutt, J. A. Synthesis of platinum nanowheels using a bicellar template. *Journal of American Chemical Society*, 2008, *130*, 12602–12603.
10. Lal, S.; Link, S.; Halas, N. J. Nano-optics from sensing to waveguiding. *Nature Photonics*, 2007, *1* (November), 641–648.
11. Lassiter, J. B.; Knight, M. W.; Mirin, N. A.; Halas, N. J. Reshaping the plasmonic properties of an individual nanoparticle. *Nano Letters*, 2009, *9* (12), 4326–4332.
12. Halas, N. J.; Wang, H.; Nordlander, P. J.; Wu, Y. Nanoeggs with silica core and asymmetric conductive nanoshell produced using electroless plating using a particle support with optical properties for sensors. U.S. Patent Application Publication US 2010028680, 2010, 13 pp.
13. Knight, M. W.; Halas, N. J. Nanoshells to nanoeggs to nanocups: Optical properties of reduced symmetry core-shell nanoparticles beyond the quasistatic limit. *New Journal of Physics*, 2008, *10* (October).

14. Wang, H.; Wu, Y.; Lassiter, B.; Nehl, C. L.; Hafner, J. H.; Nordlander, P.; Halas, N. J. Symmetry breaking in individual plasmonic nanoparticles. *Proceedings of the National Academy of Sciences of the United States of America*, 2006, *103* (29), 10856–10860.

15. Huang, W.-C.; Tsai, P.-J.; Chen, Y.-C. Multifunctional Fe_3O_4@Au nanoeggs as photothermal agents for selective killing of nosocomial and antibiotic-resistant bacteria. *Small*, 2009, *5* (1), 51–56.

16. Yamaguchi, Y.; Igarashi, R. Novel drug delivery system achieved by nanotechnology. Nanoegg capsule. *Rigaku Janaru*, 2006, *37* (1), 27–33.

17. Yamaguchi, Y.; Nakamura, N.; Nagasawa, T.; Kitagawa, A.; Matsumoto, K.; Soma, Y.; Matsuda, T.; Mizoguchi, M.; Igarashi, R. Enhanced skin regeneration by Nanoegg formulation of all-trans retinoic acid. *Pharmazie*, 2006, *61* (2), 117–121.

18. Tonigold, M.; Hitzbleck, J.; Bahnmueller, S.; Langstein, G.; Volkmer, D. Copper (II) nanoballs as monomers for polyurethane coatings: Synthesis, urethane derivatization and kinetic stability. *Dalton Transactions*, 2009, (8), 1363–1371.

19. Iyuke, S. E.; Mamvura, T. A.; Liu, K.; Sibanda, V.; Meyyappan, M.; Varadan, V. K. Process synthesis and optimization for the production of carbon nanostructures. *Nanotechnology*, 2009, *20* (37), 375602/1–375602/10.

20. Toriyabe, Y.; Watanabe, S.; Yatsu, S.; Shibayama, T.; Mizuno, T. Controlled formation of metallic nanoballs during plasma electrolysis. *Applied Physics Letters*, 2007, *91* (4), 041501/1–041501/3.

21. Kuo, C.-G.; Chen, C.-C.; Hsieh, S.-J.; Say, W. C. Fabrication of bismuth nanoballs using thermal oil reflow. *Journal of the Ceramic Society of Japan*, 2008, *116* (November), 1193–1198.

22. Surendran, G; Ramos, L.; Pansu, B.; Prouzet, E.; Beaunier, P.; Audonnet, F.; Remita, H. Synthesis of porous platinum nanoballs in soft templates. *Chemistry of Materials*, 2007, *19* (21), 5045–5048.

23. Zhou, W.; Zheng, K.; He, L.; Wang, R.; Guo, L.; Chen, C.; Han, X; Zhang, Z. Ni/Ni$_3$C core–shell nanochains and its magnetic properties: One-step synthesis at low temperature. *Nano Letters*, 2008, *8* (4), 1147–1152.

24. Surendran, G.; Ksar, F.; Ramos, L.; Keita, B.; Nadjo, L.; Prouzet, E.; Beaunier, P.; Dieudonne, P.; Audonnet, F.; Remita, H. Palladium nanoballs synthesized in hexagonal mesophases. *Journal of Physical Chemistry C*, 2008, *112* (29), 10740–10744.

25. Farrukh, M. A.; Heng, B.-T.; Adnan, R. Surfactant-controlled aqueous synthesis of SnO_2 nanoparticles via the hydrothermal and conventional heating methods. *Turkish Journal of Chemistry*, 2010, *34* (4), 537–550.

26. Fauteux, C.; El Khakani, M. A.; Pegna, J.; Therriault, D. Influence of solution parameters for the fast growth of ZnO nanostructures by laser-induced chemical liquid deposition. *Applied Physics A: Materials Science and Processing*, 2009, *94* (4), 819–829.

27. Li, C.-p.; Zhang, J.; Pu, C.-y.; Li, J.-h.; Wang, X.-h. Preparation and optical properties of Fe-doped ZnO nanoballs. *Faguang Xuebao*, 2010, *31* (2), 265–268.

28. Sharda, J. K.; Chawla, S. Synthesis of Mn doped ZnO nanoparticles with biocompatible capping. *Applied Surface Science*, 2010, *256* (8), 2630–2635.

29. Dabbagh, H. A.; Zamani, M.; Farrokhpour, H.; Namazian, M.; Habibabadi, H. E. Influence of B, Ga and In impurities in the structure and electronic properties of alumina nanoball. *Chemical Physics Letters*, 2010, *485* (1–3), 176–182.

30. Fang, J.; Saunders, M.; Guo, Y.; Lu, G.; Raston, C. L.; Iyer, K. S. Green light-emitting $LaPO_4$:Ce^{3+}:Tb^{3+} koosh nanoballs assembled by p-sulfonato-calix[6]arene coated superparamagnetic Fe_3O_4. *Chemical Communications*, 2010, *46* (18), 3074–3076.

31. Hu, K. H.; Wang, J.; Schraube, S.; Xu, Y. F.; Hu, X. G.; Stengler, R. Tribological properties of MoS_2 nano-balls as filler in polyoxymethylene-based composite layer of three-layer self-lubrication bearing materials. *Wear*, 2009, *266* (11–12), 1198–1207.

32. Hu, K. H.; Hu, X. G. Formation, exfoliation and restacking of MoS_2 nanostructures. *Materials Science and Technology*, 2009, *25* (3), 407–414.

33. Wang, Z.; Chen, J.; Xue, X.; Hu, Y. Synthesis of monodispersed CdS nanoballs through γ-irradiation route and building core-shell structure CdS@SiO$_2$. *Materials Research Bulletin*, 2007, *42* (12), 2211–2218.

34. Hu, Y.; Chen, J.-F. Synthesis and characterization of semiconductor nanomaterials and micromaterials via gamma-irradiation route. *Journal of Cluster Science*, 2007, *18* (2), 371–387.

35. Hsieh, S.-M. Hydrogen storage apparatus containing several hydrogen storage materials. Patent CN 201140984, 2008, 7 pp.

36. Baek, K.; Hota, R.; Yun, G.; Kim, Y.; Jung, H.; Park, K. M.; Yoon, E. et al. Phthalocyanine nanoballs. *Abstracts of Papers*, 240th ACS National Meeting, Boston, MA, August 22–26, 2010, COLL-467.

37. Larsen, R.; Vetromile, C.; Lazano, A. Feola, S. Solution stability of Cu(II) hydroxy nanoballs. *Abstracts, 61st Southeast Regional Meeting of the American Chemical Society*, San Juan, Puerto Rico, October 21–24, 2009, SRM-684.

38. Larsen, R. W.; McManus, G. J.; Perry, J. J., IV; Rivera-Otero, E.; Zaworotko, M. J. Spectroscopic characterization of hydroxylated nanoballs in methanol. *Inorganic Chemistry*, 2007, *46* (15), 5904–5910.

39. Duriska, M. B.; Neville, S. M.; Lu, J.; Iremonger, S. S.; Boas, J. F.; Kepert, C. J.; Batten, S. R. Systematic metal variation and solvent and hydrogen-gas storage in supramolecular nanoballs. *Angewandte Chemie, International Edition*, 2009, *48* (47), 8919–8922, S8919/1–S8919/33.

40. Anderson, J. P.; Reynolds, B. L.; Baum, K.; Williams, J. G. Fluorescent structural DNA nanoballs functionalized with phosphate-linked nucleotide triphosphates. *Nano Letters*, 2010, *10* (3), 788–792.

41. Xie, Y.; Huang, Q.; Huang, B.; Xie, X. Low temperature synthesis of high quality carbon nanospheres through the chemical reactions between calcium carbide and oxalic acid. *Materials Chemistry and Physics*, 2010, *124* (1), 482–487.

42. Li, J.; Yuan, R.; Chai, Y.; Zhang, T.; Che, X. Direct electrocatalytic reduction of hydrogen peroxide at a glassy carbon electrode modified with polypyrrole nanowires and platinum hollow nanospheres. *Microchimica Acta*, 2010, *171* (1–2), 125–131.

43. Song, Y.; Steen, W. A.; Pena, D.; Jiang, Y.-B.; Medforth, C. J.; Huo, Q.; Pincus, J.-L. et al. Foamlike nanostructures created from dendritic platinum sheets on liposomes. *Chemistry of Materials*, 2006, *18*, 2335–2346.

44. Liu, Z.; Zhao, B.; Guo, C.; Sun, Y.; Shi, Y.; Yang, H.; Li, Z. Carbon nanotube/raspberry hollow Pd nanosphere hybrids for methanol, ethanol, and formic acid electro-oxidation in alkaline media. *Journal of Colloid and Interface Science*, 2010, *351* (1), 233–238.

45. Meng, Q.; Li, H.; Li, H. Method for preparation of metallic palladium nanosphere catalyst with multi-shell-layer structure. CN 101829789, 2010, 7 pp.

46. Feng, X. Synthesis of Ag/polypyrrole core-shell nanospheres by a seeding method. *Chinese Journal of Chemistry*, 2010, *28* (8), 1359–1362.

47. Wang, Y.; Lee, K.; Irudayaraj, J. Silver nanosphere SERS probes for sensitive identification of pathogens. *Journal of Physical Chemistry C*, 2010, *114* (39), 16122–16128.

48. Ma, D.; Zhao, J.; Li, Y.; Su, X.; Hou, S.; Zhao, Y.; Hao, X. L.; Li, L. Organic molecule directed synthesis of bismuth nanostructures with varied shapes in aqueous solution and their optical characterization. *Colloids and Surfaces, A: Physicochemical and Engineering Aspects*, 2010, *368* (1–3), 105–111.

49. Wheeler, D. A.; Newhouse, R. J.; Wang, H.; Zou, S.; Zhang, J. Z. Optical properties and persistent spectral hole burning of near infrared-absorbing hollow gold nanospheres. *Journal of Physical Chemistry C*, 2010, *114* (42), 18126–18133.

50. Morant, C.; Marquez, F.; Campo, T.; Sanz, J. M.; Elizalde, E. Niobium and hafnium grown on porous membranes. *Thin Solid Films*, 2010, *518* (23), 6799–6803.

51. Sun, J.; Li, X.; Wang, Y. Study on hydrothermal synthesis of nanometer ZnO powders with PEG as a soft-template. *Taoci Xuebao*, 2010, *31* (2), 331–335.

52. Qian, H.-S.; Hu, Y.; Li, Z.-Q.; Yang, X.-Y.; Li, L.-C.; Zhang, X.-T.; Xu, R. ZnO/ZnFe$_2$O$_4$ magnetic fluorescent bifunctional hollow nanospheres: Synthesis, characterization, and their optical/magnetic properties. *Journal of Physical Chemistry C*, 2010, *114*, 17455–17459.

53. Park, J. T.; Roh, D. K.; Patel, R.; Kim, E.; Ryu, D. Y.; Kim, J. H. Preparation of TiO$_2$ spheres with hierarchical pores via grafting polymerization and sol-gel process for dye-sensitized solar cells. *Journal of Materials Chemistry*, 2010, *20* (39), 8521–8530.

54. Wang, Z.; Chen, J. S.; Zhu, T.; Madhavi, S.; Lou, X. W. One-pot synthesis of uniform carbon-coated MoO$_2$ nanospheres for high-rate reversible lithium storage. *Chemical Communications*, 2010, *46* (37), 6906–6908.

55. Ahmed, A.; Clowes, R.; Willneff, E.; Myers, P.; Zhang, H. Porous silica spheres in macroporous structures and on nanofibres. *Philosophical Transactions of the Royal Society, A: Mathematical, Physical and Engineering Sciences*, 2010, *368* (1927), 4351–4370.

56. Umegaki, T.; Yan, J.-M.; Zhang, X.-B.; Shioyama, H.; Kuriyama, N.; Xu, Q. Co-SiO$_2$ nanosphere-catalyzed hydrolytic dehydrogenation of ammonia borane for chemical hydrogen storage. *Journal of Power Sources*, 2010, *195* (24), 8209–8214.

57. He, Q.; Zeng, L.; Wu, W.; Hu, R.; Huang, J. Preparation and magnetic comparison of silane-functionalized magnetite nanoparticles. *Sensors and Materials*, 2010, *22* (6), 285–295.

58. Wang, W.; Howe, J. Y.; Li, Y.; Qiu, X.; Joy, D. C.; Paranthaman, M. P.; Doktycz, M. J.; Gu, B. A surfactant and template-free route for synthesizing ceria nanocrystals with tunable morphologies. *Journal of Materials Chemistry*, 2010, *20* (36), 7776–7781.

59. Aghazadeh, M.; Nozad, A.; Adelkhani, H.; Ghaemi, M. Synthesis of Y_2O_3 nanospheres via heat-treatment of cathodically grown $Y(OH)_3$ in chloride medium. *Journal of the Electrochemical Society*, 2010, *157* (10), D519–D522.
60. Wu, D.; Zhu, H.; Zhang, C.; Chen, L. Novel synthesis of bismuth tungstate hollow nanospheres in water-ethanol mixed solvent. *Chemical Communications*, 2010, *46* (38), 7250–7252.
61. Wang, K.-W.; Zhu, Y.-J.; Chen, F.; Cao, S.-W. Calcium phosphate/block copolymer hybrid porous nanospheres: Preparation and application in drug delivery. *Materials Letters*, 2010, *64* (21), 2299–2301.
62. Huang, Y.; You, H.; Song, Y.; Jia, G.; Yang, M.; Zheng, Y.; Zhang, L.; Liu, K. Half opened microtubes of $NaYF_4$:Yb,Er synthesized in reverse microemulsion under solvothermal condition. *Journal of Crystal Growth*, 2010, *312* (21), 3214–3218.
63. Vlasnikov, A. K.; Kuz'menko, N. K.; Mikhailov, V. M.; Smirnov, A. S. Investigation of weak pair correlations in aluminum nanograins. *Izvestiya Rossiiskoi Akademii Nauk, Seriya Fizicheskaya*, 2010, *74* (4), 611–616.
64. Zhang, B.; Shim, V. P. W. On the generation of nanograins in pure copper through uniaxial single compression. *Philosophical Magazine*, 2010, *90* (23–24), 3293–3311.
65. Kowalska, E.; Czerwosz, E.; Kozlowski, M.; Surga, W.; Radomska, J.; Wronka, H. Structural, thermal, and electrical properties of carbonaceous films containing palladium nanocrystals. *Journal of Thermal Analysis and Calorimetry*, 2010, *101* (2), 737–742.
66. Younes, A.; Bensalem, R.; Alleg, S.; Hamouda, A.; Azzaza, S.; Sunol, J. J.; Greneche, J. M. Solid state amorphisation of a Fe-Co-Nb-B powder mixture by mechanical alloying. *Annales de Chimie*, 2010, *35* (3), 169–176.
67. An, C.; Peng, S.; Sun, Y. Facile synthesis of sunlight-driven AgCl:Ag plasmonic nanophotocatalyst. *Advanced Materials*, 2010, *22* (23), 2570–2574.
68. Yamamoto, T. A.; Nakagawa, T.; Seino, S.; Nitani, H. Bimetallic nanoparticles of PtM (M = Au, Cu, Ni) supported on iron oxide: Radiolytic synthesis and CO oxidation catalysis. *Applied Catalysis, A: General*, 2010, *387* (1–2), 195–202.
69. Suchanska, M.; Czerwosz, E.; Diduszko, R.; Dluzewski, P.; Keczkowska, J.; Kowalska, E.; Rymarczyk, J. Study of molecular and nanocrystalline structure of Pd-C films for applications as hydrogen sensors. *Elektronika*, 2010, *51* (6), 54–57.
70. Miller, J. B.; Hsieh, H.-J.; Howard, B. H.; Broitman, E. Microstructural evolution of sol-gel derived ZnO thin films. *Thin Solid Films*, 2010, *518* (23), 6792–6798.
71. Krithiga, R.; Chandrasekaran, G. Synthesis, structural and optical properties of vanadium doped zinc oxide nanograins. *Journal of Crystal Growth*, 2009, *311* (21), 4610–4614.
72. Karami, H. Investigation of sol-gel synthesized CdO-ZnO nanocomposite for CO gas sensing. *International Journal of Electrochemical Science*, 2010, *5* (5), 720–730.
73. Chen, L.; Yang, X.; Chen, J.; Liu, J.; Wu, H.; Zhan, H.; Liang, C.; Wu, Mi. Continuous shape- and spectroscopy-tuning of hematite nanocrystals. *Inorganic Chemistry*, 2010, *49* (18), 8411–8420.
74. Akhavan, O.; Azimirad, R. Photocatalytic property of Fe_2O_3 nanograin chains coated by TiO_2 nanolayer in visible light irradiation. *Applied Catalysis, A: General*, 2009, *369* (1–2), 77–82.
75. Gujar, T. P.; Kim, W.-Y.; Puspitasari, I.; Jung, K.-D.; Joo, O.-S. Electrochemically deposited nanograin ruthenium oxide as a pseudocapacitive electrode. *International Journal of Electrochemical Science*, 2007, *2*, 666–673.
76. Chan, P.-H.; Lu, F.-Hg. Low temperature hydrothermal synthesis and the growth kinetics of $BaTiO_3$ films on TiN/Si, Ti/Si, and bulk-Ti substrates. *Journal of the Electrochemical Society*, 2010, *157* (6), G130–G135.
77. Gong, J.; Li, Y.; Hu, Z.; Zhou, Z.; Deng, Y. Ultrasensitive NH_3 gas sensor from polyaniline nanograin enchased TiO_2 fibers. *Journal of Physical Chemistry C*, 2010, *114* (21), 9970–9974.
78. Li, X.-G.; Feng, H.; Huang, M.-R. Redox sorption and recovery of silver ions as silver nanocrystals on poly(aniline-co-5-sulfo-2-anisidine) nanosorbents. *Chemistry—A European Journal*, 2010, *16* (33), 10113–10123.
79. Wang, H.; Brandl, D.; Nordlander, P.; Halas, N. J. Nanorice particles: Hybrid plasmonic nanostructures which combines local fields of nanorods with plasmon resonances of nanoshells. WO 2007103802, 2007, 24 pp.
80. Wiley, B. J.; Chen, Y.; McLellan, J.; Xiong, Y.; Li, Z.-Y.; Ginger, D.; Xia, Y. Synthesis and optical properties of silver nanobars and nanorice. *Nano Letters*, 2007, *7* (4), 1032–1036.
81. Brinson, B. E.; Lassiter, J. B.; Levin, C. S.; Bardhan, R.; Mirin, N.; Halas, N. J. Nanoshells made easy: Improving Au layer growth on nanoparticle surfaces. *Langmuir*, 2008, *24* (24), 14166–14171.

82. Kealley, C. S.; Cortie, M. B. A computational exploration of the color gamut of nanoscale hollow scalene ellipsoids of Ag and Au. *Plasmonics*, 2010, *5* (1), 37–43.

83. Rebolledo, A. F.; Bomati-Miguel, O.; Marco, J. F.; Tartaj, P. A facile synthetic route for the preparation of superparamagnetic iron oxide nanorods and nanorices with tunable surface functionality. *Advanced Materials*, 2008, *20* (9), 1760–1765.

84. Srivastava, D.; Lee, I. Novel fabrication of anisotropic polymer nanoparticles using solvent-aided nanoinjection molding process. *PMSE Preprints*, 2007, *96*, 869–870.

85. Wang, H.; Brandl, D. W.; Le, F.; Nordlander, F.; Halas, N. J. Nanorice: A hybrid plasmonic nanostructure. *Nano Letters*, 2006, *6* (4), 827–832. http://pubs.acs.org/doi/pdf/10.1021/nl060209w

86. Bardhan, R.; Neumann, O.; Mirin, N.; Wang, H.; Halas, N. J. Au nanorice assemble electrolytically into mesostars. *ACS Nano*, 2009, *3* (2), 266–272.

87. Hong, X.; Chu, X.; Zou, P.; Liu, Y.; Yang, G. Magnetic-field-assisted rapid ultrasensitive immunoassays using Fe_3O_4/ZnO/Au nanorices as Raman probes. *Biosensors and Bioelectronics*, 2010, *26* (2), 918–922.

88. Naja, G.; Bouvrette, P.; Hrapovich, S.; Liu, Y.; Luong, J. H. T. Detection of bacteria aided by immuno-nanoparticles. *Journal of Raman Spectroscopy*, 2007, *38* (11), 1383–1389.

89. Wei, H.; Reyes-Coronado, A.; Nordlander, P.; Aizpurua, J.; Xu, H. Multipolar plasmon resonances in individual Ag nanorice. *ACS Nano*, 2010, *4* (5), 2649–2654.

90. Cha, S. I.; Mo, C. B.; Kim, K. T.; Hong, S. H. Ferromagnetic cobalt nanodots, nanorices, nanowires and nanoflowers by polyol process. *Journal of Materials Research*, 2005, *20* (8), 2148–2153.

91. Wang, L.; Zhang, Y.; Zhu, Y. One-pot synthesis and strong near-infrared upconversion luminescence of poly(acrylic acid)-functionalized YF_3:Yb^{3+}/Er^{3+} nanocrystals. *Nano Research*, 2010, *3* (5), 317–325.

92. Mohanty, P.; Landskron, K. Periodic mesoporous organosilica nanorice. *Nanoscale Research Letters*, 2009, *4* (2), 169–172.

93. Yan, Y.; Wang, R.; Qiu, X.; Wei, Z. Hexagonal superlattice of chiral conducting polymers self-assembled by mimicking β-sheet proteins with anisotropic electrical transport. *Journal of the American Chemical Society*, 2010, *132* (34), 12006–12012.

6 Nanocage-Type Structures

6.1 NANOCAGES, NANOBOXES, AND NANOCUBES

Nanostructures on the basis of cages, boxes, and cubes are closely related without a definite and clear distinction between them, formed by the same elements or compounds, and sometimes confused with one another in different articles. However, the number of reports on nanocubes is currently considerably higher (more than 1200) than those with the two first structures. The cages and boxes are generally hollow, and nanocubes can be hollow or filled. Serious attention is paid to elemental metal (predominantly noble metals such as Au, Pt, and Pd) nanostructures including bimetallic species; in particular, metal hollow nanostructures were extensively reviewed.[1] The main metals, obtained in nanocube form, are Au,[2] Au/Cu,[3] Pt,[4] Pt_3Co,[5] Pd,[6] and Rh (Figure 6.1)[7] and can be formed competitively with other nanostructures depending on reaction conditions. Nanocube metallic structures can be in equilibrium with boxes and cages (as well as with other nanostructures); thus, Pd nanocubes were converted into nanoboxes and nanocages (NCs) in a one-pot synthesis without the involvement of exotic templates in a solution containing ethylene glycol, water, and poly(vinylpyrrolidone) with the addition of Na_3PdCl_4 as a precursor to Pd.[8] Another example corresponds to silver nanocubes, which, being dispersed in water, were transformed into Pd-Ag or Pt-Ag nanoboxes (Figure 6.2) by adding either Na_2PdCl_4 or Na_2PtCl_4.[9] Replacement of Ag with Pd resulted in the formation of a nanobox composed of a Pd-Ag alloy single crystal, but the nanobox formed after the replacement of Ag with Pt was instead composed of distinct Pt nanoparticles.

Five different hollow cubic nanoparticles-nanocatalysts with a wall length of 75 nm were synthesized from platinum and/or palladium elements[10]: pure platinum nanocages (PtNCs), pure palladium nanocages (PdNCs), Pt/Pd hollow shell-shell NCs (where Pd was defined as the inner shell around the cavity), Pd/Pt shell-shell NCs, and Pt-Pd alloy NCs. These nanoforms were used to catalyze the reduction of 4-nitrophenol with sodium borohydride; it was suggested that the catalytic reaction took place inside the cavity of the hollow nanoparticles. Electrochemical synthesis of ultrafine Pt cubic nanoboxes (Figure 6.3, with edge length of ~6 nm and a wall thickness of 1.5 nm) was carried out from Pt-on-Ag heteronanostructures.[11] A strong shape-dependent catalytic property was observed for these Pt cubic nanoboxes, which was 1.5 times more active than hollow spheres in terms of turnover frequency for catalytic oxidation of MeOH. It was intriguing that the mode of formation of cubic nanoboxes was the result of the most stable structures, which were controlled by the reaction conditions rather than the shapes of original templates. These Pt hollow nanostructures should be of interest in the development of facet-specific catalysts, which are often required for achieving the maximum selectivity and high activity. Additionally, the synthesis and size control of a cubic PAA-Pt (PAA, polyacrylic acid sodium salt) nanoparticles (Figure 6.4) were carried out, utilizing the additive effect of NaI during the particle growth process.[12] It was found that the cubic size of PAA-Pt could be changed by adjusting the reaction temperature between 20°C and 60°C, while retaining high shape selectivity. Also, the carbon nanotube (CNT)-supported Pt nanocubes with high active and selective {100} surfaces, prepared through an electroless deposition route, displayed excellent electrocatalytic activities toward the oxygen reduction reaction, which can replace the current state-of-the art cathode catalysts in fuel cells and thereby improve the catalytic performance and utilization efficiency.[13] Mn-Pt nanocubes were synthesized[14] from Pt acetylacetonate and Mn carbonyl in the presence of oleic acid and oleylamine. These Mn-Pt nanocubes were converted into an ordered $MnPt_3$ intermetallic phase upon annealing. The Mn-Pt nanocubes, which were found

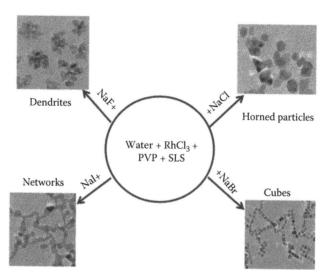

FIGURE 6.1 Summary of Rh nanostructures controlled by halogen anions (F⁻, Cl⁻, Br⁻, I⁻ in the form of sodium salts, RhCl₃ was used as precursor). (Reproduced with permission from Yuan, Q. et al., Tunable aqueous phase synthesis and shape-dependent electrochemical properties of rhodium nanostructures. *Inorg. Chem.*, 49(12), 5515–5521. Copyright 2010 American Chemical Society.)

to be enclosed by (100) surfaces, showed better activities than their spherical counterparts. These materials are promising new candidates as cathode and anode catalysts in fuel cells.

Au, Ag, and Au/Ag[15,16] square- and triangular-form nanoparticles are widely described. Thus, enclosed triangular AuAg nanostructures (triangular nanoboxes) were synthesized *via* galvanic replacement reactions from Ag nanoprisms.[17] The nanostructures were found to be hollow and did not consist of an Ag core surrounded by a Au shell. These triangular nanoboxes are attractive candidates for the encapsulation and transport of materials of interest such as drugs, radioisotopes, or magnetic materials. Ag nanocubes of 30–70 nm in edge length were synthesized with the use of CF₃COOAg as a precursor to elemental silver.[18] By adding a trace amount of NaSH and HCl to the polyol synthesis, Ag nanocubes were obtained with good quality, high reproducibility, and on a scale up to 0.19 g per batch for the 70 nm Ag nanocubes. The Ag nanocubes were found to grow in size at a controllable pace over the course of the synthesis. In addition, Ag nanocubes (edge length of 55 ± 5 nm) were prepared in water according to Reactions 6.1 and 6.2.[19] Their high stability in water would be favorable for further chemical modification and systematic investigations. Uniform Au NCs (54.6 ± 13.3 nm in outer-edge length and about 12 nm in wall thickness) were prepared by the reduction of HAuCl₄ in an aqueous mixture of hexamethylenetetramine (HMT), poly(N-vinyl-2-pyrrolidone) (PVP), and AgNO₃.[20] The electrocatalytic activity of the AuNCs toward the oxidation of Glc was explored, resulting in the fabrication of a nonenzymic Glc sensor with high sensitivity and good stability. In a related work,[21] Au-based NC (Figure 6.5) with controlled wall thickness, porosity, and optical properties were generated by dissolving Ag from Au-Ag alloy nanoboxes with H₂O₂. The process involved two steps: (a) formation of Au-Ag alloy nanoboxes with some pure Ag left behind by titrating Ag nanocubes with aqueous HAuCl₄ and (b) removal of Ag atoms from both the pure Ag remaining in the nanoboxes and the alloy walls *via* H₂O₂ etching. Due to the changes to the optical spectra, the Au-Ag alloy nanoboxes can be employed to detect H₂O₂. Gold NC, reviewed in Ref. [22] can be applied as a contrast agent for optical imaging in early stage cancer detection and as a therapeutic agent for photothermal cancer treatment.[23]

$$[Ag(NH_3)_2]^+{}_{(aq.)} + Br^- \rightleftharpoons AgBr_{(s)} + 2NH_3{}_{(aq.)} \qquad (6.1)$$

$$[Ag(NH_3)_2]^+{}_{(aq.)} + RCHO^- \ (glucose)_{(aq.)} \longrightarrow Ag \ (NPs) + RCOO^-{}_{(aq.)} + NH_4^+{}_{(aq.)} \qquad (6.2)$$

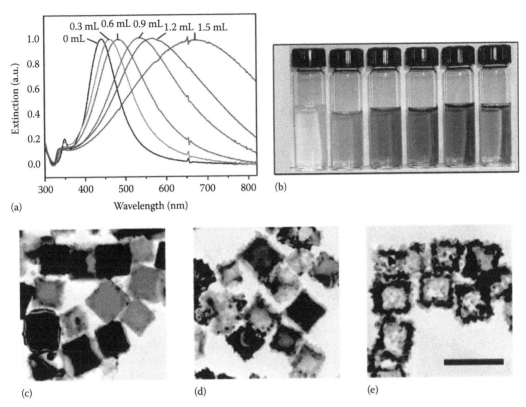

FIGURE 6.2 (a) UV–visible extinction spectra of products from galvanic replacement of Ag nanocubes with Na$_2$PtCl$_4$. The amount of 0.5 mM Na$_2$PtCl$_4$ aqueous solution added to the initial silver nanocubes is indicated on each curve. (b) Corresponding photographs of the aqueous Ag-Pd nanostructures analyzed in (a). (c through e) Transmission electron microscopy (TEM) images of the Ag-Pt nanostructures formed after the addition of (c) 0.3, (d) 0.9, and (e) 1.5 mL of a 0.5 mM Na$_2$PtCl$_4$ aqueous solution. The scale bar is 100 nm and applicable to all three images. (Reproduced with permission from Chen, J. et al., Optical properties of Pd-Ag and Pt-Ag nanoboxes synthesized *via* galvanic replacement reactions, *Nano Lett.*, 5(10), 2058–2062. Copyright 2005 American Chemical Society.)

Catalytic properties of Au-based nanostructures (including NCs, nanoboxes, and solid nanoparticles) were studied using a model reaction based on the reduction of *p*-nitrophenol by NaBH$_4$.[24] It was revealed that Au-based NCs were catalytically more active than both the nanoboxes and nanoparticles probably due to their extremely thin but continuous walls, the high content of Au, and the accessibility of both inner and outer surfaces through the pores in the walls. In addition, an interesting transformation was described: the CNT-supported Pt nanoparticles could be converted into hollow Au nanoboxes by galvanic displacement of Pt with Au.[25] These CNT-supported metal nanoparticles were shown to possess interesting optical and electrocatalytic properties. Also, Au nanocubes with controllable size (a side length of 83 ± 3 nm), synthesized by using a wet chemical method, were then coated with controllable Pd shell thickness by controlling the amount of H$_2$PdCl$_4$ salt in the Au nanocube solution.[26] These Au@Pd nanocubes (Figure 6.6) were found to be of core-shell structure with single crystal facets {100}. The surface-enhanced Raman spectroscopy (SERS) activity of carbon monoxide adsorbed at the core-shell Au@Pd nanocubes was much higher than that of the massive roughened Pd electrode. Selected useful examples of non-noble metal nanocube correspond to copper and germanium. Thus, a nonenzymic glucose (Glc) sensor was developed by potentiostatically electrodepositing metallic Cu nanocubes from a precursor solution onto vertically well-aligned multi-walled carbon nanotube (MWCNT) arrays.[27] The sensor showed significantly higher electrocatalytic activity to the oxidation of Glc in 0.1 M NaOH alkaline solution after

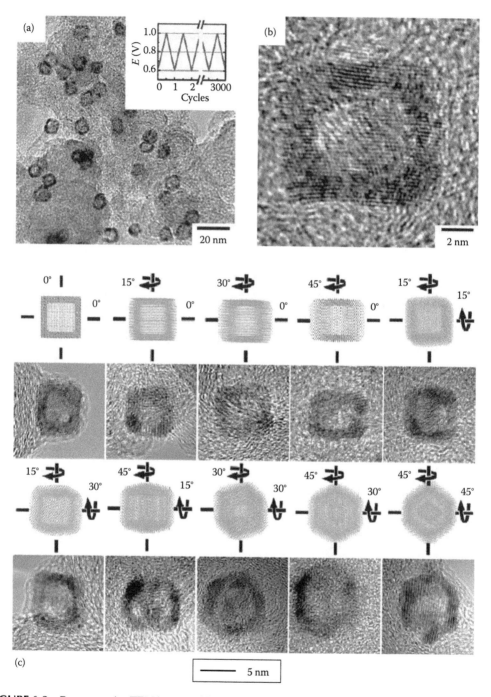

FIGURE 6.3 Representative TEM images of Pt hollow nanocubes at (a) low and (b) high magnifications and (c) individual cubes imaged under various tilting angles with respect to the direction of imaging beam. Inset in (a) shows the corresponding potential cycling profile. (Reproduced with permission from Peng, Z. et al., Electrochemical synthesis and catalytic property of Sub-10 nm platinum cubic nanoboxes. *Nano Lett.*, 10(4), 1492–1496. Copyright 2010 American Chemical Society.)

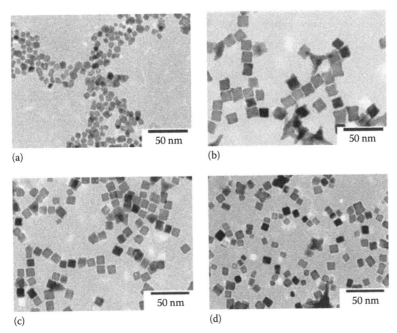

FIGURE 6.4 TEM images of PAA-Pt prepared with reaction temperatures: (a) 5°C, (b) 20°C, (c) 40°C, and (d) 60°C. (Reproduced from Yamada, M. et al., *Chem. Lett.*, 34(7), 1050. Copyright 2005 Chemistry Society of Japan.)

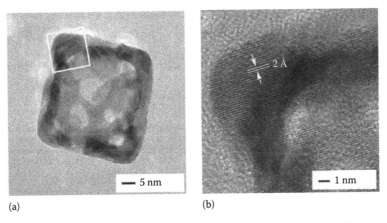

FIGURE 6.5 (a) TEM image of a representative nanocage synthesized by partially dissolving Ag from the Au-Ag nanobox with excess H_2O_2. (b) High-resolution TEM image of the area boxed in panel (a). (Reproduced with permission from Zhang, Q. et al., Dissolving Ag from Au-Ag alloy nanoboxes with H_2O_2: A method for both tailoring the optical properties and measuring the H_2O_2 concentration, *J. Phys. Chem. C*, 114(14), 6396–6400. Copyright 2010 American Chemical Society.)

modification of Cu nanocubes than before. The sensor response was found to be rapid (<1 s) and highly sensitive (1096 µA/mM/cm^2) with a wide linear range (up to 7.5 mM) and a low detection limit (1.0 µM at signal/noise ratio [S/N] = 3). In *ab initio* electronic-structure calculations of hydrogenated germanium cages Ge_nH_nM (M = Cu and Zn, n = 12–24) using density functional theory with polarized basis set (SDD) nanoclusters,[28] it was found that though the doping with Cu can be taken favorably in the cages, Zn was not.

Carbon and silicon cages or boxes are common. Thus, a simple approach for spontaneous, catalyst-free formation of highly graphitic nitrogen-containing carbon NCs (well-ordered graphitic shells with more compact graphite layer structure than that of conventional bulk graphite) was

(a)   (b)   (c)

(d)   (e)

FIGURE 6.6 Scanning electron microscope (SEM) images of 83 nm Au nanocubes with different Pd shell thickness. (a) 0.28 nm, (b) 0.56 nm, (c) 1.4 nm, (d) 2.8 nm, and (e) 4.2 nm. (From Sheng, J.-J. et al., *Can. J. Anal. Sci. Spectrosc.*, 52(3), 178–185. Copyright 2007. Reproduced with permission.)

demonstrated by using commercially available graphite rods as the initial materials.[29] The incorporation of nitrogen into the graphitic backbone of carbon NCs opens the potential for metal-free catalysis of oxygen reduction reaction in fuel cells. A thin-walled graphitic NC material with well-developed graphitic structure, a large specific surface area, and pronounced mesoporosity was synthesized and used to construct a sensing interface for an amperometric Glc biosensor, showing a high and reproducible sensitivity.[30] In addition, luminescence spectra of crystalline silicon nanoboxes etched in a silicon-on-insulator substrate showed a 125 times emission enhancement compared with the unpatterned silicon-on-insulator emission.[31] It was demonstrated that the emission enhancement partially resulted from the coupling between the electron-hole recombination inside the silicon boxes and the low group velocity optical modes of the array.

Various metal oxides form box-, cage-, or cube-like nanostructures. Among them, copper oxides are popular research objects, fabricated by distinct methods. Thus, box-shaped copper(II) oxide particles with an average diameter of ~100 nm were synthesized by warming $Cu(OH)_2$ in unmodified cyclodextrin as a capping agent at ~80°C to −90°C.[32] The synthesized CuO nanoparticles were exploited as a catalyst for the reduction of 4-nitrophenol. Nanoboxes, nanocubes, and nanospheres of cuprous oxides were readily synthesized by reducing $Cu(CH_3COO)_2 \cdot H_2O$ with ethylene glycol at different concentrations of poly(vinylpyrrolidone).[33] A highly selective adsorption characteristic of as-synthesized Cu_2O nanostructures to anionic dyes was observed due to the suggested mechanism of electrostatic adsorption. Also, copper oxide hollow nanoboxes (edge size of 195 nm and an average wall thickness of 26 nm, $(CuO)_{0.75}(Cu_2O)_{0.25}$) containing a platinum nanocluster (Figure 6.7) were fabricated in mild conditions at room temperature (r.t.) (system: chloroplatinic acid hydrate, sodium borohydride, hydrazine monohydrate, L-ascorbic acid, and sodium citrate tribasic dihydrate) by dropping a laser-irradiated aqueous platinum colloidal solution onto a copper substrate.[34] It was noted that the ionic compound of copper oxide often forms nanocubes or nanoboxes. Hollow nanoboxes instead of filled nanocubes were suggested to form as outside copper and oxygen ions migrate to photooxidized platinum nanoclusters. Because reduced platinum nanoclusters remained inside, they were trapped inside copper oxide hollow nanoboxes. Additionally, a local ordered structure constructed from solid Cu_2O nanocubes was obtained by the radiolytic reduction (γ-radiation, ^{60}Co source) of $Cu(NO_3)_2$ in a water-in-oil microemulsion composed of Triton X-100 (**1**), *n*-hexanol,

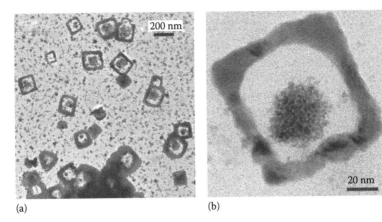

(a) (b)

FIGURE 6.7 (a) TEM and (b) HRTEM images of copper oxide nanoboxes containing a platinum nanoclu-ster. Nanoboxes were prepared after irradiation for 30 min with laser pulses of 1064 nm having a temporal width of 25 ps and a pulse energy of 0.5 mJ. (Reproduced with permission from Kim, M.R. et al., Fabrication of copper oxide nanoboxes containing a platinum nanocluster *via* an optical and galvanic route, *Cryst. Growth Des.*, 10(1), 257–261. Copyright 2010 American Chemical Society.)

cyclohexane, and water in the presence of ethylene glycol.[35] Changing Triton X-100 with Brij 56 (**2**) in the microemulsion, hollow Cu_2O nanocubes were synthesized. It was suggested that the bal-ance between the reduction rate of Cu^{2+} depending on the yield of hydrated electrons (e_{aq}^-) and the escape rate of the mixed solvent determined their final morphologies. The reaction mechanism and competitive reactions are shown by reactions 6.3 through 6.7. Finally, for composite nanocubes on copper(I) oxide basis (Cu_2O-C_{60} core-shell nanostructures), the Cu_2O cores in them could be removed, yielding monodispersed C_{60} nanoboxes.[36]

1 Triton X-100

$$CH_3(CH_2)_{14}CH_2O(CH_2CH_2O)_{10}H$$

2 Brij 56

$$\textit{Reduction: } Cu^{2+} \text{ to } Cu^+: Cu^{2+} + e_{aq}^- \rightarrow Cu^+ \tag{6.3}$$

$$\textit{Generation} \text{ of } Cu_2O \text{ through the hydrolysis of } Cu^+: Cu^+ + H_2O \rightarrow CuOH + H^+ \tag{6.4}$$

$$2CuOH \rightarrow Cu_2O + H_2O \tag{6.5}$$

$$\textit{Competing reactions: } \text{disproportionation } 2Cu^+ \rightarrow Cu + Cu^{2+} \tag{6.6}$$

$$\textit{Further reduction: } Cu^+ + e_{aq}^- \rightarrow Cu \tag{6.7}$$

Other oxides are widespread, possessing a series of useful applications. Thus, Co_3O_4 nanoboxes, prepared by a template-synthesis process and composed of compactly assembled cuboid Co_3O_4 nanocrystals, were characterized by their unexpected magnetic behavior and may lead to the possibility of novel metal oxide hierarchical structures.[37] MgO polyhedral NCs and nanocrystals, synthesized by a noncatalytic simple thermal evaporation process, were used to fabricate a highly sensitive amperometric Glc biosensor, which showed a high and reproducible sensitivity.[38] Alternatively, the nanocube of MgO was prepared in a short time of 30 s using Mg chip and steel-wool as starting materials, by means of a domestic microwave oven operated at 2.45 GHz and 1000 W, without any postannealing treatment.[39] ZnO nanoplates and nanoboxes were produced via a catalyst-free combust-oxidized (CFCO) process[40,41] (vaporization of molten Zn and instantaneous oxidation in normal atmosphere to produce Zn oxide crystals that were transported into a 100–180 m cooling duct). It was proposed that growth started from two kinds of nucleation planes parallel to the six-sided facets of nanorods that provided the nucleation sites for the nucleation planes. Alternatively, large-scale uniform ZnO dumbbells and ZnO/ZnS hollow NCs were synthesized via a facile hydrothermal route combined with subsequent etching treatment.[42] The NCs were found to be formed through preferential dissolution of the twinned (0001) plane of ZnO dumbbells. The hollow NCs showed better sensing properties to ethanol than ZnO dumbbells. In addition, the cathodoluminescence of cubic well-crystalline ZnO nanostructures (200 nm in size) showed a strong UV emission and a poor blue emission peak at r.t.[43] Sol-gel-derived In_2O_3 thin and thick films were studied as gas sensors for the detection of very low concentrations of CO and NO_2.[44] Well-aligned arrays of nano- and microtubes of SnO_2 with square or rectangular cross sections (box beams) were synthesized on quartz substrates via a simple vapor deposition process in an open atmosphere.[45] Hollow silica nanoboxes were synthesized via a simple hard-template method at r.t., employing $MnCO_3$ nanocubes as the hard template.[46] The shell thickness of nanoboxes can be controlled by the amount of tetraethylorthosilicate (TEOS). These silica nanoboxes were found to be capable of being loaded and releasing Rhodamine B, thus showing a great potential in the controlled delivery applications.

Uniform-sized cerium dioxide nanocubes and other nanostructures were synthesized by the microwave-assisted method.[47] The shape of CeO_2 nanomaterials could be controlled simply by varying the irradiation time. Pd-, Pt-, and Au-doped ceria with different crystal shapes (nanorods and nanocubes) were studied as catalysts for methanol decomposition and methanol steam reforming.[48] Ceria nanorods were found to have strong interactions with methanol compared with ceria nanocubes. Bismuth-TiO_2 nanocubes were synthesized via a facile sol-gel hydrothermal method with titanium tetraisopropoxide as the precursor.[49] The photovoltaic behavior of dye-sensitized solar cells fabricated using Bi-TiO_2 nanocubes was studied, resulting in an open-circuit voltage (Voc) of 590 mV, a short-circuit current density (Jsc) of 7.71 mA/cm^2, and the conversion efficiency (η) of 2.11% under AM 1.5 illumination, a 77% increment compared with pure TiO_2 nanocubes. Magnetic properties of 2D paddy field–like superlattices of Mn_3O_4 cubic 6 nm nanoparticles were investigated by magnetization measurements.[50] The apparent magnetic anisotropy was found to be seriously reduced with the presence of dipolar interaction. Three distinct Co_3O_4 nanostructures of nanoparticles, nanocubes (Reactions 6.8 through 6.11), and hierarchical pompon-like microspheres were prepared by facile solution routes.[51] A direct comparison of electrochemical behaviors between these three nanostructures showed interesting "nanostructure effect," which was reasonably discussed by the authors in terms of how particle sizes, nanostructures, and crystallinity of Co_3O_4 materials function in tuning their electrochemistry. As an application, the gas responses of Co_3O_4[52] nanosheets, nanorods, and nanocubes to 100 ppm C_2H_5OH at 300°C were 10.5, 4.7, and 4.5 times higher than those of the Co_3O_4 agglomerated nanopowders, respectively. These enhanced gas-sensing characteristics were attributed to the less agglomerated nanostructures of the sensing materials. In addition, CO oxidation was performed on Co_3O_4 nanobelts and nanocubes as model catalysts.[53] Single-crystalline α-Fe_2O_3 nanocubes were obtained in large quantities through a facile one-step hydrothermal synthetic route under mild conditions.[54] Aqueous iron (III) nitrate ($Fe(NO_3)_3 \cdot 9H_2O$) served as the iron source and triethylamine served as the precipitant. By prolonging the reaction time from 1 to 24 h, the evolution process of α-Fe_2O_3, from nanorhombohedra to nanohexahedron, and finally nanocube, was observed. Fe_3O_4[55] nanocubes (12 nm in size) were also obtained by a facile solvothermal method at 260°C in the presence of oleic acid and oleylamine according to the scheme shown in Figure 6.8. The nanocubes were found to be superparamagnetic at 300 K but ferromagnetic at 2 K, and the saturation magnetization is 60.3 emu/g at 300 K. Single crystalline $In(OH)_3$ architectures (3D nanocubic, microcubic,

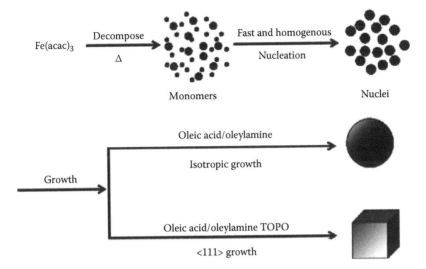

FIGURE 6.8 Schematic illustration of the mode for the growth of Fe_3O_4 nanocrystals. (Reproduced with permission from Gao, G. et al., Shape-controlled synthesis and magnetic properties of monodisperse Fe_3O_4 nanocubes, *Cryst. Growth Des.*, 10(7), 2888–2894. Copyright 2010 American Chemical Society.)

and irregular structures of $\approx 70\,nm$ to $5\,\mu m$ in size) were prepared by a rapid and efficient microwave-assisted hydrothermal (MAH) method using $InCl_3$ as the precursor at 140°C for 1 min[56]:

$$Precipitation\colon Co(NO_3)_2 + NH_3 \cdot H_2O \rightarrow Co(OH)_2 \qquad (6.8)$$

$$Co(OH)_2 + H_2O + O_2 \rightarrow Co(OH)_3 \qquad (6.9)$$

$$Hydrothermal\ process\colon Co(OH)_2 \rightarrow CoO \qquad (6.10)$$

$$Hydrothermal\ process\colon Co(OH)_3 + Organic\ species \rightarrow Co_3O_4 \qquad (6.11)$$

Simple and complex sulfides in NC and related structures are also objects of elevated attention. For instance, hollow CdS nanoboxes, with paper-thin walls of well-defined facets, were synthesized at 170°C *via* a simple reaction using Na_2SeO_3 for interior quasitemplates and ethylenediamine for exterior molecular templates.[57] Well-defined and uniform double-walled Cu_7S_4 nanoboxes with an average edge length of about 400 nm were synthesized using Cu_2O nanocubes as a sacrificial template based on an inward replacement or etching method.[58] Their formation became possible due to the repeated formation of a Cu_7S_4 layer in Na_2S solution and dissolution of the Cu_2O core in ammonia solution for two consecutive cycles. It was revealed that the double-walled Cu_7S_4 nanobox sensor exhibited enhanced performances such as higher sensitivity and shorter response time in ammonia gas sensing compared with the single-walled one. PbSe hollow single-crystalline nanoboxes were fabricated with the presence of trioctylphosphine as the structure-directing agent and stabilizer.[59] A gas bubble–assisted Ostwald ripening process was proposed to explain their formation. In a related work,[60] face-open nanoboxes (among other nanoforms such as microflowers) of lead telluride and selenide were synthesized by a simple hydrothermal method, in particular (for PbSe) from $NaSeO_3$ and PbAc as precursors.[61] It was found that the most suitable preparation condition was 10 h heating time. Inorganic fullerene-like nanoparticles consisting of WS_2 were synthesized by spray pyrolysis of $(NH_4)_2WS_4$ solutions at high temperature, presenting the unique structure of a closed, nested nanobox.[62] The nanoparticles exhibited the shoebox shape of a rectangular parallelepiped and possessed a hollow core surrounded by walls made up of stacked WS_2 layers. Ag_3CuS_2 NCs (400 nm average size and 30 nm shell thickness) were fabricated *via* a convenient ion-exchange route by Ag^+ reacting with Cu_7S_4 18-facet hollow nanopolyhedra.[63] These hollow-structured Ag_3CuS_2 exhibited better performances including higher sensitivity to ammonia and shorter response and recovery time than their rod-shape counterparts. Some other binary compounds are known; for instance, boron nitride. Thus, interaction between $B_{36}N_{36}$ fullerene-like NC and glycine amino acid was studied from the first principles.[64] It was indicated that glycine can form stable bindings with $B_{36}N_{36}$ NC *via* their carbonyl oxygen (O) active site, while the C_{60} fullerene might be unable to form stable bindings to glycine amino acid *via* their active sites. The authors predicted that the $B_{36}N_{36}$ NC can be implemented as a novel material for drug delivery applications.

Oxygen-containing metal salts represent another large group of cage-like nanostructures, among which attention is paid to titanates, for instance $BiFeO_3$-$BaTiO_3$.[65] Thus, barium titanate nanocube particles ($BaTiO_3$),[66] strontium titanate ($SrTiO_3$) (Figure 6.9),[67,68] and $SrTiO_3$ coated with platinum nanoparticles (Figure 6.10)[69] were prepared by a solvothermal method. Studies of the

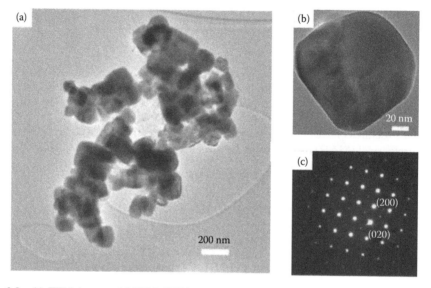

FIGURE 6.9 (a) TEM image of MSS-1 $SrTiO_3$ nanocubes, (b) TEM image of a single nanocube, and (c) corresponding SAED pattern along the [001] zone axis. The weaker spots arise from the difference in atomic scattering factors of Sr and Ti. (Reproduced with permission from Rabuffetti, F.A. et al., Synthesis-dependent first-order Raman scattering in $SrTiO_3$ nanocubes at room temperature. *Chem. Mater.*, 20, 5628–5635. Copyright 2008 American Chemical Society.)

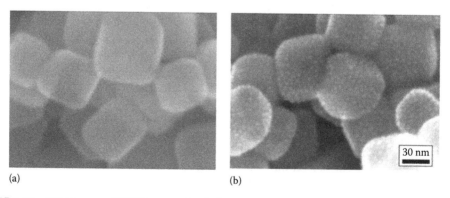

FIGURE 6.10 SEM images of $SrTiO_3$ nanocubes before (a) and after (b) coating with Pt nanoparticles using three Pt ALD growth cycles. (Christensen, S. et al.: Controlled growth of platinum nanoparticles on strontium titanate nanocubes by atomic layer deposition. *Small.* 2009, 5(6), 750–757. Copyright Wiley-VCH Verlag GmbH & Co. KGaA. Reproduced with permission.)

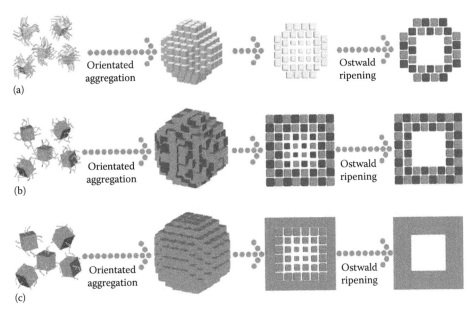

FIGURE 6.11 Schematic illustration showing three different growth mechanisms in (a) water-free, (b) 1.25 vol.% water, and (c) 5 vol.% water systems. (Reproduced with permission from Yang, X. et al., Formation mechanism of CaTiO$_3$ hollow crystals with different microstructures, *J. Am. Chem. Soc.*, 132(40), 14279–14287. Copyright 2010 American Chemical Society.)

crystal growth of CaTiO$_3$ hollow crystals with different microstructures showed that in a water-free poly(ethylene glycol) 200 (PEG-200) solution, CaTiO$_3$ nanocubes formed first,[70] undergoing an oriented self-assembly into spherical particles, enhanced by the surface-adsorbed polymer molecules. The disappearance of the small nanocubes in the cores of the spheres during an Ostwald ripening process led to spherical hollow crystals (Figure 6.11). The formation process gives a good example of the reversed crystal growth route in ceramic materials. In addition, the MnCO$_3$ hollow microspheres and nanocubes were synthesized *via* an ionic liquid-assisted hydrothermal synthetic method from MnCl$_2 \cdot 4H_2O$, NH$_4$HCO$_3$, and [bmim][Cl] (1-*n*-butyl-3-methylimidazolium chloride) **3** as precursors (Reactions 6.12 through 6.14).[71] In particular, MnCO$_3$ nanocubes were formed by the ordered aggregation of nanoplates with a high concentration of [bmim][Cl] (Figure 6.12).

1-*n*-butyl-3-methylimidazolium chloride

3

$$NH_4^+ + H_2O \rightarrow NH_3 \cdot H_2O + H^+ \tag{6.12}$$

$$HCO_3^- \leftrightarrow CO_3^{2-} + H^+ \tag{6.13}$$

$$Mn^{2+} + CO_3^{2-} \rightarrow MnCO_3 \tag{6.14}$$

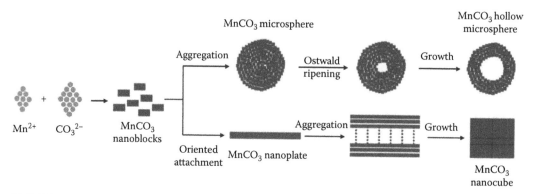

FIGURE 6.12 Schematic illustration for the growth process of the MnCO$_3$ hollow microspheres and MnCO$_3$ nanocubes. (Reproduced with permission from Duan, X. et al., Shape-controlled synthesis of metal carbonate nanostructure *via* ionic liquid-assisted hydrothermal route: The case of manganese carbonate, *Cryst. Growth Des.*, 10(10), 4449–4455. Copyright 2010 American Chemical Society.)

Zinc cobaltite hollow nanomaterials (20 nm size) with magnetic properties, which could be effectively modulated by adjusting reaction parameters, were controllably prepared under low-temperature hydrothermal condition *via* using cubic Co$_3$O$_4$ nanoparticles as precursors.[72] The formation mechanism of zinc cobaltite nanoboxes was suggested by the combination of ion diffusion and ion exchange self-generated in the reaction system. Hollow ZnSnO$_3$ nanocubes (side length of 200–400 nm) with peculiar cage- and skeleton-like architectures were prepared by a simple hydrothermal process at 180°C for 12 h.[73] The gas sensor based on these ZnSnO$_3$ nanostructures exhibited high response and quick response recovery to ethanol and HCHO. The unusual crystal structure of 12CaO · 7Al$_2$O$_3$ was found to be composed by a framework of positively charged NCs, which enable the accommodation of various negative ions (and even electrons) inside these cages.[74] La$_{0.5}$Ba$_{0.5}$MnO$_3$ nanocubes were synthesized by the hydrothermal method.[75] These nanocubes showed much higher catalytic activity and thermal stability in CO and CH$_4$ oxidation compared with the nanoparticles, maintaining the nanocube surface area well. The much higher catalytic activity and thermal stability of La$_{0.5}$Ba$_{0.5}$MnO$_3$ nanocubes were mainly predominated by the perfect single crystal structure. Finally, zeolite ZSM-5 nanoboxes with very regular pore geometry and wall thickness were obtained by pseudomorphic transformation of silicalite-1 crystals.[76] The authors noted that the coexistence of thin walls, significant mesoporosity, and high aluminum contents in the crystals is a real opportunity to improve the catalytic activity of the zeolite in reactions limited by diffusion.

Coordination compounds are rare as cage-like nanostructures. Thus, the metallosupramolecular assembly of macrocyclic and linear bisphenanthrolines (**4** [R = hexyl] and **5** [R^1 = H, X = Me and R^1 = Me, X = Br], respectively) in the presence of Cu$^+$ provided[77] nanobox structures with internal volumes of >5000 Å^3 when the ligands were designed along the HETPHEN (heteroleptic bisphenanthroline metal complexes) approach. Molecular assembly led to the formation of [Cu$_4$L$_2$L$^1_2$]$_4^+$ (L = 1,4-bis(3-phenanthrolinylethynyl)-2,5-bis(dodecyloxy)benzene; L^1 = **4**), [Cu$_n$L$_n$]$^{n+}$ (*n* = 3, 4), among other products. Several reports are dedicated to Prussian blue composites, for example, cobalt-iron Prussian blue analogs with cesium cation, tuned in the tetrahedral sites of the structure.[78] Also, graphene oxide sheets (GOs)-prussian blue (PB) nanocomposites were fabricated *via* a spontaneous redox reaction in an aqueous solution containing FeCl$_3$, K$_3$[Fe(CN)$_6$] and GOs.[79] The products represented PB nanocubes formed on the surface of GOs retaining their excellent electrochemical activity; moreover, the obtained nanocomposites even showed a higher sensitivity toward the electrocatalytical reduction of H$_2$O$_2$ than that of MWCNT/PB nanocomposites. The main applications for these nanocomposites could be electrochemical sensor and biofuel cell.

4   5

6.2 NANOCAPSULES

In contrast to closely related NCs or nanoboxes, capsule-like nanostructures as a term are widely reported (almost 4000 reports are currently available in this area, mainly organic or polymeric examples) due to a host of applications related first of all with encapsulation (e.g., protein nanoencapsulation using layered zirconium phosphate nanoparticles[80]) and drug delivery (review on medical uses[81]), as well as other uses (food,[82] biocatalysis,[83] herbicide delivery[84]). These nanostructures cannot be called as "less-common." Therefore, here we briefly examine selected examples of inorganic nanocapsules and their applications. Thus, decomposition of hydrocarbons (methane, ethane, and ethylene) on the nickel catalyst led to the formation of many disordered carbon structures and Ni nanocapsules coated with a graphite layer.[85] Electromagnetic-wave absorption properties of FeNiMo/C nanocapsules with FeNiMo nanoparticles as cores and carbon as shells were studied.[86] These nanocapsules, synthesized by arc-discharging $Fe_{11}Ni_{79}Mo_{10}$ (at.%) alloy in ethanol, exhibited outstanding reflection loss (RL < −20 dB) in the range of 13–17.8 GHz with an absorber thickness of 1.7 mm, and an optimal RL of −64.0 dB was obtained at 13.2 GHz with an absorber thickness of 1.9 mm. The excellent electromagnetic properties were ascribed to the good match between high permeability from the core of FeNiMo nanoparticles and low permittivity from the special core-shell microstructure in the FeNiMo/C nanocapsules. Controlled synthesis of uniformly shaped

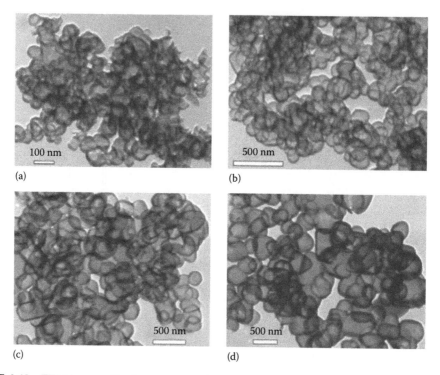

FIGURE 6.13 TEM images of carbon nanocapsules prepared by ethanol-assisted thermolysis of $Zn(Ac)_2$ at 600°C for different times: (a) 3 h, (b) 6 h, (c) 24 h, and (d) 48 h. (Reproduced with permission from Zheng, M. et al., Simple shape-controlled synthesis of carbon hollow structures. *Inorg. Chem.*, 49(19), 8674–8683. Copyright 2010 American Chemical Society.)

carbon hollow structures was carried out by an ethanol-assisted thermolysis of zinc acetate.[87] It was revealed that the generated zinc oxide nanostructures acted as *in situ* templates to form (Reactions 6.15 through 6.17) the carbon hollow structures (nanospheres, nanocapsules [Figures 6.13 through 6.15], nanorods, and microtubes). These as-synthesized carbon hollow structures exhibited excellent thermal and structural stability to temperatures as high as 1200°C:

$$CH_3COOZnOOCCH_3 \rightarrow ZnCO_3 + CH_3COCH_3 \qquad (6.15)$$

$$ZnCO_3 \rightarrow ZnO + CO_2 \qquad (6.16)$$

In the presence of ZnO: $mCH_3COCH_3 + nCH_3CH_2OH$

$$\rightarrow ZnO@C \text{ structures} + (2m + n)H_2 + (m - n)H_2O \qquad (6.17)$$

Uniform hollow inorganic core-shell-structured multifunctional mesoporous nanocapsules (Figures 6.16 and 6.17), composed of functional inorganic (Fe_3O_4, Au, etc.) nanocrystals as cores, a thin mesoporous silica shell, and a huge cavity in between, were described.[88] The excellent biocompatibility of the obtained multifunctional nanocapsules ($Fe_3O_4@mSiO_2$) was demonstrated by the low cytotoxicity against various cell lines, low hemolyticity against human blood red cells, and no significant coagulation effect against blood plasma. Hollow mesoporous zirconia nanocapsules (hm-ZrO_2) with a hollow core/porous shell structure were demonstrated as effective vehicles for anticancer drug delivery.[89] A loading of 102% related to the weight of hm-ZrO_2 was achieved by the nanocapsules with an inner diameter of 385 nm. hm-ZrO_2 loaded DOX (doxorubicin) released more

FIGURE 6.14 SEM and TEM images of the samples obtained by ethanol-assisted thermolysis of Zn(Ac)$_2$, which was first heated at 400°C for different times and then heated to 600°C for 12 h. (a, b) 1 h, (c, d) 3 h, and (e, f) 6 h. (Reproduced from Zheng, M. et al., Simple shape-controlled synthesis of carbon hollow structures. *Inorg. Chem.*, 49(19), 8674–8683. Copyright 2010 American Chemical Society.)

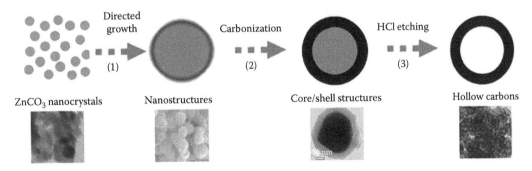

FIGURE 6.15 Schematic representation of the possible formation mechanism of the carbon hollow structures by ethanol-assisted thermolysis of Zn(Ac)$_2$. (Reproduced from Zheng, M. et al., Simple shape-controlled synthesis of carbon hollow structures. *Inorg. Chem.*, 49(19), 8674–8683. Copyright 2010 American Chemical Society.)

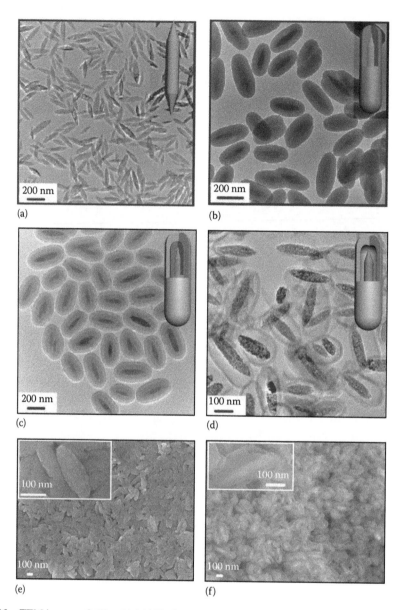

FIGURE 6.16 TEM images of ellipsoidal (a) Fe_2O_3, (b) $Fe_2O_3@SiO_2$, (c) $Fe_2O_3@SiO_2@mSiO_2$, and (d) $Fe_3O_4@mSiO_2$; secondary electron SEM image of ellipsoidal Fe_2O_3 (e, inset: SEM image at high magnification) and backscattered electron SEM (f) image of ellipsoidal $Fe_3O_4@mSiO_2$ nanocapsules (inset: purposely selected backscattered electron image of broken nanocapsules to reveal the hollow nanostructure). (Reproduced with permission from Chen, Y. et al., Core/shell structured hollow mesoporous nanocapsules: A potential platform for simultaneous cell imaging and anticancer drug delivery, *ACS Nano*, 4(10), 6001–6013. Copyright 2010 American Chemical Society.)

drugs in cancer cells than in normal cells, leading to more cytotoxicity toward tumor cells and less cytotoxicity to healthy cells than free DOX.

Among other inorganic nanocapsules, we note bioluminescent nanocapsules (Figure 6.18)[90] on CdTe quantum dot basis. Organic nanocapsules are common; here we mention a classic compound, polyaniline (PANI, standard organic nanotechnological object, mentioned also in other sections),

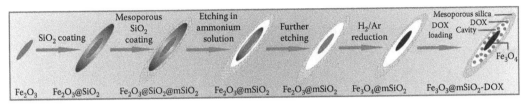

FIGURE 6.17 Schematic representation for the preparation of hollow core-shell-structured mesoporous nanocapsules with magnetic Fe_3O_4 as the core and thin mesoporous silica layer as the shell. Ellipsoidal Fe_2O_3 nanocrystals were coated by solid and mesoporous silica layer by traditional sol-gel process. Then, the fabricated composite nanoellipsoidals were etched in ammonium solution to partially or entirely remove the middle solid silica layer to prepare rattle-type (or yolk-shell type) nanocapsules using the structural difference between solid silica layer and mesoporous silica layer. After H_2/Ar reduction, the Fe_2O_3 core could be converted into magnetic Fe_3O_4. DOX molecules could be encapsulated into mesopores and cavities in the prepared magnetic nanocapsules. (Reproduced with permission from Chen, Y. et al., Core/shell structured hollow mesoporous nanocapsules: A potential platform for simultaneous cell imaging and anticancer drug delivery, *ACS Nano*, 4(10), 6001–6013. Copyright 2010 American Chemical Society.)

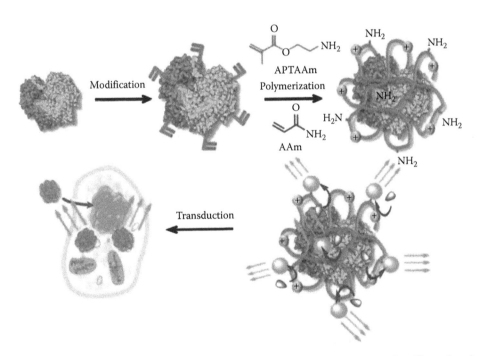

FIGURE 6.18 Scheme of forming robust, cell-permeable bioluminescent nanocapsules. (Reproduced with permission from Du, J. et al., Quantum-dot-decorated robust transductable bioluminescent nanocapsules, *J. Am. Chem. Soc.*, 132(37), 12780–12781. Copyright 2010 American Chemical Society.)

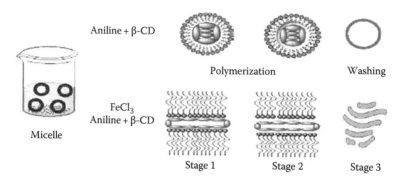

FIGURE 6.19 Models of the overall fabrication procedure of PANI nanocapsules and nanotubes. (From Prasannan, A. and P.-D. Hong, Formation mechanism of polyaniline nanotubes and nanocapsules sing β-cyclodextrin, http://ir.lib.ntust.edu.tw:8080/dspace/bitstream/987654321/14728/1/2010-Formation%20Mechanism%20 Polyaniline%20Nanotubes%20Nanocapsules.pdf. Reproduced with permission of http://ir.lib.ntust.edu.tw.)

whose nanotubes and nanocapsules (Figure 6.19) were readily fabricated through chemical oxidation polymerization in dodecyltrimethylammonium chloride (DoTAC) microemulsion polymerization using β-cyclodextrin (β-CD).[91] In the $H_2O/FeCl_3/DoTAC/\beta$-CD system, the aqueous $FeCl_3$ solution played a role in increasing the ionic strength and decreasing the second critical micelle concentration of DoTAC. As a result, DoTAC cylindrical micelles could be spontaneously formed in an aqueous medium.

REFERENCES

1. Sun, Y.; Mayers, B.; Xia, Y. Metal nanostructures with hollow interiors. *Advanced Materials*, 2003, *15* (7–8), 641–646.
2. Lu, C.-L.; Prasad, K. S.; Wu, H.-L.; Ho, J.-an A.; Huang, M. H. Au nanocube-directed fabrication of Au-Pd core-shell nanocrystals with tetrahexahedral, concave octahedral, and octahedral structures and their electrocatalytic activity. *Journal of the American Chemical Society*, 2010, *132* (41), 14546–14553.
3. Liu, Y.; Walker, A. R. H. Monodisperse gold-copper bimetallic nanocubes: Facile one-step synthesis with controllable size and composition. *Angewandte Chemie, International Edition*, 2010, *49* (38), 6781–6785.
4. Loukrakpam, R.; Chang, P.; Luo, J.; Fang, B.; Mott, D.; Bae, I.-T.; Naslund, H. R.; Engelhard, M. H.; Zhong, C.-J. Chromium-assisted synthesis of platinum nanocube electrocatalysts. *Chemical Communications*, 2010, *46* (38), 7184–7186.
5. Yang, H.; Zhang, J.; Sun, K.; Zou, S.; Fang, J. Enhancing by weakening: Electrooxidation of methanol on Pt₃Co and Pt nanocubes. *Angewandte Chemie, International Edition*, 2010, *49* (38), 6848–6851.
6. Zhao, H.; Fu, H.-G.; Tian, C.-G.; Ren, Z.-Y.; Tian, G.-H. Facile shape-controlled synthesis of palladium nanostructures on copper for promising Surface-enhanced Raman scattering. *Materials Letters*, 2010, *64* (20), 2255–2257.
7. Yuan, Q.; Zhou, Z.; Zhuang, J.; Wang, X. Tunable aqueous phase synthesis and shape-dependent electrochemical properties of rhodium nanostructures. *Inorganic Chemistry*, 2010, *49* (12), 5515–5521.
8. Xiong, Y.; Wiley, B.; Chen, J.; Li, Z.-Y.; Yin, Y.; Xia, Y. Corrosion-based synthesis of single-crystal Pd nanoboxes and nanocages and their surface plasmon properties. *Angewandte Chemie, International Edition*, 2005, *44* (48), 7913–7917.
9. Chen, J.; Wiley, B.; McLellan, J.; Xiong, Y.; Li, Z.-Y.; Xia, Y. Optical properties of Pd-Ag and Pt-Ag nanoboxes synthesized via galvanic replacement reactions. *Nano Letters*, 2005, *5* (10), 2058–2062.
10. Mahmoud, M. A.; Saira, F.; El-Sayed, M. A. Experimental evidence for the nanocage effect in catalysis with hollow nanoparticles. *Nano Letters*, 2010, *10* (9), 3764–3769.
11. Peng, Z.; You, H.; Wu, J.; Yang, H. Electrochemical synthesis and catalytic property of sub-10 nm platinum cubic nanoboxes. *Nano Letters*, 2010, *10* (4), 1492–1496.
12. Yamada, M.; Kon, S.; Miyake, M. Synthesis and size control of Pt nanocubes with high selectivity using the additive effect of NaI. *Chemistry Letters*, 2005, *34* (7), 1050–1051.
13. Wang, Q.; Geng, B.; Tao, B. A facile room temperature chemical route to Pt nanocube/carbon nanotube heterostructures with enhanced electrocatalysis. *Journal of Power Sources*, 2011, *196* (1), 191–195.

14. Kang, Y.; Murray, C. B. Synthesis and electrocatalytic properties of cubic Mn-Pt nanocrystals (nanocubes). *Journal of the American Chemical Society*, 2010, *132* (22), 7568–7569.

15. Takahashi, Y.; Zettsu, N.; Nishino, Y.; Tsutsumi, R.; Matsubara, E.; Ishikawa, T.; Yamauchi, K. Three-dimensional electron density mapping of shape-controlled nanoparticle by focused hard x-ray diffraction microscopy. *Nano Letters*, 2010, *10* (5), 1922–1926.

16. Xing, S.; Feng, Y.; Tay, Y. Y.; Chen, T.; Xu, J.; Pan, M.; He, J.; Hng, H. H.; Yan, Q.; Chen, H. Reducing the symmetry of bimetallic Au@Ag nanoparticles by exploiting eccentric polymer shells. *Journal of the American Chemical Society*, 2010, *132* (28), 9537–9539.

17. Aherne, D.; Gara, M.; Kelly, J. M.; Gun'ko, Y. K. From Ag nanoprisms to triangular AuAg nanoboxes. *Advanced Functional Materials*, 2010, *20* (8), 1329–1338.

18. Zhang, Q.; Li, W.; Wen, L.-P.; Chen, J.; Xia, Y. Facile synthesis of Ag nanocubes of 30 to 70 nm in edge length with CF_3COOAg as a precursor. *Chemistry—A European Journal*, 2010, *16* (33), 10234–10239, S10234/1–S10234/3.

19. Yu, D.; Yam, V. W.-W. Controlled synthesis of monodisperse silver nanocubes in water. *Journal of the American Chemical Society*, 2004, *126*, 13200–13201.

20. Zhang, Y.; Xu, F.; Sun, Y.; Guo, C.; Cui, K.; Shi, Y.; Wen, Z.; Li, Z. Seed-mediated synthesis of Au nanocages and their electrocatalytic activity towards glucose oxidation. *Chemistry—A European Journal*, 2010, *16* (30), 9248–9256, S9248/1–S9248/3.

21. Zhang, Q.; Cobley, C. M.; Zeng, J.; Wen, L.-P.; Chen, J.; Xia, Y. Dissolving Ag from Au-Ag alloy nanoboxes with H_2O_2: A method for both tailoring the optical properties and measuring the H_2O_2 concentration. *Journal of Physical Chemistry C*, 2010, *114* (14), 6396–6400.

22. Skrabalak, S. E.; Chen, J.; Sun, Y.; Lu, X.; Au, L.; Cobley, C. M.; Xia, Y. Gold nanocages: Synthesis, properties, and applications. *Accounts of Chemical Research*, 2008, *41* (12), 1587–1595.

23. Xia, Y. Gold nanocages: A new class of plasmonic nanostructures for biomedical applications. *Abstracts of Papers, 234th ACS National Meeting*, Boston, MA, August 19–23, 2007, COLL-528.

24. Zeng, J.; Zhang, Q.; Chen, J.; Xia, Y. A comparison study of the catalytic properties of Au-based nanocages, nanoboxes, and nanoparticles. *Nano Letters*, 2010, *10* (1), 30–35.

25. Qu, L.; Dai, L.; Osawa, E. Shape/size-controlled syntheses of metal nanoparticles for site-selective modification of carbon nanotubes. *Journal of the American Chemical Society*, 2006, *128* (16), 5523–5532.

26. Sheng, J.-J.; Li, J.-F.; Yin, B.-S.; Ren, B.; Tian, Z.-Q. A preliminary study on surface-enhanced Raman scattering from Au and Au@Pd nanocubes for electrochemical applications. *Canadian Journal of Analytical Sciences and Spectroscopy*, 2007, *52* (3), 178–185.

27. Yang, J.; Zhang, W.-D.; Gunasekaran, S. An amperometric non-enzymatic glucose sensor by electrode-positing copper nanocubes onto vertically well-aligned multi-walled carbon nanotube arrays. *Biosensors and Bioelectronics*, 2010, *26* (1), 279–284.

28. Kumar, M.; Singh, B. J.; Kajjam, S.; Bandyopadhyay, D. Effect of transition metal doping on hydrogenated germanium nanocages: A density functional investigation. *Journal of Computational and Theoretical Nanoscience*, 2010, *7* (1), 296–301.

29. Li, Y.; Zhou, C.; Xie, X.; Shi, G.; Qu, L. Spontaneous, catalyst-free formation of nitrogen-doped graphitic carbon nanocages. *Carbon*, 2010, *48* (14), 4190–4196.

30. Guo, C. X.; Sheng, Z. M.; Shen, Y. Q.; Dong, Z. L.; Li, C. M. Thin-walled graphitic nanocages as a unique platform for amperometric glucose biosensor. *ACS Applied Materials and Interfaces*, 2010, *2* (9), 2481–2484.

31. Cluzel, B.; Pauc, N.; Calvo, V.; Charvolin, T.; Hadji, E. Nanobox array for silicon-on-insulator luminescence enhancement at room temperature. *Applied Physics Letters*, 2006, *88* (13), 133120/1–133120/3.

32. Pande, S.; Pal, T. Controlled synthesis of CuO box shaped nanoparticles and their application in nitrophenol reduction. *Journal of the Indian Chemical Society*, 2010, *87* (4), 405–415.

33. Huang, L.; Peng, F.; Yu, H.; Wang, H. Synthesis of Cu_2O nanoboxes, nanocubes and nanospheres by polyol process and their adsorption characteristic. *Materials Research Bulletin*, 2008, *43* (11), 3047–3053.

34. Kim, M. R.; Kim, S. J.; Jang, D.-J. Fabrication of copper oxide nanoboxes containing a platinum nanocluster via an optical and galvanic route. *Crystal Growth and Design*, 2010, *10* (1), 257–261.

35. Chen, Q.; Shen, X.; Gao, H. Formation of solid and hollow cuprous oxide nanocubes in water-in-oil microemulsions controlled by the yield of hydrated electrons. *Journal of Colloid and Interface Science*, 2007, *312*, 272–278.

36. Liu, D.; Yang, S.; Lee, S.-T. Preparation of novel cuprous oxide-fullerene[60] core-shell nanowires and nanoparticles via a copper(I)-assisted fullerene-polymerization reaction. *Journal of Physical Chemistry C*, 2008, *112* (18), 7110–7118.

37. He, T.; Chen, D.; Jiao, X.; Wang, Y. Co_3O_4 nanoboxes: Surfactant-templated fabrication and microstructure characterization. *Advanced Materials*, 2006, *18* (8), 1078–1082.

38. Umar, A.; Rahman, M. M.; Hahn, Y.-B. MgO polyhedral nanocages and nanocrystals based glucose bio-sensor. *Electrochemistry Communications*, 2009, *11* (7), 1353–1357.
39. Takahashi, N. Simple and rapid synthesis of MgO with nano-cube shape by means of a domestic micro-wave oven. *Solid State Sciences*, 2007, *9*, 722–724.
40. Mahmud, S.; Abdullah, M. J.; Zakaria, M. Z. Growth model for nanoplates and nanoboxes of zinc oxide from a catalyst-free combust-oxidized process. *Synthesis and Reactivity in Inorganic, Metal-Organic, and Nano-Metal Chemistry*, 2006, *36* (1), 17–22.
41. Mahmud, S.; Johar Abdullah, M.; Putrus, G.; Chong, J.; Karim Mohamad, A. Nanostructure of ZnO fabri-cated via French process and its correlation to electrical properties of semiconducting varistors. *Synthesis and Reactivity in Inorganic, Metal-Organic, and Nano-Metal Chemistry*, 2006, *36* (2), 155–159.
42. Yu, X.-L.; Ji, H.-M.; Wang, H.-L.; Sun, J.; Du, X.-W. Synthesis and sensing properties of ZnO/ZnS nano-cages. *Nanoscale Research Letters*, 2010, *5* (3), 644–648.
43. Yu, L.; Qu, F.; Wu, X. Facile hydrothermal synthesis of novel ZnO nanocubes. *Journal of Alloys and Compounds*, 2010, *504* (1), L1–L4.
44. Taurino, A.; Catalano, M.; Siciliano, P.; Gurlo, A.; Barsan, N.; Weimar, U.; Ivanovskaya, M. Tuning of the gas-sensing properties of self-assembled In$_2$O$_3$ nanoboxes prepared by sol gel techniques. *Sensors and Microsystems, Proceedings of the 8th Italian Conference*, Trento, Italy, February 12–14, 2003. Di Natale, C. (Ed.). pp. 191–194.
45. Liu, Y.; Dong, J.; Liu, M. Well-aligned "nano-box-beams" of SnO$_2$. *Advanced Materials*, 2004, *16* (4), 353–356.
46. Zhao, M.; Kang, W.; Zheng, L.; Gao, Y. Synthesis of silica nanoboxes via a simple hard-template method and their application in controlled release. *Materials Letters*, 2010, *64* (8), 990–992.
47. Tao, Y.; Wang, H.; Xia, Y.; Zhang, G.; Wu, H.; Tao, G. Preparation of shape-controlled CeO$_2$ nanocrystals via microwave-assisted method. *Materials Chemistry and Physics*, 2010, *124* (1), 541–546.
48. Yi, N.; Si, R.; Saltsburg, H.; Flytzani-Stephanopoulos, M. Methanol reactions over metals (Pd, Pt and Au) doped ceria. *Abstracts of Papers, 240th ACS National Meeting*, Boston, MA, August 22–26, 2010, CATL-85.
49. An'amt, M. N.; Radiman, S.; Huang, N. M.; Yarmo, M. A.; Ariyanto, N. P.; Lim, H. N.; Muhamad, M. R. Sol-gel hydrothermal synthesis of bismuth-TiO$_2$ nanocubes for dye-sensitized solar cell. *Ceramics International*, 2010, *36* (7), 2215–2220.
50. Chen, W.; Chen, C.; Guo, L. Magnetization reversal of two-dimensional superlattices of Mn3O4 nanocubes and their collective dipolar interaction effects. *Journal of Applied Physics*, 2010, *108* (4), 043912/1–043912/8.
51. Guo, B.; Li, C.; Yuan, Z.-Y. Nanostructured Co$_3$O$_4$ materials: Synthesis, characterization, and elec-trochemical behaviors as anode reactants in rechargeable lithium ion batteries. *Journal of Physical Chemistry C*, 2010, *114* (29), 12805–12817.
52. Choi, K.-I; Kim, H.-R.; Kim, K.-M.; Liu, D.; Cao, G.; Lee, J.-H. C$_2$H$_5$OH sensing characteristics of vari-ous Co$_3$O$_4$ nanostructures prepared by solvothermal reaction. *Sensors and Actuators, B: Chemical*, 2010, *B146* (1), 183–189.
53. Hu, L.; Sun, K.; Peng. Q.; Xu, B.; Li, Y. Surface active sites on Co$_3$O$_4$ nanobelt and nanocube model catalysts for CO oxidation. *Nano Research*, 2010, *3*, 363–368.
54. Qin, W.; Yang, C.; Yi, R.; Gao, G. Hydrothermal synthesis and characterization of single-crystalline α-Fe$_2$O$_3$ nanocubes. *Journal of Nanomaterials*, 2011, Article ID 159259, 5 pp.
55. Gao, G.; Liu, X.; Shi, R.; Zhou, K.; Shi, Y.; Ma, R.; Takayama-Muromachi, E.; Qiu, G. Shape-controlled synthesis and magnetic properties of monodisperse Fe$_3$O$_4$ nanocubes. *Cryst. Growth and Design*, 2010, *10* (7), 2888–2894.
56. Motta, F. V.; Lima, R. C.; Marques, A. P. A.; Li, M. S.; Leite, E. R.; Varela, J. A.; Longo, E. Indium hydroxide nanocubes and microcubes obtained by microwave-assisted hydrothermal method. *Journal of Alloys and Compounds,* 2010, *497* (1–2), L25–L28.
57. Kim, M. R.; Jang, D.-J. One-step fabrication of well-defined hollow CdS nanoboxes. *Chemical Communications*, 2008, (41), 5218–5220.
58. Zhang, W.; Chen, Z.; Yang, Z. An inward replacement/etching route to synthesize double-walled Cu$_7$S$_4$ nanoboxes and their enhanced performances in ammonia gas sensing. *Physical Chemistry Chemical Physics*, 2009, *11* (29), 6263–6268.
59. Chen, S.; Zhang, X.; Hou, X.; Zhou, Q.; Tan, W. One-pot synthesis of hollow PbSe single-crystalline nano-boxes via gas bubble assisted Ostwald ripening. *Crystal Growth and Design*, 2010, *10* (3), 1257–1262.
60. Poudel, B.; Wang, W. Z.; Wang, D. Z.; Huang, J. Y.; Ren, Z. F. Shape evolution of lead telluride and selenide nanostructures under different hydrothermal synthesis conditions. *Journal of Nanoscience and Nanotechnology*, 2006, *6* (4), 1050–1053.

61. Terra, F. S.; Mahmoud, G. M.; Nasr, Mahmoud; El Okr, M. M. Preparation and structural properties of PbSe nanomaterial. *Surface and Interface Analysis*, 2010, *42* (6–7), 1239–1243.

62. Bastide, S.; Duphil, D.; Borra, J.-P.; Levy-Clement, C. WS_2 closed nanoboxes synthesized by spray pyrolysis. *Advanced Materials*, 2006, *18* (1), 106–109.

63. Chen, Z.; Zhang, W.; Yang, Z. Synthesis and ammonia sensing property of Ag_3CuS_2 nanocages obtained from Cu_7S_4 18-facet hollow nanopolyhedra. *Journal of Crystal Growth*, 2009, *311* (12), 3347–3351.

64. Ganji, M. D.; Yazdani, H.; Mirnejad, A. $B_{36}N_{36}$ fullerene-like nanocages: A novel material for drug delivery. *Physica E: Low-Dimensional Systems and Nanostructures*, 2010, *42* (9), 2184–2189.

65. Park, T.-J.; Papaefthymiou, G. C.; Viescas, A. J.; Lee, Y.; Zhou, H.; Wong, S. S. Composition-dependent magnetic properties of $BiFeO_3$-$BaTiO_3$ solid solution nanostructures. *Physical Review B: Condensed Matter and Materials Physics*, 2010, *82* (2), 024431/1–024431/10.

66. Wada, S.; Nozawa, A.; Iwatsuki, S.; Kuwabara, T.; Takei, T.; Kumada, N.; Pulpan, P.; Uchida, H. Dispersion of barium titanate and strontium titanate nanocubes and their selective accumulations. *Key Engineering Materials*, 2010, *445 (Electroceramics in Japan XIII)*, 183–186.

67. Rabuffetti, F. A.; Kim, H.-S.; Enterkin, J. A.; Wang, Y.M.; Lanier, C. H.; Marks, L. D.; Poeppelmeier, K. R.; Stair, P. C. Synthesis-dependent first-order Raman scattering in $SrTiO_3$ nanocubes at room temperature. *Chemistry of Materials*, 2008, *20*, 5628–5635.

68. Rabuffetti, F. A.; Stair, P. C.; Poeppelmeier, K. R. Synthesis-dependent surface acidity and structure of $SrTiO_3$ nanoparticles. *Journal of Physical Chemistry C*, 2010, *114* (25), 11056–11067.

69. Christensen, S. T.; Elam, J. W.; Rabuffetti, F. A.; Ma, Q.; Weigand, S. J.; Lee, B.; Seifert, S. et al., Controlled growth of platinum nanoparticles on strontium titanate nanocubes by atomic layer deposition. *Small*, 2009, *5* (6), 750–757.

70. Yang, X.; Fu, J.; Jin, C.; Chen, J.; Liang, C.; Wu, M.; Zhou, W. Formation mechanism of $CaTiO_3$ hollow crystals with different microstructures. *Journal of the American Chemical Society*, 2010, *132* (40), 14279–14287.

71. Duan, X.; Lian, J.; Ma, J.; Kim, T.; Zheng, W. Shape-controlled synthesis of metal carbonate nanostructure via ionic liquid-assisted hydrothermal route: The case of manganese carbonate. *Crystal Growth and Design*, 2010, *10* (10), 4449–4455.

72. Tian, L.; Chen, L.; Yi, L.-H. Fabrication and magnetic properties of hollow zinc cobaltite nanomaterials through ion-exchange. *Wuji Huaxue Xuebao*, 2010, *26* (1), 49–54.

73. Zeng, Y.; Zhang, T.; Fan, H.; Lu, G.; Kang, M. Synthesis and gas-sensing properties of $ZnSnO_3$ cubic nanocages and nanoskeletons. *Sensors and Actuators, B: Chemical*, 2009, *B143* (1), 449–453.

74. Feldbach, E.; Denks, V. P.; Kirm, M.; Kunnus, K.; Maaroos, A. Intrinsic excitons in $12CaO.7Al_2O_3$. *Radiation Measurements*, 2010, *45* (3–6), 281–283.

75. Liang, S.; Xu, T.; Teng, F.; Zong, R.; Zhu, Y. The high activity and stability of $La_{0.5}Ba_{0.5}MnO_3$ nanocubes in the oxidation of CO and CH_4. *Applied Catalysis, B: Environmental*, 2010, *96* (3–4), 267–275.

76. Wang, Y; Tuel, A. Nanoporous zeolite single crystals: ZSM-5 nanoboxes with uniform intracrystalline hollow structures. *Microporous and Mesoporous Materials*, 2008, *113* (1–3), 286–295.

77. Schmittel, M.; Ammon, H.; Kalsani, V.; Wiegrefe, A.; Michel, C. Quantitative formation and clean metal exchange processes of large void (>5000 $Å^3$) nanobox structures. *Chemical Communications*, 2002, (21), 2566–2567.

78. Xu, J. F.; Liu, H.; Liu, P.; Liang, C. H.; Wang, Q.; Fang, J.; Zhao, J. H.; Shen, W. G. Magnetism study of $CI_xCO_y[Fe(CN)_6] \cdot zH_2O$ (CI = Rb,Cs) Prussian blue nanoparticles. *Journal of the Iranian Chemical Society*, 2010, *7* (Suppl.), S123–S129.

79. Liu, X.-W.; Yao, Z.-J.; Wang, Y.-F.; Wei, X.-W. Graphene oxide sheet-prussian blue nanocomposites: Green synthesis and their extraordinary electrochemical properties. *Colloids and Surfaces, B: Biointerfaces*, 2010, *81* (2), 508–512.

80. Diaz, A.; David, A.; Perez, R.; Gonzalez, M. L.; Baez, A.; Wark, S. E.; Zhang, P.; Clearfield, A.; Colon, J. L. Nanoencapsulation of insulin into zirconium phosphate for oral delivery applications. *Biomacromolecules*, 2010, *11* (9), 2465–2470.

81. Van Hest, J. C. M. (Ed.). Micro-and nanocapsules for biological and biomedical applications. *Macromolecular Bioscience*, 2010, *10* (5), 96 pp.

82. Nakagawa, K.; Surassmo, S.; Min, S.-G.; Choi, M.-J. Dispersibility of freeze-dried poly(epsilon-caprolactone) nanocapsules stabilized by gelatin and the effect of freezing. *Journal of Food Engineering*, 2011, *102* (2), 177–188.

83. Phuoc, L.-T.; Laveille, P.; Chamouleau, F.; Renard, G.; Drone, J.; Coq, B.; Fajula, F.; Galarneau, A. Phospholipid-templated silica nanocapsules as efficient polyenzymatic biocatalysts. *Dalton Transactions*, 2010, *39* (36), 8511–8520.

84. Bhattacharyya, A.; Bhaumik, A.; Pathipati, U. R.; Mandal, S.; Epidi, T. T. Nanoparticles—A recent approach to insect pest control. *African Journal of Biotechnology*, 2010, *9* (24), 3489–3493.

85. Narkiewicz, U.; Podsiadly, M.; Jedrzejewski, R.; Pelech, I. Catalytic decomposition of hydrocarbons on cobalt, nickel and iron catalysts to obtain carbon nanomaterials. *Applied Catalysis, A: General*, 2010, *384* (1–2), 27–35.

86. Liu, X. G.; Ou, Z. Q.; Geng, D. Y.; Han, Z.; Wang, H.; Li, B.; Brueck, E.; Zhang, Z. D. Enhanced absorption bandwidth in carbon-coated supermalloy FeNiMo nanocapsules for a thin absorb thickness. *Journal of Alloys and Compounds*, 2010, *506* (2), 826–830.

87. Zheng, M.; Liu, Y.; Zhao, S.; He, W.; Xiao, Y.; Yuan, D. Simple shape-controlled synthesis of carbon hollow structures. *Inorganic Chemistry*, 2010, *49* (19), 8674–8683.

88. Chen, Y.; Chen, H.; Zeng, D.; Tian, Y.; Chen, F.; Feng, J.; Shi, J. Core/shell structured hollow mesoporous nanocapsules: A potential platform for simultaneous cell imaging and anticancer drug delivery. *ACS Nano*, 2010, *4* (10), 6001–6013.

89. Tang, S.; Huang, X.; Chen, X.; Zheng, N. Hollow mesoporous zirconia nanocapsules for drug delivery. *Advanced Functional Materials*, 2010, *20* (15), 2442–2447.

90. Du, J.; Yu, C.; Pan, D.; Li, J.; Chen, W.; Yan, M.; Segura, T.; Lu, Y. Quantum-dot-decorated robust transductable bioluminescent nanocapsules. *Journal of the American Chemical Society*, 2010, *132* (37), 12780–12781.

91. Prasannan, A.; Hong, P.-D. Formation mechanism of polyaniline nanotubes and nanocapsules sing β-cyclodextrin. 2010. http://ir.lib.ntust.edu.tw/bitstream/987654321/14728/1/2010-Formation%20Mechanism%20Polyaniline%20Nanotubes%20Nanocapsules.pdf, http://ir.lib.ntust.edu.tw/handle/987654321/14728?locale=en-us

7 "Nanovegetation" World

7.1 NANOTREES

7.1.1 NANOTREES IN GENERAL

Tree-like is a relatively frequent nano- and microstructural type, obtained for a series of inorganic compounds, mostly for silicon, silica, and carbon. Generally, they are closely related to nanowires, which are modified with branches; sometimes they are produced from them[1] and can appear together with other nanostructures. In the case of silicon, the structure and electronic properties of its nanotrees using large-scale quantum mechanical simulations were investigated.[2] Electronic structure analysis showed that the formation of nanotrees was accompanied by a narrowing of the HOMO-LUMO gap; this fact could lead to their application in molecular devices due to the predicted enhancement in the conducting properties. Si nanotrees were found to be suitable also for thermoelectrical applications.[3] The effects of trimethylaluminum (TMA) on silicon nanowires grown by chemical vapor deposition (CVD) were investigated in the 650°C–850°C growth temperature range using Au as the growth catalyst and SiH_4 in H_2 carrier gas as the Si precursor.[4] Depending on the substrate temperature and TMA partial pressure, the structure's morphology evolved from wires to tapered needles, pyramids, or nanotrees (the TMA presence was linked, in particular, to a branched growth leading to Si nanotrees).

Hierarchical silicon oxycarbide tree-like nanostructures (Figure 7.1), which consisted of trunks and abundant branches, were fabricated by electron irradiation of organosilicon polymers.[5] The nanotree structures had two different morphologies, cluster-assembled and nanowire-assembled, whose formation can be controlled by the irradiation parameters. Superhydrophobic silica nanotrees were obtained by the sol-gel method with hybrid silica sol and jelly-like resorcinol formaldehyde resin Ref. [6]. The surface roughness of the silica nanotree film is about 20 nm, and it is transparent and superhydrophobic with a water contact angle higher than 150°. Single crystal silicon was found to be self-organized into ammonium silicon hexafluoride {$(NH_4)_2SiF_6$}-based complex structures under exposure to vapors of an HF-rich $HF:HNO_3$ mixture[7] forming structures like pillars or micro- and nanotrees perpendicular to the wafer plane. It was shown that the reaction of HNO_3 with $(NH_4)_2SiF_6$ and the H_2O vapor plays an important role in the formation of these features.

Beautiful porous carbon tree-like nanostructures (Figure 7.2) were obtained[8] (as well as SiC) by the electron-beam (E-beam)-induced deposition method using a transmission electron microscopy (TEM) in poor vacuum conditions where hydrocarbons (being attracted to the substrates by the local electric fields) were present in the chamber. The authors found that the adequate ion dose to create well-defined saw-tooth nanopatterns was between 8 and 10 nC/mm². A high concentration of sp^2 sites in nanostructures was revealed, suggesting that they were made of graphite-like hydrogenated amorphous carbon. Y-junction hollow carbon nanotrees (with close-ended branches but open-ended roots, length ~5 μm, an average external trunk diameter of ~500 nm, branch diameters ranging from 70 to 100 nm, and wall thickness ranging from 24 to 30 nm) were obtained in a 90% yield by the reaction of ferrocene with 1,2-dichlorobenzene at 180°C.[9] A simple and effective method based on Coulomb effect for separating SWNTs from their bundles was developed.[10] The separated SWNTs were radially distributed at the end of the original bundle and finally formed a nanotree (Figure 7.3). The formation mechanism was proposed to be Coulomb explosion. As an important application of carbon nanotrees, the fuel cell, equipped with an anode or a cathode containing carbon-metal nanotree catalysts consisting of CNT trunks with a plurality of highly oriented

FIGURE 7.1 Silicon oxycarbide tree-like nanostructures. (Cho, S.O. et al.: Controlled synthesis of abundantly branched, hierarchical nanotrees by electron irradiation of polymers. *J. Adv. Mater.* 2006. 18(1). 60–65. Copyright Wiley-VCH Verlag GmbH & Co. KGaA. Reproduced with permission.)

FIGURE 7.2 Carbon tree-like nanostructures, obtained in a porous silicon sample. (Reproduced from *Micron*, 40, Sola, F. et al., Growth and characterization of branched carbon nanostructures arrays in nano-patterned surfaces from porous silicon substrates, 80–84, Copyright 2009. Reproduced with permission from Elsevier; http://www.ifn.upr.edu/people/17-luis-fonseca, accessed January 15, 2011.)

carbon nanotube (CNT) branches, where a metal is attached on inner or outer surfaces of the tubes,[11] provided high power generation efficiency by accelerating catalytic functions.

Tree-like nanostructures have also been obtained for several transition and noble elemental metals as well as their oxides and salts. Thus, a metallic tungsten nanotree was formed on an Al_2O_3 substrate by using $W(CO)_6$ gas by E-beam technique.[14] Platinum metal nanotrees were prepared by assembling Pt nanoparticles onto a DNA network template and reducing metal ions by photoinduced method.[15] The concentration of DNA network and the reaction time played important roles in forming platinum nanotrees. An environmentally friendly one-step method (H_2O as a benign solvent and vitamin B1 as a reducing agent) to synthesize Pd nanobelts, nanoplates, and nanotrees using vitamin B1 without using any special capping agents at room temperature (r.t.) was elaborated,[16] and it can be extended to prepare other noble nanomaterials such as Au and Pt. The Pd nanoparticles showed excellent

FIGURE 7.3   Typical nanotrees of SWNTs formed at the breaking points of SWNT bundle. (Reproduced with permission from Liu, G. et al., Coulomb explosion: A novel approach to separate single-walled carbon nanotubes from their bundle, *Nano Lett.*, 9(1), 239–244. Copyright 2009 American Chemical Society.)

catalytic activity for several C–C bond-forming reactions such as Suzuki, Heck, and Sonogashira reactions under microwave irradiation conditions. Palladium nanotrees can also be produced electrochemically; thus, uniform nanostructured Pd (standing Pd nanoplates and nanotrees [Figure 7.4]) was electrodeposited directly on an Au substrate through a facile template-free approach, which involved a cyclic electrochemical deposition or dissolution of Pd.[17] Improved and superior electrocatalytic activities toward the oxidation of EtOH were observed from these electrodes coated with nanostructured Pd. A simple route for the synthesis of such highly catalytic Pd nanostructures was proposed.

An interesting example of the use of TEM grids for nanostructure preparation was reported. Thus, the copper–carbon substrate of a TEM grid reacted with an aqueous silver nitrate solution within minutes to yield spectacular tree-like silver dendrites, without using any added capping or reducing reagents.[18] The authors noted that these results demonstrated a facile, aqueous, room-temperature synthesis of a range of noble metal nano- and mesostructures that have widespread technological potential in the design and development of next-generation fuel cells, catalysts, and

FIGURE 7.4   Pd nanotrees (about 1 μm high and 100 nm thick). (Reproduced with permission from Jia, F. et al., Electrochemical synthesis of nanostructured palladium of different morphology directly on gold substrate through a cyclic deposition/dissolution route, *J. Phys. Chem. C*, 113(17), 7200–7206. Copyright 2009 American Chemical Society.)

antimicrobial coatings. In addition, a tree-like nanostructured Ag crystal, with stems, branches, and leaves, was synthesized by preforming Au seeds, soaking, and annealing, based on monolithic mesoporous silica.[19] The obtained Ag nanotrees were of single-crystal nature and statistically symmetric in geometry. A novel lab-on-a-chip nanotree enzyme reactor, constructed from gold-tipped branched nanorod structures grown on SiN_x-covered wafers, was demonstrated for the detection of acetylcholine.[20] Significantly higher enzymic activity was found for the nanotree reactors than for the nanorod reactors, most likely due to the increased gold surface area and thereby higher enzyme binding capacity. The morphology of Au nanotrees was discussed in Ref. [21]. Unusual biaxial textures in Cu nanorod films, branching into nanotrees, were observed.[22]

The first successful attempt (one-pot synthesis approach involving the reaction of a titanium surface with the vapor generated from a hydrochloric acid solution in a Teflon-lined autoclave) to grow ordered tree-like titania TiO_2 nanoarrays was reported.[23] By adjusting the initial HCl concentration, films of different rutile structures including nanotrees (Figure 7.5), dendritic nanobundles, and nanorods can be selectively obtained. A possible formation mechanism (Figure 7.6) for the interesting architectures was proposed based on series of time-dependent experiments. Thin films of aligned WO_3 nanotrees were synthesized using a simple hydrothermal reaction again starting from elemental metal (in this case, tungsten).[24] These WO_3 nanotrees, composed of trunks and branches, were single crystals oriented in the (001) direction. The thin films exhibited efficient electrochromism due to their large tunnels in the crystal and nanochannels between the nanotrees.[25]

FIGURE 7.5 Branched nanotrees grown from and rooted on titanium sputtered substrate: (a, b) top and side view of a nanotree "forest"; (c, d) a typical individual nanotree. (Reproduced with permission from Yang, X. et al., Hierarchically nanostructured rutile arrays: Acid vapor oxidation growth and tunable morphologies, *ACS Nano*, 3(5), 1212–1218. Copyright 2009 American Chemical Society.)

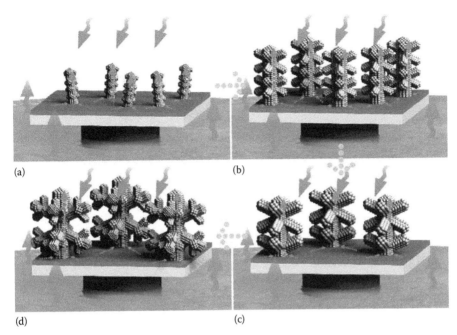

(a)   (b)

(d)   (c)

FIGURE 7.6 Schematic illustration of the growth process: (a) no densely distributed rod-like trunks grow from titanium, and the side surfaces are eroded; (b) obvious branches appear and the trunk tends to be more orientated due to space limitation; (c) both branches and trunks tend to be larger in dimension because of Ostwald ripening; (d) ternary growth of the branches. (Reproduced with permission from Yang, X. et al., Hierarchically nanostructured rutile arrays: Acid vapor oxidation growth and tunable morphologies, *ACS Nano*, 3(5), 1212–1218. Copyright 2009 American Chemical Society.)

FIGURE 7.7 Highly branched Fe_3O_4 nanostructures. (Reproduced with permission from Chu, Y. et al., Growth and characterization of highly branched nanostructures of magnetic nanoparticles, *J. Phys. Chem. B*, 110(7), 3135–3139. Copyright 2006 American Chemical Society.)

Magnetite nanoparticles of Fe_3O_4 were obtained by Reaction 7.1 and grew into large highly branched nanostructures including nanochains and highly branched nanotrees (Figure 7.7) in the solid state through a postannealing process.[26] The driving force for the formation of the highly ordered nanostructures includes interaction between the nanoparticles and interaction through surface-capping molecules.

$$Fe(acac)_3 \; + \; \text{(vinylpyrrolidone)} \quad \xrightarrow{\text{Refluxing}} \quad \text{Fe}_3\text{O}_4 \text{ complex} \qquad\qquad (7.1)$$

Gallium oxide (Ga_2O_3) nanotrees, as well as nanobelts and nanotubes (depending on the temperature zone), were obtained on Si(111) substrates by thermal evaporation of gallium in the ammonia atmosphere.[27] Photoluminescence spectrum under excitation at 396 nm showed a green emission, which is probably attributed to vacancies in Ga_2O_3. ZnO nanotrees (ZnO is one of the most important wide band gap semiconductors with a wide and direct band gap of 3.37 eV and high binding exciton energy of 60 meV[28]) were obtained by different methods; thus, complex ZnO nanotrees were observed, among nanosaws, nanobelts, and nanowires, in the products of the spray-pyrolysis-assisted thermal evaporation method of ZnO.[29] The unique nanotree ZnO arrays (Figure 7.8), obtained in large scales at low cost by direct hydrothermal oxidation of zinc metal to complex ZnO nano-/ microcrystallite forest, exhibited high photocatalytic activity; this fact may help develop a new suite of branched nanostructures for wurtzite semiconductors in general for a wide variety of important applications.[30] Microwave absorption properties of ZnO nanotrees were studied.[31] Among other oxides, obtained in nanotree forms, we note Al_2O_3[32,33] and MgO.[34]

FIGURE 7.8 ZnO nanopine tree. (Reproduced with permission from Zhao, F. et al., ZnO pine-nanotree arrays grown from facile metal chemical corrosion and oxidation, *Chem. Mater.*, 20(4), 1197–1199. Copyright 2008 American Chemical Society.)

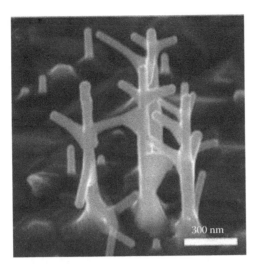

FIGURE 7.9 Four interconnected InAs nanotrees, viewed at an angle of 45° to the surface normal. (Reproduced with permission from Dick, K.A. et al., Position-controlled interconnected InAs nanowire networks, *Nano Lett.*, 6(12), 2842–2847. Copyright 2006 American Chemical Society.)

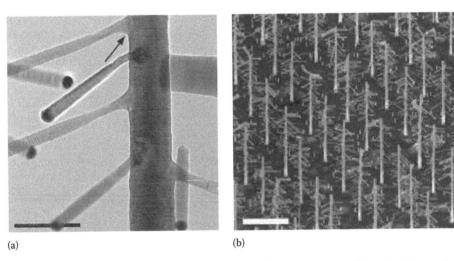

(a)   (b)

FIGURE 7.10 (a) TEM shows a tree trunk of gallium phosphide grown on a gold seed, with branches grown from smaller gold particles; (b) scanning electron microscopy (SEM) shows an ordered array of nanotrees viewed at 45° from the normal. (Reproduced with permission from Lerner, E.J., Growing nanotrees, *Ind. Phys.*, 10(5), 16–19. http://www.aip.org/tip/INPHFA/vol-10/iss-5/p16.pdf. Copyright 2004 American Institute of Physics.)

Metal salts, synthesized as tree-like structures, belong mainly to sulfides and selenides (CdS and CdSe[35]), phosphides (GaP,[36] GaP/InP[37]), and arsenides (InAs) (Figure 7.9), that is, having important applications as semiconductors. Thus, the growth of GaP nanotrees (Figure 7.10) took place when Au aerosol particles were deposited on a substrate with the aid of an electrostatic field, acting as seeds in the tree growth process.[38] On each Au seed with a size of 40–70 nm, a tree trunk of GaP was then grown by a technique called vapor-liquid-solid (VLS) growth. The nanotree arrays may be applied as light emitters or photovoltaic converters. When the Au-In particles were used to seed a second generation of InAs nanowires, producing nanotrees, the branches exhibited a—two to three times higher growth rate and more regular shape than those seeded by pure Au particles.[39,40] This result was attributed by the authors to the decreased interaction between the seed particle and the trunk nanowires when Au-In particles were used. Nanoscale tree-like $FeCl_3$-S structures were

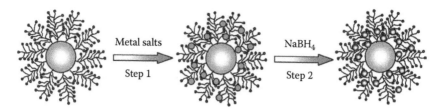

FIGURE 7.11   Preparation of PS–PEGMA–Ag composite particles: Ag ions are adsorbed by the brush particles in aqueous phase. Subsequent reduction leads to well-defined Ag nanoparticles immobilized in the dense surface layer of the particles. (Reproduced from *Polymer*, 47, Lu, Y. et al., 'Nano-tree'—type spherical polymer brush particles as templates for metallic nanoparticles, 4985–4995, Copyright 2006, with permission from Elsevier.)

assembled[41] and their unique characteristics, based on the high reactivity of $FeCl_3$ moiety and the branched thread morphology of sulfur moiety, endowed them with multiple excellent functions, particularly acting as reactive seeds or templates to directly induce organic polymerization reaction and guide the growth of polycyclopentadiene nanofibers.

A few organic nanotrees are known. For example, spherical polymer brush particles with a nanotree-type morphology, which consisted of a solid poly(styrene) core (diameter ~100 nm) onto which chains of a bottlebrush polymer were densely grafted, were prepared in aqueous dispersion by photo emulsion polymerization using the macromonomer poly(ethylene glycol) methacrylate (PEGMA) (Figure 7.11).[42] These particles can be used as nanoreactors for the generation and immobilization of well-defined silver nanoparticles exhibiting an excellent colloidal stability; their catalytic activity was investigated by monitoring the reduction of 4-nitrophenol by $NaBH_4$ in the presence of these silver nanocomposite particles.

7.1.2   NANOPINES

An array of Au nanopine trees was obtained by E-beam evaporation of a few nanometers of Au on bare alumina templates with pores (50 nm wide and 500 nm long).[43] This synthetic approach for producing nanoscale field emitters could lead to a versatile and inexpensive technology for synthesizing flexible arrays of nanoscale cold cathode emitters. Fe_3N nanodendrites were synthesized[44] directly by reduction nitriding of α-Fe_2O_3 nanopine dendrites (obtained from $K_3[Fe(CN)_6]$ by the hydrothermal method[45]) in a mixed stream of H_2-NH_3. Fe_3N basically retained the dendritic morphology of the starting material α-Fe_2O_3 (Figure 7.12).

In a related investigation, carried out starting from similar precursors, but by a different method, the nanostructured metal (Fe, Co, Mn, Cr, Mo) oxides were fabricated under microwave irradiation conditions in pure water from inexpensive compounds [$K_4Fe(CN)_6$, $K_3Co(CN)_6$, $K_3Mn(CN)_6$, $K_3Mo(CN)_8$, and $K_3Cr(CN)_6$] without using any reducing or capping reagent. The metal oxides self-assembled into pine morphology (Figure 7.13), among many others (i.e., octahedra, spheres, triangular rods, and hexagonal snowflake-like 3D morphologies).[46] After being functionalized and coated with Pd metal (Figure 7.14), pine-structured nanoiron oxides were studied as a support for various catalytic organic transformations, catalyzing various C–C coupling and hydrogenation reactions with high yields.

Wurtzite CdS and CdSe nanostructures with complex morphologies including fractal nanotrees (which are a buildup of nanopines) can be produced *via* a solvothermal approach in a mixed solution made of diethylenetriamine (DETA) and deionized water (DIW).[47] The produced CdS nanotrees (Figure 7.15) were found to be made up of many branched nanopines with different growth directions and displayed high hyperbranched nanostructures. In a related report,[48] the controllable synthesis of hexagonal pine-like Zn-doped CdSe ($Cd_{1-x}Zn_xSe$, $x=0.1$–0.3) nanotrees from the

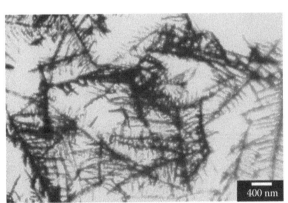

FIGURE 7.12 α-Fe_2O_3 (precursor) and Fe_3N (final product) pine-like dendrites. (Reproduced from *J. Solid State Chem.*, 178, Cao, M. et al., Magnetic iron nitride nanodendrites, 2390–2393, Copyright 2005, with permission from Elsevier.)

self-prepared mixed-metal precursors was achieved by two facile steps: first, mixed-metal precursors ($x = 0.1$–0.3) were prepared directly through precipitation reactions of stoichiometric cadmium acetate, zinc acetate, and sodium selenite in distilled water under ambient condition; second, pure hexagonal phase pine-like $Cd_{1-x}Zn_xSe$ ($x = 0.1$–0.3) nanotrees with different ratios were produced *via* a solvothermal treatment of the precursor compounds in hydrazine hydrate at 180°C for 12 h.

The nanostructures of PbS (Figure 7.16) were synthesized *via* CVD with $PbCl_2$ and elemental sulfur[49] or H_2S[50] as precursors under argon flow with a co-flow of H_2 at atmospheric pressure and with temperatures between 600°C and 650°C (Reactions 7.2 and 7.3; the Reaction 7.4 is thermodynamically unfavorable). The effects of various growth parameters, such as hydrogen

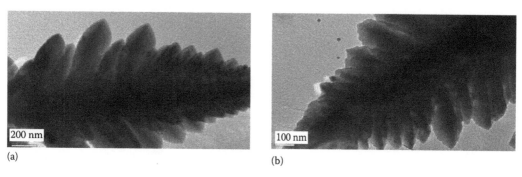

(a)   (b)

FIGURE 7.13 α-Fe$_2$O$_3$ nanopine. (Reproduced with permission from Polshettiwar, V. et al., Self-assembly of metal oxides into 3D nanostructures: Synthesis and application in catalysis, *ACS Nano*, 3(3), 728–736. Copyright 2009 American Chemical Society.)

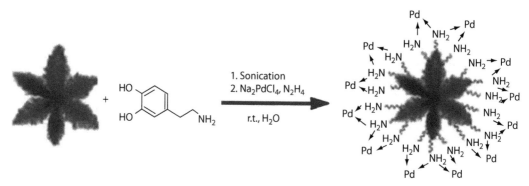

FIGURE 7.14 Synthesis of functionalized ferrites with Pd coating. (Reproduced with permission from Polshettiwar, V. et al., Self-assembly of metal oxides into 3D nanostructures: Synthesis and application in catalysis, *ACS Nano*, 3(3), 728–736. Copyright 2009 American Chemical Society.)

FIGURE 7.15 CdS microtrees. (Reproduced with permission from Yao, W.-T. et al., Nanotrees *via* a solvothermal approach in a mixed solution and their photocatalytic property, *J. Phys. Chem. B*, 110(24), 11704–11710. Copyright 2006 American Chemical Society.)

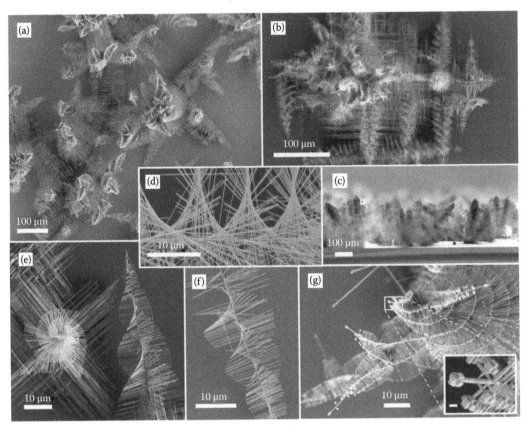

FIGURE 7.16 SEM micrographs of PbS pine tree nanowires. (a) Overview of a dense forest of many nanowire trees. (b) Tree clusters showing epitaxial growth along <100> directions. (c) Side view of growth substrate showing forest growth. (d through f) High-magnification views of trees highlighting the twisting (Eshelby twist) of the central trunk and helical rotating branches, with (e) further illustration of branch epitaxy on the tree trunk and (f) showing a tree with fewer branches. (g) An example of "tree-on-tree" morphology that can be occasionally observed. (Inset) A magnified view of the tips of nanowires after synthesis, highlighting the cubes that sometimes decorate the tips. The inset scale bar is 200 nm. The images are false colored. (Reproduced from Bierman, M. J. et al., *Science*, 320(5879), 1060–1063. Copyright 2008. With permission.)

flow, temperature, pressure, reaction time (Figure 7.17), and the growth substrates employed were studied. An illustration of the growth process of a PbS nanowire pine tree nanostructure is found in Figure 7.18.

$$H_2(g) + S(s) \rightarrow H_2S(g) \tag{7.2}$$

$$PbCl_2(s) + H_2S(g) \rightarrow PbS(s) + 2HCl(g) \tag{7.3}$$

$$PbCl_2(s) + 1/2S_2(g) \rightarrow PbS(s) + Cl_2(g) \quad \text{(thermodynamically unfavorable)} \tag{7.4}$$

7.1.3 Nanopalms

Nanopalm-like trees are lesser represented. Silica nanowires (diameters in the range of 15–49 nm and lengths up to 500 nm), obtained by E-beam irradiation on porous silicon films, were exposed to poor vacuum conditions in which carbon aggregation from the surrounding gas was promoted

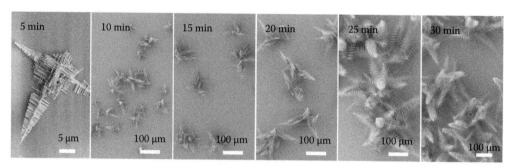

FIGURE 7.17 Progression of PbS nanowire pine trees over reaction time. (Reproduced with permission from Albert Lau, Y.K. et al., Formation of PbS nanowire pine trees driven by screw dislocations, *J. Am. Chem. Soc.*, 131(45), 16461–16471. Copyright 2009 American Chemical Society.)

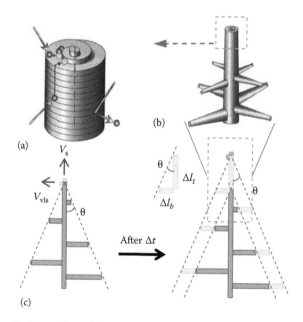

FIGURE 7.18 Schematic illustration of the growth process of a PbS nanowire pine tree nanostructure. (a) Anisotropic nanowire growth driven by screw dislocation growth spiral at the magnified trunk tip. (b) 3D representation of the tip region of a tree highlighting that the fast-growing trunk is driven by a screw dislocation, while the branches are driven by vapor-liquid-solid (VLS) mechanism. (c) The tree evolves progressively, leading to the formation of a cone angle (θ) of the outer envelope that represents the relative growth rate (trunk versus branch, or V_s/V_{vls}) cot θ. (Reproduced with permission from Albert Lau, Y.K. et al., Formation of PbS nanowire pine trees driven by screw dislocations, *J. Am. Chem. Soc.*, 131(45), 16461–16471. Copyright 2009 American Chemical Society.)

by the local electric fields enhanced at the tip of the silica wires, leading to heterostructures showing a nanopalm-like shape.[51,52] ZnO nanopalms were obtained,[53] among other CuO and ZnO nanostructures, by directly heating a CuZn alloy (brass) on a hotplate in ambient conditions. An individual ZnO nanopalm (Figure 7.19) had a width of about 350 nm at the root part and separated into several nanowires with tip diameters of about 20 nm on the upper region. Figure 7.20 briefly generalizes the morphological dependence of products on the Zn concentrations in the brasses and growth temperatures.

FIGURE 7.19 ZnO nanopalm. (Zhu, Y. et al.: Co-synthesis of ZnO–CuO nanostructures by directly heating brass in air. *Adv. Funct. Mater.* 2006. 16. 2415–2422. Copyright Wiley-VCH Verlag GmbH & Co. KGaA. Reproduced with permission.)

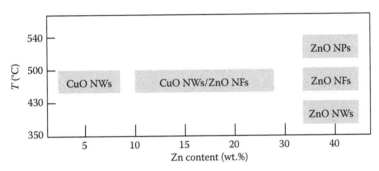

FIGURE 7.20 Morphological dependence of products on Zn contents in brass and growth temperatures. The following dominant morphologies are shown: CuO nanowires (CuO NWs), ZnO nanowires (ZnO NWs), ZnO nanoflakes (ZnO NFs), and ZnO nanopalms (ZnO NPs). (Zhu, Y. et al.: Co-synthesis of ZnO–CuO nanostructures by directly heating brass in air. *Adv. Funct. Mater.* 2006. 16. 2415–2422. Copyright Wiley-VCH Verlag GmbH & Co. KGaA. Reproduced with permission.)

7.2 NANOLEAVES

Leaf-like nanostructures are known mainly for metal oxides, although carbon, some metals, intermetallic compounds, and metal salts have been also reported. Thus, immobilization of Ag nanoleaves on γ-mercaptopropyltrimethoxysilane-modified chemical vapor deposited diamond film was carried out.[54] Ferromagnetic MnSb films, containing nanoleaves that had a width and thickness of ~100 and 20 nm, respectively, were synthesized on Si wafers by physical vapor deposition.[55] 2D and 3D carbon nanoflowers, prepared on silicon (111) substrates by plasma-enhanced CVD, using CH_4, H_2 and Ar as reactive gases in the presence of Fe catalyst, were formed by various nanoleaves (2D, related closely to the flux ratio of gas and the reaction pressure) or hundreds of nanofibers (3D, depended mainly on the growth temperature).[56]

Cu(OH)$_2$ nanoleaves were converted into to 2D CuO nanoleaves (Figure 7.21) on a large scale by a reconstructive transformation, which consisted of the nucleation of CuO followed by a two-step oriented attachment of the CuO particles to 1D CuO nanoribbons and then to 2D CuO

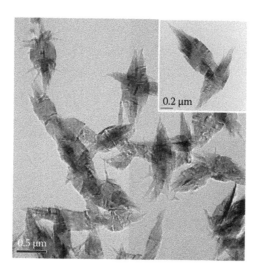

FIGURE 7.21   CuO nanoleaves; the upper inset is two single CuO nanoleaves. (Reproduced with permission from Xu, H. et al., Hierarchical-oriented attachment: From 1D Cu(OH)$_2$ nanowires to 2D CuO nanoleaves, *Cryst. Growth Des.*, 7(12), 2720–2724. Copyright 2007 American Chemical Society.)

nanoleaves, in which a hierarchical-oriented attachment is involved.[57] In a related publication,[58] single-crystalline CuO nanoleaves with an average thickness of ~10 nm and with lateral sizes of hundreds of nanometers to several micrometers were synthesized by microwave heating of an aqueous solution containing copper salt and sodium hydroxide. The band gap of CuO nanoleaves was estimated to be ~2.13 eV, which showed significant blue shift compared with that of the bulk. ZnO nanoleaves were synthesized by using a CVD method[59] and also by pulsed laser ablation (Nd:YAG 532 nm),[60] in the last case at room temperature and pressure using a Zn target in an aqueous solution of Na dodecyl sulfate. The growth mechanism appeared to involve an increase in the structural complexity from 0D nanoparticles to 1D nanorods and then broadening of these into 2D nanoleaf structures. Alternatively, using a mixture of ZnO and Te powders as the source material, ZnO nanoleaves, constructed with a nanowire and a nanodisk on one side of the nanowire near the top, with high yield and uniform morphology were fabricated by thermal evaporation.[61] These ZnO nanoleaves could be used in nanolasers, sensors, and photoelectronic nanodevices. SnO$_2$ nanoleaves and nanopencils were synthesized[62] on single silicon substrates using Au-Ag alloying catalyst-assisted carbothermal evaporation. In comparison with SnO$_2$ nanopencils, a new peak at 456 nm in the measured PL spectra of SnO$_2$ nanoleaves was observed, implying that more luminescence centers existed in these nanostructures. A freezing ferromagnetic moment model was proposed to explain the exchange bias discovered in α-Fe$_2$O$_3$ antiferromagnetic nanoleaves,[63] synthesized by oxygenating pure iron.[64] The small exchange bias may possibly originate from a different magnetic order on the surface of the nanoleaves or the coexistence of a tiny amount of Fe$_3$O$_4$. Single-crystalline TiO$_2$ nanoleaves were functionalized with catechol-group-terminated Zn(II)-porphyrin, coordinated with ZnP with trans-2,2′-ethylene 4,4′-bipyridyl, showing the self-assembly (Figures 7.22 and 7.23) of TiO$_2$ nanoleaves based on a supramolecular interaction and formation of a facet-selective, self-assembled, 2D stacking structure.[65]

Uniform single-crystal boehmite leaf-like nanosheets with high anisotropy (with a lateral size of 4.5 μm × 9.0 μm and a thickness of 60–90 nm) and flower-like superstructures consisting of single-crystal petals were prepared by a simple hydrothermal method and transformed by calcination into single-crystal gamma-alumina nanostructures while keeping their morphology.[66] Highly crystalline CePO$_4$, CePO$_4$:Tb^{3+}, and CePO$_4$:Dy^{3+} nanoleaves with monoclinic structure and dispersible in solvents such as water and methanol were prepared[67] by a relatively low temperature (at 140°C) synthesis in an ethylene glycol medium from Ce$_2$(CO$_3$)$_3$ · 5H$_2$O, Tb$_4$O$_7$, and Dy$_2$O$_3$ as

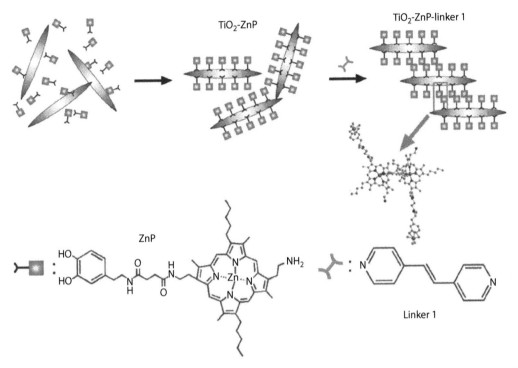

FIGURE 7.22 Illustration of coordination-bond-assisted self-assembly of TiO$_2$ nanoleaves. (Reproduced with permission from Yang, C. et al., Facet-selective 2D self-assembly of TiO$_2$ nanoleaves *via* supramolecular interactions, *Chem. Mater.*, 20(24), 7514–7520. Copyright 2008 American Chemical Society.)

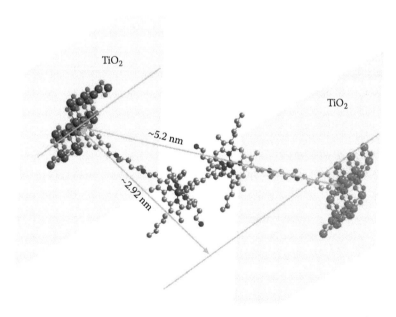

FIGURE 7.23 Molecular modeling of ZnP and *trans*-2,2'-ethylene 4,4'-bipyridyl on the surface of TiO$_2$ nanoleaves. (Reproduced with permission from Yang, C. et al., Facet-selective 2D self-assembly of TiO$_2$ nanoleaves *via* supramolecular interactions, *Chem. Mater.*, 20(24), 7514–7520. Copyright 2008 American Chemical Society.)

(a)   (b)

(c)

FIGURE 7.24   (a) Low-magnification FESEM image of the GaN leaves; the scale bar is 20 μm. (b) High-magnification FESEM image of the GaN leaves. (c) TEM image of a GaN nanoleaf. (Inset) HRTEM image of the catalyst tip; the scale bar is 5 nm. (Reproduced from Li, J. et al., Physical and electrical properties of chemical vapor grown GaN nano-/microstructures, *Inorg. Chem.*, 47(22), 10325–10329. Copyright 2008 American Chemical Society.)

precursors and incorporated into silica sols by the sol-gel method, exhibiting improved luminescence properties compared with silica sols directly doped with lanthanide ions. These systems can be useful for developing biosensors for studying enzyme and protein activities based on luminescent silica sols. Industrially important gallium arsenide and phosphide were also obtained as nanoleaves and characterized. Thus, wurtzitic Ga nitride nano- and microleaves (Figure 7.24) were obtained grown through a CVD method from β-Ga$_2$O$_3$ as precursor according to the reactions 7.5 and 7.6.[68] Field effect transistors based on individual GaN nanoleaves were fabricated, and the electric transport results revealed a pronounced n-type gating effect of the GaN nanostructures. GaAs nanowires and nanoleaves were grown by molecular beam epitaxy using Mn as a growth catalyst.[69] Bi$_2$S$_3$ nanoleaves (Figure 7.25), together with other morphologies, including nanorods, dandelion-like nanostructures, nanoflowers, and nanocabbages, were synthesized from the single-source precursor Bi(SCOPh)$_3$ or multiple-source precursors by using a colloidal solution method or hydrothermal method.[70] The nanoleaves had an average diameter of 89 ± 11 nm and a length of 471 ± 52 nm and resembled some natural long leaves with a wider middle part and two tapering points.

$$Ga_2O_3(s) + 4Ga(l) \rightarrow 3Ga_2O(g) \qquad\qquad (7.5)$$

$$Ga_2O(g) + 2NH_3(g) \rightarrow 2GaN(s) + H_2O(g) + 2H_2(g) \qquad\qquad (7.6)$$

(a)   (b)

FIGURE 7.25 SEM images of Bi_2S_3 nanoleaves. (Reproduced with permission from Lu, T. et al., Morphology-Controlled synthesis of Bi_2S_3 nanomaterials *via* single- and multiple-source approaches, *Cryst. Growth Des.*, 8(2), 734–738. Copyright 2008 American Chemical Society.)

7.3 NANOFORESTS

Nanoforests (Figures 7.5a, 7.10b, and 7.26), as a conglomerate of nanotrees and nanobushes and other "nanovegetation" described here, as well as simple nanorods and nanowires, have been reported mainly for CNTs.[71] A key growth precursor for SWCNTs is acetylene,[72] although ethylene mixtures with H_2[73] (the equipment is shown in Figure 7.27) or $C_2H_4/H_2/H_2O/Ar$[74] were used, mainly by CVD technique. Thus, a CNT nanoforest was prepared[75] by biased thermal CVD on Co-containing amorphous C composite films. These nanoforest multiwalled CNTs (MWCNTs) had thin diameters between 10 and 20 nm. Vertically aligned small diameter (single- and few-walled) CNT forests were also grown by thermal CVD over the temperature range 560°C–800°C and 10^{-5} to 14 mbar partial pressure range, using acetylene as the feedstock and Al_2O_3-supported Fe nanoparticles as the catalyst.[76] The mechanism of their formation (Figure 7.28) is described by Reactions 5.7 through 5.11; an alternative mechanism is shown in Figure 7.29.[77] Effects of aluminum buffer layer on the synthesis of highly vertically aligned CNT forests were systematically studied.[78] These CNT forests were

FIGURE 7.26 Nanoforest (carbon nanotubes). (Reproduced from Zhou, J. N., Multi-scale flow interaction with nanofabric structures, PhD thesis (partial fulfillment), California Institute of Technology, Pasadena, CA, http://etd.caltech.edu/etd/available/etd-04202005-172426/unrestricted/Jijie_ZHOU_dissertation.pdf. Copyright 2005. With permission of the *Caltech Library*.)

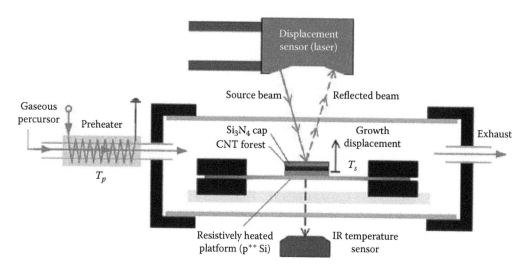

FIGURE 7.27 Obtaining a CNT forest by the CVD method. (Reproduced with permission from Meshot, E.R. et al., Engineering vertically aligned carbon nanotube growth by decoupled thermal treatment of precursor and catalyst, *ACS Nano*, 3(9), 2477–2486. Copyright 2009 American Chemical Society.)

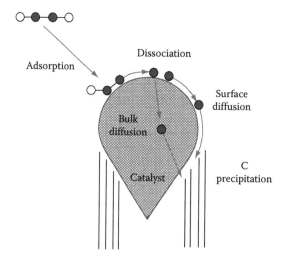

FIGURE 7.28 Schematic of CNT growth process. (Reproduced with permission from Wirth, C.T. et al., Diffusion- and reaction-limited growth of carbon nanotube forests, *ACS Nano*, 3(11), 3560–3566. Copyright 2009 American Chemical Society.)

synthesized using the CVD system on an iron nanoparticle catalyzed substrate supported by a thin aluminum buffer layer with an Fe/Al thickness of 0.5 and 40 nm, respectively. Yarned CNT fibers were spun from the grown CNT forests for applications to reinforced materials.

$$C_2H_{2\,gas} \leftrightarrow C_2H_{2\,ads} \tag{7.7}$$

$$C_2H_{2\,ads} \leftrightarrow 2CH_{ads} \tag{7.8}$$

$$CH_{ads} \leftrightarrow C_{ads} + H_{ads} \tag{7.9}$$

$$2H_{ads} \rightarrow H_{2\,gas} \tag{7.10}$$

$$C_{ads} \rightarrow C_{CNT} \tag{7.11}$$

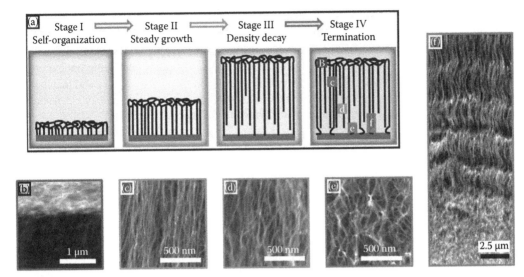

FIGURE 7.29 Collective growth mechanism of a CNT forest: (a) growth stages and SEM images of (b) tangled crust at the top of forest, (c) aligned and dense morphology near the top of a self-terminated forest, (d) less aligned and less dense morphology in the lower region of a self-terminated forest, and (e, f) randomly oriented morphology at the bottom, induced by the loss of the self-supporting forest structure. (Reproduced with permission from Bedewy, M. et al., Collective mechanism for the evolution and self-termination of vertically aligned carbon nanotube growth, *J. Phys. Chem. C*, 113(48), 20576–20582. Copyright 2009 American Chemical Society.)

Introduction of CO_2 was found to be a facile way to tune the growth of vertically aligned double- or single-walled CNT forests on wafers.[80] In the absence of CO_2, a double-walled CNT convexity was obtained; with increasing concentration of CO_2, the morphologies of the forests transformed first into radial blocks and finally into bowl-shaped forests. Nanomanipulation of bismuth nanowires, CNTs, and organic nanofibers was demonstrated[81] on CNT forests using a sharp tungsten tip. The remarkable suitability of CNT forests as nonstick surfaces for nanomanipulation was found to be indeed the strong contact-line pinning in combination with the nanostructured surface, which allowed homogeneous dispersion and easy manipulation of individual particles. Other information on CNT forests is discussed in Refs. [82–85].

In addition to CNTs, nanodiamonds were also grown as nanoforests. Thus, a boron-doped diamond nanorod forest (BDDNF) electrode was fabricated by the hot filament CVD (HFCVD) method (Figure 7.30).[86] This BDDNF electrode, with the possibility of fabricating a sensitive biosensor for glucose without any catalyst or mediators, showed good activity in direct detection of glucose by simply putting the bare BDDNF electrode into the glucose solution. Among metallic nanoforests, noble metals have been obtained in this form. Thus, studies of ultralow-loss optical diamagnetism in silver nanoforests (a composite metamaterial in which silver nanowires are aligned inside a finite-thickness dielectric host medium) revealed a frequency region where the nanoforest exhibited a strong diamagnetic response while simultaneously allowing for high transmission of

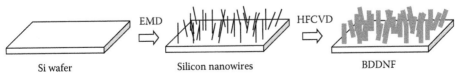

FIGURE 7.30 Plots of fabrication of the BDDNF. (Reproduced with permission from Luo, D. et al., Fabrication of boron-doped diamond nanorod forest electrodes and their application in nonenzymatic amperometric glucose biosensing, *ACS Nano*, 3(8), 2121–2128. Copyright 2009 American Chemical Society.)

incident electromagnetic waves.[87] The optical reflectivity of polished silicon covered by a forest of vertically standing gold nanowires (grown inside the pores of an anodic alumina matrix) was found to be considerably smaller than that of bare polished silicon and strongly influenced by scattering of light by the nanowires.[88] The formed gold nanowire or alumina film has interesting optical properties, such as transparency in the mid-IR and high absorbance in the visible regions and may be used as a low-reflectivity coating and a low-pass filter. Au nanowire forests, fabricated by electrodeposition into porous Al_2O_3 templates,[89] had a mean diameter 30, 60, and 70 nm.

A few metal oxides in nanoforest form are known. Thus, ZnO nanorod forests were grown by wrapping nylon fibers using a two-step process:[90] the formation of ZnO seeds at nylon fiber surfaces, induced by the dip coating of ZnO nanosols and the growth of the ZnO seeds into nanorod forests, carried out *via* a wet chemical route from an equimolar solution of zinc nitrate hexahydrate and hexamethylenetetramine (HMT). The TiO_2 nanotubes were found to grow in HCl dissolved in ethylene glycol and 2-propanol by anodization of Ti with locally rapid breakdown of the passive TiO_2 film forming a forest of nanotube-bearing microtowers with the background of the passive TiO_2 film.[91] These bundles of assembled groups of titania nanotubes look like pillar corals. The author stated that the low relative permittivity of 2-propanol led to lowering of the dissociation of HCl and hence lowering the activity of H^+ and Cl^- ions, which in turn led to the suppression of dissolution of titania and an increase in the growth rate of the titania nanotubes. Organic nanoforest structures are poorly represented; among them we note those of peptide nanotubes.[92]

At present, applications for CNTs and other nanoforests are intensively being searched, in particular as biosensors.[93] Especially, electrochemiluminescent immunosensors combining single-wall carbon nanotube forests with $[Ru(bpy)_3]^{2+}$-silica-secondary antibody nanoparticles for sensitive detection of cancer biomarker prostate-specific antigen were described.[94] CNT forests offer attractive anisotropic mechanical, thermal, optical, and electric properties, and their anisotropic structure is enabled by the self-organization of a large number of CNTs.[95] This process is governed by individual CNT diameter, spacing, and the CNT-to-CNT interaction. Among other uses, optically enhanced nanoforest cathodes are electron sources developed for high-power devices,[96] whose operation principle was based on coherent light coupling photon energy into resonantly tuned, nanometer sized antennae, able to generate large current densities with little applied energy.

7.4 NANOBUSHES

Nanobush structures are still rare; a few reports used ZnO as a model compound. Thus, divergent micro-/nanorod assemblies such as hemispherical dandelion, rice plant type bush of ZnO were obtained in an ethylenediamine-water solvent in solvothermal conditions.[97] Increase in the percentage of ethylenediamine resulted in the formation of smaller assemblies of relatively thin nanorods. The authors noted that the possibility of formation of architectural assemblies (paintbrush-like nano-/micro rod assembly, double-sided brush, windmill type, nanobush, or nanorods) depended on the pH and solvent mixture used. ZnO nanoneedles can also assemble into bush-like nanostructures.[98] For the same compound, nanoflowers with two typical morphologies—plate-like and bush-like—were described.[99] The CNTs were grown with bush-like nanostructures covered around the micron fibers or web-like nanostructures crossing between the fibers at different synthetic conditions.[100,101] The developed CNT metal filter had a higher filtration efficiency without a significant difference in the pressure drop compared with the conventional metal filter, which is because the CNTs function as the trap of pollutant nanoparticles. A bush-like morphology of mesostructured silica with stems, branches, and leaves was grown under mild basic conditions *via* an auxiliary solvent evaporation induced self-assembly process.[102] The Er_2SiO_5 nanobush (Figure 7.31), obtained from $ErCl_3 \cdot 6H_2O$ on Si nanowire arrays as templates, is also known.[103] Other nanobushes were described in Refs. [104,105].

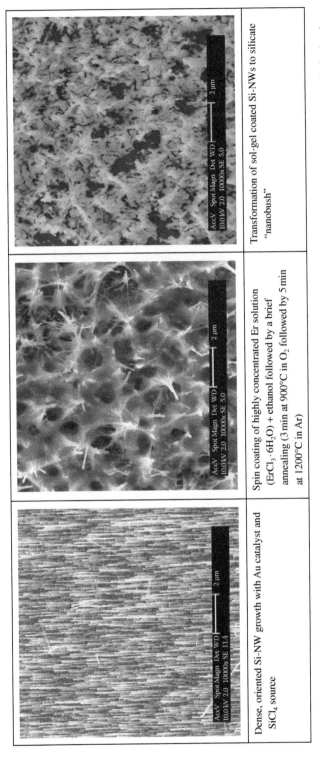

| Dense, oriented Si-NW growth with Au catalyst and SiCl₄ source | Spin coating of highly concentrated Er solution (ErCl₃·6H₂O) + ethanol followed by a brief annealing (3 min at 900°C in O₂ followed by 5 min at 1200°C in Ar) | Transformation of sol-gel coated Si-NWs to silicate "nanobush" |

FIGURE 7.31 Er₂SiO₅ nanobush formation. (Reproduced with permission from Suh, K. et al., Large-scale fabrication of single-phase Er₂SiO₅ nanocrystal aggregates using Si nanowires, *Appl. Phys. Lett.*, 89, 223102. Copyright 2006 American Institute of Physics; from http://www.sciencetouch.net/html/guide/download.php?code=ppts&filename=PPT53.pdf, accessed on January 31, 2011.)

7.5 NANOMUSHROOMS

Mushroom-like nanostructures are known for a series of elemental substances, their oxides, nitrides, salts, polymers, as well as their combinations. Carbon, traditionally obtained in a grand variety of nanostructures, can also exist as nanomushrooms. Thus, two types of nanocarbons with a mushroom shape (with a conical tip, made from helically stacked graphitic cones, and pyramidal tip, made from graphitic cups stacking to each other), having ~2 μm length and ~0.5 μm diameter, were found in common glassy carbon powder.[107] It was established that their diamagnetic behavior strongly depends on both the temperature and the magnetic field. A facile thermal CVD method for fabricating dense and vertically well-aligned bamboo-like CNT arrays on a Cu substrate from acetone was developed[108] resulting in mushroom-like carbon nanostructures, formed at 800°C, and well-aligned CNTs, produced at 850°C. Position-specified fabrication (Figures 7.32 and 7.33) of epitaxial Si grains on an amorphous thermal oxide layer on Si(001) substrates was described.[109] The growth was initiated from Si nanodots that were isolated from the substrate by the oxidation of mushroom-shaped overgrown structures. The present technique is applicable to the integration of various single-crystal materials onto Si substrates.

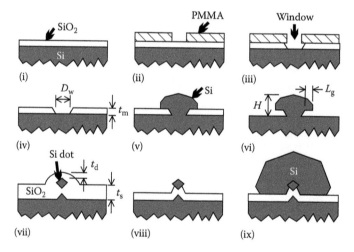

FIGURE 7.32 Process flow for forming epitaxial Si grains on a thermal oxide layer on Si: epitaxial growth of a mushroom-shaped overgrown structure (steps i–vi), formation of an isolated Si nanodot by thermal oxidation (step vii), and epitaxial regrowth to form the final structure (steps viii and ix). (Reproduced with permission from Yasuda, T. et al., Specified formation of epitaxial Si grains on thermally oxidized Si(001) surfaces *via* isolated nanodots, *Chem. Mater.*, 16(18), 3518–3523. Copyright 2004 American Chemical Society.)

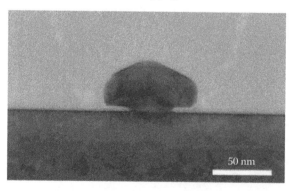

FIGURE 7.33 Mushroom-shaped structure of Si formed by selective-area epitaxial growth and the subsequent oxide mask removal (cross-sectional TEM image). (Reproduced with permission from Yasuda, T. et al., Specified formation of epitaxial Si grains on thermally oxidized Si(001) surfaces *via* isolated nanodots, *Chem. Mater.*, 16(18), 3518–3523. Copyright 2004 American Chemical Society.)

Mushroom-like nanostructures have been described for elemental gold and its composites, for example, thiolated poly(ethylene glycol) with linker molecule *p*-mercaptoaniline (pMA) on Au nanoshells.[110] Figures 7.34 and 7.35 show the appearance and formation mechanism and procedure of Au-aniline nanomushrooms.[111] Nanomushroom-type nanostructures on Au basis[112] can be used as biomolecular sensors.

Hybrid nanostructures consisting of tungsten oxide nanorods with mushroom-shaped carbon caps were grown on electrochemically etched tungsten tips by thermal CVD with methane and argon.[113] The presence of crystalline monoclinic $W_{18}O_{49}$ in the nanorods and the cap consisting of entirely amorphous

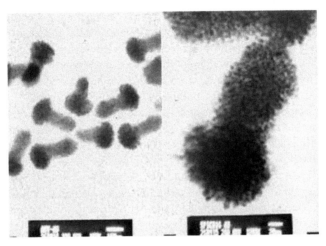

FIGURE 7.34 Nanomushroom (aniline-Au). (Shiigi, H. et al., Self-organization of an organic-inorganic hybrid nanomushroom by a simple synthetic route at the organic/water interface, *Chem. Commun.*, (24), 3615–3617, 2009. Reproduced by permission of The Royal Society of Chemistry.)

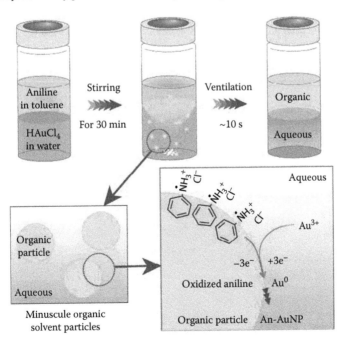

FIGURE 7.35 Formation of aniline-Au nanomushroom-type nanoparticles. (Shiigi, H. et al., Self-organization of an organic-inorganic hybrid nanomushroom by a simple synthetic route at the organic/water interface, *Chem. Commun.*, (24), 3615–3617, 2009. Reproduced by permission of The Royal Society of Chemistry.)

carbon was revealed, and a growth mechanism was offered involving the reduction of tungsten oxide WO_3, present on the tungsten surface, by methane at 900°C. A ZnO complex nanostructure with mushroom-like morphology was prepared by hydrolysis of Zn acetate dihydrate {$Zn(AcO)_2 \cdot 2H_2O$} in H_2O-MeOH mixed solvent at 60°C.[114] It was shown that the mushroom-like particles were transformed from cauliflower-like layered basic Zn acetate, $Zn_5(OH)_8(AcO)_2 \cdot 2H_2O$, and composed of ZnO sub-units with average size <10 nm. In a related work, $Zn(NO_3)_2 \cdot 6H_2O$, sodium dodecylbenzenesulfonate [$CH_3(CH_2)_{11}C_6H_4SO_3Na$], and NaOH in ethanol were used as precursors for obtaining ZnO morphologies with distinctive shapes ranging from mushrooms, bihemispheres, dumbbells, bilayer hexagonal disks, and prisms to flower-like nanosheet aggregations.[115] Other useful information on the morphology and wettability of ZnO 1D nanostructures is given in Ref. [116]. Strontium tungstate nanomushrooms (average particle size 20–24 nm) were synthesized by controlled precipitation in an aqueous medium.[117] In the case of polyvinyl alcohol (PVA) use, the morphology resembled a cauliflower. Studying luminescence properties, it was shown that both forms showed strong emissions around 425 nm. For single-crystalline mushroom-like AlN nanorod arrays, the low turn-on field and high current stability demonstrated that this mushroom-like AlN nanorod array is a promising field emission material.[118]

Among organic compounds and polymers, mushroom-like nanostructures have been reported for polypyrrole,[119] isoprene-styrene diblock copolymers,[120] and miniaturized rod-coil triblock copolymers BPEIS {(biphenyl ester)3-(isoprene)9-(styrene)9}.[121] Several intriguing nanostructures based on poly(3-trimethoxysilylpropyl methacrylate-polyethylene glycol Me ether methacrylate) [poly(TMSMA-r-PEGMA)], such as mushroom-like nanopillars, vertical nanopillars, and nanospheres, were manufactured using capillary lithography and a UV-curable mold consisting of polyurethane acrylate.[122] The 2D patterns formed by supramolecular materials deposited from solutions of precursors **1–5** on oxidized silicon substrates were composed of mushroom-shaped nanostructures measuring 2–5 nm in cross section and approximately 7–8 nm in height.[123] Other intriguing nanomushroom structures are described in Refs. [124,125].

Chart 1

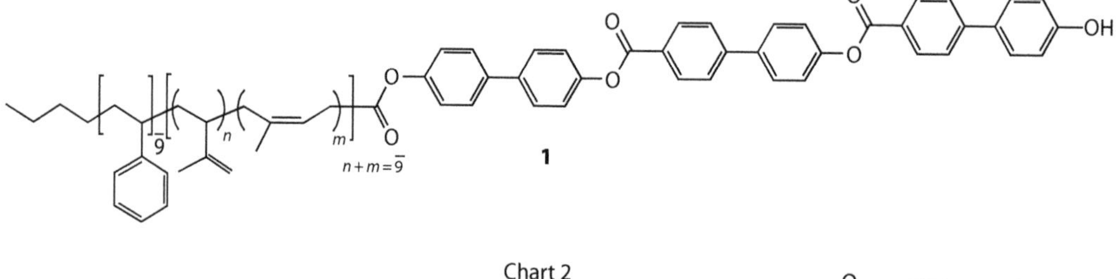

1

Chart 2

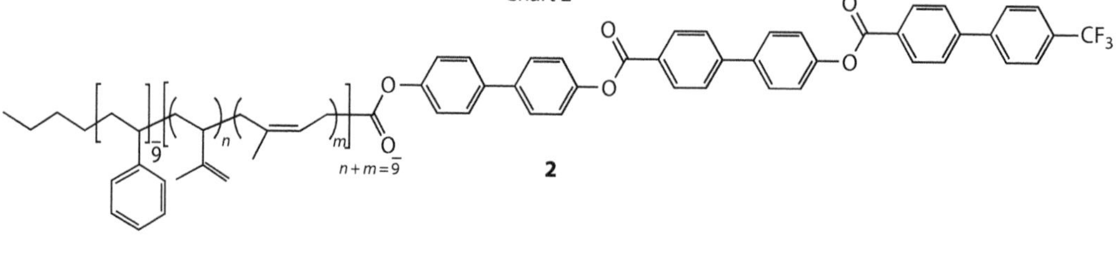

2

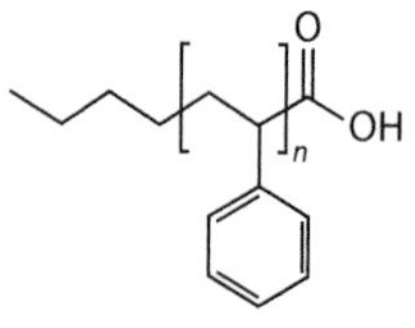

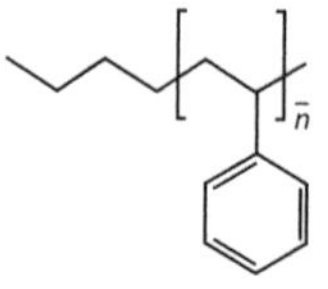

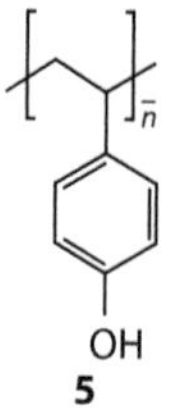

3   **4**   **5**

7.6 NANOFLOWERS

7.6.1 Nanoflowers in General

Nanoflowers (flower-like nanoparticles) are not currently reported so frequently, as, for example, nanowires[126] or one of the hottest nanotechnology area—well-known CNTs. However, in the last decade the number of related literature has increased. Last year, a series of various nanoflower- and nanoflower-like structures have been obtained, frequently together or in equilibrium with other nanoforms, depending on reaction conditions. Current and possible applications of nano-flowers as optoelectronic devices or sensors, in catalysis, and solar cells caused a definite interest in them.

7.6.1.1 Elemental Nanoflowers (Metals and Carbon)

There are several elemental metals (mainly Au[127,128]), for which the forms of nanoflower and flower-like nanoparticles have been obtained by the reduction of metal salts or other compounds. In the case of gold, $HAuCl_4$ was always used as a precursor. Thus, the sheet-like gold nanoparticles, pre-pared in colloidal form under UV light irradiation of the mixture of $HAuCl_4$ aqueous solution and poly(vinylpyrrolidone) (PVP) ethanol solution in the presence of silver ions, were found to self-assemble into nanoflowers by a centrifuging process.[129] Ag ions and PVP served as structure-directing agents to produce the sheet-like particles with a suitable size, which could assemble into the flower-like structures due to the combination of Van der Waals force and the anisotropic hydro-phobic attraction between the nanoparticles. Gold nanoflowers were also obtained by a one-pot synthesis using N-2-hydroxyethylpiperazine-N-2-ethanesulphonic acid as a reducing or stabilizing agent and $HAuCl_4$ as a gold source.[130] The obtained Au nanoflowers showed excellent electrocata-lytic activity toward the oxidation of methanol and the reduction of oxygen to H_2O_2, better than that of the spherically shaped citrate-stabilized Au nanoparticles. Citrate anion is one of the common reductants for obtaining nanostructure. Thus, $HAuCl_4$ and trisodium citrate were used as precur-sors to obtain various gold nanoforms (flower-like nanoparticle arrays, nanoflowers (Figure 7.36), nanowire networks, and nanosheets of gold on the solid substrates).[131] The first two nanostructures were formed at trisodium citrate/$HAuCl_4$ in a ratio of 1:1 and 2.6:1, respectively. Au nanoflowers (300–400 nm diameter) and Fe_3O_4/Au composite nanomaterials were also prepared in similar con-ditions in the presence of citrate, poly(ethylene glycol), and sodium acetate.[132] Additionally, the for-mation and stability of the Au nanoflowers, obtained from a mixture of $HAuCl_4$ and hydroxylamine

FIGURE 7.36 SEM images of gold nanoflowers (30 min reaction time). (Reproduced with permission from Wang, T. et al., Surfactantless synthesis of multiple shapes of gold nanostructures and their shape-dependent SERS spectroscopy, *J. Phys. Chem. B*, 110(34), 16930–16936. Copyright 2006 American Chemical Society.)

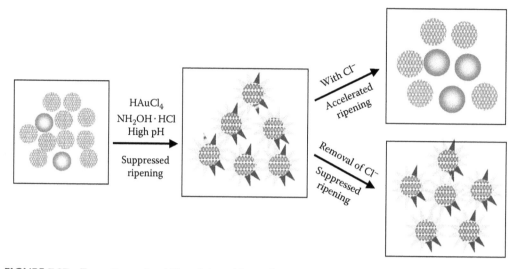

FIGURE 7.37 Formation and stability of the gold nanoflowers and change of their morphology in the presence or absence of chlorine ions. (Reproduced with permission from Zhao, L. et al., Formation and stability of gold nanoflowers by the seeding approach: The effect of intraparticle ripening, *J. Phys. Chem. C*, 113(38), 16645–16651. Copyright 2009 American Chemical Society.)

as growth solution, were shown to be affected greatly by the intraparticle ripening induced by the chlorine ions that existed in the reaction system (Figure 7.37).[133]

In addition to monometallic gold nanoflowers,[134] novel Au-Pt bimetallic flower nanostructures, including two parts (the "light" and the "pale" part) and consisting of many small bimetallic nanoparticles, were fabricated on a polyamidoamine dendrimers-modified surface by electrodeposition using H_2PtCl_6 and $HAuCl_4$ as sources of metal atoms.[135] Stable polyamidoamine dendrimers assisted the formation of nanoflowers during the electrodeposition process; the morphologies of bimetallic nanoflowers depended on the electrodeposition time and potential and the layer number of assembled dendrimers. The content of Au element in the nanoflowers was higher than that of Pt element.

Uniform nickel nanoflowers, consisting of hundreds of smaller primary nanoparticles with an average dimension of about 6.3 nm, were prepared by the reduction of nickel chloride with hydrazine hydrate in the presence of PVP in ethylene glycol in the presence of an appropriate amount of Na_2CO_3 under microwave irradiation.[136] The authors noted that a small amount of NaOH and a higher concentration of hydrazine were beneficial to the formation of Ni nanoflowers with a smaller diameter and a narrower distribution. Among nanoflowers of other elemental metals, Pt nanoflowers are known;[137] those prepared (Figure 7.38)[138] by the reduction of $H_2PtCl_6 \cdot 7H_2O$ with $NaBH_4$, showed good performance (both high sensitivity and high molecular weight) for surface-assisted laser desorption or ionization mass spectrometry (SALDI-MS) of various biomolecules, including peptides and phospholipids. Additionally, zinc electrode material, containing nanoflowers each provided with a 40–300 nm core and 3–20 radial branches with diameter of 10–50 nm and length of 0.3–1.0 μm[139] and tin nanoflowers[140] were also recently reported. Various controlled shapes and sizes of colloidal ferromagnetic cobalt nanoparticles were fabricated by the polyol process using $Co(acac)_3$, 1,2-hexadecanediol, oleylamine, and oleic acid within octylether.[141] Additionally, the composites of flowery hexagonal close-packed cobalt crystallites covered with multiwalled carbon nanotubes (MWCNTs) were fabricated in high yield *via* a catalytic pyrolysis method.[142]

In respect of carbon itself, in addition to thousands of patents, books, experimental and review articles, dedicated to intensively studied CNTs and fullerenes, this element was also obtained in the nanoflower form. Thus, carbon nanoflowers with graphitic features were prepared at 650°C with high yields using a reduction–pyrolysis–catalysis route using glycerin as a carbon source and magnesium and ferrocene as reductants and catalysts.[143] The obtained nanoflowers with a

FIGURE 7.38 Platinum nanoflowers. (Reproduced with permission from Kawasaki, H. et al., Platinum nanoflowers for surface-assisted laser desorption/ionization mass spectrometry of biomolecules, *J. Phys. Chem. C*, 111(44), 16278–16283. Copyright 2007 American Chemical Society.)

hollow core had diameters ranging from 200 to 600 nm. In another report,[144] iron was used as a catalyst for CNT flower formation from a gas mixture of Ar and C_2H_2 at a temperature range of 700–900 K for 300 s (5 min). Iron wire was sparked for 1, 2, 10, and 100 times to form iron dots or islands on the glass slides, where CNTs, composed of carbon with hexagonal structure, were further grown. Also, carbon nanoflowers, among other nanostructures (fullerene, carbon nanopowders, nanofibers, and nanotubes) were produced by carrying out a gas membrane micro-arc discharge on a cathode in an aqueous solution (basic electrolyte, 0.01–2 mol/L solutions of mineral acids, bases, or chlorides).[145]

7.6.1.2 Metal Oxide Nanoflowers

In line with elemental metals, their oxides, together with sulfides, have been obtained in the nano-flower form much more frequently. Their synthesis is based mainly on zero-valent metal oxidation or decomposition of hydroxides or other compounds. Among the reported metal oxides, ZnO[146] (Figure 7.39) has been mostly characterized and its several applications, for example, for solar cells,[147] have been proposed. Its morphologies usually include nanowires, nanorods, nanobelts, nanorings, nanosheets, nanodisks, nanoflowers, nanoneedles, nanonails, nanopencils, and nano-flakes.[148–150] As another important application of ZnO nanoflowers, a new amperometric biosen-sor for hydrogen peroxide was developed based on the adsorption of horseradish peroxidase at the glassy carbon electrode modified with zinc oxide nanoflowers, produced by electrodeposition onto MWCNT film.[151] Rapid response, expanded linear response range, and excellent stability were found for this biosensor.

ZnO nanoflowers with two typical morphologies—plate-like and bush-like—and composed of pyramidal nanorods, growing from a central point, were synthesized on AlN films by the solution method.[99] ZnO nanostructures, including single-crystal nanowires, nanoneedles, nanoflowers, and tubular whiskers, their morphology dependent on the temperature zone and the pressure of oxygen, were fabricated (Figure 7.40) at 550°C *via* the oxidation of metallic Zn powder without a metal cata-lyst according to simple Reaction 7.12.[152] Nanoflowers were composed of hundreds of "nanopetals" which have a tadpole-like structure, and the tubular whisker has a hexagonal hollow cone structure. The growth process of ZnO nanostructures was attributed by the authors to the vapor-solid process, rather than the more common catalyst-assisted VLS process.

$$Zn(g) + \tfrac{1}{2}O_2(g) \rightarrow ZnO \qquad (7.12)$$

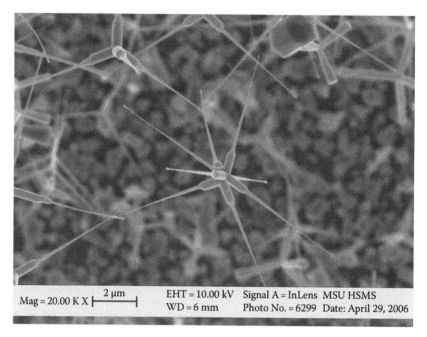

FIGURE 7.39 ZnO tetrapods (snowflakes). (Reproduced from Tretyakov, Yu. D. (Ed.), *Nanotechnologies. The Alphabet for Everyone*, FIZMATLIT, Moscow, Russia, 344–345, 2008. With permission of authors of http://www.nanometer.ru, Moscow State University, Russia.)

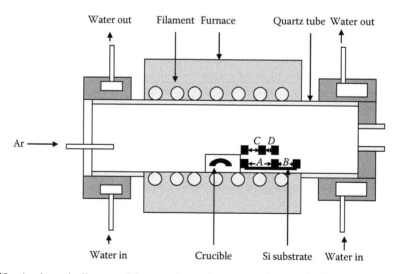

FIGURE 7.40 A schematic diagram of the experimental apparatus for growth of ZnO nanostructures by the solid-vapor-phase process. (Reproduced with permission from Sun, X.H. et al., Synthesis and synchrotron light-induced luminescence of ZnO nanostructures: Nanowires, nanoneedles, nanoflowers, and tubular whiskers, *J. Phys. Chem. B*, 109(8), 3120–3125. Copyright 2005 American Chemical Society.)

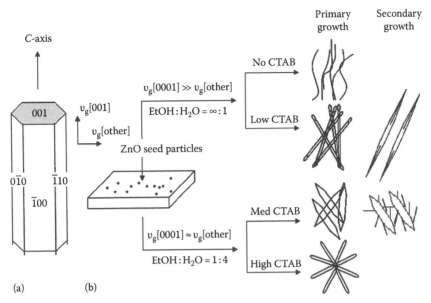

FIGURE 7.41 (a) Illustration of a zincite crystal with various planes marked; (b) growth schematic diagram of ZnO nanostructures in different experimental conditions. (Reproduced with permission from Wen, B. et al., Controllable growth of ZnO nanostructures by a simple solvothermal process, *J. Phys. Chem. C*, 112(1), 106–111. Copyright 2008 American Chemical Society.)

ZnO nanoflowers, among many other nanostructures (nanotowers, nanovolcanoes, nanorods, nanotubes, nanocorns, nanoshuttles, nanoworms, and nanowires), formed depending on the reaction conditions (Figure 7.41), were prepared using the hydrothermal technique[153,154] starting from zinc nitrate, cetyltrimethylammoniumbromide (CTAB), and HMT at 80°C–90° for 0.5–6h (Reaction 7.13). Other related information on ZnO nanoflowers is reported in Refs. [155–159] and shown in Table 7.1:

$$Zn^{2+} + CH_3(CH_2)_{15}N(Br)(CH_3)_3/C_6H_{12}N_4 \rightarrow Zn^{2+}\text{-amino complex} \rightarrow Zn(OH)_2 \rightarrow ZnO \quad (7.13)$$

Adjusting experimental parameters in MgO nanostructure production by chemical vapor transport and condensation,[34] its distinct nanoforms could be obtained. Thus, single crystalline MgO nanoflowers, consisting of MgO nanofibers (20–80 nm), were synthesized *via* a conventional evaporation method using the high-purity magnesium powders and distilled water as starting materials.[160] The obtained MgO nanoflowers have a much higher relative dielectric constant as compared with MgO micropowders and may be useful in providing insight into the formation of microbiological systems and reinforcing composite materials. The formation mechanism of MgO nanoflowers was discussed in Ref. [161] and includes Mg particle formation on the Si substrate, formation of MgO clusters as nucleation centers on the magnesium melt surface, and the nucleation of short MgO nanofibers; then growth of the MgO nanofibers occurred, and finally MgO nanoflower formation. The authors noted that the nucleation and growth process of the nanoflowers seem to be a vapor-solid mechanism, similar to that shown for obtaining ZnO nanoflowers from zinc and oxygen[149] and that the total heating time during the reaction process is a critical factor for the development of MgO nanoflowers. This is not a unique example of nanofiber-containing nanoflowers: thus, nanoflowers weakly ferromagnetic at ~34 K Co_3O_4, consisting of numerous Co_3O_4 nanofibers (diameters of 20–40 nm and lengths 100–500 nm), were prepared through a sequential process of a hydrothermal reaction and heat treatment.[162]

TABLE 7.1

Representative Examples of Nanoflower Fabrication

Compound	Synthesis Method	Variety of Formed Nanostructures, Including Nanoflowers	References
Ni	Nonaqueous sol-gel route involving the reaction of nickel acetylacetonate with benzyl alcohol (solvent and reductant) at 200°C in the presence of varying magnetic fields	Nanospheres, nanowires, and nanoflowers	[212]
Ag	Reduction of silver nitrate in solution of sodium acetate, L-ascorbic acid, sodium citrate, and poly(ethylene glycol)	Rose-, spike-, and snowflake-shaped silver nanostructures	[213]
ZnO	CVD	Nanosleeve fishes, radial nanowire arrays, nanocombs, and nanoflowers (adjusting the source temperature and the gas flow rate). Sisal-like nanoflowers	[214,215]
	Thermal treatment of $Zn(NH_3)_4^{2+}$ precursor in aqueous solvent, using ammonia as the structure directing agent	Flower-like ZnO nano-/microstructures	[216]
	Solution deposition method	Nanorods and nanoflowers (nanoflowers were synthesized on the nanorods)	[217]
	Ultrasonic pyrolysis of $Zn(CH_3COO)_2 \cdot 2H_2O$ at 380°C–500°C	Nanoblades and nanoflowers	[218]
	Solvothermal method	Nanoflowers, assembled by many thin and uniform hexagonal-structured ZnO nanosheets, with a thickness of around 6 nm	[219]
	At 90°C for 5 h without surfactant assistance, varying concentrations of NaOH, and different amounts of $N_2H_4 \cdot H_2O$	Rod-like and chrysanthemum-like ZnO nanostructures contained many radial nanorods	[220]
SnO$_2$	Controlled shape-preserving thermal oxidation process (precursor—3D Sn nanoflowers)	Superhydrophobic 3D flowers with nanoporous petals	[221]
CoO	Decomposition of Co(II) oleate complex at 280–320°C in noncoordinating solvent octadecene containing dodecanol/oleic acid	Cubic nanocrystals with various morphologies and sizes	[222]
Bi$_2$O$_3$	Oxidative metal vapor-phase deposition technique with controlled flow of oxygen and constant working pressures	The large-area arrays of 1D nanowires and nanoflowers	[223]
α-Fe$_2$O$_3$ (hematite)	Stirring Fe(NO$_3$)$_3$, urea, and ethylene glycol as precursors at 170°C and further calcination at 450°C	Flower-like micrometer-sized hematite particles	[224]
	Solvothermal reactions at 140°C using ethanol solution of FeCl$_3 \cdot 6H_2O$ as a precursor and sequential calcinations	The flower-like nanostructures, composed of nanosheets with a thickness of ~20 nm	[225]
Y$_2$O$_3$:Eu^{3+}	Hydrothermal reaction and sequential calcinations	Flower-like nanoarchitectures	[226]
CeO$_2$	Rapid thermolysis of $(NH_4)_2Ce(NO_3)_6$ in oleic acid/oleylamine	Nanoflowers with controlled shape (cubic, four-petaled, and star-like) and tunable size (10–40 nm)	[227]

TABLE 7.1 (continued)
Representative Examples of Nanoflower Fabrication

Compound	Synthesis Method	Variety of Formed Nanostructures, Including Nanoflowers	References
WS_2	Atmospheric pressure chemical vapor deposition (APCVD) process	MoS_2 and WS_2 inorganic fullerene-like nanostructures (onion-like nanoparticles, nanotubes) and elegant 3D nanoflowers	[228]
Bi_2S_3	Synthesis *via* different surfactants such as Triton X-100+OP-10, TX-10, Triton X-100 by refluxing with bismuth nitrate and thiourea as reactants (at 85°C–110°C, 3 h)	Well-crystallized nanoflowers with different morphologies	[229]
AlN	Direct nitridation of aluminum precursor	AlN conic nanoflowers, nanowires, quasi-aligned nanocones, and polycrystalline thin film	[230]
EuF_3	Ultrasonic irradiation; reaction in solution of $Eu(NO_3)_3$ and KBF_4 under ambient conditions without any template or surfactant	Single-crystalline nanoflower with a novel 3D nanostructure	[231]
$CoWO_4$	Alcohol–thermal process at 180°C without any surfactants and structure-directing agents	Flower-like hollow nanostructures, which consist of $CoWO_4$ nanorods	[232]
$MgCO_3$	Interaction in solution of $MgCl_2 \cdot 6H_2O$, urea, dispersant at 80°C–99°C in water bath, 3–8 h, aging for 1–3 h, filtering precipitate, water-washing, and drying at 100°C for 3 h	3D nanoflower structure	[233]
$Pb(Zr_{0.52}Ti_{0.48})O_3$	Pulsed-laser deposition	Flower-like nanostructures on the $CoFe_2O_4$ seeds	[234]
$CaTiO_3$	Environmentally friendly solvothermal technique	$CaTiO_3$ nanoflowers (or also $SrTiO_3$ nanocubes, and $BaTiO_3$ nanospheres). Red fluorescence originating from intra 4f 1D2–3H4 transition of Pr^{3+} is observed by doping Pr^{3+} ions in $CaTiO_3$ nanoflowers	[235]

CuO nanoflowers (containing a very large number of nanometer-scaled flakes [petals]) and each petal further branched into tips at its end) were synthesized directly on Cu plates in KOH solution at room temperature[163] as a result of a series of steps, including oxidation, complex formation, condensation, Ostwald ripening, and dissolution. The CuO nanoflower film, directly formed by chemical oxidation of copper foil under hydrothermal condition, was used as active electrode material of nonenzymic electrochemical sensors for H_2O_2 detection,[164] showing excellent electrocatalytic activity, large surface-to-volume ratio, and efficient electron transport property. Hierarchical porous NiO nanoflowers with porous petals were fabricated by calcining the isomorphological β-$Ni(OH)_2$ precursor, preliminary prepared from nickel dimethylglyoximate and NaOH.[165] In comparison with that of the standard nanoparticles, the higher electrochemical lithium intercalation could be ascribed to their special nanoporous structure.

The α-MnO_2 nanocrystal nanoflowers, together with nanowires and nanoplates, have been successfully synthesized by a common hydrothermal treatment of different solutions containing $KMnO_4$ and NH_4X (X = Cl^-, Ac^-, NO_3^-, SO_4^{2-}, and PO_4^{3-}) at 140°C for 24 h.[166] The patterned SnO_2

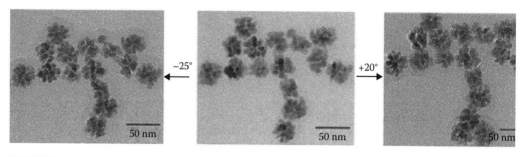

FIGURE 7.42   Rotation of a group of In$_2$O$_3$ nanoflowers. (Reproduced with permission from Narayanaswamy, A. et al., Formation of nearly monodisperse In$_2$O$_3$ nanodots and oriented-attached nanoflowers: Hydrolysis and alcoholysis vs pyrolysis, *J. Am. Chem. Soc.*, 128(31), 10310–10319. Copyright 2006 American Chemical Society.)

nanoflower arrays, with a unit diameter of ~50 μm and composed of nanorods, were synthesized *via* vapor-phase transport method using a patterned Au catalyst film, prepared on a silicon wafer by radio frequency magnetron sputtering and photolithographic patterning processes.[167] Evolution processes for MoO$_3$ nanoflowers and their other morphologies (nanosheets and nanobelts), obtained from metallic molybdenum and H$_2$O$_2$ with further exposition in autoclave to decompose the formed H$_2$Mo$_2$O$_3$(O$_2$)$_4$ (Reaction 7.14), were studied in Ref. [168]. It was proposed that the environmentally friendly synthesis process may be extended for obtaining nanostructures of other metal (W, Ti, and Cr) oxides:

$$2Mo + 10H_2O_2 \rightarrow [Mo_2O_3(O_2)_4]^{2-} + 2H^+ + 9H_2O \qquad (7.14)$$

Single crystalline and nearly monodisperse In$_2$O$_3$ nanocrystals with both dot and flower shapes were prepared using indium carboxylates as the precursors with or without alcohol as the activating reagents in a hydrocarbon solvent under elevated temperatures.[169] The 3D nature of these nanoflowers was confirmed by rotation experiments (Figure 7.42). When the specimen was rotated either way (left and right image), the 2D projection of each nanoflower roughly remained the same, but the structural detail of each flower changed substantially. This indicated that the nanoflowers are true 3D structures instead of plates with their thickness different from their diameter. Application of a new approach, the so-called *limited ligand protection* (LLP) method, resulted in 3D nanoflowers for In$_2$O$_3$, ZnO, CoO, and MnO; meanwhile, when the system had sufficient ligand protection for the nanocrystals, nanodots were found to be the stable products.[170,171] Their function is that LLP destabilizes the primary nanoparticles and promotes their three-dimensionally oriented attachment into complex nanostructures. Other reported oxide-based nanoflowers include those of amorphous silica flower-like nanowire[172,173] or porous silica microflowers (obtained by adding supercritical CO$_2$—reactant and modifier to the morphology and porosity of silica—into the sodium silicate aqueous solutions).[174]

7.6.1.3   Nanoflowers of Hydroxides and Oxo-Salts

Only a few reported examples of nanoflowers are known up to date for metal hydroxides and oxo-salts. Thus, uniform Mg(OH)$_2$ nanoflowers were prepared by a simple hydrothermal reaction of MgCl$_2$ and CO(NH$_2$)$_2$, without any organic additive or catalyst.[175] The influence of temperature on the formation of Mg(OH)$_2$ nanoflowers was studied. Nickel hydroxide nanoflowers (together with nanosheets) have been hydrothermally synthesized using Ni(CH$_3$COO)$_2$·4H$_2$O in mixed solvents of ethylene glycol or ethanol and DIW at 200°C.[176] The biomineral botallackite Cu$_2$(OH)$_3$Cl nanoflowers (Figure 7.43) were obtained with the aid of Au$_3$Cu hollow nanostructures at room temperature,

100 nm

FIGURE 7.43 TEM image of a single $Cu_2(OH)_3Cl$ nanoflower. (Reproduced with permission from Hsiao, M.-T. et al., One-pot synthesis of hollow Au_3Cu_1 spherical-like and biomineral botallackite $Cu_2(OH)_3Cl$ flowerlike architectures exhibiting antimicrobial activity, *J. Phys. Chem. B*, 110(1), 205–210. Copyright 2006 American Chemical Society.)

prepared by the reaction (Reactions 7.15 through 7.17) of Cu nanoparticles (synthesized by laser ablation of CuO powder in propanol) with $HAuCl_4$.[177] Both hollow nanospheres and nanoflowers presented antimicrobial activity toward *Streptococcus aureus*.

$$3Cu(s) + 2AuCl_4^- \rightarrow 2Au(s) + 3Cu^{2+} + 8Cl^- \qquad (7.15)$$

$$H_2O \leftrightarrow H^+ + OH^- \qquad (7.16)$$

$$2Cu^{2+} + Cl^- + 3OH^- \rightarrow Cu_2(OH)_3Cl \qquad (7.17)$$

3D nanostructured $Cd_4Cl_3(OH)_5$ nanoflower arrays were controllably synthesized through a low-temperature hydrothermal process.[178] Source of cadmium ions, concentration in solution, temperature, and reaction time are the factors influencing the possibility of formation of the flower-like structure. The $Cd_4Cl_3(OH)_5$ nanoflowers as template compounds were further transformed into $Cd(OH)_2$ and CdS nanoflowers by an anion-exchange reaction. Some other metal salts in the form of nanoflowers are shown in Table 7.1.

7.6.1.4 Sulfide, Selenide, and Telluride Nanoflowers

The first two groups of title nanoflower structures are widely represented and possess a series of useful applications. Thus, highly ordered hierarchical structures wurtzite urchin-like CdS nanoflowers, made of CdS nanorods, were produced *via* the solvothermal approach from $Cd(Ac)_2 \cdot 2H_2O$, $CS(NH_2)_2$ as precursors in a mixed solution made of DETA and DIW.[35] The possibility of obtaining different morphologies of CdS(Se) nanocrystals can be easily controlled *via* tuning the volume ratio of DETA and DIW; a solvothermal reaction in a mixed amine/water can access a variety of complex morphologies of semiconductor materials. Various CdS nanostructures possess a photocatalytic activity, demonstrated in the photodegradation of acid fuchsine at ambient temperature and related due to the unique structural features of CdS. A microwave-induced semisolvothermal reaction (i.e., involving simultaneous usage of nonaqueous and aqueous solvents) between the same reagents was also used to produce CdS nanoflowers among other obtained nanostructures.[179] Simultaneous application of microwave and ultrasound treatment allowed obtaining CdS nanoflowers too. Thus, well-defined flower-like CdS nanostructures, consisting of hexagonal nanopyramids and nanoplates depending on different sulfur sources, were synthesized by applying ultrasound and microwave at the same time.[180] It is shown that the synergistic effect of microwave and sonochemistry is the main mechanism for the formation of nanoflowers. Due to the induced shift in optical properties, this nanostructure may have potential applications in optoelectronics devices, catalysis, and solar cells.

Visually striking nanoflowers composed of $ZnS:Mn^{2+}$ nanoparticles of 2–5 nm size were prepared and characterized.[181] The observed blue-green emissions were attributed to structural defects and are the dominant luminescence from the nanoflowers. ZnSe nanoflowers with a zinc blende structure were synthesized using a nontoxic, simple, cheap, and reproducible strategy, which meets the standard of green chemistry, from ZnO, Se, and olive oil at 330°C.[182] The mechanism of ZnSe nanoflower formation was offered, based on the formation of mononuclear complexes due to the reaction of ZnO and olive oil, their further conversion to multinuclear compounds and reaction of Zn ions with injected Se forming closely connected ZnSe particles, thus looking like nanoflowers. ZnTe nanocrystals in different forms, including nanoflowers, were obtained using a series of amines as activation agents for the zinc precursor (zinc stearate), octylamine (OA), dodecylamine (DDA), octadecylamine (ODA), and trioctylamine (TOA).[183] The steric effect of alkyl chains plays an important role in the formation of nanoflowers: they are produced only when no amine or TOA is used. Nonstoichiometric copper sulfides (Cu_7S_4), having uniform hexapetalous snowflake-like morphology, were synthesized in high yields by a one-step solvothermal method at 150°C.[184] CoS nanowires, assembled with nanoflowers, were isolated by a biomolecule-assisted hydrothermal process, in which L-cysteine was used as the sulfide source and directing molecule.[185]

Bi_2S_3 nanoflowers were prepared from different precursors: $Bi(NO_3)_3$ (on an alumina template by a photochemical method in the presence of thioacetamide (TAA) and nitrilotriacetic acid at room temperature),[186] Bi-thiol complexes $Bi(SCl_2)_3$ (nanorods, evolution further to nanoflowers over time, were formed by reaction with TAA in a coordinating solvent, dodecanethiol, at 95°C),[187] and the single-source precursor $Bi(SCOPh)_3$ or multiple-source precursors (nanoflowers and other morphologies [nanorods, dandelion-like nanostructures, nanoleaves, and nanocabbages] were formed by using a colloidal solution or hydrothermal methods).[188]

Preparation of GaS and GaSe nanowalls or nanoflowers by thermal exfoliation around 900°C and their transformation to Ga_2O_3 and GaN nanowalls was reported in Ref. [189]. Elegant 3D MoS_2 nanoflowers (Figure 7.44), consisting of tens to hundreds of hexagonal petals (100–300 nm wide and several nanometers thick) and exhibiting excellent field emitter properties, were uniformly formed *via* heating a MoO_2 thin film formed on a Mo foil in a vapor sulfur atmosphere at 950°C–1000°C.[190] Hydrothermal synthesis led to different MoS_2 morphologies, in particular nanoflowers at much longer aging period.[191] SnS nanoflowers (product composed of well-crystallized orthorhombic SnS nanoflowers) were synthesized *via* thioglycolic acid (TGA)-assisted hydrothermal process at 250°C.[192] Critical factors (e.g., TGA, $SnCl_2/Na_2S$ mol ratio of the reactants, hydrothermal temperature, and sulfur source) for hydrothermal formation of the SnS nanoflowers were discussed. PbS downy-velvet-flower-like nanostructures, together with other morphologies (rod-like, belt-like, and dendrite-like), were fabricated in aqueous solution by the assistance of a classic surfactant CTAB,[193] already mentioned here for ZnO nanoflower preparation.[27] In this method, basic acetate of lead, formed at an initial reaction step, was a precursor to control the crystal nucleation rate. PbSe and PbTe flower-like structures were also described.[194]

A few double sulfide nanoflowers were reported; for example, $CuInS_2$ flower vase-like nanostructure arrays were successfully prepared on Cu-tape substrates by a so-called "copper indium sulfide on Cu-tape" (CISCuT) method without using any template or catalyst.[195] $CoMoS_4$ nanoflowers were synthesized by a precipitation method from Na_2MoO_4, $CoCl_2 \cdot 6H_2O$ and CH_3CSNH_2 as starting materials.[196]

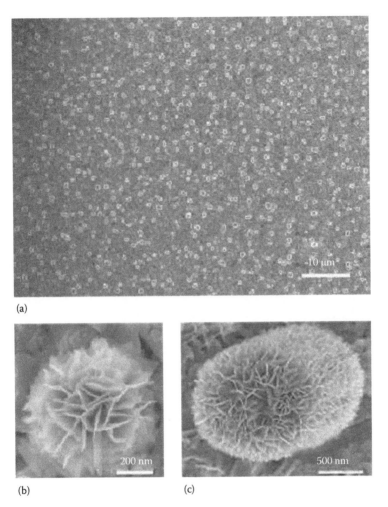

FIGURE 7.44 SEM images of MoS$_2$ flower-like nanostructures: (a) low-magnification view; (b, c) high-magnification view. (Reproduced with permission from Li, Y.B. et al., MoS$_2$ nanoflowers and their field-emission properties, *Appl. Phys. Lett.*, 82(12), 1962–1964. Copyright 2003 American Institute of Physics.)

7.6.1.5 Nitride and Phosphide Nanoflowers

A considerably lesser number of nanoflowers, based on title compounds, is known in comparison with metal compounds with VI Group elements. Thus, a flower-like Si-doped AlN nanoneedle array (The nanoneedles are several microns in length, and their base and tip diameters are in the range 50–150 and 5–30 nm respectively) was grown from cobalt particles seeded on an Si substrate by evaporating AlCl$_3$ and SiCl$_4$ in NH$_3$ medium.[197] This method may facilitate the development of efficient AlN nanostructure field emission devices. Among other structures, AlN nanoflowers were studied by Auger electron spectroscopy.[198] A hexangular InN nanoflower pattern (Figure 7.45)[199,200] was grown on a c-plane (0001) sapphire by metal organic chemical vapor deposition (MOCVD) using trimethylindium and ammonia as precursors with intentional introduction of hydrogen gas. A stable existence of metallic indium is achieved with the aid of hydrogen. This will induce the growth of InN nanoflowers *via* self-catalysis VLS process. The impact of metallic indium clusters on the optical properties of InN was analyzed.[201]

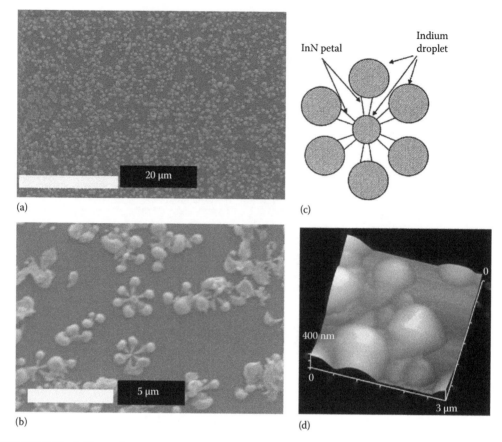

FIGURE 7.45 InN nanoflowers: (a) large-area SEM, (b) high-magnification SEM top view, (c) schematic illustration, and (d) AFM stereo view. (Reproduced with permission from Kang, T.T. et al., InN nanoflowers grown by metal organic chemical vapor deposition, *Appl. Phys. Lett.*, 89(7), 071113. Copyright 2006 American Institute of Physics.)

Crystalline GaP nanoflowers (Figure 7.46) with cubic structure composed of numerous GaP nanowires were synthesized through heating InP and Ga_2O_3 powders.[202] Cathodoluminescence measurements showed that GaP nanoflowers emit at ~600 nm and additional low-intensity emission peaks were observed at ~450 nm. The authors proposed that GaP nanoflowers may be valuable for future nanodevice design. Similar nanoflower-like GaP nanostructures, constituted also by numerous nanowires (diameters of 80–300 nm; lengths varying from several to tens of mcm), were fabricated by another route: through the close-spaced vapor transport technique (CSVT) by growing on crystalline GaAs, using a GaP powder source in the absence of any catalyst.[203] Co_2P flower-like nanocrystals, together with nanorods, were synthesized by a polymer-assisted hydrothermal method at 190°C–220°C using $CoCl_2 \cdot 6H_2O$ as Co-source, yellow phosphorous as P-source, and polyacrylamide as surfactant.[204]

7.6.1.6 Nanoflowers Formed by Organic and Coordination Compounds

According to the available bibliography, nanoflowers of organic and related compounds are currently a practically nondeveloped area; only a few compounds are reported. Thus, nanorods and nanoflowers (bundles of nanorods) of bis(8-hydroxyquinolinato)cadmium (CdQ_2) complex were fabricated in an oleic acid-Na oleate-EtOH-hexane-H_2O system at 55°C–100°C.[205] It was revealed that a longer time and a higher temperature would result in nanoflowers, while higher

FIGURE 7.46 GaP nanoflowers. (Reproduced with permission from Liu, B.D. et al., Synthesis and optical study of crystalline GaP nanoflowers, *Appl. Phys. Lett.*, 86, 083107. Copyright 2005 American Institute of Physics.)

concentrations of the reactants and the surfactant with a lower temperature and a shorter reaction time would be appropriate for the formation of nanorods. The obtained CdQ_2 nanorods could be introduced as the building blocks for novel optoelectronic devices. Field emission scanning electron microscopy (FESEM) images showed a nanoflower-like structure (among other structures observed: nanoparticles, nanocabbages, and nanoribbons) for the gold-coated quartz substrates of copper phthalocyanine (CuPc) thin films deposited at room temperature (30°C) on quartz and annealed at 750°C (Figure 7.47).[206,207] A more extensive study of this and other MPc (M = Cu, Ni, Sn, Mg, and Zn) was reported in Ref. [208]. Nanoflowers, nanosheets, and nanoribbons of the alkali earth phenylphosphonates were prepared *via* a solvothermal approach by the reaction of alkali earth metal salt with phenylphosphonate in ethylene glycol.[209] Uniform self-doped poly(*o*-aminobenzenesulfonic-acid-co-aniline) nanoflowers, exhibiting a remarkable electroactivity at an extended pH range from 3 to 13.5, were prepared by an electrochemical preparation without any

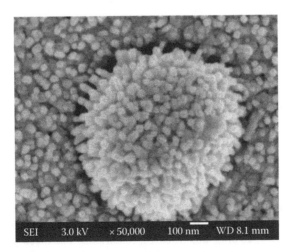

FIGURE 7.47 Copper phthalocyanine nanoflower. (Reproduced with permission from Karan, S. et al., Templating effects and optical characterization of copper (II) phthalocyanine nanocrystallites thin film: Nanoparticles, nanoflowers, nanocabbages, and nanoribbons, *J. Phys. Chem. C*, 111(20), 7352–7365. Copyright 2007 American Chemical Society.)

other supporting electrolytes.[210] Nanoflowers of 1,1,2,3,4,5-hexaphenylsilole (HPS), together with nanoglobules and microglobules, were obtained on reprecipitation with THF as a solvent and water as a nonsolvent.[211]

Selected representative examples of obtained nanoflowers are shown in Table 7.1.

7.6.2 DANDELION-TYPE NANOFLOWERS

Dandelion-type nanoflowers, in our opinion, should be discussed as part of the "traditional" nanoflowers mentioned earlier, since they present a special nanoflower type, related to nanoflowers and nanourchins at the same time. Dandelion-like nanostructures are represented by a series of distinct types of mainly inorganic compounds. Thus, a template-free method was used to prepare 3D dandelion-like nickel nanostructures *via* reduction of $NiCl_2$ with hydrazine hydrate in ethylene glycol solution at 100°C.[236] As-prepared Ni samples had obvious shape anisotropy and were composed of fine nanocrystallites, while they had significantly enhanced ferromagnetic properties than those of bulk Ni and Ni nanoparticles. Core-shell PtRu clusters resembling dandelions were formed on thiolated CNTs by the difference in bond strength with surface thiol groups between Pt and Ru single atoms.[237]

Metal oxides frequently form nanodandelions. Thus, the ZnO nanorod arrays with dandelion-like morphology were directly grown on copper substrates by a hydrothermal synthesis process at 80°C.[238] It was believed that the unique dandelion-like binary structure played an important role in the electrochemical performance of the array electrodes, opening the possibility to fabricate micrometer or nanometer hierarchical ZnO films that might be applied in lithium-ion batteries. In addition, ZnO nanoneedles, constructed on zinc layers by immersing in an aqueous NH_4OH solution at 80°C,[239] were used to fabricate the ZnO films based on dandelion-like ZnO microspheres, exhibiting excellent superhydrophilicity. Doped dandelion-like TiO_2 microsphere-assembled nanorods were synthesized from rutile powders using concentrated NaOH and either urea or thiourea under hydrothermal conditions (200°C for 24 h) leading to N- or S-doped TiO_2[240] exhibiting high photocatalytic activity in the photodegradation of aqueous Methylene Blue solution. The authors showed that concentrated urea (or thiourea) and NaOH served as additives that help in the construction of the dandelion-like structures.

Pure rutile phase crystalline TiO_2 powder with 3D dandelion-like structure was synthesized by using a hydrothermal method with $TiCl_3$ as the main starting material.[241] The typical 3D dandelion structure had an average diameter of 1.5–2 μm, was packed radially by nanorods with [001] preference growth direction, and possessed a high photocatalytic activity. The large size, low melting point bimetal Ga-Sn can be used as an effective catalyst for the large-scale growth of highly aligned, closely packed silicon oxide nanowire bunches, in particular prickly spheres, whisk-like, echinus-like, hedgehog-like, and dandelions-like silicon oxide nanowires, formed under different atomic ratio of Ga and Sn in the alloy ball.[242] The rod-like and dandelion-like CuO nanomaterials were prepared by the decomposition of a copper complex $[Cu(pbbt)Cl_2]_2 \cdot CH_3OH$ (pbbt = 1,1′-(1,3-propylene)-bis-1H-benzotriazole) in the presence of suitable surfactants and alkalies under hydrothermal conditions.[243] The average diameter of the dandelion-like CuO microspheres is 2 μm. It was established that the sensitivity of the CuO nanorods was better than that of dandelion-like CuO particles. Alternatively, CuO dandelion-like nanostructures were prepared from $[Cu(CH_3COO)_2 \cdot H_2O]$ and NaOH as precursors under hydrothermal conditions too.[244] The schematic diagram of the growth process of these CuO dandelion structures is shown in Figure 7.48. A very high surface area ~325 m²/g was revealed; remarkably enhanced photoconductivity under white light irradiation of the CuO dandelions, observed compared with the nanocrystals, was attributed to the presence of oxygen-related hole-trap states at the large surface area of the dandelions. The fast response (τ = 24 s) of the photocurrent holds promise for the fast photosensing device applications.

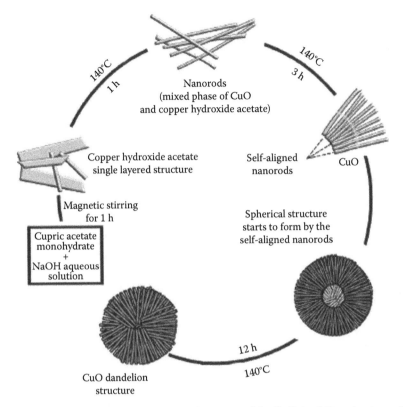

FIGURE 7.48 The schematic diagram of the growth process of the CuO dandelion structures. (Reproduced with permission from Manna, S. et al., Template free synthesis of mesoporous CuO Dandelion structures for optoelectronic applications, *ACS Appl. Mater. Interfaces.* Copyright 2010 American Chemical Society.)

SnO_2 dandelion-like architectures, composed of numerous 1D tetragonal prism nanorods with diameter <50 nm, were synthesized by a hydrothermal method with the help of the surfactant PVP. The sensors fabricated from the SnO_2 nanorods exhibited good sensitivity, high selectivity, and rapid response and recovery times to ethanol vapors at 280°C. Nanostructured α-MnO_2 thin films with different morphologies were grown on the platinum substrates from $MnSO_4$, $(NH_4)_2S_2O_8$, and Na_2SO_4 as precursors by a solution method without any assistance of a template or surfactant (Reaction 7.18).[245] It was revealed that morphological evolution from dandelion-like (Figure 7.49) spheres to nanoflakes of the as-grown MnO_2 was controlled by synthesis temperature. The α-MnO_2 thin films composed of dandelion-like spheres exhibited high specific capacitance, good rate capability, and excellent long-term cycling stability.

$$MnSO_4 + (NH_4)_2S_2O_8 + 2H_2O \rightarrow MnO_2 + (NH_4)_2SO_4 + 2H_2SO_4 \qquad (7.18)$$

α- and β-$Ni(OH)_2$ with 3D nanostructures were synthesized[246] in a H_2O-in-oil reverse micelle/microemulsion system from nickel nitrate and CTAB as precursors. α-$Ni(OH)_2$ (more exactly $[Ni(OH)_{1.38}]$ $(CO_3)_{0.21}(OCN)_{0.12}(NO_3)_{0.08}(H_2O)_{0.44}$) with dandelion-like nanostructures (Figure 7.50) was obtained in the reverse microemulsion of CTAB/H_2O/cyclohexane/n-pentanol with a molar ratio of H_2O to surfactant of 40, meanwhile the β-$Ni(OH)_2$ phase with flower-like nanostructures was formed in the same reverse micelles system with a molar ratio of H_2O to surfactant of 10. VOOH dandelion-like nanostructures are also known.[247]

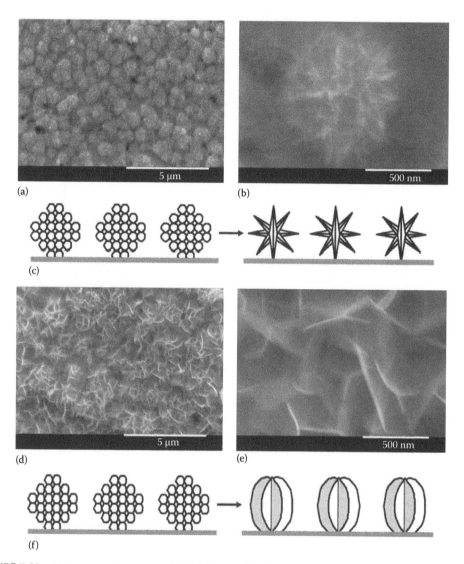

FIGURE 7.49 (a) Low-magnification and (b) high-magnification and (c) schematic illustration for the possible formation mechanism of dandelion-like MnO_2 microspheres, FESEM images for 80°C synthesized MnO_2: (d) low-magnification and (e) high-magnification and (f) schematic illustration for the possible formation mechanism of MnO_2 nanoflakes. (Reproduced with kind permission from Springer Science+Business Media: *Nanoscale Res. Lett.*, Facile synthesis of novel nanostructured MnO_2 thin films and their application in supercapacitors, 4(9), 2009, 1035–1040, Xia, H. et al.)

Self-assembled dandelion-like hydroxyapatite (HAp) nanostructures (5–8 μm in diameter, composed of radially oriented nanorods with an average diameter of ~200 nm) were successfully synthesized *via* a mild template-free hydrothermal process, using EDTA as the surfactant.[248] These nanostructures had appropriate thermal stability up to 1200°C. $LnPO_4$ nanostructures (Figure 7.51) were prepared at r.t. from lanthanide nitrates and $Na_2HPO_4 \cdot 12H_2O$ or $NaH_2PO_4 \cdot 2H_2O$ on solid substrates by the layer-by-layer adsorption and reaction method comprising repetitive adsorption of lanthanide ions and subsequent reaction with phosphate ions.[249] The product morphology can be controlled between 3D dandelion-like nanoarchitectures composed of single-crystalline 1D nanoneedles and 1D nanoneedles lying on the substrates by tailoring the concentration of the deposition solutions.

(a)

(b)

(c)

(d)

FIGURE 7.50 SEM images of an α-Ni(OH)₂ sample: (a) overview of sample; (b) a single dandelion-like α-Ni(OH)₂ microsphere; (c) detailed sphere structure assembled by nanowires; and (d) α-Ni(OH)₂ nanospheres. (Reproduced with permission from Cao, M. et al., Self-assembled nickel hydroxide 3D nanostructures: A nanomaterial for alkaline rechargeable batteries, *Cryst. Growth Des.*, 7(1), 170–174. Copyright 2007 American Chemical Society.)

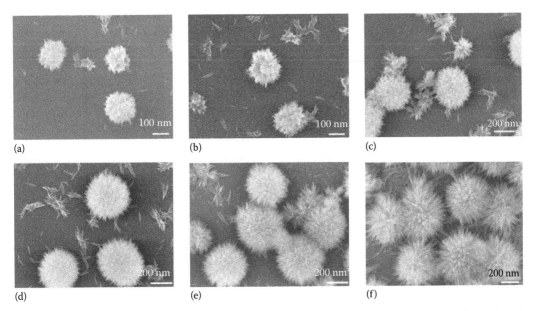

(a)

(b)

(c)

(d)

(e)

(f)

FIGURE 7.51 LaPO₄ nanoarchitecture with different deposition cycles: (a) 3, (b) 5, (c) 7, (d) 10, (e) 15, and (f) 30. (Reproduced with permission from Liu, X. et al., Fabrication of lanthanide phosphate nanocrystals with well-controlled morphologies by layer-by-layer adsorption and reaction method at room temperature, *Cryst. Growth Des.*, 9(8), 3707–3713. Copyright 2009 American Chemical Society.)

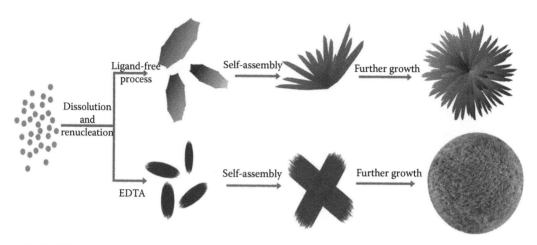

FIGURE 7.52 Possible morphological evolution of the $Y_2(WO_4)_3$:Eu 3D hierarchical architectures with different shapes in the absence and presence of EDTA, respectively. (Reproduced with permission from Xu, L. et al., Self-assembled 3D architectures of $Y_2(WO_4)_3$:Eu: Controlled synthesis, growth mechanism, and shape-dependent luminescence properties, *Cryst. Growth Des.*, 9(7), 3129–3136. Copyright 2009 American Chemical Society.)

Monoclinic $Y_2(WO_4)_3$:Eu with 3D hierarchical architectures (Figure 7.52) were synthesized by a hydrothermal method in ligand-free and chelating ligand-mediated processes, respectively, using Eu_2O_3, Y_2O_3, and $Na_2WO_4 \cdot 2H_2O$ as precursors.[250] The authors established that microflowers assembled from 2D nanoflakes were obtained in a surfactant- and template-free process, whereas microspheres with dandelion-like appearance assembled from 1D nanoplates were observed upon the introduction of the appropriate amount of EDTA to the precursor. The optical properties of these phosphors were found to be strongly dependent on the morphology and size. Additionally, the dandelion-like structure exhibited the strongest red emission with high color purity.

Dandelion-like microcrystallites and long belt-like nanostructures of Sr-V-O materials (β-SrV_2O_6) were synthesized under mild hydrothermal reaction conditions in the presence of mineralizer adipic acid at 220°C for 60h and 180°C for 60h, respectively.[251] $SrCO_3$ dandelion-like architectures with 5–10 µm were prepared by an aqueous solution route at r.t. using $SrCl_2$, Na_2CrO_4, and NaOH as the starting reaction reagents and distilled water as the solvent.[252] The growth process of $SrCO_3$ architectures, suggested by the authors as "rod to dumbbell to sphere" growth mechanism, involved the growth of single-crystalline $SrCrO_4$ nanowires (Reaction 7.19), the formation of CO_3^{2-} ions through the reaction between CO_2 gas from air and aqueous NaOH solution (Reaction 7.20), and finally the formation of $SrCO_3$ architectures through the reaction between $SrCrO_4$ and CO_3^{2-} ions (Reactions 7.21 and 7.22):

$$Sr^{2+} + CrO_4^{2-} \rightarrow SrCrO_4 \tag{7.19}$$

$$CO_2 + 2OH^- \rightarrow CO_3^{2-} \tag{7.20}$$

$$SrCrO_4 \rightarrow Sr^{2+} + CrO_4^{2-} \tag{7.21}$$

$$Sr^{2+} + CO_3^{2-} \rightarrow SrCO_3 \tag{7.22}$$

Other dandelion-like inorganic nanostructures are represented by sulfides. Thus, Bi_2S_3 nanorod bundles as well as 3D dandelion-like nanostructures were prepared in high yield at r.t. in a very simple system composed only of $Bi(NO_3)_3$, TAA, HCl, and distilled water.[253] It was suggested that a splitting crystal growth and self-assembly mechanism should be responsible for the formation of these structures. This kind of Bi_2S_3 nanostructures may find potential applications in hydrogen storage, high-energy batteries, as well as luminescence and catalysis fields. Phase-controlled synthesis of hierarchical nanostructured β-In_2S_3 dandelion flowers was realized by a rapid microwave solvothermal process using indium metal, nitric acid, and thiourea as precursors.[254]

Finally, a few organic nanodandelions are known. Thus, superhydrophobic dandelion-like 3D microstructures (about 5 μm) self-assembled from 1D nanofibers of polyaniline (PANI) were prepared by a self-assembly process in the presence of perfluorosebacic acid (PFSEA) as a dopant.[255]

As a conclusion to the nanoflower section, nanoflower-type nanostructures are currently known for several main groups of mostly inorganic compounds (carbon, elemental metals, their alloys, and compounds with the elements of V and VI Groups of the Periodic Table) and a few coordination and organic compounds. Their main methods of fabrication include oxidation of elemental metals, reduction of metal salts, and thermal decomposition of relatively unstable compounds, hydrothermally or electrochemically. Frequently, the formation of nanoflowers is competitive or in equilibrium with other nanoforms, depending on reagents ratio, temperature, and other conditions; this nanoform can be changed over time. Nanoflower structure may consist of such simple nanostructures as nanorods, nanowalls, or nanowires. Nanoflowers obviously are at the initial stage of their research in comparison with, for instance, nanowires or nanotubes. Further intensive development is expected in this area, and we hope that the organic nanoflowers, practically unknown at present, will be represented by a considerably higher number of intriguing examples.

7.7 NANOBOUQUETS

The term *nanobouquets* is a blurred (diffuse) one: corresponding nanostructures may be called quite differently. Nanobouquets, which may be formed together with other nanostructures in varying synthesis conditions, can consist not only of nanoflowers but also of nanowires, nanorods, nanotrees, and nanoplates. Thus, shuttle-like fusiform bundles of L-cysteine-Pb nanowires split and open out to form dandelion-like flowers with well-aligned architectures (nanobouquets, Figure 7.53).[256] The authors offered a cost-effective method by which ordered nanostructures can be produced on a large scale, at room temperature, and under atmospheric pressure. An aqueous solution of cysteine and lead acetate were used as precursors, leading to spindly bundles of nanowires forming dandelion-like structures with a highly oriented morphology, which, being heated under hydrothermal conditions, decomposed resulting in hierarchical PbS microstructures with various attractive shapes, including spherical, needle-like, and different flower-like structures.

Using a mixture of In_2O_3 and carbon as the starting material, In_2O_3 nanobouquets were obtained along with nanotrees and nanowires.[257] The growth of the nanowires occurs by the VLS mechanism. Starting with In metal, only the oxide nanowires were obtained. In addition, among a variety of cubic phase indium oxide based submicron or nanostructures including arrow-like, bead-like, comb-like wires, and regular polyhedra (octahedrons and triangular slab-like structures), In_2O_3 bouquets (Figure 7.54) were successfully synthesized at different zones of an Si substrate using an improved CVD system.[258] The concentration gradient of indium oxide vapor at different regions of

FIGURE 7.53 L-cysteine-Pb nanobouquets. (Shen, X.-F., Yan, X.-P.: Facile shape-controlled synthesis of well-aligned nanowire architectures in binary aqueous solution. *Angew. Chem., Int. Ed.* 2007. 46(40). 7659–7663. Copyright Wiley-VCH Verlag GmbH & Co. KGaA. Reproduced with permission.)

FIGURE 7.54 In$_2$O$_3$ nanowire bouquets. (Reproduced with permission from Yin, W. et al., Controllable synthesis of various In$_2$O$_3$ submicron/nanostructures using chemical vapor deposition, *Cryst. Growth Des.*, 9(5), 2173–2178. Copyright 2009 American Chemical Society.)

the Si substrate and the growth kinetics of different In$_2$O$_3$ crystalline facets were considered by the authors as the main factors for the formation of different structures in a propositional mechanism (Figure 7.55).

Sphere-shaped bouquet-like microscale aggregates (Figure 7.56) of bismuth telluride nanoplates were obtained by reducing bismuth chloride and orthotelluric acid with hydrazine in the presence of TGA (which serves as the shape- and size-directing agent) followed by room-temperature aging (which promotes nanoplate aggregation).[259] Adaptation of a scalable approach to synthesize and hierarchically assemble nanostructures with controlled doping could be attractive for tailoring novel thermoelectric materials for applications in high-efficiency refrigeration and harvesting electricity from heat.

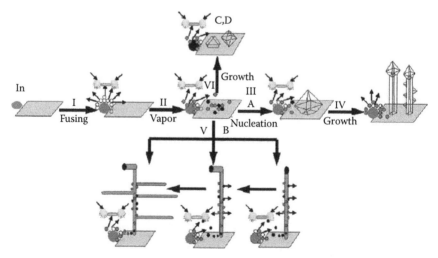

FIGURE 7.55 The formation process and structural models of the different submicron/nanostructures. (I) Formation of nanosized In droplets. (II) Conversion of In into In$_x$O (x = 1, 2) and In$_2$O$_3$ vapors (mainly In$_x$O). (III) Nucleation of In$_2$O$_3$ particles into octahedra and (IV) continuation of 1D growth along [100] until the formation of arrow-like wires. Also, indicating the bead-like wire growth with some octahedron attached on In$_2$O$_3$ submicron wires in zone A. (V) Schematic diagram of the formation of bouquets and comb-like nanowires in zone B. (VI) Schematic diagram of the formation of regular polyhedra in zones C and D. ○, Nanosized in; ●, In(v); ●, In$_x$O(v); ●, In$_2$O$_3$(v); ⟳, Oxygen. (Reproduced with permission from Yin, W. et al., Controllable synthesis of various In$_2$O$_3$ submicron/nanostructures using chemical vapor deposition, *Cryst. Growth Des.*, 9(5), 2173–2178. Copyright 2009 American Chemical Society.)

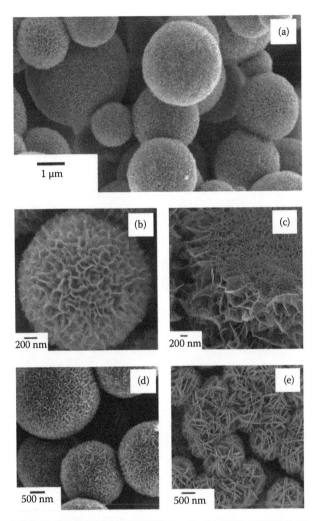

FIGURE 7.56 SEM images of bouquets of Bi_2Te_3 microspheres. (a) Low- and, (b) high-magnification SEM images of Bi_2Te_3 microspheres, (c) a cross-section showing that the microspheres are comprised of interconnected 4 ± 2 nm nanoplates. SEM micrographs from bismuth telluride microspheres synthesized using (d) 0.25 and (e) 2.50 vol% TGA capture the effect of the TGA concentration on the nanoplate thickness. (Reproduced with permission from Wang, T. et al., Microsphere bouquets of bismuth telluride nanoplates: Room-temperature synthesis and thermoelectric properties, *J. Phys. Chem. C*, 114(4), 1796–1799. Copyright 2010 American Chemical Society.)

7.8 NANODEWDROPS

Elemental metal dewdrop-like nanostructures are only available. Thus, the Ar^+-sputter-induced cone evolution on Mo-seeded Cu led to the growth of hillock-like crystallites on the cone surface, identified as Cu, and it grew selectively on the Mo lattice of seed layers.[260] It was proposed that a viscous overlayer of Cu on the growing Mo lattice condensed into dewdrop-like crystallites upon switching off the ion beam. A bimetallic Au-Ag heterostructured material (2 μm in diameter; 50 nm Ag nanoparticles) was prepared by a selective growing strategy of an Ag nanodewdrop on the petal tip of a Au flower using an electrochemical method with $AgNO_3$ and $HAuCl_4$ as precursors.[261] It was established that the presence of Ag nanodewdrops could facilitate the oxidation of $Ru(bpy)_3^{2+}$ complex in electrogenerated chemiluminescence measurements and dramatically enhance the emission intensity. The formed heterogeneous bimetallic alloy material (Figure 7.57) can be used for potential applications in chemical and biological sensors.

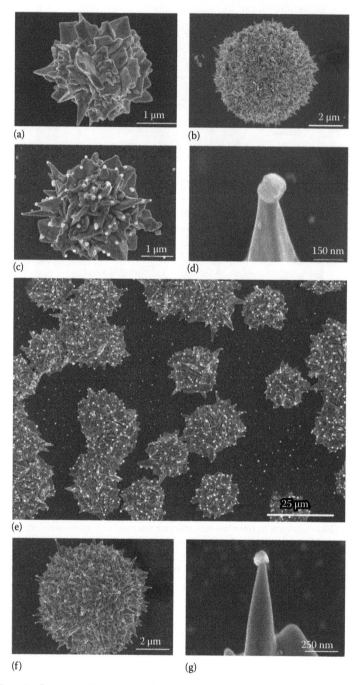

FIGURE 7.57 Bare Au flowers (a, b) and Au-Ag heterostructured flowers (HSFs) (c through g). The deposition times for (a) and (b) are 2 and 10 min, respectively. (c) Au-Ag HSF with (a) as the building block; (d) a detailed Au-Ag heterostructure of (c); (e) large-scale Au-Ag HSFs; (f) Au-Ag HSF with (b) as the building block; and (g) a detailed Au-Ag heterostructure of (f). (Reproduced with permission from Gao, L. et al., Selective growth of Ag nanodewdrops on Au nanostructures: A new type of bimetallic heterostructure, *Langmuir*, 25(19), 11844–11848. Copyright 2009 American Chemical Society.)

7.9 NANOACORNS

Nanoacorns belong to a nanoparticle type, consisting of two chemical species (such as CoPd sulfide, see the following), where the distribution of chemical species inside the particle becomes a determinant for its properties. Such nanoparticles are of standing interest since they can exhibit catalytic, electronic, optical, and magnetic properties distinct from those of nanoparticles comprising the corresponding single-chemical species. Acorn-like nanostructures are not well known and have been reported for bimetallic inorganic compounds, for instance Cu-InS (Figure 7.58), which were synthesized from the thermal decomposition of a mixture of a Cu-oleate and In-oleate complex in dodecanethiol.[262] By varying the reaction temperature and time, it was possible to synthesize Cu-In sulfide nanocrystals with acorn, bottle, and larva shapes. The anisotropically phase-segregated CoPd sulfide nanoparticles, named by the authors as "CoPd nanoacorns" (Figure 7.59) and consisting of crystalline Co_9S_8 and amorphous PdS_x phases with the Co_9S_8 (001) plane at their interface, were spontaneously

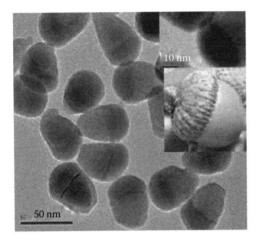

FIGURE 7.58 Cu-InS nanoacorns. (Reproduced with permission from Choi, S.-H. et al., One-pot synthesis of copper-indium sulfide nanocrystal heterostructures with acorn, bottle, and larva shapes, *J. Am. Chem. Soc.*, 128, 2520–2521. Copyright 2006 American Chemical Society.)

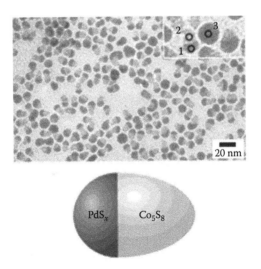

FIGURE 7.59 CoPd sulfide nanoacorns. (Reproduced with permission from Teranishi, T. et al., Nanoacorns: Anisotropically phase-segregated CoPd sulfide nanoparticles, *J. Am. Chem. Soc.*, 126(32), 9914–9915. Copyright 2004 American Chemical Society.)

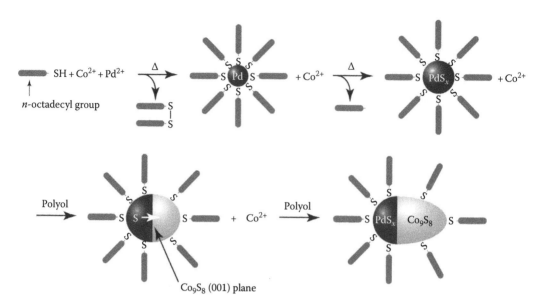

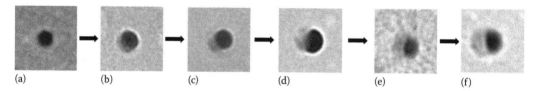

Co$_9$S$_8$ (001) plane

FIGURE 7.60 Schematic illustration of the speculated formation mechanism of the CoPd nanoacorns. (Reproduced with permission from Teranishi, T. et al., Nanoacorns: Anisotropically phase-segregated CoPd sulfide nanoparticles, *J. Am. Chem. Soc.*, 126(32), 9914–9915. Copyright 2004 American Chemical Society.)

FIGURE 7.61 Polymerization process: the primary seed particles (a), initially form an eye-ball-like morphology (b, c), followed by the formation of a moon-like (d), shape and the final acorn-shaped morphologies (e, f). (Misra, A. and Urban, M.W.: Acorn-shape polymeric nano-colloids: Synthesis and self-assembled films. *Macromol. Rapid Commun.* 2010. 31(2). 119–127. Copyright Wiley-VCH Verlag GmbH & Co. KGaA. Reproduced with permission.)

generated by reducing the corresponding metal precursors with 1,2-hexadecanediol in the presence of various alkanethiols.[263–265] Their formation mechanism is presented in Figure 7.60. It was found to spontaneously form through the anisotropic growth of the Co$_9$S$_8$ phase after the generation of PdS$_x$ nanoparticles.

As an organic example of acorn-like nanostructure, two distinct phase-separated copolymers within one colloidal particle, that is, poly(Me methacrylate) (PMMA)/*n*-butylacrylate (nBA) and poly(nBA)/pentafluorostyrene (p-PFS) phases, resulted in unique acorn-shaped morphologies (Figure 7.61), capable of coalescence.[266] The formed colloidal particles were found to be stable and able to self-assemble during coalescence, depending on the surface energy of a substrate.

7.10 NANOMULBERRY

Mulberry-like nanostructures are known for silicon and silica, as well as for some metal oxides and salts. Thus, nanocrystalline hydrogenated Si films, produced in the same reactor by hot wire and hot-wire plasma-assisted CVD, showed a mulberry-like grain agglomerate structure.[267] Modified silica (SiO$_2$) nanoparticles, prepared by the mechanochemical method,[268] were used to encapsulate a calcium carbonate (CaCO$_3$) particle surface to form mulberry-like composite particles through violent stirring, preventing agglomeration of SiO$_2$ nanoparticles. CaCO$_3$/SiO$_2$ composite particles

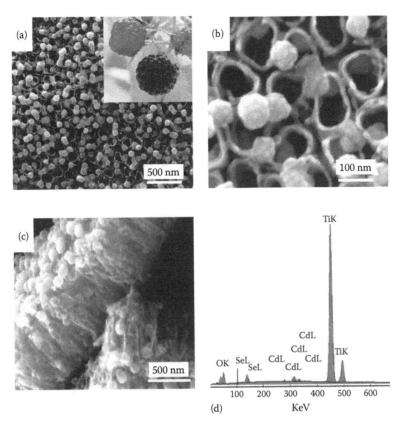

FIGURE 7.62 "Mulberry-like" CdSe nanocluster anchored TiO_2: (a) and (b) are top views obtained at two different magnifications; inset of (a) is the actual mulberry; (c) is the cross-section view; and (d) is the EDX spectrum. (Reproduced with permission from Zhang, H. et al., "Mulberry-like" CdSe nanoclusters anchored on TiO_2 nanotube arrays: A novel architecture for remarkable photo-energy conversion efficiency, *Chem. Mater.*, 21(14), 3090–3095. Copyright 2009 American Chemical Society.)

maintained the nanoeffect of nano-SiO_2 and had multiple surface structures, which led to vast potential applications in reinforcing filler of rubber and superhydrophobic surface preparation.[269] By simple mixed solvents, the morphologies (spherical, mulberry-like, nanospherical, and top-like products) of hematite (α-Fe_2O_3) particles could be manipulated by systematically adjusting the polarity of the mixed solvent *via* the ratio control.[270] It was established that the magnetic properties of the products were critically affected by their grain size and assembly morphologies. Mulberry-like crystalline CdSe nanoclusters (Figures 7.62 and 7.63) were assembled into vertically aligned TiO_2 nanotubes by photoassisted electrodeposition method to form the architecture for fabricating semiconductor nanocrystal sensitized photovoltaic cell.[271] The remarkable photovoltaic performances should be ascribed to the high-quality of 3D multijunction structure and the driving force for electron transfer in mulberry-like nanoclusters.

7.11 BROCCOLI-LIKE ARCHITECTURES

A few reports have been dedicated to clearly expressed broccoli-like nano- and microparticles. Thus, the multimorphological coordination polymer La(1,3,5-BTC)(H_2O)$_6$ was prepared at r.t. on a large scale (1,3,5-BTC = 1,3,5-benzenetricarboxylate), in particular in a broccoli-like (Figure 7.64) structural nanoform (together with sheaf-like, urchin-like, and fan-like hierarchical architectures made of uniform nanorods).[272] These 3D architectures can exhibit tunable white-light emission by co-doping Tb^{3+} and Eu^{3+}. Broccoli-like CNTs[273] on polycarbonate (PC) substrates were fabricated

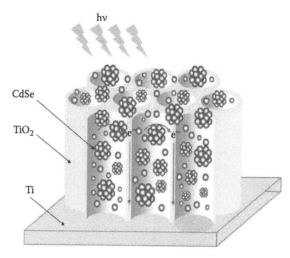

FIGURE 7.63 Structure of the "mulberry-like" CdSe nanocluster-anchored TiO$_2$ nanotube electrode. (Reproduced with permission from Zhang, H. et al., "Mulberry-like" CdSe nanoclusters anchored on TiO$_2$ nanotube arrays: A novel architecture for remarkable photo-energy conversion efficiency, *Chem. Mater.*, 21(14), 3090–3095. Copyright 2009 American Chemical Society.)

FIGURE 7.64 SEM images of the La(1,3,5-BTC)(H$_2$O)$_6$ broccoli-like architecture. (Reproduced with permission from Liu, K. et al., Room-temperature synthesis of multi-morphological coordination polymer and tunable white-light emission, *Cryst. Growth Des.*, 10(1), 16–19. Copyright 2010 American Chemical Society.)

according to the scheme in Figure 7.65 using Mo sputtering to polystyrene spheres (PS) on an Si substrate, ultrasonic removal of PS, and transferring to PC substrate. The products can be applied in various flexible devices, such as solar cells and displays.

7.12 NANOGRASSES

Nanograss (as well as *nanofur*) is an example of dynamically tunable nanostructured surfaces, whose behavior can be reversibly switched between superhydrophobic and hydrophilic states either actively, by the application of electric voltage, or passively, in response to the change in environmental conditions such as humidity.[274] Such "superlyophobic" surfaces that repel virtually any liquid even if the surface material by itself is highly wettable have potential applications in an exceptionally broad range, including areas as diverse as self-cleaning coatings, controllable permeability membranes for portable power generation, substrates for directed biological cell migration and differentiation, and hydrodynamic drag reduction surfaces.

A considerable number of "grass" reports is devoted to elemental silicon (Figure 7.66) and more rarely silica nanograsses, obtained by various methods, for example, plasma fabrication techniques,[275] in particular through hydrogen plasma etching alone, that is, without the use of a mask or a catalyst,[276]

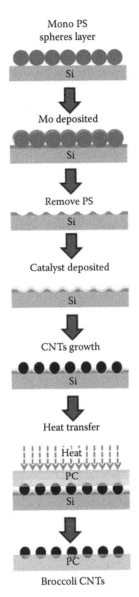

FIGURE 7.65 Schematic representation of broccoli-shaped CNTs on flexible substrates. (Reproduced with permission from Hsieh, K.-C. et al., Iridescence of patterned carbon nanotube forests on flexible substrates: From darkest materials to colorful films, *ACS Nano*, 4(3), 1327–1336. Copyright 2010 American Chemical Society.)

which led to a well-aligned silicon nanograss. Chemically modified silicon nanograss, combining nanoscale topography with lithographically defined chemical patterns, exhibited a high wettability gradient that allowed tailoring of complex droplet shapes.[277] The wetting properties of the silicon nanograss surfaces, adjusted by the photochemical immobilization of polymeric thin films,[278] can range from superwetting to impaled drops and superhydrophobicity ("condensation resistant" nanograss). The oxidized nanograss is completely wetting, while the polymer-coated nanograss has a contact angle of ca. 170° and is ultrahydrophobic. Highly sensitive SERS substrates were fabricated by metalizing (Ag or Au) Si nanograss fabricated by a Bosch process on single crystalline silicon.[279,280] It was revealed that the sensitivity of the substrates depends on the target molecules, the excitation laser wavelengths, and the metal coating on the silicon nanograss. Coating the nanostructurized surface of silicon nanograss with a thin layer of alumina (20–200 nm), subsequently released by sacrificial

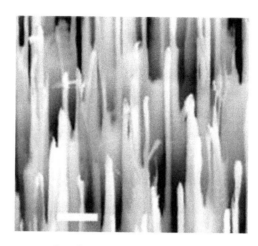

FIGURE 7.66 Silicon nanograss surface. The scale bar is 1 μm. (Reproduced with permission from Kumar, R.T.R. et al., Simple approach to superamphiphobic overhanging silicon nanostructures, *J. Phys. Chem. C*, 114(7), 2936–2940. Copyright 2010 American Chemical Society.)

plasma etching, suspended thin Al_2O_3 membranes with exceptional thermal properties offering several advantages over the traditional silicon nitride membranes were fabricated.[281] The effect of clamping ring materials and chuck temperature on the formation of silicon nanograss was studied.[282] Self-organized nanostructures in fused silica, interestingly termed as *glass grass*, which were produced by plasma dry etching methods, appeared as "grass," "needles," "pillars," or even "tubes" depending on etching conditions.[283] The bulk SiC/SiO_2 nanocomposites, synthesized without the presence of catalyst by high-frequency induction heating of SiO and activated carbon fibers at 1400°C for 20 min, exhibited the morphologies of self-assembly nanograss, nanocolumn, and pine-tree-branch-like nanostructure.[284] A simple fabrication of a superhydrophobic nanosurface consisted of a grass-like silica thin film on the inner wall of a glass tube, and its characteristics in water motion and water movement were reported.[285] Aqueous solutions containing a high content of metal ions were cross-moved without washing the tube used, and no cross-contamination occurred after cross-movement.

In addition to silicon, for other elemental nanograsses, carbon nanoforms in nanograss forms are known. Thus, a BDD nanograss array was prepared[287] simply on a heavily doped BDD film by reactive ion etching for use as an electrochemical sensor, which improved the reactive site, promoted the electrocatalytic activity, accelerated the electron transfer, and enhanced the selectivity. In particular, detection of catechol was performed on the as-grown BDD electrodes and the nanograss array BDD.[288] The nanograss array BDD showed higher electrocatalytic activity toward the catechol detection than did the as-grown BDD. Good linearity was observed for a concentration range from 5 to 100 μM with a sensitivity of 719.71 mA/M/cm^2 and a detection limit of 1.3 μM on the nanograss array BDD. Various morphologies of CNTs such as individuals, random networks parallel to the surface of the substrate ("grasses"), and vertically aligned forests of single- and MWCNTs were CVD-grown by varying the nominal thickness of the cobalt catalyst under the same reaction condition.[289] Longer growth time increased the CNT length, which caused further change in CNT morphologies from individuals to grasses and grasses to forests. Additionally, CNTs nanograss is applied as a gamma detector.[290]

In addition to SiO_2 discussed here, a few other oxide nanograsses are known, for example, for β-Ga_2O_3, which, together with nanowires, nanobelts, and nanosheets, were synthesized through microwave plasma of liquid phase Ga-containing H_2O in Ar atmosphere using Si as the substrate.[291] 2D SnO_2 nanograsses were prepared on single-crystal Si substrates by catalyst-assisted thermal evaporation.[292] The large field emission current from these products was observed at a high turn-on voltage, which was attributed to a shorter length and a wide emitter radius. Undesired TiO_2 etching-induced nanograss formation was observed[293] during extended exposure of the oxide tubes to a fluoride-containing electrolyte under obtaining anodic self-organized TiO_2 nanotube layers. To prevent this effect, on optimally pretreated surfaces thin rutile-type oxide layers can be formed in the initial stage of anodization; these layers can in the subsequent tube growth process efficiently protect the tube tops and thus allow the growth of nanotube layers with highly ordered and defined morphology.[294] Organic nanograsses are represented by a polymer system: a simple fabrication method for a moth-eye antireflective structure by using plasma-induced polymer nanograss was suggested on the basis of UV-curable photosetting polymer (SU8) and thermoplastic polymer (PMMA), exposed to plasma etching.[295]

7.13 NANOKELPS

Six kinds of SiC nanoarchitectures, including SiC nanowires, nanopyramids, and nanobones, were obtained by the thermal evaporation method, while nanokelps (Figures 7.67 and 7.68), nanoflowers, and nanocombs were achieved in the solution route.[296] Similar SiC kelp-like nanobelts were obtained through direct pyrolysis of methyltrichlorosilane at 600°C in an autoclave.[297] These SiC nanoarchitectures may have potential applications in the nanoelectronics, nanooptics, nanocomposites, catalysts, and other areas of nanoscience and nanotechnology.

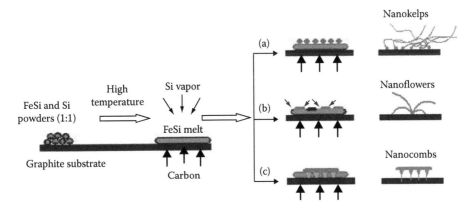

FIGURE 7.67 Schematic view of the formation of different SiC nanoarchitectures using FeSi and Si powders as source materials. (Reproduced with kind permission from Springer Science+Business Media: *Appl. Phys. A: Mater. Sci. Process.*, Thermal evaporation and solution strategies to novel nanoarchitectures of silicon carbide, 88(4), 2007, 679–685, Wu, R.B. et al.)

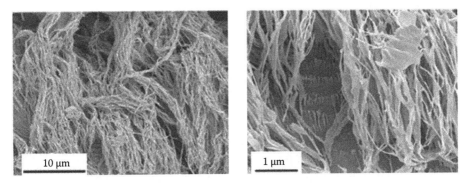

FIGURE 7.68 SiC nanokelps. (Reproduced with kind permission from Springer Science+Business Media: *Appl. Phys. A: Mater. Sci. Process.*, Thermal evaporation and solution strategies to novel nanoarchitectures of silicon carbide, 88(4), 2007, 679–685, Wu, R.B. et al.)

FIGURE 7.69 ZnO corn-like nanostructures. (Reproduced with permission from Wen, B. et al., Controllable growth of ZnO nanostructures by a simple solvothermal process, *J. Phys. Chem. C*, 112(1), 106–111. Copyright 2008 American Chemical Society.)

7.14 NANOCORNS

Nanocorns are not widespread structures and are known mainly for zinc compounds, for example, $ZnC_2O_4 \cdot 2H_2O$ nanorods, which were obtained in the form of nanocorn-like-rod, among other structures (nanosheet and nanochain), by a solvothermal-assisted heat treatment method.[298] The products decompose to ZnO by heating, resulting in nanorods adhibited to form nanosheets, and then rolled to form a nanocorn rod. Finally, nanocorn rod was dispersed to form nanochains. The solvothermal method starting from zinc nitrate, CTAB, and HMT at 80°C–90°C for 0.5–6 h for growing ZnO nanostructures with different morphologies, including nanocorns (Figure 7.69), is based on the following Reaction 7.23, combining two different coordination agents and adjusting the ratio of solvents and the reaction time.[154] Additionally, barium tungstate nanocorns with lengths of 200–800 nm and with diameters of 20–50 nm in the middle section were synthesized by a stepwise solution-phase method,[299] showing that several experimental parameters (the volume ratio of DMF/H$_2$O, and the quantity of urea and CTAB) played important roles for the morphological control of BaWO$_4$ nanostructures.

$$Zn^{2+} + CH_3(CH_2)_{15}N(Br)(CH_3)_3/C_6H_{12}N_4 \rightarrow Zn^{2+}\text{-amino complex} \rightarrow Zn(OH)_2 \rightarrow ZnO$$

$$(7.23)$$

7.15 NANOCACTUS

Cactus-like formations, made of nano- microrods or wires, are known for several metal oxides and salts. Thus, 1D CdS nanostructures, including flower- and cactus-like micro-/ nanorods and nanostructures, were synthesized by the electrochemical template deposition technique, using polycarbonate membranes, by controlling various reaction parameters.[300] It was found that apart from the dimensions of the pores of the templates, the geometrical morphologies of the CdS 1D nanostructures were significantly influenced by the synthesizing parameters also. A related work[301] described CdS cactus-like, hierarchical nanostructures, consisting of tiny nanoneedles projecting out from the outer surfaces of parent CdS nanotube motifs. Cactus-like β-Ga$_2$O$_3$ nanostructures (Figure 7.70), composed of a hollow microsphere and numerous β-Ga$_2$O$_3$ nanowires grown from the surface, were synthesized by a carbon thermal reduction process.[302] The field emission properties of the products revealed the turn-on fields of 12.6 V/μm and the threshold fields of 23.2 V/μm of the product. A schematic illustration of the formation of hollow cactus-like β-Ga$_2$O$_3$ nanostructures is shown in Figure 7.71. The growth mechanism, proposed by the authors, includes the reduction of initial Ga$_2$O$_3$ by graphite and

(a) (b) (c) (d)

FIGURE 7.70 (a, b) Low-magnification SEM images of the cactus-like β-Ga$_2$O$_3$ nanostructures. (c) SEM image of high-density Ga$_2$O$_3$ nanowires aligned on the sphere surface. (d) Local-magnification SEM image of the tip of the nanowire. (Reproduced with permission from Cao, C. et al., Growth and field emission properties of cactus-like gallium oxide nanostructures, *J. Phys. Chem. C*, 112, 95–98. Copyright 2008 American Chemical Society.)

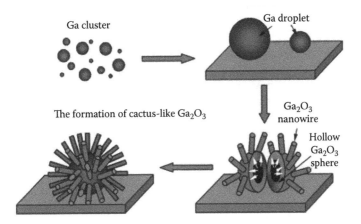

FIGURE 7.71 Schematic illustration of the formation of hollow cactus-like β-Ga$_2$O$_3$ nanostructures. The whole process includes the formation of Ga spheres, oxidation of the spheres, evaporation of the inner Ga in the spheres and nanowire growth from the surface of spheres, and finally the formation of hollow cactus-like β-Ga$_2$O$_3$ nanostructures. (Reproduced with permission from Cao, C. et al., Growth and field emission properties of cactus-like gallium oxide nanostructures, *J. Phys. Chem. C*, 112, 95–98. Copyright 2008 American Chemical Society.)

FIGURE 7.72 GeO$_2$ cactus-like formations. (From Hidalgo, P. et al., *Nanotechnology*, 16, 2521, 2005. Reproduced with permission from IOP Science.)

CO to form Ga$_2$O vapor (Reactions 7.24 and 7.25), interaction of Ga$_2$O with CO (Reaction 7.26) resulting in Ga microspheres, further oxidation of Ga by residual oxygen yielding a thin Ga$_2$O$_3$ layer on the Ga sphere surface, evaporation of inner Ga producing hollow Ga$_2$O$_3$ microspheres, and, finally, epitaxial growth of nanowires from the formed nuclei on the surface of these microspheres:

$$Ga_2O_3 + 2C \rightarrow Ga_2O + 2CO \tag{7.24}$$

$$Ga_2O_3 + 2CO \rightarrow Ga_2O + 2CO_2 \tag{7.25}$$

$$Ga_2O + 3CO \rightarrow 2Ga + C + 2CO_2 \tag{7.26}$$

Mysterious micrometer-size GeO$_2$ cactus-like formations (Figure 7.72),[303] which were obtained from a compacted GeO$_2$ powder as a precursor by its thermal treatment at 1100°C under argon flow, showed enhanced CL emission and a slight spectral blue shift as compared with the starting material. Also, ZnO cactus-like (among many others) nanostructures (Figure 7.73), representing a typical wonderful "nano-north-Mexican" landscape, were obtained from Zn(NO$_3$)$_2$ and hexamethyltetramine as precursors.[304] The scheme of their hierarchical growth is shown in Figure 7.74 and, as it is seen, this is different in comparison with the aforementioned mechanism for Ga$_2$O$_3$ nanocactus formation. In a related report, ZnO thin films, nanorods, nanotowers, and nanocactuses were grown on Si substrates in MOCVD process as a function of growth conditions including source flow rates and growth temperatures.[305] This variety of nanostructures was explained by the authors due to the 3D growth behavior of ZnO. It is well known that ZnO is very promising for applications in optical devices such as light-emitting diodes and laser diodes with high efficiency, covering blue and ultraviolet ranges, due to its wide band gap of 3.37 eV and large exciton binding energy of 60 meV. In this respect, studies of novel ZnO architectures could expand the range for these and other useful applications.

7.16 NANOSPINES

A few nanospine reports are available. Thus, layered titanate nanosheets and anatase-type titania nanospines were selectively grown from a molecular titanium precursor, TiF$_4$, by a reaction with various cationic species in an agar gel matrix.[306] A spherical architecture of anatase nanospines was produced through a moderate reaction with ammonia provided by decomposition

(a)

(b)

FIGURE 7.73 Micropatterned array of tertiary ZnO "cactus" structures (SEM images), produced with four growth stages on a microstamped Ag substrate: (a) micropatterned array viewed at an oblique angle and (b) dense array viewed normal to the substrate. (Sounart, T.L. et al.: Sequential nucleation and growth of complex nanostructured films. *Adv. Funct. Mater.* 2006. 16. 335–344. Copyright Wiley-VCH Verlag GmbH & Co. KGaA. Reproduced with permission.)

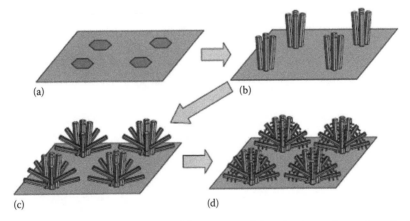

(a)   (b)

(c)   (d)

FIGURE 7.74 Schematic illustration of the multistep, sequential nucleation and growth method leading to truly hierarchical structures: (a) creation of nucleation centers on a substrate by, for example, micropatterning or microstamping; (b) growth of patterned nanorods on a substrate; (c) secondary growth from the patterned nanorods; and (d) tertiary growth from the secondary rods. (Sounart, T.L. et al.: Sequential nucleation and growth of complex nanostructured films. *Adv. Funct. Mater.* 2006. 16. 335–344. Copyright Wiley-VCH Verlag GmbH & Co. KGaA. Reproduced with permission.)

FIGURE 7.75 Arrays of BN sheathed ZnS nanospines. (Reproduced with kind permission from Springer Science+Business Media: *Nano Res.*, Crystallographically-oriented nanoassembly by ZnS tricrystals and subsequent 3D epitaxy, 2(9), 2009, 688–694, Zhu, Y. et al.)

of urea. Other spine-like nanostructures are represented by Au,[307] graphite,[308] and ZnS-BN[309] (important as semiconductors) (Figure 7.75).

7.17 NANOSHEAFS

Sheaf-like nanostructures are relatively widespread among less-common nanoforms; curiously, a considerable part of reports has been dedicated to lanthanide compounds, in particular cerium. Among elemental metals, their oxides, and hydroxides, Ni wire was synthesized by microwave heating (MW, playing a crucial role in stabilizing the pure Ni phase) of the preorganized rod microstructure of crystals consisting of Ni^{2+} ions and triethanolamine (TEA).[310] It was established that, for an Ni-TEA crystal, the TEA molecules give submicrometer-sized needle bundles that appeared as a sheaf of straw tied in the middle (Figure 7.76). The authors showed that the organized structure stemmed from the H bonding and that TEA is a potent organic ligand due to its ability to form an organized structure through hydrogen bonding for synthesizing the Ni wire. The applied synthesis method could be extended to synthesize other magnetic metals and alloys in air and aqueous medium. Sheaf-like CuO nanostructures were prepared by a hydrothermal process conducted at 120°C for 24 h.[311] It was found that the sheaf-like CuO anode material for lithium-ion batteries can exhibit a high initial discharge

(a)

(b)

FIGURE 7.76 (a) Optical micrograph and (b) FESEM image of the Ni-TEA crystal. (Reproduced with permission from Sahu, R.K. et al., Microwave-assisted synthesis of magnetic Ni wire from a metal-organic precursor containing Ni(II) and triethanolamine, *Cryst. Growth Des.*, 8(10), 3754–3760. Copyright 2008 American Chemical Society.)

FIGURE 7.77 Representative SEM images of the nanostructures constituted by CeO_2 nanowires. The individual nanowires are about 10 nm thick and over 15 μm long. The scale bars represent 5 μm. (Reproduced with permission from Kim, S. et al., Anomalous electrical conductivity of nanosheaves of CeO_2, *Chem. Mater.*, 21(7), 1182–1186. Copyright 2009 American Chemical Society.)

capacity of 965 mAh/g. After 41 cycles, the electrode can deliver a capacity of 580 mAh/g. In a related research,[312] sheaf-like CuO consisting of nanoplatelets were also prepared by a hydrothermal method at 100°C from $Cu(NO_3)_2 \cdot 3H_2O$ and NaOH in the presence of an ionic liquid 1-butyl-3-Me imidazolium tetrafluoroborate ($[BMIM]BF_4$), served as a cosolvent and modifiers in the reaction system. β-FeO(OH) (akaganeite-Q) nanostructures with a sheaf-like morphology were obtained *via* the hydrolysis of $FeCl_3$ solutions at different temperatures.[313] The individual filaments had an average diameter of ~60 nm, and the sheaves are ~3.2 μm in length at 80°C for 10 h.

A 1D CeO_2 nanostructure with a sheaf-like morphology (Figure 7.77a), exhibiting the O-ionic conductivity distinctively higher than that of conventional CeO_2 electrolyte, was obtained on Si substrates from $Ce(NO_2)_3 \cdot 6H_2O$, and HMT ($C_6H_{12}N_4$ as precursors.[314] It was established that the O nonstoichiometry in the CeO_2 nanowires constituting the sheaf was very small, which was attributed to enhanced O-ion mobility at the interfaces between CeO_2 nanowires in the sheaf rather than to increased charge-carrier concentration, which is often responsible for enhanced ionic conductivity in nanostructured ionic conductors. The sheaf (Figure 7.77a) of CeO_2 nanowires, being about 20 μm long and less than 10 nm thick, initially grew and developed to form the structure with a "pom-pom-ball"-like morphology (Figure 7.77h). According to the authors' opinion, the shown sequential images of splitting indeed suggested that the splitting mechanism can possibly explain the crystal developments in CeO_2 nanostructures.

Hierarchically nanostructured coordination polymer $Ce(1,3,5-BTC)(H_2O)_6$ architectures with tunable morphologies were prepared on a large scale from $Ce(NO_3)_3 \cdot 6H_2O$ and 1,3,5-benzentricarboxylic acid ($1,3,5-H_3BTC$) *via* a rapid solution phase method at r.t.[315] By rationally adjusting the synthetic parameters such as concentration, reaction temperature, surfactant, solvent, static, and ultrasonic treatment, $Ce(1,3,5-BTC)(H_2O)_6$ with straw-sheaf-like (Figure 7.78), flower-like, wheatear-like, straw-like, bundle-like, and urchin-like architectures, and nanorods can be selectively obtained. It was established that the presence of ultrasonic treatment could interrupt the crystal growth and crystal splitting and most importantly result in the morphology evolution and size modulation.

Sheaf-like bundles of $CePO_4$ nanostructures were reported, indicating that the synthesis mechanism likely involved a crystal-splitting step.[316] The products demonstrated an intense redox-sensitive green photoluminescence, which was exploited, in addition to their inherently high biocompatibility and low toxicity, for potential applications in biological imaging and labeling of cells. Sheaf-like Tb phosphate hydrate $TbPO_4 \cdot H_2O$ hierarchical architectures (Figure 7.79), composed of filamentary nanorods and useful for optoelectronic devices, were fabricated by a hydrothermal method from $Tb(NO_3)_3 \cdot 6H_2O$,

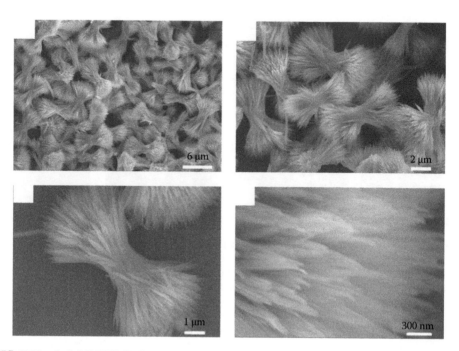

FIGURE 7.78 $Ce(1,3,5\text{-BTC})(H_2O)_6$ strawsheaf-like architectures composed of nanowires at different magnifications. (Reproduced with permission from Liu, K. et al., Hierarchically nanostructured coordination polymer: Facile and rapid fabrication and tunable morphologies, *Cryst. Growth Des.*, 10(2), 790–797. Copyright 2010 American Chemical Society.)

disodium ethylenediamine tetraacetate, CTAB, and $(NH_4)_2HPO_4 \cdot 12H_2O$.[317] Shape evolution of the half sheaf-like structure of $TbPO_4 \cdot H_2O$ is presented in Figure 7.80. It was revealed that the formation of the sheaf-like $TbPO_4 \cdot H_2O$ architectures followed the crystal-splitting theory.

EDTA was shown to mediate the splitting of $LnVO_4$ (Ln = Ce and Nd) nanocrystals from rods to sheaves and to spherulites, depending on reaction conditions (Table 7.2) under hydrothermal conditions with remarkable controllability and uniformity.[318] The roles of EDTA probably were to chelate the Ln^{3+} ions in solution, thus decreasing the nanocrystal nucleation rate but increasing the growth rate, and interfere with the nanocrystal growth by capping the nanocrystal surfaces.

Complex Sb_2S_3 and Sb_2Se_3 nanostructures with a sheaf-like hierarchical morphology were prepared[319] on a large scale at 180°C by a hydrothermal method in the presence of PVP playing a key role in their formation. According to the proposed mechanism (Figure 7.81), the PVP-stabilized nanorod seeds, formed at the initial step, developed in subsequent growth stages to a dumbbell structure, and completed their development as a closed sphere with an equatorial notch. The obtained materials had a band gap of 1.56 eV for Sb_2S_3 and 1.13 eV for Sb_2Se_3. Bi_2S_3 nanostructures with a sheaf-like morphology were obtained by the colloidal solution method *via* reaction of Bi acetate-oleic acid complex with elemental sulfur in 1-octadecene.[320] These structures may form by the splitting crystal growth mechanism (Figure 7.82), known for the formation of some mineral crystals in nature. In a surfactant system of TOA and oleic acid, $H_2Fe_3(CO)_9P(t\text{-Bu})$ reacted to form $Fe_4(CO)_{12}\{P(t\text{-Bu})\}_2$, which decomposed to give Fe_2P nanorods and "bundles"; addition of increasing amounts of oleic acid resulted in crystal splitting, while the addition of microliter amounts of an alkane enhanced the crystal splitting to give sheaf-like structures (Figure 7.83).[321] Other salts are represented by olivine-type $LiFePO_4$ crystals, prepared from a precursor phase $Fe_3(PO_4)_2(H_2O)_8$ by a reaction with LiOH under a hydrothermal condition[322] using L-(+)-ascorbic acid as a reducing and capping agent resulting in, in particular, sheaf-like

FIGURE 7.79 SEM images of the TbPO$_4$·H$_2$O products annealed at different temperatures for 2 h: (a) 90°C; (b) 250°C; (c) 450°C; (d) 650°C; (e) 700°C; (f) 750°C (inset is the partial enlarged drawing); (g) 800°C; (h) 1150°C. (Reproduced with permission from Yang, M. et al., Synthesis and luminescence properties of sheaflike TbPO$_4$ hierarchical architectures with different phase structures, *J. Phys. Chem. C*, 113(47), 20173–20177. Copyright 2009 American Chemical Society.)

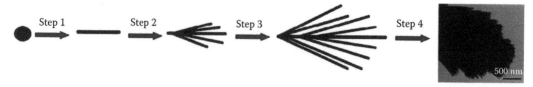

FIGURE 7.80 Schematic illustration of the formation and shape evolution of half sheaf-like structure of $TbPO_4 \cdot H_2O$. (Reproduced with permission from Yang, M. et al., Synthesis and luminescence properties of sheaflike $TbPO_4$ hierarchical architectures with different phase structures, *J. Phys. Chem. C*, 113(47), 20173–20177. Copyright 2009 American Chemical Society.)

TABLE 7.2
Morphologies of the $CeVO_4$ Samples Obtained with Different Reactant Ratios[a]

$Ce(NO_3)_3$	Na_3VO_4	EDTA	Morphology of $CeVO_4$
1	1	0	Nanorods
1	1	1	Nanowires
1	1	1.5	Straw-sheaf
2	2	3	Straw-sheaf with larger fantails
3	3	4.5	Double-head broccoli

Source: Reproduced with permission from Deng, H. et al., Additive-mediated splitting of lanthanide orthovanadate nanocrystals in water: Morphological evolution from rods to sheaves and to spherulites, *Cryst. Growth Des.*, 8(12), 4432–4439. Copyright 2008 American Chemical Society.

[a] The hydrothermal reactions were at 180°C and lasted for 24 h. The amounts of the reactants are in units of mmol.

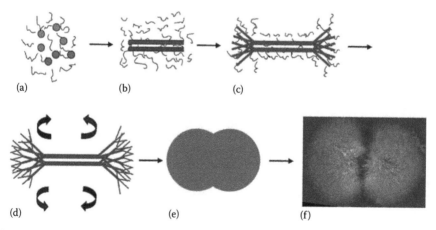

(a)   (b)   (c)

(d)   (e)   (f)

FIGURE 7.81 The proposed formation mechanism of hierarchical Sb_2S_3 nanostructures made of nanorods: (a) Nucleation of nanoparticles. (b) Growth of nanorod crystals controlled and stabilized by PVP molecules. (c, d) Branching at the ends of the primary rods and formation of dumbbell intermediates. (e, f) Final sphere with an equatorial notch. (Reproduced with permission from Chen, G.-Y. et al., The fractal splitting growth of Sb_2S_3 and Sb_2Se_3 hierarchical nanostructures, *J. Phys. Chem. C*, 112(3), 672–679. Copyright 2008 American Chemical Society.)

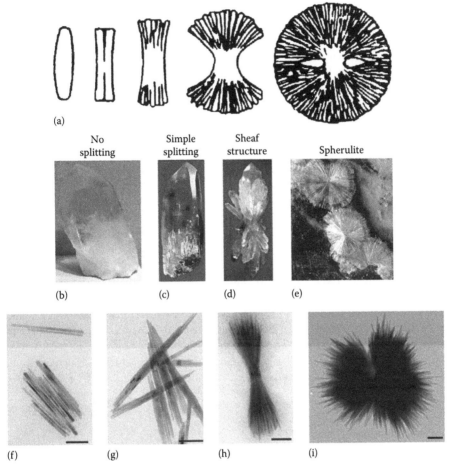

FIGURE 7.82 Schematic illustration of crystal splitting and examples of different forms of splitting in minerals and our synthesized Bi_2S_3 nanostructures. Part (a) shows the successive stages of splitting during crystal growth. Panels (b) through (e) show photographs of minerals that take on different forms of crystal splitting. Panel (b) shows a picture of quartz without splitting, while panels (c, d) show the late stage of simple splitting and early stage of sheaf splitting of quartz, respectively. Panel (e) is a wavellite spherulite. Panels (f) through (i) show representative TEM images of Bi_2S_3 nanostructures with different forms of splitting. All the scale bars in panels (f) through (i) denote 100 nm. (Reproduced with permission from Tang, J. et al., Crystal splitting in the Growth of Bi_2S_3, *Nano Lett.*, 6(12), 2701–2706. Copyright 2006 American Chemical Society.)

FIGURE 7.83 Effect of addition of alkanes on the formation of Fe_2P sheaf-like structures. All scale bars represent 200 nm. (Reproduced with permission from Kelly, A.T. et al., Iron phosphide nanostructures produced from a single-source organometallic precursor: Nanorods, bundles, crosses, and spherulites, *Nano Lett.*, 7(9), 2920–2925. Copyright 2007 American Chemical Society.)

mesocrystals of LiFePO$_4$ nanorods with high crystallinity and porous structure. Additionally, BaWO$_4$ sheaf-like nanostructures with different degrees of crystal splitting were synthesized and characterized.[323]

In the case of organic compounds, reported in the available literature, racemic random polylactide (PDLLA), which is completely amorphous in nature and mainly prepared to be of spherical form for the use of sustained drug release, was fabricated in the forms of lath-, sheaf-like morphologies of PDLLA microarchitectures by the double emulsion-solvent evaporation in the presence of glycerol or epirubicin.[324] It was found that the formation of lath-like microarchitectures was due to glycerol, while the formation of sheaf-like microarchitectures was owing to both glycerol and epirubicin.

7.18 NANOONIONS

Onion-like nanostructures[325] are known for a variety of elements and compounds, first of all for carbon. Thus, preparation, functionalization, and characterization of carbon nanoonions were generalized.[326,327] Generally, CNOs are fabricated by catalytic thermal decomposition of hydrocarbons and their derivatives. For example, nanocarbon materials were synthesized[328] by the catalytic decomposition of acetylene at 400°C by using Fe/Al$_2$O$_3$ as a catalyst. It was shown that nanoonion-like fullerenes (NOLF) (Figure 7.84) encapsulating a Fe$_3$C core were obtained, with a structure of stacked graphitic fragments, with diameters ranging from 15 to 50 nm. When the product was further heat-treated at 1100°C for 2 h, NOLF with a clear concentric graphitic layer structure were obtained. The growth mechanism of NOLF encapsulating metal cores was suggested (Figure 7.85) to follow a vapor-solid growth model.

Onion-like carbon nanoparticles (30–100 nm in diameter) were achieved to be prepared by thermolysis of an NaN$_3$-C$_6$Cl$_6$ mixture in Ar and air atmosphere (Reaction 7.27).[329] Their formation

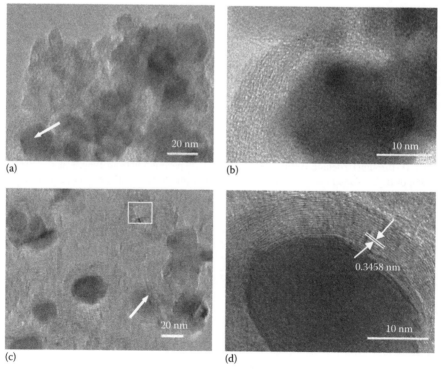

FIGURE 7.84 HRTEM images of the product. (a) Low-magnification image; (b) an individual NOLF; (c) low-magnification image of the product after heat treatment; (d) an individual NOLF. (Reproduced from Liu, X. et al., *Chin. Sci. Bull.*, 54(1), 137, 2009. With permission.)

C$_2$H$_2$

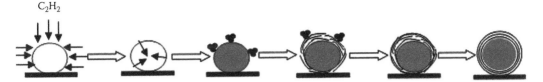

FIGURE 7.85 Suggested VS growth model. (Reproduced from Liu, X. et al., *Chin. Sci. Bull.*, 54(1), 137, 2009. With permission.)

was likely caused by a shock wave and a rapid increase in pressure during thermolysis. CNO were also efficiently produced[330] by reductive carbonization of C$_2$Cl$_6$ using sodium azide under the presence of a ferrocene. This self-induced growth process could produce nanofibers and nanotubes ca. 20–100 nm in diameter and several microns long. Additionally, high-yield synthesis of CNTs and nanoonions (CNOs) on a catalytic nickel substrate using counterflow diffusion flames was investigated.[331] When oxygen was increased to 50% in the lower flow, only CNOs were synthesized. An increase in methane concentration from 15% to 45% led to a higher yield and a greater diameter (ranging from 5 to 60 nm) of CNOs. Additionally, it was found[332] that hydrocarbon nanotubes (HCNTs) and nanoonions (HCNOs) (hereafter collectively referred to as HCNT(O)s) can be converted to CNTs and CNOs (hereafter collectively referred to as CNT(O)s) upon prolonged sonication *via* hybrid CNT ⊂ HCNT and CNO ⊂ HCNO nanostructures as intermediates, respectively:

$$6NaN_3 + C_6Cl_6 \rightarrow 6C + 6NaCl + 9N_2 \tag{7.27}$$

The surface of CNO were functionalized[333] (reaction 7.28) with pyridyl groups by first oxidizing the surface by reaction with H$_2$SO$_4$/HNO$_3$ and then reacting the carboxylated nanoonions with 4-aminopyridine to give the amidated product. In order to demonstrate that these soluble functionalized CNOs could be used to generate self-assembled nanoporous networks, they were shown to complex with zinc tetraphenylporphyrin forming **6**. In addition, a ferrocene-carbon onion derivative, where ferrocene acted as an electron-donating moiety, while the CNO serves as the electron acceptor, was synthesized.[334] On average, the CNOs had a spherical appearance with six shells. It was established that, after absorption of a photon at low energy, there is emission from CNOs characterized by larger external shells and a lower degree of functionalization.

$$\tag{7.28}$$

6

Onion-like fullerenes (OLFs) are known; they can be generated by arc discharge in water between pure graphite electrodes.[335] It was revealed that the OLFs with diameter about 25 nm were of a high degree of graphitization. The structural stability of CNO $C_{20}@C_{60}@C_{240}$ (Figure 7.86) was investigated by performing molecular dynamics computer simulations.[336] It was found that CNO is not so resistive against heat treatment, nor is it as strong as isolated single carbon nanoballs. Although single nanoballs resist heat treatment up to 4300 K, nanoonion disintegrates after 2600 K. Additionally, a fullerene-like material ($C_{48}N_{12}$ aza-fullerene) consisting of cross-linked nanoonions of C and N was discovered.[337] Growth of the onion shells takes place atom by atom on a substrate surface and yields thin solid films during magnetron sputter deposition.

Carbon-onion-supported platinum nanoparticles were synthesized using a simplified arc-discharge-in-solution method.[338] Pt nanoparticles with diameters of 5 nm were found to form during reduction of chloroplatinic acid and were simultaneously supported on carbon onions formed in the arc discharge process. Additionally, CNOs can be hydrogenated[339] or form composites, such as onion-like carbon-PMMA composite films.[340]

Other onion-like nanostructures are rare; the examples are nanoonions of threefold Fe and $CoFe_2O_4$,[341] MoS_2,[342] and Au:Fe:Au.[343]

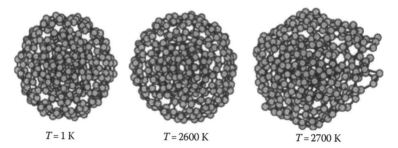

| $T = 1$ K | $T = 2600$ K | $T = 2700$ K |

FIGURE 7.86 The structures of carbon nanoonion $C_{20}@C_{60}@C_{240}$ at various temperatures. Relaxed structures at 1, 2600, and 2700 K. (Reproduced with permission from Erkoc, S., Stability of carbon nanoonion C20@C60@ C240: Molecular dynamics simulations, *Nano Lett.*, 2(3), 215–217. Copyright 2002 American Chemical Society.)

REFERENCES

1. Gentile, P.; David, T.; Dhalluin, F.; Buttard, D.; Pauc, N.; Den Hertog, M.; Ferret, P.; Baron, T. The growth of small diameter silicon nanowires to nanotrees. *Nanotechnology*, 2008, *19* (12), 125608/1–125608/5.
2. Menon, M.; Richter, E.; Lee, I.; Raghavan, P. Si-nanotrees: Structure and electronic properties. *Journal of Computational and Theoretical Nanoscience*, 2007, *4* (2), 252–256.
3. Markussen, T.; Jauho, A.-P.; Brandbyge, M. Surface-decorated silicon nanowires. A route to high-ZT thermoelectrics. *Physical Review Letters*, 2009, *103* (5), 055502/1–055502/4.
4. Oehler, F.; Gentile, P.; Baron, T.; Den Hertog, M.; Rouviere, J.; Ferret, P. The morphology of silicon nanowires grown in the presence of trimethylaluminium. *Nanotechnology*, 2009, *20* (24), 245602/1–245602/7.
5. Cho, S. O.; Lee, E. J.; Lee, H. M.; Kim, J. G.; Kim, Y. J. Controlled synthesis of abundantly branched, hierarchical nanotrees by electron irradiation of polymers. *Advanced Materials*, 2006, *18* (1), 60–65.
6. Tian, H.; Gao, X.; Yang, T.; Li, D.; Chen, Y. Fabrication and characterization of superhydrophobic silica nanotrees. *Journal of Sol-Gel Science and Technology*, 2008, *48* (3), 277–282.
7. Kalem, S. Self-organization of ammonium silicon hexafluoride complex low-dimensional structures on Silicon. *Superlattices and Microstructures*, 2008, *44* (4–5), 705–713.
8. Sola, F.; Resto, O.; Biaggi-Labiosa, A.; Fonseca, L. F. Growth and characterization of branched carbon nanostructures arrays in nano-patterned surfaces from porous silicon substrates. *Micron*, 2009, *40*, 80–84.
9. Yao, Z.; Zhu, X.; Li, X.; Xie, Y. Synthesis of novel Y-junction hollow carbon nanotrees. *Carbon*, 2007, *45* (7), 1566–1570.
10. Liu, G.; Zhao, Y.; Zheng, K; Liu, Z.; Ma, W.; Ren, Y.; Xie, S.; Sun, L. Coulomb explosion: A novel approach to separate single-walled carbon nanotubes from their bundle. *Nano Letters*, 2009, *9* (1), 239–244.
11. Haba, M. Fuel cell using carbon-metal nanotree electrocatalyst. JP 2006294493, 2006, 11 pp.
12. Sola, F.; Resto, O.; Biaggi-Labiosa, A.; Fonseca, L. F. *Micron*, 2009, *40*, 80.
13. http://www.ifn.upr.edu/people/17-luis-fonseca, accessed January 31, 2011.
14. Hasegawa, A.; Mitsuishi, K.; Furuya, K.; Nanotree structures and their formation by vapor-phase deposition process with electron-beam irradiation. JP 2005111645 A 20050428, 2005, 7 pp.
15. Yang, L.; Sun, B.; Meng, F.; Zhang, M.; Chen, X.; Li, M.; Liu, J. One-step synthesis of UV-induced Pt nanotrees on the surface of DNA network. *Materials Research Bulletin*, 2009, *44* (6), 1270–1274.
16. Nadagouda, M. N.; Polshettiwar, V.; Varma, R. S. Self-assembly of palladium nanoparticles: Synthesis of nanobelts, nanoplates and nanotrees using vitamin B1, and their application in carbon-carbon coupling reactions. *Journal of Materials Chemistry*, 2009, *19* (14), 2026–2031.
17. Jia, F.; Wong, K.-W.; Zhang, L. Electrochemical synthesis of nanostructured palladium of different morphology directly on gold substrate through a cyclic deposition/dissolution route. *Journal of Physical Chemistry C*, 2009, *113* (17), 7200–7206.
18. Nadagouda, M. N.; Varma, R. S. Silver trees: Chemistry on a TEM grid. *Australian Journal of Chemistry*, 2009, *62*, 260.
19. Kan, C.; Cai, W.; Hofmeister, H. Tree-like Ag nanostructures based on monolithic mesoporous silica. *Journal of Materials Research*, 2004, *19* (5), 1328–1332.
20. Risveden, K.; Dick, K. A.; Bhand, S.; Rydberg, P.; Samuelson, L.; Danielsson, B. Branched nanotrees with immobilized acetylcholine esterase for nanobiosensor applications. *Nanotechnology*, 2010, *21* (5), 055102/1–055102/8.
21. Dick, K. A.; Kodambaka, S.; Reuter, M. C.; Deppert, K.; Samuelson, L.; Seifert, W.; Wallenberg, L. R.; Ross, F. M. The morphology of axial and branched nanowire heterostructures. *Nano Letters*, 2007, *7* (6), 1817–1822.
22. Li, H. F.; Kar, A. K.; Parker, T.; Wang, G.-C.; Lu, T. M. The morphology and texture of Cu nanorod films grown by controlling the directional flux in physical vapor deposition. *Nanotechnology*, 2008, *19* (33), 335708/1–335708/8.
23. Yang, X.; Zhuang, J.; Li, X.; Chen, D.; Ouyang, G.; Mao, Z.; Han, Y.; He, Z.; Liang, C.; Wu, M.; Yu, J. C. Hierarchically nanostructured rutile arrays: Acid vapor oxidation growth and tunable morphologies. *ACS Nano*, 2009, *3* (5), 1212–1218.
24. Shibuya, M.; Miyauchi, M. Site-selective deposition of metal nanoparticles on aligned WO$_3$ nanotrees for super-hydrophilic thin films. *Advanced Materials*, 2009, *21* (13), 1373–1376.
25. Shibuya, M.; Miyauchi, M. Efficient electrochemical reaction in hexagonal WO$_3$ forests with a hierarchical nanostructure. *Chemical Physics Letters*, 2009, *473* (1–3), 126–130.
26. Chu, Y.; Hu, J.; Yang, W.; Wang, C.; Zhang, J. Z. Growth and characterization of highly branched nanostructures of magnetic nanoparticles. *Journal of Physical Chemistry B*, 2006, *110* (7), 3135–3139.

27. Jiang, H.; Chen, Y.; Zhou, Q.; Su, Y.; Xiao, H.; Zhu, L.-A. Temperature dependence of Ga_2O_3 micro/nanostructures via vapor phase growth. *Materials Chemistry and Physics*, 2007, *103* (1), 14–18.

28. Kumari, L.; Li, W. Z. Synthesis, structure and optical properties of zinc oxide hexagonal microprisms. *Crystal Research and Technology*, 2010, *45* (3), 311–315.

29. Liu, W.-C.; Cai, W. One-dimensional and quasi-one-dimensional ZnO nanostructures prepared by spray-pyrolysis-assisted thermal evaporation. *Applied Surface Science*, 2008, *254* (10), 3162–3166.

30. Zhao, F.; Li, X; Zheng, J.-G.; Yang, X.; Zhao, F.; Wong, K. S.; Wang, J.; Lin, W.; Wu, M.; Su, Q. ZnO pine-nanotree arrays grown from facile metal chemical corrosion and oxidation. *Chemistry of Materials*, 2008, *20* (4), 1197–1199.

31. Zhuo, R. F.; Qiao, L.; Feng, H. T.; Chen, J. T.; Yan, D.; Wu, Z. G.; Yan, P. X. Microwave absorption properties and the isotropic antenna mechanism of ZnO nanotrees. *Journal of Applied Physics*, 2008, *104* (9), 094101/1–094101/5.

32. Zhou, J.; Deng, S. Z.; Chen, J.; She, J. C.; Xu, N. S. Synthesis of crystalline alumina nanowires and nanotrees. *Chemical Physics Letters*, 2002, *365* (5–6), 505–508.

33. Cheng, W.; Steinhart, M.; Gösele, U.; Wehrspohn, R. B. Tree-like alumina nanopores generated in a non-steady-state anodization. *Journal of Materials Chemistry*, 2007, *17*, 3493–3495.

34. Hao, Y.; Meng, G.; Zhou, Y.; Kong, M.; Wei, Q.; Ye, M.; Zhang, L. Tuning the architecture of MgO nanostructures by chemical vapour transport and condensation. *Nanotechnology*, 2006, *17* (19), 5006–5012.

35. Yao, W.-T.; Yu, S.-H.; Liu, S.-J.; Chen, J.-P.; Liu, X.-M.; Li, F.-Q. Architectural control syntheses of CdS and CdSe nanoflowers, branched nanowires, and nanotrees via a solvothermal approach in a mixed solution and their photocatalytic property. *Journal of Physical Chemistry B*, 2006, *110* (24), 11704–11710.

36. Dick, K. A.; Deppert, K.; Martensson, T.; Seifert, W.; Samuelson, L. Growth of GaP nanotree structures by sequential seeding of 1D nanowires. *Journal of Crystal Growth*, 2004, *272* (1–4), 131–137.

37. Karlsson, L. S.; Larsson, M. W.; Malm, J.-O.; Wallenberg, L. R.; Dick, K. A.; Deppert, K.; Seifert, W.; Samuelson, L. Crystal structure of branched epitaxial III-V nanotrees. *NANO*, 2006, *1* (2), 139–151.

38. Lerner, E. J. Growing nanotrees. *Industrial Physicist*, 2004, *10* (5), 16–19. http://www.aip.org/tip/INPHFA/vol-10/iss-5/p16.pdf

39. Dick, K. A.; Geretovszky, Z.; Mikkelsen, A.; Karlsson, L. S.; Lundgren, E.; Malm, J.-O.; Andersen, J. N. et al. Improving InAs nanotree growth with composition-controlled Au-In nanoparticles. *Nanotechnology*, 2006, *17* (5), 1344–1350.

40. Dick, K. A.; Deppert, K.; Karlsson, L. S.; Seifert, W.; Wallenberg, L. R.; Samuelson, L. Position-controlled interconnected InAs nanowire networks. *Nano Letters*, 2006, *6* (12), 2842–2847.

41. Zheng, J.; Song, J.; Zhao, J.; Li, L.; Zhu, Z. Self-assembly of $FeCl_3$-S nanotrees and their application in directed growth of nanofibers. *Materials Letters*, 2008, *2* (25), 4069–4071.

42. Lu, Y.; Mei, Y.; Walker, R.; Ballauff, M.; Drechsler, M. 'Nano-tree'—Type spherical polymer brush particles as templates for metallic nanoparticles. *Polymer*, 47, 2006, 4985–4995.

43. Cahay, M.; Garre, K.; Fraser, J. W.; Lockwood, D. J.; Semet, V.; Thien Binh, V.; Kanchibotla, B.; Bandyopadhyay, S.; Grazulis, L.; Das, B. Field emission properties of metallic nanostructures self-assembled on nano-porous alumina and silicon templates. *Journal of Vacuum Science and Technology, B: Microelectronics and Nanometer Structures—Processing, Measurement, and Phenomena*, 2008, *26* (2), 885–890.

44. Cao, M.; Liu, T.; Sun, G.; Wu, X.; He, X.; Hu, C. Magnetic iron nitride nanodendrites. *Journal of Solid State Chemistry*, 2005, *178*, 2390–2393.

45. Cao, M.; Liu, T.; Gao, S.; Sun, G.; Wu, X.; Hu, C.; Wang, Z. L. Single-crystal dendritic micro-pines of magnetic α-Fe_2O_3: Large-scale synthesis, formation mechanism, and properties. *Angewandte Chemie*, 2005, *117* (27), 4269–4273.

46. Polshettiwar, V.; Baruwati, B.; Varma, R. S. Self-assembly of metal oxides into three-dimensional nanostructures: Synthesis and application in catalysis. *ACS Nano*, 2009, *3* (3), 728–736.

47. Yao, W.-T.; Yu, S.-H.; Liu, S.-J.; Chen, J.-P.; Liu, X.-M.; Li, F.-Q. Nanotrees via a solvothermal approach in a mixed solution and their photocatalytic property. *Journal of Physical Chemistry B*, 2006, *110* (24), 11704–11710.

48. Yang, Y.; Chai, Y.; Li, B.; Du, F. Controllable synthesis of hexagonal pine-like $Cd_{1-x}Zn_xSe$ nanotrees using the self-prepared precursors. *Materials Chemistry and Physics*, 2009, *116* (2–3), 335–338.

49. Bierman, M. J.; Albert Lau, Y. K.; Kvit, A. V.; Schmitt, A. L.; Jin, S. Dislocation-driven nanowire growth and Eshelby twist. *Science*, 2008, *320* (5879), 1060–1063.

50. Albert Lau, Y. K.; Chernak, D. J.; Bierman, M. J.; Jin, S. Formation of PbS nanowire pine trees driven by screw dislocations. *Journal of the American Chemical Society*, 2009, *131* (45), 16461–16471.

51. Sola, F.; Resto, O.; Biaggi-Labiosa, A. M.; Fonseca, L. F. Electron-beam induced growth of silica nanowires and silica/carbon heterostructures. *Materials Research Society Symposium Proceedings (Low-Dimensional Materials—Synthesis, Assembly, Property Scaling, and Modeling)*, San Francisco, CA, 2007, Vol. 1017E, Paper #: 1017-DD12-31.

52. Sola, F.; Resto, O.; Biaggi-Labiosa, A.; Fonseca, L. F. Electron-beam induced growth of silica nanorods and heterostructures in porous silicon. *Nanotechnology*, 2007, *18*, 405308.

53. Zhu, Y.; Sow, C.-H.; Yu, T.; Zhao, Q.; Li, P.; Shen, Z.; Yu, D.; Thong, J. T.-L. Co-synthesis of ZnO–CuO nanostructures by directly heating brass in air. *Advanced Functional Materials*, 2006, *16*, 2415–2422.

54. Zhao, J.; Tian, R.; Zhi, J. Deposition of silver nanoleaf film onto chemical vapor deposited diamond substrate and its application in surface-enhanced Raman scattering. *Thin Solid Films*, 2008, *516* (12), 4047–4052.

55. Dai, R.; Chen, N.; Zhang, X.; Peng, C.; Wu, J. Ferromagnetic MnSb films consisting of nanorods and nanoleaves. *Bandaoti Xuebao*, 2007, *28* (5), 661–664.

56. Ma, X.; Yuan, B. Fabrication of carbon nanoflowers by plasma-enhanced chemical vapor deposition. *Applied Surface Science*, 2009, *255* (18), 7846–7850.

57. Xu, H.; Wang, W.; Zhu, W.; Zhou, L.; Ruan, M. Hierarchical-oriented attachment: From one-dimensional $Cu(OH)_2$ nanowires to two-dimensional CuO nanoleaves. *Crystal Growth and Design*, 2007, *7* (12), 2720–2724.

58. Liang, Z.-H.; Zhu, Y.-J. Microwave-assisted synthesis of single-crystalline CuO nanoleaves. *Chemistry Letters*, 2004, *33* (10), 1314–1315.

59. Yang, Y.; Liao, Q.; Qi, J.; Guo, W.; Zhang, Y. Synthesis and transverse electromechanical characterization of single crystalline ZnO nanoleaves. *Physical Chemistry Chemical Physics*, 2010, *12* (3), 552–555.

60. Yang, L.; May, P. W.; Yin, L.; Scott, T. B. Growth of self-assembled ZnO nanoleaf from aqueous solution by pulsed laser ablation. *Nanotechnology*, 2007, *18* (21), 215602/1–215602/5.

61. Zhang, C.-Z.; Gao, H.; Zhang, D.; Zhang, X.-T. Local homoepitaxial growth and optical properties of ZnO polar nanoleaves. *Chinese Physics Letters*, 2008, *25* (1), 302–305.

62. Wang, B.; Ouyang, G.; Li, I. L.; Xu, P. Growth and photoluminescence of SnO_2 nanoleaves and nanopencils synthesized by Au-Ag alloying catalyst assisted carbothermal evaporation. *Nano Science and Nano Technology: An Indian Journal*, 2009, *3* (2), 21–27.

63. Xu, Y.-Y.; Dong, Z.; Zhang, X.-J.; Jin, W.-T.; Kashkarov, P.; Zhang, H. A freezing ferromagnetic moment model for exchange bias in α-Fe_2O_3 nanoleaves. *International Journal of Modern Physics C: Computational Physics, Physical Computation*, 2009, *20* (5), 761–768.

64. Xu, Y. Y.; Zhao, D.; Zhang, X. J.; Jin, W. T.; Kashkarov, P.; Zhang, H. Synthesis and characterization of single-crystalline α-Fe_2O_3 nanoleaves. *Physica E: Low-Dimensional Systems and Nanostructures*, 2009, *41* (5), 806–811.

65. Yang, C.; Yang, Z.; Gu, H.; Chang, C. K.; Gao, P.; Xu, B. Facet-selective 2D self-assembly of TiO_2 nanoleaves via supramolecular interactions. *Chemistry of Materials*, 2008, *20* (24), 7514–7520.

66. Liu, Y.; Ma, D.; Han, X.; Bao, X.; Frandsen, W.; Wang, D.; Su, D. Hydrothermal synthesis of microscale boehmite and gamma nanoleaves alumina. *Materials Letters*, 2008, *62* (8–9), 1297–1301.

67. Gulnar, A. K.; Sudarsan, V.; Vatsa, R. K.; Hubli, R. C.; Gautam, U. K.; Vinu, A., Tyagi, A.K. $CePO_4$:Ln (Ln = Tb^{3+} and Dy^{3+}) nanoleaves incorporated in silica sols. *Crystal Growth and Design*, 2009, *9* (5), 2451–2456.

68. Li, J.; Liu, J.; Wang, L.-S.; Chang, R. P. H. Physical and electrical properties of chemical vapor grown GaN nano/microstructures. *Inorganic Chemistry*, 2008, *47* (22), 10325–10329.

69. Martelli, F.; Piccin, M.; Bais, G.; Jabeen, F.; Ambrosini, S.; Rubini, S.; Franciosi, A. Photoluminescence of Mn-catalyzed GaAs nanowires grown by molecular beam epitaxy. *Nanotechnology*, 2007, *18* (12), 125603/1–125603/4.

70. Lu, T.; Han, Y. T.; Vittal, J. J. Morphology-controlled synthesis of Bi_2S_3 nanomaterials via single- and multiple-source approaches. *Crystal Growth and Design*, 2008, *8* (2), 734–738.

71. Zhang, Y. B.; Lau, S. P.; Li, H. F. Field emission from nanoforest carbon nanotubes grown on cobalt-containing amorphous carbon composite films. *Journal of Applied Physics*, 2007, *101* (3), 033524/1–033524/5.

72. Zhong, G.; Hofmann, S.; Yan, F.; Telg, H.; Warner, J. H.; Eder, D.; Thomsen, C.; Milne, W. I.; Robertson, J. Acetylene: A key growth precursor for single-walled carbon nanotube forests. *Journal of Physical Chemistry C*, 2009, *113* (40), 17321–17325.

73. Meshot, E. R.; Plata, D. L.; Tawfick, S.; Zhang, Y.; Verploegen, E. A.; Hart, A. J. Engineering vertically aligned carbon nanotube growth by decoupled thermal treatment of precursor and catalyst. *ACS Nano*, 2009, *3* (9), 2477–2486.

74. Hasegawa, K.; Noda, S.; Sugime, H.; Kakehi, K.; Maruyama, S.; Yamaguchi, Y. Growth window and possible mechanism of millimeter-thick single-walled carbon nanotube forests. *Journal of Nanoscience and Nanotechnology*, 2008, *8* (11), 6123–6128.

75. Zhang, Y. B.; Lau, S. P.; Huang, L.; Tay, B. K. Carbon nanotubes grown on cobalt-containing amorphous carbon composite films. *Diamond and Related Materials*, 2006, *15* (1), 171–175.

76. Wirth, C. T.; Zhang, C.; Zhong, G.; Hofmann, S.; Robertson, J. Diffusion- and reaction-limited growth of carbon nanotube forests. *ACS Nano*, 2009, *3* (11), 3560–3566.

77. Bedewy, M.; Meshot, E. R.; Guo, H.; Verploegen, E. A.; Lu, W.; Hart, A. J. Collective mechanism for the evolution and self-termination of vertically aligned carbon nanotube growth. *Journal of Physical Chemistry C*, 2009, *113* (48), 20576–20582.

78. Choi, B. H.; Yoo, H.; Kim, Y. B.; Lee, J. H. Effects of Al buffer layer on growth of highly vertically aligned carbon nanotube forests for in situ yarning. *Microelectronic Engineering*, 2010, *87* (5–8), 1500–1505.

79. Zhou, J. N. Multi-scale flow interaction with nanofabric structures. PhD thesis (partial fulfillment). California Institute of Technology, Pasadena, CA, 2005. http://etd.caltech.edu/etd/available/etd-04202005-172426/unrestricted/Jijie_ZHOU_dissertation.pdf

80. Huang, J.; Zhang, Q.; Zhao, M.; Wei, F. Process intensification by CO_2 for high quality carbon nanotube forest growth: Double-walled carbon nanotube convexity or single-walled carbon nanotube bowls? *Nano Research*, 2009, *2* (11), 872–881.

81. Gjerde, K.; Kumar, R. T. R.; Andersen, K. N.; Kjelstrup-Hansen, J.; Teo, K. B. K.; Milne, W. I.; Persson, C.; Molhave, K.; Rubahn, H.-G.; Boegild, P. On the suitability of carbon nanotube forests as non-stick surfaces for nanomanipulation. *Soft Matter*, 2008, *4* (3), 392–399.

82. Huynh, C. P.; Hawkins, S. C. Understanding the synthesis of directly spinnable carbon nanotube forests. *Carbon*, 2010, *48* (4), 1105–1115.

83. Vinten, P.; Marshall, P.; Lefebvre, J.; Finnie, P. Distinct termination morphologies for vertically aligned carbon nanotube forests. *Nanotechnology*, 2010, *21* (3), 035603/1–035603/7.

84. Yasuda, S.; Futaba, D. N.; Yamada, T.; Satou, J.; Shibuya, A.; Takai, H.; Arakawa, K.; Yumura, M.; Hata, K. Improved and large area single-walled carbon nanotube forest growth by controlling the gas flow direction. *ACS Nano*, 2009, *3* (12), 4164–4170.

85. Guo, Ha.; Lu, W. Effect of areal density on the growth morphology of nanotube forests. *Journal of Computational and Theoretical Nanoscience*, 2009, *6* (10), 2152–2155.

86. Luo, D.; Wu, L.; Zhi, J. Fabrication of boron-doped diamond nanorod forest electrodes and their application in nonenzymatic amperometric glucose biosensing. *ACS Nano*, 2009, *3* (8), 2121–2128.

87. Cook, J. J. H.; Tsakmakidis, K. L.; Hess, O. Ultralow-loss optical diamagnetism in silver nanoforests. *Journal of Optics A: Pure and Applied Optics*, 2009, *11* (11), 114026/1–114026/9.

88. Fernandes, G. E.; Kim, J. H.; Liu, Z.; Shainline, J.; Osgood, R.; Xu, J. Using a forest of gold nanowires to reduce the reflectivity of silicon. *Optical Materials*, 2010, *32* (5), 623–626.

89. Dou, R.; Derby, B. The growth and mechanical properties of gold nanowires. *Materials Research Society Symposium Proceedings (Mechanics of Nanoscale Materials)*, Boston. 2008, Vol. 1086E, Paper No. 1086-U08-01.

90. Xue, C.-H.; Wang, R.-L.; Zhang, J.; Jia, S.-T.; Tian, L.-Q. Growth of ZnO nanorod forests and characterization of ZnO-coated nylon fibers. *Materials Letters*, 2010, *64* (3), 327–330.

91. Hassan, F. M. B.; Nanjo, H.; Venkatachalam, S.; Kanakubo, M.; Ebina, T. Effect of the solvent on growth of titania nanotubes prepared by anodization of Ti in HCl. *Electrochimica Acta*, 2010, *55* (9), 3130–3137.

92. Reches, M.; Gazit, E. Controlled patterning of aligned self-assembled peptide nanotubes. *Nature Nanotechnology*, 2006, *1* (3), 195–200.

93. Rusling, J. F.; Yu, X.; Munge, B. S.; Kim, S. N.; Papadimitrakopoulos, F. Single-wall carbon nanotube forests in biosensors. In *Engineering the Bioelectronic Interface*. Davis, J. J. (Ed.). RSC Publishing, Cambridge, U.K., 2009, pp. 94–118.

94. Sardesai, N.; Pan, S.; Rusling, J. Electrochemiluminescent immunosensor for detection of protein cancer biomarkers using carbon nanotube forests and $[Ru-(bpy)_3]^{2+}$-doped silica nanoparticles. *Chemical Communications*, 2009, (33), 4968–4970.

95. De Volder, M. F. L.; Vidaud, D. O.; Meshot, E. R.; Tawfick, S.; John Hart, A. Self-similar organization of arrays of individual carbon nanotubes and carbon nanotube micropillars. *Microelectronic Engineering*, 2010, *87* (5–8), 1233–1238.

96. Zeier, W. A.; Kovaleski, S. D.; McDonald, K. F. Electron emission from thin film optically enhanced nanoforest cathodes. *IEEE Transactions on Dielectrics and Electrical Insulation*, 2009, *16* (4), 1088–1092.

97. Dev, A.; Kar, S.; Chaudhuri, S. ZnO hierarchical nanostructures: Simple solvothermal synthesis and growth mechanism. *Journal of Nanoscience and Nanotechnology*, 2008, *8* (9), 4506–4513.

98. Zhang, P.; Lee, T.; Xu, F.; Navrotsky, A.; Rock, P.A. Energetics of ZnO nanoneedles. Surface enthalpy, stability, and growth. *Journal of Materials Research*, 2008, *23* (6), 1652–1657.

99. Gao, H.-Y.; Yan, F.-W.; Zhang, Y.; Li, J.-M.; Zeng, Y.-P. Synthesis and characterization of ZnO nanoflowers grown on AlN films by solution deposition. *Chinese Physics Letters*, 2008, *25* (2), 640–643.

100. Lee, D. G.; Park, S. J.; Park, Y. O.; Ryu, E. I. Synthesis of nanostructures by direct growth of carbon nanotubes on micron-sized metal fiber filter and its filtration performance. *Hwahak Konghak*, 2007, *45* (3), 264–268.

101. Park, S. J.; Lee, D. G. Performance improvement of micron-sized fibrous metal filters by direct growth of carbon nanotubes. *Carbon*, 2006, *44* (10), 1930–1935.

102. Kapoor, M. P.; Kasama, Y.; Yanagi, M.; Yokoyama, T.; Nanbu, H.; Juneja, L. R. Mesostructured silica with bush-like morphology and its transformation into nanometer-sized mesoporous silica particles. *Chemistry Letters*, 2007, *36* (5), 626–627.

103. Suh, K.; Shina, J. H.; Seo, S.-J.; Bae, B.-S. Large-scale fabrication of single-phase Er_2SiO_5 nanocrystal aggregates using Si nanowires. *Applied Physics Letters*, 2006, *89*, 223102.

104. Song, M.; Mitsuishi, K.; Furuya, K. Controlled fabrication of nanometer-sized bushes on insulator substrates with assistance of electron beam irradiation. *Ceramic Transactions*, 2005, *159 (Ceramic Nanomaterials and Nanotechnology III)*, 31–38.

105. Guo, S.; Rzayev, J.; Bailey, T. S.; Zalusky, A. S.; Olayo-Valles, R.; Hillmyer, M. A. Nanopore and nanobushing arrays from ABC triblock thin films containing two etchable blocks. *Chemistry of Materials*, 2006, *18* (7), 1719–1721.

106. http://www.sciencetouch.net/html/guide/download.php?code=ppts&filename=PPT53.pdf, accessed on January 31, 2011.

107. Ohtani, T.; Nishikawa, T.; Harada, K.; Ikeda, K.; Takayama, N. Novel nanocarbons with a mushroom shape found in glassy carbon powder. *Journal of Alloys and Compounds*, 2009, *483* (1–2), 491–494.

108. Peng, X.; Koczkur, K.; Chen, A. Synthesis of well-aligned bamboo-like carbon nanotube arrays from ethanol and acetone. *Journal of Physics D: Applied Physics*, 2008, *41*(9), 095409/1–095409/6.

109. Yasuda, T.; Tada, T.; Yamasaki, S.; Gwo, S.; Hong, L.-S. Position-specified formation of epitaxial Si grains on thermally oxidized Si(001) surfaces via isolated nanodots. *Chemistry of Materials*, 2004, *16* (18), 3518–3523.

110. Levin, C. S.; Bishnoi, S.W.; Grady, N. K.; Halas, N. J. Determining the conformation of thiolated poly(ethylene glycol) on Au nanoshells by surface-enhanced Raman scattering spectroscopic assay. *Analytical Chemistry*, 2006, *78* (10), 3277–3281.

111. Shiigi, H.; Morita, R.; Yamamoto, Y.; Tokonami, S.; Nakao, H.; Nagaoka, T. Self-organization of an organic-inorganic hybrid nanomushroom by a simple synthetic route at the organic/water interface. *Chemical Communications*, 2009, (24), 3615–3617.

112. Naya, M.; Tani, T.; Tomaru, Y.; Li, J.; Murakami, N. Nanophotonics bio-sensor using gold nanostructure. *Proceedings of SPIE (Plasmonics: Metallic Nanostructures and Their Optical Properties VI)*, 2008, Vol. 7032, pp. 70321Q/1–70321Q/9.

113. Kichambare, P.; Hii, K.-F.; Vallance, R. R.; Sadanadan, B.; Rao, A. M.; Javed, K.; Menguc, M. P. Growth of tungsten oxide nanorods with carbon caps. *Journal of Nanoscience and Nanotechnology*, 2006, *6* (2), 536–540.

114. Wang, H.; Xie, C.; Zeng, D.; Yang, Z. Controlled organization of ZnO building blocks into complex nanostructures. *Journal of Colloid and Interface Science*, 2006, *297* (2), 570–577.

115. Zhang, X. L.; Qiao, R.; Qiu, R.; Kim, J. C.; Kang, Y. S. Fabrication of hierarchical ZnO nanostructures via a surfactant-directed process. *Crystal Growth and Design*, 2009, *9* (6), 2906–2910.

116. Wu, X.; Cai, W.; Qu, F.-Y. Tailoring the morphology and wettability of ZnO one-dimensional nanostructures. *Wuli Xuebao*, 2009, *58* (11), 8044–8049.

117. Joseph, S.; George, T.; George, K. C.; Sunny, A. T.; Mathew, S. Morphology tuning of strontium tungstate nanoparticles. *AIP Conference Proceedings*, 2007, *929 (Nanotechnology and Its Applications)*, 95–99.

118. Tang, Y. B.; Cong, H. T.; Zhao, Z. G.; Cheng, H. M. Field emission from AlN nanorod array. *Applied Physics Letters*, 2005, *86* (15), 153104/1–153104/3.

119. Yang, J.; Xiao, Y.; Martin, D. C. Electrochemical polymerization of conducting polymer coatings on neural prosthetic devices: Nanomushrooms of polypyrrole using block copolymer thin films as templates. *Materials Research Society Symposium Proceedings*, 2002, *734 (Polymer/Metal Interfaces and Defect Mediated Phenomena in Ordered Polymers)*, 187–197.

120. Higuchi, T.; Tajima, A.; Motoyoshi, K.; Yabu, H.; Shimomura, M. Frustrated phases of block copolymers in nanoparticles. *Angewandte Chemie, International Edition*, 2008, *47* (42), 8044–8046.

121. Yang, H.-C.; Luh, T.-Y.; Chen, C.-L. Multiscale simulation of the mushroom-shaped nanoscale supramolecular system: from self-assembly process to functional properties. *Lecture Series on Computer and Computational Sciences*, 2005, *4B (Advances in Computational Methods in Sciences and Engineering)*, 1828–1831.

122. Suh, K.Y.; Choi, S.-J.; Baek, S. J.; Kim, T. W.; Langer, R. Observation of high-aspect-ratio nanostructures using capillary lithography. *Advanced Materials*, 2005, *17* (5), 560–564.

123. Gunther, J.; Stupp, S. I. Surface patterns of supramolecular materials. *Langmuir*, 2001, *17* (21), 6530–6539.

124. Chen, H. M.; Pang, L.; Kher, A.; Fainman, Y. Three-dimensional composite metallodielectric nanostructure for enhanced surface plasmon resonance sensing. *Applied Physics Letters*, 2009, *94* (7), 073117/1–073117/3.

125. Sayar, M.; Stupp, S. I. Self-organization of rod-coil molecules into nanoaggregates: A coarse grained model. *Macromolecules*, 2001, *34* (20), 7135–7139.

126. Rao C. N. R.; Deepak, F. L.; Gundiah, G.; Govindaraj, A. Inorganic nanowires. *Progress in Solid State Chemistry*, 2003, *31* (1), 5–147.

127. Chang, J.; Lee, J.-H.; Najeeb, C. K.; Kang, W.-S.; Nam, G.-H.; Kim, J.-H. Fabrication of gold nanoflowers by template-assisted electrodeposition. *Journal of Nanoscience and Nanotechnology*, 2009, *9* (12), 7002–7006.

128. Zhong, L.; Zhai, X.; Zhu, X.; Yao, P.; Liu, M. Vesicle-directed generation of gold nanoflowers by gemini amphiphiles and the spacer-controlled morphology and optical property. *Langmuir*, 26 (8), 5876–5881.

129. Wang, L.; Wei, G.; Guo, C.; Sun, L.; Sun, Y.; Song, Y.; Yang, T.; Li, Z. Photochemical synthesis and self-assembly of gold nanoparticles. *Colloids and Surfaces A: Physicochemical and Engineering Aspects*, 2008, *312* (2), 148–153.

130. Jena, B. K.; Raj, C. R. Synthesis of flower-like gold nanoparticles and their electrocatalytic activity towards the oxidation of methanol and the reduction of oxygen. *Langmuir*, 2007, *23* (7), 4064–4070.

131. Wang, T.; Hu, X.; Dong, S. Surfactantless synthesis of multiple shapes of gold nanostructures and their shape-dependent SERS spectroscopy. *Journal of Physical Chemistry B*, 2006; *110* (34): 16930–16936.

132. Yang, Z.; Lin, Z. H.; Tang, C. Y.; Chang, H. T. Preparation and characterization of flower-like gold nanomaterials and iron oxide/gold composite nanomaterials. *Nanotechnology*, 2007, *18* (25), 255606.

133. Zhao, L.; Ji, X.; Sun, X.; Li, J.; Yang, W.; Peng, X. Formation and stability of gold nanoflowers by the seeding approach: The effect of intraparticle ripening. *Journal of Physical Chemistry C*, 2009, *113* (38), 16645–16651.

134. Jena, B. K.; Raj, C. R. Seedless. Surfactantless room temperature synthesis of single crystalline fluorescent gold nanoflowers with pronounced SERS and electrocatalytic activity. *Chemistry of Materials*, 2008, *20* (11), 3546–3548.

135. Qian, L.; Yang, X. Polyamidoamine dendrimers-assisted electrodeposition of gold-platinum bimetallic nanoflowers. *Journal of Physical Chemistry B.*, 2006, *110* (33), 16672–16678.

136. Xu, W.; Liew, K.Y.; Liu, H.; Huang, T.; Sun, C.; Zhao, Y. Microwave-assisted synthesis of nickel nanoparticles. *Materials Letters*, 2008, *62* (17), 2571–2573.

137. Sun, S.; Yang, D.; Villers, D.; Zhang, G.; Sacher, E.; Dodelet, J. Template- and surfactant-free room temperature synthesis of self-assembled 3D Pt nanoflowers from single-crystal nanowires. *Advanced Materials*, 2008, *20* (3), 571–574.

138. Kawasaki, H.; Yonezawa, T.; Watanabe, T.; Arakawa, R. Platinum nanoflowers for surface-assisted laser desorption/ionization mass spectrometry of biomolecules. *Journal of Physical Chemistry C*, 2007, *111* (44), 16278–16283.

139. Chen, J.; Li, C.; Zhang, S.; Ma, H.; Gao, F.; Tao, Z.; Liang, J. CN 101000954, 2007.

140. Sun, X.; Li, R.; Zhou, Y.; Cai, M.; Liu, H. One-dimensional metal and metal oxide nanostructures. WO/2008/031005, 2008.

141. Cha, S. I.; Mo, C. B.; Kim, K. T.; Hong, S. H. Ferromagnetic cobalt nanodots, nanorices, nanowires and nanoflowers by polyol process. *MRS Website*, 2007, *20* (8), 2148–2153. http://www.mrs.org/s_mrs/sec_subscribe.asp?CID=2225&DID=159765&action=detail

142. Zheng, Z.; Xu, B.; Huang, L.; He, L.; Ni, X. Novel composite of Co/carbon nanotubes: Synthesis, magnetism and microwave absorption properties. *Solid State Science*, 2008, *10* (3), 316–320.

143. Du, J.; Liu, Z.; Li, Z.; Han, B.; Sun, Z.; Huang, Y. Carbon nanoflowers synthesized by a reduction–pyrolysis–catalysis route. *Materials Letters*, 2005, *59* (4), 456–458.

144. Thongtem, S.; Singjai, P.; Thongtem, T. Preyachoti S. Growth of carbon nanoflowers on glass slides using sparked iron as a catalyst. *Materials Science and Engineering A*, 2006, *423* (1), 209–213.

145. He, Y.; Zhao, H.; Kong, X. CN 1962431 A 20070516, 2007.

146. Tretyakov, Yu. D (Ed.). *Nanotechnologies. The Alphabet for Everyone.* FIZMATLIT, Moscow, Russia, 2008, pp. 344–345.

147. Jiang, C. Y.; Sun, X. W.; Lo, G. Q.; Kwong, D. L.; Wang, J. X. Improved dye-sensitized solar cells with a ZnO-nanoflower photoanode. *Applied Physics Letters,* 2007, *90,* 263501.

148. Hsueh, T. J.; Hsu, C. L. Fabrication of gas sensing devices with ZnO nanostructure by the low-temperature oxidation of zinc particles. *Sensors and Actuators: B. Chemical,* 2008, *131* (2), 572–576.

149. Ge, C.; Bai, Z.; Hu, M.; Zeng, D.; Cai, S.; Xie, C. Preparation and gas-sensing property of ZnO nanorod-bundle thin films. *Materials Letters,* 2008, *62* (15), 2307–2310.

150. Li, C; Fang, G.; Liu, N.; Ren, Y.; Huang, H.; Zhao, X. Snowflake-like ZnO structures: Self-assembled growth and characterization. *Materials Letters,* 2008, *62* (12), 1761–1764.

151. Baia, H. P.; Lua, X. X.; Yanga, G. M.; Yang, Y. H. Hydrogen peroxide biosensor based on electrodeposition of zinc oxide nanoflowers onto carbon nanotubes film electrode. *Chinese Chemical Letters,* 2008, *19* (3), 314–318.

152. Sun, X. H.; Lam, S.; Sham, T. K.; Heigl, F.; Jürgensen, A.; Wong, N. B. Synthesis and synchrotron light-induced luminescence of ZnO nanostructures: Nanowires, nanoneedles, nanoflowers, and tubular whiskers. *Journal of Physical Chemistry B,* 2005, *109* (8), 3120–3125.

153. Tong, Y.; Liu, Y.; Dong, L.; Zhao, D.; Zhang, J.; Lu, Y.; Shen, D.; Fan, X. Growth of ZnO nanostructures with different morphologies by using hydrothermal technique. *Journal of Physical Chemistry B,* 2006, *110* (41), 20263–20267.

154. Wen, B.; Huang, Y.; Boland, J. J. Controllable growth of ZnO nanostructures by a simple solvothermal process. *Journal of Physical Chemistry C,* 2008, *112* (1), 106–111.

155. Liu, N.; Wu, D.; Wu, H.; Liu, C.; Luo, F. A versatile and "green" electrochemical method for synthesis of copper and other transition metal oxide and hydroxide nanostructures. *Materials Chemistry and Physics,* 2008, *107* (2), 511–517.

156. Fang, Z.; Tang, K.; Shen, G.; Chen, D.; Kong, R.; Lei, S. Self-assembled ZnO 3D flowerlike nanostructures. *Materials Letters,* 2006, *60* (20), 2530–2533.

157. Zhang, J.; Yang, Y.; Xu, B.; Jiang, F.; Li, J. Shape-controlled synthesis of ZnO nano- and micro-structures. *Journal of Crystal Growth,* 2005, *280* (3), 509–515.

158. Jung, S.-H.; Oh, Em.; Lee, K.-H.; Yang, Y.; Park, C. G.; Park, W.; Jeong, S.-H. Sonochemical preparation of shape-selective ZnO nanostructures. *Crystal Growth and Design,* 2008, *8* (1), 265–269.

159. Wahab, R.; Ansari, S. G.; Kim, Y. S.; Seo, H. K.; Kim, G. S.; Khang, G.; Shin, H. S. Low temperature solution synthesis and characterization of ZnO nano-flowers. *Materials Research Bulletin,* 2007, *42* (9), 1640–1648.

160. Fang, X. S.; Ye, C. H.; Xie, T.; Wang, Z. Y.; Zhao, J.; Zhang, L. Regular MgO nanoflowers and their enhanced dielectric responses. *Applied Physics Letters,* 2006, 88, 013101.

161. Fang, X.-S.; Ye, C.-H.; Zhang, L.-D.; Zhang, J.-X.; Zhao, J. W., Yan, P. Direct observation of the growth process of MgO nanoflowers by a simple chemical route. *Small,* 2005, *1* (4), 422–428.

162. Zhang, Y.-G.; Chen, Y.-C.; Zhao, Y.-G. Synthesis and magnetic properties of Co_3O_4 nanoflowers. *Chinese Journal of Chemical Physics,* 2007, *20* (5), 601–606.

163. Yu, L.; Zhang, G.; Wua, Y.; Baia, X.; Guo, D. Cupric oxide nanoflowers synthesized with a simple solution route and their field emission. *Journal of Crystal Growth,* 2008, *310* (12), 3125–3130.

164. Song, M.-J.; Hwang, S. W.; Whang, D. Non-enzymatic electrochemical CuO nanoflowers sensor for hydrogen peroxide detection. *Talanta,* 2010, *80* (5), 1648–1652.

165. Ni, X.; Zhang, Y.; Tian, D.; Zheng, H.; Wang, X. Synthesis and characterization of hierarchical NiO nanoflowers with porous structure. *Journal of Crystal Growth,* 2007, *306* (2), 418–421.

166. Gao, Y; Wang, Z.; Liu, S. X. Y.; Qian, Y. Influence of anions on the morphology of nanophase alpha-MnO_2 crystal via hydrothermal process. *Journal of Nanoscience Nanotechnology,* 2006, *6* (8), 2576–2579.

167. Zhang, Y.; Yu, K.; Li, G.; Peng, D.; Zhang, Q.; Xu, F.; Bai, W.; Ouyang, S.; Zhu, Z. Synthesis and field emission of patterned SnO_2 nanoflowers. *Materials Letters,* 2006, *60* (25), 3109–3112.

168. Li, G.; Jiang, L.; Pang, S.; Peng, H.; Zhang Z. Molybdenum trioxide nanostructures: The evolution from helical nanosheets to crosslike nanoflowers to nanobelts. *Journal of Physical Chemistry B,* 2006, *110* (48), 24472–24475.

169. Narayanaswamy, A.; Xu, H.; Pradhan, N.; Kim, M.; Peng, X. Formation of nearly monodisperse In_2O_3 nanodots and oriented-attached nanoflowers: hydrolysis and alcoholysis vs pyrolysis. *Journal of the American Chemical Society,* 2006, *128* (31), 10310–10319.

170. Narayanaswamy, A.; Xu, H.; Pradhan, N.; Peng, X. Crystalline nanoflowers with different chemical compositions and physical properties grown by limited ligand protection. *Angewandte Chemie, International Edition (English),* 2006, *45* (32), 5361–5364.

171. Peng, X.; Narayanaswamy, A.; Pradhan, N. Preparation of multi-dimensional complex nanocrystal structures. U.S. Patent 2008081016, 2008.

172. Jin, L.; Wang, J.; Cao, G.; Choy, W. C. H. Fabrication and characterization of amorphous silica nanostructures. *Physics Letters A*, 2008, *372* (25), 4622–4626.

173. Wei, Q.; Meng, G.; An, X.; Hao, Y., Zhang, L. D. Synthesis and photoluminescence of aligned straight silica nanowires on Si substrate. *Solid State Communications*, 2006, *138* (7), 325–330.

174. Zhang, J.; Liu, Z.; Han, B.; Wang, Y.; Li, Z. Yang, G. A simple and inexpensive route to synthesize porous silica microflowers by supercritical CO_2. *Microporous and Mesoporous Materials*, 2005, *87* (1), 10–14.

175. Yan, C.; Xue, D.; Zou, L.; Yan, X.; Wang, W. Preparation of magnesium hydroxide nanoflowers. *Journal of Crystal Growth*, 2005, *282* (3), 448–454.

176. Yang, L.-X.; Zhu, Y.-J.; Tong, H.; Liang, Z.-H.; Li, L.; Zhang, L. Hydrothermal synthesis of nickel hydroxide nanostructures in mixed solvents of water and alcohol. *Journal of Solid State Chemistry*, 2007, *180* (7), 2095–2101.

177. Hsiao, M.-T.; Chen, S.-F.; Shieh, D.-B.; Yeh, C.-S. One-pot synthesis of hollow Au_3Cu_1 spherical-like and biomineral botallackite $Cu_2(OH)_3Cl$ flowerlike architectures exhibiting antimicrobial activity. *Journal of Physical Chemistry B*, 2006, *110* (1), 205–210.

178. Zhong, H.; Li, Y.; Zhou, Y.; Yang, C.; Li, Y. Controlled synthesis of 3D nanostructured $Cd_4Cl_3(OH)_5$ templates and their transformation into $Cd(OH)_2$ and CdS nanomaterials. *Nanotechnology*, 2006, *17* (3), 772–777.

179. Amalnerkar, D. P.; Lee, H.-Y.; Hwang, Y. K.; Kim, D.-P.; Chang, J. S. Swift morphosynthesis of hierarchical nanostructures of CdS via microwave-induced semisolvothermal route. *Journal of Nanoscience and Nanotechnology*, 2007, *7* (12), 4412–4420.

180. Taia, G.; Guo, W. Sonochemistry-assisted microwave synthesis and optical study of single-crystalline CdS nanoflowers. *Ultrasonics Sonochemistry*, 2008, *15* (4), 350–356.

181. Chen, W.; Bovin, J.-O.; Wang, S.; Joly, A.G.; Wang, Y.; Sherwood, P.M.A. Fabrication and luminescence of $ZnS:Mn^{2+}$ nanoflowers. *Journal of Nanoscience Nanotechnology*, 2005, *5* (9), 1309–1322.

182. Dai, Q.; Xiao, N.; Ning, J.; Li, C.; Li, D.; Zou, B.; Yu, W. W. Synthesis and mechanism of particle- and flower-shaped ZnSe nanocrystals: Green chemical approaches toward green nanoproducts. *Journal of Physical Chemistry C*, 2008, *112* (20), 7567–7571.

183. Lee, S.H.; Kim, Y.J.; Park, J. Shape evolution of ZnTe nanocrystals: Nanoflowers, nanodots, and nanorods. *Chemistry of Materials*, 2007, *19* (19), 4670–4675.

184. Cao, X.; Lu, Q.; Xu, X.; Yan, J.; Zeng, H. Single-crystal snowflake of Cu_7S_4: Low temperature, large scale synthesis and growth mechanism. *Materials Letters*, 2008, *62* (17), 2567–2570.

185. Bao, S. J.; Li, C. M.; Guo, C. X.; Qiao, Y. Biomolecule-assisted synthesis of cobalt sulfide nanowires for application in supercapacitors. *Journals of Power Sources*, 2008, *180* (1), 676–681.

186. Zhao, W.-B.; Zhu, J.-J.; Xu, J.-Z.; Chen, H.-Y. Photochemical synthesis of Bi_2S_3 nanoflowers on an alumina template. *Inorganic Chemical Communications*, 2004, *7* (7), 847–850.

187. Liu, Z.; Xu, D.; Liang, J.; Lin, W.; Yu, W.; Qian; Y. Low-temperature synthesis and growth mechanism of uniform nanorods of bismuth sulfide. *Journal of Solid State Chemistry*, 2005, *178* (3), 950–955.

188. Tian, L.; Han, Y. T.; Vittal J. J. Morphology-controlled synthesis of Bi_2S_3 nanomaterials via single- and multiple-source approaches. *Crystal Growth and Design*, 2008, *8* (2), 734–738.

189. Gautam, U. K.; Vivekchand, S. R. C.; Govindaraja, A.; Rao, C. N. R. GaS and GaSe nanowalls and their transformation to Ga_2O_3 and GaN nanowalls. *Chemical Communications*, 2005, 3995–3997.

190. Li, Y. B.; Bando, Y.; Golberg, D. MoS_2 nanoflowers and their field-emission properties. *Applied Physics Letters*, 2003, *82* (12), 1962–1964.

191. Wei, R.; Yang, H.; Du, K.; Fu, W. Jian, Y.; Yu, Q.; Liu, S.; Li, M.; Zou, G. A facile method to prepare MoS_2 with nanoflower-like morphology. *Materials Chemistry and Physics*, 2008, *108* (2–3), 188–191.

192. Zhu, H.; Yang, D.; Zhang, H. Hydrothermal synthesis, characterization and properties of SnS nanoflowers. *Materials Letters*, 2006, *60* (21–22), 2686–2689.

193. Dong, L.; Chu, Y.; Liu, Y.; Li, M.; Zou, G. Surfactant-assisted fabrication PbS nanorods, nanobelts, nanovelvet-flowers and dendritic nanostructures at lower temperature in aqueous solution. *Journal of Colloid and Interface Science*, 2006, *301* (2), 503–510.

194. Ren, Z.; Chen, G.; Poudel, B.; Kumar, S.; Wang, W. Dresselhaus. M. U.S. Patent 7,255,846, 2007.

195. Das, K.; Datta, A.; Chaudhuri, S. $CuInS_2$ flower vaselike nanostructure arrays on a Cu tape substrate by the copper indium sulfide on Cu-Tape (CISCuT) method: Growth and characterization. *Crystal Growth and Design*, 2007, *7* (8), 1547–1552.

196. Guo, G.; Song, Z.; Cong, C.; Zhang, K. $CoMoS_4$ nanoflowers as anode for secondary lithium batteries. *Journal of Nanoparticle Research*, 2007, *9* (4), 653–656.

197. Tang, Y. B.; Cong, H. T.; Wang, Z. M.; Cheng, H.-M. Catalyst-seeded synthesis and field emission properties of flowerlike Si-doped AlN nanoneedle array. *Applied Physics Letters*, 2006, *89*, 253112.

198. Ecke, G.; Cimalla, V.; Tonisch, K.; Lebedev, V.; Romanus, H.; Ambacher, O.; Liday, J. Analysis of nanostructures by means of Auger electron spectroscopy. *Journal of Electrical Engineering*, 2007, *58* (6), 301–306.

199. Kang, T.-T.; Liu, X.; Zhang, R. Q.; Hu, W. G. Cong, G.; Zhao, F.; Zhu, Q. InN nanoflowers grown by metal organic chemical vapor deposition. *Applied Physics Letters*, 2006, *89* (7), 071113.

200. Sandhu, A. Nanostructures: Say it with flowers. *Nature Nanotechnology*, 2006. http://www.nature.com/nnano/reshigh/2006/0906/full/nnano.2006.48.html

201. Kang, T. T.; Hashimoto, A.; Yamamoto, A. Optical properties of InN containing metallic indium. *Applied Physics Letters*, 2008, *92* (11), 111902.

202. Liu, B. D.; Bando, Y.; Tang, C. C.; Golberg, D.; Xie, R. G.; Sekiguchi, T. Synthesis and optical study of crystalline GaP nanoflowers. *Applied Physics Letters*, 2005, *86*, 083107.

203. Felipe, C.; Chavez, F.; Angeles-Chavez, C.; Lima, E. Goly, O.; Penna-Serra, R. Morphology of nanostructured GaP on GaAs: Synthesis by the close-spaced vapor transport technique. *Chemical Physics Letters*, 2007, *439* (1), 127–131.

204. Liu, S.; Qian, Y.; Ma, X. Polymer-assisted synthesis of Co_2P nanocrystals. *Materials Letters*, 2008, *62* (1), 11–14.

205. Chen, W.; Peng, Q.; Li, Y. Luminescent bis(8-hydroxyquinoline)cadmium complex nanorods. *Crystal Growth and Design*, 2008, *8* (2), 564–567.

206. Karan, S.; Basak, D.; Mallik, B. Copper phthalocyanine nanoparticles and nanoflowers. *Chemical Physics Letters*, 2007, *434* (4), 265–270.

207. Karan, S.; Mallik, B. Templating effects and optical characterization of copper (II) phthalocyanine nanocrystallites thin film: Nanoparticles, nanoflowers, nanocabbages, and nanoribbons. *Journal of Physical Chemistry C*, 2007, *111* (20), 7352–7365.

208. Karan, S.; Mallik, B. Nanoflowers grown from phthalocyanine seeds: Organic nanorectifiers. *Journal of Physical Chemistry C*, 2008, *112* (7), 2436–2447.

209. Xu, X. J.; Zhou, L. H.; Lu, C. Z. Mixed-solvothermal synthesis of hybrid nanostructures of alkaline earth phenylphosphonates. *Materials Letters*, 2007, *61* (28), 4980–4983.

210. Wang, Z.; Jiao, L.; You, T.; Niu, L.; Donga, S.; Ivaska, A. Electrochemical preparation of self-doped poly(*o*-aminobenzenesulfonic acid-co-aniline) microflowers. *Electrochemistry Communications*, 2005, *7* (9), 875–878.

211. Bhongale, C. J.; Chang, C. W.; Wei-Guang, D. E.; Hsu, C. S.; Dong, Y.; Tang, B. Z. Formation of nanostructures of hexaphenylsilole with enhanced color-tunable emissions. *Chemical Physics Letters*, 2006, *419* (4), 444–449.

212. Jia, F.; Zhang, L.; Shang, X.; Yang, Y. Non-aqueous sol-gel approach towards the controllable synthesis of nickel nanospheres, nanowires, and nanoflowers. *Advanced Materials*, 2008, *20* (5), 1050–1054.

213. Yang, Z.; Chiu, T.-C.; Chang, H.-T. Preparation and characterization of different shapes of silver nanostructures in aqueous solution. *Open Nanoscience Journal*, 2007, *1*, 5–12.

214. Xu, F.; Yu, K.; Li, G.; Li, Q.; Zhu, Z. Synthesis and field emission of four kinds of ZnO nanostructures: Nanosleeve-fishes, radial nanowire arrays, nanocombs and nanoflowers. *Nanotechnology*, 2006, *17* (12), 2855–2859.

215. Geng, B.; Liu, X.; Wei, X.; Wang, S. Synthesis, characterization and optical properties of regularly shaped, single-crystalline sisal-like ZnO nanostructures. *Materials Letters*, 2005, *59* (28), 3572–3576.

216. Wu, C.; Qiao, X.; Luo, L.; Li, H. Synthesis of ZnO flowers and their photoluminescence properties. *Materials Research Bulletin*, 2008, *43* (7), 1883–1891.

217. Gao, H.; Yan, F.; Li, J.; Zeng, Y.; Wang, J. Synthesis and characterization of ZnO nanorods and nanoflowers grown on GaN-based LED epiwafer using a solution deposition method. *Journal of Physics D: Applied Physics*, 2007, *40* (12), 3654–3659.

218. Suh, H.-W.; Kim, G.-Y.; Jung, Y.-S.; Choi, W.-K.; Byun, D. Growth and properties of ZnO nanoblade and nanoflower prepared by ultrasonic pyrolysis. *Journal of Applied Physics*, 2005, *97* (4), 044305.1–044305.6.

219. Pan, A.; Yu, R.; Xie, S.; Zhang, Z.; Jin, C.; Zou, B. ZnO flowers made up of thin nanosheets and their optical properties. *Journal of Crystal Growth*, 2005, *282* (1), 165–172.

220. Wang, Y.; Li, X.; Lu, G.; Chen, G.; Chen, Y. Synthesis and photo-catalytic degradation property of nanostructured-ZnO with different morphology. *Materials Letters*, 2008, *62* (15), 2359–2362.

221. Chen, A.; Peng, X.; Koczkur, K.; Miller, B. Super-hydrophobic tin oxide nanoflowers. *Chemical Communications*, 2004, (17), 1964–1965.

222. Zhang, Y.; Zhu, J.; Song, X.; Zhong, X. Controlling the synthesis of CoO nanocrystals with various morphologies. *Journal of Physical Chemistry C*, 2008, *112* (14), 5322–5327.

223. Kumari, L.; Lin, J.-H.; Ma, Y.-R. Synthesis of bismuth oxide nanostructures by an oxidative metal vapour phase deposition technique. *Nanotechnology*, 2007, *18* (29), 295605.

224. Cao, A.; Cao, L.; Gao, D. Fabrication of nonaging superhydrophobic surfaces by packing flowerlike hematite particles. *Applied Physics Letters*, 2007, 91, 034102.

225. Zeng, S.; Tang, K.; Li, T.; Liang, Z.; Wang, D.; Wang, Y.; Qi, Y.; Zhou, W. Facile route for the fabrication of porous hematite nanoflowers: Its synthesis, growth mechanism, application in the lithium ion battery, and magnetic and photocatalytic properties. *Journal of Physical Chemistry C*, 2008, *112* (13), 4836–4843.

226. Zeng, S.; Tang, K.; Li, T.; Liang, Z. 3D flower-like Y_2O_3:Eu^{3+} nanostructures: Template-free synthesis and its luminescence properties. *Journal of Colloid and Interface Science*, 2007, *316* (2), 921–929.

227. Zhou, H.-P.; Zhang, Y.-W.; Mai, H.-X.; Sun, X.; Liu, Q.; Song, W. G.; Yan, C. H. Spontaneous organization of uniform CeO_2 nanoflowers by 3D oriented attachment in hot surfactant solutions monitored with an in situ electrical conductance technique. *Chemistry—A European Journal*, 2008, *14* (11), 3380–3390.

228. Li, X.-L.; Ge, J.-P.; Li, Y.-D. Atmospheric pressure chemical vapor deposition: An alternative route to large-scale MoS_2 and WS_2 inorganic fullerene-like nanostructures and nanoflowers. *Chemistry*, 2004, *10* (23), 6163–6171.

229. Zhang, Q.; Zhu, Q.-A.; Sun, X.-F.; Gong, M. et al. Controlled synthesis of Bi_2S_3 nanomaterials with different morphologies by surfactants. *Wuji Huaxue Xuebao*, 2008, *24* (4), 547–552.

230. Yu, L.; Hu, Z.; Ma, Y.; Huo, K.; Chen, Y.; Sang, H.; Lin, W.; Lu, Y. Evolution of aluminum nitride nanostructures from nanoflower to thin film on silicon substrate by direct nitridation of aluminum precursor. *Diamond and Related Materials*, 2007, *16* (8), 1636–1642.

231. Zhu, L.; Liu, X.; Meng, J.; Cao, X. Facile sonochemical synthesis of single-crystalline europium fluorine with novel nanostructure. *Crystal Growth and Design*, 2007, *7* (12), 2505–2511.

232. You T.; Cao, G.; Song, X.; Fan, C.; Zhao, W.; Yin, Z.; Sun, S. Alcohol-thermal synthesis of flowerlike hollow cobalt tungstate nanostructures. *Materials Letters*, 2008, *62* (8), 1169–1172.

233. Wang, B.; Yu, C.; Jing, D. Method for manufacturing basic magnesium carbonate with 3D nanoflower structure. CN 101058430, 2007, 5 pp.

234. Zhou J.-P.; Wang, P.-F.; Qiu, Z.-C.; Zhu, G.-Q.; Liu, P. Flower-like $Pb(Zr_{0.52}Ti_{0.48})O_3$ nanoparticles on the $CoFe_2O_4$ seeds. *Journal of Crystal Growth*, 2008, *310* (2), 508–512.

235. Zhang, X.; Zhang, J.; Jin, Y.; Zhao, H.; Wang, X.-J. Large-scale fabrication of Pr^{3+} doped or undoped nanosized $ATiO_3$ (A = Ca, Sr, Ba) with different shapes *via* a facile solvothermal technique. *Crystal Growth and Design*, 2008, *8* (3), 779–781.

236. Tian, J. T.; Gong, C. H.; Yu, L. G.; Wu, Z. S.; Zhang, Z. J. Synthesis of dandelion-like three-dimensional nickel nanostructures via solvothermal route. *Chinese Chemical Letters*, 2008, *19* (9), 1123–1126.

237. Kim, Y.-T.; Lee, H.; Kim, H.-J.; Lim, T.-H. PtRu nano-dandelions on thiolated carbon nanotubes: a new synthetic strategy for supported bimetallic core-shell clusters on the atomic scale. *Chemical Communications*, 2010, *46* (12), 2085–2087.

238. Wang, H.; Pan, Q.; Cheng, Y.; Zhao, J.; Yin, G. Evaluation of ZnO nanorod arrays with dandelion-like morphology as negative electrodes for lithium-ion batteries. *Electrochimica Acta*, 2009, *54* (10), 2851–2855.

239. Pan, Q.; Cheng, Y. Superhydrophobic surfaces based on dandelion-like ZnO microspheres. *Applied Surface Science*, 2009, *255* (6), 3904–3907.

240. Ren, L.; Huang, X.; Sun, F.; He, X. Preparation and characterization of doped TiO_2 nanodandelion. *Materials Letters*, 2007, *61* (2), 427–431.

241. Bai, X.; Xie, B.; Pan, N.; Wang, X.; Wang, H.; Bai, X.; Xie, B.; Pan, N.; Wang, X.; Wang, H. Novel three-dimensional dandelion-like TiO_2 structure with high photocatalytic activity. *Journal of Solid State Chemistry*, 2008, *181* (3), 450–456.

242. Zhang, J.; Yang, Y.; Ding, S.; Li, J.; Wang, X. Bimetal Ga-Sn catalyzed growth for the novel morphologies of silicon oxide nanowires. *Materials Science & Engineering, B: Advanced Functional Solid-State Materials*, 2008, *150* (3), 180–186.

243. Mu, Y.; Yang, J.; Han, S.; Hou, H.; Fan, Y. Syntheses and gas-sensing properties of CuO nanostructures by using $[Cu(pbbt)Cl_2]_2 \cdot CH_3OH$ as a precursor. *Materials Letters*, 2010, *64* (11), 1287–1290.

244. Manna, S.; Das, K.; De, S. K. Template free synthesis of mesoporous CuO Dandelion structures for optoelectronic applications. *ACS Applied Materials and Interfaces*, 2010, 2(5), 1536–1542.

245. Xia, H.; Xiao, W.; Lai, M. O.; Lu, L. Facile synthesis of novel nanostructured MnO_2 thin films and their application in supercapacitors. *Nanoscale Research Letters*, 2009, *4* (9), 1035–1040.

246. Cao, M.; He, X.; Chen, J.; Hu, C. Self-assembled nickel hydroxide three-dimensional nanostructures: A nanomaterial for alkaline rechargeable batteries. *Crystal Growth and Design*, 2007, *7* (1), 170–174.

247. Xie, Y.; Du, Y.-F.; Xiao, X.; Wu, C.-Z. The combination of self-produced template methodology and self-limitation properties of target materials towards three-dimensional hierarchical nanostructures. *Zhongguo Kexue Jishu Daxue Xuebao*, 2008, *38* (6), 569–575.

248. Lak, A.; Mazloumi, M.; Mohajerani, M.; Kajbafvala, A.; Zanganeh, S.; Arami, H.; Sadrnezhaad, S. K. Self-assembly of dandelion-like hydroxyapatite nanostructures via hydrothermal method. *Journal of the American Ceramic Society*, 2008, *91* (10), 3292–3297.

249. Liu, X.; Wang, Q.; Gao, Z.; Sun, J.; Shen, J. Fabrication of lanthanide phosphate nanocrystals with well-controlled morphologies by layer-by-layer adsorption and reaction method at room temperature. *Crystal Growth and Design*, 2009, *9* (8), 3707–3713.

250. Xu, L.; Shen, J.; Lu, C.; Chen, Y.; Hou, W. Self-assembled three-dimensional architectures of $Y_2(WO_4)_3$:Eu: Controlled synthesis, growth mechanism, and shape-dependent luminescence properties. *Crystal Growth and Design*, 2009, *9* (7), 3129–3136.

251. Li, Z.-A.; Yang, H.-X.; Tian, H.-F.; Zhang, Y.; Li, J.-Q. Fabrication and characterization of micro-pattern dandelion-like and nanobelts of β-SrV_2O_6 via hydrothermal process. *Chinese Journal of Chemical Physics*, 2007, *20* (6), 727–732.

252. Wang, W.-S.; Zhen, L.; Xu, C.-Y.; Yang, L.; Shao, W.-Z. Room temperature synthesis of hierarchical $SrCO_3$ architectures by a surfactant-free aqueous solution route. *Crystal Growth and Design*, 2008, *8* (5), 1734–1740.

253. Dong, L.; Chu, Y.; Zhang, W. A very simple and low cost route to Bi_2S_3 nanorods bundles and dandelion-like nanostructures. *Materials Letters*, 2008, *62* (27), 4269–4272.

254. Naik, S. D.; Jagadale, T. C.; Apte, S. K.; Sonawane, R. S.; Kulkarni, M. V.; Patil, S. I.; Ogale, S. B.; Kale, B. B. Rapid phase-controlled microwave synthesis of nanostructured hierarchical tetragonal and cubic β-In_2S_3 dandelion flowers. *Chemical Physics Letters*, 2008, *452* (4–6), 301–305.

255. Zhu, Y.; Li, J.; Wan, M.; Jiang, L. Superhydrophobic 3D microstructures assembled from 1D nanofibers of polyaniline. *Macromolecular Rapid Communications*, 2008, *29* (3), 239–243.

256. Shen, X.-F.; Yan, X.-P. Facile shape-controlled synthesis of well-aligned nanowire architectures in binary aqueous solution. *Angewandte Chemie, International Edition*, 2007, *46* (40), 7659–7663.

257. Kam, K. C.; Deepak, F. L.; Cheetham, A. K.; Rao, C. N. R. In_2O_3 nanowires, nanobouquets and nanotrees. *Chemical Physics Letters*, 2004, *397* (4–6), 329–334.

258. Yin, W.; Cao, M.; Luo, S.; Hu, C.; Wei, B. Controllable synthesis of various In_2O_3 submicron/nanostructures using chemical vapor deposition. *Crystal Growth and Design*, 2009, *9* (5), 2173–2178.

259. Wang, T.; Mehta, R.; Karthik, C.; Ganesan, P. G.; Singh, B.; Jiang, W.; Ravishankar, N.; Borca-Tasciuc, T.; Ramanath, G. Microsphere bouquets of bismuth telluride nanoplates: Room-temperature synthesis and thermoelectric properties. *Journal of Physical Chemistry C*, 2010, *114* (4), 1796–1799.

260. Okuyama, F. Growth of copper nanocrystallites on copper seed cones: Evidence for the presence of a liquid phase on an ion-impacting cone surface. *Journal of Vacuum Science and Technology, B: Microelectronics and Nanometer Structures*, 1996, *14* (2), 768–771.

261. Gao, L.; Fan, L.; Zhang, J. Selective growth of Ag nanodewdrops on Au nanostructures: A new type of bimetallic heterostructure. *Langmuir*, 2009, *25* (19), 11844–11848.

262. Choi, S.-H.; Kim, E.-G.; Hyeon, T. One-pot synthesis of copper-indium sulfide nanocrystal heterostructures with acorn, bottle, and larva shapes. *Journal of the American Chemical Society*, 2006, *128*, 2520–2521.

263. Teranishi, T.; Inoue, Y.; Nakaya, M.; Oumi, Y.; Sano, T. Nanoacorns: Anisotropically phase-segregated CoPd sulfide nanoparticles. *Journal of the American Chemical Society*, 2004, *126* (32), 9914–9915.

264. Teranishi, T.; Inoue, Y.; Saruyama, M.; Nakaya, M.; Kanehara, M. Anisotropically phase-segregated Co_9S_8/PdS_x nanoacorns: Stability improvement and new heterostructures. *Chemistry Letters*, 2007, *36* (4), 490–491.

265. Teranishi, T.; Saruyama, M.; Nakaya, M.; Kanehara, M. Anisotropically phase-segregated Pd-Co-Pd sulfide nanoparticles formed by fusing two Co-Pd sulfide nanoparticles. *Angewandte Chemie, International Edition*, 2007, *46* (10), 1713–1715.

266. Misra, A.; Urban, M. W. Acorn-shape polymeric nano-colloids: Synthesis and self-assembled films. *Macromolecular Rapid Communications*, 2010, *31* (2), 119–127.

267. Ferreira, I.; Martins, R.; Cabrita, A.; Fortunato, E.; Vilarinho, P. A comparison study of nanocrystalline undoped silicon films produced by hot wire and hot wire plasma assisted technique, *Proceedings of the 16th European Photovoltaic Solar Energy Conference*, Glasgow, U.K., May 1–5, 2000. Scheer, H. (Ed.), Vol. 1, pp. 421–424.

268. Yang, J.; Zhou, Z.; Wen, X.; He, B.; Yang, Z. Mechano-chemical surface modification of nano silica. *Guisuanyan Xuebao*, 2010, *38* (2), 320–326.

269. Yang, J.; Pi, P.; Wen, X.; Zheng, D.; Xu, M.; Cheng, J.; Yang, Z. A novel method to fabricate superhydrophobic surfaces based on well-defined mulberry-like particles and self-assembly of polydimethylsiloxane. *Applied Surface Science*, 2009, *255* (6), 3507–3512.

270. Yu, R.; Li, Z.; Wang, D.; Lai, X.; Xing, C.; Xing, X. Morphology manipulation of α-Fe$_2$O$_3$ in the mixed solvent system. *Solid State Sciences*, 2009, *11* (12), 2056–2059.

271. Zhang, H.; Quan, X.; Chen, S.; Yu, H.; Ma, N. "Mulberry-like" CdSe nanoclusters anchored on TiO$_2$ nanotube arrays: A novel architecture for remarkable photo-energy conversion efficiency. *Chemistry of Materials*, 2009, *21* (14), 3090–3095.

272. Liu, K.; You, H.; Zheng, Y.; Jia, G.; Huang, Y.; Yang, M.; Song, Y.; Zhang, L.; Zhang, H. Room-temperature synthesis of multi-morphological coordination polymer and tunable white-light emission. *Crystal Growth and Design*, 2010, *10* (1), 16–19.

273. Hsieh, K.-C.; Tsai, T.-Y.; Wan, D.; Chen, H.-L.; Tai, N.-H. Iridescence of patterned carbon nanotube forests on flexible substrates: From darkest materials to colorful films. *ACS Nano*, 2010, *4* (3), 1327–1336.

274. Krupenkin, T. Nanograss, nanobricks, nanonails, and other things useful in your nanolandscaping. *Abstracts of Papers, 237th ACS National Meeting*, Salt Lake City, UT, March 22–26, 2009, POLY-333.

275. Shieh, J.; Lin, C. H.; Yang, M. C. Plasma nanofabrications and antireflection applications. *Journal of Physics D: Applied Physics*, 2007, *40* (8), 2242–2246.

276. Yang, M.-C.; Shieh, J.; Hsu, C.-C.; Cheng, T.-C. Well-aligned silicon nanograss fabricated by hydrogen plasma dry etching. *Electrochemical and Solid-State Letters*, 2005, *8* (10), C131–C133.

277. Jokinen, V.; Sainiemi, L.; Franssila, S. Complex droplets on chemically modified silicon nanograss. *Advanced Materials*, 2008, *20* (18), 3453–3456.

278. Dorrer, C.; Ruehe, J. Wetting of silicon nanograss. From superhydrophilic to superhydrophobic surfaces. *Advanced Materials*, 2008, *20* (1), 159–163.

279. Hu, M.; Tang, J.; Ou, F. S.; Kuo, H. P.; Wang, S.-Y.; Li, Z.; Williams, R. S. Metal-coated Si nanograss as highly sensitive SERS sensors. *Proceedings of SPIE*, 2009, *7312 (Advanced Environmental, Chemical, and Biological Sensing Technologies VI)*, 73120I/1–73120I/6.

280. Tang, J.; Ou, F. S.; Kuo, H. P.; Hu, M.; Stickle, W. F.; Li, Z.; Williams, R. S. Silver-coated Si nanograss as highly sensitive surface-enhanced Raman spectroscopy substrates. *Applied Physics A: Materials Science and Processing*, 2009, *96* (4), 793–797.

281. Sainiemi, L.; Grigoras, K.; Franssila, S. Suspended nanostructured alumina membranes. *Nanotechnology*, 2009, *20* (7), 075306/1–075306/6.

282. Dixit, P.; Miao, J. Effect of clamping ring materials and chuck temperature on the formation of silicon nanograss in deep RIE. *Journal of the Electrochemical Society*, 2006, *153* (8), G771–G777.

283. Lilienthal, K.; Stubenrauch, M.; Fischer, M.; Schober, A. Fused silica "glass grass": fabrication and utilization. *Journal of Micromechanics and Microengineering*, 2010, *20* (2), 025017/1–025017/11.

284. Zhou, W. M.; Liu, X.; Zhang, Y. F.; Lai, Y. J.; Guo, X. Q. Synthesis and characterization of Si-SiO$_2$ nanocomposites. *Physica E: Low-Dimensional Systems and Nanostructure*, 2007, *36* (1), 128–131.

285. Yuan, J.-J.; Jin, R.-H. Water motion and movement without sticking, weight loss and cross-contaminant in superhydrophobic glass tube. *Nanotechnology*, 2010, *21* (6), 065704/1–065704/4.

286. Kumar, R. T. R.; Mogensen, K. B.; Boggild, P. Simple approach to superamphiphobic overhanging silicon nanostructures. *Journal of Physical Chemistry C*, 2010, *114* (7), 2936–2940.

287. Wei, M.; Terashima, C.; Lv, M.; Fujishima, A.; Gu, Z.-Z. Boron-doped diamond nanograss array for electrochemical sensors. *Chemical Communications*, 2009, (24), 3624–3626.

288. Lv, M.; Wei, M.; Rong, F.; Terashima, C.; Fujishima, A.; Gu, Z.-Z. Electrochemical detection of catechol based on as-grown and nanograss array boron-doped diamond electrodes. *Electroanalysis*, 2010, *22* (2), 199–203.

289. Kakehi, K.; Noda, S.; Maruyama, S.; Yamaguchi, Y. Individuals, grasses, and forests of single- and multi-walled carbon nanotubes grown by supported Co catalysts of different nominal thicknesses. *Applied Surface Science*, 2008, *254* (21), 6710–6714.

290. Luling, M.; Matthieu, G.; Veneruso, A. Nanograss gamma detector. European Patent Application *EP* 2007-103888, 2008, 13 pp.

291. Zhu, F.; Yang, Z. X.; Zhou, W. M.; Zhang, Y. F. Direct synthesis of beta gallium oxide nanowires, nanobelts, nanosheets and nanograsses by microwave plasma. *Solid State Communications*, 2006, *137* (4), 177–181.

292. Wang, B., Yang, Y. H., Wang, C. X., Xu, N. S., Yang, G. W. Field emission and photoluminescence of SnO$_2$ nanograss. *Journal of Applied Physics*, 2005, *98* (12), 124303/1–124303/4.

293. Song, Y.-Y.; Lynch, R.; Kim, D.; Roy, P.; Schmuki, P. TiO₂ Nanotubes: Efficient suppression of top etching during anodic growth. *Electrochemical and Solid-State Letters*, 2009, *12* (7), C17–C20.

294. Kim, D.; Ghicov, A.; Schmuki, P. TiO₂ nanotube arrays: Elimination of disordered top layers ("nanograss") for improved photoconversion efficiency in dye-sensitized solar cells. *Electrochemistry Communications*, 2008, *10* (12), 1835–1838.

295. Lee, K.-J.; Jeong, J.-H.; Kim, K.-D.; Lee, J.; Choi, D.-G. Fabrication of plasma-induced polymer nanograss for a synthetic moth-eye antireflection nanostructure. *Journal of the Korean Physical Society*, 2009, *55* (2), 566–571.

296. Wu, R. B.; Yang, G. Y.; Pan, Y.; Wu, L. L.; Chen, J. J.; Gao, M. X.; Zhai, R.; Lin, J. Thermal evaporation and solution strategies to novel nanoarchitectures of silicon carbide. *Applied Physics A: Materials Science & Processing*, 2007, *88* (4), 679–685.

297. Dong, C.; Zou, G.; Liu, E.; Xi, B.; Huang, T.; Qian, Y. Synthesis of kelp-like crystalline β-SiC nanobelts and their apical growth mechanism. *Journal of the American Ceramic Society*, 2007, *90* (2), 653–656.

298. Yang, L.; Tang, Y.-H.; Zhao, S.-H.; Chen, J.-H.; Hu, A.-P.; Wang, Q.; Wang, G.-Z. Synthesis, controlling and photoluminescence study of ZnO nanostructures under solvothermal-assisted heat treatment. *Hunan Daxue Xuebao, Ziran Kexueban*, 2009, *36* (5), 57–62.

299. Luo, Y.; Tu, Y.; Yu, B.; Liu, J.; Li, J.; Jia, Z. Synthesis of hierarchical barium tungstate corns and their shape evolution process. *Materials Letters*, 2007, *61* (30), 5250–5254.

300. Jindal, Z.; Verma, N. K. Electrochemical template-assisted fabrication of CdS micro/nanostructures. *Physica E: Low-dimensional Systems and Nanostructures*, 2009, *41* (10), 1752–1756.

301. Wong, S. S. Template-based synthesis of nanostructures. *Abstracts of Papers, 239th ACS National Meeting*, San Francisco, CA, March 21–25, 2010, IEC-27.

302. Cao, C., Chen, Z.; An, X.; Zhu, H. Growth and field emission properties of cactus-like gallium oxide nanostructures. *Journal of Physical Chemistry C*, 2008, *112*, 95–98.

303. Hidalgo, P.; Mendez, B.; Piqueras, J. GeO₂ nanowires and nanoneedles grown by thermal deposition without a catalyst. *Nanotechnology*, 2005, *16*, 2521–2524.

304. Sounart, T. L.; Liu, J.; Voigt, J. A.; Hsu, J. W. P.; Spoerke, E. D.; Tian, Z. R.; Jiang, Y. Sequential nucleation and growth of complex nanostructured films. *Advanced Functional Materials*, 2006, 16, 335–344.

305. Kim, S.-W.; Kim, H.-K.; Jeong, S.-W.; Fujita, S.; Kim, K.-K. Route from ZnO thin films to nanostructures on Si substrates by metal organic chemical vapor deposition. *Journal of the Korean Physical Society*, 2007, *51* (Suppl. 3, *Proceedings of the 14th Korean Conference on Semiconductors*, Jeju, Korea, February 8–9, 2007), S207–S211.

306. Takezawa, Y.; Imai, H. Structural control on crystal growth of titanate in aqueous system: Selective production of nanostructures of layered titanate and anatase-type titania. *Journal of Crystal Growth*, 2007, *308* (1), 117–121.

307. Liu, Y.-J.; Zhang, Z.-Y.; Zhao, Q.; Dluhy, R. A.; Zhao, Y.-P. The role of the nanospine in the nanocomb arrays for surface enhanced Raman scattering. *Applied Physics Letters*, 2009, *94* (3), 033103/1–033103/3.

308. Jyouzuka, A.; Nakamura, T.; Onizuka, Y.; Mimura, H.; Matsumoto, T.; Kume, H. Emission characteristics and application of graphite nanospine cathode. *Journal of Vacuum Science and Technology, B: Microelectronics and Nanometer Structures—Processing, Measurement, and Phenomena*, 2010, *28* (2), C2C31–C2C36.

309. Zhu, Y.; Ruan, Q.; Xu, F. Crystallographically-oriented nanoassembly by ZnS tricrystals and subsequent three-dimensional epitaxy. *Nano Research*, 2009, *2* (9), 688–694.

310. Sahu, R. K.; Ray, A. K.; Mishra, T.; Pathak, L. C. Microwave-assisted synthesis of magnetic Ni wire from a metal-organic precursor containing Ni(II) and triethanolamine. *Crystal Growth and Design*, 2008, *8* (10), 3754–3760.

311. Pan, Q.; Huang, K.; Ni, S.; Yang, F.; Lin, S.; He, D. Synthesis of sheaf-like CuO from aqueous solution and their application in lithium-ion batteries. *Journal of Alloys and Compounds*, 2009, *484* (1–2), 322–326.

312. Zhang, M.; Xu, X.; Zhang, M. Hydrothermal synthesis of sheaf-like CuO via ionic liquids. *Materials Letters*, 2008, *62* (3), 385–388.

313. Hu, Y.; Chen, K. Crystal splitting in the growth of β-FeO(OH). *Journal of Crystal Growth*, 2007, *308* (1), 185–188.

314. Kim, S.; Lee, J. S.; Mitterbauer, C.; Ramasse, Q. M.; Sarahan, M. C.; Browning, N. D.; Park, H. J. Anomalous electrical conductivity of nanosheaves of CeO₂. *Chemistry of Materials*, 2009, *21* (7), 1182–1186.

315. Liu, K.; You, H.; Jia, G.; Zheng, Y.; Huang, Y.; Song, Y.; Yang, M.; Zhang, L.; Zhang, H. Hierarchically nanostructured coordination polymer: Facile and rapid fabrication and tunable morphologies. *Crystal Growth and Design*, 2010, *10* (2), 790–797.

316. Zhang, F.; Wong, S. S. Ambient large-scale template-mediated synthesis of high-aspect ratio single-crystalline, chemically doped rare-earth phosphate nanowires for bioimaging. *ACS Nano*, 2010, *4* (1), 99–112.

317. Yang, M.; You, H.; Song, Y.; Huang, Y.; Jia, G.; Liu, K.; Zheng, Y.; Zhang, L.; Zhang, H. Synthesis and luminescence properties of sheaflike $TbPO_4$ hierarchical architectures with different phase structures. *Journal of Physical Chemistry C*, 2009, *113* (47), 20173–20177.

318. Deng, H.; Liu, C.; Yang, S.; Xiao, S.; Zhou, Z.-K.; Wang, Q.-Q. Additive-mediated splitting of lanthanide orthovanadate nanocrystals in water: Morphological evolution from rods to sheaves and to spherulites. *Crystal Growth and Design*, 2008, *8* (12), 4432–4439.

319. Chen, G.-Y.; Bin, D.; Cai, G.-B.; Zhang, T.-K.; Dong, W.-F.; Zhang, W.-X.; Xu, A.-W. The fractal splitting growth of Sb_2S_3 and Sb_2Se_3 hierarchical nanostructures. *Journal of Physical Chemistry C*, 2008, *112* (3), 672–679.

320. Tang, J.; Alivisatos, A. P. Crystal splitting in the growth of Bi_2S_3. *Nano Letters*, 2006, *6* (12), 2701–2706.

321. Kelly, A. T.; Rusakova, I.; Ould-Ely, T.; Hofmann, C.; Luettge, A.; Whitmire, K. H. Iron phosphide nanostructures produced from a single-source organometallic precursor: Nanorods, bundles, crosses, and spherulites. *Nano Letters*, 2007, *7* (9), 2920–2925.

322. Uchiyama, H.; Imai, H. Preparation of $LiFePO_4$ mesocrystals consisting of nanorods through organic-mediated parallel growth from a precursor phase. *Crystal Growth and Design*, 2010, *10* (4), 1777–1781.

323. He, J.; Han, M.; Shen, X.; Xu, Z. Crystal hierarchically splitting in growth of $BaWO_4$ in positive cat-anionic microemulsion. *Journal of Crystal Growth*, 2008, *310* (21), 4581–4586.

324. Zhou, Z.; Xu, J.; Liu, X.; Li, X.; Li, S.; Yang, K.; Wang, X.; Liu, M.; Zhang, Q. Non-spherical racemic polylactide microarchitectures formation via solvent evaporation method. *Polymer*, 2009, *50* (15), 3841–3850.

325. Sano, N. Onion-like nanoparticles and their synthesis by arc in water. *Funtai Kogaku Kaishi*, 2003, *40* (1), 1924.

326. Rivera-De Leon, M.; Melin, F.; Palkar, A.; Echegoyen, L. Preparation, functionalization, characterization of carbon nanoonions. *Abstracts of Papers, 233rd ACS National Meeting*, Chicago, IL, March 25–29, 2007, CHED–1433.

327. Liu, Y. Chemistry of novel nanoscale carbon materials: Nanodiamond and carbon nano-onions. *Dissertation Abstracts International, B*, 2006, *67* (5), 2548, Avail. UMI, Order No. DA3216745, 2006, 139 pp.

328. Liu, X.; Wang, C.; Yang, Y. Z.; Guo, X.-M.; Wen, H.-R.; Xu, B.-S. Synthesis of nano onion-like fullerenes by using Fe/Al_2O_3 as catalyst by chemical vapor deposition. *Chinese Science Bulletin*, 2009, *54* (1), 137–141.

329. Bystrzejewski, P.; Rummeli, M. H.; Gemming, T.; Lange, H.; Huczco, A. Catalyst-free synthesis of onion-like carbon nanoparticles. *New Carbon Materials*, 2010, *25* (1), 1–8.

330. Huczko, A.; Cudzilo, S.; Bystrzejewski, M.; Lange, H.; Szala, M. Preparation of nanomaterials by combustion synthesis. *Biuletyn Wojskowej Akademii Technicznej*, 2005, *54* (10), 57–71.

331. Hou, S.-S.; Chung, D.-H.; Lin, T.-H. High-yield synthesis of carbon nano-onions in counterflow diffusion flames. *Carbon*, 2009, *47* (4), 938–947.

332. Teo, B. K.; Sun, X. H.; Li, C. P.; Wong, N. B.; Lee, S. T. Observation of hybrid carbon nanostructures as intermediates in the transformation from hydrocarbon nanotubes and nano-onions to carbon nanotubes and nano-onions via sonolysis on silicon nanowires and nanodots, respectively. *Chemistry of Materials*, 2010, *22* (4), 1297–1308.

333. Palkar, A.; Kumbhar, A.; Athans, A. J.; Echegoyen, L. Pyridyl-functionalized and water-soluble carbon nano onions: First supramolecular complexes of carbon nano onions. *Chemistry of Materials*, 2008, *20* (5), 1685–1687.

334. Cioffi, C. T.; Palkar, A.; Melin, F.; Kumbhar, A.; Echegoyen, L.; Melle-Franco, M.; Zerbetto, F. et al. A carbon nano-onion-ferrocene donor-acceptor system: Synthesis, characterization and properties, *Chemistry—A European Journal*, 2009, *15* (17), 4419–4427, S4419/1–S4419/3.

335. Yao, Y.; Wang, X.; Guo, J.; Yang, X.; Xu, B. Tribological property of onion-like fullerenes as lubricant additive. *Materials Letters*, 2008, *62* (16), 2524–2527.

336. Erkoc, S. Stability of carbon nanoonion $C_{20}@C_{60}@C_{240}$: Molecular dynamics simulations. *Nano Letters*, 2002, *2* (3), 215–217.

337. Hultman, L.; Stafstrom, S.; Czigany, Z.; Neidhardt, J.; Hellgren, N.; Brunell, I. F.; Suenaga, K.; Colliex, C. Cross-linked nano-onions of carbon nitride in the solid phase: Existence of a novel $C_{48}N_{12}$ aza-fullerene. *Physical Review Letters*, 2001, *87* (22), 225503/1–225503/4.

338. Guo, J.; Wang, X.; Xu, B. One-step synthesis of carbon-onion-supported platinum nanoparticles by arc discharge in an aqueous solution. *Materials Chemistry and Physics*, 2009, *113* (1), 179–182.

339. Kintigh, J.; Miller, G. P. Hydrogenation of fullerenes, single-walled nanotubes and nanoonions. *Abstracts, 34th Northeast Regional Meeting of the American Chemical Society*, Binghamton, NY, October 5–7, 2006, NRM-021.

340. Macutkevic, J.; Adomavicius, R.; Krotkus, A.; Seliuta, D.; Valusis, G.; Maksimenko, S.; Kuzhir, P. Batrakov, K.; Kuznetsov, V.; Moseenkov, S.; Shenderova, O.; Okotrub, A.V.; Langlet, R.; Lambin, Ph. Terahertz probing of onion-like carbon-PMMA composite films. *Diamond and Related Materials*, 2008, *17* (710), 1608–1612.

341. Abe, M.; Takeshi, S. Magneto-optical enhancement by surface plasmon resonance in magnetic "nano-onions" with multicore-shell structures. *Journal of Applied Physics*, 2005, *97* (10, Pt. 3), 10M514/1–10M514/2.

342. Remskar, M.; Virsek, M.; Mrzel, A. The MoS_2 nanotube hybrids. *Applied Physics Letters*, 2009, *95* (13), 133122/1–133122/3.

343. Wiggins, J.; Carpenter, Everett, E.; O'Connor, Charles, J. Phenomenological magnetic modeling of Au:Fe:Au nano-onions. *Journal of Applied Physics*, 2000, *87* (9, Pt. 2), 5651–5653.

8 "Nanoanimal" World

8.1 NANOLARVAE

Only a few reports are available in the area of nanolarvae. Thus, the shape-controlled Cu-In sulfide heterostructured nanocrystals were synthesized by thermal decomposition of a mixture of Cu-oleate and In-oleate complex in dodecanethiol.[1,2] Varying the reaction temperature and time, it was possible to prepare Cu-InS nanocrystals not only with larva (Figure 8.1) but also with acorn and bottle shapes. Precisely when the reaction was carried out for 1 h at 250°C, the *Fannia canicularis* larva-shaped nanocrystals (nanolarvae) of 45 nm × 185 nm were produced (composition $CuIn_{1.30}S_{1.24}$). The following Reaction 8.1 was established to be responsible for the production of Cu_2S nanocrystals; the formation mechanism was discussed including generation of Cu_2S seeds, and subsequent growth of indium sulfide (In_2S_3) occurred on these seeds via a so-called seed-mediated growth mechanism. The authors proved that the initially formed polydisperse Cu_2S nanoparticles were aggregated via a process similar to the oriented attachment mechanism, and subsequent incorporation of In_2S_3 species-generated larva-shaped nanostructures with a constant Cu/In atomic ratio:

$$C_{12}H_{25}SH + 2Cu\text{-}(OC(=O)\text{-}C_{17}H_{33}) \rightarrow Cu_2S + 2C_{17}H_{33}C(=O)OH + C_{12}H_{24} \tag{8.1}$$

8.2 NANOWORMS

In contrast to larva-like nanostructures, "nanoworms"[3] (or nanostructures with worm-like pores) are much more widespread in nanotechnological reports and represented, as well as other less-common nanoforms, mainly by inorganic metal oxides and salts, although *elemental* nanoworms (metals or nonmetals, alone or supported on composites) have also been obtained. Among them, gold nanostructures with different morphologies including spheres, rods, and worm-like nanostructures (Figure 8.2) were prepared by a three-step process treating an aqueous solution of chloroauric acid with sodium citrate and poly(vinyl pyrrolidone) (PVP).[4] Two possible growth mechanisms were proposed in the formation of these worm-like particles: the first is related to the stage of formation of nanorods from spherical colloids, and the second described the formation of worm-like particles from nanorods. Au catalysts were supported on wormhole hexagonal mesoporous silica (HMS) for the CO oxidation reaction, and the effect of support modification with cerium on the catalysts for this reaction was studied.[5] The highest activity of the Au/Ce-HMS catalyst in CO oxidation was associated with its higher gold dispersion and larger degree of coverage of HMS by CeO_2, thereby increasing the effectiveness of oxygen mobility. The influence of 20 kHz ultrasound waves of high intensity (40 W/cm²) on preformed citrate-protected gold nanoparticles (GNPs) (25 nm) in water and in the presence of surfactants[6] after 60 min of sonication led to a worm-like or ring-like structure. This effect could be of interest for ultrasonic melting of inorganic materials on a nanoscale to produce metal structures with different morphologies and properties.

An interesting worm-like nanostructure was reported for cubic Pd crystals inside giant carbon clusters, forming worm-like carbon nanoworms.[7] The head of the worm, typically 20–50 nm, consisted of Pd encapsulated in carbon, and the body of the worm, several hundred nanometers in length, consisted of many sections of carbon tubes with cone-shaped voids. Graphite was found to be a superior support for Pd (worm-like small Pd nanostructures) catalyst for the direct synthesis of H_2O_2 from H_2 and O_2.[8] The formed composite showed high H_2O_2 formation activity. Worm-like

FIGURE 8.1 Nanolarva (Cu-InS). (Reproduced with permission from Choi, S.-H. et al., One-pot synthesis of copper-indium sulfide nanocrystal heterostructures with acorn, bottle, and larva shapes, *J. Am. Chem. Soc.*, 128, 2520–2521. Copyright 2006 American Chemical Society.)

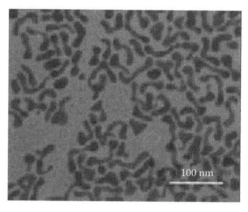

FIGURE 8.2 TEM image of gold worm-like nanostructures. (Reproduced with permission from Kemal, L. et al., Experiment and theoretical study of poly(vinyl pyrrolidone)-controlled gold nanoparticles, *J. Phys. Chem. C*, 112(40), 15656–15664. Copyright 2008 American Chemical Society.)

defective nanorods and nanospheres of silver were synthesized by photochemical decomposition of silver oxalate in water by UV irradiation in the presence of cetyltrimethylammoniumbromide (CTAB) and PVP, respectively.[9] The formation of defective Ag nanocrystals was attributed to the heating effect of UV-visible irradiation. The synthesis and formation of PtAg alloy nanowires in the presence of oleylamine and oleic acid through the oriented attachment was carried out, resulting in wormlike nanowires at the composition $Pt_{53}Ag_{47}$.[10] It was proposed that the formation of alloy nanowires is mostly driven by the interplay between the binding energy of capping agents on alloy surfaces and the diffusion of atoms at the interface upon the collision of primary nanoparticles. Worm-like Ge nanostructures (diameters 10–80 nm and lengths up to 1000 nm) can be directly prepared in an aqueous medium under ambient conditions by using widely available GeO_2 (in the form of germanate ions) as a precursor and $NaBH_4$ for 24 h reaction.[11] This synthesis route may be a good candidate for synthesizing a wide variety of crystallized Ge nanomaterials and devices due to its low cost, low safety risk, facileness, high yield (above 70% and in gram scale), and convenience for adding other chemicals (i.e., dopants or morphology modifying agents) into the reaction system.

Several *carbon* nanostructures are also known in worm-like forms, which are frequently formed together with various other structures at the same time depending on reaction conditions. Thus, high-purity (99.21 wt.%) helical carbon nanotubes (HCNTs) were prepared in a large quantity over

Fe nanoparticles (fabricated using a coprecipitation/H_2 reduction method) by acetylene decomposition at 450°C, together with worm-like carbon nanotubes (CNTs) as well as carbon nanocoils (CNCs, produced in large quantities), if H_2 was present throughout the acetylene decomposition.[12] Since the HCNTs and worm-like CNTs were attached to Fe nanoparticles, the nanomaterials were found to be high in magnetization. Pyrolysis of ruthenocene carried out in an atmosphere of argon or hydrogen was found to give rise to spherical carbon nanostructures, in particular worm-like carbon structures (Figure 8.3).[13] Worm-like expanded structures (graphite flakes) (Figure 8.4) are commonly observed in expanded graphite intercalated compounds. In addition, properties of graphene wormholes in which a short nanotube acts as a bridge between two graphene sheets, where the honeycomb carbon lattice is curved from the presence of 12 heptagonal defects, were studied.[14]

As for predominant nanostructures of various types, worm-like structures have also been reported for *silicon* and *silica*. Among them, silicon nanowires and nanoworms were selectively grown at temperatures below 400°C by plasma-enhanced CVD using silane as the Si source and

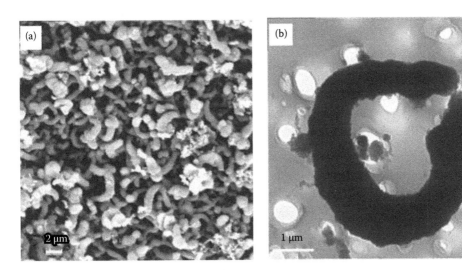

FIGURE 8.3 (a) SEM image of carbon structures prepared by pyrolysis of ruthenocene with a mixture of argon (150 sccm) and ethylene (50 sccm) bubbled through thiophene at 950°C and (b) the TEM image of these structures. (Reproduced with permission from Springer Science+Business Media: *Bull. Mater. Sci.*, Carbon nanostructures and graphite-coated metal nanostructures obtained by pyrolysis of ruthenocene and ruthenocene–ferrocene mixtures, 30(1), 2007, 23–29, Panchakarla, L.S. and Govindaraj, A.)

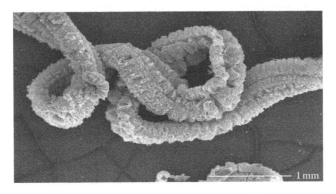

FIGURE 8.4 Worm-like expanded structures (graphite flakes) commonly observed in expanded graphite intercalated compounds. (From Wong, S.-C. et al., Materials processes of graphite nanostructured composites using ball milling. *Mater. Manuf. Processes*, 21, 159–166, 2006. Reproduced with permission from Taylor & Francis.)

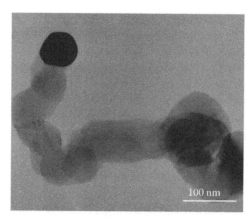

FIGURE 8.5 Amorphous, worm-like Si structures. (Reproduced with permission from Hofmann, S. et al., Gold catalyzed growth of silicon nanowires by plasma enhanced chemical vapor deposition, *J. Appl. Phys.*, 94(9), 6005–6012. Copyright 2003, American Institute of Physics.)

gold as the catalyst.[16] Si worm-like wires (80–300 nm diameter) were shown to be randomly oriented, with a rapidly varying growth direction (Figure 8.5). Sometimes, silicon-containing nanoworms were fabricated with the use of further removed or sacrificial templates. For instance, a surfactant-free synthesis of mesoporous and hollow silica nanoparticles, in which boron acted as the templating agent and then selectively removed, afforded mesoporous pure silica nanoparticles with wormhole-like pores or, depending on the synthetic conditions, silica nanoshells.[17] In a related work, exquisite hierarchical worm-like silica nanotubes were fabricated by a simple sol-gel method[18] using poly(2-(dimethylamino)ethyl methacrylate)-grafted multiwall carbon nanotubes (MWCNT-g-PDMAEMA) as a sacrificial template. Subsequent hydrolysis and polycondensation of tetraethoxysilane produced MWCNT-silica nanocomposites with numerous silica nanowires arranged on their surfaces. After the removal of templates by calcination, hierarchical worm-like silica nanotubes with some of the characteristics of mesoporous materials, such as large surface area, multiple pore distribution, and large pore volume, were obtained. Hollow silica nanoworms with circular pore channels parallel to the shell surfaces and holes at the terminals were prepared using its self-assemblies as templates via a single-templating approach.[19] In-catalyzed Si nanostructures (root-like, worm-like, and tapered) were synthesized by the H radical-assisted deposition method.[20] The shapes of Si nanostructure strongly depended on the hydrogen radical treatment temperatures and the growth temperatures. The silica nanosphere–sucrose nanocomposite, with a wormhole-like mesostructure, was readily formed by the simple addition of sucrose into the colloidal silica solution.[21] In addition, mesoporous silicalite-1 (zeolite with a high ratio of silica to alumina) nanospheres with diameters of 300–500 nm and possessing worm-like porous walls were prepared using silicalite-1 seeds as silica source, CTAB as the surfactant, and diethyl ether as the cosolvent.[22] This composite exhibited high surface areas and large total pore volumes resulting in high dynamic adsorptive capacities and super-hydrophobic properties and can be used to conduct the adsorption experiments under static and dynamic conditions.

Discussing *metal oxides* as nanoworms, we note that the main interest in *iron*-containing worm-like nanostructures is caused by the fact that segmented nanoworms composed of magnetic iron oxide and coated with a polymer are able to find and attach to tumors. Thus, magnetic nanoworms on the basis of iron oxide core in a dextran coating, developed by Park et al.,[23,24] are used for amplified tumor targeting and therapy (Figure 8.6). Such nanostructures, as it is seen, possess major contact with a cell surface (Figure 8.7) and, therefore, more effective interactions take place than those with nanospherical particles.

Fe$_3$O$_4$-polydivinylbenzene (PDVB) nanoworms, possessing superparamagnetic properties at room temperature, but ferromagnetism at 5 K, were synthesized by precipitation–polymerization of divinylbenzene in the presence of oleic acid coated iron oxide nanoparticles.[25] The superparamagnetic

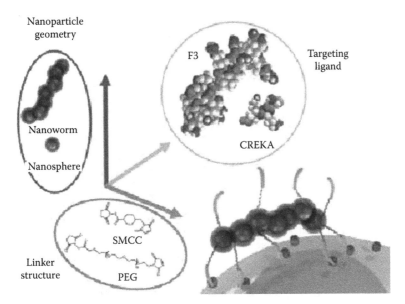

FIGURE 8.6 Use of magnetic nanoworms on iron oxide basis for tumor targeting. (Park, J.-H. et al.: Systematic surface engineering of magnetic nanoworms for in vivo tumor targeting. *Small.* 2009. 5(6). 694–700. Copyright Wiley-VCH Verlag GmbH & Co. KGaA. Reproduced with permission.)

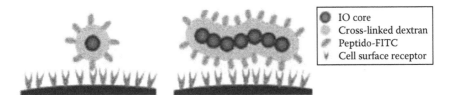

FIGURE 8.7 Conceptual scheme illustrating the increased multivalent interactions expected between receptors on a cell surface and targeting ligands on a nanoworm compared with a nanosphere. (Park, J.-H. et al.: Magnetic iron oxide nanoworms for tumor targeting and imaging. *Adv. Mater.* 2008. 20. 1630–1635. Copyright Wiley-VCH Verlag GmbH & Co. KGaA. Reproduced with permission.)

nanoworms could be well dispersed in ethanol, were capable of easy separation by an external magnetic field, and have potential applications in drug delivery/targeting, magnetic resonance imaging, and nanoprobes for diagnosis and disease treatment. Uncoated single crystalline α-Fe_2O_3 nanoflakes and nanowires were controllably synthesized simply by heating iron foil at different temperatures.[26] When the temperature was above 800°C, the worm-like structures were formed on the surface shown in Figure 8.8, because the vapor pressure might be too high and causes 3D growth. Additionally, Si-doped α-Fe_2O_3 worm-like nanostructures are known (Figure 8.9).

Several other *p*- and transition metal oxides, discovered in a nanoworm form, possess intriguing physicochemical, structural, electrical, and mechanical properties. Thus, γ-Al_2O_3 sol with a wormhole-like pore structure was prepared from aluminum alkoxide as a precursor by a hydrothermal process, and γ-Al_2O_3/polyimide nanocomposite films with a network structure were synthesized by an *in situ* synthesis process. A continuous network of inorganic phase was clearly observed in the nanocomposite films when the γ-Al_2O_3 mass fraction exceeded 12%. A strong material X-VO_x was formulated by nanocasting a conformal 4 nm thin layer of an isocyanate-derived polymer on the entangled worm-like skeletal framework of typical vanadia aerogels of 100–200 nm length and 30–40 nm thickness.[28,29] It was shown that X-VO_x remains ductile even at −180°C, a characteristic not found in most materials; this unusual ductility is derived from interlocking and sintering-like fusion of nanoworms during compression. X-VO_x emerges as an ideal material for force protection under impact. Interconnected flacks and

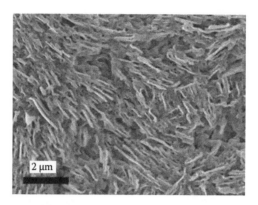

FIGURE 8.8 α-Fe$_2$O$_3$ worm-like nanostructures on Fe foils after heating it at 900°C for 10 h. (Reproduced with permission from Liao, L. et al., Morphology controllable synthesis of α-Fe$_2$O$_3$ 1D nanostructures: Growth mechanism and nanodevice based on single nanowires, *J. Phys. Chem. C*, 112, 10784–10788. Copyright 2008 American Chemical Society.)

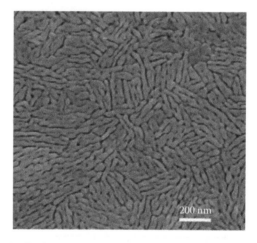

FIGURE 8.9 0.5% Si-doped α-Fe$_2$O$_3$ worm-like nanostructures, obtained from citric acid [C$_6$H$_8$O$_7$], iron(III) nitrate hydrate [Fe(NO$_3$)$_3 \cdot$ 9H$_2$O], and ethyleneglycol (C$_2$H$_6$O$_2$) as precursors. (Reproduced from *Sol. Energy Mater. Sol. Cells*, 93, Souza, F.L. et al., Nanostructured hematite thin films produced by spin-coating deposition solution: Application in water splitting, 362–368, Copyright 2009, with permission from Elsevier.)

nanoworm structures of cobalt oxide on glass and copper substrates were deposited using the chemical bath deposition method from aqueous cobalt chloride (CoCl$_2 \cdot$ 6H$_2$O) solution.[30] The electrical resistivity exhibited the semiconducting behavior of cobalt oxide thin film.

A synthesis process of mesoporous titanium dioxide (TiO$_2$) materials with tetra-*n*-butyl titanate (titanium butoxide Ti[O(*n*-Bu)]$_4$) as the precursor at ambient conditions resulted in the disordered wormhole-like mesostructure without discernible long-range order, formed by the agglomeration of TiO$_2$ nanoparticles.[31] These TiO$_2$ materials exhibited good performance (up to 98% removal) in dibenzylthiophene oxidation. The same precursor was used to obtain Fe-doped mesoporous TiO$_2$ microspheres, suitable for photodegradation of methyl orange, which were fabricated by an ultrasonic-hydrothermal method also using octadecylamine as a structure-directing agent.[32] The formed disordered wormhole-like mesostructure of Fe-doped TiO$_2$ microspheres (diameters ~300 to 400 nm) was found to be formed by the agglomeration of nanoparticles with an average size of about 10 nm. Additionally, metal oxide nanoworm-like structures have been reported for Ta$_2$O$_5$-Pt (Figure 8.10)[33] thin films and ZnO[34] (Figures 8.11 and 8.12). In the case of ZnO, the applied

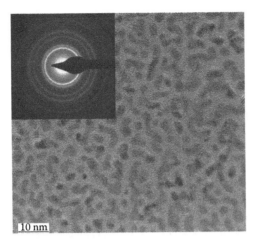

FIGURE 8.10 Nanoworm (Ta$_2$O$_5$-Pt electrode). (Reproduced from *Electrochem. Commun.*, 7(2), Park, K.-W. and Toney, M.F., Electrochemical and electrochromic properties of nanoworm-shaped Ta$_2$O$_5$-Pt thin-films, 151–155, http://www.slac.stanford.edu/cgi-wrap/getdoc/slac-pub-10454.pdf, Copyright 2005, with permission from Elsevier.)

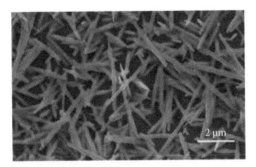

FIGURES 8.11 ZnO nanoworms. (Reproduced with permission from Wen, B. et al., Controllable growth of ZnO nanostructures by a simple solvothermal process, *J. Phys. Chem. C*, 112(1), 106–111. Copyright 2008 American Chemical Society.)

FIGURES 8.12 ZnO nanoworms. (Reproduced with permission from Wen, B. et al., Controllable growth of ZnO nanostructures by a simple solvothermal process, *J. Phys. Chem. C*, 112(1), 106–111. Copyright 2008 American Chemical Society.)

solvothermal method starting from zinc nitrate, CTAB, and hexamethylenetetramine (HMT) at 80°C–90°C for 0.5–6 h for growing ZnO nanostructures with different morphologies, including nanoworms, is based on the following Reaction 8.2, combining two different coordination agents and adjusting the ratio of solvents and the reaction time:

$$Zn^{2+} + CH_3(CH_2)_{15}N(Br)(CH_3)_3/C_6H_{12}N_4 \rightarrow Zn^{2+}\text{-amino complex} \rightarrow Zn(OH)_2 \rightarrow ZnO \quad (8.2)$$

In the case of *metal salt* nanoworm structures, we note a hydrothermal LiOH treatment of a nanostructured δ-MnO$_2$ precursor, which produced a Li-rich Li$_{1+x}$MnO$_{3-\delta}$ phase with nanoworm-like hierarchically assembled 2D nanoplate morphology.[35] It was established that tetravalent Mn ions are stabilized in octahedral sites of a Li$_2$MnO$_3$-type layered structure composed of edge-shared MnO$_6$/LiO$_6$ octahedra. The formed product showed superior electrode performance over the precursor Mn oxides and bulk Li-rich manganate. Near-monodisperse KCeF$_4$ worm-like nanowires were synthesized via co-thermolysis of K(CF$_3$COO) and Ce(CF$_3$COO)$_3$ in a hot oleic acid/oleylamine/1-octadecene solution.[36] Additionally, copper selenide CuSe nanomaterial with worm shape was prepared using a protein solution.[37]

Among worm-like nanostructures of *organic compounds* and *polymers*, we emphasize first of all polyaniline (PANI), frequently used for nanotechnological purposes. Thus, two different aniline dimers, *N*-phenyl-1,2-phenylenediamine (2-PPD) and *N*-phenyl-1,4-phenylenediamine (4-PPD), were used as starting monomers in PANI synthesis.[38] It was found that 2-PPD dimer alone produced only an amorphous PANI oligomer with a flaky morphology, while the 4-PPD provided either linear nanofiber or a spaghetti-like hollow nanofiber structures comprising of worm-like fibril subunits. By adjusting the molar fed ratio of 4-PPD to 2-PPD in the copolymerization, long PANI nanofibers with length up to tens of microns, bundled together by single PANI fibrils with a diameter ~3 to 5 nm, were formed. A possible formation mechanism was proposed, taking into account the reactivity difference at positions 4 and 2 on the 4-PPD and 2-PPD, respectively. Uniform worm-shaped polymeric nanostructures in aqueous solution were fabricated by dissolving water-soluble sacrificial layer poly(vinyl alcohol) (PVA),[39] making this protocol highly compatible with biomaterials. Compared with the worm-shaped nanostructures made by self-assembly, these lithographically defined nanoworms had much better controllability and uniformity on the shape, size, and aspect ratio.

Ligand-stabilized platinum nanoparticles were self-assembled with poly(isoprene-*block*-dimethylaminoethyl methacrylate) (PI-b-PDMAEMA, **1**) block copolymers generating organic–inorganic hybrid materials with spherical micellar, worm-like micellar, lamellar, and inverse hexagonal morphologies, whose disassembly generated isolated metal NP-based nanospheres, cylinders, and sheets, respectively.[40] Nanostructured noncrystalline cresol-formaldehyde material NCF-1, synthesized through hydrothermal condensation of *m*-cresol and formaldehyde at 363 K in the presence of supramolecular assembly of cationic surfactant, CTAB as structure directing agent, revealed a nanorod morphology with the diameter of the rods 30–50 nm and a disordered wormhole-like nanostructure with pores of ~2.5 nm.[41] This composite material exhibited photoluminescence property at room temperature, which could be utilized for the fabrication of novel organic optical devices. Long, highly stable, and densely packed edge-on worm-like nanocolumns of hexa-*peri*-hexabenzocoronenes (two HBC derivatives bearing five 3,7-dimethyloctyl solubilizing groups, the sixth peripheral substituent being terminated by either a thioacetate group [HBC-Sac, **2**] or a 1,2-dithiolate ring [HBC-SS, **3**]) (Figures 8.13 and 8.14) spontaneously grew from solution as self-assembled monolayers chemisorbed on gold.[42] The requirements for the self-assembly of 2,3-annulated core-substituted naphthalene diimide into discrete worm-like nanostructures were explored in MeOH-CHCl₃ solutions.[43] Additionally, a systematic study of the thermodynamics, structure, and rheology of mixtures of cationic worm-like micelles and like-charged nanostructures was carried out.[44]

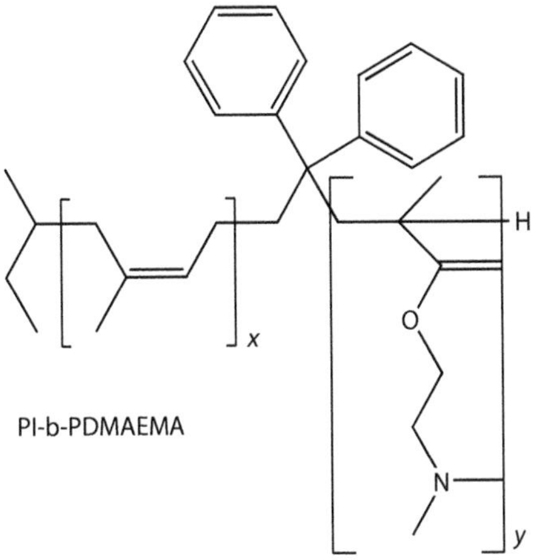

PI-b-PDMAEMA

1

HBC-sac   R =

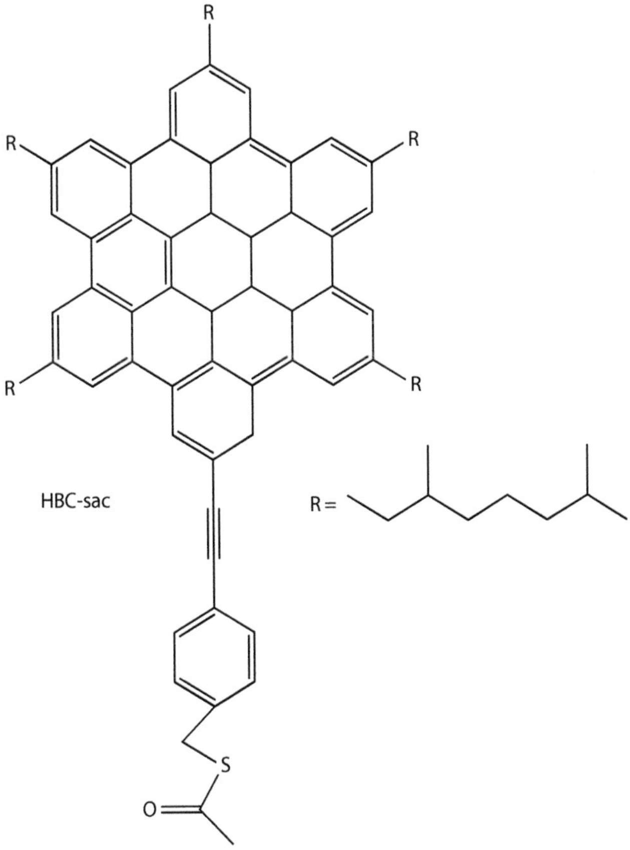

2

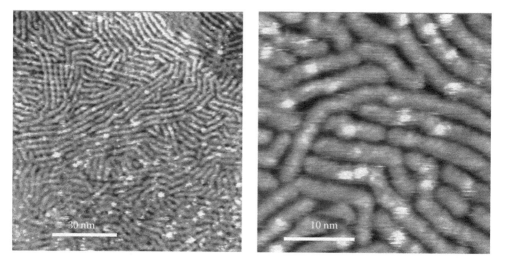

FIGURE 8.13 STM images of HBC-Sac nanocolumns chemisorbed on Au(111) after deprotection of the thioacetate terminal group. (Reproduced with permission from Piot, L. et al., Growth of long, highly stable, and densely packed worm-like nanocolumns of hexa-peri-hexabenzocoronenes via chemisorption on Au(111), *J. Am. Chem. Soc.*, 131(4), 1378–1379. Copyright 2009 American Chemical Society.)

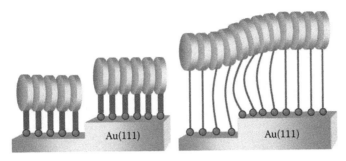

FIGURE 8.14 Schematic structure of HBC nanocolumns chemisorbed on Au(111) via sulfur atoms (as small circles) showing the "disruptive" step crossing of HBC-Sac bearing a short and rigid grafting chain (left) and the "soft" step crossing of HBC-SS bearing a long and flexible chain (right). (Reproduced with permission from Piot, L. et al., Growth of long, highly stable, and densely packed worm-like nanocolumns of hexa-peri-hexabenzocoronenes via chemisorption on Au(111), *J. Am. Chem. Soc.*, 131(4), 1378–1379. Copyright 2009 American Chemical Society.)

8.3 NANOSQUAMAE

Squama-like nanostructures do not belong to a well-known nanoform; only a few publications are available in this field and almost all correspond to metal oxides. Thus, various diversified morphologically modulated ZnO nanostructures including nanorods, nanotetrahedrons, nano-fans, nanodumbbells, and nanosquamas (Figure 8.15) were prepared via an effective aminolytic reaction of zinc carboxylates with oleylamine in noncoordinating and coordinating solvents.[45] A series of hierarchical squama-like nanostructured/porous titania materials (Figure 8.16) (with wormhole-like mesopores of nanoparticle assembly in each squama) doped with different contents of cerium (Ce/TiO$_2$, with a mixture of Ce$^{3+/4+}$ oxidation states) were synthesized by utilizing the oil-in-water emulsion technique (the formation mechanism is represented by Figure 8.17).[46] The samples were found to exhibit a pure anatase crystalline phase, to have catalytic activity in the photodegradation of Rhodamine B (Reactions 8.3 through 8.6), and to be a good support of GNPs to remove CO by catalytic oxidation. The nanocomposites (CaDHCAp/BC) of excellent nanomaterial hydroxyapatites (Hap, with outstanding bioactivity) and bacterial cellulose (BC, a remarkably versatile biomaterial) were prepared by alkaline treatment, Ca^{2+} activation, and biomimetic mineralization.[47] The product, which may have potential application as an orthopedic biomaterial, consisted of calcium-deficient carbonate-containing hydroxyapatite (CaDHCAp) in the 3D network of BC nanofibers; the apatite crystals deposited along BC nanofibers were partially substituted with calcium carbonate, and the uniform spherical apatite particles were composed of squama-shaped nanosized apatite crystals. Additionally, 3D squama-like macroaggregates of Co^{II}-doped Y$_2$O$_3$ optical functional nanoparticles (4–10 nm), synthesized by the plasma arc-discharge method (Figure 8.18), were self-assembled by disordered nanoparticles.[48] The formation mechanism of the aggregates was ascribed to the periodic coagulation of Y nanoparticles.

$$Ce^{4+} + e^- \rightarrow Ce^{3+} \tag{8.3}$$

$$Ce^{3+} + O_2 \rightarrow Ce^{4+} + \cdot O_2^- \tag{8.4}$$

$$\cdot O_2^- + H^+ \rightarrow \cdot HO_2 \tag{8.5}$$

$$\cdot HO_2 + RhB \rightarrow \text{Achromatous organic species} \tag{8.6}$$

FIGURE 8.15 Various ZnO nanosquamas prepared in TOP from different zinc alkylcarboxylates including (a) hexanoic acid, (b) octanoic acid, and (c) oleic acid. (Reproduced with permission from Zhang, Z. et al., Modulation of the morphology of ZnO nanostructures via aminolytic reaction: From nanorods to nanosquamas, *Langmuir*, 22(14), 6335–6340. Copyright 2006 American Chemical Society.)

8.4 NANOURCHINS

Nanourchins, which belong to a similar structural type as nanoflowers (indeed, nanourchins are frequently confused with flower-like nanostructures), are known for a variety of inorganic compounds and a few organics and have many applications, first of all in catalysis due to the high surface area of this nanostructure. In the case of *carbon*, sea-urchin-shaped nanostructured hollow carbon microspheres (e.g., hollow carbon microspheres containing CNTs extending outward from the central microsphere) were fabricated by formation of the carbon microsphere followed by incorporation of iron nanospheres (as growth catalysts) and then CVD of carbon (as nanotubes) from ethylene as the carbon source.[49] After incorporation of platinum on such a carbon support, the formed nanocomposite could be used for methanol electrochemical oxidation in fuel cells. Hollow, urchin-like ferromagnetic carbon spheres with electromagnetic function and high conductivity at room temperature could be used as a reversible dye adsorbent. These nanostructures were prepared from template-free-synthesized, urchin-like, hollow spheres of PANI containing FeCl$_3$ as

FIGURE 8.16 Cerium-doped titania materials of the anatase crystalline phase with a Squama-like morphology. (Reproduced with permission from Ma, T.-Y. et al., Hierarchically structured squama-like cerium-doped titania: Synthesis, photoactivity, and catalytic CO oxidation, *J. Phys. Chem. C*, 113(38), 16658–16667. Copyright 2009 American Chemical Society.)

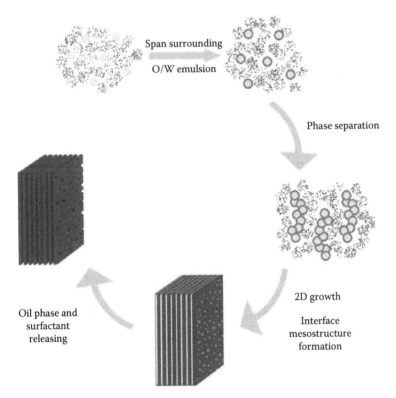

FIGURE 8.17 Formation mechanism of squama-like Ce/TiO$_2$ scheme, where the small and blue spots stand for Ce/TiO$_2$ in the water phase and Span-60 at the O/W interface, respectively, and the oil phase of cyclohexane is painted yellow. (Reproduced with permission from Ma, T.-Y. et al., Hierarchically structured squama-like cerium-doped titania: Synthesis, photoactivity, and catalytic CO oxidation, *J. Phys. Chem. C*, 113(38), 16658–16667. Copyright 2009 American Chemical Society.)

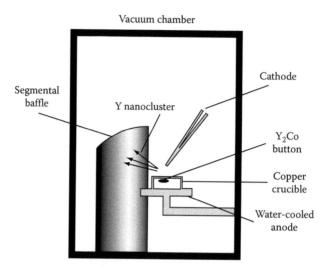

FIGURE 8.18 Schematic diagram of the arc-discharge furnace employed to synthesize the Co^{II}-doped Y_2O_3 nanoparticles. (Reproduced from *J. Alloys Compd.*, 457(1–2), Liu, X.G. et al., Co-doped Y_2O_3 optical functional nanoparticles and novel self-assembly squama-like aggregates, 517–521, Copyright 2008, with permission from Elsevier.)

the precursors by a carbonization process at 1200°C under an argon atmosphere.[50] The high conductivity at room temperature resulted from the graphite-like structure, whereas α-Fe or γ-Fe_2O_3 nanoparticles produced by $FeCl_3$ during the carbonization were attributed to the ferromagnetic properties. An urchin-structured multiwall carbon nanotube-hard carbon spherule (MWNTs/HCS) composite, where HCS was thoroughly coated by intercrossed MWNTs, was synthesized through acetylene pyrolysis[51] and showed an outstanding lithium uptake/release property: reversible charge capacity of 445 mAh/g can be delivered by this composite after 40 cycles. The hybrid sphere/CNT nanostructures were grown in the aerosol phase by an on-the-fly process by spray pyrolysis, followed by CNT catalytic growth (Figure 8.19) at the surface of the spherical nanoparticles yielding urchin-like nanostructures consisting of numerous CNTs attached to an alumina/iron oxide sphere.[52] These hybrid nanoparticles were dispersed to poly-alpha-olefin with sonication and a small amount of surfactants to form stable nanofluids, whose effective thermal conductivity was increased by about 21% at room temperature for particle volume fractions of 0.2%. The role of sulfur in the synthesis of novel carbon morphologies was studied.[53] For the materials, obtained by sulfur-assisted CVD, it was demonstrated that S not only acted on the catalyst but also can be

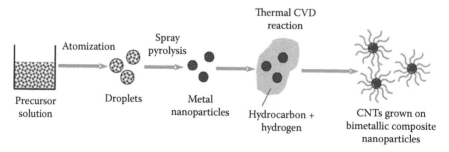

FIGURE 8.19 Schematic of gas phase growth pathways of CNTs grown on metal oxide nanoparticles. The precursor solution was prepared with 3 wt.% $Fe(NO)_3$ + $Al(NO)_3$ solution with the $Fe(NO)_3$: $Al(NO)_3$ ratio of 1:1 for hybrid sphere/CNT particles. (From Han, Z.H. et al., *Nanotechnology*, 18, 105701, 4 pp., 2007. Reproduced with permission from IOP Science.)

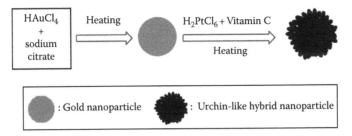

FIGURE 8.20 Procedure to design the urchin-like hybrid Au/Pt nanoparticles. (Reproduced with permission from Guo, S. et al., A novel urchin-like gold/platinum hybrid nanocatalyst with controlled size, *J. Phys. Chem. C*, 112(35), 13510–13515. Copyright 2008 American Chemical Society.)

detected in the carbon lattice of the nanostructures. Sulfur is responsible for inducing curvature and therefore influencing the final carbon nanostructure morphology, in particular sea-urchin-like nanostructures. Carbon nanourchins are reviewed, among other nanostructures, as nanostructured materials for batteries.[54]

Due to its biocompatibility and ability for functionalization, gold is one of the main and common *metals*, obtained in the nanostructural forms, so gold urchin-like nanostructures are not an exception. Thus, the excitation of localized surface plasmons of GNPs of various shapes, including urchins with average diameters of 53–72 nm, for live imaging of cells of the central nervous system, was exploited.[55] In relation to the biocompatibility of urchin (as well as spherical and rod-like) GNPs, within the examined concentration range (0.1–250 nM), they were found to be harmless toward neurons or glia, regardless of their shape and surface chemistry and promised to be good candidates for the development of nanosensors and drug delivery systems for the central nervous system. A simple strategy (Figure 8.20) to obtain quasi-monodisperse Au/Pt hybrid nanoparticles with urchin-like morphology and controlled size and Pt shell thickness was offered using $HAuCl_4$, H_2PtCl_6, and ascorbic acid.[56,57] The Pt shell thickness and size of Au/Pt urchin-like nanoparticles (from 3 to 70 nm) can be easily controlled by changing the molar ratios of Au to Pt and heating of $HAuCl_4$-citrate aqueous solution, respectively. Au/Pt hybrid nanoparticles exhibited higher catalytic activities than those of Pt NPs with similar size. The authors expected that the as-prepared Au/Pt hybrid NPs, combined with supporting materials such as CNTs and mesoporous carbon, could be used as advanced cathode material for fuel cells in the future. In the case of other noble metals, solvent-stabilized urchin-like self-assemblies of Pd nanoparticles with a small diameter, uniform size, and good dispersion were obtained by thermal decomposition of palladium acetate with Me *i*-Bu ketone as a solvent in the presence of a small amount of ethylene glycol and KOH under microwave irradiation for 60 s.[58] A simple and effective wet chemical route to direct synthesis (Reaction 8.7) of well-dispersed Pt nanoparticles with an urchin-like morphology was proposed and carried out by simply mixing H_2PtCl_6 aqueous solution and PVP with the initial molar ratio of 1:3.5 kept constant at 30°C for 3 days in the presence of formic acid.[59] These urchin-like Pt nanostructures showed excellent electrocatalytic activity in the reduction of dioxygen and oxidation of methanol and could be used as a promising nanoelectrocatalyst.

$$H_2PtCl_6 + 2HCOOH \rightarrow Pt + 6Cl^- + 6H^+ + 2CO_2 \tag{8.7}$$

Nickel, traditionally occupying an important place in nanotechnology due to its magnetic and catalytic properties, has been reported in nanourchin forms, among others. Thus, nickel nanomaterials with a wide range of morphologies (Figure 8.21) and sizes, such as superfine nanoparticles, urchin-like chains, smooth chains, rings, and hexagonal $Ni/Ni(OH)_2$ heterogeneous structure plates, were synthesized in a single reaction system (PVP K30, $NiCl_2 \cdot 6H_2O$, ethylene glycol, $N_2H_4 \cdot H_2O$, 60°C) by simply adjusting the reaction conditions.[60] The urchin-like nickel chains exhibited the best

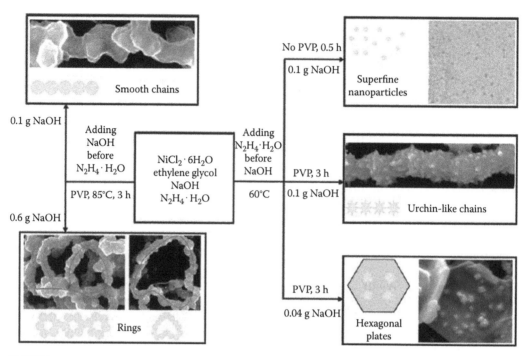

FIGURE 8.21 Flowchart for the synthesis of differently shaped nickel samples. (Reproduced with permission from Wang, C. et al., Controlled synthesis of hierarchical nickel and morphology-dependent electromagnetic properties, *J. Phys. Chem. C*, 114(7), 3196–3203. Copyright 2010 American Chemical Society.)

absorption property in contrast with other as-synthesized samples and other reported nickel structures, which can be attributed to the geometrical effect, high initial permeability, point discharge effect, and multiple absorption; thus, prepared nickel nanomaterials can be applied as promising absorbing materials. In a related report,[61] magnetic chains self-assembled from urchin-like hierarchical Ni nanostructures (average diameter 2–4 μm, composed of well-aligned sword-like nanopetals growing radially from the surfaces of the spherical particles) were synthesized by a simple hydrothermal method for 4 h at 115°C without any template or surfactant. These nickel chain-like architectures displayed ferromagnetic behavior. Urchin-like core-shell composite hollow spheres were fabricated by the assembly of nickel nanocones on the surface of hollow glass spheres.[62] FeNi$_3$ alloy nanostructures were synthesized by hydrothermal methods in the surfactant/*n*-octane/*n*-hexanol/water quaternary reverse microemulsion systems.[63] Sea-urchin-like particles, obtained when CTAB was used as the surfactant, were composed of many nanorods with diameters of ~42 nm and lengths of 0.4–1.2 μm. Both the spherical and sea-urchin-like FeNi$_3$ samples exhibited typical ferromagnetic behavior at room temperature.

For other elemental substances, only a few "nanourchin" reports are available. Thus, a solution-phase process was demonstrated for the preparation of urchin-like bismuth nanostructures by reducing bismuth nitrate with ethylene glycol at 180°C with an 85% yield.[64] The Mg/air batteries made from the vapor-deposited Mg structures, especially the battery made from the Mg sea-urchin-like nanostructures, displayed superior electrochemical properties to those of the various existing power sources.[65] A kind of interesting urchin-like selenium nanostructure consisting of Se nanorods with a diameter of ~100 nm and a length of 4–5 μm was synthesized based on a dismutation reaction in a buffer system at ambient conditions with the use of CTAB as a shape-directing agent.[66]

Among nanourchin structures, the metal oxides are common, especially vanadium oxides (VO$_x$)[67,68] (Figure 8.22). They possess spherical structures composed of a radially oriented array of VO$_x$ nanotubes (VO$_x$-NTs) and vanadium oxide nanorods (VO$_x$-NRDs).[69] The Raman scattering

FIGURE 8.22   SEM image of nanourchin (V_2O_5). (Reproduced with permission from O'Dwyer, C. et al., Nano-urchin: The formation and structure of high density spherical clusters of vanadium oxide nanotubes, *Chem. Mater.*, 18(13), 3016–3022. Copyright 2006 American Chemical Society.)

spectrum of the nanourchin exhibited a band at $1014\,\text{cm}^{-1}$ related to the distorted gamma conformation of the vanadium pentoxide (γ-V^{5+}). The structure of nanourchins, nanotubes, and nanorods of vanadium oxide nanocomposite was found to be strongly dependent on the valency of the vanadium, its associated interactions with the organic surfactant template, and on the packing mechanism and arrangement of the surfactant between vanadate layers.[70] The nanourchins contained vanadate layers in the nanotubes (γ-V^{5+} mentioned earlier), whereas the nanorods, by comparison, showed evidence for V^{5+} and V^{4+} species-containing ordered VO_x lamina.

Two isostructural materials (one planar and the other curved), ethylenediammonium (enH_2) intercalated V_7O_{16} and V_2O_5 nanourchins (VONUs), were synthesized employing pH control using *n*-dodecylamine as an amine template.[72] The structure of the vanadium oxide layer in these compounds was similar to that of vanadium oxide nanotubes (VONTs). The vanadium oxidation state in tubular structures appeared to be higher than in planar compounds such as (enH_2)V_7O_{16} and BaV_7O_{16}. Urchin-like VO_2(B) nanostructures (Figure 8.23) composed of radially aligned nanobelts were synthesized by the homogeneous reduction reaction between peroxovanadic acid and oxalic acid under hydrothermal conditions (Reactions 8.8 through 8.10):[73]

$$V_2O_5 + 4H_2O_2 \rightarrow 2[VO(O_2)_2]^- + 2H^+ + 3H_2O \tag{8.8}$$

$$10[VO(O_2)_2]^- + 10H^+ + H_2C_2O_4 + 6H_2O \rightarrow V_{10}O_{24}\cdot 12H_2O + 2CO_2 + 10O_2 \tag{8.9}$$

$$V_{10}O_{24}\cdot 12H_2O + 4H_2C_2O_4 \rightarrow 10\,VO_2(B) + 8CO_2 + 16H_2O \tag{8.10}$$

Urchin-like nanostructures are known for CuO and some copper hydroxy salts, in particular malachite. Thus, urchin-like CuO (sphere diameter about $1\,\mu\text{m}$), consisting of closely packed nanorods with a diameter of $10\,\text{nm}$, were synthesized by a PEG-assisted hydrothermal route at a relatively low temperature of $100°C^{74}$ (or, alternatively, via a simple water–ethylene glycol mixed-solvothermal

(a)   (b)

(c)

FIGURE 8.23   Urchin-like $VO_2(B)$ nanostructures composed of radially aligned nanobelts synthesized with 0.05 mol/L oxalic acid at 180°C for 12 h. (a) Low-magnification SEM image, (b) high-magnification SEM image, and (c) TEM image. The inset in part C shows a typical SAD (selected area diffraction) pattern taken from an individual nanobelt. (Reproduced with permission from Li, G. et al., Synthesis of urchin-like VO_2 nanostructures composed of radially aligned nanobelts and their disassembly, *Inorg. Chem.*, 48, 1168–1172. Copyright 2009 American Chemical Society.)

route at the same temperature of 100°C for 12 h, employing $CuCl_2$ and KOH as starting reactants in the absence of any surfactant or template[75]). No $Cu^{II} \rightarrow Cu^{I}$ reduction by PEG was observed; this was attributed by the authors to high PEG concentration. Additionally, the sensor based on the urchin-like CuO nanostructures exhibited excellent ethanol-sensing properties at a reduced working temperature (200°C), which showed a sensitivity two times higher than that of CuO particles (about 100 nm, made from calcinations of $Cu(NO_3)_2$ at 400°C), possibly contributing to the fancy 3D nanostructures.

The Cu_2O nanocubes were subsequently oxidized to form CuO hollow cubes, spheres, and urchin-like particles, through a sequential dissolution-precipitation process.[76] The CuO urchin-like particles exhibited excellent electrochemical performance and stability, superior to those of hollow structures, for lithium-ion battery anode materials. $Cu_2(OH)_2CO_3$ and CuO hierarchical nanostructures with variable morphologies were also synthesized by controlled heating hydrated nanoparticles.[77] The change mechanism from one morphology to others was established, starting the growth of nanostructures with nanoparticles, forming loose aggregates; the nanoparticles within aggregates reorganized to form urchin-like structures, which consisted of dense nanorods. Adsorbed water played an important role during the formation of malachite nanostructures. Additionally, urchin-like copper salts can serve as precursors of other nanostructure types. Thus, for example, as a result of homogeneous reaction between peroxovanadic acid and cupric acetate, the disassembly of

$Cu_3(OH)_2V_2O_7 \cdot 2H_2O$ nanourchins, composed of radially aligned nanobelts, led[78] to ultralong single crystalline CuV_2O_6 nanobelts after the growth of this compound along the direction [010].

Urchin-like ZnO nanostructures possessing distinct morphologies can be fabricated by elemental oxidation as by decomposition of zinc salts. Thus, ZnO nanostructures were grown on Si(100) substrates by direct oxidation of metallic Zn powder at 600°C, resulting in sea-urchin-like formations (Figure 8.24), consisting of straight nanowires of ZnO with blunt faceted ends with a sudden reduction in diameter projecting out, having diameters of 30–60 nm and lengths of 2–4 μm.[79] The mechanism of their formation is shown in Figure 8.25. On the contrary, sea-urchin-like ZnO nanomaterials were alternatively prepared by decomposition of a zinc acetate precursor in the presence

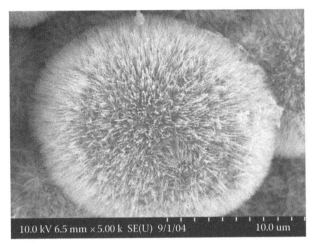

FIGURE 8.24 Sea-urchin-like ZnO nanostructure. (Reproduced from *J. Cryst. Growth*, 277, Sekar, A. et al., Catalyst-free synthesis of ZnO nanowires on Si by oxidation of Zn powders, 471–478, Copyright 2005, with permission from Elsevier.)

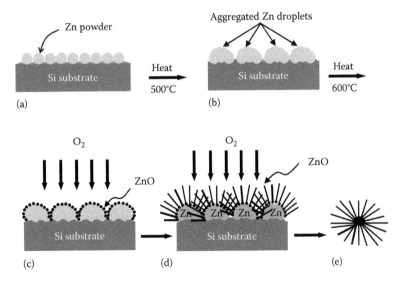

FIGURE 8.25 Schematic illustration of the growth process of the sea-urchin-like ZnO nanostructures: (a) Zn powders dispersed on Si substrate, (b) formation of aggregated Zn droplets, (c) formation of ZnO nuclei on the surface of Zn droplets, (d) growth of nanowires out of the droplets, and (e) formation of sea-urchin-like ZnO nanostructure. (Reproduced from *J. Cryst. Growth*, 277, Sekar, A. et al., Catalyst-free synthesis of ZnO nanowires on Si by oxidation of Zn powders, 471–478, Copyright 2005, with permission from Elsevier.)

of sodium hydroxide and ethylene glycol in an ethanol solution using a solvothermal method at 180°C for 12 h[80] or, alternatively, on Si(100) by simple oxidation of ZnO films at 600°C consisting of ZnO nanorods.[81] The ZnO urchin was constructed of well-assembled nanorods of length ~3 μm range and diameter ~20 nm.[82] An urchin-like Zn/ZnO core-shell structure, composed of a micrometer-scale sphere-shaped metallic Zn core and a shell made up of numerous radially protruding single-crystalline ZnO nanorods, was synthesized and deposited directly on an indium tin oxide (ITO) glass substrate by using a thermal evaporation method.[83] In this hybrid composite, the Zn core made direct contact with the ITO layer, which enhanced the interface bonding and conductance between the Zn/ZnO structure and the substrate. The urchin-like ZnO nanostructures showed a good field emission performance, comparable to the best value of ZnO nanostructures, and good stability; it can be a promising candidate for anode material in solar energy conversion devices.[84] Additionally, different doping phases can influence the formation of ZnO morphology; thus, ZnO doped with Ga_2O_3 was found to possess urchin-like morphology, in contrast to doping with In_2O_3, when nanorods were observed.[85]

Several reports are dedicated to tungsten oxide nanourchins, obtained mainly by the hydrothermal method. Thus, hierarchical WO_3 urchin-like structures were produced by a hydrothermal method at 180°C using sodium tungstate, ammonium EDTA salt, and Na_2SO_4.[86] The formed product possessed electrochemical response and luminescence properties. It was established that, using Na^+ EDTA salt, WO_3 nanowire bundles were formed instead of urchins. Large-scale vertically aligned and double-sided Co-doped hexagonal tungsten oxide nanorod arrays, synthesized by a facile hydrothermal method without using any template, catalyst, or substrate, were possibly formed from urchin-like microspheres via a self-assembly and fusion process.[87] It was found that these arrays showed excellent gas sensing performance in 1-butanol vapor, with rapid response and high sensitivity. A composite hierarchical hollow structure, consisting of discrete WO_2 hollow core spheres with $W_{18}O_{49}$ nanorod shells (hollow urchins), showed unusual magnetic behavior.[88]

Among other urchin-like oxides, Zn-doped SnO_2 nanourchins, assembled by the nanocones with diameters ranging from 20 to 60 nm and having high photocatalytic activity efficiency, were synthesized (Reactions 8.11 through 8.16) from $SnCl_4 \cdot 5H_2O$ and $Zn(NO_3)_2 \cdot 6H_2O$ as precursors in the mixed solvents of ethylenediamine (EDA), ethanol, and deionized water.[89] The photocatalytic activity of these Zn-doped SnO_2 urchin samples was evaluated by the degradation of RhB aqueous solution, revealing high photocatalytic activity efficiency:

$$Sn^{4+} + 6OH^- \rightarrow Sn(OH)_6^{2-} \tag{8.11}$$

$$Zn^{2+} + Sn(OH)_6^{2-} \rightarrow ZnSn(OH)_6 \tag{8.12}$$

$$ZnSn(OH)_6 + 4OH^- \rightarrow Sn(OH)_6^{2-} + ZnSn(OH)_4^{2-} \tag{8.13}$$

$$Sn(OH)_6^{2-} + nEn \rightarrow \{[Sn(En)_n](OH)_6\}^{2-} \tag{8.14}$$

$$Zn(OH)_4^{2-} + mEn \rightarrow \{[Zn(En)_m](OH)_4\}^{2-} \tag{8.15}$$

$$(1-x)[Sn(En)_n](OH)_6^{2-} + x[Zn(En)](OH)_4^{2-}$$
$$\rightarrow Sn_{1-x}Zn_xO_2 + 2H_2O + 2(1-x)OH^- + (n - nx + mx)En \quad (0 < x < 1) \tag{8.16}$$

A hydrothermal reaction (hydrolysis of $FeCl_3$ in the solution containing different anions [SO_4^{2-}, Cl^-, NO_3^-, ClO_3^-, ClO_4^-, $C_2O_4^-$, Br^-]) and sequential calcinations were carried out for the selective synthesis of 3D α-Fe_2O_3 urchin-like and other nanostructures without employing templates or matrixes.[90] The as-obtained α-Fe_2O_3 nanostructures showed photocatalytic activity on the decomposition of RhB upon irradiation by the UV light. An electrorheological (ER) suspension

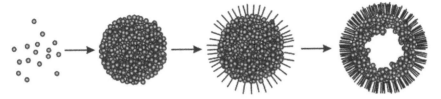

FIGURE 8.26 Formation of the α-MnO$_2$ urchin-like structures. (Reproduced with permission from Yu, P. et al., Shape-controlled synthesis of 3D hierarchical MnO$_2$ nanostructures for electrochemical supercapacitors, *Cryst. Growth Des.*, 9(1), 528–533. Copyright 2009 American Chemical Society.)

composed of Cr-doped titania particles with a sea-urchin-like hierarchical morphology (consisting of high-density rutile Cr-doped titania nanorods assembled radially on the surfaces of particles) was developed, demonstrating a distinct enhancement in ER properties.[91] It was found that the suspension of hierarchical Cr-doped titania particles (surface area of 65 m^2/g) possesses a stronger ER effect compared with the corresponding suspension of smooth nonhierarchical Cr-doped titania particles (surface area of 5 m^2/g). A simple hydrothermal route for the synthesis of various MnO$_2$ nanostructures using Mn$_3$O$_4$ powder as raw material in H$_2$SO$_4$ solution led to urchin-like γ-MnO$_2$ nanostructures and single-crystal β-MnO$_2$ nanorods, obtained at 80°C and 180°C, respectively, as well as to MnOOH nanowires prepared in a diluted acid solution.[92] Alternatively, urchin-like nano-/micro-hybrid α-MnO$_2$ balls, composed of single crystalline α-MnO$_2$ nanorods, were obtained from KMnO$_4$ and H$_2$SO$_4$ as precursors,[93] or from MnSO$_4 \cdot$H$_2$O, K$_2$S$_2$O$_8$, and concentrated sulfuric acid without using any template and surfactant.[94] Synthesis mechanism is shown in Figure 8.26, including formation of a large number of nuclei (a), their aggregation to microspheres (b) through the reaction between MnSO$_4$ and K$_2$S$_2$O$_8$ in acid solution, slow crystal growth (c), and formation of an interior cavity (d), resulting in the core-shell α-MnO$_2$ structures. Synthesis of urchin-shaped 3D CeO$_2$ via a surfactant-free route with water, ethanol, and ethylene glycol solvents was reported.[95] Both 3D CeO$_2$ and 5 wt.% CuO supported by 3D CeO$_2$ exhibited high catalytic activity for CO conversion.

Aluminum, iron, and indium oxyhydroxides were reported to be obtained in nanourchin forms. Thus, InOOH with urchin-like nanostructures, composed of nanorods with a diameter of several nanometers and a length <100 nm, was prepared through polymer-assisted hydrolysis of the In^{3+} cation in water/ethanol mixed solvent.[96] γ-AlOOH architectures with hollow and self-encapsulated structures were selectively prepared by a facile one-step wet-chemical route.[97] It was established that the condensed cores could be self-encapsulated into the hollow shells and that the hollow shells were constructed with γ-AlOOH nanoflakes and showed urchin-like morphologies. Alternatively, urchin-shaped hollow particles consisting of boehmite (AlOOH) nanorods (Figure 8.27) were prepared by the slow addition of ethyl alcohol into an aqueous sodium

(a)   (b)   (c)

FIGURE 8.27 Urchin-shaped boehmite particles prepared at 60°C after (a) 1 h, (b) 24 h, and (c) 72 h. (Reproduced from Chang, T.S. et al., *J. Ceram. Process. Res.*, 2009, 10(6), 832–839. With permission.)

aluminate solution.[98] Further heat treatment at 900°C transformed the amorphous spherical particles into γ-Al$_2$O$_3$ particles, maintaining their spherical morphologies. In addition, α-FeOOH nanocrystals (rod-like, bundle-like, and urchin-like) were synthesized in high yield via a template-free hydrothermal method at low temperature.[99] The morphology and composition of the samples were controlled by slowly releasing the SO$_4^{2-}$ ions from ammonium persulfate.

Aluminum nitride (AlN) has been prepared in nanourchin form by different routes. First, AlN branched nanostructures with tree shapes and sea-urchin shapes were synthesized via a one-step improved DC arc discharge plasma method without any catalyst and template.[100] It was shown that the branches of tree-shaped nanostructures grew in a sequence of nanowires, nanomultipeds, and nanocombs. Alternatively, sixfold-symmetrical AlN hierarchical nanostructures including urchin-like and flower-like ones assembled by AlN nanoneedles were obtained through the chemical reaction between AlCl$_3$ and NH$_3$.[101] The urchin-like nanostructures showed better optical and field-emission properties in comparison with the flower-like ones, indicating the potential applications of the sixfold-symmetrical AlN hierarchical nanostructures in optoelectronic and field-emission devices. Several kinds of TiO$_2$/TiB$_2$ hybrid materials with different morphologies, including a hollow bipyramid structure with truncations, pineapple structure, urchin structure, and nanowall structure, were synthesized by a solvothermal approach from TiB$_2$ and HF in the aqueous solution of EDA.[102,103] In respect of the effect of ethylene diamine on the shape change of the final products, it was established that, with the increase in EDA, anatase TiO$_2$ on the TiB$_2$ core was gradually evolved from nanoparticles, nanorods to nanosheets. The schematic illustration of a three-step sequential growth model for the formation of the TiB$_2$/TiO$_2$ heterostructure is shown in Figure 8.28.

Sulfides, selenides, tellurides, phosphides, and related metal salts, which have great importance as semiconductors and parts of electronics, have been fabricated as urchin-like nanostructures and have found potential applications. Thus, semiconductor ZnS with complex 3D architectures, such as nanorod (or nanowire) networks, urchin-like nanostructures, nearly monodisperse nanospheres self-assembled from nanorods, and 1D nanostructures (rods and wires) was synthesized[104] and exhibited excellent photocatalytic activity for the degradation of *Acid fuchsine*. Bi$_2$S$_3$ microcrystallites with 3D superstructures were prepared by the reaction of Bi(NO$_3$)$_3$ with sodium *O*-isopropyldithiocarbonate (*i*-Pr xanthate, C$_3$H$_7$OCS$_2$Na) at 80°C in DMF solution without any surfactants, aggregated to

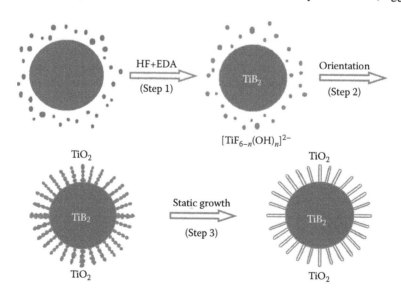

FIGURE 8.28 The schematic illustration of a three-step sequential growth model for the formation of the TiB$_2$/TiO$_2$ heterostructure. (Reproduced with permission from Huang, F. et al., Facile synthesis, growth mechanism, and UV-Vis spectroscopy of novel urchin-like TiO$_2$/TiB$_2$ heterostructures, *Cryst. Growth Des.*, 9(9), 4017–4022. Copyright 2009 American Chemical Society.)

urchin-like globules in the microscale, if $C_3H_7OCS_2Na$ was slowly added to $Bi(NO_3)_3$ in DMF solution at 80°C.[105] In the case of mixing two precursors at room temperature and then heating to 80°C, Bi_2S_3 nanobelts were formed. The Bi_2S_3 urchin-like nanosphere product was found to be of promising application value in the field of electrochemical DNA detection analysis.[106] Without the assistance of any surfactant and template, urchin-like In_2S_3 microspheres (Figure 8.29) constructed with nanoflakes of 15–30 nm thickness were prepared via the hydrothermal reaction of $InCl_3$ and thioacetamide in an acetic acid aqueous solution at 80°C for 6–24 h.[107] These In_2S_3 nanourchins could be completely transformed into In_2O_3 by heating in air at 600°C for 5 h, resulting in In_2O_3 consisting of urchin-like microsphere built up by 20–30 nm thickness nanoflakes and 20–40 nm nanoparticles. The described synthesis method is simple, mild, and cheap, which makes it suitable to be scaled up for industrial production of multifunctional In_2S_3 and In_2O_3 nanomaterials.

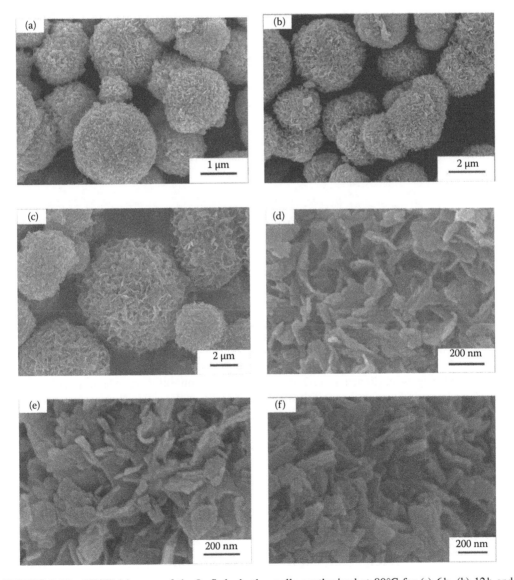

FIGURE 8.29 FESEM images of the In_2S_3 hydrothermally synthesized at 80°C for (a) 6 h, (b) 12 h and (c) 24 h; and (d through f) are the higher magnification FESEM images of the products in (a through c). (Reproduced from *Mater. Lett.*, 63(9–10), Bai, H.X. et al., Simple synthesis of urchin-like In_2S_3 and In_2O_3 nanostructures, 823–825, Copyright 2009, with permission from Elsevier.)

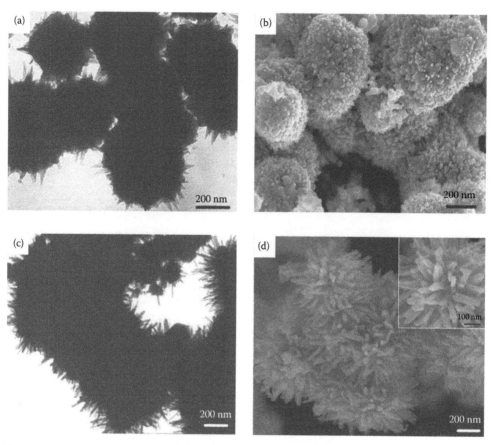

FIGURE 8.30 TEM and SEM images of sea-urchin-like $Ni_{0.85}Se$ nanocrystals obtained at 100°C for different reaction times (a, b) 12 h; (c, d) 24 h. (Reproduced from *Mater. Sci. Eng. B*, 140, Liu, X. et al., Hydrothermal synthesis and characterization of sea urchin-like nickel and cobalt selenides nanocrystals, 38–43, Copyright 2007, with permission from Elsevier.)

Sea-urchin-like nanorod-based nickel and cobalt selenides nanocrystals (Figure 8.30) were selectively synthesized via a hydrothermal reduction route in which hydrated nickel chloride and hydrated cobalt chloride were employed to supply Ni and Co source and aqueous hydrazine ($N_2H_4 \cdot H_2O$) was used as a reducing agent (Reactions 8.17 and 8.18).[108] The composition, morphology, and structure of final products could be easily controlled by adjusting the molar ratios of reactants and process parameters such as hydrothermal time. Nanoscale CdTe urchins, discovered in a tri-*n*-octylphosphine oxide (TOPO) system, consisted of a core and several attached arms 3 nm wide (their lengths could be controlled with the reaction time),[109] whose lengths could be tuned into CdTe nanourchins, which led to a change in their photophysical properties. Cobalt phosphide (Co_2P) nanocrystals with urchin-like structures were synthesized via a water–ethanol mixed-solvothermal route, employing white phosphorus and cobalt dichloride as starting reactants, sodium acetate as the pH adjustor, and sodium dodecyl benzene sulfonate as the surfactant.[110] Sodium acetate was found to play an important role in the formation of Co_2P nanocrystals with urchin-like structures.

$$2MCl_2 + N_2H_4 + 4OH^- \rightarrow 2M\downarrow + N_2\uparrow + 4H_2O + 4Cl^- \qquad (8.17)$$

$$x\,M + y\,Se \rightarrow M_xSe_y \qquad (8.18)$$

$$(x = 0.85, y = 1 \text{ or } x = 1, y = 2)$$

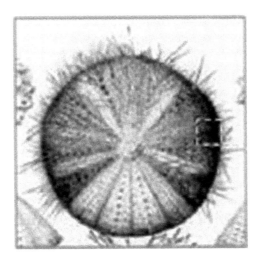

FIGURE 8.31 $CaCO_3$ urchin-like nanostructure with $MgCO_3$ precipitates. (From Tamerler, C. and Sarikaya, M., *Philos. Trans. R. Soc. London, Ser. A*, 367, 1705–1726, 2009. Reproduced with permission from The Royal Society.)

Among oxygen-containing salts, we note $CaCO_3$ sea-urchin spine (Figure 8.31), which is a single-crystal calcite with complex architecture, containing internal nanometer-scale $MgCO_3$ precipitates.[111] Hierarchical titanate nanostructures, in particular urchin-like (10 μm) ones, were hydrothermally synthesized in NaOH solutions (2 M in the case of urchins) using common titania powders as starting materials.[112] Since both base concentration and reaction temperature affected the reaction rate, the formation of various titanate nanostructures was proposed as a growth speed controlled process. 3D urchin-like $MnWO_4$ microspheres with a diameter of ~1 to 1.2 μm assembled by nanorods with a length of 240 nm were fabricated by a cationic surfactant CTAB assisted hydrothermal method (Figure 8.32).[113] Magnetic measurement indicates that urchin-like $MnWO_4$ microspheres showed a weak ferromagnetic ordering at low temperature due to spin-canting and surface spins of microspheres, whereas much shorter $MnWO_4$ nanorods showed antiferromagnetism at low temperature. An urchin-like $BiPO_4$ structure composed of nanorods,[114] synthesized from $Bi(NO_3)_3$ and tetraphosphoric acid ($H_6P_4O_{13}$) in a Teflon stainless steel autoclave at 100°C for 12 h, was found to be a useful host for rare-earth ions; thus, Ln^{3+} was doped in $BiPO_4$ and an efficient energy transfer from Bi^{3+} to Ln^{3+} took place, making $BiPO_4$:Ln (Ln = Eu, Tb, Dy) an emitter of strong luminescence in the visible region. Additionally, hierarchically structured coatings on glass substrates from soda lime glass were fabricated by the one-step hydrothermal method.[115] The surfaces of the coatings

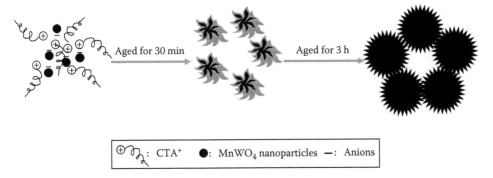

FIGURE 8.32 Schematic illustration of the growth process of urchin-like $MnWO_4$ microspheres. (Reproduced with permission from Zhou, Y.-X. et al., Surfactant-assisted hydrothermal synthesis and magnetic properties of urchin-like $MnWO_4$ microspheres, *J. Phys. Chem. C*, 112(35), 13383–13389. Copyright 2008 American Chemical Society.)

FIGURE 8.33 Schematic illustration of the PVD (physical vapor deposition) method for preparing and analyzing the growth process of the MOEP nanowires. (Reproduced with permission from Hu, J.-S. et al., Metal octaethylporphyrin nanowire array and network toward electric/photoelectric devices, *J. Phys. Chem. C*, 113(36), 16259–16265. Copyright 2009 American Chemical Society.)

were roughly composed of flower-like particles assembled by nanoflakes or urchin-like particles constructed by nanowires, exhibiting superhydrophilicity and, after surface modification by 1H, 1H, 2H, 2H-perfluorooctyltriethoxysilane, superhydrophobicity.

A few organic and coordination compounds, as well polymers and their composites, are reported as nanourchins. Thus, the multimorphological coordination polymer La(1,3,5-BTC) $(H_2O)_6$ was prepared through splitting crystal growth at room temperature on a large scale (1,3,5-BTC = 1,3,5-benzenetricarboxylate)[116] in the forms of sheaf-like, broccoli-like, urchin-like, and fan-like hierarchical architectures made of uniform nanorods obtained through splitting crystal growth. These 3D architectures can exhibit tunable white-light emission by co-doping Tb^{3+} and Eu^{3+}. A vapor transfer deposition method (Figure 8.33) was developed to fabricate metal (metal = Co, Ni, Cu, Zn, Mg) octaethylporphyrin (MOEP, **4**) nanowire arrays in a large area on a variety of substrates.[117] A vaporization-condensation-recrystallization (VCR) mechanism was proposed to understand the formation (Figure 8.34) of nanowires and thus guide the synthesis of 3D sea-urchin-like nanowire assemblies and 2D nanowire networks. These porphyrin nanowires demonstrated a good field emission property, resulting in the fabrication of the photoelectrical device that showed a good light-induced signal amplification behavior.

4

Heterogeneous molybdenum catalysts, applied for efficient epoxidation of olefins using *t*-Bu hydroperoxide as oxidant, were synthesized using sea-urchin-like PANI hollow microspheres constructed with their own oriented nanofiber arrays as support.[118] The catalytic activity of the PANI microsphere-supported catalysts (95% conversion) was found to be higher than that observed for its corresponding homogeneous catalyst (85% conversion) and the conventional PANI-supported catalyst (65% conversion). A hard template method was developed to synthesize uniform urchin-like polystyrene (PS)/α-Fe_2O_3 composite hollow microspheres under hydrothermal conditions using

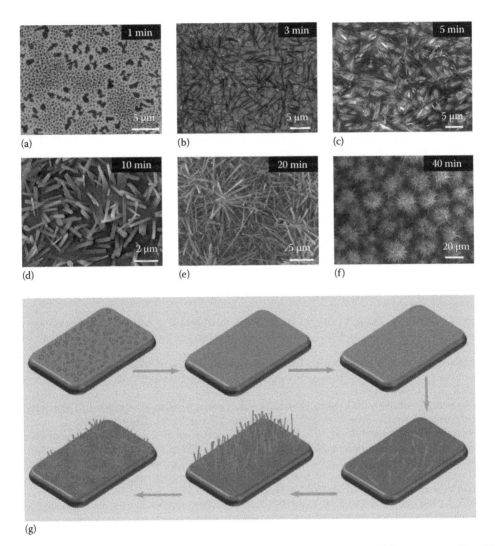

FIGURE 8.34 (a through f) Morphology evolution of CoOEP nanowires at different stages. The SEM images were taken from the samples collected by terminating deposition at 1, 3, 5, 10, 20, and 40 min, respectively, after deposition temperature reached 380°C. (g) Schematic illustration of the formation of CoOEP nanostructures. (Reproduced with permission from Hu, J.-S. et al., Metal octaethylporphyrin nanowire array and network toward electric/photoelectric devices, *J. Phys. Chem. C*, 113(36), 16259–16265. Copyright 2009 American Chemical Society.)

$FeSO_4$ and $KClO_3$ as precursors for Fe_2O_3.[119] In comparison with the Fe_2O_3 of other structures, the composite hollow microspheres had good photocatalytic activity and a large surface area. Among other organic nanourchins, octaalkynylsilsesquioxanes[120] and chitosan[121] have been reported.

REFERENCES

1. Choi, S. H.; Hyeon, T. Synthesis of shaped-controlled ZnO and copper indium sulfide nanocrystals: Nanostructured materials. *Abstracts of Papers, 235th ACS National Meeting*, New Orleans, LA, April 6–10, 2008, PHYS-435.
2. Choi, S.-H.; Kim, E.-G.; and Hyeon, T. One-pot synthesis of copper-indium sulfide nanocrystal heterostructures with acorn, bottle, and larva shapes. *Journal of the American Chemical Society*, 2006, *128*, 2520–2521.

3. Sasaki, T.; Tour, J. M. Synthesis of a new photoactive nanovehicle: A nanoworm. *Organic Letters*, 2008, *10* (5), 897–900.

4. Kemal, L.; Jiang, X. C.; Wong, K.; Yu, A. B. Experiment and theoretical study of poly(vinyl pyrrolidone)-controlled gold nanoparticles. *Journal of Physical Chemistry C*, 2008, *112* (40), 15656–15664.

5. Hernandez, J. A.; Gomez, S.; Pawelec, B.; Zepeda, T. A. CO oxidation on Au nanoparticles supported on wormhole HMS material: Effect of support modification with CeO_2. *Applied Catalysis, B: Environmental*, 2009, *89* (1–2), 128–136.

6. Radziuk, D.; Grigoriev, D.; Zhang, W.; Su, D.; Moehwald, H.; Shchukin, D. Ultrasound-assisted fusion of preformed gold nanoparticles. *Journal of Physical Chemistry C*, 2010, *114* (4), 1835–1843.

7. Wang, Y. Encapsulation of palladium crystallites in carbon and the formation of wormlike nanostructures. *Journal of the American Chemical Society*, 1994, *116* (1), 397–398.

8. Hu, B.; Zhang, Q.; Wang, Y. Pd/graphite as a superior catalyst for the direct synthesis of hydrogen peroxide from H_2 and O_2. *Chemistry Letters*, 2009, *38* (3), 256–257.

9. Navaladian, S.; Viswanathan, B.; Varadarajan, T. K.; Viswanath, R. P. Fabrication of worm-like nanorods and ultrafine nanospheres of silver via solid-state photochemical decomposition. *Nanoscale Research Letters*, 2009, *4* (5), 471–479.

10. Peng, Z.; You, H.; Yang, H. Composition-dependent formation of platinum silver nanowires. *ACS Nano*, 2010, *4* (3), 1501–1510.

11. Jing, C.; Zang, X.; Bai, W.; Chu, J.; Liu, A. Aqueous germanate ion solution promoted synthesis of worm-like crystallized Ge nanostructures under ambient conditions. *Nanotechnology*, 2009, *20* (50), 505607/1–505607/8.

12. Qi, X.; Zhong, W.; Deng, Y.; Au, C.; Du, Y. Synthesis of helical carbon nanotubes, worm-like carbon nanotubes and nanocoils at 450°C and their magnetic properties. *Carbon*, 2009, Volume Date 2010, *48* (2), 365–376.

13. Panchakarla, L. S.; Govindaraj, A. Carbon nanostructures and graphite-coated metal nanostructures obtained by pyrolysis of ruthenocene and ruthenocene–ferrocene mixtures. *Bulletin of Materials Science*, 2007, *30* (1), 23–29.

14. Gonzalez, J.; Herrero, J. Graphene wormholes: A condensed matter illustration of Dirac fermions in curved space. *Nuclear Physics B*, 2009, Volume Date 2010, *B825* (3), 426–443.

15. Wong, S.-C.; Sutherland, E. M.; Uhl, F. M. Materials processes of graphite nanostructured composites using ball milling. *Materials and Manufacturing Processes*, 2006, *21*, 159–166.

16. Hofmann, S.; Ducati, C.; Neill, R. J.; Piscanec, S.; Ferrari, A. C.; Geng, J.; Dunin-Borkowski, R. E.; Robertson, J. Gold catalyzed growth of silicon nanowires by plasma enhanced chemical vapor deposition. *Journal of Applied Physics*, 2003, *94* (9), 6005–6012.

17. Bau, L.; Bartova, B.; Arduini, M.; Mancin, F. Surfactant-free synthesis of mesoporous and hollow silica nanoparticles with an inorganic template. *Chemical Communications*, 2009, (48), 7584–7586.

18. Yang, L.-P.; Zou, P.; Pan, C.-Y. Preparation of hierarchical worm-like silica nanotubes. *Journal of Materials Chemistry*, 2009, *19* (13), 1843–1849.

19. Li, B.; Pei, X.; Wang, S.; Chen, Y.; Zhang, M.; Li, Y.; Yang, Y. Formation of hollow mesoporous silica nanoworm with two holes at the terminals. *Nanotechnology*, 2010, *21* (2), 025601/1–025601/7.

20. Jeon, M.; Kamisako, K. Growth of in-catalyzed silicon nanostructures by hydrogen radical-assisted deposition method. Effect of substrate temperatures. *Journal of Nanoscience and Nanotechnology*, 2008, *8* (10), 5188–5192.

21. Lee, D.-W.; Yu, C.-Y.; Lee, K.-H. Facile synthesis of mesoporous carbon and silica from a silica nanosphere-sucrose nanocomposite. *Journal of Materials Chemistry*, 2009, *19* (2), 299–304.

22. Hu, Q.; Dou, B. J.; Tian, H.; Li, J. J.; Li, P.; Hao, Z. P. Mesoporous silicalite-1 nanospheres and their properties of adsorption and hydrophobicity. *Microporous and Mesoporous Materials*, 2010, *129* (1–2), 30–36.

23. Park, J.-H.; von Maltzahn, G.; Zhang, L.; Derfus, A. M.; Simberg, D.; Harris, T. J.; Ruoslahti, E.; Bhatia, S. N.; Sailor, M. J. Systematic surface engineering of magnetic nanoworms for in vivo tumor targeting. *Small*, 2009, *5* (6), 694–700.

24. Park, J.-H.; von Maltzahn, G.; Zhang, L.; Schwartz, M. P.; Ruoslahti, E.; Bhatia, S. N.; Sailor, M. J. Magnetic iron oxide nanoworms for tumor targeting and imaging. *Advanced Materials*, 2008, *20*, 1630–1635.

25. Liu, Q.; Shen, S.; Zhou, Z.; Tian, L. A facile route to synthesis of superparamagnetic Fe_3O_4-PDVB nanoworms. *Materials Letters*, 2009, *63* (30), 2625–2627.

26. Liao, L.; Zheng, Z.; Yan, B.; Zhang, J. X.; Gong, H.; Li, J. C.; Liu, C.; Shen, Z. X.; Yu, T. Morphology controllable synthesis of α-Fe_2O_3 1D nanostructures: Growth mechanism and nanodevice based on single nanowire. *Journal of Physical Chemistry C*, 2008, *112*, 10784–10788.

27. Souza, F. L.; Lopes, K. P.; Nascente, P. A. P.; Leite, E. R. Nanostructured hematite thin films produced by spin-coating deposition solution: Application in water splitting. *Solar Energy Materials and Solar Cells*, 2009, *93*, 362–368.

28. Luo, H.; Churu, G.; Fabrizio, E. F.; Schnobrich, J.; Hobbs, A.; Dass, A.; Mulik, S. et al. Synthesis and characterization of the physical, chemical and mechanical properties of isocyanate-crosslinked vanadia aerogels. *Journal of Sol-Gel Science and Technology*, 2008, *48* (1–2), 113–134.

29. Leventis, N.; Sotiriou-Leventis, C.; Mulik, S.; Dass, A.; Schnobrich, J.; Hobbs, A.; Fabrizio, E. F. et al. Polymer nanoencapsulated mesoporous vanadia with unusual ductility at cryogenic temperatures. *Journal of Materials Chemistry*, 2008, *18* (21), 2475–2482.

30. Kandalkar, S. G.; Gunjakar, J. L.; Lokhande, C. D.; Joo, O.-S. Synthesis of cobalt oxide interconnected flacks and nano-worms structures using low temperature chemical bath deposition. *Journal of Alloys and Compounds*, 2009, *478* (1–2), 594–598.

31. Huang, D.; Wang, Y. J.; Cui, Y. C.; Luo, G. S. Direct synthesis of mesoporous TiO_2 and its catalytic performance in DBT oxidative desulfurization. *Microporous and Mesoporous Materials*, 2008, *116* (1–3), 378–385.

32. Li, H.; Liu, G.; Chen, S.; Liu, Q. Novel Fe doped mesoporous TiO_2 microspheres: Ultrasonic-hydrothermal synthesis, characterization, and photocatalytic properties. *Physica E: Low-Dimensional Systems and Nanostructures*, 2010, *42* (6), 1844–1849.

33. Park, K.-W.; Toney, M. F. Electrochemical and electrochromic properties of nanoworm-shaped Ta_2O_5-Pt thin-films. *Electrochemistry Communications*, 2005, *7* (2), 151–155.

34. Wen, B., Huang, Y., and Boland, J. J. Controllable growth of ZnO nanostructures by a simple solvothermal process. *Journal of Physical Chemistry C*, 2008, *112* (1), 106–111.

35. Baek, J. Y.; Ha, H.-W.; Kim, I.-Y.; Hwang, S.-J. Hierarchically assembled 2D nanoplates and 0D nanoparticles of lithium-rich layered lithium manganates applicable to lithium ion batteries. *Journal of Physical Chemistry C*, 2009, *113* (40), 17392–17398.

36. Du, Y.-P.; Zhang, Y.-W.; Sun, L.-D.; Yan, C.-H. Optically active uniform potassium and lithium rare earth fluoride nanocrystals derived from metal trifluoroacetate precursors. *Dalton Transactions*, 2009, (40), 8574–8581.

37. Huang, P.; Kong, Y.; Gao, F.; Cui, D. Method for preparation of copper selenide nanomaterial with worm shape by use of protein solution. CN 101591008, 2009, 7 pp.

38. Sun, Q.; Deng, Y. Morphology studies of polyaniline lengthy nanofibers formed via dimers copolymerization approach. *European Polymer Journal*, 2008, *44* (11), 3402–3408.

39. Tao, L.; Zhao, X. M.; Gao, J. M.; Hu, W. Lithographically defined uniform worm-shaped polymeric nanoparticles. *Nanotechnology*, 2010, *21* (9), 095301/1–095301/6.

40. Li, Z.; Sai, H.; Warren, S. C.; Kamperman, M.; Arora, H.; Gruner, S. M.; Wiesner, U. Metal nanoparticle-block copolymer composite assembly and disassembly. *Chemistry of Materials*, 2009, *21* (23), 5578–5584.

41. Nandi, M.; Bhaumik, A. Nanorods of all organic porous m-cresol-formaldehyde having photoluminescence at room temperature. *Materials Chemistry and Physics*, 2009, *114* (2–3), 785–788.

42. Piot, L.; Marie, C.; Dou, X.; Feng, X.; Mullen, K.; Fichou, D. Growth of long, highly stable, and densely packed worm-like nanocolumns of hexa-peri-hexabenzocoronenes via chemisorption on Au(111). *Journal of the American Chemical Society*, 2009, *131* (4), 1378–1379.

43. Bhosale, S. V.; Jani, C.; Lalander, C. H.; Langford, S. J. Solvophobic control of core-substituted naphthalene diimide nanostructures. *Chemical Communications*, 2010, *46* (6), 973–975.

44. Helgeson, M. E.; Hodgdon, T. K.; Kaler, E. W.; Wagner, N. J.; Vethamuthu, M.; Ananthapadmanabhan, K. P. Formation and rheology of viscoelastic "double networks" in wormlike micelle-nanoparticle mixtures. *Langmuir*, 2010, *26* (11), 8049–8060.

45. Zhang, Z.; Liu, S.; Chow, S.; Han, M.-Y. Modulation of the morphology of ZnO nanostructures via aminolytic reaction: From nanorods to nanosquamas. *Langmuir*, 2006, *22* (14), 6335–6340.

46. Ma, T.-Y.; Cao, J.-L.; Shao, G.-S.; Zhang, X.-J.; Yuan, Z.-Y. Hierarchically structured squama-like cerium-doped titania: Synthesis, photoactivity, and catalytic CO oxidation. *Journal of Physical Chemistry C*, 2009, *113* (38), 16658–16667.

47. Shi, S.; Chen, S.; Zhang, X.; Shen, W.; Li, X.; Hu, W.; Wang, H. Biomimetic mineralization synthesis of calcium-deficient carbonate-containing hydroxyapatite in a three-dimensional network of bacterial cellulose. *Journal of Chemical Technology and Biotechnology*, 2009, *84* (2), 285–290.

48. Liu, X. G.; Du, J.; Geng, D. Y.; Ma, S.; Liang, J. M.; Tong, M.; Zhang, Z. D. Co-doped Y_2O_3 optical functional nanoparticles and novel self-assembly squama-like aggregates. *Journal of Alloys and Compounds*, 2008, *457* (1–2), 517–521.

49. Piao, Y., An, K., Kim, J., Yu, T., and Hyeon, T. Sea urchin shaped carbon nanostructured materials: Carbon nanotubes immobilized on hollow carbon spheres. *Journal of Materials Chemistry*, 2006, *16* (29), 2984–2989.

50. Zhu, Y.; Li, J.; Wan, M.; Jiang, L. Electromagnetic functional urchin-like hollow carbon spheres carbonized by polyaniline micro/nanostructures containing $FeCl_3$ as a precursor. http://onlinelibrary.wiley.com/doi/10.1002/ejic.200900040/abstract. *European Journal of Inorganic Chemistry*, 2009, (19), 2860–2864. July 2009.

51. Shu, J. Urchin-structured MWNTs/HCS composite as anode material for high-capacity and high-power lithium-ion batteries. *Electrochemical and Solid-State Letters*, 2008, *11* (12), A219–A222.

52. Han, Z. H.; Yang, B.; Kim, S. H.; Zachariah, M. R. Application of hybrid sphere/carbon nanotube particles in nanofluids. *Nanotechnology*, 2007, *18*, 105701, 4 pp.

53. Romo-Herrera, J. M.; Cullen, D. A.; Cruz-Silva, E.; Ramirez, D.; Sumpter, B. G.; Meunier, V.; Terrones, H.; Smith, D. J.; Terrones, M. The role of sulfur in the synthesis of novel carbon morphologies: From covalent Y-junctions to sea-urchin-like structures. *Advanced Functional Materials*, 2009, *19* (8), 1193–1199.

54. Chen, J.; Cheng, F. Combination of lightweight elements and nanostructured materials for batteries. *Accounts of Chemical Research*, 2009, *42* (6), 713–723.

55. Hutter, E.; Boridy, S.; Labrecque, S.; Maysinger, D.; Winnik, F. M. Imaging of gold nanourchins, nanospheres, and nanorods in glia and neurons. *Abstracts of Papers, 238th ACS National Meeting*, Washington, DC, August 16–20, 2009, PMSE-441.

56. Guo, S.; Wang, L.; Dong, S.; Wang, E. A novel urchinlike gold/platinum hybrid nanocatalyst with controlled size. *Journal of Physical Chemistry C*, 2008, *112* (35), 13510–13515.

57. Guo, S.; Dong, S.; Wang, E. A general method for the rapid synthesis of hollow metallic or bimetallic nanoelectrocatalysts with urchinlike morphology. *Chemistry—A European Journal*, 2008, *14* (15), 4689–4695.

58. Tong, X.; Wang, F.; Yu, M.; Diao, X.; Huang, T.; Liu, H. Synthesis of solvent-stabilized self-assemblies of Pd nanoparticles. *Zhongnan Minzu Daxue Xuebao, Ziran Kexueban*, 2009, *28* (1), 4–7.

59. Wang, L.; Guo, S.; Zhai, J.; Dong, S. Facile synthesis of platinum nanoelectrocatalyst with urchinlike morphology. *Journal of Physical Chemistry C*, 2008, *112* (35), 13372–13377.

60. Wang, C.; Han, X.; Xu, P.; Wang, J.; Du, Y.; Wang, X.; Qin, W.; Zhang, T. Controlled synthesis of hierarchical nickel and morphology-dependent electromagnetic properties. *Journal of Physical Chemistry C*, 2010, *114* (7), 3196–3203.

61. Zhu, L.-P.; Liao, G.-H.; Zhang, W.-D.; Yang, Y.; Wang, L.-L.; Xie, H.-Y. Template-free synthesis of magnetic chains self-assembled from urchin-like hierarchical Ni nanostructures. *European Journal of Inorganic Chemistry*, 2010, (8), 1283–1288.

62. An, Z.; Pan, S.; Zhang, J.; Song, G. Facile synthesis of urchin-like glass/nickel core/shell composite hollow spheres. *Dalton Transactions*, 2008, (38), 5155–5158.

63. Wang, R.-H.; Jiang, J.-S.; Hu, M. Controlled syntheses of $FeNi_3$ alloy nanostructures via reverse microemulsion-directed hydrothermal methods. *Wuli Huaxue Xuebao*, 2009, *25* (10), 2167–2172.

64. Tang, C.; Zhang, Y. X.; Wang, G.; Wang, H. Q.; Li, G. Fabrication of urchin-like bismuth nanostructures via a facile solvothermal route. *Chemistry Letters*, 2008, *37* (7), 722–723.

65. Li, C.-s.; Li, W.-y.; Zhou, C.-y.; Ma, H.; Chen, J. Magnesium nano/micro structures: Their shape-controlled synthesis and Mg/air cell application. *Proceedings of International Forum on Green Chemical Science and Engineering and Process Systems Engineering*, Tianjin, China, October 8–10, 2006. Vol. 2, pp. 934–935.

66. Fan, S.; Li, G.; Zhang, X.; Mu, H.; Zhou, B.; Gong, L.; Liang, H.; Guo, L.; Guo, J. Facile synthesis of urchin-like selenium nanostructures in a buffer system at ambient conditions. *Crystal Growth and Design*, 2009, *9* (1), 95–99.

67. O'Dwyer, C.; Lavayen, V.; Newcomb, S. B.; Benavente, E.; Ana, M. A. S.; Gonzalez, G.; Torres Sotomayor, C. M. Atomic layer structure of vanadium oxide nanotubes grown on nanourchin structures. *Electrochemical and Solid-State Letters*, 2007, *10* (4), A111–A114.

68. Hu, C.-C.; Chang, K.-H.; Huang, C.-M.; Li, J.-M. Anodic deposition of vanadium oxides for thermal-induced growth of vanadium oxide nanowires. *Journal of the Electrochemical Society*, 2009, *156* (11), D485–D489.

69. Lavayen, V.; O'Dwyer, C.; Ana, M. A. S.; Newcomb, S. B.; Benavente, E.; Gonzalez, G.; Torres Sotomayor, C. M. Comparative structural-vibrational study of nano-urchin and nanorods of vanadium oxide. *Physica Status Solidi B: Basic Solid State Physics*, 2006, *243* (13), 3285–3289.

70. O'Dwyer, C.; Lavayen, V.; Newcomb, S. B.; Santa Ana, M. A.; Benavente, E.; Gonzalez, G.; Sotomayor Torres, C. M. Vanadate conformation variations in vanadium pentoxide nanostructures. *Journal of Electrochemical Science*, 2007, *154* (8), K29–K35.

71. O'Dwyer, C.; Navas, D., Lavayen, V., Benavente, E.; Santa Ana, M. A.; Gonzalez, G.; Schmidt, M.; Newcomb, S. B.; Sotomayor Torres, C. Nano-urchin: The formation and structure of high density spherical clusters of vanadium oxide nanotubes. *Chemistry of Materials*, 2006, *18* (13), 3016–3022.

72. Roppolo, M.; Jacobs, C. B.; Upreti, S.; Chernova, N. A.; Whittingham, M. S. Synthesis and characterization of layered and scrolled amine-templated vanadium oxides. *Journal of Materials Science*, 2008, *43* (14), 4742–4748.

73. Li, G.; Chao, K.; Zhang, C.; Zhang, Q.; Peng, H.; Chen, K. Synthesis of urchin-like VO_2 nanostructures composed of radially aligned nanobelts and their disassembly. *Inorganic Chemistry*, 2009, *48*, 1168–1172.

74. Li, J.-Y.; Xi, B.; Pan, J.; Qian, Y. Synthesis and gas sensing properties of urchin-like CuO self-assembled by nanorods through a poly(ethylene glycol)-assisted hydrothermal process. *Advanced Materials Research*, 2009, *79–82* (Pt. 1, Multi-Functional Materials and Structures II), 1059–1062.

75. Hong, J.; Li, J.; Ni, Y. Urchin-like CuO microspheres: Synthesis, characterization, and properties. *Journal of Alloys and Compounds*, 2009, *481* (1–2), 610–615.

76. Park, J. C.; Kim, J.; Kwon, H.; Song, H. Gram-scale synthesis of Cu_2O nanocubes and subsequent oxidation to CuO hollow nanostructures for lithium-ion battery anode materials. *Advanced Materials*, 2009, *21* (7), 803–807.

77. Sun, J.; Jia, Y.; Jing, Y.; Yao, Y.; Ma, J.; Gao, F.; Xia, C. Formation process of $Cu_2(OH)_2CO_3$ and CuO hierarchical nanostructures by assembly of hydrated nanoparticles. *Journal of Nanoscience and Nanotechnology*, 2009, *9* (10), 5903–5909.

78. Li, G.; Wu, W.; Zhang, C.; Peng, H.; Chen, K. Synthesis of ultra-long single crystalline CuV_2O_6 nanobelts. *Materials Letters*, 2010, *64* (7), 820–823.

79. Sekar, A.; Kim, S. H.; Umar, A.; Hahn, Y. B. Catalyst-free synthesis of ZnO nanowires on Si by oxidation of Zn powders. *Journal of Crystal Growth*, 2005, *277*, 471–478.

80. Du, J.-M.; Chen, Z.-Q.; Guo, W. Controlled synthesis of sea-urchin-like ZnO nanomaterials with the aid of ethylene glycol using a solvothermal method. *Jiegou Huaxue*, 2010, *29* (1), 126–133.

81. Tripathi, K.; Zulfequar, M.; Husain, M.; Khan, Z. H. Synthesis and characterisation of sea urchin-like nanostructures of ZnO on Si (100). *International Journal of Nanoparticles*, 2009, *2* (1/2/3/4/5/6), 111–118.

82. Yadav, R. S.; Pandey, A. C. Micro-Raman and photoluminescence study of urchin-like ZnO structure assembled with nanorods synthesized by hydrothermal method. *Structural Chemistry*, 2009, *20* (6), 1093–1097.

83. Tang, D.-M.; Liu, G.; Li, F.; Tan, J.; Liu, C.; Lu, G. Q.; Cheng, H.-M. Synthesis and photoelectrochemical property of urchin-like Zn/ZnO core-shell structures. *Journal of Physical Chemistry C*, 2009, *113* (25), 11035–11040.

84. Jiang, H.; Hu, J.; Gu, F.; Li, C. Stable field emission performance from urchin-like ZnO nanostructures. *Nanotechnology*, 2009, *20* (5), 055706/1–055706/5.

85. Yang, S.-H.; Tsai, C.-H.; Li, S.-X. Growth mechanism and characteristics of ZnO nanostructures with In and Ga dopings. http://www.electrochem.org/meetings/scheduler/abstracts/216/0187.pdf. Accessed April 15, 2010.

86. Ha, J.-H.; Muralidharan, P.; Kim, D. K. Hydrothermal synthesis and characterization of self-assembled WO_3 nanowires/nanorods using EDTA salts. *Journal of Alloys and Compounds*, 2009, *475* (1–2), 446–451.

87. Shen, X.; Wang, G.; Wexler, D. Large-scale synthesis and gas sensing application of vertically aligned and double-sided tungsten oxide nanorod arrays. *Sensors and Actuators, B: Chemical*, 2009, *B143* (1), 325–332.

88. Jeon, S. o.; Yong, K. A novel composite hierarchical hollow structure: One-pot synthesis and magnetic properties of $W_{18}O_{49}$-WO_2 hollow nanourchins. *Chemical Communications*, 2009, (45), 7042–7044.

89. Jia, T.; Wang, W., Long, F.; Fu, Z.; Wang, H.; Zhang, Q. Synthesis, characterization, and photocatalytic activity of Zn-doped SnO_2 hierarchical architectures assembled by nanocones. *Journal of Physical Chemistry C*, 2009, *113* (21), 9071–9077.

90. Zeng, S.; Tang, K.; Li, T.; Liang, Z. Hematite with the urchinlike structure: Its shape-selective synthesis, magnetism, and enhanced photocatalytic performance after TiO_2 encapsulation. *Journal of Physical Chemistry C*, 2010, *114* (1), 274–283.

91. Yin, J.; Zhao, X.; Xiang, L.; Xia, X.; Zhang, Z. Enhanced electrorheology of suspensions containing sea-urchin-like hierarchical Cr-doped titania particles. *Soft Matter*, 2009, *5* (23), 4687–4697.

92. Wang, F.; Wang, Y.-M.; Wen, Y.-X.; Su, H.-F.; Li, B. Structural and morphological transformation of MnO_2 nanostructures from Mn_3O_4 precursor. *Wuli Huaxue Xuebao*, 2010, *26* (2), 521–526.

93. Chen, Y.; Hong, Y.; Ma, Y.; Li, J. Synthesis and formation mechanism of urchin-like nano/micro-hybrid α-MnO$_2$. *Journal of Alloys and Compounds*, 2009, Volume Date 2010, *490* (1–2), 331–335.

94. Yu, P.; Zhang, X.; Wang, D.; Wang, L.; Ma, Y. Shape-controlled synthesis of 3D hierarchical MnO$_2$ nanostructures for electrochemical supercapacitors. *Crystal Growth and Design*, 2009, *9* (1), 528–533.

95. Shan, W.; Dong, X.; Ma, N.; Yao, S.; Feng, Z. The synthesis of three-dimensional CeO$_2$ and their catalytic activities for CO oxidation. *Catalysis Letters*, 2009, *131* (3–4), 350–355.

96. Chen, L.-Y.; Liang, Y.; Zhang, Z.-D. Corundum-type In$_2$O$_3$ urchin-like nanostructures: Synthesis derived from orthorhombic InOOH and application in photocatalysis. *European Journal of Inorganic Chemistry*, 2009, (7), 903–909.

97. Wu, X.; Wang, D.; Hu, Z.; Gu, G. Synthesis of γ-AlOOH (γ-Al$_2$O$_3$) self-encapsulated and hollow architectures. *Materials Chemistry and Physics*, 2008, *109* (2–3), 560–564.

98. Chang, T. S.; Na, J. H.; Jung, C. Y.; Koo, S. M. An easy one-pot synthesis of structurally controlled aluminum hydroxide particles from an aqueous sodium aluminate solution. *Journal of Ceramic Processing Research*, 2009, *10* (6), 832–839.

99. Sun, Z.; Feng, X.; Hou, W. Morphology-controlled synthesis of α-FeOOH and its derivatives. *Nanotechnology*, 2007, *18* (45), 455607/1–455607/9.

100. Lei, W.; Liu, D.; Zhu, P.; Chen, X.; Hao, J.; Wang, Q.; Cui, Q.; Zou, G. One-step synthesis of AlN branched nanostructures by an improved DC arc discharge plasma method. *Crystal Engineering Communications*, 2010, *12* (2), 511–516.

101. Zhang, F.; Wu, Q.; Wang, X.; Liu, N.; Yang, J.; Hu, Y.; Yu, L.; Wang, X.; Hu, Z.; Zhu, J. 6-Fold-symmetrical AlN hierarchical nanostructures: Synthesis and field-emission properties. *Journal of Physical Chemistry C*, 2009, *113* (10), 4053–4058.

102. Huang, F.; Fu, Z.; Yan, A.; Wang, W.; Wang, H.; Wang, Y.; Zhang, J.; Zhang, Q. Several shape-controlled TiO$_2$/TiB$_2$ hybrid materials with a combined growth mechanism. *Materials Letters*, 2009, *63* (30), 2655–2658.

103. Huang, F.; Fu, Z.; Yan, A.; Wang, W.; Wang, H.; Wang, Y.; Zhang, J.; Cheng, Y.; Zhang, Q. Facile synthesis, growth mechanism, and UV-Vis spectroscopy of novel urchin-like TiO$_2$/TiB$_2$ heterostructures. *Crystal Growth and Design*, 2009, *9* (9), 4017–4022.

104. Xiong, S.; Xi, B.; Wang, C.; Xu, D.; Feng, X.; Zhu, Z.; Qian, Y. Tunable synthesis of various wurtzite ZnS architectural structures and their photocatalytic properties. *Advanced Functional Materials*, 2007, *17* (15), 2728–2738.

105. Han, Q.; Sun, Y.; Wang, X.; Chen, L.; Yang, X.; Lu, L. Controllable synthesis of Bi$_2$S$_3$ hierarchical nanostructures: Effect of addition method on structures. *Journal of Alloys and Compounds*, 2009, *481* (1–2), 520–525.

106. Zhou, X.; Shi, H.; Zhang, B.; Fu, X.; Jiao, K. Facile synthesis and electrochemical application of surface-modified Bi$_2$S$_3$ urchin-like nano-spheres at room temperature. *Materials Letters*, 2008, *62* (17–18), 3201–3204.

107. Bai, H. X.; Zhang, L. X.; Zhang, Y. C. Simple synthesis of urchin-like In$_2$S$_3$ and In$_2$O$_3$ nanostructures. *Materials Letters*, 2009, *63* (9–10), 823–825.

108. Liu, X.; Zhang, N.; Yi, R.; Qiu, G.; Yan, A.; Wu, H.; Menga, D.; Tang, M. Hydrothermal synthesis and characterization of sea urchin-like nickel and cobalt selenides nanocrystals. *Materials Science and Engineering B*, 2007, *140*, 38–43.

109. Bao, J.; Shen, Y.; Sun, Y.; Yue, Y.; Chen, X.; Dai, N. Controlled synthesis of nanoscale CdTe urchins. *Chemical Research in Chinese Universities*, 2009, *25* (2), 147–150.

110. Ni, Y.; Li, J.; Zhang, L.; Yang, S.; Wei, X. Urchin-like Co$_2$P nanocrystals: Synthesis, characterization, influencing factors and photocatalytic degradation property. *Materials Research Bulletin*, 2009, *44* (5), 1166–1172.

111. Tamerler, C.; Sarikaya, M. Molecular biomimetics: Nanotechnology and bionanotechnology using genetically engineered peptides. *Philosophical Transactions of the Royal Society of London, Series A*, 2009, *367*, 1705–1726.

112. Dong, Z.; Zhao, Y.; Su, H.; Yu, J.; Li, L. Hierarchical titanate nanostructures through hydrothermal treatment of commercial titania powders. *Zeitschrift fuer Anorganische und Allgemeine Chemie*, 2009, *635* (3), 417–419.

113. Zhou, Y.-X.; Zhang, Q.; Gong, J.-Y.; Yu, S.-H. Surfactant-assisted hydrothermal synthesis and magnetic properties of urchin-like MnWO$_4$ microspheres. *Journal of Physical Chemistry C*, 2008, *112* (35), 13383–13389.

114. Guan, M.; Sun, J.; Tao, F.; Xu, Z. A host crystal for the rare-earth ion dopants: Synthesis of pure and Ln-doped urchinlike $BiPO_4$ structure and its photoluminescence. *Crystal Growth and Design*, 2008, *8* (8), 2694–2697.

115. Liu, X.; He, J. One-step hydrothermal creation of hierarchical microstructures toward superhydrophilic and superhydrophobic surfaces. *Langmuir*, 2009, *25* (19), 11822–11826.

116. Liu, K.; You, H.; Zheng, Y.; Jia, G.; Huang, Y.; Yang, M.; Song, Y.; Zhang, L.; Zhang, H.. Room-temperature synthesis of multi-morphological coordination polymer and tunable white-light emission. *Crystal Growth and Design*, 2010, *10* (1), 16–19.

117. Hu, J.-S.; Ji, H.-X.; Wan, L.-J. Metal octaethylporphyrin nanowire array and network toward electric/photoelectric devices. *Journal of Physical Chemistry C*, 2009, *113* (36), 16259–16265.

118. Ding, H.; Wang, G.; Yang, M.; Luan, Y.; Wang, Y.; Yao, X. Novel sea urchin-like polyaniline micro-spheres-supported molybdenum catalyst: Preparation, characteristic and functionality. *Journal of Molecular Catalysis A: Chemical*, 2009, *308* (1–2), 25–31.

119. Zhang, Y.; Chu, Y.; Dong, L. One-step synthesis and properties of urchin-like PS/α-Fe_2O_3 composite hollow microspheres. *Nanotechnology*, 2007, *18* (43), 435608/1–435608/5.

120. Asuncion, M. Z.; Roll, M. F.; Laine, R. M. Octaalkynylsilsesquioxanes, nano sea urchin molecular building blocks for 3D-nanostructures. *Macromolecules*, 2008, *41* (21), 8047–8052.

121. Qi, L.; Pal, S.; Dutta, P.; Seehra, M.; Pei, M. Morphology controllable nanostructured chitosan matrix and its cytocompatibility. *Journal of Biomedical Materials Research, Part A*, 2008, *87A* (1), 236–244.

9 "Home"-Like Nanostructures

9.1 NANOBOWLS

Nanobowls, closely related to nanocups (see later), are known mainly for some elemental metals and oxides. As in the case of several other classic and less-common nanostructures, the indisputable leadership in their formation belongs to zinc oxide due to the great interest for photonic applications due to its wide band gap (3.37 eV), large exciton binding energy (60 meV), and its use as a promising material for light-emitting devices. ZnO in contact with noble metals may undergo charge transfer, a phenomenon essential to photocatalysis, photovoltaics, and next generation nanodevices.[1] In this respect, the development of ZnO nanostructures (nanobowls and nanobagels) as well as core–shell ZnO structures with Au nanoshell (silica core–Au shell) core and ZnO epilayer coating with unique light emission properties, which may be useful in nanophotonics applications, has been extensively carried out by various techniques, frequently yielding doped or undoped different structures at the same time or depending on reaction conditions. Thus, ZnO and Cu_2O were deposited by the electrochemical method *via* the colloidal crystal template.[2] Not only 3D inverse opal structures and 2D nanobowls fabricated but also nanoparticles with a controlled shape were prepared. Mn-doped ZnO diluted magnetic semiconductor (DMS) nanostructures (nanorod, bowl, and cage, located at different deposition temperature zones) were synthesized by direct reaction of zinc metal and manganese chloride powder under oxygen environment using CVD method.[3] It was revealed that the doped nanorods exhibited low-temperature ferromagnetism at 5 K with Curie temperature around 37 K, whereas the DMS nanocrystalline bowl/cage structures had room-temperature (r.t.) ferromagnetic behavior. Unusual bowl-, trough-, and ring-shaped structures resulted from the temperature-induced self-assembly of ZnO nanoparticles, leading to bowls and rings also serving as a template to make metal or metal oxide replicas.[4] The tiny bowls were envisaged by authors not only to hold fluids of ultralow volume but also to grow nanoparticles, immobilize biomolecules, and screen submicrometer-sized particles.

With different catalysts, zinc oxide nanobowls, among other nanostructures, were prepared on silicon substrate by CVD using zinc acetylacetonate hydrate as precursor[5,6] or polymer films with spherical nanowell arrays as sacrificial templates to form the bowl-like ZnO nanoparticle structures (surface-patterned and free-standing ZnO nanobowls).[7] We note that the use of polymer patterns, for instance polystyrene (PS), is common for inorganic nanostructure formation, in particular nanobowls; PS template had a multilayer structure consisting of a periodically hexagonal array of PS spheres in three dimensions, and the zinc oxide film had a surface morphology arrayed by ZnO hollow hemispheres like nanobowls with high smoothness and density. By using PS opal template as substrate, zinc oxide film samples were obtained by a magnetron sputtering process.[8] It was established that the zinc oxide produced by magnetron sputtering duplicated the surface structure of the opal template. Additionally, Zn and other metal oxide micro-/nanobowls (Fe, Cu, and Co) were fabricated by combining the unique advantages of two simple techniques: template-assisted self-assembly technique by which PS micro-/nanospheres could be readily patterned and positioned on a large scale and hot-plate technique.[9]

Among other metal oxides, great attention has been paid to a classic nanostructure object, TiO_2. Thus, highly dense, ordered arrays of a titania (TiO_2) nanomaterial with a bowl-shape morphology were generated by using poly(styrene-block-ethylene oxide) (PS-b-PEO) block copolymers as templates combined with a sol–gel process.[10] The addition of a second inorganic precursor into

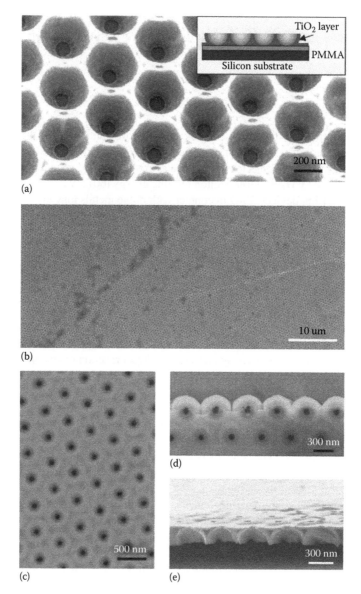

FIGURE 9.1 (a) An SEM image of TiO₂ nanobowl arrays on a PMMA layer. Inset: schematic of the modified configuration for fabricating free-standing nanobowl sheets. (b) A low-magnification SEM image of a bowl-side-down nanobowl sheet on a silicon substrate. (c) A high-magnification SEM image of the backside of a nanobowl sheet. (d) Edge of a bowl-side-down nanobowl sheet. (e) Cross section of a bowl-side-down nanobowl sheet. (Reproduced with permission from Wang, X. et al., Large-size liftable inverted-nanobowl sheets as reusable masks for nanolithography, *Nano Lett.*, 5(9), 1784–1788. Copyright 2005 American Chemical Society.)

the common solution led to arrays of composite Au-TiO₂ nanobowls with controlled shape and size, exhibiting unique photophysical properties. Large-area, liftable, ordered TiO₂ nanobowl sheets (Figure 9.1) were fabricated using the template of self-assembled PS spheres, followed by atomic layer deposition (ALD), ion milling, and etching.[11,12] By introducing a thin organic layer between the nanobowls and the substrate, the extremely thin (20–30 nm) nanobowl sheet was made as big as a few square centimeters. The mechanism (Figure 9.2) included removal of the top layer of TiO₂

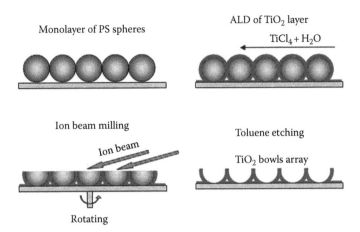

FIGURE 9.2 Schematic of the synthesis strategy for TiO$_2$ nanobowl arrays. (Reproduced with permission Wang, X.D. et al., Large-scale fabrication of ordered nanobowl arrays, *Nano Lett.*, 4(11), 2223–2226. Copyright 2004 American Chemical Society.)

with an ion beam followed by the removal of the polymer beads, yielding a micrometer-scaled surface with TiO$_2$-nanobowls.[13,14] These and a variety of other titanium dioxide nanostructures and nanomaterials were generalized in an excellent review.[15] Additionally, photocatalytic activities of the TiO$_2$ nanobowls were studied in terms of the degradation of a typical dye.[16] Large-area periodical bowl-like cobalt oxide (Co$_3$O$_4$) array films were prepared by a self-assembled monolayer PS sphere template and electrodeposition (Figure 9.3, Reactions 9.1 through 9.3).[17] The resulting Co$_3$O$_4$ films consisted of periodic, interconnected networks of monodisperse submicrometer pores

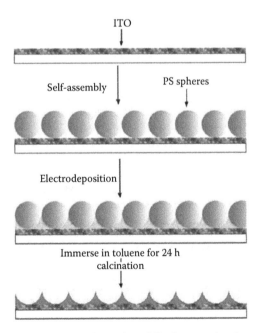

FIGURE 9.3 Schematic illustration for the formation of Co$_3$O$_4$ macrobowl array films. (Reproduced with permission from Xia, X.-H. et al., Cobalt oxide ordered bowl-like array films prepared by electrodeposition through monolayer polystyrene sphere template and electrochromic properties, *ACS Appl. Mater. Interfaces*, 2(1), 186–192. Copyright 2010 American Chemical Society.)

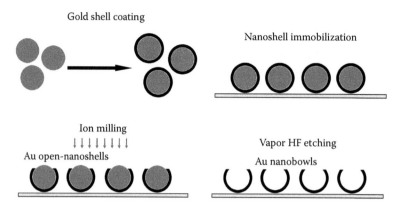

FIGURE 9.4 Experimental procedure for fabricating Au nanobowls. (Reproduced with permission Ye, J. et al., Fabrication, characterization, and optical properties of gold nanobowl submonolayer structures, *Langmuir*, 25(3), 1822–1827. Copyright 2009 American Chemical Society.)

with a diameter of 1 μm; the individual bowl contained a large number of pores with a diameter of 50 ± 20 nm, and the interstices between the bowls were filled with Co_3O_4 nanoflakes:

$$NO_3^- + H_2O + 2e^- \rightarrow NO_2^- + 2OH^- \tag{9.1}$$

$$Co^{2+} + 2OH^- \rightarrow Co(OH)_2 \tag{9.2}$$

$$3Co(OH)_2 + 2OH^- \rightarrow Co_3O_4 + 4H_2O + 2e^- \tag{9.3}$$

Elemental metals and nonmetals in bowl-like nanostructures are mainly represented by gold, although silver, nickel, cobalt, and carbon nanotubes (CNTs) have also been reported. Thus, a versatile method to fabricate (Figure 9.4) hollow Au nanobowls and complex Au nanobowls (with an Au nanoparticle core offering the capability to create plasmon hybridized nanostructures) (Figure 9.5) based on an ion milling and a vapor HF etching technique was proposed.[18] As a result, two different-sized hollow Au nanobowls were fabricated by milling and etching submonolayers of Au nanoshells deposited on a substrate. Optical properties of hollow Au nanobowls with different sizes showed highly tunable plasmon resonance ranging from the visible to the near-IR region. In addition, particle immobilization, ion milling, and vapor HF etching were subsequently applied to the Au@SiO_2@Au particles on the substrate, yielding complex Au nanobowls (Figure 9.6). The authors showed that there was one Au nanoparticle core inside each nanobowl structure and the locations of the cores were random; Au cores may sit in the middle of nanobowl bottoms or attach to the inner sidewalls of nanobowls. Another route to form an Au nanobowl array (Figure 9.7) was offered by PMMA-mediated nanotransfer printing technique (Figure 9.8).[19] The use of PMMA film as a mediator introduced several features to this transfer approach, such as high efficiency, fidelity, universality, controllability, and multilevel transferability. Additionally, the finite difference time domain method was applied to predict how optical plasmon properties are modified if the symmetry geometry of gold shell nanostructures is broken.[20] The simulations included three kinds of gold open shell nanostructures of nanobowls, open nanocages, and open eggshells. The optical transitions of gold open shell nanostructures were explained by the plasmon hybridization theory combined with numerical calculations.

Silver bowl-like nanostructures were fabricated from the elemental metal as by decomposition of metal salts. Thus, a low-cost and high-throughput method to fabricate large-area silver nanobowl arrays *via* thermal evaporation of silver on a self-assembled monolayer of nanospheres (Figure 9.9) was reported.[21] The nanobowl array possessed a hierarchical structure, composed of

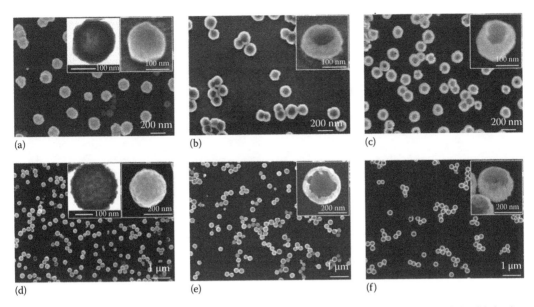

FIGURE 9.5 SEM images of different sized Au nanobowl submonolayers at each step of the fabrication process. (a, d) Self-assembled Au nanoshells. (b, e) Ion milled Au open nanoshells. (c, f) Vapor HF etched Au nanobowls. All insets are the corresponding highly magnified SEM/TEM images. (Reproduced with permission from Ye, J. et al., Fabrication, characterization, and optical properties of gold nanobowl submonolayer structures, *Langmuir*, 25(3), 1822–1827. Copyright 2009 American Chemical Society.)

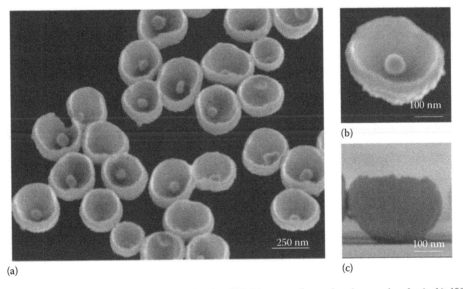

FIGURE 9.6 (a) Low- and (b, c) high-magnification SEM images of complex Au nanobowls: (a, b) 40° tilt; (c) 89° tilt. (Reproduced with permission from Ye, J. et al., Fabrication, characterization, and optical properties of gold nanobowl submonolayer structures, *Langmuir*, 25(3), 1822–1827. Copyright 2009 American Chemical Society.)

silver nanoparticles with average diameter size of ~10 nm, which can serve as a reaction container and catalyst. It was found that the surface plasmon resonance of silver nanoparticles existed on the nanobowl array. In addition, an Ag bowl-like array film with hierarchical structures on glass substrate was prepared by another route using the colloidal monolayer as a template.[22] These micro-structures were provided by a colloidal template of PS latex spheres and nanostructures resulting

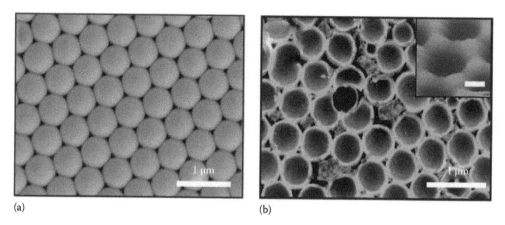

(a)   (b)

FIGURE 9.7   SEM images of (a) gold-coated PS nanosphere array and (b) gold nanobowl array. Inset, 3D AFM image. Scale bar, 200 nm. (Reproduced with permission from Jiao, L. et al., Creation of nanostructures with poly(methyl methacrylate)-mediated nanotransfer printing, *J. Am. Chem. Soc.*, 130(38), 12612–12613. Copyright 2008 American Chemical Society.)

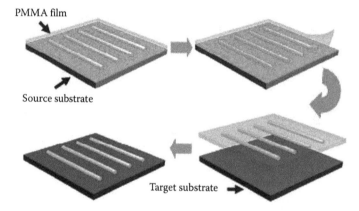

FIGURE 9.8   Illustration of the procedures of PMMA-mediated nanotransfer printing technique. (Reproduced with permission from Jiao, L. et al., Creation of nanostructures with poly(methyl methacrylate)-mediated nanotransfer printing, *J. Am. Chem. Soc.*, 130(38), 12612–12613. Copyright 2008 American Chemical Society.)

from the thermal decomposition of Ag acetate (Figure 9.10). Due to the lotus leaf-like morphology with hierarchical micro/nanostructures, the film displayed an extraordinary superhydrophobicity after chemical modification. The nanoscale aluminum bowls were derived from the porous alumina and were used as flexible nanoscale reactors for the preparation of nanoparticles.[23] A single nanoparticle or just a small number of metal (e.g., Pt) nanoparticles or semiconductor nanoparticles (e.g., CdSe or CdSe/ZnS core–shell nanostructures) in the nanobowls were thus prepared. Ordered arrays of hybrid nickel nanostructures (both nanobowl and pillar arrays) (Figure 9.11), exhibiting uniform sizes, were fabricated according to the scheme in Figure 9.12.[24] A vertically aligned CNT film was grown on a metal-coated PS submicron sphere array and then transferred onto a polycarbonate (PC) substrate by microwave heating, leading to a new architecture of a 2D metallic nanobowl array on a thermoplastic substrate.[25] In the case of CNT nanobowls, the introduction of CO_2 was found to be a facile way to tune the growth of vertically aligned double- or single-walled CNT forests on wafers.[26] In the absence of CO_2, a double-walled CNT convexity was obtained; meanwhile, with increasing concentration of CO_2, the morphologies of the forests transformed first into radial blocks and finally into bowl-shaped forests. The addition of CO_2 was speculated to generate water and serve as a weak oxidant for high-quality CNT growth.

A comparison of two techniques for silver nanobowl fabrication

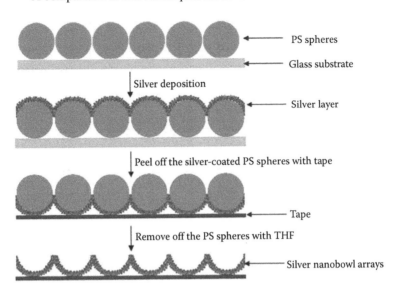

FIGURE 9.9 Schematic illustration of the procedure for fabricating a silver nanobowl array from metallic silver. (Reproduced with permission from Xu, M. et al., Fabrication of functional silver nanobowl arrays via sphere lithography, *Langmuir*, 25(19), 11216–11220. Copyright 2009 American Chemical Society.)

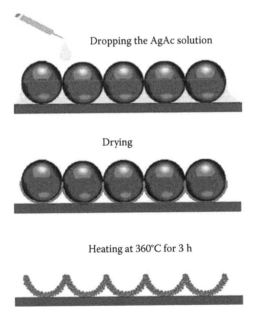

FIGURE 9.10 Schematic illustration of fabrication process of the hierarchical silver bowl-like ordered array film from AgAc solution. (Reproduced with permission from Li, Y. et al., Silver hierarchical bowl-like array: Synthesis, superhydrophobicity, and optical properties, *Langmuir*, 23(19), 9802–9807. Copyright 2007 American Chemical Society.)

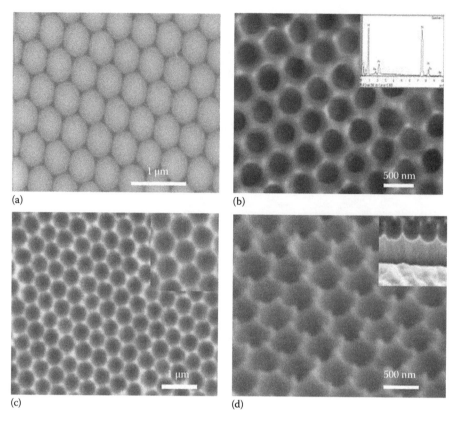

FIGURE 9.11 SEM images of nickel nanostructures during the fabrication process. (a) Monolayer self-assembly of PS spheres on silicon substrates. (b) Nanobowl arrays after PS spheres are removed, and the inset shows the EDS spectra recorded from the nanostructures. (c) Nickel nanobowl array after gold film is removed, and the inset is the high-magnification image. (d) Nickel nanostructure arrays with a tilt angle of 45°; hexagonal distributed nickel nanopillars on the top of nanobowls can be clearly seen. The inset is the cross section with a tilt angle of 45°. (Reproduced from Chen, X. et al., *Opt. Express*, 16(16), 11888, 2008. With permission.)

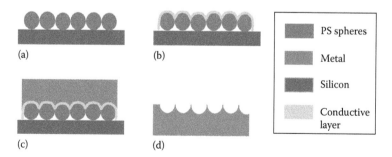

FIGURE 9.12 Process flow chart of the fabrication of large-area ordered metallic nanostructure arrays. (a) The silicon substrate is coated with a monolayer of PS spheres. (b) A thin conductive layer is deposited on top of the PS spheres by thermal evaporation. (c) A thick metal sheet is electrochemically deposited on top of the conductive layer. (d) After removal of PS spheres and the conductive layer, periodic nanostructure arrays of the desired material are obtained. (Reproduced from Chen, X. et al., *Opt. Express*, 16(16), 11888, 2008. With permission.)

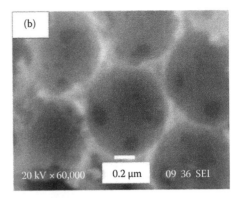

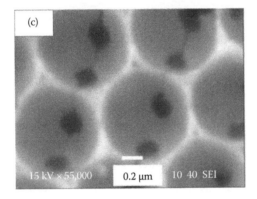

FIGURE 9.13 SEM micrographs of cobalt (a) nanobowl structure annealed at 500°C for 3 h in N_2 atmosphere, (b) nanobowl structure without annealing and only washed in toluene, (c) nanobowl structure annealed at 500°C for 3 h in N_2 atmosphere followed by toluene etching. (Srivastava, A.K. et al., Template assisted assembly of cobalt nanobowl arrays, *J. Mater. Chem.*, 15(41), 4424–4428, 2005. Reproduced by permission of The Royal Society of Chemistry.)

Magnetic cobalt[27] (wall size of <100 nm) (Figure 9.13) and cobalt ferrite ($CoFe_2O_4$)[28] (Figure 9.14) nanostructures with curved surfaces (i.e., nanobowls and hollow nanospheres) in a periodic array were fabricated by *in situ* reduction of a cobalt and iron salt solution mixture in the interstitial spaces of 3D, close-packed, PS sphere templates (latex spheres with 1 μm diameter).[29] It was established that the coercivity of cobalt nanobowls was found to be greater than the coercivity of cobalt nanoparticles of the same size range; the coercivity of $CoFe_2O_4$ nanobowls was greater than that of cobalt nanobowls. Ordered Pt nanobowls (bowl-shaped shells) and, alternatively, nanocups (cup-shaped shells) were prepared by a combination of a porous polymer template and a nanocrystal-seeded electroless plating technique.[30] The formed materials showed such intriguing properties as reduced symmetry of the building blocks, a well-ordered structure, and a high ratio of surface area to volume, useful in catalysts, sensors, and photonic crystals. Si nanobowls in elemental form were also reported.[31]

A few polymers have been fabricated as nanobowl structures (not as patterns for inorganic bowl-like nanostructures). Thus, "bowl" structures were produced for a broad material set of functional initiated chemical vapor deposition (iCVD) films, including organic polymers (pBA, pHEMA), fluoropolymers (pPFDA, pPFM), and organosilicones (pV4D4).[32] The properties of these materials ranged from hydrophilic (pHEMA) to hydrophobic (pPFDA); from soluble linear polymers (pBA) to heavily cross-linked networks (pV4D4); and to films with highly reactive pendent groups that are readily biofunctionalized (pPFM). A monolayer of PS nanospheres on a substrate was used as a template for replicating the poly(dimethylsiloxane) (PDMS) nanobowl mold,[33] resulting in good

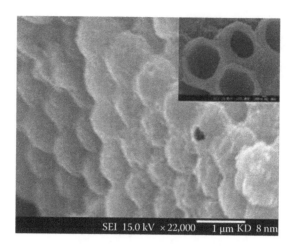

FIGURE 9.14 SEM micrograph of cobalt ferrite ($CoFe_2O_4$) nanobowl network annealed at 400°C for 4 h in air atmosphere followed by toluene etching. Micrograph shown in the inset represents the morphology of the top layer of the network. (Reproduced from *Thin Solid Films*, 505(1–2), Srivastava, A.K. et al., The processing and characterization of magnetic nanobowls, 93–96, Copyright 2006, with permission from Elsevier.)

uniformity and smooth internal surface observed in the nanobowl arrays. A 2D, ordered, large-area, and liftable conducting polymer-nanobowl sheet was fabricated *via* chemical polymerization using the monolayer self-assembled from PS spheres at the aqueous/air interface as a template.[34] A 2D, ordered, large-area, liftable, and patterned polyaniline (PANI) nanobowl monolayer containing Au nanoparticles was generated with the monolayer self-assembled PS spheres at the aqueous/air interface as a template[35] using $HAuCl_4$ as an oxidizing agent for the polymerization of aniline in an aqueous solution. The obtained product can be used in practical gas detectors. Additionally, anomalous polyimide (PI) nanoparticles with controllable morphology, that is, porous hollow structures, hollow structures, or bowl-like structures, were fabricated by blending a second polymer with poly(amic acid) (PAA, the precursor of PI) through the reprecipitation method and subsequent imidization.[36]

9.2 NANOCUPS

Nanocups, closely related to nanobowls and nanocaps, are represented by gold, platinum, stacked CNTs, ZnO, hematite, silica, and some metal salts. Thus, an individual Au nanoshell was controllably reshaped into a reduced-symmetry nanoegg, then a semishell or nanocup by an electron-beam (E-beam)-induced ablation method (Figure 9.15), transforming its plasmonic properties.[37] The splitting of plasmon modes and the onset of electroinductive plasmons upon controlled, incremental opening of the outer metallic layer of the nanoparticle were observed. In addition, Au nanocups at their magnetoinductive resonance were found to have the unique ability to redirect scattered light in a direction dependent on cup orientation, as a true 3D nanoantenna.[38] Additionally, SiO_2 nanoparticles with a Au cup-shaped shell and, alternatively, a Au cap, were obtained by a combination of nanoscale masking techniques and nanoparticle-seeded electroless plating.[39] A combination of a porous polymer template and a nanocrystal-seeded electroless plating technique was applied to prepare ordered Pt nanobowls (bowl-shaped shells) and, alternatively, nanocups (cup-shaped shells).[30] The authors noted that these materials showed some intriguing properties, such as reduced symmetry of the building blocks, a well-ordered structure, and a high ratio of surface area to volume, all of which are useful in many areas such as catalysts, sensors, and photonic crystals.

The electrochemical activity of stacked N-doped carbon nanotube cups (NCNCs, Figure 9.16) was explored in comparison with commercial Pt-decorated CNTs,[40] demonstrating for the nanocup catalyst comparable performance to that of the Pt catalyst in an oxygen reduction reaction. The NCNC electrodes were used for H_2O_2 oxidation and consequently for glucose detection. In addition,

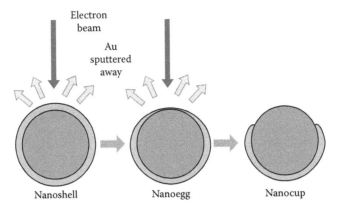

FIGURE 9.15 Schematic illustrating Au nanoshell E-beam ablation process resulting in the transformation of a nanoshell to nanoegg to nanocup. (Reproduced from *Nano Lett.*, 9(12), Lassiter, J.B. et al., Reshaping the plasmonic properties of an individual nanoparticle, 4326–4332, Copyright 2009, with permission from Elsevier.)

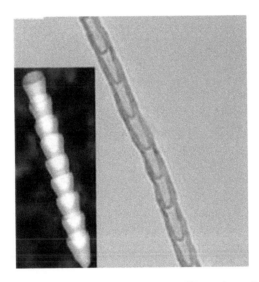

FIGURE 9.16 TEM image of stacked NCNCs. Inset: cartoon illustrating orientation of the nanocups in stacked NCNC. (Reproduced with permission from Tang, Y. et al., Electrocatalytic activity of nitrogen-doped carbon nanotube cups, *J. Am. Chem. Soc.*, 131(37), 13200–13201. Copyright 2009 American Chemical Society.)

highly engineered hollow CNT nanocup arrays were fabricated (Figure 9.17) using precisely controlled short nanopores inside AAO templates.[41] The nanocups were effectively used to hold and contain metal nanoparticles, leading to the formation of multicomponent hybrid nanostructures with unusual morphologies. Additionally, cup-shaped nanocarbons generated by the electron-transfer reduction of cup-stacked CNTs were functionalized with porphyrins (H_2P) as light-capturing chromophores.[42]

Semiconductor ZnO nanocups of hexagonal wurtzite single-crystalline structure were obtained through ultrasound irradiation of the precursor aqueous solution of $Zn(NO_3)_{23} \cdot 6H_2O$ and hexamethylenetetramine $\{(CH_2)_6N_4, HMT\}$ for 20 min at r.t.[43] We note that such nanostructural type, among others (nanorods, nanodisks, nanoflowers, and nanospheres, Figures 9.18 and 9.19) was obtained also by sonochemical route.[44] The sonochemical growth mechanism of ZnO nanostructures was considered according to the following Reactions 9.4 through 9.8, including participation of free radicals $\cdot H$, $\cdot OH$, $\cdot O_2^-$, and $\cdot HO_2$ generated by sonolysis. A bottom-up approach was used to tune

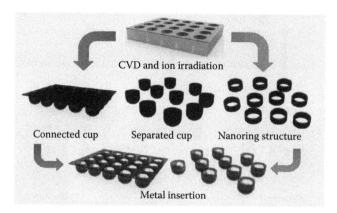

FIGURE 9.17 Schematic illustrating the fabrication process and resulting architectures of connected arrays of CNTs nanocup film, individually separated nanocups, nanorings, and metal nanoparticle–nanocup heterostructures. (Reproduced with permission from Chun, H. et al., Engineering low-aspect ratio carbon nanostructures: Nanocups, nanorings, and nanocontainers, *ACS Nano*, 3(5), 1274–1278. Copyright 2009 American Chemical Society.)

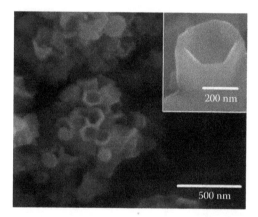

FIGURE 9.18 ZnO nanocups. (Reproduced with permission from Jung, S.-H. et al., Sonochemical preparation of shape-selective ZnO nanostructures, *Cryst. Growth Des.*, 8(1), 265–269. Copyright 2008 American Chemical Society.)

the morphology of single-crystalline hematite from hollow spheres to nanocups.[45] The mechanism involved the formation of nanocups through buckling of the spheres, similar to a deflated ball. As the shape changes, there is a drastic change in magnetic properties.

$$(CH_2)_6N_4 + 6H_2O \rightarrow 4NH_3 + 6HCHO \tag{9.4}$$

$$NH_3 + H_2O \rightarrow NH_4^+ + OH^- \tag{9.5}$$

$$Zn^{2+} + 4OH^- \rightarrow Zn(OH)_4^{2-} \tag{9.6}$$

$$Zn(OH)_4^{2-}))) \rightarrow ZnO + H_2O + 2OH^- \tag{9.7}$$

$$Zn^{2+} + 2{\bullet}O_2^-))) \rightarrow ZnO + 1.5O_2 \tag{9.8}$$

An interesting technique (Figure 9.20), able to sort nanoparticles, based on their dimensions due to the interactions between charged droplets and a nonlinear electrostatic field, allowed sorting silica

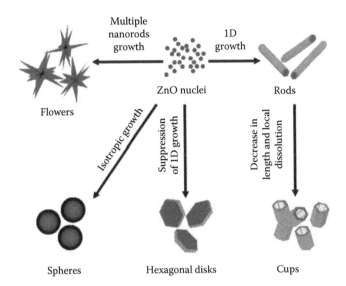

FIGURE 9.19 Schematic of the shape-selective synthesis of ZnO nanorods, nanocups, nanodisks, nano-flowers, and nanospheres *via* a sonochemical route. (Reproduced with permission from Jung, S.-H. et al., Sonochemical preparation of shape-selective ZnO nanostructures, *Cryst. Growth Des.*, 8(1), 265–269. Copyright 2008 American Chemical Society.)

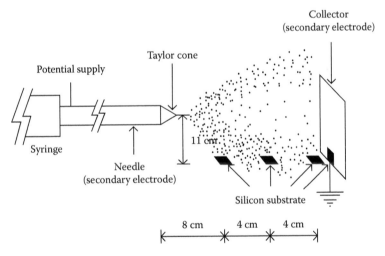

FIGURE 9.20 Experimental setup showing three floating electrodes, one counter electrode, and an elec-trospraying source that is connected to the pumping system. (Reproduced from Deotare, P. and Kameoka, J., *J. Nanomater.*, 2007, DOI Bookmark: 10.1155/2007/71259. Hindawi Publishing Corporation.)

nano- and microcups (Figure 9.21) into three groups with mean diameters of 0.31, 0.7, and 1.1 μm.[46] This method improved the nanoparticle fabrication process not only by decreasing the standard deviation of its dimensions but also by increasing its yield, since nanoparticles with different mean diameters can be generated at the same time. The fabrication of silica nanocomposite cups by elec-trospraying blended polymer–sol–gel solutions followed by calcinations was also reported.[47]

CdS nanostructures with different morphologies including nanocups were fabricated through an effective C-assisted thermal evaporation method.[48] CdS multipods and nanocups showed mainly green emission centered at ~496 nm, in contrast to nanobrushes exhibiting predominant red emis-sion band peaking at ~711 nm. Thus, carbon not only affected the growth process but also influenced the properties of CdS nanostructures. The charge separation between excited CdSe semiconductor

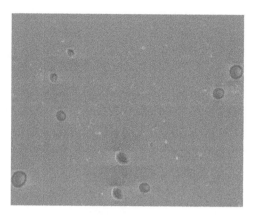

FIGURE 9.21 Silica nanocups. (Reproduced from Deotare, P. and Kameoka, J., *J. Nanomater.*, 2007, DOI Bookmark: 10.1155/2007/71259. Hindawi Publishing Corporation.)

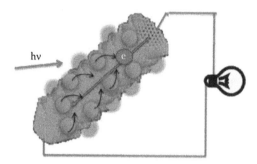

FIGURE 9.22 Illustration of SCCNT based solar cell. The "open-cup" structure provides a tubular carbon structure with many open functionalization sites. (Reproduced with permission from Farrow, B. and Kamat, P. V., CdSe quantum dot sensitized solar cells. Shuttling electrons through stacked carbon nanocups, *J. Am. Chem. Soc.*, 131(31), 11124–11131. Copyright 2009 American Chemical Society.)

quantum dots and stacked-cup carbon nanotubes (SCCNTs) generated photocurrent in a quantum dot-sensitized solar cell (QDSC) (Figure 9.22).[49] The ability of SCCNTs to collect and transport electrons from excited CdSe was established from photocurrent measurements; the morphology and excited state properties of SCCNT-CdSe composites demonstrated their usefulness in energy conversion devices. Boron nitride nanotubes (BNNTs) comprising both coaxial and cup-stacking (Figure 9.23) tubular structures (similar to stacking CNTs, see earlier discussion) were selectively synthesized at 900°C on patterned bilayer (Fe/Al) catalysts by the plasma-assisted CVD (PACVD) method.[50] The unstable dangling bonds had a unique structure, which in turn led to the highly reactive properties of the tube's surface. The authors noted that promising applications may be found in hydrogen storage and sensing devices.

9.3 NANOPLATES

The plate-like nanostructures are relatively widespread and represented practically by inorganic compounds, mainly by noble metals (gold, silver, palladium, rhodium, sometimes as composites with polymers), metal oxides, and oxygen-containing metal salts. In the case of elemental metals, a review[51] was dedicated to the use of different metal nanoparticles (including nanoplates) or their compounds in the organic transformations and syntheses of biologically important compounds. Additionally, energy transport in plasmon waveguides on chains of metal nanoplates was reviewed in Ref. [52]. Hydrothermal route has been applied very frequently, although other physicochemical,

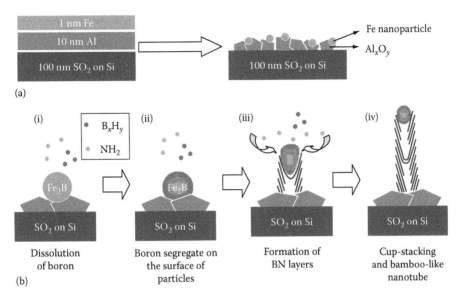

(a)

(i)

B_xH_y

NH_2

(ii)

(iii)

(iv)

Fe_2B

Fe_2B

SO_2 on Si

SO_2 on Si

SO_2 on Si

SO_2 on Si

Dissolution of boron

Boron segregate on the surface of particles

Formation of BN layers

Cup-stacking and bamboo-like nanotube

(b)

FIGURE 9.23 Schematic models for the growth of BNNTs: (a) The initial bilayer is transformed during the pretreatment process into an Al_xO_y island, where nanosized Fe particles were uniformly distributed on the surface of the Al_xO_y clusters. (b) A four-step model depicts the growth of the BN nanotube as follows: (i) Boron atoms dissolve in the Fe catalyst. (ii) Boron segregates onto the particle's surface. (iii) Boron reacts with nitrogen to form BN layers. (iv) The dynamic growth of catalyst reshaping combined with precipitation of BN layers forms the cup-stacking and bamboo-like BN nanotube. (Reproduced with permission from Su, C.-Y. et al., Selective growth of boron nitride nanotubes by the plasma-assisted and iron-catalytic CVD methods, *J. Phys. Chem. C*, 113(33), 14681–14688. Copyright 2009 American Chemical Society.)

chemical, and biological techniques were also used. Nanoplate shapes can vary (see the following): pentagons, hexagons, triangles, and squares/rectangles, and other forms were observed. The formed particles can be plate-like or can form such more complex structures as nanoflowers. Main attention is paid to gold (Au and Ag nanoplates were reviewed in Ref. [53]), whose varieties of nanostructures are very rich, in particular those with plate-like shapes, sometimes unique, such as, for example, corolla- (Figure 9.24) and propeller-like architectures made of nanoplates.[54] Their preparation

FIGURE 9.24 Gold nanocorolla. (Reproduced with permission from Soejima, T. and Kimizuka, N., One-pot room-temperature synthesis of single-crystalline gold nanocorolla in water. *J. Am. Chem. Soc.*, 131(40), 14407–14412. Copyright 2009 American Chemical Society.)

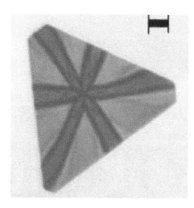

FIGURE 9.25 TEM images of gold nanoplates in PVA films, formed with stress patterns. Scale bar = 50 nm. (Porel, S. et al., Polygonal gold nanoplates in a polymer matrix, *Chem. Commun.*, 2387–2389, 2005. Reproduced by permission of The Royal Society of Chemistry.)

process was based on the simultaneous growth and etching of gold nanoplates in aqueous solution, which occurred in the course of photoreduction of $Au(OH)_4^-$ ions. The presence of bromide ion, poly(vinylpyrrolidone) (PVP), and molecular oxygen for obtaining these nanostructures was indispensable. The authors envisaged that complex metal nanoarchitecture may find unique applications in which their intrinsic higher energy surfaces play decisive roles.

Polygonal gold nanoplates (Figure 9.25) were generated *in situ* in poly(vinyl alcohol) film through thermal treatment, the polymer serving as the reducing agent and stabilizer for the nanoparticle formation and enforcing preferential orientation of the plates.[55] Regularity was observed (Figure 9.26) in the evolution of the shape; pentagons, hexagons, triangles, and squares/rectangles were formed with increasing concentration, decreasing temperature, and increasing heating time. Additionally, a single tree-type multiple-head surfactant, bis-(amidoethyl-carbamoylethyl)octadecylamine ($C_{18}N_3$), which functions as both the reducing and capping agent in the reaction system, was used to fabricate gold nano- and microplates (Figure 9.27) from $HAuCl_4 \cdot H_2O$ as a precursor.[56] These gold nano- and microplates greatly enhanced the surface-enhanced Raman scattering (SERS) of AA molecules compared with that of other gold nanoparticle morphologies. In addition to chemical reduction routes mentioned earlier, biological routes are also known; thus, it was demonstrated that the bacterium *Rhodopseudomonas capsulata* was capable of producing gold nanoparticles extracellularly and the as-formed gold nanoparticles were quite stable in solution.[57] The shape of the gold nanoparticles was controlled by pH. In addition, a single-step r.t. biosynthetic route for producing triangular and hexagonal gold nanoplates using pear fruit was reported.[58] Additional information of Au nanoplates was reported in Refs. [59,60].

A considerable number of reports are devoted to silver nanoplates; here we present only a brief description of selected representative works, where silver plate-like nanostructures were obtained by different routes. Thus, direct synthesis of Ag nanoplates on GaAs wafers was developed by solution/solid interfacial reaction strategy, in which aqueous solutions of pure $AgNO_3$ reacted with the GaAs wafers at r.t. according to Reactions 9.9 and 9.10 (h^+ is a hole released by Ag^+, which is injected into the GaAs lattice through the seed).[61] Regular silver triangular nanoplates with a thickness of 5–20 nm and a size tunable from 60 to 400 nm were obtained by the reduction of silver nitrate with sodium borohydride in the presence of sodium citrate and dioctyl sulfosuccinate sodium salt {bis(2-ethylhexyl) sulfosuccinate sodium salt} and heat treatment at 60°C for 36–50 h.[62] Additionally, Ag triangular nanoplates were synthesized by photoirradiation to $AgNO_3$ ethanol solution in the presence of polyvinylpyrrolidone (PVP) under anaerobic conditions at r.t.[63] Also, laser irradiation (laser pulses) truncated the vertices of Ag triangular nanoplates, with an average edge size of 110 nm and

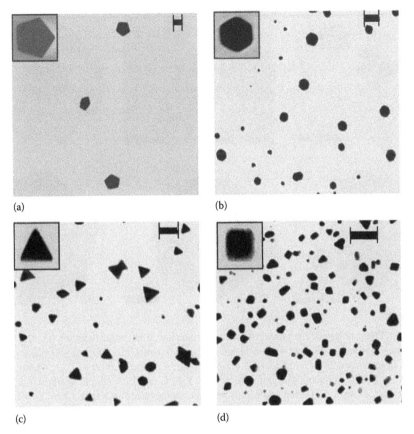

FIGURE 9.26 TEM images of Au-PVA films with polygonal nanoplates generated under different conditions. The Au/PVA ratio, temperature of heating (°C), and time of heating (min) are indicated in that order in parenthesis: (a) pentagons (0.04, 170, 5); (b) hexagons (0.08, 130, 30); (c) triangles (0.12, 100, 60); and (d) squares/rectangles (0.18, 100, 60). Scale bar, 50 nm. Inset shows enlarged view of a single nanoplate with the dominant shape in each case. (Porel, S. et al., Polygonal gold nanoplates in a polymer matrix, *Chem. Commun.*, 2387–2389, 2005. Reproduced by permission of The Royal Society of Chemistry.)

an average thickness of 14 nm, to form nearly regular hexagonal and spherical silver nanoplates gradually.[64] Additional information on Ag nanoplates can be obtained in Refs. [65,66]:

$$2GaAs + 12h^+ + 6H_2O \rightarrow 12H^+ + Ga_2O_3 + As_2O_3 \tag{9.9}$$

$$2GaAs + 12Ag^+ + 6H_2O \rightarrow 12Ag + 12H^+ + Ga_2O_3 + As_2O_3 \tag{9.10}$$

A series of useful parameters that can be tuned to control the formation of Pd nanostructures (including nanoplates, Figure 9.28) with a specific shape in a solution-phase synthesis were reviewed.[67] Both the crystallinity of seeds and the growth rates of different crystallographic facets were found to be important in determining the shape of resultant nanostructures. More specifically, the crystallinity of a seed can be controlled by manipulating the reduction rate. Additionally, a series of applications of palladium nanostructures was emphasized (high sensitivity SERS detection, plasmonic detection of hydrogen gas, use in catalysis, hydrogen storage, photothermal cancer treatment, and optical imaging contrast enhancement, among others). Unprecedented ultrathin rhodium nanoplates (1.3 nm thickness) were prepared at 50°C *via* metal–metal interactions between rhodium(I)

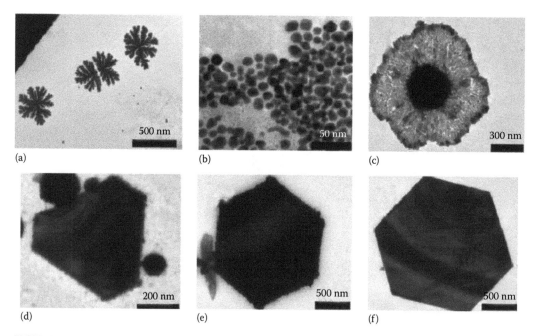

FIGURE 9.27 TEM images of the products collected at various times after reaction, (a) 30 s, (b) high magnification of the gold nanostructures shown in panel (a), (c) 1.5 min, (d) 5 min, (e) 20 min, and (f) 30 min. The standard reaction condition is 0.5 M KCl, 1 mM $C_{18}N_3$, and a molar ratio of $C_{18}N_3/HAuCl_4 = 6.9$ were maintained at 80°C. (Reproduced with permission from Lin, G. et al., A simple synthesis method for gold nano- and microplate fabrication using a tree-type multiple-amine head surfactant, *Cryst. Growth Des.*, 10(3), 1118–1123. Copyright 2010 American Chemical Society.)

precursors formed from the reaction of rhodium carbonyl chloride dimer complex with oleylamine.[68] The formation of rhodium nanoplates was observed only by using $[Rh(CO)_2Cl]_2$ and not using $[Rh(COD)Cl]_2$, $\{Rh[P(OPh)_3]_2Cl\}_2$, and $RhCl_3$. Considering the suggested formation mechanism in Figure 9.29, 2D shape evolution, according to authors' opinion, would be quite dependent on the van der Waals interaction between the coordinated oleylamines.

Bimetallic plate-like nanostructures of noble metals are represented by porous triangular Ag-Pd nanoplates of various ratios M^1/M^2, prepared by a galvanic displacement reaction in which $Pd(OAc)_2$ reacted with Ag nanoplates.[69] Studies of the kinetics of electroless copper deposition catalyzed by these bimetallic nanostructures revealed that $Ag_{18}Pd_1$ nanoplates showed better catalytic activity than that of $Ag_{18}Pd_{1.5}$ and $Ag_{18}Pd_2$ nanoplates. Also, nanoplates consisting of a Pd core and a Pt shell were obtained by the reduction of a Pt precursor with citric acid (CA) as a reducing agent in an aqueous solution, leading to a thin, uniform Pt shell around the Pd nanoplate, demonstrating the layer-by-layer epitaxial growth of Pt on Pd.[70]

A few other metals were reported in plate-like nanoforms. Thus, tensile/compressive deformation processes of a single crystalline copper nanoplate were simulated by molecular dynamics.[71] The extreme tensile or compressive elastic strains were shown as 0.08 and −0.03, respectively, and the stress–strain relation was nearly linear between them. Also, amorphous iron nanoplates with a large diameter-to-thickness aspect ratio were prepared by a reduction process and used for cyclohexane oxidation to yield cyclohexanol and cyclohexanone with a higher conversion rate, possessing a higher selectivity to yield cyclohexanol.[72]

A wide variety of oxides have been found as plate-like nanostructures; different methods (hydrothermal, microwave, or wet-chemical routes) were used. Thus, the formation mechanism of silica nanoplates (elemental silicon nanoplates were described in Refs. [73,74] by exfoliation of a phyllosilicate magnesium containing clay, Lucentite, in an aqueous solution of poly(acrylic acid) (PAAc) was studied.[75]

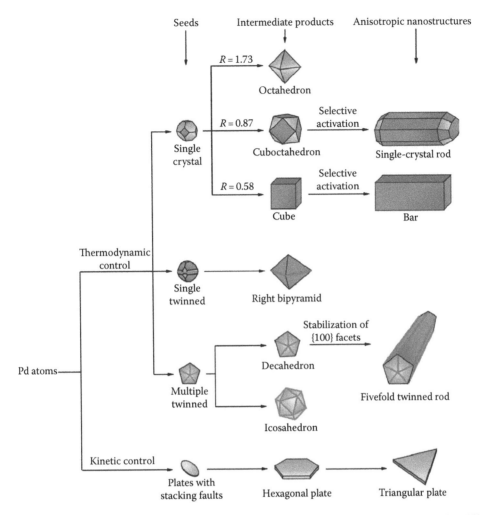

Seeds   Intermediate products   Anisotropic nanostructures

FIGURE 9.28 A schematic illustration of the reaction pathways that lead to Pd nanostructures with different shapes. As the essence of a synthesis, a palladium precursor is reduced to produce Pd atoms, which subsequently aggregate to form nuclei. Once the nuclei have grown past a certain size, they become seeds with a single-crystal, single-twinned, or multiple-twinned structure. If stacking faults are involved, the seeds will grow into plate-like nanostructures. (Xiong, Y. and Xia, Y.: Shape-controlled synthesis of metal nanostructures: The case of palladium. *Adv. Mater.* 2007. 19, 3385–3391. Copyright Wiley-VCH Verlag GmbH & Co. KGaA. Reproduced with permission.)

It was shown that non-surface (bulk) Mg ions were not chemically involved in the PAAc/clay intercalation (although MgO nanoplates are also known[76]) but were substantially involved in the exfoliation, resulting in the silica nanoplates. Heating a mixture of ZnO powder and graphite powder at 1000°C–1200°C for 10–40min produced a nanostructure binding ZnO nanoplate with a thickness of ~50nm and ZnO nanorod with a diameter of ~40nm.[77] The product was found to be suitable for piezoelectric elements, varistors, fluorescent materials, sensors, and transparent conductive films. Alternatively, ZnO films were grown on polycrystalline Zn foil by cathodic electrodeposition in an aqueous Zn chloride/Ca chloride solution at 80°C.[78] Varying electrochemical parameters, nanorods, nanoplates (50nm in thickness and several microns in diameter), and 3D crystals were obtained. Titanates with various sizes and shapes (including nanoplates) doped with Li[+] were prepared using rutile-phase TiO$_2$ nanopowders in strong basic solution of NaOH by hydrothermal process.[79] Doping with other metal ions yielded other

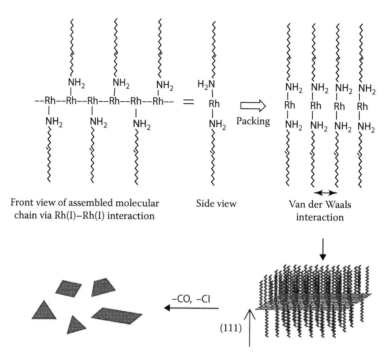

Front view of assembled molecular   Side view   Van der Waals
chain via Rh(I)–Rh(I) interaction   interaction

FIGURE 9.29 Suggested growth mechanism of rhodium nanoplates. (Reproduced with permission from Jang, K. et al., Low-temperature synthesis of ultrathin rhodium nanoplates via molecular orbital symmetry interaction between rhodium precursors, *Chem. Mater.*, 22(4), 1273–1275. Copyright 2010 American Chemical Society.)

nanostructures (nanobelts or nanotubes), so it was suggested that the particle shape of titanates can be controlled only by a small amount of doping elements in NaOH aqueous solutions. In the case of TiO_2 itself, it was established that a mesoporous nanoplate-like structure for Sr^{2+}-doped TiO_2, obtained by sol–gel method from $Ti[OCH(CH_3)_2]_4$ and strontium nitrate $Sr(NO_3)_2 \cdot 6H_2O$ as precursors, exhibited a higher photocatalytic activity than both TiO_2 nanoparticles and commercial TiO_2 (Degussa P-25) due to increases in the band gap energy and surface area.[80]

Two kinds of WO_3 square nanoplates were prepared by a hydrothermal method using (+)-tartaric acid or CA as supporting agents.[81] It was established that WO_3 square nanostructures prepared in the presence of (+)-tartaric acid had a hexagonal phase, length of ~200 nm, and thickness of ~100 nm, while WO_3 nanostructures synthesized in the presence of CA had an orthorhombic phase, length of ~500 nm, and thickness of ~100 nm. Alternatively, hexagonal and truncated hexagonal-shaped MoO_3 nanoplates (Figure 9.30) with a large surface area were synthesized through a simple catalyst-free vapor-deposition method in Ar atmosphere under ambient pressure.[82] Alcohol response sensors using 2D WO_3 nanoplates as active elements were investigated in Ref. [83].

Among other transition metal oxides, PANI–vanadium oxide hybrid hierarchical architectures assembled from nanoscale building blocks, such as nanoplates and nanobelts, were synthesized by a one-step hydrothermal homogeneous reaction between aniline and peroxovanadic acid without the aid of any surfactant or template.[84] It was established that with increasing reaction time or temperature, the alignment of nanoplates in the hierarchical architectures became gradually dense. Hexagonal magnetite (Fe_3O_4) nanoplates with an average edge length of 80 nm were prepared in large quantities by a facile microwave-assisted route.[85] Mn_3O_4 hexagonal nanoplates were synthesized *via* a solvent-assisted hydrothermal oxidation process at low temperature.[86] Their high capability of catalytic oxidation of formaldehyde to formic acid at r.t. and atmospheric pressure was established. Alternatively, manganese oxide nanoplates with different shapes (from disk-shaped to hexagonal nanoplates by introducing different organic additives) were prepared based on an

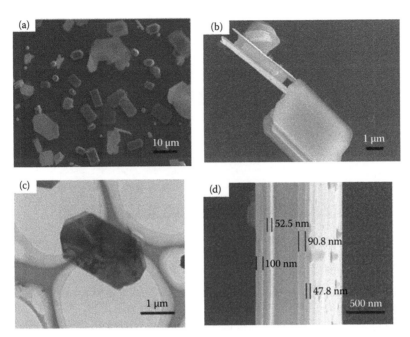

FIGURE 9.30 (a) SEM image of as-synthesized MoO_3 hexagonal nanoplate at low magnification. (b) High-magnification SEM image of MoO_3 hexagonal nanoplate. (c) TEM image of a typical layered HNP. (d) Lateral section of a typical layered MoO_3 hexagonal nanoplate. (Reproduced with permission from Chen, X. et al., Synthesis and characterization of hexagonal and truncated hexagonal shaped MoO_3 nanoplates, *J. Phys. Chem. C*, 113(52), 21582–21585. Copyright 2009 American Chemical Society.)

ethylene glycol-mediated route,[87] whose first step consisted of precipitation of manganese alkoxide precursor in a polyol process from the reaction of manganese acetate with ethylene glycol. Rare-earth oxides are represented by cerium oxide, whose nanoplates and nanorods were obtained *via* the thermal decomposition of a mixture of cerium acetate, oleic acid (OA), oleylamine, and 1-octadecene under controlled atmospheres.[88] Also, the synthesis of its binary oxides as $CuO-CeO_2$ with a plate-like morphology, prepared by a solvothermal method and calcined at 400°C, exhibited an excellent catalytic activity for benzene oxidation despite its relatively low surface area and could catalyze the complete oxidation of benzene at a temperature as low as 240°C.[89]

Among metal hydroxides, a biomolecular-assisted hydrothermal method was applied to synthesize β-phase $Ni(OH)_2$ peony-like complex nanostructures with second-order nanoplate structure exhibiting high-power Ni/MH battery performance, close to the theoretical capacity of $Ni(OH)_2$, as well as controlled wetting behavior.[90] Oxyhalides are represented by ultrathin BiOCl nanoflakes, nanoplate arrays, and curved nanoplates, synthesized *via* an ionothermal synthetic route by using $Bi(NO_3)_3 \cdot 5H_2O$ as a precursor and an ionic liquid 1-hexadecyl-3-methylimidazolium chloride ([C16 Mim]Cl) as all-in-one solvent, simply adjusting reaction temperature.[91] This ionothermal synthetic route using long-chain ionic liquid has potential application in fabricating polar nanomaterials with novel morphologies and improved properties in aspects of optics, electrochemistry, and catalysis. The excellent adsorption performance of the as-prepared BiOCl nanoplates makes them useful with potential applications in wastewater treatment. Alternatively, BiOCl nanoplates, among other structures, were synthesized through a sonochemical route.[92] Hierarchical bismuth oxybromide (BiOBr) nanoplate microspheres, used to remove NO from indoor air under visible light irradiation (λ > 420 nm), were synthesized[93] with a nonaqueous, sol–gel method using bismuth nitrate and cetyltrimethylammonium bromide (CTAB) as precursors. The excellent catalytic activity and long-term activity of these microspheres were attributed to their special hierarchical structure, favorable for diffusion of NO oxidation intermediate compounds and final products. Among non-O-containing

FIGURE 9.31   Star-like $BiVO_4$ products made of nanoplates. (Reproduced with permission from Sun, S. et al., Efficient methylene blue removal over hydrothermally synthesized starlike $BiVO_4$, *Ind. Eng. Chem. Res.*, 48(4), 1735–1739. Copyright 2009 American Chemical Society.)

halides, lanthanum fluoride was prepared starting from NH_4F and $La(NO_3)_3 \cdot 6H_2O$ as precursors by molten salt route leading to $LaF_3:Eu^{3+}$ nanoplates with tunable size.[94]

A series of oxygen-containing salts have been reported as nanoplates, for instance porous hierarchical assemblies of Li-rich $Li_{1+x}MnO_{3-\delta}$ 2D nanoplates as well as isolated 0D nanocrystalline homologs, synthesized *via* lithiation of nanostructured Mn oxides under hydrothermal conditions.[95] It was demonstrated that tetravalent Mn ions were stabilized in octahedral sites of a Li_2MnO_3-type layered structure composed of edge-shared MnO_6/LiO_6 octahedra. A chemical method with small polymer templates of polyvinyl alcohol was explored to obtain nanoplates (15–35 nm thickness) of another related nonstoichiometric compound $La_{0.67}Ca_{0.33}MnO_3$ of single magnetic domains.[96] Nanoplate structured, stacked, star-like bismuth vanadate ($BiVO_4$) products (Figure 9.31) were synthesized using a hydrothermal method, with a water/ethanol mixed solvent and EDTA chelating agent.[97] These samples exhibited high visible light-driven photocatalytic efficiency: thus, for methylene blue (MB) degradation under visible light irradiation ($\lambda > 420$ nm), ~91% of MB was degraded within 25 min. Alternatively, well-defined m-$BiVO_4$ nanoplates with exposed {001} facets were synthesized by a hydrothermal route too, but in this case without the use of any template or organic surfactant.[98] The product exhibited greatly enhanced activity in the visible-light photocatalytic degradation of organic contaminants and photocatalytic oxidation of water for O_2 generation. Another vanadate is represented by porous $YVO_4:Sm$ nanoplates, obtained by a low-temperature synthesis route of fabricating *via* a co-precipitation method using commercially available Y_2O_3, NH_4VO_3, Sm_2O_3, and ethylene glycol as the reacting precursors.[99] Additionally, single-crystal HAP nanoplates were synthesized by hydrothermal method.[100]

Nitrogen-doped $CaNb_2O_6$ nanoplates with an ellipsoid-like morphology were prepared using the hydrothermal method, followed by heat treatment at various temperatures in an NH_3 atmosphere.[101] It was found that the nitrogen doping in the $CaNb_2O_6$ nanoplates led to the formation of nanoparticles with a size of 10 nm at the surface and the red shift of the light absorption edge into the visible light region. Compared with the undoped powder, the N-doped $CaNb_2O_6$ nanoplate powder exhibited higher photocatalytic activities for the degradation of rhodamine B dye solution under visible light irradiation (>420 nm). 2D free-standing single-crystalline $SrMoO_4$ nanoplates of obvious quasi-square shape were controllably synthesized using a liquid approach (Figure 9.32) at r.t.[102] It was noted that these quasi-square nanoplates can be formed through the oriented connection and self-assembly process. Such $SrMoO_4$ nanoplates presented a strong and broad blue PL emission, which exhibited a noted blue shift, indicating that the square-shaped $SrMoO_4$ nanostructures have great potential to be applied in luminescent and optoelectronic devices. A layered inorganic perovskite submicrometer-scale material, nanoplated bismuth titanate ($Bi_4Ti_3O_{12}$) submicrospheres (NBTSMs) constructed with

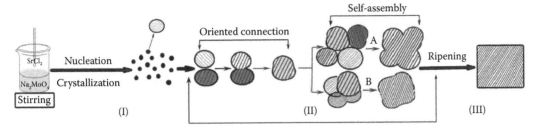

FIGURE 9.32 Schematic illustration of the formation and shape evolution of quasi-square SrMoO$_4$ nanoplates in the whole synthetic process: (I) Nucleation and crystallization; (II) Oriented connection and self-assembly (A: side-by-side, B: layer-by-layer); (III) Ripening. (Reproduced with permission from Mi, Y. et al., Room-temperature synthesis and luminescent properties of single-crystalline SrMoO$_4$ nanoplates, *J. Phys. Chem. C*, 113(49), 20795–20799. Copyright 2009 American Chemical Society.)

tens of Bi$_4$Ti$_3$O$_{12}$ nanoplates, was synthesized by a hydrothermal synthesis strategy.[103] It was revealed that the NBTSM-based composite was a satisfying matrix for proteins to effectively retain their native structure and bioactivity and may find potential applications in biomedical, food, and environmental analysis and detection. The production method of nanoplates of the same compound was also patented by another research group.[104] Other polymetallic O-containing salts are represented by calcium cobaltite (Ca$_3$Co$_4$O$_9$) powders (thickness of the order of 10–50 nm and a diameter of around 300 nm).[105] The authors noted that this is a potentially new anode material with high capacity for Li-ion batteries.

Nitrides are represented by Si$_3$N$_4$, prepared by the following route. Growth of single-crystalline Si$_3$N$_4$ nanoplates (Figure 9.33), obtained[106] *via* catalyst-assisted pyrolysis of polymeric precursor (polyaluminasilazane) in the presence of a catalyst (FeCl$_2$), involved three basic steps: formation and aggregation of the nanoparticles, grain coalescence within selective areas, and growth of one coarsened grain at the expense of the rest of the nanoparticles *via* an Oswald ripening process (Figure 9.34) assisted by the oriented attachment mechanism, in which Si$_3$N$_4$ nanoparticles were first formed due to the decomposition of the precursor and aggregated into large-sized nanostructures. Final reaction with the participation of SiO, formed from an intermediate compound "SiAlCNTs," is shown by reaction 9.11:

$$3SiO + 3CO + 2N_2 \rightarrow Si_3N_4 + 3CO_2 \tag{9.11}$$

Chalcogen salts, simple or mixed with oxides or other metal, are also common, as oxide mentioned earlier. Thus, CuS architectures including nanoplates (Figure 9.35), representing a hexagonal-structured phase and composed of intersectional nanoplates, were prepared from Cu(NO$_3$)$_2\cdot$2H$_2$O, sulfur, and ethylene glycol according to Reactions 9.12 through 9.14. They were found to have good photocatalytic activity in the degradation of MB under the irradiation of solar light.[107] Mixed-valent copper (II) oxide/copper(I) sulfide superhydrophobic CuO@Cu$_2$S nanoplate vertical arrays were synthesized from the hierarchical CuO precursors *via* a solution-immersion process.[108] It was found that the wettability of this composite film could be easily changed from hydrophilic to superhydrophobic with simple fluorination modification. Single crystalline Cu$_9$BiS$_6$ nanoplates were synthesized *via* a hydrothermal route.[109] Good quality gallium oxide/gallium sulfide nanoplates with ultrathin thickness were prepared *via* the reaction of gallium oxide nanoribbons with sulfur solution.[110] Additionally, Zn:In(OH)$_y$S$_z$ was described[111]:

$$HOCH_2CH_2OH \rightarrow CH_3CHO + H_2O \tag{9.12}$$

$$S + 2CH_3CHO \rightarrow CH_3CO\text{-}OCCH_3 + S^{2-} + 2H^+ \tag{9.13}$$

$$Cu^{2+} + S^{2-} \rightarrow CuS(s) \tag{9.14}$$

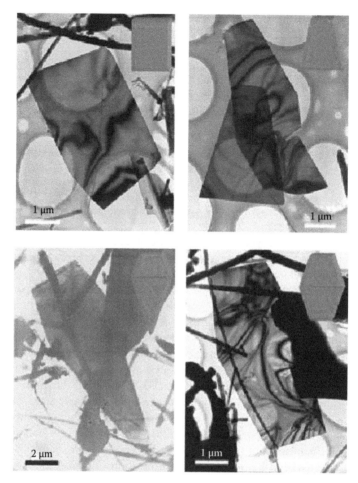

FIGURE 9.33 Typical types of the obtained Si_3N_4 nanoplates with various shapes. (Reproduced with permission from Yang, W. et al., Ostwald ripening growth of silicon nitride nanoplates, *Cryst. Growth Des.*, 10(1), 29–31. Copyright 2010 American Chemical Society.)

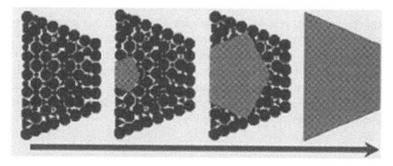

FIGURE 9.34 Schematic model for Oswald ripening growth of the Si_3N_4 nanoplate. (Reproduced with permission from Yang, W. et al., Ostwald ripening growth of silicon nitride nanoplates, *Cryst. Growth Des.*, 10(1), 29–31. Copyright 2010 American Chemical Society.)

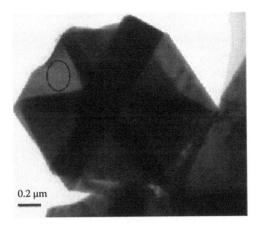

In the case of other chalcogenides, the synthesis of 2D single-crystal berzelianite ($Cu_{2-x}Se$) nanosheets (in-plane diameter-to-thickness ratio ~100) and nanoplates (in-plane diameter-to-thickness ratio ~10) was carried out *via* a simple, "green," and environmentally benign method of injecting a Cu(I)-complex precursor into Se solution in paraffin.[112] This process can be used for the creation of materials for solar energy conversion and also in a wide range of photonic devices operating in the near-IR. Ultrathin Bi_2Te_3 and Bi_2Se_3 nanoplates (Figure 9.36) with thickness down to 3 nm (three quintuple layers), were obtained *via* catalyst-free vapor–solid growth mechanism.[113] These nanoplates had an extremely large surface-to-volume ratio and can be electrically gated more effectively than the bulk form, potentially enhancing surface state effects in transport measurements.

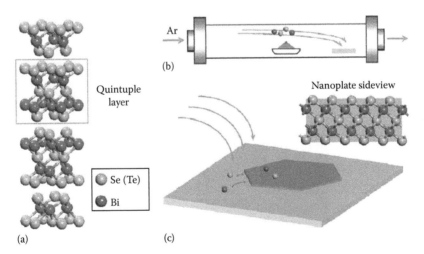

FIGURE 9.36 (a) Layered crystal structure of Bi_2Se_3 and Bi_2Te_3, with each quintuple layer (QL) formed by five Bi and Se (or Te) atomic sheets. (b) A schematic drawing of a vapor–solid (VS) growth process for few-layer nanoplates (NPs) of Bi_2Se_3 and Bi_2Te_3 in a horizontal tube furnace. (c) A schematic drawing of NPs growth mechanism, involving gas-phase atoms diffused and attached to the side surface. A side view of the NPs (1QL) shows the Se-terminated top and bottom surfaces with saturated bonds and side surfaces with dangling bonds ready to bind with incoming atoms. (Reproduced with permission from Kong, D. et al., Few-layer nanoplates of Bi_2Se_3 and Bi_2Te_3 with highly tunable chemical potential, *Nano Lett.*, 2010, 10(6), 2245–2250. Copyright American Chemical Society.)

9.4 NANOFORKS

Fork-like nanostructures are indeed rare and are known for Au, B, some oxides, and salts. Thus, nanosized Au particles were prepared by deposition on MgO from $HAuCl_4 \cdot 4H_2O$ as a precursor, yielding fork-shaped crystals (Figure 9.37), among other nanoforms, of the nanometer size.[114] The growth of these formations was explained as probably directed by the cracks in the layers of the support, but it must also be the inherent property of the deposited metal. Catalyst-free growth of α-tetragonal boron nanoribbons (consisting of boron and small amounts of oxygen and carbon), some of which seemed as nanoforks (Figure 9.38), was observed by pyrolysis of diborane at 630°C–750°C and pressure 200 mTorr in a quartz tube furnace.[115] The growth of nanoribbons was believed to be largely determined by growth kinetics. The obtained nanoribbons may find application in composites and as components in nanoelectromechanical systems.

ZnO forks, brooms, and spheres, possessing self-assembled nanowalls and nanocones of ZnO and exhibiting visible green defect-related emission on UV excitation, were prepared *via* direct thermal evaporation of metallic Zn flakes.[116] A series of different morphologies of β-Ga_2O_3 including fork-like structures (Figure 9.39) were grown *in situ* on the surface of gallium grains and films by heating gallium substrates at 750°C–1000°C for 2 h in air.[117] Their formation mechanism is shown in Figure 9.40. The products may have enormous potential applications in nanoscale photoelectron devices, catalysts, and chemical sensors. Additionally, two kinds of TiO_2 nanoforks with defined

FIGURE 9.37 Au fork-shaped crystal. (Reproduced with kind permission from Springer Science+Business Media: *React. Kinet. Catal. Lett.*, Formation of Au nanorods and nanoforks over MgO support, 87(2), 2006, 263–268, Fasi, A. et al.)

FIGURE 9.38 Boron nanoribbons, having slots in the middle forming a fork-like nanostructure. (Reproduced with permission from Xu, T.T. et al., Crystalline boron nanoribbons: Synthesis and characterization, *Nano Lett.*, 4(5), 963–968. Copyright 2004 American Chemical Society.)

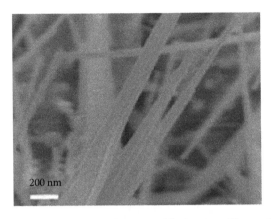
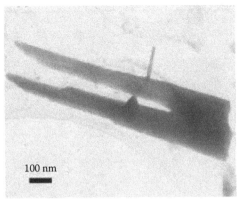

FIGURE 9.39 β-Ga₂O₃ prepared by heating silicon substrates coated with a layer of gallium metal in air at 950°C–1000°C. (Reproduced with kind permission from Springer Science+Business Media: *Sci. China Ser. E Tech. Sci.*, In-situ growth and photoluminescence of β-Ga₂O₃ cone-like nanowires on the surface of Ga substrates, 52(6), 2009, 1712–1721, Liu, R.N. et al.)

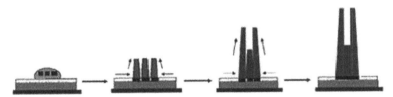

FIGURE 9.40 Schematic illustration of the growth mechanism for β-Ga₂O₃ fork-like nanobundles. (Reproduced with kind permission from Springer Science+Business Media: *Sci. China Ser. E Tech. Sci.*, In-situ growth and photoluminescence of β-Ga₂O₃ cone-like nanowires on the surface of Ga substrates, 52(6), 2009, 1712–1721, Liu, R.N. et al.).

boundary structures were observed.[118] One is a bent wire composed of two straight whiskers related by twinning on a (101) plane with an angle of 114° between the two legs, and the other by twinning on a (301) plane with an angle of 55° between the legs (Figure 9.41). It is known that TiO_2 surfaces play an important role in heterogeneous catalysis, photocatalysis, gas sensing, and electrode–electrolyte interactions. Therefore, the authors noted that the materials composed of its nanostructures can not only be employed as a prototype in the study of surface-related properties, and made into nanosensors with a very high sensitivity, but can also be used to weave a network and fabricate ceramic membranes with high mechanical strength and high temperature stability.

Single crystalline Sb_2Te_3 nanoforks (Figure 9.42) were prepared[119] in the ethylene glycol system under solvothermal conditions from Sb_2O_3, Te powder, HCl, and HNO_3 as precursors. Ethylene glycol was used not only as a solvent but also as a reducing agent. The nitric acid used played an important role in the formation of such a unique shape. The proposed growth mechanism for formation (Reactions 9.15 and 9.16) of this nanoform is shown in Figure 9.43. In addition, electrospun nanofibers of the (PVA)/LaCl₃ composite were employed to prepare the LaOCl nanofibers by calcination.[120] The addition of LaCl₃ led to the formation of fork segments in the structure of electrospun PVA/LaCl₃ composite nanofibers, thereby changing the decomposition behavior of the fibers. The resultant LaOCl nanofibers showed a good sensing behavior for CO_2 gas.

$$Te + 2e \rightarrow Te^{2-} \tag{9.15}$$

$$2Sb^{3+} + 3Te^{2-} \rightarrow Sb_2Te_3 \tag{9.16}$$

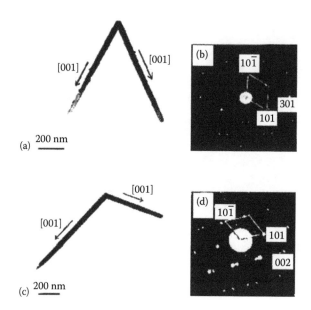

FIGURE 9.41 (a) TEM image of a TiO_2 nanofork with (301) interface and (b) selected area diffraction pattern of the twined boundary between the two legs. (c) TEM image of a nanofork and (d) the SADP of its (101) twined boundary. (Reproduced with kind permission from Springer Science+Business Media: *Eur. Phys. J. D Atom. Mol. Opt. Phys.*, Titania from nanoclusters to nanowires and nanoforks, 24(1–3), 2003, 355–360, Wang, G. and Li, G.)

FIGURE 9.42 Nanofork (Sb_2Te_3). (Reproduced with permission from Shi, S. et al., Controlled solvothermal synthesis and structural characterization of antimony telluride nanoforks, *Cryst. Growth Des.*, 9(5), 2057–2060. Copyright 2009 American Chemical Society.)

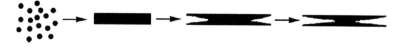

FIGURE 9.43 Proposed growth mechanism for the formation of Sb_2Te_3 nanoforks. (Reproduced with permission from Shi, S. et al., Controlled solvothermal synthesis and structural characterization of antimony telluride nanoforks, *Cryst. Growth Des.*, 9(5), 2057–2060. Copyright 2009 American Chemical Society.)

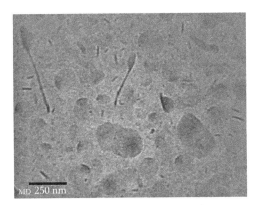

FIGURE 9.44 Nanospoon (solid lipid nanoparticles). (From Jores, K. et al., From solid lipid nanoparticles (SLN) to nanospoons. Visions and reality of colloidal lipid dispersions, *30th International Symposium on Controlled Release of Bioactive Materials*, Glasgow, U.K., Vol. 29, Controlled Release Society, St. Paul, MN, 2003, p. 181. Reproduced with permission from Controlled Release Society.)

9.5 NANOSPOONS

According to the available literature, "spoon-like nanostructure" (Figure 9.44), composed of large medium-chain triglycerides spot sticking on the solid lipid nanoparticle (SLN) surface, was reported.[121,122]

9.6 NANOBROOMS

Several studies have reported relatively rare broom-like nanostructures belonging to distinct types of mainly inorganic compounds. Thus, the formation process of TiO_2-based nanorods during hydrothermal synthesis starting from an amorphous $TiO_2 \cdot H_2O$ gel and NaOH was studied,[123] resulting in sodium tri-titanate ($Na_2Ti_3O_7$) and, after the ion-exchange reaction with HCl, $H_2Ti_3O_7$ particles with a rod-like and relatively unusual broom-like morphology (Figure 9.45). The presence of planar defects can be attributed to the exfoliation of the zigzag ribbon layers into 2D titanates and to the condensation of the layers of TiO_6 octahedra into 3D frameworks. The TiO_2 itself was also obtained

FIGURE 9.45 TEM image and corresponding electron diffraction pattern showing the existence of exfoliation at the end of thin rods of $H_2Ti_3O_7$. (Reproduced with permission from Kolen'ko, Yu.V. et al., Hydrothermal synthesis and characterization of nanorods of various titanates and titanium dioxide, *J. Phys. Chem. B*, 110, 4030–4038. Copyright 2006 American Chemical Society.)

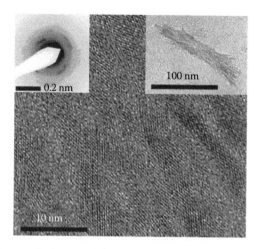

FIGURE 9.46 HRTEM image of a bunched TiO$_2$ nanostructure composed of anatase nanorods, collected after quenching in dry ice. Insets display the selected area electron diffraction (SAED) pattern and a low-resolution image of the nanostructure. (Cottam, B.F. et al., Accelerated synthesis of titanium oxide nanostructures using microfluidic chips, *Lab Chip*, 7, 167–169, 2007. Reproduced by permission of The Royal Society of Chemistry.)

FIGURE 9.47 Small broom-like TiO$_2$ particles are visible after about 90 days of hydrothermal treatment at 45°C. (Reproduced from Klein, S.M. et al., *J. Mater. Res.*, 18(6), 1457, 2003. With permission from Materials Research Society.)

in nanobroom form by performing the reaction in a microfluidic environment (Figure 9.46)[124] or by the hydrothermal method under highly acidic conditions (Figure 9.47).[125]

Cobalt spinel Co$_3$O$_4$ superstructures with broom-like, dandelion-like, and rose-like morphologies were obtained by thermal conversion of the cobalt hydroxide carbonate.[126] The superstructures were semiconducting with transitions corresponding to 775 and 530 nm in the UV–vis spectroscopy. The regular and homogeneous single-crystal CoMoO$_4$·3/4H$_2$O nanorods (diameters ~100 to 300 nm and lengths ~8 to 15 μm), prepared by a precipitation method from cobalt nitrate and sodium molybdate solutions as precursors,[127] can be converted from a broom-like to cage-like structure by controlling the reaction temperature. Thus, the broom-like microbunches were obtained at 50°C, while at 80°C dispersive nanorods can be prepared and at 90°C the morphology of the products converted to cage-like microspheres (Figures 9.48 and 9.49). The as-prepared products may have potential applications in optics, catalysis, and grating materials.

Other inorganic nanobroom structures are represented by nitrides and sulfides. Thus, broom-like single crystalline GaN nanomaterials, grown on Si(100) substrates through ammoniating the complex precursor [(NH$_4$)$_3$GaF$_6$], under a constant flow of NH$_3$ at 900°C in a quartz tube, were obtained upon

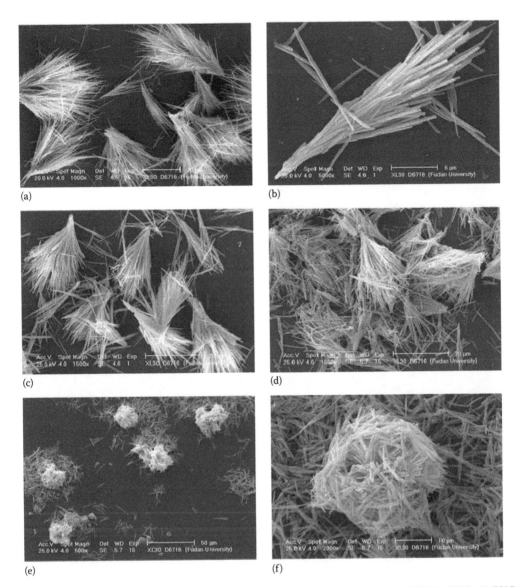

FIGURE 9.48 Typical SEM images of CoMoO$_4 \cdot$ 3/4H$_2$O products prepared at (a, b) 50°C; (c) 60°C; (d) 70°C; (e, f) 90°C for 3 h. (Reproduced with kind permission from Springer Science+Business Media: *J. Mater. Sci.*, Temperature-controlled assembly and morphology conversion of CoMoO$_4 \cdot$ 3/4H$_2$O nano-superstructured grating materials, 44(23), 2009, 6356–6362, Zhao, J. et al.)

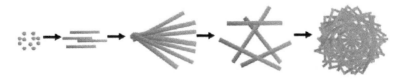

FIGURE 9.49 Schematic illustration of the formation process of the CoMoO$_4 \cdot$ 3/4H$_2$O products. (Reproduced with kind permission from Springer Science+Business Media: *J. Mater. Sci.*, Temperature-controlled assembly and morphology conversion of CoMoO$_4 \cdot$ 3/4H$_2$O nano-superstructured grating materials, 44(23), 2009, 6356–6362, Zhao, J. et al.)

the influence of a very fine ZnO nanoparticle interlayer in between the silicon substrate and the Ga source complex materials.[128] It was established that in the absence of ZnO interlayer, well-organized micron size hexagonal GaN particles resulted. The CdS nanorods and broom-like nanoparticles were synthesized by using gelatin solution as stabilizer.[129] The last nanostructural type was formed after 15 days, when fist synthesized nanorods (150–710 nm length, the diameter from 50 to 140 nm) turned into

(a)

(b)

(c)   (d)

(e)   (f)

(g)

(h)

FIGURE 9.50   (a) Typical SEM image with low magnification revealing the overall morphology of several bunches of radial arrays composed of whiskers. (b) EDAX pattern of the Sb_2S_3 whiskers shown in (a). (c through f) Several SEM images showing individual bunches of radial arrays. The length of the bars equals 10 mm. (g, h) SEM images with higher magnifications showing sectional images of radial arrays. (Reproduced with permission from Wang, H. et al., Novel microwave-assisted solution-phase approach to radial arrays composed of prismatic antimony trisulfide whiskers, *Langmuir*, 19, 10993–10996. Copyright 2003 American Chemical Society.)

the broom-like nanoparticles. The CdS samples showed obvious quantum size effect on UV–visible and fluorescence spectra. Additionally, a microwave-assisted solution-phase approach to produce radially aligned prismatic Sb_2S_3 whisker arrays was established. Broom-like arrays (Figure 9.50) were obtained from anhydrous $SbCl_3$ and thiourea as precursors in DMF under ambient air for 20 min under MW heating.[130]

CNTs were also obtained in the said form. Thus, a dual-porosity CNT material of seamlessly connected highly porous aligned nanotubes and lowly porous closely packed nanotubes was created[131] by using the capillary action of liquids. Millimeter-tall SWNT forests synthesized by water-assisted chemical vapor deposition were used as a model system. As a result, diverse structures using toothpicks, broom-like objects (Figure 9.51b and c), tepees (Figure 9.51f) (see also other "building-like" nanostructures in Figures 9.130 and 9.131), as well as liquid thin films, bubbles, vapors, and superink jet printing were obtained.

Bioorganic nanobrooms also exist; thus, highly bioactive alkylated peptide amphiphiles, capable of self-assembling into cylindrical nanofibers, were discovered as sequences for these molecules that can eliminate all curvature from the nanostructures. They formed in water and generated completely flat nanobelts with giant dimensions.[132] The sequences were found to have an alternating sequence with hydrophobic and hydrophilic side chains, and variations in monomer concentration generate a "broom" morphology (Figure 9.52) with twisted ribbons, which reveals the mechanism through which giant nanobelts form. The authors noted that, with proper functionalization, these nanostructures can offer a novel architecture to present epitopes to cells for therapeutic applications.

9.7 NANOBRUSHES

This nanostructural type, as well as some others reported here, is mainly represented by zinc oxide, prepared by distinct routes. Thus, ZnO nanobrushes on quartz substrate were prepared by a direct atmospheric evaporation method using Zn metal flakes using activated charcoal as a catalyst, which facilitated the formation of nanobrushes.[133] These aligned nanobrushes can find potential applications as nanopower generators and high aspect ratio AFM probes by virtue of the piezoelectric property of zinc oxide. ZnO nanobrushes and other morphologies, such as nanorods and, nanowires, were synthesized by a hydrothermal process without using any structure-directing reagent.[134] The results revealed that all the prepared nanostructured ZnO powders showed high response to ethanol, among which, the 3D nanobrushes showed the highest and fastest (less than 10 s) response (sensitivity), demonstrating excellent potential for ethanol sensors. The main advantages of brush-like hierarchical ZnO nanostructures (Figure 9.53 through 9.55) in ethanol sensing[135] were their excellent selectivity and low detection limit (with detectable ethanol concentration in ppm). It was noted that the response of brush-like hierarchical nanostructures was greater than that of initial ZnO nanowires.

Mixed and doped ZnO nanobrushes also possessed useful properties. Thus, ternary nanostructures with the composition of ZnO-Ga_2O_3 nanobrushes (Figure 9.56) (Ga_2O_3 as the core and ZnO as the branches of self-assembling symmetry in six equiangular directions around the core), among others, were found to be excellent materials for gas sensing applications due to their large surface areas and structural defects.[138,139] Their formation mechanism is shown in Figure 9.57. Vertically aligned Cu-doped ZnO nanonails and nanoneedles (Figure 9.58), obtained from Zn and $CuCl_2$ powders as source materials in the equipment shown in Figure 9.59, were observed as a result of continuous evolution between various morphologies, including nanobrushes (without $CuCl_2$ in the source, under the same experimental conditions, the pure ZnO samples always showed the commonly reported nanowire morphology).[140] It was shown that the offered chloride-based thermal evaporation method appeared to be a general approach to fabricate transition-metal-doped ZnO nanomaterials with novel morphologies. A controlled doping had extensive implications and affected the crystalline and electronic structures of ZnO. High-density hierarchical brush-like nanostructures

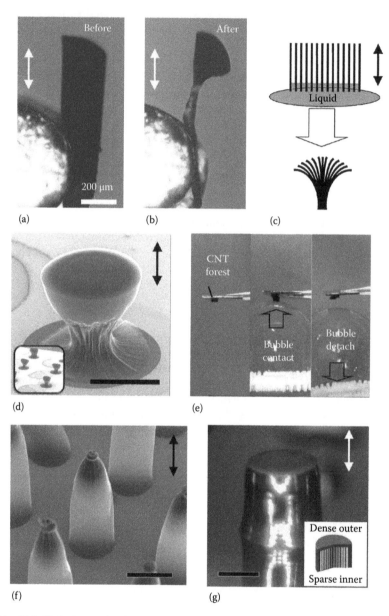

FIGURE 9.51 Hybrid dual-porosity SWNT material. (a, b) Before and after images of a high-porosity SWNT forest material (a) and hybrid dual-porosity material (b) made by partial collapse of the forest material. White lines indicate the tube alignment direction. (c) Model of the partial collapse process. (d) Mushroom brush structure created by bottom contact using thin liquid layer. Scale bar, 250 μm. Inset, array of mushrooms. (e) Photograph of bubble process. (f) Tepee structures created from top contact using bubble process. Scale bar: 500 μm. (g) Hard-coated SWNT structure created by complete liquid bubble envelopment. Inset, diagram of the hard-coated sparse SWNTs. Scale bar, 250 μm. (Reproduced with permission from Don, N.F. et al., Dual porosity single-walled carbon nanotube material, *Nano Lett.*, 9(9), 3302–3307. Copyright 2009 American Chemical Society.)

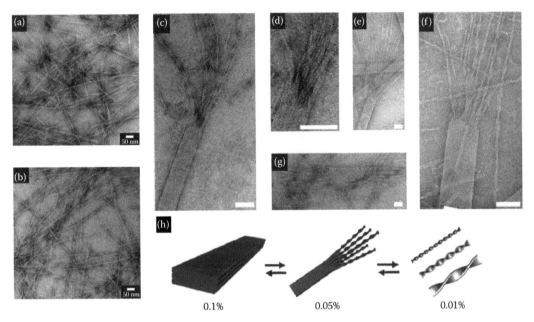

FIGURE 9.52 Twisted nanoribbons at 0.01 wt.% aqueous solution and intermediate structures (broom morphology) of nanobelts transforming into twisted nanoribbons at 0.05 wt.% solution. (a, b) Narrower nanobelts and twisted nanoribbons are observed at a concentration of 0.01 wt.%. The twist pitch increases with an increase in nanoribbon width. (c through f) Twisted nanoribbons sprouting from one nanobelt end. (d) A closer view of (c). There is a gradual transition from flat nanobelt to twisted nanoribbons. The longer the distance from the wide nanobelt, the more likely the nanoribbons will twist in their natural states. (g) Nanobelts split from both ends into narrower nanobelts. The split ribbons are too short to be twisted because they are too close to the wide nanobelt. (h) Schematic representation of the morphological transitions with a change in concentration. Scale bars of panels (c through g): 100 nm. All the TEM samples were negatively stained with 2% (w/v) uranyl acetate aqueous solution. (Reproduced with permission from Cui, H. et al., Self-assembly of giant peptide nanobelts. *Nano Lett.*, 9(3), 945–951. Copyright 2009 American Chemical Society.)

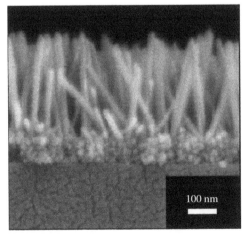

FIGURE 9.53 ZnO bulk nanocrystalline columns, or nanowires, grown in a brush conformation. (Reproduced from Coakley, K.M. et al., *MRS Bull.*, 30, 37, 2005. With permission.)

FIGURE 9.54 ZnO nanobrushes. (Reproduced from Singh, S., Electrical transport and optical studies of transition metal ion doped ZnO and synthesis of ZnO based nanostructures by chemical route, thermal evaporation and pulsed laser deposition, PhD thesis, 2009, Indian Institute of Technology Madras, Chennai, India.)

FIGURE 9.55 SEM image of the brush-like hierarchical ZnO nanostructure after sintering at 600°C for about 2 h (prepared from sequential nucleation and growth following a hydrothermal process). (Reproduced with permission from Zhang, Y. et al., Brush-like hierarchical ZnO nanostructures: Synthesis, photoluminescence and gas sensor properties, *J. Phys. Chem. C*, 113, 3430–3435. Copyright 2009 American Chemical Society.)

of Ce-doped ZnO, synthesized by a two-step approach (sol–gel and CVD),[141] were composed of two parts: a micron-sized prism-like base and nanosized vertical nanorod arrays. The nanobrush was shown to be enclosed by the (0001) and {0110} facets and grew toward the [0001] direction. Additionally, useful information on ZnO nanobrushes was presented in Ref. [142]. An interesting procedure, using amorphous carbon nanotube (a-CNT) brushes (Figure 9.60) (prepared hydrothermally in PC membranes) as the starting material, led to nanorod brushes of α-Al_2O_3 (Figure 9.61), MoO_3, and ZnO (Figure 9.62).[143] The brushes of α-Al_2O_3 and MoO_3 were shown to be made up of single crystalline nanorods; in the case of ZnO brushes, the nanorod bristles were made by the fusion of 15–25 nm size nanoparticles and were porous in nature. The authors stated that the presence of carboxylic and phenolic functional groups presented on the surface of a-CNT helped to hydrolyze the metal precursors within the voids of the membrane that would subsequently form a polymer network (Figure 9.63). Further calcinations at high temperatures, coupled with the exothermic heat generated by the combustion of carbon, facilitated the formation of crystalline nanorods of metal oxides.

Hierarchically organized micro/nanostructures of TiO_2, suitable for photocatalytic applications and consisting of small TiO_2 nanotubes that form brush-type shells around long central oxide cores, were fabricated by means of anodic oxidation of Ti nanorods prepared by *glancing angle deposition*.[144] The same compound was used as a brush-type support for metal particles. Thus, nanocomposites

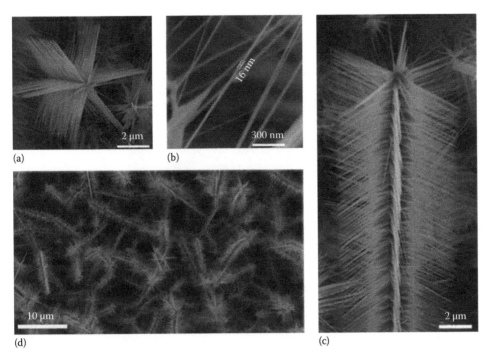

(a)   (b)   (d)   (c)

FIGURE 9.56 SEM micrograph of ZnO-Ga$_2$O$_3$ nanobrushes with long needles. (a) Sixfold orientation of needles growing in six equiangular directions around the core. (b) High-magnification image of needles showing the size in the middle of the needle. (c) Parallel growth of needles along the length of the NB. (d) Low-magnification SEM image of the ZnO-Ga$_2$O$_3$ nanobrushes to show the abundance. (Reproduced with permission from Mazeina, L. et al., Controlled growth of parallel oriented ZnO nanostructural arrays on Ga$_2$O$_3$ nanowires. *Cryst. Growth Des.*, 9(2), 1164–1169. Copyright 2009 American Chemical Society.)

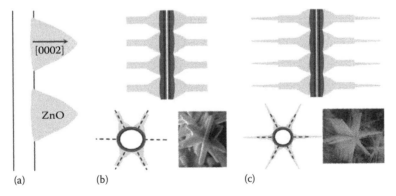

(a)   (b)   (c)

FIGURE 9.57 Schematic illustrating the formation of ZnO-Ga$_2$O$_3$ nanostructures. (a) ZnO bristles nucleates on Ga$_2$O$_3$ core in [0001] direction. (b) Several rows of ZnO bristles form around the core to produce a fourfold nanobrush with twofold symmetry. ZnGa$_2$O$_4$ is considered to be already present at this stage. (c) Additional rows of bristles grow in between previously formed four; bristles elongate forming a nanobrush with long needles. (Reproduced with permission from Mazeina, L. et al., Controlled growth of parallel oriented ZnO nanostructural arrays on Ga$_2$O$_3$ nanowires. *Cryst. Growth Des.*, 9(2), 1164–1169. Copyright 2009 American Chemical Society.)

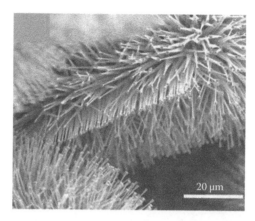

FIGURE 9.58 Cu-doped ZnO nanobrushes with a high density of side branches. (Reproduced with permission from Zhang, Z. et al., Cu-doped ZnO nanoneedles and nanonails: Morphological evolution and physical properties, *J. Phys. Chem. C*, 112(26), 9579–9585. Copyright 2008 American Chemical Society.)

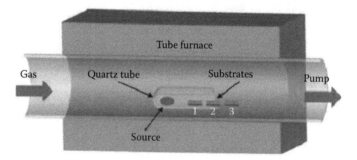

FIGURE 9.59 A schematic drawing of the experimental setup to obtain Cu-doped ZnO nanostructures. (Reproduced with permission from Zhang, Z. et al., Cu-doped ZnO nanoneedles and nanonails: Morphological evolution and physical properties, *J. Phys. Chem. C*, 112(26), 9579–9585. Copyright 2008 American Chemical Society.)

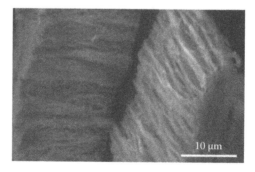

FIGURE 9.60 FESEM image of a-CNTs brushes. (Reproduced from Raidongia, K. and Eswaramoorthy, M., *Bull. Mater. Sci.*, 31(1), 87, 2008. Indian Academy of Sciences.)

consisting of 10 nm Au, Pt, Pd nanoparticles and nanoparticles of TiO_2 (anatase) were synthesized (Figure 9.64) by reduction of the metal ions adsorbed on the surface of as-prepared TiO_2 nanoparticles, which are immobilized on spherical polyelectrolyte brush particles [SPB, consisting of a PS core from which long chains of poly(styrene sodium sulfonate) are grafted] as a carrier system.[145] These metal $NP/TiO_2@SPB$ composite particles exhibited a high colloidal stability and were found to be excellent heterogeneous photocatalysts for the degradation of the dye Rhodamine B under UV irradiation. Additionally, α-MnO_2 brush-like nanostructure was described.[146]

FIGURE 9.61 FESEM image of α-Al₂O₃ nanorod brushes. (Reproduced from Raidongia, K. and Eswaramoorthy, M., *Bull. Mater. Sci.*, 31(1), 87, 2008. Indian Academy of Sciences.)

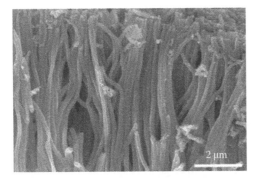

FIGURE 9.62 FESEM image of ZnO nanorod brushes. (Reproduced from Raidongia, K. and Eswaramoorthy, M., *Bull. Mater. Sci.*, 31(1), 87, 2008. Indian Academy of Sciences.)

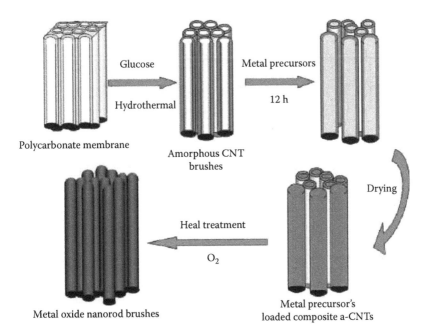

FIGURE 9.63 Synthesis of metal oxide nanorod brushes. (Reproduced from Raidongia, K. and Eswaramoorthy, M., *Bull. Mater. Sci.*, 31(1), 87, 2008. Indian Academy of Sciences.)

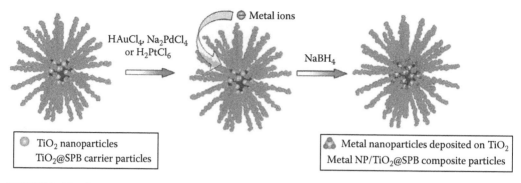

FIGURE 9.64 Synthesis of metal NP/TiO$_2$@SPB composite particles. (Reproduced with permission from Lu, Y. et al., Composites of metal nanoparticles and TiO$_2$ immobilized in spherical polyelectrolyte brushes, *Langmuir*, 26(6), 4176–4183. Copyright 2010 American Chemical Society.)

Some metallic brush-like nanostructures have been reported, mainly those of gold and generally fabricated by an anodic alumina template method. Thus, Au nanobrush membranes were prepared using a modified template method and were immobilized by poly(3,4-ethylenedioxythiophene) (PEDOT)[147] or polypyrrole (PPy)[148] films by an electropolymerization. It was established that the PEDOT film on an as-prepared working nanobrush electrode showed higher electrochromic coloration compared with the PEDOT film on an Au planar electrode. In the case of the PPy film, this film on the nanobrush membrane electrode showed higher electrochemical stability under continuous cyclic polarization. Gold nanowire arrays and gold nanobrushes were also prepared[149] by electrodeposition of gold into an alumina template coated with gold film and then chemical etching alumina directly or rubbing gold film and chemical etching alumina. In the gold nanobrushes, local nanowires collect in bundles after complete dissolution of alumina matrix due to the transverse tension during drying in the air. An Ag nanobrush material, also prepared[150] by the alumina template method, was used as the working electrode in an Li/Ag nanobrush cell to study the agglomeration phenomena of nanosized alloy particles as anode material in Li ion batteries. Heterogeneous magnetic nanobrushes, consisting of Co nanowire arrays and ferromagnetic Fe$_{70}$Co$_{30}$ nanofilm, were fabricated (Figure 9.65) using an AAO template method combined with sputtering technology.[151] It was established that the

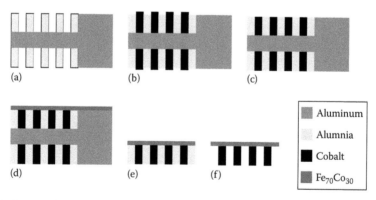

FIGURE 9.65 Preparation scheme of magnetic nanobrush: (a) The anodic aluminum oxide (AAO) template covered with Al was fabricated by anodization of Al film in an acidic solution; (b) Cobalt nanowires were electrodeposited into the nanoporous channels; (c) The surface of film was smoothed by dilute nitric acid solution; (d) The surface of film was covered with a Fe$_{70}$Co$_{30}$ layer by the sputtering technique; (e) Al substrate and another cobalt nanowire arrays were removed by HgCl$_2$ solution; (f) After eroding the AAO template using NaOH solution, a nanobrush was obtained. (Reproduced with kind permission from Springer Science+Business Media: *Nanoscale Res. Lett.*, Tunable magnetic properties of heterogeneous nanobrush: From nanowire to nanofilm, 5, 2010, 853–858, Ren, Y. et al.)

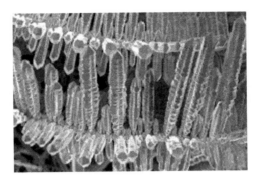

FIGURE 9.66 Lead nanobrushes, obtained at the reduction potential −1.9 V for solution containing 5 mM lead acetate and 0.1 M boric acid. The scale bar length is 500 nm. The growth time is 60 s. (Reproduced with permission from Xiao, Z.-L. et al., Tuning the architecture of mesostructures by electrodeposition, *J. Am. Chem. Soc.*, 126(8), 2316–2317. Copyright 2004 American Chemical Society.)

magnetic anisotropy of nanobrush depended on the thickness of $Fe_{70}Co_{30}$ layer, and its total anisotropy originated from the competition between the shape anisotropy of nanowire arrays and nanofilm. At last, a grand variety of lead mesostructures, including nanobrushes (Figure 9.66), were synthesized through systematically exploring electrodeposition parameters.[152]

In addition to metallic nanobrushes, a multifunctional nanobrush (attributed to nanodevices) using CNTs was described.[153] In a related report, super long (7 mm) aligned, mainly double-walled, high-density CNTs (Figure 9.67) were synthesized from ethylene, used as the source gas of carbon, with high reproducibility using CVD technique (typical CVD growth was carried out at 750°C for 30 min as the standard growth time).[154] Thin layers of Fe were used as the catalyst, and an Al_2O_3 buffer layer was used between the Si/SiO_2 substrate and the Fe catalyst. The relative levels of ethylene and water, as well as ethylene and H_2 during the CVD, were found to be the most important factor for the growth of such super long aligned CNTs. In addition, amorphous CNT brushes were prepared using glucose as the carbon precursor.[155] The functional surfaces of these nanotubes were covered with gallium ions (Figure 9.68) and then calcined to get gallium oxide nanotube brushes, which were further converted to crystalline GaN nanotube brushes (Figure 9.69) by treatment with ammonia at 800°C (there is additional related information on GaN[156] (Figure 9.70), InN,[157] and AlN[158] nanobrushes).

Calcogenides and other salts are very poorly represented by brush-like nanostructures; here we note ZnSe nanobrushes[159] and CdS nanostructures[160] with different morphologies, varying from multipods to nanobrushes to nanocups, and sizes, which were fabricated through an effective C-assisted thermal evaporation method. CdS nanobrushes exhibited a predominant red emission band peaking at ~711 nm, explained by the influence of carbon on the properties of CdS nanostructures. Strontium stannate ($SrSnO_3$) nanostructures (nanosticks and nanobrushes) were obtained by microwave-assisted calcination of a $SrSn(OH)_6$ precursor powder.[161] Compared with other conventional calcination methods mentioned in the literature, this procedure led to a remarkable decrease in the reaction time and the synthesis temperature owing to direct interaction of radiation with the material.

Organic and coordination brush-like compounds are represented by polymers and phthalocyanines, respectively; frequently, hybrid materials and composites on their basis were created. Thus, iron phthalocyanine (FePc) films of different thicknesses were deposited by molecular beam epitaxy (MBE) as a function of substrate temperature (25°C–300°C) and deposition rate (0.02–0.07 nm/s), yielding the morphology of a 60 nm alpha-phase film tuning from nanobrush (nearly parallel nanorods aligned normal to the substrate plane) to nanoweb (nanowires forming a web-like structure in the plane of the substrate) (see the Section 9.13 on web-like nanostructures later) by changing the deposition rate from 0.02 to 0.07 nm/s.[162,163] Biodegradable inorganic/organic hybrid materials were prepared by grafting poly(-lactide) (PLLA) from porous silicon (pSi) films and microparticles using tin(II) 2-ethylhexanoate catalyzed ring opening polymerization with pSi surface-bound

FIGURE 9.67 Images of super long aligned brush-like CNTs: (a) digital photographic image of 7 mm long aligned CNTs, (b) an SEM image of the top surface of the 7 mm long aligned CNTs, (c) a low-magnification SEM image of the upper part of the aligned nanotubes, (d) a high-magnification SEM image of the upper part, (e) a low-magnification SEM image of the lower part of the aligned nanotubes, and (f) a high-magnification SEM image of the nanotubes. (Reproduced from Chakrabarti, S. et al., *Jpn. J. Appl. Phys.*, 45(28), L720, 2006.)

hydroxyl groups as initiators.[164] The formation of PLLA nanobrushes on the pSi surface was confirmed. These biodegradable hybrid materials can find uses in tissue engineering and drug delivery, for example, in applications where complex degradation profiles are required, which cannot be achieved with one type of material alone. The reaction between two nonconducting chemicals, aniline and silver nitrate, in solutions of acetic acid yielded a composite of two conducting components, PANI and metallic silver, combining the electric properties of metals and the materials properties of polymers.[165] PANI was found to be present as nanotubes or nanobrushes composed of thin nanowires. Polymethylmethacrylates are represented by poly(methylmethacrylate) (PMMA) brush arrays, formed on the functionalized silicon surface[166] and nanopatterned brushes of thermoresponsive poly(2-(2-methoxyethoxy)ethyl methacrylate) (PMEO$_2$MA), displaying collapse temperature in the physiological range, synthesized (Figure 9.71) with grafting diameter from a few micrometers down to 35 nm.[167,168] Also, biocompatible and biodegradable nanobrushes as composite ingredients and reactive precursors for reinforced hydrogels were prepared from cellulose and HAP nanoparticles *via* ring-opening polymerization of dl-lactide and stannous octoate as the catalyst.[169] More information on polymer brush-like nanostructures is available in Refs. [170,171].

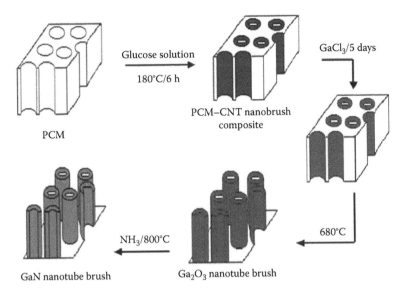

PCM

Glucose solution
180°C/6 h

PCM–CNT nanobrush
composite

GaCl₃/5 days

680°C

NH₃/800°C

GaN nanotube brush

Ga₂O₃ nanotube brush

FIGURE 9.68 Schematic showing the formation of nanotube brushes of carbon, Ga₂O₃, and GaN. (Reproduced with permission from Dinesh, J. et al., Use of amorphous carbon nanotube brushes as templates to fabricate GaN nanotube brushes and related materials, *J. Phys. Chem. C*, 111(2), 510–513. Copyright 2007 American Chemical Society.)

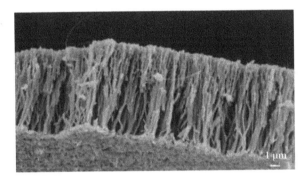

FIGURE 9.69 GaN nanotube brush. (Reproduced with permission from Dinesh, J. et al., Use of amorphous carbon nanotube brushes as templates to fabricate GaN nanotube brushes and related materials, *J. Phys. Chem. C*, 111(2), 510–513. Copyright 2007 American Chemical Society.)

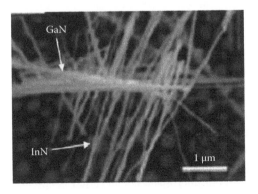

FIGURE 9.70 FESEM image of GaN/InN nanobrushes. (Mieszawska, A.J. et al.: The synthesis and fabrication of one-dimensional nanoscale heterojunctions. *Small*. 2007. 3(5). 722–756. Copyright Wiley-VCH Verlag GmbH & Co. KGaA. Reproduced with permission.)

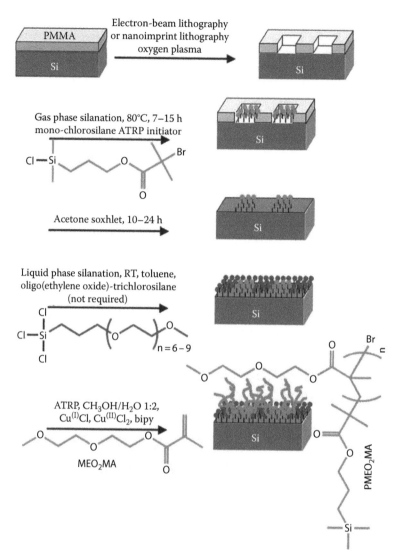

FIGURE 9.71 Schematic description of the fabrication of nanopatterned PMEO$_2$MA brushes. (Reproduced with permission from Jonas, A.M. et al., Effect of nanoconfinement on the collapse transition of responsive polymer brushes, *Nano Lett.*, 8(11), 3819–3824. Copyright 2008 American Chemical Society.)

9.8 NANOCARPETS

Relatively rare carpet-like nanostructures, related to nanobrushes, are formed mainly by nanotubes, nanowires or nanorods, belong to different types of compounds, and possess special properties (in particular, superhigh hydrophobicity) in comparison with a majority of other less-common nanostructures. For slender nanostructures, the nanostructure may change after coming into contact with liquid (this phenomenon is known as the "nanocarpet effect"[172–174]). The newly formed nanostructures may have a great impact on the hydrophobicity of a surface. The nanocarpet effect is a simple and effective method to apply to a variety of nanowire, nanotube, or nanorod structures to generate a hierarchical structure with both microscale domains and nanoscale features (nanorod tips) for superhydrophobicity and other surface property studies. As an example, by coating a fluorocarbon monolayer on a bundled Si nanorod array substrate, a superhydrophobic surface with a contact angle ~167° and sliding angle ~2° was created due to the nanocarpet effect.[175] It was noted that without

FIGURE 9.72 Nickel carpet-like structure. (Neto, C. et al., On the superhydrophobic properties of nickel nanocarpets, *Phys. Chem. Chem. Phys.*, 11(41), 9537–9544, 2009. Reproduced by permission of The Royal Society of Chemistry.)

forming the nanocarpet, only a moderately hydrophobic surface with contact angle <151° and sliding angle >17° can be obtained. The main reason for the superhydrophobicity was found to be the formation of sharp pyramidal bundles, which effectively reduced the area of solid–liquid contact. A surface formed by dense, aligned nickel nanowires in a nanocarpet form (Figure 9.72) was prepared by electrodeposition through an alumina membrane template followed by dissolution of the membrane.[176] The nickel nanowires were found to be highly rigid, perpendicularly aligned in the nanocarpet with respect to the substrate, and they touch each other at the tips, forming microscale "tepee"-shaped aggregates. Being coated with a hydrophobic surfactant (stearic acid), this nanocarpet got superhydrophobic properties (advancing contact angle ~158°) and retained its superhydrophobicity after periods of immersion in water.

CNT carpet-like nanostructures have been obtained mainly by CVD. Thus, massive carpets of well-packed, vertically aligned, and very long multiwalled carbon nanotubes (MWCNTs) were synthesized by this classic technique.[177] Ferrocene was frequently used as a precursor in such syntheses, for instance in a CVD-assisted growth of MWCNTs (Figure 9.73) in 60:40 toluene/1,2-diazine at 760°C in argon[178] or in the pyrolysis of aerosols obtained from C_6H_6/NiCp$_2$/FeCp$_2$ mixtures generating aligned Invar (Fe-Ni alloy, $Fe_{65}Ni_{35}$)-filled CNT carpet-like samples of high purity.[179] Radio-frequency hot-filament CVD (RFHFCVD), where the reactant gases (methane and hydrogen gas used as the carbon source and diluted gas, respectively) were first dissociated by the MW field before entering the chamber, thus, ensuring that there was a dense growth of CNTs from a large

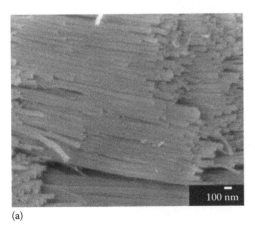
(a)

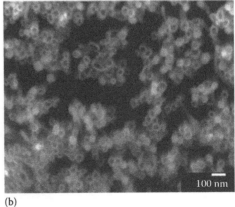
(b)

FIGURE 9.73 MWCNT carpet: (a) side view and (b) view looking down on the top of the carpet. (Krzysztof, K. et al.: Three-dimensional internal order in multiwall carbon nanotubes grown by chemical vapor deposition. *Adv. Mater.* 2005, 17(6), 760–763. Copyright Wiley-VCH Verlag GmbH & Co. KGaA. Reproduced with permission.)

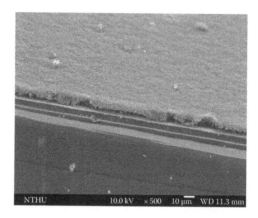

FIGURE 9.74   Carpet-like CNT films. (Chen, S.Y. et al., *J. Phys. D Appl. Phys.*, 37, 273, 2004. Reproduced with permission from Institute of Physics Publishing.)

Ni catalyst particle, was proposed for the fabrication of CNT films and carpet-like nanostructures (Figure 9.74).[180] The growth mechanisms were found to be either "tip growth" or "base growth" depending on the size of the catalyst metal particles involved. It was concluded that the oxide iron group alloys such as Ni/NiO_x, $CuNi/CuNiO_x$, and $AgNi/AgNiO_x$ were essential to yield successful growth of CNTs. Additionally, CVD method for anchoring densely packed, vertically aligned arrays of CNTs (Figure 9.75) into silicone layers using spin coating, CNT insertion, curing, and growth substrate removal was proposed.[181] CNT arrays of 51 and 120 μm height were anchored into silicone layers of thickness 26 and 36 μm, respectively. Alternatively to CVD techniques, the pulse electrodeposition technique was utilized to deposit nanosized (≤10 nm) Ni catalysts on carbon

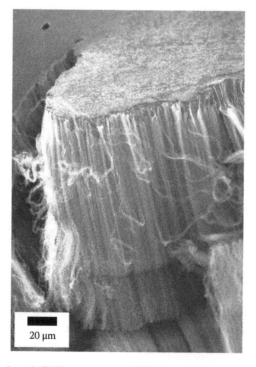

FIGURE 9.75   Double-anchored CNT nanocarpet. RTV (curable elastomeric material (RTV615, GE Silicones) is not present in the center of the nanocarpet, only at the top and bottom. (Sansom, E.B. et al., *Nanotechnology*, 19, 035302, 6 pp., 2008. Reproduced from Institute of Physics Publishing.)

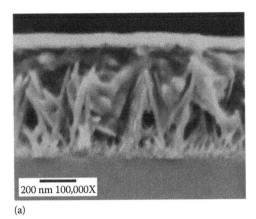
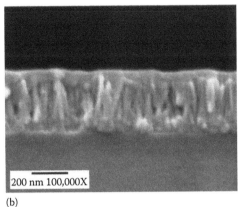

(a)   (b)

FIGURE 9.76   (a) Cross-sectional SEM image of ZnO fiber/P3HT device without complete intercalation of polymer. (b) SEM image of P3HT completely intercalated into the nanocarpet structure treated with phenyltrichlorosilane. (Reproduced with permission from Olson, D.C. et al., The effect of atmosphere and ZnO morphology on the performance of hybrid poly(3-hexylthiophene)/ZnO nanofiber photovoltaic devices, *J. Phys. Chem. C*, 111, 16670–16678. Copyright 2007 American Chemical Society.)

fabric (CF)[182] allowing the growth of carbon nanofibers (CNFs) on CF. The preparation of CNF-coated CF (carpet-like CF) was carried out in a thermal CVD system with an optimum loading of Ni catalysts deposited in the PED pulse range from 20 to 320 cycles. CNFs that grew at 813 K had a hydrophobic surface, like lotus leaves.

Inorganic oxides and salts as nanocarpets are represented very poorly, in contrast with a majority of classic and less-common nanostructures. Thus, photovoltaic devices consisting of poly(3-hexylthiophene) (P3HT) intercalated into a carpet-like mesoporous structure of ZnO nanofibers (grown *via* a low-temperature hydrothermal route from a solution of zinc nitrate precursor) (Figure 9.76), were fabricated.[183] Silicon-deficient mullite ($Al_{5.65}Si_{0.35}O_{9.175}$) single crystal nanowires (50–100 nm), forming a carpet-like structure (Figure 9.77), were synthesized[184] in large quantities on mica substrates assisted by the intermediate fluoride species. The supposed reactions are shown by reactions 9.17 through 9.21. The nanowires had strong photoluminescence emission bands at 310, 397, 452, and 468 nm.

$$2NH_3(g) \rightarrow N_2(g) + 3H_2(g) \tag{9.17}$$

$$H_2(g) + O_2(g)(residual) \rightarrow H_2O(g) \tag{9.18}$$

$$6NaF + Al_2O_3(s) + H_2O(g) \rightarrow AlF_3(g) + AlOF(g) + 3Na_2O(s) + 2HF(g) \tag{9.19}$$

$$HF(g) + mica(Al–Si–O) \rightarrow AlF_3(g) + AlOF(g) + SiF_4(g) + H_2O(g) \tag{9.20}$$

$$AlF_3(g) + 4.65AlOF(g) + 0.35SiF_4(g) + 4.525H_2O(g)$$

$$\rightarrow Al_{5.65}Si_{0.35}O_{9.175} \text{ (nanowires)} + 9.05HF(g) \tag{9.21}$$

Two-step synthesis of nanotubes of organic compounds that form pure, well-defined, nanostructured materials was developed using the strategy shown in Figure 9.78[185] using hydrazide derivatives of single-chain diacetylene lipids. In a family of synthesized amine salt derivatives **1–7** (seven individual amine, amine salt, and quaternary ammonium salt derivatives of diacetylene), the only precursor that was found to have the potential to form nanotubes (in a single step and with 100% yield) was the secondary amine salt (compound **3**). The obtained nanocarpets made of nanotubes

(a)   (b)

(c)   (d)

FIGURE 9.77 (a) SEM images of the $Al_{5.65}Si_{0.35}O_{9.175}$ carpet-like nanowire film grown on the mica substrate. (b) Enlarged view of the nanowires. The inset shows the facet tip, smooth surface of the nanowires with a diameter of approximately 50–100 nm. (c, d) SEM images of the aligned nanowires grown at the edge of the substrate, exhibiting similar diameters and lengths as those grown in the center of the substrate. (Chen, Y. et al., Fluoride-assisted synthesis of mullite ($Al5.65Si0.35O9.175$) nanowires, *Chem. Commun.*, 2780–2782, 2006. Reproduced by permission of The Royal Society of Chemistry.)

(Figure 9.79) possessed antimicrobial properties. The versatility of these materials as a platform for further nanostructure design and synthesis is enhanced by its biocidal activity.

9.9   NANOCOMBS

Comb-like nanostructures can be considered as rare nanoforms for inorganic compounds except ZnO (see the following), for which a host of reports have been published in the last decade. We note that this compound is known in a variety of other nanostructures. At the same time, other nanocombs are represented by gold, several oxides, nitrides, sulfides, silicides, and carbides, as well as some coordination compounds and polymers. Thus, tungsten oxide comb-like nanostructures were synthesized using a two-step thermal evaporation method.[186] The teeth of the comb structure were found to be well aligned and vertical to the side surfaces of the cores. A variety of cubic-phase indium oxide (In_2O_3) based submicron/nanostructures including arrow-like, bead-like, bouquet, comb-like wires, and regular polyhedra (octahedrons and triangular slab-like structures) were synthesized from In grains as precursors (Reactions 9.22 through 9.25) at different zones of a Si substrate using an improved CVD system.[187] The concentration gradient of In_2O_3 vapor at different regions of the Si substrate and the growth kinetics of different In_2O_3 crystalline facets were considered as main factors for the formation of different structures in a propositional mechanism. Anatase-type TiO_2 single nanocrystals with comb-like morphology, among other types (boat-like, sheet-like, leaf-like, quadrate, rhombic, and wire-like) of particle morphologies, were prepared[188]

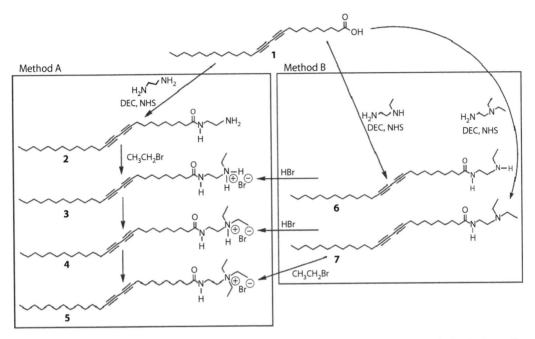

FIGURE 9.78 Strategy of the two-step synthesis of organic nanotubes. DEC, 1-[3-(dimethylamino)propyl]-3-ethylcarbodiimide; NHS, *N*-hydroxysuccinimide. (Reproduced with permission from Lee, S.B. et al., Self-assembly of biocidal nanotubes from a single-chain diacetylene amine salt, *J. Am. Chem. Soc.*, 126(41), 13400–13405. Copyright 2004 American Chemical Society.)

FIGURE 9.79 Nanocarpet (different views) of the compound **3**. (Reproduced with permission from Lee, S.B. et al., Self-assembly of biocidal nanotubes from a single-chain diacetylene amine salt, *J. Am. Chem. Soc.*, 126(41), 13400–13405. Copyright 2004 American Chemical Society.)

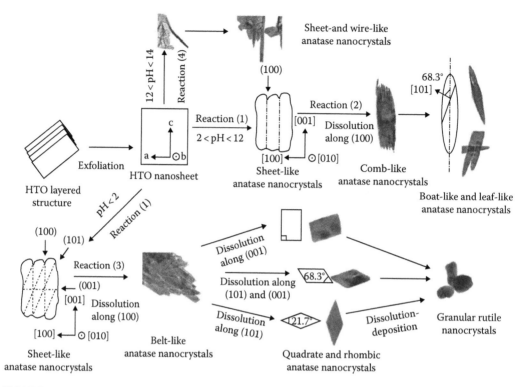

FIGURE 9.80 Scheme of the reaction mechanism for the formation of TiO$_2$ nanocrystals from titanate nanosheets under hydrothermal reaction conditions. (Reproduced with permission from Wen, P. et al., Single nanocrystals of anatase-type TiO$_2$ prepared from layered titanate nanosheets: Formation mechanism and characterization of surface properties, *Langmuir*, 23, 11782–11790. Copyright 2007 American Chemical Society.)

by hydrothermal treatment of a layered titanate (K$_{0.8}$Ti$_{1.73}$Li$_{0.27}$O$_4$) nanosheet colloidal solution. Reaction mechanism for various TiO$_2$ nanostructures is shown in Figure 9.80:

$$\text{In} \rightarrow \text{Nanosized In} \tag{9.22}$$

$$\text{Nanosized In} \rightarrow \text{In}_{(vapor)} \tag{9.23}$$

$$\text{In}_{(v)} + O_2 = 2In_2O_{(v)} \quad \text{or} \quad In_{(v)} + O_2 = 2InO_{(v)} \tag{9.24}$$

$$4InO_{(v)} + 2O_2 = In_2O_3 \quad \text{or} \quad In_2O_{(v)} + O_2 = In_2O_3 \tag{9.25}$$

The AlN branches of tree-shaped nanostructures, synthesized from aluminum and NH$_3$ as precursors *via* a one-step improved DC arc discharge plasma method without any catalyst and template, grew in a sequence of nanowires, nanomultipeds, and nanocombs.[189] The double-sided aluminum nitride nanocomb had a double-sided comb-like structure formed by nanowire array, wherein the nanowire with a diameter of 70–100 nm and length of 1–2 μm was formed by hexagonal wurtzite crystal structured aluminum nitride.[190] The product can be used in laser interference/coupling, nanolaser array, and NEMS. GaN nanocombs, together with nanorods, were grown on Si(111) substrates by plasma-assisted MBE[191] In another work, a regular self-organized 2D nanocomb structure was created *in situ* in an ultrahigh vacuum on the n-GaN(0001) surface as a result of multilayer adsorptions of Cs and Ba.[192] The structure was found to be highly regular in the microrange, arranged as combs 60–70 nm in diameter with a wall height of ~7 nm. A self-organization model was proposed, which implies the formation of a surface 2D long period incommensurate phase interacting with the superstructure of the Cs$^+$ and Ba^{2+} ion clusters with allowance for the polaron compensation on the GaN surface.

The silicon nitride nanocomb containing Si_3N_4 nanowires was prepared pressing Si, C, and silica powders by a discharging reaction at 10–30 kPa for 5–10 min.[193] The product can have applications in quantum component, photoluminescence, nanoelectromechanical systems, etc. SiC nanocombs, among a variety of other nanostructures, were obtained by solution method.[194] It was demonstrated that these SiC nanoarchitectures all have a face-centered cubic structure. The study on these nano-architectures may be helpful in the further research toward controllable formation of nanostructures and in finding potential applications for nanodevices. In addition to FeSi nanowires, obtained *via* chemical vapor transport using $FeSi_2$ as the source material and I_2 as the transport agent, other morphologies including nanocombs, nanoflowers, and micron-sized crystals were also observed during the synthesis at various temperature zones of the growth substrates.[195] Sulfides are mainly represented by ZnS- and CdS-containing nanostructures. Thus, comb nanostructures, among belt-, saw, and windmill-like ones, of wurtzite-structured ZnS were prepared by a simple catalyst-free thermal evaporation technique,[196] whereas $Cd_xZn_{1-x}S$ nanostructures were prepared by one-step MOCVD.[197] CdS arrays, nanowires, and nanocombs were selectively prepared through an atmospheric pressure CVD (APCVD) process with $CdCl_2$ and S as sources.[198] The morphologies could be controlled by adjusting the deposition position, the temperature, and the flux of the carrier gas. Air-stable solid molecular precursor $Cd[(SeP(i\text{-}Pr)_2)N]_2$ was used to synthesize CdSe comb-like (Figure 9.81) and other nanostructures by two approaches (Figure 9.82).[199] These results might enable a novel nanoscale field effect phototransistor operation scheme enabled by CdSe tripod nanostructures.

FIGURE 9.81   SEM image of CdSe nanocomb structure. Scale bar, 1 µm. (Reproduced with permission from Zhang, Y. et al., Catalytic and catalyst-free synthesis of CdSe nanostructures with single-source molecular precursor and related device application, *Nano Lett.*, 9(1), 437–441. Copyright 2009 American Chemical Society.)

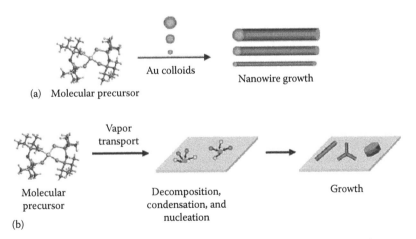

FIGURE 9.82   Schematics of two synthetic approaches of CdSe nanostructures *via* single-source molecular precursor. (a) Catalyst-assisted VLS-based process; (b) catalyst-free chemical vapor transport and condensation (CVTC) based process. (Reproduced with permission from Zhang, Y. et al., Catalytic and catalyst-free synthesis of CdSe nanostructures with single-source molecular precursor and related device application, *Nano Lett.*, 9(1), 437–441. Copyright 2009 American Chemical Society.)

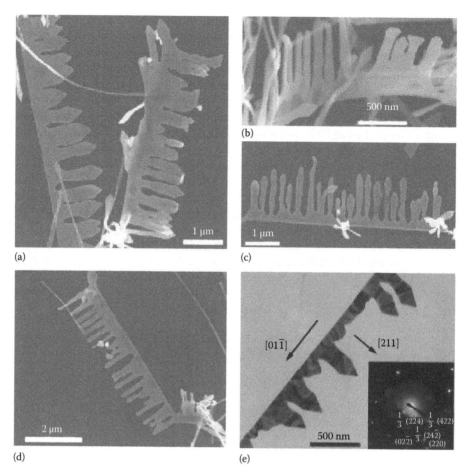

FIGURE 9.83 SEM (a through d) and TEM (e) images of gold nanocombs obtained by a two-step process with temperature changing from 4°C to 27°C. (Inset) ED pattern corresponding to the whole region of (e). (Reproduced with permission from Zhao, N. et al., Controlled synthesis of gold nanobelts and nanocombs in aqueous mixed surfactant solutions, *Langmuir*, 24(3), 991–998. Copyright 2008 American Chemical Society.)

Elemental nanocombs are known for gold only. Thus, well-defined gold nanobelts and unique gold nanocombs (Figure 9.83) made of nanobelts were readily synthesized by the reduction of $HAuCl_4$ with AA in aqueous mixed solutions of the cationic surfactant CTAB and the anionic surfactant sodium dodecylsulfonate (SDSn).[200] Their formation mechanism is shown in Figure 9.84. Single-crystalline gold nanocombs consisting of a <110>-oriented stem nanobelt and numerous <211>-oriented nanobelts grown perpendicularly on one side of the stem were fabricated by a two-step process with the temperature changing from 4°C to 27°C. The SERS from Au nanocombs and nanorods under different excitation conditions was studied,[201] showing that the SERS intensity from nanocombs is always larger than that from nanorods, but the polarized SERS dependence is similar for the two nanostructures. These results agreed quantitatively well with the local E-field calculations, and the nanospine in the nanocomb increases the local E-field over all surfaces of the nanocomb structure.

The $HgCl_2$-mediated interfacial self-assembly of coordination polymer nanocombs composed of (Zn, Pd) tetrapyridylporphyrin and 4,4′-bipyridyl ligands was reported.[202] It was shown that lengths and widths of the core stems of the nanocombs were ~10 and 1 μm, respectively. It was confirmed that, when the 4,4′-bipyridyl ligand was replaced by its derivatives, 2,2′-bipyridyl or 4,4′-trimethylenedipyridine, coordination polymer nanocombs could not be obtained. Therefore, the formation of such nanocombs was closely dependent on the structure of the bipyridyl, the mixed molar ratios of porphyrin

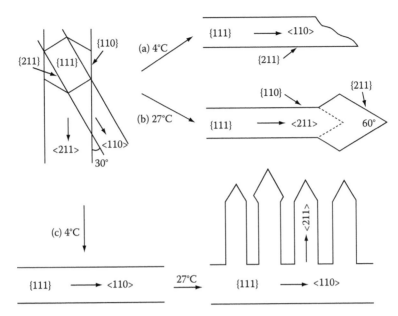

FIGURE 9.84 Schematic illustration of the formation mechanism of gold nanobelts and nanocombs. In the upper left corner is presented a projection perpendicular to the top surface of the {111} plane including a hexagonal prism to show the relative angles between different growth directions and side faces of the two kinds of nanobelts. (Reproduced with permission from Zhao, N. et al., Controlled synthesis of gold nanobelts and nanocombs in aqueous mixed surfactant solutions, *Langmuir*, 24(3), 991–998. Copyright 2008 American Chemical Society.)

and 4,4′-bipyridyl, and the interfacial reaction time. Also, the synthesis, spectroscopic characterization, and antimicrobial efficiency of gold and silver nanoparticles (prepared by reducing solutions of the salts of silver or gold and the copolymer in THF) embedded in novel amphiphilic comb-type graft copolymers with good film-forming properties were described.[203] The formed amphiphilic comb-type graft copolymers, synthesized by the reaction of chlorinated polypropylene (PP) (Mw = 140,000 Da) with polyethylene glycol (PEG) (Mn = 2000 Da) at different molar ratios, were found to be highly antimicrobial by virtue of their antiseptic properties to *Escherichia coli* and *Staphylococcus aureus*.

Zinc oxide, reviewed in Ref. [204], an important semiconducting and piezoelectric material, has three key characteristics. First, it is a semiconductor, with a direct band gap of 3.37 eV and a large excitation binding energy (60 meV), exhibiting near-UV emission. Second, due to its noncentrosymmetrical symmetry, it is piezoelectric, which is a key phenomenon in building electro-mechanical coupled sensors and transducers. Finally, ZnO is biosafe and biocompatible and can be used for biomedical applications without coating. With these unique advantages, ZnO is one of the most important nanomaterials for integration with microsystems and biotechnology. The authors suggested that ZnO could be the next most important nanomaterial after CNTs. Therefore, it is not surprising that its nanocombs, as well as a series of other nanostructures, are the object of permanent interest. Thus, unique ZnO nanocombs with *cuboid nanobranches* are synthesized by vaporizing zinc and graphite powders at 700°C on anodic aluminum oxide (AAO) template substrates (Figure 9.85).[205] Their formation mechanism is shown in Figure 9.86. Nanocombs of ZnO with *double-sided teeth* (Figure 9.87) were observed.[206] The data showed that the Zn-terminated (0001) surface was responsible for the formation of the teeth, whereas the oxygen-terminated (0001⁻) surface is chemically inactive and does not grow teeth. Nanocombs with a *disc cap* structure of ZnO (Figure 9.88) were synthesized[207] on Si substrates by using pure zinc powders as the source materials based on a vapor-phase transport process. Single crystalline Fe-doped ZnO *nanocantilever* arrays (Figure 9.89), closely related to nanocombs, were synthesized by thermal evaporating amorphous Zn-Fe-C-O composite powder.[208] *Ultralong* ZnO nanocombs were synthesized on silicon substrates with

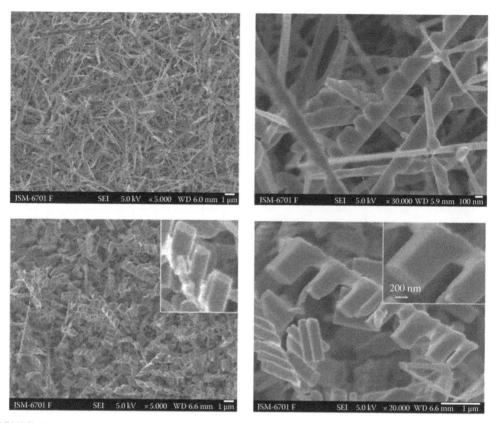

FIGURE 9.85 ZnO nanocombs with cuboid nanobranches. (Zhuo, R.F. et al., *J. Phys. D Appl. Phys.*, 41, 185405, 13 pp., 2008. Reproduced with permission from Institute of Physics Publishing.)

a high growth rate of ~7 μm/s using a thermal evaporation and condensation method promoted by Cu catalysts.[209] The lengths of the ZnO nanocombs ranged from several millimeters to more than 1 cm and the diameters of the branches were about 300 nm. It was found that the ultralong ZnO nanocombs can act as effective optical components in miniaturized integrated optics systems. Moreover, ZnO nanocombs were used as templates to form SiC nanocombs (Figure 9.90), replicating ZnO-based nanostructures by SiC at low temperatures.[210] Subsequent dissolution of ZnO left a SiC shell (Figure 9.91) that preserved the same shape as the original ZnO nanostructure. One of the many applications of ZnO single crystal nanocombs, synthesized in bulk quantity by vapor-phase transport, was their use as a basis for a glucose biosensor for glucose oxidase (GOx) loading.[211] The zinc oxide nanocomb glucose biosensor showed a high sensitivity for glucose detection and high affinity of GOx to glucose.

9.10 NANOFANS

It was observed that, for the same specific compound, fan-like nanostructures have usually been obtained together with a large series of other nanostructures changing reaction conditions. For elemental metals, the nanofans are known for Pt[212] and Au. Thus, HAuCl$_4$ when reduced by Fe^{2+} ions in aqueous solution in the presence of sodium dodecyl sulfate (SDS) produced multibranched Au nanoparticles including a nanofan shape.[213] Conglomerates of CNTs were also obtained in a fan-like form (Figure 9.92) (which seems, in our opinion, also as a flying bird) from polycyclic aromatic hydrocarbons (PAH) as precursors by template-assisted pyrolysis in alumina pores (Figure 9.93).[214] The length of the structures was ~5 mm, thus much smaller than the total thickness of the alumina

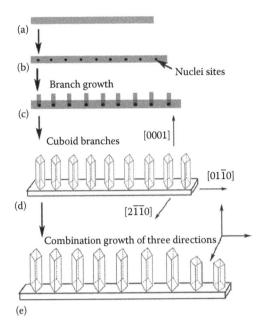

FIGURE 9.86 Schematic illustration of the proposed growth mechanism of ZnO comb-like structures with cuboid nanobranches. (a) Stem forming in the initial stage. (b) ZnO nuclei forming on the (0 0 0 1) surface of the comb stem. (c) Branching rods growing along the [0 0 0 1] direction. (d) Growing along three directions forming a rectangular cross section. (e) A trapezium comb-like structure with a triangular profile forming. (Zhuo, R.F. et al., *J. Phys. D Appl. Phys.*, 41, 185405, 13 pp., 2008. Reproduced with permission from Institute of Physics Publishing.)

FIGURE 9.87 Double-side teethed ZnO comb structures. (a) A low-magnification SEM image showing the high yield of the symmetric combs. (b, c) double-sided ZnO nanocombs with different morphologies. (Reproduced from *Chem. Phys. Lett.*, 417, Lao, C.S. et al., Formation of double-side teethed nanocombs of ZnO and self-catalysis of Zn-terminated polar surface, 359–363, Copyright 2005, with permission from Elsevier.)

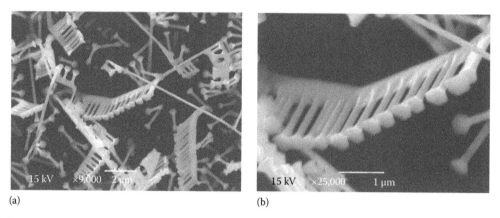

(a)   (b)

FIGURE 9.88 Disc-capped ZnO nanocomb with low (a) and high (b) magnifications. (Li, X. et al., *Chin. Phys. Lett.*, 24(12), 3495, 2007. Reproduced with permission from Institute of Physics Publishing.)

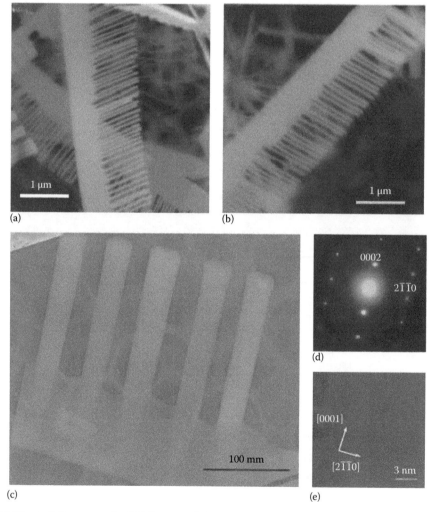

FIGURE 9.89 (a, b) Representative SEM images of the ZnO nanocantilevers (NCs); (c) TEM image of Fe-doped ZnO NCs; (d) the corresponding electron diffraction pattern, showing that the NC-stem direction is [2110], while the teeth direction is [0001]; (e) HRTEM image of the Fe-doped ZnO NC. (Reproduced from Zhang, B. et al., *Chin. Sci. Bull.*, 53(11), 1639, 2008. With permission from Science in China Press.)

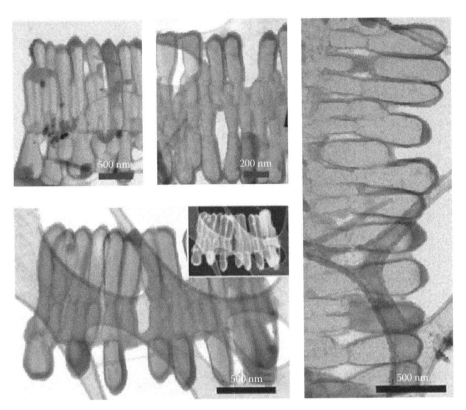

FIGURE 9.90 SiC-shell nanocombs. Inset in (d) is a darkfield TEM image of a SiC hollow nanocomb. (Zhou, J. et al.: SiC-shell nanostructures fabricated by replicating ZnO nano-objects: A technique for producing hollow nanostructures of desired shape. *Small.* 2006, 2(11), 1344–1347. Copyright Wiley-VCH Verlag GmbH & Co. KGaA. Reproduced with permission.)

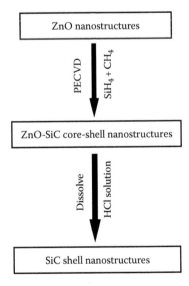

FIGURE 9.91 Schematic diagram showing the fabrication process of SiC shell nanostructures using ZnO nanoobjects as templates. (Zhou, J. et al.: SiC-shell nanostructures fabricated by replicating ZnO nano-objects: A technique for producing hollow nanostructures of desired shape. *Small.* 2006, 2(11), 1344–1347. Copyright Wiley-VCH Verlag GmbH & Co. KGaA. Reproduced with permission.)

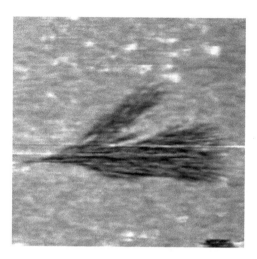

FIGURE 9.92 Carbon nanotube fan-like structure. (Reproduced from Palermo, V. et al., Unconventional nanotubes self-assembled in alumina channels: Morphology and surface potential of isolated nanostructures at surfaces, *Phil. Trans. R. Soc. A*, 365, 1577–1588. Copyright 2007 American Scientific Publishers.)

template (60 μm), suggesting that only the pores closer to the membrane side, where anodic etch was initiated, possessed this fan-like structure.

Among metal oxides, ZnO nanofans (Figure 9.94) and other structures including nanorods, nanotetrahedrons, nanodumbbells, and nanosquamas were prepared *via* an effective aminolytic reaction of zinc carboxylates with oleylamine in noncoordinating and coordinating solvents.[215] The growth of ZnO nanofans (Figure 9.95) and other structures was controlled by the thermal evaporation of Zn and a mixture of In and In_2S_3.[216] The ZnO nanofan was found to be composed of a set of ZnO nanonail arrays evenly distributed at the main stem. Both the morphologies of the products and their construction units could be efficiently controlled by simple adjustment of the weight ratio of In/In_2S_3. These novel hierarchical ZnO nanoarchitectures may be attractive building blocks for creating optical or other nanodevices. Large-scale fan-shaped rutile TiO_2 nanostructures, composed of several TiO_2 nanorods with diameters of ~5 nm and lengths of 300–350 nm, were prepared by a hydrothermal method using only $TiCl_4$ and chloroform/water as solvents.[217] Optical adsorption investigation showed that the fan-shaped TiO_2 nanostructures possess optical band gap energy of 3.11 eV. Under appropriate conditions such as nanobelt concentration and controlled solvent evaporation, $\beta\text{-}Ga_2O_3$ nanobelts with diameters of 50–100 nm and lengths of tens to hundreds of microns assembled into a fan-like structure on the substrate.[218]

Rare earth carbonate nanostructures $M_2(CO_3)_3$, in particular those with a fan-like shape (Figure 9.96), were synthesized in the reverse micelle system using rare earth nitrates $M(NO_3)_3 \cdot nH_2O$ and ammonium carbonate or bicarbonate as precursors.[219] The influence of the experimental conditions on the morphology of fan-like products was studied on $Eu_2(CO_3)_3$, revealing that, with the increase in aging temperature, the morphologies of the products changed from complex flower-like superstructures to double fan-like structures. The morphology of rare earth carbonates can be controlled through regulating the interaction between the corresponding ion or nanocrystal and the template molecule.

The coordination polymer $La(1,3,5\text{-}BTC)(H_2O)_6$ was prepared through simple and mild 1-step approaches at r.t. on a large scale (1,3,5-BTC = 1,3,5-benzenetricarboxylate) in various nanoforms including fan-like (Figure 9.97).[220] These 3D architectures can exhibit tunable white-light emission by co-doping Tb^{3+} and Eu^{3+}. Additionally, rectangular-microtube-based fan-like PANI structures were readily fabricated by oxidation polymerization of low-concentration aniline in dilute HCl aqueous solution at r.t.[221]

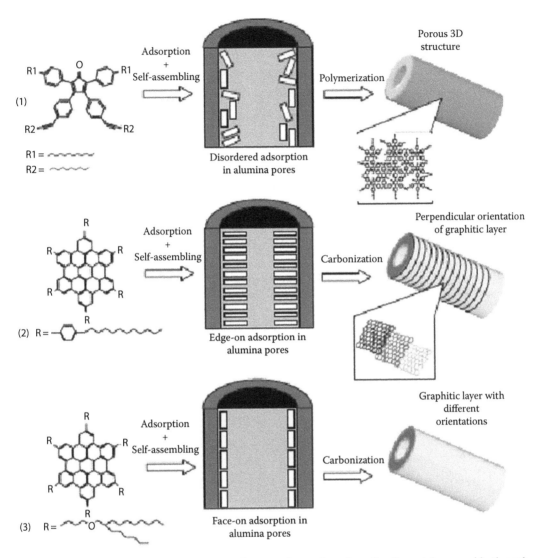

(1)

R1 =
R2 =

Adsorption
+
Self-assembling

Disordered adsorption
in alumina pores

Polymerization

Porous 3D
structure

(2) R =

Adsorption
+
Self-assembling

Edge-on adsorption in
alumina pores

Carbonization

Perpendicular orientation
of graphitic layer

(3) R =

Adsorption
+
Self-assembling

Face-on adsorption in
alumina pores

Carbonization

Graphitic layer with
different
orientations

FIGURE 9.93 The molecules used to create the nanotubes and a schematic of nanotube assembly through surface adsorption and pyrolitic, high-temperature graphitization. (Reproduced from Palermo, V. et al., Unconventional nanotubes self-assembled in alumina channels: morphology and surface potential of isolated nanostructures at surfaces, *Phil. Trans. R. Soc. A*, 365, 1577–1588. Copyright 2007 American Scientific Publishers.)

9.11 NANOSPINDLES

Spindle-like nanostructures have been found for a series of elements and inorganic compounds; we note that a certain number of reports are devoted to oxides and salts of rare-earth elements. Frequently, this nanostructural type was formed together with a variety of other structures, such as, nanorods or nanoflowers. In the case of elemental metals, gold, silver, copper, iron, and its alloys were obtained in nanospindle form. Thus, gold nanoparticles with different morphologies, such as spindle, octahedron, and decahedron, were obtained from $HAuCl_4$ as a precursor in the presence and absence of surfactants at r.t.[222] The surfactants played a crucial role in the size- and shape-controlled synthesis of gold nanoparticles. In a related work, spindle-shaped gold nanoparticles were prepared in high yield by a wet chemical approach using L-ascorbic acid (AA) as a

Varieties of ZnO nanofans

FIGURE 9.94 ZnO nanofans obtained from zinc carboxylates with oleylamine. (Reproduced with permission from Zhang, Z. et al., Modulation of the morphology of ZnO nanostructures via aminolytic reaction: From nanorods to nanosquamas, *Langmuir*, 22(14), 6335–6340. Copyright 2006 American Chemical Society.)

FIGURE 9.95 ZnO nanofan obtained by the atmospheric pressure thermal evaporation of Zn powder and the mixture of In and In_2S_3 powders in a two-heating-zone furnace system. (Reproduced with permission from Shen, G. et al., Growth of self-organized hierarchical ZnO nanoarchitectures by a simple In/In_2S_3 controlled thermal evaporation process, *J. Phys. Chem. B*, 109(21), 10779–10785. Copyright 2005 American Chemical Society.)

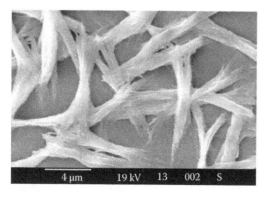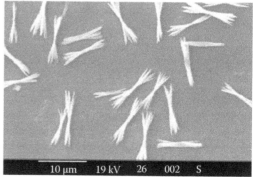

FIGURE 9.96 SEM images of fan-like $Nd_2(CO_3)_3$ (left) and $Sm_2(CO_3)_3$ (right) nanostructures. (Reproduced with permission from Gu, F. et al., Crystallization of rare earth carbonate nanostructures in the reverse micelle system, *Cryst. Growth Des.*, 7(8), 1452–1458. Copyright 2007 American Chemical Society.)

FIGURE 9.97 SEM images of the La(1,3,5-BTC)(H₂O)₆. (Reproduced with permission from Liu, K. et al., Room-temperature synthesis of multi-morphological coordination polymer and tunable white-light emission, *Cryst. Growth Des.*, 10(1), 16–19. Copyright 2010 American Chemical Society.)

reductant in the presence of CTAB at r.t.[223] It was found that the spindle-shaped architecture of gold nanoparticles was drastically influenced by the mass ratio of CTAB/AA, CTAB/HAuCl₄, HAuCl₄/AA, and the concentration of CTAB, AA, HAuCl₄. Almost the same route was applied for obtaining spindle-shaped silver nanoparticles, prepared in high yield using CA as a reductant in the presence of SDS at r.t.[224] As well as in the previous case, the spindle-shaped architecture of silver nanoparticles was drastically influenced by the mass ratio of SDS to CA and the concentration of silver nitrate. Cu nanoparticles (Figure 9.98) with a mean diameter of 10–15 nm were prepared and self-assembled *via* the discharge of bulk copper rods in a CTAB/AA solution.[225] The last compound was used as a protective agent to prevent the nascent Cu nanoparticles from oxidation in the solution. The authors noted that such a low-temperature and nonvacuum method, exhibiting the characters of both physical and chemical processes, provides a versatile choice for the economical preparation and assembly of various metal nanostructures.

Spindle porous iron nanoparticles were synthesized by reducing the presynthesized hematite (α-Fe₂O₃) spindle particles with hydrogen gas.[226] N₂ adsorption/desorption results showed a Brunauer–Emmett–Teller (BET) surface area of 29.7 m²/g and a continuous pore size distribution from 2 to 100 nm. Dispersion behavior of spindle-type iron nanoparticles with thin oxide surface layer was controlled by adding various dispersants to organic solvents.[227] Among the dispersions of Fe nanoparticles, the methylethyl ketone suspension with OA as a dispersant showed the lowest viscosity. The dispersion was found to be promoted by the chemisorption of OA on the particles surface, which resulted in the steric repulsion. Highly dispersed ethanol suspensions of spindle-type

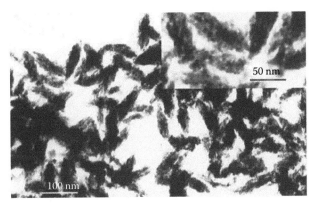

FIGURE 9.98 Spindle-like nanostructures produced from Cu arc-discharge in deionized water. (Reproduced from *J. Solid State Chem.*, 177, Xie, S.Y. et al., Preparation and self-assembly of copper nanoparticles via discharge of copper rod electrodes in a surfactant solution: A combination of physical and chemical processes, 3743–3747, Copyright 2004, with permission from Elsevier.)

Fe-Co particles were obtained by the addition of OA with the application of ultrasonication.[228] By using this suspension, spindle-type Fe-Co nanoparticles were coated by a uniform silica layer without agglomeration by a sol–gel method. Magnetically recyclable Co–B hollow nanospindles were prepared by using poly(styrene-comethacrylic acid) nanospindles as scarified templates.[229] Their catalytic performance is comparable to that of the reported Pt black concerning hydrogen yield. In addition, Co–B hollow nanospindles were found to have good reusability in the application due to their unique magnetic properties.

As well as in the majority of other nanostructures, the leadership in metal oxide nanospindle structures belongs to zinc oxide and less to the TiO_2 and hematite, obtained by different techniques including the hydrothermal method. Thus, different shapes of hierarchical ZnO nanostructures such as flower-like, spindle-like, and spherical were obtained at 90°C using the ascorbate ion as shape-directing/capping agent.[230] It was established that the spherical/quasi-spherical ZnO nanoparticles might aggregate through oriented attachment to produce spindle-like and flower-like nanostructures. The growth unit (Figure 9.99) for the formation of ZnO nanostructures by the alkaline hydrolysis of Zn^{2+} is the $[Zn(OH)_4]^{2-}$ ion, which further undergoes dehydration according to reactions 9.26 and 9.27 to produce ZnO. Almost the same combination of nanostructures, spindle/flower-like ZnO nanostructured arrays, was directly grown on glass substrates using triethanolamine (TEA) as a complexing agent by chemical bath deposition (CBD).[231] The morphology of the ZnO crystallites with star- or needle-like spindles was altered to flower-like nanostructures by adjusting the concentration of the complexing agent:

$$Zn^{2+} + 4OH^- = Zn(OH)_4^{2-} \tag{9.26}$$

$$Zn(OH)_4^{2-} = ZnO + H_2O + 2OH^- \tag{9.27}$$

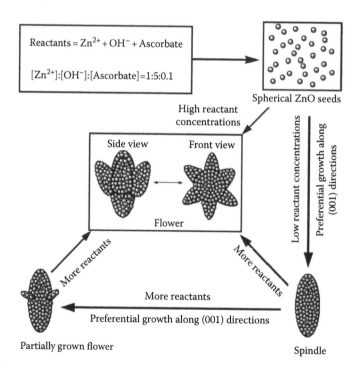

FIGURE 9.99 Representation of the probable growth mechanism of flower- and spindle-like ZnO nanostructures. (Reproduced with permission from Raula, M. et al., Ascorbate-assisted growth of hierarchical ZnO nanostructures: Sphere, spindle, and flower and their catalytic properties, *Langmuir.* 2010, 26(11), 8769–8782. Copyright American Chemical Society.)

A spindle-shaped nanoscale zinc oxide monocrystal was also obtained from zinc nitrate and NaOH as precursors under hydrothermal conditions.[232] The same method was used for obtaining several nano/microscale structures of ZnO including nanospindles, microbundles, microrods, and nanoflowers in diethanolamine from $Zn(NO_3)_2 \cdot 6H_2O$ as precursor without a surfactant in the presence of ammonium molybdate $(NH_4)_6Mo_7O_{24} \cdot 4H_2O$, whose addition had great effects on the final morphologies of the products.[233] The proposed reaction mechanism is described by Reactions 9.28 through 9.31. Additionally, ZnO/ZnS core–shell composites (having possible applications in the fields of luminescence, electronics, and sensors) and pure ZnS microspindles (Figures 9.100 through 9.103) were fabricated by a low-temperature hydrothermal growth *via* the reaction of ZnO nanospindles (obtained from $Zn(CH_3COO)_2 \cdot 2H_2O$ as precursor) and thioacetamide (TAA) according to the Reactions 9.32 and 9.33[234]:

$$HN(CH_2CH_2OH)_2 + H_2O \rightarrow H_2N(CH_2CH_2OH)_2^+ + OH^- \tag{9.28}$$

$$Zn^{2+} + 2OH^- \rightarrow Zn(OH)_2 \tag{9.29}$$

$$Zn(OH)_2 + 2OH^- \rightarrow Zn(OH)_4^{2-} \tag{9.30}$$

$$Zn(OH)_4^{2-} \rightarrow ZnO + 2OH^- + H_2O \tag{9.31}$$

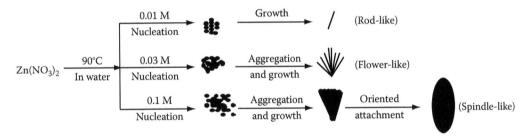

FIGURE 9.100 Schematic illustration of the possible formation process for rod-like, flower-like, and spindle-like ZnO structures from zinc nitrate and hexamethyltetramine. (Reproduced with permission from Peng, W. et al., Synthesis and structures of morphology-controlled ZnO nano- and microcrystals, *Cryst. Growth Des.*, 6(6), 1518–1522. Copyright 2006 American Chemical Society.)

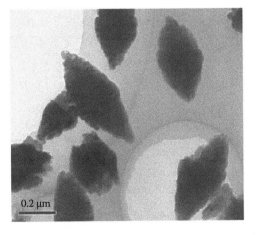

FIGURE 9.101 Typical TEM images of pure ZnO nanospindles. (Fei, L. et al.: Conversion from ZnO nanospindles into ZnO/ZnS core/shell composites and ZnS microspindles. *Cryst. Res. Technol.* 2009, 402–408, 44(4). Copyright Wiley-VCH Verlag GmbH & Co. KGaA. Reproduced with permission.)

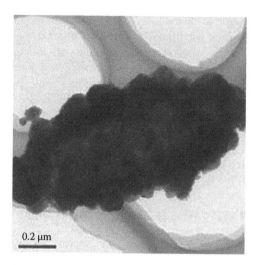

FIGURE 9.102 Pure ZnS microspindles. (Fei, L. et al.: Conversion from ZnO nanospindles into ZnO/ZnS core/shell composites and ZnS microspindles. *Cryst. Res. Technol.* 2009, 402–408, 44(4). Copyright Wiley-VCH Verlag GmbH & Co. KGaA. Reproduced with permission.)

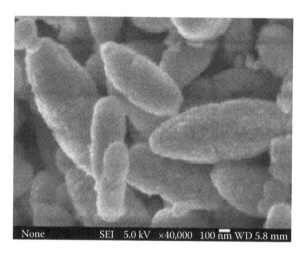

FIGURE 9.103 ZnO/ZnS core–shell microspindles. (Fei, L. et al.: Conversion from ZnO nanospindles into ZnO/ZnS core/shell composites and ZnS microspindles. *Cryst. Res. Technol.* 2009, 402–408, 44(4). Copyright Wiley-VCH Verlag GmbH & Co. KGaA. Reproduced with permission.)

$$CH_3CSNH_2 + H_2O \rightarrow CH_3CONH_2 + H_2S \tag{9.32}$$

$$ZnO(s) + H_2S \rightarrow ZnS(s) + H_2O \tag{9.33}$$

Anatase TiO_2 nanospindles were prepared from peroxotitanium complex TiO_2 {may be $Ti_2(O_2)$ $(OOH)(OH)_a(H_2O)_b$, $Ti_2(O_2)_2(OH)_a(H_2O)_b$ and even $[(Ti_2O_5)(OH)_2]_\infty$} as a precursor in aqueous solution by a low-temperature process at pH = 6.[236] Also, spindle-like mesoporous anatase TiO_2 particles (Figure 9.104) were directly synthesized at a low temperature (95°C) by using the same aqueous peroxotitanium solution with polyacrylamide (PAM).[237] Formation mechanism is shown in Figure 9.105. The mesoporous TiO_2 had a BET-specific surface area of 89.6 m²/g and showed high crystallinity, thermal stability, and good photocatalytic activity in the degradation of Rhodamine B. A related investigation using peroxotitanates was also carried out.[238] Other information on TiO_2 nanospindles was reported in Ref. [239].

FIGURE 9.104 TiO$_2$ spindle-like nanoparticles. (Reproduced with permission from Liu, X. et al., Highly crystalline spindle-shaped mesoporous anatase titania particles: Solution-phase synthesis, characterization, and photocatalytic properties, *Langmuir*, 6(11), 7671–7674. Copyright 2010 American Chemical Society.)

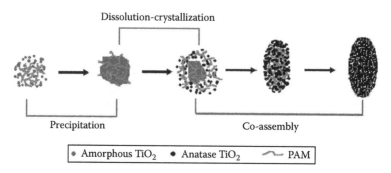

FIGURE 9.105 Schematic illustration of the formation mechanism of mesoporous TiO_2. (Reproduced with permission from Liu, X. et al., Highly crystalline spindle-shaped mesoporous anatase titania particles: Solution-phase synthesis, characterization, and photocatalytic properties, *Langmuir*, 6(11), 7671–7674. Copyright 2010 American Chemical Society.)

Single-crystalline porous hematite nanostructures with a weak ferromagnetic behavior were synthesized by a low temperature reflux condensation method from two different iron sources, namely, $FeCl_3 \cdot 6H_2O$ and $Fe(NO_3)_3 \cdot 9H_2O$, which were hydrolyzed in the presence of urea to selectively prepare nanorods and spindle-like nanostructures.[240] The photocatalytic activity of the prepared nanostructures toward RhB reflected this variation in the pore size distribution and specific surface area, by showing a higher activity for the nanorods sample. Magnetic studies by VSM showed a weak ferromagnetic behavior in both the samples due to shape anisotropy. Form factor and magnetic properties of silica-coated spindle-type hematite nanoparticles were detected from SAXS measurements with applied magnetic field and magnetometry measurements.[241] The particles were found to align with their long axis perpendicular to the applied field. Additionally, the silica coating reduced the effective magnetic moment of the particles; this effect was found to be enhanced with field strength and can be explained by superparamagnetic relaxation in the highly porous particles. A series of reports are dedicated to gold-hematite composites; thus, spindle-shaped Au/α-Fe_2O_3, obtained by an amino acid-assisted hydrothermal method, had higher catalytic activity than catalysts based on rhombohedral Fe_2O_3.[242] It was established that large hematite crystals induced higher catalytic activity than smaller crystals; for rhombohedral Au/α-Fe_2O_3 catalysts, medium-sized α-Fe_2O_3 nanocrystals exhibited high catalytic activity for CO oxidation. In a related research, spindle Fe_2O_3@Au core–shell particles were prepared by attaching the Au nanoparticles of 2 nm onto the Fe_2O_3 particles through organosilane molecules.[243]

Different methods have been used to obtain other metal oxide spindle-like nanoparticles. Thus, spindles of GeO_2 were prepared by the electric field–assisted laser ablation in liquids.[244,245] A shape-dependent red shift of emission wavelength was observed when the shape of GeO_2 nanostructures transformed into spindle from cube. Tungsten trioxide (*n*-type semiconductor material, which has found promising applications in information display devices and highly sensitive optical memory materials) powder with the spindle-like morphology (Figure 9.106), possessing high photochromic properties, was prepared *via* the hydrothermal method from sodium tungstate as a precursor with oxalic acid as the organic inducer. The induced product was still hexagonal WO_3, made up of regularly spindle particles with 200–300 nm length and 30–50 nm width.[246] Additionally, orthorhombic-phase and cubic-phase Sb_2O_3 microcrystals with a variety of morphologies and structures, such as microspindles, nanoplates, and octahedral, were selectively synthesized in high yield.[247] Three different nanostructures of CuO (wires, platelets, and spindles [Figure 9.107]) were synthesized from $Cu(NO_3)_2 \cdot 3H_2O$ as a precursor by its treatment with $NH_3 \cdot H_2O$ and NaOH and further heating the formed $Cu(OH)_2$.[248] A comparison of these three nanostructures showed an attractive phenomenon, that is, the electron transfer ability of CuO nanospindles was stronger than that of CuO nanowires or nanoplatelets. The application of these nanostructures in electrochemical detection of glucose was exhibited.

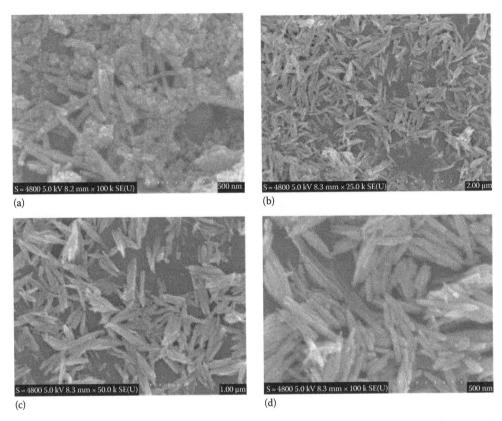

(a)

(b)

(c)

(d)

FIGURE 9.106 SEM images of pure WO$_3$ (a) and oxalic acid-induced WO$_3$ (b through d) powders. (Reproduced with kind permission from Springer Science+Business Media: *Sci. China Ser. B Chem.*, Synthesis and photochromic properties of WO$_3$ powder induced by oxalic acid, 52(5), 2009, 609–614, Shen, Y. et al.)

(a)

(b)

FIGURE 9.107 Images of CuO nanospindles: (a) low-magnification view, (b) high-magnification view. (Reproduced with permission from Zhang, X. et al., Different CuO nanostructures: Synthesis, characterization, and applications for glucose sensors, *J. Phys. Chem. C*, 112(43), 16845–16849. Copyright 2008 American Chemical Society.)

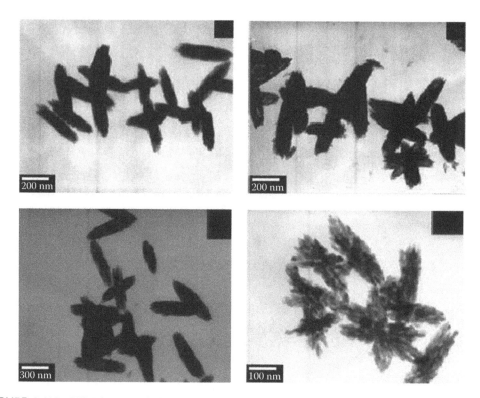

FIGURE 9.108 TEM images of the BaMoO$_4$ obtained with hydrothermal reaction for 10 min, 20 min, 30 min, and 1 h, respectively, after r.t. stirring for 4 h. (Tian, G. and Sun, S.: Hydrothermal synthesis of nano-crystalline BaMoO$_4$ under mild conditions using simple additive, *Cryst. Res. Technol.* 2010, 45(2), 188–194. Copyright Wiley-VCH Verlag GmbH & Co. KGaA. Reproduced with permission.)

Different morphologies of β-FeOOH including rod- and spindle-shaped crystals were synthesized *via* a hydrothermal reaction at low temperature from FeCl$_3 \cdot$6H$_2$O and CO(NH$_2$)$_2$ as precursors.[249] Laser ablation of gallium metal was performed to prepare spindle-like GaOOH particles in aqueous solution.[250] The formation of well-defined GaOOH was strongly associated with the addition of a cationic CTAB surfactant. It was found that spindle-like GaOOH could be grown in below, near, or above critical micelle concentrations of CTAB *via* an aging process.

Large-scale high-quality BaMoO$_4$ bone-like, spindle-like (Figure 9.108), and wheatear-like nanocrystals were obtained in aqueous solutions under mild conditions from Na$_2$MoO$_4 \cdot$2H$_2$O as a precursor with Na$_3$C$_6$H$_5$O$_7 \cdot$2H$_2$O (trisodium citrate dihydrate) as an additive.[251] Their formation mechanism is shown in Figure 9.109. The as-prepared BaMoO$_4$ nanocrystals had a strong blue emission peak at 481.5 nm. CaMoO$_4$ nanostructures with different morphologies, such as ellipsoid-like, spindle-like, and sphere-like, were synthesized in a simple cationic surfactant-CTAB-microemulsion system at r.t.[252] The CaMoO$_4$ nanostructures exhibited excellent photoluminescence properties with the same new green emission peaks at 495 nm.

FIGURE 9.109 The propositional formation process of BaMoO$_4$ nanocrystals. (Tian, G. and Sun, S.: Hydrothermal synthesis of nanocrystalline BaMoO$_4$ under mild conditions using simple additive, *Cryst. Res. Technol.* 2010, 45(2), 188–194. Copyright Wiley-VCH Verlag GmbH & Co. KGaA. Reproduced with permission.)

Several carbonates were reported in nanospindle forms. Thus, research status of different shapes of the nanocalcium carbonate, such as spindly, cubiform, catenoid, flaky, and bulbiform, and their applications in coatings were reviewed.[253] Modified calcium carbonate nanoparticles with cubic- and spindle-like configuration were synthesized *in situ* by the typical bobbling (gas-liquid-solid) method.[254] The modifiers, such as sodium stearate, octadecyl dihydrogen phosphate (ODP) and OA, were used to obtain hydrophobic nanoparticles. According to the results, the active ratio of $CaCO_3$ modified by ODP was ca. 99.9% and the value of whiteness was 97.3% when the dosage of modifiers reached 2%. When modified $CaCO_3$ was filled into PVC, the mechanical properties of products such as rupture intensity, pull intensity, and fuse temperature were improved greatly. A grand variety of $MnCO_3$ nanostructures including cocoons, prisms, spindles, and anamorphic rhombohedra were obtained using hydrazine hydrate as the complexing agent in a hydrothermal system.[255] The morphology of the products could be simply controlled by changing the reactant, the reaction temperature, and the reaction time. Spindle $FeCO_3$ was used as sacrificial templates *via* direct thermal decomposition and sealed thermal decomposition to obtain ellipsoidal Fe_2O_3 and Fe_3O_4 microparticles according to Reaction 9.34[256]:

$$6FeCO_3 + O_2 \rightarrow 2Fe_3O_4 + 6CO_2 \tag{9.34}$$

Large-scale silicon carbide nanowires and bamboo-like nanowires and spindle SiC nanochains with lengths of up to several millimeters were synthesized by a coat-mix, molding, carbonization, and high-temperature sintering process, using silicon powder and phenolic resin as the starting materials.[257] Both of the bamboo-like SiC nanowires and spindle SiC nanochains exhibited uniform periodic structures. The spindle-shaped nanosilver sulfide with a length of 70–120 nm was obtained[258] from sodium sulfide and silver nitrate as precursors. The obtained spindle-shaped or rod-like nanosilver sulfide has a good dispersibility, uniform shape, and high stability. Cadmium telluride spindle- and rod-like micrometer- and nanometer-scaled crystals were synthesized through a two-step route: CdTe nanoparticles were synthesized first *via* hydrothermal method and followed by diethylenetriamine (DETA) treatment.[259] The advantages of this synthetic route, when compared with others, were mild conditions in which the reaction was carried out and without any prerequisite on the particles in the second step. Additionally, $CuIn(WO_4)_2$ porous nanospindles (Figure 9.110) and nanorods were synthesized through a low-cost hydrothermal method without introducing any

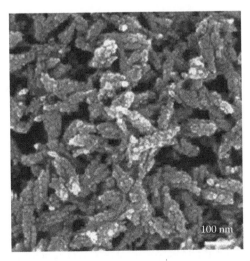
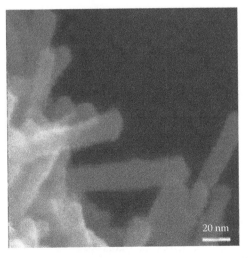

FIGURE 9.110 $CuIn(WO_4)_2$ porous nanospindles. (Song, S. et al., $CuIn(WO_4)_2$ nanospindles and nanorods: Controlled synthesis and host for lanthanide near-infrared luminescence properties, *Cryst. Eng. Comm.*, 11, 1987–1993, 2009. Reproduced by permission of The Royal Society of Chemistry.)

template or surfactants.[260] The near-infrared luminescence of lanthanide ions (Er, Nd, Yb, and Ho) doped $CuIn(WO_4)_2$ nanostructures, especially in the 1300–1600 nm region, was discussed and of particular interest in applications for telecommunications.

A series of publications are devoted to spindle-like nanostructures of lanthanide compounds. Thus, a general ultrasonic irradiation method was established for the synthesis of the lanthanide orthovanadate $LnVO_4$ (Ln = La, Ce, Pr, Nd, Sm, Eu, Gd, Tb, Dy, Ho, Er, Tm, Yb, Lu) nanoparticles from an aqueous solution of $Ln(NO_3)_3$ and NH_4VO_3 without any surfactant or template.[261] The as-formed $LnVO_4$ particles had a spindle-like shape (Figure 9.111) with an equatorial diameter of 30–70 nm and a length of 100–200 nm, which were the aggregates of even smaller nanoparticles of 10–20 nm. Their formation mechanism is shown in Figure 9.112. Bismuth vanadate ($BiVO_4$) spindle particles with a monoclinic scheelite structure were synthesized *via* a sonochemical method.[262]

FIGURE 9.111 Typical TEM images of the lanthanide orthovanadate samples (a) m-$LaVO_4$, (b) t-$LaVO_4$, (c) $CeVO_4$, (d) $PrVO_4$, (e) $NdVO_4$, (f) $SmVO_4$, (g) $EuVO_4$, (h) $GdVO_4$, (i) $TbVO_4$, (j) $DyVO_4$, (k) $HoVO_4$, (l) $ErVO_4$, (m) $TmVO_4$, (n) $YbVO_4$, and (o) $LuVO_4$ together with the SAED pattern of (p) $GdVO_4$ nanoparticles. (Reproduced with permission from Yu, C. et al., Spindle-like lanthanide orthovanadate nanoparticles: Facile synthesis by ultrasonic irradiation, characterization, and luminescent properties, *Cryst. Growth Des.*, 9(2), 783–791. Copyright 2009 American Chemical Society.)

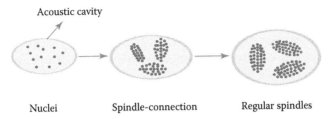

Nuclei   Spindle-connection   Regular spindles

FIGURE 9.112 Schematic illustration for the formation process of spindle-like $LnVO_4$ nanoparticles by the sonochemical process. (Reproduced with permission from Yu, C. et al., Spindle-like lanthanide orthovanadate nanoparticles: Facile synthesis by ultrasonic irradiation, characterization, and luminescent properties, *Cryst. Growth Des.*, 9(2), 783–791. Copyright 2009 American Chemical Society.)

The as-prepared $BiVO_4$ photocatalyst exhibited a hollow interior structure constructed from the self-assembly of cone-shaped primary nanocrystals. A much higher photocatalytic activity of these spindle particles was found in comparison with the SSR-$BiVO_4$ material for degradation of rhodamine-B under visible light irradiation, which may be ascribed to its special single-crystalline nanostructure.

Spindle hexagonal-phase $TbPO_4 \cdot H_2O$ nanostructured luminescent materials composed of nanorods were fabricated from $Tb(NO_3)_3 \cdot 6H_2O$ and phosphoric acid at 100°C for 6–8 h.[263,264] The advantages of this reaction route were the recyclable raw material phosphoric acid, no need of expensive surfactant as template, no use of organic solvent, environmental protection, low reaction temperature, high conversion, saving energy, and being easy for industrialization. These spindle-like hierarchical nanostructures, composed of ordered nanorods of 80–90 nm in diameter and lengths of up to 200–300 nm, displayed better photoluminescence than $TbPO_4 \cdot H_2O$ that was synthesized on both the nano- and microscale. High-efficiency green-luminescent $LaPO_4$:Ce,Tb hierarchical nanostructures with uniform spindle shape were constructed by a Pluronic P123-assisted hydrothermal approach *via* self-assembling from single-crystalline nanowires.[265] The resulting uniform spindle-shaped microstructures were composed of several tens of aligned single-crystalline nanowires. These $LaPO_4$:Ce,Tb spindle-shaped microstructures showed efficient photoluminescence.

Spherical, microsized, rod-shaped, and spindle-like CeO_2 particles were synthesized[266] and the catalysts on their basis were prepared by loading palladium chloride onto the CeO_2 support matrix. It was shown that the catalytic activities prepared by spherical cerium were superior to the catalysts prepared by spindle-like or rod-shaped cerium and the catalyst $PdCl_2/CeO_2$ (nanospheres) showed good activity, high yield, and good stability. Alternatively, spherical cerium oxide nanoparticles and microsized, rod-shaped, and spindle-like CeO_2 structures were synthesized using a polyol method from $[(NH_4)_2Ce(NO_3)_6]$ as a precursor (Reaction 9.35).[267] The CeO_2 nanospheres, formed from the decomposition of the cerium precursor, took part in further reactions with ethylene glycol to yield the rods/spindle-like cerium formiate particles and finally CeO_2 after calcination. The higher catalytic activity on CO conversion in a spindle-like sample can be explained from the extent of Ce(IV) reduction and the oxygen vacancy:

$$(NH_4)_2Ce(NO_3)_6 \rightarrow 4H_2O + 2N_2O + 4NO_2 + O_2 + CeO_2 \text{ (above 185°C) (nanospheres)} \quad (9.35)$$

Monodisperse hexagonal $Ln(OH)_3$ (Ln = Eu, Sm) submicrospindles (spindle-like shape with an equatorial diameter of 80–200 nm and a length of 500–900 nm, which were found to be aggregates of even smaller nanoparticles) with uniform morphology and size were synthesized in a large scale *via* a aqueous solution route from the mixture of aqueous solutions of $LnCl_3$ and NaOH at 5°C without using any surfactant or template.[268] Through calcining the $Ln(OH)_3$ submicrospindles, the Ln_2O_3 particles with the same morphology could be obtained. Uniform spindle-like $Y(OH)_3$ nanorod bundles with a length of about 11 μm and a diameter of about 2 μm in the middle part were successfully prepared *via* a hydrothermal method at 200°C for 12 h in the presence of EDTA.[269] The nanorod bundles were shown to be composed of numerous nanorods, and all these nanorods were orientationally aligned and grow uniformly along the bundles. A strong red emission centering at 613 nm is realized in the Eu^{3+} doped Y_2O_3 nanorod bundles, which may find potential application in the field of color display and solid-state lasers.

Among other lanthanide complexes, yttrium fluoride nanoparticles with spindle structure were prepared from yttrium chloride, selenium dioxide, and sodium fluoride at a molar ratio of 1:2:3 by hydrothermal synthesis.[270] The spindle-like yttrium fluoride nanoparticles had a thickness of 30 nm. The single-crystalline perovskite-like oxide $La_{2-x}Sr_xCuO_4$ ($x = 0$, 1) nano-/microparticles with spindle-, rod-, and short chain-like morphologies were synthesized with spindle-like single-crystalline CuO, plate-like single-crystalline La_2O_3, and nitrates as metal source by hydrothermal treatment and calcination.[271] It was shown that the partial doping of Sr to the La_2CuO_4 lattice increased the surface-adsorbed oxygen amount, Cu^{3+} content, and reducibility. The $LaSrCuO_4$ exhibited high catalytic activity, giving a methane consumption rate of up to 40.9 mmol/(g/h) at 675°C. Cerium oxycarbonate $(Ce^{3+})_2O(CO_3)_2 \cdot H_2O$, prepared in the form of elongated crystalline spindles (5–10 μm long) by heating the solution of nitrates, urea, and polyvinyl-pyrrolidine in aqueous medium at 80°C, was thermally decomposed to give rise to smaller elongated rods of nanostructured ceria.[272] Additionally, large-scale tetragonal $NaLaMo_2O_8$ nanocrystallines with spindle morphology were prepared through a facile method by a low-temperature hydrothermal treatment of Na_2MoO_4, $LaCl_3$, and EDTA in an ethanolamine/water mixed solution.[273] High-quality and uniform nanospindles with a mean length of 260 nm and a mean width of 100 nm can be easily obtained.

A few organic-inorganic spindle-like composites are known. Thus, hydroxyapatite (HAp), widely used as biomaterials (bone replacement, implant, etc.) thanks to its good biocompatibility, was used as a basis to obtain its spindle-type hydroxyapatite/chitosan (HAp/CS) nanocomposite (long axis is ~41 nm, short axis is ~22 nm) by precipitation method from $CaCl_2$ and Na_2HPO_4.[274] By using $FeCl_3 \cdot 6H_2O$ as an oxidant, an organic/inorganic complex nanostructure, poly(3,4-ethylenedioxythiophene)-/β-akaganeite {PEDOT/β-Fe^{3+}O(OH,Cl)} nanospindle, was synthesized in aqueous solution in the presence of CTAB and PAAc.[275] The product had the dual properties of both a conducting polymer (PEDOT) and an inorganic single crystal (β-akaganeite). The formation of PEDOT/β-Fe^{3+}O(OH,Cl) nanospindles may be due to the concurrence of the polymerization of EDOT and the hydrolyzation of $FeCl_3$ in aqueous solution at 50°C and the result of their interactions (Figure 9.113). Such composite nanomaterials may be anticipated to show good photocatalysis, chemical adsorption, and UV screen properties.

9.12　NANOTROUGHS

These nanostructures are very rare and represented by ZnO and NiO. Thus, post-thermal decomposition of single-crystalline β-nickel hydroxide {β-$Ni(OH)_2$} nanoplates of hexagonal structure, synthesized through hydrothermal process,[276,277] led to the formation of single-crystalline NiO nanostructures (Figure 9.114) with landscape dimension of 25–120 nm including nanorolls, nanotroughs and nanoplates. The fractions of nanorolls, nanotroughs, and nanoplates were estimated to be about 75%, 10%, and 15%, respectively. The sizes of the central hole in NiO nanorolls and the low-lying ground in NiO nanotroughs were in the range of 10–24 nm. Two photoluminescence emission peaks appear at 390.5 and 467 nm in the photoluminescence spectrum of NiO nanostructures and were assigned to the $^1T_{1g}$ (G) → $^3A_{2g}$ and $^1T_{2g}$ (D) → $^3A_{2g}$ transitions of Ni^{2+} in oxygen octahedral sites, respectively. In addition, bowl-, trough-, and ring-shaped structures resulted from the temperature-induced self-assembly of ZnO nanoparticles.[4]

9.13　NANOWEBS

Nanowebs[278] (Figure 9.115) are of great technological interest since they contain nanowire densities on the order of $10^9/cm^2$. Nanowebs are beautiful objects, reported for some compounds nonrelative to each other. First of all, we note that a series of porous polymers,[279] frequently obtained by

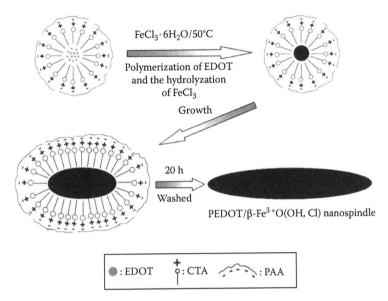

FIGURE 9.113 Growth process of PEDOT/β-Fe³⁺O(OH,Cl) nanospindles in aqueous solution: [EDOT]) 0.025 M, [EDOT]/[FeCl₃·6H₂O]) 1:3, [CTAB]) 0.0075 M, [PAA]) 0.02 mg/mL, 50°C, 20 h. (Reproduced with permission from Mao, H. et al., Preparation and characterization of PEDOT/β-Fe₃+O(OH,Cl) nanospindles with controllable sizes in aqueous solution, *J. Phys. Chem. C*, 112, 20469–20480. Copyright 2008 American Chemical Society.)

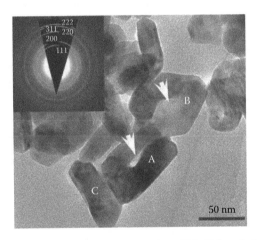

FIGURE 9.114 NiO nanostructures with landscape dimension of 25–120 nm include nanorolls, nanotroughs, and nanoplates, as marked with A, B, and C, respectively. (From Qi, Y. et al., *J. Mater. Sci. Mater. Electron.*, 20(5), 479–483, 2009. With permission.)

electrospinning technique, have been reported in web-like nanoform; here we note only some of them. Nanoweb porous structures are currently applied in filters,[280–282] fibers,[283] and composites,[284,285] in particular those containing CNTs, for example, MWCNT/poly(ethylene terephthalate) (PET) nanowebs, obtained by electrospinning.[286] It was established that tensile strength, tensile modulus, thermal stability, and the degree of crystallinity increased with increasing MWCNT concentration. Another example of CNT-polymer composite nanowebs is polyurethane (PU) and PU/MWCNT nanocomposite nanofibers and nanowebs, both with diameters of 350 nm, which were prepared by an electrospinning process from PU DMF solutions.[287] Multifunctional nanowebs were prepared using a cyclodextrin-based inclusion complex (CD-IC) to disperse MWNTs within electrospun

FIGURE 9.115 Nanoweb of cellulose acetate. (http://www.ntcresearch.org/projectapp/project_pages/ M04-CD01/pdf_files/Figure%201.pdf. Accessed January 15, 2011.)

polyvinylidene fluoride nanofibrous membranes (PVdF-NFM).[288] Subsequently, MWNT(CD-IC)/ PVdF-NFM was loaded with gold particles. The Au/MWNT(CD-IC)/PVdF-NFM was found to be electroactive and showed excellent electrocatalytic activity in the oxidation of AA. As an example of non-CNT-containing web-like polymers, we note the octaphenyl-POSS (polyhedral oligomeric silsesquioxane OPPOSS), which was blended with PMMA to prepare inorganic-organic hybrid nanofibers/nanowebs using the electrospinning method.[289] Also, a poly(dimethylsiloxane) perylene bisimide (PDMS-perylene bisimide) web-like nanostructure (Figure 9.116) was synthesized by Reactions 9.36 of perylene-3,4,9,10-tetracarboxylic dianhydride with aminoterminated polysiloxane in *m*-cresol at elevated temperatures.[290] The surface roughness of polyacrylonitrile (PAN) nanowebs was studied.[291] Four roughness parameters such as maximum height, ten point height, arithmetic mean of roughness (AMR), and root mean square were measured, and AMR parameter was used as surface roughness. It was established that increasing the fiber diameters of nanowebs led to the enhancement of the surface roughness of samples. Additionally, nanofibers of phenylsilsesquiazanes (SSQZ), as a precursor of inorganic silicon nitride nanoweb, were prepared using the electrospinning process by ammonolysis of Ph trichlorosilane and condensation of the resulting aminosilanes.[292]

$$(9.36)$$

Web-like CNT (Figure 9.117) were formed by laser ablation (Nd:YAG laser with 532 nm wavelength, 10.54 W power, 30 min) using a graphite target containing Ni and Co catalysts, each with weight percentage of 10%.[294] It was shown that the diameters of the CNTs formed using Ni and Co catalysts were between 50 and 150 nm in size. Carbon nanostructures may be precursors of other nanoweb structures. Thus, when carbon nanofilms were deposited on the Si(100) wafers at

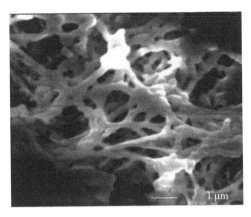
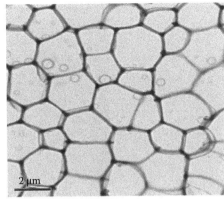

FIGURE 9.116 SEM (left) and TEM (right) images of the PDSM samples with molecular weight 900 of PDSM segment (from CHCl₃ solution). (Reproduced with permission from Yao, D. et al., Pigment-mediated nanoweb morphology of poly(dimethylsiloxane)-substituted perylene bisimides. *Macromolecules*, 39(23), 7786–7788. Copyright 2006 American Chemical Society.)

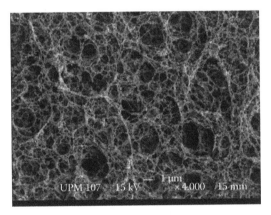

FIGURE 9.117 SEM image of CNTs collected from the PLAD process by using graphite-Ni pellet. (Reproduced from Yahya, N., *Solid State Sci. Technol.*, 15(1), 22, 2007.)

substrate temperatures from r.t. to 700°C by ultrahigh-vacuum ion beam sputtering,[295] a nanoweb-like morphology of the crystalline SiC formation was also observed on the surface of film. The much enhanced hardness and Young's modulus at 700°C were attributed to the SiC formation and nanoweb-like morphology. Additionally, web-like SWNT structures on the surface of zeolite template (Figure 9.118) were reported.

In addition to CNTs, boron nitride web-like nanostructures are known. Thus, a modification of arc-discharge method was used for the synthesis of web-like nanomaterial (Figure 9.119) made of BN double-walled nanotubes.[297] Detailed HRTEM examination of the ends of BN nanotubes indicates continuity between the graphitic BN layers that coat boron nanoparticles, that is, nanococoon, and the nanotube. The hierarchical structure of SiC-SiO₂ core–shell nanowebs in the form of intersecting nanowires and nanocables, augmented by variable amounts of SiO₂ membranes, was reported.[298] Such types of hierarchical nanostructures, together with the controllability, may offer super mechanical properties in composite applications.

Among elemental metals in nanoweb forms, platinum nanoparticles dispersed in water were soldered by gold into higher-order structures such as nanowebs (Figure 9.120).[299] The originality of this method was as follows: gold nanoparticles in water containing the platinum nanoparticles were melted by irradiation by a pulsed 532 nm laser, which selectively excited the surface

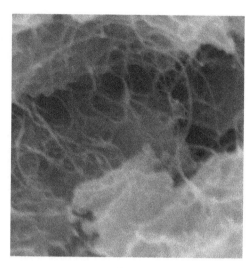

FIGURE 9.118 Web-like SWNT structures on the surface of zeolite template. (Endo, M. et al., Applications of carbon nanotubes in the twenty-first century, *Philos. Trans. R. Soc. Lond. A*, 362, 2223–2238, 2004. Reproduced by The Royal Society.)

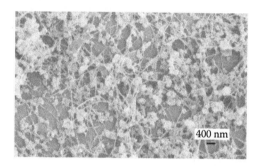

FIGURE 9.119 Web-like material: BN nanotubes can easily reach several microns in length and are systematically associated with nanoparticles. (Reproduced from Altoe, M.V.P., *J. Mater. Sci.*, 38, 4805, 2003. With permission from Kluwer Academic Publishers.)

plasmon band of the gold nanoparticles, and the melted gold nanoparticles soldered the platinum nanoparticles together into nanowebs. Other noble metal nanowebs are also known.[300] The authors noted that one of the unique advantages of this type of metallic nanowebs is that they offer a hybrid semisolid–semisolution interface for the immobilized molecules, thus allowing the surface-immobilized molecules, esp. biomacromolecules, to reside in conformations close to their native states. The inexpensive combination of cryogenically milled Cu_3Ge powders sonochemically processed in an ultrasonic cleaner led to a nanostructured composite composed of 4.5 nm diameter Cu nanocrystals embedded in a 3D amorphous $CuGeO_3$ polyhedron web matrix[301] with wires of 5–15 nm.

Metal nanoweb-like oxides are represented by few examples. Thus, deposition of RuO_2 onto porous SiO_2 substrates *via* subambient decomposition of ruthenium tetroxide from a nonaqueous solution led to a self-wired formation of nanowebs and ultimately nanoscale films of RuO_2.[302] A method of producing networks of low melting metal oxides such as crystalline Ga oxide composed of 1D nanostructures was patented.[303] These crystalline networks are defined as nanowebs, nanowire networks, and/or 2D nanowires. The individual segments of the polygonal network consisted of both nanowires and nanotubules of β-Ga oxide (Figure 9.121).[304] A possible mechanism for the fast self-assembly of crystalline metal oxide nanowires involved multiple nucleation and

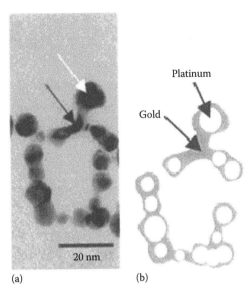

(a) (b)

FIGURE 9.120 (a) Electron micrograph of platinum and gold nanowebs produced by laser irradiation at 532 nm onto a mixed solution of platinum and gold nanoparticles with a molar ratio of 0.2. (b) Schematic view of the nanoweb shown in panel (a). (Reproduced with permission from Mafune, F. et al., Nanoscale soldering of metal nanoparticles for construction of higher-order structures, *J. Am. Chem. Soc.*, 125(7), 1686–1687. Copyright 2003 American Chemical Society.)

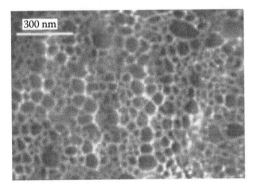

FIGURE 9.121 Ga_2O_3 nanoweb. (From Graham, U.M. et al., *Adv. Funct. Mater.*, 13(7), 576, 2003. Reproduced with permission from University of Kentucky.)

coalescence *via* oxidation-reduction reactions at the molecular level. The method of producing nanowebs can be extended to other low melting metals and their oxides such as SnO_2, ZnO, and In_2O_3. Multilayer nanowire webs of aluminum borate ($Al_{18}B_4O_{33}$) were synthesized by a solid-state reaction of boron, boron oxide, and aluminum oxide at high temperature without using any template or substrate.[305] $Al_{18}B_4O_{33}$ nanowebs mainly consisted of BO_3, AlO_4, AlO_5, and AlO_6 units. A series of Reactions 9.37 through 9.39 were found to take place after formation of the first product $Al_4B_2O_9$. The addition of boron should be responsible for the generation of the web morphology. Other salts are represented by sulfides; thus, different nanoscaled hierarchical CuS superstructures, such as tree-like superstructures, nanowires, and nanowebs, were prepared by an ion-tailoring method.[306] Also, 2D sulfide nanowebs (Figures 9.122 and 9.123) were prepared using a hyposulfite self-decomposition route (Reaction 9.40).[307] This route included four necessary steps: (1) preparation of the metastable metal hyposulfite nanoflats; (2) decomposition of the metal

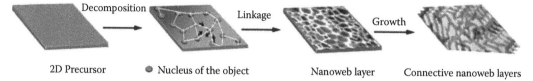

2D Precursor   ● Nucleus of the object   Nanoweb layer   Connective nanoweb layers

FIGURE 9.122 Growth process involved in the FeS$_2$ nanoweb formation. (Reproduced with permission from Gao, P. et al., From 2D nanoflats to 2D nanowire networks: A novel hyposulfite self-decomposition route to semiconductor FeS$_2$ nanowebs, *Cryst. Growth Des.*, 6(2), 583–587. Copyright 2006 American Chemical Society.)

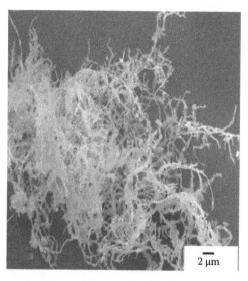

FIGURE 9.123 FESEM image of FeS$_2$ nanowebs. (Reproduced with permission from Gao, P. et al., From 2D nanoflats to 2D nanowire networks: A novel hyposulfite self-decomposition route to semiconductor FeS$_2$ nanowebs, *Cryst. Growth Des.*, 6(2), 583–587. Copyright 2006 American Chemical Society.)

hyposulfite nanoflat precursors and nucleation of metal sulfides; (3) coalescence and growth of the initial metal sulfide nanocrystals; (4) formation of the metal sulfide nanowebs and elimination of the metal sulfate by-product. As an example, semiconductor FeS$_2$ monocrystalline 2D nanowebs were obtained. The band gap of FeS$_2$ nanowires was found to be 5.64 eV by UV–visible absorption spectrum, showing its promising application for the reversible conversion between solar energy and electrical or chemical energy.

$$2Al_2O_3 + B_2O_3 \rightarrow Al_4B_2O_9 \tag{9.37}$$

$$2B + Al_2O_3 \rightarrow 2Al + B_2O_3 \tag{9.38}$$

$$9Al_4B_2O_9 \rightarrow 2Al_{18}B_4O_{33} + 5B_2O_3 \tag{9.39}$$

$$4S_2O_3{}^{2-} \rightarrow S_2{}^{2-} + 3SO_4{}^{2-} + 3S \tag{9.40}$$

Coordination compounds are represented by phthalocyanines. Thus, iron phthalocyaninate with nanobrush and nanoweb shapes was obtained by MBE as a function of substrate temperature (25°C–300°C) and deposition rate (0.02–0.07 nm/s). The morphology of a 60 nm α-phase film was tuned from nanobrush (nearly parallel nanorods aligned normal to the substrate plane) to nanoweb (nanowires forming a web-like structure in the plane of the substrate) by changing the deposition rate from 0.02 to 0.07 nm/s.[163,308] Meso- and nanoscopic Cu phthalocyanine quasi-1D structures, single crystalline whiskers, and nanowebs were used to fabricate organic conductometric gas-sensing devices for NO_x detection.[309] The obtained results demonstrated the excellent sensitivity and selectivity of these quasi-1D chemiresistors.

9.14 OTHER RARE "HOME" NANOSTRUCTURES

9.14.1 NANOSOMBRERO

Mexican sombrero-type nanostructures (Figure 9.124) were produced using silica nanospheres upon irradiation with laser (Figure 9.125).[31] An interesting phenomenon was observed: at the lower laser energy only bowl-type dents were produced on the silicon surface, while at the higher

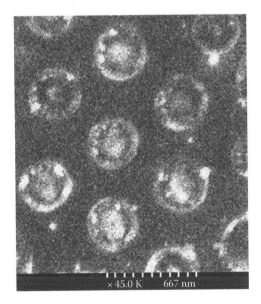

FIGURE 9.124 Sombrero type, produced using silica nanospheres upon irradiation with a 248 nm KrF excimer laser. The laser fluence delivered was 10 mJ/cm². (Lu, Y. and Chen, S.C., *Nanotechnology*, 14, 505, 2003. Reproduced with permission from Institute of Physics Publishing.)

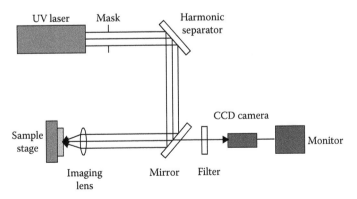

FIGURE 9.125 Experimental setup for laser nanopatterning of a silicon surface. (Lu, Y. and Chen, S.C., *Nanotechnology*, 14, 505, 2003. Reproduced with permission from Institute of Physics Publishing.)

laser intensity, sombrero-type bumps were formed. Although produced nanostructures were of the order of 300 nm, the authors expected that finer patterns can be produced by the use of smaller nanospheres.

9.14.2 NANOBOWKNOT

Ordered aggregation of scattered single spherical polystyrene-b-poly(acrylic acid) (PS-b-PAA) micelles led to various structures, including bowknot arrays (Figure 9.126), according to the mechanism shown in Figure 9.127.[310] These nanostructures are related to broom- and sheaf-like nanoforms.

FIGURE 9.126 PS-b-PAA bowknot. (Reproduced with permission from Gao, L. et al., Formation of spindle-like aggregates and flowerlike arrays of polystyrene-b-poly(acrylic acid) micelles, *Langmuir*, 20(12), 4787–4790. Copyright 2004 American Chemical Society.)

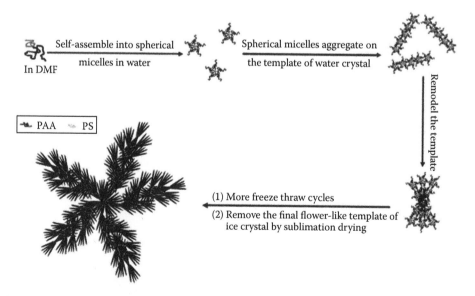

FIGURE 9.127 Formation of spindle-like, bowknot-like, and flower-like arrays. (Reproduced with permission from Gao, L. et al., Formation of spindlelike aggregates and flowerlike arrays of polystyrene-b-poly(acrylic acid) micelles, *Langmuir*, 20(12), 4787–4790. Copyright 2004 American Chemical Society.)

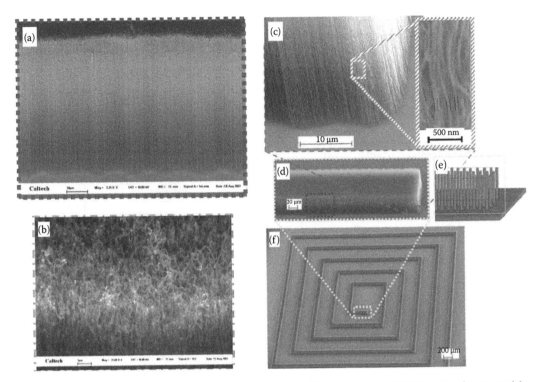

FIGURE 9.128 Hierarchical structure of a nanowick in 45° tilted views. (a) Sidewall of a nanowick; (b) the nanowick top surface; (c) the nanotube roots on the substrate; inset: 20 nm thick carbon nanotubes as the building block for the nanowick; three nanotubes are sketched out to compare with the deeper ones behind them (note: marginal nanotubes look thicker than inside tubes because of the charging effect by the E-beam); (d) a segment of a 60 μm high 100 μm wide nanowick, comprising the nanotubes; (e) diagram of the structure in (d); (f) a nanowick with arbitrary micropatterns can be centimeters long. (Zhou, J.J. et al., *Nanotechnology*, 17, 4845, 2006. Reproduced with permission from Institute of Physics Publishing.)

9.14.3 NANOWICK*

A nanowick prototype, consisting of dense arrays of aligned CNTs (Figure 9.128), for liquid transport and chemical analysis on microfluidic devices, was introduced.[311] Its fabrication process is shown in Figure 9.129.

9.14.4 EIFFEL TOWER–LIKE AND NANO–NEW YORK NANOSTRUCTURES

Vapor–liquid–solid (VLS) grown led to SiC nanowires with Eiffel Tower shape (Figure 9.130) by varying the pressure of the source species.[312] Commercially available polysilazane was used as the source material. The Eiffel Tower–shaped nanowires consisted of a cone, an Eiffel Tower base, and an Eiffel Tower tip (Figure 9.130b and c). The Eiffel Tower tip was shown to be about 1 μm long with a diameter of ~20 nm. The diameter of the nanowires was found to be strongly related to the pressure and pressure variation rate of the source species. Another example of building-like nanostructures is "nano–New York" (Figure 9.131) and previously mentioned nanotepee (see Figure 9.51).

---

* See also Section 3.9.

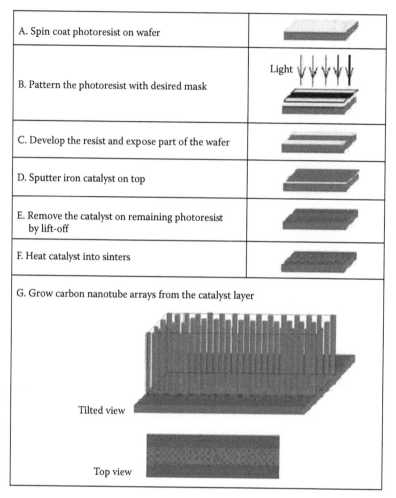

A. Spin coat photoresist on wafer	
B. Pattern the photoresist with desired mask	
C. Develop the resist and expose part of the wafer	
D. Sputter iron catalyst on top	
E. Remove the catalyst on remaining photoresist by lift-off	
F. Heat catalyst into sinters	

G. Grow carbon nanotube arrays from the catalyst layer

Tilted view

Top view

FIGURE 9.129 A diagram of nanowick fabrication and its final configuration. A–G: patterns in the catalyst layer are fabricated with photolithography. F, G: nanotubes are produced in arrays with individual nanotubes being vertically oriented with respect to the substrate. The nanotubes only exist in the designed regions. (Zhou, J.J. et al., *Nanotechnology*, 17, 4845, 2006. Reproduced with permission from Institute of Physics Publishing.)

9.14.5 NANOLADDERS

A few ladder-like nanostructures are known, for instance ZnO nanoladders, consisting of two nanobelts whose growth directions are different from each other, synthesized by a simple CVD technique.[314] A two-step growth model was speculated by authors to interpret the growth of ZnO nanoladder based on the detailed structure analyses. The supersaturation of the Zn vapor and the Coulomb repulsion between the two nanobelts were proposed as the dominating factors for the growth of ZnO nanoladders. Large-scale ear-like Si_3N_4 dendrites were prepared by the reaction of SiO_2/Fe composites and Si powders in N_2 atmosphere.[315] Changing the Fe content in the SiO_2/Fe

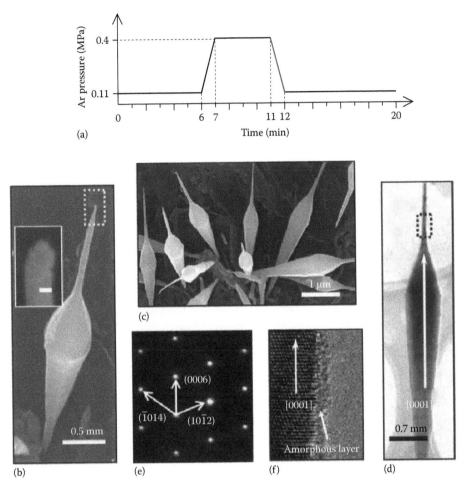

FIGURE 9.130 (a) Schematic showing a three-step processing schedule. (b) High-magnification SEM image of an Eiffel Tower–shaped SiC nanowire; the inset is an enlarged image of the tip area (the scale bar corresponds to 10 nm). (c) SEM image of Eiffel Tower–shaped SiC nanowires. (d) TEM image of a nanowire. (e, f) SAED pattern and HRTEM image, respectively, taken from the area marked by the square in (d). (Reproduced with permission from Wang, H. et al., Morphology control in the vapor-liquid-solid growth of SiC nanowires, *Cryst. Growth Des.*, 8(11), 3893–3896. Copyright 2008 American Chemical Society.)

composites, nanosized ladder-like Si_3N_4 with a length of hundreds of nanometers to several microns and a width of 100–300 nm was also obtained.

A series of reports are devoted to CNT ladder-like structures. Thus, various single-walled CNTs (SWNTs) and double-walled CNTs (DWNTs) with different diameters were used as containers for cubic octameric $H_8Si_8O_{12}$ molecules (Figure 9.132).[316] Depending on the diameter of the CNTs, two types of structures were formed inside the CNTs: in the case of CNTs with inner diameters ranging from 1.2 to 1.4 nm, an ordered self-assembled structure composed of $H_8Si_{4n}O_8n-4$ molecules was formed through the transformation of $H_8Si_8O_{12}$; in the case of CNTs with inner diameters larger than 1.7 nm, a disordered structure was formed, indicating strong interactions between the CNTs and the encapsulated $H_8Si_{4n}O_8n-4$ molecules. In addition, MWCNTs were uniformly grown directly

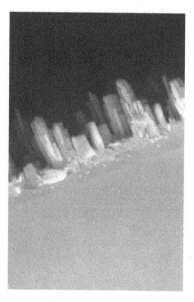

FIGURE 9.131 Nano–New York {$K_4Nb_6O_{17}$ on the monocrystalline SrTiO3(100) support}. (Reproduced from Mankevich, A.S., http://www.nanometer.ru/2007/09/14/mocvd_4265.html. Accessed June 4, 2010. With permission.)

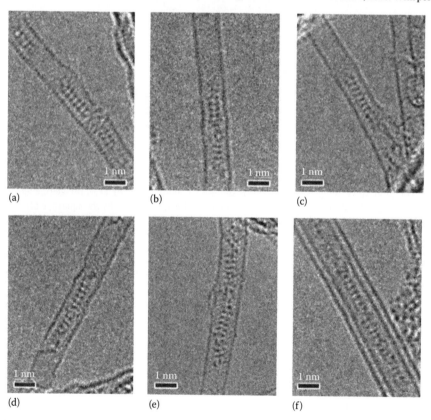

FIGURE 9.132 HRTEM images of $H_8Si_8O_{12}$ encapsulated inside the various CNTs. (a through d) HRTEM images of $H_8Si_8O_{12}$@HiPCO SWNTs, (e) HRTEM image of $H_8Si_8O_{12}$@FH-P SWNT, and (f) HRTEM image of $H_8Si_8O_{12}$@FH-P DWNT. (Reproduced with permission from Liu, Z. et al., Self-assembled double ladder structure formed inside carbon nanotubes by encapsulation of $H_8Si_8O_{12}$, *ACS Nano*, 3(5), 1160–1166, Copyright 2009 American Chemical Society.)

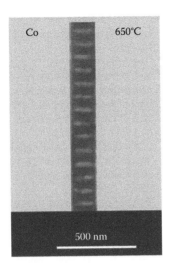

FIGURE 9.133 Carbon nanoladder. (Reproduced from *Surf. Coat. Technol.*, 203(10–11), Sharma, S. P. and Lakkad, S.C., Morphology study of carbon nanospecies grown on carbon fibers by thermal CVD technique, 1329–1335. Copyright 2009, with permission from Elsevier.)

on metal-catalyzed carbon fibers by catalytic decomposition of acetylene at ~700°C using thermal CVD process, accompanied by the formation of various carbon nanostructures such as nanofilaments, nanofibers, nanoladders (Figure 9.133), nanospirals, nanojunctions, nanocones, nanoonions, etc., accompanying nanotubes.[317] The authors noted that precise control of the process parameters was extremely important to grow a desired morphology; otherwise, it could result in a wide variety of undesired carbon nanospecies too.

REFERENCES

1. Bardhan, R.; Wang, H.; Tam, F.; Halas, N. ZnO Nanobowls, nanobagels and core-shell ZnO nanostructures. *Abstracts, 62nd Southwest Regional Meeting of the American Chemical Society*, Houston, TX, October 19–22, 2006, SRM-373.
2. Fu, M.; Zhao, A.; Zhou, J.; He, D.; Wang, Y. Morphologically controlled electrodeposition of ZnO and Cu_2O by the colloidal crystal template method. *Journal of Nonlinear Optical Physics and Materials*, 2009, *18* (4), 611–616.
3. Liu, J. J.; Yu, M. H.; Zhou, W. L. Fabrication of Mn-doped ZnO diluted magnetic semiconductor nanostructures by chemical vapor deposition. *Journal of Applied Physics*, 2006, *99* (8, Pt. 3), 08M119/1–08M119/3.
4. Krishna, K. S.; Mansoori, U.; Selvi, N. R.; Eswaramoorthy, M. Form emerges from formless entities: Temperature-induced self-assembly and growth of ZnO nanoparticles into zeptoliter bowls and troughs. *Angewandte Chemie, International Edition*, 2007, *46* (31), 5962–5965, S5962/1–S5962/7.
5. Yuan, A.; Bao, X.; Wang, P.; Zhou, H.; Yuan, Y. Characterization of zinc acetylacetonate hydrate synthesized by chemical vapor deposition. *Jiangsu Keji Daxue Xuebao, Ziran Kexueban*, 2007, *21* (1), 75–79.
6. Wang, Y.; Zhang, J.; Chen, X.; Li, X.; Sun, Z.; Zhang, K.; Wang, D.; Yang, B. Morphology-controlled fabrication of polygonal ZnO nanobowls templated from spherical polymeric nanowell arrays. *Journal of Colloid and Interface Science*, 2008, *322* (1), 327–332.
7. Wang, Y.; Chen, X.; Zhang, J.; Sun, Z.; Li, Y.; Zhang, K.; Yang, B. Fabrication of surface-patterned and free-standing ZnO nanobowls. *Colloids and Surfaces, A: Physicochemical and Engineering Aspects*, 2008, *329* (3), 184–189.
8. Zhang, L.-B.; Xiong, Y.-Y.; Chu, B.-L.; Yu, H.-H.; Xiao, H. Fabrication of zinc oxide thin films by magnetron sputtering based on opal template and its optical properties. *Wuji Cailiao Xuebao*, 2007, *22* (6), 1117–1121.
9. Yu, T.; Varghese, B.; Shen, Z.; Lim, C.-T.; Sow, C.-H. Large-scale metal oxide nanostructures on template-patterned microbowls: A simple method for growth of hierarchical structures. *Materials Letters*, 2008, *62* (3), 389–393.

10. Li, X.; Peng, J.; Kang, J.-H.; Choy, J.-H.; Steinhart, M.; Knoll, W.; Kim, D. H. One step route to the fabrication of arrays of TiO_2 nanobowls via a complementary block copolymer templating and sol-gel process. *Soft Matter*, 2008, *4* (3), 515–521.

11. Wang, X.; Lao, C.; Graugnard, E.; Summers, C. J.; Wang, Z. L. Large-area nanobowl sheet for nanodots patterning. *Abstracts of Papers, 231st ACS National Meeting*, Atlanta, GA, March 26–30, 2006, INOR-821.

12. Wang, X.; Lao, C.; Graugnard, E.; Summers, C. J.; Wang, Z. L. Large-size liftable inverted-nanobowl sheets as reusable masks for nanolithography. *Nano Letters*, 2005, *5* (9), 1784–1788.

13. Wang, X. D.; Graugnard, E.; King, J. S.; Wang, Z. L.; Summers, C. J. Large-scale fabrication of ordered nanobowl arrays. *Nano Letter*, 2004, *4* (11), 2223–2226.

14. Knez, M.; Nielsch, K.; Niinistö, L. Synthesis and surface engineering of complex nanostructures by atomic layer deposition. *Advanced Materials*, 2007, *19*, 3425–3438.

15. Chen, X.; Mao, S. S. Titanium dioxide nanomaterials: Synthesis, properties, modifications, and applications. *Chemical Reviews*, 2007, *107*, 2891–2959.

16. Peng, J.; Li, X.; Kim, D. H.; Knoll, W. Fabrication and photocatalytic activities of morphology-controlled titania nanoobject arrays by block copolymer templates. *Macromolecular Rapid Communications*, 2007, *28* (21), 2055–2061.

17. Xia, X.-H.; Tu, J.-P.; Zhang, J.; Xiang, J.-Y.; Wang, X.-L.; Zhao, X.-B. Cobalt oxide ordered bowl-like array films prepared by electrodeposition through monolayer polystyrene sphere template and electrochromic properties. *ACS Applied Materials and Interfaces*, 2010, *2* (1), 186–192.

18. Ye, J.; Van Dorpe, P.; Van Roy, W.; Borghs, G.; Maes, G. Fabrication, characterization, and optical properties of gold nanobowl submonolayer structures. *Langmuir*, 2009, *25* (3), 1822–1827.

19. Jiao, L.; Fan, B.; Xian, X.; Wu, Z.; Zhang, J.; Liu, Z. Creation of nanostructures with poly(methyl methacrylate)-mediated nanotransfer printing. *Journal of the American Chemical Society*, 2008, *130* (38), 12612–12613.

20. Ye, J.; Lagae, L.; Maes, G.; Borghs, G.; Van Dorpe, P. Symmetry breaking induced optical properties of gold open shell nanostructures. *Optics Express*, 2009, *17* (26), 23765–23771.

21. Xu, M.; Lu, N.; Xu, H.; Qi, D.; Wang, Y.; Chi, L. Fabrication of functional silver nanobowl arrays via sphere lithography. *Langmuir*, 2009, *25* (19), 11216–11220.

22. Li, Y.; Li, C.; Cho, S. O.; Duan, G.; Cai, W. Silver hierarchical bowl-like array: Synthesis, superhydrophobicity, and optical properties. *Langmuir*, 2007, *23* (19), 9802–9807.

23. Sun, Y.; Chen, X.; Yue, Y.; Zhang, R.-J.; Dai, N. Flexible aluminum nanobowls for alternative preparation of individual or a small number of nanoparticles. *Chemical Research in Chinese Universities*, 2009, *25* (2), 143–146.

24. Chen, X.; Wei, X.; Jiang, K. Large-scale fabrication of ordered metallic hybrid nanostructures. *Optics Express*, 2008, *16* (16), 11888–11893.

25. Chen, T.-H.; Tsai, T.-Y.; Hsieh, K.-C.; Chang, S.-C.; Tai, N.-H.; Chen, H.-L. Two-dimensional metallic nanobowl array transferred onto thermoplastic substrates by microwave heating of carbon nanotubes. *Nanotechnology*, 2008, *19* (46), 465303/1–465303/6.

26. Huang, J.; Zhang, Q.; Zhao, M; Wei, F. Process intensification by CO_2 for high quality carbon nanotube forest growth: Double-walled carbon nanotube convexity or single-walled carbon nanotube bowls? *Nano Research*, 2009, *2* (11), 872–881.

27. Srivastava, A. K.; Madhavi, S.; White, T. J.; Ramanujan, R. V. Template assisted assembly of cobalt nanobowl arrays. *Journal of Materials Chemistry*, 2005, *15* (41), 4424–4428.

28. Srivastava, A. K.; Madhavi, S.; White, T. J.; Ramanujan, R. V. The processing and characterization of magnetic nanobowls. *Thin Solid Films*, 2006, *505* (1–2), 93–96.

29. Srivastava, A. K.; Madhavi, S.; White, T. J.; Ramanujan, R. V. Cobalt-ferrite nanobowl arrays: Curved magnetic nanostructures. *Journal of Materials Research*, 2007, *22* (5), 1250–1254.

30. Liu, J.; Zhu, M.; Zhan, P.; Dong, H.; Dong, Y.; Qu, X.; Nie, Y.; Wang, Z. Morphology-controllable fabrication of ordered platinum nanoshells with reduced symmetry. *Nanotechnology*, 2006, *17* (16), 4191–4194.

31. Lu, Y.; Chen, S. C. Nanopatterning of a silicon surface by near-field enhanced laser irradiation. *Nanotechnology*, 2003, *14* (5), 505–508.

32. Trujillo, N. J.; Baxamusa, S. H.; Gleason, K. K. Grafted functional polymer nanostructures patterned bottom-up by colloidal lithography and initiated chemical vapor deposition (iCVD). *Chemistry of Materials*, 2009, *21* (4), 742–750.

33. Chen, X.; Wei, X.; Jiang, K. Fabrication of large-area nickel nanobump arrays. *Microelectronic Engineering*, 2009, *86* (4–6), 871–873.

34. Chen, J.; Chao, D.; Lu, X.; Zhang, W.; Manohar, S. K. General synthesis of two-dimensional patterned conducting polymer-nanobowl sheet via chemical polymerization. *Macromolecular Rapid Communications*, 2006, *27* (10), 771–775.

35. Jiang, S.; Chen, J.; Tang, J.; Jin, E.; Kong, L.; Zhang, W.; Wang, C. Au nanoparticles-functionalized two-dimensional patterned conducting PANI nanobowl monolayer for gas sensor. *Sensors and Actuators, B: Chemical*, 2009, *B140* (2), 520–524.

36. Zhao, G.; Ishizaka, T.; Kasai, H.; Hasegawa, M.; Nakanishi, H.; Oikawa, H. Anomalous polyimide nanoparticles prepared from blending of unlike polymers. *Molecular Crystals and Liquid Crystals*, 2009, *504*, 9–17.

37. Lassiter, J. B.; Knight, M. W.; Mirin, N. A.; Halas, N. J. Reshaping the plasmonic properties of an individual nanoparticle. *Nano Letters*, 2009, *9* (12), 4326–4332.

38. Mirin, N. A., Halas, N. J. Light-bending nanoparticles. *Nano Letters*, 2009, *9* (3), 255–1259.

39. Charnay, C.; Lee, A.; Man, S.-Q.; Moran, C. E.; Radloff, C.; Bradley, R. K.; Halas, N. J. Reduced symmetry metallodielectric nanoparticles: Chemical synthesis and plasmonic properties. *Journal of Physical Chemistry B*, 2003, *107* (30), 7327–7333.

40. Tang, Y.; Allen, B. L.; Kauffman, D. R.; Star, A. Electrocatalytic activity of nitrogen-doped carbon nanotube cups. *Journal of the American Chemical Society*, 2009, *131* (37), 13200–13201.

41. Chun, H.; Hahm, M. G.; Homma, Y.; Meritz, R.; Kuramochi, K.; Menon, L.; Ci, L.; Ajayan, P. M.; Jung, Y. J. Engineering low-aspect ratio carbon nanostructures: Nanocups, nanorings, and nanocontainers. *ACS Nano*, 2009, *3* (5), 1274–1278.

42. Ohtani, M.; Saito, K.; Fukuzumi, S. Synthesis, characterization, redox properties, and photodynamics of donor-acceptor nanohybrids composed of size-controlled cup-shaped nanocarbons and porphyrins. *Chemistry—A European Journal*, 2009, *15* (36), 9160–9168, S9160/1–S9160/3.

43. Deng, C.-H.; Hu, H.-M.; Huang, X.-H.; Zhu, S.-F. Sonochemical synthesis of semiconductor ZnO nanocups. *Wuji Huaxue Xuebao*, 2009, *25* (10), 1742–1746.

44. Jung, S.-H.; Oh, E.; Lee, K.-H.; Yang, Y.; Park, C. G.; Park, W.; Jeong, S.-H. Sonochemical Preparation of shape-selective ZnO nanostructures. *Crystal Growth and Design*, 2008, *8* (1), 265–269.

45. Jagadeesan, D.; Mansoori, U.; Mandal, P.; Sundaresan, A.; Eswaramoorthly, M. Hollow spheres to nanocups: Tuning the morphology and magnetic properties of single-crystalline α-Fe₂O₃ nanostructures. *Angewandte Chemie, International Edition*, 2008, *47* (40), 7685–7688.

46. Deotare, P.; Kameoka, J. Sorting of silica nanocups by diameter during fabrication process. *Journal of Nanomaterials*, 2007, DOI Bookmark: 10.1155/2007/71259 research article. http://www.hindawi.com/journals/jnm/2007/071259/abs/

47. Deotare, P. B.; Kameoka, J. Fabrication of silica nanocomposite-cups using electrospraying. *Nanotechnology*, 2006, *17* (5), 1380–1383.

48. Zhang, M.; Zhai, T.; Wang, X.; Liao, Q.; Ma, Y.; Yao, J. Carbon-assisted morphological manipulation of CdS nanostructures and their cathodoluminescence properties. *Journal of Solid State Chemistry*, 2009, *182* (11), 3188–3194.

49. Farrow, B.; Kamat, P. V. CdSe quantum dot sensitized solar cells. Shuttling electrons through stacked carbon nanocups. *Journal of the American Chemical Society*, 2009, *131* (31), 11124–11131.

50. Su, C.-Y.; Juang, Z.-Y.; Chen, K.-F.; Cheng, B.-M.; Chen, F.-R.; Leou, K.-C.; Tsai, C.-H. Selective growth of boron nitride nanotubes by the plasma-assisted and iron-catalytic CVD methods. *Journal of Physical Chemistry C*, 2009, *113* (33), 14681–14688.

51. Kidwai, M. Nanoparticles in green catalysis. In *Handbook of Green Chemistry: Green Catalysis*. Anastas, P. T.; Crabtree, R. H. (Eds.). Wiley-VCH Verlag GmbH & Co. KGaA, Weinheim, Germany, 2009, Vol. 2, pp. 81–92.

52. Saj, W. M.; Antosiewich, T. J.; Pniewski, J.; Szoplik, T. Energy transport in plasmon waveguides on chains of metal nanoplates. *Opto-Electronics Review*, 2006, *14* (3), 243–251.

53. Kawasaki, H.; Gold and silver nanoplate. In *Kinzoku Nano/Maikuro Ryushi no Keijo/Kozo Geigyo Gijutsu*. Yonezawa, T. (Ed.). 2009, pp. 53–62.

54. Soejima, T.; Kimizuka, N. One-pot room-temperature synthesis of single-crystalline gold nanocorolla in water. *Journal of the American Chemical Society*, 2009, *131* (40), 14407–14412.

55. Porel, S.; Singhb, S.; Radhakrishnan, T. P. Polygonal gold nanoplates in a polymer matrix. *Chemical Communications*, 2005, 2387–2389.

56. Lin, G.; Lu, W.; Cui, W.; Jiang, L. A simple synthesis method for gold nano- and microplate fabrication using a tree-type multiple-amine head surfactant. *Crystal Growth and Design*, 2010, *10* (3), 1118–1123.

57. He, S.; Guo, Z.; Zhang, Y.; Zhang, S.; Wang, J.; Gu, N. Biosynthesis of gold nanoparticles using the bacteria *Rhodopseudomonas capsulata*. *Materials Letters*, 2007, *61*, 2007, 3984–3987.

58. Ghodake, G. S.; Deshpande, N. G.; Lee, Y. P.; Jin, E. S. Pear fruit extract-assisted room-temperature biosynthesis of gold nanoplates. *Colloids and Surfaces, B: Biointerfaces*, 2010, *75* (2), 584–589.

59. Fan, X.; Guo, Z. R.; Hong, J. M.; Zhang, Y.; Zhang, J. N.; Gu, N. Size-controlled growth of colloidal gold nanoplates and their high-purity acquisition. *Nanotechnology*, 2010, *21* (10), 105602/1–105602/7.

60. Huang, Y.; Wang, W.; Liang, H.; Xu, H. Surfactant-promoted reductive synthesis of shape-controlled gold nanostructures. *Crystal Growth and Design*, 2009, *9* (2), 858–862.

61. Sun, Y. Synthesis of Ag nanoplates on GaAs wafers: Evidence for growth mechanism. *Journal of Physical Chemistry C*, 2010, *114* (2), 857–863.

62. Wang, Y.-H.; Zhang, Q.; Wang, T.; Zhou, J. Nanoframes directed synthesis of triangular silver nanoplates. *Wuji Huaxue Xuebao*, 2010, *26* (3), 365–373.

63. Murayama, H.; Hashimoto, N.; Tanaka, H. Ag triangular nanoplates synthesized by photo-induced reduction: Structure analysis and stability. *Chemical Physics Letters*, 2009, *482* (4–6), 291–295.

64. Kim, J.-Y.; Kim, S. J.; Jang, D.-J. Laser-induced shape transformation and electrophoretic analysis of triangular silver nanoplates. *Journal of Separation Science*, 2009, *32* (23–24), 4161–4166.

65. Washio, I.; Xiong, Y.; Yin, Y.; Xia, Y. Reduction by the end groups of poly(vinyl pyrrolidone): A new and versatile route to the kinetically controlled synthesis of Ag triangular nanoplates. *Advanced Materials*, 2006, *18*, 1745–1749.

66. Charles, D. E.; Aherne, D.; Gara, M.; Ledwith, D. M.; Gun'ko, Y. K.; Kelly, J. M.; Blau, W. J.; Brennan-Fournet, M. E. Versatile solution phase triangular silver nanoplates for highly sensitive plasmon resonance sensing. *ACS Nano*, 2010, *4* (1), 55–64.

67. Xiong, Y.; Xia, Y. Shape-controlled synthesis of metal nanostructures: The case of palladium. *Advanced Materials*, 2007, *19*, 3385–3391.

68. Jang, K.; Kim, H. J.; Son, S. U. Low-temperature synthesis of ultrathin rhodium nanoplates via molecular orbital symmetry interaction between rhodium precursors. *Chemistry of Materials*, 2010, *22* (4), 1273–1275.

69. Lee, C.-L.; Tseng, C.-M.; Wu, R.-B.; Wu, C.-C.; Syu, C.-M. Porous Ag-Pd triangle nanoplates with tunable alloy ratio for catalyzing electroless copper deposition. *Colloids and Surfaces, A: Physicochemical and Engineering Aspects*, 2009, *352* (1–3), 84–87.

70. Lim, B.; Wang, J.; Camargo, P. H. C.; Jiang, M.; Kim, M. J.; Xia, Y. Synthesis of palladium-platinum bimetallic core-shell nanoplates through seeded epitaxial growth. *Abstracts of Papers, 238th ACS National Meeting*, Washington, DC, August 16–20, 2009, INOR-230.

71. Yang, X.; Zhou, T.; Chen, C. A molecular dynamics simulation of single crystalline copper nano-plate under uniaxial tensile/compressive loading. *Huazhong Keji Daxue Xuebao, Ziran Kexueban*, 2006, *34* (12), 62–64.

72. Guan, J.; Yan, G. Iron nanoplates: Facile synthesis, formation mechanism, and selective catalytic oxidation of cyclohexane. *Abstracts of Papers, 238th ACS National Meeting*, Washington, DC, August 16–20, 2009, CATL-026.

73. Wang, J.; Huang, Q.-A.; Yu, H. Size and temperature dependence of Young's modulus of a silicon nanoplate. *Journal of Physics D: Applied Physics*, 2008, *41* (16), 165406/1–165406/5.

74. Wang, J.; Huang, Q.-A.; Yu, H. Effect of (2 × 1) surface reconstruction on elasticity of a silicon nanoplate. *Chinese Physics Letters*, 2008, *25* (4), 1403–1406.

75. Tran, N. H.; Wilson, M. A.; Milev, A. S.; Dennis, G. R.; Kannangara, G. S. K. Mechanism of silica nanoplate formation from lucentite. *Surface Review and Letters*, 2007, *14* (2), 235–239.

76. Meshkani, F.; Rezaei, M. Effect of process parameters on the synthesis of nanocrystalline magnesium oxide with high surface area and plate-like shape by surfactant assisted precipitation method. *Powder Technology*, 2010, *199* (2), 144–148.

77. Bando, Y.; Zhang, J.-H. Method to produce zinc oxide nano-plate and nano-rod binding material. JP 2006199552, 2006, 5 pp.

78. Illy, B.; Shollock, B. A.; MacManus-Driscoll, J. L.; Ryan, M. P. Electrochemical growth of ZnO nanoplates. *Nanotechnology*, 2005, *16* (2), 320–324.

79. Lee, N.-H.; Oh, H.-J.; Yoon, C.-R.; Guo, Y.; Park, K.-S.; Kim, B.-W.; Kim, S.-J. Understanding for controls of particle shape of various titanates with layered structure. *Journal of Nanoscience and Nanotechnology*, 2008, *8* (10), 5158–5161.

80. Kumaresan, L.; Mahalakshmi, M.; Palanichamy, M.; Murugesan, V. Synthesis, characterization, and photocatalytic activity of Sr^{2+} doped TiO_2 nanoplates. *Industrial and Engineering Chemistry Research*, 2010, *49* (4), 1480–1485.

81. Su, X.; Xiao, F.; Li, Y.; Jian, J.; Sun, Q.; Wang, J. Synthesis of uniform WO_3 square nanoplates via an organic acid-assisted hydrothermal process. *Materials Letters*, 2010, *64* (10), 1232–1234.

82. Chen, X.; Lei, W.; Liu, D.; Hao, J.; Cui, Q.; Zou, G. Synthesis and characterization of hexagonal and truncated hexagonal shaped MoO_3 nanoplates. *Journal of Physical Chemistry C*, 2009, *113* (52), 21582–21585.

83. Chen, D.; Hou, X.; Wen, H.; Wang, Y.; Wang, H.; Li, X.; Zhang, R. et al. The enhanced alcohol-sensing response of ultrathin WO_3 nanoplates. *Nanotechnology*, 2010, *21* (3), 035501/1–035501/12.

84. Zhang, C.; Wang, L.; Peng, H.; Chen, K.; Li, G. Polyaniline-vanadium oxide hybrid hierarchical architectures assembled from nanoscale building blocks by homogeneous polymerization. *Polymer International*, 2009, *58* (12), 1422–1426.

85. Zhou, H.-F.; Yi, R.; Li, J.-H.; Su, Y.; Liu, X.-H. Microwave-assisted synthesis and characterization of hexagonal Fe_3O_4 nanoplates. *Solid State Science*, 2010, *12* (1), 99–104.

86. Ahmed, K. A. M.; Zeng, Q.; Wu, K.; Huang, K. Mn_3O_4 nanoplates and nanoparticles: Synthesis, characterization, electrochemical and catalytic properties. *Journal of Solid State Chemistry*, 2010, *183* (3), 744–751.

87. Liu, L.; Yang, Z.; Liang, H.; Yang, H.; Yang, Y. Shape-controlled synthesis of manganese oxide nanoplates by a polyol-based precursor route. *Materials Letters*, 2010, *64* (7), 891–893.

88. Lin, H.-L.; Wu, C.-Y.; Chiang, R.-K. Facile synthesis of CeO_2 nanoplates and nanorods by [100] oriented growth. *Journal of Colloid and Interface Science*, 2009, Volume Date 2010, *341* (1), 12–17.

89. Hu, C.; Zhu, Q.; Chen, L.; Wu, R. CuO-CeO_2 binary oxide nanoplates: Synthesis, characterization, and catalytic performance for benzene oxidation. *Materials Research Bulletin*, 2009, *44* (12), 2174–2180.

90. Cao, H.; Zheng, H.; Liu, K.; Warner, J. H. Bioinspired peony-like β-$Ni(OH)_2$ nanostructures with enhanced electrochemical activity and superhydrophobicity. *ChemPhysChem*, 2010, *11* (2), 489–494.

91. Ma, J.; Liu, X.; Lian, J.; Duan, X.; Zheng, W. Ionothermal synthesis of $BiOCl$ nanostructures via a long-chain ionic liquid precursor route. *Crystal Growth and Design*, 2010, *10* (6), 2522–2527.

92. Lei, Y.; Wang, G.; Song, S.; Fan, W.; Zhang, H. Synthesis, characterization and assembly of $BiOCl$ nanostructure and their photocatalytic properties. *Crystal Engineering Communications*, 2009, *11* (9), 1857–1862.

93. Ai, Z.; Ho, W.; Lee, S.; Zhang, L. Efficient photocatalytic removal of NO in indoor air with hierarchical bismuth oxybromide nanoplate microspheres under visible light. *Environmental Science and Technology*, 2009, *43* (11), 4143–4150.

94. Tian, Y.; Jiao, X.; Zhang, J.; Sui, N.; Chen, D.; Hong, G. Molten salt synthesis of LaF_3:Eu^{3+} nanoplates with tunable size and their luminescence properties. *Journal of Nanoparticle Research*, 2010, *12* (1), 161–168.

95. Baek, J. Y.; Ha, H.-W.; Kim, I.-Y.; Hwang, S.-J. Hierarchically assembled 2D nanoplates and 0D nanoparticles of lithium-rich layered lithium manganates applicable to lithium ion batteries. *Journal of Physical Chemistry C*, 2009, *113* (40), 17392–17398.

96. De, D.; Ram, S.; Banerjee, A.; Gupta, A.; Roy, S. K. Magnetic properties of $La_{0.67}Ca_{0.33}MnO_3$ nanoplates. *IEEE Transactions on Magnetics*, 2009, *45* (10), 4352–4356.

97. Sun, S.; Wang, W.; Zhou, L.; Xu, H. Efficient methylene blue removal over hydrothermally synthesized starlike $BiVO_4$. *Industrial and Engineering Chemistry Research*, 2009, *48* (4), 1735–1739.

98. Xi, G.; Ye, J. Synthesis of bismuth vanadate nanoplates with exposed {001} facets and enhanced visible-light photocatalytic properties. *Chemical Communications*, 2010, *46* (11), 1893–1895.

99. Wang, J.; Xu, Y.; Cen, Q.; Zhang, X.; Cui, Y. Synthesis and luminescent properties of porous YVO_4:Sm nanoplates. *Materials Science Forum*, 2009, *620–622 (Eco-Materials Processing and Design X)*, 541–544.

100. Li, L.-Y.; Song, W.-H.; Chen, T.-H. Synthesis of hydroxyapatite nanoplates by the hydrothermal reaction method with anionic amino acid surfactant. *Wuli Huaxue Xuebao*, 2009, *25* (11), 2404–2408.

101. Cho, I.-S.; Kim, D. W.; Noh, T. H.; Lee, S.; Yim, D. K.; Hong, K. S. Preparation of N-doped $CaNb_2O_6$ nanoplates with ellipsoid-like morphology and their photocatalytic activities under visible-light irradiation. *Journal of Nanoscience and Nanotechnology*, 2010, *10* (2), 1196–1202.

102. Mi, Y.; Huang, Z.; Hu, F.; Li, Y.; Jiang, J. Room-Temperature synthesis and luminescent properties of single-crystalline $SrMoO_4$ nanoplates. *Journal of Physical Chemistry C*, 2009, *113* (49), 20795–20799.

103. Chen, X.; Hu, J.; Chen, Z.; Feng, X.; Li, A. Nanoplated bismuth titanate sub-microspheres for protein immobilization and their corresponding direct electrochemistry and electrocatalysis. *Biosensors and Bioelectronics*, 2009, *24* (12), 3448–3454.

104. Koinuma, H; Matsumoto, Y; Takahashi, R. Bi-layered compound nano-plate, array thereof, process for producing them, and apparatus utilizing the same. WO 2006013826, 2006, 37 pp.

105. Kim, D.-W.; Ko, Y.-D.; Park, J.-S.; Je, H.-J.; Son, J.-W.; Kim, J. Electrochemical performance of calcium cobaltite nano-plates. *Journal of Nanoscience and Nanotechnology*, 2009, *9* (7), 4056–4060.

106. Yang, W.; Gao, F.; Wei, Gg; An, L. Ostwald ripening growth of silicon nitride nanoplates. *Crystal Growth and Design*, 2010, *10* (1), 29–31.

107. Li, F.; Bi, W.; Kong, T.; Qin, Q. Optical, photocatalytic properties of novel CuS nanoplate-based architectures synthesised by a solvothermal route. *Crystal Research and Technology*, 2009, *44* (7), 729–735.

108. Zhang, X.; Guo, Y.; Zhang, P.; Wu, Z.; Zhang, Z. Superhydrophobic CuO@Cu$_2$S nanoplate vertical arrays on copper surfaces. *Materials Letters*, 2010, *64* (10), 1200–1203.

109. Zeng, Y.; Li, H.; Xiang, B.; Ma, H.; Qu, B.; Xia, M.; Wang, Y.; Zhang, Q.; Wang, Y. Synthesis and characterization of phase-purity Cu$_9$BiS$_6$ nanoplates. *Materials Letters*, 2010, *64* (9), 1091–1094.

110. Xu, J.; Jang, K.; Jung, I. G.; Kim, H. J.; Oh, D.-H.; Ahn, J. R.; Son, S. U. Cutting gallium oxide nanoribbons into ultrathin nanoplates. *Chemistry of Materials*, 2009, *21* (19), 4347–4349.

111. Zhang, L.-S.; Wong, K.-H.; Zhang, D.-Q.; Hu, C.; Yu, J. C.; Chan, C.-Y.; Wong, P.-K. Zn:In(OH)$_y$S$_z$ solid solution nanoplates: Synthesis, characterization, and photocatalytic mechanism. *Environmental Science and Technology*, 2009, *43* (20), 7883–7888.

112. Deng, Z.; Mansuripur, M.; Muscat, A. J. Synthesis of two-dimensional single-crystal berzelianite nanosheets and nanoplates with near-infrared optical absorption. *Journal of Materials Chemistry*, 2009, *19* (34), 6201–6206.

113. Kong, D.; Dang, W.; Cha, J. J.; Li, H.; Meister, S.; Peng, H.; Liu, Z.; Cui, Y. Few-layer nanoplates of Bi$_2$Se$_3$ and Bi$_2$Te$_3$ with highly tunable chemical potential. *Nano Letters*, 2010, *10* (6), 2245–2250.

114. Fasi, A.; Palinko, I.; Hernadi, K.; Kiricsi, I. Formation of Au nanorods and nanoforks over MgO support. *Reaction Kinetics and Catalysis Letters*, 2006, *87* (2), 263–268.

115. Xu, T. T.; Zheng, J. G.; Wu, N. Q.; Nicholls, A. W.; Roth, J. R.; Dikin, D. A.; Ruoff. R. S. Crystalline boron nanoribbons: Synthesis and characterization. *Nano Letters*, 2004, *4* (5), 963–968.

116. Singh, S.; Ramachandra Rao, M. S. Green light emitting oxygen deficient ZnO forks, brooms and spheres. *Scripta Materialia*, 2009, *61* (2), 169–172.

117. Liu, R. N.; Yang, H. Q.; Zhang, R. G.; Dong, H. X.; Chen, X. B.; Li, L.; Zhang, L.H.; Ma, J. H.; Zheng, H. R. In-situ growth and photoluminescence of β-Ga$_2$O$_3$ cone-like nanowires on the surface of Ga substrates. *Science in China Series E: Technological Sciences*, 2009, *52* (6), 1712–1721.

118. Wang, G.; Li, G. Titania from nanoclusters to nanowires and nanoforks. *European Physical Journal D: Atomic, Molecular and Optical Physics*, 2003, *24* (1–3), 355–360.

119. Shi, S.; Cao, M.; Hu, C. Controlled solvothermal synthesis and structural characterization of antimony telluride nanoforks. *Crystal Growth and Design*, 2009, *9* (5), 2057–2060.

120. Chen, Y.; Qian, Q.; Liu, X.; Xiao, L.; Chen, Q. LaOCl nanofibers derived from electrospun PVA/lanthanum chloride composite fibers. *Materials Letters*, 2010, *64* (1), 6–8.

121. Jores, K.; Mehnert, W.; Drechsler, M.; Bunjes, H.; Johann, C.; Mäder, K. Investigations on the structure of solid lipid nanoparticles (SLN) and oil-loaded solid lipid nanoparticles by photon correlation spectroscopy, field-flow fractionation and transmission electron microscopy. *Journal of Controlled Release*, 2004, *95* (2), 217–227.

122. Jores, K.; Mehnert, W.; Bunjes, H.; Drechsler, M.; Mäder, K. From solid lipid nanoparticles (SLN) to nanospoons. Visions and reality of colloidal lipid dispersions, *30th International Symposium on Controlled Release of Bioactive Materials*, Glasgow, U.K., Vol. 29, Controlled Release Society, St. Paul, MN, 2003, p. 181.

123. Kolen'ko, Yu. V.; Kovnir, K. A.; Gavrilov, A. I.; Garshev, A. V.; Frantti, J.; Lebedev, O. I.; Churagulov, B. R.; Van Tendeloo, O. G.; Yoshimura, M. Hydrothermal synthesis and characterization of nanorods of various titanates and titanium dioxide. *Journal of Physical Chemistry B*, 2006, *110*, 4030–4038.

124. Cottam, B. F.; Krishnadasan, S.; deMello, A. J.; deMello, J. C.; Shaffer, M. S. P. Accelerated synthesis of titanium oxide nanostructures using microfluidic chips. *Lab on a Chip*, 2007, *7*, 167–169.

125. Klein, S. M.; Choi, J. H.; Pine, D. J.; Lange, F. F. Synthesis of rutile titania powders: Agglomeration, dissolution, and reprecipitation phenomena. *Journal of Materials Research*, 2003, *18* (6), 1457–1464.

126. Li, B.; Xie, Y.; Wu, C.; Li, Z.; Zhang, J. Selective synthesis of cobalt hydroxide carbonate 3D architectures and their thermal conversion to cobalt spinel 3D superstructures. *Materials Chemistry and Physics*, 2006, *99* (2–3), 479–486.

127. Zhao, J.; Wu, Q.-S.; Wen, M. Temperature-controlled assembly and morphology conversion of CoMoO$_4$·3/4H$_2$O nano-superstructured grating materials. *Journal of Materials Science*, 2009, *44* (23), 6356–6362.

128. Uthirakumar, P.; Suh, E.-K.; Hong, C.-H. The impact of ZnO nanoparticle interlayer on the growth and morphology of broom-like single crystalline gallium nitride. *Journal of Nanoscience and Nanotechnology*, 2008, *8* (10), 5351–5355.

129. Tang, S.-H.; Huang, Z.-Y.; Huang, J.-B. Synthesis and spectral properties of broom-like CdS nanoparticles in gelatin solution. *Huaxue Xuebao*, 2007, *65* (15), 1432–1436.

130. Wang, H.; Zhu, J.-M.; Zhu, J.-J.; Yuan, L.-M.; Chen, H.-Y. Novel microwave-assisted solution-phase approach to radial arrays composed of prismatic antimony trisulfide whiskers. *Langmuir*, 2003, *19*, 10993–10996.

131. Futaba, D.N.; Miyake, K.; Murata, K.; Hayamizu, Y.; Yamada, T.; Sasaki, S.; Yumura, M.; Hata, K. Dual porosity single-walled carbon nanotube material. *Nano Letters*, 2009, *9* (9), 3302–3307.

132. Cui, H.; Muraoka, T.; Cheetham, A. G.; Stupp, S. I. Self-assembly of giant peptide nanobelts. *Nano Letters*, 2009, *9* (3), 945–951.

133. Singh, S.; Reddy, S. B.; Kottaisamy, M.; Rao, M. S. R. Formation of ZnO nanobrushes in direct atmosphere using a carbon catalyst and a Zn metal source. *Nano*, 2008, *3* (5), 361–365.

134. Zhang, W.-D.; Zhang, W.-H.; Ma, X.-Y. Tunable ZnO nanostructures for ethanol sensing. *Journal of Materials Science*, 2009, *44* (17), 4677–4682.

135. Zhang, Y.; Xu, J.; Xiang, Q.; Li, H.; Pan, Q.; Xu, P. Brush-like hierarchical ZnO nanostructures: Synthesis, photoluminescence and gas sensor properties. *Journal of Physical Chemistry C*, 2009, *113*, 3430–3435.

136. Coakley, K. M.; Liu, Y.; Goh, C.; McGehee, M. D. Ordered organic-inorganic bulk heterojunction photovoltaic cells, *MRS Bulletin*, 2005, *30*, 37–40.

137. Singh, S. Electrical transport and optical studies of transition metal ion doped ZnO and synthesis of ZnO based nanostructures by chemical route, thermal evaporation and pulsed laser deposition. PhD thesis, 2009, Indian Institute of Technology Madras, Chennai, India.

138. Mazeina, L.; Prokes, S. M.; Arnold, S. P.; Glaser, E. R.; Perkins, F. K. Incorporation of novel ternary Sn-Ga-O and Zn-Ga-O nanostructures into gas sensing devices. *Proceedings of SPIE*, 2009, *7406* (*Nanoepitaxy: Homo- and Heterogeneous Synthesis, Characterization, and Device Integration of Nanomaterials*), 740605/1–740605/12.

139. Mazeina, L.; Picard, Y. N.; Prokes, S. M. Controlled growth of parallel oriented ZnO Nanostructural arrays on Ga_2O_3 nanowires. *Crystal Growth and Design*, 2009, *9* (2), 1164–1169.

140. Zhang, Z.; Yi, J. B.; Ding, J.; Wong, L. M.; Seng, H. L.; Wang, S. J.; Tao, J. G. et al. Cu-Doped ZnO nanoneedles and nanonails: Morphological evolution and physical properties. *Journal of Physical Chemistry C*, 2008, *112* (26), 9579–9585.

141. Gong, H.-C.; Zhong, J.-F.; Zhou, S.-M.; Zhang, B.; Li, Z.-H.; Du, Z.-L. Ce-induced single-crystalline hierarchical zinc oxide nanobrushes. *Superlattices and Microstructures*, 2008, *44* (2), 183–190.

142. Xu, S.; Wei, Y.; Liu, J.; Yang, R.; Wang, Z. L. Integrated multilayer nanogenerator fabricated using paired nanotip-to-nanowire brushes. *Nano Letters*, 2008, *8* (11), 4027–4032.

143. Raidongia, K.; Eswaramoorthy, M. Synthesis and characterization of metal oxide nanorod brushes. *Bulletin of Materials Science*, 2008, *31* (1), 87–92.

144. Turkevych, I.; Pihosh, Y.; Hara, K.; Wang, Z.-S.; Kondo, M. Hierarchically organized micro/nano-structures of TiO_2. *Japanese Journal of Applied Physics*, 2009, *48* (6, Pt. 2), 06FE02/1–06FE02/4.

145. Lu, Y.; Lunkenbein, T.; Preussner, J; Proch, S.; Breu, J.; Kempe, R.; Ballauff, M. Composites of metal nanoparticles and TiO_2 immobilized in spherical polyelectrolyte brushes. *Langmuir*, 2010, *26* (6), 4176–4183.

146. Tang, N.; Tian, X.; Yang, C.; Pi, Z. Facile synthesis of α-MnO_2 nanostructures for supercapacitors. *Materials Research Bulletin*, 2009, *44* (11), 2062–2067.

147. Kimura, G.; Yamada, K. Electrochromism of poly(3,4-ethylenedioxythiophene) films on Au nano-brush electrode. *Synthetic Metals*, 2009, *159* (9–10), 914–918.

148. Yamada, K.; Seya, K.; Kimura, G. Electrochromism of poly(pyrrole) film on Au nano-brush electrode. *Synthetic Metals*, 2009, *159* (3–4), 188–193.

149. Zhai, X.-f.; Mu, C.; Xu, D.-S.; Tong, L.-M.; Zhu, T.; Du, W.-MN. Effect of the film of gold nanowire arrays on surface enhanced Raman scattering. *Guangpuxue Yu Guangpu Fenxi*, 2008, *28* (10), 2329–2332.

150. Wu, X.; Li, H.; Chen, L.; Huang, X. Agglomeration and the surface passivating film of Ag nano-brush electrode in lithium batteries. *Solid State Ionics*, 2002, *149* (3,4), 185–192.

151. Ren, Y.; Dai, Y. Y.; Zhang, B.; Liu, Q. F.; Xue, D. S.; Wang, J. B. Tunable magnetic properties of heterogeneous nanobrush: From nanowire to nanofilm. *Nanoscale Research Letters*, 2010, *5*, 853–858.

152. Xiao, Z.-L.; Han, C. Y.; Kwok, W.-K.; Wang, H.-H.; Welp, U.; Wang, J.; Crabtree, G. W. Tuning the architecture of mesostructures by electrodeposition. *Journal of the American Chemical Society*, 2004, *126* (8), 2316–2317.

153. Ghasemi-Nejhad, M. N.; Cao, A. Development of nanodevices, nanostructures, nanocomposites, and hierarchical nanocomposites at Hawaii Nanotechnology Laboratory. *NSTI Nanotech 2007, Nanotechnology Conference and Trade Show*, Santa Clara, CA, May 20–24, 2007. Laudon, M.; Romanowicz, B. (Eds.). Vol. 4, pp. 538–542.

154. Chakrabarti, S.; Nagasaka, T.; Yoshikawa, Y.; Pan, L.; Nakayama, Y. Growth of super long aligned brush-like carbon nanotubes. *Japanese Journal of Applied Physics*, 2006, *45* (28), L720–L722.

155. Dinesh, J.; Eswaramoorthy, M.; Rao, C. N. R. Use of amorphous carbon nanotube brushes as templates to fabricate GaN nanotube brushes and related materials. *Journal of Physical Chemistry C*, 2007, *111* (2), 510–513.

156. Mieszawska, A. J.; Jalilian, R.; Sumanasekera, G. U.; Zamborini, F. P. The synthesis and fabrication of one-dimensional nanoscale heterojunctions. *Small*, 2007, *3* (5), 722–756.

157. Lan, Z.-H.; Liang, C.-H.; Hsu, C.-W.; Wu, C.-T.; Lin, H.-M.; Dhara, S.; Chen, K.-H.; Chen, L.-C.; Chen, C.-C. Nanohomojunction (GaN) and nanoheterojunction (InN) nanorods on one-dimensional GaN nanowire substrates. *Advanced Functional Materials*, 2004, *14* (3), 233–237.

158. Shi, Z.; Radwan, M.; Kirihara, S.; Miyamoto, Y.; Jin, Z. Enhanced thermal conductivity of polymer composites filled with three-dimensional brushlike AlN nanowhiskers. *Applied Physics Letters*, 2009, *95* (22), 224104/1–224104/3.

159. Wang, C.; Wang, J.; Li, Q.; Yi, G.-C. ZnSe-Si bi-coaxial nanowire heterostructures. *Advanced Functional Materials*, 2005, *15* (9), 1471–1477.

160. Zhang, M.; Zhai, T.; Wang, X.; Liao, Q.; Ma, Y.; Yao, J. Carbon-assisted morphological manipulation of CdS nanostructures and their cathodoluminescence properties. *Journal of Solid State Chemistry*, 2009, *182* (11), 3188–3194.

161. Bohnemann, J.; Libanori, R.; Moreira, M. L.; Longo, E. High-efficient microwave synthesis and characterisation of $SrSnO_3$. *Chemical Engineering Journal*, 2009, *155* (3), 905–909.

162. Debnath, A. K.; Samanta, S.; Singh, A.; Aswal, D. K.; Gupta, S. K.; Yakhmi, J. V. Molecular beam epitaxy growth of iron phthalocyanine nanostructures. *AIP Conference Proceedings (Transport and Optical Properties of Nanomaterials)*, Allahabad, India, 2009, Vol. 1147, pp. 347–353.

163. Debnath, A. K.; Samanta, S.; Singh, A.; Aswal, D. K.; Gupta, S. K.; Yakhmi, J. V.; Deshpande, S. K.; Poswal, A. K.; Suergers, C. Growth of iron phthalocyanine nanoweb and nanobrush using molecular beam epitaxy. *Physica E: Low-Dimensional Systems and Nanostructures*, 2008, *41* (1), 154–163.

164. McInnes, S. J. P.; Thissen, H.; Choudhury, N. Roy; Voelcker, N. H. New biodegradable materials produced by ring opening polymerization of poly(L-lactide) on porous silicon substrates. *Journal of Colloid and Interface Science*, 2009, *332* (2), 336–344.

165. Blinova, N. V.; Bober, P.; Hromadkova, J.; Trchova, M.; Stejskal, J.; Prokes, J. Polyaniline-silver composites prepared by the oxidation of aniline with silver nitrate in acetic acid solutions. *Polymer International*, 2010, *59* (4), 437–446.

166. Hou, S.; Li, Z.; Li, Q.; Liu, Z. Fan. Poly(methyl methacrylate) nanobrushes on silicon based on localized surface-initiated polymerization. *Applied Surface Science*, 2004, *222* (1–4), 338–345.

167. Jonas, A. M.; Hu, Z.; Glinel, K.; Huck, W. T. S. Effect of nanoconfinement on the collapse transition of responsive polymer brushes. *Nano Letters*, 2008, *8* (11), 3819–3824.

168. Jonas, A. M.; Hu, Z.; Glinel, K.; Huck, W. T. S. Chain entropy and wetting energy control the shape of nanopatterned polymer brushes. *Macromolecules*, 2008, *41* (19), 6859–6863.

169. Gitsov, I.; Overgaard, A. K.; Hole, B. Preparation of biocompatible and biodegradable nanobrushes from cellulose and hydroxyapatite nanocrystals. *Abstracts, 37th Middle Atlantic Regional Meeting of the American Chemical Society*, New Brunswick, NJ, May 22–25, 2005, GENE-408.

170. Schilp, S.; Ballav, N.; Zharnikov, M. Fabrication of a full-coverage polymer nanobrush on an electron-beam-activated template. *Angewandte Chemie, International Edition*, 2008, *47* (36), 6786–6789.

171. Zhang, H.; Cao, G.; Wang, Z.; Yang, Y.; Shi, Z.; Gu, Z. Tube-covering-tube nanostructured polyaniline/carbon nanotube array composite electrode with high capacitance and superior rate performance as well as good cycling stability. *Electrochemistry Communications*, 2008, *10* (7), 1056–1059.

172. Fan, J.-G.; Fu, J.-X.; Collins, A.; Zhao, Y.-P. The effect of the shape of nanorod arrays on the nanocarpet effect. *Nanotechnology*, 2008, *19* (4), 045713/1–045713/8.

173. Zhao, Y.-P.; Fan, J.-G. Clusters of bundled nanorods in nanocarpet effect. *Applied Physics Letters*, 2006, *88* (10), 103123/1–103123/3.

174. Fan, J.-G.; Dyer, D.; Zhang, G.; Zhao, Y.-P. Nanocarpet effect: Pattern formation during the wetting of vertically aligned nanorod arrays. *Nano Letters* (2004), *4* (11), 2133–2138.

175. Fan, J.; Zhao, Y. Nanocarpet effect induced superhydrophobicity. *Langmuir*, 2010, *26* (11), 8245–8250.

176. Neto, C.; Joseph, K. R.; Brant, W. R. On the superhydrophobic properties of nickel nanocarpets. *Physical Chemistry Chemical Physics*, 2009, *11* (41), 9537–9544.

177. Castellino, M.; Tortello, M.; Bianco, S.; Musso, S.; Giorcelli, M.; Pavese, M.; Gonnelli, R. S.; Tagliaferro, A. Thermal and electronic properties of macroscopic multi-walled carbon nanotubes blocks. *Journal of Nanoscience and Nanotechnology*, 2010, *10* (6), 3828–3833.

178. Krzysztof, K.; Shaffer, M.; Windle, A. Three-dimensional internal order in multiwall carbon nanotubes grown by chemical vapor deposition. *Advanced Materials*, 2005, *17* (6), 760–763.

179. Grobert, N.; Mayne, M.; Terrones, M.; Sloan, J.; Dunin-Borkowskif, R. E.; Kamalakaran, R.; Seeger, T. et al., Metal and alloy nanowires: Iron and invar inside carbon nanotubes. In CP591, *Electronic Properties of Molecular Nanostructures*, Kuzmany, H. et al. (Eds.). American Institute of Physics, College Park, MD, 2001, pp. 287–290.

180. Chen, S. Y.; Miao, H. Y.; Lue, J. T.; Ouyang, M. S. Fabrication and field emission property studies of multiwall carbon nanotubes. *Journal of Physics D: Applied Physics*, 2004, *37*, 273–279.

181. Sansom, E. B.; Rinderknecht, D.; Gharib, M. Controlled partial embedding of carbon nanotubes within flexible transparent layers. *Nanotechnology*, 2008, *19*, 035302, 6 pp.

182. Hung, K.-H.; Tzeng, S.-S.; Kuo, W.-S.; Wei, B.; Ko, T.-H. Growth of carbon nanofibers on carbon fabric with Ni nanocatalyst prepared using pulse electrodeposition. *Nanotechnology*, 2008, *19*, 295602.

183. Olson, D. C.; Shaheen, S. E.; Collins, R. T.; Ginley, D. S. The effect of atmosphere and ZnO morphology on the performance of hybrid poly(3-hexylthiophene)/ZnO nanofiber photovoltaic devices. *Journal of Physical Chemistry C*, 2007, *111*, 16670–16678.

184. Chen, Y.; Chi, B.; Liu, Q.; Mahond, D. C.; Chen, Y. Fluoride-assisted synthesis of mullite ($Al_{5.6}5Si_{0.35}O_{9.175}$) nanowires. *Chemical Communications*, 2006, 2780–2782.

185. Lee, S. B.; Koepsel, R.; Stolz, D. B.; Warriner, H. E.; Russell, A. J. Self-assembly of biocidal nanotubes from a single-chain diacetylene amine salt. *Journal of the American Chemical Society*, 2004, *126* (41), 13400–13405.

186. Hong, K.; Xie, M.; Hu, R.; Wu H. Synthesis of tungsten oxide comblike nanostructures. *Journal of Materials Research*, 2008, *23* (10), 2657–2661.

187. Yin, W.; Cao, M.; Luo, S.; Hu, C.; Wei, B. Controllable synthesis of various In_2O_3 submicron/nanostructures using chemical vapor deposition. *Crystal Growth and Design*, 2009, *9* (5), 2173–2178.

188. Wen, P.; Itoh, H.; Tang, W.; Feng, Q. Single Nanocrystals of anatase-type TiO_2 Prepared from layered titanate nanosheets: Formation mechanism and characterization of surface properties. *Langmuir*, 2007, *23*, 11782–11790.

189. Lei, W.; Liu, D.; Zhu, P.; Chen, X.; Hao, J.; Wang, Q.; Cui, Q.; Zou, G. One-step synthesis of AlN branched nanostructures by an improved DC arc discharge plasma method. *Crystal Engineering Communications*, 2010, *12* (2), 511–516.

190. Lei, W.; Cui, Q.; Liu, D.; Zhang, J.; Chen, X.; Hao, J.; Jin, Y.; Zhan, B.; Wang, Q.; Liang, G. High-purity double-sided/single-sided and double-sided mixed aluminum nitride nanocomb and its preparation. CN 101513996, 2009, 12 pp.

191. Hsiao, C. L.; Tu, L. W.; Chi, T. W.; Seo, H. W.; Chen, Q. Y.; Chu, W. K. Buffer controlled GaN nanorods growth on Si(111) substrates by plasma-assisted molecular beam epitaxy. *Journal of Vacuum Science and Technology B*, 2006, *24* (2), 85–851.

192. Benemanskaya, G. V.; Vikhnin, V. S.; Timoshnev, S. N. Self-organization of nanostructures on the n-GaN(0001) surface in the Cs and Ba adsorption. *JETP Letters*, 2008, *87* (2), 111–114.

193. Cui, Q.; Wang, Q.; Zhang, J.; Liang, G.; Hao, J.; Liu, D.; Lei, W.; Jin, Y.; Li, M. Silicon nitride nanocomb and its preparation. CN 101531350, 2009, 13 pp.

194. Wu, R. B.; Yang, G. Y.; Pan, Y.; Wu, L. L.; Chen, J. J.; Gao, M. X.; Zhai, R.; Lin, J. Thermal evaporation and solution strategies to novel nanoarchitectures of silicon carbide. *Applied Physics A: Materials Science and Processing*, 2007, *88* (4), 679–685.

195. Szczech, J. R.; Jin, S. Epitaxially-hyperbranched FeSi nanowires exhibiting merohedral twinning. *Journal of Materials Chemistry*, 2010, *20* (7), 1375–1382.

196. Ma, C.; Moore, D.; Li, J. G.; Wang, Z. L. Nanobelts, nanocombs, and nanowindmills of wurtzite ZnS. *Advanced Materials*, 2003, *15* (3), 228–231.

197. Zhang, X. Z.; Zhai, T. Y.; Ma, Y.; Yao, J. N.; Yu, D. P. Polarity determination for the CdxZn1-xS nanocombs by EELS. *Journal of Electron Microscopy*, 2008, *57* (1), 7–11.

198. Ge, J. P.; Li, Y. D. Selective atmospheric pressure chemical vapor deposition route to CdS arrays, nanowires, and nanocombs. *Advanced Functional Materials*, 2004, *14* (2), 157–162.

199. Zhang, Y.; Tang, Y.; Lee, K.; Ouyang, M. Catalytic and catalyst-free synthesis of CdSe nanostructures with single-source molecular precursor and related device application. *Nano Letters*, 2009, *9* (1), 437–441.

200. Zhao, N.; Wei, Y.; Sun, N.; Chen, Q.; Bai, J.; Zhou, L.; Qin, Y.; Li, M.; Qi, L. Controlled synthesis of gold nanobelts and nanocombs in aqueous mixed surfactant solutions. *Langmuir*, 2008, *24* (3), 991–998.

201. Liu, Y.-J.; Zhang, Z.-Y.; Zhao, Q.; Dluhy, R. A.; Zhao, Y.-P. The role of the nanospine in the nanocomb arrays for surface enhanced Raman scattering. *Applied Physics Letters*, 2009, *94* (3), 033103/1–033103/3.

202. Liu, B.; Chen, M.; Nakamura, C.; Miyake, J.; Qian, D.-J. Coordination polymer nanocombs self-assembled at the water-chloroform interface. *New Journal of Chemistry*, 2007, *31* (6), 1007–1012.

203. Kalayci, O. A.; Comert, F. B.; Hazer, B.; Atalay, T.; Cavicchi, K. A.; Cakmak, M. Synthesis, characterization, and antibacterial activity of metal nanoparticles embedded into amphiphilic comb-type graft copolymers. *Polymer Bulletin*, 2010, *65*, 215–216.

204. Wang, Z. L.; Kong, X. Y.; Ding, Y.; Gao, P.; Hughes, W. L.; Yang, R.; Zhang, Y. Semiconducting and piezoelectric oxide nanostructures induced by polar surfaces. *Advanced Functional Materials,* 2004, *14* (10), 943–956.

205. Zhuo, R. F.; Feng, H. T.; Liang, Q.; Liu, J. Z.; Chen, J. T.; Yan, D.; Feng, J. J. et al., Morphology-controlled synthesis, growth mechanism, optical and microwave absorption properties of ZnO nanocombs. *Journal of Physics D: Applied Physics*, 2008, *41*, 185405, 13 pp.

206. Lao, C. S.; Gao, X.; Yang, R. S.; Zhang, Y.; Dai, Y.; Wang, Z. L. Formation of double-side teethed nanocombs of ZnO and self-catalysis of Zn-terminated polar surface. *Chemical Physics Letters*, 2005, *417*, 359–363.

207. Li, X.; Xu, C.-X.; Zhu, G.-P.; Yang, Y.; Liu, J.-P.; Sun, X.-W.; Cui, Y.-P. Disc-capped ZnO nanocombs. *Chinese Physics Letters*, 2007, *24* (12), 3495.

208. Zhang, B.; Zhou, S. M.; Wang, H. W.; Du, Z. L. Raman scattering and photoluminescence of Fe-doped ZnO nanocantilever arrays. *Chinese Science Bulletin*, 2008, *53* (11), 1639–1643.

209. Yu, K.; Zhang, Q.; Wu, J.; Li, L.; Xu, Y.; Huang, S.; Zhu, Z. Growth and optical applications of centimeter-long ZnO nanocombs. *Nano Research*, 2008, *1* (3), 221–228.

210. Zhou, J.; Liu, J.; Yang, R.; Lao, C.; Gao, P.; Tummala, R.; Xu, N. S.; Wang, Z. L. SiC-shell nanostructures fabricated by replicating ZnO nano-objects: A technique for producing hollow nanostructures of desired shape. *Small*, 2006, *2* (11), 1344–1347.

211. Wang, J. X.; Sun, X. W.; Wei, A.; Lei, Y.; Cai, X. P.; Li, C. M.; Dong, Z. L. Zinc oxide nanocomb biosensor for glucose detection. *Applied Physics Letters*, 2006, *88* (23), 233106/1–233106/3.

212. Gazzadi, G. C.; Frabboni, S.; Menozzi, C. Suspended nanostructures grown by electron beam-induced deposition of Pt and TEOS precursors. *Nanotechnology*, 2007, *18* (44), 445709/1–445709/7.

213. Murugadoss, A.; Chattopadhyay, A. Stabilizing agent specific synthesis of multi-branched and spherical Au nanoparticles by reduction of $AuCl_4^-$ by Fe^{2+} ions. *Journal of Nanoscience and Nanotechnology*, 2007, *7* (6), 1730–1735.

214. Palermo, V.; Liscio, A., Talarico, A. M.; Zhi, L., Müllen, K.; Samorì, P. Unconventional nanotubes self-assembled in alumina channels: Morphology and surface potential of isolated nanostructures at surfaces. *Philosophical Transactions of the Royal Society A*, 2007, *365*, 1577–1588.

215. Zhang, Z.; Liu, S.; Chow, S.; Han, M.-Y. Modulation of the morphology of ZnO nanostructures via aminolytic reaction: From nanorods to nanosquamas. *Langmuir*, 2006, *22* (14), 6335–6340.

216. Shen, G.; Bando, Y.; Lee, C.-J. Growth of self-organized hierarchical ZnO nanoarchitectures by a simple In/In_2S_3 controlled thermal evaporation process. *Journal of Physical Chemistry B*, 2005, *109* (21), 10779–10785.

217. Ye, M.; Chen, Z.; Wang, W.; Zhen, L.; Shen, J. Large-scale synthesis and characterization of fan-shaped rutile TiO_2 nanostructures. *Materials Letters*, 2008, *62* (19), 3404–3406.

218. Guo, Y.; Zhang, J.; Zhu, F.; Yang, Z. X.; Xu, J.; Yu, J. Self-assembly of β-Ga_2O_3 nanobelts. *Applied Surface Science*, 2008, *254* (16), 5124–5128.

219. Gu, F.; Wang, Z.; Han, D.; Guo, G.; Guo, H. Crystallization of rare earth carbonate nanostructures in the reverse micelle system. *Crystal Growth and Design*, 2007, *7* (8), 1452–1458.

220. Liu, K.; You, H.; Zheng, Y.; Jia, G.; Huang, Y.; Yang, M.; Song, Y.; Zhang, L.; Zhang, H. Room-temperature synthesis of multi-morphological coordination polymer and tunable white-light emission. *Crystal Growth and Design*, 2010, *10* (1), 16–19.

221. Zhou, C.; Han, J.; Guo, R. Polyaniline fan-like architectures of rectangular sub-microtubes synthesized in dilute inorganic acid solution. *Macromolecular Rapid Communications*, 2009, *30* (3), 182–187.

222. Chen, R.; Wu, J.; Li, H.; Cheng, G.; Lu, Z.; Che, C.-M. Fabrication of gold nanoparticles with different morphologies in HEPES buffer. *Rare Metals*, 2010, *29* (2), 180–186.

223. Li, Z.; Gu, A. Growth of spindle-shaped gold nanoparticles in cetyltrimethyl-ammonium bromide solutions. *Micro and Nano Letters*, 2009, *4* (3), 142–147.

224. Li, Z.; Gu, A.; Zhou, Q. Growth of spindle-shaped silver nanoparticles in SDS solutions. *Crystal Research and Technology*, 2009, *44* (8), 841–844.

225. Xie, S.-Y.; Ma, Z.-J.; Wang, C.-F.; Lin, S.-C.; Jiang, Z.-Y; Huang, R.-B.; Zheng, L.-S. Preparation and self-assembly of copper nanoparticles via discharge of copper rod electrodes in a surfactant solution: A combination of physical and chemical processes. *Journal of Solid State Chemistry*, 2004, *177*, 3743–3747.

226. Lv, B.; Xu, Y.; Wu, D.; Sun, Y. Preparation and magnetic properties of spindle porous iron nanoparticles. *Materials Research Bulletin*, 2009, *44* (5), 961–965.

227. Sato, K.; Furukawa, N.; Leng, Y.; Li, J.-G.; Kurashima, K.; Kamiya, H.; Ishigaki, T. Dispersion behavior of spindle-type iron nanoparticles in organic solvents. *Funtai Kogaku Kaishi*, 2008, *45* (11), 773–779.

228. Sato, K.; Iijima, M.; Leng, Y.; Li, J.-G.; Kurashima, K.; Yoshida, T.; Kamiya, H.; Ishigaki, T. Oxidation-resistant silica-coating on highly dispersed spindle-type Fe-Co nanoparticles. *Funtai oyobi Funmatsu Yakin*, 2009, *56* (5), 232–235.

229. Tong, D. G.; Zeng, X. L.; Chu, W.; Wang, D.; Wu, P. Magnetically recyclable hollow Co-B nanospindles as catalysts for hydrogen generation from ammonia borane. *Journal of Materials Science*, 2010, *45*, 2862–2867.

230. Raula, M.; Rashid, M. H.; Paira, T. K.; Dinda, E.; Mandal, T. K. Ascorbate-assisted growth of hierarchical ZnO nanostructures: Sphere, spindle, and flower and their catalytic properties. *Langmuir*, 2010, *26* (11), 8769–8782.

231. Saravana Kumar, R.; Sudhagar, P.; Sathyamoorthy, R.; Matheswaran, P.; Kang, Y. S. Direct assembly of ZnO nanostructures on glass substrates by chemical bath deposition through precipitation method. *Superlattices and Microstructures*, 2009, *46* (6), 917–924.

232. Huang, J.; Xia, C.; Cao, L.; Yin, L.; Wu, J.; He, H. Method for preparing spindle-shaped nanoscale zinc oxide monocrystal. CN 101319371, 2008, 8 pp.

233. Wang, S. P.; Zhong, D. S. L.; Xu, H. L. Shape tuning of ZnO with ammonium molybdate and their morphology-dependent photoluminescence properties. *Journal of Physics: Conference Series*, 2009, *188*, 012034.

234. Li, F.; Liu, X.; Kong, T.; Li, Z.; Huang, X. Conversion from ZnO nanospindles into ZnO/ZnS core/shell composites and ZnS microspindles. *Crystal Research and Technology*, 2009, *44* (4), 402–408.

235. Peng, W.; Qu, S.; Cong, G.; Wang, Z. Synthesis and structures of morphology-controlled ZnO nano- and microcrystals. *Crystal Growth and Design*, 2006, *6* (6), 1518–1522.

236. Zhang, Y.; Wu, L.; Zeng, Q.; Zhi, J. An Approach for controllable synthesis of different-phase titanium dioxide nanocomposites with peroxotitanium complex as precursor. *Journal of Physical Chemistry C*, 2008, *112* (42), 16457–16462.

237. Liu, X.; Gao, Y.; Cao, C.; Luo, H.; Wang, W. Highly crystalline spindle-shaped mesoporous anatase titania particles: Solution-phase synthesis, characterization, and photocatalytic properties. *Langmuir*, 2010, *6* (11), 7671–7674.

238. Gao, Y.; Luo, H.; Mizusugi, S.; Nagai, M. Surfactant-free synthesis of anatase TiO_2 nanorods in an aqueous peroxotitanate solution. *Crystal Growth and Design*, 2008, *8* (6), 1804–1807.

239. Li, J.; Yu, Y.; Chen, Q.; Li, J.; Xu, D. Controllable synthesis of TiO_2 single crystals with tunable shapes using ammonium-exchanged titanate nanowires as precursors. *Crystal Growth and Design*, 2010, *10* (5), 2111–2115.

240. Bharathi, S.; Nataraj, D.; Mangalaraj, D.; Masuda, Y.; Senthil, K.; Yong, K. Highly mesoporous α-Fe_2O_3 nanostructures: Preparation, characterization and improved photocatalytic performance towards Rhodamine B (RhB). *Journal of Physics D: Applied Physics*, 2010, *43* (1), 015501/1–015501/9.

241. Reufer, M.; Dietsch, H.; Gasser, U.; Hirt, A.; Menzel, A.; Schurtenberger, P. Morphology and orientational behavior of silica-coated spindle-type hematite particles in a magnetic field probed by small-angle x-ray scattering. *Journal of Physical Chemistry B*, 2010, *114* (14), 4763–4769.

242. Wang, G.-H.; Li, W.-C.; Jia, K.-M.; Spliethoff, B.; Schueth, F.; Lu, A.-H. Shape and size controlled α-Fe_2O_3 nanoparticles as supports for gold-catalysts: Synthesis and influence of support shape and size on catalytic performance. *Applied Catalysis, A: General*, 2009, *364* (1–2), 42–47.

243. Shen, H.-X.; Yao, J.-L.; Gu, R.-A. Fabrication and characteristics of spindle Fe_2O_3@Au core/shell particles. *Transactions of Nonferrous Metals Society of China*, 2009, *19* (3), 652–656.

244. Liu, P.; Cui, H.; Wang, C. X.; Yang, G. W. From nanocrystal synthesis to functional nanostructure fabrication: Laser ablation in liquid. *Physical Chemistry Chemical Physics*, 2010, *12* (16), 3942–3952.

245. Liu, P.; Wang, C. X.; Chen, X. Y.; Yang, G. W. Controllable fabrication and cathodoluminescence performance of high-index facets GeO_2 micro- and nanocubes and spindles upon electrical-field-assisted laser ablation in liquid. *Journal of Physical Chemistry C*, 2008, *112* (35), 13450–13456.

246. Shen, Y.; Zhu, H.; Huang, R.; Zhao, L.; Yan, S. N. Synthesis and photochromic properties of WO_3 powder induced by oxalic acid. *Science in China, Series B–Chemistry*, 2009, *52* (5), 609–614.

247. Wang, D.; Zhou, Y.; Song, C.; Shao, M. Phase and morphology controllable synthesis of Sb_2O_3 microcrystals. *Journal of Crystal Growth*, 2009, *311* (15), 3948–3953.

248. Zhang, X.; Wang, G.; Liu, X.; Wu, J.; Li, M.; Gu, J.; Liu, H.; Fang, B. Different CuO nanostructures: Synthesis, characterization, and applications for glucose sensors. *Journal of Physical Chemistry C*, 2008, *112* (43), 16845–16849.

249. Wei, C.; Wang, X.; Nan, Z.; Tan, Z. Thermodynamic studies of rod- and spindle-shaped β-FeOOH crystals. *Journal of Chemical and Engineering Data*, 2010, *55* (1), 366–369.

250. Huang, C.-C.; Yeh, C.-S.; Ho, C.-J. Laser ablation synthesis of spindle-like gallium oxide hydroxide nanoparticles with the presence of cationic cetyltrimethylammonium bromide. *Journal of Physical Chemistry B*, 2004, *108*, 4940–4945.

251. Tian, G.; Sun, S. Hydrothermal synthesis of nanocrystalline $BaMoO_4$ under mild conditions using simple additive. *Crystal Research and Technology*, 2010, *45* (2), 188–194.

252. Yin, Y.; Gao, Y.; Sun, Y.; Zhou, B.; Ma, L.; Wu, X.; Zhang, X. Synthesis and photoluminescent properties of $CaMoO_4$ nanostructures at room temperature. *Materials Letters*, 2010, *64* (5), 602–604.

253. Liu, X.; Zhong, H. Research progress in the preparation of different configurations of nano-calcium carbonate. *Shanghai Tuliao*, 2009, *47* (4), 22–25.

254. Zhao, L. N.; Feng, J. D.; Wang, Z. C. In situ synthesis and modification of calcium carbonate nanoparticles via a bobbling method. *Science in China, Series B: Chemistry*, 2009, *52* (7), 924–929.

255. Lei, S.; Liang, Z.; Zhou, L.; Tang, K. Synthesis and morphological control of $MnCO_3$ and $Mn(OH)_2$ by a complex homogeneous precipitation method. *Materials Chemistry and Physics*, 2009, *113* (1), 445–450.

256. Xuan, S.; Chen, M.; Hao, L.; Jiang, W.; Gong, X.; Hu, Y.; Chen, Z. Preparation and characterization of microsized $FeCO_3$, Fe_3O_4 and Fe_2O_3 with ellipsoidal morphology. *Journal of Magnetism and Magnetic Materials*, 2008, *320*, 164–170.

257. Zhao, H.; Shi, L.; Li, Z.; Tang, C. Silicon carbide nanowires synthesized with phenolic resin and silicon powders. *Physica E: Low-Dimensional Systems and Nanostructures*, 2009, *41* (4), 753–756.

258. Wang, H.; Lv, L.; Shi, Y.; Wang, F. Preparation of spindle-shaped or rodlike nano silver sulfide. CN 101654277, 2010, 11 pp.

259. Zhu, J.; Si, S.; Zuo, R.; Duan, W.; Jiang, Y. A new two-step route to CdTe micrometer-scaled spindles and nanorods. *Journal of Nanoscience and Nanotechnology*, 2010, *10* (5), 3109–3111.

260. Song, S.; Zhang, Y.; Feng, J.; Ge, X.; Dapeng, Liu; Weiqiang, Fan; Yongqian, Lei; Yan, Xing; Hongjie, Zhang. CuIn $(WO_4)_2$ nanospindles and nanorods: Controlled synthesis and host for lanthanide near-infrared luminescence properties. *Crystal Engineering Communications*, 2009, *11*, 1987–1993.

261. Yu, C.; Yu, M.; Li, C.; Zhang, C.; Yang, P.; Lin, J. Spindle-like lanthanide orthovanadate nanoparticles: Facile synthesis by ultrasonic irradiation, characterization, and luminescent properties. *Crystal Growth and Design*, 2009, *9* (2), 783–791.

262. Liu, W.; Cao, L.; Su, G.; Liu, H.; Wang, X.; Zhang, L. Ultrasound assisted synthesis of monoclinic structured spindle $BiVO_4$ particles with hollow structure and its photocatalytic property. *Ultrasonics Sonochemistry*, 2010, *17* (4), 669–674.

263. Yu, R.; Bao, J.; Zhang, J.; Yang, X.; Chen, J.; Xing, X. Method for synthesis of hexagonal phase $TbPO_4 \cdot H_2O$ nano/micro multilevel structure material. CN 101481103, 2009, 10 pp.

264. Bao, J.; Yu, R.; Zhang, J.; Yang, X.; Wang, D.; Deng, J.; Chen, J.; Xing, X. Controlled synthesis of terbium orthophosphate spindle-like hierarchical nanostructures with improved photoluminescence. *European Journal of Inorganic Chemistry*, 2009, (16), 2388–2392.

265. Fu, Z.; Bu, W. High efficiency green-luminescent $LaPO_4$:Ce,Tb hierarchical nanostructures: Synthesis, characterization, and luminescence properties. *Solid State Sciences*, 2008, *10* (8), 1062–1067.

266. Liu, J.; Wang, F.; Dewil, R. CeO_2 Nanocrystalline-supported palladium chloride: An effective catalyst for selective oxidation of alcohols by oxygen. *Catalysis Letters*, 2009, *130* (3–4), 448–454.

267. Ho, C.; Yu, J. C.; Kwong, T.; Mak, A. C; Lai, S. Morphology-controllable synthesis of mesoporous CeO_2 nano- and microstructures. *Chem. Mater.*, 2005, *17*, 4514–4522.

268. Xu, Z.; Li, C.; Yang, P.; Hou, Z.; Zhang, C.; Lin, J. Uniform $Ln(OH)_3$ and Ln_2O_3 (Ln = Eu, Sm) submicrospindles: Facile synthesis and characterization. *Crystal Growth and Design*, 2009, *9* (9), 4127–4135.

269. Zhong, S.; Wang, S.; Xu, H.; Hou, H.; Wen, Z.; Li, P.; Wang, S.; Xu, R. Spindlelike Y_2O_3:Eu^{3+} nanorod bundles: Hydrothermal synthesis and photoluminescence properties. *Journal of Materials Science*, 2009, *44* (14), 3687–3693.

270. Gao, P.; Wang, X. Method for preparing yttrium fluoride nanoparticles with spindle structure by hydrothermal synthesis. CN 101503210, 2009, 8 pp.

271. Zhang, Y.; Zhang, L.; Deng, J.; Wei, L.; Dai, H.; He, H. Hydrothermal fabrication and catalytic performance of single-crystalline $La_{2-x}Sr_xCuO_4$ (x = 0, 1) with specific morphologies for methane oxidation. *Cuihua Xuebao*, 2009, *30* (4), 347–354.

272. Bakiz, B.; Guinneton, F.; Dallas, J.-P.; Villain, S.; Gavarri, J.-Ra. From cerium oxycarbonate to nanostructured ceria: Relations between synthesis, thermal process and morphologies. *Journal of Crystal Growth*, 2008, *310* (12), 3055–3061.

273. Fan, H.; Zhou, J.; Ai, S. Large scale $NaLaMo_2O_8$ nanospindles: Synthesis, characterization, and luminescence properties. *Journal of Nanoscience and Nanotechnology*, 2010, *10* (2), 865–870.

274. Tran, V. H.; Nguyen, N. T.; Le, D. T.; Tran, D. L. Study on synthesis and characterization of biomedical hydroxyapatite/chitosan nanocomposite. *Tap Chi Hoa Hoc*, 2007, *45* (DB), 79–84.

275. Mao, H.; Lu, X.; Chao, D.; Cui, L.; Li, Y.; Zhang, W. Preparation and characterization of PEDOT/β-Fe^{3+}O(OH,Cl) nanospindles with controllable sizes in aqueous solution. *Journal of Physical Chemistry C*, 2008, *112*, 20469–20480.

276. Qi, Y., Qi, H., Lu, C., Yang, Y., Zhao, Y. Photoluminescence and magnetic properties of β-Ni(OH)$_2$ nanoplates and NiO nanostructures. *Journal of Materials Science: Materials in Electronics*, 2009, *20* (5), 479–483.

277. Qia, Y.; Qib, H.; Lia, J.; Lu, C. Synthesis, microstructures and UV–vis absorption properties of β-Ni(OH)$_2$ nanoplates and NiO nanostructures. *Journal of Crystal Growth*, 2008, *310* (18), 4221–4225.

278. Ruan, L.; Zhang, H.; Luo, H.; Liu, J.; Tang, F.; Shi, Y.-K.; Zhao, X. Designed amphiphilic peptide forms stable nanoweb, slowly releases encapsulated hydrophobic drug, and accelerates animal hemostasis. *PNAS*, 2009, *106* (13), 5105–5110. http://www.pnas.org/content/106/13/5105.full.pdf ± html

279. Borhani, S.; Ravandi, S. A. H.; Etemad, S. G. Evaluation of surface roughness of polyacrylonitrile nanowebs. *Iranian Journal of Polymer Science and Technology (Persian Edition)*, 2008, *21* (1), 61–69.

280. Chen, G.; Gommeren, H. J. C.; Knorr, L. M. Liquid filtration media. U.S. Patent Application US 2009026137, Publication 2009, 8 pp., Cont.-in-part of U.S. Ser. No. 74,164. A1 20090129 US 2008-284027.

281. Chi, C.-H.; Lim, H. S. Pleated nanoweb structures for filters. Application Publication 2009, 8 pp. US 2009064648 A1 20090312 US 2007-899803.

282. Jones, D. C.; Stone, W. H. Fuel filter. U.S. Patent Application Publication US 2008105626, 2008, 5 pp. A1 20080508 US 2006-591733.

283. Torres-Peiro, S.; Diez, A.; Cruz, J. L.; Andres, M. V.; Cordeiro, C. M. B., de Matos, C. J. Fabrication and postprocessing of Ge-doped nanoweb fibers. *SAIP Conference Proceedings (1st Workshop on Specialty Optical Fibers and Their Applications, 2008)*, 2008, Vol. 1055, pp. 50–53.

284. Hutchenson, K. W.; Page, M. A., Raghavanpillai, A.; Reinartz, S.; Stancik, C. M., Van Gorp, J. J. Method for production of nanoweb composite material containing short perfluorinated alkyl chains. U.S. Patent US 2009047498, Application Publication, 2009, 11 pp. A1 20090219 US 2007–837647.

285. Darling, K. A.; Reynolds, C. L. Jr.; Leonard, D. N., Duscher, G.; Scattergood, R. O.; Koch, C. C. Self-assembled three-dimensional Cu-Ge nanoweb composite. *Nanotechnology*, 2008, *19* (13), 135603/1–135603/6.

286. Ahn, B. W.; Chi, Y. S.; Kang, T. J. Preparation and characterization of multi-walled carbon nanotube/poly(ethylene terephthalate) nanoweb. *Journal of Applied Polymer Science*, 2008, *110* (6), 4055–4063.

287. Kimmer, D.; Slobodian, P.; Petras, D.; Zatloukal, M.; Olejnik, R.; Saha, P. Polyurethane/multiwalled carbon nanotube nanowebs prepared by an electrospinning process. *Journal of Applied Polymer Science*, 2009, *111* (6), 2711–2714.

288. Kim, T.-G.; Ragupathy, D.; Gopalan, A. I.; Lee, K.-P. Electrospun carbon nanotubes-gold nanoparticles embedded nanowebs: Prosperous multi-functional nanomaterials. *Nanotechnology*, 2010, *21* (13), 134021.

289. Kang, S.-G.; Bae, Y.-H.; Quan, S.-L.; Chin, I.-J. Electrospun PMMA/polyhedral oligomeric silsesquioxane (POSS) nanohybrid nanofibers. *PMSE Preprints*, 2009, *101*, 1293–1294.

290. Yao, D.; Tuteja, B.; Sundararajan, P. R. Pigment-mediated nanoweb morphology of poly(dimethylsiloxane)-substituted perylene bisimides. *Macromolecules*, 2006, *39* (23), 7786–7788.

291. Borhani, S.; Ravandi, S. A. Hosseini; Etemad, S. G. Evaluation of surface roughness of polyacrylonitrile nanowebs. *Iranian Journal of Polymer Science and Technology (Persian Edition)*, 2008, *21* (1), 61–69.

292. Kang, S.; Kwon, O.-M.; Choi, H.-M.; Kwark, Y.-Je. Preparation of silsesquiazane nanofibers using electrospinning techniques. *Hankook Sumyu Gonghakhoeji*, 2008, *45* (6), 353–358.

293. http://www.ntcresearch.org/projectapp/project_pages/M04-CD01/pdf_files/Figure%201.pdf. Accessed May 25, 2010.

294. Yahya, N.; Guan, B. H.; Pah, L. M. Catalytic effect of formation of a web-like carbon nanostructures. *Solid State Science and Technology*, 2007, *15* (1), 22–29.

295. Chung, C. K.; Hung, S. T.; Lai, C. W. Effect of microstructure on the mechanical properties of carbon nanofilms deposited on the Si(100) at high temperature under ultra high vacuum. *Surface and Coatings Technology*, 2009, *204* (6–7), 1066–1070.

296. Endo, M.; Hayashi, T.; Kim, Y. A.; Terrones, M.; Dresselhaus, M. S. Applications of carbon nanotubes in the twenty-first century. *Philosophical Transactions of the Royal Society of London A*, 2004, *362*, 2223–2238.

297. Altoe, M. V. P.; Sprunck, J. P.; Gabriel, J.-C. P.; Bradley, K. Nanococoon seeds for BN nanotube growth. *Journal of Materials Science*, 2003, *38*, 4805–4810.

298. Shim, H. W.; Huang, H. Nanowebs and nanocables of silicon carbide. Nanotechnology, 2007, *18* (33), 335607/1–335607/5.

299. Mafune, F.; Kohno, J.-Y.; Takeda, Y.; Kondow, T. Nanoscale soldering of metal nanoparticles for construction of higher-order structures. *Journal of the American Chemical Society*, 2003, *125* (7), 1686–1687.

300. Li, Y.; Li, X.-Y. Noble metallic nanoweb interfaces for chemo- and bio-sensing by surface-enhanced micro-Raman spectroscopy. *Abstracts of Papers, 236th ACS National Meeting*, Philadelphia, PA, August 17–21, 2008, ANYL-236.

301. Darling, K. A.; Reynolds, C. L., Jr.; Leonard, D. N.; Duscher, G.; Scattergood, R. O.; Koch, C. C. Self-assembled three-dimensional Cu-Ge nanoweb composite. *Nanotechnology*, 2008, *19* (13), 135603/1–135603/6.

302. Rolison, D. R.; Chervin, C. N.; Long, J. W.; Lubers, A. M.; Pettigrew, K. A. Making the Most of an electrochemically critical material: Self-wiring metallic nanoskins of ruthenium dioxide onto (dirt-cheap) glass filter paper. *Abstracts, Central Regional Meeting of the American Chemical Society*, Cleveland, OH, May 20–23, 2009, CRM-009.

303. Sunkara, M. K.; Sharma, S.; Davis, B. H.; Graham, U. M. Formation of metal oxide nanowire networks (nanowebs) of low-melting metals. U.S. Patent 2007209576, 2007, 29 pp.

304. Graham, U. M.; Sharma, S.; Sunkara, M. K.; Davis, B. H. Nanoweb formation: 2D self-assembly of semiconductor gallium oxide nanowires/nanotubes. *Advanced Functional Materials*, 2003, *13* (7), 576–581.

305. Song, H.; Luo, J.; Zhou, M.; Elssfah, E.; Zhang, J.; Lin, J.; Liu, S.; Huang, Y.; Ding, X.; Gao, J.; Tang, C. Multilayer quasi-aligned nanowire webs of aluminum borate. *Crystal Growth and Design*, 2007, *7* (3), 576–579.

306. Gao, P.; Wang, X.; Chen, Y. A novel tailoring route for the fabrication of different hierarchical CuS superstructures. *Chemistry Letters*, 2009, *38* (6), 554–555.

307. Gao, P.; Xie, Y.; Ye, L.; Chen, Y.; Guo, Q. From 2D nanoflats to 2D nanowire Networks: A novel hyposulfite self-decomposition route to semiconductor FeS_2 nanowebs. *Crystal Growth and Design*, 2006, *6* (2), 583–587.

308. Debnath, A. K., Samanta, S., Singh, A., Aswal, D. K., Gupta, S. K., Yakhmi, J. V., Deshpande, S. K., Poswal, A. K.; Suergers, C. Growth of iron phthalocyanine nanoweb and nanobrush using molecular beam epitaxy. *Physica E: Low-Dimensional Systems and Nanostructures*, 2008, *41* (1), 154–163.

309. Strelcov, E.; Kolmakov, A. Copper phthalocyanine quasi-1D nanostructures: Growth morphologies and gas sensing properties. *Journal of Nanoscience and Nanotechnology*, 2008, *8* (1), 212–221.

310. Gao, L.; Shi, L.; An, Y.; Zhang, W.; Shen, X.; Guo, S.; He, B. Formation of spindlelike aggregates and flowerlike arrays of polystyrene-b-poly(acrylic acid) micelles. *Langmuir*, 2004, *20* (12), 4787–4790.

311. Zhou, J. J.; Noca, F.; Gharib, M. Flow conveying and diagnosis with carbon nanotube arrays. *Nanotechnology*, 2006, *17*, 4845–4853.

312. Wang, H., Xie, Z.; Yang, W.; Fang, J.; An, L. Morphology control in the vapor-liquid-solid growth of SiC nanowires. *Crystal Growth and Design*, 2008, *8* (11), 3893–3896.

313. Mankevich, A. S. http://www.nanometer.ru/2007/09/14/mocvd_4265.html. Accessed June 4, 2010.

314. Wang, H.; Yong, B.; Wan, Y.; Chen, B.; Fang, Y.; Wang, Y.; Sha, J. Structure analyses and growth mechanism of ZnO nanoladders. *Materials Letters*, 2010, *64* (17), 1925–1928.

315. Wang, F.; Jin, G.-Q.; Wang, Y.-Y.; Guo, X.-Y. Large-scale synthesis of ear-like Si_3N_4 dendrites from SiO_2/Fe composites and Si powders. *Materials Research Bulletin*, 2008, *43* (7), 1858–1864.

316. Liu, Z.; Joung, S.-K.; Okazaki, T.; Suenaga, K.; Hagiwara, Y.; Ohsuna, T.; Kuroda, K.; Iijima, S. Self-assembled double ladder structure formed inside carbon nanotubes by encapsulation of $H_8Si_8O_{12}$. *ACS Nano*, 2009, *3* (5), 1160–1166.

317. Sharma, S. P.; Lakkad, S. C. Morphology study of carbon nanospecies grown on carbon fibers by thermal CVD technique. *Surface and Coatings Technology*, 2009, *203* (10–11), 1329–1335.

10 "Nanotechnical" Structures and Devices

10.1 NANOSPRINGS/NANOCOILS/NANOSPIRALS

Helix-shaped nanostructures are described in the available literature as nanosprings, nanocoils, and nanospirals. These structures are generally reviewed among other related nanoforms (such as nanowires, nanorods, or nanoneedles).[1,2] A computational quantum chemistry design of nanospirals was also reported.[3] As the majority of other related nanostructures, helix nanostructures are represented mainly by inorganic compounds, in particular carbon nanotubes[4] (CNTs) and other carbon nanoforms. Thus, a constant-force nanospring might be formed from a configuration of concentric CNTs (Figure 10.1), where the van der Waals force provides the extension-independent restoring force.[5] A telescoped nanotube with only one active (sliding) surface pair was expected to act as a constant-force spring. A fully telescoped MWNT (Figure 10.2) originally had nine walls, with an outer diameter of 8 nm and an inner diameter of 1.3 nm.

Carbon microcoils were prepared by catalytic CVD of acetylene, using Ni as the catalyst and thiophene as the promoter.[6] A new catalyst (ball milled nickel sulfide) was developed on purpose to avoid the introduction of noxious and unpleasant thiophene during the reaction process. Additionally, coil-in-coil[7] carbon nanocoils (CNCs, Figure 10.3) were similarly synthesized by means of acetylene decomposition using nickel nanoparticles as catalysts. It was revealed that there were often several CNCs self-assembled in one nanospring. The yield of coil-in-coil CNCs was high up to 11 g in each run at the decomposition temperature of 450°C. Among metal-carbon composite helices, tungsten-containing carbon (WC) nanosprings fabricated by focused-ion-beam CVD (FIB-CVD) using a source gas mixture of phenanthrene and $W(CO)_6$ showed unique characteristics that they could expand and contract as flexibly as macroscale springs.[8] It was found that the spring constants of the springs rose as the tungsten content increased.

Diamond-like carbon (DLC) nanowires were used to compose nanosprings fabricated by FIB-CVD.[9] The DLC nanowires of the as-grown nanosprings had elastic double structures, in which a 50 nm diameter core containing 3 at.% gallium in addition to carbon was enclosed in an outer 25 nm wide DLC shell. It was revealed that the carbon densities of the core and the shell were similar, indicating that the density of the core was higher than that of the shell owing to the incorporation of Ga into the core. However, the core density was approximately halved by 800°C annealing. This was attributed to the vaporization of Ga and the movement of C from the core to the shell. The same research group studied the mechanical characteristics of iron-containing nanosprings fabricated by FIB-CVD using a ferrocene source gas.[10] The shear and Young's moduli were found to be 34 and 92 GPa, respectively. Studying the annealing effect of Fe-containing nanosprings, it was observed that a droplet containing Fe and Ga was found on the nanospring after annealing at 600°C. It was confirmed that the decrease in the spring constant after annealing at 600°C was due to Ga removal, similar to the DLC nanosprings.

A series of publications are dedicated to silicon nanosprings. Thus, the four-turn Si nanosprings (Figure 10.4) were grown[11] using the oblique angle deposition technique with substrate rotation and were rendered conductive by coating with a 10 nm thick Co layer using CVD. The electromagnetic force was shown to lead to spring compression, which was measured with the AFM tip. Additionally, five-armed Si nanospring coatings, grown[12] by glancing-angle deposition (GLAD) on Si(111) substrates prepatterned with submicrometer-sized SiO_2 and polystyrene spheres, were

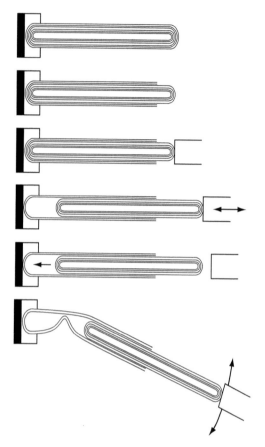

FIGURE 10.1 A schematic for MWNT processing and mechanical manipulations. (Reproduced with permission from Zettl, A. and John, C., Sharpened nanotubes, nanobearings, and nanosprings, In *CP544, Electronic Properties of Novel Materials—Molecular Nanostructures*, Kuzmany, H. et al. (Eds.), American Institute of Physics, Melville, NY, Copyright 2000, American Institute of Physics.)

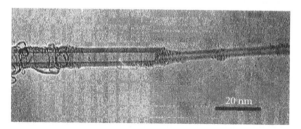

FIGURE 10.2 A fully telescoped MWNT. (Reproduced with permission from Zettl, A. and John, C., Sharpened nanotubes, nanobearings, and nanosprings, In *CP544, Electronic Properties of Novel Materials—Molecular Nanostructures*, Kuzmany, H. et al. (Eds.), American Institute of Physics, Melville, NY, Copyright 2000, American Institute of Physics.)

irradiated with 1.2 MeV Ar^{8+} ions at liquid nitrogen temperature at different ion fluences varying from 10^{15} to 10^{17} ions/cm^2. The pitch of the nanosprings shows a decrease in length and an increase in the width of its arms after the irradiation.

Si-sculptured thin films consisting of spiral-, screw-, and column-shaped nanostructures were grown by ion beam–induced GLAD of Si on rotating bare Si[001] substrates at different substrate rotational speeds and substrate temperatures.[13] Glancing-angle ion beam–assisted deposition was used for the growth of amorphous Si nanospirals (Figure 10.5) onto [001]Si substrates

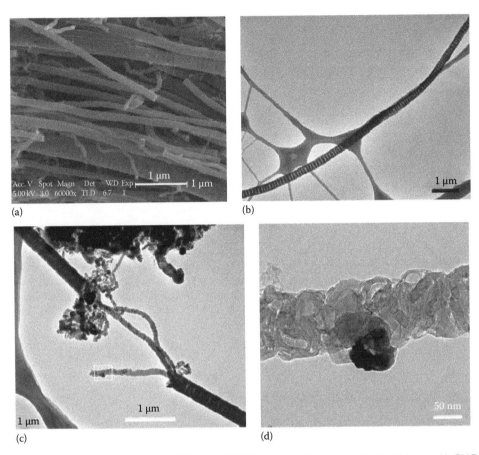

(a)

(b)

(c)

(d)

FIGURE 10.3 Microstructure of the CNCs. (a) FE-SEM image. (b through d) TEM images: (c) CNC that was ruptured after being agitated in an ultrasonic bath; (d) magnified image of the corresponding position marked in panel (c). (Reproduced with permission from Tang, N. et al., Coil-in-coil carbon nanocoils: 11 Gram-scale synthesis, single nanocoil electrical properties, and electrical contact improvement, *ACS Nano*, 4(2), 781–788. Copyright 2010 American Chemical Society.)

in a temperature range from room temperature to 475°C.[14] Amorphous Si nanocolumns, square nanospirals, and multilayer spiral/column rods were fabricated on bare Si substrates and monolayer colloid substrates by GLAD.[15] It was revealed that the size of the deposited Si columns and spirals increased with the size of colloid particles for fixed incident angle of deposition flux. The applied method is promising in the fabrication of photonic crystals.

Based on self-scrolling of Si/Cr bilayer films on Si(001) wafers, various micro- and nanostructures including microclaws, microtubes, helical nanobelts, nanorings, and nanospirals were fabricated in a controllable way.[16,17] It was found that multiple-turn Si/Cr nanorings exhibited a "sticky force" between neighboring bilayers, which is mainly attributed to the van der Waals force. Thus, multiple-turn Si/Cr or SiGe/Si/Cr tubes were self-closing and may be used as pipes for microfluidics. Si/SiGe rolled-up nanosprings were also prepared, and their electric and mechanical properties (Figures 10.6 and 10.7) were investigated.[18–20] It was found that the spring is extended to 91% of its original length; moreover, the springs could also be reproducibly extended to more than 180% of their original length. It is expected that these springs can be used as ultrasensitive force sensors. Additionally, a two-turn, eight-armed, rectangular Si/Ni heterogeneous nanospring structure on Si(100) was fabricated[21] using a multilayer GLAD technique resulting in the multilayered nanosprings with a height of ~1.98 μm composed of alternating layers of amorphous Si nanorods ~580 nm in length and fcc Ni nanorods ~420 nm in length, both with a diameter of ~35 nm.

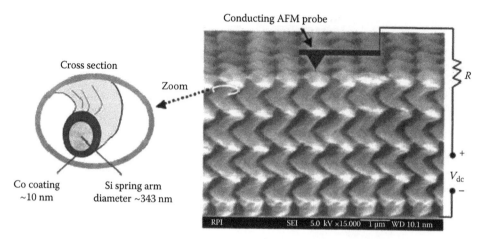

FIGURE 10.4 Cross-sectional SEM image of four-turn Si nanosprings. A schematic of the circuit used for the electromechanical measurements including a Pt (30 nm) coated AFM tip, a resistor R (200–500 Ω), and a DC power supply V_{dc} (0–24 V) is also shown. The schematic on the left shows the cross section of the conductive Co coating on the Si nanospring. (Reproduced with permission from Singh, J.P. et al., Metal-coated Si springs: Nanoelectromechanical actuators, *Appl. Phys. Lett.*, 84(18), 3657–3659, Copyright 2004, American Institute of Physics.)

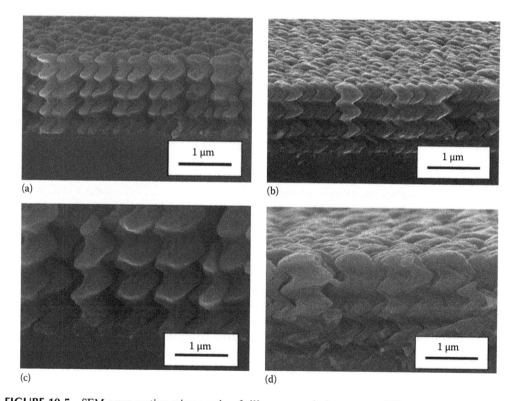

FIGURE 10.5 SEM cross section micrographs of silicon nanospirals grown at different substrate temperatures (a) 150°C, (b) 225°C, (c) 275°C, and (d) 475°C. (Reproduced from *Nucl. Instrum. Methods Phys. Res., Sect. B*, 244 (1), Schubert, E. et al., Chiral silicon nanostructures, 40–44, Copyright 2006, with permission from Elsevier.)

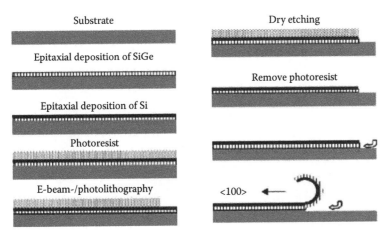

FIGURE 10.6 Basic process sequence: initial planar bilayer, patterned through conventional microfabrication techniques, assembles itself into 3D nanostructures during wet etch release. (Reproduced from *Sens. Actuators A*, 130–131, Bell, D.J. et al., Three-dimensional nanosprings for electromechanical sensors, 54–61, Copyright 2006, with permission from Elsevier.)

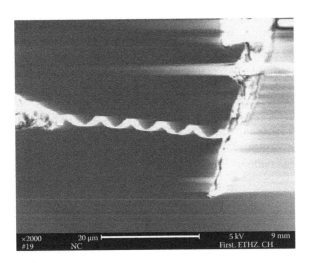

FIGURE 10.7 Nanospring attached to conductive tape. (Reproduced from *Sens. Actuators A*, 130–131, Bell, D.J. et al., Three-dimensional nanosprings for electromechanical sensors, 54–61, Copyright 2006, with permission from Elsevier.)

Several pure metal nanosprings have been reported; thus, the optical properties of gold square nanospirals were investigated numerically in the hundred terahertz range as a function of the geometrical parameters of the nanospirals.[22] Cu nanosprings on a Si substrate were obtained using GLAD technique.[23] Arrays of Cr zigzag nanosprings (Figure 10.8) and slanted nanorods, 15–55 and 40–80 nm wide, respectively, were grown on SiO_2/Si substrates by GLAD,[24] and they exhibited a reversible change in resistivity upon loading and unloading, by 50% for nanosprings and 5% for nanorods, indicating their potential as pressure sensors.

ZnO[25] helix nanostructures are also known, among other metal oxides and hydroxides. Thus, various ZnO nanostructures including nanospirals were reviewed in Ref. [26]. Uniform indium-doped ZnO nanospirals free of defects were synthesized by one-step thermal evaporation.[27] The typical radius of curvature of the nanospirals was found to be several micrometers. The In-doped nanospirals were expected to have interesting optoelectronic and mechanical properties and

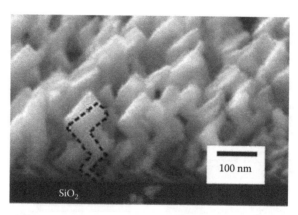

FIGURE 10.8 Cross-sectional scanning electron micrographs of Cr nanosprings. (Reproduced with permission from Kesapragada, S.V. et al., Nanospring pressure sensors grown by glancing angle deposition, *Nano Lett.*, 6(4), 854–857. Copyright 2006 American Chemical Society.)

could be potential building blocks in nanoscale optoelectronic and electromechanical systems. ZnO nanopowder, synthesized[28] by annealing the precursors in oxygen gas using the chemical precipitation method, showed various morphologies from nanorod to cobble as the annealing temperature increased from 500°C to 1000°C, while spiral structures were observed in the samples annealed at 900°C and 1000°C. Also, in a related work,[29] pregrowth pressure control in a solid–vapor process led to the formation of single-crystal ZnO nanosprings at a high yield (>50%). ZnSe nanospirals, with Zn blende structured building blocks exhibiting unconventional mosaic configuration, were fabricated *via* a two-stage growth process changing pressure.[30] It was observed that the formation of nanospirals was closely related to the pressure variation. A new strategy for the growth of spiral $Cd(OH)_2$ with 1D nanostructure, yielding an average length of 1–3 µm and width up to 30–100 nm, was developed in a one-step process *via* the microwave heating approach.[31] Different cadmium salts have more influence on the structure and morphology of the obtained products.

The fabrication, assembly, and characterization of InGaAs/GaAs piezoresistive nanosprings for creating nanoelectromechanical systems were described.[32] With their strong piezoresistive response, low stiffness, large-displacement capability, and excellent fatigue resistance, they were well suited to function as sensing elements in high-resolution, large-range electromechanical sensors. GaN nanostructure synthesis was carried out in a quartz tube furnace using NH_3 and liquid Ga as precursors and H_2 as the carrier gas.[33] The growth process gave two types of structures, straight nanowires and irregular growth sometimes resulting in nanospirals. Using the present growth system and gas flow setup, it was possible to synthesize ultralong nanowires and spirals, with overall lengths exceeding 70 µm. The formation of spirals themselves maybe related to the polarization properties of GaN. In a related work,[34] ultrathin AlN/GaN crystalline porous freestanding nanomembranes were fabricated on Si(111) by selective Si etching and self-assembled into various geometries such as tubes, spirals, and curved sheets. Nanopores with sizes from several to tens of nanometers were produced in nanomembranes of 20–35 nm nominal thickness, caused by the island growth of AlN on Si(111). Cubic structured nanosprings, InP nanosprings, were synthesized[35] *via* a simple thermochemical process using InP and ZnS as the source materials. Each InP nanospring was found to be formed by rolling up a single InP nanobelt with the growth direction along the (111) orientation. The formation of these novel nanostructures was revealed to be mainly attributed to the minimization of the electrostatic energy due to the polar charges on the ±(002) side surfaces of cubic InP.

The formation of helical nanowires (nanosprings) of boron carbide was observed, and a growth mechanism, based on the work of adhesion of the metal catalyst and the tip of the nanowire, was developed.[36]

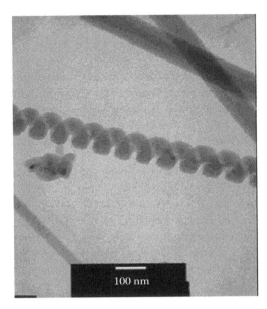

100 nm

FIGURE 10.9 Amorphous silicon carbide nanosprings. (Reproduced with permission from Zhang, D. et al., Silicon carbide nanosprings, *Nano Lett.*, 3(7), 983–987. Copyright 2003 American Chemical Society.)

The growth model demonstrates that the asymmetry necessary for helical growth is introduced when the following conditions are met: (1) the radius of the droplet is larger than the radius of the nanowire and (2) the center of mass of the metal droplet is displaced laterally from the central axis of the nanowire. Amorphous silicon carbide nanosprings (Figure 10.9), grown *via* a vapor–liquid–solid mechanism, as well as biphase (crystalline core/amorphous sheath) helical nanowires, were synthesized by plasma-enhanced CVD.[37]

Among other salts, $NaFe_4P_{12}$ nanosprings were synthesized using a hydrothermal method at 170°C for 24 h.[38] These nanosprings were 1000–5000 nm in length and 200–1000 nm in diameter, consisted of coiled nanobelts of 80–150 nm in width and 20–50 nm in thickness, and had the skutterudite crystal structure.

Organic nanospirals are rare; thus, polypyrrole nanospirals were obtained by the method called "surfactant crystallites–directed fabrication approach" which was presented for the first time to obtain conducting polymer with ordered nanostructures.[39] By using this approach, PPy with spiral nanostructures was obtained through an oxidation polymerization process using surfactant crystallites with dislocation defect as templates since the PPy preferred to grow along the step surface of the crystallites. Supramolecular assembly of the carbamate derivative, 3,3,4,4,5,5,6,6,7,7,8,8,9,9, 10,10,11,11,12,12,12-henicosa fluorododecyl 1-naphthylcarbamate ($F_{10}C_2Np$) with nanofiber or nanospiral structure, which bear partially fluorinated alkyl tails on their molecular skeletons, was reported.[40] Additionally, DNA nanospirals are known.[41]

Nanospirals possess several important applications, in particular as sensors[42] and as a medium for hydrogen storage. Thus, silica nanosprings, consisting of multiple silica nanowires approximately 10–15 nm in diameter, which coherently coil to form the nanospring, created a unique interface between neighboring nanowires, facilitating the reversible room temperature adsorption of approximately 1 wt.% molecular hydrogen at modest pressures (~20 bar).[43] The room temperature storage capacity can be further increased to ~3.5 wt.% at room temperature and 60 bar of pressure by coating the surface of the nanosprings with Pd nanoparticles. This ability of silica nanosprings to adsorb H_2 and subsequent release at 100 K was explained by the author through a combination of electrostatic interactions arising from their unique nanoscale geometry, their unique surface chemistry, and spillover upon the addition of Pd nanoparticles.

10.2   NANOAIRPLANES, NANOPROPELLERS, AND NANOWINDMILLS

Zinc oxide (ZnO), which, as it is well known, forms a lot of distinct nanostructures, was also reported[44] in the exotic nanoairplane form (Figure 10.10). Its preparation was carried out by thermal evaporation technique in a single-stage furnace (evaporation temperature 800°C–900°C, time 5–30 min). The morphologies of the nanoairplanes were found to be uniform, and no other ZnO shapes were observed in the sample. Every nanoairplane was composed of three "wings" (50 nm), having a common axis, and a ZnO nanowire with a diameter of 100–200 nm extending out from the "nose." The ZnO nanoairplane is a wurtzite structure with lattice constants of $a = 0.324$ nm and $c = 0.519$ nm.

Gold nanopropellers (Figure 10.11), which could also be considered as nanoflowers (see Chapter 7), were obtained[45] from $HAuCl_4 \cdot 4H_2O$ by UV-irradiation ($\lambda_{300-400} = 16$ mW/cm^2) of its aqueous solution in the presence of bromide ion, poly(vinylpyrrolidone) (PVP), and molecular oxygen after adjusting pH to 10 and the formation of $Au(OH)_4^-$. The corolla- or propeller-like gold nanoplates formed under these conditions were single crystalline. The authors noted that although the overall reactions proceed in one pot, the crystal growth and oxidative dissolution show dynamic interplay and occur with different kinetics at nanointerfaces that are affected

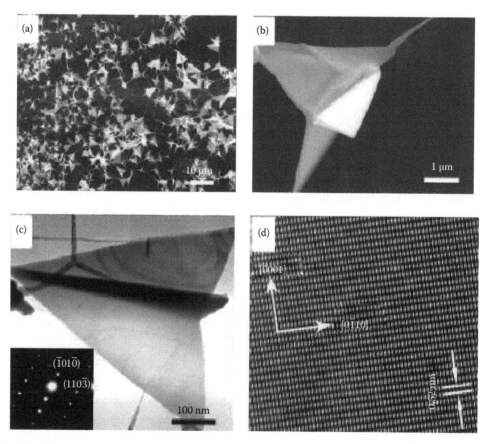

FIGURE 10.10   SEM and TEM images of the novel ZnO nanoairplanes. (a, b) Low- and high-magnification SEM images of the ZnO nanoairplanes. (c) TEM image of the nanoairplane. The inset is the SAED pattern from one wing of the nanoairplane. (d) HRTEM image of a nanoairplane wing showing the lattice image of the perfect single-crystalline structure and its crystalline orientation. (From Liu, F. et al., *Nanotechnology*, 15, 949, 2004. Reproduced with permission from Institute of Physics Science.)

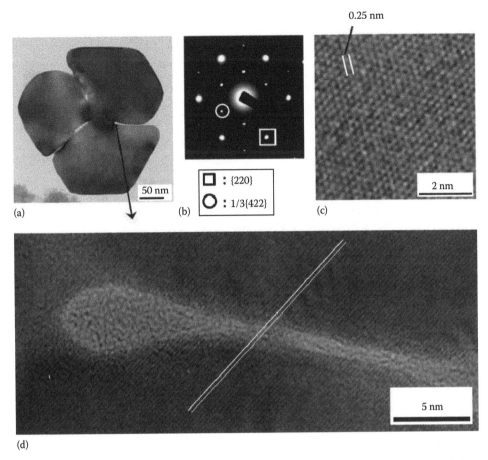

0.25 nm

☐ : {220}

◯ : 1/3{422}

50 nm

2 nm

5 nm

(a) (b) (c)

(d)

FIGURE 10.11 Transmission electron micrograph of gold nanopropeller (a), SAED pattern (b), HRTEM image of a lattice structure observed for the (111) surface of a gold nanoplate (c), and lattice structure of gold nanoplates around a crevasse (d). Nanoplates were obtained under the condition of $[Au(OH)_4^-] = 18.0$ mM; $[NaBr] = 16.7$ mM; $[PVP] = 0.25$ g/mL. Irradiation time = 6 h. The fraction of nanopropeller in nanocorolla was ca. 6%. (Reproduced with permission from Tetsuro, S. and Nobuo, K., One-pot room-temperature synthesis of single-crystalline gold nanocorolla in water, *J. Am. Chem. Soc.*, 131, 14407–14412, Copyright 2009 American Chemical Society.)

by the changes in concentration of $Au(OH)_4^-$ and the other species over time. A preliminary mechanism was proposed, which involved the concurrent crystal growth and oxidative etching on the surface of nanocrystals.

Related windmill structures are known for ZnO (prepared from ZnO thin film-coated substrates, treated solvothermally in water at pH 10),[46] ZnS,[47] and CNTs.[48] A ZnS nanostructure (Figure 10.12a) corresponds to a sixfold symmetry that closely resembles a windmill. Like the nanowires, the nanowindmills had the wurtzite crystal structure and a primary growth direction of [0001]. The six blades of the windmill grow along ±[1010], ±[0110], and ±[1100]. The mechanism (Figure 10.12b) proposed by authors is as follows.[49] First, an individual nanowire grows along the c-axis. Due to the nature of the wurtzite crystal structure, the wire is not cylindrical in cross section but hexagonal. At some point either during the synthesis or during the cooling down process, a secondary growth is triggered at the surface of the sidewalls of the nanowire. The growth continues laterally forming the blades of the windmill. At this time, it is unclear why the blades grow from the surface of the

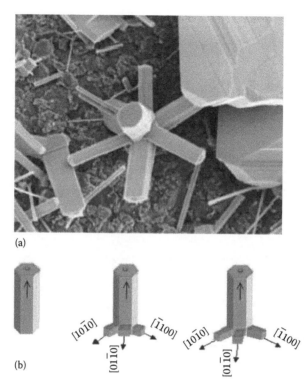

(a)

(b)

FIGURE 10.12 (a) SEM image of a ZnS nanowindmill and (b) schematic of proposed windmill growth process. (Reproduced from Christopher M., Systematic investigation on the growth of one dimensional wurtzite nanostructures, PhD dissertation, Georgia Institute of Technology, Atlanta, GA, 2005.)

wires growing laterally and increasing in thickness. The secondary growth process likely occurs at elevated temperatures and not during the cooling process. The data collected from the ZnS nanostructure study supported this since all nanowindmills were observed at the highest temperature zones of the as-deposited material.

10.3   NANOBOAT

Anatase-type TiO_2 single nanocrystals with boat-like (Figure 10.13) particle morphology, among others (comb-like, sheet-like, leaf-like, quadrate, rhombic, and wire-like shapes), were prepared[50] by hydrothermal treatment of a layered titanate nanosheet ($K_{0.8}Ti_{1.73}Li_{0.27}O_4$) colloidal solution. The titanate nanosheets were transformed to the TiO_2 nanocrystals by two types of reactions (an *in situ* topotactic structural transformation reaction and a dissolution–deposition reaction on the surface). It was revealed that the nanocrystal morphology is dependent on the reaction temperature, the pH value of the reacting solution, and the exfoliating agent (the reaction mechanism is shown in Figure 10.14).

10.4   NANOSAWS

Saw-like nanostructures were described in an excellent review.[51] This structural type is rare; only a few metal salts were found as nanosaws. Thus, a large number of ZnS nanosaws (Figures 10.15 and 10.16) were synthesized on Si substrates in the presence of Au catalyst by thermally evaporating ZnS powder at 1200°C.[52] It was shown that the temperature of the Si substrates used for collection of the products is a critical experimental parameter for the formation of ZnS nanostructures with different morphologies. The as-grown nanosaws may have potential applications in photocatalysis

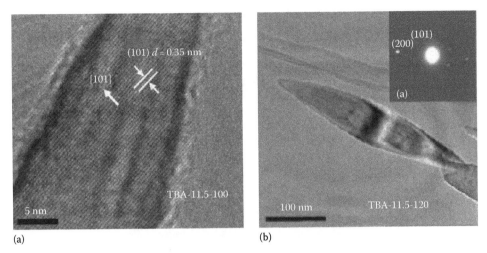

FIGURE 10.13 TiO$_2$ boat-like nanostructures obtained by hydrothermal treatment of TBA-HTO colloidal solution at pH 11.5°C and 120°C. (Reproduced with permission from Wen, P. et al., Single nanocrystals of anatase-type TiO$_2$ prepared from layered titanate nanosheets: Formation mechanism and characterization of surface properties, *Langmuir*, 23, 11782–11790. Copyright 2007 American Chemical Society.)

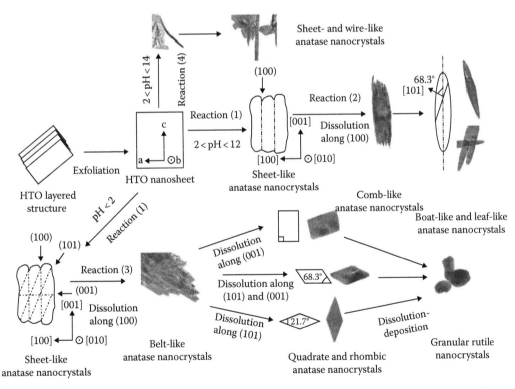

FIGURE 10.14 Scheme of the reaction mechanism for the formation of TiO$_2$ nanocrystals from titanate nanosheets under hydrothermal reaction conditions. (Reproduced with permission from Wen, P. et al., Single nanocrystals of anatase-type TiO$_2$ prepared from layered titanate nanosheets: Formation mechanism and characterization of surface properties, *Langmuir*, 23, 11782–11790. Copyright 2007 American Chemical Society.)

FIGURE 10.15 An FE-SEM image of a double-side ZnS nanosaw. (Reproduced from Xiang, W.U. et al., *Chin. Phys. Lett.*, 25(2), 737, 2008. Institute of Physics Science.)

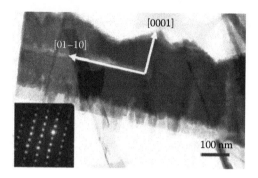

FIGURE 10.16 A TEM image of ZnS nanosaw segment, inset: an SAED pattern of the segment. (Reproduced from Xiang, W.U. et al., *Chin. Phys. Lett.*, 25(2), 737, 2008. Institute of Physics Science.)

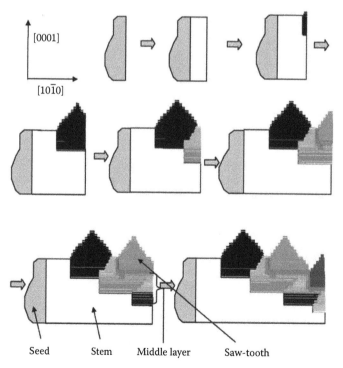

FIGURE 10.17 The schematic diagram of the growth process of ZnS nanosaws. (Reproduced with permission from Jian, Y. et al., Effect of stacking fault on the formation of the saw-teeth of ZnS nanosaws, *Cryst. Growth Des.*, 8(5), 1723–1726. Copyright 2008 American Chemical Society; Christopher, M. et al., Single-crystal CdSe nanosaws, *J. Am. Chem. Soc.*, 126, 708–709. Copyright 2004 American Chemical Society.)

FIGURE 10.18 (a) Low- and (b) high-magnification SEM images of as-grown α-Si₃N₄ nanosaws. The upper right inset in (b) shows the typical thickness of a nanosaw. (Shen, G. et al.: Systematic investigation of the formation of 1D α-Si3N4 nanostructures by using a thermal-decomposition/nitridation process. *Chem. Eur. J.* 12. 2987–2993. 2006. Copyright Wiley-VCH Verlag GmbH & Co. KGaA. Reproduced with permission.)

and nanodevices based on ZnS nanostructures. In a related article,[53] similarly obtained ZnS were reported to have been fabricated, and the growth mechanism for saw teeth is shown in Figure 10.17. CdSe nanosaws were similarly fabricated.[54]

Thermal-decomposition/nitridation method for the large-scale synthesis of 1D α-Si₃N₄ nanostructures (Figure 10.18), such as millimeter-scale microribbons, nanosaws, nanoribbons, and nanowires, was offered.[55] All these nanostructures have a single-crystalline nature and predominately grow along the [011] direction. These 1D nanostructures were found to be formed by thermal decomposition, followed by the nitridation of SiO at a high temperature in a nitrogen gas atmosphere (Reactions 10.1 and 10.2):

$$2SiO = Si + SiO_2 \tag{10.1}$$

$$3Si + 2N_2 = Si_3N_4 \tag{10.2}$$

10.5 NANOBRIDGES

Talking about nanobridges, we can mention the nanobridge as a nanostructure itself or a part of a more complex system such as a nanodevice (mainly the last variant). Bridge-like nanostructures are usually made of nanowires forming bridges in nanodevices. Thus, gold nanowires (0.8–3 nm in thickness and 5–10 nm in length)[56] were made *in situ* in a UHVTEM equipped with a field emission electron gun. Using electron-beam (E-beam) bombardment, nanoholes in a gold (001) film with 3 nm thickness were made, thereby forming a bridge such as that shown in Figures 10.19a and 10.20. The bridge was freestanding, suspended by the Au(001) film at both ends. When the bridge was irradiated further, it became thinner (Figure 10.19b). The good reproducibility and stability of the nanobridges is promising for experiments of electronic transport as compared with the previously reported nanocontact. The experimentally observed stability of nanobridges arising from the tensile deformation of (110) gold nanowires can be explained by a multishell lattice structure that forms during the plastic deformation of the nanowires.[57] In addition, the multishell structures were found to have an inherent stability that is dependent on the external loading rate applied to the nanowires.

A series of other metals have been reported in nanobridge form. Thus, body-centered pentagonal nanobridge structures were observed under high strain rate tensile loading on an initially constrained ⟨100⟩/{100} Cu nanowire at various temperatures.[58] A large inelastic deformation of ~50% was obtained under both isothermal loading and adiabatic loading. The observed interesting stability property and high strength of elongated nanowires could have various potential applications in nanomechanical and nanoelectronic devices. Superconducting quantum interference devices (SQUIDs)

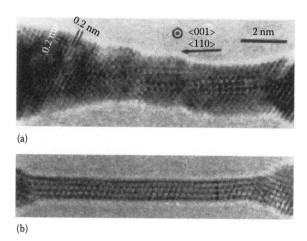

(a)

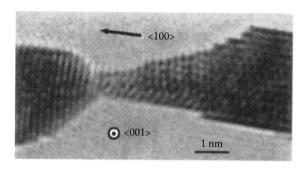

(b)

FIGURE 10.19 Transmission electron micrographs showing the formation of a nanowire: (a) an image of Au(001) film with closely neighboring nanoholes, an initial stage of the nanowire and (b) the thinnest nanobridge with four atomic rows. (Reproduced with permission from Yukihito, K. and Kunio, T., *Phys. Rev. Lett.*, 79(18), 3455–3458, 1997. Copyright 1997 by the American Physical Society.)

FIGURE 10.20 A neck-shaped bridge with an axis in the (100) direction of the Au(001) films. (Reproduced with permission from Yukihito, K. and Kunio, T., *Phys. Rev. Lett.*, 79(18), 3455–3458, 1997. Copyright 1997 by the American Physical Society.)

consisting of Al nanobridges of varying length L contacted with (2D) and (3D) banks were studied.[59] 3D nanobridge SQUIDs with $L \leq 150$ nm (~4 times the superconducting coherence length) exhibited ~70% critical current modulation with applied magnetic field; in contrast, 2D nanobridge SQUIDs exhibited significantly lower critical current modulation. Bi nanowire interconnections (diameter of 192 nm) between two prepatterned electrodes using a combination of on-film formation of nanowires and self-assembly were fabricated.[60] Bi nanowires were found to grow laterally from a multilayer structure with a Cr (or SiO_2) overlayer on top of a Bi thin film through thermal annealing to relieve vertically stored compressive stress. Such self-assembled lateral nanowire growth can be utilized as an easy means for fabricating a variety of nanowire devices without the use of catalysts or complex patterning processes. E-beam lithography and a bilayer liftoff process were used to fabricate magnetic Ni bridge-like nanostructures with constriction widths of 22–41 nm.[61] The growth of epitaxial Ge nanowires was investigated on (100), (111) B, and (110) GaAs substrates in the growth temperature range from 300°C to 380°C.[62] Using the property that, unlike epitaxial Ge nanowires on Ge or Si substrates, Ge nanowires on GaAs substrates grew predominantly along the ⟨110⟩ direction, vertical ⟨110⟩ Ge nanowires were epitaxially grown on GaAs(110) surface. Ge nanowires growing along the ⟨110⟩ directions are particularly attractive candidates for forming nanobridge devices on conventional (100) surfaces. Additionally, the transport property of a superconducting Pb nanobridge, which was carved by focus ion beam technique from an atomically flat single-crystal Pb

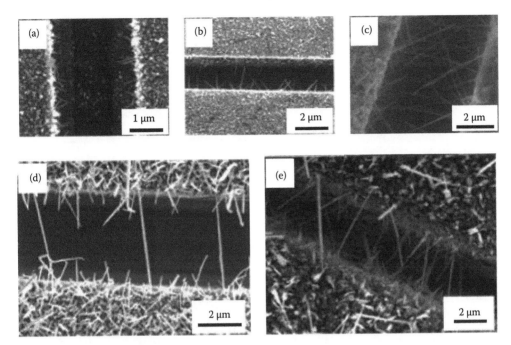

FIGURE 10.21 FE-SEM images of ZnO nanobridges formed across two electrodes at 600°C for 20 min (a), 60 min (b), and 120 min (c) and at 850°C for 60 min (d) and 120 min (e), respectively. (Reproduced with permission from Lee, J.S. et al., Direct formation of catalyst-free ZnO nanobridge devices on an etched Si substrate using a thermal evaporation method, *Nano Lett.*, 6(7), 1487–1490. Copyright 2006 American Chemical Society.)

thin film grown on Si(111) substrate, was investigated.[63] It was established that below the superconducting transition temperature Tc, the nanobridge exhibited a series of sharp voltage steps as a function of current. Just below the critical current, the voltages versus current curve showed a power-law behavior in the low temperature region but Ohmic near the Tc.

A few metal oxide nanobridge reports are known, mainly those on ZnO.[64] Thus, well-aligned single-crystalline ZnO nanobridges were synthesized selectively across the prefabricated electrodes on silicon substrates (Figure 10.21) by a single-step thermal evaporation method without using any metal catalysts or a predeposited ZnO seed layer that was a prerequisite for such synthesis.[65] Careful control of the reaction time and the substrate temperature allowed the nanobridges to form almost exclusively across the electrodes. The ZnO nanowires almost exclusively grow near the edges of the electrodes to form the nanobridges across the electrodes under precisely controlled reaction conditions. This single-step process indeed offered a simple and a cost-effective way to integrate self-assembled nanodevices based on individual and/or a large number of ZnO nanowires with conventional circuits without using E-beam lithography techniques and/or additional costly deposition processes. In addition, UV photodetectors with laterally aligned ZnO nanobridge arrays (Figure 10.22), growing upward in the face-to-face direction, thereby forming biaxial compressive stress where nanobridges intersect, were fabricated.[66] Compared with conventional thin-film photodetectors, the nanobridge devices markedly enhanced the photosensitivity and blue shift (30 nm) of the spectral response. Such nanobridge devices are a promising alternative for transforming advanced optoelectronic integration circuits with a 1D structure into miniaturized devices. In addition, wurtzite ZnO nanobridges (Figures 10.23 and 10.24) and aligned nanonails were synthesized by thermal vapor transport and condensation method from a mixture of ZnO, In_2O_3, and graphite powder.[67] The nanobridges, containing very low concentration of indium in the structure, had two rows of c-axis ZnO nanorods epitaxially grown on the edges of the {0001} plane of the ZnO nanobelt. These materials have potential in applications such as optoelectronics.

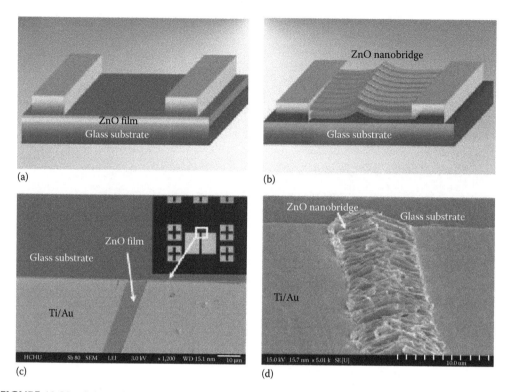

FIGURE 10.22 Schematic for the fabricated UV photodetector and the FE-SEM image with a 45° tilt angle for the devices. Panels (a) and (c) show the conventional ZnO film device. Panels (b) and (d) show the nano-bridge device. The inset of (c) shows an optical micrograph of the patterned photodetectors. (Reproduced with permission from Peng, S.-M. et al., ZnO nanobridge array UV photodetectors, *J. Phys. Chem. C*, 114(7), 3204–3208. Copyright 2010 American Chemical Society.)

FIGURE 10.23 SEM images of the ZnO nanobridges synthesized by vapor transport and condensation method. (a) Low-magnification image showing the abundance of the nanobridges. Scale bar = 4 μm. (b) Medium magnification image of side view of a nanobridge. Scale bar = 500 nm. (Reproduced with permission from Lao, J.Y. et al., ZnO nanobridges and nanonails, *Nano Lett.*, 3(2), 235–238. Copyright 2003 American Chemical Society.)

In addition to ZnO, a single-crystalline α-Fe_2O_3 nanobridge (8 nm diameter and 240 nm length) was laterally grown between two electrodes by one-step thermal oxidation of 100 nm Fe film at 350°C in air to form a nanobridge photodetector.[68] The photosensitivity of this photodetector was larger than 80% with the illumination of the visible-IR light (wavelength: 400–800 nm).

Several binary inorganic compounds are known as nanobridges. Thus, MgB_2 (this compound is widely applied in the fabrication of superconductors) grain boundary nanobridges (length of 100 nm)

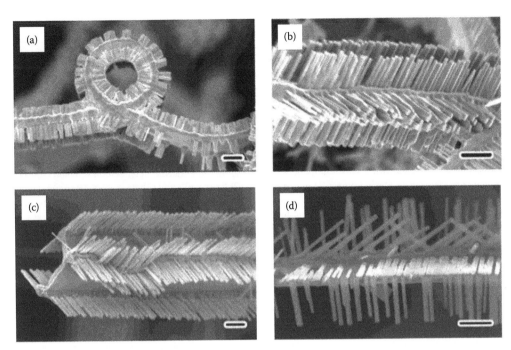

FIGURE 10.24 SEM images of the ZnO nanobridge variations: (a) roller-coaster-like and (b) joined twin-like nanobridges; (c, d) combination of nanobridge and fourfold symmetry. Scale bars = 1 μm. (Reproduced with permission from Lao, J.Y. et al., ZnO nanobridges and nanonails, *Nano Lett.*, 3(2), 235–238. Copyright 2003 American Chemical Society.)

were fabricated by FIB etch.[69] In a related work,[70] MgB_2 nanobridges were obtained with rather good electrical and transport properties both on SiN and sapphire. Such fabrication of superconducting MgB_2 nanobridges on sapphire with a thickness on the order of a few tens of nanometers represents a step forward in the field of nanodevices, such as single-photon detectors, based on this mid-temperature superconducting material. Direct integration of an ensemble of GaN nanowires onto a microchip produced a viable nanobridge device with good alignment and contact performance, the design of which demonstrated the potential of nanowires for sensor development.[71] These GaN nanobridges had strong surface-enhanced photoconductivity with ultrahigh response. Also, nickel monosilicide (NiSi) nanowires were fabricated by metal-induced growth at 575°C.[72] The solid-state reaction of Ni and Si provided linear grown nanowires. The self-assembled nanobridge can be applied to form nanocontacts at relatively low temperatures. Among other inorganics forming bridge-like structures and devices, we note some cuprates and manganates; thus, transport properties (found to be nonlinear) of an insulating submicrometric $YBa_2Cu_3O_{7-\delta}$ bridge, patterned on a thin film, were investigated.[73] The manganite (La, Pr, Ca)MnO_3 (well known for its micrometer-scale phase separation into coexisting ferromagnetic metallic and antiferromagnetic insulating [AFI] regions) was applied for fabricating bridges with widths smaller than the phase separation length scale.[74] Tunneling magnetoresistance across naturally occurring AFI tunnel barriers separating adjacent ferromagnetic regions spanning the width of the bridges was observed.

10.6 NANOTHERMOMETERS

Several reviews on nanothermometers have been published,[75–77] in particular those describing nanothermometers using liquid metals and nanotubes of oxides (e.g., In_2O_3, MgO, SiO_2).[78–80] But the first and the most important example of these nanodevices consisted of CNTs. Thus, a nanothermometer based on CNTs was synthesized in a vertical radiofrequency furnace (which differs from a one-step

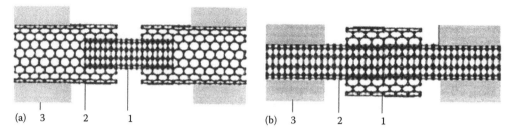

FIGURE 10.25 Schematic diagrams of the nanothermometers based on double-walled CNTs: (a) the telescopic nanothermometer with a movable inner wall and (b) the nanothermometer with a movable shuttle in the form of an outer wall. Designations: (1) movable wall, (2) fixed wall, and (3) electrodes. (Reproduced with permission from Springer Science+Business Media: *Phys. Solid State*, An electromechanical nanothermometer based on thermal vibrations of carbon nanotube walls, 51(6), 2009, 1306–1314, Popova, A.M. et al.)

arc-discharge method).[81] In a related work,[82] an interesting concept was proposed for an electromechanical nanothermometer. The temperature measurements were performed by measuring the conductivity of the nanosystem on CNT basis (shuttle nanothermometer with a movable outer wall and the telescopic nanothermometer with a movable inner wall, Figure 10.25), which depends substantially on the temperature due to the relative thermal vibrations of nanoobjects forming the nanosystem. It was shown that this nanothermometer can be used for measuring the temperature in localized regions with sizes on the order of several hundred nanometers. Also, the systems of the shuttle nanothermometer with a movable inner wall and the telescopic nanothermometer with a movable outer wall are also possible.

A homogeneous mixture of Ga_2O_3 and pure, amorphous, active carbon (weight ratio, 7.8:1) was reacted in an open carbon crucible under a flow of pure N_2 gas: at 1360°C, Reaction 10.3 occurred. However, on the inner surface of a pure graphite outlet pipe at the top of the furnace, the temperature was lower (around 800°C), causing Reaction 10.4 to occur, during which the "nanothermometers" (Figure 10.26) were created. As an improvement of this technique, the following oxidation-assisted approach was proposed.[83] When a Ga-filled CNT was heated in air for an appropriate length of time, an oxide marker was formed on the inner wall of the CNT due to partial oxidation of the gallium. Thus, the temperature to which the nanotube was exposed can be retrieved by progressively heating the CNT

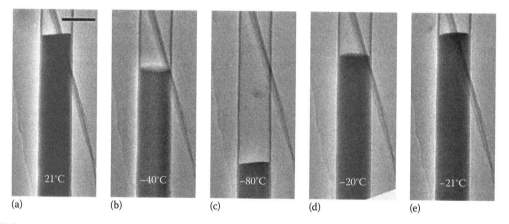

FIGURE 10.26 TEM images of a CNT confined with Ga at different temperatures: (a) Taken at room temperature, (b) taken at −40°C; (c) taken at −80°C and the confined Ga solidified at this temperature, (d) taken at −20°C and the solid Ga melted at this temperature, (e) taken at room temperature. Scale bar = 100 nm. (Liu, Z., University of Sydney, Sydney, Australia, http://www.atip.org/atip-publications/atip-news/2006/4226–060721an_carbon_nanotubes_as_nano-thermometers.html; Reproduced from Liu, Z. et al., *Nanotechnology*, 17(15), 3681, 2006. With permission from Institute of Physics Science.)

until the liquid gallium reached the oxide marker. Additionally, it was found that Ga-filled CNTs possess a unique combination of two types of conductivity: low-resistive Ga-filled nanotube segments and relatively high-resistive empty nanotube regions.[84] It was possible to produce gaps in the Ga filling of any required length by AFM indentation, thus, affecting the nanotube electrical properties:

$$Ga_2O_3(solid) + 2C(solid) \rightarrow Ga_2O(vapor) + 2CO(vapor) \quad (10.3)$$

$$Ga_2O(vapor) + 3CO(vapor) \rightarrow 2Ga(liquid) + C(solid) + 2CO_2(vapor) \quad (10.4)$$

In addition to gallium-filled CNTs, described earlier, indium-filled CNTs (diameters of 100–200 nm and length of ~10 μm) were synthesized *via* a CVD technique.[85] It was shown that the melting and expansion behavior of indium were different from that in a macroscopic state allowing to clarify the problems of indium filling usage in CNT-based nanothermometer. Fabrication of nonfilled CNT nanothermometers is also possible; thus, an interesting concept was proposed for an electromechanical nanothermometer.[82,86] In this device, the temperature measurements were performed by measuring the condition of the nanosystem (CNTs), which depends substantially on the temperature due to the relative thermal vibrations of nanoobjects forming the nanosystem. It was shown that this nanothermometer can be used for measuring the temperature in localized regions with sizes on the order of several hundred nanometers.

Inorganic nanotubes (see the corresponding section later) and nanoparticles have also been used as a support or moving force for nanothermometer devices. Thus, a nanothermometer was fabricated using silver nanoparticles with irregular shape and the size varying from 10 to 100 nm, aggregated on the carbon supporting film coated on a TEM grid.[87] After heating to a certain temperature, smaller silver nanoparticles nucleated and grew on the whole carbon film. The average diameter of the single crystal silver nanospheres increased with the increase in heating temperature while areal density decreased with the heating temperature. Thousands of such nanothermometers constitute a 2D compound macroscopic thermometer, which can record 2D temperature fields. Solid-Pb-filled ZnO nanotubes were also tested for use as nanothermometers[88]; it was established that the expansion of the filling with increasing temperature or the corresponding changes in capacitance can be measured and related to temperature. The authors emphasized such advantages of this nanothermometer as extremely low fabrication costs, superior reliability, and lower demands on structural integrity of the outer shell compared with nanothermometers based on liquid fillings. A highly effective one-step approach was developed to synthesize single-crystalline MgO nanotubes and *in situ* fill nanotubes with Ga.[89] The axes of nanotubes were in the [100] direction of cubic MgO. Linear thermal expansion behavior recorded for liquid gallium column confined in the MgO nanotube made possible the creation of a wide-temperature range nanothermometer with superior mechanic properties and environmental structural stability. Au(Si)-filled β-Ga_2O_3 nanotubes were fabricated by an effective one-step CVD method, and the Au(Si) interior was introduced by capillarity.[90] A linear thermal expansion coefficient of Au(Si) as high as 1.5×10^{-4} (1/K) within a single crystal Ga_2O_3 shell up to 800°C was observed by *in situ* transmission electron microscopy. This structure can be used as a wide-range high-temperature nanothermometer within localized regions of nanosystems. Additionally, well-aligned silica nanotubes with lengths up to 0.9–1.0 mm were prepared *via* a vapor–liquid–solid process and partially filled with indium.[91] The thermal expansion behavior of a liquid indium column inside a silica nanotube was studied as a prospective inorganic nanotube-based nanothermometer.

Nanothermometers, on the basis of CNTs or other inorganics, have also been applied for biological systems. Thus, filled CNTs (Figure 10.27) have the potential to act as sensors that might provide noninvasive temperature control in biological systems on a cellular level; in this case, the temperature can be detected by measuring NMR parameters on the filling materials.[92] The most temperature-sensitive parameter was found to be the [127]I spinlattice relaxation rate in CuI-CNT, which can provide the temperature detection with an accuracy of 2 K. The beneficial feature of a CNT was to provide protection of both (1) a human body against toxic adverse effects from the filling material and (2) a filling material against chemical and biochemical exposure. A nanothermometer, capable

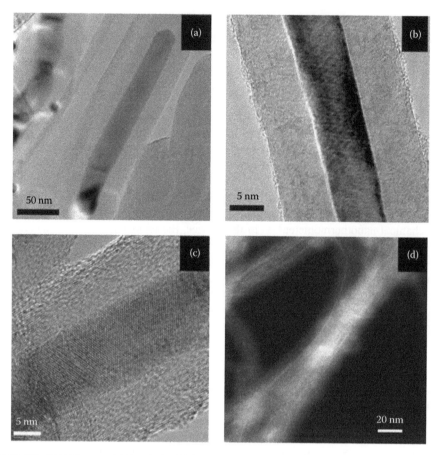

FIGURE 10.27   TEM images of (a) cobalt-, (b) iron-, (c) CuI-, and (d) AgCl-filled CNT. (Vyalikh, A. et al.: A nanoscaled contactless thermometer for biological systems. *Phys. Status Solidi B*. 244 (11). 4092–4096. 2007. Copyright Wiley-VCH Verlag GmbH & Co. KGaA. Reproduced with permission.)

of accurately determining the temperature of solutions as well as biological systems such as HeLa cancer cells and based on the temperature-sensitive fluorescence of $NaYF_4 : Er^{3+}$, Yb^{3+} nanoparticles, where the intensity ratio of the green fluorescence bands of the Er^{3+} dopant ions (2H11/2 $\rightarrow$ 4I15/2 and 4S3/2 $\rightarrow$ 4I15/2) changes with temperature, was created.[93] These fluorescent nanothermometers measured the internal temperature of the living cell from 25°C to its thermally induced death at 45°C.

10.7 NANOTWEEZERS

GaN nanotweezers (Figure 10.28) consisting of a bottom rod (~100 to 150 nm diameter and ~200 to 500 nm length) and two arms (diameters ~40 to 70 nm in the bottoms and ~15 to 30 nm in the tops; 0.8 to 1.5 μm in length) were obtained by CVD by reacting gallium and ammonia on etched cubic MgO(100) single-crystal substrates according to Reaction 10.5.[94] The authors inferred that the fabrication of the GaN nanotweezers is associated with small convex hillocks on the surface of the etched cubic MgO (100) single-crystal substrates and that the nanotweezers grow by a growth mechanism that is similar to vapor-phase heteroepitaxy. If bias voltages are applied to the nanotweezers, it will be possible to make the ends of the tweezer arms bend toward each other from their relaxed positions.

$$Ga + NH_3 \rightarrow GaN + H_2 \tag{10.5}$$

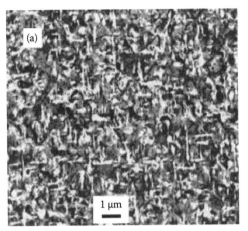
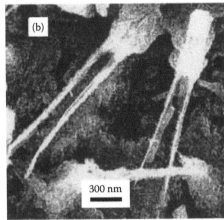

FIGURE 10.28 Field-emission scanning electron microscopy (FE-SEM) images of GaN nanotweezers formed on planar substrates of cubic MgO (100) single crystal. (a) Many nanotweezers with two growth directions normal to each other and (b) two nanotweezers. (Reproduced with permission from Springer Science+Business Media: *Appl. Phys. A*, GaN nanotweezers, 76(1), 2003, 115–118, Li, Z.J. et al.)

CNT nanotweezers are well known.[95,96] Thus, both bent-type and straight-type nanotweezers based on two carbon nanowires by means of localized CVD using a focused ion beam (FIB-CVD) were demonstrated.[97] The location, dimension, and gap between the two nanowires were precisely controlled such that the tweezing motion and the operation voltage can be easily adjusted. Potential applications of these nanotweezers include the manipulation of nanoparticles and nanoscale objects. In addition, a method to make a nanotweezer composed of two CNT arms was offered.[98] The CNT arm was fabricated by attaching a multiwall CNT on a tungsten tip *via* manual assembly. Since each CNT arm has a macroactuator, namely, a separated tweezer arm, it was possible to close and open the nanotweezer repeatedly. Electrochemical etching was used to cut the CNT in the CNT arm, and the cutting resolution was approximately a few hundred nanometers. The system consisted of a pair of electrostatically actuated silicon nanotweezers, and a differential capacitive sensor that was connected to the moving tip of the tweezers was used for achieving subnanometer displacement resolution (around 0.2 nm), detecting the trapping of DNA nanowires, and measuring the evolution of their biomechanical characteristics through the shift of the resonant frequency.[99] Other elemental nanotweezers are known; thus, a hybrid nanomaterial incorporating Te and Se components within a multisegmented nanowire morphology was synthesized through a facile aqueous-phase reaction at room temperature.[100] The obtained Se-Te-Se structures exhibited a self-organization property, thereby enabling the formation of nanotweezers at elevated temperatures. Figure 10.29a shows a typical structure of nanowires assembled in parallel ways when the Se-Te-Se nanowire solution was dried at 100°C on a copper grid and the solvent evaporated. Figure 10.29b is a TEM image of nanowires dried at 140°C. Since the melting temperature of Se (144°C) is lower than that of Te (449.8°C), Se segments of the nanowires can melt first and became a glue to fix the assembled structures (Figure 10.29c and d). At 140°C, Se parts of the nanowires melted and merged into a single head where the Te parts remained as separated legs resulting in a "nanotweezers" structure. Single-crystalline ternary $Zn_xCd_{1-x}S$ nanocombs, which had "comb-shaped" teeth on one side, were synthesized by a one-step metalorganic CVD process at 420°C.[101] Because of the uniform structure and perfect geometrical shape, these nanoteeth could be potentially used as nanocantilever arrays for nanosensors and nanotweezers.

As a representative coordination compound in the role of nanotweezer, we note a zinc porphyrine complex. Thus, nanotweezers consisting of two chiral porphyrins and phenanthrene

FIGURE 10.29 TEM images of assembled Se-Te-Se nanowires heat treated at (a) 100°C and (b) 140°C. (c, d) "Nanotweezers" formed by selective melting of Se segments. (Reproduced with permission from Vinod, T.P. et al., Multisegmented Se-Te-Se hybrid nanowires: A building unit with inbuilt block and glue functionality, *Langmuir*, 26(12), 9195–9197. Copyright 2010 American Chemical Society.)

in between were synthesized according to reaction (10.6).[102] These nanotweezers showed high selectivity toward (6,5)-SWNTs possessing the smallest diameter among the major components of CoMoCAT-SWNTs. Only the single stereoisomer of (6,5)-SWNTs was highly enriched through the extension of CoMoCAT-SWNTs with phenanthrene-bridged chiral diporphyrin nanotweezers.

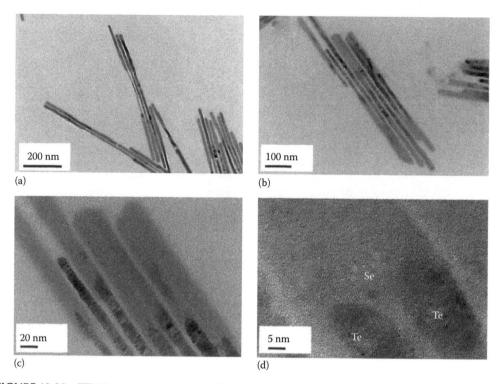

(10.6)

10.8 NANOCARS AND NANOTRUCKS

Nanocars, developed by the research group of Tour,[103–108] are "molecular machines consisting of molecular-scale chassis, axles, and wheels that can directionally roll across solid surfaces." Two complementary approaches can be considered in the fabrication of miniaturized devices[109]: (a) the top-down approach, which reduces the size of macroscopic objects to reach an equivalent microscopic entity using photolithography and related techniques, and (b) the bottom-up approach, which builds functional microscopic or nanoscopic entities from molecular building blocks. These nanocars are mainly based on carborane and fullerene wheels; additionally, a class of molecular wheels was developed based on a *trans*-alkynyl(dppe)$_2$ ruthenium complex.[110,111] The low rotation barrier around the alkyne bond in this complex allows a free rotation of the wheel, while the bulky phosphine ligands offer good interactions with metallic surfaces.

A motorized nanocar (Figure 10.30), which bears a light-activated unidirectional molecular motor and an oligo(phenylene ethynylene) chassis and axle system with four carboranes to serve as the wheels, was synthesized.[112] This proposed propulsion scheme is shown in Figure 10.31.

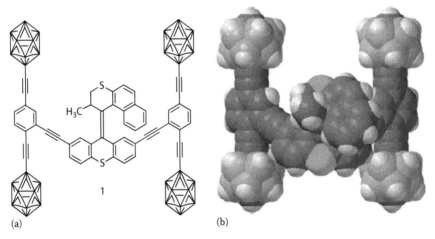

FIGURE 10.30 (a) Structure of motorized nanocar on *p*-carborane basis. The *p*-carborane wheels have BH at every intersection except at the top and bottom vertexes, which represent C and CH positions, *ipso* and *para*, respectively, relative to the alkynes. (b) The space-filling model of this nanocar. (Reproduced with permission from Morin, J.-F. et al., En route to a motorized nanocar, *Org. Lett.*, 8(8), 1713–1716. Copyright 2006 American Chemical Society.)

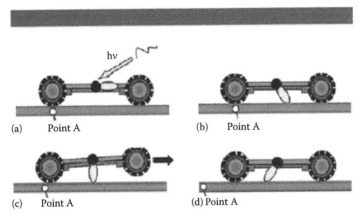

FIGURE 10.31 Proposed propulsion scheme for the motorized nanocar where (a) 365 nm light would impinge upon the motor, which in conjunction with a heated substrate (at least 65°C) (b) affords motor rotation and (c) sweeps across the surface to (d) propel the nanocar forward. (Reproduced with permission from Morin, J.-F. et al., En route to a motorized nanocar, *Org. Lett.*, 8(8), 1713–1716. Copyright 2006 American Chemical Society.)

The convergent five-step synthesis (yields 46%–49%) of inherently highly fluorescent nanocars incorporating 4,4-difluoro-4-bora-3a,4a-diaza-s-indacene (BODIPY)-containing axles and *p*-carborane wheels (compounds **1–4**) was reported.[113] The authors expected that these nanocars exhibit rolling motion with predetermined patterns over smooth surfaces, depending on their chassis.

Design, syntheses reaction (10.7) by *in situ* ethynylation of fullerenes, and testing of fullerene-wheeled single molecular nanomachines (Figure 10.32), namely, nanocars and nanotrucks, composed of three basic components that include spherical fullerene wheels, freely rotating alkynyl axles, and a molecular chassis, were also presented.[114,115] The use of spherical wheels based on C_{60} and freely rotating axles based on alkynes permitted directed nanoscale rolling of the molecular structure on gold surfaces. The obtained nanocars were stable and stationary on the gold surface at room temperature for a wide range of tunneling parameters. The authors attributed their stability to a relatively strong adhesion force between the fullerene wheels and the underlying gold.

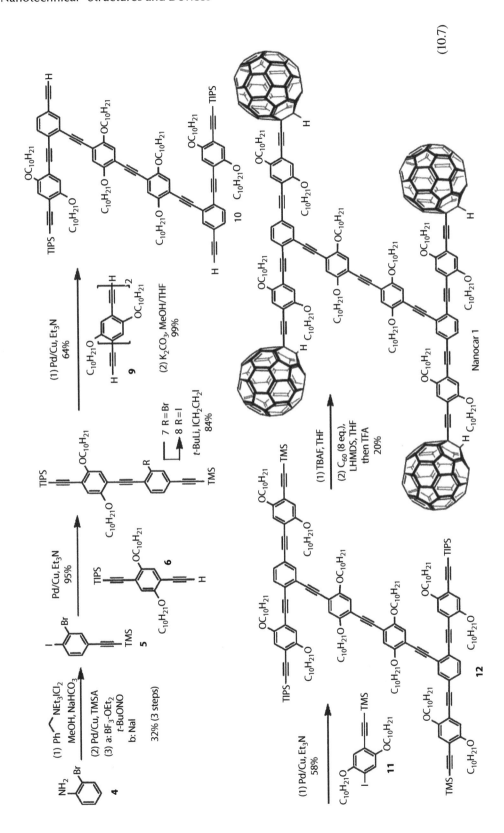

(10.7)

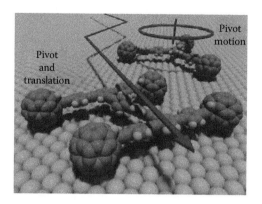

FIGURE 10.32 A summary of the two methods of motion for the different nanocar structures. (Reproduced with permission from Shirai, Y. et al., Directional control in thermally driven single-molecule nanocars, *Nano Lett.*, 5(11), 2330–2334. Copyright 2005 American Chemical Society.)

10.9 NANOBALANCES

The nanobalance for applying a voltage across the nanotube and its count electrode was constructed on the basis of CNTs (have diameters of 5–50 nm and lengths of 1–20 μm), produced by an arc-discharge technique and agglomerated into a fiber-like rod.[116] The specimen holder required the translation of the nanotube *via* either mechanical movement by a micrometer or axial directional piezo. The CNT can be charged by an externally applied voltage. The induced charge is distributed mostly at the tip of the CNT, and the electrostatic force results in the deflection of the nanotube (Figure 10.33). The nanotube is a very flexible structure, and it can be bent to 90° and still recovers its original shape. Additionally, the mass of a particle attached at the end of the spring (Figure 10.34) can be determined if the vibration frequency is measured, provided the spring constant is calibrated. This "nanobalance" (the most

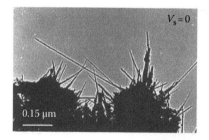

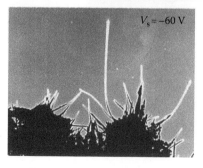

FIGURE 10.33 Electrostatic deflection of a carbon nanotube induced by a constant field across the electrodes. Quantification of the deflection gives the electrical charge on the CNT and the mechanical strain on the fiber. (Reproduced from *J. Phys. Chem. Solids*, 61, Wang, Z.L. et al., Measuring physical and mechanical properties of individual carbon nanotubes by in situ TEM, 1025–1030, Copyright 2000, with permission from Elsevier.)

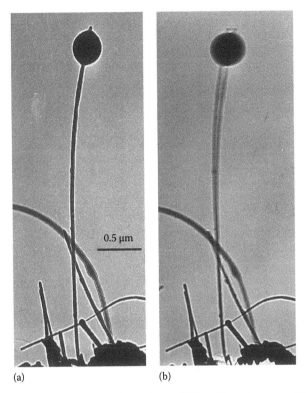

0.5 µm

(a)   (b)

FIGURE 10.34 A small particle attached at the end of a CNT at (a) stationary and (b) first harmonic resonance ($\upsilon = 0.968$ MHz). The effective mass of the particle was measured to be ~ 22 fg (1 f $= 10^{-15}$). (Reproduced from *J. Phys. Chem. Solids*, 61, Wang, Z.L. et al., Measuring physical and mechanical properties of individual carbon nanotubes by in situ TEM, 1025–1030, Copyright 2000, with permission from Elsevier.)

sensitive and smallest balance in the world) was shown to be able to measure the mass of a particle as small as 22 ± 6 fg (1 f $= 10^{-15}$). Other nanobalances have been described in Refs. [117–119].

10.10 NANOGRIDS

Several reports are dedicated to grid-like nanostructures, for instance those for DNA.[120,121] Metal complex nanogrids are also known; thus, the rational design of nanoporous 2D supramolecular structures by the hierarchical assembly of organic molecules (1,3,5-tricarboxylic benzoic acid [trimesic acid, TMA] molecules) and Fe atoms) and transition metal atoms (Fe) at surfaces was demonstrated (Figure 10.35).[122] From the primary components, secondary mononuclear chiral complexes were formed, which represented antecedents for tertiary polynuclear metalorganic nanogrids. These nanogrids represented the constituents of the eventually evolving 2D networks comprising homochiral nanocavity arrays. Also, the growth of epitaxial CaF_2 and SrF_2 thin films on single-crystalline *r*-cut sapphire, MgO(001) and biaxially textured Ni-W polycrystalline tape by low-temperature MOCVD was reported.[123] Certain growth conditions were shown to result in a unique 3D-ordered nanogrid structure (Figure 10.36) of the films, making them a perfect nanotemplated substrate for the epitaxial growth of other functional layers.

10.11 NANOMESH*

In 2003, Corso et al. from the University of Zurich, Switzerland, published in *Science* the discovery of an inorganic nanostructured 2D material, called *nanomesh* (we consider it related to nanogrid). It consisted of a highly regular mesh of hexagonal boron nitride (BN) with a 3 nm periodicity and a 2 nm

* See Chapter 21 for more details.

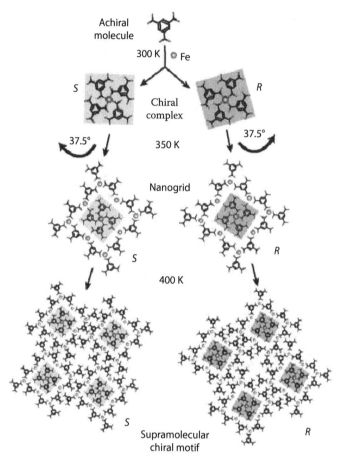

FIGURE 10.35 *Aufbau* of dissymmetric supramolecular motifs mediated by hierarchical assembly of simple achiral species. (1,3,5-tricarboxylic benzoic acid [trimesic acid, TMA]) molecules and Fe atoms represent the primary units that are employed for the formation of secondary chiral mononuclear $(Fe(TMA)_4)$ complexes. The complexes are antecedents for tertiary polynuclear nanogrids, which are in turn the supramolecular motifs for the assembly of homochiral nanocavity arrays. (Reproduced with permission from Spillmann, H. et al., Hierarchical assembly of two-dimensional homochiral nanocavity arrays, *J. Am. Chem. Soc.*, 125, 10725–10728. Copyright 2003 American Chemical Society.)

hole size and was formed by self-assembly on an Rh(111) single-crystalline surface.[124] Two layers of mesh cover the surface uniformly after high-temperature exposure of the clean rhodium surface to borazine $(HBNH)_3$. This unique structure (Figure 10.37) was found[125] to be stable under ambient atmosphere (it does not decompose up to temperatures of at least 796°C), which provides an important basis for technological applications such as templating and coating. The suggested (12 × 12) periodicity of this reconstruction was unambiguously confirmed. The nanomesh is a coincidence structure of 13 h-BN units per 12 Rh substrate units, and no major deviations from the in-plane bulk positions occur for the bulk Rh atoms. In addition, the BN nanomesh can serve as a template to organize molecules and clusters. These characteristics promise interesting applications of the nanomesh in areas such as nanocatalysis, surface functionalization, spintronics, quantum computing, and data storage media like hard drives. Among a series of BN nanomesh investigations, we note the study on the temperature-dependent microscopic structure and the dynamics of adsorbed Xe at different temperatures on single-sheet h-BN on an Rh(111) nanomesh.[126] It was shown that the site-specific adsorption arose from two different interactions of similar magnitude with respect to their lateral variations: the

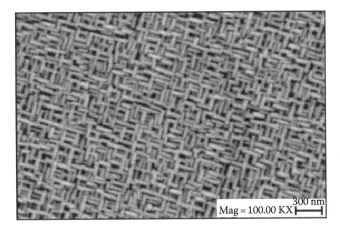

FIGURE 10.36 SEM micrograph of rapidly grown SrF_2 film on r-sapphire with (001) orientation representing a nanogrid film structure. (Reproduced with permission from Blednov, A.V. et al., Epitaxial calcium and strontium fluoride films on highly mismatched oxide and metal substrates by MOCVD: Texture and morphology, *Chem. Mater.*, 22, 175–185. Copyright 2010 American Chemical Society.)

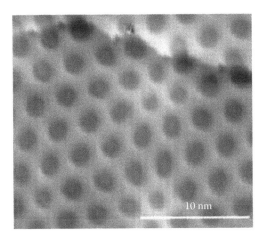
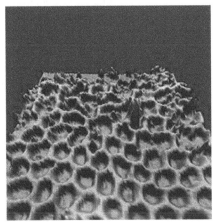

FIGURE 10.37 STM image of the h-BN on Rh(111) nanomesh taken after preparation in ultrahigh vacuum. (Reproduced from *Surf. Sci.*, 601, Bunk, O. et al., Surface x-ray diffraction study of boron-nitride nanomesh in air, L7–L10, Copyright 2007, with permission from Elsevier.)

first can be attributed to a van der Waals type interaction, whereas the second originates from lateral variation of the electrostatic surface potential and is of a polarization type. Both types are responsible for stabilizing dynamic and static Xe rings in these pores. Other aspects of h-BN are described in Refs. [127–129].

A few other mesh-like nanostructures are known, in particular those of C_{60} and graphene.[130] Thus, a new graphene nanostructure, called by authors[131] a graphene nanomesh, can open up a bandgap in a large sheet of graphene to create a semiconducting thin film. These nanomeshes were prepared using block copolymer lithography and can have variable periodicities and neck widths as low as 5 nm. Graphene nanomesh field-effect transistors can support currents nearly 100 times greater than individual graphene nanoribbon devices, and the on–off ratio, which is comparable with the values achieved in individual nanoribbon devices, can be tuned by varying the neck width. Among other related publications, hexagonal graphene nanomeshes with sub-10 nm ribbon width were fabricated.[132] Also, local photodegradation of graphene oxide sheets at the tip of ZnO nanorods was used to achieve semiconducting graphene nanomeshes (Figure 10.38).[133] These graphene

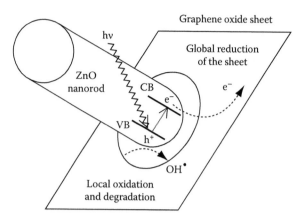

FIGURE 10.38 Schematic illustration of the mechanism describing the formation of graphene nanomeshes by using the photocatalytic property of the ZnO nanorods. (Reproduced with permission from Akhavan, O., Graphene nanomesh by ZnO nanorod photocatalysts, *ACS Nano*, 4(7), 4174–4180. Copyright 2010 American Chemical Society.)

nanomeshes contained smaller O-containing carbonaceous bonds and higher defects than those of the as-prepared graphene oxide sheets. An extended 2D C_{60} nanomesh was fabricated by controlling the binary molecular phases of C_{60} and pentacene on Ag(111).[134] The skeleton of the C_{60} nanomesh was found to be stabilized by the strong molecule–metal interfacial interactions [C_{60}-Ag(111) and pentacene-Ag(111)] and was further modified by the pentacene-C_{60} donor–acceptor intermolecular interactions. This C_{60} nanomesh can serve as an effective template to selectively accommodate guest C_{60} molecules at the nanocavities.

In addition to BN and carbon mesh-like nanostructures, metal oxides[135] are known, in particular ZnO[136] nanomesh; on its basis, AlGaInP-based light-emitting diodes with nanomesh ZnO layers were fabricated using nanosphere lithography.[137] The formation of bismuth nanodots on a well-ordered (4 × 4) vanadium oxide nanomesh on Pd(111) was investigated.[138] At very low Bi coverage, a Bi atom located in the vanadium oxide nanomesh and every Bi atom formed an isolated Bi nanodot. When the Bi coverage increased, most of the nanoholes in the nanomesh were occupied by Bi nanodots, and the remaining Bi atoms formed Bi clusters on the vanadium oxide surface. The template-directed assembly of two planar molecules (copper phthalocyanine [CuPc] and pentacene) on SiC nanomesh was studied by scanning tunneling microscopy and photoelectron spectroscopy, respectively.[139] Both molecules were trapped as single molecules in the cells of SiC nanomesh at low coverage. Metal-containing mesh-like nanostructures are represented by a semitransparent nanomesh Cu electrode on a polyethylene terephthalate (PET) substrate, obtained using metal transfer from a polydimethylsiloxane (PDMS) stamp and nanoimprint lithography.[140] It was found that a uniform pressure of 30 psi and a temperature of 100°C were needed for the transfer of the Cu mesh structure from the PDMS stamp onto the PET substrate. A fabricated semitransparent Cu electrode exhibited high transmittance in the visible range and good electrical conductivity. In addition, a regular square mesh structure with 4.5 nm periodicity was formed after annealing an adsorption layer of Ti on W(100) at 1370 K.[141] This nanomesh consisted of two self-assembled layers on top of the Ti adlayer, ~3.5 physical monolayers thick on average.

10.12 NANOFOAMS

In 2006, Dr. Bryce Tappan at Los Alamos National Laboratory discovered a technique (Figure 10.39) for producing metal nanofoams by igniting pellets of energetic metal bis(tetrazolato)amine complexes in an inert atmosphere.[142] Nanofoams of iron, cobalt, nickel, copper, silver, and palladium have been prepared through this technique. These materials exhibited densities as low as

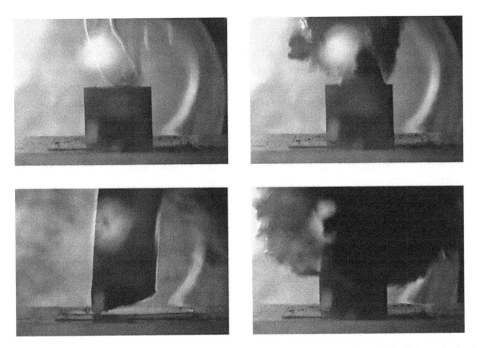

FIGURE 10.39 In the upper left frame, the slanted U-shape with the bright spot is a resistively heated wire igniting a pellet pressed from a high-nitrogen transition-metal complex. (The spot is a reflection from the window of the experimental chamber.) As the pellet rapidly burns, its volume dramatically increases as nitrogen gas released by the combustion creates nanoscopic pores in the coalescing metal particles that are also released. (Reproduced from http://www.rdmag.com/Awards/RD-100-Awards/2005/09/Exclusive. Accessed on January 30, 2011.)

11 mg/cm^3, pore diameters of 20 nm to 1 μm, and surface areas as high as 258 m^2/g.[143] These values compare favorably with those of silica aerogels, the lightest known solids. The formed foams were found to be effective catalysts.

Currently, nanofoams are known for a series of elements and compounds, mainly for carbon[144] (prepared, in particular, by laser ablation[145]). Thus, high-repetition-rate laser ablation and deposition of carbon vapors resulted in the formation of quite different carbonaceous structures depending on the pressure of the ambient Ar gas in the chamber.[146] DLC films formed at a pressure below ~0.1 Torr whereas a DLC nanofoam (Figure 10.40) was created above 0.1 Torr. The bulk density of

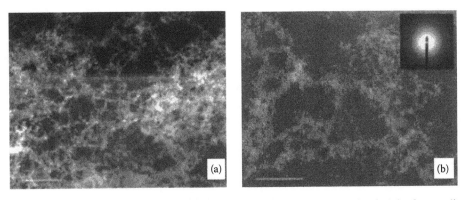

FIGURE 10.40 Scanning (a) and transmission (b) electron microscope images showing the freestanding carbon foam. The bars in the images are 1 mm (a) and 100 nm (b). (Reproduced with permission from Springer Science+Business Media: *Appl. Phys. A*, Formation of cluster-assembled carbon nano-foam by high-repetition-rate laser ablation, 70, 2000, 135–144, Rode, A.V. et al.)

various foam samples was found to be in the range $(2–10) \times 10^{-3}$ g/cm^3, and the specific surface area was 300–400 m^2/g. Carbon nanofoams (CNFs) were also prepared[147] by air cooling the pyrolysate of high purified acetylene gas. Each particle of the nanofoam was constructed by two to four mutually nested carbon balls with ~300 nm diameter. Undoubted room temperature ferromagnetism with saturation magnetization of 0.11 emu/g at 300 K in CNFs can be attributed to the carbon balls themselves instead of impurities. Additionally, within foam-like carbon, onion concentric fullerene-like structures (see details later) can be formed.[148] It was shown that an optimal annealing temperature of 4000 K is required to form well-ordered onions concentric fullerene-like spheres. The onions formed from the outer layer first, and a model was offered in which the background pressure must be sufficient to allow atoms to cluster, yet low enough to allow annealing into well-ordered onions.

Si nanofoam is a porous material with a nanometer structure produced through a sol–gel process and was used as a heat insulator.[149] It was expected that the nanofoam may work as a good acoustic matching layer of an airborne ultrasonic transducer for highly sensitive and wideband ultrasound transmission/detection since the nanofoam has an extremely low acoustic impedance. In addition, light-ion irradiation can significantly improve the elastic modulus and hardness of low-density silica nanofoams (aerogels).[150] Both elastic modulus and hardness saturation with ion fluence and the saturation level increase with the ion stopping power. In addition, organometallic dysprosium complex embedded polyhedral silica nanofoam (PNF-SiO$_2$) was prepared[151] *via* a sol–gel process. It was revealed that the particles of the organometallic complex were evenly distributed on the walls of the polyhedral cells; additionally, the characteristic meso- and macrocell structure of PNF-SiO$_2$ composed of polyhedral cells of silica joined at mesosized windows was preserved also after incorporation.

Polymeric nanofoams are also known. Thus, a route to thermally stable polyimide nanofoams was developed[152] from graft copolymers consisting of thermally stable and thermally labile blocks as a continuous and disperse phase, respectively. The copolymers were synthesized *via* the reaction of sulfone-based diamines with aromatic dianhydrides through the poly(amic acid) precursor, followed by thermal imidization. Foam formation was achieved by thermolysis of the thermally labile block, leaving pores of the size and shape corresponding to the initial copolymer morphology. The nanofoams showed good processability, high thermal stability, and low dielectric constant, which are prerequisites for application in the electronic industry. Additionally, polyimide nanofoamed films were prepared from the polyimide precursors (PMDA-ODA) and poly(ethylene oxide) (PEO) in N,N-dimethylacetamide.[153] The films with a proper amount of PEO displayed relatively low dielectric constant compared with the pure polyimide film.

10.13 NANOJUNCTIONS

Nanojunctions are known mainly for metals, although some oxides and other compounds have also been reported. Thus, 1D BN nanotube nanojunctions are of great interest for both fundamental and applied research because of their stable and excellent mechanical and physical properties.[154] Large quantities of highly pure BN nanotube multiple Y- (Figure 10.41) and T-junctions were synthesized by annealing pure boron nanowire precursors in an N$_2$ atmosphere at 1500°C. It was revealed that the products possess a concentric tubular structure and stoichiometric BN composition. Junctions of silver–copper oxide and silver–zinc oxide (Figure 10.42) were prepared by electrodeposition followed by oxidation within anodic aluminum oxide (AAO) membranes having pores with a diameter of 20 nm.[155]

Ag nanowires with a graded diameter from 8 to 32 nm in AAO membranes were fabricated by the direct-current electrodeposition.[156] It was shown that there was a transport behavior similar to that of a metal–semiconductor junction along the axial direction in the diameter-graded Ag nanowires. Such a homogeneous nanojunction will be of great fundamental and practical significance. Also, an effective photocatalytic disinfection of *Escherichia coli* K-12 using a AgBr-Ag-Bi$_2$WO$_6$

FIGURE 10.41 HRTEM image of a typical Y-junction region in the BN nanotube nanofeather. (Reproduced from Cao, L.M. et al., *Nanotechnology*, 18, 155605, 4 pp., 2007. With permission from Institute of Physics Science.)

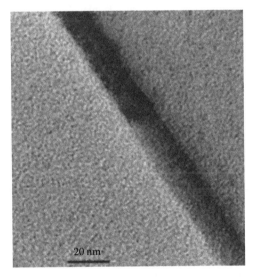

FIGURE 10.42 Transmission electron micrograph for silver–zinc oxide nanojunction. (Reproduced from Bose, A. et al., *Bull. Mater. Sci.*, 32(3), 227–230, 2009. With permission.)

nanojunction system as a catalyst under visible light ($\lambda \geq 400\,nm$) irradiation was reported.[157] This visible-light-driven nanojunction could completely inactivate $5 \times 10^7\,cfu/mL$ *E. coli* K-12 within 15 min, which was superior to other photocatalysts such as Bi_2WO_6 superstructure, $Ag\text{-}Bi_2WO_6$ and $AgBr\text{-}Ag\text{-}TiO_2$ composite. It was found that the diffusing hydroxyl radicals generated both by the oxidative pathway and the reductive pathway play an important role in the photocatalytic disinfection. In addition, 200 nm diameter Au contacts were fabricated by E-beam lithography on sputtered thin film vanadium oxide grown on conducting substrates and current perpendicular to plane electron transport measurements were performed with a conducting tip atomic force microscope.[158] Sharp jumps in electric current were observed in the I–V characteristics of the nano-VO_2 junctions and were attributed to the manifestation of the metal–insulator transition. These results

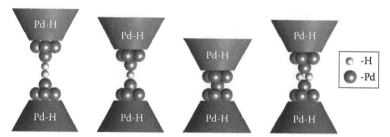

FIGURE 10.43 Possible realizations of the new hydrogen-related atomic configurations. (Reproduced from Csonka, Sz. et al., *Phys. Rev. Lett.*, 93(1), 016802-1–016802-4. Copyright 2004 by the American Physical Society.)

are of potential relevance to novel oxide electronics utilizing metal–insulator transitions. An experimental study of palladium nanojunctions (Figure 10.43) in a hydrogen environment was presented.[159] Two hydrogen-related atomic configurations were found, which had conductances of ~0.5 and ~1 quantum unit ($2e^2/h$). In hydrogen-embedded platinum junctions, a single configuration appeared, which corresponded to a bridge of a H_2 molecule between the platinum electrodes. Other important information on nanojunctions of metals and their composites is given in Au,[160] Pd,[161] Au-NiO-Ni,[162] n-Si(111)/Ir,[163] and NiO-Ni.[164]

Silicon oxide nanojunction structures with various shapes, such as X type, Y type, T type, ring like, and tree like, were fabricated in a self-assembled manner by the hydrothermal method without any metallic catalyst.[165] Both the silicon oxide nanowire part and the junction part consisted of the same chemical composition, forming homogeneous homojunctions and suitable for application in nanoscale optoelectronics devices. The authors noted that the formation of silicon oxide nanojunctions may be influenced by the surrounding environment in the reaction kettle, the growth space among the silicon oxide nanowires, and the weight of SiO droplets at the growth tip. Among other nanojunctions, those of graphene,[166,167] CNTs,[168] InGaAs/InAlAs,[169] and organometallic nanojunctions[170] are known.

10.14 NANOPAPER

Nanopapers are known for various carbon-containing[171] compounds and TiO_2. Thus, wood nanofibrils were used to prepare porous cellulose nanopaper of remarkably high toughness.[172] This nanopaper (Figure 10.44) sample showed very high toughness, $W_A = 15\,MJ/m^3$, in uniaxial tension, and this was associated with a strain-to-failure as high as 10%. Despite a porosity of 28% for the toughest nanopaper, the Young's modulus (13.2 GPa) and tensile strength (214 MPa) were found to be remarkably high. The high toughness of highly porous nanopaper was found to be related to the nanofibrillar network structure and high mechanical nanofibril performance.

The electrical and shape-memory behavior of a self-assembled carbon nanofiber (CNF) nanopaper incorporated with a shape-memory polymer (SMP) was studied.[173,174] The CNF nanopaper could have highly conductive properties. In addition, multifunctional nanocomposites were fabricated using carbon nanopaper sheets.[175] Vapor-grown CNFs were preformed as carbon nanopaper sheets, resulting in a porous structure with highly entangled CNFs, which was integrated into the laminates through the vacuum-assisted resin transfer molding process. It was revealed that the carbon nanopaper sheet was fully integrated to the laminates. Metal-coated carbon nanopapers have been reported too. For instance, after synthesizing a nanopaper (10–20 μm thick, free standing sheets of self-assembled SWNTs), it was decorated with Pt nanoparticles by electroless deposition.[176] Also, the carbon nanopaper sheet, consisting of randomly oriented single-walled nanotubes and vapor-grown CNFs, was coated with nickel by laser pulse deposition.[177] Noncarbon nanopaper

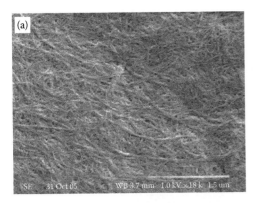

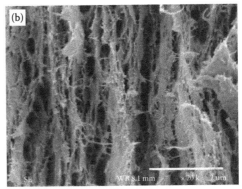

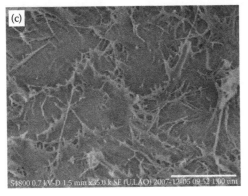

FIGURE 10.44 FE-SEM micrographs of (a) a cellulose nanofibril film surface showing a fibrous network (scale bar is 1.5 μm), (b) the cross section of a fracture surface of a film showing a layered structure (scale bar is 2 μm), and (c) a fracture surface viewed perpendicular to the film surface (scale bar is 1 μm). These films were dried from water suspension. The film in (a) is prepared from DP-1100, and the other two are prepared from DP-800. (Reproduced from *Biomacromolecules*, 9(6), Henriksson, M. et al., Cellulose nanopaper structures of high toughness, 1579–1585, Copyright 2008, with permission from Elsevier.)

is represented by TiO_2, which was fabricated from a titanium dioxide nanoribbon or its mixture with additives.[178] This nanopaper can be used in the fields such as high-temperature filtration, electronic devices, photocatalysts, etc.

10.15 NANOBATTERIES

Nanobatteries, reviewed in,[179–181] are represented by a series of combinations of elemental metals, salts, or composites. Thus, arrays of individual Li nanobatteries were constructed using alumina membranes with pores 200 nm in diameter,[182] accommodating a PEO-Li triflate electrolyte and capping with a cathode material, an ambigel of V_2O_5, making arrays of individual nanobatteries. These small Li batteries can be viable, miniaturized power sources for future nanodevices. The FIB technique was used to obtain thin cross-sectioned nanobattery samples out of a solid-state microbattery (Figure 10.45); they were studied by TEM microscopy in an *ex situ* mode to determine the origin of both the high percentage rate of faulty assembled batteries and of the microbattery capacity fading upon cycling.[183] The authors noted that these batteries cannot presently be carried out at the various states of charge or discharge, due in such special cases to the reactivity of the negative electrode materials under ambient atmosphere. CNTs, encapsulating metallic lithium, can potentially act as a miniaturized nanobattery.[184] Such a battery would be potentially useful in the next generation of

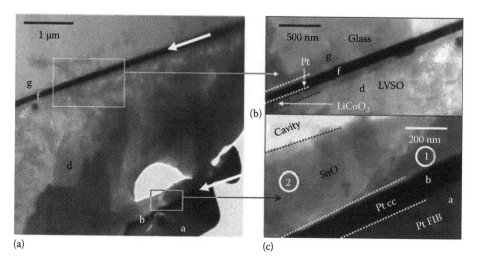

FIGURE 10.45 (a) TEM image of the cycled microbattery. The arrows indicate the positive (top) and the negative (bottom) Pt current collectors layers. (b, c) Higher-magnification images of the $LiCoO_2$ positive electrode region and SnO negative electrode region, respectively. LVSO is $Li_2O-V_2O_5-SiO_2$. (Reproduced from *Chem. Mater.*, 20, Brazier, A. et al., First cross-section observation of an all solid-state lithium-ion "nanobattery" by transmission electron microscopy, 2352–2359, Copyright 2008, with permission from Elsevier.)

communication and remote sensing devices, where a pulse of current is required for their operation. Molybdenum disulfide tubular nanostructures[185] were prepared by treating molybdenum disulfide powder with butyllithium to produce Li_xMoS_2 ($x \sim 1$), which was then fully exfoliated in water and restacked by acidifying the resulting suspension and further sonication in MeOH. These MoS_2 nanostructures can be applied in micro- and nanoscale 3D battery architectures. Among other inorganic potential nanobatteries, we note Al_2O_3[186] and AgI.[187]

Potential organic nanobatteries are represented by poly(vinyl alcohol)-modified particles synthesized in dispersion and functionalized[188] by the simple grafting of ferrocenoyl chloride on the accessible hydroxyl groups of the PVA stabilizer. It was shown that an electrochemical contribution between the semiconducting core and the outer redox system leads to new tunable nanobatteries. In addition, a bionanobattery (Figure 10.46) was developed[189] using ferritins (iron storage proteins) reconstituted with both an iron core (Fe-ferritin) and a cobalt core (Co-ferritin). The reducing capability was determined as well as the half-cell electrical potentials, indicating an electrical output of nearly 0.5 V for the battery cell. This bionanobattery can play a key role in moving to a distributed power storage system for electronic applications.

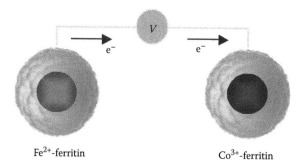

Fe^{2+}-ferritin   Co^{3+}-ferritin

FIGURE 10.46 Bionanobattery unit cell. (From King, G.C. et al., Development of a bio-nanobattery for distributed power storage systems. *Smart Electronics, MEMS, and bioMEMS, and Nanotechnology, Conference*, San Diego, CA, ETATS-UNIS, March 15, 2004, Vol. 5389, pp. 461–467.)

10.16 E-NOSE AND E-TONGUE

Electronic instruments mimicking the mammalian olfactory system are often referred to as "electronic noses" (E-noses, Figure 10.47).[190,191] The human nose is widely used as an analytical sensing tool to assess the quality of drinks, foodstuffs, perfumes, and many other household products in our daytime activities, and of many products in the food, cosmetic, and chemical industries.[192] However, practical use of the human nose is severely limited by the fact that the human sense of smell is subjective, often affected by physical and mental conditions, and tires easily. Consequently, there is considerable need for a device that could mimic the human sense of smell and could provide an objective, quantitative estimation of smell or odor. Thanks to recent nanotechnology breakthroughs, the fabrication of mesoscopic and even nanoscopic E-noses is now feasible in the size domain where miniaturization of the microanalytical systems encounters principal limitations. Electronic noses have provided a plethora of benefits to a variety of commercial industries, including the agricultural, biomedical, cosmetics, environmental, food, manufacturing, military, pharmaceutical, regulatory, and various scientific research fields.[193] Advances have improved product attributes, uniformity, and consistency as a result of increases in quality control capabilities afforded by electronic-nose monitoring of all phases of industrial manufacturing processes.

Among a series of E-noses, a portable E-nose based on hybrid CNT-SnO$_2$ gas sensors was fabricated using E-beam evaporation by powder mixing.[195] Doping of CNT improved the sensitivity of hybrid gas sensors, while the quantity of CNT had a direct effect on the selectivity to volatile organic compounds, that is, MeOH and EtOH. Based on the proposed methods, this instrument can monitor and classify 1 vol.% of MeOH contamination in whiskeys. A series of reports of Sysoev et al.[196–198] also describes the use of SnO$_2$ in the creation of E-noses (Figure 10.48). It was demonstrated that a discrimination between H$_2$ and CO can be achieved using the electronic nose approach through an analysis of the responses from the three-nanowire array. The obtained results can be used for further development of a real world "nanoscopic E-nose" based on the array of individual metal oxide quasi-1D nanostructures.

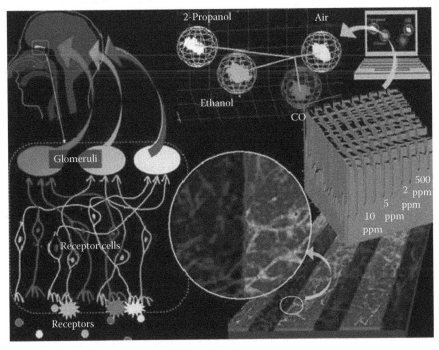

FIGURE 10.47 Function of an E-nose. (Reproduced from http://www.nanowerk.com/http://www.nanowerk.com/spotlight/spotid=3331.php, accessed on January 30, 2011.)

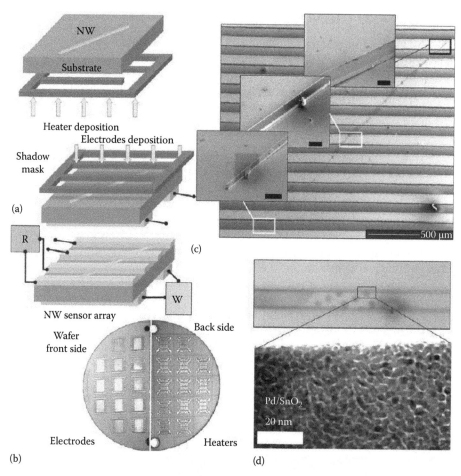

FIGURE 10.48 Fabrication and characterization of the single SnO$_2$ NB E-nose. (a, b) The deposition of Pt electrodes onto the individual NB and meander heaters onto the back of the Si/SiO$_2$ wafer. The ohmmeter R and the power source W serve to conduct resistance measurements of the NB segments and heating of the substrate correspondingly. (c) Survey SEM image of the ultralong (ca. 1.5 mm) wedge-like SnO$_2$ nanobelt. The inset SEM images were taken at different locations along the NB and demonstrate the degree of the morphological changes (scale bar corresponds to 2 μm). (d) (Top) polarized optical microscopy images (×1000) of Pd patches deposited onto the three segments using the shadow mask; (bottom) TEM image of few ML Pd deposit indicates that the growth proceeds *via* Volmer–Weber (cluster) mode. (Reproduced with permission from Sysoev, V.V. et al., Single-nanobelt electronic nose: Engineering and tests of the simplest analytical element, *ACS Nano*, 4(8), 4487–4494. Copyright 2010 American Chemical Society.)

Pure and 10% wt./wt. Au-doped ZnO nanostructure sensors were produced using a thermal oxidation technique with sintering temperature at 700°C under oxygen atmosphere at a flow rate of 500 mL/min and used as sensing devices in a portable E-nose.[199] The sensors were demonstrated to be sensitive to various volatile organic compounds, especially ethanol vapor. This portable E-nose can detect the difference between alcohol beverages and alcohol solutions and can distinguish between white and red wines with the same percentage of alcohol.

Analogously, E-tongues are used in food analysis.[200] As a representative example, we note that an innovative hybrid sensor array (E-tongue) formed by voltammetric electrodes chemically modified with three families of electroactive substances (conducting polymers, phthalocyanine complexes, and perylenes) was developed.[201] This e-tongue demonstrated a good-quality ability in discriminating and recognizing among 12 Spanish red wines based on their denomination of origin, grape variety, and vintage.

Another rapidly developing area, which could be attributed to this "nanotechnical" section, corresponds to *nanoelectromechanical systems* (NEMs). We note here only this field of nanotechnology mentioning three new types of nanoelectromechanical systems based on CNTs, reported in[202] an electromechanical *nanothermometer* (see Chapter 10), a *nanorelay*, and a *nanomotor*. The nanorelay was a prototype of a memory cell, and the nanoactuator can be used for the transformation of the forward force into the relative rotation of the walls. Relative motion of the walls in these nanosystems was defined by the shape of the interwall interaction energy surface.

REFERENCES

1. Ronning, C.; Schwen, D. One dimensional material from semiconductors. Nanowires, nanosaws, nanospirals. *Physik in Unserer Zeit*, 2006, *37* (1), 34–40.
2. Prinz, V. Ya. Three-dimensional self-shaping nanostructures based on free stressed heterofilms. *Russian Physics Journal (Translation of Izvestiya Vysshikh Uchebnykh Zavedenii, Fizika)*, 2003, *46* (6), 568–576.
3. Mezey, P. G. Computational quantum chemistry design of nanospirals and nanoneedles. *Lecture Series on Computer and Computational Sciences*, 2006, *6 (Trends and Perspectives in Modern Computational Science)*, 222–230.
4. Vaudreuil, S.; Bousmina, M. Stretchable carbon nanosprings production by a catalytic growth process. *Journal of Nanoscience and Nanotechnology*, 2009, *9* (8), 4880–4885.
5. Zettl, A.; Cumigs, J. Sharpened nanotubes, nanobearings, and nanosprings. In *CP544, Electronic Properties of Novel Materials—Molecular Nanostructures*. Kuzmany, H. et al. (Eds.). American Institute of Physics, Melville, NY, 2000. pp 526–532.
6. Mukhopadhyay, K.; Porwal, D.; Rao, K. U. B. Carbon micro/nano spring structures in the absence of sulphurous promoter by CCVD method. *Journal of Nanoscience and Nanotechnology*, 2007, *7* (6), 1851–1854.
7. Tang, N.; Kuo, W.; Jeng, C.; Wang, L.; Lin, K.; Du, Y. Coil-in-coil carbon nanocoils: 11 Gram-scale synthesis, single nanocoil electrical properties, and electrical contact improvement. *ACS Nano*, 2010, *4* (2), 781–788.
8. Nakamatsu, K.; Igaki, J.; Nagase, M.; Ichihashi, T.; Matsui, S. Mechanical characteristics of tungsten-containing carbon nanosprings grown by FIB-CVD. *Microelectronic Engineering*, 2006, *83* (4–9), 808–810.
9. Nakamatsu, K.-i.; Ichihashi, T.; Kanda, K.; Haruyama, Y.; Kaito, T.; Matsui, S. Nanostructure analysis of nanosprings fabricated by focused-ion-beam chemical vapor deposition. *Japanese Journal of Applied Physics*, 2009, *48* (10), 105001/1–105001/4.
10. Nakai, Y.; Kang, Y.; Okada, M.; Haruyama, Y.; Kanda, K.; Ichihashi, T.; Matsui, S. Mechanical characteristics of nanosprings fabricated by focused-ion-beam chemical vapor deposition using ferrocene source gas. *Japanese Journal of Applied Physics*, 2010, *49* (6, Pt. 2), 06GH07/1–06GH07/4.
11. Singh, J. P.; Liu, D.-L.; Ye, D.-X.; Picu, R. C.; Lu, T.-M.; Wang, G.-C. Metal-coated Si springs: Nanoelectromechanical actuators. *Applied Physics Letters*, 2004, *84* (18), 3657–3659.
12. Nagar, R.; Patzig, C.; Rauschenbach, B.; Sathe, V.; Kanjilal, D.; Mehta, B. R.; Singh, J. P. Ion beam induced anisotropic deformation of Si nanosprings. *Journal of Physics D: Applied Physics*, 2009, *42* (14), 145404/1–145404/8.
13. Patzig, C.; Rauschenbach, B. Temperature effect on the glancing angle deposition of Si sculptured thin films. *Journal of Vacuum Science and Technology, A: Vacuum, Surfaces, and Films*, 2008, *26* (4), 881–886.
14. Schubert, E.; Fahlteich, J.; Hoeche, Th.; Wagner, G.; Rauschenbach, B. Chiral silicon nanostructures. *Nuclear Instruments and Methods in Physics Research, Section B: Beam Interactions with Materials and Atoms*, 2006, *244* (1), 40–44.
15. Zhao, Y.-P.; Ye, D.-X.; Wang, P.-I.; Wang, G.-C.; Lu, T.-M. Fabrication of Si nanocolumns and Si square spirals on self-assembled monolayer colloid substrates. *International Journal of Nanoscience*, 2002, *1* (1), 87–97.
16. Zhang, L.; Dong, L.; Bell, D. J.; Nelson, B. J.; Schoenenberger, C.; Gruetzmacher, D. Fabrication and characterization of freestanding Si/Cr micro- and nanospirals. *Microelectronic Engineering*, 2006, *83* (4–9), 1237–1240.
17. Zhang, L.; Dong, L.; Bell, D. J.; Nelson, B. J.; Gruetzmacher, D. A. Fabrication and characterization of self-scrolling Si/Cr micro- and nanostructures. *Proceedings of the First IEEE International Conference on Nano/Micro Engineered and Molecular Systems*, Zhuhai, China, January 18–21, 2006, pp. 1268–1271.

18. Gruetzmacher, D.; Zhang, L.; Dong, L.; Bell, D.; Nelson, B.; Prinz, A.; Ruh, E. Ultra flexible SiGe/Si/Cr nanosprings. *Microelectronics Journal*, 2008, *39* (3–4), 478–481.

19. Bell, D. J.; Sun, Y.; Zhang, L.; Dong, L. X.; Nelson, B. J.; Grutzmacher, D. Three-dimensional nanosprings for electromechanical sensors. *Sensors and Actuators A: Physical*, 2006, *130–131*, 54–61.

20. Gruetzmacher, D.; Dais, C.; Zhang, L.; Mueller, E.; Solak, H. H. Templated self-organization of SiGe quantum structures for nanoelectronics. *Materials Science and Engineering, C: Biomimetic and Supramolecular Systems*, 2007, *27* (5–8), 947–953.

21. He, Y.; Fu, J.; Zhang, Y.; Zhao, Y.; Zhang, L.; Xia, A.; Cai, J. Multilayered Si/Ni nanosprings and their magnetic properties. *Small*, 2007, *3* (1), 153–160.

22. Abdeddaim, R.; Guida, G.; Priou, A.; Gallas, B.; Rivory, J. Negative permittivity and permeability of gold square nanospirals. *Applied Physics Letters*, 2009, *94* (8), 081907/1–081907/3.

23. Copper nano-springs for compliant interconnect applications. *IP.com Journal*, http://ip.com/IPCOM/000135873, 2006, *6* (5A), 29.

24. Kesapragada, S. V.; Victor, P.; Nalamasu, O.; Gall, D. Nanospring pressure sensors grown by glancing angle deposition. *Nano Letters*, 2006, *6* (4), 854–857.

25. Gao, P. X.; Wang, A. L. Nanoarchitectures of semiconducting and piezoelectric zinc oxide. *Journal of Applied Physics*, 2005, *97*, 044304.

26. Wang, Z. L. Zinc oxide nanostructures: Growth, properties and applications. *Journal of Physics: Condensed Matter*, 2004, *16*, R829–R858.

27. Gao, H.; Ji, H.; Zhang, X.; Lu, H.; Liang, Y. Indium-doped ZnO nanospirals synthesized by thermal evaporation. *Journal of Vacuum Science and Technology, B: Microelectronics and Nanometer Structures—Processing, Measurement, and Phenomena*, 2008, *26* (2), 585–588.

28. Zhong, H.-M.; Liu, Q.; Sun, Y.; Lu, W. Annealing effect on structure and green emission of ZnO nanopowder by decomposing precursors. *Chinese Physics B*, 2009, *18* (11), 5024–5028.

29. Gao, P. X.; Wang, Z. L. High-yield synthesis of single-crystal nanosprings of ZnO. *Small*, 2005, *1* (10), 945–949.

30. Jin, L.; Wang, J.; Choy, W. C. H. Growth of ZnSe nanospirals with bending mediated by Lomer-Cottrell sessile dislocations through varying pressure. *Crystal Growth and Design*, 2008, *8* (10), 3829–3833.

31. Peng, Y.; Liu, Z.-Y.; Liu, S.-H. Facile microwave heating approach to spiral $Cd(OH)_2$ nanowires with 1D nanostructure. *Gaodeng Xuexiao Huaxue Xuebao*, 2010, *31* (1), 7–10.

32. Hwang, G.; Hashimoto, H.; Bell, D. J.; Dong, L.; Nelson, B. J.; Silke, S. Piezoresistive InGaAs/GaAs nanosprings with metal connectors. *Nano Letters*, 2009, *9* (2), 554–561.

33. Koley, G.; Cai, Z. Growth of gallium nitride nanowires and nanospirals. *Materials Research Society Symposium Proceedings (Nanowires and Carbon Nanotubes—Science and Applications)*, Pittsburg, CA, 2007, Volume Date 2006, Vol. 963E, Paper # 0963-Q10–17.

34. Mei, Y.; Thurmer, D. J.; Deneke, C.; Kiravittaya, S.; Chen, Y.-F.; Dadgar, A.; Bertram, F. et al. Fabrication, self-assembly, and properties of ultrathin AlN/GaN porous crystalline nanomembranes: Tubes, spirals, and curved sheets. *ACS Nano*, 2009, *3* (7), 1663–1668.

35. Shen, G. Z.; Bando, Y.; Zhi, C. Y.; Yuan, X. L.; Sekiguchi, T.; Golberg, D. Single-crystalline cubic structured InP nanosprings. *Applied Physics Letters*, 2006, *88* (24), 243106/1–243106/3.

36. McIlroy, D. N.; Zhang, D.; Kranov, Y.; Han, H.; Alkhateeb, A.; Grant, N. M. The effects of crystallinity and catalyst dynamics on boron carbide nanospring formation. *Materials Research Society Symposium Proceedings (Three-Dimensional Nanoengineered Assemblies)*, Boston, MA, 2003, Volume Date 2002, Vol. 739, pp. 165–173.

37. Zhang, D.; Alkhateeb, A.; Han, H.; Mahmood, H.; McIlroy, D. N.; Grant, N. M. Silicon carbide nanosprings. *Nano Letters*, 2003, *3* (7), 983–987.

38. Liu, H.; Cui, H.; Wang, J.; Gao, L.; Han, F.; Boughton, R. I.; Jiang, M. Growth of $NaFe_4P_{12}$ skutterudite single crystalline nanosprings synthesized through a hydrothermal-reduction-alloying method. *Journal of Physical Chemistry B*, 2004, *108* (35), 13254–13257.

39. Chen, W.; Xue, G. Synthesis of polypyrrole nanospirals in the presence of surfactant crystallites. *Abstracts of Papers, 239th ACS National Meeting*, San Francisco, CA, March 21–25, 2010, PMSE-98.

40. Chen, P.; Ma, X.; Zhang, Y.; Hu, K.; Liu, M. Nanofibers and nanospirals fabricated through the interfacial organization of a partially fluorinated compound. *Langmuir*, 2007, *23* (22), 11100–11106.

41. Wang, C.; Huang, Z.; Lin, Y.; Ren, J.; Qu, X. Artificial DNA nano-spring powered by protons. *Advanced Materials*, 2010, *22* (25), 2792–2798.

42. Zhao, Y. Designing nanostructures for optical sensor applications. *Abstracts, 57th Southeast/61st Southwest Joint Regional Meeting of the American Chemical Society*, Memphis, TN, November 1–4, 2005, NOV04-045.

43. McIlroy, D. N. Role of nanoscale geometry and surface chemistry on hydrogen storage by silica nano-springs. *Abstracts, Joint 65th Northwest and 22nd Rocky Mountain Regional Meeting of the American Chemical Society*, Pullman, WA, June 20–23, 2010, NWRM-189.

44. Liu, F.; Cão, P. J.; Zhang, H. R.; Li, J. Q.; Gao, H. J. Controlled self-assembled. nanoaeroplanes, nano-combs, and tetrapod-like networks of zinc oxide. *Nanotechnology*, 2004, *15*, 949–952.

45. Tetsuro, S.; Nobuo, K. One-pot room-temperature synthesis of single-crystalline gold nanocorolla in water. *Journal of the American Chemical Society*, 2009, *131*, 14407–14412.

46. Dev, A.; Kar, S.; Chaudhuri, S. ZnO hierarchical nanostructures: Simple solvothermal synthesis and growth mechanism. *Journal of Nanoscience and Nanotechnology*, 2008, *8* (9), 4506–4513.

47. Ma, C.; Moore, D.; Li, Jing; W., Zhong L. Nanobelts, nanocombs, and nanowindmills of wurtzite ZnS. *Advanced Materials*, 2003, *15* (3), 228–231.

48. Lambert, C. J.; Bailey, S. W. D.; Cserti, J. Oscillating chiral currents in nanotubes: A route to nanoscale magnetic test tubes. *Condensed Matter*, 2008, 1–4, arXiv.org, e-Print Archive, arXiv:0809.4216v1 [cond-mat.mes-hall].

49. Christopher M. Systematic investigation on the growth of one dimensional wurtzite nanostructures. PhD dissertation, Georgia Institute of Technology, Atlanta, GA, 2005.

50. Wen, P.; Itoh, H.; Tang, W.; Feng, Q. Single nanocrystals of anatase-type TiO_2 prepared from layered titanate nanosheets: Formation mechanism and characterization of surface properties. *Langmuir*, 2007, *23*, 11782–11790.

51. Ma, C.; Moore, D.; Ding, Y.; Li, J.; Wang, Z. L. Nanobelt and nanosaw structures of II–VI semiconduc-tors. *International Journal of Nanotechnology*, 2004, *1* (4), 431–451.

52. Xiang, W. U.; Jie-He, S. U. I.; Wei, C. A. I.; Jiang, P. Temperature-controllable preparation of ZnS nano-saws on Si substrate. *Chinese Physics Letters*, 2008, *25* (2), 737–739.

53. Jian, Y.; Zhaoming, W.; Lide, Z. Effect of stacking fault on the formation of the saw-teeth of ZnS nano-saws. *Crystal Growth and Design*, 2008, *8* (5), 1723–1726.

54. Christopher, M.; Yong, D.; Daniel, M.; Xudong, W.; Wang, Z. L. Single-crystal CdSe nanosaws. *Journal of the American Chemical Society*, 2004, *126*, 708–709.

55. Shen, G.; Bando, Y.; Liu, B.; Tang, C.; Huang, Q.; Golberg, D. Systematic investigation of the forma-tion of 1D α-Si_3N_4 nanostructures by using a thermal-decomposition/nitridation process. *Chemistry A European Journal*, 2006, *12*, 2987–2993.

56. Yukihito, K.; Kunio, T. Gold nanobridge stabilized by surface structure. *Physical Review Letters*, 1997, *79* (18), 3455–3458.

57. Park, H. S.; Zimmerman, J. A. Stable nanobridge formation in (110) gold nanowires under tensile defor-mation. *Scripta Materialia*, 2006, *54*, 1127–1132.

58. Sutrakar, V. K.; Mahapatra, D. R. Coupled effect of size, strain rate, and temperature on the shape mem-ory of a pentagonal Cu nanowire. *Nanotechnology*, 2009, *20* (4), 045701/1–045701/10.

59. Vijay, R.; Levenson-Falk, E. M.; Slichter, D. H.; Siddiqi, I. Approaching ideal weak link behavior with three dimensional aluminum nanobridges. *Applied Physics Letters*, 2010, *96* (22), 223112/1–223112/3.

60. Ham, J.; Kang, J.; Noh, J.-S.; Lee, W. Self-assembled Bi interconnections produced by on-film formation of nanowires for in situ device fabrication. *Nanotechnology*, 2010, *21* (16), 165302/1–165302/5.

61. Claudio-Gonzalez, D.; Husain, M. K.; de Groot, C. H.; Bordignon, G.; Fischbacher, T.; Fangohr, H. Fabrication and simulation of nanostructures for domain wall magnetoresistance studies on nickel. *Journal of Magnetism and Magnetic Materials*, 2010, *322* (9–12), 1467–1470.

62. Song, M. S.; Jung, J. H.; Kim, Y.; Wang, Y.; Zou, J.; Joyce, H. J.; Gao, Q.; Tan, H. H.; Jagadish, C. Vertically standing Ge nanowires on GaAs(110) substrates. *Nanotechnology*, 2008, *19* (12), 125602/1–125602/6.

63. Wang, J.; Ma, X.-C.; Qi, Y.; Ji, S.-H.; Fu, Y.-S.; Lu, L.; Jin, A.-Z. et al. Dissipation in an ultrathin super-conducting single-crystal Pb nanobridge. *Journal of Applied Physics*, 2009, *106* (3), 034301/1–034301/4.

64. Pelatt, B. D.; Huang, C. C.; Conley, J. F. Jr. ZnO nanobridge devices fabricated using carbonized photo-resist. *Solid-State Electronics*, 2010, *54* (10), 1143–1149.

65. Lee, J. S.; Islam, M. S.; Kim, S. Direct formation of catalyst-free ZnO nanobridge devices on an etched Si substrate using a thermal evaporation method. *Nano Letters*, 2006, *6* (7), 1487–1490.

66. Peng, S.-M.; Su, Y.-K.; Ji, L.-W.; Wu, C.-Z.; Cheng, W.-B.; Chao, W.-C. ZnO nanobridge array UV pho-todetectors. *Journal of Physical Chemistry C*, 2010, *114* (7), 3204–3208.

67. Lao, J. Y.; Huang, J. Y.; Wang, D. Z.; Ren, F. ZnO nanobridges and nanonails. *Nano Letters*, 2003, *3* (2), 235–238.

68. Hsu, L.-C.; Kuo, Y.-P.; Li, Y.-Y. On-chip fabrication of an individual α-Fe_2O_3 nanobridge and application of ultrawide wavelength visible-infrared photodetector/optical switching. *Applied Physics Letters*, 2009, *94* (13), 133108/1–133108/3.

69. Lee, S.-G.; Hong, S.-H.; Kang, W. N.; Kim, D. H. MgB$_2$ grain boundary nanobridges prepared by focused ion beam. *Journal of Applied Physics*, 2009, *105* (1), 013924/1–013924/4.

70. Portesi, C.; Borini, S.; Taralli, E.; Rajteri, M.; Monticone, E. Superconducting MgB$_2$ nanobridges and meanders obtained by an electron beam lithography-based technique on different substrates. *Superconductor Science and Technology*, 2008, *21* (3), 034006/1–034006/5.

71. Chen, R.-S.; Wang, S.-W.; Lan, Z.-H.; Tsai, J. T.-H.; Wu, C.-T.; Chen, L.-C.; Chen, K.-H.; Huang, Y.-S.; Chen, C.-C. On-chip fabrication of well-aligned and contact-barrier-free GaN nanobridge devices with ultrahigh photocurrent responsivity. *Small*, 2008, *4* (7), 925–929.

72. Kim, J.; Lee, D.; Anderson, W. A. NiSi nanowires and nanobridges formed by metal-induced growth. *Materials Research Society Symposium Proceedings (Assembly at the Nanoscale—Toward Functional Nanostructured Materials)*, Boston, MA, 2006, Volume Date 2005, Vol. 901E, Paper # 0901-Ra22-03-Rb22-03.

73. Fruchter, L.; Kasumov, A. Y.; Briatico, J.; Ivanov, A. A.; Nicholaichik, V. Non linear transport properties of an insulating YBCO nano-bridge. *European Physical Journal B: Condensed Matter and Complex Systems*, 2010, *73* (3), 361–365.

74. Singh-Bhalla, G.; Biswas, A.; Hebard, A. F. Tunneling magnetoresistance in (La,Pr,Ca)MnO$_3$ nano-bridges. *Condensed Matter*, 2009, 1–13, arXiv.org, e-Print Archive, arXiv:0902.4386v1 [cond-mat. mtrl-sci].

75. Fabrication and characterization of one-dimensional nanoscale heterostructures. *Progress in Advanced Materials Research*, 2007, 185–219.

76. Bando, Y. Carbon nanotubes as use of nanothermometer. Zhan, J.; Voler, N. H. (Eds.). *Materials Integration*, 2004, *17* (6), 34–40.

77. Bando, Y. World smallest nanothermometer using carbon nanotube. *Kagaku*, 2004, *59* (6), 20–24.

78. Bando, Y. Oxide-nanotubes as use of nanothermometer. *Seramikkusu*, 2006, *41* (4), 262–266.

79. Bando, Y. Nanothermometer using oxide nanotubes. *Materials Integration*, 2004, Volume Date 2005, *18* (1), 42–47.

80. Bando, Y. Study of nanomaterials by using state-of-the-art microscopy. *Kagaku to Kogyo*, 2004, *57* (6), 595–600.

81. Yihua, G; Yoshio, B. Carbon nanothermometer containing gallium. *Nature*, 2002, *415* (7), 599–600.

82. Popova, A. M.; Lozovik, Yu. E.; Bichoutskaia, E.; Ivanchenko, G. S.; Lebedev, N. G.; Krivorotov, E. K. An electromechanical nanothermometer based on thermal vibrations of carbon nanotube walls. *Physics of the Solid State*, 2009, *51* (6), 1306–1314.

83. Liu, Z.; Bando, Y.; Hu, J.; Ratinac, K.; Ringer, S. P. A novel method for practical temperature measurement with carbon nanotube nanothermometers. *Nanotechnology*, 2006, *17* (15), 3681–3684.

84. Zhan, J.; Bando, Y.; Hu, J.; Golberg, D. Nanothermometers: Bulk synthesis and calibration. *Abstracts of Papers, 232nd ACS National Meeting*, San Francisco, CA, September 10–14, 2006, INOR-488.

85. Gao, Y.; Bando, Y.; Golberg, D. Melting and expansion behavior of indium in carbon nanotubes. *Applied Physics Letters*, 2002, *81* (22), 4133–4135.

86. Popov, A. M.; Lozovik, Y. E.; Bichoutskaia, E.; Ivanchenko, G. S.; Lebedev, N. G.; Krivorotov, E. K. Electromechanical nanothermometer based on carbon nanotubes. *Fullerenes, Nanotubes, and Carbon Nanostructures*, 2008, *16* (5–6), 352–356.

87. Lan, Y.; Wang, H.; Chen, X.; Wang, D.; Chen, G.; Ren, Z. Nanothermometer using single crystal silver nanospheres. *Advanced Materials*, 2009, *21* (47), 4839–4844.

88. Wang, C.-Y.; Gong, N.-W.; Chen, L.-J. High-sensitivity solid-state Pb(core)/ZnO(shell) nanother-mometers fabricated by a facile galvanic displacement method. *Advanced Materials*, 2008, *20* (24), 4789–4792.

89. Li, Y. B.; Bando, Y.; Golberg, D.; Liu, Z. W. Ga-filled single-crystalline MgO nanotube: Wide-temperature range nanothermometer. *Applied Physics Letters*, 2003, *83* (5), 999–1001.

90. Gong, N. W.; Lu, M. Y.; Wang, C. Y.; Chen, Y.; Chen, L. J. Au(Si)-filled β-Ga$_2$O$_3$ nanotubes as wide range high temperature nanothermometers. *Applied Physics Letters*, 2008, *92* (7), 073101/1–073101/3.

91. Li, Y.; Bando, Y.; Golberg, D. Indium-assisted growth of aligned ultra-long silica nanotubes. *Advanced Materials*, 2004, *16* (1), 37–40.

92. Vyalikh, A.; Klingeler, R.; Hampel, S.; Haase, D.; Ritschel, M.; Leonhardt, A.; Borowiak-Palen, E. et al. A nanoscaled contactless thermometer for biological systems. *Physica Status Solidi B*, 2007, *244* (11), 4092–4096.

93. Vetrone, F.; Naccache, R.; Zamarron, A.; Juarranz de la Fuente, A.; Sanz-Rodriguez, F.; Martinez Maestro, L.; Martin Rodriguez, E.; Jaque, D.; Sole, J. G.; Capobianco, J. A. Temperature sensing using fluorescent nanothermometers. *ACS Nano*, 2010, *4* (6), 3254–3258.

94. Li, Z. J.; Chen, X. L.; Dai, L.; Li, H. J.; Liu, H. W.; Gao, H. J.; Xu, Y. P. GaN nanotweezers. *Applied Physics A: Materials Science and Processing*, 2003, *76* (1), 115–118.

95. Nakayama, Y. Nanomachine "nanotweezers." *Kagaku to Kogyo*, 2003, *56* (6), 663–666.

96. Lieber, C. M.; Hafner, J. H.; Cheung, C. L.; Kim, P. Direct growth of carbon nanotubes, and their use in nanotweezers. WO 2002026624, 2002, 46 pp.

97. Chang, J.; Min, B.-K.; Kim, J.; Lee, S.-J.; Lin, L. Electrostatically actuated carbon nanowire nanotweezers. *Smart Materials and Structures*, 2009, *18* (6), 065017/1–065017/7.

98. Lee, J.; Kim, S. Manufacture of a nanotweezer using a length controlled CNT arm. *Sensors and Actuators, A: Physical*, 2005, *A120* (1), 193–198.

99. Yamahata, C.; Collard, D.; Domenget, A.; Hosogi, M.; Kumemura, M.; Hashiguchi, G.; Fujita, H. Silicon nanotweezers: A new biophysical tool for molecular experimentation. *IEEE 21st International Conference on Micro Electro Mechanical Systems*, Technical Digest, Tucson, AZ, January 13–17, 2008, Vol. 2, pp. 681–684.

100. Vinod, T. P.; Park, M.; Kim, S.-H.; Kim, J. Multisegmented Se-Te-Se hybrid nanowires: A building unit with inbuilt block and glue functionality. *Langmuir*, 2010, *26* (12), 9195–9197.

101. Zhai, T.; Zhang, X.; Yang, W.; Ma, Y.; Wang, J.; Gu, Z.; Yu, D.; Yang, H.; Yao, J. Growth of single crystalline $Zn_xCd_{1-x}S$ nanocombs by metallo-organic chemical vapor deposition. *Chemical Physics Letters*, 2006, *427*(4–6), 371–374.

102. Wang, F.; Matsuda, K.; Rahman, A. F. M. M.; Peng, X.; Kimura, T.; Komatsu, N. Simultaneous discrimination of handedness and diameter of single-walled carbon nanotubes (SWNTs) with chiral diporphyrin nanotweezers leading to enrichment of a single enantiomer of (6, 5)-SWNTs. *Journal of the American Chemical Society*, 2010, *132* (31), 10876–10881.

103. Tour, J. M. NanoCars. *Abstracts of Papers, 65th Southwest Regional Meeting of the American Chemical Society*, El Paso, TX, November 4–7, 2009, SWRM-130.

104. Sasaki, T.; Osgood, A. J.; Alemany, L. B.; Kelly, K. F.; Tour, J. M. Synthesis of a nanocar with an angled chassis. Toward circling movement. *Organic Letters*, 2008, *10* (2), 229–232

105. Sasaki, T.; Tour, J. M. Synthesis of a dipolar nanocar. *Tetrahedron Letters*, 2007, *48* (33), 5821–5824.

106. Akimov, A. V.; Nemukhin, A. V.; Moskovsky, A. A.; Kolomeisky, A. B.; Tour, J. M. Molecular dynamics of surface-moving thermally driven nanocars. *Journal of Chemical Theory and Computation*, 2008, *4* (4), 652–656.

107. Sasaki, T.; Guerrero, J. M.; Tour, J. M. The assembly line: Self-assembling nanocars. *Tetrahedron*, 2008, *64* (36), 8522–8529.

108. Khatua, S.; Guerrero, J. M.; Claytor, K.; Vives, G.; Kolomeisky, A. B.; Tour, J. M.; Link, S. Micrometer-scale translation and monitoring of individual nanocars on glass. *ACS Nano*, 2009, *3* (2), 351–356.

109. Vives, G.; Tour, J. M. Synthesis of single-molecule nanocars. *Accounts of Chemical Research*, 2009, *42* (3), 473–487.

110. Vives, G.; Tour, J. M. Synthesis of a nanocar with organometallic wheels. *Abstracts of Papers, 238th ACS National Meeting*, Washington, DC, August 16–20, 2009, INOR-346.

111. Vives, G.; Tour, J. M. Synthesis of a nanocar with organometallic wheels. *Tetrahedron Letters*, 2009, *50* (13), 1427–1430.

112. Morin, J.-F.; Shirai, Y.; Tour, J. M. En route to a motorized nanocar. *Organic Letters*, 2006, *8* (8), 1713–1716.

113. Godoy, J.; Vives, G.; Tour, J. M. Synthesis of highly fluorescent BODIPY-based nanocars. *Organic Letters*, 2010, *12* (7), 1464–1467.

114. Shirai, Y.; Osgood, A. J.; Zhao, Y.; Yao, Y.; Saudan, L.; Yang, H.; Chiu, Y.-H. et al. Surface-rolling molecules. *Journal of the American Chemical Society*, 2006, *128* (14), 4854–4864.

115. Shirai, Y.; Osgood, A. J.; Zhao, Y.; Kelly, K. F.; Tour, J. M. Directional control in thermally driven single-molecule nanocars. *Nano Letters*, 2005, 5 (11), 2330–2334.

116. Wang, Z. L.; Poncharal, P.; de Heer, W. A. Measuring physical and mechanical properties of individual carbon nanotubes by in situ TEM. *Journal of Physics and Chemistry of Solids*, 2000, *61*, 1025–1030.

117. Mirmohseni, A.; Shojaei, M.; Feizi, M. A. H.; Azhar, F. F.; Rastgouye-Houjaghan, M. Application of quartz crystal nanobalance and principal component analysis for detection and determination of nickel in solution. *Journal of Environmental Science and Health, Part A: Toxic/Hazardous Substances and Environmental Engineering*, 2010, *45* (9), 1119–1125.

118. Williams, O. A.; Mortet, V.; Daenen, M.; Haenen, K. The diamond nano-balance. *Journal of Nanoscience and Nanotechnology*, 2009, *9* (6) 3483–3486.

119. Huang, Y.; Bai, X.; Zhang, Y. In situ mechanical properties of individual ZnO nanowires and the mass measurement of nanoparticles. *Journal of Physics: Condensed Matter*, 2006, *18* (15), L179–L184.

120. Park, S. H.; Peng, Y.; Yan, L.; Reif, J. H.; LaBean, T. H.; Hao, Y. Programmable DNA self-assemblies for nanoscale organization of ligands and proteins. *Nano Letters*, 2005, *5* (4), 729–733.

121. Junping, Z.; Yan, L.; Yonggang, K.; Hao, Y. Periodic square-like gold nanoparticle arrays templated by self-assembled 2D DNA nanogrids on a surface. *Nano Letters*, 2006, *6* (2), 248–251.

122. Spillmann, H.; Dmitriev, A.; Lin, N.; Messina, P.; Barth, J. V.; Kern, K. Hierarchical assembly of two-dimensional homochiral nanocavity arrays. *Journal of the American Chemical Society*, 2003, *125*, 10725–10728.

123. Blednov, A. V.; Gorbenko, O. Y.; Samoilenkov, S. V.; Amelichev, V. A.; Lebedev, V. A.; Napolskii, K. S.; Kaul, A. R. Epitaxial calcium and strontium fluoride films on highly mismatched oxide and metal substrates by MOCVD: Texture and morphology. *Chemistry of Materials*, 2010, *22*, 175–185.

124. Corso, M.; Auwarter, W.; Muntwiler, M.; Tamai, A.; Greber, T.; Osterwalder, J. Boron nitride nanomesh. *Science*, 2004, *303*, 217–220.

125. Bunk, O.; Corso, M.; Martoccia, D.; Herger, R.; Willmott, P. R.; Patterson, B. D.; Osterwalder, J.; van der Veen, J. F.; Greber, T. Surface x-ray diffraction study of boron-nitride nanomesh in air. *Surface Science*, 2007, *601*, L7–L10.

126. Widmer, R.; Passerone, D.; Mattle, T.; Sachdev, H.; Groning, O. Probing the selectivity of a nanostructured surface by xenon adsorption. *Nanoscale*, 2010, *2* (4), 502–508.

127. Laskowski, R.; Blaha, P. Ab initio study of h-BN nanomeshes on Ru(001), Rh(111), and Pt(111). *Physical Review B: Condensed Matter and Materials Physics*, 2010, *81* (7), 075418/1–075418/6.

128. Martoccia, D.; Brugger, T.; Bjorck, M.; Schlepuetz, C. M.; Pauli, S. A.; Greber, T.; Patterson, B. D.; Willmott, P. R. h-BN/Ru(0001) nanomesh: A 14-on-13 superstructure with 3.5 nm periodicity. *Surface Science*, 2010, *604* (5–6), L16–L19.

129. Ma, H.; Brugger, T.; Berner, S.; Ding, Y.; Iannuzzi, M.; Hutter, J.; Osterwalder, J.; Greber, T. Nano-ice on boron nitride nanomesh: Accessing proton disorder. *ChemPhysChem*, 2010, *11* (2), 399–403.

130. Bai, J.; Zhong, X.; Jiang, S.; Huang, Y.; Duan, X. Graphene nanomesh. *Nature Nanotechnology*, 2010, *5* (3), 190–194.

131. Jingwei, B.; Xing, Z.; Shan, J.; Yu, H.; Xiangfeng, D. Graphene nanomesh. *Nature Nanotechnology*, 2010, *5* (3), 190–194.

132. Liang, X.; Jung, Y.-S.; Wu, S.; Ismach, A.; Olynick, D. L.; Cabrini, S.; Bokor, J. Formation of bandgap and subbands in graphene nanomeshes with sub-10 nm ribbon width fabricated via nanoimprint lithography. *Nano Letters*, 2010, *10* (7), 2454–2460.

133. Akhavan, O. Graphene nanomesh by ZnO nanorod photocatalysts. *ACS Nano*, 2010, *4* (7), 4174–4180.

134. Zhang, H. L.; Chen, W.; Huang, H.; Chen, L.; Wee, A. T. S. Preferential trapping of C_{60} in nanomesh voids. *Journal of the American Chemical Society*, 2008, *130*, 2720–2721.

135. Ke, X. B.; Zheng, Z. F.; Liu, H. W.; Zhu, H. Y.; Gao, X. P.; Zhang, L. X.; Xu, N. P. et al. High-flux ceramic membranes with a nanomesh of metal oxide nanofibers. *Journal of Physical Chemistry B*, 2008, *112* (16), 5000–5006.

136. Fu, M.; Zhou, J.; Yu, J. Flexible photonic crystal fabricated by two-dimensional free-standing ZnO nanomesh arrays. *Journal of Physical Chemistry C*, 2010, *114* (20), 9216–9220.

137. Chen, J.-J.; Su, Y.-K.; Lin, C.-L.; Kao, C.-C. Light output improvement of AlGaInP-based LEDs with nano-mesh ZnO layers by nanosphere lithography. *IEEE Photonics Technology Letters*, 2010, *22* (6), 383–385.

138. Hayazaki, S.; Matsui, T.; Zhang, H. L.; Chen, W.; Wee, A. T. S.; Yuhara, J. Formation of bismuth nanodot in (4 × 4) vanadium oxide nanomesh on Pd (111). *Surface Science*, 2008, *602* (12), 2025–2028.

139. Shi, C.; Wei, C.; Han, H.; Xingyu, G.; Dongchen, Q.; Yuzhan, W.; Wee, A. T. S. Template-directed molecular assembly on silicon carbide nanomesh: Comparison between CuPc and pentacene. *ACS Nano*, 2010, *4* (2), 849–854.

140. Kang, M.-G.; Guo, L. J. Semitransparent Cu electrode on a flexible substrate and its application in organic light emitting diodes. *Journal of Vacuum Science and Technology, B: Microelectronics and Nanometer Structures—Processing, Measurement, and Phenomena*, 2007, *25* (6), 2637–2641.

141. Trembulowicz, A.; Ciszewski, A. A square titanium nanomesh on W(100). *Nanotechnology*, 2007, *18* (34), 345303/1–345303/4.

142. http://www.rdmag.com/Awards/RD-100-Awards/2005/09/Exclusive

143. http://www.lanl.gov/orgs/tt/pdf/techs/nanofoam.pdf

144. Rode, A. V.; Gamaly, E. G.; Christy, A. G.; Fitz Gerald, J. G.; Hyde, S. T.; Elliman, R. G.; Luther-Davies, B.; Veinger, A. I.; Androulakis, J.; Giapintzakis, J. Unconventional magnetism in all-carbon nanofoam. *Physical Review B*, 2004, *70*, 054407, 9 pp.

145. Gamaly, E. G.; Rode, A. V. Nanostructures created by lasers. In *Encyclopedia of Nanoscience and Nanotechnology*. Nalwa, H. S. (Ed.). American Scientific Publishers, Stevenson Ranch, CA, 2004, Vol. 7, pp. 783–809. http://laserspark.anu.edu.au/pubs/gamaly04nanostructures.pdf

146. Rode, A. V.; Gamaly, E. G.; Luther-Davies, B. Formation of cluster-assembled carbon nano-foam by high-repetition-rate laser ablation. *Applied Physics A*, 2000, *70*, 135–144.

147. Li, S.; Ji, G.; Lü, L. Magnetic carbon nanofoams. *Journal of Nanoscience and Nanotechnology*, 2009, *9*, 1133–1136.

148. Lau, D. W. M.; McCulloch, D. G.; Marks, N. A.; Madsen, N. R.; Rode, A. V. High-temperature formation of concentric fullerene-like structures within foam-like carbon: Experiment and molecular dynamics simulation. *Physical Review B*, 2007, *75*, 233408.

149. Iino, T.; Nakamura, K. Acoustic and acousto-optic characteristics of silicon nanofoam. *Japanese Journal of Applied Physics*, 2009, *48* (7, Pt. 2), 07GE01/1–07GE01/5.

150. Kucheyev, S. O.; Hamza, A. V.; Worsley, M. A. Ion-beam-induced stiffening of nanoporous silica. *Journal of Physics D: Applied Physics*, 2009, *42* (18), 182003/1–182003/4.

151. Hussami, L. L.; Corkery, R. W.; Kloo, L. Study of [Dy(6-*p*-xylene)(GaCl$_4$)$_3$]-incorporated polyhedral silica nanofoam. *Microporous and Mesoporous Materials*, 2010, *132* (3), 480–486.

152. Mehdipour-Ataei, S.; Saidi, S. Preparation and characterization of sulfone-based polyimide nanofoams grafted with poly (propylene glycol). *e-Polymers*, 2007.

153. Zhang, Y. H.; Li, Y.; Huang, H. T.; Ke, S. M.; Zhao, L. H.; Chan, H. L. W. Synthesis and characterization of polymer nanofoams. *Key Engineering Materials*, 2007, *334–335* (Pt. 2, *Advances in Composite Materials and Structures*), 821–824.

154. Cao, L. M.; Zhang, X. Y.; Tian, H.; Zhang, Z.; Wang, W. K. Boron nitride nanotube branched nanojunctions. *Nanotechnology*, 2007, *18*, 155605, 4 pp.

155. Bose, A.; Chatterjee, K.; Chakravorty, D. Metal–semiconductor nanojunctions and their rectification characteristics. *Bulletin of Materials Science*, 2009, *32* (3), 227–230.

156. Wang, X. W.; Yuan, Z. H. Electronic transport behavior of diameter-graded Ag nanowires. *Physics Letters A*, 2010, *374* (22), 2267–2269.

157. Zhang, L.-S.; Wong, K.-H.; Yip, H.-Y.; Hu, C.; Yu, J. C.; Chan, C.-Y.; Wong, P.-K. Effective photocatalytic disinfection of *E. coli* K-12 using AgBr-Ag-Bi$_2$WO$_6$ nanojunction system irradiated by visible light: The role of diffusing hydroxyl radicals. *Environmental Science and Technology*, 2010, *44* (4), 1392–1398.

158. Ruzmetov, D.; Gopalakrishnan, G.; Deng, J.; Narayanamurti, V.; Ramanathan, S. Electrical triggering of metal-insulator transition in nanoscale vanadium oxide junctions. *Journal of Applied Physics*, 2009, *106* (8), 083702/1–083702/5.

159. Csonka, Sz.; Halbritter, A.; Mihály, G.; Shklyarevskii, O. I.; Speller, S.; van Kempen, H. Conductance of Pd-H nanojunctions. *Physical Review Letters*, 2004, *93* (1), 016802-1–016802-4.

160. Garcia-Lekue, A.; Wang, L. W. Plane-wave-based electron tunneling through Au nanojunctions: Numerical calculations. *Physical Review B: Condensed Matter and Materials Physics*, 2010, *82* (3), 035410/1–035410/9.

161. Scott, G. D.; Palacios, J. J.; Natelson, D. Anomalous transport and possible phase transition in palladium nanojunctions. *ACS Nano*, 2010, *4* (5), 2831–2837.

162. Sun, J.-L.; Zhao, X.; Zhu, J.-L. Metal-insulator transition in Au-NiO-Ni dual Schottky nanojunctions. *Nanotechnology*, 2009, *20* (45), 455203/1–455203/5.

163. Munoz, A. G.; Lewerenz, H. J. Electroplating of iridium onto single-crystal silicon: Chemical and electronic properties of n-Si(111)/Ir nanojunctions. *Journal of the Electrochemical Society*, 2009, *156* (5), D184–D187.

164. Zhao, X.; Sun, J.-L.; Zhu, J.-L. Field-induced semiconductor-metal transition in individual NiO-Ni Schottky nanojunction. *Applied Physics Letters*, 2008, *93* (15), 152107/1–152107/3.

165. Lin, L. W.; Tang, Y. H.; Chen, C. S. Self-assembled silicon oxide nanojunctions. *Nanotechnology*, 2009, *20* (17), 175601/1–175601/10.

166. Lue, X.; Zheng, Y.; Xin, H.; Jiang, L. Spin polarized electron transport through a graphene nanojunction. *Applied Physics Letters*, 2010, *96* (13), 132108/1–132108/3.

167. Chen, Y. P.; Xie, Y. E.; Wei, X. L.; Sun, L. Z.; Zhong, J. X. Resonant transmission in three-terminal triangle graphene nanojunctions with zigzag edges. *Solid State Communications*, 2010, *150* (13–14), 675–679.

168. Ren, C.; Xu, Z.; Zhang, W.; Li, Y.; Zhu, Z.; Huai, P. Theoretical study of heat conduction in carbon nanotube hetero-junctions. *Physics Letters A*, 2010, *374* (17–18), 1860–1865.

169. Sadi, T.; Thobel, J.-L. Analysis of the high-frequency performance of InGaAs/InAlAs nanojunctions using a three-dimensional Monte Carlo simulator. *Journal of Applied Physics*, 2009, *106* (8), 083709/1–083709/9.

170. Konopka, M.; Turansky, R.; Doltsinis, N. L.; Marx, D.; Stich, I. Organometallic nanojunctions probed by different chemistries: thermo-, photo-, and mechano-chemistry. *Advances in Solid State Physics*, 2009, *48*, 219–235.

171. Krivenko, A. G.; Komarova, N. S. Electrochemistry of nanostructured carbon. *Russian Chemical Reviews*, 2008, *77* (11), 927–943.

172. Henriksson, M.; Berglund, L. A.; Isaksson, P.; Lindstroem, T.; Nishino, T. Cellulose nanopaper structures of high toughness. *Biomacromolecules*, 2008, *9* (6), 1579–1585.

173. Lu, H.; Liu, Y.; Gou, J.; Leng, J.; Du, S. Electrical properties and shape-memory behavior of self-assembled carbon nanofiber nanopaper incorporated with shape-memory polymer. *Smart Materials and Structures*, 2010, *19* (7), 075021/1–075021/7.

174. Lu, H.; Liu, Y.; Gou, J.; Leng, J.; Du, S. Synergistic effect of carbon nanofiber and carbon nanopaper on shape memory polymer composite. *Applied Physics Letters*, 2010, *96* (8), 084102/1–084102/3.

175. Zhao, X.; Gou, J.; Song, G.; Ou, J. Strain monitoring in glass fiber reinforced composites embedded with carbon nanopaper sheet using Fiber Bragg Grating (FBG) sensors. *Composites, Part B: Engineering*, 2009, *40B* (2), 134–140.

176. Bromberg, M. R.; Patlolla, A.; Segal, R.; Feldman, Y.; Wang, Q.; Iqbal, Z.; Frenkel, A. I. Synthesis and characterization of platinum nanoparticles on single-walled carbon nanotube "nanopaper" support. *Journal of Physics: Conference Series*, 2009, *190*.

177. Gou, J.; Blanco, R.; Zhao, Z.; Khan, A.; Appalla, A. Synthesis of nickel-coated carbon nanopaper sheets by pulse laser deposition. *Materials Research Society Symposium Proceedings (Transport Behavior in Heterogeneous Polymeric Materials and Composites)*, San Francisco, CA, 2007, Vol. 1006E, Paper # 1006-R01–09.

178. Du, G.; Wang, Y.; Liu, H.; Liu, D.; Huang, L.; Liu, J.; Qin, S.; Tao, X.; Wang, J.; Jiang, M. Method for preparing titanium dioxide nanopaper. CN 101126213, 2008, 12 pp.

179. Lowy, D. A.; Patrut, A. Nanobatteries: Decreasing size power sources for growing technologies. *Recent Patents on Nanotechnology*, 2008, *2* (3), 208–219.

180. Anon. Fast-charging nano batteries. *American Ceramic Society Bulletin*, 2006, *85* (10), 21–22.

181. Long, J. W.; Dunn, B.; Rolison, D. R.; White, H. S. Three-dimensional battery architectures. *Chemical Reviews*, 2004, *104*, 4463–4492.

182. Vullum, F.; Teeters, D. Investigation of lithium battery nanoelectrode arrays and their component nanobatteries. *Journal of Power Sources*, 2005, *146* (1–2), 804–808.

183. Brazier, A.; Dupont, L.; Dantras-Laffont, L.; Kuwata, N.; Kawamura, J.; Tarascon, J.-M. First cross-section observation of an all solid-state lithium-ion "nanobattery" by transmission electron microscopy. *Chemistry of Materials*, 2008, *20*, 2352–2359.

184. Das, M.; Bittencourt, C.; Pireaux, J.-J.; Shivashankar, S. A. Metallic Li in carbonaceous nanotubes grown by metalorganic chemical vapor deposition from a metalorganic precursor. *Applied Organometallic Chemistry*, 2008, *22* (11), 647–658.

185. Falcao, E. H. L.; Viculis, L. M.; Mack, J. J.; Kwon, C.-W.; Kaner, R. B.; Dunn, B. S.; Wudl, F. Preparation of nanostructured molybdenum disulfide for nanobattery electrode application. *Abstracts of Papers, 227th ACS National Meeting*, Anaheim, CA, March 28–April 1, 2004, INOR-566.

186. McManus, J.; DeShazer, M.; Tetters, D.; Rowland, J.; Holmstrom, S. Growth of nanoporous aluminum oxide membranes. *Abstracts, 59th Southwest Regional Meeting of the American Chemical Society*, Oklahoma City, OK, October 25–28, 2003, p. 20.

187. Liang, C.; Terabe, K.; Hasegawa, T.; Aono, M.; Iyi, N. Anomalous phase transition and ionic conductivity of AgI nanowire grown using porous alumina template. *Journal of Applied Physics*, 2007, *102* (12), 124308/1–124308/5.

188. Costa, M.; Mumtaz, M.; Cloutet, E.; Cramail, H.; Ruiz, J.; Astruc, D. Grafting electron reservoirs at the periphery of core-shell PEDOT particles: Toward new molecular batteries. *Abstracts of Papers, 239th ACS National Meeting*, San Francisco, CA, March 21–25, 2010, PMSE-452.

189. King, G. C.; Choi, S. H.; Chu, S.-H.; Kim, J.-W.; Park, Y.; Lillehei, P.; Watt, G. D.; Davis, R.; Harb, J. N. Development of a bio-nanobattery for distributed power storage systems. *Smart Electronics, MEMS, and bioMEMS, and Nanotechnology, Conference*, San Diego, CA, ETATS-UNIS, March 15, 2004, Vol. 5389, pp. 461–467.

190. Romain, A. C.; Nicolas, J. Long term stability of metal oxide-based gas sensors for e-nose environmental applications: An overview. *Sensors and Actuators, B: Chemical*, 2010, *B146* (2), 502–506.

191. Korel, F.; Balaban, M. O. Electronic nose technology in food analysis. In *Handbook of Food Analysis Instruments*. Otles, S. (Eds.). CRC Press, Boca Raton, FL, 2009, pp. 365–378.

192. Nanto, H. Electronic nose (e-NOSE) system. *Materials Integration*, 2008, *21* (5, 6), 99–104.

193. Wilson, A. D.; Baietto, M. Applications and advances in electronic-nose technologies. *Sensors*, 2009, *9* (7), 5099–5148.
194. http://www.nanowerk.com/spotlight/spotid=3331.php
195. Wongchoosuk, C.; Wisitsoraat, A.; Tuantranont, A.; Kerdcharoen, T. Portable electronic nose based on carbon nanotube-SnO_2 gas sensors and its application for detection of methanol contamination in whiskeys. *Sensors and Actuators, B: Chemical*, 2010, *B147* (2), 392–399.
196. Sysoev, V. V.; Strelcov, E.; Sommer, M.; Bruns, M.; Kiselev, I.; Habicht, W.; Kar, S.; Gregoratti, L.; Kiskinova, M.; Kolmakov, A. Single-nanobelt electronic nose: Engineering and tests of the simplest analytical element. *ACS Nano*, 2010, *4* (8), 4487–4494.
197. Sysoev, V. V.; Goschnick, J.; Schneider, T.; Strelcov, E.; Kolmakov, A. A Gradient microarray electronic nose based on percolating SnO_2 nanowire sensing elements. *Nano Letters*, 2007, *7* (10), 3182–3188.
198. Sysoev, V. V.; Button, B. K. L.; Wepsiec, K.; Dmitriev, S.; Kolmakov, A. Toward the nanoscopic electronic nose: Hydrogen vs carbon monoxide discrimination with an array of individual metal oxide nano- and mesowire sensors. *Nano Letters*, 2006, *6* (8), 1584–1588.
199. Wongchoosuk, C.; Choopun, S.; Tuantranont, A.; Kerdcharoen, T. Au-doped zinc oxide nanostructure sensors for detection and discrimination of volatile organic compounds. *Materials Research Innovations*, 2009, *13* (3), 185–188.
200. Escuder-Gilabert, L.; Peris, M. Review: Highlights in recent applications of electronic tongues in food analysis. *Analytica Chimica Acta*, 2010, *665* (1), 15–25.
201. Parra, V.; Arrieta, A. A.; Fernandez-Escudero, J. A.; Garcia, H.; Apetrei, C.; Rodriguez-Mendez, M. L.; de Saja, J. A. E-tongue based on a hybrid array of voltammetric sensors based on phthalocyanines, perylene derivatives and conducting polymers: Discrimination capability towards red wines elaborated with different varieties of grapes. *Sensors and Actuators, B: Chemical*, 2006, *B115* (1), 54–61.
202. Popov, A. M.; Bichoutskaia, E.; Lozovik, Y. E.; Kulish, A. S. Nanoelectromechanical systems based on multi-walled nanotubes: Nanothermometer, nanorelay, and nanoactuator. *Physica Status Solidi A: Applications and Materials Science*, 2007, *204* (6), 1911–1917.

11 Nanostructures Classified as Polyhedra

11.1 NANOTRIANGLES

Triangle-like nanoparticles are predominantly represented by noble metals,[1] such as silver (Ag)[2,3] and gold[4] (frequently prepared by biogenic/bioreduction methods,[5] whose use in nanomaterial synthesis offers an environmentally benign alternative to the traditional chemical synthesis routes), although a few other inorganic compounds have been reported. Thus, biogenic gold nanotriangles and spherical Ag nanoparticles were synthesized by a simple procedure using *Aloe vera* leaf extract as the reducing agent.[6] It was revealed that multiply twinned particles play an important role in the formation of gold nanotriangles. On the contrary, reduction of Ag ions by *Aloe vera* extract led to the formation of spherical Ag nanoparticles of 15.2 ± 4.2 nm size. The effect of halide ions on the formation of biogenically prepared gold nanotriangles (Figure 11.1) using the leaf extract of lemongrass (*Cymbopogon flexuosus*) plant was studied.[7] It was shown that the presence of halide ions either during the growth of gold nanoparticles by the reduction of aqueous $AuCl_4^-$ ions using lemongrass leaf extract or after the synthesis of the particles significantly affected the morphology of the particles formed. The authors emphasized that, although F^- ions do not bind to the gold particle surface, Cl^-, Br^-, and I^- ions bind strongly to the (111) surface of gold and, by virtue of the mismatch between the halide adlayer and the (111) surface, alter the morphology of the particles formed; Cl^- ions promoted the formation of nanotriangles, whereas Br^- and I^- ions led to distortion of the triangular morphology in that order. In addition, the electric field interaction of hexagonal arrays of gold nanotriangles positioned on glass slides was studied using finite-difference time domain (FDTD) simulations.[8] It was shown that the plasmonic properties of this array of nanostructures were susceptible to modification by changing any of the studied variables: the length of the triangles ($L = 100$–300 nm), the irradiation wavelength ($\lambda = 400$–1100 nm), and the direction of the linearly polarized input field (either P_x or P_y).

Ag nanoparticles were prepared[9] by nanosphere lithography (NSL) and were subsequently released into solution (Figure 11.2). The resulting nanoparticles were found to be asymmetrically functionalized to produce either single isolated nanoparticles or dimer pairs. In addition, colloidal Ag nanoparticles were synthesized by reacting aqueous $AgNO_3$ with *Medicago sativa* seed exudates under nonphotomediated conditions.[10] In particular, the formation of single-crystalline Ag nanoplates, forming hexagonal particles and nanotriangles with edge lengths of 86–108 nm, while pH adjustment to 11 resulted in monodisperse Ag nanoparticles with an average size of 12 nm, was observed. Additionally, stabilized films of exudate-synthesized Ag nanoparticles were found to be effective antibacterial agents. In addition, by placing Ag nanotriangles or Au nanovoids on SiO_2 films containing Si nanocrystals, the photoluminescence (PL) intensity (luminescence properties) was found[11] to be enhanced. The largest PL enhancement was obtained when the excitation wavelength coincided with the absorption band of Ag nanotriangles. In the case of Au/Ag nanostructures, a simple and reproducible method[12] for the synthesis of triangular Au core-Ag shell nanoparticles included first preparation of the triangular gold core by the reduction of gold ions. Utilizing the negative charge on the gold nanotriangles, Ag ions were bound to their surface and thereafter reduced by ascorbic acid under alkali conditions.

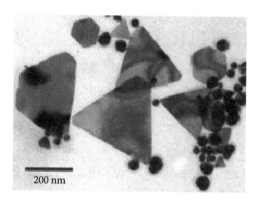

FIGURE 11.1 Gold nanotriangles. (Reproduced with permission from Rai, A. et al., Role of halide ions and temperature on the morphology of biologically synthesized gold nanotriangles, *Langmuir*, 22(2), 736–741. Copyright 2006 American Chemical Society.)

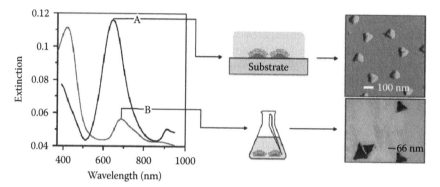

FIGURE 11.2 Fabrication, optical characterization, and structural characterization of released Ag nanoparticles (100 and 50 nm). A—Ag nanoparticles were first fabricated *via* NSL on glass substrates. A representative LSPR spectrum of hexadecanethiol-functionalized nanoparticles in ethanol had an extinction maximum located at 645.6 nm. Atomic force microscopy (AFM) revealed average nanoparticle heights of 52 nm and widths of 100 nm. B—Upon dislodging the nanoparticles from the glass substrate into bulk ethanol, a representative LSPR spectrum had two maxima located at 417.9 and 682.1 nm. TEM revealed nanoparticles with in-plane widths of ~95 nm. (Reproduced with permission from Haes, A.J. et al., Triangular Ag nanotriangles fabricated by nanosphere lithography, *J. Phys. Chem. B*, 109(22), 11158–11162. Copyright 2005 American Chemical Society.)

The sharp vertices of the triangles coupled with the core-shell structure were expected to have potential for the sensitive detection of biomolecules.

Other metal nanotriangles are rare like metal salts. Thus, the localized surface plasmon resonance (LSPR) of Al nanoparticles fabricated by NSL was examined by UV–visible extinction spectroscopy and electrodynamics theory, showing, comparing Al, Ag, Cu, and Au triangular nanoparticles, that the LSPR λ_{max} had the ordering Au > Cu > Ag > Al, while the full width at half-maximum satisfied Al > Au > Ag > Cu. Tungsten was deposited on silicon wafer and/or glass substrates by using random incidence sputtering deposition and thermal vapor deposition techniques, resulting in a 2D nanotriangle with size within 100 nm per side and 210 nm apart from each other.[13] The magnetization states in Ni triangular dots under an applied magnetic field were studied using variable-field magnetic force microscopy (VF-MFM) imaging.[14] The nanostructures presented magnetic vortices as ground states, which move under an external magnetic field. In addition, Cu nanotriangles were used for the growth of superconducting Nb films.[15] Carbon nanotriangles are also known.[16]

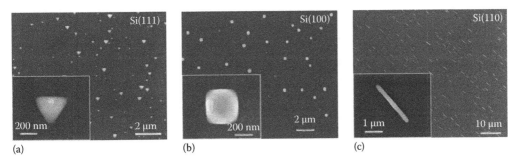

(a)   (b)   (c)

FIGURE 11.3 Scanning electron microscopy (SEM) images of shape-controlled Cu_3Si nanostructures grown at 960°C for 1 min using 30 nm Au NPs as catalyst. NTs, NS, and NWs are synthesized on substrates of (a) Si(111), (b) Si(100), and (c) Si(110), respectively. Insets show magnified views of individual nanostructures. (Reproduced with permission from Zhang, Z. et al., Self-assembled shape- and orientation-controlled synthesis of nanoscale Cu3Si triangles, squares, and wires, *Nano Lett.*, 8(10), 3205–3210. Copyright 2008 American Chemical Society.)

Nanoscale Cu_3Si triangles, squares, and wires (Figure 11.3) were grown[17] on Si(111), (100), and (110) substrates through a template-free Au-nanoparticle-assisted vapor transport method. Au nanoparticles absorbed Cu vapor and facilitated the rate-limited diffusion of Si, which was critical for the shape-controlled growth of Cu_3Si. The authors noted that such a bottom-up approach to synthesize shape- and orientation-controlled Cu_3Si nanostructures might be applicable to the tailored growth of other materials. Fe silicide nanorods on Si(111) substrates were fabricated by both deposition at high temperatures and electron-beam (E-beam)-induced deposition; the formation of either nanotriangles or nanorods was observed depending on the surface geometry of the substrates.[18] Zinc oxalate, $[Zn(C_2O_4) \cdot 2H_2O]$, was used as a precursor to prepare zinc oxide nanotriangles (Figure 11.4) by thermal decomposition.[19,20] In addition, manganese (oxyhydr)oxides, synthesized at low temperature (60°C or 95°C) from MnO_4^- reduction by $S_2O_3^{2-}$ in water, gave[21] various nanostructures, in particular, feitknechtite β-MnOOH, identified as an intermediate phase during the formation of γ-MnOOH. The transformation of feitknechtite nanotriangles into manganite nanowires (Figure 11.5) was found to occur through the dissolution of the feitknechtite particle core and recrystallization on the edge of the feitknechtite nanotriangles. For organic and biomolecules, the nanotriangle term is unknown, except DNA.[22]

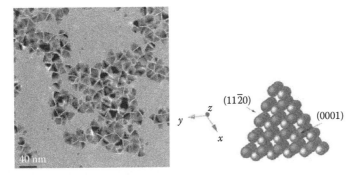

FIGURE 11.4 ZnO nanotriangles (triangular prisms), prepared in HD (1-hexadecanol 99%). (Reproduced with permission from Andelman, T. et al., Morphological control and photoluminescence of zinc oxide nanocrystals, *J. Phys. Chem. B*, 109, 14314–14318. Copyright 2005 American Chemical Society.)

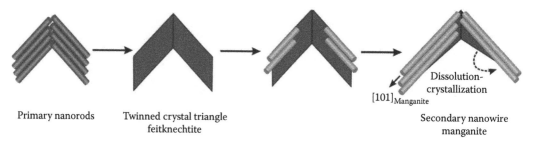

Primary nanorods   Twinned crystal triangle   Secondary nanowire
                            feitknechtite   manganite

FIGURE 11.5 Reaction scheme for the formation of feitknechtite β-MnOOH triangles and their evolution toward manganite wires. (Reproduced with permission from Portehault, D. et al., Evolution of nanostructured manganese (oxyhydr)oxides in water through MnO$_4^-$ reduction, *Cryst. Growth Des.*, 10(5), 2168–2173. Copyright 2010 American Chemical Society.)

11.2 NANOSQUARES AND NANORECTANGLES

Square-like nanostructures are not as widespread as nanotriangles mentioned earlier. Among metals, a noncatalytic and template-free vapor transport process was developed to make possible the simultaneous growth of single-crystalline tiny nanowires, nanosquares, nanodisks, and polycrystalline nanoparticles.[23] These four types of Sn nanostructures showed typical diamagnetic behavior in magnetization measurements, with the three anisotropically shaped nanostructures (nanowires, nanosquares, and nanodisks) showing one order of magnitude enhancement in the working magnetic field ranges for superconductivity, compared with bulk Sn and Sn nanoparticles. The magnetic field range was found to be the broadest for nanowires, followed by nanodisks, nanosquares, and nanoparticles. In addition, the superconducting transition temperature Sn nanosquares (Figure 11.6) were about 3.7 K, which was very close to that of bulk β-Sn.[24] Magnetization measurements showed that the critical magnetic fields for nanosquares increased significantly as compared with that of bulk Sn. Cobalt nanorods, fabricated by hydrothermal microemulsion, could spontaneously self-assemble into 2D square-shaped frames or dense multilayer nanosquares assisted by a magnetic field.[25] Also, the growth of AlN nanocolumns by ammonium nitridation of Al nanosquares embedded in SiO$_2$ on Si(111) substrates was studied.[26] As a result, selective nitridation of the Al nanosquares on the SiO$_2$ mask was obtained in the temperature window of 600°C–700°C. The well-shaped AlN nanocolumn arrays with diameters confined by the lateral size of the Al nanosquares (~100 nm) were observed.

Metal oxides are represented by monodispersed and well-aligned samples of TiO$_2$ nanosquares (single crystals characterized by slightly truncated shape bounded by {101} facets), synthesized in large quantities in the presence of Me$_4$NOH (TMAOH).[27] It was found that TMAOH accelerated the formation of crystalline anatase and played a structural template role to modify the particle shape to nanosquare. In addition, single-crystalline nanosquare sheets (with typical length of the sides of 200–600 nm and thickness varying from 30 to 100 nm) of Bi$_2$O$_3$ nanomaterial were synthesized in a large area by thermal evaporation of commercial Bi$_2$O$_3$ powder at high temperatures.[28] Metal oxide nanosquares can also be formed by nanobelts, as in the case of InO$_x$ (Figure 11.7).[29] In addition, nanoscale Mg(OH)$_2$ with different morphologies, from nanosheets, nanosquares, and nanodisks, were synthesized *via* a simple hydrothermal reaction of Mg(CH$_3$COO)$_2$ and sodium hydroxide in the presence of citrate.[30] Organic nanosquares are known for tricosanol crystals, whose growth was templated down to 60 nm by chemical nanopatterns on γ-aminopropylsilanized Si wafer, providing access to polycrystalline nanosquares and nanolines.[31]

Nanorectangles are extremely rare; thus, In(OH)$_3$ nanorectangles were solvothermally synthesized[32] at 160°C. Also, SiO$_3^{2-}$ doped TiO$_2$ films with oriented nanoneedle and nanorectangle block structure were prepared by a hydrothermal method.[33]

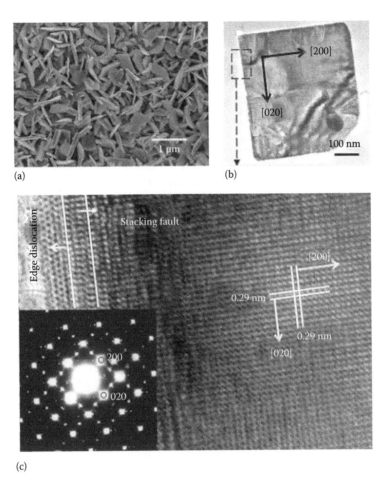

(a)

(b)

(c)

FIGURE 11.6 (a, b) SEM and TEM images, respectively, of the as-deposited square-shaped Sn nanostructure collected in the region of a lower temperature of 180°C. (c) HRTEM image in the edge region as marked in (b). The inserted SAED pattern was recorded along the [001] zone axis. (Reproduced with permission from Hsu, Y.-J. et al., Vapor-solid growth of Sn nanowires: Growth mechanism and superconductivity, *J. Phys. Chem. B*, 109(10), 4398–4403. Copyright 2005 American Chemical Society.)

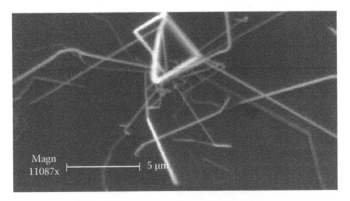

FIGURE 11.7 InO$_x$ nanobelts forming a "nanosquare" on a silicon substrate (without Au catalyst) deposited at 750°C. (From Fechete, A.C. et al., Growth of indium oxide nanostructures by thermal evaporation. Nanoscience and nanotechnology, 2006, *ICONN '06, International Conference on Nanoscience and Nanotechnology*, Brisbane, Queensland, Australia, July 3–7, 2006, ISBN: 1-4244-0452-5, © 2006 IEEE Xplore Digital Library.)

11.3 NANOTETRAHEDRA

This nanostructural type is also rare and is represented by a few elements (in particular, Si[34]), oxides, and salts, frequently formed as distinct nanostructures depending on reaction conditions. Thus, single-crystalline silicon nanoparticles bounded by (111) faces in the form of a tetrahedral were fabricated in a size range from 20 to 1000 nm side length.[35] Faces, edges, and tips can be chemically modified within certain limits, opening a possibility to facilitate the self-assembly into new supermaterials such as photonic crystals in the diamond lattice. Well-defined nanostructured Pd particles (nanocubes, nanotetrahedra, nanopolyhedra, and nanorods) were synthesized in aqueous solution using a seeding-mediated approach with cetyltrimethylammonium bromide (CTAB) acting as both capping and structure-directing agent.[36] Platinum nanocubes and tetrahedrons or octahedrons[37] are also known.

ZnO nanostructures with a controllable morphology were obtained by hot mixing reverse micelles containing $Zn(NO_3)_2$ or a monoethanol amine aqueous solution.[38] The obtained ZnO nanotetrahedrons (Figure 11.8) gave a strong blue emission arising from an interface state. AgI nanoparticles of various shapes were prepared by solution-based routes using H_2O-solution anionic or cationic polyelectrolytes as capping agents.[39] In particular, using poly(sodium acrylate) (PAS), the truncated-tetrahedron shaped γ-AgI nanoparticles (nanotetrahedra) in a Zn-blende structure (3C) were obtained. Also, uniform single-crystalline γ-CuI nanocrystals with tetrahedral morphology (Figure 11.9) were synthesized from the reaction between Cu_n and KI under ambient conditions.[40] I− performed the roles of reactant and surface stabilizing agent, thereby reducing the complexity of the reaction by eliminating the need for an external capping agent. Nanotetrahedrons

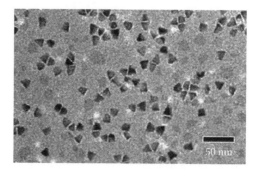

FIGURE 11.8 TEM image of ZnO nanotetrahedrons. (Reproduced with permission from Mao, J. et al., Control of the morphology and optical properties of ZnO nanostructures via hot mixing of reverse micelles, *Langmuir*, 26(17), 13755–13759. Copyright 2010 American Chemical Society.)

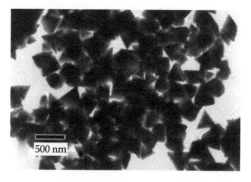

FIGURE 11.9 TEM image of γ-CuI nanotetrahedrons. (Reproduced with permission from Ng, C.H.B. and Fan, W.Y., Facile synthesis of single-crystalline γ-CuI nanotetrahedrons and their induced transformation to tetrahedral CuO nanocages, *J. Phys. Chem. C*, 111(26), 9166–9171. Copyright 2007 American Chemical Society.)

grew from single-crystalline seeds whose growth was promoted by iodide-induced oxidative etching of twinned seeds. Removal of excess free I⁻ in solution (Reactions 11.1 and 11.2) initiates the dissolution of CuI nanotetrahedrons and simultaneous oxidation to CuO nanocages (Figures 11.10 and 11.11). In addition, the formation of CdS nanotetrahedrons, pencil-shaped nanorods, tetrapods, prickly spheres, and high aspect-ratio hexagonal nanoprisms was, respectively, achieved by

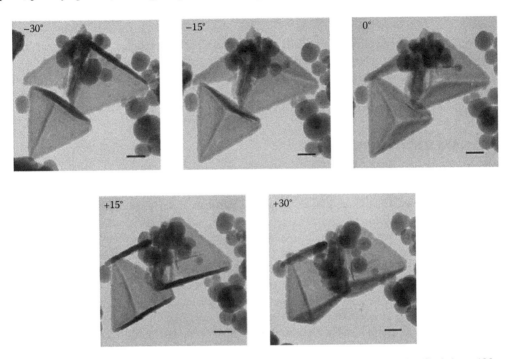

FIGURE 11.10 TEM image of tetrahedral CuO nanocages tilted by five different angles. Scale bars, 100 nm. (Reproduced with permission from Ng, C.H.B. and Fan, W.Y., Facile synthesis of single-crystalline γ-CuI nanotetrahedrons and their induced transformation to tetrahedral CuO nanocages, *J. Phys. Chem. C*, 111(26), 9166–9171. Copyright 2007 American Chemical Society.)

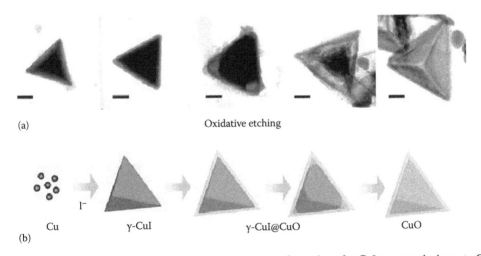

FIGURE 11.11 (a) TEM images and (b) scheme for the transformation of γ-CuI nanotetrahedrons to CuO nanocages *via* Kirkendall effect. Scale bars, 200 nm. (Reproduced with permission from Ng, C.H.B. and Fan, W.Y., Facile synthesis of single-crystalline γ-CuI nanotetrahedrons and their induced transformation to tetrahedral CuO nanocages, *J. Phys. Chem. C*, 111(26), 9166–9171. Copyright 2007 American Chemical Society.)

adjusting the ratio of two solvents ethylenediamine and ethylene glycol under the solvothermal condition.[41] None of the surfactants or other templates was needed in the process.

$$CuI \rightarrow Cu^+ + I^- \tag{11.1}$$

$$2CuI + O_2 + I^- \rightarrow 2CuO + I_3^- \tag{11.2}$$

Additionally, it was predicted[42] theoretically that nanoparticles of octupolar symmetry (nanotriangles and nanotetrahedra), whose orientation cannot be affected by means of linear optics, subjected to a coherent mixture of fundamental and second harmonic fields will rotate and orient controlled by the relative phase between these fields. This is due to the generation of the second-harmonic polarization and its interaction with the second-harmonic field.

11.4 NANOOCTAHEDRA

Nanooctahedra are represented by Au,[43] its core-shell nanoparticles (for instance, Au@Pd[44]), a series of oxides (e.g., Co_3O_4[45]), and some salts; these nanostructures have been frequently obtained together with other nanoforms. Thus, three kinds of gold nanoparticles, namely nanorods, nanospheres, and nanooctahedrons with different sizes were fabricated.[46] Single-crystal Au octahedra ~85% of the product with an edge length of 32.4 ± 2.3 nm and singly twinned truncated bipyramids ~15% were synthesized by reducing $HAuCl_4$ with N-vinyl pyrrolidone in an aqueous solution in the presence of a proper amount of cetyltrimethylammonium chloride.[47] The mechanistic study indicated that the formation of Au nanooctahedra could be explained by oxidative etching. Core-shell nanocrystals composed of Pd and Au were composed using both Pd nanocubes and Au nanooctahedrons as cores,[48] yielding single-crystalline Pd@Au@Pd@Au and Au@Pd@Au@Pd core-trishell nanocrystals of narrow size distributions. The thickness of each shell can be readily varied by changing the amounts of metal salts. In addition to gold, the fabrication of large-area Cu nanooctahedra with well-defined shape, good monodispersity, and uniform distribution on a gold film substrate was carried out by a very simple, rapid, and cost-effective low-potential electrodeposition technique.[49] The size of the octahedra depended on deposition time and was adjustable over a wide range. The shape evolution was found to be remarkably different from that of octahedral particles formed in solution.

Among metal oxides, faceted iron oxide nanoparticles with octahedral shape (Figure 11.12) were synthesized through controlled modification of the iron oleate decomposition method.[50] Both metallic iron and iron oxide were found to be presented within the nanooctahedra. Also, monosized octahedron-shaped magnetite (Fe_3O_4) nanoparticles (Figure 11.13) with average sizes ranging from 8 to ~430 nm were fabricated.[51] A simple solvent-evaporation assembly process was applied to obtain either 2D monolayer or 3D microrod superstructures made of 21 nm-sized nanooctahedra by applying a weak magnetic field (~0.06 T) in the horizontal or vertical direction, respectively.

Reduction of copper nitrate in Triton X-100 water-in-oil microemulsions by γ-irradiation (Reactions 11.3 through 11.5) resulted in cuprous oxide octahedron nanocrystals (Figure 11.14) smaller than 100 nm.[52] The average edge length of these octahedron-shaped nanocrystals varied from 45 to 95 nm as a function of the dose rate. Also, Cu-Cu_2O core-shell nanoparticles of different shapes (cubic, cuboctahedral, and octahedral) over an extended nanosize regime of 5–400 nm were deposited on a hydrogen-terminated Si(100) substrate by using a capping-agent-free electrochemical method from $CuSO_4 \cdot 5H_2O$ as precursor.[53] These nanoparticles had a crystalline core-shell structure, with a face-centered cubic (FCC) metallic Cu core and a simple cubic Cu_2O shell with a CuO outer layer:

$$H_2O \ (\gamma\text{-irradiation}) \rightarrow e_{aq.}^-, H_3O, \ ^+H, H_2, OH, H_2O_2 \tag{11.3}$$

$$Cu^{2+} + e_{aq.}^- \rightarrow Cu^+ \tag{11.4}$$

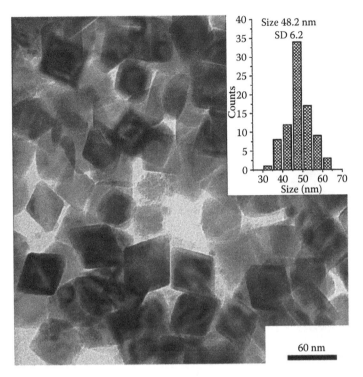

FIGURE 11.12   Representative TEM image of iron oxide nanooctahedra. (Reproduced with permission from Shavel, A. et al., Shape control in iron oxide nanocrystal synthesis, induced by trioctylammonium ions, *Chem. Mater.*, 21(7), 1326–1332. Copyright 2009 American Chemical Society.)

$$Cu^+ + OH^- \rightarrow Cu(OH) \rightarrow Cu_2O + 0.5H_2O \tag{11.5}$$

The hydrothermal method was used to fabricate monodisperse CeO_2 hollow spheres assembled by nanooctahedra.[54] These CeO_2 hollow spheres, assembled by nanooctahedra with an average edge length of ~20 nm, exhibited higher catalytic properties than those of commercial CeO_2 powder. In a related report,[55] uniform single-crystalline CeO_2 nanooctahedrons and nanorods were synthesized by a hydrothermal synthesis process using only $Ce(NO_3)_3 \cdot 6H_2O$ as a cerium resource and $Na_3PO_4 \cdot 6H_2O$ as a mineralizer (Reactions 11.6 through 11.8), into which no surfactant or template was introduced. By tuning the hydrothermal treatment time, the morphology evolution between the nanooctahedron and nanorod was observed (Figure 11.15). The morphological evolution mechanism suggested that the shape variation from nanooctahedron to nanorod might be a phosphate ion-related dissolution-recrystallization process:

$$Na_3PO_4 + 2H_2O \leftrightarrow Na^+ + 2OH^- + H_2PO_4^- \tag{11.6}$$

$$Ce^{3+} + OH^- + H_2O + O_2 \rightarrow Ce(H_2O)_x(OH^-)_y^{(4-x)+} \tag{11.7}$$

$$Ce(H_2O)_x(OH^-)_y^{(4-x)+} + H_2O \rightarrow CeO_2 \cdot 2nH_2O + H_3O^+ \tag{11.8}$$

Uniform single-crystal $CdMoO_4$ nanooctahedra in the hemline regime of 30 nm were synthesized *via* a reverse-microemulsion route at room temperature (r.t.).[56] Room-temperature PL measurement exhibited a blue emission peaking at 438 nm under the 360 nm excitation, which was mainly attributed to the existence of intrinsic distortions into the $[MoO_4]$ tetrahedron moiety in the lattices.

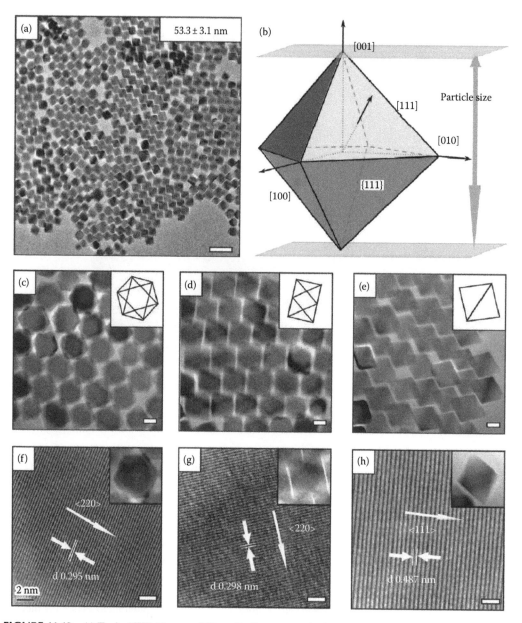

FIGURE 11.13   (a) Typical TEM image of 53 nm Fe$_3$O$_4$ nanooctahedra. (b) Schematic 3D model of one octahedron-shaped nanoparticle. TEM and HRTEM images of 53 nm Fe$_3$O$_4$ nanooctahedra with different projection shapes: (c, f) hexagonal (zone axis: $\langle 111 \rangle$), (d, g) rectangle (zone axis: $\langle 112 \rangle$), and (e, h) parallelogram (zone axis: $\langle 110 \rangle$). Scale bars: (a) 100 nm; (c) through (e) 20 nm; and (f) through (h) 2 nm. (Reproduced with permission from Li, L. et al., Synthesis of magnetite nanooctahedra and their magnetic field-induced two-/three-dimensional superstructure, *Chem. Mater.*, 22(10), 3183–3191. Copyright 2010 American Chemical Society.)

While studying pitting corrosion of austenitic stainless steels, a number of nanosized octahedral MnCr$_2$O$_4$ crystals (Figure 11.16) (with a spinel structure) were found to be embedded in the MnS medium, generating local MnCr$_2$O$_4$/MnS nanogalvanic cells.[57] High yield of nickel ferrite nanooctahedra with size distribution from 40 to 90 nm was achieved through a hydrothermal method.[58] Magnetic measurements indicated that the sample was soft-magnetic material with much lower coercivity and much higher saturation magnetization compared with the nickel ferrite nanocrystals with similar size distribution but irregular shapes. Chalcogenides are represented by molybdenum

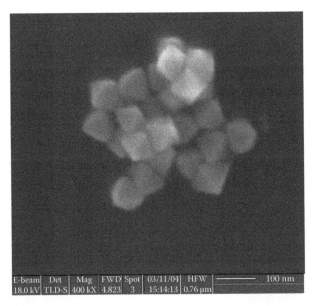

FIGURE 11.14 Cu$_2$O nanooctahedra. (Reproduced from *J. Colloid Interface Sci.*, 284, Ping, H. et al., Size-controlled preparation of Cu$_2$O octahedron nanocrystals and studies on their optical absorption, 510–515, Copyright 2005, with permission from Elsevier.)

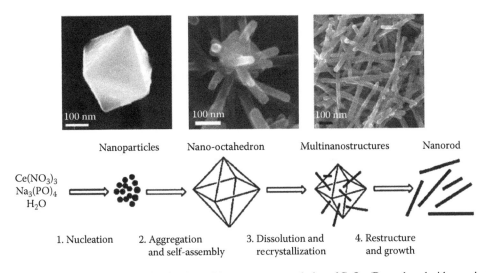

FIGURE 11.15 Schematic illustration for the multinanostructure evolution of CeO$_2$. (Reproduced with permission from Lai, Y. et al., Template-free hydrothermal synthesis of CeO$_2$ nano-octahedrons and nanorods: Investigation of the morphology evolution, *Cryst. Growth Des.*, 8(5), 1474–1477. Copyright 2008 American Chemical Society.)

salts; thus, nanooctahedra of MoS$_2$ are considered to be the true inorganic fullerenes, exhibiting different properties from the bulk and also other closed-cage morphologies of the same material.[59] In addition, pulsed laser vaporization was used to produce nanooctahedra of MoSe$_2$.[60]

In addition to "standard" octahedra, a series of reports have been devoted to *truncated* octahedral nanostructures belonging to distinct types of compounds, for example, Pt,[61] PtAu,[62] Fe$_3$O$_4$,[63] TiO$_2$,[64] Cu$_2$O,[65] CeO$_2$[66] (its morphology changes during growth are shown in Figure 11.17), tantalite-niobate (KTa$_{1-x}$Nb$_x$O$_3$, KTN, Figure 11.18),[67] TiN,[68] PbS (Figure 11.19),[69] and PbSe,[70] forming among other nanostructures (nanocubes).

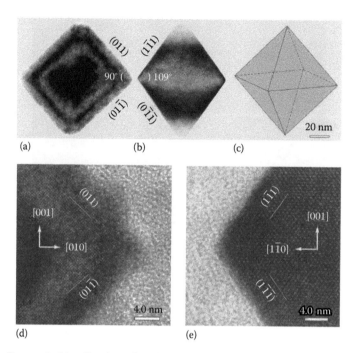

FIGURE 11.16 Geometric identification of $MnCr_2O_4$ nanooctahedron by conventional TEM techniques. (a) A bright-field TEM micrograph of an $MnCr_2O_4$ octahedron imaged along the [100] direction. (b) TEM micrograph of the same $MnCr_2O_4$ octahedron as that in (a) but imaged along the [110] direction. (c) Schematic illustration of an ideal octahedron. (d) HREM images of an octahedron taken along the [100] direction. (e) HREM images of an octahedron taken along the [110] direction. The amorphous-like area around $MnCr_2O_4$ octahedron results from the dissolution of MnS. (Reproduced from *Acta Mater.*, 58, Zheng, S.J. et al., Identification of $MnCr_2O_4$ nanooctahedron in catalyzing pitting corrosion of austenitic stainless steels 5070–5085, Copyright 2010, with permission from Elsevier.)

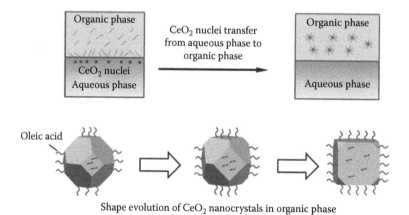

FIGURE 11.17 Schematic illustration of the growth mechanism for ceria nanocrystals with different morphologies. (Reproduced with permission from Dang, F. et al., Characteristics of CeO_2 nanocubes and related polyhedra prepared by using a liquid-liquid interface, *Cryst. Growth Des.*, 10(10), 4537–4541. Copyright 2010 American Chemical Society.)

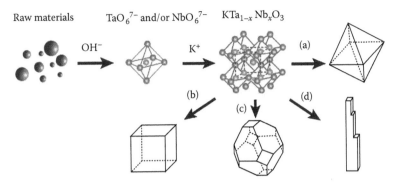

FIGURE 11.18 A schematic illustration of the possible formation mechanisms of KTN nanostructures with various morphologies synthesized by hydrothermal treatment; Ta and/or Nb atoms are located in the polyhedrons of oxygen atoms. (a) $KTaO_3$ octahedrons; (b) $KTa_{0.77}Nb_{0.23}O_3$, $KTa_{0.50}Nb_{0.50}O_3$ and $KTa_{0.35}Nb_{0.65}O_3$ cube; (c) $KNbO_3$ truncated octahedron nanostructures; (d) $KTa_{0.25}Nb_{0.75}O_3$ tower-like nanostructures. (Reproduced with permission from Hu, Y. et al., Controllable hydrothermal synthesis of $KTa_{1-x}Nb_xO_3$ nanostructures with various morphologies and their growth mechanisms, *Cryst. Growth Des.*, 8(3), 832–837. Copyright 2008 American Chemical Society.)

FIGURE 11.19 PbS truncated octahedrons. (Reproduced with permission from Wang, N. et al., Truncated octahedron crystals with high symmetry and their large-scale assembly into regular patterns by a simple solution route, *ACS Nano*, 2(2), 184–190. Copyright 2008 American Chemical Society.)

11.5 NANOPYRAMIDS

Along with a majority of other nanoforms, Au[71–76] nanopyramids are being intensively studied. Thus, a bottom-up approach (Figure 11.20) for fabricating periodic arrays of gold nanopyramids with nanoscale sharp tips was developed.[77] This templating technology is scalable and compatible with standard microfabrication, enabling large-scale production of surface-enhanced Raman spectroscopy (SERS)-active electrodes for *in situ* electrochemical studies and sensitive electroanalysis. Alternatively, a wafer-scale 3D plasmonic crystal consisting of gold nanopyramid arrays using a colloidal templating technique was developed.[78,79] These surface plasmons in the arrays allowed obtaining real-time, label-free, and high-sensitivity biomolecular binding measurements. In addition, direct and reversible electron transfer of myoglobin (Mb) was achieved at a nanopyramidal gold surface, which was fabricated by one-step electrodeposition.[80] Mb was found to be stably confined on the nanopyramidal gold surface and maintained electrocatalytic activity toward hydrogen peroxide. Among other elemental nanopyramids, those for Si were reported.[81]

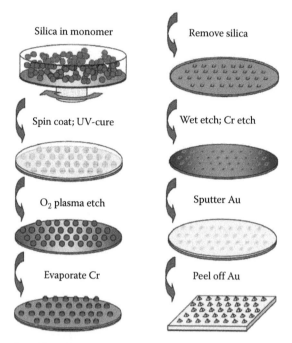

FIGURE 11.20 Schematic outline of the templating procedures for fabricating gold nanopyramid array by using a spin-coated monolayer colloidal crystal as template. (Reproduced with permission from Tzung-Hua, L. et al., Electrochemical SERS at periodic metallic nanopyramid arrays, *J. Phys. Chem. C*, 113(4), 1367–1372. Copyright 2009 American Chemical Society.)

Well-dispersed ZnO nanomaterials were prepared by direct calcination of zinc stearate.[82] It was revealed that the morphology of ZnO transformed from nanosheets to hexagonal nanopyramids and then to nanoparticles at 573, 673, and 773 K, respectively. Doped ZnO nanopyramids were also developed; thus, monodisperse ZnO:Eu nanocrystals (mainly composed of nanopyramids and dot-shaped nanocrystals and having an average size in the range of 16–32 nm) were prepared by the code composition of metal acetylacetonate precursors in a mixture of oleylamine and 1-octadecene.[83] With increasing dopant concentration of Eu^{3+} ions, Eu_2O_3 species tended to segregate on the surfaces of nanocrystals. Plasma-assisted synthesis of nanostructures (one of the most precise and effective approaches used in nanodevice fabrication) can lead to arrays of crystalline CdO nano/micropyramids.[84] These nanostructures grew *via* unconventional plasma-assisted oxidation of a cadmium foil exposed to inductively coupled plasmas with a narrow range of process parameters. It was shown that the size of the pyramidal structures can be effectively controlled by the fluxes of oxygen atoms and ions impinging on the cadmium surface. Mn-doped In_2O_3 nanopyramids (Figure 11.21, Mn content was below 1%) were grown by a catalyst-free thermal process at 700°C using InN and Mn_2O_3 powders as precursors.[85] Manganese doping of In_2O_3 inhibited the growth of elongated structures at the surface and reduced the grain size of the resulting nanostructured In_2O_3 ceramic when the treatment temperatures were below 700°C. Among other oxides, SnO_2[86] and $(La,Sr)_xO_y$[87] nanopyramids (in the latter case, sitting on a $La_{0.7}Sr_{0.3}MnO_3$ epitaxial wetting layer) are known.

A few oxygen-containing salts in nanopyramid form are known. As an example, monoclinic $SrAl_2O_4$ monocrystals, with a 3D dendritic structure, were found to be represented by microstructures selected from nanopyramids, nanorods, nanowires, and nanobelts.[88] The nanopyramid was found to be composed of a polyhedral pyramid or prism and a top ball, wherein the polyhedral pyramid or prism has a length of 100 nm to 500 μm and a section diameter of 20 nm to 10 μm, and the top ball was a monocrystalline Al ball with a face-centered crystalline structure and a

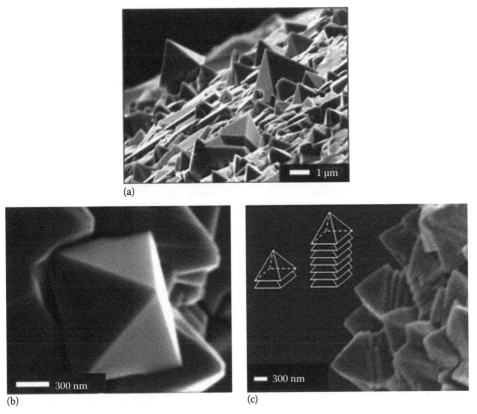

FIGURE 11.21 (a, b) SEM images showing Mn-doped In_2O_3 pyramids obtained on a sample treated at 700°C during 15 h. (c) Stacked structure of pyramids corresponding to a sample treated at 800°C for 15 h. (Reproduced with permission from Maestre, D. et al., Micro- and nanopyramids of manganese-doped indium oxide, *J. Phys. Chem. C*, 114(27), 11748–11752. Copyright 2010 American Chemical Society.)

diameter of 50 nm to 30 μm. On the contrary, binary compounds, such as carbides (SiC[89]), simple and double nitrides (InGaN[90]), phosphides, or sulfides, are more common. Thus, the impact of base size and shape on the evolution control of multifaceted InP (100) nanopyramids was reported.[91] $In_xGa_{1-x}N/GaN$ quantum wells were grown on dense arrays of self-assembled GaN nanopyramids (Figure 11.22) formed by selective area growth.[92] The pyramids were shown to have significantly reduced defect (green-yellow) band emission, and the quantum well luminescence was correspondingly intense. Single-crystalline wurtzite Al nitride (AlN) tetragonal nanopyramids were fabricated through thermal evaporation of Al in diluted NH_3 flux.[93] The observation of the polar surface-induced anisotropic growth in AlN nanostructures was expected to provide an insightful sample to study the microscopic crystal growth mechanism of AlN and other Group III nitrides. In addition, AlGaN hexagonal nanopyramids were formed on GaN/Al_2O_3, AlN/Al_2O_3, and $Al_{0.5}Ga_{0.5}N/Al_2O_3$ using selective area heteroepitaxy.[94] The growth rates of nanostructures were found to be much higher when grown on an AlN template in comparison to that on a GaN template under the same growth conditions.

PbS triangular nanopyramids (Figure 11.23), among other nanostructures, were synthesized at the air/$PbCl_2$ aqueous solution interface *via* reaction between Pb^{2+} and H_2S gas under poly(9-vinyl-carbazole) (PVK) thin films at 20°C.[95] The formation of the nanopyramids should be related to lattice matching between the 2D arrays of N atoms in PVK Langmuir monolayers at 20°C and the (111) crystal face of PbS. In a related report,[96] the PbS nanorings were transformed from triangular PbS nanopyramids that were formed under Langmuir monolayers of arachidic acid through interfacial

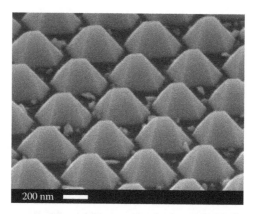

FIGURE 11.22 GaN nanopyramids formed by selective area growth self-assembly with InGaN/GaN quantum wells on the facets. (Reproduced with permission from Liu, C. et al. *Appl. Phys. Exp.*, 2, 121002. Copyright 2009 The Japan Society of Applied Physics.)

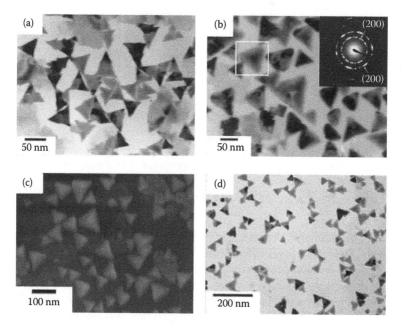

FIGURE 11.23 TEM (a, b, d) and SEM (c) micrographs of PbS triangular nanopyramids formed at the air/water interface at 20°C. The concentration of PbCl$_2$ aqueous solution is 1×10^{-4} mol/L, and the volumes of H$_2$S gas are 0.02, 0.10, 0.1, and 1.0 mL for the images of (a) through (d), respectively. (Reproduced with permission from Wang, C.-W. et al., Triangular PbS nano-pyramids, square nanoplates, and nanorods formed at the air/water interface, *Cryst. Growth Des.*, 8(8), 2660–2664. Copyright 2008 American Chemical Society.)

reactions between Pb^{2+} ions in the subphases and H$_2$S in the gaseous phase. Well-defined flower-like CdS nanostructures, which consisted of hexagonal nanopyramids and nanoplates depending on different sulfur sources, were synthesized by applying ultrasound and microwave simultaneously.[97] This structure-induced shift in optical properties may have potential applications in optoelectronic devices, catalysis, and solar cells. Additionally, nanostructured CdIn$_2$S$_4$ with a fascinating marigold flower' morphology were synthesized using a hydrothermal method and mixed morphologies (flowers, spheres, and pyramids) using a microwave method.[98] In the microwave synthesis, the product was formed within 15 min, whereas by the hydrothermal method >24 h was required. The marigold flowers, nanoparticle spheres, and nanopyramids of $CdIn_2S_4$ synthesized by the microwave method gave almost 30% enhancement in the degradation of methylene blue (MB) as compared with CdS under direct sunlight. $CdIn_2S_4$ has potential for applications in solar energy conversion and optoelectronic devices.

11.6 NANOICOSAHEDRA

Important observations on icosahedron and related structures were generalized in a review of Jose-Yacaman et al.[99] The authors emphasized that classical theory of nanoparticle stability predicts that for sizes <1.5–2 nm the icosahedral (I_h) structure should be the most stable, and then between around 2–5 nm, the decahedral (D_h) shape should be the most stable. Beyond that, FCC structures will be the predominant phase. However, in the experimental side, I_h and D_h particles can be observed (Figure 11.24) much beyond the 5 nm limit. In fact, it is possible to find I_h and D_h particles even in the mesoscopic range. Conversely, it is possible to find FCC particles with a size <1.5 nm. Mechanisms to stabilize fivefold nanoparticles are as follows: internal strain, introduction of planar defects (twins, stacking faults), the introduction of steps, and kinks on the surface, external modification of shape by facet truncation, displacement of the fivefold axis, which reduces the elastic energy, and splitting of the fivefold axis.

Iron nanoparticles (highly desirable for their potential applications in magnetic and catalytic industry) were synthesized in the form of I_h FCC Fe nanoparticles (Figure 11.25) with a size of 5–13 nm by a specifically designed thermodynamic-governed synthetic route, which is facile but highly efficient and reproducible.[100] It is expected that as-synthesized Fe nanoparticles with sharp corners and edges would be beneficial for tailoring chemical and physical properties at the nanoscale. In addition, for Fe clusters, their structure and magnetism of iron clusters with up to 641 atoms were investigated by means of density functional theory calculations including full geometric optimizations.[101] In addition to body centered cubic (BCC) isomers, found to be lowest in energy when the clusters contain more than about one hundred atoms, another stable conformation with a close-packed particle core and an I_h surface, while intermediate shells are partially transformed along the Mackay path between I_h and cuboctahedral geometry, was identified.

Self-assembled Ag nanocrystals supported on a $SrTiO_3$ (001) – (2 × 1) substrate were studied using scanning tunneling microscopy, showing that Ag formed nanocrystals with fivefold symmetry that had an icosahedral shape (Figure 11.26).[102] The authors believed that the nanocrystals are a growth shape that evolved from small icosahedral nuclei. Nanocrystals of this type are expected to have

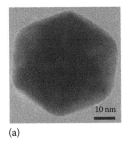
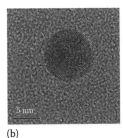

(a)   (b)   (c)   (d)

FIGURE 11.24 Electron microscopy images of an (a) icosahedron of ~80 nm, (b) an icosahedron of about 7 nm, (c) a decahedron of ~300 nm, and (d) a decahedron of few micrometers. (Mayoral, A. et al., Nanoparticle stability from the nano to the meso interval, *Nanoscale*, 2, 335–342. Copyright 2010. Reproduced by permission of The Royal Society of Chemistry.)

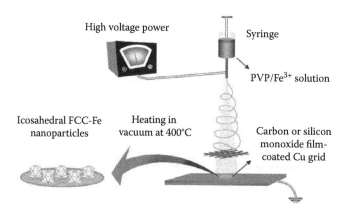

FIGURE 11.25 Schematic diagram of the synthetic procedure of icosahedral FCC-Fe nanoparticles. PVP/Fe^{3+} fluid jet was sprayed out from the needle and uniformly dispersed onto an amorphous carbon or silicon monoxide film-coated Cu grid using standard electrospinning process. Afterward, the Cu grid was heated in vacuum at 400°C for the fabrication of icosahedral FCC-Fe nanoparticles. (Reproduced with permission from Tao, L. et al., Icosahedral face-centered cubic Fe nanoparticles: Facile synthesis and characterization with aberration-corrected TEM, *Nano Lett.*, 9(4), 1572–1576. Copyright 2009 American Chemical Society.)

different properties from their single-crystal counterparts and may therefore be of use in novel chemical and optical application.

Shape-controlled synthesis of Pd nanostructures (Figures 11.27 and 11.28), including icosahedron-like nanoforms, was described.[103] It was generalized that both the crystallinity of seeds and the growth rates of different crystallographic facets are important in determining the shape of resultant nanostructures. More specifically, the crystallinity of a seed can be controlled by manipulating the reduction rate. In addition, star polyhedral Au nanocrystals (multiple-twinned crystals with fivefold symmetry and single crystals) were synthesized[104] by colloidal reduction with ascorbic acid in H_2O at r.t. These respective classes correspond to icosahedra and cuboctahedra, two Archimedean solids, with preferential growth of their {111} surfaces. It was proposed that the driving force for star nanocrystal formation could be the reduction in surface energy that the crystals experience. Icosahedrally derived star nanocrystals possess a geometric morphology closely resembling the great stellated dodecahedron, a Kepler–Poinsot solid. In addition, 55-atom "crown-gold" alloy nanocluster $Au_{43}Cu_{12}$, in contrast to the pure Au_{55} nanocluster, exhibiting a low-symmetry C_1 structure, showed a multishell structure, denoted by $Au@Cu_{12}@Au_{42}$, with the highest icosahedral group symmetry.[105] Its density functional calculations suggested that this geometric magic-number nanocluster possesses comparable catalytic capability as a small-sized Au_{10} cluster for the CO oxidation (Figure 11.29), due in part to its low-coordinated Au atoms on vertexes.

Nanoscale icosahedral particles are also known for other alloy systems. Thus, the study of tensile fracture structure in the as-quenched $Zr_{80}Pt_{20}$ alloy hardened by the homogeneous precipitation of nanoicosahedral phase indicated that the tensile fracture occurred in the residual amorphous phase in the intergranular region and was obtained in the nanocomposite alloy consisting of icosahedral and amorphous phases.[106] Similar $Zr_{70}Pd_{30}$ nanoicosahedral phase was also described.[107,108] An application field of Au was presented as an alloying element in Cu-based amorphous alloys with a tendency to form nanoscale icosahedral particles having quasicrystalline symmetry upon devitrification.[109,110] Nanoicosahedral particles with a size below 10 nm were found to be formed in the amorphous matrix of $Cu_{55}Zr_{30}Ti_{10}Au_5$ amorphous alloy in the initial stage of the devitrification process. A similar composition of a nanoscale icosahedral phase in a $Cu_{55}Zr_{30}Ti_{10}Pd_5$ amorphous alloy was studied by 3D atom probe analysis.[111] It was established that the composition of the nanoicosahedral phase that precipitated from the amorphous alloy was only slightly different from that of the matrix, that is, the icosahedral phase was Cu-based containing >50 at.% Cu.

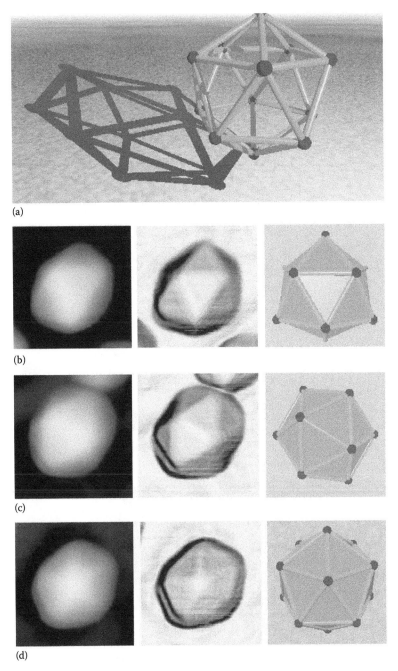

(a)

(b)

(c)

(d)

FIGURE 11.26 (a) 3D representation of an icosahedron. Supported Ag icosahedral nanocrystals in the (b) face orientation, (c) edge orientation, and (d) point orientation. The panels on the left show topographic STM images (image sizes $15 \times 15\,\text{nm}^2$; $V_s = +4.0\,\text{V}$, $I_t = 30\,\text{pA}$). The central panels show derivative STM images to better illustrate their shapes. The right-hand panels show models of the various supported icosahedral orientations. (Reproduced with permission from Silly, F. and Castell, M.R., Growth of Ag icosahedral nanocrystals on a SrTiO$_3$ (001) support, *Appl. Phys. Lett.*, 87, 213107, Copyright 2005, American Institute of Physics.)

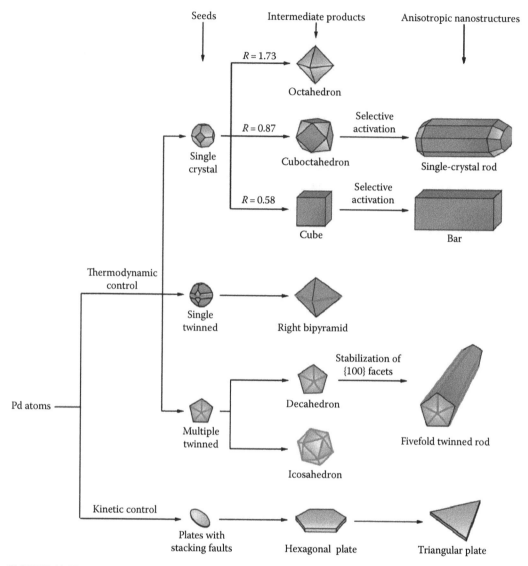

FIGURE 11.27 A schematic illustration of the reaction pathways that lead to Pd nanostructures with different shapes. (Yujie, X. and Younan, X.: Shape-controlled synthesis of metal nanostructures: The case of palladium. *Adv. Mater.* 2007. 19. 3385–3391. Copyright. Wiley-VCH Verlag GmbH & Co. KGaA. Reproduced with permission.)

In addition, the nanoicosahedral phase was formed as a primary phase in the $Zr_{65}Al_{7.5}Ni_{10}Cu_{17.5-x}$ Pd_x ($x = 5$, 10, and 17.5) glassy alloys.[112] The nanoicosahedral grains were found to be distributed homogeneously, indicating the homogeneous nucleation mode. The formation of the nanoicosahedral phase by the addition of Pd implies the existence of icosahedral short-range order in the glassy state of the Zr-Al-Ni-Cu alloy.

11.7 NANODODECAHEDRA

Dodecahedron-like nanoparticles are relatively rare and are mainly represented by gold (Ag^{113}) and its alloys ($AuPd^{114}$), as well as some other metals. Thus, gold nanododecahedrons[115] (Figure 11.30) were prepared at 100°C–155°C from N,N-a dimethylformamide (N,N-DMF was

FIGURE 11.28 (a) SEM and (b) high-resolution TEM images of palladium icosahedrons prepared with citric acid as a reductant in water at 90°C at $t = 26$ h. The inset of (a) shows the blow-up SEM image of a single Pd icosahedron. (Yujie, X. and Younan, X.: Shape-controlled synthesis of metal nanostructures: The case of palladium. *Adv. Mater.* 2007. 19. 3385–3391. Copyright. Wiley-VCH Verlag GmbH & Co. KGaA. Reproduced with permission.)

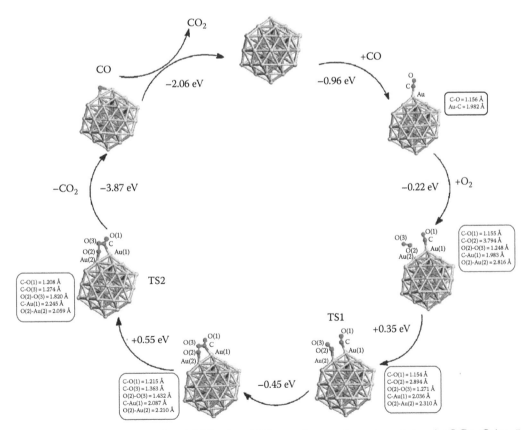

FIGURE 11.29 Catalytic cycle of $2CO + O_2 \rightarrow 2CO_2$ reaction on the anion cluster $Au@Cu_{12}@Au_{42}^-$ Calculation is done at the PBE/DNP level of theory. TS1 and TS2 denote the first and second transition states, respectively. (Reproduced with permission from Yi, G. et al., Icosahedral crown gold nanocluster $Au_{43}Cu_{12}$ with high catalytic activity, *Nano Lett.*, 10, 1055–1062. Copyright 2010 American Chemical Society.)

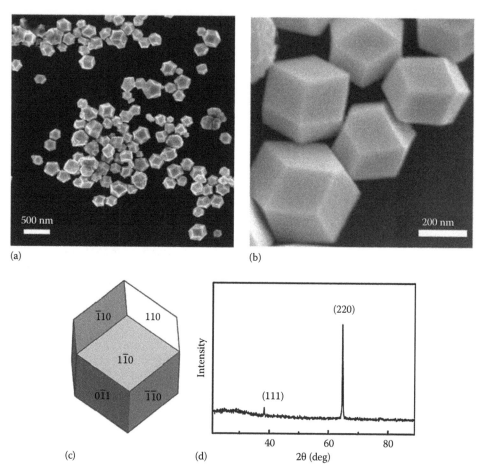

FIGURE 11.30 (a) Low- and (b) high-magnification SEM images of the rhombic dodecahedral Au nanocrystals. (c) An ideal rhombic dodecahedron enclosed by 12 {110} facets. (d) XRD pattern of the rhombic dodecahedral Au nanocrystals. (Reproduced with permission from Jeong, G.H. et al., Polyhedral Au nanocrystals exclusively bound by {110} facets: The rhombic dodecahedron, *J. Am. Chem. Soc.*, 131(5), 1672–1673. Copyright 2009 American Chemical Society.)

used as both reductant and solvent[116]) solution of chloroauric acid as a precursor. The quashed dodecahedron of the gold nanocrystals is composed of six trapezoidal surfaces and six rhombohedral surfaces. The trapezoidal surface is a twin crystal surface, and the rhombohedral face is a single-crystal surface. The obtained quashed dodecahedral gold nanocrystals had high surface energy and high catalytic performance. Pd single-crystal nanothorns (Pd NTs, Figure 11.31) were electrochemically prepared (Reactions 11.9 and 11.10) at r.t. without template or surfactant.[117] The nanothorn was found to be made by a succession of epitaxic dodecahedrons of decreasing sizes aligned in the direction of the (111) plane. Pd NTs have higher catalytic activity than commercial Pd black for the oxidation of formic acid. The morphology alteration from δ- to α-Cr nanoparticles and the phase transition temperature were directly observed by an *in situ* experiment using a transmission electron microscopy (TEM).[118] The icositetrahedral nanoparticle of δ-Cr transformed into a rhombic dodecahedral nanoparticle of α-Cr upon heating at 550°C. The most stable shape of the α-Cr nanoparticle was found to be the rhombic dodecahedron.

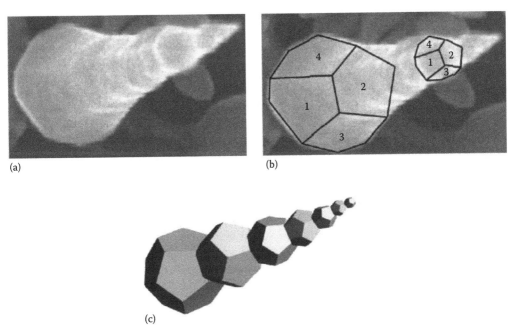

FIGURE 11.31 SEM image of (a) a Pd nanothorn, (b) the same Pd nanothorn showing its dodecahedron basis, and (c) a representation of the nanothorn with its sequence of epitaxic dodecahedrons of decreasing volumes as a model for the actual Pd nanothorn shown in (a). (Reproduced with permission from Meng, H. et al., Electrosynthesis of Pd single-crystal nanothorns and their application in the oxidation of formic acid, *Chem. Mater.*, 20(22), 6998–7002. Copyright 2008 American Chemical Society.)

$$PdCl_6^{2-} + 2e^- \rightarrow PdCl_4^{2-} + 2Cl^- \tag{11.9}$$

$$PdCl_4^{2-} + 2e^- \rightarrow Pd + 4Cl^- \tag{11.10}$$

Among other compounds, magnetite dodecahedral nanocrystals (two different ribs were ~300 and 200 nm) were fabricated using an ethylenediaminetetraacetic acid (EDTA)-mediated hydrothermal route.[119] It was shown that the products were of the cubic inverse spinel structure. Nanoparticle ammonium zinc phosphate with dodecahedron morphology was synthesized *via* the solid-state reaction at r.t.[120] It was shown that the obtained product was nanoparticle NH_4ZnPO_4 with size distribution between 100 and 300 nm, which belongs to hexagonal crystal system. In addition, nano-structured Ni_3S_2 thin films, NiS nanowhiskers, and NiS_2 single crystals with a dodecahedron [101] unit polytype shape were directly synthesized in ethylenediamine.[121] It was established that a new molecular precursor $Ni(en)_3S_4$ derived in ethylenediamine with an en = ethylenediamine ligand was an excellent precursor for the synthesis of nickel sulfide nanoparticles. Nickel sulfide with different phases and special shapes may find applications in catalysis, semiconductors, and magnetic devices. Also, a unique magnetic nanosystem, constructed from Ge_{20} dodecahedrons and Mn, showed a ferromagnetic transition at ~10 K.[122] In this system $Ba_8Mn_xGe_{46-x}$ ($x = 1$–2), the Mn atoms can be incorporated with accurate control in the position of the crystal lattice. It was revealed that the d electrons are almost localized on Mn atoms but also affected by conduction electrons spreading over the clathrate network.

11.8 NANOCUBOCTAHEDRA

Cuboctahedron-like nanostructures are known mainly for metals (Ag,[123] Au,[124] Pt[125]), alloys (PtCo[126]) and some binary compounds. Thus, rhodium nanoparticles of different shapes (triangular, spherical, cuboctahedron, rhombohedral) and sizes (5–15 nm) were synthesized by a modified polyol method.[127] Structural phase transition of an Ni nanocluster from cuboctahedron to icosahedron at 1100 K was investigated by means of molecular dynamics simulation.[128] It was established that minimizing surface energy enables the Ni nanocluster to drive the transition; the interior energy is increased to as much as that of icosahedron prior to the transition. The introduction of metallic traces into the synthesis of platinum nanocrystals (Pt NCs) was investigated as a surfactant-independent means of controlling shape.[129] In the presence of metallic cobalt (a strong reducer for Pt cations) cubic Pt NCs were obtained, while cobalt ions or gold NCs have no effect on the synthesis, and as a result, polypods were obtained (Figure 11.32). Intermediate shapes such as cemented cubes or cuboctahedron NCs are also obtained under similar conditions.

Binary metal compounds are mainly represented by platinum combinations. Thus, FePt octapod, cuboctahedron, truncated cube, and nanocube were prepared from cuboctahedral seed (Figure 11.33).[130] The formations of FePt nanostructures were mainly attributed to the differences in the growth rate between the {111} and {100} planes of cuboctahedral seeds. The magnetic measurements showed that the order of volume, V (nanocube) > V (octapod) > V (cuboctahedron), obviously reflected the order of saturated magnetization (Ms), Ms (nanocube) > Ms (octapod) > Ms (cuboctahedron). In addition, based on a first-principle calculation combined with the cluster expansion technique and Monte Carlo statistical simulation, the segregation behavior in 55-atom $Pt_{28}Rh_{27}$ nanoparticles in a cuboctahedron shape was examined.[131]

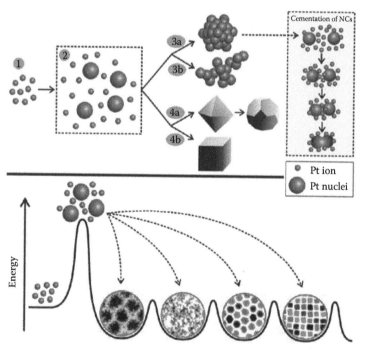

FIGURE 11.32 Schematic diagrams illustrating the growth mechanism of the Pt NCs and the degeneracy of this system. Top panel. Key: Pt ions in solution (1); nucleation and growth of Pt nuclei (2); the formation of multibranched nanostars (3a) and polypods (3b); and the formation of cuboctahedron (4a) and cubic (4b) nanocrystals. Inset: cementation of two nanocrystals in the presence of monomers. Bottom panel. Scheme not drawn to scale. (Reproduced with permission from Lim, S.I. et al., Synthesis of platinum cubes, polypods, cuboctahedrons, and raspberries assisted by cobalt nanocrystals, *Nano Lett.*, 10(3), 964–973. Copyright 2010 American Chemical Society.)

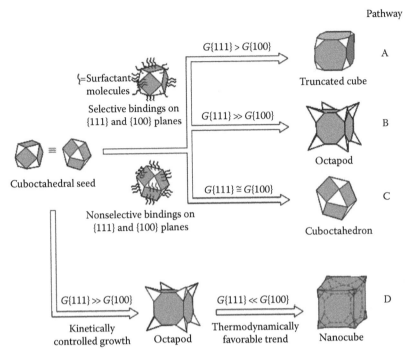

FIGURE 11.33 Schematic illustration of the formations of truncated cube (pathway A), octapod (pathway B), cuboctahedron (pathway C), and nanocube (pathway D). All the nanocrystals were formed originally from the cuboctahedral seeds (left). The white and gray colors represent the {111} and {100} facets of the FePt nanocrystals, respectively. (Reproduced with permission from Chou, S.-W., Controlled growth and magnetic property of FePt nanostructure: Cuboctahedron, octapod, truncated cube, and cube, *Chem. Mater.*, 21(20), 4955–4961. Copyright 2009 American Chemical Society.)

Among other compounds, niobium carbide nanoparticles were synthesized by flowing methane, ethylene, or acetylene gas through a plasma generated from an arc discharge between two niobium electrodes.[132] It was found that these nanoparticles adopt cubic morphology with methane gas, a mixture of cubes and cuboctahedron morphology with ethylene gas, and solely a cuboctahedron morphology with acetylene gas (Figure 11.34). The authors supposed that a change in particle morphology might be attributed to either the ethylene and acetylene free radicals or the increase in carbon concentration effecting the relative growth rates of the {111} and {100} facets on an NbC seed crystal. In a related research,[133] it was found that TiC nanoparticles (Figure 11.35) prefer a cubic morphology at low concentrations of methane and a cuboctahedron morphology at high concentrations of methane. Copper sulfide crystal represents a concaved cuboctahedron structure with 14 faces comprising of 6 squares and 8 triangles.[134] In a related report,[135] a wonderful composite structure made of unique CuS cuboctahedron crystals and coating silica spheres (Figure 11.36) with diameters ranged from 30 to 730 nm was demonstrated. 3-Aminopropyltrimethoxysilane acted as a linker, which provided hydrogen bonding and metal–ligand bonding interaction with the facets of CuS crystals and covalent bonding of its functional silane groups onto the surface of hydrophilic silica spheres, resulting in the formation of such a CuS/silica composite structure.

11.9 NANOCUBES

Cube-like nanoparticles are widespread in comparison with other polyhedron-like nanostructures, so here we present the most representative examples. A host of reports are devoted to metals (especially Au,[136] Ag,[137] Pd, and Pt), their alloys, or core-shell structures, whose nanocube structures were obtained chemically or by microwave heating. Thus, monodispersed Ag nanocubes

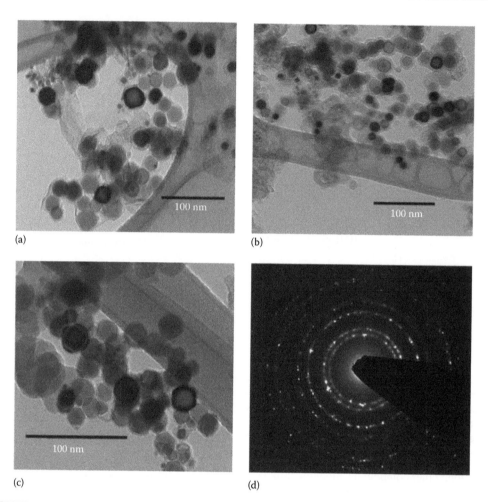

FIGURE 11.34 (a) Niobium carbide nanoparticles synthesized with 10% acetylene. (b) Niobium carbide synthesized with 50% acetylene. (c) Niobium carbide nanoparticles synthesized with 100% acetylene. (d) SADP of the nanoparticles confirming that they are NbC with a stoichiometry of 1 metal to 1 carbon. (Reproduced with permission from Grove, D.E. et al., Effect of hydrocarbons on the morphology of synthesized niobium carbide nanoparticles, *Langmuir*, 26(21), 16517–16521. Copyright 2010 American Chemical Society.)

(Figure 11.37, 25–45 nm) were prepared by adding a trace amount of sodium sulfide (Na_2S) or sodium hydrosulfide (NaHS) to the conventional polyol synthesis (Reactions 11.11 and 11.12); the reaction time was significantly shortened from 16–26 h to 3–8 min.[138] These small nanocubes are of great interest for biomedical applications. Additionally, these Ag nanocubes were further used as a sacrificial template to generate gold nanocages (Reaction 11.13). The microwave technique was employed in the shape-controlled synthesis of palladium nanoparticle H_2PdCl_4 as a precursor and tetraethylene glycol (TEG) as both a solvent and a reducing agent in the presence of PVP and CTAB in 80 s,[139] resulting in Pd nanocubes and nanobars with a mean size of ~23.8 nm. The formation of $PdBr_4^{2-}$ due to the coordination replacement of the ligand Cl^- ions in $PdCl_4^{2-}$ ions by Br^- ions in the presence of bromide was responsible for the synthesis of Pd nanocubes and nanobars. Monodispersed gold nanocubes of highly uniform size were fabricated by a simple electrochemical method[140] (Figure 11.38). The lengths of the edges of the gold nanocubes were about 30 nm. The growth solution was prepared from two cationic surfactant solutions as micelle templates with added acetone solvent.

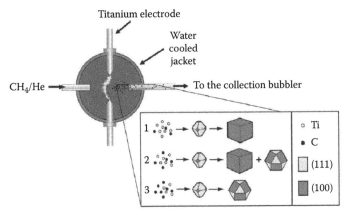

FIGURE 11.35 Principal components of the arc discharge vessel and the different nanoparticle morphologies with varying carbon concentrations. (Reproduced with permission from Grove, D.E., Effect of carbon concentration on changing the morphology of titanium carbide nanoparticles from cubic to cuboctahedron, *ACS Nano*, 4(1), 49–54. Copyright 2010 American Chemical Society.)

Conventional polyol synthesis:

$$2HOCH_2CH_2OH \rightarrow 2CH_3CHO + 2H_2O \qquad (11.11)$$

$$2Ag^+ + 2CH_3CHO \rightarrow CH_3CO-OCCH_3 + 2Ag + 2H^+ \qquad (11.12)$$

$$3Ag + HAuCl_4 \rightarrow Au + HCl + 3AgCl \qquad (11.13)$$

By carefully adjusting the reaction parameters in the synthesis of gold–copper bimetallic nanocubes, the edge length of the resulting nanocubes was controlled to form 3.4, 5, 23, 45, and 85 nm nanocubes, and also their composition could be varied from $AuCu_3$ to Au_3Cu.[141] Networks of single-wall carbon nanotubes (SWCNTs) decorated with Au-coated Pd (Au/Pd) nanocubes were employed as electrochemical biosensors, which exhibit excellent sensitivity (2.6 mA/mM/cm²) and a low estimated detection limit (2.3 nM) at a signal-to-noise ratio of 3 ($S/N = 3$) in the amperometric sensing of hydrogen peroxide.[142] Similarly, glucose oxidase (GOx) was linked to the surface of the nanocubes for amperometric glucose sensing. The exhibited glucose detection limit of 1.3 M ($S/N = 3$) and the linear range spanning from 10 µM to 50 mM substantially surpass other carbon nanotube (CNT)-based biosensors. Another useful application of mixed-metal nanocubes is methanol oxidation on Pt_3Co nanocubes.[143] We emphasize that platinum-containing nanocubes (as well as other nanostructures) are of extreme importance due to these catalytic applications. As examples, we note Mn-Pt nanocubes, synthesized from Pt acetylacetonate and Mn carbonyl in the presence of oleic acid and oleylamine.[144] These Mn-Pt nanocubes were converted into an ordered $MnPt_3$ intermetallic phase upon annealing. These materials are promising new candidates as cathode and anode catalysts in fuel cells. In addition, monodisperse, crystalline Pt_9Co nanocubes were prepared by using a one-pot colloidal method.[145] The nanocubes showed significantly enhanced electrocatalytic activity toward oxygen reduction. FePt nanocubes[146] are also known, among other platinum-metal cube-like arrays.

Metal oxides (for instance, Mn_3O_4[147]) are also common as nanocubes or their precursors and can be prepared by a variety of methods. Thus, the CuO hollow nanostructures were prepared by adding aqueous ammonia solutions with Cu_2O nanocube colloidal solutions.[148] In a related study,[149] highly uniform size-controlled Cu_2O nanocubes were formed by means of pulse electrodeposition. Cu_2O (cuprite) nanocubes with high catalytic properties were synthesized by Cu

FIGURE 11.36 FESEM images of the alignment of silica spheres on CuS crystal outer surfaces with different diameters: (a, b) 30 nm, (c, d) 130 nm, and (e) 730 nm. (Reproduced with permission from Wan, Y. et al., Loading of silica spheres on concaved CuS cuboctahedrons using 3-aminopropyltrimethoxysilane as a linker, *Langmuir*, 23(16), 8526–8530. Copyright 2007 American Chemical Society.)

nanopowder in distilled water.[150] The single phase of Cu_2O nanocubes showed uniform size with an average edge length of 30 nm. A part of Cu_2O was changed to CuO phase in the air. The freshly prepared Cu_2O shows the highest catalytic activity, and it decreases with increasing CuO phase. Iron oxides in distinct oxidation states were fabricated as nanocubes. Thus, single-crystalline $\alpha\text{-}Fe_2O_3$ nanocubes were obtained in large quantities through a facile one-step hydrothermal synthetic route under mild conditions from aqueous iron(III) nitrate $\{Fe(NO_3)_3 \cdot 9H_2O\}$ as iron source and triethylamine as precipitant.[151] By prolonging the reaction time from 1 to 24 h, the evolution process of $\alpha\text{-}Fe_2O_3$, from nanorhombohedra to nanohexahedron, and finally

26 µM 28 µM 30 µM 32 µM

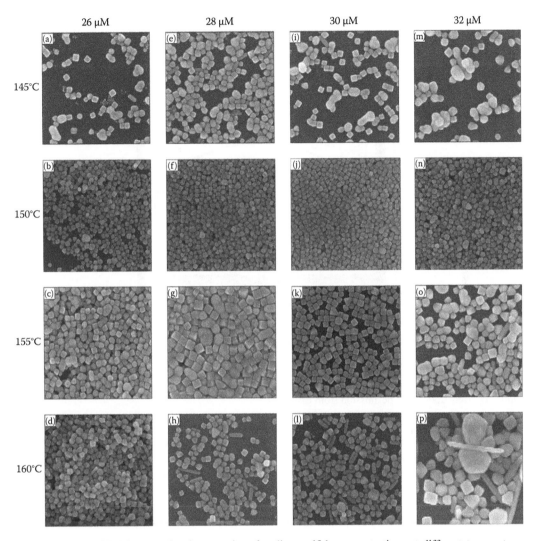

FIGURE 11.37 SEM images showing a series of sodium sulfide concentrations at different temperatures (145°C–160°C). Each row shows reactions that were conducted in the same bath at the same temperature, side by side, and with the same precursor solutions. Each column was performed with the same sodium sulfide concentration (26–32 µM). ((a)-(d): At the concentration of sulfide species in the solution under 28 µM; (m)-(p): at their concentration higher than 30 µM.) (Reproduced from *Chem. Phys. Lett.*, 432, Siekkinen, A.R. et al., Rapid synthesis of small silver nanocubes by mediating polyol reduction with a trace amount of sodium sulfide or sodium hydrosulfide, 491–496, http://dipts.washington.edu/gemsec/publications/pdfs/SiekkinenXiaChem-Phys-Lett.pdf, Copyright 2006, with permission from Elsevier.)

nanocube, was observed. Alternatively, iron oxide nanocubes were prepared by thermal decomposition of iron oleate complex in the presence of oleic acid *via* microwave-assisted solvothermal method, followed by Ostwald ripening procedures.[152] The primary nanoparticles synthesized by microwave heating were low crystalline spheres with an average diameter of about 6 nm. After aging at 180°C, these iron oxides transform to crystalline α-Fe_2O_3 and trace amount of Fe_3O_4. Monodisperse Fe_3O_4 nanocubes were synthesized by a facile solvothermal method at 260°C in the presence of oleic acid and oleylamine.[153] The as-synthesized nanocubes were found to be ferromagnetic at 2 K, whereas they were superparamagnetic at 300 K. A phase transformation induced by reduction in the γ-Fe_2O_3 nanocubes synthesized by oxidizing the corresponding FeO/Fe_3O_4 nanocubes was achieved in solution.[154] The γ-Fe_2O_3 nanocubes were transformed to Fe_3O_4

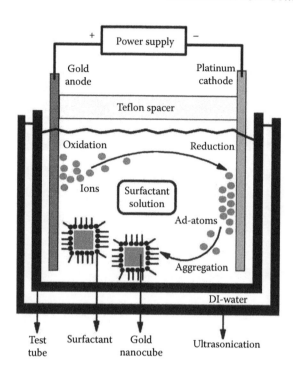

FIGURE 11.38 Schematic diagram of the electrochemical apparatus for the synthesis of gold nanocubes. (Reproduced from *Mater. Lett.*, 60, Chien-Jung, H. et al., Electrochemical synthesis of gold nanocubes, 1896–1900, Copyright 2006, with permission from Elsevier.)

nanocubes within 260°C–300°C range and the FeO/Fe$_3$O$_4$ nanospheres at 350°C. CO oxidation was performed on Co$_3$O$_4$ nanobelts and nanocubes as model catalysts.[155] The Co$_3$O$_4$ nanobelts that have a predominance of exposed {011} planes were found to be more active than Co$_3$O$_4$ nanocubes with exposed {001} planes. The essence of shape and crystal plane effect was revealed by the fact that the turnover frequency of Co^{3+} sites of {011} planes on Co$_3$O$_4$ nanobelts was far higher than that of {001} planes on Co$_3$O$_4$ nanocubes. In addition, monodispersed CeO$_2$ nanocubes were prepared from Ce(NO$_3$)$_3 \cdot$H$_2$O and NaOH (Reaction 11.14)[156] or alternatively by using a liquid–liquid interface.[66] The morphology of CeO$_2$ nanocrystals from the aqueous phase changed from truncated octahedron to nanocube in the toluene phase at the H$_2$O–toluene interface. In addition, well-defined ceria nanocubes covered by oleic acid with exposed {100} facets were synthesized and exhibited exclusive selectivity for the oxidation of toluene to benzaldehydes in liquid phase by O$_2$.[157] The ceria nanocubes exhibited excellent reducibility and high oxygen storage capacity, indicating that they are potential novel catalytic materials. Metal hydroxides are represented by single-crystalline In(OH)$_3$ architectures, prepared by a rapid (1 min) and efficient microwave-assisted hydrothermal method using InCl$_3$ as the precursor at 140°C.[158] These samples were found to be composed of 3D nanocubic, microcubic, and irregular structures of ~70 nm to 5 μm size:

$$2Ce(NO_3)_3 + 6OH^- + 0.5O_2 \text{ (g)} \rightarrow 2CeO_2 \text{ (s)} + 6NO_3^- + 3H_2O \qquad (11.14)$$

A considerable number of reports are dedicated to niobates, tantalates, and titanates, for instance SrTiO$_3$.[159] Thus, the crystal growth of CaTiO$_3$ hollow crystals with different microstructures was investigated, revealing that in a water-free poly(ethylene glycol) 200 (PEG-200) solution, CaTiO$_3$ nanocubes formed first (Figure 11.39).[160] Then the nanocubes underwent an oriented self-assembly into spherical particles, enhanced by the surface-adsorbed polymer molecules. The study of

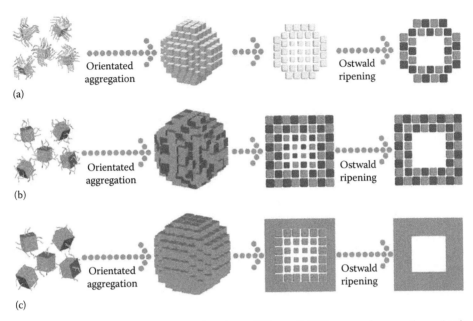

FIGURE 11.39 Schematic illustration showing three different $CaTiO_3$ nanocube growth mechanisms in (a) water-free, (b) 1.25 vol.% water, and (c) 5 vol.% water systems. (Reproduced with permission from Yang, X. et al., Formation mechanism of $CaTiO_3$ hollow crystals with different microstructures, *J. Am. Chem. Soc.*, 132(40), 14279–14287. Copyright (2010) American Chemical Society.)

hydrothermally grown $KNbO_3$ and $KTaO_3$ nanocrystals showed that the $KNbO_3$ crystal had a nanorod shape while the $KTaO_3$ crystal had a nanocube shape.[161] The existence of a hydroxyl group in both nanocrystals was revealed, whose presence seems to be related to the enlarged lattice constants, the lowered structural phase transition temperatures, the broad phonon width, and the increased number of Raman-active phonon modes in $KNbO_3$ and $KTaO_3$ nanocrystals. Truncated nanocubes of barium titanate were synthesized using a rapid, facile microwave-assisted hydrothermal route (Reaction 11.15).[162] A hydrothermal synthesis of single-crystalline nanocubes (50–100 nm) with a pseudo-cubic perovskite structure composed of lanthanum barium manganite (Figure 11.40, $La_{1-x}Ba_xMnO_3$) with three different doping levels (x = 0.3, 0.5, and 0.6) was also

100 nm

FIGURE 11.40 TEM image of $La_{0.5}Ba_{0.5}MnO_3$ nanocubes, illustrating that the reaction produces isolated nanocubes ranging from 20 to 500 nm in size. (Reproduced with permission from Urban, J.J. et al., Synthesis of single-crystalline $La_{1-x}Ba_xMnO_3$ nanocubes with adjustable doping levels, *Nano Lett.*, 4(8), 1547–1550. Copyright 2004 American Chemical Society.)

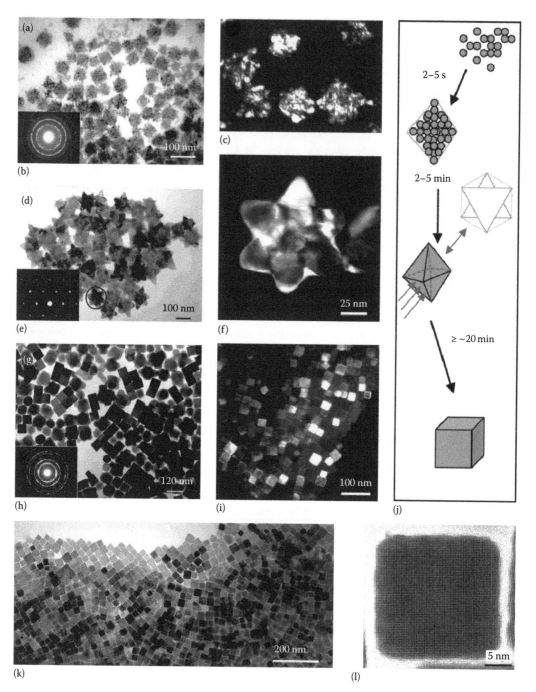

FIGURE 11.41 (a, c, b) TEM images and diffraction pattern of 5-s-growth PbSe clusters at 230°C. (d, f, e) TEM images and diffraction pattern of a selected single PbSe octahedron grown for 2–5 min at 230°C. (g, h) TEM image and diffraction pattern of PbSe nanocubes after five injections within a total aging time of ~25 min at a growth temperature of 230°C; (i, k, l) TEM images of a size-selected PbSe nanocube assembly and HR-TEM image of a single PbSe nanocube (the sample was grown through the same procedure at 180°C for ~25 min). (j) A scheme illustrating the formation process of PbSe nanocubes; (a), (d), (g), (k), and (l) are bright-field images, whereas (c), (f), and (i) are dark-field images. (Reproduced with permission from Weigang, L. et al., Formation of PbSe nanocrystals: A growth toward nanocubes, *J. Phys. Chem. B*, 109, 19219–19222. Copyright 2005 American Chemical Society.)

carried out.[163] Magnetic measurements performed on nanocube ensembles showed that the magnetic properties depended on the doping level:

$$2Ba(NO_3)_{3(aq.)} + 2TiO(NO_3)_{2(aq.)} + 8C_2H_5NO_{2(aq.)} + 5O_{2(g)}$$

$$\rightarrow 2BaTiO_{3(s)} + 9N_{2(g)} + 16CO_{2(g)} + 20H_2O_{(g)} \tag{11.15}$$

Chalcogenides are also of a certain interest. Thus, PbS nanocubes were prepared by a solvothermal process with ethylenediamine,[164] meanwhile monodispersed PbSe nanocubes were synthesized through a hydrothermal method by Pb^{2+}-EDTA complex and Na_2SeSO_3[165] or by rapidly injecting a phenyl ether solution of lead acetate and trioctylphosphine selenide ([TOP]-Se) into a vigorously stirred hot phenyl ether at 180°C or 230°C, in the presence of oleic acid.[166] In the last case, a two-step evolution mechanism was proposed, indicating that the shape evolution of PbSe nanocrystals (Figure 11.41) is dependent on the growth time, whereas the crystalline size can be tuned by varying the growth temperature under the studied conditions. Alternatively, the pure FCC lead chalcogenide nanocubes were synthesized in hydrazine hydrate saturated alkaline solution at r.t. and for a short growth time.[167] The size of PbS, PbSe, and PbTe nanocubes was 200–300, 50–120, and 30–60 nm, respectively. It was found that the growth steps of lead chalcogenides (especially PbTe) nanostructures could be controlled in the strong hydrazine hydrate alkaline environment. In addition, $AgPb_{10}SbTe_{12}$ crystals with nanocubic and flower-like morphologies were fabricated (Figure 11.42) by a facile solution route.[168]

For the synthesis of silver chloride nanocubes (Figure 11.43), silver nitrate and hydrochloric acid were used as precursors in ethylene glycol and poly(vinyl pyrrolidone) as a surfactant.[169] These AgCl-based nanoparticles with controlled morphologies may play an important role in various applications such as photocatalyst, SERS, and antiseptic biomedical devices. Among other compounds, we note that uniform multicolor upconversion luminescent Re^{3+} doped $NaYF_4$ nanocubes

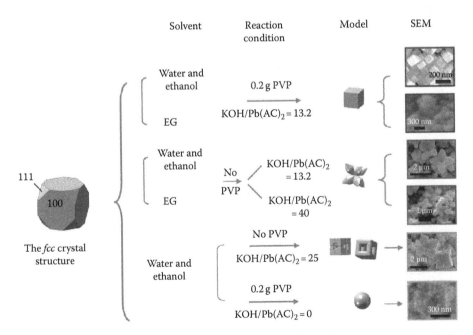

FIGURE 11.42 The summary schematic illustration of the experimental parameters and morphology evolution for the $AgPb_{10}SbTe_{12}$ samples. (Reproduced with permission from Yun, B.K. et al., Possible role of hydroxyl group on local structure and phase transition of $KNbO_3$ and $KTaO_3$ nanocrystals, *Phys. B Condensed Matter*, 405(23), 4866–4870. Copyright 2010 American Chemical Society.)

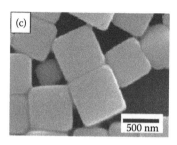

FIGURE 11.43 SEM images of (a) cube-shaped Ag-grain aggregates, (b) AgCl nanocubes with Ag nanoparticles at corners, and (c) AgCl nanocubes with truncated corners obtained by reducing AgCl nanocubes with $NaBH_4$, L-ascorbic acid, and citric acid, respectively. (Reproduced from Seungwook, K. et al., *Bull. Korean Chem. Soc.*, 31(10), 2918, 2010. With permission.)

were fabricated through a facile ethylene glycol (EG)/ionic liquid interfacial synthesis route at 80°C, with the ionic liquids acting as both reagents and templates.[170] In addition, a series of reports are devoted to various Prussian blue analogs, in particular those of cobalt-iron.[171]

11.10 NANOPRISMS

Prism-like nanoparticles, as well as nanocubes, are widespread in comparison with other polyhedron-like nanostructures, but the peculiarity is that a predominant number of reports have been devoted to Ag[172–178] and less to gold[179–182] (Figure 11.44) nanoprisms, reviewed in.[183] Thus, the surface plasmon resonance spectra of Ag nanoprisms in the presence of halide ions change gradually with reaction time.[184] The changes in the spectra correspond to the shape transformation of Ag nanoprisms. Optical properties and sensing performance of Ag/SiO$_2$/Ag triangular nanoprisms were investigated.[185] It was revealed that when the thickness of the SiO$_2$ dielectric layer increases, the plasmon coupling between the top and the bottom Ag layers becomes weaker and the red shift speed becomes slower. When the thickness of SiO$_2$ reaches 60 nm, the coupling disappears. A photoinduced method for converting large quantities of Ag nanospheres into triangular nanoprisms (Figure 11.45) by irradiation with a conventional 40 W fluorescent light was reported.[186] This light-driven process results in a colloid with distinctive optical properties that directly relate to the nanoprism shape of the particles. Gold nanoprisms with an average edge size of ∼140 nm and thickness of ∼8 nm were achieved in high-purity (∼97%) by exploiting the electrostatic aggregation and shape effects through a modified seed-mediated approach.[187] The gold nanoprisms spontaneously aggregated into precipitate, whereas most of the spherical ones were still kept in the solution. These high-purity colloidal gold nanoprisms exhibited remarkably enhanced surface plasmon resonance and can be a promising nanostructured system for plasmonic sensor applications. Very fast synthesis of gold nanoprisms (110 nm) under microwave heating for 60–90 s in the presence of 2-naphthol (2-N) as a reducing agent was carried out.[188] The growth of the particles with different shapes (spherical, polygonal, or triangular/prisms, etc.) were found to be functions of the surfactant-to-metal-ion molar ratio and the concentration of 2-N. The offered method might find a variety of applications in many areas such as catalysis, clinical and diagnostic medicine, and nanoelectronics. In addition, DNA-conjugated triangular gold nanoprisms were synthesized and characterized to determine the factors that allow adsorption of alkylthiol-modified oligonucleotides on different facets of an anisotropic gold nanoparticle.[189] Face-selective DNA–ligand adsorption processes were found to be time dependent and can be exploited to selectively immobilize DNA on the different edges and faces of the particle.

Unique Au$_{core}$-Ag$_{shell}$ triangular bifrustum nanocrystals (Figures 11.46 and 11.47) were synthesized in an aqueous solution using a seed-mediated approach.[190] The formation of the Ag layer on the Au nanoprism seeds leads to structures with highly tunable dipole and quadrupole surface plasmon resonances. The authors noted that the use of Ag to modulate the geometry of Au seeds

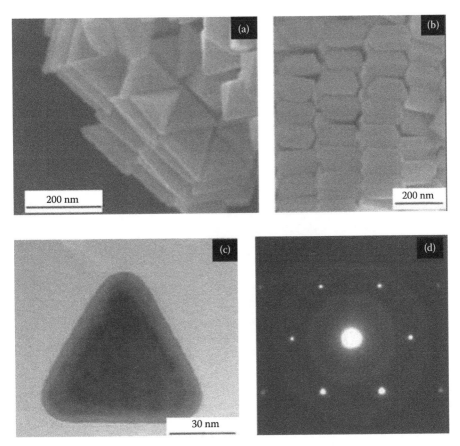

FIGURE 11.44 (a) SEM image of the Au nanoprisms with a 3D (3D) pattern. (b) Side-view SEM image of the Au nanoprisms with a brick-like wall structure. (c) TEM image of a single nanoprism. (d) SAED pattern taken from a single nanoprism as shown in (c). (Reproduced with permission from Yingzhou, H. et al., Surfactant-promoted reductive synthesis of shape-controlled gold nanostructures, *Cryst. Growth Des.*, 9(2), 858–862. Copyright 2009 American Chemical Society.)

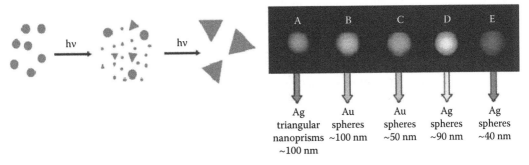

FIGURE 11.45 Transformation Ag nanospheres-nanoprisms (left). Rayleigh light-scattering of particles deposited on a microscope glass slide (right). (Reproduced from Jin, R. et al., *Science*, 294, 1901–1903. Copyright 2001. With permission.)

is a facile way to potentially extend the utility of these structures in many areas where plasmonic wavelengths are important, including diagnostic labels, energy harvesting, optical transport, and therapeutics.

Other elements and compounds in nanoprism form are rare. Thus, vertically aligned carbon hexagonal nanoprism arrays were grown on molybdenum substrates by the catalyst-assisted pulsed

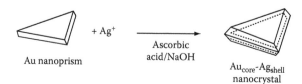

FIGURE 11.46 Formation of Au_{core}-Ag_{shell} triangular bifrustum nanocrystals. (Reproduced with permission from Yoo, H. et al., Core-shell triangular bifrustums, *Nano Lett.*, 9(8), 3038–3041. Copyright 2009 American Chemical Society.)

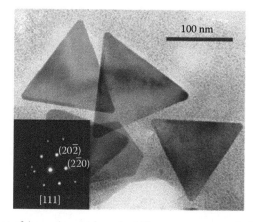

FIGURE 11.47 TEM image of Au_{core}-Ag_{shell} triangular bifrustum nanocrystals (10 μL of 10 mM $AgNO_3$ solution). Inset: Electron diffraction pattern of the top of a nanocrystal. (Reproduced with permission from Yoo, H. et al., Core-shell triangular bifrustums, *Nano Lett.*, 9(8), 3038–3041. Copyright 2009 American Chemical Society.)

laser deposition techniques.[191] The carbon hexagonal nanoprisms had uniform shape and length, almost aligned vertically on the substrate, and the average diameters were about 30 nm. The sample with vertically aligned carbon hexagonal nanoprism arrays exhibited better field emission behaviors than that with aligned carbon nanorod arrays. Radial ZnO clusters were synthesized in a large scale by a simple microemulsion method using sulfonate-polystyrene (S-PS) as a template.[192] The clusters were found to be formed by the radial growth of ZnO hexagonal nanorods. These radial ZnO structures possess hexagonal prism structures. Alternatively, ZnO nanoprisms were synthesized from zinc sheet with a purity of 99.99% by CO_2 laser irradiation and coaxially transporting O_2.[193] It was established that high-purity ZnO nanoprisms can be grown from a zinc sheet, ZnO nanoprisms had single-crystalline characteristics with [001] direction growth, and its appearance depends predominantly on the supersaturation of O_2 in the reacting area. Fe_3O_4 nanoprisms, synthesized by a solvothermal process,[194] exhibited crystal plane-dependent electrochemical activities, which opens an approach to nanoscale materials with controlled electrochemical properties. Ceria materials with sphere, rod, and prism (edge length of about 600 nm) morphologies were synthesized *via* the hydrolysis of solutions containing $Ce(NO_3)_3$ and urea under hydrothermal conditions.[195] These ceria materials exhibited a higher catalytic activity in the oxidation of CO than conventionally prepared CeO_2. A highly oriented, well-aligned hierarchical Zn-In-O nanobelt-nanoprism array was synthesized by vapor-phase transport and condensation using GaN epilayer as the substrate.[196] The upper nanobelts were found to be ZnO:In with an average Zn/In molar ratio of ~9:1, and the subjacent nanoprisms were $In_2O_3(ZnO)_m$ (*m* = 2, 3, 4, and 5) structures. This bottom-up, self-assembled, highly oriented, and well-aligned hierarchical nanoarray may find applications in newly emerging vertically integrated nanoarray circuits. Hollow hexagonal fluorapatite nanoprisms (an outer length of about 1 μm, an average outer diameter of 150 nm, and a shell thickness of about

50 nm) were prepared *via* a template-free method with the assistance of Na$_2$EDTA/citric acid mixed chelating reagent.[197] SiC hexagonal nanoprisms with a diameter of ~100 nm and 2 μm length were prepared by a reaction of multiwall CNTs and Si vapor in an Astro furnace at 1450°C for 3 h.[198] The PL spectrum of the nanoprisms exhibited a significant blue shift relative to bulk 3C-SiC and other nanostructured SiC.

REFERENCES

1. Haes, A. J.; Zou, S.; Schatz, G. C.; Van Duyne, R. P. Nanoscale optical biosensor: Short range distance dependence of the localized surface plasmon resonance of noble metal nanoparticles. *Journal of Physical Chemistry B*, 2004, *108* (22), 6961–6968.
2. Murray, W. A.; Suckling, J. R.; Barnes, W. L. Overlayers on silver nanotriangles: Field confinement and spectral position of localized surface plasmon resonances. *Nano Letters*, 2006, *6* (8), 1772–1777.
3. Zhang, X.; Hicks, E. M.; Zhao, J.; Schatz, G. C.; Van Duyne, R. P. Electrochemical tuning of silver nanoparticles fabricated by nanosphere lithography. *Nano Letters*, 2005, *5* (7), 1503–1507.
4. Sajanlal, P. R.; Pradeep, T. Reply to "comments on electric-field-assisted growth of highly uniform and oriented gold nanotriangles on conducting glass substrates." *Advanced Materials*, 2009, *21* (13), 1320–1321.
5. Sinha, S.; Pan, I.; Chanda, P.; Sen, S. K. Nanoparticles fabrication using ambient biological resources. *Journal of Applied Bioscience*, 2009, *19*, 1113–1130.
6. Chandran, S. P.; Chaudhary, M.; Pasricha, R.; Ahmad, A.; Sastry, M. Synthesis of gold nanotriangles and silver nanoparticles using *Aloe vera* plant extract. *Biotechnology Progress*, 2006, *22* (2), 577–583.
7. Rai, A.; Singh, A.; Ahmad, A.; Sastry, M. Role of halide ions and temperature on the morphology of biologically synthesized gold nanotriangles. *Langmuir*, 2006, *22* (2), 736–741.
8. Galarreta, B. C.; Norton, P. R.; Lagugne-Labarthet, F. Hexagonal array of gold nanotriangles: Modeling the electric field distribution. *Journal of Physical Chemistry C*, 2010, *114* (47), 19952–19957.
9. Haes, A. J.; Zhao, J.; Zou, S.; Own, C. S.; Marks, L. D.; Schatz, G. C.; Van Duyne, R. P. Solution-phase, triangular Ag nanotriangles fabricated by nanosphere lithography. *Journal of Physical Chemistry B*, 2005, *109* (22), 11158–11162.
10. Lukman, A. I.; Gong, B.; Marjo, C. E.; Roessner, U.; Harris, A. T. Facile synthesis, stabilization, and anti-bacterial performance of discrete Ag nanoparticles using *Medicago sativa* seed exudates. *Journal of Colloid and Interface Science*, 2010, Volume Date 2011, *353* (2), 433–444.
11. Mochizuki, Y.; Fujii, M.; Hayashi, S.; Tsuruoka, T.; Akamatsu, K. Enhancement of photoluminescence from silicon nanocrystals by metal nanostructures made by nanosphere lithography. *Journal of Applied Physics*, 2009, *106* (1), 013517/1–013517/5.
12. Rai, A.; Chaudhary, M.; Ahmad, A.; Bhargava, S.; Sastry, M. Synthesis of triangular Au core-Ag shell nanoparticles. *Materials Research Bulletin*, 2007, *42* (7), 1212–1220.
13. Jia, D.; Goonewardene, A. Two-dimensional nanotriangle and nanoring arrays on silicon wafer. *Applied Physics Letters*, 2006, *88* (5), 053105/1–053105/3.
14. Jaafar, M.; Yanes, R.; Asenjo, A.; Chubykalo-Fesenko, O.; Vazquez, M.; Gonzalez, E. M.; Vicent, J. L. Field induced vortex dynamics in magnetic Ni nanotriangles. *Nanotechnology*, 2008, *19* (28), 285717/1–285717/8.
15. Gonzalez, E. M.; Nunez, N. O.; Anguita, J. V.; Vicent, J. L. Transverse rectification in superconducting thin films with arrays of asymmetric defects. *Applied Physics Letters*, 2007, *91* (6), 062505/1–062505/3.
16. Merkulov, V. I.; Melechko, A. V.; Guillorn, M. A.; Lowndes, D. H.; Simpson, M. L. Effects of spatial separation on the growth of vertically aligned carbon nanofibers produced by plasma-enhanced chemical vapor deposition. *Applied Physics Letters*, 2002, *80* (3), 476–478.
17. Zhang, Z.; Wong, L. M.; Ong, H. G.; Wang, X. J.; Wang, J. L.; Wang, S. J.; Chen, H.; Wu, T. Self-assembled shape- and orientation-controlled synthesis of nanoscale Cu$_3$Si triangles, squares, and wires. *Nano Letters*, 2008, *8* (10), 3205–3210.
18. Tanaka, M.; Han, M.; Takeguchi, M.; Chu, F.; Shimojo, M.; Mitsuishi, K.; Furuya, K. Morphology of iron silicide nanorods formed by electron-beam-induced deposition using ultrahigh-vacuum transmission electron microscope. *Japanese Journal of Applied Physics, Part 1: Regular Papers, Brief Communications and Review Papers*, 2005, *44* (7B), 5635–5638.
19. Salavati-Niasari, M.; Mir, N.; Davar, F. ZnO nanotriangles: Synthesis, characterization and optical properties. *Journal of Alloys and Compounds*, 2009, *476* (1–2), 908–912.

20. Andelman, T.; Gong, Y.; Polking, M.; Yin, M.; Kuskovsky, I.; Neumark, G.; O'Brien, S. Morphological control and photoluminescence of zinc oxide nanocrystals. *Journal of Physical Chemistry B*, 2005, *109*, 14314–14318.

21. Portehault, D.; Cassaignon, S.; Baudrin, E.; Jolivet, J.-P. Evolution of nanostructured manganese (oxy-hydr)oxides in water through MnO_4^- reduction. *Crystal Growth and Design*, 2010, *10* (5), 2168–2173.

22. Wang, W.; Yang, Y.; Cheng, E.; Zhao, M.; Meng, H.; Liu, D.; Zhou, D. A pH-driven, reconfigurable DNA nanotriangle. *Chemical Communications*, 2009, (7), 824–826.

23. Hsu, Y.-J.; Lu, S.-Y.; Lin, Y.-F. Nanostructures of Sn and their enhanced, shape-dependent superconducting properties. *Small*, 2006, *2* (2), 268–273.

24. Hsu, Y.-J.; Lu, S.-Y. Vapor-solid growth of Sn nanowires: Growth mechanism and superconductivity. *Journal of Physical Chemistry B*, 2005, *109* (10), 4398–4403.

25. Liu, W.; Zhong, W.; Wu, X.; Tang, N.; Du, Y. Hydrothermal microemulsion synthesis of cobalt nano-rods and self-assembly into square-shaped nanostructures. *Journal of Crystal Growth*, 2005, *284* (3–4), 446–452.

26. Qi, B.; Agnarsson, B.; Goethelid, M.; Olafsson, S.; Gislason, H. P. High-resolution x-ray photoemission spectroscopy study of AlN nano-columns grown by nitridation of Al nano-squares on Si(111) substrates with ammonia. *Thin Solid Films*, 2010, *518* (14), 3632–3639.

27. Chen, Y.; He, X.; Zhao, X.; Yuan, Q.; Gu, X. Preparation, characterization, and growth mechanism of a novel aligned nanosquare anatase in large quantities in the presence of TMAOH. *Journal of Colloid and Interface Science*, 2007, *310* (1), 171–177.

28. Zhao, M.; Chen, X. L.; Ma, Y. J.; Jian, J. K.; Dai, L.; Xu, Y. P. Bi_2O_3 nanosquare sheets grown on Si substrate. *Applied Physics A: Materials Science and Processing*, 2004, *78* (3), 291–293.

29. Fechete, A. C.; Wlodarski, W.; Holland, A. S.; Kalantar-zadeh, K. Growth of indium oxide nano-structures by thermal evaporation. Nanoscience and nanotechnology, 2006. *ICONN '06. International Conference on Nanoscience and Nanotechnology*, Brisbane, Queensland, Australia, July 3–7, 2006, ISBN: 1-4244-0452-5.

30. Jia, B.; Gao, L. Morphology transformation of nanoscale magnesium hydroxide: From nanosheets to nanodisks. *Journal of the American Ceramic Society*, 2006, *89* (12), 3881–3884.

31. Plain, J.; Pallandre, A.; Nysten, B.; Jonas, A. M. Nanotemplated crystallization of organic molecules. *Small*, 2006, *2* (7), 892–897.

32. Wang, X.; Zhang, M.; Liu, J.; Luo, T.; Qian, Y. Shape- and phase-controlled synthesis of In_2O_3 with various morphologies and their gas-sensing properties. *Sensors and Actuators, B: Chemical*, 2009, *B137* (1), 103–110.

33. Xu, X.; Yin, D.; Wu, S.; Wang, J.; Lu, J. Preparation of oriented SiO_3^{2-} doped TiO_2 film and degradation of methylene blue under visible light irradiation. *Ceramics International*, 2010, *36* (2), 443–450.

34. Ramirez, F.; Heyliger, P. R.; Rappe, A. K.; Leisure, R. G. Breakdown of frequency-spectra scaling of Si nanoparticles. *Physical Review B: Condensed Matter and Materials Physics*, 2007, *76* (8), 085415/1–085415/6.

35. Berenschot, J. W.; Tas, N. R.; Jansen, H. V.; Elwenspoek, M. Chemically anisotropic single-crystalline silicon nanotetrahedra. *Nanotechnology*, 2009, *20* (47), 475302/1–475302/7.

36. Berhault, G.; Bisson, L.; Thomazeau, C.; Verdon, C.; Uzio, D. Preparation of nanostructured Pd particles using a seeding synthesis approach—Application to the selective hydrogenation of buta-1,3-diene. *Applied Catalysis, A: General*, 2007, *327* (1), 32–43.

37. Urchaga, P.; Baranton, S.; Napporn, T. W.; Coutanceau, C. Selective syntheses and electrochemical characterization of platinum nanocubes and nanotetrahedrons/octahedrons. *Electrocatalysis*, 2010, *1* (1), 3–6.

38. Mao, J.; Li, X.-L.; Qin, W.-J.; Niu, K.-Y.; Yang, J.; Ling, T.; Du, X.-W. Control of the morphology and optical properties of ZnO nanostructures via hot mixing of reverse micelles. *Langmuir*, 2010, *26* (17), 13755–13759.

39. Guo, Y.-G.; Lee, J.-S.; Maier, J. Preparation and characterization of AgI nanoparticles with controlled size, morphology and crystal structure. *Solid State Ionics*, 2006, *177* (26–32), 2467–2471.

40. Ng, C. H. B.; Fan, W. Y. Facile synthesis of single-crystalline γ-CuI nanotetrahedrons and their induced transformation to tetrahedral CuO nanocages. *Journal of Physical Chemistry C*, 2007, *111* (26), 9166–9171.

41. Chu, H.; Li, X.; Chen, G.; Zhou, W.; Zhang, Y.; Jin, Z.; Xu, J.; Li, Y. Shape-controlled synthesis of CdS nanocrystals in mixed solvents. *Crystal Growth and Design*, 2005, *5* (5), 1801–1806.

42. Stockman, M. I.; Li, K.; Brasselet, S.; Zyss, J. Octupolar metal nanoparticles as optically driven, coherently controlled nanorotors. *Chemical Physics Letters*, 2006, *433* (1–3), 130–135.

43. Kim, D.; Heo, J.; Kim, M.; Lee, Y. W.; Han, S. W. Size-controlled synthesis of monodisperse gold nanooctahedrons and their surface-enhanced Raman scattering properties. *Chemical Physics Letters*, 2009, *468* (4–6), 245–248.

44. Lee, Y.-W.; Kim, M.; Kim, Z.-H.; Han, S.-W. One-step synthesis of Au@Pd core-shell nanooctahedron. *Journal of the American Chemical Society*, 2009, *131* (47), 17036–17037.

45. Geng, B.; Zhan, F.; Fang, C.; Yu, N. A facile coordination compound precursor route to controlled synthesis of Co_3O_4 nanostructures and their room-temperature gas sensing properties. *Journal of Materials Chemistry*, 2008, *18* (41), 4977–4984.

46. Mustafa, D. E.; Yang, T.; Xuan, Z.; Chen, S.; Tu, H.; Zhang, A. Surface plasmon coupling effect of gold nanoparticles with different shape and size on conventional surface plasmon resonance signal. *Plasmonics*, 2010, *5* (3), 221–231.

47. Li, W.; Xia, Y. Facile synthesis of gold octahedra by direct reduction of HAuCl4 in an aqueous solution. *Chemistry—An Asian Journal*, 2010, *5* (6), 1312–1316.

48. Wang, F.; Wang, J. Step-by-step heteroepitaxial growth of Pd@Au@Pd@Au and Au@Pd@Au@Pd core-multishell nanocrystals. *Abstracts of Papers, 240th ACS National Meeting*, Boston, MA, August 22–26, 2010, INOR-749.

49. Tang, S. C.; Meng, X. K.; Vongehr, S. An additive-free electrochemical route to rapid synthesis of large-area copper nano-octahedra on gold film substrates. *Electrochemistry Communications*, 2009, *11* (4), 867–870.

50. Shavel, A.; Rodriguez-Gonzalez, B.; Pacifico, J.; Spasova, M.; Farle, M.; Liz-Marzan, L. M. Shape control in iron oxide nanocrystal synthesis, induced by trioctylammonium ions. *Chemistry of Materials*, 2009, *21* (7), 1326–1332.

51. Li, L.; Yang, Y.; Ding, J.; Xue, J. Synthesis of magnetite nanooctahedra and their magnetic field-induced two-/three-dimensional superstructure. *Chemistry of Materials*, 2010, *22* (10), 3183–3191.

52. Ping, H.; Xinghai, S.; Hongcheng, G. Size-controlled preparation of Cu_2O octahedron nanocrystals and studies on their optical absorption. *Journal of Colloid and Interface Science*, 2005, *284*, 510–515.

53. Radi, A.; Pradhan, D.; Sohn, Y.; Leung, K. T. Nanoscale shape and size control of cubic, cuboctahedral, and octahedral Cu-Cu_2O core-shell nanoparticles on Si(100) by one-step, templateless, capping-agent-free electrodeposition. *ACS Nano*, 2010, *4* (3), 1553–1560.

54. Yang, Z.; Han, D.; Ma, D.; Liang, H.; Liu, L.; Yang, Y. Fabrication of monodisperse CeO_2 hollow spheres assembled by nano-octahedra. *Crystal Growth and Design*, 2010, *10* (1), 291–295.

55. Lai, Y.; Ranbo, Y.; Jun, C.; Xianran, X. Template-free hydrothermal synthesis of CeO_2 nano-octahedrons and nanorods: Investigation of the morphology evolution. *Crystal Growth and Design*, 2008, *8* (5), 1474–1477.

56. Li, Y.; Mi, Y.; Jiang, J.; Huang, Z. Room-temperature synthesis of $CdMoO_4$ nanooctahedra in the hemline length of 30 nm. *Chemistry Letters*, 2010, *39* (7), 760–761.

57. Zheng, S. J.; Wang, Y. J.; Zhang, B.; Zhu, Y. L.; Liu, C.; Hub, P.; Ma, X. L. Identification of $MnCr_2O_4$ nano-octahedron in catalyzing pitting corrosion of austenitic stainless steels. *Acta Materialia*, 2010, *58*, 5070–5085.

58. Yao, C.; Yuanhui, Z.; Yuansheng, W.; Feng, B.; Yong, Q. Synthesis and magnetic properties of nickel ferrite nano-octahedra. *Journal of Solid State Chemistry*, 2005, *178*, 2394–2397.

59. Enyashin, A. N.; Bar-Sadan, M.; Sloan, J.; Houben, L.; Seifert, G. Nanoseashells and nanooctahedra of MoS_2: Routes to inorganic fullerenes. *Chemistry of Materials*, 2009, *21* (23), 5627–5636.

60. Parilla, P. A.; Dillon, A. C.; Parkinson, B. A.; Jones, K. M.; Alleman, J.; Riker, G.; Ginley, D. S.; Heben, M. J. Formation of nanooctahedra in molybdenum disulfide and molybdenum diselenide using pulsed laser vaporization. *Journal of Physical Chemistry B*, 2004, *108*, 6197–6207.

61. Ferreira, P. J.; Yang, S.-H. Formation mechanism of Pt single-crystal nanoparticles in proton exchange membrane fuel cells. *Electrochemical and Solid-State Letters*, 2007, *10* (3), B60–B63.

62. Ren, R.; Xu, J.; Ren, D.-N. Study on the synthesis of Pd/Au nanostructure and spectral, characteristics of particle size composition and spins. *Guangpuxue Yu Guangpu Fenxi*, 2010, *30* (7), 1858–1861.

63. Song, Q.; Ding, Y.; Wang, Z. L.; Zhang, Z. J. Formation of orientation-ordered superlattices of magnetite magnetic nanocrystals from shape-segregated self-assemblies. *Journal of Physical Chemistry B*, 2006, *110* (50), 25547–25550.

64. Shan, G.-B.; Demopoulos, G. P. The synthesis of aqueous-dispersible anatase TiO_2 nanoplatelets. *Nanotechnology*, 2010, *21* (2), 025604/1–025604/9.

65. Guo, L.; Zhang, H.; Zhang, D. Preparation method of polyhedron-like cuprous oxide nanoparticle. 2009, Patent CN 101348275, 10 pp.

66. Dang, F.; Kato, K.; Imai, H.; Wada, S.; Haneda, H.; Kuwabara, M. Characteristics of CeO_2 nanocubes and related polyhedra prepared by using a liquid-liquid interface. *Crystal Growth and Design*, 2010, *10* (10), 4537–4541.

67. Hu, Y.; Gu, H.; Hu, Z.; Di, W.; Yuan, Y.; You, J.; Cao, W.; Wang, Y.; Chan, H. L. W. Controllable hydrothermal synthesis of $KTa_{1-x}Nb_xO_3$ nanostructures with various morphologies and their growth mechanisms. *Crystal Growth and Design*, 2008, *8* (3), 832–837.

68. Yang, L.; Yu, H.; Xu, L.; Ma, Q.; Qian, Y. Sulfur-assisted synthesis of nitride nanocrystals. *Dalton Transactions*, 2010, *39* (11), 2855–2860.

69. Wang, N.; Cao, X.; Guo, L.; Yang, S.; Wu, Z. Facile synthesis of PbS truncated octahedron crystals with high symmetry and their large-scale assembly into regular patterns by a simple solution route. *ACS Nano*, 2008, *2* (2), 184–190.

70. Cheng, C.; Xu, G.; Zhang, H. Facile solvothermal synthesis of nanostructured PbSe with anisotropic shape: Nanocubes, submicrometer cubes and truncated octahedron. *Journal of Crystal Growth*, 2009, *311* (5), 1285–1290.

71. Stoerzinger, K. A.; Hasan, W.; Lin, J. Y.; Robles, A.; Odom, T. W. Screening nanopyramid assemblies to optimize surface enhanced Raman scattering. *Journal of Physical Chemistry Letters*, 2010, *1* (7), 1046–1050.

72. Hasan, W.; Lee, J.; Henzie, J.; Odom, T. W. Selective functionalization and spectral identification of gold nanopyramids. *Journal of Physical Chemistry C*, 2007, *111* (46), 17176–17179.

73. Tian, Y.; Liu, H.; Deng, Z. Electrochemical growth of gold pyramidal nanostructures: Toward super-amphiphobic surfaces. *Chemistry of Materials*, 2006, *18* (25), 5820–5822.

74. Tian, Y.; Liu, H.; Zhao, G.; Tatsuma, T. Shape-controlled electrodeposition of gold nanostructures. *Journal of Physical Chemistry B*, 2006, *110* (46), 23478–23481.

75. Burgin, J.; Liu, M.; Guyot-Sionnest, P. Dielectric sensing with deposited gold bipyramids. *Journal of Physical Chemistry C*, 2008, *112* (49), 19279–19282.

76. Liu, H.; Tian, Y.; Xia, P. Pyramidal, rodlike, spherical gold nanostructures for direct electron transfer of copper, zinc-superoxide dismutase: Application to superoxide anion biosensors. *Langmuir*, 2008, *24* (12), 6359–6366.

77. Tzung-Hua, L.; Linn, N. C.; Tarajano, L.; Bin, J.; Peng, J. Electrochemical SERS at periodic metallic nanopyramid arrays. *Journal of Physical Chemistry C*, 2009, *113* (4), 1367–1372.

78. Chung, P.-Y.; Lin, T.-H.; Schultz, G.; Jiang, P.; Batich, C. Gold nanopyramidal plasmonic crystals for label-free biosensing. *Proceedings of SPIE*, 2010, *7759* (Biosensing III), 775908/1–775908/7.

79. Chung, P.-Y.; Lin, T.-H.; Schultz, G.; Batich, C.; Jiang, P. Nanopyramid surface plasmon resonance sensors. *Applied Physics Letters*, 2010, *96* (26), 261108/1–261108/3.

80. Xia, P.; Liu, H.; Tian, Y. Cathodic detection of H_2O_2 based on nanopyramidal gold surface with enhanced electron transfer of myoglobin. *Biosensors and Bioelectronics*, 2009, *24* (8), 2470–2474.

81. Gong-Ru, L.; Chi-Kuan, L.; Li-Jen, C.; Yu-Lun, C. Synthesis of Si nanopyramids at SiO_x/Si interface for enhancing electroluminescence of Si-rich SiO_x. *Applied Physics Letters*, 2006, *89*, 093126.

82. Guo, G.; Shi, C.; Tao, D.; Qian, W.; Han, D. Synthesis of well-dispersed ZnO nanomaterials by directly calcining zinc stearate. *Journal of Alloys and Compounds*, 2009, *472* (1–2), 343–346.

83. Du, Y.-P.; Zhang, Y.-W.; Sun, L.-D.; Yan, C.-H. Efficient energy transfer in monodisperse Eu-doped ZnO nanocrystals synthesized from metal acetylacetonates in high-boiling solvents. *Journal of Physical Chemistry C*, 2008, *112* (32), 12234–12241.

84. Cvelbar, U.; Ostrikov, K.; Mozetic, M. Reactive oxygen plasma-enabled synthesis of nanostructured CdO: Tailoring nanostructures through plasma-surface interactions. *Nanotechnology*, 2008, *19* (40), 405605/1–405605/7.

85. Maestre, D.; Martinez de Velasco, I.; Cremades, A.; Amati, M.; Piqueras, J. Micro- and nano-pyramids of manganese-doped indium oxide. *Journal of Physical Chemistry C*, 2010, *114* (27), 11748–11752.

86. Das, S.; Chaudhuri, S.; Maji, S. Ethanol-water mediated solvothermal synthesis of cube and pyramid shaped nanostructured tin oxide. *Journal of Physical Chemistry C*, 2008, *112* (16), 6213–6219.

87. Carretero-Genevrier, A.; Gazquez, J.; Puig, T.; Mestres, N.; Sandiumenge, F.; Obradors, X.; Ferain, E. Vertical (La,Sr)MnO₃ nanorods from track-etched polymers directly buffering substrates. *Advanced Functional Materials*, 2010, *20* (6), 892–897.

88. He, L.; Liang, J.; Shen, Z. Method for preparing $SrAl_2O_4$ nanomaterial. CN 101319377, 2008, 16 pp.

89. Sun, Y.; Cui, H.; Yang, G. Z.; Huang, H.; Jiang, D.; Wang, C. X. The synthesis and mechanism investigations of morphology controllable 1-D SiC nanostructures via a novel approach. *Crystal Engineering Communications*, 2010, *12* (4), 1134–1138.

90. Zakharov, D. N.; Colby, R.; Wildeson, I. H.; Ewoldt, D. A.; Liang, Z.; Zaluzec, N.; Garcia, R. E.; Sands, T. D.; Stach, E. A. Transmission electron microscopy study of defects in nanopyramidal InGaN LEDs structures. *Microscopy and Microanalysis*, 2010, *16* (Suppl. 2), 1508–1509.

91. Yuan, J.; Wang, H.; Van Veldhoven, P. J.; Noetzel, R. Impact of base size and shape on formation control of multifaceted InP nanopyramids by selective area metal organic vapor phase epitaxy. *Journal of Applied Physics*, 2009, *106* (12), 124304/1–124304/4.

92. Liu, C.; Satka, A.; Krishnan Jagadamma, L.; Edwards, P. R.; Allsopp, D.; Martin, R. W.; Shields, P.; Kovac, J.; Uherek, F.; Wang, W. Light emission from InGaN quantum wells grown on the facets of closely spaced GaN nano-pyramids formed by nano-imprinting. *Applied Physics Express*, 2009, *2*, 121002.

93. Zheng, J.; Song, X.; Yu, B.; Li, X. Asymmetrical AlN nanopyramids induced by polar surfaces. *Applied Physics Letters*, 2007, *90* (19), 193121/1–193121/3.

94. Jindal, V.; Grandusky, J.; Jamil, M.; Tripathi, N.; Thiel, B.; Shahedipour-Sandvik, F.; Balch, J.; LeBoeuf, S. Effect of interfacial strain on the formation of AlGaN nanostructures by selective area heteroepitaxy. *Physica E: Low-Dimensional Systems and Nanostructures*, 2008, *40* (3), 478–483.

95. Wang, C.-W.; Liu, H.-G.; Bai, X.-T.; Xue, Q.; Chen, X.; Lee, Y.-I.; Hao, J.; Jiang, J. Triangular PbS nano-pyramids, square nanoplates, and nanorods formed at the air/water interface. *Crystal Growth and Design*, 2008, *8* (8), 2660–2664.

96. Xin, G.-Q.; Ding, H.-P.; Yang, Y.-G.; Shen, S.-L.; Xiong, Z.-C.; Chen, X.; Hao, J.; Liu, H.-G. Triangular single-crystalline nanorings of PbS formed at the air/water interface. *Crystal Growth and Design*, 2009, *9* (4), 2008–2012.

97. Tai, G.; Guo, W. Sonochemistry-assisted microwave synthesis and optical study of single-crystalline CdS nanoflowers. *Ultrasonics Sonochemistry*, 2008, *15* (4), 350–356.

98. Apte, S. K.; Garaje, S. N.; Bolade, R. D.; Ambekar, J. D.; Kulkarni, M. V.; Naik, S. D.; Gosavi, S. W.; Baeg, J. O.; Kale, B. B. Hierarchical nanostructures of $CdIn_2S_4$ via hydrothermal and microwave methods: Efficient solar-light-driven photocatalysts. *Journal of Materials Chemistry*, 2010, *20* (29), 6095–6102.

99. Mayoral, A.; Barron, H.; Estrada-Salas, R.; Vazquez-Duran, A.; Jose-Yacaman, M. Nanoparticle stability from the nano to the meso interval. *Nanoscale*, 2010, *2*, 335–342.

100. Tao, L.; Lin, X.; Jing, Z.; Huimin, Y.; Hengqiang, Y.; Rong, Y.; Zhiying, C. et al. Icosahedral face-centered cubic Fe nanoparticles: Facile synthesis and characterization with aberration-corrected TEM. *Nano Letters*, 2009, *9* (4), 1572–1576.

101. Rollmann, G.; Gruner, M. E.; Hucht, A.; Meyer, R.; Entel, P.; Tiago, M. L.; Chelikowsky, J. R. Shell-wise Mackay transformation in iron nano-clusters. *Physical Review Letters*, 2007, *99* (8), 083402, 4 pp.

102. Silly, F.; Castell, M. R. Growth of Ag icosahedral nanocrystals on a $SrTiO_3$ (001) support. *Applied Physics Letters*, 2005, *87*, 213107.

103. Yujie, X.; Younan, X. Shape-controlled synthesis of metal nanostructures: The case of palladium. *Advanced Materials*, 2007, *19*, 3385–3391.

104. Burt, J. L.; Elechiguerra, J. L.; Reyes-Gasga, J.; Martin Montejano-Carrizales, J.; Jose-Yacaman, M. Beyond Archimedean solids: Star polyhedral gold nanocrystals. *Journal of Crystal Growth*, 2005, *285* (4), 681–691.

105. Yi, G.; Nan, S.; Yong, P.; Zeng, X. C. Icosahedral crown gold nanocluster $Au_{43}Cu_{12}$ with high catalytic activity. *Nano Letters*, 2010, *10*, 1055–1062.

106. Saida, J.; Inoue, A. Microstructure of tensile fracture in nanoicosahedral quasicrystal dispersed $Zr_{80}Pt_{20}$ amorphous alloy. *Scripta Materialia*, 2004, *50* (10), 1297–1301.

107. Saida, J.; Matsushita, M.; Inoue, A. Nanoicosahedral quasicrystalline phase in Zr-Pd and Zr-Pt binary alloys. *Journal of Applied Physics*, 2001, *90* (9), 4717–4724.

108. Saida, J.; Matsushita, M.; Inoue, A. Direct observation of icosahedral cluster in $Zr_{70}Pd_{30}$ binary glassy alloy. *Applied Physics Letters*, 2001, *79* (3), 412–414.

109. Louzguine, D. V.; Inoue, A. Gold as an alloying element promoting formation of a nanoicosahedral phase in a Cu-based alloy. *Journal of Alloys and Compounds*, 2003, *361* (1–2), 153–156.

110. Louzguine, D. V.; Kato, H.; Inoue, A. Investigation of mechanical properties and devitrification of Cu-based bulk glass formers alloyed with noble metals. *Science and Technology of Advanced Materials*, 2003, *4* (4), 327–331.

111. Louzguine-Luzgin, D. V.; Inoue, A.; Nagahama, D.; Hono, K. Composition and structure of Cu-based nanoicosahedral phase in Cu-Zr-Ti-Pd alloy. *Applied Physics Letters*, 2005, *87* (21), 211918/1–211918/3.

112. Saida, J.; Matsushita, M.; Inoue, A. Nucleation and grain growth kinetics of nano icosahedral quasicrystalline phase in $Zr_{65}Al_{7.5}Ni_{10}Cu_{17.5-x}Pd_x$ (x = 5, 10 and 17.5) glassy alloys. *Materials Transactions, JIM*, 2000, *41* (11), 1505–1510.

113. Tytik, D. L.; Belashchenko, D. K.; Sirenko, A. N. Structural transformations in silver nanoparticles. *Journal of Structural Chemistry*, 2008, *49* (1), 109–116.

114. Montejano-Carrizales, J. M.; Rodríguez-López, J. L.; Pal, U.; Miki-Yoshida, M.; Jose-Yacaman, M. The completion of the Platonic atomic polyhedra: The dodecahedron. *Small*, 2006, *2* (3), 351–355.

115. You, T.; Wang, D. Method for preparing quashed dodecahedral gold nanocrystals with high surface energy and high catalytic performance. Patent CN 101775647, 2010, 13 pp.

116. Jeong, G. H.; Kim, M.; Lee, Y. W.; Choi, W.; Oh, W. T.; Park, Q.-H.; Han, S. W. Polyhedral Au nanocrystals exclusively bound by {110} facets: The rhombic dodecahedron. *Journal of the American Chemical Society*, 2009, *131* (5), 1672–1673.

117. Meng, H.; Sun, S.; Masse, J.-P.; Dodelet, J.-P. Electrosynthesis of Pd single-crystal nanothorns and their application in the oxidation of formic acid. *Chemistry of Materials*, 2008, *20* (22), 6998–7002.

118. Kido, O.; Kamitsuji, K.; Kurumada, M.; Sato, T.; Kimura, Y.; Suzuki, H.; Saito, Y.; Kaito, C. Morphological alteration upon phase transition and effects of oxygen impurities of chromium nanoparticles. *Journal of Crystal Growth*, 2005, *275* (1–2), e1745–e1750.

119. Chen, F.; Gao, Q.; Hong, G.; Ni, J. Synthesis and characterization of magnetite dodecahedron nanostructure by hydrothermal method. *Journal of Magnetism and Magnetic Materials*, 2008, *320* (11), 1775–1780.

120. Yuan, A.-Q.; Wu, J.; Chen, J.; Liang, R.-L.; Huang, Z.-Y.; Tong, Z.-F. Thermochemical properties and decomposition kinetics of nanoparticle NH_4ZnPO_4 with dodecahedron morphology. *Yingyong Huaxue*, 2007, *24* (1), 12–16.

121. Yu, S.-H.; Yoshimura, M. Fabrication of powders and thin films of various nickel sulfides by soft solution-processing routes. *Advanced Functional Materials*, 2002, *12* (4), 277–285.

122. Kawaguchi, T.; Tanigaki, K.; Yasukawa, M. Ferromagnetism in germanium clathrate: $Ba_8Mn_2Ge_{44}$. *Applied Physics Letters*, 2000, *77* (21), 3438–3440.

123. Lu, X.; Tuan, H.-Y.; Chen, J.; Li, Z.-Y.; Korgel, B. A.; Xia, Y. Mechanistic Studies on the galvanic replacement reaction between multiply twinned particles of Ag and $HAuCl_4$ in an organic medium. *Journal of the American Chemical Society*, 2007, *129* (6), 1733–1742.

124. Barnard, A. S.; Curtiss, L. A. Predicting the shape and structure of face-centered cubic gold nanocrystals smaller than 3 nm. *Chemical Physics Letters*, 2006, *7* (7), 1544–1553.

125. Qi, W. H.; Huang, B. Y.; Wang, M. P.; Yin, Z. M.; Li, J. Molecular dynamic simulation of the size- and shape-dependent lattice parameter of small platinum nanoparticles. *Journal of Nanoparticle Research*, 2009, *11* (3), 575–580.

126. Lim, S. I.; Varon, M.; Ojea-Jimenez, I.; Arbiol, J.; Puntes, V. Exploring the limitations of the use of competing reducers to control the morphology and composition of Pt and Pt-Co nanocrystals. *Chemistry of Materials*, 2010, *22* (15), 4495–4504.

127. Obuya, E.; Harrigan, W.; Jones, W. E. Electronic structure modification of electrospun titania by Rhodium nanoparticles. *Abstracts of Papers*, 240th ACS National Meeting, Boston, MA, August 22–26, 2010, INOR-509.

128. Joe, M.; Kim, S.-P.; Lee, K.-R. Study on the phase transition behavior of Ni nano-clusters using molecular dynamics simulation. *Journal of Computational and Theoretical Nanoscience*, 2009, *6* (11), 2442–2445.

129. Lim, S. I.; Ojea-Jimenez, I.; Varon, M.; Casals, E.; Arbiol, J.; Puntes, V. Synthesis of platinum cubes, polypods, cuboctahedrons, and raspberries assisted by cobalt nanocrystals. *Nano Letters*, 2010, *10* (3), 964–973.

130. Chou, S.-W.; Zhu, C.-L.; Neeleshwar, S.; Chen, C.-L.; Chen, Y.-Y.; Chen, C.-C. Controlled growth and magnetic property of FePt nanostructure: Cuboctahedron, octapod, truncated cube, and cube. *Chemistry of Materials*, 2009, *21* (20), 4955–4961.

131. Yuge, K. Segregation of $Pt_{28}Rh_{27}$ bimetallic nanoparticles: A first-principles study. *Journal of Physics: Condensed Matter*, 2010, *22* (24), 245401/1–245401/6.

132. Grove, D. E.; Gupta, U.; Castleman, A. W., Jr. Effect of hydrocarbons on the morphology of synthesized niobium carbide nanoparticles. *Langmuir*, 2010, *26* (21), 16517–16521.

133. Grove, D. E.; Gupta, U.; Castleman, A. W., Jr. Effect of carbon concentration on changing the morphology of titanium carbide nanoparticles from cubic to cuboctahedron. *ACS Nano*, 2010, *4* (1), 49–54.

134. Chen, J.; Wu, C.; Tian, J.; Li, W.; Yu, S.; Tian, Y. Three-dimensional imaging of a complex concaved cuboctahedron copper sulfide crystal by x-ray nanotomography. *Applied Physics Letters*, 2008, *92* (23), 233104/1–233104/3.

135. Wan, Y.; Wu, C.; Min, Y.; Yu, S.-H. Loading of silica spheres on concaved CuS cuboctahedrons using 3-aminopropyltrimethoxysilane as a linker. *Langmuir*, 2007, *23* (16), 8526–8530.

136. Ringe, E.; McMahon, J. M.; Sohn, K.; Cobley, C.; Xia, Y.; Huang, J.; Schatz, G. C.; Marks, L. D.; Van Duyne, R. P. Unraveling the effects of size, composition, and substrate on the localized surface plasmon resonance frequencies of gold and silver nanocubes: A systematic single-particle approach. *Journal of Physical Chemistry C*, 2010, *114* (29), 12511–12516.

137. Lee, S. Y.; Hung, L.; Lang, G. S.; Cornett, J. E.; Mayergoyz, I. D.; Rabin, O. Dispersion in the SERS enhancement with silver nanocube dimers. *ACS Nano*, 2010, *4* (10), 5763–5772.

138. Siekkinen, A. R.; McLellan, J. M.; Chen, J.; Xia, Y. Rapid synthesis of small silver nanocubes by mediating polyol reduction with a trace amount of sodium sulfide or sodium hydrosulfide. *Chemical Physics Letters*, 2006, *432*, 491–496.

139. Yu, Y.; Zhao, Y.; Huang, T.; Liu, H. Microwave-assisted synthesis of palladium nanocubes and nanobars. *Materials Research Bulletin*, 2010, *45* (2), 159–164.

140. Chien-Jung, H.; Yeong-Her, W.; Pin-Hsiang, C.; Ming-Chang, S.; Teen-Hang, M. Electrochemical synthesis of gold nanocubes. *Materials Letters*, 2006, *60*, 1896–1900.

141. Liu, Y.; Hight Walker, A. R. Monodisperse gold-copper bimetallic nanocubes: Facile one-step synthesis with controllable size and composition. *Angewandte Chemie, International Edition*, 2010, *49* (38), 6781–6785, S6781/1–S6781/24.

142. Claussen, J. C.; Franklin, A. D.; Fisher, T. S.; Porterfield, D. M. Electrochemical biosensor. U.S. Patent US 2010285514, 2010, 47 pp.

143. Yang, H.; Zhang, J.; Sun, K.; Zou, S.; Fang, Je. Enhancing by weakening: Electrooxidation of methanol on Pt3Co and Pt nanocubes. *Angewandte Chemie, International Edition*, 2010, *49* (38), 6848–6851, S6848/1–S6848/8.

144. Kang, Y.; Murray, C. B. Synthesis and electrocatalytic properties of cubic Mn-Pt nanocrystals (nanocubes). *Journal of the American Chemical Society*, 2010, *132* (22), 7568–7569.

145. Choi, S.-I.; Choi, R.; Han, S. W.; Park, J. T. Synthesis and characterization of Pt9Co nanocubes with high activity for oxygen reduction. *Chemical Communications*, 2010, *46* (27), 4950–4952.

146. Min, C.; Jaemin, K.; Liu, J. P.; Hongyou, F.; Shouheng, S. Synthesis of FePt nanocubes and their oriented self-assembly. *Journal of the American Chemical Society*, 2006, *128*, 7132–7133

147. Chen, W; Chen, C.; Guo, L. Magnetization reversal of two-dimensional superlattices of Mn3O4 nanocubes and their collective dipolar interaction effects. *Journal of Applied Physics*, 2010, *108* (4), 043912/1–043912/8.

148. Kang, H.; Jung, H. S.; Kim, J. Y.; Park, J. C.; Kim, M.; Song, H.; Park, K. H. Immobilized CuO hollow nanospheres catalyzed alkyne-azide cycloadditions. *Journal of Nanoscience and Nanotechnology*, 2010, *10* (10), 6504–6509.

149. Song, Y.-J.; Han, S.-B.; Lee, H.-H.; Park, K.-W. Size-controlled Cu2O nanocubes by pulse electrodeposition. *Journal of the Korean Electrochemical Society*, 2010, *13* (1), 40–44.

150. Uhm, Y. R.; Park, J. H.; Kim, W. W.; Lee, M. K.; Rhee, C. K. Novel synthesis of Cu2O nano cubes derived from hydrolysis of Cu nano powder and its high catalytic effects. *Materials Science and Engineering A*, 2007, *449–451*, 817–820.

151. Qin, W.; Yang, C.; Yi, R.; Gao, G. Hydrothermal synthesis and characterization of single-crystalline α-Fe2O3 nanocubes. *Journal of Nanomaterials*, 2011, Article ID 159259, 5 pp.

152. Jiang, F. Y.; Wang, Ch. M.; Fu, Y.; Liu, R. C. Synthesis of iron oxide nanocubes via microwave-assisted solvothermal method. *Journal of Alloys and Compounds*, 2010, *503* (2), L31–L33.

153. Gao, G.; Liu, X.; Shi, R.; Zhou, K.; Shi, Y.; Ma, R.; Takayama-Muromachi, E.; Qiu, G. Shape-controlled synthesis and magnetic properties of monodisperse Fe3O4 nanocubes. *Crystal Growth and Design*, 2010, *10* (7), 2888–2894.

154. Hai, H. T.; Kura, H.; Takahashi, M.; Ogawa, T. Phase transformation of FeO/Fe3O4 core/shell nanocubes and facile synthesis of Fe3O4 nanocubes. *Journal of Applied Physics*, 2010, *107* (9, Pt. 2), 09E301/1–09E301/3.

155. Hu, L.; Sun, K.; Peng, Q.; Xu, B.; Li, Y. Surface active sites on Co3O4 nanobelt and nanocube model catalysts for CO oxidation. *Nano Research*, 2010, *3* (5), 363–368.

156. Zhiqiang, Y.; Kebin, Z.; Xiangwen, L.; Qun, T.; Deyi, L.; Sen, Y. Single-crystalline ceria nanocubes: Size-controlled synthesis, characterization and redox property. *Nanotechnology*, 2007, *18*, 185606, 4 pp.

157. Lv, J.; Shen, Y.; Peng, L.; Guo, X.; Ding, W. Exclusively selective oxidation of toluene to benzaldehyde on ceria nanocubes by molecular oxygen. *Chemical Communications*, 2010, *46* (32), 5909–5911.

158. Motta, F. V.; Lima, R. C.; Marques, A. P. A.; Li, M. S.; Leite, E. R.; Varela, J. A.; Longo, E. Indium hydroxide nanocubes and microcubes obtained by microwave-assisted hydrothermal method. *Journal of Alloys and Compounds*, 2010, *497* (1–2), L25–L28.

159. Rabuffetti, F. A.; Kim, H. S.; Enterkin, J. A.; Wang, Y.; Lanier, C. H.; Marks, L. D.; Poeppelmeier, K. R.; Stair, P. C. Synthesis-dependent first-order Raman scattering in SrTiO₃ nanocubes at room temperature. *Chemistry of Materials*, 2008, *20*, 5628–5635.

160. Yang, X.; Fu, J.; Jin, C.; Chen, J.; Liang, C.; Wu, M.; Zhou, W. Formation mechanism of CaTiO₃ hollow crystals with different microstructures. *Journal of the American Chemical Society*, 2010, *132* (40), 14279–14287.

161. Yun, B. K.; Koo, Y. S.; Jung, J. H.; Song, M.; Yoon, S. Possible role of hydroxyl group on local structure and phase transition of KNbO₃ and KTaO₃ nanocrystals. *Physica B: Condensed Matter*, 2010, *405* (23), 4866–4870.

162. Swaminathan, V.; Pramana, Stevin, S.; White, T. J.; Chen, L.; Chukka, R.; Ramanujan, R. V. Microwave synthesis of noncentrosymmetric BaTiO₃ truncated nanocubes for charge storage applications. *ACS Applied Materials and Interfaces*, 2010, *2* (11), 3037–3042.

163. Urban, J. J.; Ouyang, L.; Jo, M.-H.; Wang, D. S.; Park, H. Synthesis of single-crystalline La₁₋ₓBaₓMnO₃ nanocubes with adjustable doping levels. *Nano Letters*, 2004, *4* (8), 1547–1550.

164. Paul, G. S.; Agarwal, P. Structural, optical and thermal studies on PbS nanocubes. *Physica Status Solidi C: Current Topics in Solid State Physics*, 2010, *7* (3–4), 905–908.

165. Wang, X.; Li, K.; Dong, Y.; Jiang, K. Preparation and characterization of monodispersed PbSe nanocubes. *Crystal Research and Technology*, 2010, *45* (1), 94–98.

166. Weigang, L.; Jiye, F.; Yong, D.; Zhong Lin, W. Formation of PbSe nanocrystals: A growth toward nanocubes. *Journal of Physical Chemistry B*, 2005, *109*, 19219–19222.

167. Wan, B.-Y.; Hu, C.-G.; Xi, Y.; Xu, J.; He, X.-S. Room-temperature synthesis and seebeck effect of lead chalcogenide nanocubes. *Solid State Sciences*, 2010, *12* (1), 123–127.

168. Wang, L.; Chen, G.; Wang, Q.; Zhang, H.; Jin, R.; Chen, D.; Meng, X. Shape-controlled synthesis and electrical conductivities of AgPb₁₀SbTe₁₂ materials. *Journal of Physical Chemistry C*, 2010, *114* (13), 5827–5834.

169. Seungwook, K.; Haegeun, C.; Kwon, J. H.; Yoon, H. G.; Kim, W. Facile synthesis of silver chloride nanocubes and their derivatives. *Bulletin of the Korean Chemical Society*, 2010, *31* (10), 2918–2922.

170. Zhang, C.; Chen, J. Facile EG/ionic liquid interfacial synthesis of uniform Re³⁺ doped NaYF₄ nanocubes. *Chemical Communications*, 2010, *46* (4), 592–594.

171. Xu, J. F.; Liu, H.; Liu, P.; Liang, C. H.; Wang, Q.; Fang, J.; Zhao, J. H.; Shen, W. G. Magnetism study of Cl$_x$Co$_y$[Fe(CN)₆]. *z*H₂O (CI=Rb,Cs) Prussian blue nanoparticles. *Journal of the Iranian Chemical Society*, 2010, *7* (Suppl.), S123–S129.

172. Dong, X.; Ji, X.; Jing, J.; Li, M.; Li, J.; Yang, W. Synthesis of triangular silver nanoprisms by stepwise reduction of sodium borohydride and trisodium citrate. *Journal of Physical Chemistry C*, 2010, *114* (5), 2070–2074.

173. Ciou, S.-H.; Cao, Y.-W.; Huang, H.-C.; Su, D.-Y.; Huang, C.-L. SERS enhancement factors studies of silver nanoprism and spherical nanoparticle colloids in the presence of bromide ions. *Journal of Physical Chemistry C*, 2009, *113* (22), 9520–9525.

174. Aherne, D.; Charles, D. E.; Brennan-Fournet, M. E.; Kelly, J. M.; Gun'ko, Y. K. Etching-resistant silver nanoprisms by epitaxial deposition of a protecting layer of gold at the edges. *Langmuir*, 2009, *25* (17), 10165–10173.

175. Nelayah, J.; Kociak, M.; Stephan, O.; Geuquet, N.; Henrard, L.; Garcia de Abajo, F. J.; Pastoriza-Santos, I.; Liz-Marzan, L. M.; Colliex, C. Two-dimensional quasistatic stationary short range surface plasmons in flat nanoprisms. *Nano Letters*, 2010, *10* (3), 902–907.

176. Yang, P.; Portales, H.; Pileni, M.-P. Identification of multipolar surface plasmon resonances in triangular silver nanoprisms with very high aspect ratios using the DDA method. *Journal of Physical Chemistry C*, 2009, *113* (27), 11597–11604.

177. Xue, C.; Metraux, G. S.; Millstone, J. E.; Mirkin, C. A. Mechanistic study of photomediated triangular silver nanoprism growth. *Journal of the American Chemical Society*, 2008, *130* (26), 8337–8344.

178. Lee, B.-H.; Hsu, M.-S.; Hsu, Y.-C.; Lo, C.-W.; Huang, C.-L. A facile method to obtain highly stable silver nanoplate colloids with desired surface plasmon resonance wavelengths. *Journal of Physical Chemistry C*, 2010, *114* (14), 6222–6227.

179. Banholzer, M. J.; Harris, N.; Millstone, J. E.; Schatz, G. C.; Mirkin, C. A. Abnormally large plasmonic shifts in silica-protected gold triangular nanoprisms. *Journal of Physical Chemistry C*, 2010, *114* (16), 7521–7526.

180. Kundu, S.; Lau, S.; Liang, H. Shape-controlled catalysis by cetyltrimethylammonium bromide terminated gold nanospheres, nanorods, and nanoprisms. *Journal of Physical Chemistry C*, 2009, *113* (13), 5150–5156.

181. Hill, H. D.; Hurst, S. J.; Mirkin, C. A. Curvature-induced base pair "slipping" effects in DNA-nanoparticle hybridization. *Nano Letters*, 2009, *9* (1), 317–321.

182. Yingzhou, H.; Wenzhong, W.; Hongyan, L.; Hongxing, X. Surfactant-promoted reductive synthesis of shape-controlled gold nanostructures. *Crystal Growth and Design*, 2009, *9* (2), 858–862.

183. Millstone, J. E.; Hurst, S. J.; Metraux, G. S.; Cutler, J. I.; Mirkin, C.d.A. Colloidal gold and silver triangular nanoprisms. *Small*, 2009, *5* (6), 646–664.

184. Hsu, M.-S.; Cao, Y.-W.; Wang, H.-W.; Pan, Y.-S.; Lee, B.-H.; Huang, C.-L. Time-dependent surface plasmon resonance spectroscopy of silver nanoprisms in the presence of halide ions. *Chemical Physics Letters*, 2010, *11* (8), 1742–1748.

185. Ma, W.; Yao, J.; Yang, H.; Liu, J. Optical properties and sensing performance of $Ag/SiO_2/Ag$ triangular nanoprism. *Nami Jishu Yu Jingmi Gongcheng*, 2010, *8* (3), 240–244.

186. Jin, R.; Cao, Y. W.; Mirkin, C. A.; Kelly, K. L.; Schatz, G. C.; Zheng, J. G. Photoinduced conversion of silver nanospheres to nanoprisms. *Science*, 2001, *294*, 1901–1903.

187. Guo, Z.; Fan, X.; Liu, L.; Bian, Z.; Gu, C.; Zhang, Y.; Gu, N.; Yang, D.; Zhang, J. Achieving high-purity colloidal gold nanoprisms and their application as biosensing platforms. *Journal of Colloid and Interface Science*, 2010, *348* (1), 29–36.

188. Kundu, S.; Liang, H. Shape-controlled synthesis of triangular gold nanoprisms using microwave irradiation. *Journal of Nanoscience and Nanotechnology*, 2010, *10* (2), 746–754.

189. Millstone, J. E.; Georganopoulou, D. G.; Xu W.; Li, S.; Mirkin, C. A. DNA-gold triangular nanoprism conjugates. *Small*, 2008, *4* (12), 2176–2180.

190. Yoo, H.; Millstone, J. E.; Li, S.; Jang, J.-W.; Wei, W.; Wu, J.; Schatz, G. C.; Mirkin, C. A. Core-shell triangular bifrustums. *Nano Letters*, 2009, *9* (8), 3038–3041.

191. Zhang, H. X.; Feng, P. X. Synthesis of the vertically aligned carbon hexagonal nanoprism arrays and their application for field emission. *Applied Surface Science*, 2009, *255* (11), 5939–5942.

192. Yuan, D.; Wang, G.-S.; Xiang, Y.; Chen, Y.; Gao, X.-Q.; Lin, G. Optical properties and formation mechanism of radial ZnO hexagonal nanoprism clusters. *Journal of Alloys and Compounds*, 2009, *478* (1–2), 489–492.

193. Luo, K.-y.; Zhang, H.-y.; Li, B.-q. Morphology and optical properties of ZnO nanoprisms fabricated by laser irradiation. *Cailiao Rechuli Xuebao*, 2008, *29* (6), 21–24.

194. Zeng, Y.; Hao, R.; Xing, B.; Hou, Y.; Xu, Z. One-pot synthesis of Fe_3O_4 nanoprisms with controlled electrochemical properties. *Chemical Communications*, 2010, *46* (22), 3920–3922.

195. Guan, Y.; Hensen, E. J. M.; Liu, Y.; Zhang, H.; Feng, Z.; Li, C. Template-free synthesis of sphere, rod and prism morphologies of CeO_2 oxidation catalysts. *Catalysis Letters*, 2010, *137* (1–2), 28–34.

196. Pan, N.; Xue, H.; Huang, J.; Zhang, G.; Wu, Y.; Li, M.; Wang, X.; Hou, J. Self-assembly and the properties of a highly oriented hierarchical nanobelt-nanoprism array of ternary oxide Zn-In-O. *European Journal of Inorganic Chemistry*, September, 2010, (27), 4344–4350.

197. Chen, M.; Jiang, D.; Xie, J.; Zhu, J.; Wu, Y. Template-free preparation of hollow hexagonal fluorapatite nanoprisms. *Chemistry Letters*, 2008, *37* (12), 1286–1287.

198. Wu, R. B.; Yang, G. Y.; Pan, Y.; Chen, J. J. Synthesis of silicon carbide hexagonal nanoprisms. *Applied Physics A: Materials Science and Processing*, 2007, *86* (2), 271–274.

12 Other Rare Nanostructures

12.1 NANOVOLCANOES

Nanovolcanoes are very rare. Thus, ZnO nanovolcanoes (Figure 12.1), among other nanostructures, were grown[1] (Reactions 12.1 through 12.4) using the hydrothermal technique from zinc nitrate and hexamethylenetetramine (HMT) as precursors. It was established that, by increasing the reaction time, the volcano-like and tube-like ZnO structures were formed due to the Ostwald ripening process and the selective adsorption of the complexes. In addition, the formation of large features at the edges of nanopores in freestanding silicon nitride membranes, called by authors "nanovolcanoes," was studied.[2] It was established that the rate at which the nanopores open or close was strongly influenced by sample temperature. In addition, volcano size and closing rates were found to be dependent on the initial pore size.

$$+ \text{Zn}^{2+} \longleftrightarrow \text{Zn}^{2+} \text{ amino complex} \tag{12.1}$$

$$+ 4\text{H}_3\text{O}^+ \longleftrightarrow \quad + 4\text{H}_2\text{O} \tag{12.2}$$

$$\text{Zn}^{2+} + 2\text{H}_2\text{O} \longrightarrow 2\text{H}^+ + \text{Zn(OH)}_2 \longrightarrow \text{ZnO} + \text{H}_2\text{O} \tag{12.3}$$

$$\text{Zn}^{2+} \text{ amino complex} + \text{H}_2\text{O} \longrightarrow \quad + \text{Zn(OH)}_2 \tag{12.4}$$

$$\text{ZnO} + \text{H}_2\text{O}$$

12.2 NANOSPONGES

Elemental metals (especially noble) and alloys are common as sponge-like nanostructures with distinct applications.[3] Thus, in the search of applicability as surfaces for the analyses of peptides and proteins using surface-assisted laser desorption/ionization mass spectrometry (SALDI-MS), Pt nanosponges were found to be efficient nanomaterials for proteins, with an upper detectable mass limit of ca. 25 kDa.[4] They have several advantages over organic matrixes, including lower limits

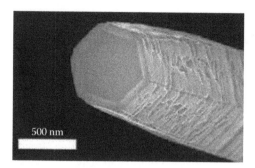 

FIGURE 12.1 Scanning electron microscopy (SEM) images of ZnO nanovolcanoes grown on an Si substrate for 6 h. (Reproduced with permission from Tong, Y. et al., Growth of ZnO nanostructures with different morphologies by using hydrothermal technique, *J. Phys. Chem. B*, 110(41), 20263–20267. Copyright 2006 American Chemical Society.)

of detection for small analytes and lower batch-to-batch variations. In addition, Pt nanosponges, nanonetworks, and nanodendrites were synthesized through a unique galvanic replacement reaction between Te nanowires and $PtCl_6^{2-}$ ions in the presence of Na dodecyl sulfate.[5] These Pt nanomaterials had large active surface areas and highly electrocatalytic activities for the oxidation of MeOH. Templateless synthesis of nanoporous Au sponge was carried out by chemical reduction of Au species in aqueous solution.[6] These obtained Au nanosponges exhibited surface-enhanced Raman scattering activity. High surface area noble metal nanosponges of Au, Ag, Pd (Figure 12.2), and Pt were synthesized by kinetically controlling the simple borohydride reduction process.[7] These 3D porous nanostructures exhibited efficient broadband optical limiting behavior. Nanosponges of Ag and Au were shown to be good substrates for surface-enhanced Raman spectroscopy (SERS) activity. Additionally, the antimicrobial activity of the self-supported nanoporous silver discs or nanoporous silver network incorporated filter papers could be utilized for water purification, particularly to screen and destroy harmful pathogens. Also, a simple approach was reported for prepared Pd/Ag and Pd/Ag/Au nanosponge alloys, which comprise network nanowires:[8] self-regulated reduction of sodium dodecyl sulfate and adding the second or third metal salt in the synthesis period, without an additional reduction agent. These alloy nanosponges with network nanowires were found to exhibit a high activity in the reduction of oxygen.

By carefully controlling the experimental parameters, large-scale nanosponges of group II-B metal (Zn and Cd) on Si substrates were fabricated through a vacuum vapor deposition (VVD) route. It was shown that nanosponges were composed of network nanowires, which are 20–250 nm in diameter and hundreds of micrometers in length. The direct synthesis of a sponge-like nanostructure, which uses magnetic nanoparticles (FePt) as the joints and semiconducting nanowires (ZnS) as the branches, in a solution-phase reaction without the use of a template was reported.[9] It was shown that a network with two components can be easily made by the sequential chemical synthesis of the two components. These materials may serve as magnetic/electronic materials to help investigate the electron transfer processes that occur between magnetic nanomaterials and nanosemiconductors.

Carbon[10] was also obtained in a nanosponge form. Thus, CVD synthesis of carbon nanotubes and nanofibers on the surface of expanded vermiculite was used to produce a highly hydrophobic floatable absorbent to remove oil spilled on water.[11] It was shown that the carbon nanotubes and nanofibers grew on FeMo catalyst impregnated on the vermiculite surface to form a "sponge structure." As a result of these carbonaceous nanosponges, the absorption of different oils remarkably increased ca. 600% with a concomitant strong decrease in undesirable water absorption. Silicon nanosponge particles were prepared from a metallurgical grade silicon powder.[12] Antireflective nanosponges were fabricated on polycrystalline Si (poly-Si) thin films using Ag-nanoparticle-assisted etching.[13] It was revealed that a 400 nm thick poly-Si nanosponge reduced an effective optical reflection of the

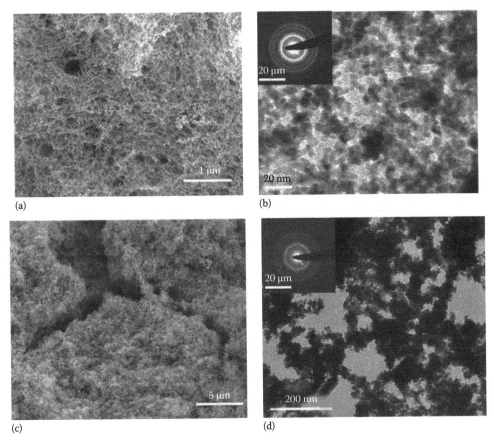

FIGURE 12.2 (a) Low-magnification field emission scanning electron microscopy (FESEM) image of palladium nanosponge. (b) TEM image showing palladium nanoparticles of 5–10 nm size fused to form a network structure. Inset shows its electron diffraction (ED) pattern. (c) Low-magnification FESEM image showing the platinum nanosponge. (d) TEM image showing the platinum networks made up of nanoparticles of size 5–10 nm. Inset shows its ED pattern. (Reproduced with permission from Krishna, K.S. et al., Mixing does the magic: A rapid synthesis of high surface area noble metal nanosponges showing broadband nonlinear optical response, *ACS Nano*, 4(5), 2681–2688. Copyright 2010 American Chemical Society.)

poly-Si thin film with substrate crystal orientation of (110) and average grain size of 250 nm from 26% to 3% at the wavelengths ranging from 400 to 1000 nm.

Nanosponge titania, integrated into a prototype device, was used for ultrasensitive detection of hydrogen.[14] This material has potential applications in multiplex sensing systems such as electronic noses and tongues, and three-dimensionally interconnected nanostructured metal oxides hold great promise as platforms for ultrasensitive sensors. TiO_2 nanosponge was also used as a sensor to molecular oxygen.[15] In the case of titania composites, a procedure was developed to coat functionalized polystyrene spheres with a well-defined layer of amorphous titanium dioxide.[16] These core-shell particles can be turned into TiO_2 nanosponge by calcining the dried particles in a furnace. Nanocomposite materials consisting of $CoFe_2O_4$ FePt and a polyethylene glycol-acrylamide gel matrix were synthesized.[17] It can be used as an active sponge-like nanomagnetic container for water-based formulations as oil-in-water microemulsions. In addition, using corresponding metal elemental nanowires as a self-sacrificed template, large-scale nanosponges consisting of curving CdS nanotubes were fabricated through a gas–solid sulfurization reaction.[18] The CdS nanotubes exhibited high photocatalytic activity for the degradation of nonbiodegradable *Rhodamine B* under visible irradiations probably due to the accessibility of both the inner and outer surfaces through the pores in the walls.

A definite number of nanosponge reports are dedicated to cyclodextrin.[19,20] Thus, cyclodextrin-based nanosponges are defined as a novel class of cross-linked derivatives of cyclodextrins.[21] They have been used to increase the solubility of poorly soluble actives, to protect the labile groups and control the release. Nanosponges were prepared from β-cyclodextrins as nanoporous materials for possible use as carriers for drug delivery.[22] The nanosponges were found capable of carrying both lipophilic and hydrophilic drugs and of improving the solubility of poorly water-soluble molecules.

12.3 NANOSTARS

Nanostars are well developed first of all for gold. Thus, gold nanostars (multibranched nanoparticles with sharp tips with extremely interesting plasmonic properties) were obtained *via* seeded growth in concentrated solutions of poly(vinylpyrrolidone) in DMF.[23] Varying the size of the seeds (2–30 nm) (Figure 12.3), a clear influence on the final nanostar dimensions as well as on the number of spikes was found, while the synthesis temperature notably affected the morphology of the particles, with more rounded morphologies formed at >60°C. Au nanostar dispersions displayed a well-defined optical response, which was found (through theoretical modeling) to comprise a main mode confined within the tips and a secondary mode confined in the central body.[24]

Elongated penta-branched gold nanocrystals with a shape resembling that of a star fruit but with sharp ends (Figure 12.4) were prepared by a seeding growth approach,[25] using cetyltrimethylammonium bromide (CTAB) as a capping surfactant and AgNO$_3$, added to the last growth solution to promote the formation of the five side branches. Smaller penta-branched nanocrystals with sizes of 70–110 nm and more fully developed larger nanocrystals with sizes of 200–350 nm can be readily prepared. These gold nanostructures should offer opportunities for detailed analysis of their optical properties and applications in molecular sensing.

The controlled synthesis of high-yield gold nanostars of varying sizes (from 45 to 116 nm in size) was carried out with high yields.[26] Their transmission electron microscope (TEM) images (Figure 12.5)

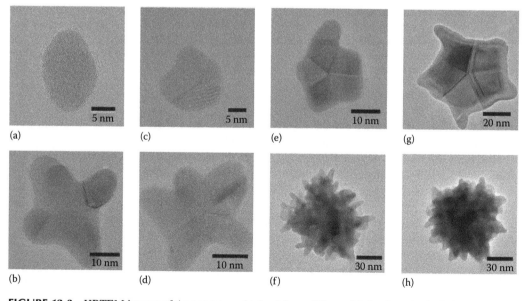

FIGURE 12.3 HRTEM images of Au nanostars obtained from different kinds of seeds: (a, b) 2 nm Pt; (c, d) 2.3 nm Au; (e, f) 15 nm Au; (g, h) 30 nm Au. [HAuCl$_4$]/[seed] ratios are 45 (a), 675 (b), 67.5 (c), 880 (d), 1.5 (e), 90 (f), 1.5 (g), and 11.5 (h). (Reproduced with permission from Barbosa, S. et al., Tuning size and sensing properties in colloidal gold nanostars, *Langmuir*, 26(18), 14943–14950. Copyright 2010 American Chemical Society.)

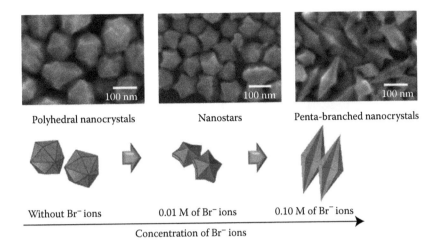

FIGURE 12.4 Shape evolution from polyhedral gold nanocrystals to nanostars and penta-branched nano-crystals with increasing bromide ion concentration in the reaction solution. For the synthesis of gold nanostars, 100 μL of 0.01 M AgNO$_3$ was also added in the last step of the reaction. (Reproduced with permission from Wu, H.-L. et al., Seed-mediated synthesis of branched gold nanocrystals derived from the side growth of pentagonal bipyramids and the formation of gold nanostars, *Chem. Mater.*, 21(1), 110–114. Copyright 2009 American Chemical Society.)

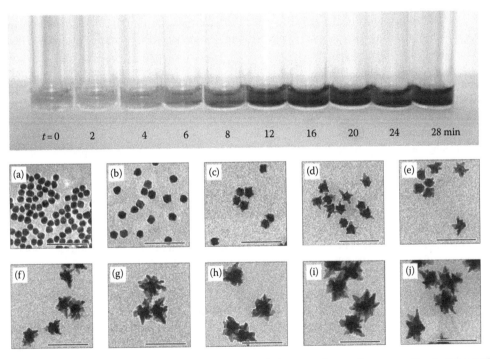

FIGURE 12.5 TEM images monitoring the nanostar evolution over time and corresponding to the samples in the figure: t = (a) 0, (b) 2, (c) 4, (d) 6, (e) 8, (f) 12, (g) 16, (h) 20, (i) 24, and (j) 28 min (Reproduced with permission from Khoury, C.G. et al., Gold nanostars for surface-enhanced Raman scattering: Synthesis, characterization and optimization, *J. Phys. Chem. C*, 112(48), 18849–18859. Copyright 2008 American Chemical Society.)

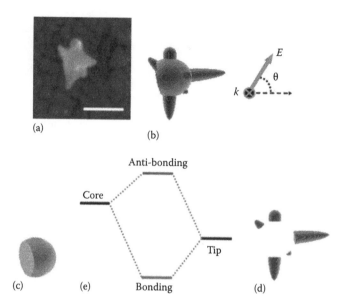

FIGURE 12.6 Structure of the gold nanostar. Panel (a) shows an experimental scanning electron micrograph. The scale bar is 100 nm. Panel (b) shows the theoretical model, consisting of a truncated spherical core (c) and tips, consisting of truncated prolate spheroids (d). Panel (e) illustrates the concept of plasmon hybridization in the nanostar. The core plasmons interact with the tip plasmons and form bonding and anti-bonding nanostar plasmons. The polarization angle is defined in the upper right corner. (Reproduced with permission from Hao, F. et al., Plasmon resonances of a gold nanostar, *Nano Lett.*, 7(3), 729–732. Copyright 2007 American Chemical Society.)

confirmed the observations of the growth mechanism: the nanostar core increases rapidly between $t = 0$ and $t = 8$ min, during which the surface protrusions that appeared at $t = 2$ min develop into distinct star branches by $t = 6$ min.

The plasmons of a gold nanostar (it is known that the plasmon resonances of a nanoparticle depend strongly on its composition and its shape) were found to be resulting from the hybridization of the plasmons of the core and tips of the nanoparticle (Figure 12.6).[27] The nanostar core serves as a nanoscale antenna, dramatically increasing the excitation cross section and the electromagnetic field enhancements of the tip plasmons.

Discussing gold alloy nanostars (e.g., Au/Pd[28]), we note a star-like morphology of Au/Ag nanoparticles (Figure 12.7), synthesized using Ag seeds followed by the addition and reduction of $HAuCl_4 \cdot 4H_2O$ with L-ascorbic acid in the presence of CTAB.[29]

In the case of other noble metals, attention has been paid to Pt.[30] Thus, a very active platinum nanocatalyst—the multiarmed nanostar single crystal (Figure 12.8)—was prepared[31] using a seed-mediated method using tetrahedral nanoparticles. The nanostars with many arms, varying from a few to over 30, were found to be formed by a growth mechanism of the seed crystals and not by the aggregation of seed crystals, which should produce twinning planes.

Among metal oxides, we note briefly V_2O_5 nanostars, obtained by combining chemical vapor transport with hydrothermal synthesis,[32] and TiO_2.[33] Also, the field emission properties of MoO_2 nanostars (length 1 µm, thickness 50 nm, width 500–700 nm; MoO_2 nanostars over a crystalline thin film containing Mo_4O_{11} nanoparticles) grown on a silicon substrate and their emission performance in various vacuum gaps were reported.[34] The grown nanostructures were found to be composed of both crystalline Mo_4O_{11} and crystalline MoO_2 structures containing 21.2% Mo^{6+}, 16.2% Mo^{5+}, 39.8% Mo^{4+}, and 22.8% $Mo^{\delta+}$ (where $0 < \delta < 4$). Due to their excellent emission properties, these nanostars can be used in vacuum microelectronic applications. Metal salts are mainly represented by chalcogenides, for instance Mn-doped PbSe.[35] Another example is PbS, for which a controllable self-assembly (Figure 12.9) of uniform star-shaped PbS nanocrystals into highly ordered structures

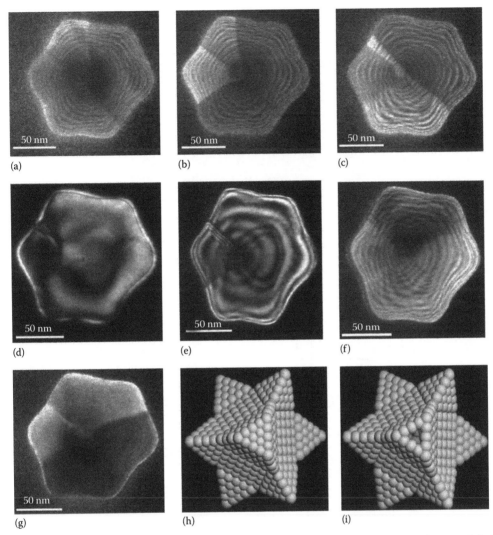

FIGURE 12.7 Weak beam dark field transmission electron microscopy (WBDF-TEM) images of Au/Ag nanoparticles when they were tilted along α angle at different angles: (a) 0°, (b) −2.6°, (c) −4.4°, (d) −11.8°, (e) −14.8°, (f) −17.8°, and (g) 23.6°; (h) simulated image; (i) artificially "hollowed" nanoparticle showing the Ag seed. (Reproduced with kind permission from Springer Science + Business Media: *Appl. Phys. A Mater. Sci. Process.*, Polyhedral shaped gold nanoparticles with outstanding near-infrared light absorption, 97(1), 2009, 11–18, Mayoral, A. et al.)

including close-packed arrays and patterned arrays was realized by evaporation-induced assembly routes.[36] The obtained PbS close-packed and patterned arrays may exhibit novel physical properties by virtue of the peculiar six-horn star shape of PbS nanocrystals as well as their well-defined locations and orientations and could find potential applications including photoelectronic and photonic nanodevices.

In the H_2O-in-oil microemulsion stabilized by the cationic gemini surfactant alkane-diyl-α,ω (dimethydodecyl-ammonium bromide), CuS nanorod, tube-like, and star sheet-like structures were prepared.[37] Also, Ni diselenide nanocrystals in well-defined star shape were synthesized *via* an improved solvothermal route with oleic acid as a capping ligand.[38] The mechanism of their formation is shown in Figure 12.10. Cubic $NiSe_2$ nanocrystals were in a star-like shape, and each nanostar consisted of a central core and six symmetric horns (the length of each horn was 85 nm, the diameter of the central core was 90 nm, and the overall length between two most distance vertexes was

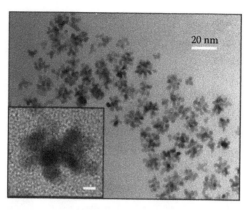

FIGURE 12.8 TEM image of new multiarmed nanostar particles. Inset is an high resolution transmission electron microscopy. (HRTEM) image of a nanostar displaying multiple arms branching in various directions from the initial seed particle. The scale bar in the insert is 2 nm. (Reproduced with permission from Mahmoud, M.A. et al., A new catalytically active colloidal platinum nanocatalyst: The multiarmed nanostar single crystal, *J. Am. Chem. Soc.*, 130(14), 4590–4591. Copyright 2008 American Chemical Society.)

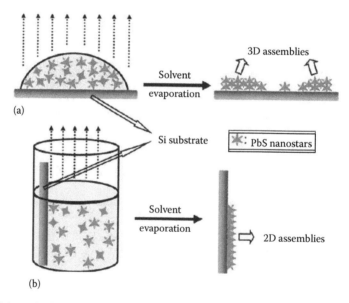

FIGURE 12.9 Schematic illustration of two methods of evaporation-induced assembly: (a) drop coating and (b) vertical deposition. (Reproduced with permission from Huang, T. et al., Controllable self-assembly of PbS nanostars into ordered structures: Close-packed arrays and patterned arrays, *ACS Nano*, 4(8), 4707–4716. Copyright 2010 American Chemical Society.)

260 nm). In addition, nearly monodispersed and monocrystal cubic-phase HgTe-ssDNA (ss, single-stranded) nanostars were obtained in a self-assembling process and have exhibited a blue shifted single narrow PL of line width 8 nm at 548.4 nm.[39]

Organometallics are represented by arene and isonitrile bridged metal carbonyls and phosphines.[40] In addition, covalently bonded organometallic oligomers and stars were designed[41] to have rigid and thermally and chemically stable organometallic repeating units such as [*trans*-Mo(Ph$_2$PCH$_2$CH$_2$Ph$_2$)$_2$(μ-CN-1,4-C$_6$H$_4$-NC)]. Ferrocenyl isonitrile was used to prepare a series of linear oligomers and star-shaped nanomaterials.

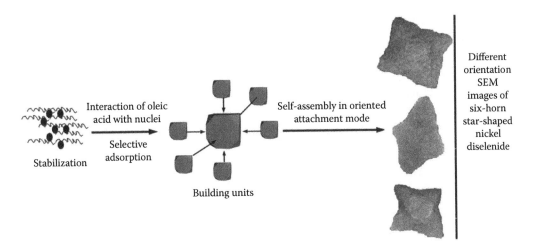

FIGURE 12.10   A schematic illustration of the formation process of the six-horn star-shaped NiSe$_2$ nano-crystals. (ᴧᴧᴧ, oleic acid; •, nuclei) (Reproduced with permission from Du, W. et al., Symmetrical six-horn nickel diselenide nanostars growth from oriented attachment mechanism, *Cryst. Growth Des.*, 7(12), 2733–2737. Copyright 2007 American Chemical Society.)

12.4   NANOGLASS CERAMICS

The feature of a special class of materials—called nanoglasses—with a glassy structure is that the atomic structure in the entire volume of the material as well as the density of the material can be tuned.[42] Nanoglasses can be prepared by a series of methods, in particular sol-gel (thus, a nanometer-sized CeO$_2$-doped nanoglass was prepared by this route[43]) or, more rarely, from polymers and their transition metal complexes.[44] Their optical properties as important characteristics are permanently studied; thus, reversible changes in the refractive index of nanoglass thin films induced by laser beams were revealed.[45] Additionally, nanostructures and optical properties of nanoglass thin films containing Co$_3$O$_4$ grains (10 nm)[46] surrounded by amorphous grain boundaries with 1 nm thickness were examined. In addition, nanoglasses can be generated, in particular, by introducing interfaces into metallic glasses on a nanometer scale. Interfaces in these nanoglasses delocalize upon annealing, so that the free volume associated with these interfaces spreads throughout the volume of the glass. The composition of nanoglasses can vary from complex metal oxides to their mixed compounds with salts, frequently giving nonstoichiometric composites. Nanoglasses have a series of applications, for instance an Fe$_2$O$_3$ base nanoglass thin film was used for optical recording disks.[47] Nanoglass was applied as a protective layer for elastomeric substrates to avoid the diffusion of aggressive components into the material to protect, such as in motor vehicles (e.g., brake components, car wheel rims, engine parts, exhaust fume ducts, etc.).[48] Also, magnetic semiconductor nanoglass Cd$_{1-x}$Mn$_x$Se was fabricated.[49]

Transparent oxyfluoride nanoglass ceramics with highly efficient up-conversion and adjustable color luminescence were developed in the 28SiO$_2$·17Al$_2$O$_3$·28PbF$_2$·22CdF$_2$·0.1NdF$_3$·xYbF$_3$·yHoF$_3$·zTmF$_3$·(4.9 − x − y − z)GdF$_3$ composition (mol.%).[50] It was revealed that heat treatments of the oxy-fluoride glasses caused the homogeneous precipitation of rare-earth ions co-doped with fluorite-type Pb$_x$Cd$_{1-x}$F$_2$ nanocrystals of about 10 nm in diameter in the glass matrix. Various colors of luminescence, including bright perfect white light, can be tuned by adjusting the concentrations of the Tm^{3+} ions in the material. Similar transparent oxyfluoride nanoglass-ceramics with a more simple composition 90(SiO$_2$)10(PbF$_2$) co-doped with 0.3 Yb^{3+} and 0.1 Er^{3+} (mol.%) were prepared by thermal treatment of precursor sol-gel glasses.[51] A precipitation of cubic PbF$_2$ nanocrystals of certain diameter in nanoglass ceramics varying from 10 to 20 nm depending on heat treatment conditions was pointed out. The incorporation of Yb^{3+} and Er^{3+} dopants in these nanocrystals has been confirmed by signatures of luminescence spectroscopy. Analogously, 95SiO$_2$-5SnO$_2$ nanoglass ceramics doped with 0.4 Sm^{3+} were fabricated.[52]

12.5   OTHER NANOOBJECTS

Nanodrugs[53] are heterogeneous structures, which capitalize on their small size to target human disease.[54] New materials, such as nanocrystals, offer a general approach to improve formulations sufficiently to achieve multiple FDA approvals. Many others, such as dendrimers, polymeric micelles, quantum dots, and inorganic nanoshells, are under active development for both incremental and revolutionary improvements in therapy and drug delivery.

Finally, several nanostructures are represented by single examples. Thus, new nanostructures were proposed based on hybrids of fullerenes and carbon nanotubes $A^+@C_n@tubeC_m$, called *nano-autoclaves*.[55] This autoclave was used to dimerize C_{20}. It was shown that $(C_{20})_2$ dimer can be synthesized this way. The advantage of such nanoautoclave is that it guarantees the absence of any impurities inside the capsule $tube@C_n$, so the final dimerization product $(C_{20})_2$ does not contain atoms of other chemical elements. ZnSe/CdS heterostructured nanocrystals (Figure 12.11) were fabricated *via* colloidal routes and comprised a *barbell*-like arrangement of ZnSe tips and CdS nanorods.[56] The resonant excitation of ZnSe tips resulted in an unprecedented fast transfer of excited electrons into CdS domains of nanobarbells (<0.35 ps), whereas selective pumping of CdS components leads to a relatively slow injection of photoinduced holes into ZnSe tips. These nanostructures possess potential photocatalytic applications.

The interesting core-shell structure is a concentric nanoshell composed of a dielectric core with alternating layers of metal, dielectric, and metal, essentially a nanoshell encased within another nanoshell, inspiring, as the authors[57] stated, its alternative Russian name of *"nanomatryushka"**

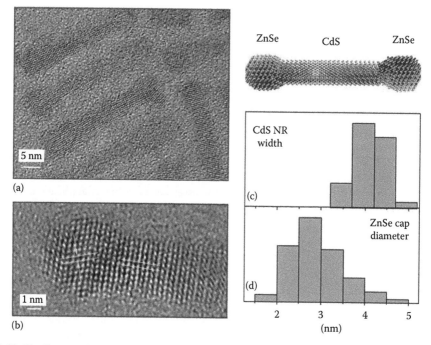

FIGURE 12.11   Structural analysis of ZnSe/CdS barbells: (a) high-resolution TEM images of ZnSe/CdS heterostructures, (b) typical ZnSe/CdS barbell showing unique directions of lattice planes for each of the material domains, (c) statistical distributions of barbell widths, and (d) diameters of ZnSe tips. (Reproduced with permission from Hewa-Kasakarage, N.N. et al., Ultrafast carrier dynamics in type II ZnSe/CdS/ZnSe nanobarbells, *ACS Nano*, 4(4), 1837–1844. Copyright 2010 American Chemical Society.)

---

* More correct spelling is "nanomatryoshka" (наноматрёшка in Russian): the matryoshka is a set of dolls of decreasing sizes placed one inside the other.

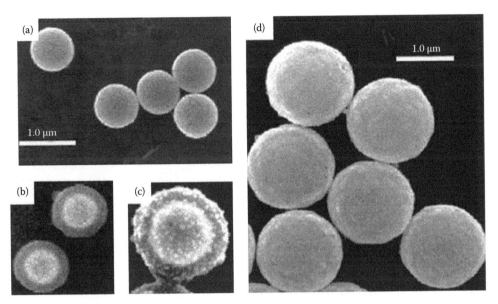

FIGURE 12.12 SEM images of (a) inner nanoshell, consisting of a silica core layer and a gold shell layer of 800 nm total diameter, (b) silica encapsulated core (235 nm silica layer), (c) seeded silica-coated nanoshell with Au colloid, and (d) complete concentric nanoshell with total particle diameter 1.4 μm. (Reproduced with permission from Wang, H. et al., Plasmonic nanostructures: Artificial molecules, *Acc. Chem. Res.*, 40, 53–62. Copyright 2007 American Chemical Society.)

(Figure 12.12). The plasmon response of this structure can be understood as a hybridization of the plasmon resonances of the inner and outer nanoshells.

Post-thermal decomposition of single-crystalline β-nickel hydroxide {β-Ni(OH)$_2$} nanoplates of hexagonal structure, synthesized through a hydrothermal process, led to the formation of single-crystalline NiO nanostructures with a landscape dimension of 25–120 nm including *nanorolls* (Figure 12.13a), nanotroughs, and nanoplates.[58] The sizes of the central hole in NiO nanorolls and the low-lying ground in NiO nanotroughs were in the range of 10–24 nm.

 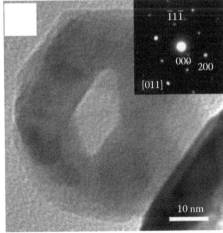

FIGURE 12.13 TEM images and corresponding selected area electron diffraction (SAED) patterns (insets) of NiO nanostructures (nanorolls, nanotroughs, and nanoplates, as marked with A, B, and C, respectively). (Reproduced with kind permission from Springer Science + Business Media: *J. Mater. Sci. Mater Electron*, Photoluminescence and magnetic properties of β-Ni(OH)$_2$ nanoplates and NiO nanostructures, 20, 2009, 479–483, Yajun, Q. et al.)

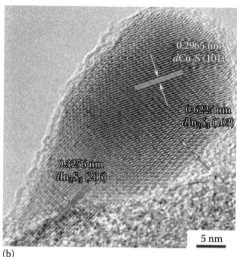

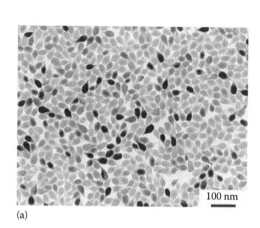

(a)   (b)

FIGURE 12.14   (a) Representative TEM image of teardrop-like nanocrystals; (b) HRTEM image of a single teardrop like nanocrystal overlaid with identifications of crystalline phases. (Reproduced with permission from Wei, H. et al., Synthesis and shape-tailoring of copper sulfide/indium sulfide-based nanocrystals, *J. Am. Chem. Soc.*, 130, 13152–13161. Copyright 2008 American Chemical Society.)

Teardrop-like quasi-core-shell $Cu_2S@In_2S_3$ nanocrystals (Figure 12.14) were obtained by a high-temperature precursor-injection method using the semiconductor nanocrystal $Cu_{1.94}S$ as a catalyst.[59]

REFERENCES

1. Tong, Y.; Liu, Y.; Dong, L.; Zhao, D.; Zhang, J.; Lu, Y.; Shen, D.; Fan, X. Growth of ZnO nanostructures with different morphologies by using hydrothermal technique. *Journal of Physical Chemistry B*, 2006, *110* (41), 20263–20267.
2. Hoogerheide, D. P.; Golovchenko, J. A. Dynamics of ion beam stimulated surface mass transport to nanopores. *Materials Research Society Symposium Proceedings* (*Ion-Beam-Based Nanofabrication*), San Francisco, CA, 2007, Vol. 1020, pp. 29–38.
3. Nanosponges go commercial. *Environmental Science and Technology*, 2006, *40* (15), 4535.
4. Chiang, C.-K.; Chiang, N.-C.; Lin, Z.-H.; Lan, G.-Y.; Lin, Y.-W.; Chang, H.-T. Nanomaterial-based surface-assisted laser desorption/ionization mass spectrometry of peptides and proteins. *Journal of the American Society for Mass Spectrometry*, 2010, *21* (7), 1204–1207.
5. Lin, Z.-H.; Lin, M.-H.; Chang, H.-T. Facile synthesis of catalytically active platinum nanosponges, nanonetworks, and nanodendrites. *Chemistry—A European Journal*, 2009, *15* (18), 4656–4662.
6. Ataee-Esfahani, H.; Fukata, N.; Yamauchi, Y. Templateless synthesis of nanoporous gold sponge with surface-enhanced Raman scattering activity. *Chemistry Letters*, 2010, *39* (4), 372–373.
7. Krishna, K. S.; Sandeep, C. S. S.; Philip, R.; Eswaramoorthy, M. Mixing does the magic: A rapid synthesis of high surface area noble metal nanosponges showing broadband nonlinear optical response. *ACS Nano*, 2010, *4* (5), 2681–2688.
8. Lee, C.-L.; Huang, Y.-C.; Kuo, L.-C.; Oung, J.-C.; Wu, F.-C. Preparation and characterization of Pd/ Ag and Pd/Ag/Au nanosponges with network nanowires and their high electroactivities toward oxygen reduction. *Nanotechnology*, 2006, *17* (9), 2390–2395.
9. Gu, H.; Zheng, R.; Liu, H.; Zhang, X.; Xu, B. Direct synthesis of a bimodal nanosponge based on FePt and ZnS. *Small*, 2005, *1* (4), 402–406.
10. Benedek, G.; Vahedi-Tafreshi, H.; Barborini, E.; Piseri, P.; Milani, P.; Ducati, C.; Robertson, J. The structure of negatively curved spongy carbon. *Diamond and Related Materials*, 2003, *12* (3–7), 768–773.
11. Moura, F. C. C.; Lago, R. M. Catalytic growth of carbon nanotubes and nanofibers on vermiculite to produce floatable hydrophobic "nanosponges" for oil spill remediation. *Applied Catalysis, B: Environmental*, 2009, *90* (3–4), 436–440.

12. Limaye, S.; Farrell, D.; Subramanian, S. Silicon nanosponge particles. Patent IE 20060360, 2006, 27 pp.

13. Chyan, J. Y.; Hsu, W. C.; Yeh, J. A. Broadband antireflective poly-Si nanosponge for thin film solar cells. *Optics Express*, 2009, *17* (6), 4646–4651.

14. Zuruzi, A. S.; MacDonald, N. C.; Moskovits, M.; Kolmakov, A. Metal oxide "nanosponges" as chemical sensors: Highly sensitive detection of hydrogen with nanosponge titania. *Angewandte Chemie, International Edition*, 2007, *46* (23), 4298–4301.

15. Zuruzi, A. S.; Kolmakov, A.; MacDonald, N. C.; Moskovits, M. Highly sensitive gas sensor based on integrated titania nanosponge arrays. *Applied Physics Letters*, 2006, *88* (10), 102904/1–102904/3.

16. Guo, L.; Gao, G.; Liu, X.; Liu, F. Preparation and characterization of TiO_2 nanosponge. *Materials Chemistry and Physics*, 2008, *111* (2–3), 322–325.

17. Bonini, M.; Lenz, S.; Falletta, E.; Ridi, F.; Carretti, E.; Fratini, E.; Wiedenmann, A.; Baglioni, P. Acrylamide-based magnetic nanosponges: A new smart nanocomposite material. *Langmuir*, 2008, *24* (21), 12644–12650.

18. Wang, Q.; Chen, G.; Zhou, C.; Jin, R.; Wang, L. Sacrificial template method for the synthesis of CdS nanosponges and their photocatalytic properties. *Journal of Alloys and Compounds*, 2010, *503* (2), 485–489.

19. Alongi, J.; Poskovic, M.; Frache, A.; Trotta, F. Novel flame retardants containing cyclodextrin nanosponges and phosphorus compounds to enhance EVA combustion properties. *Polymer Degradation and Stability*, 2010, *95* (10), 2093–2100.

20. Cyclodextrin nanosponges as carriers for biocatalysts and in the delivery and release of enzymes, proteins, vaccines, and antibodies. Patent IT 2008MI1056, 2008, 33 pp.

21. Swaminathan, S.; Pastero, L.; Serpe, L.; Trotta, F.; Vavia, P.; Aquilano, D.; Trotta, M.; Zara, G. P.; Cavalli, R. Cyclodextrin-based nanosponges encapsulating camptothecin: Physicochemical characterization, stability and cytotoxicity. *European Journal of Pharmaceutics and Biopharmaceutics*, 2010, *74* (2), 193–201.

22. Cavalli, R.; Trotta, F.; Tumiatti, W. Cyclodextrin-based nanosponges for drug delivery. *Journal of Inclusion Phenomena and Macrocyclic Chemistry*, 2006, *56* (1–2), 209–213.

23. Barbosa, S.; Agrawal, A.; Rodriguez-Lorenzo, L.; Pastoriza-Santos, I.; Alvarez-Puebla, R. A.; Kornowski, A.; Weller, H.; Liz-Marzan, L. M. Tuning size and sensing properties in colloidal gold nanostars. *Langmuir*, 2010, *26* (18), 14943–14950.

24. Kumar, P. S.; Pastoriza-Santos, I.; Rodriguez-Gonzalez, B.; García de Abajo, F. J.; Liz-Marzan, L. M. High-yield synthesis and optical response of gold nanostars. *Nanotechnology*, 2008, *19* (1), 015606/1–015606/6.

25. Wu, H.-L.; Chen, C.-H.; Huang, M. H. Seed-mediated synthesis of branched gold nanocrystals derived from the side growth of pentagonal bipyramids and the formation of gold nanostars. *Chemistry of Materials*, 2009, *21* (1), 110–114.

26. Khoury, C. G.; Vo-Dinh, T. Gold nanostars for surface-enhanced Raman scattering: Synthesis, characterization and optimization. *Journal of Physical Chemistry C*, 2008, *112* (48), 18849–18859.

27. Hao, F; Nehl, C. L.; Hafner, J. H.; Nordlander, P. Plasmon resonances of a gold nanostar. *Nano Letters*, 2007, *7* (3), 729–732.

28. Krichevski, O.; Markovich, G. Growth of colloidal gold nanostars and nanowires induced by palladium doping. *Langmuir*, 2007, *23* (3), 1496–1499.

29. Mayoral, A.; Vazquez-Duran, A.; Barron, H.; Jose-Yacaman, M. Polyhedral shaped gold nanoparticles with outstanding near-infrared light absorption. *Applied Physics A: Materials Science and Processing*, 2009, *97* (1), 11–18.

30. Formo, E.; Camargo, P. H. C.; Lim, B.; Jiang, M.; Xia, Y. Functionalization of ZrO_2 nanofibers with Pt nanostructures: The effect of surface roughness on nucleation mechanism and morphology control. *Chemical Physics Letters*, 2009, *476* (1–3), 56–61.

31. Mahmoud, M. A.; Tabor, C. E.; El-Sayed, M. A.; Ding, Y.; Wang, Z. Lin. A new catalytically active colloidal platinum nanocatalyst: The multiarmed nanostar single crystal. *Journal of the American Chemical Society*, 2008, *130* (14), 4590–4591.

32. Velazquez, J. M.; Jaye, C.; Fischer, D. A.; Banerjee, S. Nanostructured vanadium oxides by chemical vapor transport and hydrothermal methods. *Abstracts of Papers, 240th ACS National Meeting*, Boston, MA, August 22–26, 2010, INOR-622.

33. Uyanik, M. Synthesis and characterization of TiO_2 nanostars. Avail. Metadata on Internet Documents, Order No. 396036, 2008, From: Metadata Internet Doc. [German dissertation] 2008, (D1020-1), http://www.meind.de/search.py?recid=396036.

34. Khademi, A.; Azimirad, R.; Zavarian, A. A.; Moshfegh, A. Z. Growth and field emission study of molybdenum oxide nanostars. *Journal of Physical Chemistry C*, 2009, *113* (44), 19298–19304.

35. Ji, T.; Fang, J.; O'Connor, C. Preparation and characterization of Mn-doped PbSe nanocrystals with interesting shapes. *Abstracts of Papers, 225th ACS National Meeting*, New Orleans, LA, March 23–27, 2003, INOR-235.

36. Huang, T.; Zhao, Q.; Xiao, J.; Qi, L. Controllable self-assembly of PbS nanostars into ordered structures: Close-packed arrays and patterned arrays. *ACS Nano*, 2010, *4* (8), 4707–4716.

37. Chen, L.; Shang, Y.; Liu, H.; Hu, Y. Synthesis of CuS nanocrystal in cationic gemini surfactant W/O microemulsion. *Materials and Design*, 2010, *31* (4), 1661–1665.

38. Du, W.; Qian, X.; Niu, X.; Gong, Q. Symmetrical six-horn nickel diselenide nanostars growth from oriented attachment mechanism. *Crystal Growth and Design*, 2007, *7* (12), 2733–2737.

39. Rath, S.; Sarangi, S. N.; Sahu, S. N.; Nozaki, S. HgTe nanoparticles and HgTe–ssDNA nanostars: Novel properties due to size quantization effect. In *Nano-Scale Materials: From Science to Technology*. Sahu, S. N.; Choudhury, R. K.; Jena, P. (Eds.), *Proceedings of the Workshop on Nano-Scale Materials: From Science to Technology*, Puri, India, April 5–8, 2004, 2006, pp. 291–295.

40. Hunter, A. D.; Zeller, M. Towards organometallic nanowires and nanostars. *Abstracts of Papers, 229th ACS National Meeting*, San Diego, CA, March 13–17, 2005, INOR-180.

41. Hunter, A. D.; Zeller, M.; Perrine, C.; Payton, J. L.; Woolcock, J. L.; Westcott, B. L. Organometallic nanostars from ferrocenyl isonitrile to metal phosphine building blocks. *Abstracts of Papers, 228th ACS National Meeting*, Philadelphia, PA, August 22–26, 2004, INOR-569.

42. Gleiter, H. Nanoglasses: A way to solid materials with tunable atomic structures and properties. *Materials Science Forum*, 2008, *584–586* (Pt. 1, Nanomaterials by Severe Plastic Deformation IV), 41–48.

43. Wang, J.; Dong, X.; Yan, J.; Feng, X; Liu, Z.; Hong, G. Preparation of nanometer-sized CeO_2-doped nanoglass. *Xiyou Jinshu Cailiao Yu Gongcheng*, 2004, *33* (12), 1358–1361.

44. Konishi, G. Preparation of functional nano glass from soluble assembled metal complexes as precursor. *Asahi Garasu Zaidan Josei Kenkyu Seika Hokoku*, 2005, 01.03/70–01.03/77.

45. Yamamoto, H.; Tanaka, S.; Hirano, K. Reversible change of refractive index of nanoglass thin films induced by laser irradiation and applications thereof. *Materials Integration*, 2003, *16* (8), 17–22.

46. Yamamoto, H. Nanoglass thin film for optical disk with high recording density. *Optronics*, 2003, *260*, 111–115.

47. Yamamoto, H. Application of Fe_2O_3 base nanoglass thin film for optical recording disk. *Sentan Garasu no Sangyo Oyo to Atarashii Kako*, 2009, 196–205.

48. Wessels, M. Use of nanoglass as protective coating material. Patent DE 102007024042, 2008, 4 pp.

49. Murayama, A.; Oka, Y. Preparation of magnetic semiconductor nano-glass. *Kinosei arasu/Nanogarasu no Saishin Gijutsu*, 2006, 226–236.

50. Song, Z.; Zhou, D.; Qiu, J. Adjustable up-conversion luminescence color in rare earth co-doped transparent oxyfluoride nano-glass-ceramics. *Journal of Nanoscience and Nanotechnology*, 2010, *10* (3), 1969–1973.

51. del-Castillo, J.; Yanes, A. C.; Mendez-Ramos, J.; Tikhomirov, V. K.; Rodriguez, V. D. Structure and up-conversion luminescence in sol-gel derived Er^{3+}-Yb^{3+} co-doped SiO_2:PbF_2 nano-glass-ceramics. *Optical Materials*, 2009, *32* (1), 104–107.

52. Yanes, A. C.; Velazquez, J. J.; del-Castillo, J.; Mendez-Ramos, J.; Rodriguez, V. D. Site-selective spectroscopy in Sm^{3+}-doped sol-gel-derived nano-glass ceramics containing SnO_2 quantum dots. *Nanotechnology*, 2008, *19* (29), 295707/1–295707/5.

53. Yokoyama, M. Nanodrugs. *Ensho to Men'eki*, 2005, *13* (2), 123–130.

54. Vine, W.; Gao, K.; Zegelman, J. L.; Helsel, S. K. Nanodrugs: Fact, fiction & fantasy. *Drug Delivery Technology*, 2006, *6* (5), 34, 36, 38–39.

55. Glukhova, O. E.; Salii, I. N.; Meshchanov, V. P. Nano-autoclave on the basis of carbon nanopeapod. *Nano- i Mikrosistemnaya Tekhnika*, 2007, (10), 47–52.

56. Hewa-Kasakarage, N. N.; El-Khoury, P. Z.; Tarnovsky, A. N.; Kirsanova, M.; Nemitz, I.; Nemchinov, A.; Zamkov, M. Ultrafast carrier dynamics in type II ZnSe/CdS/ZnSe nanobarbells. *ACS Nano*, 2010, *4* (4), 1837–1844.

57. Wang, H.; Brandl, D. W.; Nordlander, P.; Halas, N. J. Plasmonic Nanostructures: Artificial molecules. *Accounts of Chemical Research*, 2007, *40*, 53–62.

58. Yajun, Q.; Hongyan, Q.; Chaojing, L.; Ye, Y.; Yong, Z. Photoluminescence and magnetic properties of β-$Ni(OH)_2$ nanoplates and NiO nanostructures. *Journal of Materials Science: Materials in Electronics*, 2009, *20*, 479–483.

59. Wei, H.; Luoxin, Y.; Nan, Z.; Aiwei, T.; Mingyuan, G.; Zhiyong, T. Synthesis and shape-tailoring of copper sulfide/indium sulfide-based nanocrystals. *Journal of the American Chemical Society*, 2008, *130*, 13152–13161.

Part III

Selected Intriguing Topics
in Nanotechnology

13 Coordination and Organometallic Nanomaterials

13.1 INTRODUCTION

Nanomaterials are studied by a research field, which is devoted to materials with morphological features on the nanoscale, having special properties based on their nanoscale dimensions. The advantages of nanomaterials are the large surface area or volume ratio exhibited by them, leading to a very high surface reactivity with the surrounding surface, ideal for catalysis or sensor applications, and the ability of varying their fundamental properties (e.g., magnetization, optical properties [color], melting point, hardness, etc.), relative to bulk materials without a change in chemical composition.[1] According to classic sources, materials referred to as "nanomaterials" generally fall into two categories: fullerenes and inorganic nanoparticles. At the same time, during the last decade, the number of reports, dedicated to nanomaterials based on metal-containing complexes with σ- or π-metal-ligand bonds (coordination and organometallic compounds), has dramatically increased in comparison with last years of the previous century. The recent achievements in nanomaterials in particular including metal-organic composites have been generalized in a series of monographs[2–5] and reviews on various particular aspects of metal-complex nanomaterials.[6–18] Additionally, metal complexes are frequently used as useful precursors for obtaining inorganic nanomaterials and composites, for instance nanostructures of metals and their oxides and chalcogenides, which are of extensive interest for their widespread use in catalysis, optics, electronics, optoelectronics, information storage, and biological and chemical sensing. Techniques of their synthesis from coordination compounds were reviewed.[19] The advantages for the use of metal complexes as precursors are as follows: simple process, mild synthetic conditions, clean reactions, excellent reproducibility, and high product quality make the authors' strategies significant. It is believed that the introduction of coordination structures into the synthesis of metal-oxide and chalcogenide nanocrystals may provide a versatile tool to control their growth and assembly into well-defined nanostructures. This field is out of scope of the present section, where we try to comprehend briefly the main directions of recent development in the coordination nanomaterial synthesis, properties, and applications. Attention is paid to coordination polymers, an emerging new class of hybrid nanomaterials with promising characteristics for a number of practical applications, such as gas storage and heterogeneous catalysis.[20]

13.2 COORDINATION NANOMATERIALS AND NANOCOMPOSITES

13.2.1 Nanomaterials Based on Nitrogen-Containing Ligands

Among N-containing ligand complexes forming nanomaterials, a considerable number of reports have been devoted to complex *cyanides*. Thus, the complex metal cyanides $Co_3[Co(CN)_6]_2$, $Ni(CN)_2$, $Cu(CN)$, and $M_2[Fe(CN)_6]$ (M = Co^{2+}, Ni^{2+}, Cu^{2+}) were prepared as nanomaterials from metal chlorides via a reverse microemulsion method {CTAB (cetyltrimethylammonium)/*n*-dodecane/1-hexanol/ water} followed by diethylene glycol (DEG) treatment.[21] The as-prepared complex cyanides were also used as precursors for thermal decomposition to generate the pertinent oxide materials, namely, Co_3O_4, NiO, CuO, and $CoFe_2O_4$. An approach to the synthesis of multifunctional nanoparticles was developed by using covalent anchoring of cyanobridged coordination polymer $Ni^{2+}/[Fe(CN)_6]^{3-}$ to the surface of two-photon dye-doped mesoporous silica nanoparticles,[22] leading to homogeneously dispersed unishaped nanoparticles of around 100 nm length that were coated with cyanobridged metallic coordination polymer nanoparticles. As an example of application of this type of complexes as precursors, an insertion of the polynuclear nickel(II) and copper(II) complexes into the pores of mesoporous silica sieve MCM-41, followed by the decomposition or reduction of the complexes, and a synthesis of MCM-41 or microporous coordination polymer of $Na_2Zn_3[Fe(CN)_6]_2$ in the presence of Fe_3O_4 nanoparticles allowed fabrication of a series of porous materials containing polynuclear complexes and nanoparticles of oxides or metals[23] and preserving the ferromagnetic properties of the nanoparticles and the porous properties of silica component.

A series of recent original reports[24,25] are devoted to *Prussian blue* (PB), one of the first synthetic pigments with the idealized formula $Fe_7(CN)_{18}$ with 14 to $16H_2O$, and its analogues, which are representative of coordination polymers and have played important roles in the field of molecular magnets. Thus, the dissociation of apoferritin into subunits at pH 2 followed by its re-formation at pH 8.5 in the presence of hexacyanoferrate(III) gave rise to a solution containing hexacyanoferrate(III) trapped within the apoferritin and hexacyanoferrate(III) outside it; after the addition of Fe^{II} (Figure 13.1) to the dialyzed solution, the appearance of the characteristic PB color showing discrete spherical electron dense iron particles with an average size of about 5 nm was observed.[26] The authors noted that this represented a new route for preparing metallic nanoparticles that offers control over the size and protection against aggregation. Single system bifunctional inorganic–organic hybrid nanocomposites, PB@SiO_2@BTC@Ln {BTC = benzene tricarboxylate; Ln = Tb(III)/Sm(III)} with a PB magnetic core and a luminescent lanthanide probe, showed superparamagnetic behavior and significant enhancement in luminescence intensities.[27] PB nanoparticles protected by poly(vinylpyrrolidone) were also characterized.[28] Additionally, when PB analog, Co hexacyanoferrate (Fe-CN-Co) metal coordination nanopolymers (MCNPs) were synthesized in reverse micelles of cationic surfactants, cetyltrimethylammonium halides [CTAX, X = Br, Cl], their color dramatically changed from red to green with increasing the reaction time.[29] It was established that the coordination geometry of Co(II) ions was changed from a 6-coordinate octahedral to a 4-coordinate tetrahedral with clear crystal distortion.

Coordination nanomaterials on the *porphyrin* and *phthalocyanine* basis are also known. Thus, vanadium pentoxide nanocomposites containing supramolecular iron or manganese tetrapyridyl porphyrins coordinated to four $[Ru(bipy)_2Cl]^+$ complexes were synthesized.[30] The films generated by direct deposition of these nanocomposite aqueous suspensions onto interdigitated gold electrodes

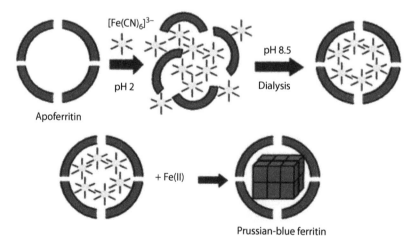

FIGURE 13.1 Schematic representation of the pH-induced dissociation-reformation process of apoferritin in the presence of hexacyanoferrate(III) and the subsequent reaction with iron(II) giving rise to Prussian blue complex within the Apoferritin cavity. (Reproduced with permission from Domínguez-Vera, J.M. and Colacio, E., Nanoparticles of Prussian blue ferritin: A new route for obtaining nanomaterials, *Inorg. Chem.*, 42(22), 6983–6985. Copyright 2003 American Chemical Society.)

exhibited good electrical and electrochemical performance for application in amperometric sensors. The 1D, 2D, or 3D nanomaterials based on metal *meso*-tetraarylporphyrins **1** (R^1 = H and R^2 = H, Me; or R^1 = OCH_3, OC_8H_{17} and R^2 = H; M = Sn, Ge, Si) alternately arranged with dicyclopropa-fullerenedicarboxylates or tricyclopropafullerenetricarboxylate were reported.[31] The nanomaterials had a high photoabsorption in the UV and visible spectrum and can be used for photoelectrical conversion purposes. The preparation of *t*-dodecanethiol-protected Au nanoparticles amenable to very rapid surface functionalization by associative mechanism was described.[32] Their reported properties were found to be ideal for the preparation of a variety of new organic–inorganic hybrid nanomaterials by coordinative self-assembly, for instance with *meso*-tetra(4-pyridyl)porphyrin, which promptly bonded to the reactive protected Au nanoparticles generating a solid that can be grown as thin films with layer-by-layer control on a suitably modified surface. Porphyrin nanomaterials were obtained by an electrostatic layer by layer assembly of [μ_4-5,10,15,20-tetra(3- or 4-pyridylporphyrin) tetrakis(bisbipyridyl)chlororuthenium(II)] and metal tetrasulfonatophthalocyaninates of transition metal ions, generating films constituted as column structures of about 30 nm high and for use in diverse fields of applications, in particular, active materials for electrochemical/electric detectors for sulfite/SO_2, nitrite/nitrate, and ascorbic acid.[33]

1

Among nanomaterials based on other related N-containing heterocycles, we note coordination polymer nanotubes, together with nanoparticles and nanowires (depending on ligand ratios), which were prepared by using the Hg^{2+}-mediated co-assembly of two ligands, tetrakis(4-pyridyl)porphine (TPyP) and tris(4-pyridyl)-1,3,5-triazine (TPyTa), at the water–$CHCl_3$ interface.[34] Nanotubular structure was also observed in the case of materials based on [{[Cd(apab)$_2$(H$_2$O)]$_3$(MOH)·G}$_n$] (apab, 4-amino-3-(pyridin-4-ylmethyleneamino)benzoate; G, guest molecules), whose structure consists of a large single-walled metal-organic nanotube of [{Cd(apab)$_2$(H$_2$O)}$_{3n}$] with an exterior wall diameter of up to 3.2 nm and an interior channel diameter of 1.4 nm, holding together by alkali metal cations to form 3D nanotubular supramolecular arrays.[35] Nanoparticles of a Pb^{II} polymer, [Pb(μ-pyr)(μ-I)$_2$]$_n$, with a net-like morphology were synthesized by the reaction of pyrazine (pyr) with Pb(NO$_3$)$_2$ and NaI via sonochemical irradiation[36] and can be used as precursors for obtaining nanostructured PbI$_2$ and PbO by calcinations. An aqueous solution of Zn(NO$_3$)$_2$·6H$_2$O and an EtOH solution of 1,4-bis(imidazol-1-ylmethyl)benzene were mixed with Fe$_2$O$_4$ nanoparticles to afford a colloidal system with encapsulated Fe$_2$O$_4$ nanoparticles.[37] A coordination polymer [Fe^{II}(4,4'-bpy)$_3$(H$_2$O)$_2$]

$(PF_6)_2 \cdot (4,4'\text{-bpy})_2 \cdot 5H_2O$ (bpy = bipyridine) was shown to be a three-dimensional network with nanosized channels constructed from 1D coordination chains via noncovalent interactions between the pyridyl rings.[38] The channels clathrate large molecules of free 4,4'-bpy. Two-layered nanocomposites, $[Co(1,10\text{-phen})_3]_yMoS_2$ and $[Co(2,2'\text{-bpy})_3]_yMoS_2$, were prepared from the interaction of MoS_2 single-layer dispersions with aqueous solutions of corresponding coordination cationic complexes.[39] In the presence of O_2, the composites exhibited good catalytic activities (enhanced upon irradiation with visible light) in the oxidization of sulfide ions into thiosulfate ions, due to their large surface areas. Other reports on N-containing ligands include the guest copper(II)-methyladenine complex, which was entrapped into beta zeolite (BEA) as a host by a process of sequential introduction of the components in the liquid phase followed by assembly inside the void space of the zeolite.[40] Also, second-generation diazadibenzoperylene dendrimers (yellow units) underwent supramolecular polymerization with Ag^+ ions to afford rigid-rod cylindrical coordination polymers, constituting a new class of π-conjugated materials linked by metal centers, which may find application in nanoscale electronic devices.[41]

13.2.2 NANOMATERIALS BASED ON N,O-CONTAINING LIGANDS

Oxygen-containing *bipyridine derivatives* and related heterocycles have also been obtained as nanomaterials. For instance, reaction of $AgPF_6$ with the asymmetric ligand 1,6-dihydro-2-methyl-6-oxo-(3,4'-bipyridine)-5-carbonitrile (L) afforded a significant Ag coordination polymer $\{[Ag_2(L)_3]_2 \cdot (MeOH)_3 \cdot (PF_6)_4\}_n$ with unique 1D twofold interpenetrating metal-organic frameworks constructed by 1D triple helical chains with nanosized cages hosting counterions as guests.[42] These compounds serve also as precursors for nanocomposites, for example, ruthenium complex $[Ru(Bpy)_2L]CL_2$ (L = 5-(trimethoxysilylpropylaminocarbonyl)-2,2'-bipyridine), which was prepared and treated with $Si(OMe)_4$ to give wet gels that were dried to form a nanocomposite based on the Ru complex supported on SiO_2.[43] Lanthanide/transition metal coordination polymers $[Ln_2Cu(\mu_6\text{-}O)_{1/6}(\mu_3\text{-}OH)_{2/3}(\mu_3\text{-}OH)_2(hpzc)_2(H_2O)_2] \cdot 2H_2O$ [Ln = Dy, Ho, Er; Hhpzc = 3-hydroxypyrazine-2-carboxylic acid], based on nanosized Ln-O-Ln inorganic rods and *in situ* obtained 3-hydroxypyrazine-2-carboxylate, were synthesized.[44] It was found by the authors to be the first example when nanorods are incorporated into metal-organic frameworks.

In the case of classic *EDTA* and related compounds, the hybrid mesoporous nanostructured material, containing *p*-amino benzoic acid bound to titanium oxide, reacted with EDTA monoanhydride or DTPA dianhydride to yield the materials[45] suitable for the complexation (chemisorption) and separation of the lanthanide cations. It was shown that the DTPA-based one was stronger at the chemisorption capacity but weaker at the selectivity. Such nanomaterials can have medical applications, for instance, $Ru(bpy)_3^{2+}$-doped Gd-Si-DTPA-functionalized silica nanoparticles, prepared having about $63,200 Gd^{3+}$ ions per nanoparticle, can be used to diagnose diseases, including cancer, cardiovascular disease, and diseases related to inflammation.[46]

Different *Schiff base complexes* have been reported as a consistent part of nanomaterials, for instance the Schiff bases of *N,N'*-bis(salicylidene)-1,4-diaminobenzene, 4-[2-(hydroxy-1-naphthalen-1-yl-methylene)-amino]-benzoic acid and 4-(benzylidene-amino)-benzoic acid, which, synthesized and encapsulated in the non-poplar channels of $(CH_3)_3Si$-MCM-41, revealed stronger luminescent intensities than that of the composites encapsulated in the polar channels of MCM-41, resulting in the lipophilic channels of host more favorable to fluorescence of the guest Schiff base molecules than the hydrophilic channels.[47] The monomer transition metal complexes [ML] (M = Mn(II), Co(II), Ni(II) and Cu(II)), synthesized from the reaction of metal acetate with bis(salicylaldehyde) oxaloyldihydrazone, H_2L; in 1:1 molar ratio in ethanol under reflux, were entrapped in the nanocavity of zeolite-Y in the liquid phase and showed catalytic activities for oxidation of cyclohexane.[48] A series of iron(III) complexes derived from a variety of pentadentate Schiff base ligands, namely, salten [H_2salten = bis(3-salicylideneaminopropyl)amine] derivatives, were synthesized and further used to stabilize and functionalize gold nanoparticles (Au-NPs).[49] The gold nanocomposites were

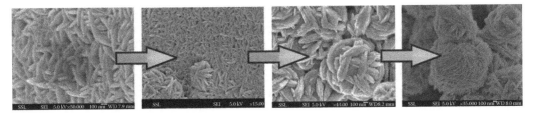

FIGURE 13.2 Probable formation mechanism for the nanomaterial CuS. (Reproduced with permission from Nagarathinam, M. et al., From self-assembled Cu(II) coordination polymer to shape-controlled CuS nanocrystals, *Cryst. Growth Design*, 9(5), 2457–2463. Copyright 2009 American Chemical Society.)

synthesized either by direct reduction of Au^{III} in the presence of iron complexes or by coordination chemistry on prefunctionalized gold nanoparticles. We emphasize that a comprehensive review, dedicated to core-shell Fe/Fe_xO_y-Au nanoparticles and their applications, has been recently published.[50]

Among other nanomaterials and their precursors, based on N,O-ligands, a hydrogen-bonded 1D coordination polymer, $\{[Cu(Hsglu)(H_2O)] \cdot H_2O\}_n$ {$H_3sglu = N$-(2-hydroxybenzyl)-L-glutamic acid}, was used as a precursor (Equations 13.1 and 13.2) as well as a sacrificial template to synthesize covellite CuS nanomaterials, whose probable formation mechanism is presented in Figure 13.2.[51] A remarkable correlation between the repeating patterns of the metal atoms in the coordination polymer with the preliminary shape of the obtained CuS was revealed. A novel 2D porous coordination polymer $[Cd(pda)_2]_n$ {pda = 3-(3-pyridyl)acrylate acid} was synthesized by hydrothermal reaction of 3-(3-pyridyl)acrylate acid and $Cd(CH_3COO)_2 \cdot 2H_2O$.[52] The nanoporous heterometalorganic 3D coordination polymer is $[Zn_7(L)_6Cd_2(H_2O)_6] \cdot 28H_2O$ (L = 1-phosphonomethylproline), belonging to a monoclinic system with a highly ordered and heterometal nanoporous structure, was obtained by the self-assembly of multinuclear clusters $[Zn_7(L)_6]$ and CdO_6 and used as a hydrogen storage material.[53] Coordination polymer chains assembled from an Ni^{II} hexaazamacrocycle and sodium 4,4′-biphenyldicarboxylate were found to be packed forming a porous framework being precursors of nanocomposites of silver particles and matrix.[54] These nanocomposites can be prepared by immersing this framework in $AgNO_3$ solution and the matrix-free silver nanoparticles (~3 nm) isolated by treatment of the solid with boiling dioctyl ether containing oleic acid. The assembly of cavitand ligands H_4L (R = CH_2CH_2Ph) and Zn^{2+} ions yielded[55] a 1D polymer composed of hexameric, chiral, closed-surface nanocapsules, $[Zn_{16}(L)_6(\mu$-OH$)(H_2O)_{14}](NO_3)_7 \cdot 12Me_2CO \cdot 47H_2O$.

$$NH_2CSNH_2 + 3H_2O \rightarrow H_2S + 2NH_4^+ + CO_3^{2-} \qquad (13.1)$$

$$[Cu(HSglu)(H_2O)] \cdot H_2O + H_2S \rightarrow CuS + H_3Sglu + 2H_2O \qquad (13.2)$$

13.2.3 NANOMATERIALS BASED ON OXYGEN- AND/OR SULFUR-CONTAINING LIGANDS

β-*Diketones*, classic ligands in coordination chemistry, are widely represented in coordination nanomaterials, containing mainly rare earth elements as complex formers. Thus, a route for coordination of Eu^{3+} by dibenzoylmethane (DBM) covalently bonded inside hexagonal mesoporous SiO_2 was established to produce the highly luminescent nanomaterial Si(DBM)-Eu(DBM)$_2$ **2**,[56] formed by nanoaggregates with 200 nm diameter and showing an efficient DBM to Eu^{3+} intramolecular energy process, namely, *antenna effect*, which favored a highly luminescent behavior in this modified hexagonal mesoporous SiO_2. Morphological studies of vacuum-deposited thin films of the Eu complex of Eu(dibenzoylmethanato)$_3$(bathophenanthroline) (Eu(DBM)$_3$ bath) from metal-organic planar light-emitting nanocomposites consisting of Au island films and vacuum-deposited Eu(DBM)$_3$ bath film showed that thermal evaporation in vacuum did not cause chemical decomposition of Eu(DBM)$_3$ bath, and the obtained films were essentially amorphous.[57] Additionally, the spectral characteristics of europium β-diketonates are controlled by the Eu^{3+} ion fluorescence.[58] Structure, photo-, electroluminescence, and electron transport properties of a planar nanoscale light-emitting device consisting of Au nanoislands and Tb(thd)$_3$ were studied by the same authors.[59] In particular, electroluminescence from the nanocomposite was observed at voltages 15–30 V, showing the emission features both of Tb(thd)$_3$ and Au nanoislands; the results can be useful for the design of planar light-emitting devices with controllable spectral characteristics. Similar studies were carried out for *8-hydroxyquinoline* complexes; thus, molecular organization and morphology of vacuum-deposited thin films of tris(8-hydroxyquinoline) aluminum (Alq$_3$) from metal-organic planar light-emitting nanocomposites consisting of gold island films and Alq$_3$ film were studied showing that thermal evaporation in vacuum did not cause dissociation of Alq$_3$ and that the obtained films were essentially amorphous.[60] The study of dielectric properties and mechanical relaxation of nanocomposites prepared by extruding at 240°C polyamide 6 and cobalt acetylacetonate, nickel acetylacetonate, Mo-carbonyl Mo(CO)$_6$, or W-carbonyl W(CO)$_6$ revealed that removing water from extruded samples by thermo-treatment led to a total fading of a dielectric loss maximum for all composites.[61]

Si(DBM)-Eu(DBM)$_2$

2

Carboxylate nanomaterials can be used for adsorption of CO molecules, such as the Ru@MOF-5 nanocomposite, obtained from [Zn$_4$O(bdc)$_3$] (MOF-5; bdc = 1,4-benzenedicarboxylate) and [Ru(cod)(cot)] (cod = 1,5-cyclooctadiene, cot = 1,3,5-cyclooctatriene) as precursors.[62] The nanopowders,[63] with improved rheological and surface mechanical properties and useful in coatings, having the structure M$_i$O$_j$(OH)$_k$[(OOC)$_l$R$_m$]$_n$ (M = Group 5 or Group 8 metal or lanthanide; R = H or organic group; i = 1–20, j = 0–40, k = 0–80, l = 1–80, m = 0–80, n = 1–80; the molar ratio of M to OOCR being 20:1 to 1:4), were obtained and used in a transparent composite coating. Surface-modified Fe$_3$O$_4$ nanoparticles were obtained by substituting [(η^5-semiquinone)Mn(CO)$_3$] for oleylamine surface protecting groups[64] and can function as a nucleus or template to generate crystalline coordination polymers that contain superparamagnetic Fe$_3$O$_4$ nanoparticles. Introducing paramagnetic metal nodes, such as Mn^{2+}, into these polymers led to hybridized magnetic properties.

Sulfur-containing coordination nanomaterials are represented mainly by thiophene derivatives and N,S-heterocycles. Thus, diodes composed of a nanoparticulate composite of poly(3,4-ethylenedioxythiophene) and a Cu-Cu^{2+} redox couple in a poly(ethylene oxide)-LiBF$_4$ polymer–electrolyte matrix between Ag and Zr electrodes showed rectifications in excess of 50,000 at applied fields of 4 V.[65] Two bilayered metal-organic frameworks with nanoporous channels, [Cd(TPT)(BDC)]·6H$_2$O and [Cd(TPT)(BDC)]·H$_2$BDC (TPT = tetrakis(4-pyridyl)thiophene, H$_2$BDC = 1,3-benzenedicarboxylic acid), were synthesized at different ligand-to-metal ratios, which demonstrated an interesting crystal-to-crystal transformation property and a special fluorescent response to the different guest molecules included.[66] Among other S-containing heterocycles, we note direct hydrothermal reactions between 6-mercaptopurine (6-MPH) and the analogous 6-thioguanine (6-ThioGH) with $NiSO_4 \cdot 6H_2O$ yielding the conductive coordination polymers $[Ni(6-MP)_2]_n \cdot 2nH_2O$ and $[Ni(6-ThioG)_2]_n \cdot 2nH_2O$, whose structures are based on 1D chains in which the deprotonated nucleobases act as the bridging ligands connecting the metal ions by short distances.[67] The potential use of these compound coordination polymers on nanomaterials for molecular electronics was suggested. Additionally, the bis(thioether)silanes $Me_2Si(CH_2SR)_2$ (R = Me, Pr-i, But) were introduced in reaction with copper(I) halides CuX to yield products with dimeric structures (X = Br, I) or complex cation $[Cu\{Me_2Si(CH_2SMe)_2\}_2]^+$ and the copper(I) dichloride anion $[CuCl_2]^-$ (X = Cl).[68] Some of these or related Ag(I), Au(I), and Pd(II) species can be applied as nanomaterials.

13.3 ORGANOMETALLIC NANOMATERIALS

Organometallic nanomaterials[69] are mainly based on ferrocene and other bis-arene π-complexes. Thus, ferrocene nanocrystals (average diameter 40 ± 5 nm) were successfully prepared by the ultrasonic-solvent-substitution method.[70] It was established that the ferrocene-nanosphere-modified glassy carbon electrode exhibited a better electrocatalytic effect than that modified by the ferrocene bulk material in phosphate buffer (pH 5.00). The method could be potentially applied to the preparation of nanomaterials of other organometallic compounds. A facile synthesis of spherical ZnS nanocrystals modified with dianion of 3-ferrocenyl-2-crotonic acid (FCA) from a 1D coordination polymer was reported.[71] This Zn(II) coordination polymer was demonstrated as both precursor and template in the preparation and surface modifications of nanomaterials. The 1D polymeric chain structure was found to be an excellent template for synthesizing organized assemblies of ZnS nanocrystals at low temperatures and ambient atmosphere. Nonlinear optical polymeric film-producing nanocomposites based on bis(arene)chromium complexes were described.[72] Syndiotactic polystyrene (sPS)/montmorillonite nanocomposites were prepared via *in situ* intercalative coordination polymerization using a mono(η^5-pentamethylcyclopentadienyl)trisbenzyloxytitanium [Cp*Ti(Obz)$_3$] complex activated by methylaluminoxanes and triisobutylaluminum.[73] It was shown that the degree of crystallinity of the sPS nanocomposite increased with increasing montmorillonite content and with higher Tg and thermal decomposition temperature than pure sPS.

13.4 NANOMATERIALS BASED ON POLYMERS

The chitosan, widely applied as a polymer support for nanomaterials, was used in the preparation of magnetic water-soluble cyanobridged metallic coordination polymer nanoparticles mentioned earlier (see complex cyanides) of controlled size forming water-soluble chitosan beads, consisting of M^{2+}/$[Fe(CN)_6]^{3-}$/chitosan (M = Ni, Cu, Co and Fe).[74] In the case of M = Gd, metallic coordination polymer nanoparticles with a chitosan shell showed high nuclear relaxivity in acidic water, which was up to six times higher than that of the actually used Gd-chelates. Methacrylic acid (MA) was used to modify the surface of TiO_2 using a Ti-carboxylic coordination bond[75] with further copolymerization with methyl methacrylate (MMA) to form a TiO_2-PMMA nanocomposite (Figure 13.3). The resulting nanocomposites exhibited improved elastic properties and have potential application in dental composites and bone cements. Soft polymeric nanomaterials in the form of nanogels were synthesized by a template-assisted method involving condensation of the poly(ethylene oxide)-polycarboxylate anions by metal ions into core-shell block ionomer complex micelles followed by chemical crosslinking of the polyion chains in the micelle cores.[76] The application of these materials for loading and release of a drug, *cis*-platin, was evaluated showing their suitability for delivery of pharmaceutical agents. Novel nonlinear optical polymeric film-producing nanocomposites based on bis(arene)chromium complexes incorporated into a CN-containing matrix were developed.[77] Polymeric nanocomposite precursors were prepared by the reaction of $Cr(Et_nC_6H_{6-n})_2$ (n = 1, 2, 3) with CN-containing vinyl monomers (acrylonitrile, crotononitrile, or Et 2-cyanopropenoate). Metal-polymer nanomaterials based on $CdCl_2$, $AgNO_3$, $HgCl_2$, K_2PtCl_4, or K_2PdCl_4 as metal ion source and the vinylpyridine polymer precursor

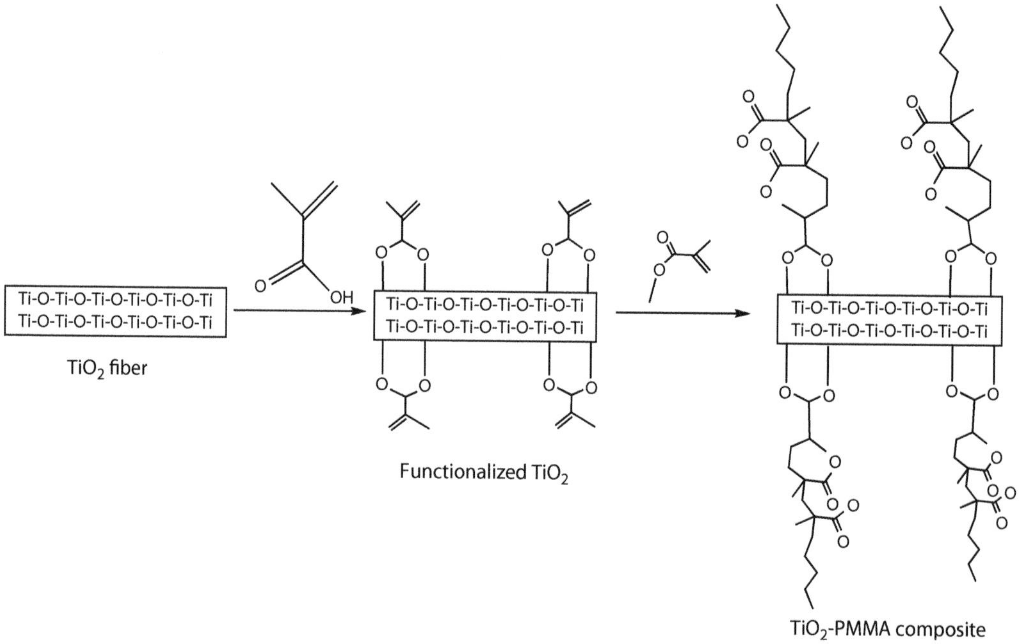

FIGURE 13.3 Schematic of functionalization of TiO_2 and formation of TiO_2-PMMA nanocomposite. (Reproduced with permission from Khaled, S.M., Synthesis of TiO_2-PMMA nanocomposite: Using methacrylic acid as a coupling agent, *Langmuir*, 3(7), 3988–3995. Copyright 2007 American Chemical Society.)

derivatives were described.[78] Among other nanocomposites with the same noble metals, palladium and platinum bis-acetylide polymers were prepared and characterized; the palladium vapor deposition yielded their hybrids with Pd(0) nanoclusters.[79] Pd(0) nanoclusters were dispersed by using the metal vapor synthesis technique in organometallic polymers $[-M(Pbu_3)_2C \equiv CC_6H_4C_6H_4C \equiv C-]_n$ (Pd-DEBP, Pt-DEBP; M = Pd, Pt). A class of composite nanomaterial CdSe quantum dots functionalized with oligo-(phenylene vinylene) (OPV) ligands (CdSe-OPV nanostructures) represented significantly modified photophysics relative to bulk blends or isolated components.[80] Single-molecule spectroscopy on these species revealed novel photophysics such as enhanced energy transfer, spectral stability, and strongly modified excited state lifetimes and blinking statistics.

Among other polymer-based nanomaterials of metal complexes, we note Au nanoparticle-Pt(II) diethynylbiphenyl polymer nanocomposites[81] and silver nanoparticles in fluoropolymer matrixes.[82] In the case of the latter, the obtained polymer matrix nanocomposite had optical properties that were dominated by the response of the nanoparticles owing to the broadbanded transparency of the fluoropolymer matrix. Other current and possible applications of polymer nanocomposites are related to catalysis and drug delivery. Thus, poly(styrene) brushes containing Co(III)-salen or piperazine side chains were prepared via atom-transfer radical polymerization from Fe_3O_4 nanoparticles modified with appropriate initiator molecules.[83] These catalysts were suitable for ring opening of epoxides and promoting the Knoevenagel condensation of benzaldehyde and malononitrile, respectively, and can be easily removed from solution via application of a magnetic field, allowing straightforward recovery and reuse. Nanoparticles with a magnetic core comprising an iron carbonyl ethyl cellulose complex with a latex coating forming a colloidal system can be used for drug delivery purposes.[84]

13.5 COORDINATION NANOMATERIALS AS PRECURSORS

As mentioned earlier, metal complexes are currently widely and successfully used as precursors for obtaining inorganic nanocomposites and nanoparticles.[85] Here we briefly pay attention to some representative examples of such processes. The synthesized coordination compounds or their composites with/on inert supports (SiO_2, TiO_2, etc.) are generally decomposed thermally (pyrolysis), ultrasonically, via laser and microwave irradiation, MO CVD, and by a series of other modern techniques forming elemental metals, their oxides, or more rarely salts inside the support structure. Conventional β-diketonates[86] or alcoholates as more complex ligand systems can serve as precursors. Thus, a series of novel germanium(II) precursors $[Ge(\mu c\text{-DBED})]_2$ (μc = bridging chelating), $[Ge(\mu\text{-DMP})(DMP)]_2$, $Ge(DPP)_2$, $[Ge(\mu\text{-OtBu})(DMBS)]_2$, $[Ge(\mu\text{-DMBS})(DMBS)]_2$, $Ge(TPS)_3(H)$, $[Ge(\mu\text{-TPST})(TPST)]_2$, and $Ge(PS)_4$ were synthesized from the reaction of $Ge[N(SiMe_3)_2]_2$ or $[Ge(OtBu)_2]_2$ and ligands N,N''-dibenzylethylenediamine (H_2-DBED), t-Bu alcohol {H-O(t-Bu)}, 2,6-dimethylphenol (H-DMP), 2,6-diphenylphenol (H-DPP), t-butyldimethylsilanol (H-DMBS), triphenylsilanol (H-TPS), triphenylsilanethiol (H-TPST), and benzenethiol (H-PS). Using a simple solution precipitation methodology, Ge^0 nanomaterials were isolated as dots and wires for the majority of these precursors.[87] A congener series using the neo-pentoxide (ONep) as $[Ti(\mu\text{-ONep})(ONep)_3]_2$, $M_2(\mu\text{-ONep})_2(\mu\text{-HONep})(ONep)_6$ (M = Zr and Hf), and related $Hf(OR)_4$ family were developed, characterized, and used for the production of nanometal oxide materials due to the importance of Group 4 (Ti, Zr, and Hf) cations, widely used in ceramic materials production.[88] The preparation of nanostructured particles of metal/metal oxides (M = Cr, Fe, Ru, and Mn) by using a pyrolysis at 800°C of organometallic polyphosphazenes with the metal anchored to the polymeric chain was described.[89]

A coordination polymer $[C_8H_{10}CdO_7]_n \cdot 4H_2O$ was prepared via a hydrothermal procedure by using 1,4-benzenedicarboxylic acid (p-BDC) and CdI(I) salt as starting materials.[90] At heating, it was converted to uniform CdO nanowires, suggesting an effective and reasonable complex-precursor procedure for the preparation of 1D crystalline nanomaterials. A limited number of organometallic complexes have been used as catalysts {typically ferrocene or $Fe(CO)_5$} to make carbon materials that have distinctive shapes.[17] Depending on the reaction conditions employed, ferrocene can be used to synthesize single-walled (SWCNTs), double-walled (DWCNTs), and multiwalled nanotubes (MWCNTs) as well as fibers and other SCNMs. The use of a 2D coordination polymer generated from a trinuclear building block $[Cu_3(HSser)_3(H_2O)_2] \cdot 2H_2O$ (H_3Sser = N-(2-hydroxybenzyl)-L-serine) (similar to the earlier described {$[Cu(Hsglu)(H_2O)] \cdot H_2O\}_n$) as a precursor in the synthesis of copper sulfide resulted in CuS nanospheres with hollow interiors.[91] The electrochemical behavior of these nanosized CuS mesoassemblies as a cathode material for Li ion batteries revealed that the reaction proceeds through an insertion and deinsertion mechanism and the CuS with a hollow interior is more efficient and has good cyclability compared with that with a flower-like morphology.

REFERENCES

1. Fahlman, B. D. *Materials Chemistry*. Springer, Berlin, Germany, 2007, 485 pp.
2. Schodek, D. L.; Ferreira, P.; Ashby, M. F. Nanomaterials. In *Nanotechnologies and Design: An Introduction for Engineers and Architects*. Butterworth-Heinemann, Oxford, U.K., 2009, 560 pp.
3. Vollath, D. *Nanomaterials: An Introduction to Synthesis, Properties and Applications*. Wiley-VCH, Weinheim, Germany, 2008, 362 pp.
4. Ozin, G. A.; Arsenault, A. C.; Cademartiri, L. *Nanochemistry: A Chemical Approach to Nanomaterials*, 2nd edn., Royal Society of Chemistry, Cambridge, U.K., 2009, 820 pp.
5. Ramesh, K. T. *Nanomaterials: Mechanics and Mechanisms*, 1st edn., Springer, Berlin, Germany, 2009, 316 pp.
6. Yamada, M. Synthesis of organic shell-inorganic core hybrid nanoparticles by wet process and investigation of their advanced functions. *Bulletin of the Chemical Society of Japan*, 2009, *82* (2), 152–170.
7. He, Z.; Ishizuka, T.; Jiang, D. Dendritic architectures for design of photo- and spin-functional nanomaterials. *Polymer Journal*, 2007, *39* (9), 889–922.
8. Haga, M.; Kobayashi, K.; Terada, K. Fabrication and functions of surface nanomaterials based on multilayered or nanoarrayed assembly of metal complexes. *Coordination Chemistry Reviews*, 2007, *251* (21–24), 2688–2701.
9. Corriu, R.; Mehdi, A.; Reye, C. Nanoporous materials: A good opportunity for nanosciences. *Journal of Organometallic Chemistry*, 2004, *689* (24), 4437–4450.

10. Braunstein, P. Functional ligands and complexes for new structures, homogeneous catalysts and nanomaterials. *Journal of Organometallic Chemistry*, 2004, *689* (24), 3953–3967.

11. Dai, J.-C.; Fu, Z.-Y.; Wu, X.-T. Supramolecular coordination polymers. In *Encyclopedia of Nanoscience and Nanotechnology*. Nalwa, H. S. (Ed.). American Scientific Publishers, Valencia, CA, 2004, Vol. 10, pp. 247–266.

12. Maekawa, H. Recent advances in structure direction of nanospace. 3. Organics, polymers, metals and semiconductors. *Materia*, 2006, *45* (7), 540–546.

13. Korczagin, I.; Lammertink, R. G. H.; Hempenius, M. A.; Golze, S.; Vancso, G. J. Surface nano- and microstructuring with organometallic polymers. *Advances in Polymer Science*, 2006, *200 (Ordered Polymeric Nanostructures at Surfaces)*, 91–117.

14. Shimomura, S.; Horike, S.; Kitagawa, S. Chemistry and application of porous coordination polymers. *Studies in Surface Science and Catalysis*, 2007, *170B (From Zeolites to Porous MOF Materials)*, 1983–1990.

15. Chan, W. K.; Cheng, K. W. Metal coordination polymers for nanofabrication. In *Frontiers in Transition Metal-Containing Polymers*. Abd-El-Aziz, A. S.; Manners, I. (Eds.). Wiley-Interscience, Hoboken, NJ, 2007, pp. 217–246.

16. Cho, C.-P.; Perng, T.-P. One-dimensional organic and organometallic nanostructured materials. *Journal of Nanoscience and Nanotechnology*, 2008, *8* (1), 69–87.

17. Nyamori, V. O.; Mhlanga, S. D.; Coville, N. J. The use of organometallic transition metal complexes in the synthesis of shaped carbon nanomaterials. *Journal of Organometallic Chemistry*, 2008, *693* (13), 2205–2222.

18. Sergeev, G. B.; Nemukhin, A. V.; Sergeev, B. M.; Shabatina, T. I.; Zagorskii, V. V. Cryosynthesis and properties of metal-organic nanomaterials. *Nanostructured Materials*, 1999, *12* (5–8), 1113–1116.

19. Xie, Y.; Xiong, Y.; Li, Z. Application of coordination chemistry in the fabrication of inorganic nanostructures. In *Frontal Nanotechnology Research*. Berg, M. V. (Ed.). Nova Science Pub Inc., Hauppauge, NY, 2007, pp. 41–59.

20. Lin, W.; Rieter, J. W.; Taylor, K. M. L. Modular synthesis of functional nanoscale coordination polymers. *Angewandte Chemie, International Edition*, 2009, *48* (4), 650–658.

21. Buchold, D. H. M., Feldmann, C. Nanoscale complex metal cyanides and thermolysis thereof. *Solid State Sciences*, 2008, *10* (10), 1305–1313.

22. Chelebaeva, E.; Raehm, L.; Durand, J.-O.; Guari, Y.; Larionova, J.; Guerin, C.; Trifonov, A. et al. Mesoporous silica nanoparticles combining two-photon excited fluorescence and magnetic properties. *Journal of Materials Chemistry*, 2010, *20* (10), 1877–1884.

23. Kolotilov, S. V.; Shvets, A. V.; Kas'yan, N. V.; Il'in, V. G.; Pavlishchuk, V. V. Development of new type of magnetic materials by insertion of ferrimagnetic Fe_3O_4 nanoparticles or heteronuclear complexes of transition metals into silica and coordination porous matrixes. *Nanosistemi, Nanomateriali, Nanotekhnologii*, 2005, *3* (4), 1033–1045.

24. Ding, Y.; Hu, Y.-L.; Gu, G.; Xia, X.-H. Controllable synthesis and formation mechanism investigation of Prussian blue nanocrystals by using the polysaccharide hydrolysis method. *Journal of Physical Chemistry C*, 2009, *113* (33), 14838–14843.

25. Liang, G.; Xu, J.; Wang, X. Synthesis and characterization of organometallic coordination polymer nanoshells of Prussian blue using miniemulsion periphery polymerization (MEPP). *Journal of the American Chemical Society*, 2009, *131* (15), 5378–5379.

26. Domínguez-Vera, J. M.; Colacio, E. Nanoparticles of Prussian blue ferritin: A new route for obtaining nanomaterials. *Inorganic Chemistry*, 2003, *42* (22), 6983–6985.

27. Pal, S.; Jagadeesan, D.; Gurunatha, K. L.; Eswaramoorthy, M.; Maji, T. K. Construction of bi-functional inorganic-organic hybrid nanocomposites. *Journal of Materials Chemistry*, 2008, *18* (45), 5448–5451.

28. Uemura, T.; Kitagawa, S. Prussian blue nanoparticles protected by poly(vinylpyrrolidone). *Journal of the American Chemical Society*, 2003, *125* (26), 7814–7815.

29. Yamada, M.; Sato, T.; Miyake, M.; Kobayashi, Y. Temporal evolution of composition and crystal structure of cobalt hexacyanoferrate nano-polymers synthesized in reversed micelles. *Journal of Colloid and Interface Science*, 2007, *315* (1), 369–375.

30. Toma, H. E.; Timm, R. A.; Gonzalez, M. F.; Araki, K. Vanadium(V) oxide—Metal organic nanocomposites as electrochemical sensing materials. *Materials Science Forum*, 2010, *636–637* (Pt. 1, Advanced Materials Forum V), 729–736.

31. Zheng, J.; Xu, H.; Ruan, H.; Zhu, Y. Nanomaterial composed of alternately arranged metalloporphyrin and fullerene and its preparation. CN 1966546, 2007, 30 pp.

32. Araki, K.; Mizuguchi, E.; Tanaka, H.; Ogawa, T. Preparation of very reactive thiol-protected gold nanoparticles: Revisiting the Brust-Schiffrin method. *Journal of Nanoscience and Nanotechnology*, 2006, *6* (3), 708–712.

33. Araki, K.; Toma, H. E.; Mayer, I. Process for obtaining porphyrin nanomaterials, such as, films and electrodes, for simultaneous analytical determinations. BR 2007000803, 2008, 44 pp.

34. Liu, B.; Qian, D.-J.; Chen, M.; Wakayama, T.; Nakamura, C.; Miyake, J. Metal-mediated coordination polymer nanotubes of 5,10,15,20-tetrapyridylporphine and tris(4-pyridyl)-1,3,5-triazine at the water-chloroform interface. *Chemical Communications*, 2006, (30), 3175–3177.

35. Luo, T.-T.; Wu, H.-C.; Jao, Y.-C.; Huang, S.-M.; Tseng, T.-W.; Wen, Y.-S.; Lee, G.-H.; Peng, S.-M.; Lu, K.-L. Self-assembled arrays of single-walled metal-organic nanotubes. *Angewandte Chemie, International Edition*, 2009, *48* (50), 9461–9464, S9461/1–S9461/12.

36. Aslani, A.; Morsali, A. Sonochemical synthesis of nano-sized metal-organic lead(II) polymer: A precursor for the preparation of nano-structured lead(II) iodide and lead(II) oxide. *Inorganica Chimica Acta*, 2009, *362* (14), 5012–5016.

37. Ruiz Molina, D.; Maspoch Comamala, D.; Imaz Gabilondo, I. Metallo-organic systems for the encapsulation and release of compounds of interest, method for their preparation, and their uses. WO 2009133229, 2009, 44 pp.

38. Izarova, N. V.; Sokolov, M. N.; Rothenberger, A.; Ponikiewski, L.; Fenske, D.; Fedin, V. P. Synthesis and crystal structure of a new metal-organic coordination polymer [Fe(4,4'-bpy)$_3$(H$_2$O)$_2$](PF$_6$)$_2$·2·(4,4'-bpy)·5H$_2$O with nanosized channels clathrate large organic molecules. *Comptes Rendus Chimie*, 2005, *8* (6–7), 1005–1010.

39. Lin, B.-Z.; Pei, X.-K.; Zhang, J.-F.; Han, G.-H.; Li, Z.; Liu, P.-D.; Wu, J.-H. Preparation and characterization of nanocomposite materials consisting of molybdenum disulfide and cobalt(II) coordination complexes. *Journal of Materials Chemistry*, 2004, *4* (13), 2001–2005.

40. Teixeira, C.; Pescarmona, P.; Carvalho, M. A.; Fonseca, A. M.; Neves, I. C. Host(beta zeolite)-guest (copper(II)-methyladenine complex) nanomaterials: Synthesis and characterization. *New Journal of Chemistry*, 2008, *32* (12), 2263–2269.

41. Wuerthner, F.; Stepanenko, V.; Sautter, A. Rigid-rod metallosupramolecular polymers of dendronized diazadibenzoperylene dyes. *Angewandte Chemie, International Edition*, 2006, *45* (12), 1939–1942.

42. Niu, C.-Y.; Wu, B.-L.; Zheng, X.-F.; Zhang, H.-Y.; Li, Z.-J.; Hou, H.-W. The first 1D twofold interpenetrating metal-organic network generated by 1D triple helical chains with nanosized cages. *Dalton Transactions*, 2007, (48), 5710–5713.

43. Khimich, N. N.; Semov, M. P.; Chepik, L. F. Nanocomposites in the Ru^{2+} organic complex-SiO$_2$ system as a new class of metal-polymer complexes. *Doklady Chemistry (Translation of the Chemistry Section of Doklady Akademii Nauk)*, 2004, *394* (2), 31–33.

44. Chen, L.-F.; Zhang, J.; Ren, G.-Q.; Li, Z.-J.; Qin, Y.-Y.; Yin, P.-X.; Cheng, J.-K.; Yao, Y.-G. Nanosized lanthanide oxide rods in I1O3 hybrid organic-inorganic frameworks involving in situ ligand synthesis. *Crystal Engineering Communications*, 2008, *10* (8), 1088–1092.

45. Rahal, R.; Daniele, S.; Pellet-Rostaing, S.; Lemaire, M. New hybrid TiO$_2$ nano-structured materials for lanthanides separation. *Chemistry Letters*, 2007, *36* (11), 1364–1365.

46. Lin, W.; Rieter, W.; Taylor, K.; Kim, J. Nanomaterials as multimodal imaging contrast agents. WO 2007-US9796, 2007, 127 pp.

47. Yin, W. Luminescence of nanosized supramolecular materials composed of host mesoporous molecular sieves and guest Schiff bases. *Faguang Xuebao*, 2005, *26* (3), 349–353.

48. Salavati-Niasari, M., Sobhani, A. Ship-in-a-bottle synthesis, characterization and catalytic oxidation of cyclohexane by host (nanopores of zeolite-Y)/guest (Mn(II), Co(II), Ni(II) and Cu(II) complexes of bis(salicylaldehyde)oxaloyldihydrazone) nanocomposite materials. *Journal of Molecular Catalysis A: Chemical*, 2008, *285* (1–2), 58–67.

49. Mayer, C. R.; Cucchiaro, G.; Jullien, J.; Dumur, F.; Marrot, J.; Dumas, E.; Secheresse, F. Functionalization of gold nanoparticles by iron(III) complexes derived from Schiff base ligands. *European Journal of Inorganic Chemistry*, (23), 3614–3623, August 2008.

50. Kharisov, B. I.; Kharissova, O. V.; José Yacamán, M.; Ortiz Mendez, U. State of the art of the bi- and trimetallic nanoparticles on the basis of gold and iron. *Recent Patents on Nanotechnology*, 2009, *3* (2), 81–98.

51. Nagarathinam, M.; Chen, J.; Vittal, J. J. From self-assembled Cu(II) coordination polymer to shape-controlled CuS nanocrystals. *Crystal Growth and Design*, 2009, *9* (5), 2457–2463.

52. Zhou, Q.-x.; Wang, H.-f.; Zhao, X.-q.; Yue, L.; Wang, Y.-j. Synthesis and crystal structure of porous coordination polymer with nanometer-sized [Cd(pda)$_2$]$_n$. *Jixie Gongcheng Cailiao*, 2004, *28* (4), 36–38, 54.

53. Cheng, P.; Liu, H.; Chen, Y. Hepta-zinc hexa(1-phosphonomethylproline) dicadmium hexaaqua octaco-sahydrate ([Zn$_7$(L)$_6$Cd$_2$(H$_2$O)$_6$]·28H$_2$O) heterometal-organic framework nanoporous three dimensional coordination polymer, and its preparation and application in hydrogen storage material. CN 101428755, 2009, 9 pp.

54. Moon, H. R.; Kim, J. H.; Suh, M. P. Redox-active porous metal-organic framework producing silver nanoparticles from AgI ions at room temperature. *Angewandte Chemie, International Edition*, 2005, *44* (8), 1261–1265.

55. Ugono, O.; Moran, J. P.; Holman, K. T. Closed-surface hexameric metal-organic nanocapsules derived from cavitand ligands. *Chemical Communications*, 2008, (12), 1404–1406.

56. DeOliveira, E.; Neri, C. R.; Serra, O. A.; Prado, A. G. S. Antenna effect in highly luminescent Eu^{3+} anchored in hexagonal mesoporous silica. *Chemistry of Materials*, 2007, *19* (22), 5437–5442.

57. Dovbeshko, G.; Fesenko, O.; Fedorovich, R.; Gavrilko, T.; Marchenko, A.; Puchkovska, G.; Viduta, L. et al. FTIR spectroscopic analysis and STM studies of electroluminescent Eu(DBM)$_3$ bath thin films vacuum deposited onto Au surface. *Journal of Molecular Structure*, 2006, *792–793*, 115–120.

58. Chubich, D. A.; Fedorovich, R. D.; Vitukhnovsky, A. G. Electrical conductivity and luminescence of metal-organic nanocomposites. *Journal of Russian Laser Research*, 2008, *29* (4), 368–376.

59. Chubich, D.; Dovbeshko, G.; Fesenko, O.; Fedorovich, R.; Gavrilko, T.; Cherepanov, V.; Marchenko, A. et al. Light-emitting diode of planar type based on nanocomposites consisting of island Au film and organic luminofore Tb(thd)$_3$. *Molecular Crystals and Liquid Crystals*, 2008, *497*, 186–195.

60. Gavrilko, T.; Fedorovich, R.; Dovbeshko, G.; Marchenko, A.; Naumovets, A.; Nechytaylo, V.; Puchkovska, G.; Viduta, L.; Baran, J.; Ratajczak, H. FTIR spectroscopic and STM studies of vacuum deposited aluminium(III) 8-hydroxyquinoline thin films. *Journal of Molecular Structure*, 2004, *704* (1–3), 163–168.

61. Voilov, D. N.; Novikov, G. F.; Pesetskii, S. S.; Efremova, A. I.; Ivanova, L. L. Dielectric properties of nano-composites based on polyamide-6 and metal containing compounds. *Plasticheskie Massy*, 2008, (3), 15–18.

62. Schroeder, F.; Esken, D.; Cokoja, M.; van den Berg, M. W. E.; Lebedev, O. I.; Van Tendeloo, G.; Walaszek, B. et al. Ruthenium nanoparticles inside porous [Zn$_4$O(bdc)$_3$] by hydrogenolysis of adsorbed [Ru(cod)(cot)]: A solid-state reference system for surfactant-stabilized ruthenium colloids. *Journal of the American Chemical Society*, 2008, *130* (19), 6119–6130.

63. Organometallic nano-powders. 2006, DE 202005014332, 31 pp.

64. Kim, S. B.; Cai, C.; Sun, S.; Sweigart, D. A. Incorporation of Fe$_3$O$_4$ nanoparticles into organometallic coordination polymers by nanoparticle surface modification. *Angewandte Chemie, International Edition*, 2009, *48* (16), 2907–2910.

65. Zhang, R.; Barnes, A.; Wang, Y.; Chambers, B.; Wright, P. V. Organo-metal diodes based on a nanoparticulate poly(3,4-ethylenedioxythiophene) composite. *Advanced Functional Materials*, 2006, *16* (9), 1161–1165.

66. Lin, J.-G.; Xu, Y.-Y.; Qiu, L.; Zang, S.-Q.; Lu, C.-S.; Duan, C.-Y.; Li, Y.-Z.; Gao, S.; Meng, Q.-J. Ligand-to-metal ratio controlled assembly of nanoporous metal-organic frameworks. *Chemical Communications*, 2008, (23), 2659–2661.

67. Amo-Ochoa, P.; Castillo, O.; Alexandre, S. S.; Welte, L.; de Pablo, P. J.; Rodriguez-Tapiador, I.; Gomez-Herrero, J.; Zamora, F. Synthesis of designed conductive one-dimensional coordination polymers of Ni(II) with 6-mercaptopurine and 6-thioguanine. *Inorganic Chemistry*, 2009, *48* (16), 7931–7936.

68. Rabinovich, D. PolythioetherSilanes: Unexpected coordination complexes and applications. *Abstracts, 59th Southeast Regional Meeting of the American Chemical Society*, Greenville, SC, October 24–27, 2007, GEN-218.

69. Zeller, M.; Hunter, A. D.; DiMuzio, S. J.; Lazich, E.; Payton, J. L.; Perrine, C.; Updegraff, J. B.; Hoff, R. Organometallic nanomaterials. *Abstracts of Papers, 231st ACS National Meeting*, Atlanta, GA, March 26–30, 2006, INOR-356.

70. Chen, P.; Wu, Q.-S.; Ding, Y.-P. Preparation of ferrocene nanocrystals by the ultrasonic-solvent-substitution method and their electrochemical properties. *Small*, 2007, *3* (4), 644–649.

71. Yang, J. X.; Wang, S. M.; Zhao, X. L.; Tian, Y. P.; Zhang, S. Y.; Jin, B. K.; Hao, X. P.; Xu, X. Y.; Tao, X. T.; Jiang, M. H. Preparation and characterization of ZnS nanocrystal from Zn(II) coordination polymer and ionic liquid. *Journal of Crystal Growth*, 2008, *310* (19), 4358–4361.

72. Klapshina, L. G., Grigoryev, I. S., Semenov, V. V., Douglas, W. E., Bushuk, B. A., Bushuk, S. B., Lukianov, A. Yu. et al. Chromium-containing organometallic self-organized materials for nanophotonics. *Northern Optics Conference Proceedings*, Bergen, Norway, June 14–16, 2006. Sudbo, A. S.; Arisholm, G. (Eds.), pp. 29–34.

73. Shen, Z.; Zhu, F.; Liu, D.; Zeng, X.; Lin, S. Preparation of syndiotactic polystyrene/montmorillonite nanocomposites via in situ intercalative polymerization of styrene with monotitanocene catalyst. *Journal of Applied Polymer Science*, 2005, *95* (6), 1412–1417.

74. Guari, Y.; Larionova, J.; Molvinger, K.; Folch, B.; Guerin, C. Magnetic water-soluble cyano-bridged metal coordination nano-polymers. *Chemical Communications*, 2006, (24), 2613–2615.

75. Khaled, S. M.; Sui, R.; Charpentier, P. A.; Rizkalla, A. S. Synthesis of TiO_2-PMMA nanocomposite: Using methacrylic acid as a coupling agent. *Langmuir*, 2007, *3* (7), 3988–3995.

76. Kim, J. O.; Nukolova, N. V.; Oberoi, H. S.; Kabanov, A. V.; Bronich, T. K. Block ionomer complex micelles with cross-linked cores for drug delivery. *Vysokomolekulyarnye Soedineniya, Seriya A i Seriya B*, 2009, *51* (6), 1033–1042.

77. Klapshina, L. G.; Grigoryev, I. S.; Lopatina, T. I.; Semenov, V. V.; Domrachev, G. A.; Douglas, W. E.; Bushuk, B. A. et al. Chromium-containing organometallic nanomaterials for non-linear optics. *New Journal of Chemistry*, 2006, *30* (4), 615–628.

78. Qian, D.; Liu, B.; Huang, W. Preparation of metal-coordinated polymeric nanomaterial. CN 1673258, 2005, 7 pp.

79. Belotti, F.; Fratoddi, I.; La Groia, A.; Martra, G.; Mustarelli, P.; Panziera, N.; Pertici, P.; Russo, M. V. Preparation of nanostructured organometallic polymer/palladium hybrids by metal vapour synthesis: Structure and morphology. *Nanotechnology*, 2005, *16* (11), 2575–2581.

80. Hammer, N. I.; Emrick, T.; Barnes, M. D. Quantum dots coordinated with conjugated organic ligands: New nanomaterials with novel photophysics. *Nanoscale Research Letters*, 2007, *2* (6), 282–290.

81. Vitale, F.; Mirenghi, L.; Piscopiello, E.; Pellegrini, G.; Trave, E.; Mattei, G.; Fratoddi, I.; Russo, M. V.; Tapfer, L.; Mazzoldi, P. Gold nanoclusters-organometallic polymer nanocomposites: Synthesis and characterization. *Materials Science and Engineering, C: Biomimetic and Supramolecular Systems*, 2007, *27* (5–8), 1300–1304.

82. See, K. C.; Spicer, J. B.; Brupbacher, J.; Zhang, D.; Vargo, T. G. Modeling interband transitions in silver nanoparticle-fluoropolymer composites. *Journal of Physical Chemistry B*, 2005, *109* (7), 2693–2698.

83. Gill, C. S.; Long, W.; Jones, C. W. Magnetic nanoparticle polymer brush catalysts: Alternative hybrid organic/inorganic structures to obtain high, local catalyst loadings for use in organic transformations. *Catalysis Letters*, 2009, *131* (3–4), 425–431.

84. Arias Mediano, J. L.; Gallardo Lara, V.; Ruiz Martinez, M. A.; Lopez-Viota Gallardo, M. Nanoparticles with magnetic core comprising iron carbonyl ethyl cellulose complex with latex coating forming colloidal system for drug delivery. ES 2324003, 2009, 28 pp.

85. Revaprasadu, N.; Mlondo, S. N. Use of metal complexes to synthesize semiconductor nanoparticles. *Pure and Applied Chemistry*, 2006, *78* (9), 1691–1702.

86. Orimoto, Y.; Toyota, A.; Furuya, T.; Nakamura, H.; Uehara, M.; Yamashita, K.; Maeda, H. Computational method for efficient screening of metal precursors for nanomaterial syntheses. *Industrial and Engineering Chemistry Research*, 2009, *48* (7), 3389–3397.

87. Boyle, T. J.; Tribby, L. J.; Ottley, L. A. M.; Han, S. M. Synthesis and characterization of germanium coordination compounds for production of germanium nanomaterials. *European Journal of Inorganic Chemistry*, (36), 5550–5560, December 2009.

88. Boyle, T. J.; Pratt, H. D., III; Burton, P. D.; Raymond, R.; Ottley, L. A. M. Group 4 coordination precursors for ceramic nanomaterials. *Abstracts of Papers*, 237th ACS National Meeting, Salt Lake City, UT, March 22–26, 2009, INOR-311.

89. Diaz, C.; Valenzuela, M. L. Synthesis of nanostructured materials by a new solid state pyrolysis organometallic polymer method. *Journal of the Chilean Chemical Society*, 2005, *50* (1), 417–419.

90. Zhang, F.; Bei, F.-L.; Cao, J.-M.; Wang, X. The preparation of CdO nanowires from solid-state transformation of a layered metal-organic framework. *Journal of Solid State Chemistry*, 2008, *181* (1), 143–149.

91. Nagarathinam, M.; Saravanan, K.; Leong, W. L.; Balaya, P.; Vittal, J. Hollow nanospheres and flowers of CuS from self-assembled Cu(II) coordination polymer and hydrogen-bonded complexes of N-(2-Hydroxybenzyl)-L-serine. *Journal of Crystal Growth and Design*, 2009, *9* (10), 4461–4470.

14 Application of Ultrasound for Obtaining Nanostructures and Nanomaterials

Ultrasound (US) is currently a common laboratory tool used to nebulize solutions into fine mixtures, emulsify mixtures, drive chemical reactions, and to disperse nanoparticles and colloids.[1] It consists of acoustic waves with frequencies of more than 20 kHz. Interacting with a species can cause structural changes and accelerate chemical reactions, disrupting the weak noncovalent interactions or disintegrating aggregated particles, but seldom favors the assembly formation.[2] On the basis of ultrasound, *sonochemistry* (synthesis of materials under nonequilibrium conditions appearing under cavitation induced by acoustic waves resulting in the creation and collapse of microbubbles, providing extreme synthesis conditions such as temperatures of ~5000 K, pressures of ~1 GPa, and cooling rates of ~1010 K/s) studies the synthesis of metals, alloys, oxides, polymers, and a variety of other classes of chemical compounds both in isolated and composite form. Since the process duration is extremely short, the resulting particles will be of nanosize.

Recently, several nanochemistry- and nanotechnology-related applications of ultrasound have been extensively reviewed. Thus, some preparation methods of nanosized materials involving ultrasonic precipitation, ultrasonic pyrolysis, ultrasonic reduction, and sonoelectrochemistry were highlighted in a series of reviews.[3–9] The use of ultrasound for obtaining and functionalization of traditional carbon materials (e.g., carbon black and activated carbon), pigments, adsorbents, and composite components on its basis, as well as the carbon nanotubes, graphene, and meso- and macroporous carbons, which are objects of a permanent interest, was described by *Skrabalak*[10] and *Xing*,[11] in particular their preparation by both ultrasonic spray pyrolysis and high-intensity ultrasound or ultrasonic modification of their properties. Additionally, latest advances in the application of ultrasonic technology in catalytic chemistry, including its application for nanomaterial preparation, active component loading, and heterogeneous chemical reactions, were generalized by Yan et al.[12] Among other important ultrasound-assisted applications in materials chemistry, we note the synthesis of ceramic nanoparticles[13] and polymer nanocomposites.[14]

In the case of medical, biomedical, and related areas, new technologies that combine the use of nanoparticles with acoustic power both in drug and gene delivery were summarized by Husseini et al.[15] It was shown that ultrasonic drug and gene delivery from nanocarriers has tremendous potential because of the wide variety of drugs and genes that could be delivered to targeted tissues by fairly noninvasive means. Other related publications include discussion on the revolution in cancer treatment, potential uses of nanoparticles in oncology,[16] and use of micelles and nanoparticles for ultrasonic drug and gene delivery.[15,17] Pharmaceutical applications of ultrasound were described by Ishtiaq et al.[18]

Nanomaterials, in particular those obtained using ultrasound, have unusual properties not found in the bulk materials, which can be exploited in numerous applications[19] such as biosensing, electronics, scaffolds for tissue engineering, and diagnostics. Ultrasound could also act as a stimulus to induce the gelation of organic liquids with low-molecular-weight gelators,[20] forming micro- and nanoemulsions,[21] in particular in cosmetic and pharmaceutical products,[22] or for insertion of nanomaterials into mesostructures.[23] A variety of ultrasonic equipment have been developed for nanotechnological purposes, for instance, an ultrasound reactor for the fabrication of nanoparticles from the gas phase[24] or an ultrasound flow reactor.[25]

In relation to the ultrasound, the most popular objects used were carbon nanotubes; platinum, gold, ZnO, and iron oxides; Al_2O_3, SiO_2, and TiO_2 polymers; and core-shell nanoparticles based on the said compounds among many others. We emphasize that ultrasonic applications in nanochemistry and nanotechnology are a very extensive area, so, in the present section we pay attention to the main recent achievements of ultrasound-assisted preparation of compounds and materials as well as certain medical applications.

14.1 METAL/ALLOY-CONTAINING NANOSTRUCTURES

The manufacture of metal nanoparticles by reduction[26] is now a common synthesis procedure leading to different metal-containing nanostructures and nanocomposites. Ultrasonically prepared elemental *metal/alloy*-containing nanocomposites on polymer basis are widespread, in particular those of noble metals, especially Au. Thus, gold-deposited iron oxide/glycol chitosan nanocomposites[27] (chitosan is a classic polysaccharide support in nanotechnology) were obtained from $FeCl_2 \cdot 4H_2O$, $FeCl_3 \cdot 6H_2O$, and $HAuCl_4$ as precursor under ultrasonication. Hybrid nanocomposites of carboxyl-terminated generation 4 (G4) poly(amidoamine) dendrimers with gold nanoparticles encapsulated inside them were described by Wei et al.[28] These US-obtained nanocomposites were used to fabricate highly sensitive amperometric glucose biosensors, which exhibited a high and reproducible sensitivity and response time less than 5 s. The combined application of polyethylene glycol (PEG) and ultrasonic irradiation was investigated as a possible method to synthesize intercalated Au/clay nanocomposites (spherical-shaped Au nanoparticles of 6–8 nm diameter spread homogeneously within the nanocomposites).[29] The intercalation method consisted of two steps: (1) intercalation of PEG into the clay matrix by ultrasonic irradiation and (2) replacement of PEG by gold nanoparticles under ultrasonic treatment. The incorporation (Figure 14.1) of Au nanoparticles in the clay matrix under ultrasound was twice as efficient with PEG as without it. These new nanocomposites can find practical application in different areas such as in electronic (computer chips, optoelectronics, information storage, sensors, catalysis, micro-/nanoelectronic devices) and medical fields. Biosynthesis of size-controlled gold nanoparticles was reported[30] using fungus *Penicillium* sp., which could successfully bioreduce and nucleate $AuCl_4^-$ ions, and lead to the assembly and formation of intracellular Au nanoparticles with spherical morphology and good monodispersity after exposure to $HAuCl_4$ solution. The intracellular gold nanoparticles could be easily separated from the fungal cell lysate by ultrasonication and centrifugation. A new type of core-shell nanostructure Au_{core}/$Ag\text{-}PVP_{shell}$ consisting of Au nanoparticles embedded in shells of Ag nanoparticles-filled polymer was ultrasonically fabricated in a two-step procedure, in which Ag nanoparticles served as an initial

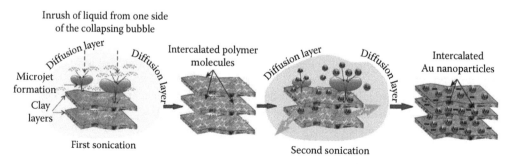

FIGURE 14.1 Scheme of ultrasonic intercalation of gold nanoparticles into the clay matrix in the presence of PEG. (Reproduced with permission from Belova, V. et al., Ultrasonic intercalation of gold nanoparticles into clay matrix in the presence of surface-active materials. Part 1. Neutral polyethylene glycol, *J. Phys. Chem. C*, 113(14), 5381–5389. Copyright 2009 American Chemical Society.)

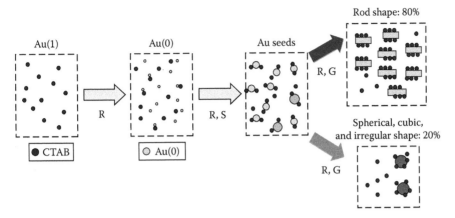

FIGURE 14.2 Schematic mechanism for the sonochemical formation of gold nanorods and nanoparticles at pH 3.5. R, reduction of Au(I); S, formation of Au seeds; G, growth of Au seeds. (Reproduced with permission from Okitsu, K. et al., One-pot synthesis of gold nanorods by ultrasonic irradiation: The effect of pH on the shape of the gold nanorods and nanoparticles, *Langmuir*, 25(14), 7786–7790. Copyright 2009 American Chemical Society.)

reductive agent for Au^{3+} ions with the existence of polyvinylpyrrolidone (PVP).[31] Solidification and spherulite growth of the polymer was induced around the as-formed Au nanoparticles such that very small Ag nanoparticles were embedded in the crystalline polymer spherulites. The surfactant- and reducer-free synthesis of gold nanoparticles from an aqueous $HAuCl_4 \cdot 4H_2O$ solution using a high-frequency (950 kHz) ultrasound in the absence of any stabilizing, capping, and reducing agents was described.[32] It was found that higher $AuCl_4^-$ concentration promoted particle growth (size increase) and plate formation, enhanced with the addition of NaCl or HCl (but not NaOH). A one-pot synthesis method to prepare gold nanorods was developed by using sonochemical reduction of gold ions in aqueous solution.[33] The size and shape of these gold nanoparticles were greatly dependent on the solution pH (3.5–7.7). The mechanism for the sonochemical formation of gold nanorods and nanoparticles is shown in Figure 14.2.

Composites of noble metal nanocrystals and titanium dioxide nanotubes were prepared by ultrasound-assisted pulsed electrodeposition with titanium dioxide nanotubes erectly grown on a titanium substrate as the carrier, based on the high dispersibility of Pd.[34] The composite catalytic electrode had stable structure, high catalytic activity, many surface catalytic active sites, and a large specific surface area for catalytic reaction. Dispersed and aggregated palladium nanoparticles with a large size distribution (centered at 20 nm) were obtained[35] by ultrasonic irradiation of $Pd(NO_3)_2$ solution in the presence of ethylene glycol and PVP. It was shown that sonochemical reduction of Pd(II) ions to Pd(0) atoms depended on sonication time and the use of a low quantity of PVP involved in obtaining aggregates of palladium nanoparticles. The $Sn(II \rightarrow IV)/Pd(II \rightarrow 0)$ redox couple was used for decorating CNTs with Pd nanoparticles by displacement reaction with the aid of ultrasonication,[36] producing nanosized particles uniformly dispersed and tightly anchored on the surfaces of CNTs. The constructed heterostructure exhibited a synergistic effect in its hydrogen storage performance. The production of Pt nanoparticles (11–15 nm) from aqueous chloroplatinic solutions in the presence of low-frequency high-power ultrasound (20 kHz) on Ti alloy electrodes performed galvanostatically at 298 K using a sonoelectrode producing ultrasonic pulses triggered and followed immediately by short applied current pulses.[37] Among other noble metals used, Ag nanoparticles,[38,39] an Ag/C nanocomposite,[40] Ag/TiO_2 nanotube composite material,[41] hydrogel[42] and solution[43] on the basis of nanoscale Ag, possessing antibacterial properties, are known. As an example, the antibacterial nanopowder, composed of shell powder carriers 80–98 wt.%, and nanoscale antibacterial active component 2–20 wt.% (on the basis of Ag^+, Cu^{2+}, Ni^{2+}, or Zn^{2+} ions) adsorbed in micropores of carriers, was reported.[44] This nanopowder has the advantages of small particle size, good compatibility with materials, good safety, no toxicity, wide antibacterial spectrum, and good stability and can be applied in plastic, rubber, fiber, coating material, ceramic, etc. Sonochemical irradiation of Fe(II) acetate aqueous solution in the presence of Ag nanopowder resulted in the deposition of magnetite nanoparticles on Ag nanocrystals and imparted them with magnetic properties.[45] The Ag-Fe_3O_4 nanocomposite (for other magnetite composites, see the following) was well attracted to a permanent magnet and demonstrated superparamagnetic behavior typical of nanomaterials in a magnetic field. The strong anchoring of magnetite to the nanosilver surface was explained by authors as a result of local melting of Ag when the magnetite nucleus is thrown at the Ag surface by high-speed sonochemical microjets.

SiC nanoparticles reinforced with magnesium and its alloys including pure Mg, Mg-(2, 4)Al-1Si, Mg-6Zn, and Mg-4Zn were fabricated by ultrasonic cavitation-based dispersion of SiC nanoparticles in magnesium melt.[46,47] As compared with an un-reinforced magnesium alloy matrix, the mechanical properties including tensile strength and yield strength were improved significantly while the ductility was retained or even improved. In a related research, magnesium matrix composites reinforced with nanosized SiC particles (n-SiCp/AZ91D) were fabricated by high-intensity ultrasonic-assisted casting.[48] The dispersion and distribution of n-SiCp in magnesium alloy melts were significantly improved by ultrasonic processing, yield strengths were remarkably improved, and the yield strength increased by 117% after gravity permanent mold casting. The room-temperature (r.t.) preparation of metallic aluminum nanoparticles (10–20 nm) by the pulsed sonoelectrochemical method has been described.[49] The authors noted that the sonoelectrochemical technique is a promising method for the fabrication of air-sensitive metallic nanoparticles that have a high, negative reduction potential. On the contrary, using Al nanoparticles (80 nm), nanothermite powders were prepared by ultrasonic mixing such aluminum powders with a variety of metal iodates (bismuth iodate, copper iodate, zinc iodate, molybdenum oxide, and iodine oxide [particle sizes 50 nm–5 μm]) and metal oxides.[50] An interesting feature of these materials (primary explosives) was the production of gaseous products rapidly converting to condensed phases upon cooling. The major products of interaction of components were as follows: for $AgIO_3$ and Al—nanoparticulate AgI and Al_2O_3, for Bi_2O_3 and Al—the bismuth vapor product condensing as Bi metal and/or reacting with oxygen from air to re-form Bi_2O_3.

Using an ultrasonic aerosol pyrolysis approach, diamond cubic Ge nanocrystals with dense, spherical morphologies and sizes ranging from 3 to 14 nm are synthesized at 700°C from an ultrasonically generated aerosol of tetrapropylgermane (TPG) precursor and toluene solvent.[51] Ni-CeO_2 nanocomposites with high microhardness were prepared in an ultrasonic field by means of cavitation effects of ultrasonic wave and selecting suitable processing parameters of pulse electrodeposition.[52] The co-deposited CeO_2 content in the composites was shown to be markedly affected by the CeO_2 particle concentration in the electrolyte. The Ni-TiN composite layer was prepared by ultrasonic-electrodeposited technology.[53] The surface of the Ni-TiN composite layer was found to be relatively flat and smooth with the ultrasound-assisted action and TiN nanoparticles added. The basic reason of improving the Ni-TiN composite layer wear resistance was that TiN nanoparticles have the dispersion strengthening effect, the fine crystal strengthening effect, and the bearing and lubricating effect. Particle (carbon nanotubes, diamond, graphite, Cu, Al, Fe)-containing liquid metals (Hg, Ga, Pb) with high heat transfer performance were prepared by mechanical stirring and ultrasonic dispersion of 1:(0.1–99) wt. ratio particle–metal mixtures.[54]

Bimetallic and alloy nanoparticles and composites on their basis are widely represented. Thus, the ultrasonic irradiation technique was employed to prepare Cu-Ga/polymethyl methacrylate nanoparticles,[55] in which there were chemical actions between the Cu-Ga alloy and PMMA. The synthesis of FeCr alloy nanoparticles using a method that couples electrodeposition of metals with the employment of high-power ultrasound was described.[56] The final product was a suspension of nanoparticles with high purity and high surface/volume ratio, which could be controlled by varying process parameters. Shape-controllable metal nanocrystal/carbon nanotube heterostructures are also known.[57] The synthesis of FeCo nanoparticles using a method that couples electrodeposition of metals with the employment of high-power ultrasound was described using a titanium alloy horn ultrasound generator.[58] The primary role of ultrasound in this process was to induce cavitation phenomenon in the electrolyte and the ablation of the metallic nuclei from the cathodic surface. The final product was a suspension of nanoparticles with high purity and surface/volume ratio, which can be controlled by varying process parameters. Platinum-cobalt (PtCo) alloy nanoparticles were successfully fabricated by the ultrasonic-electrodeposition method, using an inclusion complex film of functionalized cyclodextrin-ionic liquid as support.[59] The resulting modified glassy carbon electrode showed excellent catalytic activity for glucose oxidation, so it is promising as a nonenzymic glucose sensor. A nonenzymatic glucose sensor based on highly dispersed PtM (M = Ru, Pd, and Au) nanoparticles on composite film of multiwalled carbon nanotubes (MWNTs)-ionic liquid (trihexyltetradecylphosphonium bis(trifluoromethylsulfonyl)imide) was fabricated by ultrasonic-electrodeposition method.[60,61] This novel nonenzyme sensor has potential application in glucose detection. Among other metal pairs, FeNi[62] alloy nanoparticles and nanocomposites Cu-Zn-Al-ZSM-5[63] have been ultrasonically prepared and characterized.

Concluding this section, we note that ultrasonic treatment is widely used to produce various metal nanoparticles[64] (in particular, *Rieke* metals[65]) nanoalloys, and metal-containing nanocomposites on different supports starting from metals and their salts as precursors in combination with reduction and electrodeposition methods. At the same time, ultrasound has been applied for the activation of elemental metals in the synthesis and coordination of organometallic compounds.[66]

14.2 CARBON NANOTUBES, GRAPHENE, DIAMOND, AND FULLERENES

MWCNTs with minimal defects as templates and facilely fabricated carbon nanotube-polyaniline (PANI) nanocomposites with uniform core-shell structures were prepared by ultrasonic assisted *in situ* polymerization.[67] By varying the ratio of aniline monomers and carbon nanotubes, the thickness of PANI layers can be effectively controlled. The effective site-selective interaction between the π-bonds in the aromatic ring of the PANI and the graphitic structure of carbon nanotubes should strongly facilitate the charge-transfer reaction between the two components. CNTs can be ultrasonically filled with magnetic metal iron nanoparticles.[68] A general, rapid, template-free, one-step, and continuous approach was designed to obtain rattle-type hollow carbon spheres (M@carbon, M = multiple Sn, Pt, Ag, or Fe-FeO nanoparticles) via ultrasonic spray pyrolysis of aqueous solutions containing Na citrate and the corresponding inorganic metal salts,[69] controlling the content of encapsulated nanoparticles in M@carbon via tuning the concentration of metal salts. Sn@carbon exhibited high capacity and good cycle performance when it was used as anode material for a Li battery. Wet chemical technique to produce carbon nanoscrolls at low temperature, based on the use of readily available acceptor-type graphite intercalation compounds, was reported[70] using the initial graphite intercalation compound, first exfoliated to produce a suspension of graphene monolayers in ethanol and subsequently sonicated yielding a suspension of carbon nanoscrolls. The C_{70} nanoscale monocrystal material, composed of granules, rods, or tubes, was prepared from saturated C_{70} m-xylene solution (as the mother liquor) and linear saturated mono alcohol (as the shape regulator) by ultrasonic vibration, standing, and a series of consequent steps yielding C_{70} nanoscale crystals.[71] Ultrasonic stability of C_{70}-gold nanoparticle multilayer films was studied by exposing them to ultrasonic irradiated surroundings, which resulted in partial desorption and a little aggregation of nanoparticles on solid surfaces.[72] Al last, ultrasonication at 150 kHz was applied to disperse diamond powders with a primary particle size of 5 nm.[73]

The transformation of 2D graphene oxide (GO) nanosheets into carbon nanotubes was achieved by sonicating GO in 70% nitric acid.[74] Main steps of these reactions are shown in Figure 14.3. This net chemical process (from open-face carbon sheets to curled carbon nanostructures) is, according to the authors' opinion, almost magical in its "one-pot" transformation: the open, 2D GO sheets can be readily decomposed into polyaromatics in minutes, which are further reconstituted by acid dehydration reactions to form larger carbon nanoparticles and nanotubes.

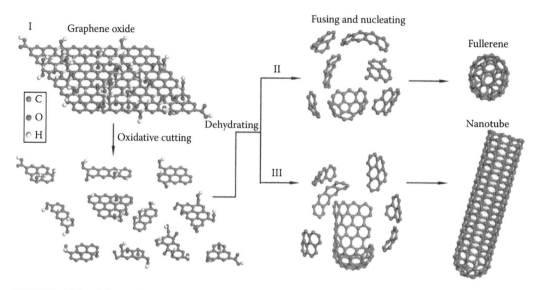

FIGURE 14.3 Schematic illustration of the mechanism for transforming GO nanosheets into carbon nanoparticles and nanotubes following their ultrasonication in acid. (I) Oxidative cutting of GO produces PAH molecules in concentrated HNO_3. In the dehydrating acidic medium, the polyaromatic fragments fuse and nucleate into (II) carbon nanoparticles or (III) nanotubes via acid-catalyzed intramolecular or intermolecular dehydration reactions. (Reproduced with permission from Wang, S. et al., Room-temperature synthesis of soluble carbon nanotubes by the sonication of graphene oxide nanosheets, *J. Am. Chem. Soc.*, 131(46), 16832–16837. Copyright 2009 American Chemical Society.)

14.3 OXIDE- AND HYDROXIDE-CONTAINING NANOSTRUCTURES

Among oxide-containing nanoparticles and nanocomposites, the most attention has been paid, undoubtedly, to iron oxides and TiO_2. Thus, talc $[Mg_3Si_4O_{10}(OH)_2]$ microparticles were coated with hydrophilic and hydrophobic TiO_2 nanoparticles in liquid CO_2.[75] The talc was found to be completely coated with TiO_2 nanoparticles, and a smooth coating surface was obtained; the flowability and wettability of the talc was improved by TiO_2 nanoparticle coating. A commercially available dry titania nanopowder with a mean primary particle diameter of approximately 30 nm was mixed into an epoxy resin/hardener system to produce nanocomposite samples.[76] Processing techniques such as ultrasonication and particle surface modification were used to produce nanocomposites with varying degrees of particle mixture homogeneity. Highly crystalline metal oxide nanoparticles of TiO_2, WO_3, and V_2O_5 were synthesized in just a few minutes by reacting transition metal chloride with benzyl alcohol using ultrasonic irradiation under argon atmosphere in a nonaqueous solvent at 363 K.[77] The particles' size and shape revealed: (a) "quasi" 0D, spherical TiO_2 particles (3–7 nm); (b) the V_2O_5 particles with a "quasi" 1D ellipsoidal morphology (lengths of 150–200 nm and widths of 40–60 nm); (c) the WO_3 particles as "quasi" 2D platelets with square shapes (facets 30–50 nm). Among other TiO_2-containing nanocomposites, we note a fluoridated hydroxyapatite/titanium dioxide nanocomposite,[78] an enzyme electrode on the basis of MWCNTs-TiO_2/Nafion composite,[79] and TiO_2 nanotubes, prepared using anodization of Ti foils in H_3PO_4 and ethylene glycol by mechanical stirring and ultrasonic method, which were found to demonstrate potential in the photoelectrocatalytic degradation (see also Section 14.6) of methyl orange dye[80] and higher activity as compared with the stirring method. The addition of oxidants such as oxygen and H_2O_2 demonstrated improvement in the methyl orange degradation. For TiO_2 nanofilms, prepared using the ultrasonic atomization technology, the analytic results showed that the products were uniform TiO_2 nanofilms with anatase 2 crystal phase and a particle diameter of 20–80 nm.[81] The sol-gel synthesis technique was suitably modified by incorporating ultrasound to study the effect of cavitation on the phase transformation, crystallite size, crystallinity, and morphology to synthesize nanostructured TiO_2 (95% yield vs. 86% by classic sol-gel process) via sol-gel technique to obtain a 100% rutile polymorph of nanostructured TiO_2.[82]

An enormous number of publications on the nanoparticles and nanocomposites on the basis of iron oxides (mainly Fe_3O_4), in particular recently reviewed Au/Fe_xO_y core-shell nanoparticles,[83] testify about a permanent interest of nanotechnologists to these nonexpensive magnetic nanomaterials with unusual properties (e.g., shape memory effect) having a series of important applications, for instance for drug delivery systems.[84] Thus, biocompatible PDLLA/magnetite nanocomposites were reported by Zheng et al.[85] Fe_3O_4 nanoparticles with an average size of 20 nm were synthesized by chemical co-precipitation and mixed uniformly with a PDLLA matrix, in particular under ultrasonication. These nanocomposites displayed a desirable shape memory effect. Among other nanocomposites, we note magnetic Fe_3O_4/polyphosphazene nanofibers (several microns in length and 50–100 nm in diameter with Fe_3O_4 nanoparticles of 5–10 nm attached on the surface through coordination behavior), prepared via a facile approach by ultrasonic irradiation.[86] Magnetic studies showed that the magnetic nanofibers exhibited good superparamagnetic properties with a high magnetization saturation value of about 36 emu/g. In the case of closely related poly (cyclotriphosphazene-co-4,4'-sulfonyldiphenol) (PZS) use, the magnetic PZS nanotubes, with good thermal stability and superparamagnetic properties, were formed.[87] They were found to be of 50–100 nm in outer diameter and 5–10 nm in inner diameter; the Fe_3O_4 nanoparticles with a diameter of 5–10 nm were embedded in the walls of the nanotubes.

Fe_3O_4 MNPs with improved peroxidase-like activity were prepared through an advanced reverse co-precipitation method under the assistance of ultrasound irradiation.[88] The H_2O_2-activating ability of Fe_3O_4 MNPs was evaluated using RhB as a model compound of organic pollutants to be degraded showing ability to activate H_2O_2 and removing ca. 90% of RhB in 60 min. Additionally, magnetite was ultrasonically prepared not only in nanoparticles form[89] but also as nanocrystals and nanofilms. Thus, magnetite thin film was prepared by one-liquid ultrasonic spray plating using an aqueous solution containing $FeCl_2$, $NaNO_2$, and dextran.[90] Using this method, Fe_3O_4 film was deposited at 90°C; after postdeposition annealing at 850°C in N_2, NaCl, coexisting in the film, was evaporated, and crystallinity and magnetic property of the film were improved. Fe_3O_4 magnetic nanocrystals were prepared by coprecipitation of Fe^{2+} and Fe^{3+} ions in an ammonia solution with ultrasonic enhancement and modification by the anionic surfactant sodium dodecyl sulfate.[91] It was shown that the Fe_3O_4 magnetic nanoparticles (~10 nm) are well-crystallized particles with good dispersivity and thermostability, superparamagnetic, and might be applied to biological and medical fields such as cell or enzyme immobilization. Cobalt and manganese analogues of magnetite, Co_3O_4 and Mn_3O_4, were synthesized as uniform sphere-like or cubic spinel nanocrystals from acetate salts and NaOH or tetramethylammonium hydroxide (TMAH) as precursors.[92]

Among a high number of recent reports on the ultrasonic production of SiO_2 nanostructures, we note polypropylene/silica nanocomposites,[93] $Fe_3O_4@SiO_2$ core-shell nanoparticles for plasmid DNA purification,[94] and Silicalite-1 (colloidal zeolite on the SiO_2 basis) nanocrystals with organic functional groups, prepared by a simple ultrasonic treatment in methanol.[95] The organic functionalization enhanced the hydrophobicity of Silicalite-1 nanocrystals, which has proven to be very useful for fabricating monolayer zeolite films. Other oxides are lesser presented. The study of sonochemical reactions with MSU-X mesoporous alumina in aqueous solutions revealed[96] that sonication ($f = 20$ kHz, $I = 30$ W/cm^2, $W_{aq.} = 0.67$ W/mL, $T = 36°C–38°C$, Ar) caused significant acceleration of m-Al_2O_3 dissolution in the pH range of 4–11, forming nanorods and nanofibers of boehmite, {AlO(OH)} at short-time sonication (pH = 4) or, at prolonged ultrasonic treatment, aggregated nanosheets in weakly acid solutions or plated nanocrystals in alkaline solutions (boehmite and small amounts of bayerite, $Al(OH)_3$). The authors concluded that the effect of ultrasound on the textural properties of mesoporous alumina as well as on the transformation of nanosized bayerite to boehmite can be consistently attributed to the transient strong heating of the liquid shell surrounding the cavitation bubble, which caused the chemical processes similar to those that occurred during hydrothermal treatment. The ultrasonic vibration and conventional diamond grinding of

Al_2O_3/ZrO_2 nanoceramics were performed to investigate the effects of workpiece ultrasonic vibration on the brittle-ductile transition mechanism and the effects of grit size, worktable speed, and grinding depth on the critical depth of cut.[97]

Another classic research object in nanotechnology, ZnO, has been ultrasonically prepared in a variety of nanoforms, from nanowires to nanofilms. Thus, ultrasonic irradiation of a mixture of ZnO nanorods, $Ag(NH_3)^{2+}$, and formaldehyde in an aqueous medium yielded ZnO nanorod/Ag nanoparticle composites,[98] in which the ZnO nanorods were coated with Ag nanoparticles with a mean size of several tens of nanometers. Doping ZnO nanowires with transition metal[99] was carried out by forming a Zn thin film on a substrate, impregnating this film with an aqueous solution of Zn salt, reductant, and transition metal and applying ultrasonication to the aqueous solution to grow transition metal-doped ZnO nanowires on the Zn thin film. Oriented zinc nanoparticles in zinc oxide matrix were obtained by Lu et al.,[100] and a nanoscale-laminated ZnO nanofilm was synthesized by Cao et al.[101] Zinc oxide nanoparticles (300 nm) were synthesized and deposited on the surface of glass slides using ultrasound irradiation.[102] The antibacterial activities of these ZnO-glass composites were tested against *Escherichia coli* (Gram neg.) and *Staphylococcus aureus* (Gram pos.) cultures, demonstrating a significant bactericidal effect, even in a 0.13 wt.% coated glass. (We note that the bactericidal effect was obviously obtained for Ag nanoparticle-based composites; see the previous section). A similar effect, even in a 1 wt.% coated fabric, was found by the same authors for copper oxide nanoparticles, synthesized and subsequently deposited on the surface of cotton fabrics using ultrasound irradiation.[103] Ultrasonically synthesized monodispersed Cu_2O nanoparticles were also reported.[104]

The porous WO_3 (pore size 2–5 nm) nanoparticles were synthesized using a high-intensity ultrasound irradiation of commercially available WO_3 nanoparticles (80 nm) in ethanol.[105] These sonochemically modified porous WO_3 nanoparticles dispersed more uniformly over the entire volume of the epoxy (without any settlement or agglomeration) as compared with the unmodified WO_3/epoxy nanocomposites. At the same time, WO_3 was also ultrasonically prepared as an oriented nanofilm.[106] V_2O_5 nanomaterials including nanoribbons, nanowires, and microflakes (Figure 14.4)

(a)   (b)   (c)

(d)   (e)   (f)   (g)

FIGURE 14.4 Scanning electron microscopy (SEM) images of V_2O_5 (a) nanoribbons, (b) nanowires, and (c) microflakes; transmission electron microscopy (TEM) and HRTEM images of (d, e) nanoribbons and (f, g) nanowires. (Reproduced with permission from Chou, S.-L. et al., High capacity, safety, and enhanced cyclability of lithium metal battery using a V_2O_5 nanomaterial cathode and room temperature ionic liquid electrolyte, *Chem. Mater.*, 20(22), 7044–7051. Copyright 2008 American Chemical Society.)

were fabricated by an ultrasonic-assisted hydrothermal method and combined with a postannealing process.[107] The rechargeable lithium battery using obtained V_2O_5 nanoribbons as cathode materials could be the next generation lithium battery with high capacity, safety, and long cycle life.

Among other oxides, delamination of layered manganese oxide into colloidal nanosheets occurred[108] when manganese oxide intercalated with tetramethylammonium ions was ultrasonically dispersed in acetonitrile. Using $SnCl_4 \cdot 5H_2O$ and ammonia as raw materials, SnO_2 nanoparticles were synthesized by ultrasonic irradiation-assisted sol-gel method.[109] The shape of the synthesized 20 nm SnO_2 nanoparticles was round, and their dispersivity was greatly improved by the anionic surfactant, that is, citric acid. Room-temperature ferromagnetism was revealed in $Sn_{1-x}Mn_xO_2$ nanocrystalline thin films prepared by ultrasonic spray pyrolysis.[110] Two methods of compacting dry poly- and nanodisperse powders (such as yttria-stabilized zirconia) into compacts of a complicated shape with uniform density distribution in the volume were developed: pressing under powerful ultrasonic action and collector pressing by the control of friction force redistribution.[111] Among lanthanide oxides, CeO_2 nanoparticles were prepared by an ultrasonic atomization process using low-cost $Ce(NO_3)_3 \cdot 6H_2O$, NH_4HCO_3, and $NaOH$ as starting materials.[112] Mist drops of solutions were generated and used as space-confined microreactors for the nucleation, growth, and crystallization of CeO_2 nanoparticles (3 nm) under r.t. conditions. Nanosized (15 nm) lanthanum oxide, belonging to a hexagonal crystal system, was ultrasonically prepared with $LaCl_3$ and $CO(NH_2)_2$ as raw materials by hydrolyzation of urea.[113]

Hydroxides and related compounds are represented considerably lesser in available literature. Thus, nano γ-Ni oxyhydroxide (nano γ-NiOOH), a new cathode material for alkaline Zn/Ni batteries, was synthesized by a sonochemical intercalation method[114] using $NiCl_2$ solution, NaOH, NaClO as oxidant, allowing all four elementary reactions (precipitation, oxidation, cation exchange, and H_2O molecule intercalation). $Ni(OH)_2$ was fabricated as a thin film and also used as active electrode material.[115] $CaSn(OH)_6$ nanotubes were fabricated in high yield and at low cost by the sonochemical precipitation method at r.t.[116] revealing a direct rolling process from nanosheets to nanotubes. The transient $CaSn(OH)_6$ nanosheets were formed as intermediates produced by the spontaneous self-assembly and transformation of amorphous colloid clusters. Possible applications of the products belong to medicine and to pharmaceuticals through to materials science.

14.4 METAL SALTS AND COMPLEXES

A high number of publications are dedicated to ultrasonic preparation of metal sulfide and selenide nanostructures, very popular and important research objects in the current nanotechnology, mainly ZnS,[117] object of thousands of publications. Thus, zinc sulfide nanorods of wurtzite structure were grown using a simple sol-gel method via ultrasonication, in the presence of a capping agent.[118] The photoluminescent spectrum of ZnS nanorods exhibited green emission, which may find applications in optoelectronic devices. The same compound, but in the form of nanocrystallite powder, was synthesized[119] by a sonochemical technique using zinc chloride and thioacetamide as raw materials. The crystal size of the powder was about 10 nm, and it decreased slightly with the increase in ultrasonic irradiation power, whereas its reaction rate increased in a reaction time of <110 min, and the synthesis reaction rate of the powder increased with the increase in ultrasonic irradiation power. Metallic Ag-doped polycrystalline CdS nanoparticles were obtained by an ultrasound-assisted microwave synthesis method.[120] A structural evolution from cubic to hexagonal with increasing molar ratios of Ag^+/Cd^{2+} from 0% to 5% was confirmed. The photocatalytic activity of different Ag-doped samples showed a reasonable change due to different ratios of Ag, which doped into CdS. We note that such a combination of microwave heating and ultrasound treatment sometimes is applied in chemical processes and corresponding equipment has been elaborated.[121] CdS/polyacrylonitrile nanocomposites with nonlinear optical properties are also known,[122] among others. A simple ultrasonic method was developed to synthesize rod-like SnS nanocrystals, using tin chloride and thioacetamide as starting materials and ethanolamine and water as solvents.[123] The ultrasonic condition was believed to be critical for the formation of SnS with pure phase, providing the energy to form rod-like nanostructures and help preventing the hydrolysis of Sn^{2+} to form tin oxides and hydrates and oxidation of the final products. Nanofilms are also common for metal sulfides; for instance Bi_2S_3 nanofilm was ultrasonically fabricated[124] and nanophase Gd_2S_3 thin films were deposited by a so-called asynchronous pulse ultrasonic spray pyrolysis method on glass substrates.[125]

A certain number of quantum dots (QD) on metal sulfide and selenide basis, in particular GaAs/AlGaAs,[126] CdSe and CdSe-ZnSe,[127] and PbS,[128] have been reported. It was noted that ultrasound triggered the formation of a peptide-based supramolecular gel (stable for months), readily doped with CdSe/ZnS QD affording highly luminescent gels.[129] The hollow spherical CdSe QD assemblies, synthesized via a sonochemical approach that utilizes β-cyclodextrin as a template reagent,[130] had an average diameter of 70 nm and were found to consist of an assembly of monodispersed 5 nm sized CdSe QD. The observed unique electrogenerated chemiluminescence intensity and stability of the synthesized spherical nanoassemblies could allow for potential sensor applications of CdSe QD in water. 3D arrays of close-packed $AgBiS_2$ QD (average QD radius 4.2 nm, twice as small compared with the QD solid obtained without ultrasonic treatment) in thin film form were synthesized using sonochemical approach.[131]

Carbonate and silicate nanostructures are also common objects for ultrasonic applications, especially $CaCO_3$.[132] Thus, the combined effect of surfactant and ultrasound in the synthesis of calcium carbonate nanoparticles (nanocalcite with cubic structure) was studied by Sonawane et al.[133] The effect of different surfactants (polyacrylic acid, steric acid, sodium tripolyphosphate, and myristic acid) on the synthesis of nano $CaCO_3$ was investigated, demonstrating that nanocalcite crystals can be synthesized, for high potential industrial applications such as filler in the paper and polymer industry. Primary nanoparticles of vaterite (a metastable form of Ca carbonate) precipitated in aqueous solutions under 2.64 MHz sonication experienced layered agglomeration.[134] It was concluded that, under the experimental conditions used, heterogeneous nucleation of primary vaterite particles takes place. A 20 nm inorganic mineral particle, mainly including $Mg_3Si_2O_5(OH)_4$, $KAlSiO_4$, and $CaMg(CO_3)_2$, was prepared by the ultrasonic nanometer grinder.[135] It was shown that the inorganic nanomineral material as an oil additive had excellent stable-dispersion property. The $Sr(OH)_2$ and $SrCO_3$ nanostructures were synthesized by reaction of strontium(II) acetate and sodium hydroxide or TMAH via the ultrasonic method.[136]

Nanocrystalline hydroxyapatite was prepared by a synthetic method with the aid of ultrasonic mixing using $Ca(NO_3)_2$ and $NH_4H_2PO_4$ as raw materials.[137] The results indicated that the best power and time combination is 80 W and 1.5 h, respectively. Bioactive monetite (anhydrous calcium hydrogen phosphate, $CaHPO_4$) with an orderly layered structure assembled by nanosheets was successfully synthesized by a sonochemical-assisted method in the presence of CTAB.[138] The assembled process of monetite nanosheets under the effects of CTAB and ultrasound is shown in Figure 14.5. The authors noted that, with the assistance of ultrasound, the disordered nanosheets can be assembled into closely attached oriented stacking to obtain minimum energy, demonstrating the potential application of sonochemistry to obtain an oriented-assembled architecture. Among other phosphates, we note the layered iron Ph phosphate ($Fe(OH)(C_6H_5PO_4H)_{1.6}(H_2PO_4)_{0.4}\cdot 5.1H_2O$), which is composed of a multilayer alternating bilayer of Ph groups of the phosphates and amorphous iron phosphate phase and was exfoliated in ethanol under ultrasonic irradiation.[139] The small crystallites of quasi sphere aggregates $NaNO_2$ nanoparticles prepared by normal ice sublimation method in ethanol were reassembled by ultrasonic vibration to form more stable square aggregates.[140] Microwave and ultrasound, two methodologies for combined rapid synthesis, mentioned earlier for obtaining Ag/CdS nanoparticles, were both applied (50–50 W) to fabricate other nanostructures, in particular Pb(OH)Br nanowires (45% yield compared with 23% by conventional heating).[141]

Such generally hard and refractory compounds in their normal state as nitrides (TiN nanopowder[142]) and carbides (Fe_3C and χ-$Fe_{2.5}C$)[143] have also been ultrasonically obtained as nanoforms. An air-passivated hafnium subhydride subcarbide nanopowder (10 nm crystallites agglomerated into larger porous structures) $Hf_{3.47}C_{1.00}H_{2.40}Li_{0.10}O_{0.12}$ was prepared via a slurry sonochemical reaction

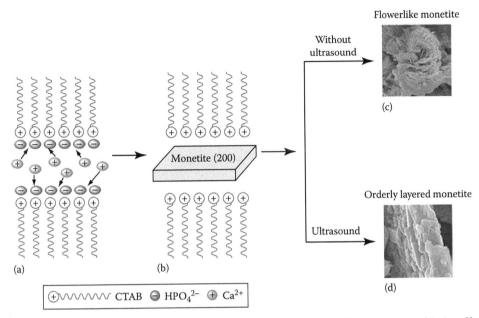

FIGURE 14.5 Schematic graph of the orderly layered assembly of monetite nanosheets with the effect of CTAB and ultrasonic radiation: (a) the CTAB polar groups absorb PO_4^{3-} ions and act as active sites for nucleation of monetite, (b) monetite nanosheets form on the surface of the CTAB micelles, (c) nanosheets assemble in a disordered way without ultrasound, and (d) ordered stacking of nanosheets happens under ultrasonic radiation. (Reproduced with permission from Ruan, Q. et al., Ultrasonic-irradiation-assisted oriented assembly of ordered monetite nanosheets stacking, *J. Phys. Chem. B*, 113(4), 1100–1106. Copyright 2009 American Chemical Society.)

of $HfCl_4$ and LiH, followed by a 900°C vacuum annealing.[144] Once air-passivated, the hafnium subcarbide subhydride nanopowder was found to be indefinitely air-stable and produced 6.7 kJ/g of heat when oxidized by O_2.

Oxygen-containing salts are well represented in the available literature. Thus, lanthanide orthovanadate $LnVO_4$ (Ln = La, Ce, Pr, Nd, Sm, Eu, Gd, Tb, Dy, Ho, Er, Tm, Yb, Lu) nanoparticles were ultrasonically synthesized from an aqueous solution of $Ln(NO_3)_3$ and NH_4VO_3 without any surfactant or template.[145] It was established that ultrasonic irradiation had a strong effect on the morphology of the $LnVO_4$ nanoparticles: $LnVO_4$ particles had a spindle-like shape with an equatorial diameter of 30–70 nm and a length of 100–200 nm, which were the aggregates of even smaller nanoparticles of 10–20 nm. Nanosized $BiVO_4$ with high visible-light-induced photocatalytic activity was synthesized via an ultrasonic-assisted method for 30 min with PEG.[146] This essentially stable product exhibited excellent visible-light-driven photocatalytic efficiency for degrading organic dye, which was increased to nearly 12 times that of the products prepared by traditional solid-state reaction. A new effect of ultrasonic irradiation on the formation of $BaTiO_3$ particles was identified.[147] Ultrasonication caused the aggregation of the original 5–10 nm $BaTiO_3$ particles in the same crystal axis and accelerated their formation significantly. As a result, narrow size distribution was obtained for the aggregated particles under ultrasonic irradiation. A simple ultrasonic-assisted ion-exchange/intercalation process (Figure 14.6) that enables unwrapping 1D titanate nanotubes, obtained from TiO_2 nanopowder and NaOH as precursors, into 2D titanate nanosheets in the presence of Tetrabutylammonium hydroxide (TBAOH) was reported.[148] The resulting titanate nanosheets possess larger band gap energy (~3.75 eV) than that of the original nanotubes (~3.30 eV), which might be attributed to the quantum size effect within the 2D titanate nanosheets with small thickness.

A luminescent nanomaterial, monoclinic wolframite-type $HgWO_4$ nanorod (diameter ~200 nm; length ~2000 nm) was prepared by hydrothermal method together with ultrasonic technique.[149]

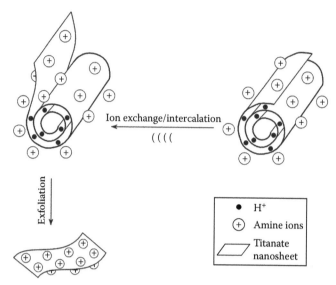

FIGURE 14.6 Irreversible topotactic transformation from titanate nanotubes to titanate nanosheets. (Reproduced with permission from Gao, T. et al., Topological properties of titanate nanotubes, *J. Phys. Chem. C*, 112(23), 8548–8552. Copyright 2008 American Chemical Society.)

Such a combination changed greatly HgWO$_4$ optical behaviors, shifting the fluorescent emitting peaks (365 and 495 nm) to the central region and finally forming a wider one at 435 nm. These unique optical performances might result from both small sizes caused by the ultrasonic irradiation procedure and the involvement of incompact d^{10} electrons. Visible-light-induced photocatalyst Bi$_2$MO$_6$ (M = W, Mo) nanocrystals (nanoplates 100 nm [M = W] and nanoparticles 150 nm [M = Mo]) were synthesized via an ultrasonic-assisted method[150] showing the photocatalytic activities about four to six times higher than that of the products prepared by traditional solid-state reaction due to an important role of ultrasonic irradiation in the formation of the nanomaterials with a smaller crystal size and larger surface area, which was beneficial to their photocatalytic activities. A mixed ionic-electronic conducting nanocomposite La$_{0.8}$Sr$_{0.2}$Ni$_{0.4}$Fe$_{0.6}$O$_3$ (LSNF)-Ce$_{0.8}$Gd$_{0.2}$O$_{2-\sigma}$ (GDC) was prepared via ultrasonic dispersion of nanocrystalline powders of perovskite and fluorite oxides in water with the addition of surfactant, followed by drying and sintering up to 1300°C.[151]

In the case of ultrasonic applications to prepare nanoforms of coordination compounds, microporous Prussian blue analogue Zn$_3$[Co(CN)$_6$]$_2 \cdot x$H$_2$O microspheres and micropolyhedrons self-assembled by nanoparticles (Figure 14.7) were synthesized under ultrasonic conditions using PVP as a surfactant.[152] N$_2$ adsorption properties confirmed the existence of micropores in both Zn$_3$[Co(CN)$_6$]$_2$ microspheres and micropolyhedrons, which may have very important applications in future gas adsorption. A gelation mechanism was proposed based on a readily available coordination polymer

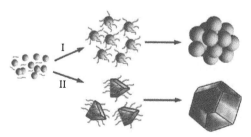

FIGURE 14.7 Schematic illustration of the growth mechanism for the formation of Zn$_3$[Co(CN)$_6$]$_2$ microspheres and micropolyhedrons. (Reproduced with permission from Du, D. et al., Morphology-controllable synthesis of microporous Prussian blue analogue Zn$_3$[Co(CN)$_6$]2·xH$_2$O microstructures, *Langmuir*, 25(12), 7057–7062. Copyright 2009 American Chemical Society.)

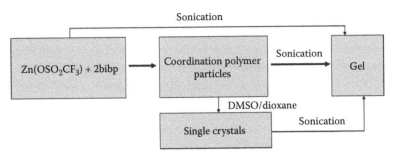

FIGURE 14.8   Morphological transformations of {Zn(bibp)$_2$(OSO$_2$CF$_3$)$_2$}$_n$. (Reproduced with permission from Zhang, S. et al., Ultrasound-induced switching of sheetlike coordination polymer microparticles to nanofibers capable of gelating solvents, *J. Am. Chem. Soc.*, 131(5), 1689–1691. Copyright 2009 American Chemical Society.)

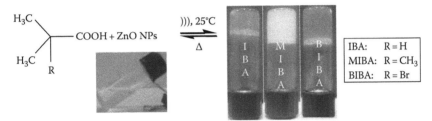

FIGURE 14.9   Dispersion of ZnO and isobutyric acids in organic solvents (left) and resulting organogelation (right). (Reproduced with permission from Kotal, A. et al., Ultrasound-induced *in situ* formation of coordination organogels from isobutyric acids and zinc oxide nanoparticles, *Langmuir*, 26(9), 6576–6582. Copyright 2010 American Chemical Society.)

{Zn(bibp)$_2$(OSO$_2$CF$_3$)$_2$}$_n$, in which ultrasound changed dramatically the morphology of the material from sheet-like microparticles into nanofibers, which further entangled each other to form a 3D fibrillar network, thereby resulting in the immobilization of organic fluids (Figure 14.8).[153]

The discovery of ultrasound-induced *in situ* formation of coordination organogels using various isobutyric acids (such as 2-methylisobutyric acid or 2-bromoisobutyric acid) and zinc oxide nanoparticles (obtained from Zn(NO$_3$)$_2$·6H$_2$O) was described (Figure 14.9).[154] It was established that the ultrasound irradiation triggered the quick dissolution of zinc oxide nanoparticles by isobutyric acids, resulting in the *in situ* formation of zinc isobutyrate complexes that underwent fast sonocrystallization into gel fibers, which is probably a very rare case of ultrasound-induced organogelation where metal oxide nanoparticles are used as the precursor.

A variety of low-dimensional nano- and microscale coordination polymer materials of {[Cu(en)$_2$][KFe(CN)$_6$]}$_n$ including short nanorods, nanocube, and hollow micro-arrowhead were prepared by a self-assembly method in aqueous solution.[155] It was suggested that PVP and ultrasonic radiation played a key role in the morphological development of the nano-/microcrystals. At near r.t. and under ultrasonication, nanocrystalline lanthanum 8-quinolinolate was synthesized by a one-step solid-state chemical reaction.[156] It was noted that nanocrystallites of quality can be produced with a high selectivity in high yields and little or no agglomeration normally found in reactions in liquid phase.

14.5   POLYMERIC AND MACROCYCLIC NANOSTRUCTURES/NANOCOMPOSITES

Encapsulation of inorganic nanoparticles (as a core) by polymers (as a shell) is one of the most interesting research subjects, in our point of view, that lead to the synthesis of nanocomposites,[157] whose properties include properties of not only the organic polymer (e.g., optical properties, toughness, processability, flexibility, etc.) but also the inorganic nanoparticles (e.g. mechanical strength, thermal stability, etc.). Some of the applied preparative methods are dry spray, dispersion, suspension, emulsion, and miniemulsion polymerization techniques, frequently under high-shear ultrasonic irradiation. Thus, nanosilver/[P(AMPS-co-MMA)] composite materials were prepared with the silver nitrate solution containing AMPS and MMA monomers without initiator or reducer, in which Ag$^+$ ion was reduced to nanosilver particles and the monomers were copolymerized by ultrasonic simultaneously.[158] The authors concluded that there is a kind of interaction phenomena of nanometal silver with an effective polymer matrix in the nanocomposite materials. Related polymer grafted carbon nanotubes (single-walled carbon nanotubes [SWCNTs]) were easily prepared by *in situ* sonochemically initiated radical polymerization of Me methacrylate.[159] The surface-modified SWCNTs were then dispersible in good solvents for PMMA, such as dichloromethane and chloroform. Vinylated magnesium hydroxide (MH) nanosheets were prepared with 3-(trimethoxysilyl) Pr methacrylate (γ-MPS) and pristine MH nanosheets, and then the MH/polystyrene (PS) hybrid nanoparticles with MH-cores and PS-shell were prepared by ultrasonic wave-assisted *in situ* copolymerization of vinylated MH nanosheets and styrene.[160] It was shown that the covalent interaction between PS and MH improved the thermal stability of PS. The graphite nanosheet/polyaniline (GN/PANI) nanorod composites were fabricated via ultrasonic polymerization of the aniline monomer in the presence of GN, which was used as electric filling.[161] It was noted that ultrasonic can effectively restrain the agglomerate of the aniline and come to uniform nanorod composites; The thermal stability of GN/PANI nanorod composites was superior to pure PANI. In a related work, an ultrasonic irradiation-assisted polymerization method was used to prepare PANI nanotubes (at low doping concentration) and nanofibers (at high doping concentration) by using different doping concentrations of mineral or organic acids.[162] The authors stated that the mechanism of the formation of the nanotubes and nanofibers may be due to the formation of different kinds of micelles and the prevention of the second growth of the PANI by ultrasonic irradiation.

Curcumin nanoparticles were ultrasonically prepared by chitosan-graft-vinyl acetate copolymers to evaluate the *in vitro* release of curcumin.[163] The encapsulation efficiency of nanoparticles was up to 91.6%. The *in vitro* release profile showed the slower release rate of curcumin. A continuous ultrasound-assisted process using a single screw compounding extruder with an ultrasonic attachment was developed to prepare polyolefin/clay nanocomposites.[164] High-density polyethylene and isotactic polypropylene were compared. A layer of a polyethylene-silver nanoparticle composite was deposited on a five-layer barrier film structure using different layer deposition methods (laminating, casting, and spraying over the multilayer structure).[165] For the casting and spraying methods, the silver nanoparticles were previously ultrasonically dispersed in the polymer solution. The biocide effect of the multilayer films was found to be dependent on the silver nanoparticle content and on the deposition method, indicating possible use in the design of industrial films for packaging. A combination of acid hydrolysis and ultrasound was applied to obtain a high yield of cellulose nanofibers from flax fibers and microcrystalline cellulose.[166] A significant enhancement in thermal and mechanical properties was achieved with a low addition of cellulose nanofibers to the polymer matrix. In a related work, rod-like cellulose nanofibrils were prepared from poly(lactic) acid by sulfuric acid hydrolysis and incorporated into dimethylformamide (DMF) by ultrasonication to obtain a stable suspension.[167] Nanocrystalline cellulose, prepared by hydrolysis of cellulose, was described by Jiang et al.[168]

Some syntheses of well-known porphyrins and phthalocyanines have been reported in respect of ultrasonic-assisted treatment. Thus, porphyrin nanoparticles were produced by a combination of the reprecipitation and sonication method.[169] No self-aggregation of the constituent porphyrin chromophores was confirmed; additionally, the nanoparticles exhibited interesting optical properties, a large bathochromic shift in the absorption spectra. In addition, the ultrasound was used for *in situ* synthesis of metal phthalocyanine/carbon nanotube composite[170] and copper phthalocyanine nanofilms.[171]

14.6 APPLICATIONS IN CATALYSIS AND DEGRADATION OF TOXIC SUBSTANCES

Ultrasound has been widely applied in the creation and activation of nanocatalysts or in catalytical processes with their use, for instance, in ultrasonic degradation of toxic substances. Thus, supported platinum or platinum-metal (Au, Ag, Co, etc.) nanoparticle catalyst was prepared by ultrasonic electrodeposition.[172] The effect of ultrasonic treatment on the crystallinity and activity of platinum nanoparticles was demonstrated,[173] showing that, after 1 h of ultrasonic treatment in all solutions (water, PVP, and ethylene glycol), Pt nanoparticles were found to be more crystalline and ordered. The fastest catalysis by these nanoparticles (in the reaction of the hexacyanoferrate(III) reduction by thiosulfate ions) was enabled after sonication in PVP solution for 1 h, whereas Pt nanoparticles with low crystallinity formed after 20 min of sonication in ethylene glycol were found to be the least efficient catalysts. Pt nanoparticles were supported on mesoporous TiO_2 (Pt/TiO_2) by an ultrasound-assisted reduction method.[174] The Pt/TiO_2 showed a higher catalytic activity for the oxygen reduction reaction than the Pt supported on C; the enhanced activity could be attributed to the better dispersion and interaction of Pt nanoparticles with the support. Another known catalyst on TiO_2 basis is visible-light-activated cuprous oxide/titanium dioxide nanocomposite photocatalyst.[175] Among other nanocatalysts supported on oxides, we note the dual-metal Rh_xAg_{1-x}/Y ($x = 0.1–0.9$, Y is a metal oxide) nanocatalyst for purification of automobile exhaust gas[176] and highly active catalyst Au/MnO_x-CeO_2 used for the preferential oxidation of CO in H_2-rich stream.[177]

As a valuable work in continuation of obtaining a series of Fe_3O_4@inorganic@shell type three-component composite particles, such as, for example, Fe_3O_4@SiO_2@Au or Fe_3O_4@SiO_2@CdTe, and more rare Fe_3O_4@polymer@shell composites (attributed to the synthesis difficulties of the Fe_3O_4@organic shell composite particles), we can consider the preparation of the superparamagnetic Fe_3O_4@PANI with well-defined core/shell nanostructures (Figure 14.10), synthesized via an ultrasound-assisted *in situ* surface polymerization method.[178] The negatively charged Au nanoparticles (4 nm) were effectively assembled onto the positively charged surface of the as-synthesized Fe_3O_4@polyaniline core/shell microspheres via electrostatic attraction to form the Fe_3O_4@PANI@Au composite microspheres. As-prepared inorganic/organic nanocomposite can be used as a magnetically recoverable nanocatalyst for the reduction of a selected substrate. These possible catalytic and biomedical applications are derived from the rational combination of magnetic properties with surface plasmon resonance and luminescence.

An efficient and green procedure was developed for the synthesis of 1,8-dioxo-octahydroxanthene derivatives in water under ultrasound irradiation, using nanosized MCM-41-SO_3H.[179] Using this method, several types of aromatic aldehydes with electron-withdrawing groups as well as electron-donating groups were rapidly converted to the corresponding 1,8-dioxo-octahydroxanthenes in good to excellent yields and for short reaction times.

A series of sonochemically synthesized catalysts have been widely used for degradation of organic pollutants and dyes, especially TiO_2 and its composites. Thus, photocatalysts the photocatalytic and sonophotocatalytic efficiencies of as-prepared $Au-TiO_2$ were evaluated by studying the degradation of a representative organic contaminant, nonylphenol ethoxylate surfactant in aqueous solutions.[180] The catalytic activities of these nanomaterials were compared for the degradation of a polydisperse nonylphenol ethoxylate, Teric GN9, by photocatalysis and sonophotocatalysis under visible light/high-frequency ultrasound irradiation. In a related work, TiO_2 and $Ag-TiO_2$ catalysts, prepared by an ultrasonic-assisted sol-gel method, were used for photodegradation and mineralization of the chlorophenols.[181] It was established that the rate of a mineralization trend was observed in the order of P < 2-CP < 2,4-DCP < 2,6-DCP < 4-CP < 2,4,6-TCP by using the $Ag-TiO_2$ catalyst, and the concentration of the main aromatic intermediate products was considerably lower for the $Ag-TiO_2$ photocatalysts than for pure TiO_2. The 2,4-dichlorophenol was also destroyed by ultrasound cooperating nanoscale iron,[182] showing a remarkable improvement of degradation (nanoscale iron alone: 11.3%, ultrasound alone: 36%, ultrasound and nanoscale iron: 88.3%). $Na_5PV_2Mo_{10}O_{40}$

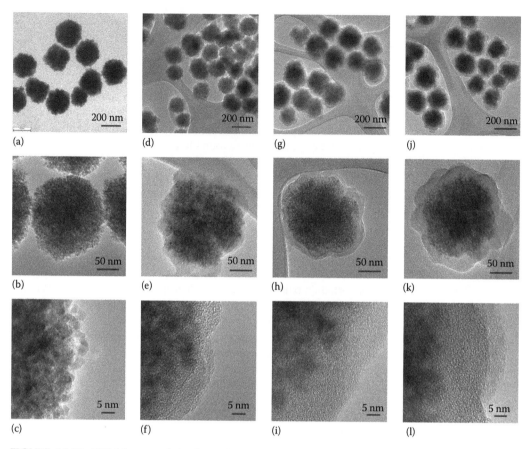

FIGURE 14.10 TEM images of the Fe_3O_4@PANI core/shell nanocomposite with the PANI shell thickness of 0 nm (a through c), 7 nm (d through f), 15 nm (g through i), and 25 nm (j through l), which were prepared at a concentration of aniline in a reaction system with 0, 1.2×10^{-3}, 1.6×10^{-3}, and 2.0×10^{-3} M, respectively. (Reproduced with permission from Xuan, S. et al., Preparation, characterization, and catalytic activity of core/shell Fe_3O_4@Polyaniline@Au nanocomposites, *Langmuir*, 25(19), 11835–11843. Copyright 2009 American Chemical Society.)

supported on nanoporous anatase TiO_2 particles, TiO_2-PVMo, was used as an efficient photocatalyst for photocatalytic degradation of different dyes by visible light using O_2 as oxidant.[183] The TiO_2-PVMo composite showed higher photocatalytic and sonocatalytic activity than pure polyoxometalate or pure TiO_2. Degradation of acid red B in the presence of nanosized ZnO powder under ultrasonic irradiation[184] indicated that the effects of sonocatalytic degradation were more obvious in the presence of nanometer ZnO. A sono-Fenton system based on Fe@Fe_2O_3 core-shell nanowires was used to degrade a recalcitrant pollutant, pentachlorophenol (PCP) (Figure 14.11), resulting in an economic and environmentally friendly technique to eliminate persistent organic pollutants in contaminated water.[185]

14.7 APPLICATIONS IN DRUG DELIVERY AND TUMOR TREATMENTS

A wide variety of nanoobjects, from nanometals to nanofilms and polymer nanocomposites, are currently used in combination with ultrasound for the purpose of drug delivery. Thus, a method/system utilizing the interaction of electromagnetic pulses or ultrasonic radiation with nano- and microparticles for enhancement of drug delivery in solid cancer tumors was described in a comprehensive review.[186] These particles can be attached to antibodies directed against antigens in

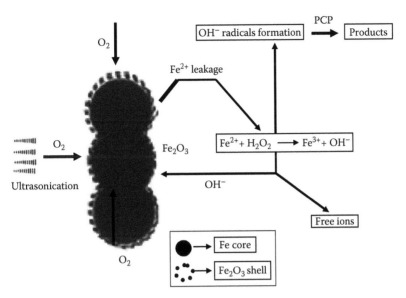

FIGURE 14.11   The possible degradation pathway of PCP in the sono-Fenton system based on the $Fe@Fe_2O_3$ core-shell nanowires. (Reproduced with permission from Luo, T. et al., $Fe@Fe_2O_3$ core-shell nanowires as iron reagent. 4. Sono-Fenton degradation of pentachlorophenol and the mechanism analysis, *Phys. Chem. C*, 112(23), 8675–8681. Copyright 2008 American Chemical Society.)

tumor vasculature and selectively delivered to tumor blood vessel wall. Silica nanotubes (SNTs) were demonstrated as a versatile host for controlled drug delivery and biosensing.[187] In the case of controlled drug delivery triggered by ultrasound, drug yield as function of time was found to be heavily dependent on the ultrasound impulse protocol: impulses of shorter duration (~0.5 min) and shorter time intervals between successive impulses resulted in higher drug yields. A nano-porous (20–40 nm pores) polymer thin film fabricated by a light-induced polymerization process was developed as a potential platform for drug delivery.[188] Ultrasound-enhanced cumulative and pulsatile release revealed the advantages of ultrasound in controlled drug delivery. On the basis of the specific physicochemical properties of the functional branched poly-ε-caprolactones, the drug delivery systems, consisting of nano- and microspheres with regular spherical shapes and narrow size distributions, were fabricated through a "melting ultrasonic dispersion method" and a "melting emulsion method," without involving toxic organic solvents.[189] The drug release study showed the release of the drug, which could be effectively sustained by entrapment in the nano- and micro-spheres. A process to obtain nano-/microcapsules with a single core of liquid perfluorocarbons within a biodegradable polymeric shell of homogeneous thickness was designed[190] to encapsulate several liquid perfluorocarbons: perfluorohexane, perfluorodecalin, and perfluorooctyl bromide. The obtained system was rather versatile: the mean size of the capsules can be adjusted between 70 nm and 25 µm and the thickness-to-radius ratio can be easily modulated by simply modifying the polymer-to-perfluorocarbon ratio. The ability of oxygen-loaded chitosan bubbles to exchange oxygen in the presence or in the absence of ultrasound was studied, showing that oxygen delivery was enhanced by sonication.[191]

Stable nanocolloids of insoluble drugs with very high drug content (up to 90 wt.%) were easily prepared through the application of the layer-by-layer technology alternate adsorption of oppo-sitely charged polyelectrolytes on the surface of drug nanoparticles produced by ultrasonication of larger drug crystals.[192] Such polymeric coating prevented drug nanoparticle aggregation and cre-ated a firm polymeric shell on their surface. This may represent a novel approach to prepare con-venient dosage forms of poorly soluble drugs. The study of targeted chemotherapeutic intervention on solid tumors by means of ultrasound and doxorubicin- or paclitaxel-loaded perfluoropentane nanoemulsions was carried out.[193] Under the action of a tumor-directed therapeutic ultrasound, nanodroplets of the emulsions, accumulated in a tumor by passive targeting, are converted into vapor microbubbles. *In vivo*, the nanodroplets strongly retained the loaded drugs; yet under ultrasound-mediated vaporization they released the drugs into the tumor tissue, thereby implementing effective targeting into the tumor.

REFERENCES

1. Forney, M. W.; Poler, J. C. Sonochemical formation of methyl hydroperoxide in polar aprotic solvents and its effect on single-walled carbon nanotube dispersion stability. *Journal of the American Chemical Society*, 2010, *32* (2), 791–797.
2. Hasobe, T.; Oki, H.; Sandanayaka, A. S. D.; Murata, H. Sonication-assisted supramolecular nanorods of meso-diaryl-substituted porphyrins. *Chemical Communications*, 2008, 724–726.
3. He, S.; Ha, J.; Wang, Y.; Hou, Z. Applications of sonochemistry in preparing nanosized materials. *Huaxue Tongbao*, 2008, *71* (11), 846–851.
4. Mastai, Y.; Gedanken, A. Sonochemistry and other novel methods developed for the synthesis of nanoparticles. In *Chemistry of Nanomaterials*. Rao, C. N. R.; Mueller, A.; Cheetham, A. K. (Eds.). Wiley-VCH, Weinheim, Germany, 2004, Vol. 1, pp. 113–169.
5. Manoharan, S. S.; Rao, M. L. Sonochemical synthesis of nanomaterials. In *Encyclopedia of Nanoscience and Nanotechnology*. Nalwa, H. S. (Ed.). American Scientific Publishers, Valencia, CA, 2004, Vol. 10, pp. 67–82.
6. Okitsu, K.; Maeda, Y. Sonochemistry for materials creation and preparation. *Kagaku to Kogyo*, 2004, *57* (8), 819–822.
7. Gedanken, A. Using sonochemistry for the fabrication of nanomaterials. *Ultrasonics Sonochemistry*, 2004, *11* (2), 47–55.
8. Saez, V.; Mason, T. J. Sonoelectrochemical synthesis of nanoparticles. *Molecules*, 2009, *14* (10), 4284–4299.
9. Theerdhala, S.; Vitta, S.; Bahadur, D. Magnetic nanoparticles through sonochemistry. *Materials Technology*, 2008, *23* (2), 88–93.
10. Skrabalak, S. E. Ultrasound-assisted synthesis of carbon materials. *Physical Chemistry Chemical Physics*, 2009, *11* (25), 4930–4942.
11. Xing, Y. Sonochemistry of carbon nanotubes. In *Chemistry of Carbon Nanotubes*. Basiuk, V. A.; Basiuk, E. V. (Eds.). American Scientific Publishers, CA, 2008, *1*, 159–168.
12. Yan, G.; Huo, C.; Liu, H. Application of ultrasonic technologies in catalytic chemistry. Application of ultrasonic technologies in catalytic chemistry. *Gongye Cuihua*, 2007, *15* (2), 1–5.
13. Enomoto, N.; Hojo, J.; Nakagawa, Z.-E. Sonochemistry and ceramic nanoparticles synthesis. *Materials Integration*, 2005, *18* (4), 39–46.
14. Zhang, K.; Park, B.-J.; Fang, F.-F.; Choi, H. J. Sonochemical preparation of polymer nanocomposites. *Molecules*, 2009, *14* (6), 2095–2110.
15. Husseini, G. A.; Pitt, W. G. Micelles and nanoparticles for ultrasonic drug and gene delivery. *Advanced Drug Delivery Reviews*, 2008, *60* (10), 1137–1152.
16. Wilson, B. C. Photonic and non-photonic based nanoparticles in cancer imaging and therapeutics. *NATO Science Series, II: Mathematics, Physics and Chemistry*, 2006, *239 (Photon-Based Nanoscience and Nanobiotechnology)*, 121–157.
17. Reddy, L. H. Drug delivery to tumors: Recent strategies. *Journal of Pharmacy and Pharmacology*, 2005, *57* (10), 1231–1242.
18. Ishtiaq, F.; Farooq, R.; Farooq, U.; Farooq, A.; Siddique, M.; Shah, H.; Mukhatar-Ul-Hassan; Shaheen, M. A. Application of ultrasound in pharmaceutics. *World Applied Sciences Journal*, 2009, *6* (7), 886–893.
19. Moreira, S.; Silva, N. B.; Almeida-Lima, J.; Rocha, H. A. O.; Medeiros, S. R. B.; Alves, C. Jr.; Gama, F. M. BC nanofibres: In vitro study of genotoxicity and cell proliferation. *Toxicology Letters*, 2009, *189* (3), 235–241.
20. Naota, T.; Koori, H. Molecules that assemble by sound: An application to the instant gelation of stable organic fluids. *Journal of the American Chemical Society* 2005, *127*, 9324–9325.
21. Kentish, S.; Wooster, T. J.; Ashokkumar, M.; Balachandran, S.; Mawson, R.; Simons, L. The use of ultrasonics for nanoemulsion preparation. *Innovative Food Science and Emerging Technologies*, 2008, *9* (2), 170–175.
22. Tal-Figiel, B.; Figiel, W. Micro- and nanoemulsions in cosmetic and pharmaceutical products. *Journal of Dispersion Science and Technology*, 2008, *29* (4), 611–616.

23. Gedanken, A.; Koltypin, Y.; Perkas, N.; Besson, M.; Vradman, L.; Herskowitz, M.; Landau, M. V. The sonochemical insertion of nanomaterials into mesostructures. *Transactions of the Indian Ceramic Society*, 2004, *63* (3), 137–144.

24. Geipel-Kern, A. Big system for small particles. Evonik industries produces nanoparticles in an ultrasound reactor. *Process*, 2009, *16* (10), 12–13.

25. Pohl, B.; Oezyilmaz, N.; Brenner, G.; Peuker, U. A. Characterization of a conical geometry for an ultrasonic flow reactor. *Chemie Ingenieur Technik*, 2009, *81* (10), 1613–1622.

26. Abe, M.; Sakai, H.; Sakai, T.; Enomoto, H. Manufacture of metal nanoparticles by reduction. 2009, JP 2009057594, 22 pp.

27. Yuk, S. H.; Oh, G. S. Method for fabricating gold deposited iron oxide/glycol chitosan nanocomposite, and MRI contrast agent containing it. KR 2009119324, 2009, 13 pp.

28. Wei, Y.; Li, Y.; Zhang, N.; Shi, G.; Jin, L. Ultrasound-radiated synthesis of PAMAM-Au nanocomposites and its application on glucose biosensor. *Ultrasonics Sonochemistry*, 2009, Volume Date 2010, *17* (1), 17–20.

29. Belova, V.; Andreeva, D. V.; Mohwald, H.; Shchukin, D. G.; Belova, V.; Andreeva, D. V.; Mohwald, H. Ultrasonic intercalation of gold nanoparticles into clay matrix in the presence of surface-active materials. Part 1. Neutral polyethylene glycol. *Journal of Physical Chemistry C*, 2009, *113* (14), 5381–5389.

30. Zhang, X.; He, X.; Wang, K.; Wang, Y.; Li, H.; Tan, W. Biosynthesis of size-controlled gold nanoparticles using fungus, *Penicillium* sp. *Journal of Nanoscience and Nanotechnology*, 2009, *9* (10), 5738–5744.

31. Kan, C.; Zhu, J.; Wang, C. Ag nanoparticle-filled polymer shell formed around Au nanoparticle core via ultrasound-assisted spherulite growth. *Journal of Crystal Growth*, 2009, *311* (6), 1565–1570.

32. Sakai, T.; Enomoto, H.; Torigoe, K.; Sakai, H.; Abe, M. Surfactant- and reducer-free synthesis of gold nanoparticles in aqueous solutions. *Colloids and Surfaces, A: Physicochemical and Engineering Aspects*, 2009, *347* (1–3), 18–26.

33. Okitsu, K.; Sharyo, K.; Nishimura, R. One-pot synthesis of gold nanorods by ultrasonic irradiation: The effect of pH on the shape of the gold nanorods and nanoparticles. *Langmuir*, 2009, *25* (14), 7786–7790.

34. Zhao, G.; Lei, Y.; Tong, X. Method for chemically assembling noble metal nanocrystals based on titania nanotube array. CN 101560669, 2009, 8 pp.

35. Nemamcha, A.; Rehspringer, J. L. Morphology of dispersed and aggregated PVP-Pd nanoparticles prepared by ultrasonic irradiation of Pd(NO$_3$)$_2$ solution in ethylene glycol. *Reviews on Advanced Materials Science*, 2008, *18* (8), 687–690.

36. Chang, J.-K.; Chen, C.-Y.; Tsai, W.-T. Decorating carbon nanotubes with nanoparticles using a facile redox displacement reaction and an evaluation of synergistic hydrogen storage performance. *Nanotechnology*, 2009, *20* (49), 495603/1–495603/7.

37. Zin, V.; Pollet, B. G.; Dabala, M. Sonoelectrochemical (20 kHz) production of platinum nanoparticles from aqueous solutions. *Electrochimica Acta*, 2009, *54* (28), 7201–7206.

38. Zhang, W.; Qiao, X.; Chen, J. Synthesis of silver nanoparticles—Effects of concerned parameters in water/oil microemulsion. *Materials Science and Engineering, B: Solid-State Materials for Advanced Technology*, 2007, *142* (1), 1–15.

39. Hayashi, Y.; Sekino, T.; Niihara, K. New phenomena and applications by combination of metal oxide and ultrasound. *Transactions of the Materials Research Society of Japan*, 2002, *27* (1), 121–124.

40. Fu, H.; Wang, B.; Tian, C.; Wang, L.; Tian, G. Method for preparing Ag/C nanocomposite with good stability. CN 101480604, 2009, 12 pp.

41. Duan, X.; Li, H.; Liu, Y. Method for preparing inorganic antibacterial Ag/TiO$_2$ nanotube composite material. CN 101485981, 2009, 8 pp.

42. Ma, D.; Xie, X.; Zhang, L. Antibacterial supramolecular hydrogel containing nanoscale silver, preparation and application thereof. CN 101564400, 2009, 16 pp.

43. Situ, J. Nano-silver disinfection solution. CN 101341883, 2009, 5 pp.

44. Li, Y.; Song, W.; Zhang, Z.; Lin, H.; Guo, X. Antibacterial nanopowder with shell powder carriers. CN 101151967, 2008, 9 pp.

45. Perkas, N.; Amirian, G.; Rottman, C.; de la Vega, F.; Gedanken, A. Sonochemical deposition of magnetite on silver nanocrystals. *Ultrasonics Sonochemistry*, 2008, Volume Date 2009, *16* (1), 132–135.

46. Cao, G.; Konishi, H.; Li, X. Recent developments on ultrasonic cavitation based solidification processing of bulk magnesium nanocomposites. *International Journal of Metalcasting*, 2008, *2* (1), 57–65, 67–68.

47. Cao, G.; Choi, H.; Konishi, H.; Kou, S.; Lakes, R.; Li, X. Mg-6Zn/1.5%SiC nanocomposites fabricated by ultrasonic cavitation-based solidification processing. *Journal of Materials Science*, 2008, *43* (16), 5521–5526.

48. Jia, X. Y.; Liu, S. Y.; Gao, F. P.; Zhang, Q. Y.; Li, W. Z. Magnesium matrix nanocomposites fabricated by ultrasonic assisted casting. *International Journal of Cast Metals Research*, 2009, *22* (1–4), 196–199.

49. Mahendiran, C.; Ganesan, R.; Gedanken, A. Sonoelectrochemical synthesis of metallic aluminum nanoparticles. *European Journal of Inorganic Chemistry*, 2009, (14), 2050–2053.

50. Johnson, C. E.; Higa, K. T.; Albro, W. R. Nanothermites with condensable gas products. *Proceedings of the 35th International Pyrotechnics Seminar*, Fort Collins, CO, 2008, pp. 159–168.

51. Stoldt, C. R.; Haag, M. A.; Larsen, B. A. Preparation of freestanding germanium nanocrystals by ultrasonic aerosol pyrolysis. *Applied Physics Letters*, 2008, *93* (4), 043125/1–043125/3.

52. Xue, Y.; Lan, M.; Li, J.; Liu, H.; Ma, W. Pulse electrodeposited $Ni-CeO_2$ nanocomposites in an ultrasonic field. *Tezhong Zhuzao Ji Youse Hejin*, 2009, *29* (4), 313–316.

53. Xue, J.; Wu, Me.; Xia, F. Study on microstructure and properties of ultrasonic-electrodeposited Ni- TiN composite layer. *Biaomian Jishu*, 2009, *38* (4), 13–15, 38.

54. Xie, K.; Liu, J. Method for preparing particle-containing liquid metal with high heat transfer performance by mechanical stirring and ultrasonic dispersion. CN 101418210, 2009, 13 pp.

55. Du, H.; Zhao, S.-X.; Xu, G.-C. Preparation of Cu-Ga/PMMA nanocomposites by ultrasonic. *Shanxi Daxue Xuebao, Ziran Kexueban*, 2009, *32* (3), 432–435.

56. Zin, V.; Dabala, M. Iron-chromium alloy nanoparticles produced by pulsed sonoelectrochemistry: Synthesis and characterization. *Acta Materialia*, 2009, Volume Date 2010, *58* (1), 311–319.

57. Zhang, X.; Sun, H. Method for preparing shape-controllable metal nanocrystal/carbon nanotube heterostructure by electrochemical deposition. CN 101538007, 2009, 15 pp.

58. Zin, V.; Zanella, A.; Agnoli, A.; Brunelli, K.; Dabala, M. Sonoelectrochemical synthesis of FeCo nanoparticles: study of the effects of bath's composition on process efficiency and particles features. *Current Nanoscience*, 2009, *5* (2), 232–239.

59. Zhao, F.; Xiao, F.; Zeng, B. Electrodeposition of PtCo alloy nanoparticles on inclusion complex film of functionalized cyclodextrin-ionic liquid and their application in glucose sensing. *Electrochemistry Communications*, 2010, *12* (1), 168–171.

60. Xiao, F.; Zhao, F.; Mei, D.; Mo, Z.; Zeng, B. Nonenzymatic glucose sensor based on ultrasonic-electrodeposition of bimetallic PtM (M = Ru, Pd and Au) nanoparticles on carbon nanotubes-ionic liquid composite film. *Biosensors and Bioelectronics*, 2009, *24* (12), 3481–3486.

61. Xiao, F.; Zhao, F.; Zhang, Y.; Guo, G.; Zeng, B. Ultrasonic electrodeposition of gold-platinum alloy nanoparticles on ionic liquid-chitosan composite film and their application in fabricating nonenzyme hydrogen peroxide sensors. *Journal of Physical Chemistry C*, 2009, *113* (3), 849–855.

62. Gurmen, S.; Ebin, B.; Stopic, S.; Friedrich, B. Nanocrystalline spherical iron-nickel (Fe-Ni) alloy particles prepared by ultrasonic spray pyrolysis and hydrogen reduction (USP-HR). *Journal of Alloys and Compounds*, 2009, *480* (2), 529–533.

63. Liu, Y.; Huang, W.; Zhao, Y.; Dou, T. Ultrasound promoted direct synthesis of nano Cu-Zn-Al-ZSM-5 in acid medium. *Reaction Kinetics and Catalysis Letters*, 2009, *96* (1), 157–163.

64. Kharissova, O.V.; Kharisov, B. I. Synthetic techniques and applications of activated nanostructurized metals: Highlights up to 2008. *Recent Patents on Nanotechnology*, 2008, *2* (2), 103–119.

65. Garza-Rodríguez, L. A.; Kharisov, B. I.; Kharissova, O. V. Overview on the synthesis of activated micro- and nanostructurized Rieke metals: History and present state. *Synthesis and Reactivity in Inorganic and Metal-Organic Chemistry*, 2009, *39* (5), 270–290.

66. Garnovskii, A. D.; Kharisov, B. I. (Eds.). *Direct Synthesis of Coordination and Organometallic Compounds*. Marcel Dekker, Inc., New York, 1999, 244 pp.

67. Li, L.; Qin, Z.-Y.; Liang, X.; Fan, Q.-Q.; Lu, Y.-Q.; Wu, W.-H.; Zhu, M.-F. Facile fabrication of uniform core-shell structured carbon nanotube-polyaniline nanocomposites. *Journal of Physical Chemistry C*, 2009, *113* (14), 5502–5507.

68. Zhao, D.; Li, X.; Shen, Z. Method for filling carbon nanotubes with magnetic metal iron nanoparticles. CN 101456074, 2009, 9 pp.

69. Zheng, R.; Meng, X.; Tang, F.; Zhang, L.; Ren, J. A general, one-step and template-free route to rattle-type hollow carbon spheres and their application in lithium battery anodes. *Journal of Physical Chemistry C*, 2009, *113* (30), 13065–13069.

70. Savosin, M. V.; Mochalin, V. N.; Yaroshenko, A. P.; Lazareva, N. I.; Konstantinova, T. E.; Barsukov, I. V.; Prokofiev, I. G. Carbon nanoscrolls produced from acceptor-type graphite intercalation compounds. *Carbon*, 2007, *45* (14), 2797–2800.

71. Liu, B.; Liu, D.; Cui, W.; Zou, G. Method for preparing monodisperse C_{70} nanoscale monocrystal material. CN 101550591, 2009, 12 pp.

72. Ko, W.-B.; Shon, Y.-S. Ultrasonic, chemical stability and preparation of self-assembled fullerene[C70]-gold nanoparticle films. *Elastomer*, 2005, *40* (4), 272–276.
73. Takeuchi, S.; Uchida, T. Decentralized improvement of nano diamond corpuscle by supersonic wave irradiation system. *Materials Integration*, 2009, *22* (6), 35–40.
74. Wang, S.; Tang, L. A.-l.; Bao, Q.; Lin, M.; Deng, S.; Goh, B.-M.; Loh, K.-P. Room-temperature synthesis of soluble carbon nanotubes by the sonication of graphene oxide nanosheets. *Journal of the American Chemical Society*, 2009, *131* (46), 16832–16837.
75. Matsuyama, K.; Mishima, K. Particle coating of talc with TiO_2 nanoparticles using ultrasonic irradiation in liquid CO_2. *Industrial and Engineering Chemistry Research*, 2010, *49* (3), 1289–1296.
76. Kane, M. C.; Londono, J. D.; Beyer, F. L.; Brennan, A. B. Characterization of the hierarchical structures of a dry nanopowder in a polymer matrix by X-ray scattering techniques. *Journal of Applied Crystallography*, 2009, *42* (5), 925–931.
77. Ohayon, E.; Gedanken, A. The application of ultrasound radiation to the synthesis of nanocrystalline metal oxide in a non-aqueous solvent. *Ultrasonics Sonochemistry*, 2009, Volume Date 2010, *17* (1), 173–178.
78. Wang, J.; Chao, Y.; Wan, Q.; Yan, K.; Meng, Y. Fluoridated hydroxyapatite/titanium dioxide nanocomposite coating fabricated by a modified electrochemical deposition. *Journal of Materials Science: Materials in Medicine*, 2009, *20* (5), 1047–1055.
79. Tian, D.; Ma, W.; Yang, X.; Xu, J. Method for preparing MWCNTs-TiO_2/Nafion composite enzyme electrode. CN 101603940, 2009, 18 pp.
80. Sohn, Y. S.; Smith, Y. R.; Misra, M.; Subramanian, V. Electrochemically assisted photocatalytic degradation of methyl orange using anodized titanium dioxide nanotubes. *Applied Catalysis, B: Environmental*, 2008, *84* (3–4), 372–378.
81. Zhang, J.-Q.; Zheng, S.-H.; Huang, J.-T. Preparation and characterization of TiO_2 nano-films based on ultrasonic atomization technology. *Wuji Cailiao Xuebao*, 2006, *21* (3), 719–724.
82. Prasad, K.; Pinjari, D. V.; Pandit, A. B.; Mhaske, S. T. Phase transformation of nanostructured titanium dioxide from anatase-to-rutile via combined ultrasound assisted sol-gel technique. *Ultrasonics Sonochemistry*, 2010, *17* (2), 409–415.
83. Kharisov, B. I.; Kharissova, O. V.; José Yacamán, M.; Ortiz Mendez, U. State of the art of the bi- and trimetallic nanoparticles on the basis of gold and iron. *Recent Patents on Nanotechnology*, 2009, *3* (2), 81–98.
84. Bolden, N. W.; Rangari, V. K.; Jeelani, S. Synthesis of magnetic nanoparticles and its application in drug delivery systems. *NSTI Nanotech, Nanotechnology Conference and Trade Show, Technical Proceedings*, Boston, MA, June 1–5, 2008. Laudon, M.; Romanowicz, B. (Eds.). Vol. 2, pp. 390–393.
85. Zheng, X.; Zhou, S.; Xiao, Y.; Yu, X.; Li, X.; Wu, P. Shape memory effect of poly(D,L-lactide)/Fe_3O_4 nanocomposites by inductive heating of magnetite particles. *Colloids and Surfaces, B: Biointerfaces*, 2009, *71* (1), 67–72.
86. Zhang, X.; Dai, Q.; Huang, X.; Tang, X. Synthesis and characterization of novel magnetic Fe_3O_4/polyphosphazene nanofibers. *Solid State Sciences*, 2009, *11* (11), 1861–1865.
87. Zhang, X.; Huang, X.; Tang, X. Novel synthesis of magnetic poly(cyclotriphosphazene-co-4,4′-sulfonyldiphenol) nanotubes with magnetic phases embedded in the walls. *New Journal of Chemistry*, 2009, *33* (12), 2426–2430.
88. Wang, N.; Zhu, L.; Wang, D.; Wang, M.; Lin, Z.; Tang, H. Sono-assisted preparation of highly-efficient peroxidase-like Fe_3O_4 magnetic nanoparticles for catalytic removal of organic pollutants with H_2O_2. *Ultrasonics Sonochemistry*, 2010, *17* (3), 526–533.
89. Dang, F.; Enomoto, N.; Hojo, J.; Enpuku, K. Sonochemical synthesis of monodispersed magnetite nanoparticles by using an ethanol-water mixed solvent. *Ultrasonics Sonochemistry*, 2009, *16* (5), 649–654.
90. Wakiya, N.; Takai, M.; Sakamoto, N.; Sakurai, O.; Shinozaki, K.; Suzuki, H. Preparation of magnetite thin film using one-liquid ultrasonic spray plating. *Transactions of the Materials Research Society of Japan*, 2008, *33* (4), 977–979.
91. Wang, B.; Zhang, F.; Qiu, J.; Zhang, X.; Chen, H.; Du, Y.; Xu, P. Preparation of Fe_3O_4 superparamagnetic nanocrystals by coprecipitation with ultrasonic enhancement and their characterization. *Huaxue Xuebao*, 2009, *67* (11), 1211–1216.
92. Askarinejad, A.; Morsali, A. Direct ultrasonic-assisted synthesis of sphere-like nanocrystals of spinel Co_3O_4 and Mn_3O_4. *Ultrasonics Sonochemistry*, 2008, Volume Date 2009, *16* (1), 124–131.
93. Tan, X.; Xu, Y.; Cai, N.; Jia, G. Polypropylene/silica nanocomposites prepared by in-situ melt ultrasonication. *Polymer Composites*, 2009, *30* (6), 835–840.

94. Wang, X.; Wang, S.; Yang, W.; Shan, Z. Ultrasonic synthesis of $Fe_3O_4@SiO_2$ nanocomposite magnetic particles for plasmid DNA purification. *Huaxue Xuebao*, 2009, *67* (1), 54–58.

95. Wee, L. H.; Wang, Z.; Mihailova, B.; Doyle, A. M. Organic functionalization of Silicalite-1 nanocrystals by ultrasonic treatment in methanol. *Microporous and Mesoporous Materials*, 2008, *116* (1–3), 59–62.

96. Chave, T.; Nikitenko, S. I.; Granier, D.; Zemb, T. Sonochemical reactions with mesoporous alumina. *Ultrasonics Sonochemistry*, 2008, Volume Date 2009, *16* (4), 481–487.

97. Wu, Y.; Zhao, B.; Zhu, X. Brittle-ductile transition in the two-dimensional ultrasonic vibration grinding of nanocomposite ceramics. *Key Engineering Materials*, 2009, *416 (Advances in Grinding and Abrasive Technology XV)*, 477–481.

98. Li, F.; Liu, X.; Qin, Q.; Wu, J.; Li, Z.; Huang, X. Sonochemical synthesis and characterization of ZnO nanorod/Ag nanoparticle composites. *Crystal Research and Technology*, 2009, *44* (11), 1249–1254.

99. Kang, S. U.; Yoon, J. Y.; Sung, D. J.; Shin, Y. H.; Jung, S. H.; Oh, Y. J. Method for doping ZnO nanowires with transition metal by ultrasonication. KR 2009095861, 2009, 10 pp.

100. Lu, W.; Li, Y.; Zhang, B.; Chen, P.; Li, T.; Li, Z.; Li, N.; Chen, X. Transition metal ion injection method for preparing oriented zinc nanoparticles in zinc oxide matrix. CN 101333686, 2008, 12 pp.

101. Cao, C.; Yang, H. Method for preparing nanoscale laminated ZnO. CN 1899969, 2007, 6 pp.

102. Applerot, G.; Perkas, N.; Amirian, G.; Girshevitz, O.; Gedanken, A. Coating of glass with ZnO via ultrasonic irradiation and a study of its antibacterial properties. *Applied Surface Science*, 2009, *256* (Suppl.), S3–S8.

103. Perelshtein, I.; Applerot, G.; Perkas, N.; Wehrschuetz-Sigl, E.; Hasmann, A.; Guebitz, G.; Gedanken, A. CuO-cotton nanocomposite: Formation, morphology, and antibacterial activity. *Surface and Coatings Technology*, 2009, *204* (1–2), 54–57.

104. Yu, W.; Xie, H.; Chen, L.; Li, Y. Preparation of monodispersed cuprous oxide nanoparticles under ultrasonic radiation. *Asian Journal of Chemistry*, 2009, *21* (9), 6927–6932.

105. Rangari, V. K.; Hassan, T. A.; Mayo, Q.; Jeelani, S. Size reduction of WO_3 nanoparticles by ultrasound irradiation and its applications in structural nanocomposites. *Composites Science and Technology*, 2009, *69* (14), 2293–2300.

106. Chen, D.; Zhang, R.; Wang, H.; Lu, H.; Xu, H. Method for preparation of oriented tungsten trioxide nanofilm having high crystallinity and good orientation. CN 101318705, 2008, 12 pp.

107. Chou, S.-L.; Wang, J.-Z.; Sun, J.-Z.; Wexler, D.; Forsyth, M.; Liu, H.-K.; MacFarlane, D. R.; Dou, S.-X. High capacity, safety, and enhanced cyclability of lithium metal battery using a V_2O_5 nanomaterial cathode and room temperature ionic liquid electrolyte. *Chemistry of Materials*, 2008, *20* (22), 7044–7051.

108. Zhang, X.; Yang, W. Electrophoretic deposition of a thick film of layered manganese oxide. *Chemistry Letters*, 2007, *36* (10), 1228–1229.

109. Liu, X.-L.; Guo, Y.; Li, Y.; Chen, L.-Q. Synthesis of SnO_2 nanoparticles by ultrasonic irradiation assisted sol-gel method. *Huaxue Shijie*, 2007, *48* (7), 385–387, 446.

110. Kaushal, A.; Bansal, P.; Vishnoi, R.; Choudhary, N.; Kaur, D. Room-temperature ferromagnetism in $Sn_{1-x}Mn_xO_2$ nanocrystalline thin films prepared by ultrasonic spray pyrolysis. *Physica B: Condensed Matter*, 2009, *404* (20), 3732–3738.

111. Khasanov, O. L.; Dvilis, E. S. Net shaping nanopowders with powerful ultrasonic action and methods of density distribution control. *Advances in Applied Ceramics*, 2008, *107* (3), 135–141.

112. Chen, Z.; Liu, Q.; Zhao, X.; Fu, M.; Chen, Y.; Sun, L. Effect of different precipitants on synthesis of CeO_2 nanoparticles by ultrasonic atomization process. *Guisuanyan Xuebao*, 2008, *36* (10), 1450–1453.

113. Han, C.; Liao, J.; Cao, X.; Ren, S.; Wang, X.; Zong, J. Preparation of nano-sized lanthanum oxide by ultrasonic homogeneous precipitation method. *Wujiyan Gongye*, 2008, *40* (10), 18–20.

114. Liu, L.; Zhou, Z.; Peng, C. Sonochemical intercalation synthesis of nano γ-nickel oxyhydroxide: Structure and electrochemical properties. *Electrochimica Acta*, 2008, *54* (2), 434–441.

115. Nie, Z.; Li, Q.; Wang, R.; Wei, Q.; Wang, Z. Manufacture of a nanoscale nickel hydroxide thin film and its use as active electrode material. CN 1986431, 2007, 12 pp.

116. Jia, Z.; Tang, Y.; Luo, L.; Li, B.; Chen, Z.; Wang, J.; Zheng, H. Room temperature fabrication of single crystal nanotubes of $CaSn(OH)_6$ through sonochemical precipitation. *Journal of Colloid and Interface Science*, 2009, *334* (2), 202–207.

117. Bang, J. H.; Helmich, R. J.; Suslick, K. S. Nanostructured $ZnS:Ni^{2+}$ photocatalysts prepared by ultrasonic spray pyrolysis. *Advanced Materials*, 2008, *20* (13), 2599–2603.

118. Senthilkumaar, S.; Selvi, R. T. Formation and photoluminescence of zinc sulfide nanorods. *Journal of Applied Sciences*, 2008, *8* (12), 2306–2311.

119. Li, J.; Huang, J.; Cao, L.; Wu, J.; He, H. Influence of ultrasonic irradiation power on the synthesis of ZnS nanocrystallites. *Guisuanyan Xuebao*, 2009, *37* (11), 1843–1846.

120. Ma, J.; Tai, G.; Guo, W. Ultrasound-assisted microwave preparation of Ag-doped CdS nanoparticles. *Ultrasonics Sonochemistry*, 2010, *17* (3), 534–540.

121. Garnovskii A. D.; Kharisov, B. I. (Eds.). *Synthetic Coordination & Organometallic Chemistry*. Marcel Dekker, New York, 2003, 513 pp.

122. Feng, M.; Chen, Y.; Gu, L.; He, N.; Bai, J.; Lin, Y.; Zhan, H. CdS nanoparticles chemically modified PAN functional materials: Preparation and nonlinear optical properties. *European Polymer Journal*, 2009, *45* (4), 1058–1064.

123. Pan, J.; Li, J.; Xiong, S.; Qian, Y. Ultrasonically assisted synthesis of Tin sulfide nanorods at room temperature. *Advanced Materials Research*, 2009, *79–82* (Pt. 1, *Multi-Functional Materials and Structures II*), 313–316.

124. Huang, J.; Wang, Y.; Cao, L.; Zhu, H.; Yin, L.; Wu, J. Method for preparing Bi_2S_3 nanofilm. CN 101407377, 2009, 8 pp.

125. Zhou, Z. Characters of χ-Gd_2S_3 film deposited by spray pyrolysis. *Huagong Shikan*, 2006, *20* (9), 14–16.

126. Yang, Q.-Z.; Jones, G. A. C.; Kelly, M. J.; Beere, H.; Farrer, I. Fabrication and characterization of GaAs/AlGaAs lateral quantum dot developed by ultrasonic agitation. *Semiconductor Science and Technology*, 2008, *23* (5), 055018/1–055018/7.

127. Huang, J.; Zhu, W.; Xiao, H. Method for preparing CdSe and CdSe-ZnSe core-shell quantum dot. CN 101585516, 2009, 9 pp.

128. Wang, Z.; Qu, S.; Liu, J.; Wang, Z. Preparation method of poly[2-methoxy-5-(3,7-dimethyloctyloxy)-1,4-phenylene vinylene]-coated PbS quantum dot and nanorod material and battery. CN 101497784, 2009, 16 pp.

129. Bardelang, D.; Zaman, Md. B.; Moudrakovski, I. L.; Pawsey, S.; Margeson, J. C.; Wang, D.; Wu, X.; Ripmeester, J. A.; Ratcliffe, C. I.; Yu, K. Interfacing supramolecular gels and quantum dots with ultrasound: Smart photoluminescent dipeptide gels. *Advanced Materials*, 2008, *20* (23), 4517–4520.

130. Liu, B.; Ren, T.; Zhang, J.-R.; Chen, H.-Y.; Zhu, J.-J.; Burda, C. Spectroelectrochemistry of hollow spherical CdSe quantum dot assemblies in water. *Electrochemistry Communications*, 2007, *9* (4), 551–557.

131. Pejova, B.; Grozdanov, I.; Nesheva, D.; Petrova, A. Size-dependent properties of sonochemically synthesized three-dimensional arrays of close-packed semiconducting $AgBiS_2$ quantum dots. *Chemistry of Materials*, 2008, *20* (7), 2551–2565.

132. Kow, K. W.; Abdullah, E. C.; Aziz, A. R. Effects of ultrasound in coating nano-precipitated $CaCO_3$ with stearic acid. *Asia-Pacific Journal of Chemical Engineering*, 2009, *4* (5), 807–813.

133. Sonawane, S. H.; Gumfekar, S. P.; Meshram, S.; Deosarkar, M. P.; Mahajan, C. M.; Khanna, P. Combined effect of surfactant and ultrasound on nano calcium carbonate synthesized by crystallization process. *International Journal of Chemical Reactor Engineering*, 2009, *7*. http://www.bepress.com/ijcre/vol7/A47.

134. Berdonosov, S. S.; Melikhov, I. V.; Znamenskaya, I. V. Layered agglomeration of primary vaterite nanoparticles during ultrasonic stirring. *Inorganic Materials*, 2005, *41* (4), 397–401.

135. Wang, P.; Zheng, S.; Yin, Y.; Su, D. Preparation and properties of nano-inorganic mineral material as lubricating oil additive. *Advanced Materials Research*, 2009, *79–82* (Pt. 2, *Multi-Functional Materials and Structures II*), 1847–1850.

136. Alavi, M. A.; Morsali, A. Syntheses and characterization of $Sr(OH)_2$ and $SrCO_3$ nanostructures by ultrasonic method. *Ultrasonics Sonochemistry*, 2009, Volume Date 2010, *17* (1), 132–138.

137. Ren, C.-J.; Liu, D.-M.; Qu, L.-J.; Tan, R.-M. Synthesis of hydroxyapatite nanoparticles in ultrasonic precipitation. *Jiamusi Daxue Xuebao, Ziran Kexueban*, 2009, *27* (3), 402–404, 407.

138. Ruan, Q.; Zhu, Y.; Zeng, Y.; Qian, H.; Xiao, J.; Xu, F.; Zhang, L.; Zhao, D. Ultrasonic-irradiation-assisted oriented assembly of ordered monetite nanosheets stacking. *Journal of Physical Chemistry B*, 2009, *113* (4), 1100–1106.

139. Tanaka, H.; Okumiya, T.; Ueda, S.-K.; Taketani, Y.; Murakami, M. Preparation of nanosheet by exfoliation of layered iron phenyl phosphate under ultrasonic irradiation. *Materials Research Bulletin*, 2009, *44* (2), 328–333.

140. Zuo, G.-H.; Li, D.-M.; Mu, J.-J.; Chen, H.-Y. Influences of ultrasonic vibration and electric field on preparing $NaNO_2$ nano-particles by ice sublimation method. *Jilin Daxue Xuebao, Lixueban*, 2008, *46* (4), 751–754.

141. Shen, X.-F. Combining microwave and ultrasound irradiation for rapid synthesis of nanowires: A case study on Pb(OH)Br. *Journal of Chemical Technology and Biotechnology*, 2009, *84* (12), 1811–1817.

142. Xiong, J.; Guo, Z.; Xiong, S.; Zuo, L.; Chen, J.; Wu, Y.; Fan, H. Method for dispersing nanoscale TiN powder. CN 101565176, 2009, 6 pp.

143. Sergiienko, R.; Shibata, E.; Akase, Z.; Suwa, H.; Shindo, D.; Nakamura, T. Synthesis of Fe-filled carbon nanocapsules by an electric plasma discharge in an ultrasonic cavitation field of liquid ethanol. *Journal of Materials Research*, 2006, *21* (10), 2524–2533.

144. Epshteyn, A.; Purdy, A. P.; Pettigrew, K. A.; Miller, J. B.; Stroud, R. M. Sonochemical synthesis of air-insensitive carbide-stabilized hafnium subhydride nanopowder. *Chemistry of Materials*, 2009, *21* (15), 3469–3472.

145. Yu, C.; Yu, M.; Li, C.; Zhang, C.; Yang, P.; Lin, J. Spindle-like lanthanide orthovanadate nanoparticles: Facile synthesis by ultrasonic irradiation, characterization, and luminescent properties. *Crystal Growth and Design*, 2009, *9* (2), 783–791.

146. Shang, M.; Wang, W.; Zhou, L.; Sun, S.; Yin, W. Nanosized $BiVO_4$ with high visible-light-induced photocatalytic activity: Ultrasonic-assisted synthesis and protective effect of surfactant. *Journal of Hazardous Materials*, 2009, *172* (1), 338–344.

147. Dang, F.; Kato, K.; Imai, H.; Wada, S.; Haneda, H.; Kuwabara, M. A new effect of ultrasonication on the formation of $BaTiO_3$ nanoparticles. *Ultrasonics Sonochemistry*, 2010, *17* (2), 310–314.

148. Gao, T.; Wu, Q.; Fjellvag, H.; Norby, P. Topological properties of titanate nanotubes. *Journal of Physical Chemistry C*, 2008, *112* (23), 8548–8552.

149. Jia, R.-P.; Ou-Yang, C.-F.; Li, Y.-S.; Yang, J.-H.; Xia, W. Preparation and optical properties of $HgWO_4$ nanorods by hydrothermal method coupled with ultrasonic technique. *Journal of Nanoparticle Research*, 2008, *10* (1), 215–219.

150. Zhou, L.; Wang, W.; Zhang, L. Ultrasonic-assisted synthesis of visible-light-induced Bi_2MO_6 (M = W, Mo) photocatalysts. *Journal of Molecular Catalysis A: Chemical*, 2007, *268* (1–2), 195–200.

151. Sadykov, V.; Kharlamova, T.; Batuev, L.; Muzykantov, V.; Mezentseva, N.; Krieger, T.; Alikina, G. et al. $La_{0.8}Sr_{0.2}Ni_{0.4}Fe_{0.6}O_3$ (LSNF)-$Ce_{0.8}Gd_{0.2}O_{2-\sigma}$ nanocomposite as mixed ionic-electronic conducting material for SOFC cathode and oxygen permeable membranes: synthesis and properties. *Composite Interfaces*, 2009, *16* (4–6), 407–431.

152. Du, D.; Cao, M.; He, X.; Liu, Y.; Hu, C. Morphology-controllable synthesis of microporous Prussian blue analogue $Zn_3[Co(CN)_6]2 \cdot xH_2O$ microstructures. *Langmuir*, 2009, *25* (12), 7057–7062.

153. Zhang, S.; Yang, S.; Lan, J.; Tang, Y.; Xue, Y.; You, J. Ultrasound-induced switching of sheetlike coordination polymer microparticles to nanofibers capable of gelating solvents. *Journal of the American Chemical Society*, 2009, *131* (5), 1689–1691.

154. Kotal, A.; Paira, T. K.; Banerjee, S.; Mandal, T. K. Ultrasound-induced in situ formation of coordination organogels from isobutyric acids and zinc oxide nanoparticles. *Langmuir*, 2010, *26* (9), 6576–6582.

155. Wu, S.; Shen, X.; Cao, B.; Lin, L.; Shen, K.; Liu, W. Shape- and size-controlled synthesis of coordination polymer $\{[Cu(en)_2][KFe(CN)_6]\}_n$ nano/micro-crystals. *Journal of Materials Science*, 2009, *44* (23), 6447–6450.

156. Li, D.-H.; Xin, X.-Q. One-step solid-state synthesis and characterization of nanocrystalline lanthanum 8-quinolinolate. *Yingyong Huaxue*, 2008, *25* (12), 1421–1424.

157. Mahdavian, A. R.; Sarrafi, Y.; Shabankareh, M. Nanocomposite particles with core-shell morphology III: Preparation and characterization of nano Al_2O_3-poly(styrene-methyl methacrylate) particles via miniemulsion polymerization. *Polymer Bulletin*, 2009, *63* (3), 329–340.

158. Xu, G. C.; Shi, J. J.; Li, D. J.; Xing, H. L. On Interaction between nano-Ag and P(AMPS-co-MMA) copolymer synthesized by ultrasonic. *Journal of Polymer Research*, 2009, *16* (3), 295–299.

159. Chen, W.; Tao, X. Ultrasound-induced functionalization and solubilization of carbon nanotubes for potential nanotextiles applications. *Materials Research Society Symposium Proceedings*, 2006, *920* (*Smart Nanotextiles*), 27–32.

160. Liu, S.; Ying, J.; Zhou, X.; Xie, X. Core-shell magnesium hydroxide/polystyrene hybrid nanoparticles prepared by ultrasonic wave-assisted in-situ copolymerization. *Hubei Materials Letters*, 2009, *63* (11), 911–913.

161. Mo, Z.; Shi, H.; Chen, H.; Niu, G.; Zhao, Z.; Wu, Y. Synthesis of graphite nanosheets/polyaniline nanorods composites with ultrasonic and conductivity. *Journal of Applied Polymer Science*, 2009, *112* (2), 573–578.

162. Lu, X.; Mao, H.; Chao, D.; Zhang, W.; Wei, Y. Fabrication of polyaniline nanostructures under ultrasonic irradiation: from nanotubes to nanofibers. *Macromolecular Chemistry and Physics*, 2006, *207* (22), 2142–2152.

163. Liu, Z.; Han, G.; Yu, J.; Dai, H. Preparation and drug release property of curcumin nanoparticles. *Zhongyaocai*, 2009, *32* (2), 277–279.

164. Lapshin, S.; Swain, S. K.; Isayev, A. I. Ultrasound aided extrusion process for preparation of polyolefin-clay nanocomposites. *Polymer Engineering and Science*, 2008, *48* (8), 1584–1591.

165. Sanchez-Valdes, S.; Ortega-Ortiz, H.; Ramos-de Valle, L. F.; Medellin-Rodriguez, F. J.; Guedea-Miranda, R. Mechanical and antimicrobial properties of multilayer films with a polyethylene/silver nanocomposite layer. *Journal of Applied Polymer Science*, 2009, *111* (2), 953–962.

166. Qua, E. H.; Hornsby, P. R.; McNally, G. M.; Sharma, S.; Lyons, G.; McCall, D. Preparation and characterisation of polyvinyl alcohol nanocomposites made from cellulose nanofibres. *66th Annual Technical Conference—Society of Plastics Engineers*, Milwaukee, WI, 2008, pp. 1609–1613.

167. Xiang, C.; Frey, M. W. Nanocomposite fibers electrospun from biodegradable polymers. *Abstracts of Papers, 235th ACS National Meeting*, New Orleans, LA, April 6–10, 2008, CELL-148.

168. Jiang, L.; Chen, X.; Li, Z. Preparation of nano-crystalline cellulose from hydrolysis by cellulase. *Huaxue Yu Shengwu Gongcheng*, 2008, *25* (12), 63–66.

169. Motlagh, M. M.; Rahimi, R.; Kachousangi, M. J. Ultrasonic method for the preparation of organic nanoparticles of porphyrin. *12th International Electronic Conference on Synthetic Organic Chemistry*, November 1–30, 2008. Seijas, J. A.; Vazquez Tato, M. P. (Eds.). MOTL/1–MOTL/7.

170. Hu, X.; Zuo, X.; Xia, D. Method for in-situ synthesis of metal phthalocyanine/carbon nanotube composite. CN 101254916, 2008, 10 pp.

171. Xue, M.; Su, J.; Ma, N.; Sheng, Q.; Zhang, Q.; Liu, Y. Production of copper phthalocyanine nanofilms. CN 101372757, 2009, 9 pp.

172. Tang, Y.; Qi, J.; Chen, Y. Preparation method for supported platinum or platinum-metal nanoparticle catalyst by ultrasonic electrodeposition. CN 101352683, 2009, 8 pp.

173. Radziuk, D.; Mohwald, H.; Shchukin, D. Ultrasonic activation of platinum catalysts. *Journal of Physical Chemistry C*, 2008, *112* (49), 19257–19262.

174. Shanmugam, S.; Gedanken, A. Synthesis and electrochemical oxygen reduction of platinum nanoparticles supported on mesoporous TiO_2. *Journal of Physical Chemistry C*, 2009, *113* (43), 18707–18712.

175. Li, X.; Hou, Y.; Zhao, Q.; Quan, X.; Chen, G. Preparation method and application of visible-light-activated cuprous oxide/titanium dioxide nanocomposite photocatalyst. CN 101537354, 2009, 12 pp.

176. He, H.; Li, Z.; Zi, X.; Liu, L.; Dai, H. Supported dual-metal rhxag1-x/y nano catalyst for purification of automobile exhaust gas and its preparation method by ultrasound-assisted membrane reaction. CN 101590406, 2009, 10 pp.

177. Tu, Y.-B.; Luo, J.-Y.; Meng, M.; Wang, G.; He, J.-J. Ultrasonic-assisted synthesis of highly active catalyst Au/MnO_x-CeO_2 used for the preferential oxidation of CO in H_2-rich stream. *International Journal of Hydrogen Energy*, 2009, *34* (9), 3743–3754.

178. Xuan, S.; Wang, Y.-X. J.; Yu, J. C.; Leung, K. C.-F. Preparation, characterization, and catalytic activity of core/shell Fe_3O_4@Polyaniline@Au nanocomposites. *Langmuir*, 2009, *25* (19), 11835–11843.

179. Rostamizadeh, S.; Amani, A. M.; Mahdavinia, G. H.; Amiri, G.; Sepehrian, H. Ultrasound promoted rapid and green synthesis of 1,8-dioxo-octahydroxanthenes derivatives using nanosized MCM-41-SO_3H as a nanoreactor, nanocatalyst in aqueous media. *Ultrasonics Sonochemistry*, 2010, *17* (2), 306–309.

180. Anandan, S.; Ashokkumar, M. Sonochemical synthesis of Au-TiO_2 nanoparticles for the sonophotocatalytic degradation of organic pollutants in aqueous environment. *Ultrasonics Sonochemistry*, 2008, Volume Date 2009, *16* (3), 316–320.

181. Rengaraj, S.; Yeon, J.-W.; Li, X.-Z.; Jung, Y.; Kim, W.-H. Heterogeneous photocatalytic degradation of chlorophenols over Ag-TiO_2 nanoparticles in an aqueous suspension. Diffusion and defect data—Solid state data, Pt. B: Solid state phenomena, 2007, *124–126* (Pt. 2, *Advances in Nanomaterials and Processing, Part* 2), 1745–1748.

182. Zhang, X.-J.; Dai, You.-Z.; Song, Y.; Yang, S.-Y.; Zhang, H. Studies on the degradation of 2,4-dichlorophenol by ultrasound cooperating with nanoscale iron. *Xiangtan Daxue Ziran Kexue Xuebao*, 2005, *27* (3), 87–90.

183. Tangestaninejad, S.; Moghadam, M.; Mirkhani, V.; Mohammadpoor-Baltork, I.; Salavati, H. Sonochemical and visible light induced photochemical and sonophotochemical degradation of dyes catalyzed by recoverable vanadium-containing polyphosphomolybdate immobilized on TiO_2 nanoparticles. *Ultrasonics Sonochemistry*, 2008, *15* (5), 815–822.

184. Jiang, Y.-C.; Tian, P.; Jiang, Z.; Han, Z.-X.; Cao, C.-Y.; Xu, R.; Xing, Z.-Q. Degradation of acid red B in the presence of nano-sized ZnO powder under ultrasonic irradiation. *Shenyang Shifan Daxue Xuebao, Ziran Kexueban*, 2008, *26* (2), 202–205.

185. Luo, T.; Ai, Z.; Zhang, L. J. Fe@Fe_2O_3 core-shell nanowires as iron reagent. 4. Sono-Fenton degradation of pentachlorophenol and the mechanism analysis. *Physical Chemistry C*, 2008, *112* (23), 8675–8681.

186. Shah, S. H.; Patel, V. R.; Shah, V. R.; Vaghani, Z. H.; Thakkar, A. Y.; Shah, S. H. Nanotechnology: A review on revolution in cancer treatment. *Pharmaceutical Reviews*, 2008, *6* (6). http://www.pharmainfo.net/reviews/nanotechnology-review-revolution-cancer-treatment

187. Kapoor, S.; Bhattacharyya, A. J. Ultrasound-triggered controlled drug delivery and biosensing using silica nanotubes. *Journal of Physical Chemistry C*, 2009, *113* (17), 7155–7163.

188. Yan, W.; Hsiao, V. K. S.; Zheng, Y. B.; Shariff, Y. M.; Gao, T.; Huang, T. J. Towards nanoporous polymer thin film-based drug delivery systems. *Thin Solid Films*, 2009, *517* (5), 1794–1798.

189. Zhang, H.; Tong, S.-Y.; Zhang, X.-Z.; Cheng, S.-X.; Zhuo, R.-X.; Li, H. Novel solvent-free methods for fabrication of nano- and microsphere drug delivery systems from functional biodegradable polymers. *Journal of Physical Chemistry C*, 2007, *111* (34), 12681–12685.

190. Pisani, E.; Tsapis, N.; Paris, J.; Nicolas, V.; Cattel, L.; Fattal, E. Polymeric nano/microcapsules of liquid perfluorocarbons for ultrasonic imaging: Physical characterization. *Langmuir*, 2006, *22* (9), 4397–4402.

191. Cavalli, R.; Bisazza, A.; Rolfo, A.; Balbis, S.; Madonnaripa, D.; Caniggia, I.; Guiot, C. Ultrasound-mediated oxygen delivery from chitosan nanobubbles. *International Journal of Pharmaceutics*, 2009, *378* (1–2), 215–217.

192. Agarwal, A.; Lvov, Y.; Sawant, R.; Torchilin, V. Stable nanocolloids of poorly soluble drugs with high drug content prepared using the combination of sonication and layer-by-layer technology. *Journal of Controlled Release*, 2008, *128* (3), 255–260.

193. Rapoport, N. Ya.; Nam, K.-H.; Gao, Z.; Kennedy, A. Application of ultrasound for targeted nanotherapy of malignant tumors. *Acoustical Physics*, 2009, *55* (4–5), 594–601.

15 Inorganic Non-Carbon Nanotubes

In addition to well-known carbon nanotubes (CNTs), described in hundreds of thousands of publications, the micro- and nanotubes (including core-shell inorganic nanotubes[1]) of a series of other inorganic compounds have been developed, mainly by the research groups of Tenne[2–6] and Rao.[7,8] Among extensive reviews,[9–13] we note an excellent discussion of TiO_2 nanostructures, including its nanotubes.[14] Generally, the formation of tubular nanostructures requires layered, anisotropic, or pseudo-layered crystal structures.[15] Inorganic nanotubes, which typically do not possess such structures, are usually synthesized using template-based methods, leading, however, to either amorphous, polycrystalline products or existing only in ultrahigh vacuum. Additionally, although initially the method of synthesis for the formation of nanotubular structures, such as, for example, WS_2 and MoS_2, involved starting from the respective oxides, it was well established that the gas-phase synthetic route (using metal chlorides, carbonyls, etc.) provides an alternative that is suitable for the synthesis of many closed caged structures and nanotubes hitherto unknown.[16]

A considerable number of reports are devoted to metal chalcogenide nanotubes, such as, for instance, NbS_2[17] or superconducting $NbSe_2$ (prepared by the decomposition of $NbSe_3$) or $PbSe$,[18] extensively discussed in a review among other inorganic nanotubes.[19] Among chalcogenides, major attention has been given to MoS_2 and WS_2 nanotubes,[20–22] fabricated by various techniques. Thus, a simple synthesis (Figure 15.1) of faceted MoS_2 nanotubes (Figure 15.2), possessing high internal surface area, and nanoflowers starting from molybdenum oxide and thiourea as the sulfur source was reported.[23] The use of thiourea enabled a one-step synthesis involving both the reduction and sulfidization in a facile manner to obtain these nanotubes. In some of the nanotubes, the presence of the oxide core was observed. The authors expected that these MoS_2 nanostructures will find important applications in energy storage, catalysis, and field emission. Mixed-phase sulfide compounds (for instance, $W_xMo_yC_zS_2$ nanotubes) are also known.[24]

WS_2 belongs to a family of layered metal dichalcogenide compounds that are known to form cylindrical (inorganic nanotubes) and polyhedral nanostructures—onion or nested fullerene-like particles.[25] Multiwall WS_2 nanotube templates were used as hosts to prepare core-shell $PbI_2@WS_2$ nanotubes (Figure 15.3) by a capillary-wetting method.[26] Conformal growth of PbI_2 layers on the inner wall of the relatively wide WS_2 nanotubes (internal diameter ~10 nm) led to nanotubular structures, not observed in narrow CNT templates. In addition, two other techniques for obtaining such core-shell nanotubes were offered,[27] including electron-beam irradiation, that is, *in situ* synthesis within a transmission electron microscope (TEM), resulting in SbI_3 nanotubes (observed either in a hollow core of WS_2 ones [$SbI_3@WS_2$ nanotubes] or atop them [$WS_2@SbI_3$ nanotubes]). The second method involved a gaseous-phase reaction, where the layered product employed WS_2 nanotubes as nucleation sites (the MoS_2 layers most often covered the WS_2 nanotube, resulting in $WS_2@MoS_2$ core-shell nanotubes). WS_2 nanotubes were also filled and intercalated by molten-phase cesium iodide (Figure 15.4).[28] It was established that the intercalation of CsI into the host concentric WS_2 lattices resulted in an increase in the interplanar spacing.

In addition to metal sulfides, boron nitride (BN)[29] is also known in nanotube form. Thus, a modification in the conventional arc-discharge method for the synthesis of nanotubes was presented.[30] By injecting pure nitrogen gas directly into the plasma, the amount of BN nanotubes produced (commonly double-walled with outer diameter around 3 nm) was greatly increased. The predominance of double-walled BN nanotubes was attributed by the authors to a direct result of the distribution of the number of

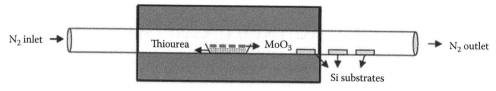

Horizontal furnace

FIGURE 15.1 Experimental setup used in the synthesis of the MoS_2 nanotubes and nanoflowers. (Reproduced from *Mater. Chem. Phys.*, 118, Deepak, F.L. et al., Faceted MoS_2 nanotubes and nanoflowers, 392–397, Copyright 2009, with permission from Elsevier.)

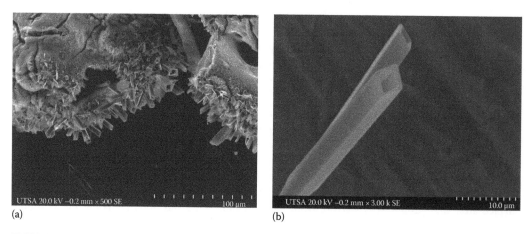

FIGURE 15.2 (a) SEM images of the as-obtained MoS_2 nanotubes. (b) Close-up view of the nanotubes showing the faceted morphology (square or rhomboidal shapes) of the nanotubes. (Reproduced from *Mater. Chem. Phys.*, 118, Deepak, F.L. et al., Faceted MoS_2 nanotubes and nanoflowers, 392–397, Copyright 2009, with permission from Elsevier.)

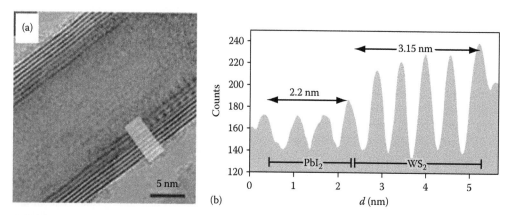

FIGURE 15.3 (a) HRTEM micrograph showing a core-shell $PbI_2@WS_2$ composite nanotube. (b) Line profile obtained from the indicated region in (a) showing two types of nanotube layers: five outer WS_2 layers with sharper contrast and an average spacing of 0.63 nm and three inner layers with more complex contrast and an average spacing of 0.73 nm, corresponding to three concentric PbI_2 nanotubes.

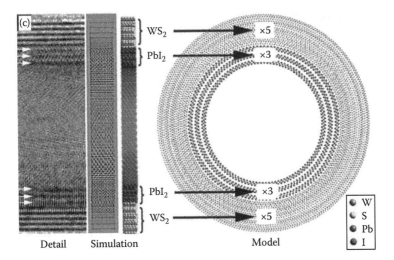

FIGURE 15.3 (continued)   (c) Detail from (a) showing the complex contrast of the inner PbI$_2$ layers (arrowed) relative to the outer WS$_2$ layers. To the right of the detail is a simulation and a cutaway space-filling model (left) and cross-sectional structure model (right) with both WS$_2$ (*aba* stacking) and PbI$_2$ layers (*abc* stacking) indicated. (Kreizman, R. et al.: Core-shell PbI$_2$@WS$_2$ inorganic nanotubes from capillary wetting. *Angew. Chem., Int. Ed.* 2009. 48(7). 1230–1233. Copyright Wiley-VCH Verlag GmbH & Co. KGaA. Reproduced with permission.)

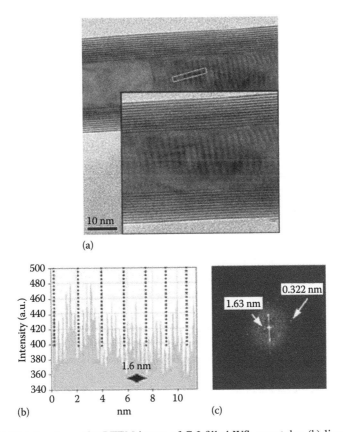

FIGURE 15.4   (a) Moiré pattern of HRTEM image of CsI filled WS$_2$ nanotube, (b) line profile integrated along the region enclosed in the rectangle, (c) fast Fourier transform (FFT) pattern. (Reproduced with kind permission from Springer Science+Business Media: *Nano Res.*, Synthesis and characterization of WS$_2$ inorganic nanotubes with encapsulated/intercalated CsI, 3(3), 2010, 170–173, Hong, S.-Y. et al.)

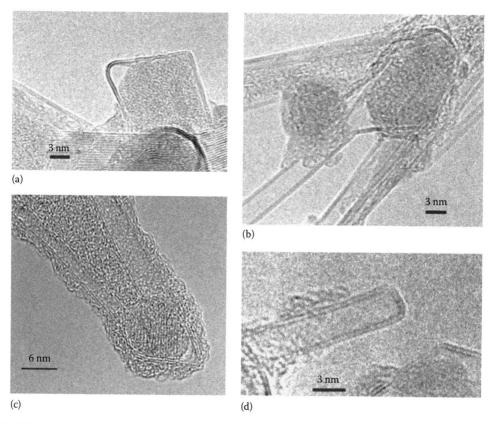

FIGURE 15.5 HRTEM images: (a) sharp facets of a double layer of graphitic BN coating a boron particle. The coating layers do not wet completely the boron particle surface, suggesting that the formation of the coating is posterior to the boron particle formation, (b) a double-wall BN nanotube with a double-layer nanococoon end, (c) two double-wall nanotubes with a common nanococoon end, and (d) a flat angular cap end. (Reproduced with kind permission from Springer Science+Business Media: *J. Mater. Sci.*, Nanococoon seeds for BN nanotube growth, 38, 2003, 4805–4810, Altoe, M.V.P. et al.)

graphitic BN layers for the nanococoons, the second major product of the synthesis. Detailed HRTEM examination (Figure 15.5) of the ends of BN nanotubes indicated continuity between the graphitic BN layers that coat boron nanoparticles, that is, nanococoon, and the nanotube.

In addition to TiO$_2$ nanotubes mentioned here,[31] a family of functional metal oxide nanotubes such as Fe$_2$O$_3$, SnO$_2$, ZrO$_2$, and SnO$_2$@Fe$_2$O$_3$ composites were fabricated using highly active carbonaceous nanofibers as templates, which were synthesized via a hydrothermal approach.[32] This general method can be further developed to synthesize uniform ternary oxide nanotubes such as BaTiO$_3$ and metal oxide composite nanotubes. The obtained uniform nanotubes might find potential applications in fields such as catalysis, chemical/biological separation, and sensing. A sol-gel deposition method was applied to synthesize high-purity SiO$_2$ and ZrO$_2$ nanotubes and nanorods using track-etched membrane templates.[33] Structures, stability, and electronic properties of inorganic dioxides MO$_2$ (M = Si, Ge, Sn) nanotubes with quasi-quadrone atomic lattice were studied by the DFT method.[34] K-type CeO$_2$ nanotubes (Figure 15.6) (based on the Kirkendall effect) were fabricated through a solid–liquid interface reaction between Ce(OH)CO$_3$ nanorods and NaOH solutions (Reactions 15.1 through 15.4).[35] It was indicated that NaOH and the reaction temperature were two key factors responsible for the formation of this type of nanotubes:

$$Ce(OH)CO_3 \rightarrow Ce^{3+} + OH^- + CO_3^{2-} \tag{15.1}$$

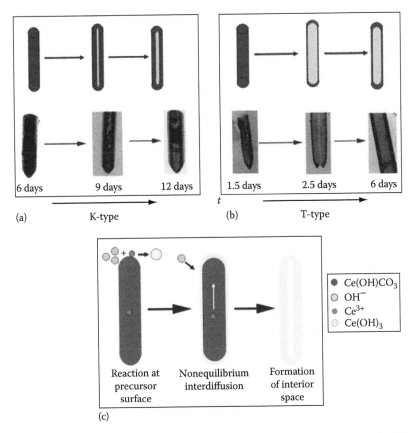

FIGURE 15.6 Sketch map and corresponding TEM images of evolution processes for K-type (a) and T-type (b) CeO_2 nanotubes; (c) schematic illustration of the procedure used to fabricate K-type nanotubes. (Reproduced with permission from Chen, G. et al., Formation of CeO_2 nanotubes from Ce(OH) CO_3 nanorods through Kirkendall diffusion, *Inorg. Chem.*, 48(4), 1334–1338. Copyright 2009 American Chemical Society.)

$$Ce^{3+} + 3OH^- \rightarrow Ce(OH)_3 \qquad (15.2)$$

$$\text{In presence of } O_2: Ce(OH)_3 \rightarrow Ce(OH)_4 \qquad (15.3)$$

$$\text{Calcination: } Ce(OH)_4 \rightarrow CeO_2 + H_2O \qquad (15.4)$$

Nanotubular oxygen-containing metal salts are represented by two U(VI) compounds, $K_5[(UO_2)_3(SeO_4)_5](NO_3) \cdot (H_2O)_{3.5}$ and $(C_4H_{12}N)_{14}[(UO_2)_{10}(SeO_4)_{17} \cdot (H_2O)]$,[36] and hydroxyapatite (HAp). Thus, uniform F-substituted HAp nanotubes (Figure 15.7) with different aspect ratios and surface properties were prepared via a hydrothermal synthetic route.[37] It was noted that a small amount of 3% of fluor doping into HAp can lead to the formation of nanotubes. Naturally occurring nanotubes, halloysite $\{Al_2Si_2O_5(OH)_4\}$ nanotubes (HNTs), were modified by silane and incorporated into epoxy resin to form nanocomposites.[38] Compared with the neat resin, about 40% increase in storage modulus at glassy state and 133% at rubbery state were achieved by incorporating 12% modified HNTs into the epoxy matrix. In addition, the nanocomposites exhibited improved flexural strength, char yield, and dimensional stability. Additionally, a quantitative correlation between the composition, diameter, and internal energy of a class of single-walled mixed oxide aluminosilico-germanate (AlSiGeOH) nanotubes was analyzed.[39]

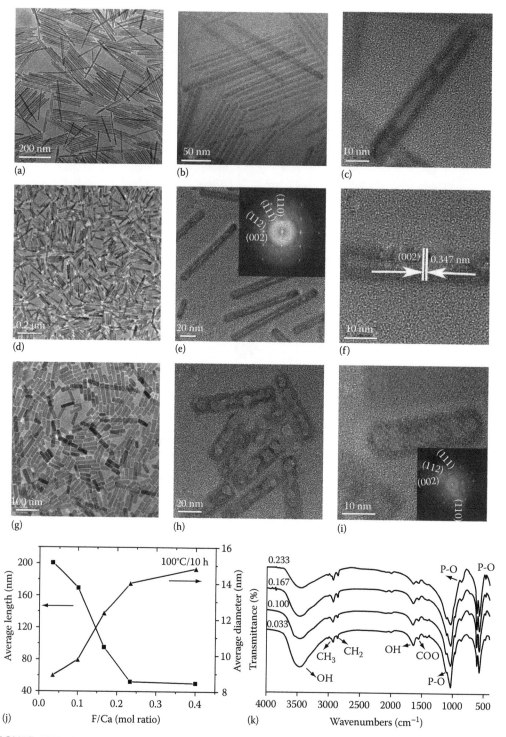

FIGURE 15.7 (a through i) Representative TEM and HRTEM images of F-HAp nanotubes with different doping levels: (a through c) F/Ca = 0.033, 100°C; (d through f) F/Ca = 0.05, 100°C; (g through i) F/Ca = 0.25, 100°C. (j) Variance of the lengths and diameters of nanotubes with differences in the F/Ca ratio. (k) Infrared spectra taken from F-HAp nanotubes with different doping levels. (Reproduced with permission from Hui, J. et al., Monodisperse F-substituted hydroxyapatite single-crystal nanotubes with amphiphilic surface properties, *Inorg. Chem.*, 48(13), 5614–5616. Copyright 2009 American Chemical Society.)

TABLE 15.1
Correlations between Bi Nanoparticle Form, Size, and Yields for the Samples, Obtained *via* MW Heating in Vacuum

Heating Time (min)	Nanoform Size (nm)	Yield (%)	Product Form
5	50–90	80	Nanoparticles
10	50–90	65	Nanoparticles and nanotubes
	630–990 (length)	30	
15	1144 (length)	95	Nanotubes with branches/protrusions
	95 (nanotube diameter)		
	63 (diameter of branches)		

Elemental metal nanotubes are represented by platinum[40] and bismuth.[41] Thus, MW-heating Bi powder in vacuum in a domestic MW oven led to the formation of bismuth nanoparticles and nanotubes without any impurities (bismuth oxide is not observed). As seen in Table 15.1, the nanoparticle size is increased with increasing heating time (5, 10, and 15 min). Heating for 5 min is already sufficient for the formation of the nanoparticles (Figure 15.8a); increasing heating time to 10 min (Figure 15.8b and c), nanoparticles and nanotubes are present in higher quantities, and at 15 min a huge number of nanotubes are observed (Figure 15.8d). High-resolution

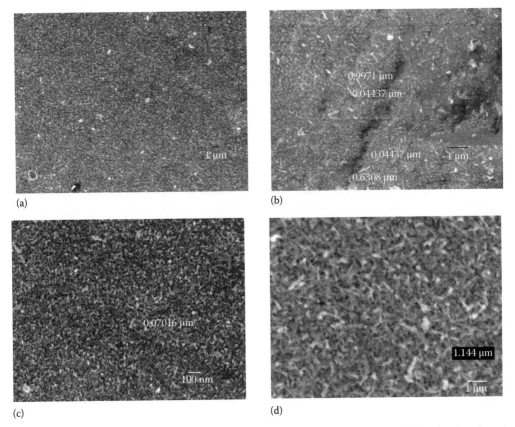

(a)   (b)   (c)   (d)

FIGURE 15.8 SEM images of bismuth nanoparticles and nanotubes, obtained at distinct heating times in vacuum: (a) 5 min (×4,000), (b) 10 min (×10,000), (c) 10 min (×10,000), (d) 15 min (×10,000),

(continued)

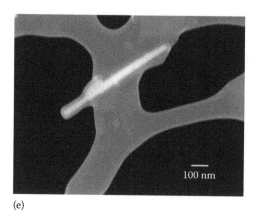

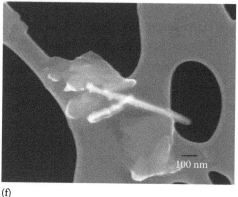

(e)   (f)

FIGURE 15.8 (continued)   (e) 10 min (×80,000), and (f) 15 min (×70,000).

scanning electron microscope (SEM) studies of the last samples (15 min heating) revealed Bi nanotubes with diameters of approximately 100 nm. In addition, it was observed that the nanotube length is increased in the row 5–10–15 min of MW-heating. In contrast to the results on the MW heating in vacuum, Bi nanoparticles (60–70 nm) with small impurities of Bi_2O_3 are formed by MW heating in air. High-resolution SEM analysis data showed for 10 and 15 min heated samples the growth of the nanotubes with the formation of small branches and protrusions (Figure 15.8e and f).

In order to verify the nanotube formation by MW heating in vacuum for 15 min, TEM studies were carried out. Figure 15.9 shows bismuth nanotubes with several easily observed structural defects and irregularities. As seen, 15 min heating in vacuum leads to the formation of long, well-shaped, nonstraight nanotubes, in whose surface a nonlinear growth of a metal layer takes place. Easy formation of bismuth nanotubes in vacuum and the shortened heating time in comparison with metal heating in air can be explained by the absence of the traces of bismuth oxide formed (preventing the bismuth nanotube formation), which is MW heated at higher temperatures in comparison with pure metal, and faster and easier mass transfer in the absence of air during MW evaporation of metal.

Among other interesting synthesis aspects related to inorganic nanotubes, we note a general method to assemble nanoparticle arrays inside inorganic nanotubes by introducing a nanoparticle solution into the nanotube nanochannels.[42] The size matching between nanotubes and nanoparticles may be used to form 1D nanoparticle assemblies of variable sizes and configurations. In addition, the *in situ* atom transfer radical polymerization "grafting from" approach was applied to the surfaces of multiwall carbon nanotubes (MWCNTs) and titania nanotubes.[43] Hybrid core-shell

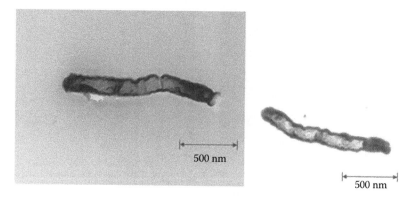

FIGURE 15.9   TEM image of the sample, MW-heated in vacuum for 15 min.

nanomaterials with the nanotubes as the core and polymer, such as polymethyl methacrylate and polystyrene, as the shell were fabricated using the functionalized nanotubes.

Among simulation models and other theoretical calculations for inorganic nanotubes, we emphasize a model[44] for polygonal nanotubes with no defects, where the chirality of the nanotube determines the shape of the cross section. Circular and polygonal nanotubes were compared based on their strain energy and interfacial energy. First-principles calculations were used to parameterize strain and interfacial energy for TiS_2 nanotubes, showing that the polygonal model is energetically favorable to the circular model when the inner radius is above a critical radius, 6.2 Å for a TiS_2 nanotube with 10 layers. These results allow computational studies to more accurately predict nanotube properties. Additionally, a method of calculation of the Coulomb contribution to the strain energy of inorganic nanotubes composed of oppositely charged atomic cylinders was proposed.[45] This method made it possible to describe the differences in the stability of inorganic nanotubes as a function of the ionicity of bonding and the atomic structure of a system. It was demonstrated that the stability of inorganic nanotubes generally decreases with an increase in the ionicity of their bonds. Also, the energetics of inorganic nanotubes were investigated using both atomistic and continuum models.[46] The CNTs and hexagonal inorganic nanotubes were found to display folding energetics essentially consistent with a continuum elastic model.

Several coordination compounds are known as nanotubes, for instance [{[Cd(apab)$_2$ (H$_2$O)]$_3$(MOH)·G}$_n$] (MI = CsI, KI, NaI, respectively; apab = 4-amino-3-(pyridin-4-ylmethyleneamino)benzoate; G = guest molecules). Their structures consist of a large single-walled metalorganic nanotube of [{Cd(apab)$_2$(H$_2$O)}$_{3n}$] with an exterior wall diameter of up to 3.2 nm and an interior channel diameter of 1.4 nm and are held together by alkali metal cations to form 3D nanotubular supramolecular arrays. Hydrothermal reactions of $CdCl_2$, oxalic acid, 1,3,5-benzenetricarboxylic acid and o-phenylenediamine (o-PD) yielded three coordination polymers [NaCd$_4$(btc)$_3$(H$_2$bbim)$_4$(H$_2$O)$_2$]·2H$_2$O, [KCd$_4$(btc)$_3$(H$_2$bbim)$_4$(H$_2$O)$_2$]·2H$_2$O, and [Cd(H$_2$bbim)(H$_2$bimbdc)(HbimbdcH)]·H$_2$O [btc = 1,3,5-benzenetricarboxylate, H$_2$bbim = 2,2′-bibenzimidazole, HbimbdcH$_2$ = 1-(2-benzimidazolyl)-3,5-benzenedicarboxylic acid].[47] They have single-walled nanotubular coordination structures and showed strong red photoluminescence at room temperature. Additionally, nanotubes of Zn(ATIBDC)(bpy)·3H$_2$O (H$_2$ATIBDC = 5-amino-2,4,6-triiodoisophthalic acid)[48] and {[Cd$_3$(phen)$_3$(HL)$_2$(H$_2$O)$_2$]·4.25H$_2$O}$_n$ {H$_4$L = p-terphenyl-type 4,4′-(1,4-phenylene)bis(2,6-dimethyl-3,5-pyridinedicarboxylic acid)}[49] were fabricated and characterized.

A host of reports and reviews are devoted to organic nanotubes.[50,51] Here we present as a typical example poly(3,4-ethylenedioxythiophene) (PEDOT) nanotubes (Figure 15.10, 500–100 nm

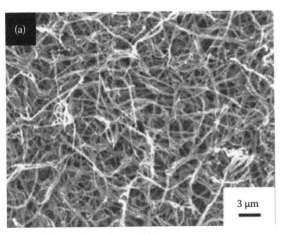
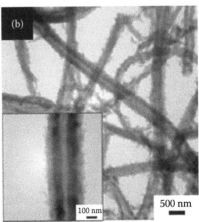

FIGURE 15.10 Microscopy images of PEDOT nanotubes: (a) SEM; (b) TEM. Inset: magnified image. (Reproduced with permission from Zhang, X.Y. et al., Chemical synthesis of PEDOT nanotubes, *Macromolecules*, 39, 470–472. Copyright 2006 American Chemical Society.)

in diameter), obtained by a rapid, room temperature, reverse emulsion polymerization method.[52] Composites of PEDOT nanotubes with noble metals, metal oxides, etc., can be readily synthesized using postsynthesis and *in situ* polymerization methods.

15.1 APPLICATIONS OF INORGANIC NANOTUBES

Present and potential applications of inorganic nanotubes have been reported in the fields of catalysis, microelectronics, Li rechargeable batteries, medical, and optoelectronics and discussed in a series of reviews, in particular in Refs. [53,54]. Thus, it was indicated that MoS_2 and WS_2 nanotubes, heading for large-scale applications in the automotive, machining, aerospace, electronics, defense, medical, and numerous other kinds of industries, formed a basis of a few products, which were recently commercialized by "ApNano Materials, Inc." ("NanoMaterials, Ltd.," see www.apnano. com).[55] Additionally, MoS_2 nanopods and spherical MoS_2 nanoparticles grow in a confined geometry of MoS_2 nanotube reactors, which subsequently serve as nanocontainers and enable safe handling and simple release of the particles on stimulation.[56] Silica nanotubes can be integrated into metal-oxide-solution field effect transistors (MOSolFETs), which exhibit rapid field effect modulation of ionic conductance.[57,58] These nanofluidic devices were further demonstrated to be useful in single molecular sensing and have potential implications in subfemtoliter analytical technology and large-scale nanofluidic integration. Semiconductor GaN nanotubes were reported to have similar properties.[59] In addition, hydrogen and Li storage in inorganic nanotubes and nanowires such as MoS_2, WS_2, TiS_2, BN, TiO_2, MnO_2, V_2O_5, Fe_2O_3, Co_3O_4, NiO, and SnO_2 was reviewed.[60] H and Li can be stored in different 1D nanostructures in various ways, including physical and chemical sorption, intercalation, and electrochemical reactions.

REFERENCES

1. Kreizman, R.; Hong, S. Y.; Deepak, F. L.; Enyashin, A. N.; Popovitz-Biro, R.; Albu-Yaron, A.; Seifert, G.; Tenne, R. Routes for core-shell inorganic nanotubes. *Abstracts of Papers*, *240th ACS National Meeting*, Boston, MA, August 22–26, 2010, FUEL-30.
2. Tenne, R. Inorganic nanotubes and fullerene-like materials of metal dichalcogenide and related layered compounds. In *Nanomaterials Handbook*. Gogotsi, Y. (Ed.). CRC, Boca Raton, FL, 2006, pp. 317–337.
3. Tenne, R.; Redlich, M. Recent progress in the research of inorganic fullerene-like nanoparticles and inorganic nanotubes. *Chemical Society Reviews*, 2010, *39* (5), 1423–1434.
4. Tenne, R. Inorganic nanotubes and fullerene-like materials of metal dichalcogenide and related layered compounds. In *Nanotubes and Nanofibers*. Gogotsi, Y. (Ed.). CRC, Boca Raton, FL, 2006, pp. 135–155.
5. Tenne, R.; Rao, C. N. R. Inorganic nanotubes, *Philosophical Transactions of the Royal Society of London, Series A: Mathematical, Physical and Engineering Sciences*, 2004, *362* (1823), 2099–2125.
6. Tenne, R.; Seifert, G. Recent progress in the study of inorganic nanotubes and fullerene-like structures. *Annual Review of Materials Research*, 2009, *39*, 387–413.
7. Rao, C. N. R.; Govindaraj, A. Synthesis of inorganic nanotubes. In *Advanced Nanomaterials*. Geckeler, K. E.; Nishide, H. (Eds.). Wiley-VCH Verlag GmbH & Co. KGaA, Weinheim, Germany, 2010, Vol. 1, pp. 195–247.
8. Rao, C. N. R.; Govindaraj, A. Synthesis of inorganic nanotubes. *Advanced Materials*, 2009, *21* (42), 4208–4233.
9. Wang, Z. L. Theme issue: inorganic nanotubes and nanowires. *Journal of Materials Chemistry*, 2009, *19* (7), 826–827.
10. Xiong, Y.; Mayers, B. T.; Xia, Y. Some recent developments in the chemical synthesis of inorganic nanotubes. *Chemical Communications*, 2005, (40), 5013–5022.
11. Remskar, M. Inorganic nanotubes. *Advanced Materials*, 2004, *16* (17), 1497–1504.
12. Remskar, M.; Mrzel, A. High-temperature fibers composed of transition metal inorganic nanotubes. *Current Opinion in Solid State and Materials Science*, 2004, *8* (2), 121–125.
13. Enyashin, A. N.; Gemming, S.; Seifert, G. Simulation of inorganic nanotubes. Springer Series in *Materials Science*, 2007, *93* (*Materials for Tomorrow*), 33–57.

14. Xiaobo, C.; Mao, S. S. Titanium dioxide nanomaterials: Synthesis, properties, modifications, and applications. *Chemical Review*, 2007, *107* (7), 2891–2959.

15. Shen, G.; Bando, Y.; Golberg, D. Recent developments in single-crystal inorganic nanotubes synthesised from removable templates. *International Journal of Nanotechnology*, 2007, *4* (6), 730–749.

16. Deepak, F. L.; Tenne, R. Gas-phase synthesis of inorganic fullerene-like structures and inorganic nanotubes. *Central European Journal of Chemistry*, 2008, *6* (3), 373–389.

17. Zhu, Y. Q.; Hsu, W. K.; Kroto, H. W.; Walton, D. R. M. An alternative route to Nbs$_2$ nanotubes. *Journal of Physical Chemistry B*, 2002, *106* (31), 7623–7626.

18. Sima, M.; Enculescu, I.; Vasile, E. Preparation of metal chalcogenides, semiconductor nanowires, and microtubes. *Romanian Reports in Physics*, 2006, *58* (2), 183–188.

19. Rao, C. N. R.; Nath, M. Inorganic nanotubes. *Journal of the Chemical Society Dalton Transactions*, 2003, 1–24.

20. Whitby, R. L. D.; Hsu, W. K.; Fearon, P. K.; Billingham, N. C.; Maurin, I.; Kroto, H.W.; Walton, D. R. M. et al. Multiwalled carbon nanotubes coated with tungsten disulfide. *Chemistry of Materials*, 2002, *14* (5), 2209–2217.

21. Kuang, H. W.; Chang, B. H.; Zhu, Y. Q.; Han, W. Q.; Terrones, H.; Terrones, M.; Grobert, N.; Cheetham, A. K.; Kroto, H. W.; Walton, D. R. M. An alternative route to molybdenum disulfide nanotubes. *Journal of the American Chemical Society*, 2000, *122* (41), 10155–10158.

22. Zhu, Y. Q.; Sekine, T.; Brigatti, K. S.; Firth, S.; Tenne, R.; Rosentsveig, R.; Kroto, H. W.; Walton, D. R. M. Shock-wave resistance of WS$_2$ nanotubes. *Journal of the American Chemical Society*, 2003, *125* (5), 1329–1333.

23. Deepak, F. L.; Mayoralb, A.; Yacaman, M. J. Faceted MoS$_2$ nanotubes and nanoflowers. *Materials Chemistry and Physics*, 2009, *118*, 392–397.

24. Hsu, W. K.; Zhu, Y. Q.; Boothroyd, C. B.; Kinloch, I.; Trasobares, S.; Terrones, H.; Grobert, N. et al. Mixed-phase W$_x$Mo$_y$C$_z$S$_2$ nanotubes. *Chemistry of Materials*, 2000, *12* (12), 3541–3546.

25. Shahar, C.; Zbaida, D.; Rapoport, L.; Cohen, H.; Bendikov, T.; Tannous, J.; Dassenoy, F.; Tenne, R. Surface functionalization of WS$_2$ fullerene-like nanoparticles. *Langmuir*, 2010, *26* (6), 4409–4414.

26. Kreizman, R.; Hong, S. Y.; Sloan, J.; Popovitz-Biro, R.; Albu-Yaron, A.; Tobias, G.; Ballesteros, B.; Davis, B. G.; Green, M. L. H.; Tenne, R. Core-shell PbI$_2$@WS$_2$ inorganic nanotubes from capillary wetting. *Angewandte Chemie, International Edition*, 2009, *48* (7), 1230–1233.

27. Kreizman, R.; Enyashin, A. N.; Deepak, F. L.; Albu-Yaron, A.; Popovitz-Biro, R.; Seifert, G.; Tenne, R. Synthesis of core-shell inorganic nanotubes. *Advanced Functional Materials*, 2010, *20* (15), 2459–2468.

28. Hong, S.-Y.; Popovitz-Biro, R.; Tobias, G.; Ballesteros, B.; Davis, B. G.; Green, M. L. H.; Tenne, R. Synthesis and characterization of WS$_2$ inorganic nanotubes with encapsulated/intercalated CsI. *Nano Research*, 2010, *3* (3), 170–173.

29. Golberg, D.; Costa, P. M. F. J.; Mitome, M.; Bando, Y. Properties and engineering of individual inorganic nanotubes in a transmission electron microscope. *Journal of Materials Chemistry*, 2009, *19* (7), 909–920.

30. Altoe, M. V. P.; Sprunck, J. P.; Gabriel, J.-C. P.; K. Bradley, K. Nanococoon seeds for BN nanotube growth. *Journal of Material Science*, 2003, *38*, 4805–4810.

31. Vásquez, J.; Lozano, H.; Lavayen, V.; Lira-Cantú, M.; Gómez-Romero, P.; Santa Ana, M. A.; Benavente, E.; González, G. High-yield preparation of titanium dioxide nanostructures by hydrothermal conditions. *Journal of Nanoscience and Nanotechnology*, 2009, *9* (2), 1103–7.

32. Gong, J.-Y.; Guo, S.-R.; Qian, H.-S.; Xu, W.-H.; Yu, S.-H. A general approach for synthesis of a family of functional inorganic nanotubes using highly active carbonaceous nanofibres as templates. *Journal of Materials Chemistry*, 2009, *19* (7), 1037–1042.

33. Chen, H.; Elabd, Y. A.; Palmese, G. R. Plasma-aided template synthesis of inorganic nanotubes and nanorods. *Journal of Materials Chemistry*, 2007, *17* (16), 1593–1596.

34. Chernozatonskii, L. A.; Sorokin, P. B.; Fedorov, A. S. New inorganic nanotubes of dioxides MO$_2$ (M = Si, Ge, Sn). *AIP Conference Proceedings (Electronic Properties of Novel Nanostructures)*, Salt Lake City, UT, 2005, Vol. 786, pp. 357–360.

35. Chen, G.; Sun, S.; Sun, X.; Fan, W.; You, T. Formation of CeO$_2$ nanotubes from Ce(OH)CO$_3$ nanorods through Kirkendall diffusion. *Inorganic Chemistry*, 2009, *48* (4), 1334–1338.

36. Krivovichev, S. V.; Tananaev, I. G.; Kahlenberg, V.; Kaindl, R.; Myasoedov, B. F. Synthesis, structure, and properties of inorganic nanotubes based on uranyl selenates. *Radiochemistry*, 2005, *47* (6), 525–536.

37. Hui, J.; Xiang, G.; Xu, X.; Zhuang, J.; Wang, X. Monodisperse F-substituted hydroxyapatite single-crystal nanotubes with amphiphilic surface properties. *Inorganic Chemistry*, 2009, *48* (13), 5614–5616.

38. Liu, M.; Guo, B.; Du, M.; Lei, Y.; Jia, D. Natural inorganic nanotubes reinforced epoxy resin nanocomposites. *Journal of Polymer Research*, 2008, *15* (3), 205–212.
39. Konduri, S.; Mukherjee, S.; Nair, S. Controlling nanotube dimensions: Correlation between composition, diameter, and internal energy of single-walled mixed oxide nanotubes. *ACS Nano*, 2007, *1* (5), 393–402.
40. Xu, Y.-X.; Zhao, X.; Jiang, X.-K.; Li, Z.-T. Organic nanotubes assembled from isophthalamides and their application as templates to fabricate Pt nanotubes. *Chemical Communications*, 2009, (28), 4212–4214.
41. Kharissova, O. V.; Osorio, M.; Kharisov, B. I.; José-Yacamán, M.; Ortiz Méndez, U. A comparison of bismuth nanoforms obtained in vacuum and air by microwave heating of bismuth powder. *Materials Chemistry and Physics*, 2010, *121*, 489–496.
42. Bai, J.; Huang, S.; Wang, L.; Chen, Y.; Huang, Y. Fluid assisted assembly of one-dimensional nanoparticle array inside inorganic nanotubes. *Journal of Materials Chemistry*, 2009, *19* (7), 921–923.
43. Gao, Y.; Yan, D. Surface modification of inorganic nanotubes by atom transfer radical polymerization. *ACS Symposium Series (Controlled/Living Radical Polymerization)*, 2006, American Chemical Society, Washington, DC, Vol. 944, pp. 279–294.
44. Tibbetts, K.; Doe, R.; Ceder, G. Polygonal model for layered inorganic nanotubes. *Physical Review B: Condensed Matter and Materials Physics*, 2009, *80* (1), 014102/1–014102/10.
45. Enyashin, A. N.; Makurin, Yu. N.; Ivanovskii, A. L. Coulomb interactions and the problem of stability of inorganic nanotubes. *Doklady Physical Chemistry*, 2004, *399* (2), 293–297.
46. Bishop, C. L.; Wilson, M. The energetics of inorganic nanotubes. *Molecular Physics*, 2008, *106* (12–13), 1665–1674.
47. Liu, W.-T.; Ou, Y.-C.; Xie, Y.-L.; Lin, Z.; Tong, M.-L. Photoluminescent metal-organic nanotubes via hydrothermal in situ ligand reactions. *European Journal of Inorganic Chemistry*, 2009, (28), 4213–4218.
48. Dai, F.; He, H.; Sun, D. Polymorphism in high-crystalline-stability metal-organic nanotubes. *Inorganic Chemistry*, 2009, *48* (11), 4613–4615.
49. Huang, K.-L.; Liu, X.; Chen, X.; Wang, D.-Q. Spontaneous assembly of 63 topological metal-organic nanotubes with distinct asymmetric subunits for the construction of hydrophilic intertube channels encapsulating rare helical water-chains. *Crystal Growth and Design*, 2009, *9* (4), 1646–1650.
50. Asakawa, M. Organic nanotubes: Molecular assemblies shape nano-sized tubes. *Kagaku to Kyoiku*, 2010, *58* (4), 180–181.
51. Masuda, M.; Kameta, N. Organic nanotubes with different inner and outer surfaces and their applications. *Nettowaku Porima*, 2010, *31* (4), 191–200.
52. Zhang, X. Y.; Lee, J.-S; Lee, G. S.; Cha, D.-K.; Kim, M. J.; Yang, D. J.; Manohar, S. K. Chemical synthesis of PEDOT nanotubes. *Macromolecules*, 2006, *39*, 470–472.
53. Tenne, R. Inorganic nanotubes and fullerene-like nanoparticles. *Journal of Materials Research*, 2006, *21* (11), 2726–2743.
54. Rapoport, L.; Fleischer, N.; Tenne, R. Applications of WS_2 (MoS_2) inorganic nanotubes and fullerene-like nanoparticles for solid lubrication and for structural nanocomposites. *Journal of Materials Chemistry*, 2005, *15* (18), 1782–1788.
55. Tenne, R.; Remskar, M.; Enyashin, A.; Seifert, G. Inorganic nanotubes and fullerene-like structures (IF). *Topics in Applied Physics*, 2008, *111* (*Carbon Nanotubes*), 631–671.
56. Remskar, M.; Mrzel, A.; Virsek, M.; Jesih, A. Inorganic nanotubes as nanoreactors: The first MoS_2 nanopods. *Advanced Materials*, 2007, *19* (23), 4276–4278.
57. Yang, P.; Goldberger, J.; Fan, R. Inorganic nanotubes and nanofluidic transistors. *Abstracts of Papers, 233rd ACS National Meeting*, Chicago, IL, March 25–29, 2007, INOR-1089.
58. Yang, P. Inorganic nanotubes, nanofluidic transistors and DNA translocation. *PMSE Preprints*, 2006, *95*, 386.
59. Goldberger, J.; Fan, R.; Yang, P. Inorganic nanotubes: A novel platform for nanofluidics. *Accounts of Chemical Research*, 2006, *39* (4), 239–248.
60. Cheng, F.; Chen, J. Storage of hydrogen and lithium in inorganic nanotubes and nanowires. *Journal of Materials Research*, 2006, *21* (11), 2744–2757.

16 Soluble Carbon Nanotubes

The functionalization of carbon nanotubes (CNTs) is a very actively discussed topic in contemporary nanotube literature because the planned modification of CNTs properties is believed to open the road to real nanotechnology applications.[1] Due to their exceptional combination of mechanical, thermal, chemical, and electronic properties, single-walled carbon nanotubes (SWNTs or SWCNTs) and multiwalled carbon nanotubes (MWNTs or MWCNTs) are considered as unique materials, with very promising future applications, especially in the field of nanotechnology, nanoelectronics, and composite materials. Additionally, CNTs are becoming highly attractive molecules for applications in medicinal chemistry. At present, potential biological applications of CNTs have been little explored. The main difficulty in integrating such materials into biological systems derives from their lack of solubility in physiological solutions. The functionalization of CNTs with the assistance of biological molecules remarkably improves the solubility of nanotubes in an aqueous or organic environment and, thus, facilitates the development of novel biotechnology, biomedicine, and bioengineering.

Nonfunctionalized CNTs are difficult to dissolve or disperse in most organic or inorganic solvents because of their long-structured features, large molecular size, or severe aggregation. The common agents used to help disperse CNTs are surfactants, which, however, can only increase the dispersibility to a limited extent, and surfactants do not affect the solubility of CNTs. Current chemical methods for water suspended SWNTs require harsh sonochemical treatments to effectively disperse nanotubes. However, these methods are currently incapable of conferring thermodynamically stable water-based dissolutions of carbon structures since surfacted SWNT solutions are simply metastable colloidal suspensions, where they transiently individualize but always reaggregate over time since this is their thermodynamically favorable state. Therefore, true water-soluble nanotube solutions are those solutions that entropically favor individualized nanotubes,[2] where the reaggregation of CNTs in a solvent is less favored, on a thermodynamic basis, than their continued solvated state.[3] In some embodiments, the extent of functionalization is dependent upon a number of factors, for example, the reactivity of the CNTs, the reactivity of the functionalizing agent, steric factors, etc. In some such embodiments, as a result of such dependencies, the extent of functionalization can be in the range of from at least about 1 functional group per every 1000 CNT carbons to at most about 1 functional group per every 2 CNT carbons.

In recent years, the CNTs, soluble in water or organic solvents, have been mentioned in a series of monographs, including book chapters on CNTs.[4–8] Some recent reviews were dedicated to the strategic approaches toward the solubilization of CNTs using chemical and physical modifications [9–11]; environmental, toxicological, and pharmacological studies related to the use of CNTs[12,13]; the main methods for the modification of CNTs with polymers[14]; application of functionalized CNTs as biosensors[15]; and discussions on the possibility of the existence of SWNTs in organic solvents in the form of clusters.[16] A fundamental review[17] was devoted to electronic and vibrational properties of the SWCNTs.

Among many others, currently used functionalization agents include aryl diazonium moieties, halogen, oxy, carbon {alkyl, alkenyl, alkynyl, aryl, and acyl(RCO′)}, and metal-based radicals. CNTs can be functionalized by oxidation (peroxyacids, metal oxidants, such as osmium tetroxide, potassium permanganate, chromates; ozone, oxygen, superoxides) or reduction reactions, interactions with thiols, carbenes, dienes, etc. In this section, we generalize the most recent advances in this area, leading to CNTs, soluble in water or common organic solvents.

16.1 FUNCTIONALIZATION BY THE USE OF ELEMENTAL METALS, INORGANICS, AND GRIGNARD REAGENTS

Generally, elemental metals have been used in the form of nanoparticles (Au, Fe, Co) or solutions in liquid ammonia (alkali metals) with organic linkers. Thus, water-soluble hybrids of MWNTs and *gold* nanoparticles (Au@MWNTs) were fabricated via the *in situ* solution method using an optoelectronic-active compound of *N,N'*-bi(2-mercaptoethyl)-perylene-3,4,9,10-tetracarboxylic diimide as interlinker and stabilizer.[18] It was found that the formed hybrid exhibited strong visible luminescence under UV lamp irradiation, which might extend its potential applications to biological labeling. In another report, dedicated to gold-CNTs hybrids, the efficient aqueous dispersion of pristine HiPco SWNTs with ionic liquid surfactants 1-dodecyl-3-methylimidazolium bromide (D) and 1-(12-mercaptododecyl)-3-methylimidazolium bromide (M), the thiolation of nanotube sidewalls, and the controlled self-assembly of positively charged SWNT-D,M composites on gold were studied.[19] *Iron*-filled Fe@MWCNTs were surface modified with various functionalities via a rapid, single-step process involving ultrasonication-assisted and microwave-induced radical polymerization reactions.[20] The offered process is universal for both hydrophobic (e.g., polystyrenes and polymethyl methacrylate) and hydrophilic (e.g., polyacrylamide, polyacrylic acids, and polyallyl alcohols) polymer chains (see also Section 16.8), which can be chemically grafted onto the surface of MWCNTs in the same conditions within ~10 min. The solubilities of the formed functionalized MWCNTs are in the range of 1200–2800 mg/L in solutions. An intriguing method for CNT derivatization with magnetic nanoparticles is described in Ref. [21]. Capped iron oxide or *cobalt and cobalt/platinum* magnetic nanoparticles were attached to CNTs by means of an interlinker molecule, a carboxylic derivative of pyrene (Figure 16.1) (see also Section 16.3 on the functionalization with aromatic molecules). The available carboxylic groups of pyrene derivative can be further linked to metal or metal oxide nanoparticles. The formed composites were highly soluble in organic solvents, such as chloroform, toluene, and hexane.

Sidewall-functionalized nanotubes, soluble in organic solvents, were prepared by alkylation of nanotube salts obtained using either *lithium, sodium, or potassium* in liquid ammonia (*Billups* reaction).[22] Such reactions can produce different types of derivatized CNTs. It was shown that the alkali metal intercalates into the SWNT ropes (in case of further reaction of nanotube salts Na-CNTs with PhI, the arylated CNTs can be produced); alkali metals used behave differently requiring distinct temperature ranges (the least range corresponds to Li). The *Billups* reaction protocol involving dissolving metal reduction of MWNTs and their subsequent alkylation or arylation was shown to produce functionalized MWNTs that were soluble in either organic or aqueous solvents. This method allows for the attachment of alkyl or aryl pendent groups, using either lithium or sodium, and has been used to produce gram quantities of alkylated MWNTs[23] (Figure 16.2).

The use of Li as the most used alkali metal in these processes was studied in detail using the Raman technique.[24] It was shown that addition of 1-iodododecane to the lithiated SWNTs resulted in the covalent attachment of dodecyl groups. The intercalation of lithium throughout the SWNT ropes led to complete dodecylation of all individual SWNTs. Lithium was also applied to yield a

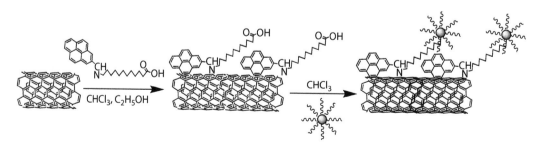

FIGURE 16.1 Modification of CNTs by capped magnetic nanoparticles.

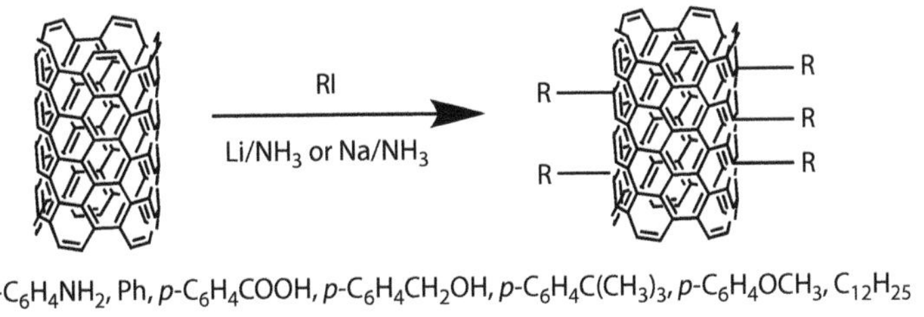

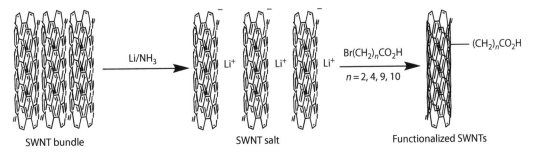

$R = p\text{-}C_6H_4NH_2, Ph, p\text{-}C_6H_4COOH, p\text{-}C_6H_4CH_2OH, p\text{-}C_6H_4C(CH_3)_3, p\text{-}C_6H_4OCH_3, C_{12}H_{25}$

FIGURE 16.2 Functionalization of MWCNTs by *Billups* reaction.

SWNT bundle   SWNT salt   Functionalized SWNTs

FIGURE 16.3 General procedure for the synthesis of carboxylic acid derivatized SWNTs.

water-soluble polyethyleneglycol poly(ethylene glycol) (PEG)-CNT conjugate in a two-step process, including the reaction of lithium, SWCNTs, and 11-bromoundecanoic acid in liquid ammonia forming carboxylic acid derivatized SWCNTs (Figure 16.3) and further interaction of the obtained product with H_2N-PEG-OMe in a mixture of solvents.[25,26]

Simple inorganic compounds are capable of solubilizing CNTs even without the formation of direct chemical bonds "CNTs-substrate." Thus, MWCNTs were dispersed in aqueous solutions through *alumina-coated silica* (ACS) nanoparticle halos.[27] MWCNTs were directly dispersed into a highly charged ACS nanoparticle aqueous solution, which was stable for weeks after ultrasonication, without functionalization of their surfaces. A stable, water-soluble suspension of CdS/aligned-MWCNT heterostructures with electron transfer from CdS nanoparticles to aligned-MWCNTs, implying potential applications in photovoltaic cells, photocatalysis, and solar energy conversion, was reported.[28] CdSe quantum dots were noncovalently attached to SWNTs through an intermediary 1-pyrenebutyric acid *N*-hydroxy-succinimide ester molecule.[29] The use of *Grignard reagents* RC(=S)SMX (wherein R is alkyl, alkenyl, or aryl, wherein any alkyl, alkenyl, or aryl may be optionally substituted; M is Be, Mg, Ca, Sr, or Ba; and X is Cl, Br, or I) in functionalization of CNTs is also known.[30]

16.2 FUNCTIONALIZATION OF CNTs IN STRONG ACIDIC MEDIA AND WITH OXYGEN-CONTAINING MOIETIES

Generally, suitable acid media include any acidic medium capable of dispersing CNTs in a substantially individualized state. By first dispersing CNTs in an acidic medium, bundled CNTs can be separated as individual CNTs, affording exposure of the CNT sidewalls, and thereby facilitating the functionalization of such CNTs, wherein functional groups are attached to the subsequently exposed sidewalls of these individualized CNTs. Strong acids, oxidizing CNT surface, lead to the formation of COOH groups, so the use of sulfuric or nitric acid can be a first step in CNT functionalization, followed by further reaction of COOH groups with, for example, $SOCl_2$ or other substances that are present in solution.

A few recent patents and articles are dedicated to use oleum to functionalize CNTs. Thus, the functionalization of SWNTs in oleum via *diazonium species* generated *in situ* (Figure 16.4) was

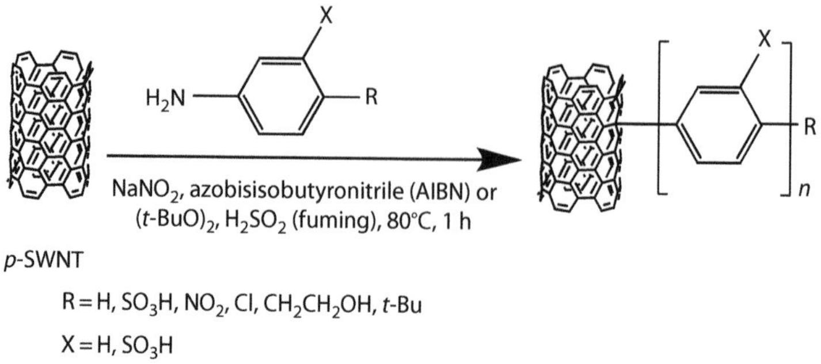

FIGURE 16.4 Functionalization of SWNTs in oleum via diazonium species generated *in situ*.

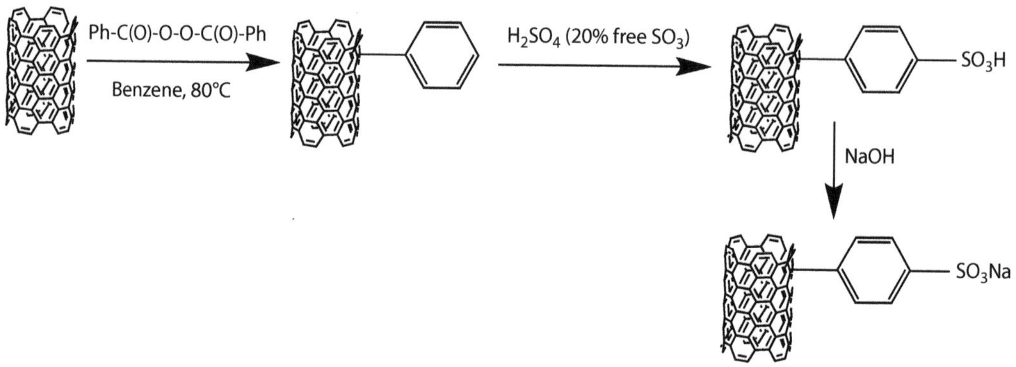

FIGURE 16.5 Synthesis of sulfonated SWNTs.

reported in Ref. [31]. In a typical example, purified SWCNTs were dispersed in oleum (20% free SO$_3$) at 80°C; then sulfanilic acid was added to the dispersion followed by sodium nitrite and 2,2'-azo-bis-isobutyrylnitrile (AIBN) and the mixture was stirred for 1 h and further filtered and washed. In other reports,[32,33] the phenylated SWNTs, obtained from SWNT and distinct quantities of benzoyl peroxide to reach most, medium, or least functionalized level, were further sulfonated by the reaction of phenylated SWNTs with sulfuric acid (simply dispersing the phenylated SWNTs in oleum), wherein –SO$_3$H substitutes for –H on the phenyl groups at a position parallel to their attachment to the SWNT (Figure 16.5). Phenyl sulfonated SWNTs are found to be true water-soluble CNTs and can serve as a platform for the development of SWNTs for several industries including pharmaceutical, energy, and electronics. The degree of functionalization was determined both qualitatively using Raman spectroscopy and quantitatively using thermogravimetric analysis and x-ray photoelectron spectroscopy. Similar derivatization of SWCNTs using the diazonium-species technique was reported in Refs. [34,35].

Aqueous solutions of SWNTs (0.03–0.15 mg/mL), stable for more than a month, were obtained by functionalization with carboxylate groups, formed as a result of acid/oxidant treatment (mixture of 9:1 concentrated H$_2$SO$_4$/30% H$_2$O$_2$).[36] It is noted that the optical absorption of the first interband transition of as-treated water-soluble semiconducting SWNTs reversibly responds to the pH change in aqueous solutions. In another report,[37] soluble (2 wt.%), ultrashort (length <60 nm), carboxylated SWCNTs were prepared in the process, predicated on oleum's ability to intercalate between individual SWNTs inside SWNT ropes; this is a procedure that simultaneously cuts and functionalizes SWNTs using a mixture of sulfuric and nitric acids. Long-time-stable solutions of CNTs, functionalized with carboxylic acids, among other species (nitrates, hydroxyls, sulfur-containing groups, carboxylic acid salts, and phosphates), were also described in the patent.[38] It was noted that the most preferred pH ranges from 3 to 6, the most preferred level of functionalized carbons on the SWCNT is 0.5–20 at.%, and the functionalized carbons may exist anywhere on the nanotubes (open or closed

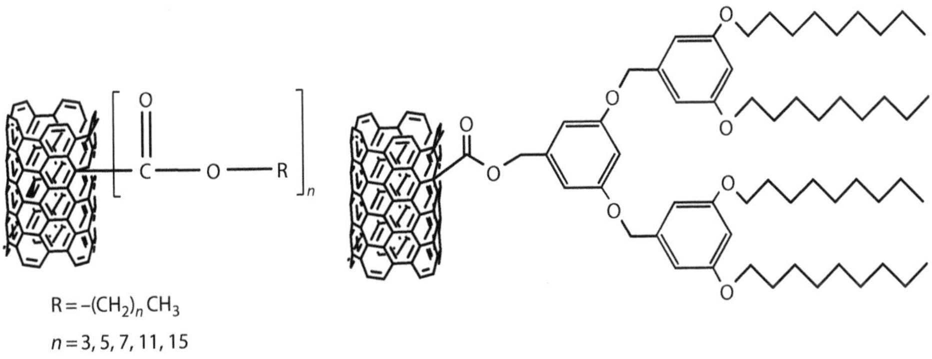

R = –(CH$_2$)$_n$ CH$_3$

n = 3, 5, 7, 11, 15

FIGURE 16.6 Ester-functionalized CNTs.

ends, external and internal sidewalls). Additionally, the behavior of CNTs, treated with acids, was studied in solvents of different polarity and in water of different pH as a function of acid treatment conditions. In contrast to untreated CNTs, soluble or dispersible in non- or low-polar solvents (acetone, alcohols), the treated CNTs were found to be soluble or dispersed in the deionized water, but not in acetone or alcohols, increasing the solubility with pH from 4 up to 10.[39]

Among O-containing organic groups, CNTs were solubilized by functionalization with *ester* moieties, for instance esters with *n*-pentyl,[40] other alkyl chains (*n*-butyl, *n*-hexyl, *n*-octyl, *n*-dodecyl, and *n*-hexadecyl),[41] or dendron-type moieties with long alkyl chains[42] (Figure 16.6), *crown ether* (4′-aminobenzo-15-crown-5-ether),[43] or *aldehydes* (anisaldehyde together with 3-methylhippuric acid).[44]

16.3 FUNCTIONALIZATION WITH COMPOUNDS CONTAINING ALKYL AND AROMATIC MOIETIES

High-speed vibration mill (HSVM) mechanochemical technique was applied to prepare SWCNTs, functionalized with some alkyl and aryl groups (Figure 16.7). As a result, SWNTs with long alkyl chains can be dissolved in many common organic solvents.[45] In another report, highly soluble pyridyl-functionalized SWCNTs were obtained by a 1,3-dipolar cycloaddition of a nitrile oxide on the SWNT walls (Figure 16.8), similar to 1,3-dipolar cycloadditions that are common for fullerene functionalization (Figure 16.9), and characterized by nuclear magnetic resonance (NMR), Fourier transform (FT)-Raman, and electron microscopy.[46] The CNTs here were *doubly functionalized*: at the tips with pentyl esters (to provide sufficient solubility in organic solvents) and on the walls by pyridyl isoxazoline groups (they are capable of coordinating the metallo-porphyrin to the pyridyl group). This composite was further used as a precursor in the synthesis of CNT-Zn-porphyrin analogue of fullerene-C$_{60}$-Zn-porphyrin. The occurrence of this complex is clearly revealed by optical spectroscopy and by the shifts in CV potentials of Zn-porphyrin in the presence of Py-SWNT, similar but larger than for the corresponding fullerene analogues.

Functionalization of SWCNTs with *phenol* groups by 1,3-dipolar cycloaddition reaction and further derivatization of the products with 2-bromoisobutyryl bromide resulted in the attachment

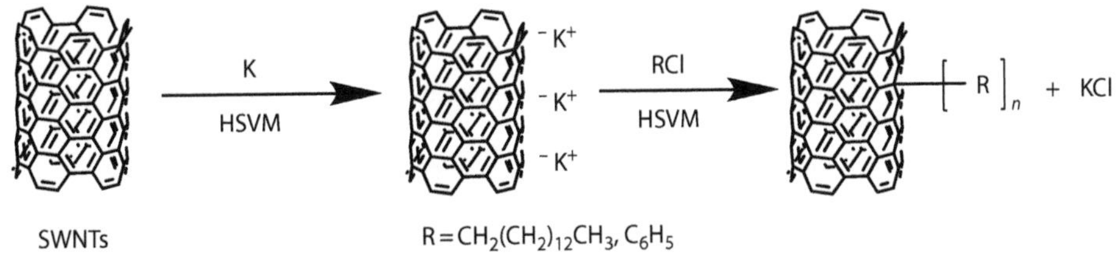

SWNTs   R = CH$_2$(CH$_2$)$_{12}$CH$_3$, C$_6$H$_5$

FIGURE 16.7 SWCNTs, functionalized with alkyl and aryl groups.

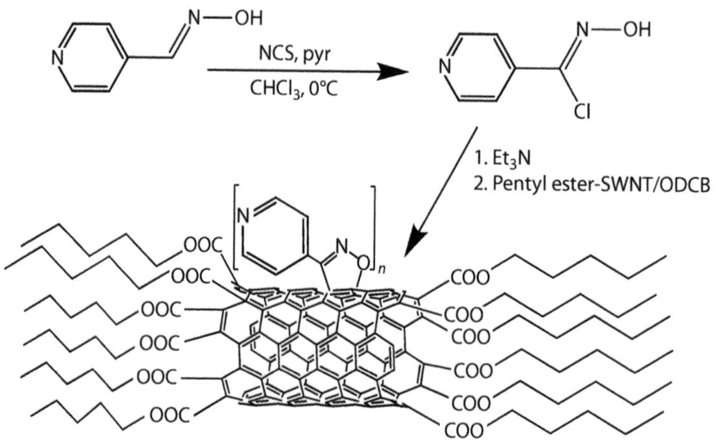

FIGURE 16.8 Synthesis of pyridyl-functionalized SWCNTs.

ZnPor-Py-C$_{60}$

ZnPor-Py-SWNT

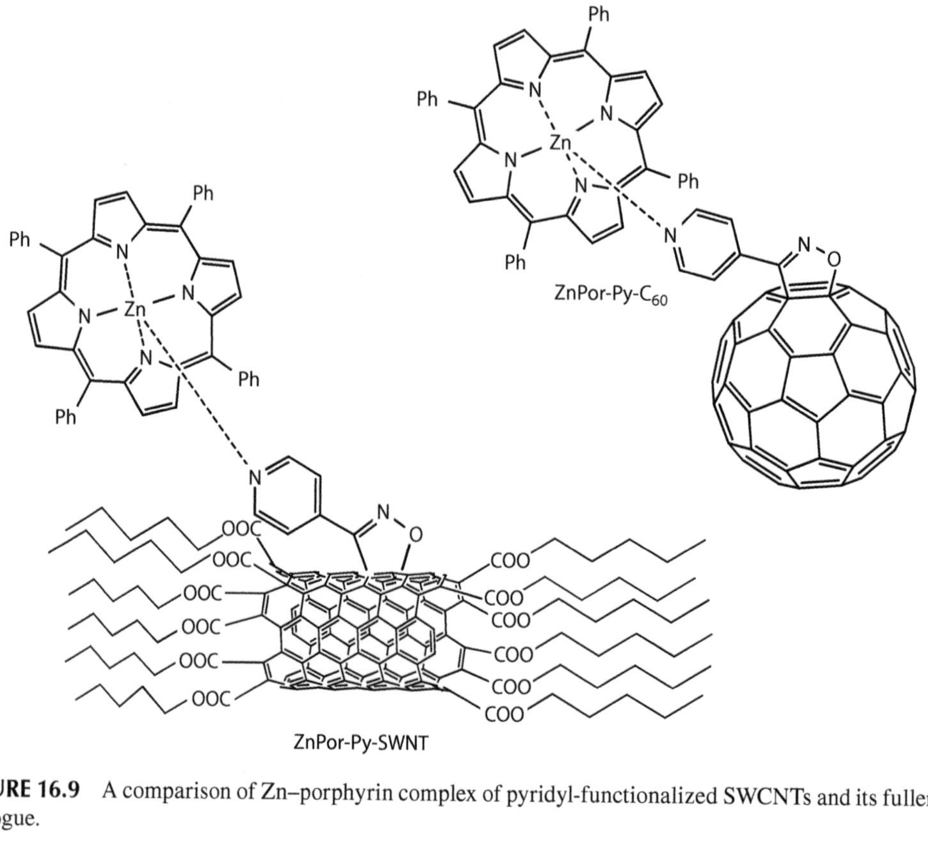

FIGURE 16.9 A comparison of Zn–porphyrin complex of pyridyl-functionalized SWCNTs and its fullerene analogue.

of atom transfer radical polymerization initiators (active in the polymerization of *t*-butyl acrylate from the surface of the nanotubes) to the sidewalls of the nanotubes (Figure 16.10).[47] The obtained SWCNTs, functionalized with poly(*t*-butyl acrylate) were soluble in a variety of organic solvents (see Section 16.8).

Polynuclear aromatics—ionic *pyrene* and *naphthalene* derivatives (Figure 16.11)—were used for obtaining water-soluble SWNT polyelectrolytes (SWNT-PEs), which are analogous to polyanions and polycations.[48] The nanotube–adsorbate interactions consist of π-π-stacking interactions between

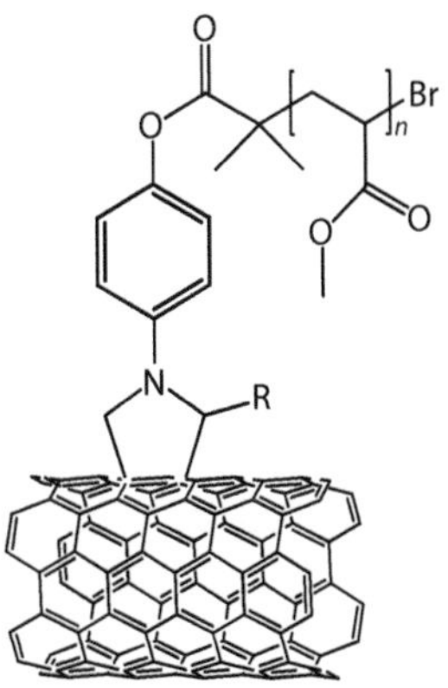

FIGURE 16.10 CNTs, obtained by 1,3-dipolar cycloaddition reaction.

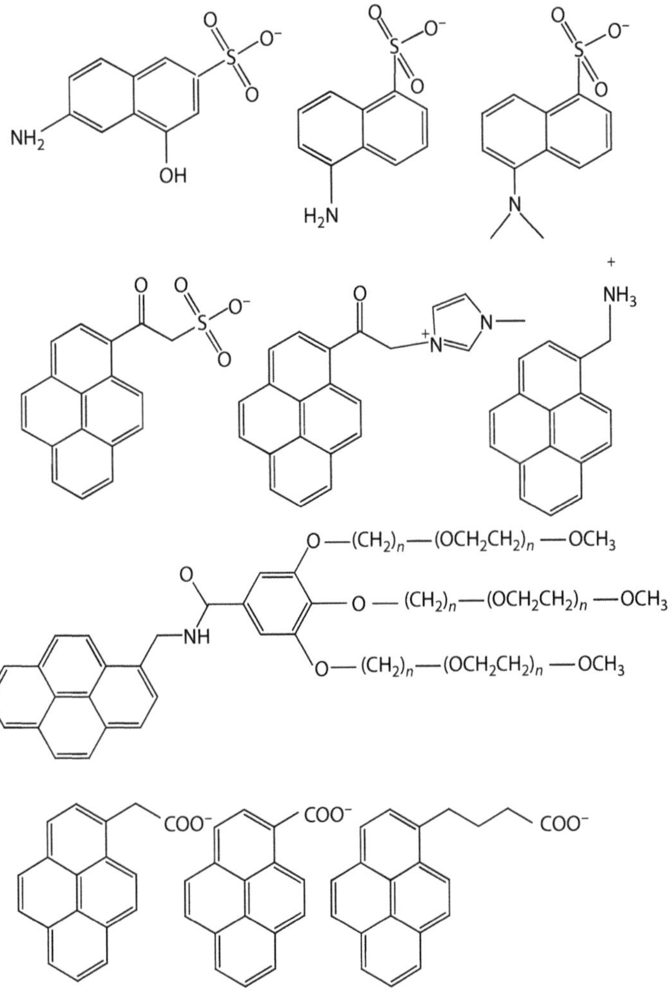

FIGURE 16.11 Polynuclear aromatics, used for obtaining water-soluble SWNT polyelectrolytes.

the aromatic core of the adsorbate and the nanotube surface and charge transfer between them. Pyrene-containing moieties in the construction of soluble CNTs were also reported in Refs. [49,50].

The SWNTs, functionalized by *ferrocene*-grafted poly(*p*-phenyleneethynylene), were found to gelate common organic solvents such as chloroform forming a freestanding CNT organogel that cannot be redispersed in any organic solvents and can be applied for obtaining an insoluble, homogeneous, electroactive SWNT film.[51] Ferrocene units can be covalently introduced onto SWNTs also by 1,3-dipolar cycloaddition of azomethine ylides, thus solubilizing them.[52]

16.4 FUNCTIONALIZATION WITH ALIPHATIC AND AROMATIC AMINE(AMIDO)-CONTAINING MOIETIES

Generally, the amino-functionalization of MWNTs can improve their dispersion in H_2O; however, other reactions may also have occurred, which influenced their dispersity in organic solvents.[53] Among other reports on the use of these moieties,[54,55] a series of amines (octadecylamine (ODA), 2-aminoanthracene, 1-H,1-H-pentadecafluorooctylamine, 4-perfluorooctylaniline, and 2,4-bis(perfluorooctyl) aniline) were reported to interact with SWCNTs, previously thermally treated in air and optionally purified with nitric acid. The formed products exhibit different solubility depending on the amines and solvents used.[56] For one of the most frequently applied amines, ODA,[57] it was revealed that the ODA chains grafted on MWNT are partially crystallized.[58] Similar functionalization of SWCNTs with octadecylamido moieties allowed creating soluble nanotubes (s-SWNT-CONH(CH$_2$)$_{17}$CH$_3$), where the mentioned groups are attached in the end groups and at defect sites; the weight percentage of the octadecylamido functionality in the s-SWNTs is about 50%.[59] Figure 16.12 shows a representative reaction between ODA y SWCNTs.[60] Additionally, a total aromatic polyimide was found[61] to have a high potential to solubilize SWNTs individually in organic solutions. When SWNT concentrations increased, the solutions became viscous and then changed to gels.

A series of *aniline derivatives* (4-pentylaniline, 4-dodecylaniline, 4-tetradocylaniline, 4-pentacosylaniline, 4-tetracontylaniline, 4-pentacontylaniline), as well as amines (ODA, nonylamine, dodecylamine, pentacosylamine, tetracontylamine, pentacontylamine), etc. and mixtures thereof were used for functionalization of CNTs.[62,63] The formed composites are soluble in carbon disulfide and common organic solvents as chlorobenzene, dichlorobenzene, trichlorobenzene, tetrahydrofuran (THF), chloroform, methylene chloride, diethylene glycol dimethyl ether, benzene, toluene, tetrachlorocarbon, pyridine, dichloroethane, diethyl ether, xylene, naphthalene, nitrobenzene, ether, and mixtures thereof. The solubilities of the CNTs in these solvents range from about 0.01–5.0 mg/mL. For a completely soluble optically active polyaniline-MWCNT composite, it was found[64] that the polymer's optical activity was retained in the presence of CNTs. Solutions were found to be easily processable into thin films, which exhibited dendritic structures only in the presence of nanotubes. Aniline was also used as a

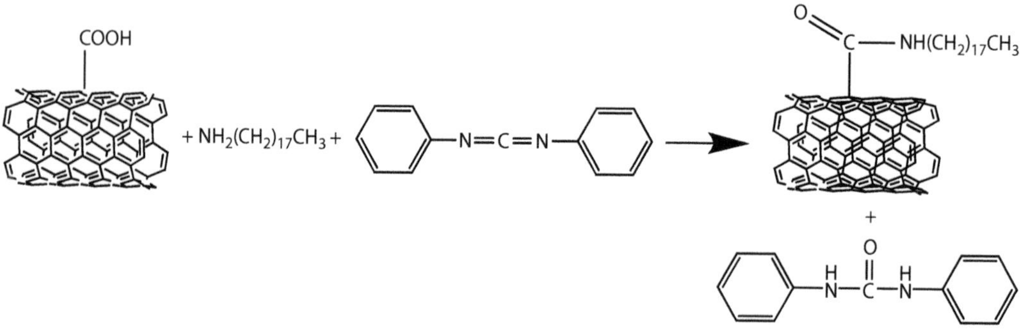

FIGURE 16.12 Amidation of SWNTs with ODA.

solvent for MWCNTs, grafted by the carboxylic acid group.[65] These nanotubes were rapidly (30 min) dissolved in aniline under microwave treatment. The solubility of SWNT in aniline is up to 8 mg/mL.[66]

16.5 FUNCTIONALIZATION WITH SULFUR-CONTAINING MOIETIES

Among *sulfur-containing* surfactants or functionalizing agents, we note the surfactants sodium dodecyl sulfate $C_{12}H_{25}OSO_3Na$ (and sometimes sodium dodecylbenzenesulfonate), whose presence together with hydroxypropyl methyl cellulose[67] or HNO_3/H_2SO_4 mixture[68] helps to assist the dispersion of CNTs. Dodecanethiol as the reaction agent was used to obtain a stable suspension of thiolated SWCNTs in toluene; the thiolation process is also observed on the exposure of the nanotubes to toluene solutions of dodecanethiol-stabilized Au nanoparticles, using them for labeling or manipulating the location of the chemical reaction sites on the tube wall.[69] Similarly, cyclic disulfides were used for introduction of sulfur-containing functional groups onto SWNTs and further treatment with gold nanoparticles.[70]

16.6 FUNCTIONALIZATION WITH MACROCYCLES

Classic macrocycles *porphyrins* and *phthalocyanines* have been successfully applied as precursors for solubilization of CNTs. Thus, the authors of Ref. [71] established that porphyrin molecules can dissolve SWNTs in organic solutions and the SWNT–porphyrin hybrid nanomaterials can be separated from the solutions; moreover, both individually dissolved nanotubes and bundled nanotubes coexisted in the solution. The solid purified SWNT (*p*-SWNT)–porphyrin nanomaterials were readily separated from the *p*-SWNT–porphyrin solution, and this nanomaterial was redissolvable in DMF. The formed porphyrin-CNT solutions can be stable for a long time; thus, water-soluble porphyrin molecules (*meso*-(tetrakis-4-sulfonatophenyl) porphine dihydrochloride) (Figure 16.13) were used to solubilize SWNTs, resulting in aqueous solutions stable for several weeks.[72] A suspension, stable more than 1 week, was prepared with the use of an anionic tetra (*p*-carboxyphenyl) porphyrin (TCPP) and MCNTs, and, as an application, a spectrofluorometric method of DNA hybridization was proposed.[73] Spectroscopic changes of tetraphenylporphyrin CNT composites in a variety of chlorinated solvents such as chloroform, dichloroethane, and dichlorobenzene as a result of sonication were studied.[74] It was established that the protonation of the porphyrin core nitrogen atoms occurs as a result of sonodegradation of the solvent molecules.

In addition to free porphyrins, their zinc-containing complexes, which are usually used for functionalization of fullerenes, are also standard solubilizing agents for CNTs. Thus, a highly soluble, conjugated Zn-porphyrin polymer was synthesized in the presence of trifluoroacetic acid in THF and found to strongly interact with the surface of SWCNTs, producing a soluble polymer–nanotube

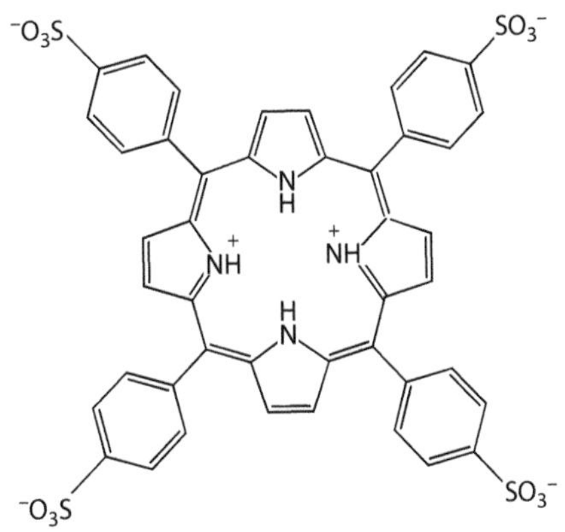

FIGURE 16.13 *meso*-(Tetrakis-4-sulfonatophenyl) porphine dihydrochloride.

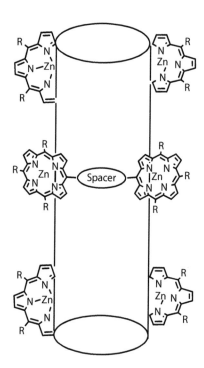

FIGURE 16.14 Zn–porphyrin complex/SWNT intermediate compound, used for CNT purification.

complex, which remains soluble after excess free polymer was removed from the solution, and could be centrifuged at high speed with no observable sedimentation.[75] The CNTs can be *purified* (see Section 16.9) from a large quantity of impurities such as carbon nanoparticles by executing a step of adding the CNTs into a solution (in THF, chloroform, dichloromethane, toluene, benzene, chlorobenzene, dimethylformamide, dimethyl sulfoxide, hexane, acetone, methanol, ethanol, isopropanol, butanol, acetonitrile, or diethylether) using Zn–porphyrin complex with their further recovering[76] (Figure 16.14). Additional details on the use of porphyrin salts in CNT functionalization were reported in Ref. [77].

In comparison with porphyrins, soluble *phthalocyanine–CNT* composites are considerably lesser represented. Thus, a water-soluble composite of oxidized MWCNTs and sulfonic acid sodium salt derivatized copper phthalocyanine for application in bilayer organic solar cells is reported.[78] CNT-molecular semiconductor thin films on the phthalocyanine basis were patented in Refs. [79,80].

Among other nonbiological molecules, used for CNTs functionalization to make them soluble, we note that *diazo dyes* were reported to be functionalizing agents for CNTs. Thus, the mixture of SWNTs and a rigid, planar, and conjugated diazo dye, Congo red (CR), can be dissolved in water with a solubility as high as 3.5 mg/mL for SWNTs.[81] The authors noted that the π-stacking interaction between adsorbed CR and SWNTs was considered responsible for the high solubility. Among other numerous functionalities, *calixarenes*,[82] *carbenes*[83] (Figure 16.15a), *carbohydrates*,[84] *carboranes*[85] (Figure 16.15b), *phenosafranin* (3,7-diamino-5-phenylphenazinium chloride),[86] and much more compounds[87–89] have been used for CNT functionalization.

16.7 FUNCTIONALIZATION WITH THE USE OF BIOMOLECULES

Soluble CNTs and especially water-soluble ones, functionalized with biomolecules, could have many applications in medicinal chemistry; therefore, a host of efforts have been dedicated to CNT treatment with biologically active species. It was shown that the high solubility of CNTs in water can be reached by functionalization with *amino acids*.[90] Thus, in the case of reaction of $NH_2(CH_2)_nCO_2H$ with fluoronanotubes, the solubility in water is controlled by the length of the hydrocarbon side chain.[91] The authors showed that the 6-aminohexanoic acid CNT-derivative

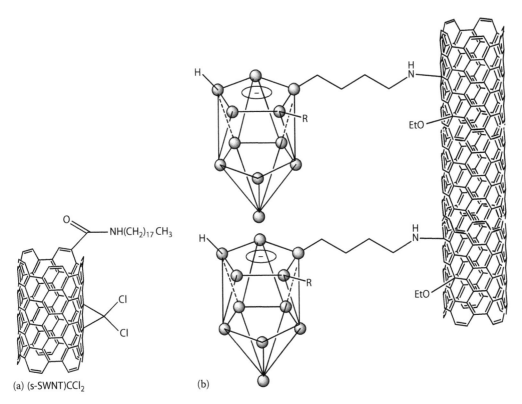

(a) (s-SWNT)CCl$_2$ (b)

FIGURE 16.15 Carbene- (a) and carborane- (b) functionalized SWNTs.

is soluble in an aqueous solution (0.5 mg/mL) between pH 4 and 11, whereas the glycine and 11-aminoundecanoic acid derivatives are insoluble across all pH values. Highly water-dispersed MWCNTs (stable concentration as high as 10 mg/mL in deionized water) were obtained by attaching the lysine HO$_2$CCH(NH$_2$)(CH$_2$)$_4$NH$_2$ onto MWCNTs by producing acyl chloride on the carboxylic groups associated with the nanotubes.[92] The functionalized MWNTs can be dispersed in water under a wide range of pH values (5–14).

Biologically active *peptides* can be easily linked through a stable covalent bond to CNTs. Thus, a CNT-bound peptide from the foot-and-mouth disease virus (FMDV) retained the structural integrity; this is an immunogenic, eliciting antibody responses of the right specificity.[93] In order to explore the utilization of CNTs in solvent and the affinities of CNTs for different peptides, the binding free energies of peptides to SWCNTs were calculated and analyzed.[94] The simulation of interactions between different peptides and SWCNTs was carried out using molecular dynamics methods, and estimation of the binding free energies of peptides onto the outer-surface of the SWCNTs was based on thermodynamics theory. A good agreement between theoretical and experimental results was observed.

Different *proteins* such as bovine serum albumin (BSA), cytochrome *c*, and horseradish peroxidase (HRP) were used to solubilize SWNTs in water aided by sonication.[95] Among other proteins used, the egg white lysozyme dispersed SWNTs, whereas papain and pepsin could not.[96] The authors concluded that the main driving force to the hydrophobic interactions between the sidewall of the SWNT and the inner hydrophobic domain exposed to the solvent during the 3D change of the protein was induced by sonication. A gonadotrophin-releasing hormone (GnRH), which was overexpressed in the plasma membrane of several types of cancer cells, was covalently anchored onto the surface of the oxidized MWCNTs via an amide linkage.[97] Sidewall coverage of MWCNTs by the GnRH was about 0.7% of the available surface area. It was also shown that the GnRH–MWCNTs entered the cells and showed toxicity in the malignant cells. The CNTs were successfully suspended in aqueous buffer solutions by their functionalization with a specific bifunctional molecule that is "sticky" to proteins 1-pyrene butanoic acid succidymidyl ester (1-pbase).[98] Among other reported proteins, attached to CNTs making them soluble, we note Cy5 labeled goat anti-rabbit IgG (anti-IgG-Cy5), chemically bonded to CNTs via a two-step process of diimide-activated amidation and observed successfully using fluorescence microscopy, obtaining the fluorescent image of highly oriented f-CNTs the first time[99]; the FMDV, leading to mono- and bis-derivatized CNTs *b* and *c* (Figure 16.16) starting from the precursor *a* possessing free amino groups[100] and more.[101–104]

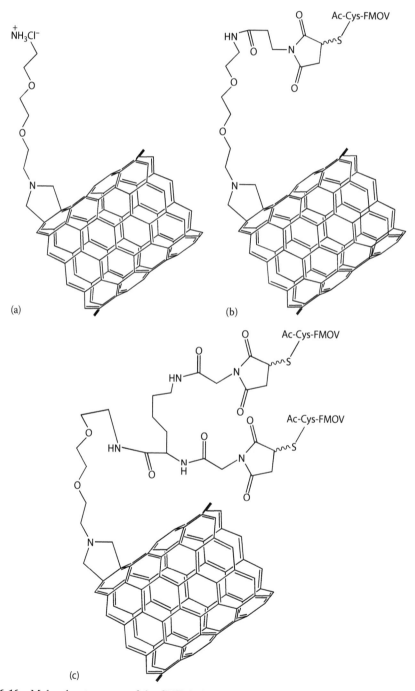

FIGURE 16.16 Molecular structures of the CNT derivative (a) and mono- (b) and bis- (c) conjugates.

Protein–CNT nanocomposites can be potentially applied for biosensor and bio-fuel cell applications. A comparison of protein-dispersing media for various engineered carbon nanoparticles is given in Ref. [105].

Enzymes, belonging to proteins, have also been extensively studied as solubilizing agents for CNTs. Thus, a variety of enzyme-MWNT conjugates in aqueous solutions, their high activity and stability, and reusability, were reported in Refs. [106,107]. The formed products are soluble in aqueous buffer, retained a high fraction of their native activity, and stable at higher temperatures relative to their solution phase counterparts. With the use of noncovalent functionalization by the surfactant Triton X-100, the hydrophobic surfaces of the CNTs are changed to hydrophilic and interact with the hydrophilic surface of Biliverdin IX beta reductase enzyme, creating a water-soluble complex with real interaction between the enzyme and CNT-Triton conjugates.[108] We would like to note glucose oxidase[109] among other enzymes[110] forming soluble composites with CNTs. Some protein conjugates possess the attributes of both soluble enzymes—high activity and low diffusional resistance—and immobilized enzymes—high stability—making them attractive choices for applications ranging from diagnostics and sensing to drug delivery.

Other recently reported biomolecules, used as dispersing agents for CNTs, are on the sugar basis. Thus, HSVM technique was applied to obtain water-soluble SWNT–*nucleotide* composites, whose solubility depends significantly on the number of phosphate groups and the kinds of bases employed.[111] The complex amylose-nanotube system was studied by molecular dynamics simulation for elucidation of the mode of interaction between the initially separated amylose and SWNT fragments, which can be either wrapping or encapsulation.[112] The authors found that amylose molecules can be used to bind with nanotubes due to the dominance of the van der Waals force and, thus, favor noncovalent functionalization of CNTs. The functionalized SWCNTs, with substantial solubility in dimethyl sulfoxide and dimethylformamide, were obtained from oxidized SWCNTs by acyl halogenation with thionyl chloride and dimethylformamide and further interaction with methyl-β-*cyclodextrin* and ODA simultaneously.[113] Such systems offer considerable advantages over polymer-based composites due to their biocompatibility and noncovalent coupling, which can potentially preserve the unique properties of the tubes.[114] The mechanism of interaction for such systems has been proposed to be dominated by hydrophobic and hydrophilic interactions along the surface of the tube.

A certain attention has been paid to composites of CNTs with a linear polysaccharide *chitosan* (CS).[115,116] Thus, a new amperometric biosensor for hydrogen peroxide was developed[117] based on cross-linking HRP by glutaraldehyde with MWNTs/CS composite film coated on a glassy carbon electrode. The biosensor had good repeatability and stability for the determination of H_2O_2. In general, the CNT–CS composite provides a suitable biosensing matrix due to its good conductivity, high stability, and good biocompatibility.[118] In addition to H_2O_2, these composites can detect different substrates (ABTS, catechol, and O_2), possessing high affinity and sensitivity, durable long-term stability, and facile preparation procedure. CS–CNT composites were also obtained in the form of fibers with a wet spinning method.[119]

DNA functionalization of CNTs holds interesting prospects in various fields including solubilization in aqueous media, nucleic acid sensing, gene-therapy, and controlled deposition on conducting or semiconducting substrates.[120] Thus, a solid-state mechanochemical reaction was used for obtaining DNA-wrapped nanotubes of both MWCNTs and SWCNTs, resulting in a high aqueous solubility of the products with a stability of >6 months.[121] It was established that the nanotubes were cut into shorter lengths and were fully covered with DNA, which in the product is intact. DNA-SWCNTs were also prepared by another method, the *layer-by-layer technique*,[122] where poly(diallyldimethylammonium) (PDDA), a positively charged PE, and DNA as a negatively charged counterpart macromolecule are alternatively deposited on the water-soluble oxidized SWCNTs (Figure 16.17). (We note that the same technique was applied also to fabricate a thin film from RNA-dissolved SWNT solutions[123]). As a representative application of the obtained DNA

FIGURE 16.17 Scheme for fabrication of DNA-modified SWCNTs.

composite, an electrode modified by DNA/PDDA/SWCNT particles exhibited larger electrocatalytic oxidation current in an aqueous solution of Ru-(bpy)$_3^{2+}$; moreover, DNA-CNTs can be used as sensors for NO$_2$ detection. Other recent achievements in DNA immobilization on the CNTs are reported in Refs. [124,125].

16.8 FUNCTIONALIZATION WITH THE USE OF POLYMERS OR THEIR PRECURSORS

In comparison with other functionalizing agents for CNTs, the polymers and their precursors, undoubtedly, have been of the most importance (together with biomolecules) during the last 10 years due to a considerable improvement in mechanical properties and stability of polymer hybrids with CNTs. A host of experimental articles and patents have been published on the basis of results using a broad spectrum of distinct polymers: polystyrene,[126–130] polyvinyl pyrrolidone,[131–133] polybutadiene,[134] poly(N-vinylcarbazole),[135] polyaniline,[136,137] polyethyleneglycol,[138,139] poly(acrylic acid) and related polymers,[140–142] polyurethane,[143] and much more.[144–153] The obtained composites (sometimes in the form of films) were soluble in water or common organic solvents, depending on the polymer used. Some representative examples of the synthesis of CNT–polymer composites and their applications are represented in Table 16.1.

16.9 SPECIAL TECHNIQUES IN THE SYNTHESIS AND PURIFICATION OF SOLUBLE CNTs

16.9.1 FUNCTIONALIZATION USING IRRADIATION OR RADIONUCLIDES

It is well known that strong sources of irradiation can produce defects and imperfections not only in biological molecules but also in inorganic materials. CNTs are not an exception to this rule: additional defects cause a higher-scale formation of attached COOH or other functional groups. Thus, MWNTs irradiated with γ-*rays* were subjected to chemical modification with thionyl chloride and decylamine.[173] The results showed that γ-radiation increased the concentration of functional groups bound to MWNTs, which arose due to the increasing number of defect sites created on the MWNTs by γ-photons. Compared with untreated MWNTs, γ-irradiation significantly enhanced the solubility of MWNTs in acetone and THF.

A series of radionuclides have been attached to CNTs with distinct purposes. Thus, *carbon-13* was used to enrich CNT compositions for improved magnetic resonance imaging.[174,175] An imaging study to determine the tissue biodistribution and pharmacokinetics of prototypical water-soluble 1,4,7,10-tetraazacyclododecane-1,4,7,10-tetraacetic acid (DOTA)-functionalized CNT labeled with *yttrium-86* (Figure 16.18) and *indium-111* in a mouse model was undertaken.[176] It was noted that the major sites of accumulation of activity resulting from the administration of [86]Y-CNT were the kidney, liver, spleen, and to a lesser extent the bone.

Water-soluble MWNTs were labeled with *technetium-99m* (this isotope ^{99m}Tc is one of the most used in medicinal chemistry[177]) to study the distribution of MWNTs modified with glucosamine in mice.[178] It was shown that MWNTs moved easily among the compartments and tissues of the body, behaving like active molecules although their apparent mean molecular weight is tremendously large. Similar results were obtained for *iodine-125* (^{125}I)-labeled CNTs.[179]

TABLE 16.1

Representative Examples of the Composites, Prepared on the Basis of Polymers and CNTs

Precursors	Product(s)	Properties and Possible Applications	References
A rigid linear polymer poly(phenyleneethynylene) (PPE) and related compounds, SWNTs	PPE/SWNTs hybrid	The water solubility of SWNTs was enhanced to 1.8 mg/mL. Used in a photovoltaic cell with the bulk heterojunction configuration	[154,155]
Poly(acrylamide) (PAM), SWNTs (*via* reversible addition–fragmentation chain transfer [reversible addition–fragmentation chain transfer, RAFT] polymerization)	PAM-g-SWNT	Relatively uniform polymer coatings present on the surface of individual, debundled nanotubes	[156]
RAFT agents, MWNTs, cationic polymer (poly(2-(dimethylamino) ethylmethacrylate)), anionic polymer (poly(acrylic acid)) and zwitterionic polymer (poly{3-[N-(3-methacrylamidopropyl)-N,N-dimethyl] ammoniopropane sulfonate})	Poly{3-[N-(3-methacrylamidopropyl)-N,N-dimethyl] ammoniopropane sulfonate} functionalized MWNTs, poly(acrylic acid) functionalized MWNTs and poly(2-(dimethylamino) ethyl methacrylate) functionalized MWNTs	Good solubility in aqueous solution	[157]
Poly(N-isopropylacrylamide) (PNIPAM), MWNTs, (*via* RAFT polymerization)	MWNT-g-PNIPAM	Good solubility in water, chloroform, and THF. Potential applications by grafting other functional polymer chains onto MWNTs	[158]

(continued)

TABLE 16.1 (continued)
Representative Examples of the Composites, Prepared on the Basis of Polymers and CNTs

Precursors	Product(s)	Properties and Possible Applications	References
Poly-*m*-aminobenzene sulfonic acid (PABS) or polyethylene glycol (PEG), SWNTs	Graft copolymers SWNT-PABS (30% loading SWNTs) and SWNT-PEG (71% loading SWNTs)	Water-soluble (5 mg/L). Fairly uniform length and diameter	[159,160]
MWNTs. *In situ* polymerization of aniline followed by sulfonation with chlorosulfonic acid in an inert solvent and by hydrolysis in water	SPAN/MWNTs SPAN, sulfonated polyaniline	SPAN/MWNTs are highly dispersible in water. Quinonoid structure of SPAN preferentially interacts with the nanotubes and is stabilized by strong pi–pi interaction between two components	[161]
3-Aminophenylboronic acid monomers, DNA, SWNTs	Poly-(anilineboronic acid)/ss-DNA/SWNT composite (PABA/ss-DNA/SWNT)	Water-soluble; the conductivity of the nanocomposite was much higher than that of the pure self-doped polyaniline in both acidic and neutral solutions	[162]

Poly(1-phenyl-1-alkyne) and poly(diphenylacetylene) derivatives carrying azido functional groups at the ends of their alkyl pendants, SWNTs

SWNTs-polyacetylene composites:

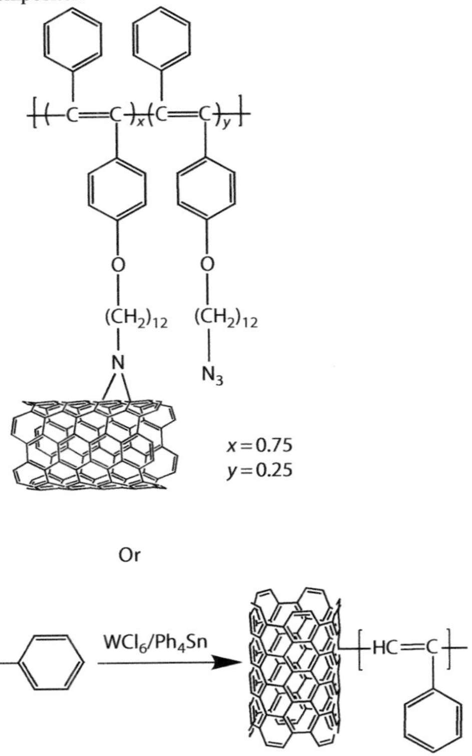

$x = 0.75$
$y = 0.25$

Soluble in common solvents

[163–165]

Double-hydrophilic block copolymer, poly(ethylene oxide)-*b*-poly [2-(*N,N*-dimethylamino)ethyl methacrylate] (PEO-*b*-PDMA), MWCNTs

(PEO-*b*-PDMA)/MWCNTs

Or

Yield 26%. Direct evidence for the individual dispersion. Possible applications: Au and Pt nanoparticles were attached on the sidewall of the modified MWCNTs by using the amino groups of PDMA segments

[166]

(*continued*)

TABLE 16.1 (continued)

Representative Examples of the Composites, Prepared on the Basis of Polymers and CNTs

Precursors	Product(s)	Properties and Possible Applications	References
Poly(ethylene-co-vinyl alcohol) (EVOH) copolymer, SWNTs, carbodiimide-activated esterification reaction conditions	EVOH-SWNT	Soluble in highly polar solvent systems such as DMSO and hot ethanol–water mixtures. A fabrication of nanocomposites in which the SWNTs are homogeneously dispersed in the polymer matrix is possible	[167]
Poly(vinyl alcohol) (PVA), MWCNTs (wet-casting method; esterification reactions)	PVA-MWCNTs films	Water-soluble. The mechanical properties of the nanocomposite films were significantly improved compared to the neat polymer film. The composite-based electrodes can be used as biosensors for glucose detection	[168–170]
Carboxylic acid-terminated hyperbranched poly(ether-ketones) (HPEKs), MWCNTs or SWCNTs	HPEK-g-SWCNT and HPEK-g-MWCNT nanocomposites	The resultant nanocomposites were homogeneously dispersed in various common polar aprotic solvents as well as in concentrated ammonium hydroxide	[171]
Thionyl chloride, SWCNTs, caprolactam, metallic sodium	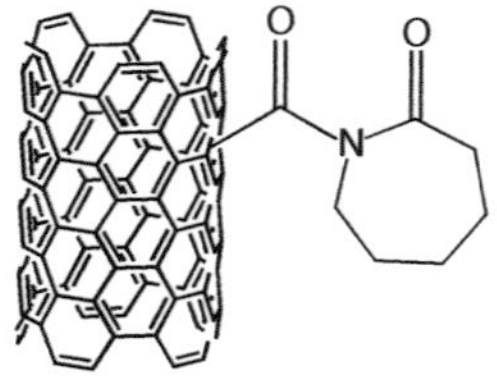 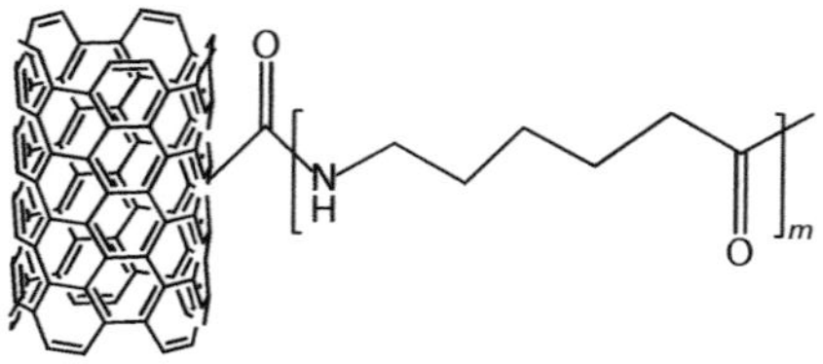	Soluble in organic solvents	[172]

Caprolactam-functionalized SWCNTs

Caprolactam- and nylon-SWCNTs

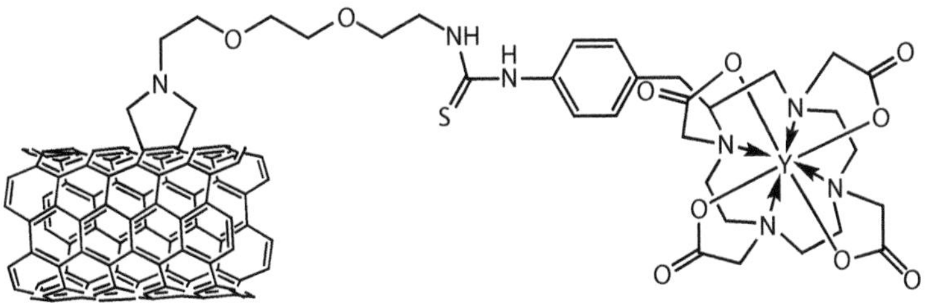

FIGURE 16.18 Yttrium-labeled CNTs.

16.9.2 OTHER SYNTHESIS METHODS

A very interesting technique called *pulsed streamer discharge*, generated in water, involving chemical reactions between radicals appearing by the pulsed streamer discharge and CNTs, allowed homogeneously dispersing them and solubilizing well in water for a month or longer.[180] A study of the mechanism revealed that –OH groups, which are known to impart a hydrophilic nature to carbon material, were introduced on the CNT surface; highly oxidative O* and H* radicals were generated in water and are responsible for the functionalization of the CNT surface by –OH groups. A great advantage of the proposed method is that there is no need for any chemical agents or additives for solubilization due to their generation from the water itself by the electrochemical reactions induced by the pulsed streamer discharge. This method was improved by the same researchers[181] by the use of gas bubbling in water. Oxygen, argon, and nitrogen were used as bubbling gas to clarify the effects of the gas species on the SWCNT solubilization efficiency, and it was established that gas bubbling has positive effects on microplasma-based SWCNT solubilization as a result of enhanced radical formation and functionalization of the SWCNT surface.

Microwave treatment, now a common technique in chemistry, was successfully applied for chemical functionalization and solubilization of SWCNTs, for instance, for their amidation and 1,3-dipolar cycloaddition[182,183] or preparation *in situ* of (MWCNT)/polystyrene or poly(methyl methacrylate) composites soluble in common organic solvents such as 1,2-dichlorobenzene, THF, and chloroform.[184] Solubility was a key feature for a successful MW-heated reaction of cycloaddition of 1,3-dipolar azomethine ylides to the sidewalls of MWNTs, resulting in MWNTs that contain 2-methylenethiol-4-(4-octadecyloxyphenyl), *N*-octyl-2-(4-octadecyloxyphenyl), or 2-(4-octadecyloxyphenyl)pyrrolidine units.[185] All these contain the 4-octadecyloxyphenyl substituent that acts as a solubilizing group. The amount of added groups after only 2 h of MW heating at 200°C was in the same range as that obtained after 100–120 h of conventional heating of soluble and insoluble MWNTs. Among other techniques, *cryogenic crushing* CNTs at liquid nitrogen temperature allowed them to be shortened and made them appreciably soluble in a solvent without any dispersant.[186] Typical lengths of less than 500 nm were obtained from 30 min crushing. A two-phase liquid–liquid *extraction process*, allowing the extraction of water-soluble SWCNTs into an organic phase, is reported in Ref. [187]. The extraction is based on electrostatic interactions between a common phase transfer agent and the sidewall functional groups on the nanotubes.

Purification of soluble CNTs from impurities (carbon nanoparticles, graphite fragments, etc.) can be carried out by several methods, in particular by *chromatography*,[188–189] flow field-flow *fractionation*,[190] and *centrifugation*.[191] Thus, through a systemic study of a series of centrifuged solutions, the authors of Ref. [192] confirmed using Raman spectroscopy that heavily functionalized amorphous carbon was fractionated into the early centrifuged solutions, whereas lightly functionalized graphite fragments as well as polyhedral carbon and metal catalysts particles were fractionated into the late centrifuged solutions and centrifuged residue, and then highly pure and well dispersed

SWNTs were collected from the middle centrifuged solutions. It is proposed that the purity, dispersibility, and aggregation state of SWNTs can be qualitatively estimated by the relative intensity of their absorption features, the fine structure, and slope of their absorption curves.

16.10 STUDY OF REACTIVITY AND PHYSICOCHEMICAL PROPERTIES OF SOLUBLE CNTs

16.10.1 SPECIAL STUDIES OF THE REACTIVITY OF SOLUBLE CNTs TOWARD ACTIVE OXYGEN SOURCES

Soluble SWCNTs were oxidized with singlet *oxygen* ($^1\Delta_g$), and the reaction progress was monitored utilizing FT-IR and UV–vis–NIR spectroscopy.[193] The results indicated reversible covalent addition of oxygen to the walls of the nanotubes, most likely producing either the [2 + 2] or [4 + 2] cycloaddition product (Figure 16.19). Dilute aqueous *ozone* solution with or without ultrasound was used to functionalize SWCNTs.[194] Both O_3 and O_3/ultrasound treatments greatly increased the stability of SWCNTs in water. The oxidation pathway was proposed as follows: at the onset of the oxidation reaction, the C=C double bond was first converted to –C–OH, which was then oxidized to –C=O and O=C-OH concurrently. Ozonating CNTs in fluorinated solvents (perfluorinated polyethers) to functionalize the sidewalls of the CNTs yielding functionalized CNTs with oxygen-containing functional moieties was reported in Refs. [195,196] Short (about 15 min) and long (about 3 h) exposures to ozone as well as cold (–78°C.), room temperature (r.t.) and hot (50°C) temperatures were tested.

The reaction of H_2O_2 with an aqueous suspension of water-soluble HiPco SWNTs encased in the surfactant sodium dodecyl sulfate was studied.[197] Preliminary studies on the mechanisms suggested that H_2O_2 withdraws electrons from the SWNT valence band by charge transfer, which suppresses the nanotube spectral intensity. CNTs were modified by oxidation with *peroxygen compounds* (inorganic peroxoacids, peroxycarboxylic acids of the formula $Q(C(O)OOH)_n$, hydroperoxides of the formula $Q(OOH)_n$, salts thereof, and combinations of any of the above, where Q is an alkyl, cycloalkylyl, aryl, or heterocyclic group of C_1 to C_{12} and n is 1 or 2).[198] The oxidized CNTs included carbon- and oxygen-containing moieties, such as carbonyl, carboxyl, aldehyde, ketone, hydroxy, phenolic, esters, lactones, quinones, and derivatives thereof. Oxidation of the nanotubes increases the degree of dispersion of aggregates of nanotubes and aids in the disassembling of such aggregates. The dispersed nanotubes are used to prepare rigid structures and can be used in electrodes and capacitors.

16.10.2 OTHER PHYSICOCHEMICAL STUDIES

The oxidation of MWCNTs in nitric acid was monitored using sample weight, *Raman spectrum*, solubility, morphology, and alignment.[199] It was noted that high solubility (20–40 mg/mL) is obtained only after prolonged exposure (24–48 h) in concentrated acid (60%) with a considerable loss of the product (60%–90%); the MWCNTs are strongly fragmented and covered by amorphous carbon after 48 h of oxidation. Moreover, it was found that the solubility correlates well with the area ratio

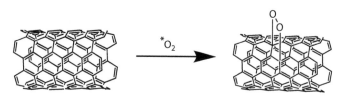

FIGURE 16.19 Products of SWCNT oxidation with oxygen.

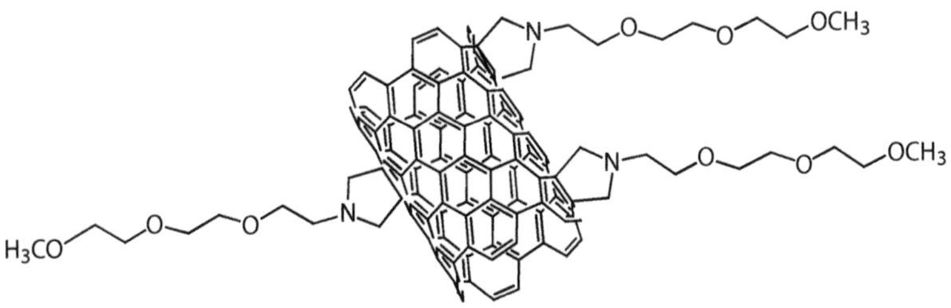

FIGURE 16.20 Pyrrolidine-functionalized nanotubes.

of the G and D bands from the Raman spectrum. In a similar investigation, aqueous dispersions of SWNTs, prepared using different dispersing agents, were also analyzed by Raman spectroscopy.[200] The influences of different dispersing agents and excitation wavelengths were discussed in comparison with UV–visible spectroscopic analysis data. The authors offered to use the most effective dispersing agent found in this study, sodium dodecylbenzene sulfonate, as a benchmark for future dispersion experiments. Electron paramagnetic resonance (EPR) studies on pristine, purified, shortened, and soluble SWNTs in various solution phases showed that the soluble SWNTs carry about one unpaired electron per 10,000 carbon atoms and give a free electron g-value.[201] *Density functional theory* calculations were carried out to study the effects of covalently binding isoniazid, an antitubercular compound to functionalized CNTs.[202] Binding energies, energies of solvation, and quantum-chemical molecular descriptors were calculated. Significant differences were observed between SWNTs and MWNTs by investigating them in depth by conventional *electrochemical techniques* in solution.[203,204] Although functionalization strongly modified the electronic properties of CNTs, the enrichment of the density of states of MWNTs with respect to SWNTs, due to larger tube diameters, is still appreciated. The bulk electronic properties of pyrrolidine-functionalized nanotubes (Figure 16.20) were obtained from cyclic voltammetry measurements and discussed in the light of quantum chemical calculations.[205]

^{13}C NMR study of highly soluble ^{13}C-enriched SWCNTs, functionalized with diamine-terminated oligomeric poly(ethylene glycol) (PEG(1500N)), allowed detection of CNTs in solution.[206] In-depth studies of CNT *solubility* in a variety of solvents were carried out by author of Ref. [207]. Dispersions of SWCNTs in various solvents and aqueous surfactant emulsions were investigated to correlate the degree of dispersion state with *Hansen solubility parameters* $\delta_t^2 = \delta_d^2 + \delta_p^2 + \delta_h^2$,[208] and it was found that the nanotubes were dispersed or suspended very well in the solvents with certain dispersive component (δ_d) values. They were precipitated in the solvents with high polar component (δ_p) values or hydrogen-bonding component (δ_h) values. The surfactants with a lipophilic group equal to and longer than decyl, containing nine methylene groups and one methyl group, contributed to the dispersion of nanotubes in water. *Molecular dynamics simulations* of liquid water near the external surface of a CNT bundle were presented in Ref. [209]. Hydrogen bonding, diffusive behavior, and rotational and vibrational motions are analyzed in the low- and high-density regimes. *Supercritical water* in the presence and absence of oxygen was used for the first time for the study of the opening and thinning of MWNTs.[210]

Among other recent investigations of CNTs, we note the study of optically sensing *sonication effects*,[211] relations between *colloidal stability and percolation* phenomena in aqueous suspensions of MWCNTs,[212] and *toxic effects*.[213] Additionally, a method, a system, and an apparatus to determine the *concentration* of CNTs in a solution were proposed.[214]

16.11 MAIN APPLICATIONS OF SOLUBLE CNTs

In addition to such applications of soluble CNTs as biosensors, composites, advanced polymers, etc., mentioned throughout the text, we would like to underline the following uses. Soluble CNTs are applied mainly in nanomedicine[215] for different purposes, the most important of which is the *drug delivery.*[216] Within the family of nanomaterials, CNTs have emerged as a new alternative and efficient tool for transporting and translocating therapeutic molecules. It has become possible after the recent discovery of their capacity to penetrate into the cells. CNTs can be loaded with active molecules by forming stable covalent bonds or supramolecular assemblies based on noncovalent interactions. As described earlier, CNTs can be functionalized with bioactive peptides, proteins, nucleic acids and drugs and used to deliver their cargos to cells and organs. Once the cargos are carried into various cells, tissues, and organs, they are able to express their biological function.[217] Because functionalized CNTs display low toxicity and are not immunogenic, such systems hold great potential in the field of nanobiotechnology and nanomedicine.[218] Thus, SWNTs and MWNTs were solubilized via the esterification of nanotube-bound carboxylic acids by oligomeric polyethylene glycol compounds.[219] The obtained water-soluble samples were used as starting materials in reactions with BSA protein in ambient aqueous solutions, yielding SWNT-BSA and MWNT-BSA conjugates. Related information is given in Ref. [220].

CNT-Alg gel (alginate hydrogel) was reported as useful scaffold material in *tissue engineering* with the sidewalls of CNTs acting as active sites for chemical functionalization.[221] Soluble CNTs were dispersed in sodium alginate solution as a cross-linker. As a result, the CNT-Alg gel showed faster gelling and higher mechanical strength than the conventional alginate gel. *Elimination of cancer cells* using SWNTs is reported in Ref. [222]. The behavior of the functionalized, water-soluble SWNTs under exposition to a noninvasive, 13.56-MHz RF field was studied. Then, human cancer cell lines were incubated with various concentrations of SWNTs and then treated in the RF field. As a result, SWNTs targeted to cancer cells may allow noninvasive RF field treatments to produce lethal thermal injury to the malignant cells. As shown in another "anticancer" report, the platinum(IV) complex c,c,t-[Pt(NH$_3$)$_2$Cl$_2$(OEt)(O$_2$CCH$_2$CH$_2$CO$_2$H)] (Figure 16.21), which is nearly nontoxic to testicular cancer cells, displays a significantly enhanced cytotoxicity profile when attached to the surface of amine functionalized soluble SWNTs.[223] The authors noted that by linking additional groups, such as cancer-cell targeting moieties, to the platinated SWNTs as longboat passengers, it may be possible to achieve highly selective constructs for use in clinical trials.

Considerable efforts have been dedicated to the preparation, characterization, and search for applications of the *films* on the basis (or starting from) CNTs.[224,225] Thus, a free-standing film made of a SWCNT–polyvinylalcohol (PVA) composite material was fabricated by pretreatment of SWCNTs with ultrasonication in water with the aid of a surfactant that promotes unbundling of aggregated SWNCTs.[226] Large-scale and homogeneous MWNT films were fabricated using the soluble ODA modified MWNT at r.t. by spin-coating method.[227] A composite film containing titania

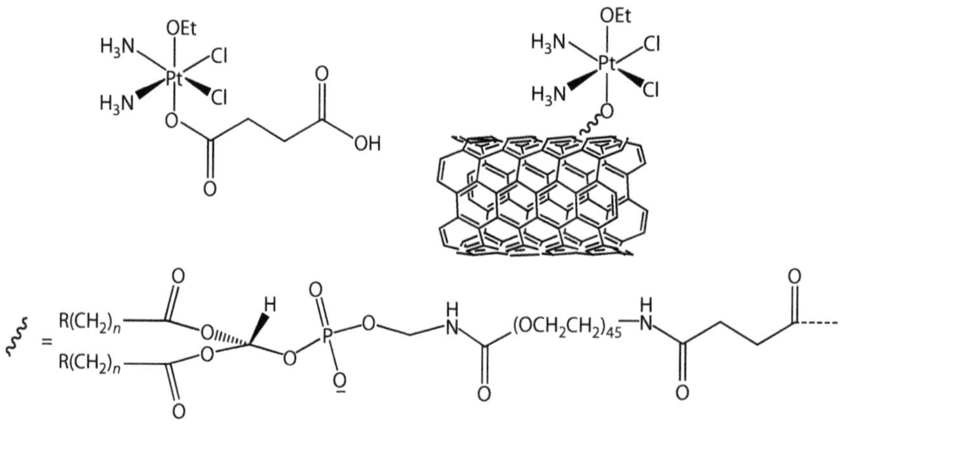

FIGURE 16.21 c,c,t-[Pt(NH$_3$)$_2$Cl$_2$(OEt)(O$_2$CCH$_2$CH$_2$CO$_2$H)] and its SWCNTs conjugate.

electrostatically linked to oxidized MWNTs (TiO_2-s-MWNTs) was prepared from a suspension of TiO_2 nanoparticles in soluble CNTs. Photoinduced charge transfer between the MWNT and TiO_2 was proposed. One representative application for such films is the modification of electrode surfaces. Thus, SEM and electrochemical studies of water-soluble SWNTs, prepared via noncovalent functionalization by CR through a physical grinding treatment and immobilized on the surface of a glassy carbon electrode, showed that SWNTs formed uniform films with porous network structures of nanosizes on the electrode surface, which were stable in neutral and acidic solutions but were unstable in basic media.[228] Examination of potential applications of the films demonstrated that the water-soluble SWNTs were the ideal materials for constructing SWNT-based electrochemical sensing films.

The directed assembly of SWCNTs at lithographically defined positions on gate oxide surfaces was reported,[229] allowing for the high yield (~90%) and parallel fabrication of SWCNT *device arrays*. The method is based on SWCNT functionalization through diazonium chemistry, further assembling with HfO_2 surface and heating to 600°C. The precision, ease, and resultant high yield of this method provide a promising route to the parallel fabrication of large-scale CNT electronics. Conducting textiles (*dyes*) can be simply prepared by incorporating CNTs through a dyeing approach, immersing textiles in an aqueous sulfonated polyaniline-CNT dispersion acting as a dye.[230] In comparison with textiles dyed with sulfonated polyaniline, the conductivity and capacitance considerably increased. The patent[231] is dedicated to hair coloring using modified CNTs. Other applications are related with the use of CNTs in *solar cells*.[232]

REFERENCES

1. Kuzmany, H.; Kukovecz, A.; Simon, F.; Holzweber, M.; Kramberger, C.; Pichler, T. Functionalization of carbon nanotubes. *Synthetic Metals*, 2004, *141* (1), 113–122.
2. Liang, F.; Billups, E. W. Water-soluble single-wall carbon nanotubes as a platform technology for biomedical applications. U.S. Patent 20070110658, 2007.
3. Tour, J. M.; Hudson, J. L.; Dyke, C.; Stephenson, J. J. Functionalization of carbon nanotubes in acidic media. WO2005113434, 2005.
4. Wiesner, M.; Bottero, J.-Y. *Environmental Nanotechnology*. McGraw-Hill Professional, New York, 2007, 540 pp.
5. Gonsalves, K.; Halberstadt, C.; Laurencin, C. T.; Nair, L. *Biomedical Nanostructures*. 2007, John Wiley & Sons, Hoboken, NJ, 507 pp.
6. Choi, S.-K. *Synthetic Multivalent Molecules: Concepts and Biomedical Applications*. Wiley-Interscience, New York, 2004, 418 pp.
7. Reich, S.; Thomsen, C.; Maultzsch, J. *Carbon Nanotubes: Basic Concepts and Physical Properties*. Wiley-VCH, Weinheim, Germany, 2004; 224 pp.
8. Jorio, A.; Dresselhaus, G.; Dresselhaus, M. S. *Carbon Nanotubes: Advanced Topics in the Synthesis, Structure, Properties and Applications*. Springer, Berlin, Germany, 2008, 720 pp.
9. Hiroto, M.; Naotoshi, N. Soluble carbon nanotubes and their applications. *Journal of Nanoscience and Nanotechnology*, 2006, *6* (1), 16–27.
10. Tasis, D.; Tagmatarchis, N.; Georgakilas, V.; Prato, M. Soluble carbon nanotubes. *Chemistry*, 2003, *9* (17), 4000–4008.
11. Nakashima, N.; Fujigaya, T. Fundamentals and applications of soluble carbon nanotubes. *Chemistry Letters*, 2007, *36* (6), 692.
12. Lacerda, L.; Bianco, A.; Prato, M.; Kostarelos, K. Carbon nanotubes as nanomedicines: From toxicology to pharmacology. *Advanced Drug Delivery Reviews*, 2006, *58* (14), 1460–1470.
13. Helland, A.; Wick, P.; Koehler, A.; Schmid, K.; Som, C. Reviewing the environmental and human health knowledge base of carbon nanotubes. *Environmental Health Perspectives*, 2007, *115* (8), 1125–1131.
14. Liu, P. Modifications of carbon nanotubes with polymers. *European Polymer Journal*, 2005, *41* (11), 2693–2703.
15. Yun, Y.; Dong, Z.; Shanov, V.; Heineman, W. R.; Halsall, H. B.; Bhattacharya, A.; Conforti, L.; Schulz, M. J. Nanotube electrodes and biosensors. *Nano Today*, 2007, *2* (6), 30–37.

16. Torrens, F.; Castellano, G. Effect of packing on the cluster nature of C nanotubes: An information entropy analysis. *Microelectronics Journal*, 2007, *38* (12), 1109–1122.

17. Burghard, M. Electronic and vibrational properties of chemically modified single-wall carbon nanotubes. *Surface Science Reports*, 2005, *58* (1), 1–109.

18. Zhou, R.; Shi, M.; Chen, X.; Wang, M.; Yang, Y.; Zhang, X.; Chen, H. Water-soluble and highly fluorescent hybrids of multi-walled carbon nanotubes with uniformly arranged gold nanoparticles. *Nanotechnology*, 2007, *18* (48), 485603.

19. Kocharova, N.; Aäritalo, T.; Leiro, J.; Kankare, J.; Lukkari, J. Aqueous dispersion, surface thiolation, and direct self-assembly of carbon nanotubes on gold. *Langmuir*, 2007, *23* (6), 3363–3371.

20. Hsin, Y. L.; Lai, J. Y.; Hwang, K. C.; Lo, S. C.; Chen, F. R.; Kai, J. J. Rapid surface functionalization of iron-filled multi-walled carbon nanotubes. *Carbon*, 2006, *44* (15), 3328–3335.

21. Georgakilas, V.; Tzitzios, V.; Gournis, D.; Petridis, D. Attachment of magnetic nanoparticles on carbon nanotubes and their soluble derivatives. *Chemistry of Materials*, 2005, *17* (7), 1613–1617.

22. Liang, F.; Alemany, L.; Beach, J. M.; Billups, E. W., Structure analyses of dodecylated single-walled carbon nanotubes. *Journal of the American Chemical Society*, 2005, *127* (40), 13941–13948.

23. Stephenson, J. J.; Sadana, A. K.; Higginbotham, A. L.; Tour., J. M. Highly functionalized and soluble multiwalled carbon nanotubes by reductive alkylation and arylation: The Billups reaction. *Chemistry of Materials*, 2006, *18*, 4658–4661.

24. Gu, Z.; Liang, F.; Chen, Z.; Sadana, A.; Kittrell, C.; Billups, W. E.; Hauge, R. H.; Smalley, R. E. In situ Raman studies on lithiated single-wall carbon nanotubes in liquid ammonia. *Chemical Physics Letters*, 2005, *410* (4), 467–470.

25. Billups, E. W.; Sadana, A.; Chattopadhyay, J. Two-step method of functionalizing carbon allotropes and PEGylated carbon allotropes made by such methods. WO07051071, 2007.

26. Chattopadhyay, J.; Cortez, F. d. J.; Chakraborty, S.; Slater, N. K. H.; Billups, W. E. Synthesis of water-soluble PEGylated single-walled carbon nanotubes. *Chemistry of Materials*, 2006, *18*, 5864–5868.

27. Tsai, Y. C.; Chiu, C. C.; Tsai, M. C.; Wu, J. Y.; Tseng, T. F.; Wu, T. M.; Hsu, S. F. Dispersion of carbon nanotubes in low pH aqueous solutions by means of alumina-coated silica nanoparticles. *Carbon*, 2007, *45* (14), 2823–2827.

28. Cai, Z.-X.; Yan, X.-P. In situ electrostatic assembly of CdS nanoparticles onto aligned multiwalled carbon nanotubes in aqueous solution. *Nanotechnology*, 2006, *17* (16), 4212–4216.

29. Landi, B. J.; Evans, C. M.; Worman, J. J.; Castro, S. L.; Bailey, S. G.; Raffaelle, R. P. Noncovalent attachment of CdSe quantum dots to single wall carbon nanotubes. *Materials Letters*, 2006, *60* (29), 3502–3506.

30. Ellis, A. V. Functionalised carbon nanotubes and methods of preparation. WO07067079, 2007.

31. Stephenson, J. J.; Hudson, J. L.; Leonard, A. D.; Price, B. K.; Tour., J. M. Repetitive functionalization of water-soluble single-walled carbon nanotubes. Addition of acid-sensitive addenda. *Chemistry of Materials*, 2007, *19*, 3491–3498.

32. Liang, F.; Billups, E. W. Water-soluble single-wall carbon nanotubes as a platform technology for biomedical applications. U.S. Patent 20070110658, 2007.

33. Liang, F.; Beach, J. M.; Rai, P. K.; Guo, W.; Hauge, R. H.; Pasquali, M.; Smalley, R. E.; Billups, W. E. Highly exfoliated water-soluble single-walled carbon nanotubes. *Chemistry of Materials*, 2006, *18*, 1520–1524.

34. Tour, J. M.; Bahr, J. L.; Yang, J. Derivatization of carbon nanotubes using diazonium species. GB2413123, 2005.

35. Hudson, J. L.; Casavant, M.; Tour, J. M. Water-soluble, exfoliated, nonroping single-wall carbon nanotubes. *Journal of the American Chemical Society*, 2004, *126* (36), 11158–11159.

36. Zhao, W.; Song, C.; Pehrsson, P. E. Water-soluble and optically pH-sensitive single-walled carbon nanotubes from surface modification. *Journal of the American Chemical Society*, 2002, *124* (42), 12418–12419.

37. Chen, Z.; Kobashi, K.; Rauwald, U.; Booker, R.; Fan, H.; Hwang, W.-F.; Tour, J. M. Soluble ultrashort single-walled carbon nanotubes. *Journal of the American Chemical Society*, 2006, *128* (32), 10568–10571.

38. Rowley, L. A.; Irvin, G. C.; Anderson, C. C.; Majumdar, D. Coating compositions containing single wall carbon nanotubes. WO07086878, 2007.

39. Shieh, Y. T.; Liu, G. L.; Wu, H. H.; Lee, C. C. Effects of polarity and pH on the solubility of acid-treated carbon nanotubes in different media. *Carbon*, 2007, *45* (9), 1880–1890.

40. Alvaro, M.; Atienzar, P.; Cruz, P. d. l.; Delgado, J. L.; Garcia, H.; Langa, F. Synthesis and photochemistry of soluble, pentyl ester-modified single wall carbon nanotube. *Chemical Physics Letters*, 2004, *386* (4), 342–245.

41. Li, S.; Qin, Y.; Shi, J.; Guo, Z.-X.; Li, Y.; Zhu., D. Electrical properties of soluble carbon nanotube/polymer composite films. *Chemistry of Materials*, 2005, *17*, 130–135.

42. Sun, Y.-P.; Huang, W.; Lin, Y.; Fu, K.; Kitaygorodskiy, A.; Riddle, L. A.; Yu, Y. J.; Carroll, D. L. Soluble dendron-functionalized carbon nanotubes: Preparation, characterization, and properties. *Chemistry of Materials*, 2001, *13*, 2864–2869.

43. Feng, L.; Li, H.; Li, F.; Shi, Z.; Gu, Z. Functionalization of carbon nanotubes with amphiphilic molecules and their Langmuir-Blodgett films. *Carbon*, 2003, *41* (12), 2385–2391.

44. Konesky, G. Cross-linked carbon nanotubes. U.S. Patent 20060275956, 2006.

45. Li, X.; Shi, J.; Qin, Y.; Wang, Q.; Luo, H.; Zhang, P.; Guo, Z. X.; Park, D. K. Alkylation and arylation of single-walled carbon nanotubes by mechanochemical method. *Chemical Physics Letters*, 2007, *444* (4), 258–262.

46. Alvaro, M.; Atienzar, P.; Cruz, P. d. l.; Delgado, J. L.; Troiani, V.; Garcia, H.; Langa, F.; Echegoyen, L. Synthesis, photochemistry, and electrochemistry of single-wall carbon nanotubes with pendent pyridyl groups and of their metal complexes with zinc porphyrin. Comparison with pyridyl-bearing fullerenes. *Journal of the American Chemical Society*, 2006, *128* (20), 6626–6635.

47. Yao, Z.; Braidy, N.; Botton, G. A.; Adronov, A. Polymerization from the surface of single-walled carbon nanotubes—Preparation and characterization of nanocomposites. *Journal of the American Chemical Society*, 2003, *125* (51), 16015–16024.

48. Paloniemi, H.; Aäritalo, T.; Laiho, T.; Liuke, H.; Kocharova, N.; Haapakka, K.; Terzi, F.; Lukkari, J. Water-soluble full-length single-wall carbon nanotube polyelectrolytes: Preparation and characterization. *Journal of Physical Chemistry B*, 2005, *109* (18), 8634–8642.

49. Shirai, H.; Kimura, M.; Miki, N. Complex of amphiphilic compound with soluble carbon nanotube. JP2006265151, 2006.

50. Guldi, D. M.; Rahman, G. M. A.; Jux, N.; Balbinot, D.; Hartnagel, U.; Tagmatarchis, N.; Prato, M. Functional single-wall carbon nanotube nanohybrids—Associating SWNTs with water-soluble enzyme model systems. *Journal of the American Chemical Society*, 2005, *127* (27), 9830–9838.

51. Chen, J.; Xue, C. A new method for the preparation of stable carbon nanotube organogels. *Carbon*, 2006, *44* (11), 2142–2146.

52. Tagmatarchis, N.; Prato, M.; Guldi, D. M. Soluble carbon nanotube ensembles for light-induced electron transfer interactions. *Physica E: Low-Dimensional Systems and Nanostructures*, 2005, *29* (3), 546–550.

53. Shen, J.; Huang, W.; Wu, L.; Hu, Y.; Ye, M. Study on amino-functionalized multiwalled carbon nanotubes. *Materials Science and Engineering A*, 2007, *464* (1), 151–156.

54. Maeda, Y.; Hasegawa, T.; Wakahara, T.; Akasaka, T.; Choi, N.; Tokumoto, H.; Kazaoui, S.; Minami, N. Chemical modification of SWNTs. *Molecular Nanostructures: XVII International Winterschool Euroconference on Electronic Properties of Novel Materials. AIP Conference Proceedings*, Kirchberg, Tirol, Austria, March 8–15, 2003, pp. 257–260.

55. Yokoi, T.; Iwamatsu, S. I.; Komai, S. I.; Hattori, T.; Murata, S., Chemical modification of carbon nanotubes with organic hydrazines. *Carbon*, 2005, *43* (14), 2869–2874.

56. Gabriel, G.; Sauthier, G.; Fraxedas, J.; Moreno-Manas, M.; Martinez, M. T.; Miravitlles, C.; Casabo, J. Preparation and characterisation of single-walled carbon nanotubes functionalised with amines. *Carbon*, 2006, *44* (10), 1891–1897.

57. Research news: Bonded to change—Nanotechnology. *Materials Today*, 2003, *6* (11), 12.

58. Xu, M.; Huang, Q.; Chen, Q.; Guo, P.; Sun, Z. Synthesis and characterization of octadecylamine grafted multi-walled carbon nanotubes. *Chemical Physics Letters*, 2003, *375* (5), 598–604.

59. Hamon, M. A.; Hu, H.; Bhowmik, P.; Niyogi, S.; Zhao, B.; Itkis, M. E.; Haddon, R. C., End-group and defect analysis of soluble single-walled carbon nanotubes. *Chemical Physics Letters*, 2001, *347* (1), 8–12.

60. Lian, Y.; Maeda, Y.; Wakahara, T.; Akasaka, T.; Kazaoui, S.; Minami, N.; Choi, N.; Tokumoto, H. Assignment of the fine structure in the optical absorption spectra of soluble single-walled carbon nanotubes. *Journal of Physical Chemistry B*, 2003, *107*, 12082–12087.

61. Shigeta, M.; Komatsu, M.; Nakashima, N. Individual solubilization of single-walled carbon nanotubes using totally aromatic polyimide. *Chemical Physics Letters*, 2006, *418* (1), 115–118.

62. Haddon, R. C.; Hamon, M. A. Method of solubilizing carbon nanotubes in organic solutions. U.S. Patent 6531513, 2003.

63. Haddon, R. C.; Chen, J.; Hamon, M. A. Method of solubilizing unshortened carbon nanotubes in organic solutions. U.S. Patent 6368569, 2002.

64. Ramasubramaniam, R.; Chen, J.; Liu, H. Homogeneous carbon nanotube/polymer composites for electrical applications. *Applied Physics Letters*, 2003, *83*, 2928.

65. Zhu, S.; Zhang, H.; Bai, R. Microwave-accelerated dissolution of MWNT in aniline. *Materials Letters*, 2007, *61* (1), 16–18.

66. Sun, Y.; Wilson, S. Method for dissolving carbon nanotubes. WO02088025, 2002.

67. Suarez, B.; Simonet, B. M.; Cardenas, S.; Valcarcel, M. Separation of carbon nanotubes in aqueous medium by capillary electrophoresis. *Journal of Chromatography A*, 2006, *1128* (1), 282–289.

68. Licea-Jimenez, L.; Henrio, P. Y.; Lund, A.; Laurie, T. M.; Perez-Garcia, S. A.; Nyborg, L.; Hassander, H.; Rychwalski, R. W. MWNT reinforced melamine-formaldehyde containing alpha-cellulose. *Composites Science and Technology*, 2007, *67* (5), 844–854.

69. Cui, J. B.; Daghlian, C. P.; Gibson, U. J. Solubility and electrical transport properties of thiolated single-walled carbon nanotubes. *Journal of Applied Physics*, 2005, *98*, 044320.

70. Nakamura, T.; Ohana, T.; Ishihara, M.; Hasegawa; Koga, Y. Chemical modification of single-walled carbon nanotubes with sulfur-containing functionalities. *Diamond and Related Materials*, 2007, *16* (4), 1091–1094.

71. Murakami, H.; Nomura, T.; Nakashima, N. Noncovalent porphyrin-functionalized single-walled carbon nanotubes in solution and the formation of porphyrin-nanotube nanocomposites. *Chemical Physics Letters*, 2003, *378* (5), 481–485.

72. Chen, J.; Collier, P. C. Noncovalent functionalization of single-walled carbon nanotubes with water-soluble porphyrins. *Journal of Physical Chemistry B*, 2005, *109* (16), 7605–7609.

73. Huang, C. Z.; Liao, Q. G.; Li, Y. F. Non-covalent anionic porphyrin functionalized multi-walled carbon nanotubes as an optical probe for specific DNA detection. *Talanta*, 2008, *75* (1), 163–166.

74. Mhuircheartaigh, E. M. N.; Blau, W. J.; Prato, M.; Giordani, S. Spectroscopic changes induced by sonication of porphyrin-carbon nanotube composites in chlorinated solvents. *Carbon*, 2007, *45* (13), 2665–2671.

75. Cheng, F.; Adronov, A. Noncovalent functionalization and solubilization of carbon nanotubes by using a conjugated Zn-porphyrin polymer. *Chemistry*, 2006, *12* (19), 5053–5059.

76. Komatsu, N.; Osuka, A.; Isoda, S.; Nakashima, N.; Murakami, H. Carbon nanotube and method of purifying the same. EP1702885, 2006.

77. Guldi, D. M.; Rahman, G. N. A.; Ramey, J.; Marcaccio, M.; Paolucci, D.; Paolucci, F.; Qin, S.; Prato, M. Donor-acceptor nanoensembles of soluble carbon nanotubes. *Chemical Communications*, 2004, (18), 2034–2035.

78. Hatton, R. A.; Blanchard, N. P.; Miller, A. J.; Silva, S. R. P. A multi-wall carbon nanotube-molecular semiconductor composite for bi-layer organic solar cells. *Physica E: Low-Dimensional Systems and Nanostructures*, 2007, *37* (1), 124–127.

79. Ross, A. H.; Ravi, S. S. Production of carbon nanotube-molecular semiconductor thin film. GB2428135, 2007.

80. Hatton, R. A.; Silva, S. R. Improvements in thin film production. WO07007061, 2007.

81. Hu, C.; Chen, Z.; Shen, A.; Shen, X.; Li, J.; Hu, S. Water-soluble single-walled carbon nanotubes via noncovalent functionalization by a rigid, planar and conjugated diazo dye. *Carbon*, 2006, *44* (3), 428–434.

82. Ogoshii, T.; Yamagishi, T.-A.; Nakamoto, Y. Supramolecular single-walled carbon nanotubes (SWCNTs) network polymer made by hybrids of SWCNTs and water-soluble calix[8]arenes. *Chemical Communications*, 2007, (45), 4776–4778.

83. Hu, H.; Zhao, B.; Hamon, M. A.; Kamaras, K.; Itkis, M. E.; Haddon, R. C. Sidewall functionalization of single-walled carbon nanotubes by addition of dichlorocarbene. *Journal of the American Chemical Society*, 2003, *125* (48), 14893–14900.

84. Dohi, H.; Kikuchi, S.; Kuwahara, S.; Sugai, T.; Shinohara, H. Synthesis and spectroscopic characterization of single-wall carbon nanotubes wrapped by glycoconjugate polymer with bioactive sugars. *Chemical Physics Letters*, 2006, *428* (1), 98–101.

85. Yinghuai, Z.; Peng, A. T.; Carpenter, K.; Maguire, J. A.; Hosmane, N. S.; Takagaki, M. Substituted carborane-appended water-soluble single-wall carbon nanotubes: New approach to boron neutron capture therapy drug delivery. *Journal of the American Chemical Society*, 2005, *127* (27), 9875–9880.

86. Curran, S. A.; Ellis, A. V.; Vijayaraghavan, A.; Ajayan, P. M. Functionalization of carbon nanotubes using phenosafranin. *Journal of Chemical Physics*, 2004, *120*, 4886.

87. Isobe, H.; Tanaka, T.; Maeda, R.; Noiri, E.; Solin, N.; Yudasaka, M.; Iijima, S.; Eiichi Nakamura, E. Preparation, purification, characterization, and cytotoxicity assessment of water-soluble, transition-metal-free carbon nanotube aggregates. *Angewandte Chemie, International Edition (English)*, 2006, *45* (40), 6676–6680.

88. Liu, Y. T.; Zhao, W.; Huang, Z. Y.; Gao, Y. F.; Xie, X. M.; Wang, X. H.; Ye., X. Y. Noncovalent surface modification of carbon nanotubes for solubility in organic solvents. *Carbon*, 2006, *44* (8), 1613–1616.

89. Arrais, A.; Diana, E.; Pezzini, D.; Rossetti, R.; Boccaleri, E. A fast effective route to pH-dependent water-dispersion of oxidized single-walled carbon nanotubes. *Carbon*, 2006, *44* (3), 587–590.

90. Georgakilas, V.; Tagmatarchis, N.; Pantarotto, D.; Bianco, A.; Briand, J.-P.; Prato, M. Amino acid functionalisation of water soluble carbon nanotubes. *Chemical Communications*, 2002, (24), 3050–3051.

91. Zeng, L.; Zhang, L.; Barron, A. R. Tailoring aqueous solubility of functionalized single-wall carbon nanotubes over a wide pH range through substituent chain length. *Nano Letters*, 2005, *5* (10), 2001–2004.

92. Hu, N.; Dang, G.; Zhou, H.; Jing, J.; Chen, C. Efficient direct water dispersion of multi-walled carbon nanotubes by functionalization with lysine. *Materials Letters*, 2007, *61* (30), 5285–5287.

93. Pantarotto, D.; Partidos, C. D.; Graff, R.; Hoebeke, J.; Briand, J.-P.; Prato, M.; Bianco, A. Synthesis, structural characterization, and immunological properties of carbon nanotubes functionalized with peptides. *Journal of the American Chemical Society*, 2003, *125* (20), 6160–6164.

94. Cheng, Y.; Liu, G. R.; Li, Z. R.; Lu, C. Computational analysis of binding free energies between peptides and single-walled carbon nanotubes. *Physica A: Statistical Mechanics and Its Applications*, 2006, *367*, 293–304.

95. Kum, M. C.; Joshi, K. A.; Chen, W.; Myung, N. V.; Mulchandani, A. Biomolecules-carbon nanotubes doped conducting polymer nanocomposites and their sensor application. *Talanta*, 2007, *74* (3), 370–375.

96. Ke, G.; Guan, W.; Tang, C.; Guan, W.; Zeng, D.; Deng, F. Covalent functionalization of multiwalled carbon nanotubes with a low molecular weight chitosan. *Biomacromolecules*, 8 (2), 322–326.

97. Yu, B. Z.; Yang, J. S.; Li, W. X. In vitro capability of multi-walled carbon nanotubes modified with gonadotrophin releasing hormone on killing cancer cells. *Carbon*, 2007, *45* (10), 1921–1927.

98. Prakash, R.; Superfine, R.; Washburn, S.; Falvo, M. R. Functionalization of carbon nanotubes with proteins and quantum dots in aqueous buffer solutions. *Applied Physics Letters*, 2006, *88*, 063102.

99. Xu, Z.; Hu, P.; Wang, S.; Wang, X. Biological functionalization and fluorescent imaging of carbon nanotubes. *Applied Surface Science*, 2008, *254* (7), 1915–1918.

100. Pantarotto, D.; Partidos, C. D.; Hoebeke, J.; Brown, F.; Kramer, E.; Briand, J.-P.; Muller, S.; Bianco, A. Immunization with peptide-functionalized carbon nanotubes enhances virus-specific neutralizing antibody responses. *Chemistry and Biology*, 2003, *10* (10), 961–966.

101. Bianco, A.; Pantarotto, D.; Prato, M. Functionalized carbon nanotubes, a process for preparing the same and their use in medicinal chemistry. WO04089819, 2004.

102. Ke, P.-C.; Wu, Y.; Rao, A. M. Lysophospholipids solubilized single-walled carbon nanotubes. WO07136404, 2007.

103. Dai, H.; Chen, R. J. Noncovalent sidewall functionalization of carbon nanotubes. U.S. Patent 20050100960, 2005.

104. Holder, P. G.; Francis, M. B. Integration of a self-assembling protein scaffold with water-soluble single-walled carbon nanotubes. *Angewandte Chemie, International Edition (English)*, 2007, *46* (23), 4370–4373.

105. Buford, M. C.; Hamilton, R. F.; Holan, A. A comparison of dispersing media for various engineered carbon nanoparticles. *Particle and Fibre Toxicology*, 2007, *4* (1), 6.

106. Asuri, P.; Karajanagi, S. S.; Sellitto, E.; Kim, D.-Y.; Kane, R. S.; Dordick, J. S. Water-soluble carbon nanotube-enzyme conjugates as functional biocatalytic formulations. *Biotechnology and Bioengineering*, 2006, *95* (5), 804–811.

107. Asuri, P.; Bale, S. S.; Pangule, R. C.; Shah, D. A.; Kane, R. S.; Dordick, J. S. Structure, function, and stability of enzymes covalently attached to single-walled carbon nanotubes. *Langmuir*, 2007, *23* (24), 12318–12321.

108. Panhuis, M. I. H.; Salvador-Morales, C.; Franklin, E.; Chambers, G.; Fonseca, A.; Nagy, J. B.; Blau, W. J.; Minett, A. Characterization of an interaction between functionalized carbon nanotubes and an enzyme. *Journal of Nanoscience and Nanotechnology*, 2003, *3* (3), 209–213.

109. Li, J.; Wang, Y.-B.; Qiu, J.-D.; Sun, D.-C.; Xia, X.-H. Biocomposites of covalently linked glucose oxidase on carbon nanotubes for glucose biosensor. *Analytical and Bioanalytical Chemistry*, 2005, *383* (6), 918–922.

110. Wang, Y.; Iqbal, Z.; Malhotra, S. V. Functionalization of carbon nanotubes with amines and enzymes. *Chemical Physics Letters*, 2005, *402* (1), 96–101.

111. Ikeda, A.; Hamano, T.; Hayashi, K.; Kikuchi, J. I. Water-solubilization of nucleotides-coated single-walled carbon nanotubes using a high-speed vibration milling technique. *Organic Letters*, 2006, *8* (6), 1153–1156.

112. Xie, Y. H.; Soh, A. K. Investigation of non-covalent association of single-walled carbon nanotube with amylose by molecular dynamics simulation. *Materials Letters*, 2005, *59* (8), 971–975.

113. Yu, J.-G.; Huang, K.-L.; Liu, S.-Q.; Tang, J.-C. Preparation and characterization of soluble methyl-β-cyclodextrin functionalized single-walled carbon nanotubes. *Physica E: Low-Dimensional Systems and Nanostructures*, 2008, *40* (3), 689–692.

114. Casey, A.; Farrell, G. F.; McNamara, M.; Byrne1, H. J.; Chambers, G. Interaction of carbon nanotubes with sugar complexes. *Synthetic Metals*, 2005, *153* (1), 357–360.

115. Rinaudo, M. Chitin and chitosan: Properties and applications. *Progress in Polymer Science*, 2006, *31* (7), 603–632.

116. Yang, H.; Wang, S. C.; Mercier, P.; Akins, D. L. Diameter-selective dispersion of single-walled carbon nanotubes using a water-soluble, biocompatible polymer. *Chemical Communications*, 2006, (13), 1425–1427.

117. Qian, L.; Yang, X. Composite film of carbon nanotubes and chitosan for preparation of amperometric hydrogen peroxide biosensor. *Talanta*, 2006, *68* (3), 721–727.

118. Liu, Y.; Qu, X.; Guo, H.; Chen, H.; Liu, B.; Dong, S. Facile preparation of amperometric laccase biosensor with multifunction based on the matrix of carbon nanotubes-chitosan composite. *Biosensors and Bioelectronics*, 2006, *21* (12), 2195–2201.

119. Spinks, G. M.; Shin, S. R.; Wallace, G. G.; Whitten, P. G.; Kim, S. I.; Kim, S. J. Mechanical properties of chitosan/CNT microfibers obtained with improved dispersion. *Sensors and Actuators: B. Chemical*, 2006, *115* (2), 678–684.

120. Daniel, S.; Rao, T. P.; Rao, K. S.; Rani, S. U.; Naidu, G. R. K.; Lee, H. Y.; Kawai, T. A review of DNA functionalized/grafted carbon nanotubes and their characterization. *Sensors and Actuators B: Chemical*, 2007, *122* (2), 672–682.

121. Nepal, D.; Sohn, J.-I.; Aicher, W. K.; Lee, S.; Geckeler, K. Supramolecular conjugates of carbon nanotubes and DNA by a solid-state reaction. *Biomacromolecules*, 2005, *6* (6), 2919–2922.

122. He, P.; Bayachou, M. Layer-by-layer fabrication and characterization of DNA-wrapped single-walled carbon nanotube particles. *Langmuir*, 2005, *21* (13), 6086–6092.

123. Ishibashi, A.; Yamaguchi, Y.; Murakami, H.; Nakashima, N. Layer-by-layer assembly of RNA/single-walled carbon nanotube nanocomposites. *Chemical Physics Letters*, 2006, *419* (4), 574–577.

124. Yang, Q. H.; Gale, N.; Oton, C. J.; Li, H.; Nandhakumar, I. S.; Tang, Z. Y.; Brown, T.; Loh, W. H. Deuterated water as super solvent for short carbon nanotubes wrapped by DNA. *Carbon*, 2007, *45* (13), 2701–2703.

125. Bianco, A.; Pantarotto, D.; Kostarelos, K.; Prato, M. Non-covalent complexes comprising carbon nanotubes. EP1605265, 2005.

126. Zhang, Z.; Zhang, J.; Chen, P.; Zhang, B.; He, J.; Hu, G. H. Enhanced interactions between multi-walled carbon nanotubes and polystyrene induced by melt mixing. *Carbon*, 2006, *44* (4), 692–698.

127. Fan, D. Q.; He, J. P.; Tang, W.; Xu, J. T.; Yang, Y. L. Synthesis of polymer grafted carbon nanotubes by nitroxide mediated radical polymerization in the presence of spin-labeled carbon nanotubes. *European Polymer Journal*, 2007, *43* (1), 26–34.

128. Sluzarenko, N.; Heurtefeu, B.; Maugey, M.; Zakri, C.; Poulin, P.; Lecommandoux, S. Diblock copolymer stabilization of multi-wall carbon nanotubes in organic solvents and their use in composites. *Carbon*, 2006, *44* (15), 3207–3212.

129. Tong, R.; Wu, H.-X.; Qiu, X.-Q.; Qian, S.-X.; Lin, Y.-H.; Cai, R.-F. Optical performance and nonlinear scattering of soluble polystyrene grafted multi-walled carbon nanotubes. *Chinese Physics Letters*, 2006, *23* (8), 2105–2108.

130. Li, H.; Adronov, A. Water-soluble SWCNTs from sulfonation of nanotube-bound polystyrene. *Carbon*, 2007, *45* (5), 984–990.

131. O'Connell, M. J.; Boul, P.; Ericson, L. M.; Huffman, C.; Wang, Y.; Haroz, E.; Kuper, C.; Smalley, R. E. Reversible water-solubilization of single-walled carbon nanotubes by polymer wrapping. *Chemical Physics Letters*, 2001, *342* (3), 265–271.

132. Sakakibara, Y.; Tokumoto, M.; Kataura, H. Saturable absorber of polyimide containing dispersed carbon nanotubes. EP1772770, 2007.

133. Li, L.-J.; Nicholas, R. J.; Chen, C.-Y.; Darton, R. C.; Baker, S. C. Comparative study of photoluminescence of single-walled carbon nanotubes wrapped with sodium dodecyl sulfate, surfactin and polyvinylpyrrolidone. *Nanotechnology*, 2005, *16* (5), S202–S205.

134. Li, C.; Liu, C.; Li, F.; Gong, Q. Optical limiting performance of two soluble multi-walled carbon nanotubes. *Chemical Physics Letters*, 2003, *380* (1), 201–205.

135. Wu, H. X.; Qiu, X. Q.; Cai, R. F.; Qian, S. X. Poly(N-vinyl carbazole)-grafted multiwalled carbon nanotubes: Synthesis via direct free radical reaction and optical limiting properties. *Applied Surface Science*, 2007, *253* (11), 5122–5128.

136. Wang, G.; Ding, Y.; Wang, F.; Li, X.; Li, C. Poly(aniline-2-sulfonic acid) modified multiwalled carbon nanotubes with good aqueous dispersibility. *Journal of Colloid and Interface Science*, 2008, *317* (1), 199–205.

137. Sainz, R.; Benito, A. M.; Martínez, M. T.; Galindo, J. F.; Sotres, J.; Baró, A. M.; Corraze, B.; Maser, W. K. A soluble and highly functional polyaniline–carbon nanotube composite. *Nanotechnology*, 2005, *16* (5), S150–S154.

138. Wang, R.; Cherukuri, P.; Duque, J. G.; Leeuw, T. K.; Lackey, M. K.; Moran, C. H.; Moore, V. C.; Schmidt, H. K. SWCNT PEG-eggs: Single-walled carbon nanotubes in biocompatible shell-crosslinked micelles. *Carbon*, 2007, *45* (12), 2388–2393.

139. Lee, J. U.; Huh, J.; Kim, K. H.; Park, C.; Jo, W. H. Aqueous suspension of carbon nanotubes via noncovalent functionalization with oligothiophene-terminated poly(ethylene glycol). *Carbon*, 2007, *45* (5), 1051–1057.

140. Liu, A.; Watanabe, T.; Honma, I.; Wang, J.; Zhou, H. Effect of solution pH and ionic strength on the stability of poly(acrylic acid)-encapsulated multiwalled carbon nanotubes aqueous dispersion and its application for NADH sensor. *Biosensors and Bioelectronics*, 2006, *22* (5), 694–699.

141. Meng, Q.-j.; Zhang, X.-x.; Bai, S.-h.; Wang, X.-c. Preparation and characterization of poly(methyl methacrylate)-functionalized carboxyl multi-walled carbon nanotubes. *Chinese Journal of Chemical Physics*, 2007, *20* (6), 660–664.

142. Liu, A.; Honma, I.; Ichihara, M.; Zhou, H. Poly(acrylic acid)-wrapped multi-walled carbon nanotubes composite solubilization in water: Definitive spectroscopic properties. *Nanotechnology*, 2006, *17* (12), 2845–2849.

143. Liu, Z.; Bai, G.; Huang, Y.; Li, F.; Ma, Y.; Guo, T.; He, X.; Lin, X.; Gao, H.; Chen, Y. Microwave absorption of single-walled carbon nanotubes/soluble cross-linked polyurethane composites. *Journal of Physical Chemistry C*, 2007, *111*, 13696–13700.

144. Hong, C. Y.; You, Y. Z.; Pan, C. Y. A new approach to functionalize multi-walled carbon nanotubes by the use of functional polymers. *Polymer*, 2006, *47* (12), 4300–4309.

145. Ford, W. E.; Wessels, J. Soluble carbon nanotubes. EP1428793, 2004.

146. Ford, W. E.; Wessels, J.; Yasuda, A. Soluble carbon nanotubes. WO04052783, 2004.

147. Ford, W. E.; Wessels, J.; Yasuda, A. Soluble carbon nanotubes. U.S. Patent 20060014375, 2006.

148. Yerushalmi-Rozen, R.; Regev, O. Method for the preparation of stable suspensions and powders of single carbon nanotubes. WO02076888, 2002.

149. Yu, H.; Cao, T.; Zhou, L.; Gu, E.; Yu, D.; Jiang, D. Layer-by-layer assembly and humidity sensitive behavior of poly(ethyleneimine)/multiwall carbon nanotube composite films. *Sensors and Actuators B: Chemical*, 2006, *119* (2), 512–515.

150. Miller, A. J.; Hatton, R. A.; Silva, S. R. P., Water-soluble multiwall-carbon-nanotube-polythiophene composite for bilayer photovoltaics. *Applied Physics Letters*, 2006, *89*, 123115.

151. Kuila, B. K.; Malik, S.; Batabyal, S. K.; Nandi, A. K. In-situ synthesis of soluble poly(3-hexylthiophene)/multiwalled carbon nanotube composite: Morphology, structure, and conductivity. *Macromolecules*, 2007, *40*, 278–287.

152. Hong, C.-Y.; You, Y.-Z.; Pan, C.-Y. Synthesis of water-soluble multiwalled carbon nanotubes with grafted temperature-responsive shells by surface RAFT polymerization. *Chemistry of Materials*, 2005, *17*, 2247–2254.

153. Cheng, F.; Imin, P.; Maunders, C.; Botton, G.; Adronov, A. Soluble, discrete supramolecular complexes of single-walled carbon nanotubes with fluorene-based conjugated polymers. *Macromolecules*, 2008, *41* (7), 2304–2308.

154. Mao, J.; Liu, Q.; Lv, X.; Liu, Z.; Huang, Y.; Ma, Y.; Chen, Y.; Yin, S. A water-soluble hybrid material of single-walled carbon nanotubes with an amphiphilic poly(phenyleneethynylene): Preparation, characterization, and photovoltaic properties. *Journal of Nanoscience and Nanotechnology*, 2007, *7* (8), 2709–2718.

155. Chen, J.; Liu, H. Polymer and method for using the polymer for solubilizing nanotubes. U.S. Patent 7244407, 2007.

156. Wang, G. J.; Huang, S. Z.; Wang, Y.; Liu, L.; Qiu, J.; Li, Y. Synthesis of water-soluble single-walled carbon nanotubes by RAFT polymerization. *Polymer*, 2007, *48* (3), 728–733.

157. You, Y.-Z.; Hong, C.-Y.; Pan, C.-Y. Directly growing ionic polymers on multi-walled carbon nanotubes via surface RAFT polymerization. *Nanotechnology*, 2006, *17* (9), 2350–2354.

158. Xu, G.; Wu, W.-T.; Wang, Y.; Pang, W.; Wang, P.; Zhu, Q.; Lu, F. Synthesis and characterization of water-soluble multiwalled carbon nanotubes grafted by a thermoresponsive polymer. *Nanotechnology*, 2006, *17* (10), 2458–2465.

159. Ni, Y.; Hu, H.; Malarkey, E. B.; Zhao, B.; Montana, V.; Haddon, R. C.; Parpura, V. Chemically functionalized water soluble single-walled carbon nanotubes modulate neurite outgrowth. *Journal of Nanoscience and Nanotechnology*, 2005, *5* (10), 1707–1712.

160. Zhao, B.; Hu, H.; Yu, A.; Perea, D.; Haddon, R. C. Synthesis and characterization of water soluble single-walled carbon nanotube graft copolymers. *Journal of the American Chemical Society*, 2005, *127* (22), 8197–8203.

161. Zhang, H.; Li, H. X.; Cheng, H. M. Water-soluble multiwalled carbon nanotubes functionalized with sulfonated polyaniline. *Journal of Physical Chemistry B*, 2006, *110* (18), 9095–9099.

162. Ma, Y.; Ali, S. R.; Wang, L.; Chiu, P. L.; Mendelsohn, R.; He, H. In situ fabrication of a water-soluble, self-doped polyaniline nanocomposite: The unique role of DNA functionalized single-walled carbon nanotubes. *Journal of the American Chemical Society*, 2006, *128* (37), 12064–12065.

163. Li, Z.; Dong, Y.; Häussler, M.; Lam, J. W. Y.; Dong, Y.; Wu, L.; Wong, K. S.; Tang, B. Z. Synthesis of, light emission from, and optical power limiting in soluble single-walled carbon nanotubes functionalized by disubstituted polyacetylenes. *Journal of Physical Chemistry B*, 2006, *110* (5), 2302–2309.

164. Tang, B.-Z.; Xu, H. Preparation, alignment, and optical properties of soluble polyphenylacetylene-wrapped carbon nanotubes. *Preprints of Second East Asian Polymer Conference*, Kowloon, Hong Kong, January 12–16, 1999, Hong Kong University of Science and Technology, Kowloon, Hong Kong, 1999, pp. 141–148.

165. Tang, B. Z.; Xu, H. Preparation, alignment, and optical properties of soluble poly(phenylacetylene)-wrapped carbon nanotubes. *Macromolecules*, 2000, *32*, 2569–2576.

166. Wang, Z.; Liu, Q.; Zhu, H.; Liu, H.; Chen, Y.; Yang, M. Dispersing multi-walled carbon nanotubes with water-soluble block copolymers and their use as supports for metal nanoparticles. *Carbon*, 2007, *45* (2), 285–292.

167. Fernando, K. A. S.; Lin, Y.; Zhou, B.; Grah, M.; Joseph, R.; Allard, L. F.; Sun, Y.-P. Poly(ethylene-co-vinyl alcohol) functionalized single-walled carbon nanotubes and related nanocomposites. *Journal of Nanoscience and Nanotechnology*, 2005, *5* (7), 1050–1054.

168. Paiva, M. C.; Zhou, B.; Fernando, K. A. S.; Lin, Y.; Kennedy, J. M.; Sun, Y.-P. Mechanical and morphological characterization of polymer-carbon nanocomposites from functionalized carbon nanotubes. *Carbon*, 2004, *42* (14), 2849–2854.

169. Zhang, N.; Xie, J.; Varadan, V. K. Soluble functionalized carbon nanotube/poly(vinyl alcohol) nanocomposite as the electrode for glucose sensing. *Smart Materials and Structures*, 2006, *15* (1), 123–128.

170. Zhou, B.; Lin, Y.; Veca, L. M.; Fernando, K. A. S.; Harruff, B. A.; Sun, Y.-P. Luminescence polarization spectroscopy study of functionalized carbon nanotubes in a polymeric matrix. *Journal of Physical Chemistry B*, 2006, *110* (7), 3001–3006.

171. Choi, J. Y.; Han, S. W.; Huh, W. S.; Tan, L. S.; Baek, J. B. In situ grafting of carboxylic acid-terminated hyperbranched poly(ether-ketone) to the surface of carbon nanotubes. *Polymer*, 2007, *48* (14), 4034–4040.

172. Qu, L.; Veca, L. M.; Lin, Y.; Kitaygorodskiy, A.; Chen, B.; McCall, A. M.; Connell, J. W.; Sun, Y.-P. Soluble nylon-functionalized carbon nanotubes from anionic ring-opening polymerization from nanotube surface. *Macromolecules*, 2005, *38*, 10328–10331.

173. Guo, J.; Li, Y.; Wu, S.; Li, W. The effects of γ-irradiation dose on chemical modification of multi-walled carbon nanotubes. *Nanotechnology*, 2005, *16* (10), 2385–2388.

174. Hurd, R. E. Magnetic resonance imaging (MRI) agents: water soluble carbon-13 enriched fullerene and carbon nanotubes for use with dynamic nuclear polarization. U.S. Patent 20070025918, 2007.

175. Hurd, R. E. Magnetic resonance imaging (MRI) agents: Water soluble carbon-13 enriched fullerene and carbon nanotubes for use with dynamic nuclear polarization. WO08008075, 2008.

176. McDevitt, M. R.; Chattopadhyay, D.; Jaggi, J. S.; Finn, R. D.; Zanzonico, P. B.; Villa, C.; Rey, D. et al. PET imaging of soluble yttrium-86-labeled carbon nanotubes in mice. *PLoS ONE*, 2007, *2* (9), e907.

177. Méndez-Rojas, M. A.; Kharisov, B. I.; Tsivadze, A. Y. Recent advances on technetium complexes: Coordination chemistry and medical applications. *Journal of Coordination Chemistry*, 2006, *59* (1), 1–63.

178. Guo, J.; Zhang, X.; Li, Q.; Li, W. Biodistribution of functionalized multiwall carbon nanotubes in mice. *Nuclear Medicine and Biology*, 2007, *34* (5), 579–583.

179. Wang, H.; Wang, J.; Deng, X.; Sun, H.; Shi, Z.; Gu, Z.; Liu, Y.; Zhao, Y. Biodistribution of carbon single-wall carbon nanotubes in mice. *Journal of Nanoscience and Nanotechnology*, 2004, *4* (8), 1019–1024.

180. Imasaka, K.; Suehiro, J.; Kanatake, Y.; Kato, Y.; Hara, M. Preparation of water-soluble carbon nanotubes using a pulsed streamer discharge in water. *Nanotechnology*, 2006, *17* (14), 3421–3427.

181. Imasaka, K.; Kato, Y.; Suehiro, J., Enhancement of microplasma-based water-solubilization of single-walled carbon nanotubes using gas bubbling in water. *Nanotechnology*, 2007, *18* (33), 335602.

182. Wang, Y.; Iqbal, Z.; Mitra, S. Microwave-induced rapid chemical functionalization of single-walled carbon nanotubes. *Carbon*, 2005, *43* (5), 1015–1020.

183. Mitra, S.; Iqbal, Z. Microwave induced functionalization of single wall carbon nanotubes and composites prepared therefrom. WO06099392, 2006.

184. Wu, H. X.; Qiu, X. Q.; Cao, W. M.; Lin, Y. H.; Cai, R. F.; Qian, S. X. Polymer-wrapped multiwalled carbon nanotubes synthesized via microwave-assisted in situ emulsion polymerization and their optical limiting properties. *Carbon*, 2007, *45* (15), 2866–2872.

185. Li, J.; Grennberg, H. Microwave-assisted covalent sidewall functionalization of multiwalled carbon nanotubes. *Chemistry*, 2006, *12* (14), 3869–3875.

186. Lee, J.; Jeong, T.; Heo, J.; Park, S. H.; Lee, D.; Park, J. B.; Han, H.; Yoon, S. M. Short carbon nanotubes produced by cryogenic crushing. *Carbon*, 2006, *44* (14), 2984–2989.

187. Ziegler, K. J.; Schmidt, D. J.; Rauwald, U.; Shah, K. N.; Flor, E. L.; Hauge, R. H.; Smalley, R. E. Length-dependent extraction of single-walled carbon nanotubes. *Nano Letters*, 2005, *5* (12), 2355–2359.

188. Zhao, B.; Hu, H.; Niyogi, S.; Itkis, M. E.; Hamon, M. A.; Bhowmik, P.; Meier, M. S.; Haddon, R. C. Chromatographic purification and properties of soluble single-walled carbon nanotubes. *Journal of the American Chemical Society*, 2001, *123* (47), 11673–11677.

189. Niyogi, S.; Hu, H.; Hamon, M. A.; Bhowmik, P.; Zhao, B.; Rozenzhak, S. M.; Chen, J.; Haddon, R. C. Chromatographic purification of soluble single-walled carbon nanotubes (s-SWNTS). *Journal of the American Chemical Society*, 2001, *123* (4), 733–734.

190. Tagmatarchis, N.; Zattoni, A.; Reschiglian, P.; Prato, M. Separation and purification of functionalised water-soluble multi-walled carbon nanotubes by flow field-flow fractionation. *Carbon*, 2005, *43* (9), 1984–1989.

191. Jia, H.; Lian, Y.; Ishitsuka, M. O.; Nakahodo, T.; Maeda, Y.; Tsuchiya, T.; Wakahara, T.; Akasaka, T. Centrifugal purification of chemically modified single-walled carbon nanotubes. *Science and Technology of Advanced Materials*, 2005, *6* (6), 571–581.

192. Lian, Y.; Maeda, Y.; Wakahara, T.; Nakahodo, T.; Akasaka, T.; Kazaoui, S.; Minami, N.; Tokumoto, H. Spectroscopic study on the centrifugal fractionation of soluble single-walled carbon nanotubes. *Carbon*, 2005, *43* (13), 2750–2759.

193. Hamon, M. A.; Stensaas, K. L.; Sugar, M. A.; Tumminello, K. C.; Allred, A. K. Reacting soluble single-walled carbon nanotubes with singlet oxygen. *Chemical Physics Letters*, 2007, *447* (1), 1–4.

194. Li, M.; Boggs, M.; Beebe, T. P.; Huang, C. P. Oxidation of single-walled carbon nanotubes in dilute aqueous solutions by ozone as affected by ultrasound. *Carbon*, 2008, *46* (3), 466–475.

195. Ziegler, K. J.; Shaver, J.; Hauge, R. H.; Smalley, R. E.; Marek, I. M. Ozonation of carbon nanotubes in fluorocarbons. U.S. Patent 20060159612, 2006.

196. Smalley, R. E.; Ziegler, K. J.; Shaver, J.; Hauge, R. H.; Marek, I. M. Ozonation of carbon nanotubes in fluorocarbons. WO07050096, 2007.

197. Song, C.; Pehrsson, P. E.; Zhao, W., Recoverable solution reaction of HiPco carbon nanotubes with hydrogen peroxide. *Journal of Physical Chemistry B*, 2005, *109* (46), 21634–21639.

198. Niu, C.; Moy, D.; Ma, J.; Chishti, A. Modification of nanotubes oxidation with peroxygen compounds. U.S. Patent 6872681, 2005.

199. Rosca, I. D.; Watari, F.; Uo, M.; Akasaka, T. Oxidation of multiwalled carbon nanotubes by nitric acid. *Carbon*, 2005, *43* (15), 3124–3131.

200. Salzmann, C. G.; Chu, B. T. T.; Tobias, G.; Llewellyn, S. A.; Green, M. L. H. Quantitative assessment of carbon nanotube dispersions by Raman spectroscopy. *Carbon*, 2007, *45* (5), 907–912.

201. Chen, Y.; Chen, J.; Hu, H.; Hamon, M. A.; Itkis, M. E.; Haddon, R. C. Solution-phase EPR studies of single-walled carbon nanotubes. *Chemical Physics Letters*, 1999, *299* (6), 532–535.

202. Gallo, M.; Favila, A.; Glossman-Mitnik, D. DFT studies of functionalized carbon nanotubes and fullerenes as nanovectors for drug delivery of antitubercular compounds. *Chemical Physics Letters*, 2007, *447* (1), 105–109.

203. Paolucci, D.; Marcaccio, M.; Bruno, C.; Paolucci, F.; Tagmatarchis, N.; Prato, M. Voltammetric quantum charging capacitance behaviour of functionalised carbon nanotubes in solution. *Electrochimica Acta*, 2008, *53* (11), 4059–4064.

204. Rivas, G. A.; Rubianes, M. D.; Rodriguez, M. C.; Ferreyra, N. F.; Luque, G. L.; Pedano, M. L.; Miscoria, S. A.; Parrado, C. Carbon nanotubes for electrochemical biosensing. *Talanta*, 2007, *74* (3), 291–307.

205. Melle-Franco, M.; Marcaccio, M.; Paolucci, D.; Paolucci, F.; Georgakilas, V.; Guldi, D. G.; Prato, M.; Zerbetto, F. Cyclic voltammetry and bulk electronic properties of soluble carbon nanotubes. *Journal of the American Chemical Society*, 2004, *126* (6), 1646–1647.

206. Kitaygorodskiy, A.; Wang, W.; Xie, S.-Y.; Lin, Y.; Fernando, K. A. S.; Wang, X.; Qu, L.; Sun, Y.-P. NMR detection of single-walled carbon nanotubes in solution. *Journal of the American Chemical Society*, 2005, *127* (20), 7517–7520.

207. Torrens, F. Calculations on solvents and co-solvents of single-wall carbon nanotubes: Cyclopyranoses. *Theochem*, 2005, *757* (1), 183–191.

208. Ham, H. T.; Choi, Y. S.; Chung, I. J. An explanation of dispersion states of single-walled carbon nanotubes in solvents and aqueous surfactant solutions using solubility parameters. *Journal of Colloid and Interface Science*, 2005, *286* (1), 216–223.

209. Martí, J.; Gordillo, M. C. Structure and dynamics of liquid water adsorbed on the external walls of carbon nanotubes. *Journal of Chemical Physics*, 2003, *119* (23), 12540–12546.

210. Chang, J.-Y.; Ghule, A.; Chang, J.-J.; Tzing, S.-H.; Ling, Y.-C. Opening and thinning of multiwall carbon nanotubes in supercritical water. *Chemical Physics Letters*, 2002, *363* (5), 583–590.

211. Benedict, B.; Pehrsson, P. E.; Zhao, W. Optically sensing additional sonication effects on dispersed HiPco nanotubes in aerated water. *Journal of Physical Chemistry B*, 2005, *109* (16), 7778–7780.

212. Lisunova, M. O.; Lebovka, N. I.; Melezhyk, O. V.; Boiko, Y. P. Stability of the aqueous suspensions of nanotubes in the presence of nonionic surfactant. *Journal of Colloid and Interface Science*, 2006, *299* (2), 740–746.

213. Dumortier, H.; Lacotte, S.; Pastorin, G.; Marega, R.; Wu, W.; Bonifazi, D.; Briand, J.-P.; Bianco, A. Functionalized carbon nanotubes are non-cytotoxic and preserve the functionality of primary immune cells. *Nano Letters*, 2006, *6* (7), 1522–1528.

214. Zhang, Y.; Tan, S.; Lopez, H. Determination of carbon nanotube concentration in a solution by fluorescence measurement. U.S. Patent 20060141634, 2006.

215. Chan, V. S. W., Nanomedicine: An unresolved regulatory issue. *Regulatory Toxicology and Pharmacology*, 2006, *46* (3), 218–224.

216. Liu, Z.; Sun, X.; Nakayama-Ratchford, N.; Dai., H. Supramolecular chemistry on water-soluble carbon nanotubes for drug loading and delivery. *ACS Nano*, 2007, *1* (1), 50–56.

217. Klumpp, C.; Kostarelos, K.; Prato, M.; Bianco, A. Functionalized carbon nanotubes as emerging nanovectors for the delivery of therapeutics. *BBA—Biomembranes*, 2006, *1758* (3), 404–412.

218. Bianco, A.; Kostarelos, K.; Prato, M. Applications of carbon nanotubes in drug delivery. *Current Opinion in Chemical Biology*, 2005, *9* (6), 674–679.

219. Fu, K.; Huang, W.; Lin, Y.; Zhang, D.; Hanks, T. W.; Rao, A. M.; Sun, Y.-P. Functionalization of carbon nanotubes with bovine serum albumin in homogeneous aqueous solution. *Journal of Nanoscience and Nanotechnology*, 2002, *2* (5), 457–461.

220. Zhang, L. W.; Zeng, L.; Barron, A. R.; Monteiro-Riviere, N. A. Biological interactions of functionalized single-wall carbon nanotubes in human epidermal keratinocytes. *International Journal of Toxicology*, 2007, *26* (2), 103–113.

221. Kawaguchi, M.; Fukushima, T.; Hayakawa, T.; Nakashima, N.; Inoue, Y.; Takeda, S.; Okamura, K.; Taniguchi, K. Preparation of carbon nanotube-alginate nanocomposite gel for tissue engineering. *Dental Materials Journal*, 2006, *25* (4), 719–725.

222. Gannon, C. J.; Cherukuri, P.; Yakobson, B. I.; Cognet, L.; Kanzius, J. S.; Kittrell, C.; Weisman, B. R.; Curley, S. A. Carbon nanotube-enhanced thermal destruction of cancer cells in a noninvasive radiofrequency field. *Cancer*, 2007, *110* (12), 2654–2665.

223. Feazell, R. P.; Nakayama-Ratchford, N.; Dai, H.; Lippard, S. J. Soluble single-walled carbon nanotubes as longboat delivery systems for platinum(IV) anticancer drug design. *Journal of the American Chemical Society*, 2007, *129*, 8438–8439.

224. Hernandez-Lopez, J. L.; Alvizo-Paez, E. R.; Moya, S. E.; Ruiz-Garcia, J. Ordered carbon nanotube thin films produced by the trapping of water-soluble single-wall carbon nanotubes at the air/water interface. *Carbon*, 2007, *45* (12), 2448–2450.

225. Li, J.; Zhang, Y. Large-scale aligned carbon nanotubes films. *Physica E: Low-Dimensional Systems and Nanostructures*, 2006, *33* (1), 235–239.

226. Rozhin, A. G.; Sakakibara, Y.; Tokumoto, M.; Kataura, H.; Y Achiba, Y. Near-infrared nonlinear optical properties of single-wall carbon nanotubes embedded in polymer film. *Thin Solid Films*, 2004, *464*, 368–372.

227. Yu, K.; Zhu, Z.; Xu, M.; Li, Q.; Lu, W.; Chen, Q. Soluble carbon nanotube films treated using a hydrogen plasma for uniform electron field emission. *Surface and Coatings Technology*, 2004, *179* (1), 63–69.

228. Hu, C.; Chen, X.; Hu, S. Water-soluble single-walled carbon nanotubes films: Preparation, characterization and applications as electrochemical sensing films. *Journal of Electroanalytical Chemistry*, 2006, *586* (1), 77–85.

229. Tulevski, G. S.; Hannon, J.; Afzali, A.; Chen, Z.; Avouris, P.; Kagan, C. R. Chemically assisted directed assembly of carbon nanotubes for the fabrication of large-scale device arrays. *Journal of the American Chemical Society*, 2007, *129* (39), 11964–11968.
230. Panhuis, M. I. H.; Wu, J.; Ashraf, S. A.; Wallace, G. G. Conducting textiles from single-walled carbon nanotubes. *Synthetic Metals*, 2007, *157* (8), 358–362.
231. Huang, X.; Kobos, R. K.; Xu, G. Hair coloring and cosmetic compositions comprising carbon nanotubes. U.S. Patent 7276088, 2007.
232. Miller, A. J.; Hatton, R. A.; Silva, S. R. P. Interpenetrating multiwall carbon nanotube electrodes for organic solar cells. *Applied Physics Letters*, 2006, *89*, 133117.

17 Graphene

Carbon is one of the most interesting elements in the periodic table. It forms a series of allotropes; some of them are known for thousands of years (3D diamond and 2D graphite) and others were discovered 10–20 years ago (0D fullerenes and 1D carbon nanotubes). Its new allotropic form, 2D graphene (nanographenes are also known as polycyclic aromatic hydrocarbons [PAHs]), discovered in 2004, is a rapidly rising star on the horizon of materials science and condensed-matter physics. This strictly 2D material exhibits exceptionally high crystal and electronic quality, and, despite its short history, has already revealed a cornucopia of new physics and potential applications, which are briefly discussed here. Graphene represents a conceptually new class of materials that are only one atom thick (it is just one layer of carbon atoms,[1] a similar structure to graphite but is a single isolated sheet of carbon), and, on this basis, offers new inroads into low-dimensional physics that has never ceased to surprise and continues to provide a fertile ground for applications.[2] Strictly 2D crystals, such as planar graphene, have for a long time been considered as a thermodynamically unstable form with respect to the formation of curved structures such as fullerenes or nanotubes. Geim and Novoselov described[2] that small size (≪1 mm) and strong interatomic bonds ensure that thermal fluctuations cannot lead to the generation of dislocations or other crystal defects even at elevated temperature. The extracted 2D crystals become intrinsically stable by gentle crumpling in the third dimension; such 3D warping (observed on a lateral scale of ≈10 nm) leads to a gain in elastic energy but suppresses thermal vibrations (anomalously large in 2D), which above a certain temperature can minimize the total free energy (see Ref. [2] and references therein).

Despite the fact that graphene was discovered recently, its structure, properties, and applications have already been generalized in a series of monographs[3–6] and reviews[7–10] (among which we note an excellent work of Mullen on graphenes as potential material for electronics[11]). Various patents are dedicated to obtaining graphenes.[12–14] Among experimental works, we note a number of investigations of Novoselov,[15–19] who precisely discovered this material. At present, the area of graphene is one of the most popular topics in physics and nanotechnology.[20]

17.1 STRUCTURE AND PROPERTIES

Graphene is a layer of carbon atoms, connected in a hexagonal 2D crystalline lattice, that is, a plane, consisting of hexagonal cells (Figure 17.1). It can be considered as one graphite plane, separated from the voluminous crystal. Two atoms A and B are situated in an elementary cell of the crystal. Each one, at a displacement on the translation vectors, forms a sublattice from the atoms, equivalent to it, that is, crystal properties are nondependent on observation points, situated in the equivalent places of the crystal. The distance between the closest carbon atoms in graphene a_0 is 0.142 nm (Figure 17.2).

An ideal graphene consists exclusively of hexagonal cells. The presence of penta- or heptagonal cells leads to various types of defects.[21] The appearance of pentagonal cells leads to a turning of the atomic plane to a cone. Examining 12 such defects simultaneously, the fullerene structure is formed. The presence of heptagonal cells leads to the formation of saddle-type distortions of the atomic plane. Combinations of these defects and normal cells could lead to the formation of distinct surface forms. Additional data of graphene forms are given in Refs. [22–24]. A series of 2D sheets of hexagonal carbon rings, with hydrogens around the edges, have been investigated computationally as models for graphene.[25]

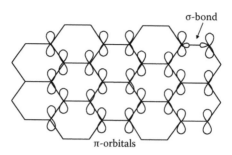

FIGURE 17.1 Structure of graphene.

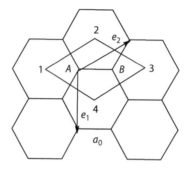

FIGURE 17.2 Hexagonal cell of graphene. e_1 and e_2 are translation vectors; the rhomb 1234 is the elemental cell.

There are a series of *graphene derivatives* reported; the basic ones are presented in Ref. [26]: $C_{62}H_{20}$ (a "flake" of graphite with hydrogen used to terminate the dangling bonds), graphene with adatom $C_{63}H_{20}$ (a twofold coordinated additional carbon atom on a graphene sheet; represents one of the stable structures of a carbon interstitial on graphite), graphene with adatom pair $C_{64}H_{20}$ (two additional carbon atoms on a graphene sheet; represents an interstitial pair on the surface of graphite), and graphene with CO_3 molecule. One important example is D_{6h} symmetric hexa-*peri*-hexabenzocoronene (HBC, "super-benzene," Figure 17.3a), containing 42 carbon atoms, its symmetric and asymmetric derivatives with a variety of functional groups (Figure 17.3b and c), and oligomers (Figure 17.4).

A series of distinct larger graphenes, containing 90, 96, 132, 150, and 222 carbon atoms in their cores and different substituents, are reviewed in Ref. [11]; some of them are as follows (Figure 17.5):

The *number of graphene layers* (sheets) is an important magnitude, which can be exactly counted by various methods. One of them is based on the quantitative analysis of electron diffraction intensity and allows detection of a single graphene sheet in a carbon nanofilm.[27] In another report,[28] the single-, bilayer-, and multiple-layer graphenes (<10 layers) are clearly discriminated on an Si substrate with a 285 nm SiO_2 capping layer by using contrast spectra, which were generated from the reflection light of a white light source. Two easy-to-use methods to determine the number of graphene layers based on contrast spectra are provided: a graphic method and an analytical method. The authors mention that the refractive index of graphene is different from that of graphite. Graphene and graphene layers were investigated on different substrates by monochromatic and white-light confocal Rayleigh scattering microscopy.[29] It was established that for a few layers (<6), the monochromatic contrast increases linearly with thickness. Rayleigh imaging is a general, simple, and quick tool to identify graphene layers, which is readily combined with Raman scattering, that provides structural identification. The intershell and interlayer interaction (complexation) energies of graphene sheets are investigated by all-electron density functional theory

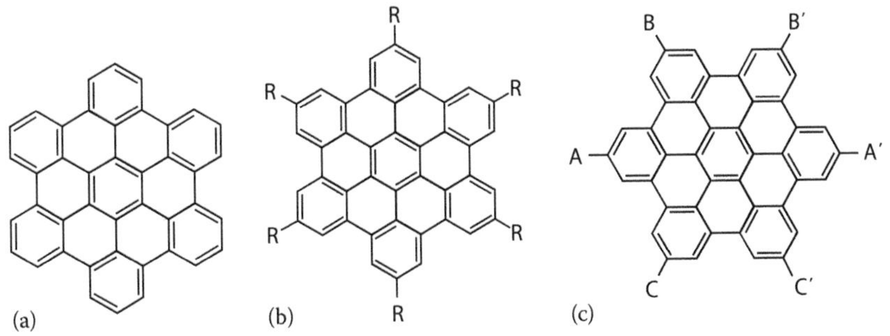

FIGURE 17.3 (a) Hexa-*peri*-hexabenzocoronene and (b,c)—its symmetric and asymmetric derivatives, respectively.

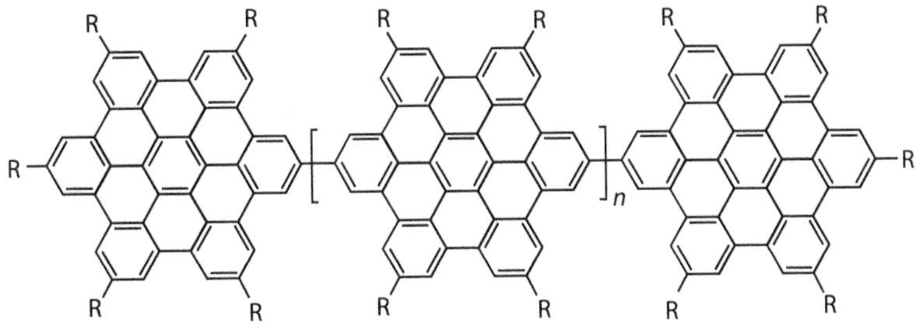

FIGURE 17.4 HBC oligomers.

(DFT) using generalized gradient approximation (GGA) functionals and an empirical correction for dispersion (van der Waals) effects (DFT-D method).[30] The theoretical approach is first applied to graphene sheet model dimers of increasing size (up to $(C_{216}H_{36})_2$). The interaction energies are extrapolated to an infinite lateral size of the sheets. The value of −66 meV/atom obtained for the interaction energy of two sheets supports the most recent experimental estimate for the exfoliation energy of graphite (−52 ± 5 meV/atom). The interlayer equilibrium distance (334 ± 3 Pm) is also obtained accurately.

One of the most frequently reported graphene materials is a graphene sheet supported by an insulating silicon dioxide substrate; its properties are intensively studied. Thus, scanning probe microscopy was employed to reveal atomic structures and nanoscale morphology of graphene-based electronic devices.[31] Atomic resolution scanning tunneling microscopy images reveal the presence of a strong spatially dependent perturbation, which breaks the hexagonal lattice symmetry of the graphitic lattice. A novel cleaning process to produce atomically clean graphene sheets was offered.

Electronic properties of graphenes are described in a series of publications.[32–35] Electrons in graphene, obeying a linear dispersion relation, behave like massless relativistic particles or quantum billiard balls. This results in the observation of a number of very peculiar electronic properties—from an anomalous quantum Hall effect to the absence of localization—in this, the first 2D material. It also provides a bridge between condensed matter physics and quantum electrodynamics and opens up new perspectives for carbon-based electronics.[36] Electrons in bilayer graphene possess an unusual property: they are chiral quasiparticles characterized by Berry phase 2π. Researchers at The University of Manchester have just found that electrons move more easily in graphene than all other materials, including gold, silicon, gallium arsenide, and carbon nanotubes,[37] and have singled graphene out as the "best possible" material for electronic applications. With a high electronic quality—measured at around 200,000 cm²/V s and more than 100 times higher than for silicon—these researchers believe graphene has the potential to improve upon the capabilities of current semiconductors and open up exciting new possibilities.

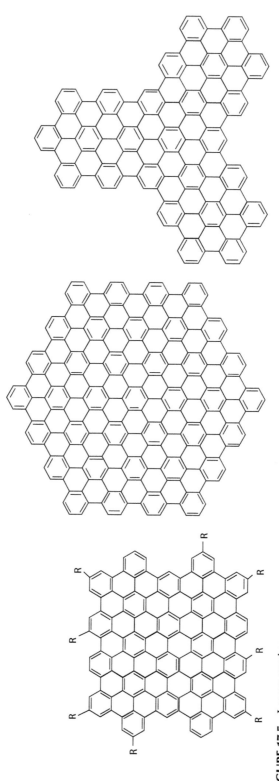

FIGURE 17.5   Larger graphenes.

The tight-binding model of bilayer graphene, which determines the band structure and low-energy quasiparticle properties of this material, is reviewed in Ref. [38]. A comprehensive theoretical study of the electronic properties and relative stabilities of edge-oxidized zigzag graphene nanoribbons (GNRs) is presented in Ref. [39]. The oxidation schemes considered include hydroxyl, lactone, ketone, and ether groups. Using screened exchange DFT, it was shown that these oxidized ribbons are more stable than hydrogen-terminated nanoribbons except for the case of the etheric groups. Electronic transport properties of individual chemically reduced graphene oxide sheets are reported in Ref. [40].

Magnetic properties of arbitrarily shaped finite graphene fragments (referred to by authors as graphene nanoflakes [GNFs]) were investigated using benzenoid graph theory and first-principles calculations.[41] It was demonstrated that the spin of a GNF depends on its shape due to topological frustration of the π-bonds. In general, the principle of topological frustration can be used to introduce large net spin and interesting spin distributions in graphene. In another report,[42] it was found that an antiferromagnetic (AFM) phase appears as the PAH reaches a certain size. This AFM phase in PAHs has the same origin as the one in infinitely long zigzag-edged GNR, namely, from the localized electronic state at the zigzag edge. The smallest PAH still having an AFM ground state is identified. Magnetism in graphene nanoislands is examined in Ref. [43].

Solution properties of graphite are reported in Ref. [44]. It was suggested that solutions of functionalized oxidized graphite in common organic solvents consist of single and few layer graphene sheets. Soluble graphene layers were also formed by reacting graphite fluoride with alkyl lithium reagents.[45] Covalent attachment of alkyl chains to the graphene layers was confirmed. It was revealed that the chemical process partially restores the sp^2 carbon network. A one-step chemical treatment of graphite fluoride allows the manipulation of a soluble form of graphene.

Optical properties of armchair-edged graphene nanoribbons (AGNRs) with many-electron effects included are reported in Ref. [46]. The characteristics of the excitons of the three distinct families of AGNRs are compared and discussed. The enhanced excitonic effects found here are expected to be of importance in optoelectronic applications of graphene-based nanostructures. A simple optical method is presented for identifying and measuring the effective optical properties of nanometer-thick, graphene-based materials, based on the use of substrates consisting of a thin dielectric layer on silicon.[47] The effective refractive index and optical absorption coefficient of graphene oxide, thermally reduced graphene oxide, and graphene are obtained by comparing the predicted and measured contrasts.

Raman spectroscopy measurements of graphenes are reported in a number of reports.[48–50] Several authors[51] investigated the influence of substrates (GaAs, sapphire, glass, and standard Si/SiO$_2$ [300 nm, served as a reference] substrates) on Raman scattering spectrum from graphene. It was found that while G peak of graphene on Si/SiO$_2$ and GaAs is positioned at 1580 cm^{-1}, it is down-shifted by ~5 cm^{-1} for graphene on sapphire and, in some cases, splits into doublets for graphene on glass with the central frequency around 1580 cm^{-1}.

Capacitance–voltage (*C–V*) characteristics are important for understanding fundamental electronic structures and device applications of nanomaterials. The *C–V* characteristics of GNRs are examined using self-consistent atomistic simulations. The results indicate the strong dependence of the GNR *C–V* characteristics on the edge shape.[52] GNR are also examined in Refs. [53–58]. It is noted that all GNR are semiconductors and they have possible applications to future quantum device. The zigzag edge of a graphene nanoribbon possesses a unique electronic state that is near the Fermi level and localized at the edge carbon atoms.[59] The chemical reactivity of these zigzag edge sites was elucidated by examining their reaction energetics with common radicals from first principles. Among other studies of this form of carbon, plasma waves in graphenes are reported in Refs. [60,61].

17.2 SYNTHESIS

The principal method for graphene production[62] is based on the mechanical exfoliation of graphite layers by micromechanical cleavage, thus obtaining high-quality samples. Thin graphite layers are put between sticky tapes and thin graphite layers are removed to get a sufficiently thin layer. Then the sample is collocated on the support of SiO_2. AFM studies are used for determining real thickness of the graphene film (0.35–0.8 nm).

Graphenes are also produced by a series of other methods, in particular by *CVD/pyrolisis techniques* as simple, economical, and reproducible methods. Thus, planer few-layer graphene films were synthesized from camphor pyrolysis on nickel substrates by a simple, cost-effective thermal CVD method.[63] Thermal decomposition of the (0001) face of a 6H–SiC wafer demonstrated the successful growth of single-crystalline films down to approximately one graphene layer.[64] A three-step process for producing nanoscale graphene plate material is reported in Ref. [65]. It includes carbonizing a polymer or heat-treating petroleum or coal tar pitch, producing micron- and nanometer-scaled graphite crystallites, exfoliating the graphite crystallites in the polymeric carbon, and subjecting the polymeric carbon containing exfoliated graphite crystallites to a mechanical attrition treatment.

Graphene films can be obtained on various supports. Thus, the synthesis of bilayer graphene thin films deposited on insulating silicon carbide is reported in Refs. [66–68]. The formed bilayer graphene can potentially possess switching functions in atomic-scale electronic devices. It is shown[69] that graphenes grown from the SiC(000$\bar{1}$) (C-terminated) surface are of higher quality than those previously grown on SiC(0001). Graphene grown on the C face can have structural domain sizes more than three times larger than those grown on the Si face while at the same time reducing SiC substrate disorder from sublimation by an order of magnitude. As noted,[70] despite the thickness of graphite layer being more than one monolayer, only the closest layer to the supporting surface takes part in conductivity, since a noncompensated charge is formed in the border SiC-C due to the differences in the output of two materials.

Graphite platelets of 1–5 μm diameter consisting of a few graphenes were generated from commercially available exfoliated graphite by *ultrasonic treatment* in benzene (1 mg material in 20 mL solvent) for 3 h. Successive oxidation of the sample was carried out at 450°C–550°C in air. AFM measurements showed that the thermal oxidation removed two to three graphenes from the platelets and it left behind single graphene layers.[71]

Chemical techniques are also used for obtaining nanometric graphene layers. Thus,[44] graphite microcrystals are treated with a mixture of HCl and H_2SO_4, forming carboxyl groups in the sample edges (Figure 17.6). Then they are transformed into chlorides using $SOCl_2$ and graphene layers

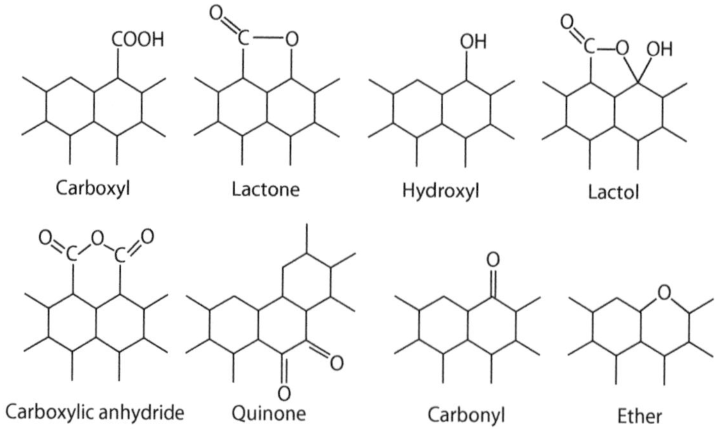

FIGURE 17.6 Surface oxygen groups on graphite.

by the action of octadecylamine in organic solvents. Novel tube-in-tube nanostructures are similarly obtained in Ref. [72]. Small graphene sheets were produced by disintegration of the graphitic nanoparticles *via* an intercalation-exfoliation process with nitric acid, during which the graphene sheets were simultaneously modified with carboxyl and hydroxyl groups at their edges. The reduction of a colloidal suspension of exfoliated graphene oxide sheets in water with hydrazine hydrate results in their aggregation and subsequent formation of a high-surface-area carbon material that consists of thin graphene-based sheets.[73] A comprehensive generalization of techniques for obtaining various, in particular larger, graphenes is given in Ref. [11].

The inclusion of heteroatoms into the graphene molecules offers the possibility of making novel PAH-based metal complexes. In the case of nitrogen as a heteroatom, oxidative cyclodehydrogenation is an important process in the formation of the new graphene of the "half-cyclized" nitrogen-heterosuperbenzene N-½HSB.[74] This heteropolyaromatic results from the FeCl$_3$-catalyzed oxidative cyclodehydrogenation of 1,2-dipyrimidyl-3,4,5,6-tetra-(4-*tert*-butylphenyl)benzene. Three new C–C bonds are formed that lock the two pyrimidines in a molecular platform comprising eight fused aromatic rings flanked by two remaining "uncyclized" phenyl rings. Heteroleptic Ru(II) complexes, [Ru(bpy)$_2$(N-½HSB)](PF$_6$)$_2$ and [Ru(bpy)$_2$(N-HSB)](PF$_6$)$_2$, which differ in the size and planarity of their aromatic ligands, were also obtained (Figure 17.7a and b). Palladium(II) complex [Pd(η^3-C$_3$H$_5$)(tetra-*peri*-(*tert*-butyl-benzo)-di-*peri*-(pyrimidino)-coronene)]PF$_6$] is also known (Figure 17.7c).[75]

Oxidation of graphite produces graphite oxide (GO, Figures 17.8 and 17.9), which is dispersible in water as individual platelets. After deposition onto Si/SiO$_2$ substrates, its chemical reduction

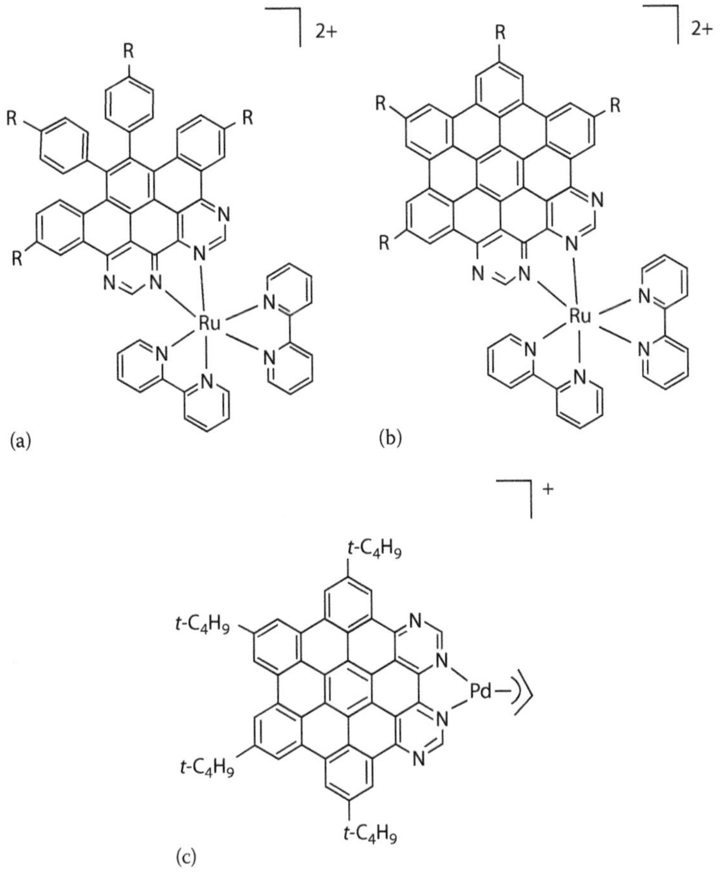

FIGURE 17.7   (a,b) Ruthenium– and (c) palladium–graphene complexes.

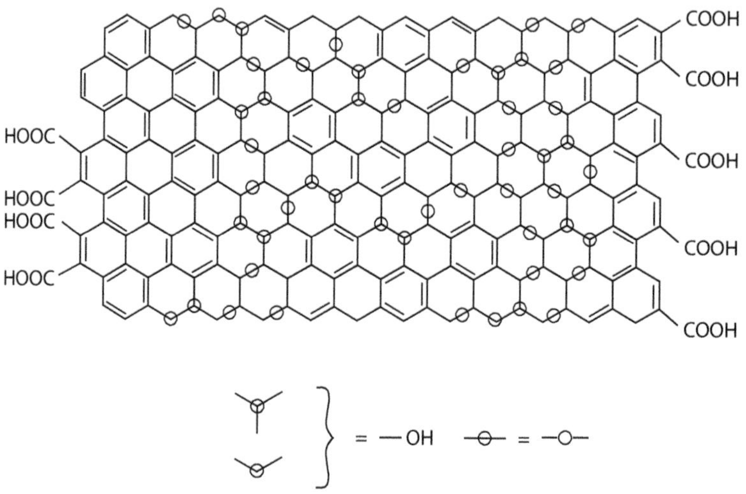

FIGURE 17.8 Structural model for graphite oxide. (From Lerf, A. et al., *J. Phys. Chem. B*, 102, 4477, 1998.)

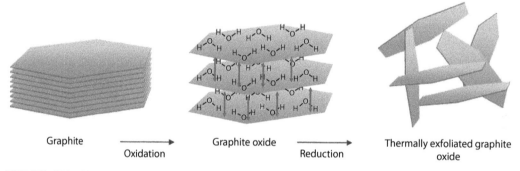

Graphite → Oxidation → Graphite oxide → Reduction → Thermally exfoliated graphite oxide

FIGURE 17.9 Formation and exfoliation of graphite oxide. (From Aksay, I. A. Developing fundamental insights pertinent to the design and processing of multifunctional nanocomposites. http://www.wtec.org/nb-france-us/aksay_i/aksay-modified.pdf, posted on March 2, 2006. With permission.)

produces graphene sheets. Electrical conductivity measurements indicate a 10,000-fold increase in conductivity after chemical reduction to graphene.[76] The thermal expansion mechanism of GO to produce functionalized graphene sheets is provided.[77] After dispersion by ultrasonication in appropriate solvents, statistical analysis by atomic force microscopy shows that 80% of the observed flakes are single sheets. Thermal exfoliation of GO can produce single sheets of functionalized graphene.[78] Although GO is an insulator, functionalized graphene produced by this method is electrically conducting. A number of functionalized GO were prepared by treatment of GO with organic isocyanates.[79] These isocyanate-treated GOs (iGOs) can then be exfoliated into functionalized graphene oxide nanoplatelets that can form a stable dispersion in polar aprotic solvents. Another route to graphenes through intercalated graphite is reported in Ref. [80] (Figure 17.10). It includes subsequent treatment of graphite with potassium, EtOH, and further exfoliation and sonication, leading to carbon nanoscrolls through graphene sheets.[81]

A straightforward technique using *electrostatic attraction* is demonstrated to transfer graphene sheets, one to a few atomic layers thick, to a selected substrate. Graphite crystal was collocated between electrodes; graphite pieces, including thin films, can move to SiO_2 support due to the electric field. Sheets from 1 to 22 layers thick have been transferred by this method.[84] The applied voltage was 1–13 kV.

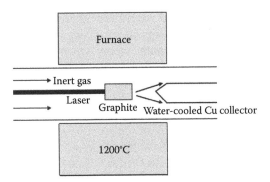

FIGURE 17.10 Nanographene growth using laser ablation method. (Reproduced with permission from http://www.wtec.org.)

Mechanisms of graphene formation and growth are examined in Ref. [85]. A reaction pathway for growth is explored in which two cyclopenta-carbon groups combine on the zigzag edge of a graphene layer.[86] The process is initiated by H addition to a five-member ring, followed by opening of that ring and the formation of a six-member ring adjacent to another five-member ring. Hydrogen abstraction reactions by methyl radicals on the zigzag and armchair edges of perylene are studied by DFT to explore various growth pathways that seem to be in line with experimental observations.[87] Notably, in the case of 3,4-dimethyl-3,4-dihydroperylene, the first two reaction steps have no or only a very low reaction barrier. The conclusion of this study is that a cascade of reactions is possible that leads to the growth of a graphene sheet on a graphite surface. The role of surface migration in the growth and structure of graphene layers is examined in Ref. [88].

Nanographene growth and texturing by Nd was investigated using pulsed *laser ablation* from a rotating graphite target operating both in vacuum ($\sim 10^{-5}$ Pa) and in He sustaining gas (~ 10 Pa).[89] As a result, thin carbon films grew on Si <100> substrates kept at temperatures from room temperature (r.t.) to 900°C (Figure 17.10). The preferential vertically oriented growth of graphene layers in vacuum and high temperature can be explained as a combined effect of different processes under a fast kinetic mode: thermal surface diffusion, in-plane growth of graphene sheets, and line source direction of activated carbon species of the laser plume. The deposited samples are characterized by a different nucleation and growth process and a more complex structure.

17.3 REACTIVITY

A certain attention is given to the interaction of graphene and simple inorganic molecules such as *hydrogen*,[90] oxygen, etc. Thus, quantum dynamics study of the Langmuir–Hinshelwood H + H recombination mechanism and H_2 formation on a graphene model surface is described in Ref. [91]. Two-dimensional semiconducting nanostructures based on single graphene sheets with lines of adsorbed hydrogen atoms were examined in Ref. [92]. It is shown that lines of adsorbed hydrogen pair atoms divide the graphene sheet into strips and form hydrogen-based superlattice structures (2HG-SL). The formation of 2HG-SL changes the electronic properties of graphene from a semimetal to semiconductor. In another research,[93] the physisorption of a hydrogen molecule on planar and curved graphene clusters is examined. The repulsive nature of a H_2–H_2 interaction and the small dependence of physisorption energies on physisorption sites favor a close-packed structure for molecular hydrogen physisorption on a planar graphene. It is shown that Li doping enhances hydrogen physisorption on graphene.[94] Hydrogen molecules are physisorbed on pure graphene with binding energies about 80–90 meV/molecule. However, the binding energies increase to 160–180 meV/molecule for many adsorption configurations of the molecule near an Li atom in the doped systems. A charge-density analysis shows that the origin of the increase in binding energy is the electronic charge transfer from the Li atom to graphene and the nanotube.

The mechanism and energy characteristics (activation energy and enthalpy) of interaction of linear and graphene carbon nanoparticles with an *oxygen* molecule are investigated by semiempirical PM3 method.[95] The oxidation activation energy depends on the structure of clusters and the interposition of the O_2 molecule and a carbon cluster. Linear clusters are oxidized mainly to CO_2; graphene clusters are oxidized to CO. Results of the reaction of *ozone* with milled graphite and different carbon black grades were based on the role played by fullerene-like carbon nanostructures, present in graphene sheets, to explain the observed gasification rates and surface functionalization.[96] A theoretical study of the sorption of *carbon dioxide* on the four-ring graphene ("unmodified" or N-, O-, and OH-substituted) structures possessing one completely unsaturated edge zigzag site is reported using the DFT (B3LYP/6-31G(d,p)) method.[97]

The reactivity of graphene in relation to complex molecules is also studied. Thus, *water* can dissociate over defective sites in graphene following many possible reaction pathways, some of which have activation barriers lower than half the value for the dissociation of bulk water.[98] This reduction is caused by spin selection rules that allow the system to remain on the same spin surface throughout the reaction. The interaction of *sulfuric acid* with graphene is studied in Ref. [99] by DFT. The results show that there is protonation of the graphene sheet by the acid, in accordance with experimental results for H_2SO_4 adsorbed onto highly oriented pyrolytic graphite and for single-wall carbon nanotubes in concentrated sulfuric acid. Nevertheless, the electronic structure of graphene is not heavily affected, and its zero-band-gap semiconducting behavior is preserved.

Using first principles DFT methods, the authors of the investigation[100] shed light on the question how *aryl groups* attach to a graphene sheet. Thus, for the basal plane, isolated phenyl groups are predicted to be weakly bonded to the graphene sheet, even though a new single C–C bond is formed between the phenyl group and the basal plane by converting a sp^2-carbon in the graphene sheet to sp^3. However, the interaction can be strengthened significantly with two phenyl groups attached to the *para* positions of the same six-member ring to form a pair on the basal plane. The strongest bonding is found at the graphene edges.

Metal–graphene interactions are also being examined. Thus, density-functional calculations were done to examine the interface between graphene and a Pt_{13} or Au_{13} cluster.[101] Introducing a carbon vacancy into a graphene sheet enhanced the interaction between graphene and the metal clusters. Five- or seven-member rings introduced into the graphene also increased the stability of the interface. Lead(II) adsorption from an aqueous solution onto a graphene layer ($C\pi$ electrons) was investigated using activated carbon and charcoal.[102] The experimental results indicate that an acidic oxygen-free graphene layer exhibits a basic character caused by $C\pi$ electrons. For a Pd-metal graphite, having a layered structure, where each Pd sheet is sandwiched between adjacent graphene sheets, superconductivity and magnetic short-range order were studied.[103]

17.4 APPLICATIONS

A series of investigations are dedicated to obtaining *carbon nanotubes* from graphenes. Depending on the concrete scheme of turning of graphite planes, the formed nanotubes can possess metallic or semiconductor properties. The single-walled nanotubes (SWNTs) are formed by folding graphene nanoribbons patterned on graphite films through adsorption of atoms of varying coverage, which introduces an external stress to drive the folding process (Figure 17.11).[104] The interaction of bare graphene nanoribbons (GNRs) was investigated by *ab initio* DFT calculations with both the local density approximation (LDA) and the GGA.[105] Remarkably, two bare 8-GNRs with zigzag-shaped edges are predicted to form an (8, 8) armchair single-wall SWCNT without any obvious activation barrier. A possible route to control the growth of specific types SWCNT via the interaction of GNRs is suggested. The use of graphene layer encapsulated catalytic metal particles for the growth of narrower multiwalled carbon nanotubes (MWCNTs) has been studied using plasma-enhanced chemical vapor deposition and conventional thermal CVD.[106] Ni–C or Fe–C composite nanoclusters were fabricated using the dc arc discharge technique with a metal–graphite

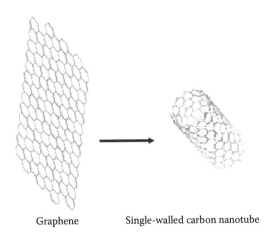

Graphene   Single-walled carbon nanotube

FIGURE 17.11   A simple route from a graphene sheet to a carbon nanotube.

composite. The results show that the diameters of the MWCNTs were reduced from 50–100 nm for a conventional Ni thin film-evaporated Si substrate to a minimum of roughly 2–4 nm in the present study. C-BN patterned SWNT synthesized by laser vaporization technique and a growth model are proposed.[107] The synthesis process consists of vaporizing, by a continuous CO_2 laser, a target made of carbon and boron mixed with a Co/Ni catalyst under N_2 atmosphere. Boron and nitrogen co-segregate with respect to carbon and form nanodomains within the hexagonal lattice of the graphene layer in a sequential manner.

Composite materials have been created on graphene basis.[108] Thus, transparent and electrically conductive composite silica films were fabricated on glass and hydrophilic SiO_x/silicon substrates by incorporation of individual graphene oxide sheets into silica sols followed by spin coating, chemical reduction, and thermal curing.[109] Ultrathin epitaxial graphite films that show remarkable 2D electron gas (2DEG) behavior were produced. These films, composed of typically three graphene sheets, were grown by thermal decomposition on the (0001) surface of 6H-SiC.[110] Interaction with a silicon carbide substrate leads to the opening of a semiconductor gap in epitaxial graphene.[111] This is an important first step toward bandgap engineering in this 2D crystal and its incorporation in electronic devices. Preparation and characterization of *graphene oxide paper*, a freestanding carbon-based membrane material made by flow-directed assembly of individual graphene oxide sheets, is reported in Ref. [112]. This new material outperforms many other paper-like materials in stiffness and strength. Its combination of macroscopic flexibility and stiffness is a result of a unique interlocking-tile arrangement of the nanoscale graphene oxide sheets.

In addition, graphene is considered as one of the most promising materials for postsilicon (carbon-based nanoelectronics) electronics, as it combines high electron mobility with atomic thickness.[113,114] For example, applications of graphenes for *transistor* production are reported in Refs. [115–117]. The transistors show a hole and electron mobility of 3735 and 795 cm²/V s, respectively, and a maximum drive-current of 1.7 mA/μm (at $V_{DS} = 1$ V), which are among the highest reported for r.t. Intrinsic current–voltage characteristics of graphene nanoribbon transistors are examined in Ref. [118].

Nanodevices[119,120] are the most prospective present and future application area for graphenes. Thus, graphenes can be potentially used in digital memory devices[121] or as building blocks for novel optoelectronic devices.[122] Microfabrication of graphene devices currently relies on the fact that graphene crystallites can be visualized using optical microscopy if prepared on top of Si wafers with a certain thickness of SiO_2.[123] Graphene's visibility depends strongly on both thickness of SiO_2 and light wavelength. It was found that by using monochromatic illumination, graphene can be isolated for any SiO_2 thickness, albeit 300 nm (the current standard) and, especially, ~100 nm are most suitable for its visual detection.

Transparent, conductive, and ultrathin graphene films are demonstrated as an alternative to the ubiquitously employed metal oxides window electrodes for solid-state dye-sensitized *solar cells*.[124] These graphene films are fabricated from exfoliated GO, followed by thermal reduction. The obtained films exhibit a high conductivity of 550 S/cm and a transparency of more than 70% over 1000–3000 nm.

Carbon-based 3D solid structures, called covalently bonded graphenes (CBGs), were investigated to elucidate *hydrogen storage* characteristics.[125] Using the density functional method and the Møller–Plesset perturbation method, it is shown that H_2 molecular binding in the CBGs is stronger than that on an isolated graphene with an increase of 20% to ~150% in binding energy, which is very promising for storage at ambient conditions. Graphene nanostructures serve as tunable storage media for molecular hydrogen, because they are lightweight, cheap, chemically inert, and environmentally benign.[126]

The possibility of chemical doping and related excellent *chemical sensor* properties of graphene have been experimentally demonstrated.[16] Thus, graphene sheets can be effective adsorbents and sensors for dioxin in the presence of calcium atoms.[127] This is due to a cooperative formation of sandwich complexes of graphene sheet or (5,5) CNT through the interaction π–Ca–π with the total binding energy of more than 3 eV. The first joint experimental and theoretical investigation of adsorbate-induced doping of graphene is presented in Ref. [128] with the NO_2 system. It is shown that the single, open-shell NO_2 molecule is found to be a strong acceptor, whereas its closed-shell dimer N_2O_4 causes only weak doping. This effect is pronounced by graphene's peculiar density of states (DOS), which provides an ideal situation for model studies of doping effects in semiconductors.

REFERENCES

1. Katsnelson, M. I.; Novoselov, K. S. Graphene: New bridge between condensed matter physics and quantum electrodynamics. *Solid State Communications*, 2007, *143* (1), 3–13.
2. Geim, A. K.; Novoselov, K. S. The rise of graphene. *Nature Materials*, 2007, *6* (3), 183–191.
3. Gogotsi, Y. (Ed.). *Carbon Nanomaterials*. CRC, Boca Raton, FL, 2006, 344 pp.
4. Jorio, A.; Dresselhaus, M. S.; Dresselhaus, G.; Gogotsi, Y. (Eds.). *Carbon Nanotubes: Advanced Topics in the Synthesis, Structure, Properties and Applications*, 1st edn. Springer, Berlin, Germany, 2008, 744 pp.
5. Haug, R. (Ed.). *Advances in Solid State Physics*, 1st edn. Springer, Berlin, Germany, Vol. 47, 2008, 363 pp.
6. Bharat Bhushan, B. (Ed.). *Springer Handbook of Nanotechnology*, 1st edn. Springer, Berlin, Germany, 2008, 363 pp.
7. Van Noorden, R. Moving towards a graphene world. *Nature*, 2006, *442*, 228–229.
8. Müller, S.; Müllen, K. Expanding benzene to giant graphenes: Towards molecular devices. *Philosophical Transactions of the Royal Society A: Mathematical, Physical and Engineering Sciences*, 2007, *365* (1855), 1453–1472.
9. Ando, T. Exotic electronic and transport properties of graphene. *Physica E: Low-Dimensional Systems and Nanostructures*, 2007, *40* (2), 213–227.
10. Aida, T.; Fukushima, T. Soft materials with graphitic nanostructures. *Philosophical Transactions of the Royal Society A: Mathematical, Physical and Engineering Sciences*, 2007, *365* (1855), 1539–1552.
11. Wu, J.; Pisula, W.; Mullen, K. Graphenes as potential material for electronics. *Chemical Reviews*, 2007, *107*, 718–747.
12. Gruner, G.; Hu, L.; Hecht, D. Graphene film as transparent and electrically conducting material. U.S. Patent 20070284557, 2007. http://www.freepatentsonline.com/20070284557.html
13. Prud'homme, R. K.; Ozbas, B.; Aksay, I. A.; Register, R. A.; Adamson, D. H. Functional graphene—Rubber nanocomposites. U.S. Patent Application Filed—Invention # 07-2323-1, 2006.
14. Jang, B. Z. Highly conductive nano-scaled graphene plate nanocomposites and products. U.S. Patent 20070158618, 2007. http://www.freshpatents.com/Bor-Z-Jang-cndirb.php
15. Abanin, D. A.; Novoselov, K. S.; Zeitler, U.; Lee, P. A.; Geim, A. K.; Levitov, L. S. Dissipative quantum hall effect in graphene near the Dirac point. *Physical Review Letters*, 2007, *98* (19), 196806.
16. Schedin, F.; Geim, A. K.; Morozov, S. V.; Hill, E. W.; Blake, P.; Katsnelson, M. I.; Novoselov, K. S. Detection of individual gas molecules adsorbed on graphene. *Nature Materials*, 2007, *6* (9), 652–655.
17. Meyer, J. C.; Geim, A. K.; Katsnelson, M. I.; Novoselov, K. S.; Obergfell, D.; Roth, S.; Girit, C.; Zettl, A. On the roughness of single- and bi-layer graphene membranes. *Solid State Communications*, 2007, *143* (1), 101–109.

18. Novoselov, K. S.; Jiang, Z.; Zhang, Y.; Morozov, S. V.; Stormer, H. L.; Zeitler, U.; Maan, J. C.; Boebinger, G. S.; Kim, P.; Geim, A. K. Room-temperature quantum Hall effect in graphene. *Science*, 2007, *315* (5817), 1379.

19. Pisana, S.; Lazzeri, M.; Casiraghi, C.; Novoselov, K. S.; Geim, A. K.; Ferrari, A. C.; Mauri, F. Breakdown of the adiabatic Born-Oppenheimer approximation in graphene. *Nature Materials*, 2007, *6* (3), 198–201.

20. van den Brink, J. Graphene: From strength to strength. *Nature Nanotechnology*, 2007, *2* (4), 199–201.

21. Hashimoto, A.; Suenaga, K.; Gloter, A.; Urita, K.; Iijima, S. Direct evidence for atomic defects in graphene layers. *Nature*, 2004, *430* (7002), 870–873.

22. Meyer, J. C.; Geim, A. K.; Katsnelson, M. I.; Novoselov, K. S.; Booth, T. J.; Roth, S. The structure of suspended graphene sheets. *Nature*, 2007, *446* (7131), 60–63.

23. Lin, C.-T.; Lee, C.-Y.; Chiu, H.-T.; Chin, T.-S. Graphene structure in carbon nanocones and nanodiscs. *Langmuir*, 2007, *23* (26), 12806–12810.

24. Chen, Y.; Lu, J.; Gao, Z. Structural and electronic study of nanoscrolls rolled up by a single graphene sheet. *Journal of Physical Chemistry C*, 2007, *111* (4), 1625–1630.

25. Peralta-Inga, Z.; Murray, J. S.; Edward Grice, M.; Boyd, S.; O'Connor, C. J.; Politzer, P. Computational characterization of surfaces of model graphene systems. *Journal of Molecular Structure: THEOCHEM*, 2001, *549* (1), 147–158.

26. Image gallery—Graphite and graphene. http://www.ewels.info/img/science/graphite/index.html (accessed January 26, 2008).

27. Shigeo, H.; Takuya, G.; Masahiro, F.; Toru, A.; Tadahiro, Y.; Yoshio, M. Single graphene sheet detected in a carbon nanofilm. *Applied Physics Letters*, 2004, *84*, 2403.

28. Ni, Z. H.; Wang, H. M.; Kasim, J.; Fan, H. M.; Yu, T.; Wu, Y. H.; Feng, Y. P.; Shen, Z. X. Graphene thickness determination using reflection and contrast spectroscopy. *Nano Letters*, 2007, *7* (9), 2758–2763.

29. Casiraghi, C.; Hartschuh, A.; Lidorikis, E.; Qian, H.; Harutyunyan, H.; Gokus, T.; Novoselov, K. S.; Ferrari, A. C. Rayleigh imaging of graphene and graphene layers. *Nano Letters*, 2007, *7* (9), 2711–2717.

30. Grimme, S.; Muck-Lichtenfeld, C.; Antony, J. Noncovalent interactions between graphene sheets and in multishell (hyper)fullerenes. *Journal of Physical Chemistry C*, 2007, *111* (30), 11199–11207.

31. Ishigami, M.; Chen, J. H.; Cullen, W. G.; Fuhrer, M. S.; Williams, E. D. Atomic structure of graphene on SiO$_2$. *Nano Letters*, 2007, *7* (6), 1643–1648.

32. Sitenko, Y. A.; Vlasii, N. D. Electronic properties of graphene with a topological defect. *Nuclear Physics, Section B*, 2007, *787* (3), 241–259.

33. Barone, V.; Hod, O.; Scuseria, G. E. Electronic structure and stability of semiconducting graphene nanoribbons. *Nano Letters*, 2006, *6* (12), 2748–2754.

34. Cortijo, A.; Vozmediano, M. A. H. Electronic properties of curved graphene sheets. *Europhysics Letters*, 2007, *77* (4), 47002.

35. Lopez-Urias, F.; Rodriguez-Manzo, J. A.; Munoz-Sandoval, E.; Terrones, M.; Terrones, H. Magnetic response in finite carbon graphene sheets and nanotubes. *Optical Materials*, 2006, *29* (1), 110–115.

36. Katsnelson, M. I. Graphene: Carbon in two dimensions. *Materials Today*, 2007, *10* (1), 20–27.

37. Graphene makes movement easy for electrons. http://www.physorg.com/news119030362.html (posted January 8, 2008).

38. McCann, E.; Abergel, D. S. L.; Fal'ko, V. I. Electrons in bilayer graphene. *Solid State Communications*, 2007, *143* (1), 110–115.

39. Hod, O.; Barone, V.; Peralta, J. E.; Scuseria, G. E. Enhanced half-metallicity in edge-oxidized zigzag graphene nanoribbons. *Nano Letters*, 2007, *7* (8), 2295–2299.

40. Gomez-Navarro, C.; Weitz, R. T.; Bittner, A. M.; Scolari, M.; Mews, A.; Burghard, M.; Kern, K. Electronic transport properties of individual chemically reduced graphene oxide sheets. *Nano Letters*, 2007, *7* (11), 3499–3503.

41. Wang, W. L.; Meng, S.; Kaxiras, E. Graphene nanoflakes with large spin. *Nano Letters*, 2008, *8* (1), 241–245.

42. Jiang, D.-E.; Sumpter, B. G.; Dai, S. First principles study of magnetism in nanographenes. *Journal of Chemical Physics*, 2007, *127* (12), 124703.

43. Fernández-Rossier, J.; Palacios, J. J. Magnetism in graphene nanoislands. *Physical Review Letters*, 2007, *99*, 177204.

44. Niyogi, S.; Bekyarova, E.; Itkis, M. E.; McWilliams, J. L.; Hamon, M. A.; Haddon, R. C. Solution properties of graphite and graphene. *Journal of the American Chemical Society*, 2006, *128* (24), 7720–7721.

45. Worsley, K. A.; Ramesh, P.; Mandal, S. K.; Niyogi, S.; Itkis, M. E.; Haddon, R. C. Soluble graphene derived from graphite fluoride. *Chemical Physics Letters*, 2007, *445* (1), 51–56.

46. Yang, L.; Cohen, M. L.; Louie, S. G. Excitonic effects in the optical spectra of graphene nanoribbons. *Nano Letters*, 2007, *7* (10), 3112–3115.

47. Jung, I.; Pelton, M.; Piner, R.; Dikin, D. A.; Stankovich, S.; Watcharotone, S.; Hausner, M.; Ruoff, R. S. Simple approach for high-contrast optical imaging and characterization of graphene-based sheets. *Nano Letters*, 2007, *7* (12), 3569–3575.

48. Graf, D.; Molitor, F.; Ensslin, K.; Stampfer, C.; Jungen, A.; Hierold, C.; Wirtz, L. Spatially resolved Raman spectroscopy of single- and few-layer graphene. *Nano Letters*, 2007, *7* (2), 238–242.

49. Calizo, I.; Miao, F.; Bao, W.; Lau, C. N.; Balandin, A. A. Variable temperature Raman microscopy as a nano-metrology tool for graphene layers and graphene-based devices. *Applied Physics Letters*, 2007, *91*, 071913.

50. Calizo, I.; Balandin, A. A.; Bao, W.; Miao, F.; Lau, C. N. Temperature dependence of the Raman spectra of graphene and graphene multilayers. *Nano Letters*, 2007, *7* (9), 2645–2649.

51. Calizo, I.; Bao, W.; Miao, F.; Lau, C. N.; Balandin, A. A. The effect of substrates on the Raman spectrum of graphene: Graphene-on-sapphire and graphene-on-glass. *Applied Physics Letters*, 2007, *91*, 201904.

52. Guo, J.; Yoon, Y.; Ouyang, Y. Gate electrostatics and quantum capacitance of graphene nanoribbons. *Nano Letters*, 2007, *7* (7), 1935–1940.

53. Son, Y.-W.; Cohen, M. L.; Louie, S. G. Energy gaps in graphene nanoribbons. *Physical Review Letters*, 2006, *97*, 216803.

54. Son, Y.-W.; Cohen, M. L.; Louie, S. G. Half-metallic graphene nanoribbons. *Nature*, 2006, *444*, 347.

55. Yang, L.; Park, C. H.; Son, Y.-W.; Cohen, M. L.; Louie, S. G. Quasiparticle energies and band gaps of graphene nanoribbons. *Physical Review Letters*, 2007, *99*, 186801.

56. Martins, T. B.; Miwa, R. H.; da Silva, A. J. R.; Fazzio, A. Electronic and transport properties of boron-doped graphene nanoribbons. *Physical Review Letters*, 2007, *98* (19), 196803.

57. Shemella, P.; Zhang, Y.; Mailman, M.; Ajayan, P. M.; Nayak, S. K. Energy gaps in zero-dimensional graphene nanoribbons. *Applied Physics Letters*, 2007, *91*, 042101.

58. Liang, G.; Neophytou, N.; Lundstrom, M. S.; Nikonov, D. E. Ballistic graphene nanoribbon metal-oxide-semiconductor field-effect transistors: A full real-space quantum transport simulation. *Journal of Applied Physics*, 2007, *102*, 054307.

59. Jiang, D.-E.; Sumpter, B. G.; Dai, S. Unique chemical reactivity of a graphene nanoribbon's zigzag edge. *Journal of Chemical Physics*, 2007, *126* (13), 134701.

60. Ryzhii, V. Terahertz plasma waves in gated graphene heterostructures. *Japanese Journal of Applied Physics*, 2006, *45*, L923. doi:10.1143/JJAP.45.L923.

61. Ryzhii, V.; Satou, A.; Otsuji, T. Plasma waves in two-dimensional electron-hole system in gated graphene heterostructures. *Journal of Applied Physics*, 2007, *101*, 024509. doi:10.1063/1.2426904.

62. Novoselov, K.; Jiang, S. D.; Schedin, F.; Booth, T. J.; Khotkevich, V. V.; Morozov, S. V.; Geim, A. K. Two-dimensional atomic crystals. *Proceedings of the National Academy of Sciences of the United States of America*, 2005, *102*, 10451. doi:10.1073/pnas.0502848102.

63. Somani, P. R.; Somani, S. P.; Umeno, M. Planer nano-graphenes from camphor by CVD. *Chemical Physics Letters*, 2006, *430* (1), 56–59.

64. Rollings, E.; Gweon, G. H.; Zhou, S. Y.; Mun, B. S.; McChesney, J. L.; Hussain, B. S.; Fedorov, A. V.; First, P. N.; de Heer, W. A.; Lanzara, A. Synthesis and characterization of atomically thin graphite films on a silicon carbide substrate. *Journal of Physics and Chemistry of Solids*, 2006, *67* (9), 2172–2177.

65. Jang, B. Z., Huang, W. C. Nano-scaled graphene plates. U.S. Patent 7071258, 2006. http://www.freepatentsonline.com/7071258.html.

66. Ohta, T.; Bostwick, A.; Seyller, T.; Horn, K.; Rotenberg, E. Controlling the electronic structure of bilayer graphene. *Science*, 2006, *313* (5789), 951–954.

67. de Heer, W. A.; Berger, C.; Wu, X.; First, P. N.; Conrad, E. H.; Li, T.; Sprinkle, M. et al. Epitaxial graphene. *Solid State Communications*, 2007, *143* (1), 92–100.

68. Riedl, C.; Starke, U. Structural properties of the graphene-SiC(0001) interface as a key for the preparation of homogeneous large-terrace graphene surfaces. *Physical Review B*, 2007, *76*, 245406.

69. Hass, J.; Feng, R.; Li, T.; Li, X.; Zong, Z.; de Heer, W. A.; First, P. N.; Conrad, E. H.; Jeffrey, C. A.; Berger, C. Highly ordered graphene for two dimensional electronics. *Applied Physics Letters*, 2006, *89*, 143106.

70. Berger, C.; Song, Z.; Li, X.; Wu, X.; Brown, N.; Naud, C.; Mayou, D. et al. Electronic confinement and coherence in patterned epitaxial graphene. *Science*, 2006, *312*, 1191. doi:10.1126/science.1125925.

71. Osvath, Z.; Darabont, Al.; Nemes-Incze, P.; Horvath, E.; Horvath, Z. E.; Biro, L. P. Graphene layers from thermal oxidation of exfoliated graphite plates. *Carbon*, 2007, *45* (15), 3022–3026.

72. Zhu, Z.; Su, D.; Weinberg, G.; Schlogl, R. Supermolecular self-assembly of graphene sheets: Formation of tube-in-tube nanostructures. *Nano Letters*, 2004, *4* (11), 2255–2259.

73. Stankovich, S.; Dikin, D. A.; Piner, R. D.; Kohlhaas, K. A.; Kleinhammes, A.; Jia, Y.; Wu, Y.; Nguyen, S. T.; Ruoff, R. S. Synthesis of graphene-based nanosheets via chemical reduction of exfoliated graphite oxide. *Carbon*, 2007, *45* (7), 1558–1565.

74. Gregg, D. J.; Bothe, E.; Höfer, P.; Passaniti, P.; Draper, S. M. Extending the nitrogen-heterosuperbenzene family: The spectroscopic, redox, and photophysical properties of "half-cyclized" N-1/2HSB and its Ru(II) complex. *Inorganic Chemistry*, 2005, *44* (16), 5654–5660.

75. Draper, M. S.; Gregg, D. J.; Schofield, E. R.; Browne, W. R.; Duati, M.; Vos, J. G.; Passaniti, P. Complexed nitrogen heterosuperbenzene: The coordinating properties of a remarkable ligand. *Journal of the American Chemical Society*, 2004, *126*, 8694.

76. Gilje, S.; Han, S.; Wang, M.; Wang, K. L.; Kaner, R. B. A chemical route to graphene for device applications. *Nano Letters*, 2007, *7* (11), 3394–3398.

77. McAllister, M. J.; Li, J.-L.; Adamson, D. H.; Schniepp, H. C.; Abdala, A. A.; Liu, J.; Herrera-Alonso, M. et al. Single sheet functionalized graphene by oxidation and thermal expansion of graphite. *Chemistry of Materials*, 2007, *19* (18), 4396–4404.

78. Schniepp, H. C.; Li, J.-L.; McAllister, M. J.; Sai, H.; Herrera-Alonso, M.; Adamson, D. H.; Prud'homme, R. K.; Car, R.; Saville, D. A.; Aksay, I. A. Functionalized single graphene sheets derived from splitting graphite oxide. *Journal of Physical Chemistry B*, 2006, *110* (17), 8535–8539.

79. Stankovich, S.; Piner, R. D.; Nguyen, S. T.; Ruoff, R. S. Synthesis and exfoliation of isocyanate-treated graphene oxide nanoplatelets. *Carbon*, 2006, *44* (15), 3342–3347.

80. Viculis, L. M.; Mack, J. J.; Kaner, R. B. A chemical route to carbon nanoscrolls. *Science*, 2003, *299*, 1361.

81. Son, Y. W. Physics of graphene and graphene nanoribbons. http://icpr.or.kr/file/vod/2007/yonsei/20071205.pdf (posted on December 5, 2007).

82. Lerf, A.; He, H.; Forster, M.; Klinowski, J. Structure of graphite oxide revisited. *Journal of Physical Chemistry B*, 1998, *102*, 4477–4482.

83. Aksay, I. A. Developing fundamental insights pertinent to the design and processing of multifunctional nanocomposites. http://www.wtec.org/nb-france-us/aksay_i/aksay-modified.pdf (posted on March 2, 2006).

84. Sidorov, A. N.; Yazdanpanah, M. M.; Jalilian, R.; Ouseph, P. J.; Cohn, R. W.; Sumanasekera, G. U. Electrostatic deposition of graphene. *Nanotechnology*, 2007, *18* (13), 135301.

85. Northrop, B. H.; Norton, J. E.; Houk, K. N. On the mechanism of peripentacene formation from pentacene: Computational studies of a prototype for graphene formation from smaller acenes. *Journal of the American Chemical Society*, 2007, *129* (20), 6536–6546.

86. Whitesides, R.; Kollias, A. C.; Domin, D.; Lester, W. A.; Frenklach, M. Graphene layer growth: Collision of migrating five-member rings. *Proceedings of the Combustion Institute*, 2007, *31* (1), 539–546.

87. Carissan, Y.; Klopper, W. Growing graphene sheets from reactions with methyl radicals: A quantum chemical study. *ChemPhysChem: A European Journal of Chemical Physics and Physical Chemistry*, 2006, *7* (8), 1770–1778.

88. Frenklach, M.; Ping, J. On the role of surface migration in the growth and structure of graphene layers. *Carbon*, 2004, *42* (7), 1209–1212.

89. Cappelli, E.; Orlando, S.; Morandi, V.; Servidori, M.; Scilletta, C. Nano-graphene growth and texturing by Nd:YAG pulsed laser ablation of graphite on silicon. *Journal of Physics: Conference Series*, 2007, *59* (1), 616–624.

90. Miura, Y.; Kasai, H.; Diño, W.; Nakanishi, H.; Sugimoto, T. First principles studies for the dissociative adsorption of H_2 on graphene. *Journal of Applied Physics*, 2003, *93*, 3395.

91. Kerkeni, B.; Clary, D. C. Quantum dynamics study of the Langmuir–Hinshelwood H + H recombination mechanism and H_2 formation on a graphene model surface. *Chemical Physics*, 2007, *338* (1), 1–10.

92. Chernozatonskii, L. A.; Sorokin, P. B. Two-dimensional semiconducting nanostructures based on single graphene sheets with lines of adsorbed hydrogen atoms. *Applied Physics Letters*, 2007, *91*, 183103.

93. Okamoto, Y.; Miyamoto, Y. Ab initio investigation of physisorption of molecular hydrogen on planar and curved graphenes. *Journal of Physical Chemistry B*, 2001, *105* (17), 3470–3474.

94. Cabria, I.; López, M. J.; Alonso, J. A. Enhancement of hydrogen physisorption on graphene and carbon nanotubes by Li doping. *Journal of Chemical Physics*, 2005, *123*, 204721.

95. Zavodinsky, V. G.; Mikhailenko, E. A. Quantum-mechanics simulation of carbon nanoclusters and their activities in reactions with molecular oxygen. *Computational Materials Science*, 2006, *36* (1), 159–165.

96. Cataldo, F. Ozone reaction with carbon nanostructures. 2: The reaction of ozone with milled graphite and different carbon black grades. *Journal of Nanoscience and Nanotechnology*, 2007, *7* (4–5), 1446–1454.

97. Gauden, P. A.; Wisniewski, M. CO$_2$ sorption on substituted carbon materials. *Applied Surface Science*, 2007, *253* (13), 5726–5731.

98. Kostov, M. K.; Santiso, E. E.; George, A. M.; Gubbins, K. E.; Nardelli, M. B. Dissociation of water on defective carbon substrates. *Physical Review Letters*, 2005, *95* (13), 136105.

99. Cordero, N. A.; Alonso, J. A. The interaction of sulfuric acid with graphene and formation of adsorbed crystals. *Nanotechnology*, 2007, *18* (48), 485705.

100. Jiang, D.-E.; Sumpter, B. G.; Dai, S. How do aryl groups attach to a graphene sheet? *Journal of Physical Chemistry B*, 2006, *110* (47), 23628–23632.

101. Okamoto, Y. Density-functional calculations of icosahedral M$_{13}$ (M = Pt and Au) clusters on graphene sheets and flakes. *Chemical Physics Letters*, 2006, *420* (4), 382–386.

102. Machida, M.; Mochimaru, T.; Tatsumoto, H. Lead(II) adsorption onto the graphene layer of carbonaceous materials in aqueous solution. *Carbon*, 2006, *44* (13), 2681–2688.

103. Suzuki, M.; Suzuki, I. S.; Walter, J. Superconductivity and magnetic short-range order in the system with a Pd sheet sandwiched between graphene sheets. *Journal of Physics: Condensed Matter*, 2004, *16* (6), 903–918.

104. Yu, D.; Liu, F. Synthesis of carbon nanotubes by rolling up patterned graphene nanoribbons using selective atomic adsorption. *Nano Letters*, 2007, *7* (10), 3046–3050.

105. Du, A. J.; Smith, S. C.; Lu, G. Q. Formation of single-walled carbon nanotube via the interaction of graphene nanoribbons: Ab initio density functional calculations. *Nano Letters*, 2007, *7* (11), 3349–3354.

106. Nagatsu, M.; Yoshida, T.; Mesko, M.; Ogino, A.; Matsuda, T.; Tanaka, T.; Tatsuoka, H.; Murakami, K. Narrow multi-walled carbon nanotubes produced by chemical vapor deposition using graphene layer encapsulated catalytic metal particles. *Carbon*, 2006, *44* (15), 3336–3341.

107. Enouz, S.; Stéphan, O.; Cochon, J.-L.; Colliex, C.; Loiseau, A. C-BN patterned single-walled nanotubes synthesized by laser vaporization. *Nano Letters*, 2007, *7* (7), 1856–1862.

108. Stankovich, S.; Dikin, D. A.; Dommett, G. H. B.; Kohlhaas, K. M.; Zimney, E. J.; Stach, E. A.; Piner, R. D.; Nguyen, S. B. T.; Ruoff, R. S. Graphene-based composite materials. *Nature*, 2006, *442*, 282–286.

109. Watcharotone, S.; Dikin, D. A.; Stankovich, S.; Piner, R.; Jung, I.; Dommett, G. H. B.; Evmenenko, G. et al. Graphene-silica composite thin films as transparent conductors. *Nano Letters*, 2007, *7* (7), 1888–1892.

110. Berger, C.; Song, Z.; Li, T.; Li, X.; Ogbazghi, A. Y.; Feng, R.; Dai, Z. et al. Ultrathin epitaxial graphite: 2D electron gas properties and a route toward graphene-based nanoelectronics. *Journal of Physical Chemistry B*, 2004, *108* (52), 19912–19916.

111. Novoselov, K. Graphene: Mind the gap. *Nature Materials*, 2007, *6* (10), 720–721.

112. Dikin, D. A.; Stankovich, S.; Zimney, E. J.; Piner, R. D.; Dommett, G. H. B.; Evmenenko, G.; Nguyen, S. B. T.; Ruoff, R. S. Preparation and characterization of graphene oxide paper. *Nature*, 2007, *448* (7152), 457–460.

113. Novoselov, K. S.; Geim, A. K.; Morozov, S. V.; Jiang, D.; Zhang, Y.; Dubonos, S. V.; Grigorieva, I. V.; Firsov, A. A. Electric field effect in atomically thin carbon films. *Science*, 2004, *306*, 666–669.

114. Areshkin, D. A.; White, C. T. Building blocks for integrated graphene circuits. *Nano Letters*, 2007, *7* (11), 3253–3259.

115. Liang, X.; Fu, Z.; Chou, S. Y.; Areshkin, D. A.; White, C. T. Graphene transistors fabricated via transfer-printing in device active-areas on large wafer. *Nano Letters*, 2007, *7* (12), 3840–3844.

116. Semenov, Y. G.; Kim, K. W.; Zavada, J. M. Spin field effect transistor with a graphene channel. *Applied Physics Letters*, 2007, *91*, 153105.

117. Chen, Z.; Lin, Y.-M.; Rooks, M. J.; Avouris, P. Graphene nano-ribbon electronics. *Physica E: Low-Dimensional Systems and Nanostructures*, 2007, *40* (2), 228–232.

118. Yan, Q.; Huang, B.; Yu, J.; Zheng, F.; Zang, J.; Wu, J.; Gu, B.-L.; Liu, F.; Duan, W. Intrinsic current-voltage characteristics of graphene nanoribbon transistors and effect of edge doping. *Nano Letters*, 2007, *7* (6), 1469–1473.

119. Jayasekera, T.; Mintmire, J. W. Transport in multiterminal graphene nanodevices. *Nanotechnology*, 2007, *18* (42), 424033.

120. Staley, N.; Wang, H.; Puls, C.; Forster, J.; Jackson, T. N.; McCarthy, K.; Clouser, B.; Liu, Y. Lithography-free fabrication of graphene devices. *Applied Physics Letters*, 2007, *90*, 143518.

121. Gunlycke, D.; Areshkin, D. A.; Li, J.; Mintmire, J. W.; White, C. T. Graphene nanostrip digital memory device. *Nano Letters*, 2007, *7* (12), 3608–3611.

122. Roddaro, S.; Pingue, P.; Piazza, V.; Pellegrini, V.; Beltram, F. The optical visibility of graphene: Interference colors of ultrathin graphite on SiO$_2$. *Nano Letters*, 2007, *7* (9), 2707–2710.

123. Blake, P.; Hill, E. W.; Castro Neto, A. H.; Novoselov, K. S.; Jiang, D.; Yang, R.; Booth, T. J.; Geim, A. K. Making graphene visible. *Applied Physics Letters*, 2007, *91*, 063124.

124. Wang, X.; Zhi, L.; Mullen, K. Transparent, conductive graphene electrodes for dye-sensitized solar cells. *Nano Letters*, 2008, *8* (1), 323–327.

125. Park, N.; Hong, S.; Kim, G.; Jhi, S.-H. Computational study of hydrogen storage characteristics of covalent-bonded graphenes. *Journal of the American Chemical Society*, 2007, *129* (29), 8999–9003.

126. Patchkovskii, S.; Tse, J. S.; Yurchenko, S. N.; Zhechkov, L.; Heine, T.; Seifert, G. Graphene nanostructures as tunable storage media for molecular hydrogen. *Proceedings of the National Academy of Sciences of the United States of America*, 2005, *102* (30), 10439–10444.

127. Kang, H. S. Theoretical study of binding of metal-doped graphene sheet and carbon nanotubes with dioxin. *Journal of the American Chemical Society*, 2005, *127* (27), 9839–9843.

128. Wehling, T. O.; Novoselov, K. S.; Morozov, S. V.; Vdovin, E. E.; Katsnelson, M. I.; Geim, A. K.; Lichtenstein, A. I. Molecular doping of graphene. *Nano Letters*, 2008, *8* (1), 173–177.

18 Nanodiamonds

Carbon takes many different forms, each with its own electronic structure, and has a fantastic range of properties. Nanodiamond (ND) is a new member of nanocarbons, which consist of nanosized tetrahedral networks. The history of the discovery of ND is unique, since ND synthesis was accidentally discovered in the USSR three times over 19 years starting from 1963 by shock compression of nondiamond carbon modifications in blast chambers.[1] The term *nanodiamond* is currently used broadly for a variety of diamond-based materials at the nanoscale ranging from single diamond clusters to bulk nanocrystalline films.[2] It is generally accepted that *nanocrystalline diamond* (NCD) consists of facets less than 100 nm in size, whereas a second term *ultrananocrystalline diamond* (UNCD) has been coined to describe material with grain sizes less than 10 nm. These differences in morphology originate in the growth process.[3]

Nanometer-sized diamond has been found in meteorites, proto-planetary nebulae, and interstellar dusts, as well as in residues of detonation and in diamond films. It is known that primitive chondritic meteorites contain up to approximately 1500 ppm of nanometer-sized diamonds, containing isotopically anomalous noble gases, nitrogen, hydrogen, and other elements. These isotopic anomalies indicate that meteoritic NDs probably formed outside our solar system before the Sun's formation (they are thus presolar grains).[4] It appears that some interstellar emission bands in the 3–4, 7–10, and approximately 21 μm spectral regions could originate from NDs.[5]

Diamond thin films have outstanding optical, electrical, mechanical, and thermal properties, which make these attractive for applications in a variety of current and future systems. In particular, the wide band gap and optical transparency of diamond thin films make them an ideal semiconductor for applications in current and future electronics. Various aspects on the NDs have been recently emphasized in a series of reviews[6–12] and books,[13,14] in particular such topics as ND preparation by chemical vapor deposition (CVD), NCD film formation from hydrogen-deficient and hydrogen-rich plasma, nanocomposite films, mechanical behavior of NCD films, and field emission characteristics,[15] diamond nanowires and their biofunctionalization of nucleic acid molecules,[16] or use of NCD to study noncovalent interaction.[17] In this section, we have tried to emphasize updated main synthesis and functionalization methods for NDs and their relationship with ND structure and properties.

18.1 SYNTHESIS METHODS FOR NANODIAMONDS

18.1.1 PRECURSORS

NDs can be produced from a series of precursors, among which the CH_4/H_2 mixtures with the addition of inert gases, O_2 or N_2, are frequently used. Thus, the synthesis of NCD films can be done using a small amount of simultaneous O_2 and N_2 addition into conventional CH_4/H_2 mixtures.[18] The NCD samples were grown in a 5 kW microwave plasma CVD (MWPCVD) system on large silicon wafers. It was shown that the morphology, microstructure, grain size, crystalline quality, and growth rate of NCD films can be tailored by simply adjusting the amount of O_2 and N_2 addition, and with increasing the ratio of addition, the crystal quality of the NCD films is significantly enhanced. Additionally, carbon nanotubes were used as ND precursors[19]; for instance multiwall carbon nanotubes (MWCNTs) served as precursor to prepare diamond in a pure hydrogen microwave discharge.[20] Also, NDs were easily formed from carbon film containing Si.[21] The growth of diamond and β-SiC was controllable by adjusting the heating temperature and the proportion of Si. A method for producing an ND, in which an ND was removed from an activated carbon containing the ND, was offered in [Ref. 22]. An ND fiber (up to 2000 nm) can also be produced by mixing a carbon source, a metal, and an acid under conditions that result in ND formation. Methods of industrial ND synthesis were generalized in a review[23] with attention to aspects of detonation decomposition of powerful mixed explosives with a negative oxygen balance (formation of ultrafine-dispersed diamonds). The most efficient technology of chemical cleaning of diamonds with nitric acid at high temperatures and pressures for producing high-purity NDs was described (for other methods, see Section 18.1.9). Other ND precursors include fullerenes,[24] graphite, β-SiC, carbide-organic systems, polycarbine, trotyl/hexogen, 1,1,1-trichloroethane, etc.

18.1.2 Hydrothermal Synthesis

Hydrothermal synthesis as any heterogeneous reactions occurring under the conditions of high-temperature/high-pressure (>100°C, >1 atm) in aqueous solutions in a closed system has a growing interest among the scientists in particular due to a possibility of synthesis of new phases or crystals growth.[25] This technique was also applied for ND synthesis[26]; thus, available hydrothermal techniques for ND synthesis were reviewed by Nickel et al.[27] The hydrothermal synthesis of diamond was frequently carried out in the silicon carbide–organic compound system.[28,29] Organic matter dissociated in a closed system to generate C–O–H supercritical fluids, known for their high dissolving power and influence on the type of elemental carbon formation. The formed carbon crystallites had typical spectra of sp^3-hybridization, clearly demonstrating the formation of nanosized diamond crystallites under subnatural conditions. Among other ND precursors in hydrothermal synthesis, chlorinated hydrocarbons such as dichloromethane and 1,1,1-trichloroethane[25,30] can be used.

18.1.3 Ion Bombardment

A successful irradiation-induced transformation of MWCNTs to diamond nanocrystals was realized with double-ion ($^{40}Ar^+$, $C_2H_6^+$) bombardment.[31] This idea of multi-ion irradiation may also be used for fabricating other nanostructures. The growth of ND under prolonged DC glow discharge plasma bombardment of 1 μm thick polycrystalline CVD diamond was studied by Gouzman et al.[32] It was established that ND formation on diamond started directly. The amorphous carbon/ND composite structure was substantial to the ND nucleation under energetic plasma bombardment. The nucleation of diamond on graphitic edges as predicted by W. R. L. Lambrecht et al. in 1993 was experimentally confirmed by Yao et al.[33] Thus, the precipitation of ND crystallites in upper layers of a film deposited by a 1 keV mass-selected carbon ion beam onto silicon held at 800°C was observed by HRTEM, selected area electron diffraction, and electron energy loss spectroscopy. Molecular dynamic simulations showed that diamond nucleation in the absence of hydrogen can occur by precipitation of diamond clusters in a dense amorphous carbon matrix generated by subplantation. After cluster formation, they can grow by thermal annealing consuming carbon atoms from the amorphous matrix.[34]

18.1.4 Laser Bombarding

Pulsed laser ablation has become an attractive method for the preparation of ND in liquids[35] or solids using carbon powders. Thus, NDs were obtained[36] by suspending carbon powders (crystalline flake graphite, microcrystalline graphite, or carbon black with particle size less than 10 μm) in a circulating liquid medium (water, alcohols, ketones, ethers, and their solutions or mixtures), bombarding the carbon powders by laser, and further purifying the product to obtain the diamond nanopowders. It was shown that only microcrystalline graphite by laser transformed into diamond (cubic diamond about 5 nm) in the three carbon materials. It also demonstrated that microcrystalline graphite was more advantageous than carbon black and crystalline flake graphite when the laser power density was 10^6 W/cm².[37] A theoretical kinetic approach to elucidate the nucleation

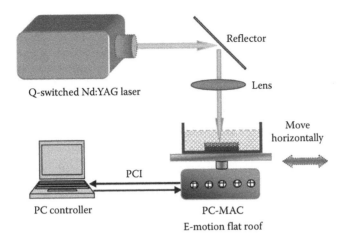

FIGURE 18.1 Schematic illustration of the experimental setup of laser irradiation in liquid. (Reproduced with permission from Liu, P. et al., Localized nanodiamond crystallization and field emission performance improvement of amorphous carbon upon laser irradiation in liquid, *J. Phys. Chem. C*, 113(28), 12154–12161. Copyright 2009 American Chemical Society.)

and growth of nanocrystals with respect to the capillary effect of the nanometer-sized curvature of crystalline nuclei was proposed by Wang et al.[38] on the example of the ND synthesis by pulsed-laser ablating a graphite target in H_2O. The authors predicted the nucleation time, growth velocity, and the grown size of NDs from the proposed kinetic model. Pulsed-laser irradiation of amorphous carbon films in a liquid phase (Figure 18.1) at room temperature and ambient pressure led to a phase transformation from amorphous carbon to ND (4–7 nm).[39] On the basis of the obtained results, it was concluded that laser irradiation in liquid actually opens a route toward self-assembly of surface micro- and nanostructures, that is, functional nanostructure manufacturing.

The diamond-like carbon films with the highest sp³ carbon bonding content were obtained at laser fluences of 850–1000 mJ/cm² by irradiation of the polycarbyne polymer films, coated on silicon substrates, with a pulsed Nd:yttrium-aluminum-garnet laser ($\lambda = 532$ nm) in argon gas atmosphere.[40] Quartz substrates were used as supports in similar experiments ($\lambda = 1064$ nm, $\tau = 20$ ns, $q = 4.9 \times 10^8$ W/cm²) under vacuum ($p = 2.6 \times 10^{-3}$ Pa).[41] The effect of low-power laser radiation on volume and surface microinclusions of graphite-like carbon during the CVD diamond film fabrication was studied.[42] The simulation calculations showed that laser irradiation accelerates the processes of graphite and nondiamond phase etching.

18.1.5 MICROWAVE PLASMA CHEMICAL VAPOR DEPOSITION TECHNIQUES

An important research and industrial method, microwave and millimeter-wave processing of materials,[43–45] in particular diamonds, was reviewed by Lewis et al.[46] CVD techniques, in particular those based on microwave-assisted approach, have become one of the most classic methods in materials production, including various carbon nanoforms, and are well generalized in a series of books.[47–49] The experimental findings providing a basic understanding of the plasma chemistry of a hydrocarbon/Ar-rich plasma environment, used for growing ND films, were discussed by Gordillo-Vazquez et al.[50] Both main ND types, NCD and UNCD, were obtained *via* this route. Among a series of related works, the NCD (40–400 nm) was grown without the help of initial nucleation sites on Ni substrates in a microwave plasma reactor (400°C–600°C) using hexane/nitrogen-based CVD.[51] In a related work, NCD thin films were prepared and characterized with MPCVD.[52] Ultrasmooth nanostructured diamond (USND) coatings were deposited by MPCVD technique using He/H₂/CH₄/N₂ gas mixture as a precursor.[53]

Various aspects of CVD applications for ND materials were studied in detail by Butler et al.,[54] in particular the growth and characteristics of NCD thin films with thicknesses from 20 nm to <5 μm containing 95 to >99.9% diamond crystallites. It was shown that the UNCD was usually grown in argon-rich, H-poor CVD environments and may contain up to 95%–98% sp[3]-bonded C. The effect of gas (mixed methane and hydrogen) flow rate on diamond deposition on mirror-polished silicon substrates in a microwave plasma reactor was studied by Chen et al.[55] Microcrystalline diamond thin films were obtained at relatively low gas flow rates (30–300 sccm), whereas NCD thin films with cauliflower-like morphology were obtained at higher gas flow rates (>300 sccm). This result may be attributed to the enhancement of diamond secondary nucleation arising from the increase in the flux rate of carbon-containing radicals reaching the diamond growth surface.

A series of patents are dedicated to various presentations of microwave equipment for fabrication of distinct diamond forms. Thus, the equipment for MWCVD processing was patented by Nanba et al.[56] The apparatus had a dielectric window for introducing microwave in a vacuum chamber and an antenna part with an edge electrode for introducing microwave in the chamber, wherein the window is sandwiched between the chamber inner surface and the electrode. The apparatus was suitable for forming a large-area high-quality semiconductor diamond film. Another microwave CVD apparatus precisely for preparation of diamond, comprising a globe-shaped discharge chamber and a coaxial antenna for feeding microwave into the chamber, with the antenna tip equipped with a work holder, positioned in the center of the globe, was proposed by Ariyada et al.[57]

18.1.6 Detonation Methods for ND Fabrication

Detonation techniques belong to the most conventional methods[58] for obtaining NDs, in particular at industrial scale. In this respect, we note a series of fundamental reviews of Dolmatov, dedicated to such various aspects of detonation-produced nanodiamonds (DND) as the structure,[59] key properties and promising fields of application of detonation-synthesis NDs,[60] and modern industrial methods for their manufacture,[61,62] etc. DND purity is an important problem, which is being continuously studied. In particular, the mechanism of prolonged water washing to remove excessive acidity was described,[63] which was offered to improve essentially the quality of NDs and the stability of their aqueous suspensions by treating them with ammonia water (to an alkalescent medium), followed by heating to 200°C–240°C under pressure (so-called thermolysis). Alternative methods to remove materials still unconverted to diamond (for details, see the following) produced[64] yields up to 60% by firing of high explosive mixtures in water confinement, avoiding the use of inert gas and preventing the oxidation and graphitization of recovered diamonds. Various selective oxidation treatments (with KNO_3/KOH, H_2O_2/HNO_3 mixtures) were carried out, leading to light gray ultra-dispersed diamond (UDD) aggregates with a yield up to 60%.

The mechanisms of various steps of DND formation have been established. Thus, problems associated with the final stages in the disintegration/purification of DND into monodisperse single-nanodiamond (DSND) particles were critically reviewed by Osawa et al.[65,66] Possible pitfalls that one might encounter during the search of industrial application of DSND were identified: low diamond–graphite transition temperature and abnormally strong tendency of the dispersed primary particles to reaggregate. The formation kinetics of detonation NDs was proposed by Titov et al.[67] As a conclusion of a series of detonation experiments with ultradisperse diamond in oxygen and oxygen-free media, the diamond cannot be produced immediately behind the wave front. The authors believed that there is a diamond-free zone and zones of diamond formation. Production of NDs with an increased colloidal stability by using an explosion synthesis was patented by Puzyr et al.[68]

18.1.7 Use of Ultrasound

Ultrasonic action during the ND manufacture or its pretreatment has got a certain use in ND fabrication and compared with other treatments. Thus, carbon materials containing C≡C were irradiated with x-rays, microwave, and/or ultrasonic waves to give varieties of carbon structures, in particular diamond thin film and fine-grain diamond.[69] Diamond microcrystals were prepared (size 6 or 9 μm, yield ~10% by mass) using ultrasonic cavitation of a suspension of hexagonal graphite in various organic solvents at ~120°C and at atmospheric pressure.[70] Thermal processing of ND powder in air at 440°C–600°C until powder weight loss reached 5%–85% led to stable suspensions in water, EtOH, and other solvents upon ultrasonic treatment.[71] Effects of pretreatment on the nuclei formation of UNCD on Si substrates were studied,[72] showing that either precoating a thin layer of titanium (~400 nm) or ultrasonication pretreatment using diamond and titanium mixed powder enhanced the nucleation process on Si substrates markedly. NCD coating using sol-gel technique as an easy coating technique for NCD films was reported by Hanada et al.,[73] making an ND sol for coating by ultrasonic dispersion in water of diamond nanoparticles (<10 nm) synthesized by a detonation method, followed by mechanical milling milled with zirconia balls. This technique allowed fabrication NCD films easily at a much lower processing temperature in comparison with conventional diamond coating. The process and apparatus for manufacture of diamond powder in the nanometer range by size reduction using ultrasonic agitation was patented by Iijima.[74]

18.1.8 Models for Nanodiamond Synthesis

In addition to the DND formation mechanisms (Section 18.1.6), various models of ND synthesis on the basis of experimental results have been offered including software. Thus, visualization techniques for modeling carbon allotropes, including NDs, were described by Adler et al.,[75,76] in particular AViz (Atomistic Visualization package), which is essential for understanding sample geometries. A cold plasma hybrid-fluid-particle approach, in agreement with experimental results, was used to describe the deposition of diamond films from pulsed MW discharges.[77] A density functional theory study was presented[78] on changes in band gap effects of NDs (hydrogen terminated diamond-like molecules, diamondoids) depending on size, shape, and the incorporation of heteroatom functionalities. The combination of increasing the size of the ND and push-pull doping is likely to make these materials highly valuable for semiconductor applications. A comparison of several theoretical models that have been proposed to describe the relative phase stability of diamonds and other forms of carbon at the nanoscale was reviewed.[79]

Simulation and bonding of dopants in NCD were reviewed by Barnard et al.[80] using quantum mechanics–based simulation methods. It is known that impurities can be introduced into diamond, resulting in materials with unusual physical and chemical properties. For electronic applications, one of the main objectives in the doping of diamond is the production of p-type and n-type semiconductor materials. The thermodynamic approach on the nanoscale to elucidate the diamond nucleation taking place in the hydrothermal synthesis and the reduction of carbide (HSRC) supercritical-fluid systems was performed by Wang et al.[81] It was shown that the nanosize-induced interior pressure of diamond nuclei could drive the metastable phase region of the diamond nucleation in HSRC into the new stable phase region of diamond in the carbon phase diagram (Figure 18.2). In the metastable phase region of diamond, 400 MPa was predicted to be the threshold pressure for the diamond synthesis by HSRC. Among other reports, the transformation of NDs into carbon-onions (and vice versa) was observed experimentally and modeled computationally,[82] outlining results of numerous computational and theoretical studies examining the phase stability of carbon nanoparticles and clarifying the complicated relationship between fullerene and diamond structures at the nanoscale.

18.1.9 Purification of Nanodiamonds

The diamonds and NDs, obtained by different methods, contain various impurities, described in the following. Small amounts of graphitic impurities and other carbon structures in diamond can alter its most important properties,[83] so a grand variety of techniques for their purification have been developed, in particular magnetic,[84] ultrasonic,[85] electrolysis[86] or electric field,[87] or laser,[88]

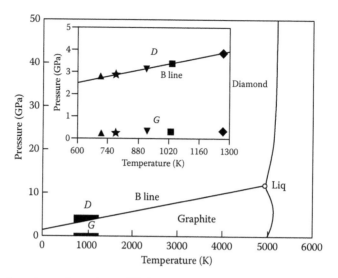

FIGURE 18.2 Carbon thermodynamic equilibrium phase diagram based on pressure and temperature. *G* region is a metastable phase region of diamond nucleation; *D* region is a new stable phase region of diamond nucleation by hydrothermal synthesis or reduction of carbide under the nanosize-induced interior pressure conditions. The inset shows the enlarged *G* and *D* region. (Reproduced with permission from Wang, C.X. et al., Thermodynamics of diamond nucleation on the nanoscale, *J. Am. Chem. Soc.*, 126(36), 11303–11306. Copyright 2004 American Chemical Society.)

as well as pure chemical methods applying generally strong oxidants (acids, potassium permanganate, chromium(VI) oxide, chromates, mixture of potassium nitrate and KOH, ozone) to oxidize and solubilize impurities. Thus, an oxidative treatment for liquid-phase purification of diamonds in an autoclave at increased temperature (215°C–250°C) and pressure under the action of MW radiation was patented: in the first stage, with a mixture of nitric acid and hydrogen peroxide, and in the second stage, with a mixture of concentrated nitric, hydrochloric, and hydrofluoric acids.[89] Milder ND-purification conditions were also reported, for instance adsorption onto a diamond nanoparticle from a bath of 10–100 g/L $(NH_4)_2Ce(NO_3)_6$ or $(NH_4)_2Ce(SO_4)_3$ solution at 30°C–70°C for 24–48 h, washing with 1.5 vol.% H_2SO_4 three to five times, washing with water, and vacuum drying at 100°C–120°C for 12–24 h.[90] A reported gas-phase technique for ND purification and modification consisted in passing a gas mixture, in the form of a dried air, oxygen, or ozone at a pressure up to 0.8 MPa, through an initial material, and simultaneously heating it from 20°C to 550°C.[91] Valuable information on various techniques for diamond purification was critically analyzed and reviewed by Lin et al.[92]

18.2   TECHNIQUES APPLIED TO STUDY STRUCTURAL AND ELECTRONIC PROPERTIES OF NANODIAMONDS

In order to study the structure of formed carbon allotropes, in particular the NDs, classic methods, usually used in materials characterization, were applied. Thus, magnetic resonance techniques (EPR and solid-state NMR[93]) as powerful nondestructive tools for studying electron-nuclear and crystalline structure, inherent electronic and magnetic properties, and transformations in carbon-based nanomaterials, in particular NDs, were reviewed by Shames et al.[94] EPR allows controlling the purity of UDD samples and studying the origin, location, and spin-lattice relaxation of radical-type carbon-inherited paramagnetic centers, among other possibilities, meanwhile solid-state NMR provides information on the crystalline quality, allows quantum estimation of the number of different allotropic forms, and reveals electron–nuclear interactions within the UDD samples under study. Raman spectroscopy of amorphous, nanostructured, diamond-like carbon and ND was reviewed by

Ferrari et al.[95] This method was useful in the determination of sp^2 carbon phases in NCD films.[96] Additionally, ND formation and its *in situ* Raman spectroscopy under the conditions of high static pressures and high temperatures were studied using carbon of different isotopic compounds in the K_2O-Na_2O-Al_2O_3-SiO_2 system and the presence of $NaAlSi_2O_6$, low-melting $K_2Si_2O_5$, and Na_2SiO_3.[97]

The sp^2/sp^3 bonding concentrations of nitrogen-doped amorphous carbon samples and ND films were determined from their soft x-ray absorption spectra.[98] The ND films were synthesized on silicon substrates in a CH_4/H_2 gas mixture by MWCVD. It was shown that ND deposition conditions, such as bias voltage and methane concentration, affect the purity of the film. The stress evolution in NCD films deposited at different temperatures (from 800°C to 400°C) was investigated,[99] showing that the intrinsic stress gradually changed from tensile to compressive with decreasing deposition temperature. A comparison study of hydrogen incorporation among nano-, micro-, and polycrystalline diamond (PCD) films grown by MWCVD was performed by Tang et al.,[100] indicating from polycrystalline to NCD film, hydrogen impurity content increases drastically with decrease in grain size. It was suggested that hydrogen incorporated not only into grain boundaries but also into other structural defects of CVD diamond. The electric conductivity and sedimentation stability of aqueous dispersions of ND agglomerates were studied by Koroleva et al.[101] The sedimentation kinetic curves showed two sections of fast and slow sedimentation. The consequences of strong and diverse electrostatic potential fields on the surface of detonation ND particles were reported by Osawa et al.[102] DNDs were also studied by time-of-flight mass spectrometry (TOF MS)[103] showing the formation of singly charged C clusters, C_n^+, with groups of clusters at $n = 1$–35, $n \sim 160$–400, and clusters with $n \sim 8000$. It was established that high C clusters consisted of an even number of carbons while the percentage of odd-numbered clusters is quite low ($\sim 5\%$–10%).

Structural and electronic properties of isolated NDs were reviewed by Raty et al.[104] Using *ab intio* calculations, it was shown that in the size range of 2–5 nm ND (so-called bucky diamond) has a fullerene-like surface and, unlike silicon and germanium, exhibits very weak quantum confinement effects. Ultrafast photoluminescence of NCD membranes and films on Si and fused SiO_2 glass, prepared by MWCVD, measured by a femtosecond upconversion technique, was reported by Preclikova et al.[105] It was established that the photoluminescence dynamics were affected by temperature and ambient pressure. In addition to the generalized report on the ND surface chemistry,[106] detailed structural investigations of patterned NCD submicrometer-tip arrays, synthesized by conformal coating of the SiO_2 nanowires with 5–10 nm sized ND grains by MWCVD, were carried out by high-resolution transmission electron microscopy.[107] Electron field emission of ND emitter arrays was observed with a threshold field of 5.5 V/μm. The structural and chemical analysis of the soot particles produced in a bell jar reactor under Ar/H_2/CH_4 microwave discharges employed for NCD film synthesis was reported by Aggadi et al.,[108] showing that the soot was mainly composed of polyhedral graphite particles of roughly 100 nm size and of smaller particles of the same carbon structure, embedded in graphite crumpled sheets. Chemical bonding study of NCD films prepared by various plasma techniques (MPWCVD and hybrid pulsed laser deposition [HPLD]) was carried out by Popov et al.[109] In the case of HPLD samples, the formation of NCD was enhanced by lower working pressures and RF powers; no such influence of the deposition conditions on the bonding structure was observed for the MWCVD films.

18.3 SOME PHYSICAL PROPERTIES OF NANODIAMONDS

The *stability* of NDs was reviewed by Barnard,[110] in particular such aspects as theoretical and computational studies on the relative stability of carbon nanoparticles, phase transitions between ND and graphitic and fullerenic forms of nano-C, coexistence of bucky diamond and other nano-C phases, stability of diamond nanowires, simulation of transitions of NDs, etc. It was distinguished between the stability of large (5–6 nm) and small (<2.5 nm) particles of nano-C. In an interesting report,[111] the stability of NDs with sizes less or more than 4 nm was studied by x-ray coherent scattering areas (CSA) and of sample specific surfaces. The authors supposed that small ND particles

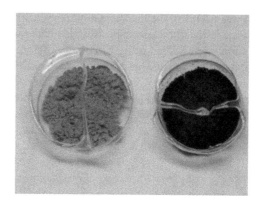

FIGURE 18.3 Initial (left) and oxidized (right) NDs. (Reproduced with permission from Springer Science+Business Media: *J. Superhard Mater.*, On the stability of small sized nanodiamonds, 29(2), 2007, 124–125, Gordeev, S.K. and Korchagina, S.B.)

exhibited higher activity when interacted with oxygen and "burned out," caused by the heat effect of the carbon oxidation with oxygen (Reactions 18.1 and 18.2). As a result of the detailed study of ND oxidation with water (Figure 18.3), it was established that it was impossible to produce NDs with particles of size below 4 nm using the oxidation processes. It was concluded that there is a critical size of the stability of ND particles, namely, about 4 nm. An ND cluster with a size below the critical one becomes unstable and rapidly oxidizes in air or transforms into a particle with a graphite-like structure. Additionally, in ND crystallites, exposed to extreme UV radiation, the surface H groups and graphite species were found to prevent graphitization, whose radiation-induced mechanism was discussed, as well as other effects influencing the stability of ND in radiation conditions[112]:

$$C + O_2 = CO_2 + 384 \text{ kJ/mol} \tag{18.1}$$

$$C + \frac{1}{2}O_2 = CO + 113 \text{ kJ/mol} \tag{18.2}$$

Nitrogen is a common *impurity* in the crystal of diamond and also one of the predominant elements to affect diamond performance[113]; nitrogen impurity location, related to the luminescent properties of DND, was detected in several detonation DND samples using a combination of TEM techniques,[114] showing that the size and morphology of DND can be modified by temperature treatment. It was suggested[115] on the basis of spectral data and elemental analysis that nitrogen can be mainly in the form of impurity N-centers of different kinds inside the diamond core of ND particles. Among other impurities in diamonds and NDs, several elements were found in the combustion residues, such as Fe, Ni, Ca, Al, Si, S, Cl, K, Na, Ti, and Mg.[116] It was observed that metal impurities more easily enter into the diamonds, especially Fe-Ni alloy. Nonuniform distribution of impurities in the diamond crystal, especially in the surface region, was the main reason for strength variation in the samples.[117] Additionally, the behavior and influence of cobalt,[118] boron,[119] beryllium, and magnesium,[120] and other dopants on diamond and ND properties and reactivity were studied.

Discussing the *solubility* of NDs, it is necessary to mention that bad dispersibility is a major cause that hinders the application of ND severely. In this respect, the techniques allowing increasing ND solubility for medical and other applications are being intensively developed. Thus, the method producing a dispersion of ND particles (clusters with a primary particle diameter of 3–5 nm and produced by a shock compaction method under a H_2 atmosphere at 300°C–800°C) in organic solvents by disaggregating an aggregate of the diamond particles was proposed by Sakai et al.[121] The formed dispersions did not cause aggregation or precipitation even after six months and maintain their stable dispersion state. Covalent functionalization of ND, leading to an increased solubility, was carried out by employing several methods,[122] involving, for example, the reaction of acid-treated ND with thionyl chloride followed by reaction with a long-chain aliphatic amine to produce the amide derivative. The products of functionalization produced excellent dispersions in CCl_4 and toluene. Interaction of ND with surfactants such as sodium bis(2-ethylhexyl) sulfosuccinate gives good dispersions in water. Other functionalization techniques resulting in enhanced solubility/dispersibility are presented later.

The *growth* of NDs has been intensively studied in whole as a mechanism of each of its steps including the pretreatment of precursors before the reaction and nucleation. Thus, effects of pretreatment on the nuclei formation of UNCD on Si substrates were studied.[73] Either precoating a thin layer of titanium (~400 nm) or ultrasonication pretreatment using diamond and titanium mixed powder enhanced the nucleation process on Si substrates markedly. Nucleation, growth, and low-temperature synthesis of diamond thin films were reviewed by Das et al.[123] The important parameters responsible for enhancing the nucleation and growth rates of diamond film were identified, and the growth of diamond thin films at low temperatures were discussed.

Various supports/substrates were investigated in respect of diamond nucleation. Thus, studies of diamond nucleation and growth on conical carbon tubular structures showed that the nucleation preferentially occurs at the tips but only occurs on the sidewalls when they are pretreated with diamond or other powder dispersions, forming an ND coating.[124] It was revealed that the diamond nucleation on the sidewalls may proceed through the formation of diamond nuclei within the walls at subsurface damage sites caused during pretreatment. MWPCVD was used to deposit a smooth NCD film on a pure titanium substrate using Ar, CH_4, and H_2 gases at moderate deposition.[125] The exceptional adhesion of a 2 μm thick diamond film to the metal substrate was observed by indentation testing ≤150 kg load. This type of materials could be used for aerospace and biomedical applications, avoiding the problem of the different thermal expansion coefficients of thin ND and titanium and its metallic alloys.[126] Nanocomposite films consisting of diamond nanoparticles of 3–5 nm diameter embedded in an amorphous carbon matrix were deposited by means of MWCVD from CH_4/N_2 gas mixtures[127] on different substrates (some of them were pretreated ultrasonically) such as Si wafers, Si coated with TiN, PCD and cubic boron nitride films, and Ti-6Al-4V alloy. The growth process seemed to be unaffected by the substrate material and the crystallinity, and the bonding environment showed no significant differences for the various substrates. A route to high-purity NCD films from C_2 dimers and related mechanisms was investigated[128] by enhancing C_2 growth chemistry in Ar-rich microwave plasmas. C_2 grows ND on diamond surfaces but rarely initiates nucleation on foreign surfaces. The phase purity can be improved by increasing the dominance of ND growth from C_2 over nondiamond growth from CH_x ($x = 0$–3) and large radicals. It was found that the growth of ND thin films from hydrocarbon/Ar-rich plasmas was very sensitive to the contribution of C_2 and C_2H species from the plasma.[129] Among other growth studies, the effect of total pressure on growth rate and quality of diamond films prepared by MWCVD was investigated by Li et al.,[130] showing that when the total pressure changed from 1.03×10^4 to 1.68×10^4 Pa the growth rate increased from 3 to 16 μm/h. Under a gas pressure lower than that used for growth, a high nucleation density of diamond films on alumina, as high as 10^8/cm^2, and creation of [100]-textured diamond films deposited on alumina were successfully achieved by MWCVD at 800°C–860°C.[131] NCD films are promising candidates for tribological applications, in particular when deposited on hard ceramic materials such as silicon nitride (Si_3N_4). Thus, MWCVD deposition of NCD was achieved using Ar/H_2/CH_4 gas mixtures on plates and ball-shaped Si_3N_4 specimens by a conventional continuous mode.[132]

18.4 CHEMICAL PROPERTIES AND FUNCTIONALIZATION OF NANODIAMONDS

The structure, solubility, and electronic/magnetic properties of ND vary depending on how the surface carbon atoms are terminated.[133] Therefore, its functionalization with active *biomolecules* is attractive for possible medical applications, among many others. In this relation, different approaches to the surface functionalization of ND, that is, the key in successful biomedical applications (in particular, biosensing), as well as modification of diamond surfaces with nucleic acids and proteins, were reviewed by Grichko et al.[134] Linear pUC19 molecules with blunt ends, prepared by restriction of the initial ring form of pUC19 DNA, and linear 0.25–10 kb DNA fragments were found to be adsorbed on NDs.[135] The amount of adsorbed linear DNA molecules depended on the size of the molecules and the size of the ND clusters. Size separation (important for passive targeting such as EPR effect in tumor imaging) of ND particles (4–25 nm) by the use of ultracentrifugation and synthesis of fluorescent ND through the surface chemical modifications for adding requisite functions to NDs such as dispersibility and visibility were reported by Komatsu.[136]

The functionalization of ND with *nonbiological organic molecules* includes the use of, in particular, amine derivatives, whose attachment led to enhanced ND solubility. Thus, ND was covalently functionalized with 1,3-propanediamine to obtain a novel derivative (ND-NH$_2$) with 29.97% of ND, in which the 1,3-propanediamine molecules were covalently attached to the NDs through amide linkages.[137] This compound dissolves not only in inorganic or organic acidic aqueous solutions but also in organic solvents such as acetone, CH$_2$Cl$_2$, NMP, DMF, DMAc, and DMSO; this might be a beneficial base for the next study and application of ND in the fields of complex electroplating and lubrication. A radio-frequency plasma discharge in the vaporized silane coupling agent *N*-(6-aminohexyl) aminopropyl trimethoxysilane was applied to successfully functionalize the NCD surface with the primary amino group NH$_2$ as confirmed by using fluorescamine-in-acetone spray as a fluorescence marker.[138] Hydrophobic blue fluorescent ND, easily dispersible in hydrophobic solvents and forming a transparent colloidal solution, was synthesized by covalent linking of octadecylamine (ODA) to the surface of ND particles (Reaction 18.3).[139] This material can be used in those applications where stable dispersions of ND in fuels, polymers, or oils are required. Due to the long hydrocarbon chains linked to its surface, the ND-ODA can be easily dispersed in hydrophobic solvents such as benzene, toluene, chloroform, dichloromethane, etc. (Figure 18.4). At the same time, it is immiscible with water and poorly miscible with polar hydrophilic organic solvents such as DMF, ethanol, methanol, and acetone.

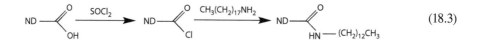

$$(18.3)$$

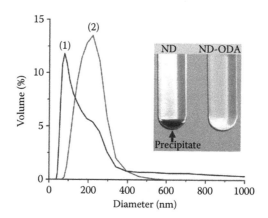

FIGURE 18.4 Particle size of ND-ODA in toluene (1) and chloroform (2); photograph of 0.01% wt. suspensions of ND and ND-ODA in toluene. (Reproduced with permission from Mochalin, V.N. and Gogotsi, Y., Wet chemistry route to hydrophobic blue fluorescent nanodiamonds, *J. Am. Chem. Soc.*, 131(13), 4594–4595. Copyright 2009 American Chemical Society.)

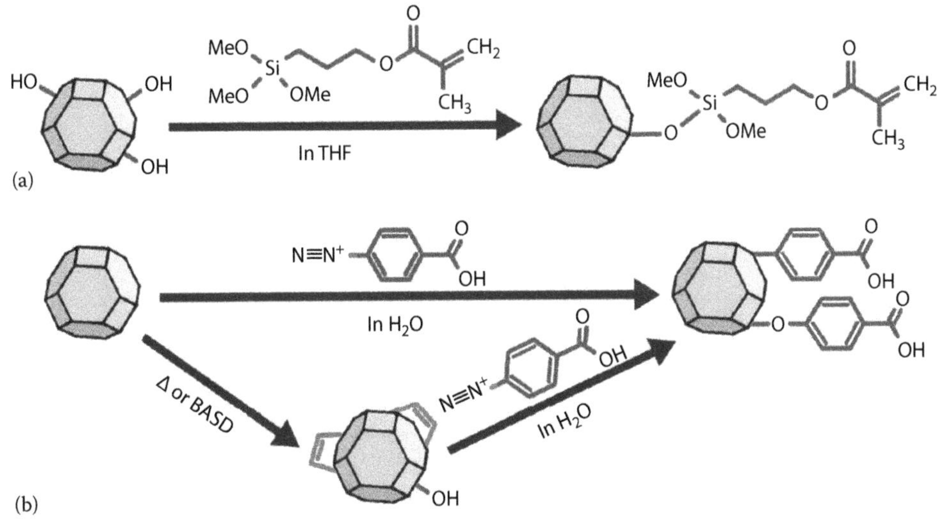

FIGURE 18.5 Surface functionalization of ND. (a) Silanization of ND surfaces. (b) Arylation of NDs by applying aromatic diazonium salts. (Reproduced with permission from Liang, Y. et al., A general procedure to functionalize agglomerating nanoparticles demonstrated on nanodiamonds, *ACS Nano*, 3(8), 2288–2296. Copyright 2009 American Chemical Society.)

A general procedure to functionalize agglomerating nanoparticles demonstrated on ND was examined by Liang et al.,[140] who reported a new technique to facilitate surface chemistry of nanoparticles in a conventional glassware system offering a beads-assisted sonication (BASD) process to break up persistent agglomerates of NDs in two different reactions (silanization with an acrylate-modified silane [Figure 18.5a] and the arylation using diazonium salts [Figure 18.5b]) for simultaneous surface functionalization. This allowed for the efficient grafting of aryl groups to the surface of primary diamond nanoparticles yielding stable, homogeneously functionalized ND particles in colloidal solution. Even in media where the original particles were flocculating, an efficient functionalization took place. The latter led to truly dispersed primary nanoparticles with a homogeneous surface termination. In a related work, functionalization of ND particles with aryl organics using *Suzuki* coupling reactions was demonstrated by Yeap et al.[141] derivatizing hydrogenated ND with aryl diazonium or with boronic acid groups followed by Suzuki cross coupling with arenediazonium tetrafluoroborate salts, in particular in a capillary microreactor (Figure 18.6). These reactions have general validity to a wide range of organic dyes that have boronic acid or halide functional groups. The dispersion of the ND in organic solvent can be thus improved by coupling hydrophobic aryl groups.

Among classic works on *"inorganic" functionalization* of ND,[142] at least at the first step, we note a suggested technology of ND surface modification,[143] which allowed separation of commercial ND powders into two fractions (differing in size characteristics), each possessing absolutely new properties as compared with the initial powder. Physical and chemical properties of these modified NDs were studied. Analysis of structural and chemical impurities in NDs, polyfunctional surface termination, agglomeration, and other features that may restrict the ND application in academic research and industrial practice were reviewed by Spitsyn et al.,[144] as well as ND purification and surface functionalization, using a high-temperature treatment in gaseous media containing hydrogen and chlorine. The methods for the covalent functionalization of nanosized (~2–10 nm) diamond powders, as well as CNTs, were reviewed by Khabashesku et al.[145] The described methods involved direct fluorination, organic free radical additions, fluorine displacement reactions in fluoro-nanotubes, and fluoro-ND producing amino, hydroxyl, and carboxyl group-terminated derivatives.[146] Thus, the reaction of nanoscale diamond powder with an elemental fluorine/hydrogen mixture at 150°C–470°C resulted in ND surface fluorination yielding a fluoro-ND with up to 8.6 at. % fluorine content,[147–149]

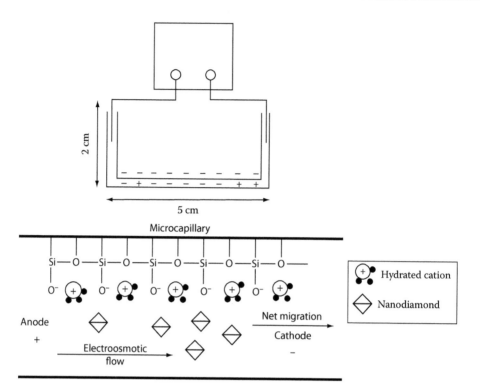

FIGURE 18.6 (Top) Glass microcapillary reactor for Suzuki coupling, (bottom) schematic showing electroosmotic flow in the glass microcapillary. (Reproduced from Yeap, W.S. et al., Detonation nanodiamond: An organic platform for the suzuki coupling of organic molecules, *Langmuir*, 25(1), 185–191. Copyright 2009 American Chemical Society.)

used as a precursor for the preparation of a series of functionalized NDs by subsequent reactions with alkyllithium reagents, diamines, and amino acids (Scheme 18.1), as well as a starting material for the diamond coating on modified glass surface (Figure 18.7). All functionalized NDs showed an improved solubility in polar organic solvents, for instance alcohols and THF, and a reduced particle agglomeration. The developed method can be extended to ND coating of various substrates, such as quartz, silicon, metals, and other materials.

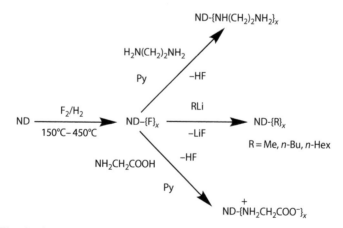

SCHEME 18.1 Fluorination and further functionalization of ND with amines and amino acids. Copyright. Reproduced with permission from Liu, Y. et al. Functionalization of nanoscale diamond powder, *Chemistry of Materials*, 2004, 16(20), 3924–3930.

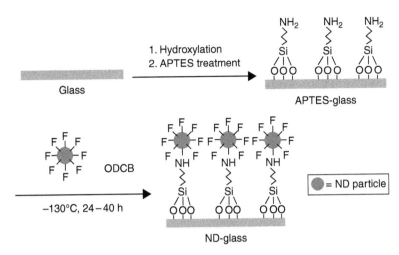

FIGURE 18.7   Reaction steps for coating glass with fluoro-ND. APTES is 3-aminopropyltriethoxysilane, ODCB is 1,2-dichlorobenzene. (Reproduced with permission from Liu, Y. et al., Fluorinated nanodiamond as a wet chemistry precursor for diamond coatings covalently bonded to glass surface, *J. Am. Chem. Soc.*, 127(11), 3712–3713. Copyright 2005 American Chemical Society.)

In addition to functionalization reactions discussed earlier, *graphitization* processes of diamond with emphasis on the low-temperature graphitization at 1370–1870 K were described and reviewed by Kunzetsov et al.[150] Some nanocarbons (such as onion-like carbon (OLC), sp^2/sp^3 nanocomposites, and nanographite) can be thus produced using controlled ND graphitization. A theoretical analysis of the graphitization of an ND in the metastable state was reported by Kwon et al.[151] An ND annealed at a relatively lower temperature suffers morphological transition into an ND-graphite core-shell structure. It was established that the ND was not completely transformed into graphite by simple annealing at a relatively lower process temperature and pressure and the highest graphitization probability decreased with increasing annealing temperature. However, according to Leyssale's data[152], the annealing of a small ND cluster at 1500 K, as studied by molecular dynamics, led to an almost fully graphitized carbon onion. The remaining 17% of sp^3 atoms were delocalized on the whole particle, both under the form of isolated point defects and of small diamond-like clusters separating large graphite-like domains. This is in accordance with the results of similar investigations, where it was found that the diamond nanoparticles begin to graphitize in the range of 1100°C–1200°C and all the particles transform into OLCs at 1400°C.[153] Other aspects of graphitization have also been discussed,[154–156] in particular ND graphitization in conditions of laser ablation[157] or stimulated by electron-hole recombination.[158]

18.5   NANODIAMOND FILMS AND COMPOSITE MATERIALS

ND films and the methods and mechanisms of their formation have been extensively covered. Thus, ND films deposited by energetic species using the direct-current glow-discharge (DC-GD) deposition technique were reviewed by Hoffman et al.[159,160] ND film and growth were explained as a subsurface process in terms of a four-step cyclic process. ND films with a large area, low stress, and smooth surface were deposited on mirror-polished Si wafers under continuous ion bombardment by a hot-filament CVD (HFCVD) chamber equipped with a substrate bias system.[161] The size of diamond clusters in the film can be controlled by a suitable selection of deposition parameters such as the ratio of CH_4 to H_2. To verify a statement that ion bombardment is the main mechanism in the bias-enhanced nucleation process during the initial stage of diamond film growth by CVD, mass-separated ion-beam deposition experiments were carried out, in which a pure C ion beam, with precisely defined low energy, was selected for investigating the ion-bombardment effect on a Si substrate.[162] The obtained results supported the ion-bombardment-enhanced nucleation mechanism. NCD films were found to be a good support for adhesion, growth, and differentiation of osteogenic cells and could be used for surface modification of bone implants to improve their integration with the surrounding bone tissue.[163]

Such substrates as carbides or silicon were used as supports for ND deposition by MPCVD techniques. Thus, the diamond coating containing elemental Si was deposited on a cemented carbide substrate with the use of H_2, CH_4, and octamethyl cyclotetrasiloxane (D4) as precursors by MWCVD.[164] The results revealed that when the flow of D4 was larger than that of CH_4, the cellular structure was obtained, and the diamond coating with good quality and good adhesion was deposited with a little Si only when the flow of D4 was equivalent to that of CH_4. NCD thin films (grain size 10–20 nm) were grown on silicon substrates at 480°C using methanol and hydrogen mixtures in MPCVD conditions.[165] The high-quality NCD film could be deposited on a 50 mm diameter Si substrate by increasing the methanol concentration and reducing the deposition temperature. NCD films are highly strained materials, in general.[166] Unstrained failure-resistant NCD films were obtained by performing bias-enhanced nucleation in an MWCVD system followed by NCD deposition in a HFCVD system. A simple method for synthesizing a diamond nanofilm with MW plasma at low temperature, patented by Zhou et al.,[167] included the steps of (a) introducing argon and hydrogen to a microwave plasma synthesis chamber, (b) ionizing at 600–750 W and 900–1100 Pa to obtain plasma, and (c) standing till the plasma is stable and introducing methane to precipitate a diamond nanofilm.

A great variety of composite materials have been manufactured on the basis of ND and *polymers* to improve properties of polymer matrixes. Thus, an efficient method was developed to achieve improved dispersion of detonation ND particles in amorphous thermoplastic matrices. It was found that the reinforcing and toughening effects of uniformly dispersed nanoparticles on polymer matrixes were pronounced at lower loading compared with traditional mixing procedure.[168] Transparent heat-resistant polymer-ND composites were obtained by Sawaguchi et al.[169] by dispersing ND in a polymerization solvent at the stage of polymerization reaction. Binary combinations of ND, few-layer graphene, and single-walled nanotubes were used to reinforce polyvinyl alcohol.[170] The mechanical properties of the resulting composites showed extraordinary synergy, improving the stiffness and hardness by as much as 400% compared with those obtained with single nanocarbon reinforcements. The effects of modifying additives of nanodispersed substances on rheological and relaxation properties of polymer-ND composite materials based on polydimethylsiloxane,[171] polyurethanes, nitrile rubber, and polycarbonates[172] were established. An effective method for introduction of dispersed filler particles of detonation ND into a polymeric matrix, based on disaggregation and uniform deposition of the filler onto the surface of polymeric particles in an inert liquid medium under the action of ultrasound, was developed.[173] Among other similar composites, the preparation and application of fluoroalkyl-containing oligomer-ND composites was reported.[174]

Inorganic-based diamond and ND composites, generally leading to hard carbide- or nitride-containing coatings, are also known. Thus, an ultrahard multilayer coating was composed of two chemically different layered nanocrystalline materials, ND and nanocubic boron nitride (nano-cBN).[175] Nano-cBN layers were directly grown on ND crystallites using ion-assisted physical vapor deposition and ion-assisted plasma enhanced CVD (PECVD), again followed by ND deposition using CVD methods in cycles until the intended number of layers of the final composite was obtained. The obtained multilayer structure possessed extreme hardness (82 GPa), high surface smoothness, significantly reduced film stress compared with a single cBN layer of equivalent thickness, and great chemical stability.[176] A study of MPCVD diamond deposition on Si_3N_4-TiN composites with different TiN amounts (0–30 vol.% TiN) was performed[177] to obtain a suitable material to be cut by electrodischarge machining, aiming at their use as substrates for cutting tools and tribological components. Composite thin films of ND and silica nanotubes were synthesized by means of MPCVD on a silica nanotube matrix that was seeded with ND particles.[178] ND grew around silica nanotubes and filled the space left between silica nanotubes to form a continuous film. In a related work, the fabrication of high-density NCD-coated silica nanofibers with diameters of 1–5 μm was reported by

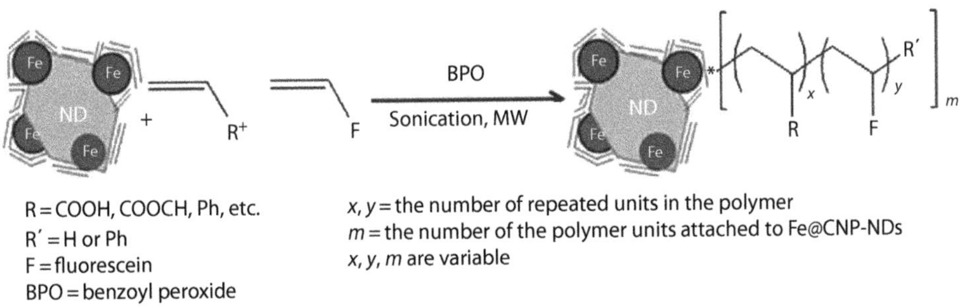

R = COOH, COOCH, Ph, etc.
R′ = H or Ph
F = fluorescein
BPO = benzoyl peroxide

x, y = the number of repeated units in the polymer
m = the number of the polymer units attached to Fe@CNP-NDs
x, y, m are variable

FIGURE 18.8 Schematic process for surface functionalization of magnetic NDs. (Reproduced with permission from Chang, I.P. et al., Preparation of fluorescent magnetic nanodiamonds and cellular imaging, *J. Am. Chem. Soc.*, 130(46), 15476–15481. Copyright 2008 American Chemical Society.)

a template technique,[179] consisting of the preparation of templates (SiO_2 nanofibers) by conventional vapor-liquid-solid method and the conformal coating of the nanofibers with ND by the MWCVD technique in hydrogen-deficient conditions. A composite with hardness over 55 GPa is obtained by cobalt infiltration (min. 6 vol.%) into an ND at a pressure of 8 GPa.[180] In the presence of cobalt, graphite-like carbon was inversely transformed into diamond with a slight time delay. The hybrid diamond-graphite nanowires, consisting of a single crystalline diamond core of 5–6 nm in diameter oriented along the [110] principal axis, and graphite shells of different thicknesses covering the core, with lengths up to a few hundred nanometers were identified as structural units in UNCD films synthesized by CVD with nitrogen gas added.[181]

Water-soluble (2.1 g/L) magnetic NDs (Figure 18.8) were prepared *via* solid-state microwave arcing of an ND-ferrocene mixed powder in a focused microwave oven.[182] It was shown that a magnetic nanodiamond (MND) was composed of iron nanoparticles encapsulated by graphene layers on the surface of NDs, had a saturation magnetization of ~10 emu/g and a coercivity field 155 G, and *via* covalent surface grafting of organic fluorescent molecules can be converted to fluorescent MNDs (FMNDs), whose color of fluorescence was readily tunable depending on the polymerizable fluorescent monomer used in the surface grafting process. The FMNDs had an aqueous solubility of ~2.1 g/L and were collectible with an external magnet, which is important in many biomedical applications. The authors stated that these NDs could serve as biocompatible cargos to transport various biologically active species, such as proteins, DNA, signal-transduction molecules, and drugs, into cells for various biomedical investigation or treatment of diseases. Also, ND composites were produced, in particular, in low-pressure microwave gas-discharge plasma,[183] laser-induced phase transition (ND on cast alloy A319),[184] or by the incorporation of ND into plated films,[185] among other techniques.[186–188]

18.6 APPLICATIONS

NDs have a series of distinct applications in various areas, in particular medicine, electrochemistry, and creation of novel materials. Thus, the sensing film for the gas sensor on the ND basis was prepared by coating semiconductor particle mixtures on a substrate and making semiconductor particles adsorb onto a metal film.[189] Nitrogen-incorporated ND electrodes were fabricated using MWCVD by N_2 incorporation achieved by the introduction of N_2 along with H_2 and CH_4 in the plasma[190] and used to study the detection of dopamine showing excellent electrochemical response. The electrochemical response of an electrode-immobilized layer of undoped, insulating diamond nanoparticles was attributed to the oxidation and reduction of surface states.[191] It was established that the potentials of these surface states were pH dependent and able to interact with solution redox species. Another electrochemistry-related achievement was the method for galvanic deposition of an ND-based coating compound by the introduction of aqueous suspension of NDs into a solution of electrolyte to initiate coating sedimentation by electrolysis, which was offered to improve the wear resistance of metal surfaces.[192] ND applications in the production of composite electrolytic coatings on the basis of Cr, Ni, Au, and Ag exhibited good service properties and savings on the use of noble metals and electrical power were exemplified,[193] as well as ND applications for the modification of plastics, antifriction lubricants, and oxide coatings grown by microarc oxidation of aluminum alloys.

The normal-temperature glass was prepared by acid- or alkali-hydrolyzing and dehydrocondensing an aqueous or alcohol solution of ND-dispersed metal alkoxide, applying on a substrate, and drying at $\leq 200°C$.[194] This glass gave a film with good deodorization and antibacterial effect owing to ND and can be applied on the substrate with any surface. The method of obtaining a polyfunctional ND sorbent for sorption and chromatographical separation, in particular of protein molecules, was offered including processing of a polysaccharide matrix, which was first cleaned from preservatives with distillate water, and then processed with 0.5–10.0 wt.% ND hydrosol.[195] The possibilities of applying DND in untraditional areas were emphasized,[196] in particular their prospective use as polyfunctional adsorbents for purification of the proteins and enterosorbent and compound carrier applied for medical purposes. The protection of documents, valuable papers, and products with diamond nanocrystals with active nitrogen-vacancy center by application or insertion of a special fluorescent mark on the protected document was offered by Zibrov et al.[197] ND was offered as a new type of lubricating additive due to its double performances of both diamond and nanomaterial.[198] The key problem was the surface modification of ND aggregates to achieve the deaggregation and stable dispersion of ND in the oil. As an example of a composition for such a lubricant, the friction material was manufactured from nanometer diamond 1.2%–14%, bronze powder 2%–25%, rare earth oxides or fluorides, among others, 1%–50%, and polytetrafluoroethylene to 100%.[199] Among other nonmedical ND applications, we note its use as an additive for a photocurable composite resin matrix, whose main filler was barium glass,[200] and as a reducing agent in energetic chlorate-based compounds.[201]

Biomedical applications of NDs are well developed and related with the recently established fact that carbon NDs are much more biocompatible than most other carbon nanomaterials, including carbon blacks, fullerenes, and carbon nanotubes.[202] Their tiny size, large surface area, and ease of functionalization with biomolecules make NDs attractive for various biomedical applications both *in vitro* and *in vivo*, for instance for single particle imaging in cells, drug delivery, protein separation, and biosensing. Some radiation-damaged NDs can emit strong and stable photoluminescence (red or green) from nitrogen-vacancy defect centers embedded in the crystal lattice.[203] Specific use of NDs in both nonconjugated and conjugated forms as enterosorbents or solid phase carriers for small molecules including lysozyme, vaccines, and drugs was reviewed.[204]

Among a series of biomedical applications of NDs, we note direct observation of a growth hormone receptor in one single cancer cell using ND–growth hormone complex as a specific probe, reported by Cheng et al.[205] The growing hormone molecules were covalently conjugated to 100 nm diameter carboxylated NDs, which can be recognized specifically by the growth hormone receptors of A549 cell. Bovine insulin was noncovalently bound to detonated NDs *via* physical adsorption in an aqueous solution and demonstrated pH-dependent desorption of sodium hydroxide.[206] NDs combined with insulin at a 4:1 ratio showed 79.8% ± 4.3% adsorption and 31.3% ± 1.6% desorption in pH-neutral and alkaline solutions, respectively. The location and distribution of 100 nm carboxylated ND particles in cell division and differentiation were investigated by Liu et al.[207] ND's clusters were carried inside of a cell but without inducing damages after long-term cell culture. Endocytic ND particles were found to be noncytotoxic in cell division and differentiation, which can be applied for the labeling and tracking of cancer and stem cells. A platform approach of water-dispersible, ND cluster-mediated interactions with several therapeutics (Purvalanol A and 4-hydroxytamoxifen) to enhance their suspension in water with preserved functionality was demonstrated by Chen et al.[208] This approach served as a facile, broadly impacting, and significant route to translate water-insoluble compounds toward treatment-relevant scenarios. For diamond nanoparticles used as fluorescent labels in cells,[209] the photoluminescence of a single color center embedded in a 30 nm diameter ND was compared with that of a single dye molecule and demonstrated perfect photostability of the color centers. Optimal parameters to achieve a high fluorescence yield were determined.

REFERENCES

1. Danilenko, V. V. On the history of the discovery of nanodiamond synthesis. *Physics of the Solid State*, 2004, *46* (4), 595–599.
2. Shenderova, O. A.; Zhirnov, V. V.; Brenner, D. W. Carbon nanostructures. *Critical Reviews in Solid State and Materials Sciences*, 2002, *27* (3–4), 227–356.
3. Williams, O. A.; Nesladek, M.; Daenen, M.; Michaelson, S.; Hoffman, A.; Osawa, E.; Haenen, K.; Jackman, R. B. Growth, electronic properties and applications of nanodiamond. *Diamond and Related Materials*, 2008, *17* (7–10), 1080–1088.
4. Huss, G. R. Meteoritic nanodiamonds: Messengers from the stars. *Elements*, 2005, *1* (2), 97–100.
5. Jones, A. P.; d'Hendecourt, L. B. Interstellar nanodiamonds. *Astronomical Society of the Pacific Conference Series*, 2004, *309*, 589–601.
6. Khan, Z. H.; Husain, M. Nanodiamond: Synthesis, transport property, field emission and applications. *Material Science Research India*, 2006, *3* (1a), 1–22.
7. Tanaka, A. Tribology of carbon composites. *Toraiborojisuto*, 2009, *54* (1), 16–21.
8. Hu, X.; Li, M.; Sun, Z.; Wang, Q.; Fan, D.; Chen, L. Research status and prospect of synthetic nano-diamonds. *Guanli Gongchengban*, 2009, *31* (2), 301–304, 317.
9. Osawa, E. Nano carbon materials. Fullerenes and diamond. *Seramikkusu*, 2004, *39* (11), 892–909.
10. Quiroz Alfaro, M. A.; Martinez Huitle, U. A.; Martinez Huitle, C. A. Nanodiamantes. *Ingenierias*, 2006, *IX* (33), 37–43.
11. Freitas Jr., R. A. A simple tool for positional diamond mechanosynthesis, and its method of manufacture. http://www.MolecularAssembler.com/Papers/DMSToolbuildProvPat.htm, 2003–2004.
12. Freitas Jr., R. A. How to make a nanodiamond. http://www.kurzweilai.net/articles/art0632.html?printable=1, 2006.
13. Doan, Ho. *Nanodiamonds: Applications in Biology and Nanoscale Medicine*. Springer, Berlin, Germany, 2009, 380 pp.
14. Shenderova, O. A.; Haber, D. M. *Ultrananocrystalline Diamond: Synthesis, Properties, and Applications*. William Andrew, Norwich, NY, 2007, 620 pp.
15. Tjong, S. C. Properties of chemical vapor deposited nanocrystalline diamond and nanodiamond/amorphous carbon composite films. *Nanocomposite Thin Films and Coatings*, 2007, 167–206.
16. Yang, N.; Uetsuka, H.; Williams, O. A.; Osawa, E.; Tokuda, N.; Nebel, C. E. Vertically aligned diamond nanowires: Fabrication, characterization, and application for DNA sensing. *Physica Status Solidi A: Applications and Materials Science*, 2009, *206* (9), 2048–2056.
17. Kong, X.-L. Nanodiamonds used as a platform for studying noncovalent interaction by MALDI-MS. *Chinese Journal of Catalysis*, 2008, *26* (10), 1811–1815.
18. Tang, C. J.; Neves, A. J.; Gracio, J.; Fernandes, A. J. S.; Carmo, M. C. A new chemical path for fabrication of nanocrystalline diamond films. *Journal of Crystal Growth*, 2008, *310* (2), 261–265.
19. Watanabe, M.; Yusa, H. Introduction of advanced materials laboratory and some topics—From repletion of basic research to development of practical materials. *Materials Integration*, 2002, *15* (9), 8–13.
20. Yang, Q.; Yang, S.; Xiao, C.; Hirose, A. Transformation of carbon nanotubes to diamond in microwave hydrogen plasma. *Materials Letters*, 2007, *61* (11–12), 2208–2211.
21. Kimura, Y.; Kaito, C. Production of nanodiamond from carbon film containing silicon. *Journal of Crystal Growth*, 2003, *255* (3–4), 282–285.
22. West, J. A.; Kennett, J. Nanodiamonds and diamond-like particles from carbonaceous material. PCT International Application, WO 2009094481 A2 20090730 Application: WO 2009-US31731 20090122. Priority: U.S. Patent 2008–62350 20080125. CAN 151:227622 AN 2009:917817, 2009, 25 pp.
23. Dolmatov, V. Yu.; Veretennikova, M. V.; Marchukov, V. A., Sushchev, V. G. Currently available methods of industrial nanodiamond synthesis. *Physics of the Solid State*, 2004, *46* (4), 611–615.
24. Gruen, D. M. Ultrananocrystalline diamond films from fullerene precursors. *Perspectives of Fullerene Nanotechnology*, 2002, *Part V*, 217–222.
25. Korablov, S.; Yokosawa, K.; Korablov, D.; Tohji, K.; Yamasaki, N. Hydrothermal formation of diamond from chlorinated organic compounds. *Materials Letters*, 2006, *60* (25–26), 3041–3044.
26. Chen, Q.; Lou, Z.; Wang, Q.; Chen, C. Recent progress in diamond synthesis. *Wuli*, 2005, *34* (3), 199–204.
27. Nickel, K. G.; Kraft, T.; Gogotsi, Y. G. Hydrothermal synthesis of diamond. In *Handbook of Ceramic Hard Materials*, Vol. 1. Wiley-VCH, Weinheim, Germany, 2000, pp. 374–389.
28. Basavalingu, B.; Byrappa, K.; Madhusudan, P. Hydrothermal synthesis of nano-sized crystals of diamond under sub-natural conditions. *Journal of Geological Society of India*, 2007, *69* (3), 665–670.

29. Basavalingu, B.; Byrappa, K.; Yoshimura, M.; Madhusudan, P.; Dayananda, A. S. Hydrothermal synthesis and characterization of micro to nano sized carbon particles. *Journal of Materials Science*, 2006, *41* (5), 1465–1469.

30. Yamasaki, N.; Yokosawa, K.; Korablov, S.; Tohjt, K. Synthesis of diamond particles under alkaline hydrothermal conditions. *Diffusion and Defect Data—Solid State Data, Part B: Solid State Phenomena*, 2006, *114* (*High Pressure Technology of Nanomaterials*), 271–276.

31. Wang, Z.-X.; Pan, Q.-Y.; Hu, J.-G.; Yong, Z.-Z.; Hu, Y.-Q.; Zhu, Z.-Y. Synthesis of diamond nanocrystals by double ions ($^{40}Ar^{+}$,$C_2H_6^{+}$) bombardment. *Wuli Xuebao*, 2007, *56* (8), 4829–4833.

32. Gouzman, I.; Fuchs, O.; Lifshitz, Y.; Michaelson, Sh.; Hoffman, A. Nanodiamond growth on diamond by energetic plasma bombardment. *Diamond and Related Materials*, 2007, *16* (4–7), 762–766.

33. Yao, Y.; Liao, M. Y.; Wang, Z. G.; Lifshitz, Y.; Lee, S. T. Nucleation of diamond by pure carbon ion bombardment—A transmission electron microscopy study. *Applied Physics Letters*, 2005, *87* (6), 063103/1–063103/3.

34. Yao, Y.; Liao, M. Y.; Kohler, Th.; Frauenheim, Th.; Zhang, R. Q.; Wang, Z. G.; Lifshitz, Y.; Lee, S. T. Diamond nucleation by energetic pure carbon bombardment. *Physical Review B: Condensed Matter and Materials Physics*, 2005, *72* (3), 035402/1–035402/5.

35. Amans, D.; Chenus, A.-C.; Ledoux, G.; Dujardin, C.; Reynaud, C.; Sublemontier, O.; Masenelli-Varlot, K.; Guillois, O. Nanodiamond synthesis by pulsed laser ablation in liquids. *Diamond and Related Materials*, 2009, *18* (2–3), 177–180.

36. Sun, J.; Zhai, Q.; Yang, X.; Lei, Y.; Du, X.; Yang, J. Synthesis of diamond nanopowders from carbon powders by laser bombarding. CN 1663909 A 20050907 Patent written in Chinese. Application: CN 2004-10093973 20041220. Priority: CAN 144:152787 AN 2005:1332207, 2005, 6 pp.

37. Sun, J.; Zhai, Q.; Du, H.; Jiang, L.; Lei, Y.; Yang, X.; Du, X. Effects of carbon material structures on the nanodiamond synthesis by laser irradiation. *Nami Jishu Yu Jingmi Gongcheng*, 2006, *4* (3), 217–220.

38. Wang, C. X.; Liu, P.; Cui, H.; Yang, G. W. Nucleation and growth kinetics of nanocrystals formed upon pulsed-laser ablation in liquid. *Applied Physics Letters*, 2005, *87* (20), 201913/1–201913/3.

39. Liu, P.; Wang, C.; Chen, J.; Xu, N.; Yang, G.; Ke, N.; Xu, J. Localized nanodiamond crystallization and field emission performance improvement of amorphous carbon upon laser irradiation in liquid. *Journal of Physical Chemistry C*, 2009, *113* (28), 12154–12161.

40. Lu, Y. F.; Huang, S. M.; Sun, Z. Raman spectroscopy of phenylcarbyne polymer films under pulsed green laser irradiation. *Journal of Applied Physics*, 2000, *87* (2), 945–951.

41. Goncharov, V. K.; Ismailov, D. R.; Lyudchik, O. R.; Petrov, S. A.; Puzyrev, M. V. Determination of the optical bandgap for diamond-like carbon films obtained by laser plasma deposition. *Journal of Applied Spectroscopy*, 2007, *74* (5), 704–709.

42. Varyukhin, V. N.; Shalaev, R. V.; Prudnikov, A. M. Properties of diamond films obtained in a glow discharge under laser irradiation. *Functional Materials*, 2002, *9* (1), 111–114.

43. Kharissova, O. V.; Osorio, M.; Garza, M.; Kharisov, B. I. Study of bismuth nanoparticles and nanotubes obtained by microwave heating. *Synthesis and Reactivity in Inorganic, Metal-Organic, and Nano-Metal Chemistry*, 2008, *38* (7), 567–572.

44. Kharissova, O. V.; Castanon, M. G.; Hernandez Pinero, J. L.; Ortiz-Mendez, U.; Kharisov, B. I. Fast production method of Fe-filled carbon nanotubes. *Mechanics of Advanced Materials and Structures*, 2009, *16* (1), 63–68.

45. Rodriguez, M. G.; Kharissova, O. V.; Ortiz-Mendez, U. Formation of boron carbide nanofibers and nanobelts from heated by microwave. *Reviews on Advanced Materials Science*, 2004, *7* (1), 55–60.

46. Lewis, D., III; Imam, M. A.; Fliflet, A. W.; Bruce, R. W.; Kurihara, L. K.; Kinkead, A. K.; Lombardi, M.; Gold, S. H. Recent advances in microwave and millimeter-wave processing of materials. *Materials Science Forum*, 2007, *539–543* (Pt. 4, *THERMEC 2006*), 3249–3254.

47. Yan, X.-T.; Xu, Y. *Chemical Vapour Deposition: An Integrated Engineering Design for Advanced Materials (Engineering Materials and Processes)*. Springer, Berlin, Germany, 2009, 327 pp.

48. Pierson, H. O. *Handbook of Chemical Vapor Deposition: Principles, Technology and Applications*, Materials Science and Process Technology Series, 2nd edn., William Andrew, Norwich, NY, 2000, 506 pp.

49. Pradeep, G. *Chemical Vapor Deposition*. V D M Verlag Dr. Mueller E.K., Saarbrücken, Germany, 2008, 112 pp.

50. Gordillo-Vazquez, F. J.; Gomez-Aleixandre, C.; Albella, J. M. Plasma chemistry in the CVD synthesis of nanodiamond films. *Proceedings—Electrochemical Society*, 2005, 2005-09(EUROCVD-15), Pennington, NJ, pp. 415–426.

51. Purohit, V. S.; Jain, D.; Sathe, V. G.; Ganesan, V.; Bhoraskar, S. V. Synthesis of nanocrystalline diamonds by microwave plasma. *Journal of Physics D: Applied Physics*, 2007, *40* (6), 1794–1800.

52. Miyake, M.; Ogino, A.; Nagatsu, M. Characteristics of nano-crystalline diamond films prepared in Ar/H₂/CH₄ microwave plasma. *Thin Solid Films*, 2007, *515* (9), 4258–4261.

53. Chowdhury, S.; Borham, J.; Catledge, S. A.; Eberhardt, A. W.; Johnson, P. S.; Vohra, Y. K. Synthesis and mechanical wear studies of ultra smooth nanostructured diamond (USND) coatings deposited by microwave plasma chemical vapor deposition with He/H₂/CH₄/N₂ mixtures. *Diamond and Related Materials*, 2008, *17* (4–5), 419–427.

54. Butler, J. E.; Sumant, A. V. The CVD of nanodiamond materials. *Chemical Vapor Deposition*, 2008, *14* (7–8), 145–160.

55. Chen, W.; Lu, X.; Yang, Q.; Xiao, C.; Sammynaiken, R.; Maley, J.; Hirose, A. Effects of gas flow rate on diamond deposition in a microwave plasma reactor. *Thin Solid Films*, 2006, *515* (4), 1970–1975.

56. Nanba, A.; Imai, T.; Nishibayashi, Y.; Yamamoto, Y.; Meguro, K. Microwave plasma CVD apparatus. JP 2005–218732 20050728. Priority: CAN 146:194557 AN 2007:143909.

57. Ariyada, O.; Sato, S.; Suzuki, H. Microwave chemical vapor deposition apparatus for preparation of diamond. JP 2006083405 A 20060330 Patent written in Japanese. Application: JP 2004-266462 20040914. Priority: CAN 144:321896 AN 2006:292991, 2006, 12 pp.

58. Pichot, V.; Comet, M.; Fousson, E.; Spitzer, D. Detonation synthesis of nanodiamonds: Their synthesis and use in pyrotechnics. *Actualite Chimique*, 2009, *329*, 8–13.

59. Dolmatov, V. Yu. The structure of a cluster of a detonation-produced nanodiamond. *Sverkhtverdye Materialy*, 2005, (1), 28–32.

60. Dolmatov, V. Yu. Detonation-synthesis nanodiamonds: Synthesis, structure, properties and applications. *Russian Chemical Reviews*, 2007, *76* (4), 339–360.

61. Dolmatov, V. Yu. Modern commercial technology for production of detonation nano-diamonds and the area of their application Report 1. *Sverkhtverdye Materialy*, 2006, (3), 10–21.

62. Dolmatov, V. Yu. Modern industrial methods for manufacture of detonation derived nanodiamonds and main areas of their use. Part 1. *Sverkhtverdye Materialy*, 2006, (2), 18–29.

63. Dolmatov, V. Yu.; Fujimura, T. Physical and chemical problems of modification of detonation nanodiamond surface properties. *Synthesis, Properties and Applications of Ultrananocrystalline Diamond*, NATO Science Series, II: Mathematics, Physics and Chemistry, Gruen, D. M.; Shenderova, O. A.; Vul, A. Ya. (Eds.). Vol. 192. Springer, New York, 2005, pp. 217–230.

64. Donnet, J. B.; Fousson, E.; Wang, T. K.; Samirant, M.; Baras, C.; Pontier Johnson, M. Dynamic synthesis of diamonds. *Diamond and Related Materials*, 2000, *9* (3–6), 887–892.

65. Osawa, E. Recent progress and perspectives in single-digit nanodiamond. *Diamond and Related Materials*, 2007, *16* (12), 2018–2022.

66. Osawa, E. Disintegration and purification of crude aggregates of detonation nanodiamond. A few remarks on nano methodology. *Synthesis, Properties and Applications of Ultrananocrystalline Diamond*, NATO Science Series, II: Mathematics, Physics and Chemistry, Gruen, D. M.; Shenderova, O. A.; Vul, A. Ya. (Eds.).Vol. 192. Springer, New York, 2005, pp. 231–240.

67. Titov, V. M.; Tolochko, B. P.; Ten, K. A.; Lukyanchikov, L. A.; Zubkov, P. I. The formation kinetics of detonation nanodiamonds. *Synthesis, Properties and Applications of Ultrananocrystalline Diamond*, NATO Science Series, II: Mathematics, Physics and Chemistry, Gruen, D. M.; Shenderova, O. A.; Vul, A. Ya. (Eds.). Vol. 192. Springer, New York, 2005, pp. 169–180.

68. Puzyr, A. P.; Bondar, V. S. Production of nanodiamonds with an increased colloidal stability by using an explosion synthesis. RU 2252192 C2 20050520 Patent written in Russian. Application: RU 2003-119416 20030626. Priority: CAN 142:448964 AN 2005:428972, 2005.

69. Nakanishi, E.; Matsui, K.; Yamaguchi, C.; Nishino, H.; Kurusu, C. Manufacture of functional carbon materials. JP 2000109310 A 20000418 Patent written in Japanese. Application: JP 99-218782 19990802. Priority: JP 98–219197 19980803. CAN 132:253156 AN 2000:247376, 2000, 5 pp.

70. Khachatryan, A. Kh.; Aloyan, S. G.; May, P. W.; Sargsyan, R.; Khachatryan, V. A.; Baghdasaryan, V. S. Graphite-to-diamond transformation induced by ultrasound cavitation. *Diamond and Related Materials*, 2008, *17* (6), 931–936.

71. Gordeev, S. K.; Korchagina, S. B. Method for preparation of nanodiamond powders for producing stable suspensions. RU 2302994 C2 20070720 Patent written in Russian. Application: RU 2004-121069 20040701. Priority: CAN 147:191800 AN 2007:789603, 2007, 3 pp.

72. Chen, L.-J.; Tai, N.-H.; Lee, C.-Y.; Lin, I.-N. Effects of pretreatment processes on improving the formation of ultrananocrystalline diamond. *Journal of Applied Physics*, 2007, *101* (6), 064308/1–064308/6.

73. Hanada, K.; Matsuzaki, K.; Sano, T. Nanocrystalline diamond films fabricated by sol-gel technique. *Surface Science*, 2007, *601* (18), 4502–4505.

74. Iijima, S. Process and apparatus for manufacture of diamond powder in nanometer range by size reduction. JP 04132606 A 19920506 Heisei. Patent written in Japanese. Application: JP 90-254415 19900925. Priority: CAN 117:153818 AN 1992:553818, 1992, 4 pp.

75. Adler, J.; Gershon, Y.; Mutat, T.; Sorkin, A.; Warszawski, E.; Kalish, R.; Yaish, Y. Visualizing nanodiamond and nanotubes with AViz. *Computer Simulation Studies in Condensed-Matter Physics XIX, Springer Proceedings in Physics*, Vol. 123. Springer, Berlin, Germany, 2009, pp. 56–60.

76. Adler, J.; Pine, P. Visualization techniques for modelling carbon allotropes. *Computer Physics Communications*, 2009, *180* (4), 580–582.

77. El Bojaddaini, M.; Chatei, H.; El Hammouti, M.; Robert, H.; Bougdira, J. Modeling of a pulsed microwave plasma discharge in view of diamond film synthesis. *Los Alamos National Laboratory, Preprint Archive, Physics*, 2007, *156*, arXiv:0711.0845v1 [physics.soc-ph].

78. Fokin, A. A.; Schreiner, P. R. Band gap tuning in nanodiamonds: First principle computational studies. *Molecular Physics*, 2009, *107* (8–12), 823–830.

79. Barnard, A. S.; Russo, S. P.; Snook, I. K. Modeling of stability and phase transformations in zero- and one-dimensional nanocarbon systems. In *Handbook of Theoretical and Computational Nanotechnology*. Rieth, M.; Schommers, W. (Eds.). American Scientific Publishers, Stevenson, CA, 2006, Vol. 9, pp. 573–622.

80. Barnard, A. S.; Russo, S. P.; Snook, I. K. Simulation and bonding of dopants in nanocrystalline diamond. *Journal of Nanoscience and Nanotechnology*, 2005, *5* (9), 1395–1407.

81. Wang, C. X.; Yang, Y. H.; Xu, N. S.; Yang, G. W. Thermodynamics of diamond nucleation on the nanoscale. *Journal of the American Chemical Society*, 2004, *126* (36), 11303–11306.

82. Barnard, A. S.; Russo, S. P.; Snook, I. K. Modeling of stability and phase transformations in quasi-zero dimensional nanocarbon systems. *Journal of Computational and Theoretical Nanoscience*, 2005, *2* (2), 180–201.

83. La Torre Riveros, L.; Tryk, D. A.; Cabrera, C. R. Chemical purification and characterization of diamond nanoparticles for electrophoretically coated electrodes. *Reviews on Advanced Materials Science*, 2005, *10* (3), 256–260.

84. Novikov, N. V.; Bogatyreva, G. P.; Nevstruev, G. F.; Il'nitskaya, G. D.; Voloshin, M. N. Magnetic methods of purification control of nanodiamond powders. *Physics of the Solid State (Translation of Fizika Tverdogo Tela (Sankt-Peterburg))*, 2004, *46* (4), 672–674.

85. Lin, K.; Hou, S.; Ma, B. Effect of ultrasonic-dispersion for diamond powder purification. *Jingangshi Yu Moliao Moju Gongcheng*, 2007, (6), 9–12.

86. Liu, S.; Liu, J.; Li, P.; Li, Y.; Shen, Y. Process for purification of diamond by electrolysis. CN 101049929 A 20071010 Patent written in Chinese. Application: CN 2007-10054424 20070518, 2007, 7 pp.

87. Ovcharenko, A. G.; Ignatchenko, A. V.; Sataev, R. R.; Brylyakov, P. M. Purification of ultradispersed diamond by removal of soluble and adsorbed impurities using electric field. SU 1815933 A1 19960620 Patent written in Russian. Application: SU 90-4855039 19900727. Priority: CAN 125:333282 AN 1996:729518, 1996.

88. Dolgaev, S. I.; Kirichenko, N. A.; Kulevskii, L. A.; Lubnin, E. N.; Simakin, A. V.; Shafeev, G. A. Laser purification of ultradispersed diamond in aqueous solution. *Quantum Electronics*, 2004, *34* (9), 860–864.

89. Doynikov, Yu. A.; Makhrachev, A. F.; Kurbatov, K. K.; Makarskii, I. V.; Adodin, E. I.; Yagupov, S. A.; Tarasova, L. G.; Kovalenko, E. G. Method for purification of diamonds. RU 2367601 C1 20090920 Patent written in Russian. Application: RU 2007-147784 20071225. Priority: CAN 151:384616 AN 2009:1153611, 2009, 7 pp.

90. Zeng, H.; An, X. Method for purifying diamond nanoparticle with Ce salt. CN 1480252 A 20040310 Patent written in Chinese. Application: CN 2003-139849 20030718. Priority: CAN 142:357372 AN 2005:20996, 2004, 11 pp.

91. Petrov, I. L.; Skryabin, Yu. A.; Shenderova, O. A. Nanodiamond material, method and device for purifying and modifying a nanodiamond. Patent written in Russian. Application: WO 2008-RU313 20080520. Priority: RU 2007–118553 20070521, 2008, 21 pp.

92. Lin, K.; Pan, Y.; Hou, S.; Xiao, H.; Ma, B. Discussion on purification techniques of synthetic diamond. *Jingangshi Yu Moliao Moju Gongcheng*, 2005, (5), 77–78, 80.

93. Dubois, M.; Guerin, K.; Petit, E.; Batisse, N.; Hamwi, A.; Komatsu, N.; Giraudet, J.; Pirotte, P.; Masin, F. Solid-state NMR study of nanodiamonds produced by the detonation technique. *Journal of Physical Chemistry C*, 2009, *113* (24), 10371–10378.

94. Shames, A. I.; Panich, A. M.; Kempinski, W.; Baidakova, M. V.; Osipov, V. Yu.; Enoki, T.; Vul, A. Ya. Magnetic resonance study of nanodiamonds. *Synthesis, Properties and Applications of Ultrananocrystalline Diamond*, NATO Science Series, II: Mathematics, Physics and Chemistry, Gruen, D. M.; Shenderova, O. A.; Vul, A. Ya. (Eds.). Vol. 192. Springer, New York, 2005, pp. 271–282.

95. Ferrari, A. C.; Robertson, J. Raman spectroscopy of amorphous, nanostructured, diamond-like carbon, and nanodiamond. *Philosophical Transactions of the Royal Society of London, Series A: Mathematical, Physical and Engineering Sciences*, 2004, *362* (1824), 2477–2512.

96. Ballutaud, D.; Jomard, F.; Pinault, M.-A.; Frangieh, G.; Simon, N. sp^2 carbon phases in nanocrystalline diamond. *ECS Transactions*, 2008, *13* (2, Dielectrics for Nanosystems 3: Materials Science, Processing, Reliability, and Manufacturing), 377–383.

97. Shushkanova, A. V.; Dubrovinsky, L.; Dubrovinskaya, N.; Litvin, Yu. A.; Urusov, V. S. Synthesis and in-situ Raman spectroscopy of nanodiamonds. *Doklady Physics*, 2008, *53* (1), 1–4.

98. Hamilton, T.; Wilks, R. G.; Yablonskikh, M. V.; Yang, Q.; Foursa, M. N.; Hirose, A.; Vasilets, V. N.; Moewes, A. Determining the sp^2/sp^3 bonding concentrations of carbon films using x-ray absorption spectroscopy. *Canadian Journal of Physics*, 2008, *86* (12), 1401–1407.

99. Xiao, X.; Sheldon, B. W.; Qi, Y.; Kothari, A. K. Intrinsic stress evolution in nanocrystalline diamond thin films with deposition temperature. *Applied Physics Letters*, 2008, *92* (13), 131908/1–131908/3.

100. Tang, C. J.; Neto, M. A.; Soares, M. J.; Fernandes, A. J. S.; Neves, A. J.; Gracio, J. A comparison study of hydrogen incorporation among nanocrystalline, microcrystalline and polycrystalline diamond films grown by chemical vapor deposition. *Thin Solid Films*, 2007, *515* (7–8), 3539–3546.

101. Koroleva, M. Yu.; Berdnikova, D. V.; Spitsyn, B. V.; Yurtov, E. V. Sedimentation stability of aqueous dispersions of nanodiamond agglomerates. *Theoretical Foundations of Chemical Engineering*, 2009, *43* (4), 478–481.

102. Osawa, E.; Ho, D.; Huang, H.; Korobov, M. V.; Rozhkova, N. N. Consequences of strong and diverse electrostatic potential fields on the surface of detonation nanodiamond particles. *Diamond and Related Materials*, 2009, *18* (5–8), 904–909.

103. Houska, J.; Panyala, N. R.; Pena-Mendez, E. M.; Havel, J. Mass spectrometry of nanodiamonds. *Rapid Communications in Mass Spectrometry*, 2009, *23* (8), 1125–1131.

104. Raty, J.-Y.; Galli, G. Structural and electronic properties of isolated nanodiamonds: A theoretical perspective. *Synthesis, Properties and Applications of Ultrananocrystalline Diamond*, NATO Science Series, II: Mathematics, Physics and Chemistry, Gruen, D. M.; Shenderova, O. A.; Vul, A. Ya. (Eds.). Vol. 192. Springer, New York, 2005, pp. 15–24.

105. Preclikova, J.; Trojanek, F.; Kromka, A.; Rezek, B.; Dzurnak, B.; Maly, P. Ultrafast photoluminescence of nanocrystalline diamond films. *Physica Status Solidi A: Applications and Materials Science*, 2008, *205* (9), 2154–2157.

106. Kulakova, I. I. Surface chemistry of nanodiamonds. *Physics of the Solid State (Translation of Fizika Tverdogo Tela (Sankt-Peterburg))*, 2004, *46* (4), 636–643.

107. Madaleno, J. C.; Singh, M. K.; Titus, E.; Cabral, G.; Gracio, J.; Pereira, L. Electron field emission from patterned nanocrystalline diamond coated a-SiO$_2$ micrometer-tip arrays. *Applied Physics Letters*, 2008, *92* (2), 023113/1–023113/3.

108. Aggadi, N.; Arnas, C.; Benedic, F.; Dominique, C.; Duten, X.; Silva, F.; Hassouni, K.; Gruen, D. M. Structural and chemical characterization of soot particles formed in Ar/H$_2$/CH$_4$ microwave discharges during nanocrystalline diamond film synthesis. *Diamond and Related Materials*, 2006, *15* (4–8), 908–912.

109. Popov, C.; Novotny, M.; Jelinek, M.; Boycheva, S.; Vorlicek, V.; Trchova, M.; Kulisch, W. Chemical bonding study of nanocrystalline diamond films prepared by plasma techniques. *Thin Solid Films*, 2006, *506–507*, 297–302.

110. Barnard, A. S. Stability of nanodiamond. In *Ultrananocrystalline Diamond: Synthesis, Properties and Applications*. Shenderova, O. A.; Gruen, D. M. (Eds.). William Andrew Publishing, New York, 2006, pp. 117–154.

111. Gordeev, S. K.; Korchagina, S. B. On the stability of small sized nanodiamonds. *Journal of Superhard Materials*, 2007, *29* (2), 124–125.

112. Butenko, Yu. V.; Coxon, P. R.; Yeganeh, M.; Brieva, A. C.; Liddell, K.; Dhanak, V. R.; Siller, L. Stability of hydrogenated nanodiamonds under extreme ultraviolet irradiation. *Diamond and Related Materials*, 2008, *17* (6), 962–966.

113. Wang, G.-z. Review on correlation of diamond performance and nitrogen. *Chaoying Cailiao Gongcheng*, 2006, *18* (2), 33–36.

114. Turner, S.; Lebedev, O. I.; Shenderova, O.; Vlasov, I. I.; Verbeeck, J.; Van Tendeloo, G. Determination of size, morphology, and nitrogen impurity location in treated detonation nanodiamond by transmission electron microscopy. *Advanced Functional Materials*, 2009, *19* (13), 2116–2124.

115. Kulakova, I. I. Chemical properties of nanodiamond. *Innovative Superhard Materials and Sustainable Coatings for Advanced Manufacturing*, NATO Science Series, II: Mathematics, Physics and Chemistry, Lee, J., Novikov, N. (Eds.). Vol. 200. Springer, New York, 2005, pp. 365–379.

116. Jiang, W.; Lu, W.-y.; Yang, B.-w.; Yao, Y.; Zhang, S.-x., Kou, Z.-l. Method for analyzing impurity elements in diamond by detecting combustion residues of diamond with EDS and SEM. *Chaoying Cailiao Gongcheng*, 2006, *18* (3), 6–11.

117. Bogatyreva, G. P.; Maevskii, V. M.; Il'nitskaya, G. D.; Nevstruev, G. F.; Tkach, V. N.; Zaitseva, I. N. Impurities and inclusions in synthetic diamond powders of the AC4 and AC6 grades. *Sverkhtverdye Materialy*, 2006, (4), 62–69.

118. Assali, L. V. C.; Machado, W. V. M.; Larico, R.; Justo, J. F. Cobalt in diamond: An ab initio investigation. *Diamond and Related Materials*, 2007, *16* (4–7), 819–822.

119. Cheng, H.-F.; Lee, Y.-C.; Lin, S.-J.; Chou, Y.-P.; Chen, T. T.; Lin, I-N. Current image tunneling spectroscopy of boron-doped nanodiamonds. *Journal of Applied Physics*, 2005, *97* (4), 044312/1–044312/5.

120. Yan, C. X.; Dai, Y.; Huang, B. B.; Long, R.; Guo, M. Shallow donors in diamond: Be and Mg. *Computational Materials Science*, 2009, *44* (4), 1286–1290.

121. Sakai, H.; Kudou, H.; Takahashi, M.; Arifuku, M. Method for producing dispersion of nanodiamond in organic solvent. WO 2008-JP3215 20081106. Priority: JP 2007-290340 20071108. CAN 150:518067 AN 2009:587231.

122. Maitra, U.; Gomathi, A.; Rao, C. N. R. Covalent and noncovalent functionalization and solubilisation of nanodiamond. *Journal of Experimental Nanoscience*, 2008, *3* (4), 271–278.

123. Das, D.; Singh, R. N. A review of nucleation, growth and low temperature synthesis of diamond thin films. *International Materials Reviews*, 2007, *52* (1), 29–64.

124. Chernomordik, B.; Dumpala, S.; Chen, Z. Q.; Sunkara, M. K. Nanodiamond tipped and coated conical carbon tubular structures. *Chemical Vapor Deposition*, 2008, *14* (7–8), 256–262.

125. Askari, S. J.; Chen, G. C.; Akhtar, F.; Lu, F. X. Adherent and low friction nano-crystalline diamond film grown on titanium using microwave CVD plasma. *Diamond and Related Materials*, 2008, *17* (3), 294–299.

126. Askari, S. J.; Chen, G. C.; Lu, F. X. Growth of polycrystalline and nanocrystalline diamond films on pure titanium by microwave plasma assisted CVD process. *Materials Research Bulletin*, 2008, *43* (5), 1086–1092.

127. Kulisch, W.; Popov, C.; Vorlicek, V.; Gibson, P. N.; Favaro, G. Nanocrystalline diamond growth on different substrates. *Thin Solid Films*, 2006, *515* (3), 1005–1010.

128. Teii, K.; Ikeda, T. Effect of enhanced C_2 growth chemistry on nanodiamond film deposition. *Applied Physics Letters*, 2007, *90* (11), 111504/1–111504/3.

129. Gordillo-Vazquez, F. J.; Albella, J. M. Distinct nonequilibrium plasma chemistry of C_2 affecting the synthesis of nanodiamond thin films from C_2H_2 (1%)/H_2/Ar-rich plasmas. *Journal of Applied Physics*, 2003, *94* (9), 6085–6090.

130. Li, X.-H.; Guo, W.-T.; Chen, X.-K.; Gan, W.; Yang, J.-P.; Wang, R.; Cao, S.-Z.; Rong, Yu. Effect of pressure on growth rate and quality of diamond films prepared by microwave plasma chemical vapor deposition. *Wuli Xuebao*, 2007, *56* (12), 7183–7187.

131. Wang, L.; Lu, J.; Su, Q.; Wu, N.; Liu, J.; Shi, W.; Xia, Y. [100]-textured growth of polycrystalline diamond films on alumina substrates by microwave plasma chemical vapor deposition. *Materials Letters*, 2006, *60* (19), 2390–2394.

132. Abreu, C. S.; Amaral, M. S.; Oliveira, F. J.; Tallaire, A.; Benedic, F.; Syll, O.; Cicala, G.; Gomes, J. R.; Silva, R. F. Tribological testing of self-mated nanocrystalline diamond coatings on Si_3N_4 ceramics. *Surface and Coatings Technology*, 2006, *200* (22–23), 6235–6239.

133. Enoki, T.; Takai, K.; Osipov, V.; Baidakova, M.; Vul, A. Nanographene and nanodiamond; new members in the nanocarbon family. *Chemistry—An Asian Journal*, 2009, *4* (6), 796–804.

134. Grichko, V. P.; Shenderova, O. A. Nanodiamond: Designing the bio-platform. In *Ultrananocrystalline Diamond*. Shenderova, O. A.; Gruen, D. M. (Eds.). William Andrew Publishing, New York, 2006, pp. 529–557.

135. Purtov, K. V.; Burakova, L. P.; Puzyr, A. P.; Bondar, V. S. The interaction of linear and ring forms of DNA molecules with nanodiamonds synthesized by detonation. *Nanotechnology*, 2008, *19* (32), 325101/1–325101/3.

136. Komatsu, N. Size separation and surface functionalization of nanodiamond particles aiming at their biomedical applications. *Hyomen Kagaku*, 2009, *30* (5), 273–278.

137. Ke, G.; Huan, S.; Huang, F.-l. Synthesis and dispersibility of derivative of 1,3-propanediamine with nanodiamond. *Gongneng Cailiao*, 2009, *40* (5), 863–866.

138. Remes, Z.; Choukourov, A.; Stuchlik, J.; Potmesil, J.; Vanecek, M. Nanocrystalline diamond surface functionalization in radio frequency plasma. *Diamond and Related Materials*, 2006, *15* (4–8), 745–748.

139. Mochalin, V. N.; Gogotsi, Y. Wet chemistry route to hydrophobic blue fluorescent nanodiamond. *Journal of the American Chemical Society*, 2009, *131* (13), 4594–4595.

140. Liang, Y.; Ozawa, M.; Krueger, A. A general procedure to functionalize agglomerating nanoparticles demonstrated on nanodiamond. *ACS Nano*, 2009, *3* (8), 2288–2296.

141. Yeap, W. S.; Chen, S.; Loh, K. P. Detonation nanodiamond: An organic platform for the suzuki coupling of organic molecules. *Langmuir*, 2009, *25* (1), 185–191.

142. Baidakova, M.; Vul, A. New prospects and frontiers of nanodiamond clusters. *Journal of Physics D: Applied Physics*, 2007, *40* (20), 6300–6311.

143. Puzyr, A. P.; Bondar, V. S.; Bukayemsky, A. A.; Selyutin, G. E.; Kargin, V. F. Physical and chemical properties of modified nanodiamonds. *Synthesis, Properties and Applications of Ultrananocrystalline Diamond*, NATO Science Series, II: Mathematics, Physics and Chemistry, Gruen, D. M.; Shenderova, O. A.; Vul, A. Ya. (Eds.). Vol. 192. Springer, New York, 2005, pp. 261–270.

144. Spitsyn, B. V.; Gradoboev, M. N.; Galushko, T. B.; Karpukhina, T. A.; Serebryakova, N. V.; Kulakova, I. I.; Melnik, N. N. Purification and functionalization of nanodiamond. *Synthesis, Properties and Applications of Ultrananocrystalline Diamond*, NATO Science Series, II: Mathematics, Physics and Chemistry, Gruen, D. M.; Shenderova, O. A.; Vul, A. Ya. (Eds.). Vol. 192. Springer, New York, 2005, pp. 241–252.

145. Khabashesku, V. N.; Margrave, J. L.; Barrera, E. V. Functionalized carbon nanotubes and nanodiamonds for engineering and biomedical applications. *Diamond and Related Materials*, 2005, *14* (3–7), 859–866.

146. Khabashesku, V. N.; Liu, Y.; Margrave, J. L.; Margrave, M. L. Functionalization of nanodiamond powder through fluorination and subsequent derivatization reactions. U.S. Patent Application Publications U.S. Patent 2005158549 A1 20050721 Patent written in English. Application: U.S. Patent 2004-996869 20041124. Priority: U.S. Patent 2003–525588 20031126, 2005, 18 pp.

147. Liu, Y.; Khabashesku, V. N.; Halas, N. J. Functionalization of nanodiamond powder and applications for glass surface diamond coatings. *Abstracts of Papers, 229th ACS National Meeting*, San Diego, CA, March 13–17, 2005, COLL-581.

148. Liu, Y.; Gu, Z.; Margrave, J. L.; Khabashesku, V. N. Functionalization of nanoscale diamond powder: Fluoro-, alkyl-, amino-, and amino acid-nanodiamond derivatives. *Chemistry of Materials*, 2004, *16* (20), 3924–3930.

149. Liu, Y.; Khabashesku, V. N.; Halas, N. J. Fluorinated nanodiamond as a wet chemistry precursor for diamond coatings covalently bonded to glass surface. *Journal of the American Chemical Society*, 2005, *127* (11), 3712–3713.

150. Kuznetsov, V. L.; Butenko, Yu. V. Nanodiamond graphitization and properties of onion-like carbon. *Synthesis, Properties and Applications of Ultrananocrystalline Diamond*, NATO Science Series, II: Mathematics, Physics and Chemistry, Gruen, D. M.; Shenderova, O. A.; Vul, A. Ya. (Eds.). Vol. 192. Springer, New York, 2005, pp. 199–216.

151. Kwon, S. J.; Park, J. G. Theoretical analysis of the graphitization of a nanodiamond. *Journal of Physics: Condensed Matter*, 2007, *19*, 386215.

152. Leyssale, J.-M.; Vignoles, G. L. Molecular dynamics evidences of the full graphitization of a nanodiamond annealed at 1500K. *Chemical Physics Letters*, 2008, *454* (4–6), 299–304.

153. Qiao, Z.; Li, J.; Zhao, N.; Shi, C.; Nash, P. Graphitization and microstructure transformation of nanodiamond to onion-like carbon. *Scripta Materialia*, 2005, Volume Date 2006, *54* (2), 225–229.

154. Narulkar, R.; Bukkapatnam, S.; Raff, L. M.; Komanduri, R. Graphitization as a precursor to wear of diamond in machining pure iron: A molecular dynamics investigation. *Computational Materials Science*, 2009, *45* (2), 358–366.

155. Brodka, A.; Hawelek, L.; Burian, A.; Tomita, S.; Honkimaeki, V. Molecular dynamics study of structure and graphitization process of nanodiamonds. *Journal of Molecular Structure*, 2008, *887* (1–3), 34–40.

156. Azevedo, A. F.; Ramos, S. C.; Baldan, M. R.; Ferreira, N. G. Graphitization effects of CH_4 addition on NCD growth by first and second Raman spectra and by x-ray diffraction measurements. *Diamond and Related Materials*, 2008, *17* (7–10), 1137–1142.

157. Kononenko, V. V.; Kononenko, T. V.; Pimenov, S. M.; Sinyavskii, M. N.; Konov, V. I.; Dausinger, F. Effect of the pulse duration on graphitisation of diamond during laser ablation. *Quantum Electronics*, 2005, *35* (3), 252–256.

158. Strekalov, V. N. Graphitization of diamond stimulated by electron-hole recombination. *Applied Physics A: Materials Science and Processing*, 2005, *80* (5), 1061–1066.

159. Gouzman, I.; Michaelson, S.; Hoffman, A. Nanodiamond films deposited from energetic species: Material characterization and mechanism of formation. In *Ultrananocrystalline Diamond*. Shenderova, O. A.; Gruen, D. M. (Eds.). William Andrew Publishing, New York, 2006, pp. 229–272.

160. Hoffman, A. Mechanism and properties of nanodiamond films deposited by the DC-GD-CVD process. *Synthesis, Properties and Applications of Ultrananocrystalline Diamond*, NATO Science Series, II: Mathematics, Physics and Chemistry, Gruen, D. M.; Shenderova, O. A.; Vul, A. Ya. (Eds.). Vol. 192. Springer, New York, 2005, pp. 125–144.

161. Liu, W.; Gu, C. The preparation and properties of nanostructured diamond films deposited by a hot-filament chemical vapor deposition method via continuous ion bombardment. *Thin Solid Films*, 2004, *467* (1–2), 4–9.

162. Liao, M.; Qin, F.; Zhang, J.; Liu, Z.; Yang, S.; Wang, Z.; Lee, S.-T. Ion bombardment as the initial stage of diamond film growth. *Journal of Applied Physics*, 2001, *89* (3), 1983–1985.

163. Grausova, L.; Bacakova, L.; Kromka, A.; Potocky, S.; Vanecek, M.; Nesladek, M.; Lisa, V. Nanodiamond as promising material for bone tissue engineering. *Journal of Nanoscience and Nanotechnology*, 2009, *9* (6), 3524–3534.

164. Liu, S.; Liu, W.; Hei, L.; Tang, W.; Lv, F. Research on preparation of diamond coatings containing Si by microwave plasma chemical vapor deposition. *Beijing Keji Daxue Xuebao*, 2007, *29* (4), 408–412, 446.

165. Man, W.; Wang, J.; Wang, C.; Ma, Z.; Wang, S.; Xiong, L. Low temperature synthesis of nanocrystalline diamond films deposited by microwave CVD. *Wuhan Huagong Xueyuan Xuebao*, 2006, *28* (4), 57–61.

166. Mujica, E. A.; Piazza, F.; De Jesus, J.; Weiner, B. R.; Wolter, S. D.; Morell, G. Synthesis of unstrained failure-resistant nanocrystalline diamond films. *Thin Solid Films*, 2007, *515* (20–21), 7906–7910.

167. Zhou, J.; Wang, L.; Liu, G.; Ouyang, S. Method for synthesizing diamond nanofilm with microwave plasma at low temperature. CN 101024893 A 20070829 Patent written in Chinese. Application: CN 2007-10051244 20070111. Priority: CAN 147:353777 AN 2007:967265, 2007, 3 pp.

168. Karbushev, V. V.; Konstantinov, I. I.; Parsamyan, I. L.; Kulichikhin, V. G.; Popov, V. A.; George, T. F. Preparation of polymer-nanodiamond composites with improved properties. *Advanced Materials Research*, 2009, *59* (*1st International Conference on New Materials for Extreme Environments, 2008*), 275–278.

169. Sawaguchi, T.; Yano, S.; Hagiwara, T.; Ito, H. Transparent heat-resistant polymer-nanodiamond composites. PCT International Application WO 2005085359 A1 20050915 Patent written in Japanese. Application: WO 2005-JP3887 20050307. Priority: JP 2004-64281 20040308, 2005, 20 pp.

170. Eswar Prasad, K.; Das, B.; Maitra, U.; Ramamurty, U.; Rao, C. N. R. Extraordinary synergy in the mechanical properties of polymer matrix composites reinforced with 2 nanocarbons. *Proceedings of the National Academy of Sciences of the United States of America*, Early Edition, 2009, 1–4, 4 pp.

171. Gavrilov, A. S.; Voznyakovskii, A. P. Rheological characteristics and relaxation properties of polymer-nanodiamond composites. *Russian Journal of Applied Chemistry*, 2009, *82* (6), 1041–1045.

172. Dolmatov, V. Yu. Polymer-diamond composites based on detonation nanodiamonds. Report 3. *Sverkhtverdye Materialy*, 2007, (4), 3–12.

173. Konstantinov, I. I.; Karbushev, V. V.; Semakov, A. V.; Kulichikhin, V. G. Combining carbon and polymeric particles in an inert fluid as a promising approach to synthesis of nanocomposites. *Russian Journal of Applied Chemistry*, 2009, *82* (3), 483–487.

174. Yoshioka, H.; Furukuwa, R.; Sawada, H. Preparation and applications of fluoroalkylated oligomeric nanoparticles. *Hyomen*, 2006, *44* (5), 167–182.

175. Zhang, W.; Lee, S.-T.; Bello, I.; Leung, K. M.; Li, H.-Q.; Zou, Y.-S.; Chong, Y. M.; Ma, K. L. Ultrahard multilayer coatings based on alternating layers of nanocrystalline diamond and nanocrystalline cubic boron nitride. U.S. Patent Application Publications. U.S. Patent 2009022969 A1 20090122 Patent written in English. Application: U.S. Patent 2007-880115 20070719. Priority: CAN 150:150364 AN 2009:93515, 2009, 14 pp.

176. Li, H. Q.; Leung, K. M.; Ma, K. L.; Ye, Q.; Chong, Y. M.; Zou, Y. S.; Zhang, W. J.; Lee, S. T.; Bello, I. Nanocubic boron nitride/nanodiamond multilayer structures. *Applied Physics Letters*, 2007, *91* (20), 201918/1–201918/3.

177. Almeida, F. A.; Belmonte, M.; Fernandes, A. J. S.; Oliveira, F. J.; Sandilva, R. F. MPCVD diamond coating of Si_3N_4-TiN electroconductive composite substrates. *Diamond and Related Materials*, 2007, *16* (4–7), 978–982.

178. Tzeng, Y.; Chen, Y.-C.; Cheng, A.-J.; Hung, Y.-T.; Yeh, C.-S.; Park, M.; Wilamowski, B. M. Chemically vapor deposited diamond-tipped one-dimensional nanostructures and nanodiamond-silica-nanotube composites. *Diamond and Related Materials*, 2009, *18* (2–3), 173–176.

179. Singh, M. K.; Titus, E.; Madaleno, J. C.; Pereira, L.; Cabral, G.; Neto, V. F.; Gracio, J. Nanocrystalline diamond on SiO_2 fiber: A new class of hybrid material. *Diamond and Related Materials*, 2008, *17* (7–10), 1106–1109.

180. Ekimov, E. A.; Zoteev, A.; Borovikov, N. F. Sintering of a nanodiamond in the presence of cobalt. *Inorganic Materials*, 2009, *45* (5), 491–494.

181. Vlasov, I. I.; Lebedev, O. I.; Ralchenko, V. G.; Goovaerts, E.; Bertoni, G.; Van Tendeloo, G.; Konov, V. I. Hybrid diamond-graphite nanowires produced by microwave plasma chemical vapor deposition. *Advanced Materials*, 2007, *19* (22), 4058–4062.

182. Chang, I. P.; Hwang, K. C.; Chiang, C.-S. Preparation of fluorescent magnetic nanodiamonds and cellular imaging. *Journal of American Chemical Society*, 2008, *130* (46), 15476–15481.

183. Yafarov, R. K. Production of nanodiamond composites in a low-pressure microwave gas-discharge plasma. *Technical Physics*, 2006, *51* (1), 40–46.

184. Blum, R.; Molian, P. Liquid-phase sintering of nanodiamond composite coatings on aluminum A319 using a focused laser beam. *Surface and Coatings Technology*, 2009, *204* (1–2), 1–14.

185. Matsubara, H. Fabrication of novel materials by the incorporation of nanodiamond into plated films. *Hyomen Kagaku*, 2009, *30* (5), 279–286.

186. Matsubara, H. Co-deposition behavior of nanodiamond with electrolessly plated nickel films. *Hyomen Gijutsu*, 2006, *57* (7), 484–488.

187. Detkov, P. Y.; Popov, V. A.; Kulichikhin, V. G.; Chukhaeva, S. I. Development of composite materials based on improved nanodiamonds. *Topics in Applied Physics*, 2007, *109* (*Molecular Building Blocks for Nanotechnology*), 29–43.

188. Uetsuka, H.; Nakamura, T.; Nebel, C. E. Nanodiamond-containing microstructure composites and method for transporting biological molecules into bodies by using them. JP 2009119561 A 20090604 Patent written in Japanese. Application: JP 2007-296378 20071115, 2009, 7 pp.

189. Lee, S. H. Gas sensor using nanodiamond and gas detection method. KR 2009066740 A 20090624 Patent written in Korean. Application: KR 2007-134421 20071220. Priority: CAN 151:92754 AN 2009:780521, 2009, 6 pp.

190. Raina, S.; Kang, W. P.; Davidson, J. L. Optimizing nitrogen incorporation in nanodiamond film for bio-analyte sensing. *Diamond and Related Materials*, 2009, *18* (5–8), 718–721.

191. Holt, K. B.; Caruana, D. J.; Millan-Barrios, E. J. Electrochemistry of undoped diamond nanoparticles: Accessing surface redox states. *Journal of American Chemical Society*, 2009, *131* (32), 11272–11273.

192. Larionova, I. S.; Belyaev, V. N.; Il'inykh, K. F.; Frolov, A. V.; Bychin, N. V.; Mitrofanov, V. M. Method for preparation and galvanic deposition of wear-resistant nanodiamonds based coating composition on metal surfaces. RU 2357017 C1 20090527 Patent written in Russian. Application: RU 2007-128703 20070725. Priority: CAN 150:565842 AN 2009:646846, 2009, 5 pp.

193. Vityaz, P. A. The state of the art and prospects of detonation-synthesis nanodiamond applications in Belarus. *Physics of the Solid State*, 2004, *46* (4), 606–610.

194. Shiozaki, S. Normal-temperature glass, its formation, and normal temperature glass coating material. JP 2009102188 A 20090514 Patent written in Japanese. Application: JP 2007-274359 20071022. Priority: CAN 150:499296 AN 2009:583008, 2009, 18 pp.

195. Purtov, K. V.; Bondar, V. S.; Puzyr, A. P. Nanodiamond sorbent and method of its obtaining. RU 2352387 C1 20090420 Patent written in Russian. Application: RU 2007-127892 20070719. Priority: CAN 150:450910 AN 2009:474289, 2009, 7 pp.

196. Bondar, V. S.; Puzyr, A. P. Possibilities and prospects for creation of new nanoprocesses based on detonation nanodiamond particles: medicobiological and technical aspects. *Konstruktsii iz Kompozitsionnykh Materialov*, 2005, (4), 80–94.

197. Zibrov, S. A.; Vasil'ev, V. V.; Velichanskii, V. L.; Pevgov, V. G.; Rudoi, V. M. Method for protection of documents, valuable papers or products with nanodiamonds with active NV centers. RU 2357866 C1 20090610 Patent written in Russian. Application: RU 2008-136466 20080910. Priority: CAN 151:7812 AN 2009:703362, 2009, 4 pp.

198. Zhang, D.; Hu, X.-g.; Tong, Y.; Huang, F.-l. The research development of nanodiamond as a lubricating additive. *Runhuayou*, 2006, *21* (1), 50–54.

199. Qu, J.; Li, X.; Song, B. Polytetrafluoroethylene friction material for ultrasonic motor. CN 1473865 A 20040211 Patent written in Chinese. Application: CN 2003-132555 20030807. Priority: CAN 142:393210 AN 2004:1024000, 2004, 4 pp.

200. Luo, J.; Liu, X.; Wang, X. Effect of proportion of nano-diamond and zirconia on color of core resin. *Xiandai Kouqiang Yixue Zazhi*, 2008, *22* (3), 251–254.

201. Comet, M.; Pichot, V.; Siegert, B.; Spitzer, D.; Moeglin, J.-P.; Boehrer, Y. Use of nanodiamonds as a reducing agent in a chlorate-based energetic composition. *Propellants, Explosives, Pyrotechnics*, 2009, *34* (2), 166–173.

202. Xing, Y.; Dai, L. Nanodiamonds for nanomedicine. *Nanomedicine*, 2009, *4* (2), 207–218.

203. Vaijayanthimala, V.; Chang, H.-C. Functionalized fluorescent nanodiamonds for biomedical applications. *Nanomedicine*, 2009, *4* (1), 47–55.

204. Schrand, A. M.; Hens, S. A. C.; Shenderova, O. A. Nanodiamond particles: Properties and perspectives for bioapplications. *Critical Reviews in Solid State and Materials Sciences*, 2009, *34* (1–2), 18–74.

205. Cheng, C.-Y.; Perevedentseva, E.; Tu, J.-S.; Chung, P.-H.; Cheng, C.-L.; Liu, K.-K.; Chao, J.-I.; Chen, P.-H.; Chang, C.-C. Direct and in vitro observation of growth hormone receptor molecules in A549 human lung epithelial cells by nanodiamond labeling. *Applied Physics Letters*, 2007, *90* (16), 163903/1–163903/3.

206. Shimkunas, R. A.; Robinson, E.; Lam, R.; Lu, S.; Xu, X.; Zhang, X.-Q.; Huang, H.; Osawa, E.; Ho, D. Nanodiamond-insulin complexes as pH-dependent protein delivery vehicles. *Biomaterials*, 2009, *30* (29), 5720–5728.

207. Liu, K.-K.; Wang, C.-C.; Cheng, C.-L.; Chao, J.-I. Endocytic carboxylated nanodiamond for the labeling and tracking of cell division and differentiation in cancer and stem cells. *Biomaterials*, 2009, *30* (26), 4249–4259.

208. Chen, M.; Pierstorff, E. D.; Lam, R.; Li, S.-Y.; Huang, H.; Osawa, E.; Ho, D. Nanodiamond-mediated delivery of water-insoluble therapeutics. *ACS Nano*, 2009, *3* (7), 2016–2022.

209. Faklaris, O.; Joshi, V.; Irinopoulou, T.; Tauc, P.; Girard, H.; Gesset, C.; Senour, M. et al. Determination of the internalization pathway of photoluminescent nanodiamonds in mammalian cells for biological labeling and optimization of the fluorescent yield. *Physics*, 2009, 1–24, arXiv.org, e-Print Archive, arXiv:0907.1148v1 [physics.optics].

19 Fulleropyrrolidines

After discovery of buckminsterfullerene C_{60} in 1985,[1] during the next three decades the organic chemistry of fullerenes was successfully developed. Among other methods of its functionalization, the widest application corresponds to 1,3-dipolar cycloaddition of azomethine ylides, leading to an important class of fullerene derivatives, pyrrolidino[3′,4′:1,2][60]fullerenes (familiarly known as *fulleropyrrolidines*), intensively developed mainly in the research groups of Prato and Martin. The main achievements in this area have been generalized in a series of monographs by Hirsch,[2,3] Guldi,[4] and other researchers,[5–11] as well as in reviews[12–20]; additionally, synthetic methods and applications of fulleropyrrolidines are examined in a series of patents.[21–31] Among the series of publications, we note an excellent recent review,[20] dedicated to new reactions in fullerene chemistry, including those of fuller-1,6-enynes (involving the fulleropyrrolidine moiety) as new and versatile blocks in fullerene chemistry and detailed discussion on retro-cycloaddition processes of fulleropyrrolidines.

The simplest compound of fulleropyrrolidine type, *N*-methylfulleropyrrolidine, is shown in Figure 19.1. The classic synthetic procedure, leading to this fullerene derivative and its numerous analogues, is the 1,3-dipolar cycloaddition of azomethine ylides[32] (reactive intermediates generated *in situ* by many different ways, in particular by decarboxylation of ammonium salts, formed as a result of condensation of α-amino acids with aldehydes) to C_{60} (*Prato* reaction). As a result of this powerful methodology for obtaining functionalized fullerene derivatives, the fulleropyrrolidines are formed, in which a pyrrolidine ring is fused to a junction between two six-member rings of a fullerene sphere[5] (Figure 19.1). When the reaction is carried out in the presence of large excesses of reagents, up to nine pyrrolidine rings can be introduced.[15] The key features and strategies of this type of reaction are summarized in Refs. [5,15,17,33]; in particular, its main advantages are as follows: (1) the reactions lead individual [6,6]-closed isomers, (2) majority of precursors are commercially available or could be easily prepared, and (3) two substituents can be simultaneously introduced into the pyrrolidine cycle. Therefore, functionalization of the fullerene sphere on the *Prato*'s reaction basis occupies a leading place in the synthesis of fullerene derivatives to get new materials and potentially biologically active compounds.

The physicochemical properties of the fulleropyrrolidines are expected to arise from the combination of the two molecular fragments, the fullerene and the pyrrolidine moiety. The electronic properties of the fulleropyrrolidines are typical of most C_{60} monoadducts.[13]

In this section, together with the description of fulleropyrrolidine derivatives, attention is paid to the functionalized carbon nanotubes (CNTs), containing pyrrolidine groups. Representative examples of the synthesis of fulleropyridines are shown in Table 19.1 and those of applications in Table 19.2.

19.1 PORPHYRIN-CONTAINING FULLEROPYRROLIDINES

Fullerene derivatives, containing porphyrins and their metallocomplexes (generally those of Zn,[62,63] Mg, or Ru), continue to be obviously one of the popular topics in fullerene organic chemistry as well due to many their applications (see the following), in particular for solar energy conversion. Among a series of publications in this area, we note the development of a new fullerene building block (*N*-pyridylfulleropyrrolidine, Figure 19.7, left), capable of forming axially symmetric complexes with metalloporphyrins, and designed with the potential for defined geometry and good electronic communication.[64] This compound was prepared by the *Prato* reaction from *N*-pyridylglycine, C_{60}, and paraformaldehyde as precursors in *o*-dichlorobenzene (31% yield) and compared with the

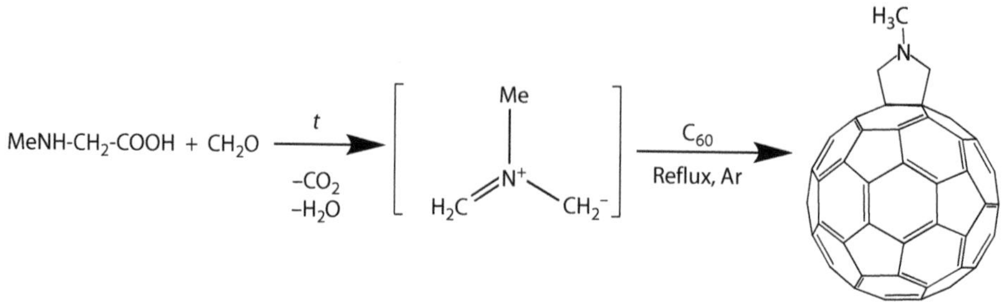

FIGURE 19.1   Classic synthesis of *N*-methylfulleropyrrolidine.

fullerene ligand, wherein the pyridine is connected *via* an insulating sp^3 carbon atom, from the point of view whether the pyridine N atom of *N*-pyridylfulleropyrrolidine really communicates with the fullerene core.

"Tail-on" and "tail-off" binding mechanisms are examined for an axial ligand coordination in porphyrin–fullerene dyads.[65] The donor–acceptor proximity is controlled either by temperature variation or by an axial ligand replacement method (Figure 19.8). It is observed that in the tail-off form the charge-separation efficiency changes to some extent in comparison with the results obtained for the tail-on form, suggesting the presence of some through-space interactions between the singlet excited zinc porphyrin and the C$_{60}$ moiety in the tail-off form.

Spectroscopic, redox, and electron transfer reactions of a self-assembled donor–acceptor dyad formed by axial coordination of magnesium *meso*-tetraphenylporphyrin (MgTPP) and fulleropyrrolidine appended with an imidazole-coordinating ligand (C$_{60}$Im) were investigated by a series of methods.[66] Spectroscopic studies revealed the formation of a 1:1 C$_{60}$Im:MgTPP supramolecular complex (Figure 19.9). Among many other porphyrin-containing fulleropyrrolidines,[53,67] the most interesting recently synthesized, in our point of view, are as follows: the supramolecular assembly containing an electron donor, zinc 5,10,15,20-*meso*-tetraferrocenylporphyrin, and a pyridine-substituted fulleropyrrolidine as an electron acceptor (Figure 19.10),[68] supramolecular ferrocene-porphyrin-fullerene constructions, in which covalently linked ferrocene-porphyrin-crown ether compounds are self-assembled with alkylammonium cation functionalized fullerenes,[69] multimodular systems, composed of three covalently linked triphenylamine, entities at the meso position of the porphyrin ring and one fulleropyrrolidine at the fourth meso position (Figure 19.11),[70] mixed assembly metal-free porphyrin (5-(3′-(2″-(3‴ or 4‴-pyridyl)fulleropyrrolidinyl-*N*)ethoxyphenyl)-10,15,20-triphenylporphyrin)—ferrocene—metal porphyrin H$_2$P-C$_{60}$Py-ZnP (Figure 19.12).[71]

In comparison with a lot of reported porphyrin-containing fulleropyrrolidines during the last decade, only a few investigations are devoted to the fulleropyrrolidines, containing phthalocyanine and its analogues,[52,72–74] having possible applications for solar energy conversion.[59,75]

19.2   FULLEROPYRROLIDINES WITH SULFUR-CONTAINING GROUPS

The synthesis of thiophene-containing fulleropyrrolidines, covalently linked to fluorescent conjugated systems bearing long alkyl chains (Figure 19.13), attached to the naphthalene or benzene moieties, has been carried out from suitably functionalized oligomers by *Prato* reaction; the obtained products were studied by cyclic voltammetry.[56]

Among other studied fulleropyrrolidines, containing thiophene moieties, we note bithiophene-fulleropyrrolidine (Figure 19.14, yield 49%)[76] and dimeric fullerene-donor-fullerene triads, containing an electron-rich pyrrole ring[57] (Figure 19.15) in the donor moiety (other multifulleropyrrolidines—see the following). A direct palladation as a pathway toward fullerene SCS-based ([C$_6$H$_2$(CH$_2$SPh)$_2$-2,6-R-4]$^-$) organometallic complexes is reported in Ref. [77]. C$_{60}$-modified polythiophene copolymers are shown in Section 19.6.

TABLE 19.1
Representative Examples of the Synthesis of Fulleropyrrolidines

Precursors and Conditions	Products	Observations	Reference
N-(Triphenylmethyl)-3,4-fulleropyrrolidine, trifluoromethanesulfonic acid, dicyclohexylcarbodiimide (DCC) coupling of 6-((biotinoyl)-amino)-hexanoic acid (biotin-X), anhydrous pyridine, CH$_2$Cl$_2$, DMF		Yield 65%	[34]

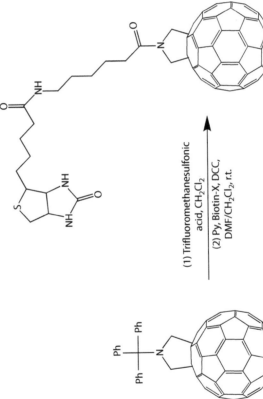

(1) Trifluoromethanesulfonic acid, CH$_2$Cl$_2$

(2) Py, Biotin-X, DCC, DMF/CH$_2$Cl$_2$, r.t.

FIGURE 19.2 Synthesis of N-6-(biotinamido)hexanoyl-3,4-fullerenopyrrolidine.

(*continued*)

TABLE 19.1 (continued)
Representative Examples of the Synthesis of Fulleropyrrolidines

Precursors and Conditions	Products	Observations	Reference
C_{60}, paraformaldehyde, 4-amino-4-carboxy-2,2,6,6-tetramethylpiperidine-1-oxyl	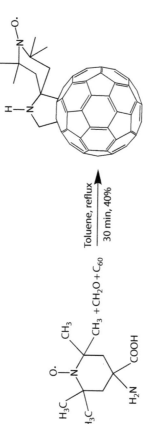	Yield 40%	[35]
C_{60}, toluene, corresponding aldehyde, amino-acid, reflux, 1 h	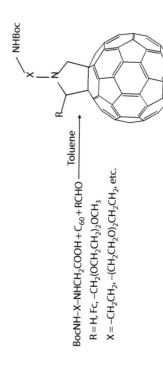	Yield 17%–44%	[36]

FIGURE 19.3 Synthesis of 3,4-fulleropyrrolidine-2-spiro-4′-[2′:2′,6,6′-tetramethyl]piperidine-1′-oxyl].

FIGURE 19.4 Obtaining N-Boc-protected amino fulleropyrrolidines.

C$_{60}$, glycine, paraformaldehyde, toluene, reflux, 2 h (first step).
Further treatment with 1-naphthoyl chloride and triethylamine, r.t., 30 min

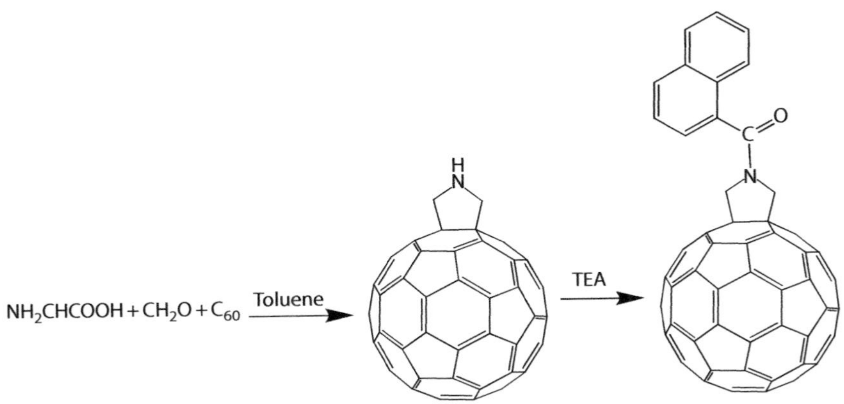

Yield 9%

[37]

NH$_2$CHCOOH + CH$_2$O + C$_{60}$ $\xrightarrow{\text{Toluene}}$ $\xrightarrow{\text{TEA}}$

FIGURE 19.5 Synthesis of *N* – 1-naphthoyl fulleropyrrolidine.

N-ethyl glycine, paraformaldehyde, Sc$_3$N@C$_{80}$ in 10 mL of *o*-dichlorobenzene, 120°C, Ar, 15 min

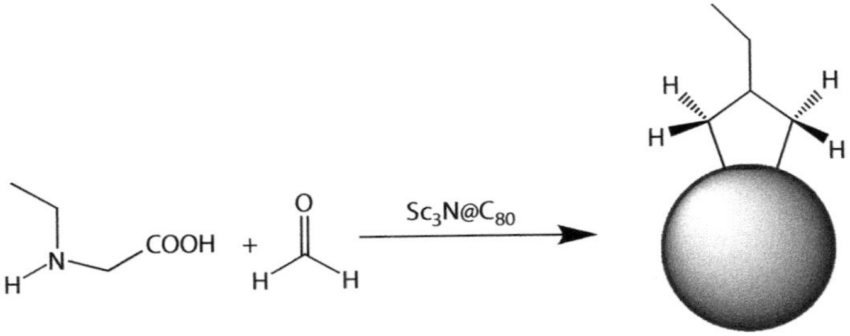

Yield 73%

[38]

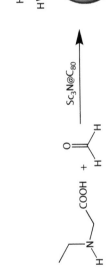

FIGURE 19.6 The first pyrrolidine adduct on Sc$_3$N@C$_{80}$.

TABLE 19.2
Examples of Academic and Industrial Applications of Fulleropyrrolidines

Structures	Properties and Characteristics	Applications	Reference
	Increasing solubility of C_{60} in organic solvents	Reduction of current in electric circuits	[39]
	Absorption in visible spectrum range	Use in Grätzel solar cells	[40]

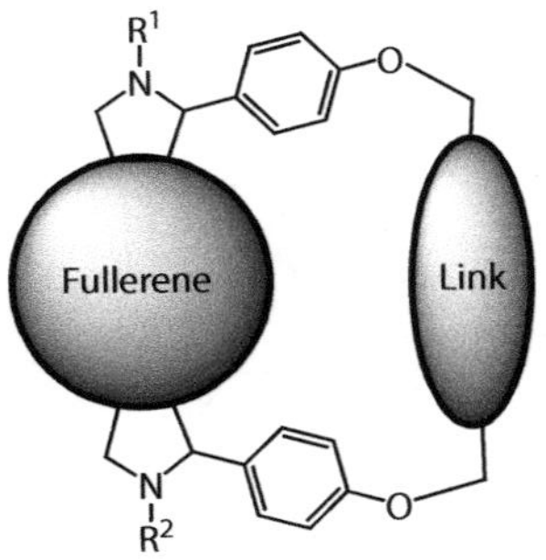

(continued)

[41]

Selective synthesis
of fulleropyrrolidine
bis-adduct

Unique absorption in the
region of 400–750 nm.
Can anchor two, three,
and even more building
blocks, such as pyridine
and polypyridine

[42]

Construction
of photoactive
multicomponent
systems containing C_{60}

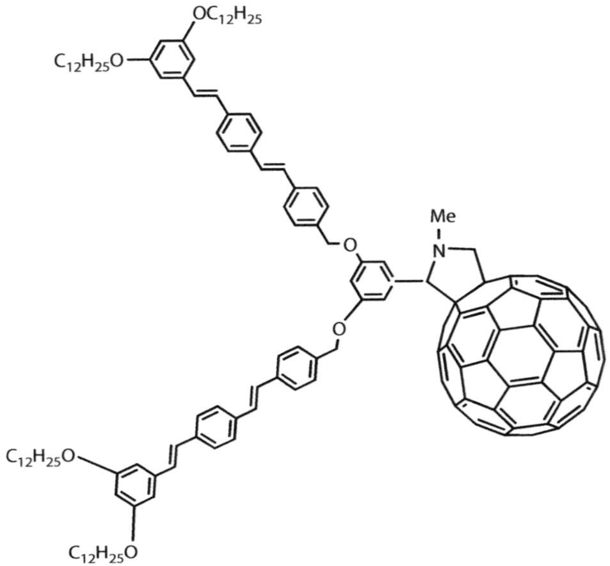

Fast photoinduced energy
transfer. Electron transfer
not possible, even in
polar solvent

TABLE 19.2 (continued)
Examples of Academic and Industrial Applications of Fulleropyrrolidines

Structures	Properties and Characteristics	Applications	Reference
	Useful for production of monomolecular layers	Obtaining controlled architectures like coplanar fullerene moieties	[43]
	In the presence of O_2, it can decay from its triplet to the ground state transferring its energy to O_2 for generation of $^1O_2$	Inside the HIV-P cavity, it can contribute to stabilization of the complexation of the enzyme	[44]
	Low flexibility	Photovoltaic cells	[12]

$R = COCH_3, CO(CH_2)_{16}CH_3, CO(CH_2)_{20}CH_3, CH_3, CO(CF_2)_6CF_6$

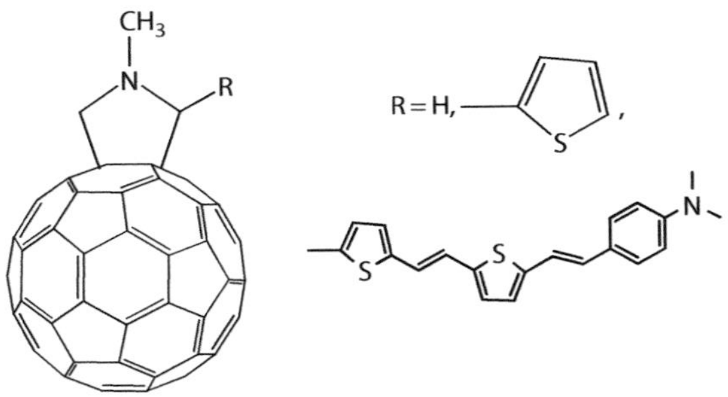

(continued)

[45] Development of optoelectronic materials

Intramolecular photoinduced charge separation in polar solvents

[46] Development of oligothiophene-based optoelectronic materials and artificial photosynthetic systems

Intramolecular photoinduced charge separation

[47] Generation of photoexcited quartet and doublet states

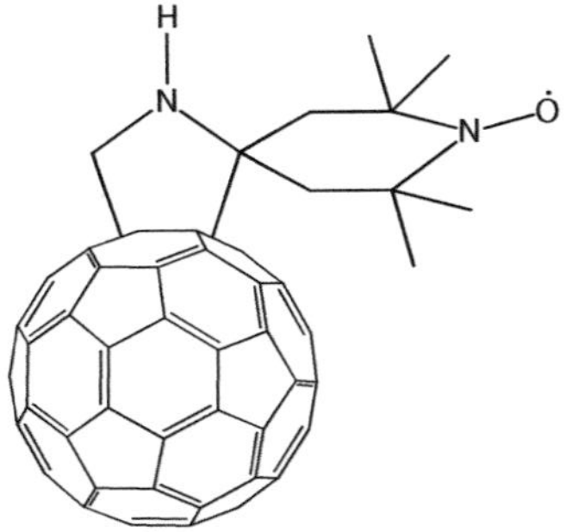

A great ability of 2D nutation spectroscopy

$R = H,$

Fe-nT-C_{60} (Figure 19.18)

TABLE 19.2 (continued)
Examples of Academic and Industrial Applications of Fulleropyrrolidines

Structures	Properties and Characteristics	Applications	Reference
	C_{60} behaves as both a stopper and a photoactive unit	Can be used to shuttle the macrocycle from close to the fullerene spheroid (in nonpolar solvents) to far away (in polar solvents)	[48]
	Precursor of other more complex structures	Preparation of metal-chelating and air-stable structures	[49]

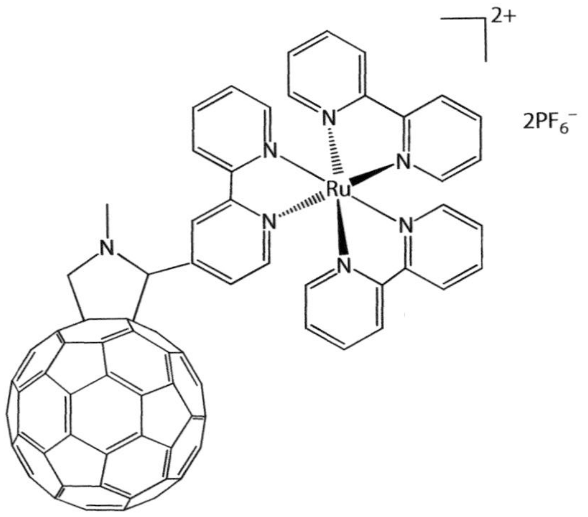

(continued)

[49]

[50]

Incorporation into new types of ruthenium–polypyridine complexes

Preparation of *N*-6-[(biotinamido)-hexanoyl]-3,4-fulleropyrrolidine

Metal-chelating, air-stable solid

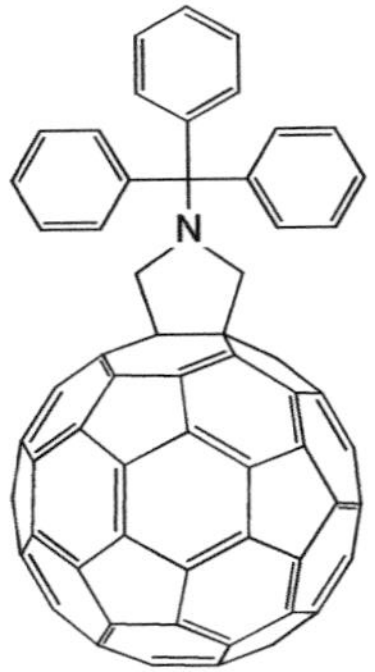

React with functional groups on proteins

2PF$_6^-$

2+

TABLE 19.2 (continued)
Examples of Academic and Industrial Applications of Fulleropyrrolidines

Structures	Properties and Characteristics	Applications	Reference
	Presence of a single biotin moiety. Unlike suspensions of C_{60} that settle within minutes in water, the fullerene-immobilized enzyme remains in suspension for extended periods of time	Use of fullerene-immobilized enzyme without any significant loss of activity	[50]

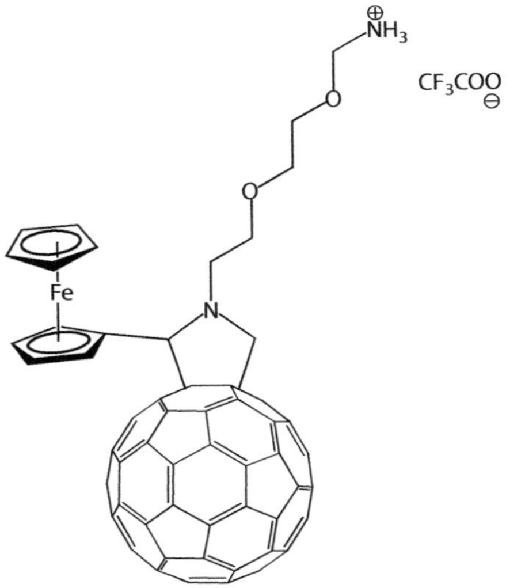

(continued)

[51]

Create a new hybrid system where C_{60} differs from its crystals or its solutions

Successful incorporation of fulleropyrrolidine into hydrophilic or modified organophilic clays by ion exchange or organic solution

[52]

Electron-transfer benefits stem for nanotubes

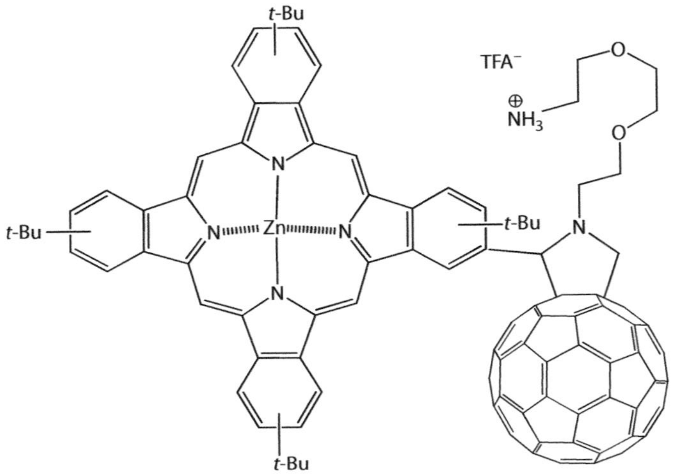

When it dissolves homogeneously, fast charge recombination occurs

TABLE 19.2 (continued)
Examples of Academic and Industrial Applications of Fulleropyrrolidines

Structures	Properties and Characteristics	Applications	Reference
	With only a slight excess of pyC$_{60}$, no fluorescence is detected at 430 and 365 nm	Construction of the assembled system ZnTFcPCpyC$_{60}$	[53]
	A rapid intramolecular deactivation of the photoexcited C$_{60}$ chromophore with decay dynamics between 570 and 340 ps	Increase the solubility of C$_{60}$ cage	[54]

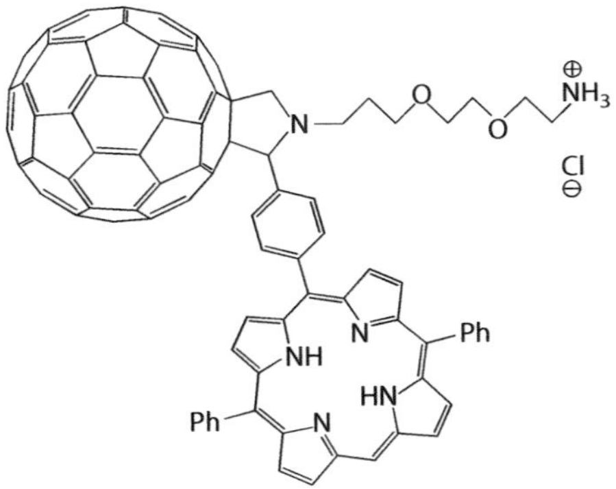

(continued)

[55]

The new electrodes give rise to improved light harvesting features in the visible

Fabrication of high-quality, robust, and photoactive ITO electrodes

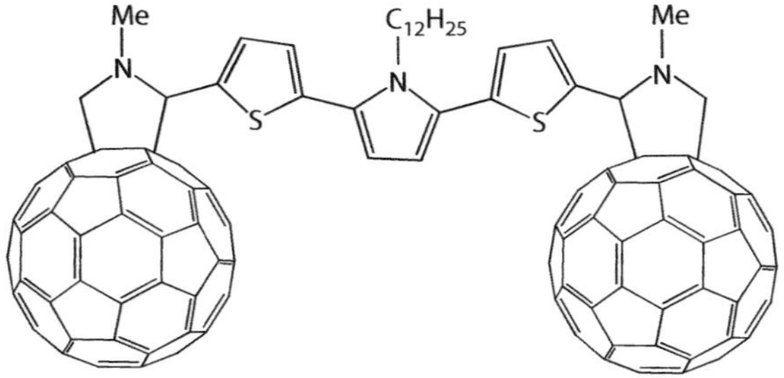

[56]

Intensity of the fullerene photoluminescence (PL) of the triads at 712 nm

Can be used to assess whether an intramolecular charge-transfer reaction occurs in a donor–acceptor system

[57]

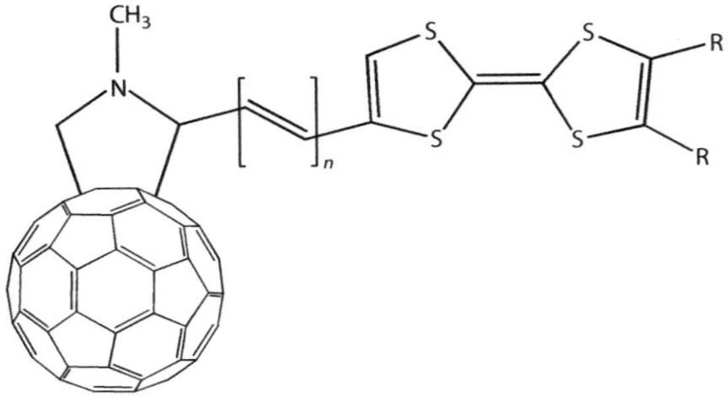

Fulleropyrrolidine is covalently linked to tetrathiafulvalene (TTF)

Create single component donor–acceptor organic compounds. Molecular electronic devices

(Figure 19.15, up)

TABLE 19.2 (continued)
Examples of Academic and Industrial Applications of Fulleropyrrolidines

Structures	Properties and Characteristics	Applications	Reference
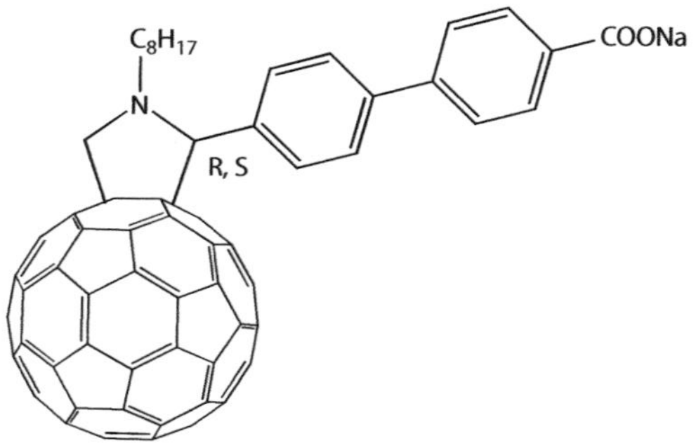	Formation of charge-separated radical pairs C_{60}^{-}—TTF^{+}	Preparation of new photovoltaic materials	[58]
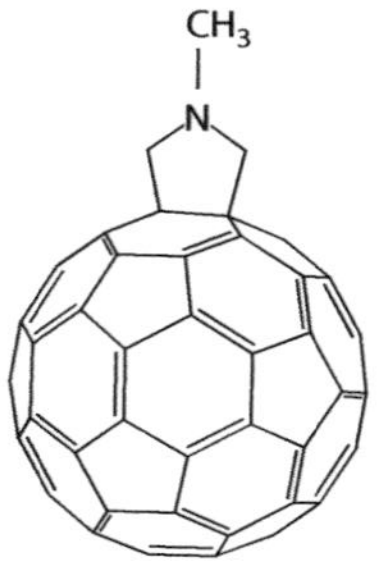	Electron acceptor	Useful in construction of new donor–acceptor systems	[59]

C$_8$H$_{17}$ N R, S COONa

CH$_3$ N

(Figure 19.1)

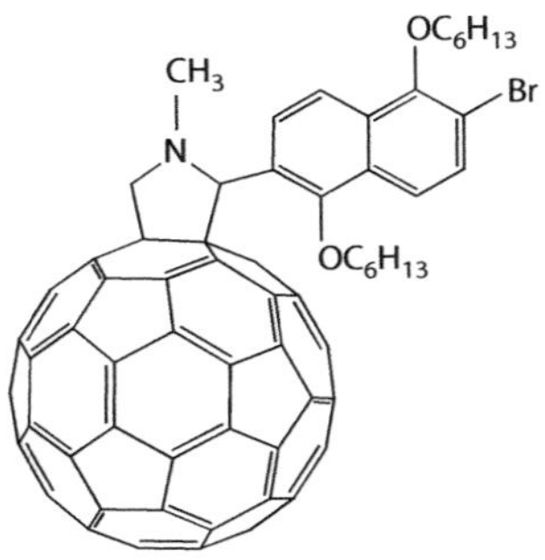

(continued)

[59]

High-performance photovoltaic devices

Fluorescence maxima at 715 nm. Cathodic side presents three one-electron reduction steps, whereas on anodic side only a single one-electron wave is detected

[60]

Electronic and optoelectronic applications as, for example, 1D electron transportation

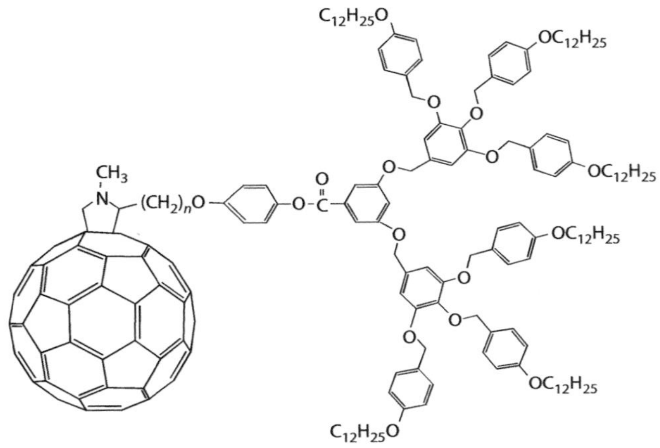

Functionalization of C_{60} with liquid-crystalline dendrimers exhibits columnar mesomorphism

TABLE 19.2 (continued)
Examples of Academic and Industrial Applications of Fulleropyrrolidines

Structures	Properties and Characteristics	Applications	Reference
	In benzonitrile, Charge separation rate (K_{CS}) $1.9 \times 10^9\,s^{-1}$. Charge recombination rate (K_{CR}) $7.7 \times 10^6\,s^{-1}$	Preparation of fulleropyrrolidinium ions with higher K_{CS} and lower K_{CR} than fulleropyrrolidine	[61]
	In benzonitrile, $K_{CS} = 2.8 \times 10^9\,s^{-1}$ $K_{CR} = 6.6 \times 10^6\,s^{-1}$	Efficient conversion of solar energy	[61]

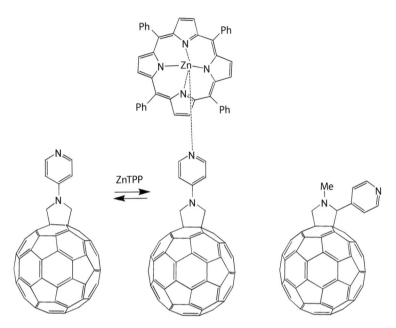

FIGURE 19.7 *N*-pyridylfulleropyrrolidine (left), its ZnTPP complex (middle), and *C*-pyridylfulleropyrrolidine (right).

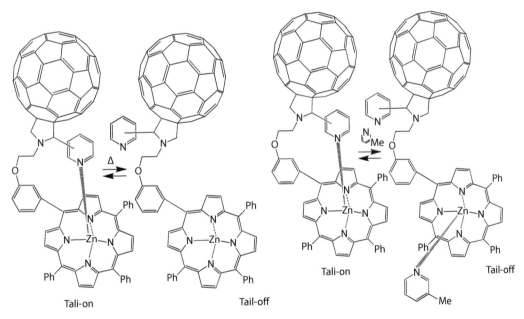

FIGURE 19.8 Temperature variation (left) or an axial ligand replacement (right) study of the donor–acceptor proximity in the porphyrin–fullerene dyad.

Highly conjugated TTF analogues with the *p*-quinodimethane structure covalently attached to C_{60} (Figure 19.16) were reported in Ref. [78]. The molecular geometry of the formed compounds was calculated from semiempirical PM-3 calculations and reveals a highly distorted electron-donor moiety with a butterfly-type structure. A series of similar TTF-containing C_{60}-pyrrolidines,[54] in particular dimeric compounds,[46,79] is known and clearly shows that the TTF

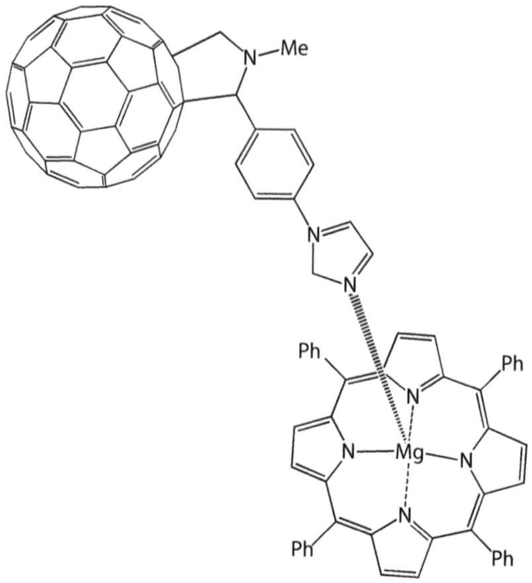

FIGURE 19.9 Magnesium *meso*-tetraphenylporphyrin–fulleropyrrolidine complex.

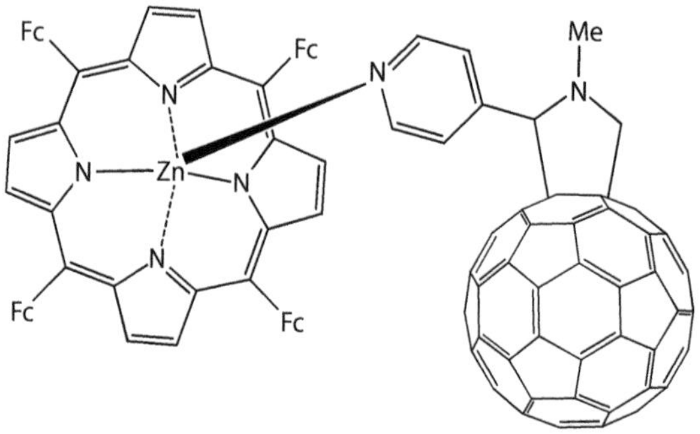

FIGURE 19.10 Supramolecular assembly of zinc 5,10,15,20-*meso*-tetraferrocenylporphyrin and pyridine-substituted fulleropyrrolidine

moiety is one of the most used S-containing ligands in the preparation of its fulleropyrrolidine dyads and triads (see also Section 19.3, dedicated to triads of ferrocene-containing fulleropyrrolidines, thiophenes, and TTF).

19.3 FERROCENE-CONTAINING FULLEROPYRROLIDINES

In addition to relatively simple C_{60}-ferrocene-containing diads (Figure 19.17), intensively developed in the last decade of the twentieth century,[80] two types of the ferrocene-oligothiophene-fullerene triads (Figure 19.18, up), Fc-nT-C_{60} directly linking the ferrocene to the oligothiophene and Fc-tm-nT-C_{60} inserting a trimethylene spacer between the ferrocene and the oligothiophene, were synthesized[81] to promote photoinduced charge separation for the oligothiophene-fullerene dyads (nT-C_{60}). The physicochemical studies of Fc-nT-C_{60} indicate conjugation between the ferrocene and oligothiophene components and markedly quenched fluorescence of the oligothiophene in comparison with that observed for the dyads nT-C_{60}. The authors stated that the additionally conjugated ferrocene evidently contributes to the stabilization of charge separation states, thus promoting intramolecular electron transfer; this fact is confirmed by the observation that the emission spectra of the nonconjugated triads Fc-tm-nT-C_{60} are essentially similar to the corresponding dyads

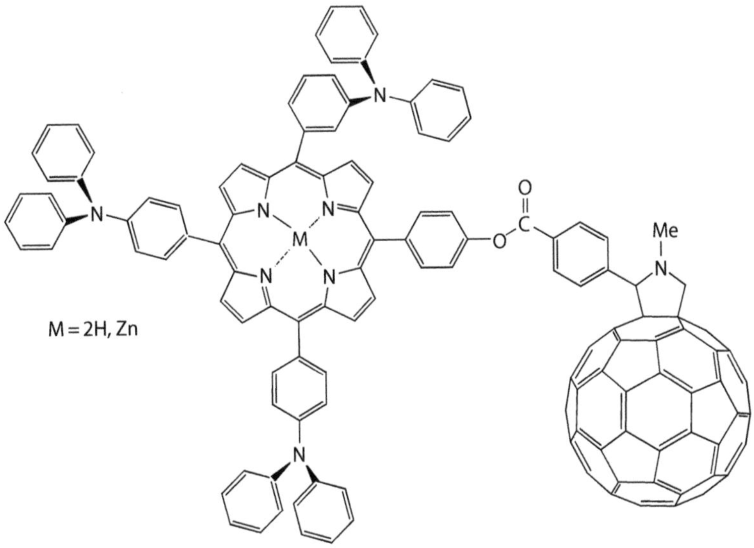

FIGURE 19.11 Multimodular systems, composed of three entities of triphenylamine, porphyrin, and fullerene.

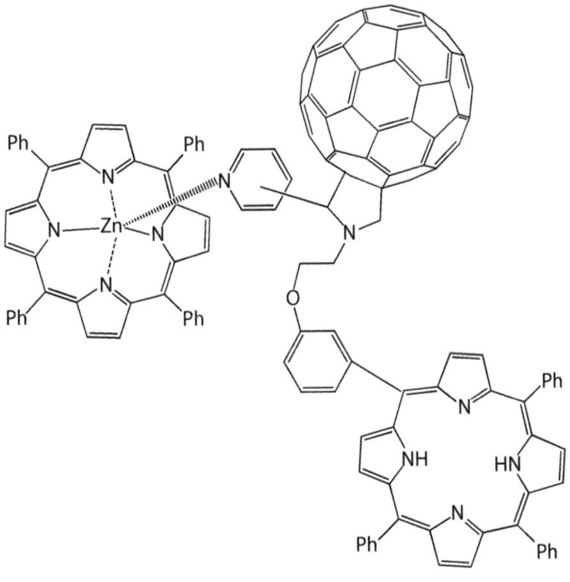

FIGURE 19.12 Supramolecular triads of the type (donor-1)-acceptor-(donor-2) composed of free-base porphyrin, fullerene, and zinc porphyrin.

nTC_{60}. In another research,[82] the ferrocene moiety is bound to N-pyrrolidine atom through C=O group, meanwhile TTF is connected to the C-pyrrolidine atom (Figure 19.19).

Among other combinations with ferrocene moieties, we note a series of molecular triads composed of ferrocene (through pyrrolidine C-binding), C_{60}, and nitroaromatic entities (N-pyrrolidine binding) R (R = $-CO-C_6H_3(3,5-NO_2)_2$, $-CH_2-C_6H_3(3,5-NO_2)_2$, $-C_6H_3(2,4-NO_2)_2$, and $-CO-C_6H_4(4-NO_2)$),[83,84] photoactive fulleropyrrolidine-perylenetetracarboxylic diimide-porphyrin triad and its zinc analogue,[85] and much more.[61,86–88] The physicochemical properties of ferrocene-containing fulleropyrrolidines are intensively studied.[41,89,90]

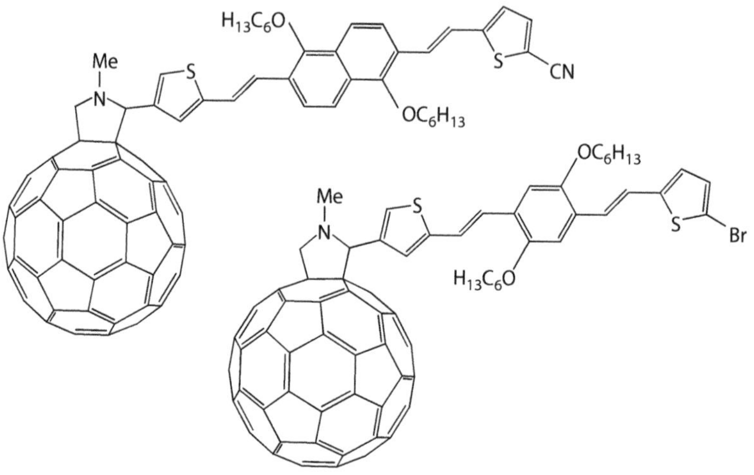

FIGURE 19.13 Fulleropyrrolidines with dihexyloxynaphthalenethiophene (left) and dihexyloxybenzene-thiophene (right) moieties.

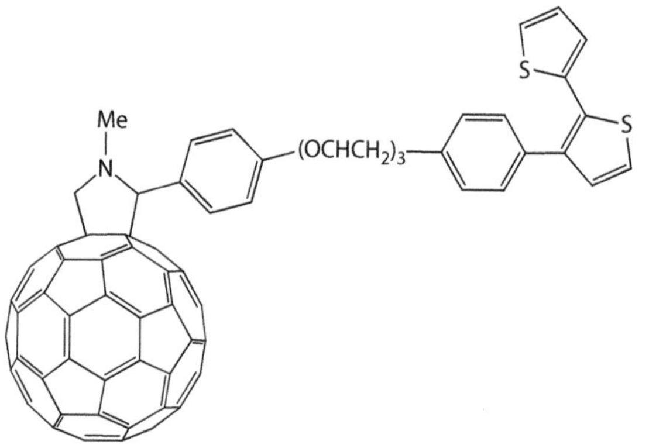

FIGURE 19.14 Bithiophene–fulleropyrrolidine.

19.4 FULLEROPYRROLIDINE BIS- AND TRIS-ADDUCTS

The synthesis, characterization, properties, and supramolecular organization of *liquid-crystalline* fullerene bis-adducts and monoadduct (Figure 19.19) used as reference compound were reported in Ref. [91]. The authors postulated that the mono- and bis-adducts formed bilayered and monolayered smectic A phases, respectively. The obtained compounds are interesting and promising supramolecular materials as they combine the self-organizing behavior of liquid crystals and retain most of the properties of C_{60}.

A series of fulleropyrrolidine bis-adducts (some of them are shown in Figure 19.20) with *trans*-1, *trans*-2, *trans*-3, *trans*-4, and equatorial bis-addition patterns were prepared from corresponding bis(benzaldehydes), sarcosine, and C_{60} by *Prato* reaction.[92]

Additionally, three chiral *N*-methylfulleropyrrolidine bis-adducts were synthesized, isolated, and completely resolved into each enantiomer using a chiral HPLC column, which were then converted to the corresponding optically active, cationic C_{60}-bis-adducts.[93] Other bis-adducts were

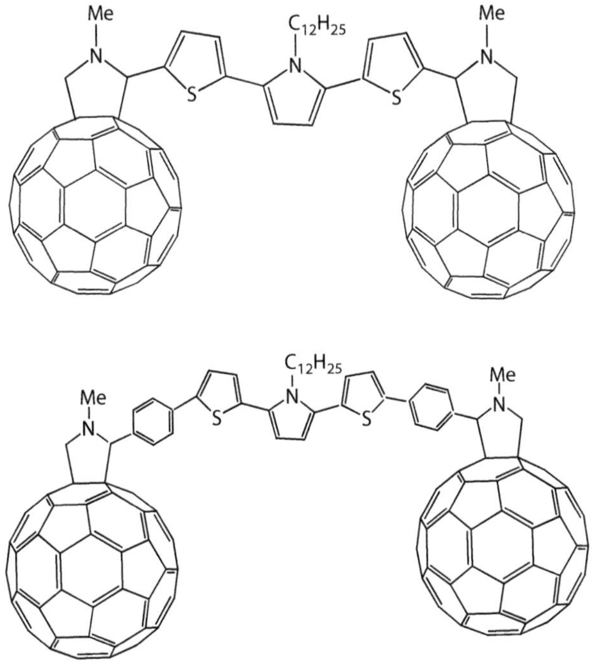

FIGURE 19.15   Dimeric fulleropyrrolidine–(S-donor)–fulleropyrrolidine triads.

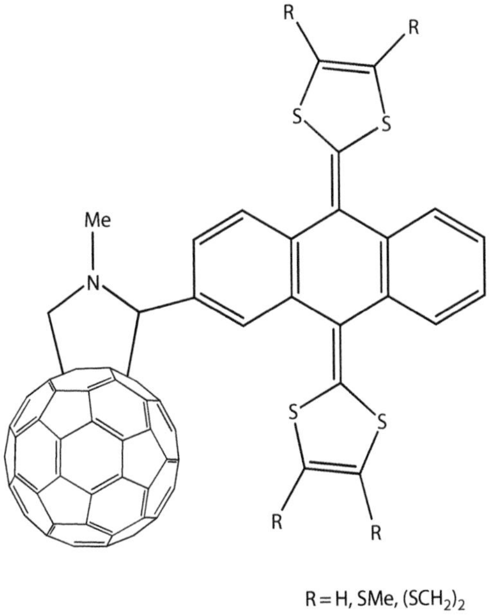

R = H, SMe, (SCH$_2$)$_2$

FIGURE 19.16   C$_{60}$–pyrrolidine–tetrathiafulvalene system.

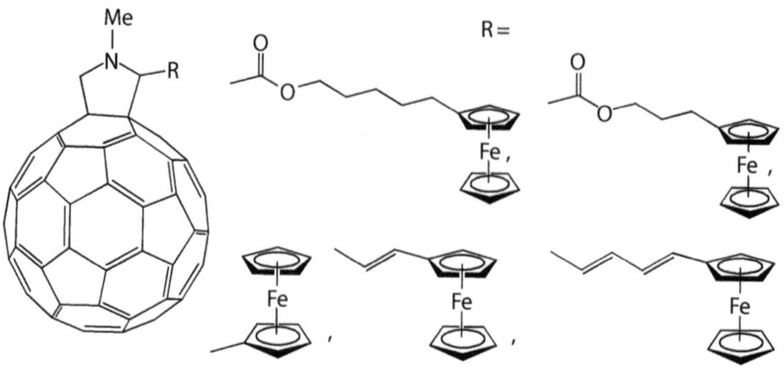

FIGURE 19.17 Fullerene–ferrocene diads.

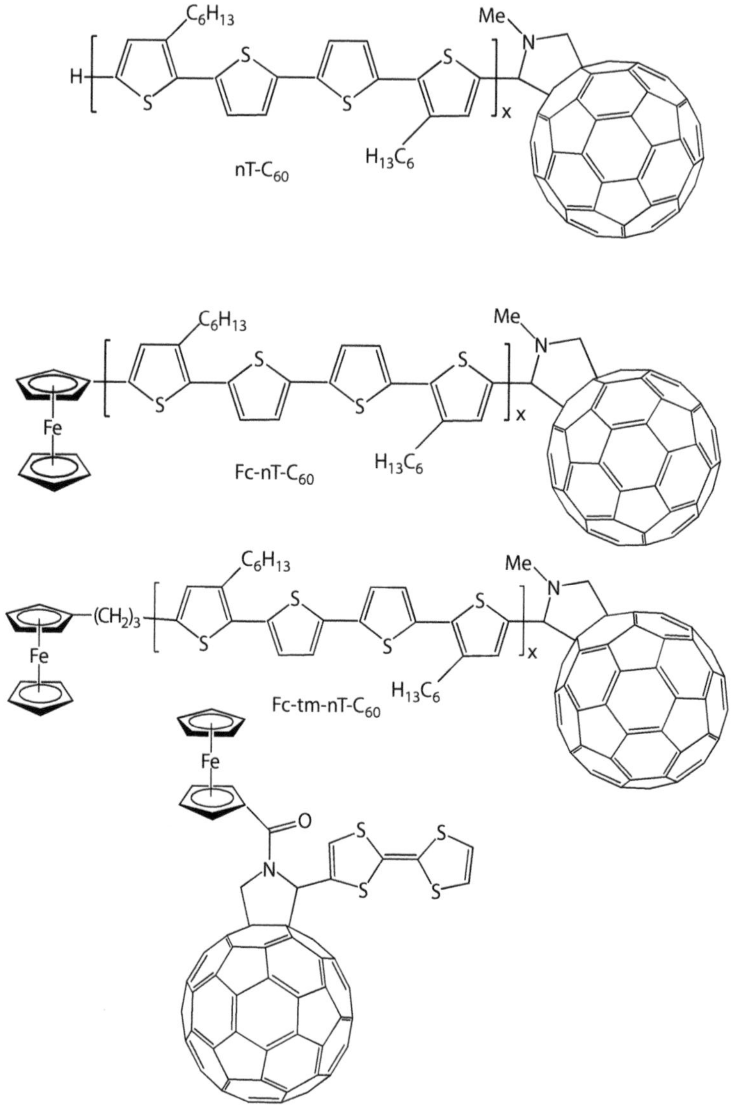

FIGURE 19.18 Ferrocene–oligothiophene–fullerene and ferrocene–tetrathiafulvalene–fullerene triads.

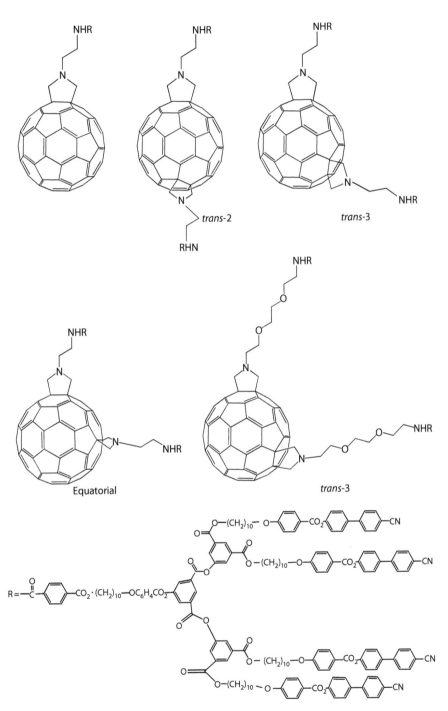

FIGURE 19.19 Fulleropyrrolidine monoadduct (reference compound, left) and liquid-crystalline fullerene bis-adducts.

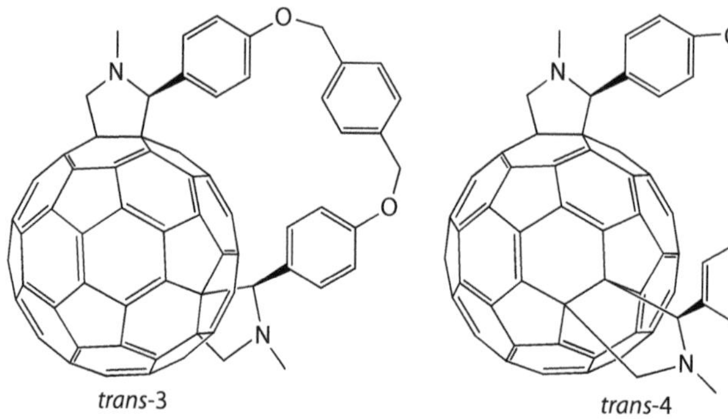

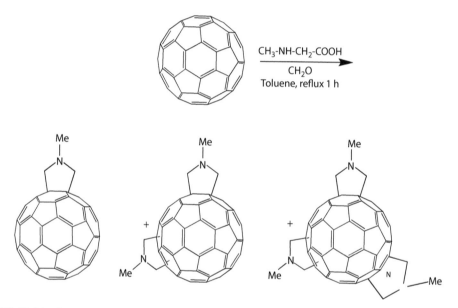

FIGURE 19.20 Examples of fulleropyrrolidine bis-adducts.

FIGURE 19.21 Synthesis of mono-, bis-, and tris-fulleropyrrolidine adducts.

reported in Ref. [94]. Tris-adducts are also known[95]; they were obtained from C_{60} as starting material (Figure 19.21). Nine fulleropyrrolidine tris-adducts, including three isomers, have been purified and unambiguously characterized. Additional information on fulleropyrrolidine bis-adducts is given in Ref. [96].

19.5 ENZYME-CONTAINING FULLEROPYRROLIDINES

N-(3-maleimidopropionyl)-3,4-fulleropyrrolidine (Figure 19.22) was successfully attached to subtilisin through site-specific immobilization[97] in a phosphate buffer containing mutant subtilisin. It was established that the nature of the immobilization support affects the catalytic properties of the enzyme. The interaction of fullerenes with biological molecules was also studied in Ref. [98].

19.6 POLYMER-CONTAINING FULLEROPYRROLIDINES

Polymer-containing fullerenes were discussed in a comprehensive review,[99] where some attention was paid to polymers with fulleropyrrolidine moieties. Thus, a series of oligophenylenevinylene (OPV) dyads (Figure 19.23) and soluble double-cable copolythiophenes were prepared by multistep

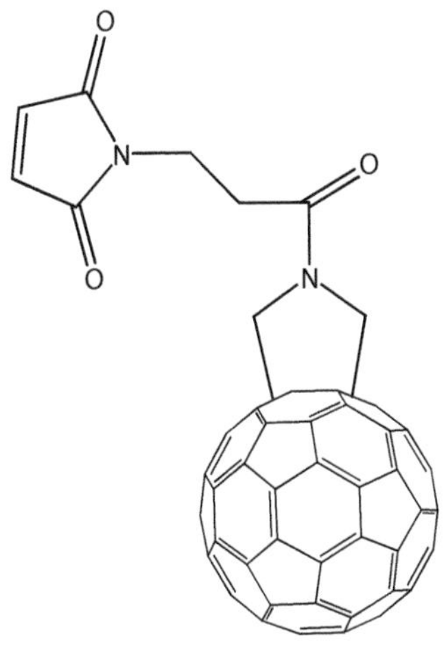

FIGURE 19.22 *N*-(3-maleimidopropionyl)-3,4-fulleropyrrolidine.

transformations from monomers or their precursors and characterized. It is noted that the copolymers contained 7% to 14% of fullerene monomer. For the second series of compounds, it was noted that photovoltaic devices incorporating these hybrids produced a photocurrent, showing that photoinduced electron transfer takes place. However, the efficiency of the devices is limited by the fact that photoinduced electron transfer from the OPV moiety to the C_{60} sphere must compete with an efficient energy transfer.

Among other polymers, containing fulleropyrrolidines, we note a series of new polymers containing porphyrin, poly(*p*-phenylenevinylene), and a pendant fullerene unit,[60] in which the aggregation superstructures of three polymers (nanobriquetting, nanofiber, and hierarchical porous structures) were revealed.

19.7 OTHER INTRIGUING FULLEROPYRROLIDINES

Liquid fullerenes on the basis of fulleropyrrolidines were synthesized by refluxing the 2,4,6-tris-(alkyloxy)benzaldehyde with *N*-methylglycine and C_{60} in dry toluene; their rheology was studied in Ref. [39]. They (Figure 19.24) are electrochemically active and have a relatively large hole mobility. It is noted that the melting points dramatically decrease from $n = 8$ to $n = 12$, a fact that is correlated with the length of alkyl chains surrounding C_{60}.

Mono-adducts of *liquid-crystalline fulleropyrrolidines*[100,101] (their bis-adducts were already discussed), were obtained from mesomorphic aldehyde-based dendrimers of first to fourth generation, sarcosine or glycine and C_{60}. Two different organizations were found depending upon the dendrimer generation: (1) the molecules are oriented in a head-to-tail fashion within the layers (for the second-generation dendrimers); (2) the mesogenic units are oriented above and below the dendritic core (for the third- and fourth-generation dendrimers).

Carbon onions, one of the most intriguing forms of carbon allotropes, large concentric (multishell) fullerenes, were prepared by arc discharge technique.[102] An onion consists of concentric shells (Figure 19.25) and has diameters between 60 and 300 nm, while the space between the internal shells has a mean value of ca. 4 nm. The authors reported new methodology that not only allows the isolation of giant fullerenes but especially renders them soluble in organic solvents. The carbon onion was found to be soluble in many organic solvents, especially chloroform and dichloromethane, with a solubility of about 10 and 7 mg/100 mL, respectively, and the solutions are stable for at least 2 weeks. It is emphasized that the functionalized onions are an example of soluble material that functions equally well in both the visible and the NIR regions.

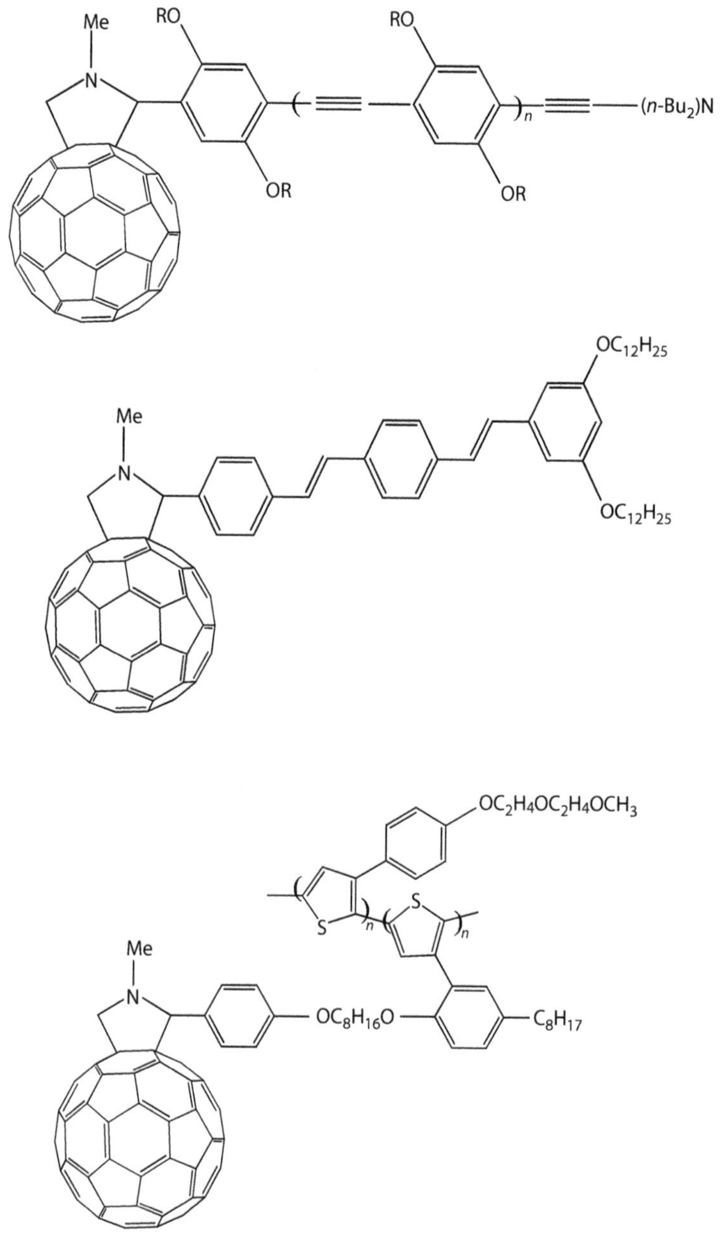

FIGURE 19.23 Oligophenylenevinylene dyads and soluble double-cable copolythiophenes.

C_{60}-*pyrrolidine-N-oxides*, in which the tertiary amine is transformed into a quaternary amine bearing an oxygen atom (Figure 19.26),[103] were prepared in moderate yields (20%–40%) by oxidation of C_{60}-pyrrolidines by a peracid (3-chloroperoxybenzoic acid). It is noted that the reaction is very selective, favoring the nitrogen atom of the pyrrolidine ring in preference to epoxidation of the fullerene cage; selective oxidation of the nitrogen atom was favored under dilute conditions, whereas concentrated solutions furnished mixtures of products. ^{15}N-labeled C_{60}-pyrrolidine-*N*-oxide molecule was also synthesized and oxidized to the *N*-oxide to confirm the formation of the final product by ^{15}N NMR. The solubility of fullerene-*N*-oxide molecules in organic solvents is different from that of their unoxidized counterparts. Stabilization of fulleropyrrolidine *N*-oxides through intrarotaxane hydrogen bonding was examined in Ref. [104].

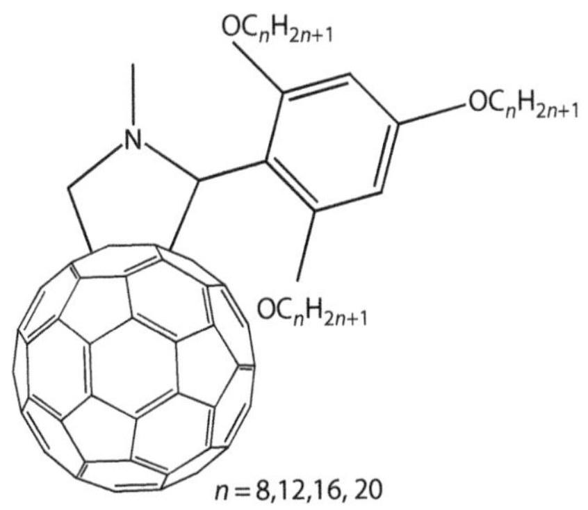

OC$_n$H$_{2n+1}$
N
OC$_n$H$_{2n+1}$
OC$_n$H$_{2n+1}$
n = 8,12,16, 20

FIGURE 19.24 Liquid fullerenes.

N
O

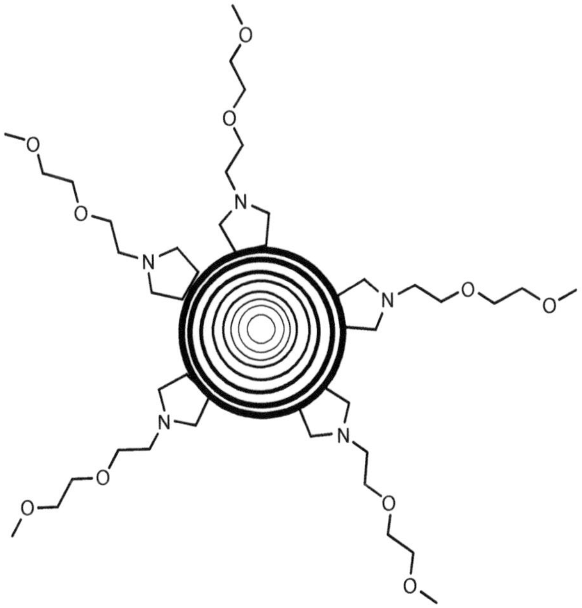

FIGURE 19.25 Carbon onions.

-O
R
N
⊕
R′

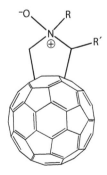

FIGURE 19.26 C_{60}-pyrrolidine-*N*-oxides.

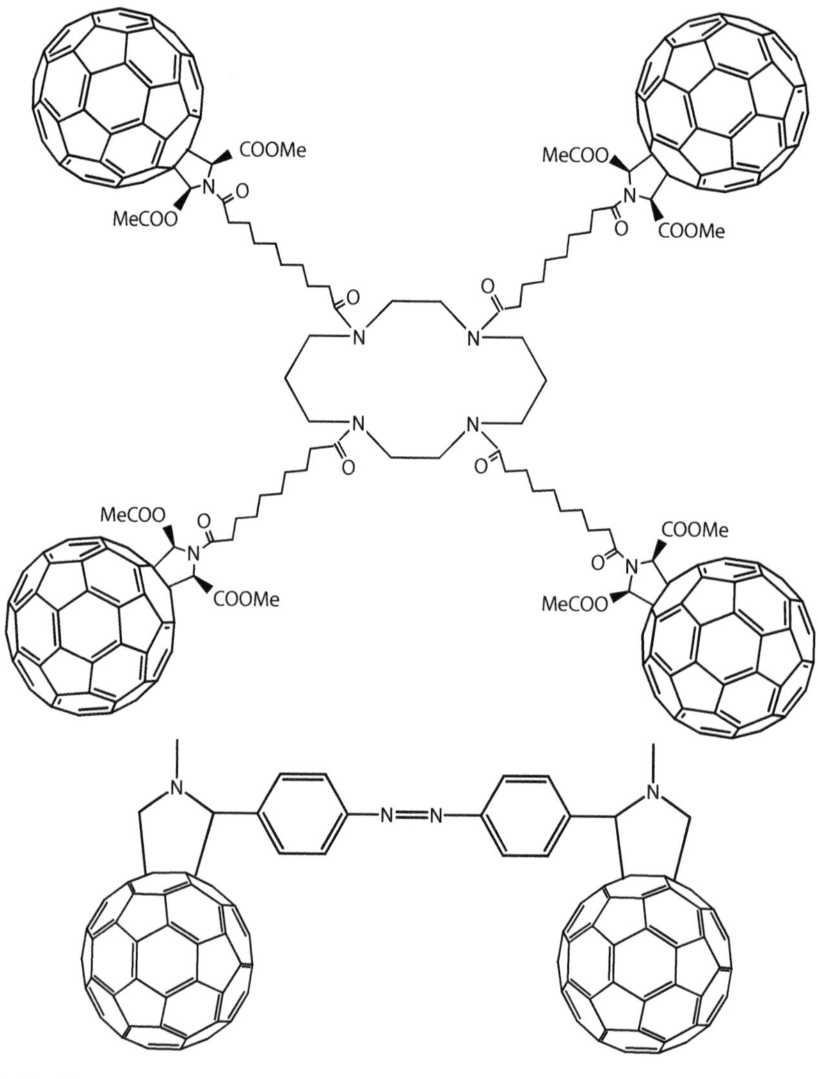

FIGURE 19.27 Examples of multifullerenopyrrolidines.

Multifulleropyrrolidines with 3, 4 (Figure 19.27) and 5 functionalized fulleropyrrolidine moieties were prepared[105] starting from 2,5-dimethoxycarbonyl[60]-fulleropyrrolidine as a starting precursor. The reaction time is 5 weeks for the hexakisfullerene. Additionally,[106] another compound (Figure 19.27, bottom), a photo-switchable fullerene dimer, one half of which comprised the endohedral N@C_{60}, was prepared by refluxing a mixture of 4,4'-azobenzaldehyde, C_{60}, and *N*-methylglycine (sarcosine) in toluene for 2 days under a nitrogen atmosphere (yield 95%). Irradiation by ultraviolet and visible light has been used to switch between the *trans* and *cis* isomers of both the C_{60}- and N@ C_{60}-based dimers. Other dimers with TTF and analogous S-donors are described in the Section 19.2 devoted to S-containing fulleropyrrolidines.

Simple fulleropyrrolidines bearing long aliphatic chains [$-C_6H_2$-3,4,5-$((OCH_2)_{15}CH_3)_3$[107,108] and $-C_6H_2$-3,4,5-$((OCH_2)_{19}CH_3)_3$[109]], showing applications as *superhydrophobic surfaces*, were prepared from the corresponding benzaldehydes, *N*-methylglycine and C_{60} in refluxing dry toluene. These robust and durable artificial nanocarbon superhydrophobic surfaces exhibit fractal morphology on both nano- and micrometer scale. The superhydrophobic films on their basis are stable toward polar organic solvents and acidic/basic media and heating. Hierarchical organization of molecular components into macroscopic objects provides a solution for fabrication of low friction, low adhesion,

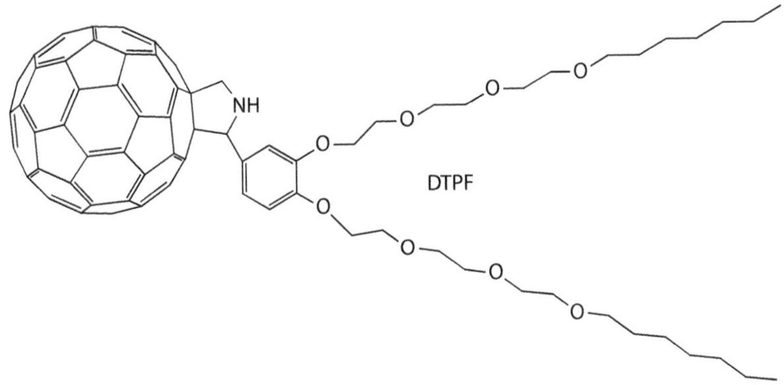

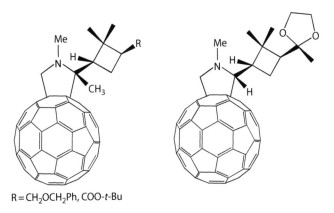

FIGURE 19.28   A C$_{60}$-pyrrolidine derivative with a hydrophobic-hydrophilic-hydrophobic structure.

and nonwetting surfaces for micro-/nano-electromechanical systems or self-cleaning surfaces. SEM and TEM images of the compounds above formed from 1,4-dioxane solutions showed flower-shaped supramolecular assemblies (for a detailed recent review of other nanoflowers of inorganic, organic, and coordination compounds, see Ref. [110]).

A C$_{60}$-pyrrolidine derivative with a *hydrophobic-hydrophilic-hydrophobic* structure (2-{3,4-di{2-[2-(2-decyloxyethoxy)ethoxy]ethoxy}}phenyl-3,4-fulleropyrrolidine [DTPF]) (Figure 19.28) has been synthesized and characterized.[111] It is noted that this compound could form stable nanospheres by simply injecting its THF solution into water and then removing THF by purging gaseous nitrogen in sequence. Moreover, nanoassemblies of DTPF nanospheres and gold nanoparticles were obtained through *in situ* photoreduction of aqueous HAuCl$_4$ in the presence of DTPF nanospheres.[49] According to the authors' opinion, the interaction between the positively charged nitrogen atom and the gold nanoparticles is the main driving force for the formation of these nanoassemblies; the elaborated synthetic technique could lead to a wide variety of super-nanostructures that show extraordinary optoelectronic properties.

Among many other reported fulleropyrrolidine derivatives, prepared and characterized in 2000–2008, we note those containing such moieties as diastereomerically pure functionalized cyclobutanes (Figure 19.29),[112] calixarenes (Figure 19.30),[113] coumarin,[114] nitrobenzene and *m*-dinitrobenzene (Figure 19.31),[115] dec-9-ynyl–(CH$_2$)$_8$C≡CH,[58] 2,2'-bipyridine (Figure 19.32)[116] and biologically active 1,4-dihydropyridines,[117] imidazolium,[48] the first fullerene derivative possessing both pyridyl and 4-imidazolylphenyl chelating groups (Figure 19.33),[118] various alkyl-,[119] phenyl-,[120] and nitroso-[121] groups, etc. A definite attention is paid to fullerene-containing rotaxanes,[50,122] sugars,[123] and crown-ethers.[124,125]

R = CH$_2$OCH$_2$Ph, COO-*t*-Bu

FIGURE 19.29   Cyclobutane-containing fulleropyrrolidines.

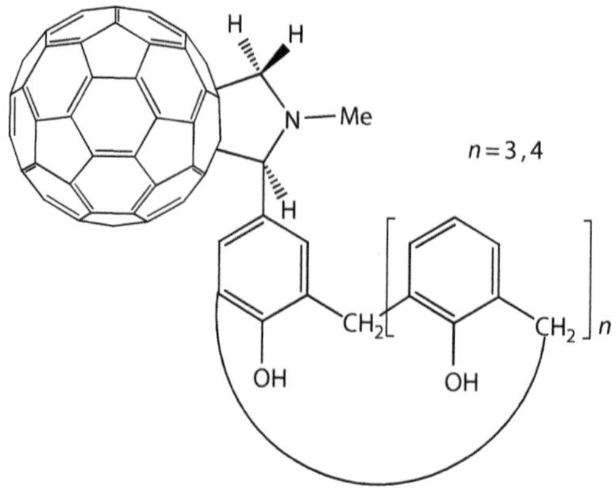

FIGURE 19.30   Calixarene-containing fulleropyrrolidines.

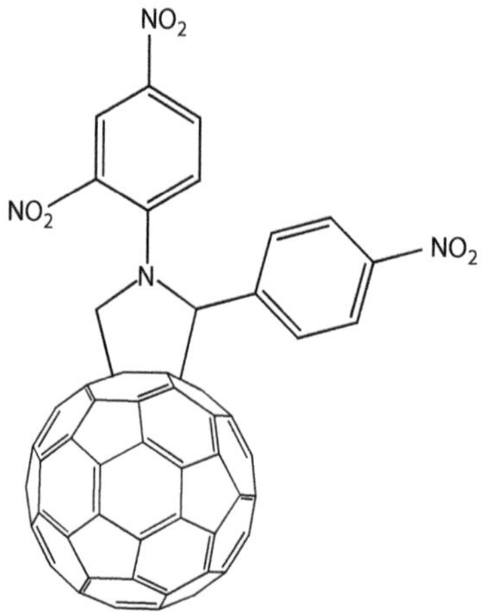

FIGURE 19.31   Nitrobenzene- and *m*-dinitrobenzene-containing fulleropyrrolidines.

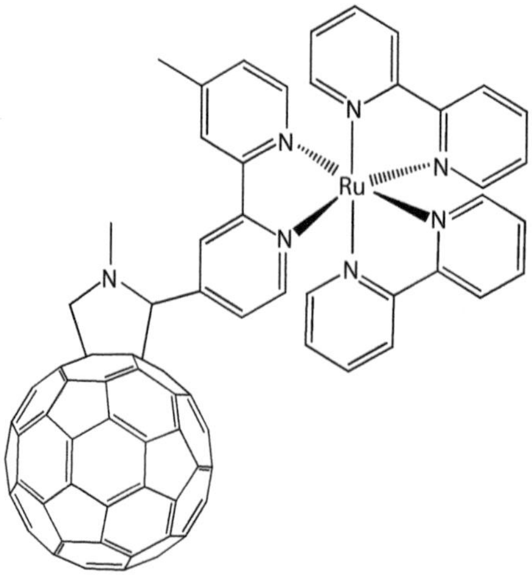

FIGURE 19.32   Bis(2,2′-bipyridine)(1′,5′-dihydro-3′-methyl-2′-(4-(4′-methyl-2,2′-bipyridinyl))-2′H-[5,6]fullereno (C_{60}-I_h)[1,9]pyrrole)ruthenium-bis-(hexafluorophosphate).

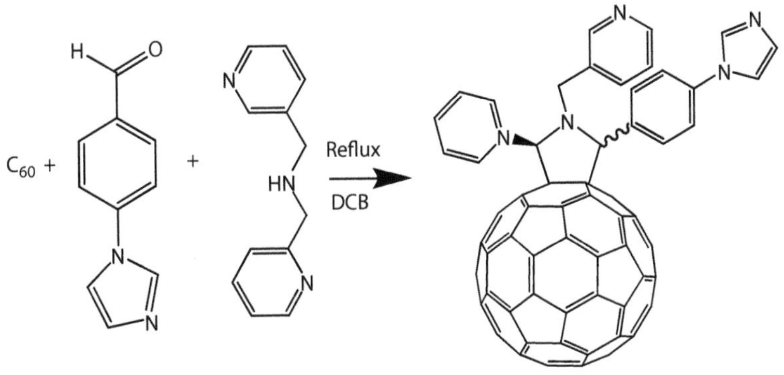

FIGURE 19.33 Synthesis of fullerene derivative possessing both pyridyl and 4-imidazolylphenyl chelating groups.

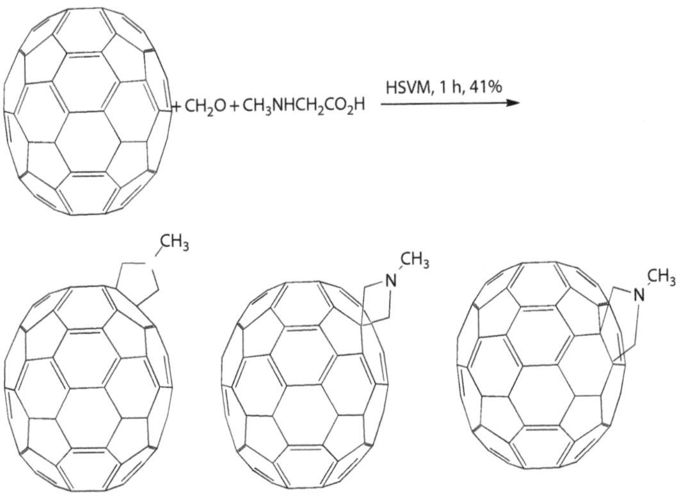

FIGURE 19.34 Formation of three C_{70}-containing pyrrolidine isomers.

19.8 C_{70}-PYRROLIDINES

In comparison with a large amount of reported C_{60}-pyrrolidines, there are a few examples of related C_{70} compounds. Thus, the reaction of C_{70} with paraformaldehyde and/or N-methylglycine under high-speed vibration milling (HSVM) solvent-free conditions was found to give three positional isomers of [70]fulleropyrrolidines in 41% yield with a ratio of 47:36:16 (Figure 19.34).[126] When C_{70} reacts only with N-methylglycine in the same conditions, only two monoadducts are formed with a total yield of 23%. (C_{80}-pyrrolidine—see example in Table 19.1.)

19.9 CARBON NANOTUBES, FUNCTIONALIZED WITH PYRROLIDINES

Prato described[15,127] the functionalization of CNTs on the basis of similar techniques for fullerenes (1,3-dipolar cycloaddition of azomethine ylides), examined earlier. The technique for CNT (either single-walled or multiwalled) solubilizing was developed on the basis of their interaction in DMF with an excess of aldehyde and α-amino acid. This way, a large number of pyrrolidine rings fused to the carbon–carbon bonds of CNTs are produced. Representative examples of typical functionalized

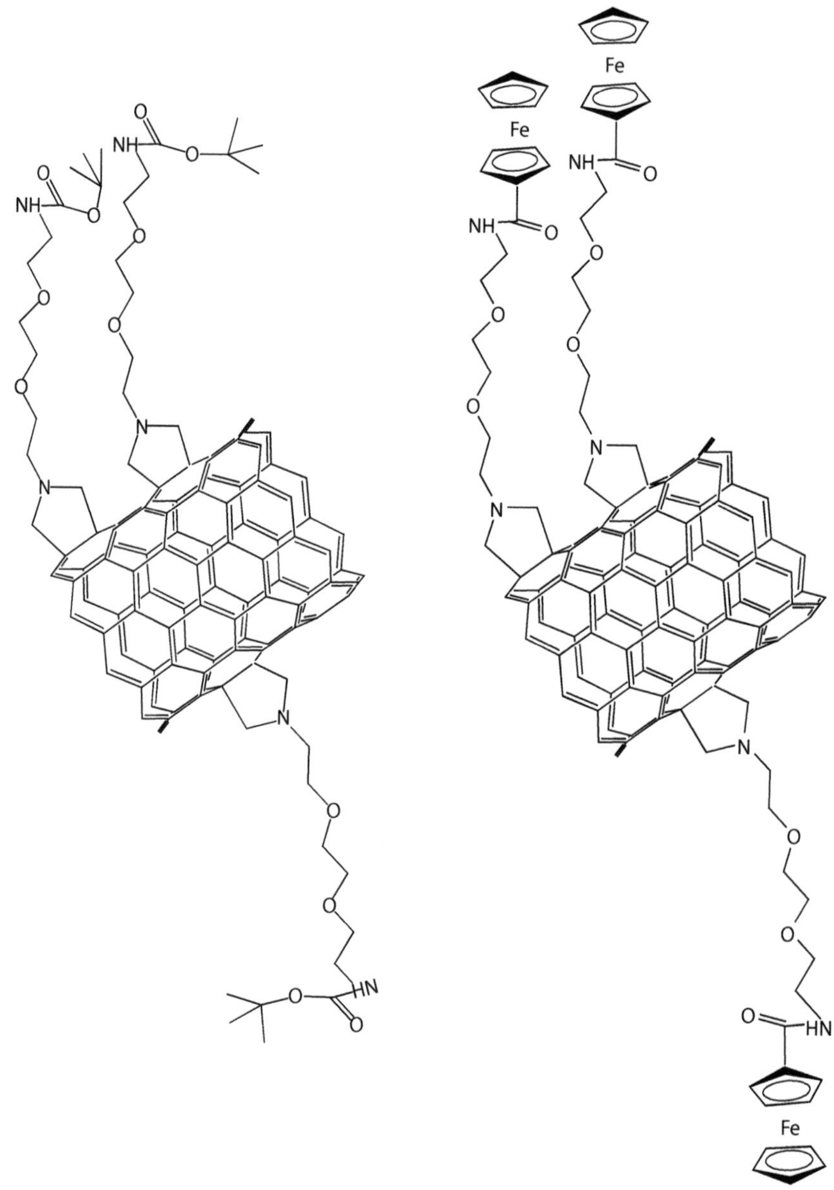

FIGURE 19.35 Functionalized CNTs.

water-soluble CTNs (*f*-NTs), in particular containing ferrocene moieties, are shown in Figure 19.35; they can serve as intermediate synthons to covalently attach peptides for future biological studies. Similar *f*-NTs with OCH_3 ending groups have solubility as high as 50 g/L through the attachment of pyrrolidines to the external wall of NT.

High filling of single-wall nanotubes (SWCNTs) with the typical exohedrally functionalized fullerene derivative of C_{60} *N*-methyl-3,4-fulleropyrrolidine C_{60}–C_3NH_7 (without excessive defects or sidewall functionalization as a result of this treatment) at the temperature of refluxing hexane is described in Ref. [128]. It was confirmed that bundles of SWCNT are highly filled with the fullero-pyrrolidine and form the $(C_{60}$–$C_3NH_7)_n$ peapods.

19.10   APPLIED TECHNIQUES TO STUDY FULLEROPYRROLIDINES

The techniques to study synthesized fulleropyrrolidines are generally the same in the majority of reports. Nevertheless, we would like to pay special attention to some of them in order to show a variety of routes to discuss the obtained experimental data. Thus, *x-ray photoemission spectroscopy* study was carried out for fulleropyrrolidine and neutral or positively charged pyrrolidine derivatives[129] in order to determine the effects of the C_{60}-cage on the pyrrolidine nitrogen. It was shown that the charge transfer from the carbon pyrrolidine ring to the C_{60}-cages is observed, and this charge redistribution influences not only the carbon atoms but also the nitrogen. The *spectral characteristics* of main excited states of the *N*-methylfulleropyrrolidine (Figure 19.1) and its several analogues with ferrocene moieties C_{60}-Fc were obtained.[130] It was noted that excited states of fullerenes and fullerene derivatives are easily generated from their ground states, due to their wide absorption window ranging from the UV to the near-IR. Among other similar investigations, we note the study of low-temperature *vibronic spectra* of two fulleropyrrolidines (1-methyl-3,4-fulleropyrrolidine and 1-methyl-2(4-pyridine)-3,4-fulleropyrrolidine) embedded in a crystalline toluene matrix,[131] and the interpretation of *continuous wave electron spin resonance* (CWESR)[132] and EPR spectra of fulleropyrrolidine bis-adducts with nitroxide addends.[133,134] Unusual luminescence of hexapyrrolidine derivatives of C_{60} with T_h and novel D_3-symmetry was reported in Ref. [135].

   Electrochemical characteristics of various synthesized fulleropyrrolidine derivatives have been extensively studied in a majority of reports.[136] Thus, the cyclic voltammetric study of a series of bisfulleropyrrolidines and bisfulleropyrrolidinium ions is presented in Ref. [137]. The eight possible stereoisomers of each series were investigated allowing the observation of up to four and five subsequent reversible reductions, respectively. Some isomers have a rather unexpectedly wide potential difference between the second and third reductions.

   Experimental and theoretical investigations of *acid-base properties* of fulleropyrrolidines were carried out[138] on the example of 2-(*n*-alkyl)fulleropyrrolidines and *N*-methyl-2-(*n*-alkyl)fulleropyrrolidines (Figure 19.36) in aqueous micellar media of a sodium dodecyl sulfate surfactant by using *ab initio* 3-21G(*) methods. A $0 < pK_a < 14$ range was covered. The obtained results indicate that the fullerene cage fused to a pyrrolidine ring increases the acidity of the protonated pyrrolidine nitrogen mainly due to the induced electronic effects and, to a smaller extent, by structural effects. Comparison of the calculated HOMO and LUMO energy levels for the protonated and free-base forms of the investigated fulleropyrrolidines indicates that electron deficiency of the C_{60} cage and the HOMO-LUMO energy gap caused by *N*-protonation are increased.

   Highly diastereoselective 1,3-dipolar cycloaddition of 1,4-dihydropyridine-containing azomethine ylides to C_{60} was theoretically investigated by means of *quantum mechanical calculations* at the semiempirical AM1 and DFT (B3LYP/6-31G*) methods,[139] taking into account that the presence of two chiral centers and one chiral axis in the resulting fulleropyrrolidines leads to four possible [6,6] cycloaddition products (Figure 19.37). It is emphasized that only the [6,6] closed regioisomer is formed from four possible regioisomers. Similar DFT studies, as well as the theoretical IR

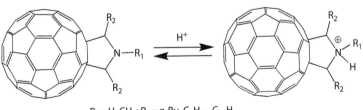

$R_1 = H, CH_3; R_2 = n\text{-Bu}, C_8H_{17}, C_{12}H_{25}$

FIGURE 19.36   Fulleropyrrolidines, used as models for determination of acid-base properties.

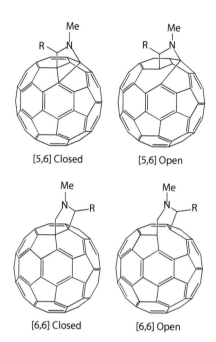

FIGURE 19.37   Four possible regioisomers from a 1,3-dipolar cycloaddition of azomethine ylides to C_{60}.

spectra, physical, chemical and thermodynamics properties, were discussed for a series of 1-(4, 5 and 6-selenenyl derivatives-3-formyl-phenyl) pyrrolidinofullerene molecules $C_{60}-C_2H_4N-[3-(CHO)C_6H_3SeX]$ (Figure 19.38)[140] and similar derivatives.[44]

Among other studies,[141] we note *semiempirical PM3* method, applied for theoretical investigation of three series of [C_{60}]fulleropyrrolidine-1-carbodithioic acid 2; 3 and 4-substituted-benzyl esters (with *o-*, *m-* and *p*-positions of phenyl ring with variable donating and acceptor substituents).[142] The optimized geometries, some of calculated energies, spatial distribution, and positions of the HOMO, LUMO, and QSAR properties of structural variables were obtained. *Synchrotron radiation photoemission* and near-edge x-ray absorption and NEXAFS were used to characterize thin films on the basis of CoTBPPs (cobalt *tetra*-butyl-phenyl porphyrins)—Py-C_{60} (*N*-methyl-2-(*p*-pyridyl)-3,4-fulleropyrrolidine), forming charge-separation complexes, which

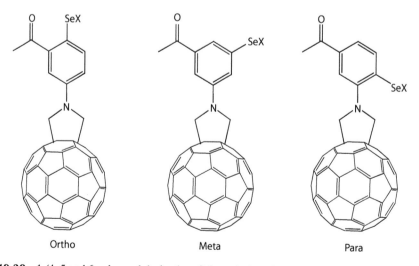

FIGURE 19.38   1-(4, 5 and 6-selenenyl derivatives-3-formyl-phenyl)pyrrolidinofullerenes.

may have applications in solar cells.[143] Excited state absorption of fullerenes was measured by the photoacoustic calorimetry technique.[144] The third-order nonlinear optical properties of fulleropyrrolidine were investigated.[145]

19.11 APPLICATIONS

According to the opinion of the authors of Ref. [13], C_{60} is a relatively *common building block* in organic synthesis and, at least in principle, any type of molecular fragment can be covalently attached to C_{60}. Facile addition of azomethine ylides to C_{60} allows the creation of a wide family of high-degree functionalized fullerene derivatives with useful applications in materials science and medicinal chemistry, for instance donor-bridge-acceptor diads, where covalently linked electro- or photoactive units are connected with C_{60} as the acceptor unit. The same authors noted thin films containing fulleropyrrolidines on the gold electrodes, which generates photocurrent upon irradiation, and obtaining a few water-soluble fulleropyrrolidines, which could possess a biological activity against different microorganisms or could be incorporated in sol-gel glasses.

Biological/medical applications of fullerene derivatives, in particular those of fulleropyrrolidines, are overviewed in Ref. [146]. Among their most interesting examined properties, we note those such as HIV-P inhibition,[147] DNA photocleavage,[148] neuroprotection, apoptosis, and antioxidant (similar to standard antioxidant BHT (2,6-di-*tert*-butyl-4-methylphenol)),[51] vitro,[149] and anti-inflammatory activity.[150] As an example, [60]fullerene amino acid was fabricated for further use in solid-phase *peptide synthesis*,[151] prepared from a fullerene derivative containing a free amino group by condensation with *N*-Fmoc-L-glutamic acid r-*tert*-butyl ester. The final peptide, very soluble in aqueous solutions, showed antimicrobial activity against two representative bacteria. Design, synthesis, and photophysical properties of C_{60}-modified proteins were reported in Ref. [152]. Thus, protoporphyrin (acid chloride form) was linked to 1′-methyl-2′-[4-(6-hydroxyhexyloxy)phenyl]-2′,3′,4′,5-tetrahydropyrrolo[3′,4′:1,2][60]fullerene in 2:3 v/v mixture CH_2Cl_2-CS_2 giving a fullerene-porphyrin conjugate, whose insertion with the use of iron and zinc salts gave corresponding metal complexes that were successfully reconstituted into apomyoglobin to produce C_{60}-myoglobins. The electrodes, modified with these final compounds, gave anodic photocurrent coupled with on-off light irradiation. These and other interesting results open possibilities for the construction of biodevices based on fullerene-protein conjugated novel materials, which promote biochemical, pharmaceutical, and medical applications. Health effects of fullerenes were also recently generalized.[153]

Nanocomposites can be successfully prepared on the basis of fulleropyrrolidine. Thus, a general method was developed to incorporate four fulleropyrrolidiniums (Figure 19.39) into the nanoscale cavities of a Nafion membrane through electrostatic interaction.[154] The modified Nafion was prepared by soaking the membrane in methanol solutions of fulleropyrrolidiniums, removing their residues, washing with methanol, and drying in vacuum. The resulting Nafion-fulleropyrrolidinium nanocomposites exhibit significant optical-limiting effects, implying their potential as broadband optical limiters to protect optical devices from damage by lasers. The fulleropyrrolidinium-incorporated Nafion membrane is expected to find a range of applications from biosensor to antibacterial materials.

The insertion and the subsequent behavior of neutral and ionic fulleropyrrolidine derivatives (Figure 19.40) into the interlayer space of an aluminosilicate-layered material (smectite clays) leads to the formation of a new hybrid system where C_{60} differs from its crystals or its solutions. It is shown that the insertion of ionic fullerene derivatives was easier because of their solubility in water.[55] In addition, not all the clay-clay platelets are intercalated by the fullerene derivatives, and a sizable amount of charge transfer takes place between the host and the guests. It is emphasized that the C_{60}-based organic/clay hybrid materials may possess interesting photophysical characteristics and could be used for the development of photofunctional devices in the form of anisotropic

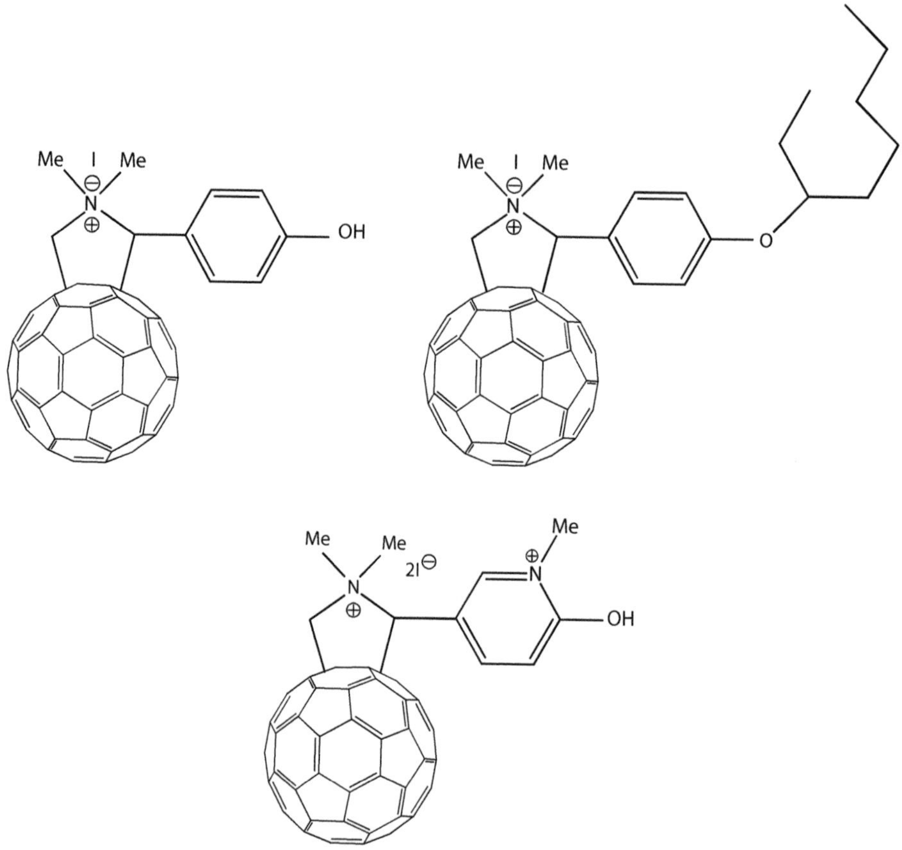

FIGURE 19.39 Some fulleropyrrolidiniums used for Nafion modification.

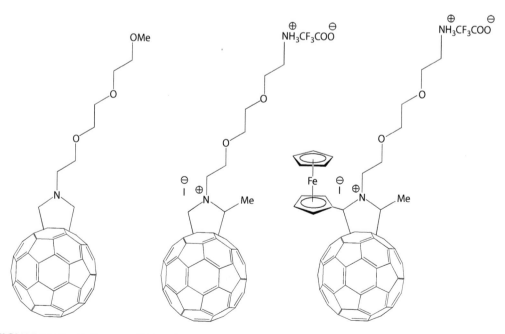

FIGURE 19.40 Fulleropyrrolidines, inserted into smectite clays.

thin films on metallic or glass surfaces. Another related application of fulleropyrrolidines is the fabrication of the layered perovskite, $(RNH_3)_2PbX_4$ (R = alkyl group, X = halogen), containing the fullerene ammonium derivative, *N*-methyl-2-(4-aminophenyl)-fulleropyrrolidine iodide, in the organic layers, in the form of a thin solid film.[155] A nanocomposite of poly*N*-vinylcarbazole (PNVC) with C_{60} was isolated from the *N*-vinylcarbazole (NVC)–C_{60} polymerization system[156]; as a result, the formation of almost spherical composite particles with average diameter in the range of 50 nm was observed.

Solar cells. Fulleropyrrolidine derivatives together with a commercial water-soluble porphyrin, bearing peripherical anionic SO_3 groups (Figure 19.41), were used to prepare thin films mentioned earlier by Langmuir–Schafer method.[157] In particular, a solution of the fulleropyrrolidine in chloroform and dimethyl sulfoxide was spread on the water surface, while the porphyrin was dissolved into the aqueous subphase. Evidence of the effective interactions between the two components at the air–water interface was obtained from the analysis of the floating layers by means of surface pressure vs. area per molecule Langmuir curves, Brewster angle microscopy, and UV–visible reflection spectroscopy. The authors suggested that this method represents a new effective tool for the fabrication of functioning devices that contain fullerene and macrocyclic derivatives as active species for photocurrent generation. Similar valuable information on the preparation of thin films of fulleropyrrolidines for organic solar cells was reported in Refs. [158,159]; new approaches toward the synthesis of [60]

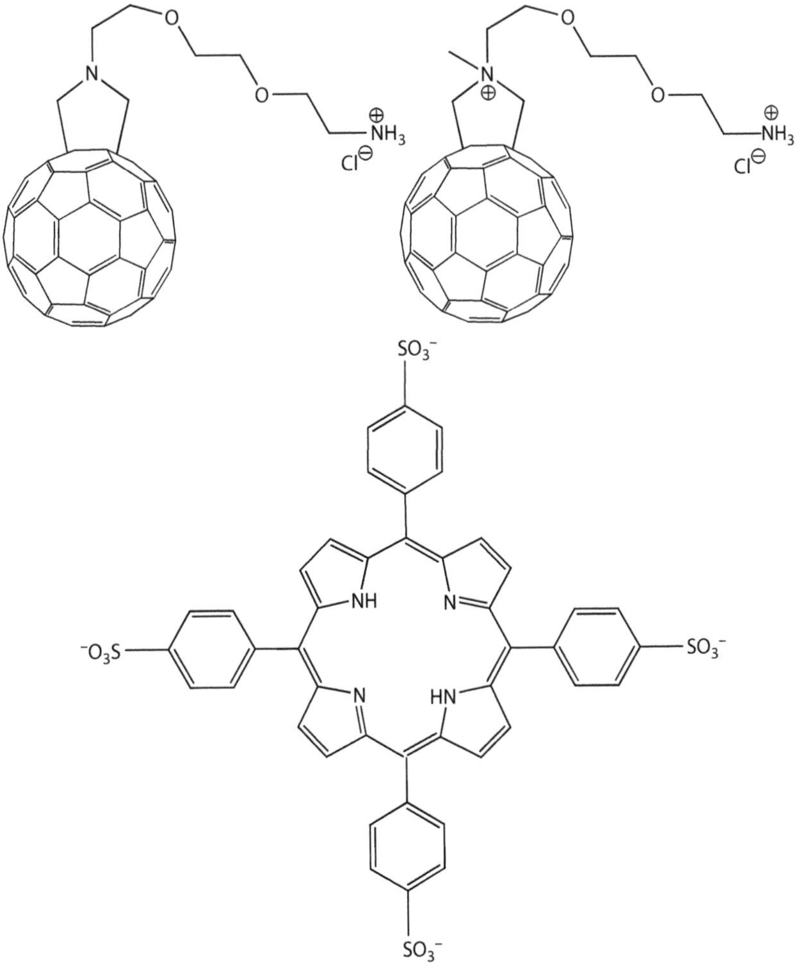

FIGURE 19.41   Precursors for preparing thin films for solar energy conversion.

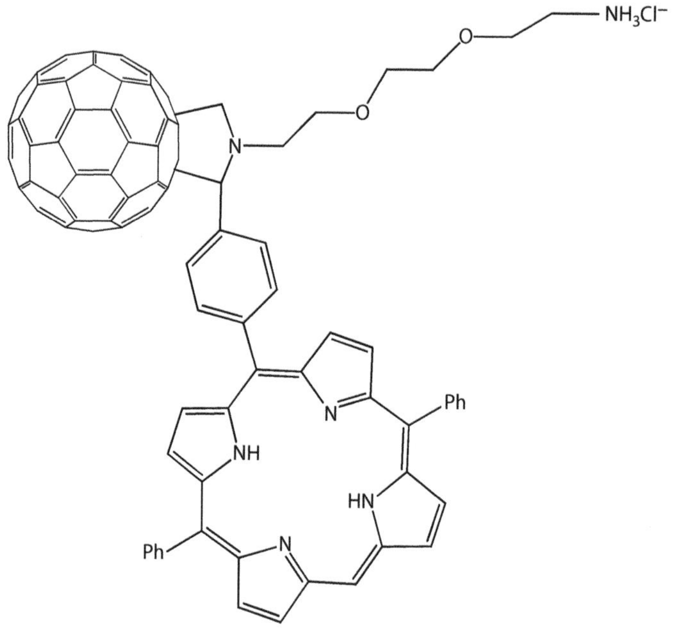

FIGURE 19.42 Fullerene-porphyrin dyad, used for electrode construction.

fullerene derivatives specifically designed to be used in the fabrication of photovoltaic devices (synthesis of functionalized C_{60} and C_{70} derivatives with enhanced solubility and systems containing a conjugated moiety antenna covalently linked to the C_{60} unit) were examined in Ref. [160].

Layer-by-layer construction of *nanostructured porphyrin-fullerene high-quality, robust, and photoactive ITO electrodes* on the basis of a fullerene–porphyrin dyad (Figure 19.42) as the active layer was reported in Ref. [161]. The new electrodes give rise to improved light harvesting features in the visible zone, with absorbances as high as 0.2 (that is, at the Soret band of H_2P) for a monolayer coverage and IPCEs of 0.38% and 0.6% under anaerobic and aerobic conditions, respectively. Among other applications of fulleropyrrolidines, we note those related with their memory effect[162] and molecular lithography.[163]

Covalent functionalization of C_{60} with liquid-crystalline dendrimers is a valuable concept for the design of fullerene-containing *liquid crystals* (see the Section 19.4 dedicated to fulleropyrrolidine bis-adducts) that display tailor-made mesomorphic properties.

APPENDIX A: SELECTED MODIFIED METHODS OF CYCLOADDITION

The *mechanochemistry* (transformation of the mechanical energy into the driving force for the chemical reactions of solid reagents, which is generally done in the absence of any solvent) of fullerenes is reviewed in Ref. [164]. Fulleropyrrolidines were also obtained in these HSVM conditions ("high-speed vibration milling," vigorous milling of the reaction mixture; see also Section 19.8) (Figure 19.43):

A few C_{60}-pyrrolidine derivatives have been obtained under simultaneous *microwave treatment* of the reaction system in a modified domestic microwave oven and a focused microwave reactor.[165] Reactions proceed within minutes (in comparison with long reaction time under classic heating) to afford the respective cycloadducts in similar or increased yields related to the described methods by conventional heating. Thus, a series of fluorinated 2,5-disubstituted C_{60}-pyrrolidine derivatives (Figure 19.44) were synthesized *via* one-pot three-component reaction of C_{60}, amino acid, and fluorinated benzaldehyde under microwave irradiation.[166]

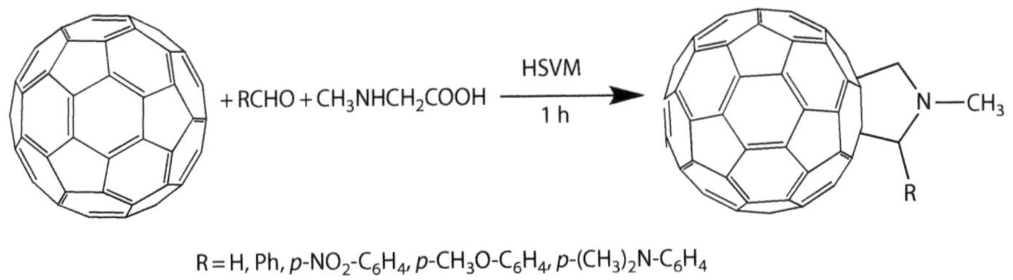
+RCHO+CH3NHCH2COOH
HSVM
1 h
N—CH3
R
R=H, Ph, p-NO2-C6H4, p-CH3O-C6H4, p-(CH3)2N-C6H4

FIGURE 19.43 HSVM synthesis of fulleropyrrolidines.

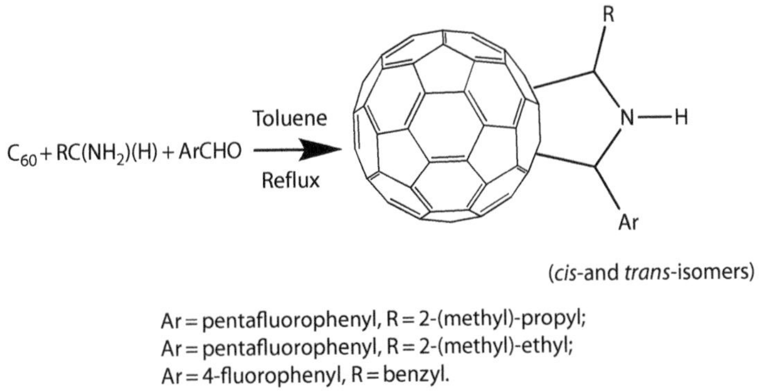
C60+RC(NH2)(H)+ArCHO
Toluene
Reflux
R
N—H
Ar
(cis-and trans-isomers)
Ar=pentafluorophenyl, R=2-(methyl)-propyl;
Ar=pentafluorophenyl, R=2-(methyl)-ethyl;
Ar=4-fluorophenyl, R=benzyl.

FIGURE 19.44 MW-obtaining C_{60}-pyrrolidine derivatives.

REFERENCES

1. Kroto, H. W.; Heath, J. R.; O'Brien, S. C.; Curl, R. F.; Smalley, R. F. C_{60}-buckminsterfullerene. *Nature*, 1985, *318* (6042), 162–163.
2. Hirsch, A. (Ed.). *Fullerenes and Related Structures*. Springer-Verlag, Berlin, Germany, 1999, 246 pp.
3. Hirsch, A.; Brettreich, M.; Wudl, F. *Fullerenes: Chemistry and Reactions*. Wiley-VCH, Weinheim, Germany, 2005, 440 pp.
4. Maggini, M.; Menna, E. Addition of azomethine ylides: Fulleropyrrolidines. In *Fullerenes: From Synthesis to Optoelectronic Properties*. Developments in Fullerene Science Series, Vol. 4. Guldi, D. M.; Martin, N. (Eds.). Springer, Berlin, Germany, 2003, 447 p.
5. Sidorov, L. N.; Yurovskaya, M. A.; Borschevskii, A. Ya.; Trushkov, I. V.; Ioffe, I. N. *Fullerenes*. Examen, Moscow, Russia, 2005, 688 pp.
6. Fowler, P. W.; Manolopoulos, D. E. *An Atlas of Fullerenes*. Dover Publications, Mineola, NY, 2007, 416 pp.
7. Margadonna, S. (Ed.). *Fullerene-Related Materials*. Springer, Berlin, Germany, 2008, 700 pp.
8. Langa, F.; Nierengarten, J.-F. (Ed.). *Fullerenes: Principles and Applications*. Royal Society of Chemistry, London, U.K., 2007, 300 pp.
9. *Fullerene Research Advances*. Kramer, C. N. (Ed.). Nova Science Pub Inc., Hauppauge, NY, 2007, 305 pp.
10. Prassides, K. *Fullerene-Based Materials: Structures and Properties*. Springer, Berlin, Germany, 2004, 294 pp.
11. Rietmeijer, F. J. M. *Natural Fullerenes and Related Structures of Elemental Carbon*. Springer, Berlin, Germany, 2006, 295 pp.
12. Hirsch, A. Principles of fullerene reactivity. *Topics in Current Chemistry*, 1999, *199*, 1–65.
13. Thompson, D. T. Platinum group metal fullerenes: Some recent studies on systems containing C_{60}. *Platinum Metals Review*, 1996, *40* (1), 23–25.
14. Prato, M. Fulleropyrrolidines: A family of full-fledged fullerene derivatives. *Accounts of Chemical Research*, 1998, *31* (9), 519–526.
15. Yurovskaya, M. A.; Trushkov, I. V. Cycloaddition to buckminsterfullerene C_{60}: Advancements and future prospects. *Russian Chemical Bulletin, International Edition*, 2002, *51* (3), 367–443.

16. Tagmatarchis, N.; Prato, M. Carbon-based materials: From fullerene nanostructures to functionalized carbon nanotubes. *Pure and Applied Chemistry*, 2005, *77* (19), 1675–1684.

17. Illescas, B. M.; Martin, N. [60]Fullerene-based electron acceptors. *Comptes Rendus Chimie*, 2006, *9* (7), 1038–1050.

18. Dirk, M.; Guldi, G. M.; Rahman, A.; Sgobba, V.; Ehli, C. Multifunctional molecular carbon materials— From fullerenes to carbon nanotubes. *Chemical Society Reviews*, 2006, *35*, 471–487.

19. Karaulova, E. N.; Bagrii, E. I. Fullerenes: Functionalization methods and perspective of applications of their derivatives. *Russian Chemical Reviews*, 1999, *68* (11), 979–998.

20. Martín, N.; Altable, M.; Filippone, S.; Martín-Domenech, A. New reactions in fullerene chemistry. *Synlett*, 2007, *18* (20), 3077–3095.

21. Xiang, X. D.; Yang, H.; Dionne, G. F. Endohedral fullerenes as spin labels and MRI contrast agents. U.S. Patent 20070048870, 2007. http://www.freepatentsonline.com/20070048870.html

22. Chiang, L. Y.; Anantharaj, V.; Haldar, M. K. Chiral (1pyrrolino) fullerene derivatives. U.S. Patent 6949660, 2005. http://www.freepatentsonline.com/6949660.html

23. Chiang, L. Y.; Anantharaj, V.; Haldar, M. K. Chiral (1pyrrolino) fullerene derivatives. U.S. Patent 20040034205, 2004. http://www.freepatentsonline.com/20040034205.html

24. Chiang, L. Y.; Anantharaj, V. E.-isomeric fullerene derivatives. U.S. Patent 6790963, 2004. http://www.freepatentsonline.com/6790963.html

25. Chiang, L. Y.; Anantharaj, V. E-isomeric fullerene derivatives. U.S. Patent 6455709, 2002. http://www.freepatentsonline.com/6455709.html

26. Bethune, D. S.; Tiwari, S. Molecular memory and logic. U.S. Patent 6472705, 2002. http://www.freepatentsonline.com/6472705.html

27. Bethune, D. S.; Tiwari, S. Molecular memory and logic. U.S. Patent 6750471, 2004. http://www.freepatentsonline.com/6750471.html

28. Bethune, D. S.; Tiwari, S. Molecular memory and logic. U.S. Patent 20030011036, 2003. http://www.freepatentsonline.com/20030011036.html

29. Chiang, L. Y.; Anantharaj, V. E-isomeric fullerene derivatives. U.S. Patent 6576655, 2003. http://www.freepatentsonline.com/6576655.html

30. Chiang, L. Y.; Anantharaj, V. E-isomeric fullerene derivatives. U.S. Patent 20030013861, 2003. http://www.freepatentsonline.com/20030013861.html

31. Chiang, L. Y.; Anantharaj, V. E-isomeric fullerene derivatives. U.S. Patent 20030009036, 2002. http://www.freepatentsonline.com/20030009036.html

32. Troshin, P. A.; Peregubov, A. S.; Muhlbacher, D.; Lyubovskaya, R. N. An efficient [2 + 3] cycloaddition approach to the synthesis of pyridyl-appended fullerene ligands. *European Journal of Organic Chemistry*, 2005, Volume date 2005, (14), 3064–3074.

33. Mateo-Alonso, A.; Sooambar, C.; Prato, M. Synthesis and applications of amphiphilic fulleropyrrolidine derivatives. *Organic and Biomolecular Chemistry*, 2006, *4* (9), 1629–1637.

34. Capaccio, M.; Gavalas, V. G.; Meier, M. S.; Anthony, J. E.; Bachas, L. G. Coupling biomolecules to fullerenes through a molecular adapter. *Bioconjugate Chemistry*, 2005, *16* (2), 241–244.

35. Corvaja, C.; Maggini, M.; Fyato, M.; Scorrano, G.; Venzin, M. C_{60} derivative covalently linked to a nitroxide radical: Time-resolved EPR evidence of electron spin polarization by intramolecular radical-triplet pair interaction. *Journal of the American Chemical Society*, 1995, *117*, 8857–8858.

36. Kordatos, K.; Da Ros, T.; Bosi, S.; Vazquez, E.; Bergamin, M.; Cusan, C.; Pellarini, F. et al. Novel versatile fullerene synthons. *Journal of Organic Chemistry*, 2001, *66*, 4915–4920.

37. Borsato, G.; Della Negra, F.; Gasparrini, F.; Misiti, D.; Lucchini, V.; Possamai, G.; Villani, C.; Zambon, A. Internal motions in a fulleropyrrolidine tertiary amide with axial chirality. *Journal of Organic Chemistry*, 2004, *69*, 5785–5788.

38. Cardona, C. M.; Kitaygorodskiy, A.; Ortiz, A.; Angeles Herranz, M.; Echegoyen, L. The first fulleropyrrolidine derivative of $Sc_3N@C_{80}$: Pronounced chemical shift differences of the geminal protons on the pyrrolidine ring. *Journal of Organic Chemistry*, 2005, *70*, 5092–5097.

39. Michinobu, T.; Nakanishi, T.; Hill, J. P.; Funahashi, M.; Ariga, K. Room temperature liquid fullerenes: An uncommon morphology of C_{60} derivatives. *Journal of the American Chemical Society*, 2006, *128*, 10384–10385.

40. Modin, J. Synthesis and evaluation of photoactive pyridine complexes for electron transfer studies and photoelectrochemical applications; Acta Universitatis Upsaliensis Uppsala, 2005. http://publications.uu.se/spikblad.xsql?dbid=6146

41. Pérez, L.; García-Martínez, J. C.; Díez-Barra, E.; Atienzar, P.; García, H.; Rodríguez-López, J.; Langa, F. Electron transfer in nonpolar solvents in fullerodendrimers with peripheral ferrocene units. *Chemistry*, 2006, *12* (19), 5149–5157.

42. Armaroli, N.; Barigelletti, F.; Ceroni, P. A fulleropyrrolidine with two oligophenyllenevinylene substituents: Synthesis, electrochemistry and photophysical properties. *International Journal of Photoenergy*, 2001, *3*, 34–40.

43. Maggini, M.; Pasimeni, L.; Prato, M.; Scorrano, G.; Valli, L. Incorporation of an acyl group in fulleropyrrolidines: Effects on langmuir monolayers. *Langmuir*, 1994, *10*, 4164–4166.

44. Jalbout, A. F.; Trzaskowski, B.; Hameed, A. J. Theoretical investigation of the electronic structure of 1-(3, 4; 3,5 and 3,6-bis-selenocyanato-phenyl) pyrrolidinofullerenes. *Journal of Organometallic Chemistry*, 2006, *691* (22), 4589–4594.

45. Liu, S.-G.; Lianhe, S.; Rivera, J.; Liu, H.; Raimundo, J. M.; Roncali, J.; Gorgues, A.; Echegoyen, L. A new dyad based on C_{60} and a conjugated dimethylaniline-substituted dithienylethylene donor. *Journal of Organic Chemistry*, 1999, *64*, 4884–4886.

46. Sanchez, L.; Sierra, M.; Martın, N.; Guldi, D. M.; Wienk, M. W.; Janssen, R. A. J. C_{60}-exTTF-C_{60} dumbbells: Cooperative effects stemming from two C_{60}s on the radical ion pair stabilization. *Organic Letters*, 2005, *7* (9), 1691–1694.

47. Mizuochi, N.; Ohba, Y.; Yamauchi, S. A two dimensional EPR nutation study on excited multiplet states of fullerene linked to nitroxide radical. *Journal of Physical Chemistry A*, 1997, *101*, 5966–5968.

48. Itoh, T.; Mishiro, M.; Matsumoto, K.; Hayase, S.; Kawatsura, M.; Morimoto, M. Synthesis of fulleropyrrolidine-imidazolium salt hybrids and their solubility in various organic solvents. *Tetrahedron*, 2008, *64* (8), 1823–1828.

49. Zhang, P.; Zhang, S.; Li, J.; Liu, D.; Guo, Z.-X.; Ye, C.; Zhu, D. The self-assembly of gold nanoparticles with C_{60} nanospheres: Fabrication and optical liming effect. *Chemical Physics Letters*, 2003, *382* (5), 599–604.

50. Mateo-Alonso, A.; Fioravanti, G.; Marcaccio, M.; Paolucci, F.; Jagesar, D. C.; Brouwer, A. M.; Prato, M. Reverse shuttling in a fullerene-stoppered rotaxane. *Organic Letters*, 2006, *8* (22), 5173–5176.

51. Bjelakovic, M. S.; Godjevac, D. M.; Milic, D. R. Synthesis and antioxidant properties of fullero-steroidal covalent conjugates. *Carbon*, 2007, *45* (11), 2260–2265.

52. Mateo-Alonso, A.; Sooambar, C.; Prato, M. Fullerene photoactive dyads assembled by axial coordination with metals. *Comptes Rendus Chimie*, 2006, *9* (7), 944–951.

53. Conoci, S.; Guldi, D. M.; Nardis, S.; Paolesse, R.; Kordatos, K.; Prato, M.; Ricciardi, G. et al. Langmuir-Shäfer transfer of fullerenes and porphyrins: Formation, deposition, and application of versatile films. *Chemistry*, 2004, *10* (24), 6523–6530.

54. Sierra, M.; Angeles Herranz, M. A.; Zhang, S.; Sanchez, L.; Martın, N.; Echegoyen, L. Self-assembly of C_{60} π-extended tetrathiafulvalene (exTTF) dyads on gold surfaces. *Langmuir*, 2006, *22*, 10619–10624.

55. Gournis, D.; Georgakilas, V.; Karakassides, M. A.; Bakas, T.; Kordatos, K.; Prato, M.; Fanti, M.; Zerbetto, F. Incorporation of fullerene derivatives into smectite clays: A new family of organic-inorganic nanocomposites. *Journal of the American Chemical Society*, 2004, *126*, 8561–8568.

56. Guldi, D. M.; Luo, C.; Swartz, A.; Gomez, R.; Segura, J. L.; Martin, N.; Brabec, C.; Sariciftci, N. S. Molecular engineering of C_{60}-based conjugated oligomer ensembles: Modulating the competition between photoinduced energy and electron transfer processes. *Journal of Organic Chemistry*, 2002, *67*, 1141–1152.

57. Beckers, E. H. A.; van Hal, P. A.; Dhanabalan, A.; Meskers, S. C. J.; Knol, J.; Hummelen, J. C.; Janssen, R. A. J. Charge transfer kinetics in fullerene-oligomer-fullerene triads containing alkylpyrrole units. *Journal of Physical Chemistry A*, 2003, *107*, 6218–6224.

58. Dattilo, D.; Armelao, L.; Maggini, M.; Fois, G.; Mistura, G. Wetting behavior of porous silicon surfaces functionalized with a fulleropyrrolidine. *Langmuir*, 2006, *22*, 8764–8769.

59. Loi, M. A.; Denk, P.; Hoppe, H.; Neugebauer, H.; Meissner, D.; Winder, C.; Brabec, C. J. et al. A fulleropyrrolidine-phthalocyanine dyad for photovoltaic applications. *Synthetic Metals*, 2003, *137* (1), 1491–1492.

60. Huang, C.; Wang, N.; Li, Y.; Li, C.; Li, J.; Liu, H.; Zhu, D. A new class of conjugated polymers having porphyrin, poly(p-phenylenevinylene), and fullerene units for efficient electron transfer. *Macromolecules*, 2006, *39*, 5319–5325.

61. Oswald, F.; Islam, D.-M. S.; Araki, Y.; Troiani, V.; de la Cruz, P.; Moreno, A.; Ito, O.; Langa, F. Synthesis and photoinduced intramolecular processes of fulleropyrrolidine-oligothienylenevinylene-ferrocene triads. *Chemistry*, 2007, *13* (14), 3924–3933.

62. D'Souza, F.; Gadde, S.; Zandler, M. E.; Itou, M.; Araki, Y.; Ito, O. Supramolecular complex composed of a covalently linked zinc porphyrin dimer and fulleropyrrolidine bearing two axially coordinating pyridine entities. *Chemical Communications*, 2004, (20), 2276–2277.

63. Xiang, Y.; Wei, X. W.; Zhang, X. M.; Wang, H. L.; Wei, X. L.; Hu, J. P.; Yin, G.; Xu, Z. Synthesis of new pyridinofullerene ligands capable of forming complexes with zinc tetraphenyl porphyrin. *Inorganic Chemistry Communications*, 2006, 9 (5), 452–455.

64. Tat, F. T.; Zhou, Z.; MacMahon, S.; Song, F.; Rheingold, A. L.; Echegoyen, L.; Schuster, D. I.; Wilson, S. R. A new fullerene complexation ligand: *N*-Pyridylfulleropyrrolidine. *Journal of Organic Chemistry*, 2004, 69, 4602–4606.

65. D'Souza, F.; Deviprasad, G. R.; El-Khouly, M. E.; Fujitsuka, M.; Ito, O. Probing the donor-acceptor proximity on the physicochemical properties of porphyrin-fullerene dyads: "Tail-on" and "tail-off" binding approach. *Journal of the American Chemical Society*, 2001, 123, 5277–5284.

66. D'Souza, F.; El-Khouly, M. E.; Gadde, S.; McCarty, A. L.; Karr, P. A.; Zandler, M. E.; Araki, Y.; Ito, O. Self-assembled via axial coordination magnesium porphyrin-imidazole appended fullerene dyad: Spectroscopic, electrochemical, computational, and photochemical Studies. *Journal of Physical Chemistry B*, 2005, 109, 10107–10114.

67. Guldi, D. M.; Da Ros, T.; Braiuca, P.; Prato, M. A topologically new ruthenium porphyrin-fullerene donor-acceptor ensemble. *Photochemical and Photobiological Sciences*, 2003, 2 (11), 1067–1073.

68. Galloni, P.; Floris, B.; De Cola, L.; Cecchetto, E.; Williams, R. M. Zinc 5, 10, 15, 20-meso-tetraferrocenylporphyrin as an efficient donor in a supramolecular fullerene C_{60} system. *Journal of Physical Chemistry C*, 2007, 111, 1517–1523.

69. D'Souza, F.; Chitta, R.; Gadde, S.; Shafiqul Islam, D.-M.; Schumacher, A. L.; Zandler, M. E.; Araki, Y.; Ito, O. Design and studies on supramolecular ferrocene-porphyrin-fullerene constructs for generating long-lived charge separated states. *Journal of Physical Chemistry B*, 2006, 110, 25240–25250.

70. D'Souza, F.; Gadde, S.; Shafiqul Islam, D.-M.; Wijesinghe, C. A.; Schumacher, A. L.; Zandler, M. E.; Araki, Y.; Ito, O. Multi-triphenylamine-substituted porphyrin-fullerene conjugates as charge stabilizing "antenna-reaction center" mimics. *Journal of Physical Chemistry A*, 2007, 111, 8552–8560.

71. D'Souza, F.; Deviprasad, G. R.; Zandler, M. E.; El-Khouly, M. E.; Fujitsuka, M.; Ito, O. Electronic interactions and photoinduced electron transfer in covalently linked porphyrin-C_{60}(pyridine) diads and supramolecular triads formed by self-assembling the diads and zinc porphyrin. *Journal of Physical Chemistry B*, 2002, 106, 4952–4962.

72. Hameed, A. J. Theoretical investigation of a phthalocyanine-fulleropyrrolidine adduct and some of its metallic complexes. *Journal of Molecular Structure: THEOCHEM*, 2006, 764 (1), 195–199.

73. El-Khouly, M. E.; Rogers, L. M.; Zandler, M. E.; Suresh, G.; Fujitsuka, M.; Ito, O.; D'Souza, F. Studies on intra-supramolecular and intermolecular electron-transfer processes between zinc naphthalocyanine and imidazole-appended fullerene. *ChemPhysChem: A European Journal of Chemical Physics and Physical Chemistry*, 2003, 4 (5), 474–481.

74. Guldi, D. M.; Gouloumis, A.; Vazquez, P.; Torres, T.; Georgakilas, V.; Prato, M. Nanoscale organization of a phthalocyanine-fullerene system: Remarkable stabilization of charges in photoactive 1-D nanotubules. *Journal of the American Chemical Society*, 2005, 127, 5811–5813.

75. Loi, M. A.; Denk, P.; Neugebauer, H.; Brabec, C.; Sariciftci, N. S.; Gouloumis, A.; Vázquez, P.; Torres, T. Fulleropyrrolidine-phthalocyanine: A new molecule for solar energy conversion. *AIP Conference Proceedings*, Paris, France, 2002, Vol. 633, pp. 488–491.

76. Cravino, A.; Zerza, G.; Neugebauer, H.; Maggini, M.; Bucella, S.; Menna, E.; Svensson, M.; Andersson, M. R.; Brabec, C. J.; Serdar Sariciftci, N. Electrochemical and photophysical properties of a novel polythiophene with pendant fulleropyrrolidine moieties: toward "double cable" polymers for optoelectronic devices. *Journal of Physical Chemistry B*, 2002, 106, 70–76.

77. Meijer, M. D.; Mulder, B.; van Klink, G. P. M.; van Koten, G. Synthesis of C_{60}-attached SCS pincer palladium(II) complexes. *Inorganica Chimica Acta*, 2003, 352, 247–252.

78. Martın, N.; Perez, I.; Sanchez, L.; Seoane, C. Synthesis and properties of the first highly conjugated tetrathiafulvalene analogues covalently attached to [60]fullerene. *Journal of Organic Chemistry*, 1997, 62, 5690–5695.

79. Segura, J. L.; Priego, E. M.; Martın, N.; Luo, C.; Guldi, D. M. A new photoactive and highly soluble C_{60}-TTF-C_{60} dimer: Charge separation and recombination. *Organic Letters*, 2000, 2 (25), 4021–4024.

80. Guldi, D. M.; Maggini, M.; Scorrano, G.; Prato, M. Intramolecular electron transfer in fullerene/ferrocene based donor-bridge-acceptor dyads. *Journal of the American Chemical Society*, 1997, 119, 974–980.

81. Kanato, H.; Takimiya, K.; Otsubo, T.; Aso, Y.; Nakamura, T.; Araki, Y.; Ito, O. Synthesis and photo-physical properties of ferrocene-oligothiophene-fullerene triads. *Journal of Organic Chemistry*, 2004, *69*, 7183–7189.

82. Angeles Herranz, M. A.; Illescas, B.; Martın, N. Donor/acceptor fulleropyrrolidine triads. *Journal of Organic Chemistry*, 2000, *65*, 5728–5738.

83. Zandler, M. E.; Smith, P. M.; Fujitsuka, M.; Ito, O.; D'Souza, F. Molecular triads composed of ferrocene, C_{60}, and nitroaromatic entities: electrochemical, computational, and photochemical investigations. *Journal of Organic Chemistry*, 2002, *67*, 9122–9129.

84. D'Souza, F.; Zandler, M. E.; Smith, P. M.; Deviprasad, G. R.; Klykov, A.; Fujitsuka, M.; Ito, O. A ferrocene-C_{60}-dinitrobenzene triad: Synthesis and computational, electrochemical, and photochemical studies. *Journal of Physical Chemistry A*, 2002, *106*, 649–656.

85. Xiao, S.; Li, Y.; Li, Y.; Zhuang, J.; Wang, N.; Liu, H.; Ning, B. et al. [60]Fullerene-based molecular triads with expanded absorptions in the visible region: Synthesis and photovoltaic properties. *Journal of Physical Chemistry B*, 2004, *108*, 16677–16685.

86. Fong II, R.; Schuster, D. I.; Wilson, S. R. Synthesis and photophysical properties of steroid-linked porphyrin-fullerene hybrids. *Organic Letters*, 1999, *1* (5), 729–732.

87. Marczak, R.; Sgobba, V.; Kutner, W.; Gadde, S.; D'Souza, F.; Guldi, D. M. Langmuir-Blodgett films of a cationic zinc porphyrin-imidazole-functionalized fullerene dyad: Formation and photoelectrochemical studies. *Langmuir*, 2007, *23*, 1917–1923.

88. Li, Y.; Wang, N.; He, X.; Wang, S.; Liu, H.; Li, Y.; Li, X. et al. Synthesis and characterization of ferrocene-perylenetetracarboxylic diimide-fullerene triad. *Tetrahedron*, 2005, *6* (6), 1563–1569.

89. Guldi, D. M.; Luo, C.; Kotov, N. A.; Da Ros, T.; Bosi, S.; Prato, M. Zwitterionic acceptor moieties: Small reorganization energy and unique stabilization of charge transfer products. *Journal of Physical Chemistry B*, 2003, *107*, 7293–7298.

90. Matino, F.; Arima, V.; Maruccio, G.; Phaneuf, R. J.; Del Sole, R.; Mele, G.; Vasapollo, G.; Cingolani, R.; Rinaldi, R. Rectifying behaviour of self assembled porphyrin/fullerene dyads on Au(111). *Journal of Physics: Conference Series*, 2007, *61*, 795–799.

91. Campidelli, S.; Vazquez, E.; Milic, D.; Lenoble, J.; Atienza Castellanos, C.; Sarova G.; Guldi, D. M.; Deschenaux, R.; Prato, M. Liquid-crystalline bisadducts of [60]fullerene. *Journal of Organic Chemistry*, 2006, *71*, 7603–7610.

92. Zhou, Z.; Schuster, D. I.; Wilson, S. R. Tether-directed selective synthesis of fulleropyrrolidine bisadducts. *Journal of Organic Chemistry*, 2006, *71*, 1545–1551.

93. Nishimura, T.; Tsuchiya, K.; Ohsawa, S.; Maeda, K.; Yashima, E.; Nakamura, Y.; Nishimura, J. Macromolecular helicity induction on a poly(phenylacetylene) with C_2-symmetric chiral [60]fullerene-bisadducts. *Journal of the American Chemical Society*, 2004, *126*, 11711–11717.

94. Kordatos, K.; Bosi, S.; Da Ros, T.; Zambon, A.; Lucchini, V.; Prato, M. Isolation and characterization of all eight bisadducts of fulleropyrrolidine derivatives. *Journal of Organic Chemistry*, 2001, *66*, 2802–2808.

95. Marchesan, S.; Da Ros, T.; Prato, M. Isolation and characterization of nine tris-adducts of *N*-methylfulleropyrrolidine derivatives. *Journal of Organic Chemistry*, 2005, *70*, 4706–4713.

96. Zoleo, A.; Bellinazzi, M.; Prato, M.; Brustolon, M.; Maniero, A. L. Multifrequency EPR study and DFT calculations of a C_{60} bisadduct anion. *Chemical Physics Letters*, 2005, *412* (4), 470–476.

97. Nednoor, P.; Capaccio, M.; Gavalas, V. G.; Meier, M. S.; Anthony, J. E.; Bachas, L. G. Hybrid nanoparticles based on organized protein immobilization on fullerenes. *Bioconjugate Chemistry*, 2004, *15*, 12–15.

98. Okada, E.; Komazawa, Y.; Kurihara, M.; Inoue, H.; Miyata, N.; Okuda, H.; Tsuchiya, T.; Yamakoshi, Y. Synthesis of C_{60} derivatives for photoaffinity labeling. *Tetrahedron Letters*, 2004, *45* (3), 527–529.

99. Wanga, C.; Guob, Z.-X.; Fua, S.; Wub, W.; Zhu, D. Polymers containing fullerene or carbon nanotube structures. *Progress in Polymer Science*, 2004, *29*, 1079–1141.

100. Lenoble, J.; Maringa, N.; Campidelli, S.; Donnio, B.; Guillon, D.; Deschenaux, R. Liquid-crystalline fullerodendrimers which display columnar phases. *Organic Letters*, 2006, *8* (9), 1851–1854.

101. Campidelli, S.; Lenoble, J.; Barbera, J.; Paolucci, F.; Marcaccio, M.; Paolucci, D.; Deschenaux, R. Supramolecular fullerene materials: Dendritic liquid-crystalline fulleropyrrolidines. *Macromolecules*, 2005, *38*, 7915–7925.

102. Georgakilas, V.; Guldi, D. M.; Signorini, R.; Bozio, R.; Prato, M. Organic functionalization and optical properties of carbon onions. *Journal of the American Chemical Society*, 2003, *125* (47), 14268–14269.

103. Brough, P.; Klumpp, C.; Bianco, A.; Campidelli, S.; Prato, M. [60]Fullerene-pyrrolidine-*N*-oxides. *Journal of Organic Chemistry*, 2006, *71*, 2014–2020.

104. Mateo-Alonso, A.; Brough, P.; Prato, M. Stabilization of fulleropyrrolidine *N*-oxides through intrarotax-ane hydrogen bonding. *Chemical Communications*, 2007, (14), 1412–1414.

105. Zhang, S.; Gan, L.; Huang, C.; Lu, M.; Pan, J.; He, X. Acylation of 2,5-dimethoxycarbonyl[60]fulleropyr-rolidine and synthesis of its multifullerene derivatives. *Journal of Organic Chemistry*, 2002, *67*, 883–891.

106. Zhang, J.; Porfyrakis, K.; Morton, J. J. L.; Sambrook, M. R.; Harmer, J.; Li, Xiao; Ardavan, A.; Briggs, G. A. D. Photoisomerization of a fullerene dimer. *Journal of Physical Chemistry C*, 2008, *112* (8), 2802–2804.

107. Nakanishi, T.; Ariga, K.; Michinobu, T.; Yoshida, K.; Takahashi, H.; Teranishi, T.; Mohwald, H.; Kurth, D. G. Flower-shaped supramolecular assemblies: Hierarchical organization of a fullerene bearing long aliphatic chains. *Small*, 2007, (12), 2019–2023.

108. Nakanishi, T.; Schmitt, W.; Michinobu, T.; Kurth, D. G.; Ariga, K. Hierarchical supramolecular fullerene architectures with controlled dimensionality. *Chemical Communications*, 2005, 5982–5984.

109. Nakanishi, T.; Michinobu, T.; Yoshida, K.; Shirahata, N.; Ariga, K.; Mohwald, H.; Kurth, D. G. Nanocarbon superhydrophobic surfaces created from fullerene-based hierarchical supramolecular assemblies. *Advanced Materials*, 2008, *20*, 443–446.

110. Kharisov, B. I. A review for synthesis of nanoflowers. *Recent Patents on Nanotechnology*, 2008, 2, 190–200.

111. Zhang, P.; Li, J.; Liu, D.; Qin, Y.; Guo, Z.-X.; Zhu, D. Self-assembly of gold nanoparticles on fullerene nanospheres. *Langmuir*, 2004, *20*, 1466–1472.

112. Illescas, B. M.; Martın, N.; Poater, J.; Sola, M.; Aguado, G. P.; Ortuno, R. M. Diastereoselective synthesis of fulleropyrrolidines from suitably functionalized chiral cyclobutanes. *Journal of Organic Chemistry*, 2005, *70*, 6929–6932.

113. Wang, J.; Gutsche, C. D. Synthesis and structure of calixarene-fullerene dyads. *Journal of Organic Chemistry*, 2000, *65*, 6273–6275.

114. Brites, M. J.; Santos, C.; Nascimento, S.; Gigante, B.; Berberan-Santos, M. N. Synthesis of [60]fullerene-coumarin polyads. *Tetrahedron Letters*, 2004, *45* (37), 6927–6930.

115. Illescas, B. M.; Martın, N. [60] Fullerene adducts with improved electron acceptor properties. *Journal of Organic Chemistry*, 2000, *65*, 5986–5995.

116. Modin, J.; Johansson, H.; Grennberg, H. New pyrazolino- and pyrrolidino[60]fullerenes with transition-metal chelating pyridine substitutents: Synthesis and complexation to Ru(II). *Organic Letters*, 2005, *7* (18), 3977–3979.

117. Suarez, M.; Verdecia, Y.; Illescas, B.; Martinez-Alvarez, R.; Alvarez, A.; Ochoa, E.; Seoane, C. et al. Synthesis and study of novel fulleropyrrolidines bearing biologically active 1,4-dihydropyridines. *Tetrahedron*, 2003, *59* (46), 9179–9186.

118. Troshin, P. A.; Troyanov, S. I.; Boiko, G. N.; Lyubovskaya, R. N.; Lapshin, A. N.; Goldshleger, N. Efficient [2 + 3] cycloaddition approach to synthesis of pyridinyl based [60]fullerene ligands. *Fullerence, Nanotubes, and Carbon Nanostructures*, 2004, *12*, 413–419.

119. Xiao, S.; Li, Y.; Li, Y.; Liu, H.; Li, H.; Zhuang, J.; Liu, Y.; Lu, F.; Zhang, D.; Zhu, D. Easy access to N-alkylation of N-unsubstituted [60]fulleropyrrolidines: Reductive amination using sodium triacetoxy-borohydride. *Tetrahedron Letters*, 2004, *45* (20), 3975–3978.

120. Segura, M.; Sanchez, L.; de Mendoza, J.; Martın, N.; Guldi, D. M. Hydrogen bonding interfaces in fullerene-TTF ensembles. *Journal of the American Chemical Society*, 2003, *125*, 15093–15100.

121. Vasapollo, G.; Mele, G.; Longo, L.; Ianne, R.; Gowenlock, B. G.; Orrell, K. G. Synthesis of novel nitroso-fulleropyrrolidines. *Tetrahedron Letters*, 2002, *43* (28), 4969–4972.

122. Da Ros, T.; Guldi, D. M.; Farran Morales, A.; Leigh, D. A.; Prato, M.; Turco, R. Hydrogen bond-assembled fullerene molecular shuttle. Organic Letters, 2003, *5* (5), 689–691.

123. Dondoni, A.; Marra, A. Synthesis of [60]fulleropyrrolidine glycoconjugates using 1,3-dipolar cycloaddi-tion with C-glycosyl azomethine ylides. *Tetrahedron Letters*, 2002, *43* (9), 1649–1652.

124. Ge, Z.; Li, Y.; Shi, Z.; Bai, F.; Zhu, D. Synthesis and photophysical characterization of a new crown ether-bearing [70]fulleropyrrolidine derivative. *Journal of Physics and Chemistry of Solids*, 2000, *61* (7), 1075–1079.

125. Ge, Z.; Li, Y.; Du, C.; Wang, S.; Zhu, D. Stable monolayers and Langmuir-Blodgett films of a new crown ether-bearing [60]fulleropyrrolidine containing benzothiazolium styryl dye. *Thin Solid Films*, 2000, *368* (1), 147–151.

126. Wang, G.-W.; Zhang, T.-H.; Hao, E.-H.; Jiao, L.-J.; Murata, Y.; Komatsu, K. Solvent-free reactions of fullerenes and *N*-alkylglycines with and without aldehydes under high-speed vibration milling. *Tetrahedron*, 2003, *59* (1), 55–60.

127. Tasis, D.; Tagmatarchis, N.; Georgakilas, V.; Gamboz, C.; Soranzo, M.-R.; Prato, M. Supramolecular organized structures of fullerene-based materials and organic functionalization of carbon nanotubes. *Comptes Rendus Chimie*, 2003, *6* (5), 597–602, 2003.

128. Mrzel, A.; Hassanien, A.; Liu, Z.; Suenaga, K.; Miyata, Y.; Yanagi, K.; Kataura, H. Effective, fast, and low temperature encapsulation of fullerene derivatives in single wall carbon nanotubes. *Surface Science*, 2007, *601* (22), 5116–5120.

129. Benne, D.; Maccallini, E.; Rudolf, P.; Sooambar, C.; Prato, M. X-ray photoemission spectroscopy study on the effects of functionalization in fulleropyrrolidine and pyrrolidine derivatives. *Carbon*, 2006, *44*, 2896–2903.

130. Guldi, D. M.; Prato, M. Excited-state properties of C_{60} fullerene derivatives. *Accounts of Chemical Research*, 2000, *33*, 695–703.

131. Razbirin, B. S.; Sheka, E. F.; Starukhin, A. N.; Korotkov, A. S.; Afanas'ev, V. A. Matrix-isolated fulleropyrrolidine molecules. Optical spectra and quantum-chemical calculations. *6th Biennial International Workshop: Fullerenes and Atomic Clusters, IWFAC'2003*. St Petersburg, Russia, June 30–July 4, 2003. p. 27. http://vs.ioffe.net/iwfac/2005/abstr/contents.pdf

132. Polimeno, A.; Zerbetto, M.; Franco, L.; Maggini, M.; Corvaja, C. Stochastic modeling of CW-ESR spectroscopy of [60]fulleropyrrolidine bisadducts with nitroxide probes. *Journal of the American Chemical Society*, 2006, *128*, 4734–4741.

133. Mazzoni, M.; Franco, L.; Corvaja, C.; Zordan, G.; Menna, E.; Scorrano, G.; Maggini, M. Synthesis, EPR and ENDOR of [60]fulleropyrrolidine bisadducts with nitroxide addends: Magnitude and sign of the exchange interaction. *ChemPhysChem: A European Journal of Chemical Physics and Physical Chemistry*, 2002, *3* (6), 527–531.

134. Corvaja, C.; Conti, F.; Franco, L.; Maggini, M. Spin-labeled fulleropyrrolidines. *Comptes Rendus Chimie*, 2006, *9* (7), 909–915.

135. Schick, G.; Levitus, M.; Kvetko, L.; Johnson, B. A.; Lamparth, I.; Lunkwitz, R.; Ma, B.; Khan, S. I.; Garcia-Garibay, M. A.; Rubin, Y. Unusual luminescence of hexapyrrolidine derivatives of C_{60} with T_h and novel D_3-symmetry. *Journal of the American Chemical Society*, 1999, *121*, 3246–3247.

136. Sandanayaka, A. S. D.; Araki, Y.; Ito, O.; Deviprasad, G. R.; Smith, P. M.; Rogers, L. M.; Zandler, M. E.; D'Souza, F. Photoinduced electron transfer in fullerene triads bearing pyrene and fluorine. *Chemical Physics*, 2006, *325* (2), 452–460.

137. Carano, M.; Da Ros, T.; Fanti, M.; Kordatos, K.; Marcaccio, M.; Paolucci, F.; Prato, M.; Roffia, S.; Zerbetto, F. Modulation of the reduction potentials of fullerene derivatives. *Journal of the American Chemical Society*, 2003, *125*, 7139–7144.

138. D'Souza, F.; Zandler, M. E.; Deviprasad, G. R.; Kutner, W. Acid-base properties of fulleropyrrolidines: Experimental and theoretical investigations. *Journal of Physical Chemistry A*, 2000, *104*, 6887–6893.

139. Alvarez, A.; Ochoa, E.; Verdecia, Y.; Suarez, M.; Sola, M.; Martın, N. Theoretical study of the highly diastereoselective 1,3-dipolar cycloaddition of 1,4-dihydropyridine-containing azomethine ylides to [60]fullerene (Prato's reaction). *Journal of Organic Chemistry*, 2005, *70*, 3256–3262.

140. Jalbout, A. F.; Hameed, A. J.; Trzaskowski, B. Study of the structural and electronic properties of 1-(4, 5 and 6-selenenyl derivatives-3-formyl-phenyl) pyrrolidinofullerenes. *Journal of Organometallic Chemistry*, 2007, *692* (5), 1039–1047.

141. Hameed, A. J. Computational note on substitution effects on the structural and electronic properties of 1-(para-substituted phenyl diazenyl)pyrrolidinofullerenes. *Journal of Molecular Structure: THEOCHEM*, 2006, *766* (2), 73–75.

142. Hameed, A. J.; Ibrahim, M.; ElHaes, H. Computational notes on structural, electronic and QSAR properties of [C_{60}]fulleropyrrolidine-1-carbodithioic acid 2; 3 and 4-substituted-benzyl esters. *Journal of Molecular Structure: THEOCHEM*, 2007, *809* (1), 131–136.

143. Arima, V.; Matino, F.; Thompson, J.; Del Sole, R.; Mele, G.; Vasapollo, G.; Cingolani, R. et al. Characterization of functionalised porphyrin films using synchrotron radiation. *Applied Surface Science*, 2005, *248* (1), 40–44.

144. Ferrante, C.; Signorini, R.; Feis, A.; Bozio, R. Excited state absorption of fullerenes measured by the photoacoustic calorimetry technique. *Photochemical and Photobiological Sciences: Official Journal of the European Photochemistry Association and the European Society for Photobiology*, 2003, *2* (7), 801–807.

145. Koudoumas, E.; Konstantaki, M.; Mavromanolakis, A.; Couris, S.; Fanti, M.; Zerbetto, F.; Kordatos, K.; Prato, M. Large enhancement of the nonlinear optical response of reduced fullerene derivatives. *Chemistry*, 2003, *9* (7), 1529–1534.

146. Da Ros, T.; Spalluto, G.; Prato, M. Biological applications of fullerene derivatives: A brief overview. *Croatica Chemica Acta*, 2001, *74* (4), 743–755.

147. Marchesan, S.; Da Ros, T.; Spalluto, G.; Balzarini, J.; Prato, M. Anti-HIV properties of cationic fullerene derivatives. *Bioorganic and Medicinal Chemistry Letters*, 2005, *15* (15), 3615–3618.

148. Da Ros, T.; Vazquez, E.; Spalluto, G.; Moro, S.; Boutorine, A.; Prato, M. Design, synthesis and biological properties of fulleropyrrolidine derivatives as potential DNA photo-probes. *Journal of Supramolecular Chemistry*, 2002, *2* (1), 327–334.

149. Illescas, B. M.; Martinez-Alvarez, R.; Fernandez-Gadea, J.; Martin, N. Synthesis of water soluble fulleropyrrolidines bearing biologically active arylpiperazines. *Tetrahedron*, 2003, *59* (34), 6569–6577.

150. Huang, S. T.; Liao, J. S.; Fang, H. W.; Lin, C. M. Synthesis and anti-inflammation evaluation of new C_{60} fulleropyrrolidines bearing biologically active xanthine. *Bioorganic and Medicinal Chemistry Letters*, 2008, *18* (1), 99–103.

151. Pellarini, F.; Pantarotto, D.; Da Ros, T.; Giangaspero, A.; Tossi, A.; Prato, M. A novel [60]fullerene amino acid for use in solid-phase peptide synthesis. *Organic Letters*, 2001, *3* (12), 1845–1848.

152. Murakami, H.; Matsumoto, R.; Okusa, Y.; Sagara, T.; Fujitsuka, M.; Ito, O.; Nakashima, N. Design, synthesis and photophysical properties of C_{60}-modified proteins. *Journal of Materials Chemistry*, 2002, *12*, 2026–2033.

153. Ostiguy, C.; Lapointe, G.; Trottier, M.; Ménard, L.; Cloutier, Y.; Boutin, M.; Antoun, M.; Normand, C. Health effects of nanoparticles. 2006, IRSST report, www.irsst.qc.ca

154. Li, J.; Zhang, S.; Zhang, P.; Liu, D.; Guo, Z.-X.; Ye, C.; Zhu, D. Nanoscale cavities for fulleropyrrolidinium in nafion membrane. *Chemistry of Materials*, 2003, *15*, 4739–4744.

155. Kikuchi, K.; Takeoka, Y.; Rikukawa, M.; Sanui, K. Fabrication and characterization of organic-inorganic perovskite films containing fullerene derivatives. *Colloids and Surfaces A: Physicochemical and Engineering Aspects*, 2005, *257*, 199–202.

156. Ballav, N. Some experimental results on poly-*N*-vinylcarbazole-buckminster-fullerene (C_{60}) nanocomposite system. *Materials Letters*, 2005, *59* (27), 3419–3422.

157. Sgobba, V.; Giancane, G.; Conoci, S.; Casilli, S.; Ricciardi, G.; Guldi, D. M.; Prato, M.; Valli, L. Growth and characterization of films containing fullerenes and water soluble porphyrins for solar energy conversion applications. *Journal of the American Chemical Society*, 2007, *129*, 3148–3156.

158. Lee, J.-K.; Fujida, K.; Tsutsui, T.; Kim, M.-R. Synthesis and photovoltaic properties of soluble fulleropyrrolidine derivatives for organic solar cells. *Solar Energy Materials and Solar Cells*, 2007, *91* (10), 892–896.

159. Guo, F.; Ogawa, K.; Kim, Y.-G.; Danilov, E. O.; Castellano, F. N.; Reynolds, J. R.; Schanze, K. S. A fulleropyrrolidine end-capped platinum-acetylide triad: The mechanism of photoinduced charge transfer in organometallic photovoltaic cells. *Physical Chemistry Chemical Physics*, 2007, *9* (21), 2724–2734.

160. Seguraa, J. L.; Giacalonea, F.; Gómeza, R.; Martín, N.; Guldi, D. M.; Luo, C.; Swartz, A. et al. Design, synthesis and photovoltaic properties of [60]fullerene based molecular materials. *Materials Science and Engineering C*, 2005, *25* (5), 835–842.

161. Guldi, D. M.; Pellarini, F.; Prato, M.; Granito, C.; Troisi, L. Layer-by-layer construction of nanostructured porphyrin-fullerene electrodes. *Nano Letters*, 2002, *2* (9), 965–968.

162. Casalbore-Miceli, G.; Camaioni, N.; Geri, A.; Ridolfi, G.; Zanelli, A.; Gallazzi, M. C.; Maggini, M.; Benincori, T. "Solid state charge trapping": Examples of polymer systems showing memory effect. *Journal of Electroanalytical Chemistry*, 2007, *603* (2), 227–234.

163. Jonas Järvholm, E. Mechanisms and development of etch resistance for highly aromatic monomolecular etch masks—Towards molecular lithography. PhD thesis, Georgia Institute of Technology, Atlanta, GA, 2007.

164. Wang, G.-W. Fullerene mechanochemistry. In *Encyclopedia of Nanoscience and Nanotechnology*. Nalva, H. S. (Ed.). American Scientific Publishers, Stevenson Ranch, CA, 2004, Vol. 3, pp. 557–565.

165. de la Cruz, P.; de la Hoz, A.; Langa, F.; Illescas, B.; Martin, N. Cycloadditions to [60] fullerene using microwave irradiation: A convenient and expeditious procedure. *Tetrahedron*, 1997, *53* (7), 2599–2608.

166. Wang, S.; Zhang, J. M.; Song, L. P.; Jiang, H.; Zhu, S. Z. One-pot three-component reaction of C_{60}, amino acid and fluorinated benzaldehyde to C_{60}-pyrrolidine derivatives. *Journal of Fluorine Chemistry*, 2005, *126* (3), 349–353.

20 Small Fullerenes C$_{20<n<60}$ and Endohedral Metallofullerenes M@C$_n$

Since the discovery of C$_{60}$ Buckminsterfullerene[1] in 1985 and its isolation and unequivocal characterization[2,3] in 1990, almost all fullerene research has focused on the stable species C$_n$ ($n = 60$ and ≥ 70), all of which conform to the *Pentagon Isolation Rule* (PIR).[4,5] The PIR maintains that the stability of C$_{60}$, C$_{70}$, and higher fullerenes is due to the fact that these fullerenes possess structures with nonabutting pentagons, whereas fullerenes with $n < 70$ (except for $n = 60$) must possess abutting pentagons. This rule applies only to the underivatized fullerenes and the addition of moieties that disrupt the sp^2 hybridization can stabilize structures with abutting pentagons.

The original cluster beam studies, which uncovered the existence of C$_{60}$, also yielded strong circumstantial evidence that many small fullerenes ($n = 20$–60) may form with varying degrees of metastability—at least in carbon vapor. Indeed, variations in intensity as a function of variations in the clustering conditions were an important factor in these original studies and in particular a critical empirical factor in identifying the special nature of C$_{60}$. Furthermore, some early mass spectrometric data were interpreted as providing compelling evidence for not only the fundamental fullerene structural concept but also the existence of metastable, small, closed-cage structures down to at least C$_{24}$.

In this section, we describe the synthesis and main properties of small fullerenes and their endohedral metal complexes.

20.1 STABILITY OF FULLERENES LOWER THAN C$_{60}$ AND THEIR SPECTRAL AND ELECTRONIC PROPERTIES

The *instability* of small fullerenes, in comparison with classic C$_{60}$, can be explained as follows.[6] It is well known that five- and six-member cycles, presenting in C$_{60}$, are the most widespread and stable in organic chemistry. The Euler theorem is a necessary and sufficient condition of polyhedra existing ($\chi = V - E + F$, where V, E, and F are, respectively, the numbers of vertices [corners], edges, and faces in the given polyhedron). For example, the C$_{22}$, containing 12 pentagonal faces and hexagonal face, cannot be constructed. The simplest fullerene could be dodecahedron C$_{20}$, containing 12 pentagonal faces and zero hexagonal face. It is obvious that, reducing fullerene size and its flexion, the structure turns more rigorous and the efficiency of aromatic conjugation decreases. The same reason causes a lower kinetic stability of small fullerenes, since the tension of double bonds and nonefficiency of their conjugation mean a comparably low position of reactive polyradical states. Moreover, in some cases main states become triplet. In high-symmetry molecules, due to the Jahn–Teller effect, a distortion of geometry of degenerated singlet states per symmetry may take place, thereby becoming main states. This is the case of the dodecahedron C$_{20}$ and more stable isomer of C$_{36}$ with D$_{6h}$ symmetry. The presence of several states, close in geometry and energy, allows expecting the dynamic Jahn–Teller effect also for these molecules.

The stability[7] of small fullerenes and its relation with their electronic structure is permanently discussed. Thus, the evolution of the properties of face-centered cubic fullerites with a variation in the number of carbon atoms (nc) in a C$_{nc}$ fullerene molecule ($15 \leq nc \leq 147$) was investigated[8] using

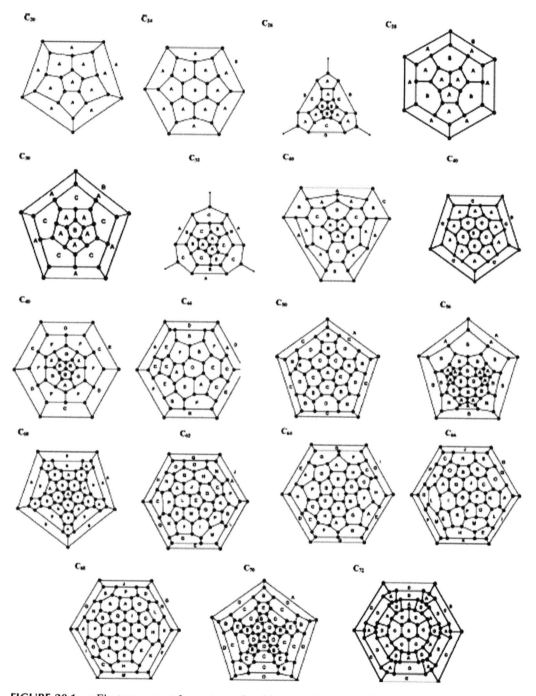

FIGURE 20.1 π-Electron content for pentagonal and hexagonal rings for 19 fullerenes. (Reproduced with permission from Randic, M. et al., Numerical Kekulé structures of fullerenes and partitioning of π-electrons to pentagonal and hexagonal rings, *J. Chem. Inf. Model.*, 47, 897–904. Copyright 2007 American Chemical Society.)

the dependence of the parameters of the interfullerene interaction in face-centered cubic fullerites on the mass of the C_{nc} fullerene molecule. It was demonstrated that, for $nc < 20$, the face-centered cubic fullerites become unstable because such light, small fullerene molecules cannot be kept by weak van der Waals forces. For $nc \geq 110$, the fullerites have anomalously low surface energies, which should lead to fragmentation of nanoclusters composed of large hollow spherical molecules C_{nc}. The range $30 < nc < 100$ was found to be optimum for the formation of stable face-centered cubic fullerites. The *ground-state structures* of small fullerenes below C_{70} were calculated using tight-binding molecular-dynamics total-energy optimization.[9] An efficient simulated-annealing scheme was used to generate closed, hollow, spheroidal cage structures for all even-numbered carbon clusters from C_{20} to C_{70}. It was concluded that, as a general trend, *fullerenes prefer geometries that separate the pentagonal rings as far apart as possible.* Except for C_{60}, C_{70}, and C_{50}, most fullerenes have relatively low symmetries.

Spectral properties of small fullerenes (as well as other related compounds, C_{60} and Si-heterofullerenes) were studied by *ab initio* calculations.[10] It was shown that C_{20}, C_{28}, C_{32}, C_{36}, and C_{50}, *the most stable small fullerenes* in the range of C_{20}–C_{50}, have characteristic features in their optical absorption spectra, originating from the geometry of the molecules. Substitutionally doped fullerenes were of interest due to their enhanced chemical reactivity. Also, structures, energies of formation, electronic spectra, and values of bulk moduli for crystalline modifications of small fullerene C_{28} with space groups Fd3m and P63/mmc were calculated at the density-functional based tight-binding level.[11] Both solid phases of C_{28} were found to have very similar characteristics, which were compared with those of crystals of other small fullerenes (C_{20} and C_{36}). Additionally, the structure and electronic properties of small fullerene structures C_{28}, C_{32}, C_{36}, C_{40}, and their C_{24} exohydrogenates, were studied by employing the quantum-chemical *ab initio* method.[12] Calculations showed that the stability order of addition of hydrogen to a carbon cage for carbon site is 5/5/5 > 5/6/5 > 6/5/6 > 6/6/6. The authors suggested that this implies the vertex of triplet pentagons of the small carbon cage is an activation site for addition. *Hueckel model calculations* were performed for small fullerene cages with 20–50 atoms.[13] The relatively high stability of C_{24}, C_{28}, C_{32}, C_{44}, and C_{50} clusters was observed in the laser vaporization experiments. These clusters have pseudo closed-shell or half-filled electronic structures with a relatively large HOMO–LUMO gap. *π-Electron content* for pentagonal and hexagonal rings was reported for the following 19 fullerenes: C_{20}, C_{24}, C_{26}, C_{28}, C_{30}, C_{32}, C_{40} (three isomers), C_{44}, C_{50}, C_{56}, C_{60}, C_{62}, C_{64}, C_{66}, C_{68}, and C_{74} (Figure 20.1)[14] by constructing their Kekulé valence structures and averaging the π-electron content of individual rings over all Kekulé valence structures. The authors noted two novel aspects arising when calculating the partitioning of π-electrons in fullerenes, which are absent when calculating the π-electron content for benzenoid or nonbenzenoid hydrocarbons: (1) The calculations are simplified due to the fact that in fullerenes all C=C bonds are shared by two adjacent rings, hence each C=C contributes always just one π-electron to a ring and (2) the search is complicated because the number of Kekulé valence structures in fullerenes is so large that one has to use a computer to construct and analyze them. Additionally, the 3D poly cycle indexes for the natural actions of the symmetry group of the small fullerene C_{24} and big fullerene C_{150} over the set of vertices, edges, and faces were computed.[15]

20.2 STRUCTURES OF SMALL FULLERENES

Among all the small fullerenes (Figure 20.2), C_{28} is the best studied.[16] The isolated molecule C_{28} is the T_d-symmetrical cage of 28 carbon atoms, connected by 42 covalent sp²-like bonds (Figure 20.3).[17] There are three essentially different kinds of bonds in molecule C_{28}:

1. Twenty-four bonds with length 1.42 Å, which are the verges of four hexagons
2. Six bonds with length 1.52 Å, which connect the vertexes of each pair of hexagons
3. Twelve bonds with length 1.46 Å, which are joined in four vertexes, common for each three neighboring pentagons

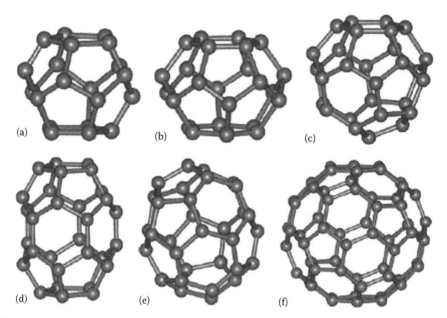

FIGURE 20.2   Some small fullerene molecules: (a) C_{20}, (b) C_{24}, (c) C_{28}, (d) C_{30}, (e) C_{32}, and, for comparison, (f) C_{60}.

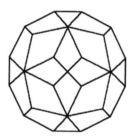

FIGURE 20.3   The structure of C_{28}.

In C_{28} structure, a carbon atom is shared by three pentagons at each of four vertices, and the associated strain of these four reactive carbons can be overcome by the addition of a monovalent atom or moiety.[18] It was suggested that the C_{28} cluster might under certain circumstances also exhibit special behavior. It was hypothesized, on the basis of intuitive chemical reasoning, that C_{28} might behave like a carbon cluster superatom because it had a tetrahedral symmetry that could exhibit "meta"-tetravalency.

Additionally, dimers $(C_{28})_2$ (the simplest polymerized structure, which is formed by two molecules of C_{28}, connected by two bridge-like bonds, directed along their common twofold axis, Figure 20.4) and cuban-like clusters $(C_{28})_8$ (cluster can be got from cuban by replacing each of carbon atoms onto fullerene molecule C_{28}; molecules C_{28} in cluster are connected by 12 pairs of bridge-like bonds directed along three perpendicular twofold axes, Figure 20.5) are known.

20.3   ELECTRONIC TRANSPORT, MECHANISM, AND GROWTH STUDIES

The *electronic transport properties* of small fullerenes sandwiched between two metallic electrodes (Figure 20.6) and modulated by a gate electrode were studied.[17] The charging energies were calculated as an important factor that influences their transport. It was revealed that the bond types

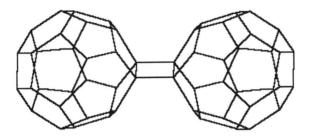

FIGURE 20.4   The structure of dimer C_{28}–C_{28}.

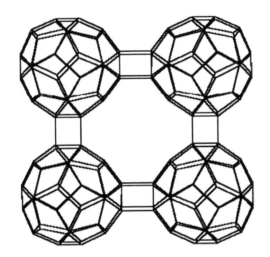

FIGURE 20.5   The structure of cubic phase of $(C_{28})_8$.

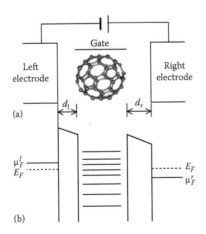

FIGURE 20.6   (a) Schematic diagram of a fullerene molecule weakly coupled between two electrodes through vacuum tunnel barriers. (b) Schematic diagram of the energy levels of the fullerene and the bias-induced energy band of the metallic electrodes. (Reproduced from Zhang, R.Q. et al., Electrical transport and electronic delocalization of small fullerenes, *J. Phys. Chem. B*, 108(43), 16636–16641. Copyright 2004 American Chemical Society.)

and molecular shapes are important in determining the charging energies and, thus, the transport properties. The calculated current–voltage and conductance characteristics were shown to relate to the electronic structures of fullerenes near the energy gaps and can be adjusted by the applied gate voltages. A ring-stacking model for the fullerene *growth mechanism* was analyzed using the semiempirical AM1 scheme.[19,20] Following different routes, small fullerenes C_{28} (T_d and D_2), C_{26} (D_{3h}), and C_{24} (D_{6d}) were constructed from monocyclic/polycyclic precursors and circumscribing them with appropriate carbon belts. Additionally, theoretical studies of the growth mechanism of small fullerene cage C_{24} were reported.[21]

To study the *mechanism of fullerene formation*, first-principle calculations of the total energies of carbon clusters with different structure and size ($n = 14$–28 atoms) were performed[22] based on the local-density approximation of the density-functional theory. For small cluster sizes ($n < 20$), the most stable structures found were the 1D ring clusters, while, for larger sizes ($n = 20$–28), clusters with different types of geometry (bowls, rings, and fullerenes) had similar energies. For $n > 28$, the fullerenes were the most stable species. Therefore, the carbon cluster regime of growth and formation was divided into three parts: (1) domination of 1D clusters, (2) fast growth of reactive small fullerene clusters, and (3) slow growth of C_{60}.

20.4 SYNTHESIS, SORPTION, AND ENCAPSULATION

In a *laser evaporation source* as the principal synthesis method, high abundances of small fullerenes, especially of C_{32}, can be achieved.[23] Under appropriate source conditions, sufficient C_{32} for fullerite preparation can also be obtained in the soot of a furnace. The electronic structure of small fullerenes determined experimentally by anion photoelectron spectroscopy was compared with calculations. A huge mass signal and a large gap of 1.3 eV comparable to the gap of C_{70} were found for C_{32}, indicating that this cluster is the most stable fullerene below C_{60}. Also C_{36}, C_{44}, and C_{50} exhibited large gaps and surprisingly high stabilities, being likely candidates for the formation of cluster materials. In addition, small fullerene derivatives can be formed from C_{60}. Thus, the main products from the reaction of $C_{60}Cl_6$ with fluorobenzene-$FeCl_3$ were found to be $C_{60}(4\text{-}FC_6H_4)_5Cl$, $1,4\text{-}(4\text{-}FC_6H_4)_2C_{60}$ and asymmetric $C_{60}(4\text{-}FC_6H_4)_4$, accompanied by a small amount of $C_{60}(4\text{-}FC_6H_4)_4H_2$.[24] Exposure of the products to light and air before processing gave O-containing components, whose fragmentation during EI mass spectrometry occurred readily to give C_{58} and derivatives, which must contain adjacent pentagons, stabilization being derived from the presence of a heptagon.

Density functional theory based on an *ab initio* method was performed to study the *chemisorption behavior* of a small fullerene molecule, C_{28}, on the Si(001)-c(2 × 1) surface.[25] The strong interactions were found to take place between C_{28} and Si(001) surface. The substrate and fullerene molecule underwent the lattice relaxation and structural deformation. The strongest chemisorption occurred at the trench site, and the adsorption energy was up to 5.00 eV. Alternatively, extensive theoretical studies on the chemisorption of a small fullerene molecule, C_{28}, on the c(4 × 4) reconstructed GaAs(001) surface were performed.[26] It was found that upon C_{28} adsorption the c(4 × 4) reconstructed GaAs(001) surface underwent considerable lattice relaxation and structural deformation of fullerene molecule also occurred. The deposition of C_{28} clusters on a semiconducting surface was simulated.[27] Additionally, using a laser vaporization source, low-energy neutral carbon clusters in the range C_{10}–C_{32} were deposited to grow thin films ($e \approx 100$ nm).[28] A large red shift (≈ 110 cm^{-1}) was measured, which can be related to the decrease in the material force constants compared with the diamond ones. Encapsulation of C_{20} and C_{30} fullerenes into semiconducting carbon nanotubes was carried out to study the possibility of bandgap engineering in such systems.[29,30] It was observed that C_{20} fullerenes behave similarly to an *n*-type dopant while C_{30} can provide *p*-type doping in some cases. The combined incorporation of both types of fullerenes (hybrid encapsulation) into the same nanotube led to a behavior similar to that found in electronic *pn*-junctions.

20.5 METAL (OR X = H, HAL, C, BN)–SMALL FULLERENE ENDOHEDRAL COMPLEXES

A few compounds of small fullerene (mainly C_{28}) with metals, H, C, and BN species have been reported. Thus, *ab initio* restricted Hartree–Fock calculations were performed on the ground and excited states of $Hf@C_{28}$ (Figure 20.7) and its positive and negative ions.[31] The calculated charge on Hf (=0.16) in $Hf@C_{28}$ indicated that Hf is covalently bound to C_{28} in $Hf@C_{28}$. The bonding between Hf and C_{28} in $Hf@C_{28}$ is mainly due to the compatibilities of the orbital energies and symmetry of the 6s and 5d orbitals of Hf (in T_d symmetry) and the higher occupied t_2, e, and a_1 orbitals of C_{28}. Stabilities of two kinds of tetravalent $M@C_{28}$ were discussed (stable complexes with M = Ti, Zr, Hf, and Ce, and M = Si, Ge, Sn, where M to C_{28} bonding is weak and mostly ionic). Similar calculations were performed on the ground and excited states of $C_{28}H_4$, $Hf@C_{28}H_4$, and their positive and negative ions.[32] Self-consistent calculations of the electronic structure of the endohedral complexes $M@C_{28}$ based on the small fullerene C_{28} (where M = Sc, Ti, V, Cr, Fe, and Cu) were performed by the nonempirical method of electron density functional in the discrete-variational scheme. The $Ti@C_{28}$ complex was found to have a maximum chemical stability, while the formation of the endohedral $Cu@C_{28}$ complex is least probable. A comparative analysis of the stability factors and electronic structure of two possible crystalline forms of small fullerene C_{28} and endohedral fullerene $Zn@C_{28}$ with diamond and lonsdaleite structures was performed using a cluster model.[33] Atoms of elements that, when placed inside C_{28} cages, had no significant effect on the stability of free small-fullerene molecules were shown to be able to dramatically change the electronic properties and reactivity of the C_{28} skeleton and to be favorable for forming small-fullerene crystal modifications, which are covalent crystals. In contrast, if the presence of foreign atoms inside C_{28} cages stabilized the isolated nanoparticles, then molecular crystals (such as C_{60} fullerites) were formed due to weak van der Waals forces. $U@C_{28}$ is also known.[34]

Among other related compounds, quasidynamic local-density simulations on a small fullerene complex (C_{28}) resulted in three endohedral complexes ($C_{28}@C$, $C_{28}@Zr$, and $C_{28}@Ti$), a carbon atom bound to the corner of the C_{28} molecule, and $C_{28}H_4$.[35] This molecule forms spontaneously from a 29-atom diamond crystallite, and, under proper circumstances, is a covalent container compound. Upon encapsulation of a Zr atom, 12.6 eV of energy is liberated leading to an unreactive closed-shell $C_{28}@Zr$ molecule. For exohedral complexes C_{28}–$C_{28}X_4$ (X = H, F, Cl, Br) on the basis of a small fullerene C_{28}, it was established that the chemical stability decreases in the series of complexes as follows: $C_{28}F_4 > C_{28}H_4 > C_{28}Br_4 > C_{28}Cl_4$.[36] The small fullerene C_{28} was supposed to display reaction

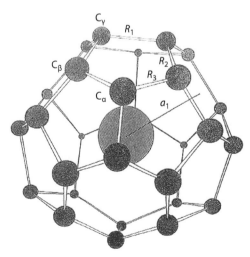

FIGURE 20.7 Structure of $Hf@C_{28}$ in T_d symmetry ($R_1 = 1.536$ Å, $R_2 = 1.41$ Å, $R_3 = 1.539$ Å, $d = 1.961$ Å).

TABLE 20.1

Parameters of Unit Cells Simulating Hybrid Nanostructures (C_{20}, C_{28})@(n,0)BN-NT

Parameter	C_{20}@(9,0)NT	C_{20}@(13,0)NT	C_{28}@(10,0)NT
Number of atoms in the cell	92	124	108
Cell composition	$C_{20}B_{36}N_{36}$	$C_{20}B_{52}N_{52}$	$C_{28}B_{40}N_{40}$
Nanotube diameter (nm)	0.709	0.102	0.787
Fullerene–nanotube wall distance (nm)	0.155	0.311	0.155
Fullerene–fullerene distance (nm)	0.454	0.454	0.206

activity in the process of the dissociative addition of complexing agents. In addition, construction of all 352 786 possible addition patterns $C_{24}H_{2m}$ and optimization with the density-functional-based tight-binding model is used to deduce that the C_{24} fullerene has an effective valence of 12.[37] The optimal structure of $C_{24}H_{12}$ has an equatorial ring of 12 sp^3 sites and is reachable by a pathway involving cumulative addition to low-energy isomers. Additionally, a quantum-chemical simulation of hybrid nanostructures consisting of regular chains of the small fullerenes C_{20} and C_{28} encapsulated into the bulk of achiral zigzag single-walled boron-nitrogen nanotubes [(C_{20}, C_{28})@BN-NT] was carried out.[38] The electronic characteristics of hybrid nanostructures were compared with those of "isolated" fullerenes and nanotubes and (C_{20}, C_{28}) + BN-NT structures (Table 20.1) simulating fullerene adsorption on tube surface as the initial stage of (C_{20}, C_{28})@BN-NT formation.

A simple *model of endohedral bonding* in small fullerene clusters was proposed.[39] It was based on a metaatom view of the cluster in which principal and angular momentum quantum numbers are assigned to the cluster eigenstates. The character of the valence electron states near the Fermi level of the cluster determines the nature of bonding to endohedral atoms. This model was found to be valid to explain the relative stability of Sn@C_{28} (hypothetical structure, not observed experimentally), Zr@C_{28} (experimentally produced structure), and U@C_{28} (complex with remarkable stability).

REFERENCES

1. Kroto, H. W.; Heath, J. R.; O'Brien, S. C.; Curl, R. F.; Smalley, R. E. C_{60}: Buckminsterfullerene, *Nature*, 1985, *318*, 162–163.
2. Haufler, R. E.; Conceicao, J.; Chibante, L. P. F.; Chai, Y.; Byrne, N. E.; Flanagan, S.; Haley, M. M. et al. Efficient production of C_{60} (buckminsterfullerene), $C_{60}H_{36}$, and the solvated buckide ion. *Journal of Physical Chemistry*, 1990, *94* (24), 8634–8636.
3. Taylor, R.; Hare, J. P.; Abdul-Sada, A. K.; Kroto, H. W. Isolation, separation and characterisation of the fullerenes C_{60} and C_{70}: The third form of carbon. *Journal of the Chemical Society, Chemical Communications*, 1990, 1423–1425.
4. Schmalz, T. G.; Seitz, W. A.; Klein, D. J.; Hite, G. E. *Journal of the American Chemical Society*, 1988, *110*, 1113–1127.
5. Dinadayalane, T. C.; Narahari Sastry, G. Isolated pentagon rule in buckybowls: A computational study on thermodynamic stabilities and bowl-to-bowl inversion barriers. *Tetrahedron*, 2003, *59*, 8347–8351.
6. Sidorov, L. N.; Yurovskaya, M. A.; Borshevskii, A. Ya.; Trushkov, I. V.; Ioffe, I. N. *Fullerenes*. Examen, Moscow, Russia, 2005, pp. 95–98.
7. Ori, O.; Cataldo, F.; Graovac, A. Topological ranking of C_{28} fullerenes reactivity. *Fullerenes, Nanotubes and Carbon Nanostructures*, 2009, *17* (3), 308–323.
8. Magomedov, M. N. On the prospects of preparing fullerites from small and large fullerenes. *Physics of the Solid State*, 2006, *48* (11), 2220–2225.
9. Zhang, B. L.; Wang, C. Z.; Ho, K. M.; Xu, C. H.; Chan, C. T. The geometry of small fullerene cages: C_{20} to C_{70}. *Journal of Chemical Physics*, 1992, *97* (7), 5007–5011.

10. Koponen, L.; Puska, M. J.; Nieminen, R. M. Photoabsorption spectra of small fullerenes and Si-heterofullerenes. *Journal of Chemical Physics*, 2008, *128* (15), 154307/1–154307/7.

11. Seifert, G.; Enyashin, A. N.; Heine, T. Hyperdiamond and hyperlonsdaleit: Possible crystalline phases of fullerene C$_{28}$. *Physical Review B: Condensed Matter and Materials Physics*, 2005, *72* (1), 012102/1–012102/4.

12. Lin, M.; Chiu, Y.-N.; Xiao, J. Theoretical study for exohydrogenates of small fullerenes C$_{28-40}$. *Journal of Molecular Structure: THEOCHEM*, 1999, *489* (2–3), 109–117.

13. Fan, M.-F.; Lin, Z.; Yang, S. Closed-shell electronic requirements for small fullerene cage structures. *Journal of Molecular Structure: THEOCHEM*, 1995, *337* (3), 231–240.

14. Randic, M.; Kroto, H. W.; Vukicevic, D. Numerical Kekulé structures of fullerenes and partitioning of π-electrons to pentagonal and hexagonal rings. *Journal of Chemical Information and Modeling*, 2007, *47*, 897–904.

15. Ghorbani, M.; Ashrafi, A. R. Cycle index of the symmetry group of fullerenes C$_{24}$ and C$_{150}$. *Asian Journal of Chemistry*, 2007, *19* (2), 1109–1114.

16. Popov, A. P.; Bazhin, I. V. Three-dimensional polymerized cubic phase of fullerenes C$_{28}$. In *Hydrogen Materials Science and Chemistry of Carbon Nanomaterials*. Veziroglu, T. N. et al. (Eds.). Kluwer Academic Publishers, Norwell, MA, 2004, pp. 329–332.

17. Zhang, R. Q.; Feng, Y. Q.; Lee, S. T.; Bai, C. L. Electrical transport and electronic delocalization of small fullerenes. *Journal of Physical Chemistry B*, 2004, *108* (43), 16636–16641.

18. Dunk, P. W.; Kaiser, N. K.; Hendrickson, C. L.; Marshall, A. G.; Kroto, H. W. Stabilization of small carbon clusters. *Abstracts of Papers, 238th ACS National Meeting*, Washington, DC, August 16–20, 2009, INOR-223.

19. Lin, W.-H.; Mishra, R. K.; Lin, Y.-T.; Lee, S.-L. Computational studies of the growth mechanism of small fullerenes: A ring-stacking model. *Journal of the Chinese Chemical Society*, 2003, *50* (3B), 575–582.

20. Mishra, R. K.; Lin, Y.-T.; Lee, S.-L. C$_{28}$ (D$_2$): Fullerene growth mechanism. *International Journal of Quantum Chemistry*, 2001, *84*, 642–648.

21. Lin, W.-H.; Tu, C.-C.; Lee, S.-L. Theoretical studies of growth mechanism of small fullerene Cage C$_{24}$ (D$_{6d}$)+. *International Journal of Quantum Chemistry*, 2005, *103* (4), 355–368.

22. Martins, J. L.; Reuse, F. A. Growth and formation of fullerene clusters. *Condensed Matter Theories*, 1998, *13*, 355–362.

23. Kietzmann, H.; Rochow, R.; Ganteför, G.; Eberhardt, W.; Vietze, K.; Seifert, G.; Fowler, P. W. Electronic structure of small fullerenes: Evidence for the high stability of C$_{32}$. *Physical Review Letters*, 1998, *81* (24), 5378–5381.

24. Darwish, A. D.; Avent, A. G.; Birkett, P. R.; Kroto, H. W.; Taylor, R.; Walton, D. R. M. Some 4-fluorophenyl derivatives of [60]fullerene; spontaneous oxidation and oxide-induced fragmentation to C$_{58}$. *Journal of the Chemical Society, Perkin Transactions 2*, 2001, (7), 1038–1044.

25. Fan, T.; Yao, S.; Zhou, C.; Han, B.; Wu, J. Adsorption of a small fullerene, C$_{28}$, on the Si(001)-c(2 × 1) surface: A density functional theory study. *Wuhan Daxue Xuebao, Lixueban*, 2007, *53* (6), 655–660.

26. Yao, S.; Zhou, C.; Ning, L.; Wu, J.; Pi, Z.; Cheng, H.; Jiang, Y. Chemisorption of C$_{28}$ fullerene on c(4 × 4) reconstructed GaAs(001) surface: A density functional theory study. *Physical Review B: Condensed Matter and Materials Physics*, 2005, *71* (19), 195316/1–195316/7.

27. Galli, G.; Canning, A.; Kim, J. Assembling small fullerenes: A molecular dynamics study. *Materials Research Society Symposium Proceedings (Covalently Bonded Disordered Thin-Film Materials)*, Denver, CO, 1998, Vol. 498, pp. 19–30.

28. Melinon, P.; Paillard, V.; Dupuis, V.; Perez, J. P.; Perez, A.; Panczer, G. Synthesis of diamond nanocrystallites using the low-energy cluster beam deposition; an indirect proof of small fullerene existence. *Carbon*, 1994, *32* (5), 1011–1013.

29. Troche, K. S.; Coluci, V. R.; Rurali, R.; Galvao, D. S. Structural and electronic properties of zigzag carbon nanotubes filled with small fullerenes. *Journal of Physics: Condensed Matter*, 2007, *19* (23), 236222/1–236222/9.

30. Troche, K. S.; Coluci, V. R.; Rurali, R.; Galvao, D. S. Doping of zigzag carbon nanotubes through the encapsulation of small fullerenes. *Los Alamos National Laboratory, Preprint Archive, Condensed Matter*, 2006, 1–17, arXiv:cond-mat/0607197.

31. Tuan, D. F. T.; Pitzer, R. M. Electronic structure of Hf@C$_{28}$ and its ions. 1. SCF calculations. *Journal of Physical Chemistry*, 1995, *99*, 9762–9767.

32. Tuan, D. F. T.; Pitzer, R. M. Electronic structures of C$_{28}$H$_4$ and Hf@C$_{28}$H$_4$ and their ions. SCF calculations. *Journal of Physical Chemistry*, 1996, *100*, 6277–6283.

33. Enyashin, A. N.; Ivanovskaya, V. V.; Makurin, Yu. N.; Ivanovskii, A. L. Modeling of the structure and electronic structure of condensed phases of small fullerenes C_{28} and $Zn@C_{28}$. *Physics of the Solid State (Translation of Fizika Tverdogo Tela (Sankt-Peterburg))*, 2004, *46* (8), 1569–1573.

34. Guo, T.; Diener, M. D.; Chai, Y.; Alford, M. J.; Haufler, R. E.; McClure, S. M.; Ohno, T.; Weaver, J. H.; Scuseria, G. E.; Smalley, R. E. Uranium stabilization of C_{28}: A tetravalent fullerene. *Science*, 1992, *257* (5077), 1661–1664.

35. Pederson, M. R.; Laouini, N. Covalent container compound: Empty, endohedral, and exohedral C_{28} fullerene complexes. *Physical Review B: Condensed Matter and Materials Physics*, 1993, *48* (4), 2733–2737.

36. Makurin, Yu. N.; Sofronov, A. A.; Ivanovskii, A. L. Electronic structure and conditions for chemical stabilization of fullerene C_{28}. Exohedral complexes $C_{28}M_4$ (M = H, Cl, Br). *Russian Journal of Coordination Chemistry*, 2000, *26* (7), 464–469.

37. Fowler, P. W.; Heine, T.; Troisi, A. Valencies of a small fullerene: Structures and energetics of $C_{24}H_{2m}$. *Chemical Physics Letters*, 1999, *312* (2–4), 77–84.

38. Ivanovskaya, V. V.; Enyashin, A. N.; Sofronov, A. A.; Makurin, Yu. N.; Ivanovskii, A. L. Quantum-chemical simulation of new hybrid nanostructures: Small fullerenes C_{20} and C_{28} in single-walled boron-nitrogen nanotubes. *Russian Journal of General Chemistry*, 2004, *74* (5), 713–720.

39. Jackson, K.; Kaxiras, E.; Pederson, M. R. Bonding of endohedral atoms in small carbon fullerenes. *Journal of Physical Chemistry*, 1994, *98* (32), 7805–7810.

21 Nanomesh and Nanohoneycomb Structures

Two nanostructural types—*nanomesh* and *nanohoneycomb*—are practically the same nanoforms, named differently by distinct researchers. Among other (mainly inorganic) compounds, a considerable number of reports are devoted to a BN nanomaterial, which was called *nanomesh* (we consider it related to nanogrid too) by Corso et al. from the University of Zurich, Switzerland, who published in 2003 in *Science* the discovery of this inorganic nanostructured 2D material. It consisted of a highly regular mesh of hexagonal boron nitride (BN) with a 3 nm periodicity and a 2 nm hole size and was formed by self-assembly on an Rh(111) single-crystalline surface.[1] Two layers of mesh cover the surface uniformly after high-temperature exposure of the clean rhodium surface to borazine $(HBNH)_3$. This unique structure (Figure 21.1) was found[2] to be stable under ambient atmosphere (it does not decompose up to temperatures of at least 796°C), which provides an important basis for technological applications like templating and coating. The suggested (12×12) periodicity of this reconstruction was unambiguously confirmed. The nanomesh is a coincidence structure of 13 h-BN units per 12 Rh substrate units, and no major deviations from the in-plane bulk positions occur for the bulk Rh atoms. In addition, the BN nanomesh can serve as a template to organize molecules and clusters. These characteristics promise interesting applications of the nanomesh in areas such as nanocatalysis, surface functionalization, spintronics, quantum computing, and data storage media like hard drives. Among a series of other BN nanomesh investigations, we note the study on the temperature-dependent microscopic structure and the dynamics of adsorbed Xe at different temperatures on single-sheet h-BN on a Rh(111) nanomesh.[3] It was shown that the site-specific adsorption arose from two different interactions of similar magnitude with respect to their lateral variations: the first can be attributed to a van der Waals type interaction, whereas the second originates from lateral variation of the electrostatic surface potential and is of polarization type. Both types are responsible for stabilizing dynamic and static Xe rings in these pores.

Graphene-like analogs of BN (Figure 21.2) were prepared by the Reactions 21.1 through 21.3 of boric acid with urea with a control on the number of layers.[4] Synthesis with a high proportion of urea yielded a product with a majority of one to four layers. It was found that the surface area of BN increased progressively with the decreasing number of layers, and the high surface area BN exhibited high CO_2 adsorption but negligible H_2 adsorption. First-principles simulations were used to determine structure, phonon dispersion, and elastic properties of BN with planar honeycomb lattice-based *n*-layer forms. BN was found to be softer than graphene and exhibited signatures of long-range ionic interactions in its optical phonons. Few-layer BN can be functionalized and solubilized by employing Lewis bases. Additionally, the elastic and plastic deformations of graphene, silicene, and BN honeycomb nanoribbons under uniaxial tension were found to determine their elastic constants.[5] Other aspects of h-BN are described in Refs. [6–8] among which we note a comprehensive review[9] on BN nanostructures, including honeycombs:

$$2B(OH)_3 \rightarrow B_2O_3 + 3H_2O \tag{21.1}$$

$$NH_2CONH_2 \rightarrow NH_3 + HNCO \tag{21.2}$$

$$B_2O_3 + 2NH_3 \rightarrow 2BN + 3H_2O \tag{21.3}$$

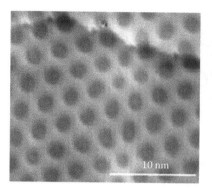
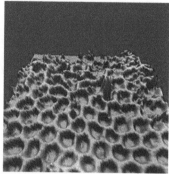

FIGURE 21.1 STM image of the h-BN on Rh(111) nanomesh taken after preparation in ultrahigh vacuum. (Reproduced from *Surface Science*, 601, Bunk, O. et al., Surface x-ray diffraction study of boron-nitride nanomesh in air, L7–L10, Copyright 2007, with permission from Elsevier.)

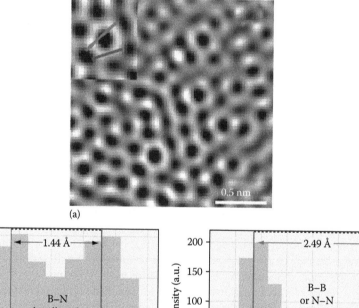

(a)

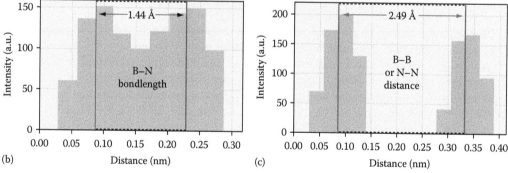

(b)

(c)

FIGURE 21.2 TEM image of few-layer BN prepared with (a) 1:12, (b) 1:24, and (c) 1:48 boric acid/urea mixture. (Reproduced with permission from Nag, A. et al., Graphene analogues of BN: Novel synthesis and properties, *ACS Nano*, 4(3), 1539–1544. Copyright 2010 American Chemical Society.)

Other *binary compounds* are known in honeycomb/mesh nanoforms as isolated examples. Thus, a study of mechanical, electronic, and magnetic properties of a 2D monolayer of SiC in honeycomb structure and its quasi-1D (quasi-1D) armchair nanoribbons using first-principles plane-wave method was presented.[10] Some of the vacancy defects, adatoms, and substitutional impurities, which were found to influence physical properties and attain magnetic moments, can be used to functionalize SiC honeycomb structures. Honeycomb-like network of vertically aligned AlN nanoplatelets was synthesized on an etched Si substrate *via* a simple vapor-phase method without a catalyst.[11] The nanoplatelets were found to be hexagonal wurtzite AlN with a thickness of 10–100 nm. This 2D AlN nanostructure is a promising field emission material. An ordered nanostructure of single-crystalline GaN nanowires in a honeycomb structure of anodic alumina was synthesized[12] through a gas reaction of Ga_2O vapor with a constant ammonia atmosphere at 1273 K in the presence of nanosized metallic indium catalysis. It was indicated that the ordered nanostructure consists of single-crystalline hexagonal wurtzite GaN nanowires in the uniform pores of anodic alumina about 20 nm in diameter and 40–50 μm in length. 2D photonic lattices, honeycomb nanostructures, were fabricated[13] by electron beam lithography with (Al,Ga)As materials. The structures were found to be stable, nonradiative surface recombination was present, and resonant coupling of light into/out of the lattice occurred at selected wavelengths satisfying a Bragg condition. In addition, using submonolayer Be deposition onto the Si(111) 7 × 7 surface under ultrahigh vacuum conditions, highly ordered honeycomb-like nanostructure arrays (Figure 21.3) were obtained.[14] It was revealed that they have composition, size, and properties similar to those theoretically predicted for the short Be-encapsulated Si nanotubes. It was indicated that the nanostructure building blocks are primarily made of silicon. Taking the $Si_{24}Be_2$ nanotube as a prototype for a "stick" element, the hypothetical structures can be suggested for "boomerang" and "propeller" elements. These are $Si_{36}Be_3$ and $Si_{48}Be_4$ clusters, respectively, shown in Figure 21.3c. In addition, the template-directed assembly of two planar molecules (copper phthalocyanine [CuPc] and pentacene) on SiC nanomesh was studied by scanning tunneling microscopy and photoelectron spectroscopy, respectively.[15] Both molecules were trapped as single molecules in the cells of SiC nanomesh at low coverage.

Elemental honeycomb nanostructures are represented by several metals (mainly nickel, due to its magnetic properties), carbon and silicon. Thus, honeycomb-like Ni composite nanostructures were prepared *via* a simple two-step solution route.[16] Homogeneous Ni nanospheres with an average diameter of ~100 nm were first obtained *via* a DMF-water mixed solvothermal route; then honeycomb-like Ni composite nanostructures were prepared through the hydrothermal carbonization of glucose solutions with suitable amounts of Ni nanospheres. It was found that the as-obtained honeycomb-like Ni composite nanostructures could be used as an electrochemical sensor for the detection of glucose molecules. Additionally, these honeycomb-like Ni composite nanostructures presented good capacities for selective adsorption of Pb^{2+}, Cd^{2+}, and Cu^{2+} ions in

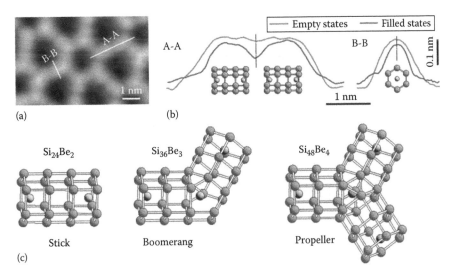

FIGURE 21.3 Atomic structure of the nanostructure building elements. (a) Fragment of the nanostructure array. (b) Line profiles along (A-A) and perpendicular (B-B) stick elements as indicated in (a). (c) Schematic diagram showing the hypothetical structure of the array building blocks: $Si_{24}Be_2$ for stick, $Si_{36}Be_3$ for boomerang, and $Si_{48}Be_4$ for propeller. (Reproduced with permission from Saranin, A.A. et al., Ordered arrays of Be-encapsulated Si nanotubes on Si (111) surface, *Nano Lett.*, 4(8), 1469–1473. Copyright 2004 American Chemical Society.)

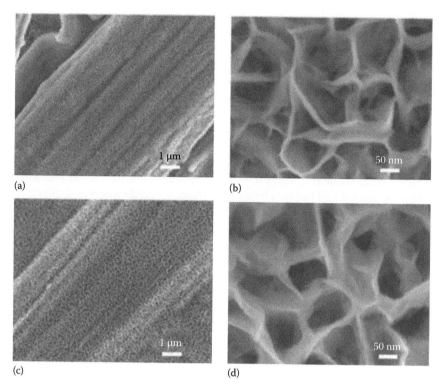

FIGURE 21.4 Morphology-controlled synthesis of Ni honeycombs on carbon paper. (a, b) SEM images of the Ni honeycombs synthesized from the reaction of 100 nm Al film on carbon paper and 0.1 M NiCl$_2$ solution. (c, d) SEM images of the Ni honeycombs synthesized from the reaction of 30 nm Al film on carbon paper and 0.1 M Ni(NO$_3$)$_2$ solution. (Reproduced with permission from Zhang, G.-X. et al., Controlled growth/patterning of Ni nanohoneycombs on various desired substrates, *Langmuir*, 26(6), 4346–4350. Copyright 2010 American Chemical Society.)

water. Alternatively, a two-step process for the growth/patterning of Ni honeycomb nanostructures on various substrates, such as carbon paper (Figure 21.4), silicon wafers, and copper grids, was offered[17] *via* the combination of a sputter-coating/patterning technique and a replacement reaction solution method. These honeycombs were found to be composed of numerous nanocells, several tens of nanometers in diameter, and with cell wall thickness of ~10 nm, randomly connecting to each other. Morphology evolution of studied Ni honeycombs is presented in Figure 21.5.

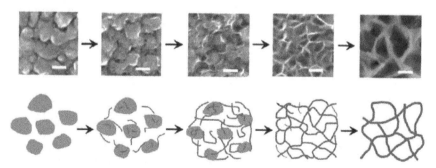

FIGURE 21.5 Schematic illustration of the formation and morphology evolution of Ni honeycombs. All the scale bars are 50 nm. (Reproduced with permission from Zhang, G.-X. et al., Controlled growth/patterning of Ni nanohoneycombs on various desired substrates, *Langmuir*, 26(6), 4346–4350. Copyright 2010 American Chemical Society.)

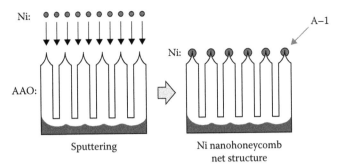

Ni:

AAO:

Ni:

A–1

Sputtering

Ni nanohoneycomb
net structure

FIGURE 21.6 Schematic diagrams for the fabrication of Ni nanohoneycomb net structure. (Reproduced from Kim, M.J. et al., Magnetic properties of Ni nanostructures fabricated using anodic aluminum oxide templates, *J. Korean Phys. Soc.*, 47(2), 313–317, 2005.)

It was demonstrated that the applied strategy is remarkably straightforward and versatile to generate metal nanostructures, with high surface areas and structural uniformity, on various substrates. The metal-substrate composites, especially with desired patterns, are expected to be ideal candidates to be used in modern electronic and optoelectronic devices, sensors, fuel cells, and energy storage systems. Also, Ni honeycomb nanostructures (Figure 21.6) can also be obtained by using anodic aluminum oxide and an Ni metal target.[18] In this case, it was confirmed that Ni formed in a polycrystalline state.

Other elements (except carbon) are poorly represented in a mesh/honeycomb-like form. Thus, metal-containing mesh-like nanostructures, represented by a semitransparent nanomesh Cu electrode on a polyethylene terephthalate (PET) substrate, were obtained using metal transfer from a polydimethylsiloxane (PDMS) stamp and nanoimprint lithography.[19] It was found that a uniform pressure of 30 psi and a temperature of 100°C were needed for the transfer of the Cu mesh structure from the PDMS stamp onto the PET substrate. A fabricated semitransparent Cu electrode exhibited high transmittance in the visible range and good electrical conductivity. In addition, a regular square mesh structure with 4.5 nm periodicity was formed after annealing an adsorption layer of Ti on W(100) at 1370 K.[20] This nanomesh consisted of two self-assembled layers on top of the Ti adlayer, ~3.5 physical monolayers thick on average. A rapid, low-cost, and sensitive monitoring of environmental contaminants (heavy metallic ions) at trace levels (<1 ng/mL) was developed[21] with the use of (SERS)-based optical sensors on the basis of a nanostructured gold honeycomb as SERS substrate and a thiol-ligand to functionalize the SERS substrate and capture Pb^{2+} and Hg^{2+} ions. Regular arrays of Si nanorods with a circular cross section in hexagonal-closed-packed and triangular cross section in honeycomb-like arrangements were grown[22] using glancing angle deposition on Si(100) and fused SiO_2 substrates that were patterned with Au dots using self-assembled mono- and double layers of polystyrene nanospheres as an evaporation mask. Different seed heights and interseed distances were found to be the main reasons for the strong distinctions between the grown nanorod arrays. Additionally, a first-principles study of bare and H-passivated armchair nanoribbons of the puckered single layer honeycomb structures of Si and Ge was reported.[23] The bandgaps of Si and Ge nanoribbons exhibited family behavior similar to those of graphene nanoribbons.

Several carbon nanostructural types (*fullerene, graphene, carbon films, nanotubes, and nanodiamonds*) are well studied in mesh/honeycomb-like forms, as in a "free" state (for instance, 2D honeycomb network, constructed from pure multiwalled (CNT [24]) as composites possessing important applications. Thus, an energy storage electrode system was fabricated *via* a template method with 1D nanostructures that were hexagonally patterned in a honeycomb-like fashion and vertically standing nanorods made of a gold-coated CNT core and a V_2O_5 shell layer.[25] The performance of this system for Li insertion and extension showed an increased capacity along with an enhanced rate of performance, which could be attributed to the aligned nanostructures having increased reaction sites, facilitated charge transport, and improved stability in the face of mechanical stress. As another example, the preparation of honeycomb structure in CNT–polymer composite films was described.[26] The CNT brush morphology can be converted into a honeycomb nanostructure on silicon substrates by evaporating a dilute polymer solution in a volatile organic solvent on the brush surface. This composite CNT honeycomb morphology showed good electrical and mechanical properties and may have potential as a robust field emitter or other thin-film, functional polymer composites. Additionally, nanostructured electrode material, which is a honeycomb carbon anode consisting of a thin carbon film containing an ordered array of monodispersed nanoscale pores (Figure 21.7), was prepared[27] using an O_2 plasma etch method in conjunction with a nanopore alumina mask. This honeycomb carbon anode showed a low-rate discharge capacity of 325 mA/hg, close to that of graphite. At high discharge rates, the honeycomb anode delivered 50 times the capacity of a thin-film control anode that did not contain the nanopore honeycomb. Improved rate capabilities were obtained because penetration of solvent and Li^+ electrolyte into the pore structure of the honeycomb anode ensured that the distance Li^+ must diffuse in the solid state is smaller than in the thin-film control electrode.

An interesting combination of two distinct carbon nanostructural types is as follows. Carbon nanotube/nanohoneycomb diamond (CNT-NANO) composite electrodes were fabricated by introducing multiwalled CNTs into the pores of nanohoneycomb diamond (400 nm diameter) using CVD technology.[28] The electrochemical behavior of these electrodes was examined by a series of techniques, showing that, at the nanohoneycomb diamond densely deposited CNTs (HD CNT-NANO), only the Li^+ intercalation process was observed. In contrast, the nanohoneycomb diamond modified

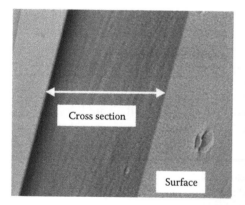
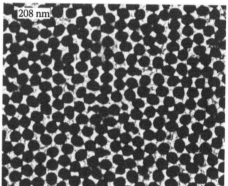

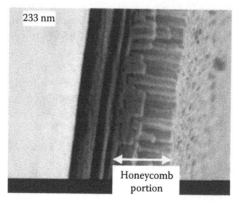

FIGURE 21.7 Electron micrographs of (top) the as-synthesized CVD carbon film, the surface (center), and cross section (bottom) of the honeycomb film. (Reproduced from Li, N. et al., Mitchell, D. T.; Lee, K.-P.; Martin, C. R. A nanostructured honeycomb carbon anode, *J. Electrochem. Soc.*, 150(7), A979–A984, 2003. With permission from Electrochemical Society.)

with CNTs in low-density (LD CNT-NANO) exhibited the combined behavior of Li$^+$ intercalation at CNTs and the electrochemical double-layer discharging on the diamond surface. Nanostructured diamond honeycomb electrodes and nanocylinder ensembles were fabricated[29] using anodic porous alumina as a mask and template, respectively. These nanostructures are expected to offer further improvements in the electrode performance of diamond for electroanalytical applications. As an example, given by the authors, nanoporous diamond electrodes modified with Pt nanoclusters in the pores were shown to have a high selectivity for small alcohol molecules. In a related article,[30] nanostructured diamond honeycomb films were fabricated using a porous Al$_2$O$_3$ membrane with an ordered hexagonal array of holes as a mask, which was laid on top of the synthetic diamond film and which was treated in an O$_2$ plasma. Possible applications of these films could be as capacitors or photonic materials.

A few examples of other carbon mesh-like nanostructures are known, in particular those of C$_{60}$ and graphene. Thus, a new graphene nanostructure, called by the authors[31] as a graphene nanomesh, can open up a bandgap in a large sheet of graphene to create a semiconducting thin film. These nanomeshes were prepared using block copolymer lithography and can have variable periodicities and neck widths as low as 5 nm. Graphene nanomesh field-effect transistors can support currents nearly 100 times greater than individual graphene nanoribbon devices, and the on–off ratio, which is comparable with the values achieved in individual nanoribbon devices, can be tuned by varying the neck width. Among other related publications, hexagonal graphene nanomeshes with sub-10 nm ribbon width were fabricated.[32] Also, local photodegradation of graphene oxide sheets at the tip of ZnO nanorods was used to achieve semiconducting graphene nanomeshes (Figure 21.8).[33] These graphene nanomeshes contained smaller O-containing carbonaceous bonds and higher defects as compared with the as-prepared graphene oxide sheets. An extended 2D C$_{60}$ nanomesh was fabricated by controlling the binary molecular phases of C$_{60}$ and pentacene on Ag(111).[34] The skeleton of the C$_{60}$ nanomesh was found to be stabilized by the strong molecule–metal interfacial interactions [C$_{60}$-Ag(111) and pentacene–Ag(111)] and was further modified by the pentacene–C$_{60}$ donor–acceptor intermolecular interactions. This C$_{60}$ nanomesh can serve as an effective template to selectively accommodate guest C$_{60}$ molecules at the nanocavities.

A series of *metal oxides*[35] *and hydroxides* have been reported in a mesh/honeycomb nanoform, in particular the most important zinc oxide.[36] Thus, a first-principles study of the atomic, electronic, and magnetic properties of 2D, single, and bilayer ZnO in honeycomb structure and its armchair and zigzag nanoribbons was carried out.[37] It was revealed that 2D ZnO in honeycomb structure and its armchair nanoribbons are nonmagnetic semiconductors but acquire net magnetic moment upon the creation of Zn–vacancy defect. Under tensile stress, the nanoribbons were found to be deformed elastically maintaining honeycomb-like structure but yield at high strains. Beyond yielding point, honeycomb-like structure undergoes a structural change and deforms plastically by forming large polygons.

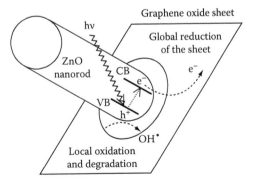

FIGURE 21.8 Schematic illustration of the mechanism describing formation of graphene nanomeshes by using the photocatalytic property of the ZnO nanorods. (Reproduced with permission from Akhavan, O., Graphene nanomesh by ZnO nanorod photocatalysts, *ACS Nano*, 4(7), 4174–4180. Copyright 2010 American Chemical Society.)

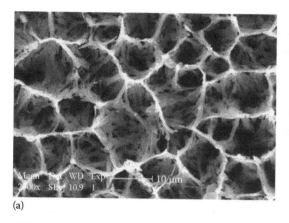
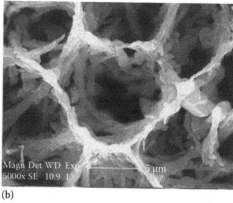

FIGURE 21.9 (a) Overview and (b) high-magnification SEM images of CuO honeycomb-like structure constructed by subplates. (Reproduced with permission from Liu, Y. et al., Anion-controlled construction of CuO honeycombs and flowerlike assemblies on copper foils, *Cryst. Growth Des.*, 7(3), 467–470. Copyright 2007 American Chemical Society.)

In addition, ZnO epitaxial layers were grown on GaN underlying films by metal-organic CVD at various temperatures.[38] An increase in growth temperature led to morphological changes from a smooth film with hexagonal-shaped surface pits to honeycomb-like nanostructures with deep hollow and additionally resulted in a decrease in dislocation density in the interfacial layers. In addition, synthesis and applications of nanostructure ZnO such as honeycomb-shaped 3D nanowalls on a ZnO/SiO$_2$/Si substrate were carried out by a chemical reaction/vapor transportation deposition technique.[39] Finally, on ZnO basis, AlGaInP-based light-emitting diodes with nanomesh ZnO layers were fabricated by using nanosphere lithography.[40]

Among other metal oxides, the honeycomb CuO nanostructure consisting of nanowires (length of 10–20 μm; diameter of 100–300 nm) was fabricated by electrodeposition of Cu(OH)$_2$ on a copper sheet and further heating.[41] Alternatively, honeycomb- and flower-like CuO nanoarchitectures (Figure 21.9) assembled from nanowires and nanoribbons were prepared (Reactions 21.4 through 21.6) on copper foils by an anion-controlled mild hydrothermal method.[42] It was established that the concentration of inorganic anions, the concentration of surfactant, and the kind of inorganic anions have direct influences on the morphology of the products. It is expected that these CuO architectures may offer exciting opportunities for potential applications in catalysis, electrochemistry, superconductivity, and superhydrophobic coating.

$$Cu + 2NaOH + (NH_4)_2S_2O_8 \rightarrow Cu(OH)_2 + Na_2SO_4 + (NH_4)_2SO_4 \tag{21.4}$$

$$Cu(OH)_2 + 2OH^- \rightarrow [Cu(OH)_4]^{2-} \tag{21.5}$$

$$[Cu(OH)_4]^{2-} \rightarrow CuO + H_2O + 2OH^- \tag{21.6}$$

In addition to nickel honeycomb-like nanostructures, mentioned earlier, their oxygen-containing compounds are also known in the same form. Thus, a simple approach was proposed to prepare a 2D NiO nanohoneycomb by thermal annealing of Ni thin film deposited onto an Si substrate by thermal evaporation.[43] Since the NiO nanohoneycomb was realized onto an Si substrate, a basic material for microelectronics and microsystem, this will probably open the door to integrate the nanohoneycomb into a microsystem, thus leading to nano-based functional devices. Nanostructured honeycomb-like nickel hydroxide thin films were synthesized *via* a simple chemical bath deposition method using Ni(NO$_3$)$_2$ as the starting material.[44] It was shown that β-Ni(OH)$_2$ has a wide optical bandgap of 3.95 eV. Several reports are devoted to alumina; for instance, the mechanical properties, including Young's modulus, the effective bending modulus, and the nominal fracture strength, of nanohoneycomb structures (anodic alumina) were measured using an AFM and a nanouniversal testing machine.[45] In a related report,[46] such anodic aluminum oxide with a nanohoneycomb structure was found to be subjected to nanoindentation along the axial direction of the honeycomb. Top-view and cross-sectional microscopic study revealed a localized mode of deformation with clear-cut elastoplastic boundaries. A crack system that is self-similar with respect to the indent size was also found, and this is thought to correspond to the discontinuous load-displacement behavior. Also, the effect of *iso*octylphenol poly(ethylene glycol) ether ($C_8H_{17}C_6H_4O[C_2H_4O]_{10}H$) on the lamellar structure (with a layer thickness of 60–80 nm) of aluminum oxyhydrate was studied.[47] Each layer of the lamellar phase consisted of hexagonal prisms with an edge length of 100–120 nm, which were assembled into a honeycomb- or a graphite-like structure. In addition, by using a μm polystyrene ball array as a template, an Al oxide honeycomb structure, consisting of the networking hexagonal Al oxide nanowall, was produced on the *n*-GaN surface (for comparison, see a related example ZnO/GaN discussed earlier) of a thin-GaN light-emitting diode (LED).[48] With the Al oxide honeycomb nanostructure, the total lighting output of thin-GaN LED was enhanced by 35%. The authors proposed that the networking nanowall of the Al oxide honeycomb structure acted as a waveguide to extend the light emitted to the outer medium effectively.

α-Fe_2O_3 hollow spheres with uniformly distributed mesoporosity on the shell were fabricated on a large scale by a smart complex precursor method, in which the composite mesoporous hollow structures were generated by utilizing different removing modes of oxalate ligands in ferric potassium oxalate (Figure 21.10).[49] The shell of the hollow spheres exhibited honeycomb-like mesoporous nanostructures (Figure 21.11) composed of single-crystal iron oxide nanoparticles, and the as-obtained α-Fe_2O_3 composite hollow structures exhibited high gas sensitivity toward formaldehyde and ethanol. Time- and frequency-resolved photoelectrochemical studies were performed[50] on a nanohoneycomb porous TiO_2 electrode prepared by photoelectrochemical etching of single crystal TiO_2. Much faster electron diffusion in the honeycomb electrode than that in colloid-based porous electrode was revealed because of the smaller electron trap density in the nanohoneycomb structure owing to the nonexistence of grain boundaries. In addition, the formation of bismuth nanodots on a well-ordered (4 × 4) vanadium oxide nanomesh on Pd(111) was investigated.[51] At very low Bi coverage, a Bi atom located in the vanadium oxide nanomesh and every Bi atom formed an isolated Bi nanodot. When the Bi coverage increased, most of the nanoholes in the nanomesh were occupied by Bi nanodots, and the remaining Bi atoms formed Bi clusters on the vanadium oxide surface.

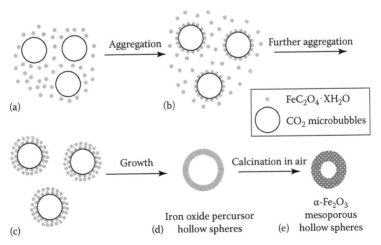

FIGURE 21.10 Schematic illustration of the evolution of α-Fe_2O_3 mesoporous hollow spheres. (Reproduced with permission from Wu, Z. et al., Hematite hollow spheres with a mesoporous shell: Controlled synthesis and applications in gas sensor and lithium ion batteries, *J. Phys. Chem. C*, 112(30), 11307–11313. Copyright 2008 American Chemical Society.)

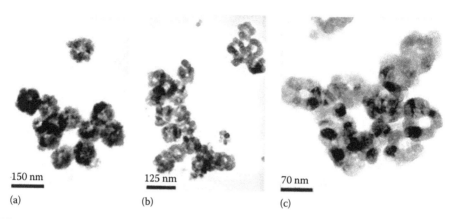

FIGURE 21.11 (a) TEM image of iron oxide precursor hollow spheres prepared with 0.2 mmol ferric potassium oxalate. (b, c) TEM images of the corresponding α-Fe$_2$O$_3$ mesoporous hollow spheres with different magnifications after calcination. (Reproduced with permission from Wu, Z. et al., Hematite hollow spheres with a mesoporous shell: Controlled synthesis and applications in gas sensor and lithium ion batteries, *J. Phys. Chem. C*, 112(30), 11307–11313. Copyright 2008 American Chemical Society.)

Some *O-containing metal salts* are known as honeycomb nanostructures. Thus, a simple emulsion-phase route was applied for the preparation of honeycomb-like basic magnesium carbonate (BMC, Mg$_5$(OH)$_2$(CO$_3$)$_4$·4H$_2$O) microspheres from magnesium salt and Na$_2$CO$_3$ at 80°C.[52] The obtained BMC samples were found to be composed of many microspheres (diameter 8–10 μm), which are interweaved by nanosized thin sheets (thickness of 20–30 nm and length >1 μm). Ceramic honeycombs (having about 50% porosity and about 10 μm mean pore size) were fabricated from clays at 1500°C *via* reactive sintering to form cordierite (2MgO$_2$·Al$_2$O$_3$·5SiO$_2$).[53] The honeycombs thus obtained have about 50% porosity and about 10 μm mean pore size. The next example is nanostructured CaCO$_3$ thin films with honeycomb-shaped nanopores, obtained[54] at the surface of urease-embedded multilayers prepared by the layer-by-layer deposition (Figure 21.12). Amorphous CaCO$_3$ (ACC) droplets were initially nucleated from the multilayer surface. Once ACC droplets successfully covered the entire surface of the urease multilayers, the ACC layers underwent the transformation

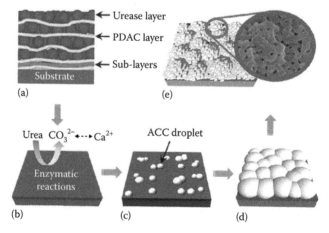

FIGURE 21.12 Schematic illustration for the preparation of a nanostructured CaCO$_3$ thin film on a urease-containing multilayer. (a) Preparation of Ur multilayers, (b) enzymatic decomposition of urea to carbonate ion, (c) nucleation and growth of ACC droplets, (d) densification and merge into ACC thin film, and (e) transformation into NCC thin film. (Reproduced with permission from Yeom, B.-J. and Char, K.-H., Nanostructured CaCO$_3$ thin films formed on the urease multilayers prepared by the layer-by-layer deposition, *Chem. Mater.*, 22(1), 101–107. Copyright 2010 American Chemical Society.)

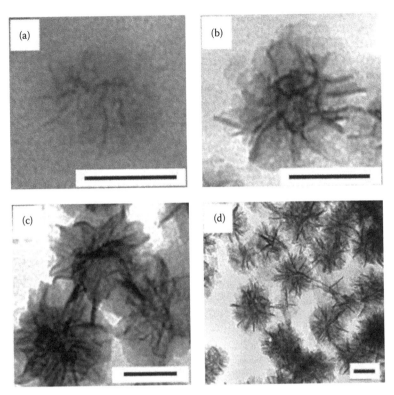

FIGURE 21.13 TEM images of honeycomb K$_x$MnO$_2$ nanospheres obtained after redox times of 0.5 h (a), 2 h (b), 5 h (c), and 20 h (d), respectively. Scale bar, 50 nm. (Reproduced with permission from Chen, H. et al., Self-assembly of novel mesoporous manganese oxide nanostructures and their application in oxidative decomposition of formaldehyde, *J. Phys. Chem. C*, 111(49), 18033–18038. Copyright 2007 American Chemical Society.)

into the crystalline {104}-faceted nanostructured calcite thin films with square pores, ~30 nm on each side, in a humidity-controlled chamber. The authors believe that their synthetic scheme to obtain biomineral thin films, combined with flexible processing methods (i.e., layer-by-layer deposition), could cast fresh insight into the manufacturing of biomineral thin films on synthetic substrates, in terms of nanostructure control and large-area fabrications for further applications.

Monodisperse K$_x$MnO$_2$ honeycomb and hollow nanospheres (Figure 21.13), stable even after ultrasonic treatment (40 kHz, 120 W) for 30 min, were prepared facilely at room temperature by varying the molar ratio of KMnO$_4$ and oleic acid (Figure 21.14).[55] The method involved a redox reaction of KMnO$_4$ and oleic acid at the O/W interface, followed by self-assembly of the formed K$_x$MnO$_2$ nanoplatelets into K$_x$MnO$_2$ nanostructures. Both the manganese oxide nanomaterials showed high catalytic activities for oxidative decomposition of formaldehyde at low temperatures. The catalytic activities of the obtained manganese oxide nanospheres were also significantly higher than those of previously reported manganese oxide octahedral molecular sieve nanorods, MnO$_x$ powders, and alumina-supported manganese-palladium oxide catalysts. Additionally, the fabricated materials could also be believed to have applications as adsorbents and separation materials, and in electrodes, electrolytes, and electromagnetic and electronic devices.

A few *organic and coordination compounds* have been also reported. Thus, nano-ordered structures, that is, 1D molecular rows and 2D honeycombs consisting of mercaptomethylthiophene derivatives were observed[56] on Au(111) by scanning tunneling microscopy at room temperature. When the samples were immersed in EtOH solution of transition metal chlorides, the nanostructures remained intact although the deposition of nanoclusters was observed in some cases. A systematic study of metal-organic honeycomb lattices (Figure 21.15) assembled from simple ditopic molecule

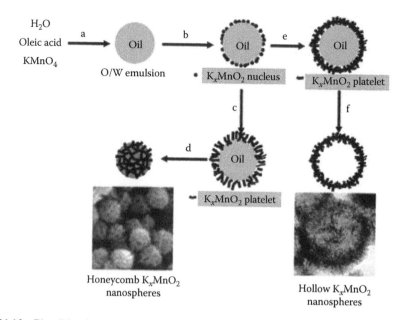

FIGURE 21.14 Plausible formation mechanism of honeycomb and hollow K_xMnO_2 nanospheres. (Reproduced with permission from Chen, H. et al., Self-assembly of novel mesoporous manganese oxide nanostructures and their application in oxidative decomposition of formaldehyde, *J. Phys. Chem. C*, 111(49), 18033–18038. Copyright 2007 American Chemical Society.)

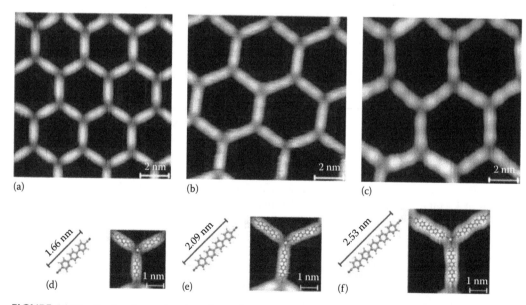

FIGURE 21.15 Tuning the cavity size of metal-organic honeycomb networks with designed linkers. The STM images show the result of Co-directed assembly of (a) NC-Ph$_3$-CN, (b) NC-Ph$_4$-CN, and (c) NC-Ph$_5$-CN, respectively. (d through f) Structure of the molecules including their length and models of the threefold Co-carbonitrile coordination motif resolved in (a through c) (yellow, cobalt center; turquoise, carbon; white, hydrogen; blue, nitrogen). The images (a through c) were taken at a tunnel current of $I = 0.1$ nA and bias voltages of 0.9, 1, and 2 V, respectively. (Reproduced with permission from Schlickum, U. et al., Metal-organic honeycomb nanomeshes with tunable cavity size, *Nano Lett.*, 7(12), 3813–3817. Copyright 2007 American Chemical Society.)

bricks (abbreviated NC-Ph$_n$-CN, whereby n can be 3, 4, or 5) and Co atoms on Ag(111) was conducted.[57] This approach enables one to fabricate size- and shape-controlled open nanomeshes with pore dimensions up to 5.7 nm. The networks were found to be shape resistant in the presence of further deposited materials and represent templates to organize guest species and realize molecular rotary systems. These nanomeshes may serve as templates for the selective organization of guest species to realize patterned media of separated, regularly distributed nanoclusters. Additionally, a highly lipophilic polyion complex [Pt(en)$_2$][PtCl$_2$(en)$_2$](1)$_4$ (1, R = CH$_3$(CH$_2$)$_7$CH:CH(CH$_2$)$_8$; en = 1,2-diaminoethane) with observed reversible thermochromism in CH$_3$OH solution was prepared[58] from 1D mixed valence Pt^{II}/Pt^{IV} complex and chiral amphiphile 1. Casting of the 0°C-purple dispersion on solid substrates affords honeycomb nanostructures in addition to the nanowires with the width of ~20 nm. More regular honeycomb structures were exclusively obtained by casting the 21°C-colorless solution. The authors indicated that the ordered honeycomb structures can be obtained on solid surfaces by the self-assembly of molecularly dispersed components [Pt(en)$_2$](I)$_2$ and $trans$-[PtCl$_2$(en)$_2$](I)$_2$. Such metal-organic honeycombs can be used as templates in the creation of other nanostructures, for instance those of regularly sized Fe and Co nanostructures, created using 2D metal-organic honeycomb lattices.[59] They consisted of dicarbonitrile–polyphenyl molecules coordinated to Co centers on Ag(111). Subsequently deposited Fe or Co atoms nucleated clusters at specific sites on top of the metal-organic lattices while leaving their hexagonal pores empty. The distance between the Fe and Co clusters and their shape can be adjusted through the lattice constant of the template by varying the number of Ph rings in the molecules.

1

REFERENCES

1. Corso, M.; Auwarter, W.; Muntwiler, M.; Tamai, A.; Greber, T.; Osterwalder, J. Boron nitride nanomesh. *Science*, 2004, *303*, 217–220.
2. Bunk, O.; Corso, M.; Martoccia, D.; Herger, R.; Willmott, P. R.; Patterson, B. D.; Osterwalder, J.; van der Veen, J. F.; Greber, T. Surface x-ray diffraction study of boron-nitride nanomesh in air. *Surface Science*, 2007, *601*, L7–L10.
3. Widmer, R.; Passerone, D.; Mattle, T.; Sachdev, H.; Groning, O. Probing the selectivity of a nanostructured surface by xenon adsorption. *Nanoscale*, 2010, *2* (4), 502–508.
4. Nag, A.; Raidongia, K.; Hembram, K. P. S. S.; Datta, R.; Waghmare, U. V.; Rao, C. N. R. Graphene analogues of BN: Novel synthesis and properties. *ACS Nano*, 2010, *4* (3), 1539–1544.
5. Topsakal, M.; Ciraci, S. Elastic and plastic deformation of graphene, silicene, and boron nitride honeycomb nanoribbons under uniaxial tension: A first-principles density-functional theory study. *Physical Review B: Condensed Matter and Materials Physics*, 2010, *81* (2), 024107/1–024107/6.
6. Laskowski, R.; Blaha, P. Ab initio study of h-BN nanomeshes on Ru(001), Rh(111), and Pt(111). *Physical Review B: Condensed Matter and Materials Physics*, 2010, *81* (7), 075418/1–075418/6.
7. Martoccia, D.; Brugger, T.; Bjorck, M.; Schlepuetz, C. M.; Pauli, S. A.; Greber, T.; Patterson, B. D.; Willmott, P. R. h-BN/Ru(0001) nanomesh: A 14-on-13 superstructure with 3.5 nm periodicity. *Surface Science*, 2010, *604* (5–6), L16–L19.
8. Ma, H.; Brugger, T.; Berner, S.; Ding, Y.; Iannuzzi, M.; Hutter, J.; Osterwalder, J.; Greber, T. Nano-ice on boron nitride nanomesh: Accessing proton disorder. *ChemPhysChem*, 2010, *11* (2), 399–403.
9. Boustani, I.; Quandt, A.; Hernández, E.; Rubio, A. New boron based nanostructured materials. *Journal of Chemical Physics*, 1999, *110* (6), 3176–3185.

10. Bekaroglu, E.; Topsakal, M.; Cahangirov, S.; Ciraci, S. First-principles study of defects and adatoms in silicon carbide honeycomb structures. *Physical Review B: Condensed Matter and Materials Physics*, 2010, *81* (7), 075433/1–075433/9.

11. Tang, Y. B.; Cong, H. T.; Cheng, H.-M. Field emission from honeycomblike network of vertically aligned AlN nanoplatelets. *Applied Physics Letters*, 2006, *89* (9), 093113/1–093113/3.

12. Cheng, G. S.; Zhang, L. D.; Chen, S. H.; Li, Y.; Li, L.; Zhu, X. G.; Zhu, Y.; Fei, G. T.; Mao, Y. Q. Ordered nanostructure of single-crystalline GaN nanowires in a honeycomb structure of anodic alumina. *Journal of Materials Research*, 2000, *15* (2), 347–350.

13. Gourley, P. L.; Wendt, J. R.; Vawter, G. A.; Brennan, T. M.; Hammons, B. E. Optical properties of two-dimensional photonic lattices fabricated as honeycomb nanostructures in compound semiconductors. *Applied Physics Letters*, 1994, *64* (6), 687–689.

14. Saranin, A. A.; Zotov, A. V.; Kotlyar, V. G.; Kasyanova, T. V.; Utas, O. A.; Okado, H.; Katayama, M.; Oura, K. Ordered arrays of Be-encapsulated Si nanotubes on Si (111) surface. *Nano Letters*, 2004, *4* (8), 1469–1473.

15. Chen, S.; Chen, W.; Huang, H.; Gao, X.; Qi, D.; Wang, Y.; Wee, A. T. S. Template-directed molecular assembly on silicon carbide nanomesh: Comparison between CuPc and pentacene. *ACS Nano*, 2010, *4* (2), 849–854.

16. Ni, Y.; Jin, L.; Zhang, L.; Hong, J. Honeycomb-like Ni composite nanostructures: Synthesis, properties and applications in the detection of glucose and the removal of heavy-metal ions. *Journal of Materials Chemistry*, 2010, *20* (31), 6430–6436.

17. Zhang, G.-X.; Sun, S.-H.; Ionescu, M. I.; Liu, H.; Zhong, Y.; Li, R.-Y.; Sun, X.-L. Controlled growth/patterning of Ni nanohoneycombs on various desired substrates. *Langmuir*, 2010, *26* (6), 4346–4350.

18. Kim, M. J.; Kim, Y. W.; Lee, J. S.; Yoo, J.-B.; Park, C. Y. Magnetic properties of Ni nanostructures fabricated using anodic aluminum oxide templates. *Journal of the Korean Physical Society*, 2005, *47* (2), 313–317.

19. Kang, M.-G.; Guo, L. Jay. Semitransparent Cu electrode on a flexible substrate and its application in organic light emitting diodes. *Journal of Vacuum Science and Technology, B: Microelectronics and Nanometer Structures—Processing, Measurement, and Phenomena*, 2007, *25* (6), 2637–2641.

20. Trembulowicz, A.; Ciszewski, A. A square titanium nanomesh on W (100). *Nanotechnology*, 2007, *18* (34), 345303/1–345303/4.

21. Leng, W.; Vikesland, P. J. Gold honeycomb based SERS sensor for ion detection in drinking water. *Abstracts of Papers, 239th ACS National Meeting*, San Francisco, CA, March 21–25, 2010, ANYL-130.

22. Patzig, C.; Rauschenbach, B.; Fuhrmann, B.; Leipner, H. S. Growth of Si nanorods in honeycomb and hexagonal-closed-packed arrays using glancing angle deposition. *Journal of Applied Physics*, 2008, *103* (2), 024313/1–024313/6.

23. Cahangirov, S.; Topsakal, M.; Ciraci, S. Armchair nanoribbons of silicon and germanium honeycomb structures. *Physical Review B: Condensed Matter and Materials Physics*, 2010, *81* (19), 195120/1–195120/6.

24. Su, J.-W.; Fu, S.-J.; Gwo, S.; Lin, K.-J. Fabrication of porous carbon nanotube network. *Chemical Communications*, 2008, (43), 5631–5632.

25. Kim, Y.-S.; Ahn, H.-J.; Nam, S. H.; Lee, S. H.; Shim, H.-S.; Kim, W. B. Honeycomb pattern array of vertically standing core-shell nanorods: Its application to Li energy electrodes. *Applied Physics Letters*, 2008, *93* (10), 103104/1–103104/3.

26. Cheng, B.; Cui, H.; Stoner, B. R.; Samulski, E. T. Solvent-induced morphology in nano-structures. In *Nanotechnology and Nano-Interface Controlled Electronic Devices* (International Workshop on Nanotechnology and NICE Devices, Nagoya, Japan, March 19–20, 2002). Iwamoto, M.; Kaneto, K.; Mashiko, S. (Eds.). 2003, pp. 399–410.

27. Li, N.; Mitchell, D. T.; Lee, K.-P.; Martin, C. R. A nanostructured honeycomb carbon anode. *Journal of the Electrochemical Society*, 2003, *150* (7), A979–A984.

28. Honda, K.; Yoshimura, M.; Kawakita, K.; Fujishima, A.; Sakamoto, Y.; Yasui, K.; Nishio, N.; Masuda, H. Electrochemical characterization of carbon nanotube/nanohoneycomb diamond composite electrodes for a hybrid anode of Li-ion battery and super capacitor. *Journal of the Electrochemical Society*, 2004, *151* (4), A532–A541.

29. Fujishima, A.; Terashima, C.; Honda, K.; Sarada, B. V.; Rao, T. N. Recent progress in electroanalytical applications of diamond electrodes. *New Diamond and Frontier Carbon Technology*, 2002, *12* (2), 73–81.

30. Masuda, H.; Watanabe, M.; Yasui, K.; Tryk, D.; Rao, T.; Fujishima, A. Fabrication of a nanostructured diamond honeycomb film. *Advanced Materials*, 2000, *12* (6), 444–447.

31. Bai, J.; Zhong, X.; Jiang, S.; Huang, Y.; Duan, X. Graphene nanomesh. *Nature Nanotechnology*, 2010, *5* (3), 190–194.

32. Liang, X.; Jung, Y.-S.; Wu, S.; Ismach, A.; Olynick, D. L.; Cabrini, S.; Bokor, J. Formation of bandgap and subbands in graphene nanomeshes with sub-10 nm ribbon width fabricated via nanoimprint lithography. *Nano Letters*, 2010, *10* (7), 2454–2460.

33. Akhavan, O. Graphene nanomesh by ZnO nanorod photocatalysts. *ACS Nano*, 2010, *4* (7), 4174–4180.

34. Zhang, H. L.; Chen, W.; Huang, H.; Chen, L.; Wee, A. T. S. Preferential trapping of C_{60} in nanomesh voids. *Journal of the American Chemical Society*, 2008, *130*, 2720–2721.

35. Ke, X. B.; Zheng, Z. F.; Liu, H. W.; Zhu, H. Y.; Gao, X. P.; Zhang, L. X.; Xu, N. P. et al. High-flux ceramic membranes with a nanomesh of metal oxide nanofibers. *Journal of Physical Chemistry B*, 2008, *112* (16), 5000–5006.

36. Fu, M.; Zhou, J.; Yu, J. Flexible photonic crystal fabricated by two-dimensional free-standing ZnO nanomesh arrays. *Journal of Physical Chemistry C*, 2010, *114* (20), 9216–9220.

37. Topsakal, M.; Cahangirov, S.; Bekaroglu, E.; Ciraci, S. First-principles study of zinc oxide honeycomb structures. *Physical Review B: Condensed Matter and Materials Physics*, 2009, *80* (23), 235119/1–235119/14.

38. Kong, B. H.; Kim, D. C.; Mohanta, S. K.; Han, W. S.; Cho, H. K.; Hong, C.-H.; Kim, H. G. Controlled growth of heteroepitaxial zinc oxide nanostructures on gallium nitride. *Journal of Nanoscience and Nanotechnology*, 2009, *9* (7), 4383–4387.

39. Miao, L.; Tanemura, S.; Yang, H. Y.; Lau, S. P. Synthesis and random laser application of ZnO nanowalls: A review. *International Journal of Nanotechnology*, 2009, *6* (7/8), 723–734.

40. Chen, J.-J.; Su, Y.-K.; Lin, C.-L.; Kao, C.-C. Light output improvement of AlGaInP-based LEDs with nano-mesh ZnO layers by nanosphere lithography. *IEEE Photonics Technology Letters*, 2010, *22* (6), 383–385.

41. Xu, J.; Yu, K.; Shang, D.; Wu, J.; Xu, Y.; Li, L.; Zhu, Z. Honeycomb CuO nanostructure and its preparation method. Patent CN101486485, 2009, 10 pp.

42. Liu, Y.; Chu, Y.; Zhuo, Y.; Li, M.; Li, L.; Dong, L. Anion-controlled construction of CuO honeycombs and flowerlike assemblies on Copper foils. *Crystal Growth and Design*, 2007, *7* (3), 467–470.

43. Zhang, K.; Rossi, C.; Alphonse, P.; Tenailleau, C. NiO nanostructured honeycomb realized by annealing Ni film deposited on silicon. *Journal of Nanoscience and Nanotechnology*, 2008, *8* (11), 5903–5907.

44. Patil, U. M.; Gurav, K. V.; Fulari, V. J.; Lokhande, C. D.; Joo, O. S. Characterization of honeycomb-like "β-Ni(OH)₂" thin films synthesized by chemical bath deposition method and their supercapacitor application. *Journal of Power Sources*, 2009, *188* (1), 338–342.

45. Jeon, J. H.; Choi, D. H.; Lee, P. S.; Lee, K. H.; Park, H. C.; Hwang, W. Measuring the tensile and bending properties of nanohoneycomb structures. *Mechanics of Composite Materials*, 2006, *42* (2), 173–186.

46. Ng, K. Y.; Lin, Y.; Ngan, A. H. W. Deformation of anodic aluminum oxide nano-honeycombs during nanoindentation. *Acta Materialia*, 2009, *57* (9), 2710–2720.

47. Zherebtsov, D. A. Template synthesis of alumina gel with a nanohoneycomb structure. *Colloid Journal*, 2009, *71* (3), 430–432.

48. Lin, C. L.; Chen, P. H.; Chan, C.-H.; Lee, C. C.; Chen, C.-C.; Chang, J.-Y.; Liu, C. Y. Light enhancement by the formation of an Al oxide honeycomb nanostructure on the *n*-GaN surface of thin-GaN light-emitting diodes. *Applied Physics Letters*, 2007, *90* (24), 242106/1–242106/3.

49. Wu, Z.; Yu, K.; Zhang, S.; Xie, Y. Hematite hollow spheres with a mesoporous shell: Controlled synthesis and applications in gas sensor and lithium ion batteries. *Journal of Physical Chemistry C*, 2008, *112* (30), 11307–11313.

50. Yoshida, T.; Shinada, A.; Oekermann, T.; Sugiura, T.; Sakai, T.; Minoura, H. Time- and frequency-resolved photoelectrochemical investigations on nano-honeycomb TiO_2 electrodes. *Electrochemistry*, 2002, *70* (6), 453–456.

51. Hayazaki, S.; Matsui, T.; Zhang, H. L.; Chen, W.; Wee, A. T. S.; Yuhara, J. Formation of bismuth nanodot in (4 × 4) vanadium oxide nanomesh on Pd(111). *Surface Science*, 2008, *602* (12), 2025–2028.

52. Gao, G.; Xiang, L. Emulsion-phase synthesis of honeycomb-like $Mg_5(OH)_2(CO_3)_4 \cdot 4H_2O$ micro-spheres and subsequent decomposition to MgO. *Journal of Alloys and Compounds*, 2010, *495* (1), 242–246.

53. Gadkaree, K. Growth of unique carbon nanostructures on ceramic substrates. *Journal of Materials Science Letters*, 2002, *21* (14), 1081–1084.

54. Yeom, B.-J.; Char, K.-H. Nanostructured $CaCO_3$ thin films formed on the urease multilayers prepared by the layer-by-layer deposition. *Chemistry of Materials*, 2010, *22* (1), 101–107.

55. Chen, H.; He, J.; Zhang, C.; He, H. Self-assembly of novel mesoporous manganese oxide nanostructures and their application in oxidative decomposition of formaldehyde. *Journal of Physical Chemistry C*, 2007, *111* (49), 18033–18038.

56. Nakamura, T.; Kondoh, H.; Matsumoto, M.; Nozoye, H. Influence of external stimuli on the nano-ordered rows and honeycombs in mercaptomethyloligothiophene monolayers on Au (111). *Molecular Crystals and Liquid Crystals Science and Technology, Section A: Molecular Crystals and Liquid Crystals*, 1998, *322*, 185–190.

57. Schlickum, U.; Decker, R.; Klappenberger, F.; Zoppellaro, G.; Klyatskaya, S.; Ruben, M.; Silanes, I. et al. Metal-organic honeycomb nanomeshes with tunable cavity size. *Nano Letters*, 2007, *7* (12), 3813–3817.

58. Lee, C.-S.; Kimizuka, N. Pillared honeycomb nanoarchitectures formed on solid surfaces by the self-assembly of lipid-packaged one-dimensional Pt complexes. *Proceedings of the National Academy of Sciences of the United States of America*, 2002, *99* (8), 4922–4926.

59. Decker, R.; Schlickum, U.; Klappenberger, F.; Zoppellaro, G.; Klyatskaya, S.; Ruben, M.; Barth, J. V.; Brune, H. Using metal-organic templates to steer the growth of Fe and Co nanoclusters. *Applied Physics Letters*, 2008, *93* (24), 243102/1–243102/3.

Part IV

Nanometals and Nanoalloys

22 Nanometals

22.1 NANOSTRUCTURED METALS IN GENERAL

22.1.1 TYPES OF ACTIVATED METALS

The area of nanometals as a part of the broader area of "nanomaterials"[1] is rapidly developing and gaining importance. During the last four decades, great efforts have been dedicated to the preparation and use of zero-valent (more exactly, *elemental* metals) in organic and organometallic reactions. Unfortunately, a majority of them are not sufficiently active due to a series of factors: high particle size (up to 359 mesh) and oxide and other films on their surface. However, these factors can be partially or completely eliminated, if metals are subjected to diverse activation techniques discussed later. Metal activation is necessitated by a host of their applications in a series of in-chemical processes, related with catalysis, organic and organometallic synthesis, etc. During the last two decades, the use of zero-valent activated and nonactivated metals as precursors in organic/organometallic processes has been described in detail in some monographs[2-7] and reviews.[8,9] In comparison with classic methods using metal salts or carbonyls, the application of metals in the elemental form frequently leads to unique compounds, which cannot be obtained by traditional routes.

In general, a nanoparticle (or nanopowder, nanocluster, or nanocrystal) is defined as a small particle with at least one dimension less than 100 nm. The physical and chemical properties of metal nanoparticles are distinct from those of both bulk metal and isolated atoms. Their main drawback is their instability and high trend for aggregation.[10] Without stabilization they fuse together, losing their special shape and properties. Activated elemental metals in nanoforms are extensively reviewed[11]; they can be divided into active metal *nanopowders*,[12] *nanowires*,[13,14] *nanolines*,[15] *nanodumbbells*,[16] *nanoclusters*,[17-19] *nanosheets*,[20] *nanorods*,[21-23] *nanoalloy*,[24,25] *nanobelts*,[26] and *nanofilms*,[27] which have a host of applications, shown later, in particular as catalysts,[28-31] magnetic materials,[32] nanocomposites,[33,34] chemical sensors,[35-37] degradation of toxic chemicals,[38] or even as possible carriers of isotopes for medical applications.[39] The present short review is dedicated to recent achievements in their obtaining and main applications in various areas of chemistry and chemical technology. Since the same type of activated metals can be frequently obtained by different techniques, the presentation of material is given according to synthetic methods.

22.1.2 METHODS FOR METAL ACTIVATION AND APPLICATIONS OF OBTAINED ACTIVE METALS

There is a series of special techniques for metal activation, which could be artificially divided into *physicochemical*, *chemical*, and *biological* methods. Of course, such a separation is conditional; any physical process is accompanied here by a chemical transformation. In this review, we consider as physicochemical methods those requiring a special equipment, for example, a ^{60}Co source for γ-irradiation of samples or the reactor for cryosynthesis. Such techniques as obtaining *Rieke* metals or decomposition of metal complexes belong to "pure" chemical methods.

22.1.2.1 Physicochemical Methods

22.1.2.1.1 Evaporation of Metals

It is the first step in the *cryosynthesis* technique followed by further interaction of a flow of formed "naked" metal atoms with frozen organic substrate.[40] This research area has been developed mainly

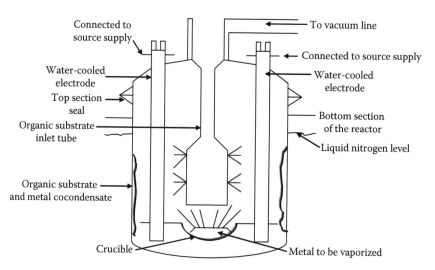

FIGURE 22.1 Macroscale stationary cocondensation apparatus for investigations of metal atom chemistry.

by *K. Klabunde*.[41] Metals {low-boiling (<1000°C) and high-boiling (>2500°C)} can be evaporated in a special high-vacuum reactor (Figure 22.1) by laser treatment, electric resistance heating, electron bombardment, and other methods. Such effective complex formers as Zn, Cd, and Pb belong to the first group of metals; the second one includes Re, Nb, Mo, W, Os, Ir, etc. The metals, which are the most frequently used in cryosynthesis (Co, Fe, Cr, Ni, Pd, La, Ce, Lu), are evaporated at 1400°C–1700°C. The interval of working temperatures usually varies from 10 to 273 K, although sometimes the syntheses are carried out as at higher (295–325 K) or as at lower (e.g., in liquid helium) temperatures. The ligands in gas-phase syntheses are represented by various inorganic and organic compounds, from simple inorganic molecules CO, NO, CO_2 to polymers.

Nowadays cryochemical procedures are applied for obtaining metal–polymer materials (by low-temperature co-condensation of metal and monomer vapors followed by low-temperature solid-state polymerization of the co-condensate under irradiation)[42–45] and various metal complexes, for instance isomeric complexes [Ni(η^3-PC$_2$Bu$_2'$)(η^5-P$_3$C$_2'$Bu$_2$)] and [Ni(η^4-P$_2$C$_2$Bu$_2'$)$_2$] (by the reaction of nickel atoms with ^{t}BuCP),[46] In(P$_3$C$_2$Bu$_2'$),[47] zerovalent bis(η-arene)lanthanide complexes of the form [Ln(η^6-C$_6$H$_3$–^{t}Bu$_3$-1,3,5)$_2$] (by codeposition of a monatomic lanthanide vapor (Sm, Eu, Tm, or Yb) and tri-*tert*-butylbenzene C$_6$H$_3$–^{t}Bu$_3$-1,3,5 onto a cold (77 K) surface),[48] and others.[49–51] Low-temperature surface chemistry and nanostructures are reviewed in Ref. [52]. In comparison with 1970s–1990s, a considerable decrease in the number of publications on the cryosynthesis/(metal vapor synthesis) of metal complexes by co-condensation is observed.

As use of related methods, we would like to note obtaining nanocrystalline copper powders, produced by the *cryogenic melting technique*.[53] The molten metal is overheated while maintained in levitation by r.f. induction in a flux of liquid nitrogen. A calefaction layer is produced at the interface between the two phases, and it is believed that the nanoparticles are formed by rapid condensation of supersaturated metal vapor. The *vapor flow condensation* technique is used for the preparation of bismuth nanoparticles by vapor condensation in tube flows.[54] Bismuth nanoparticles of average diameter 12–37 nm are made by controlling the quenching gas flow rate, carrier gas flow rate, and process pressure.

22.1.2.1.2 Chemical Vapor Deposition

It is a rapidly developing method, recently reviewed fundamentally in a comprehensive review,[55] and is an important technique for surface modification of powders through either grafting or deposition of films and coatings. In other generalizing publications,[56,57] gas-phase/chemical vapor deposition (CVD) nanoparticle production is discussed paying attention to cluster sources, formation

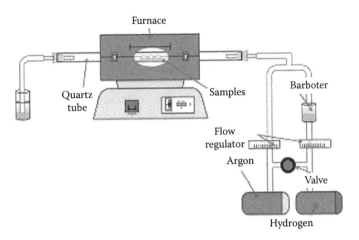

FIGURE 22.2 The diagram of the CVD equipment.

and growth mechanisms, and the use of aerodynamic focusing methods, which are coupled with supersonic expansions to obtain high-intensity cluster beams with a control on nanoparticle mass and spatial distribution.

A typical equipment for CVD is presented in Figure 22.2 and consists of a horizontal furnace with a quartz tube (with a 2.2 cm internal diameter and 2.5 cm external diameter) as the reaction chamber in which controlled flows of argon and hydrogen can be introduced directly *via* a bubbler with the desired liquid as the carbon source. The CVD has been generally used for obtaining carbon nanoparticles described elsewhere; however, some pure metallic nanoparticles can also be prepared this way, for instance Fe,[58–60] W, Mo, Al, Au, Cu, and Pt.[61] Palladium (Pd) nanoparticles are incorporated into free-standing polymer films by a one-step dry process involving simultaneous vaporization, absorption, and reduction schemes of palladium(II) bis(acetylacetonate), $Pd(acac)_2$, used as a precursor.[62]

On the other hand, metal nanoparticles can take part as *catalysts* in the formation and growth of carbon nanotubes (CNTs), obtained by the CVD method. Thus, CNTs were synthesized using hot-filament CVD on an Ni film-coated Si substrate.[63] The reported results here provide strong evidence that the metallic catalyst remains in a *liquid state* during nanotube growth. The upward-growth pulling force of the CNT layer elongates the liquid nanoparticles, which are finally broken into two parts. One part remains at the substrate surface (base of the CNTs) and is responsible for the catalytic growth of the CNTs. The other part is enclosed at the tip of the CNTs and is inactive during CNT growth. Metal (M=Ni)-containing catalysts can be removed from CNTs using molten salts, such as an LiCl–KCl eutectic molten salt, and subsequently by hydrochloric acid.[64]

A powerful but simple technique, called *cold-plasma chemical vapor deposition*, was reported[65] also for CNTs, grown directly on metallic wires. The growth occurs on an Fe catalyst supported by kanthal wires while heating under a hydrocarbon precursor gas and plasma created by a bias. A possible application for such nanotubes on metallic wires is in luminescent tubes. Another modification of the CVD, *chemical vapor reductive deposition* (CVRD) method, is used for obtaining metallic Ni nanoparticles on the surface of a titania thin-film substrate.[66] It is important that the growth of the nanoparticles here is based on the specific adsorption and heterogeneous nucleation on the surface of substrate, not *via* vapor-phase formation and subsequent sedimentation.

22.1.2.1.3 Flame (Combustion) Synthesis

It is a part of gas-phase synthesis that is examined in detail, together with other related techniques, in a recent fundamental monograph[67]; the fundamentals of particle formation in a flame environment are discussed in Ref. [68]. For nanometal production, it is reported for the production of metallic copper,[69] bismuth, and cobalt nanoparticles at production rates up to 70 g/h.[70] Optional coating of the nanoparticles by graphite further allowed the preparation of air-stabled, metal–carbon composites.

$LaCoO_3$-supported Pd is prepared similarly in a single step.[71] The reduction process involved both Pd and Co. In the 100°C–300°C range, the reduction of Co^{3+} to Co^{2+} (from $LaCoO_3$ to $La_2Co_2O_5$) and the segregation of Pd in the form of metal particles occurred. The reduction of Co was already reversible at 120°C. In contrast, Pd remained in the metallic state. The final structure of PdLCO after mild reoxidation consisted of Pd and Co particles supported on $LaCoO_3$. In contrast, reduction at 600°C led to the formation of a Pd–Co alloy. Among other achievements in the combustion synthesis of metal nanoparticles, we note a trimetallic anode electrocatalyst $Pt_{0.6}Ru_{0.3}Ni_{0.1}$,[72] metal-SnO_2 nanocomposites (prepared from metal acetates as metal nanoparticle precursors),[73] nanocrystalline Ti and TiC powders (obtained by combustion of TiO_2–Mg and TiO_2–Mg–C systems in the presence of NaCl),[74] and Ni-catalyzed nanofibers, synthesized by thermal decomposition of a nebulized Ni nitrate solution entrained into a reactive fuel mixture.[75]

A combination of reducing flame synthesis with *magnetic fields* can produce metallic nanowires. Thus, the application of an external magnetic field on a metallic cobalt nanoparticle loaded exhaust gas stream produced by reducing flame synthesis resulted in the formation of self-aligned metallic cobalt nanowires of 20–50 nm diameter, as reported in Ref. [76]. This template-free, continuous, and rapid production method afforded nanowires with an aspect ratio of over 1000 (length/diameter) at a production rate of over 30 g/h. At high magnetic fields, the nanowires formed bundles with distinctive parallel orientation. In the absence of a directing magnetic field, thermophoresis promoted an irregular growth and resulted in highly branched and porous nanomeshes on cooled, stainless steel collectors.

22.1.2.1.4 Electrochemical Activation

Electro chemical activation of metal anode allows passing its material (so-called *sacrificial metal*) to the solution of an organic ligand in a nonaqueous solvent, leading to the formation of metal ions and its further interaction with a solute. This procedure, called a *direct electrochemical synthesis*,[77] is frequently used to obtain coordination and organometallic compounds starting from elemental metals at room temperature. Figure 22.3 shows a combined equipment for electrosynthesis under

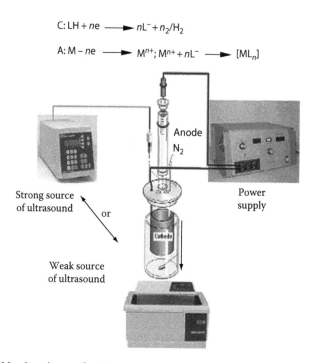

$$C: LH + ne \longrightarrow nL^- + n_2/H_2$$

$$A: M - ne \longrightarrow M^{n+}; M^{n+} + nL^- \longrightarrow [ML_n]$$

FIGURE 22.3 Combined equipment for electrosynthesis under simultaneous ultrasonic treatment.

simultaneous ultrasonic treatment. A zero-valent metal (Group I—Cu, Ag, Au; II—Mg, Ca, Zn, Cd, Hg; III—Al, In, Ga, Tl; IV—Sn, Pb, Ti, Zr, Hf; V—Sb, V, Nb, Ta; VI—Cr, Mo, W; VII—Mn, Re; VIII—Co, Ni, Pd; actinides—Th, U) is electrochemically oxidized forming corresponding metal ions, reacting with dissolved ligand and forming a coordination compound. Despite the formation of metal atoms, electrochemically oxidized to ions, this method cannot be named directly as a production method for nanoparticles. As a result, activated metals are not formed physically, as, for example, *Rieke* metals (see the following): an activated metal here is an intermediate, nonisolated product of the electrochemical reaction. Recent achievements using this tool have reported obtaining new coordination and organometallic compounds, such as, for example, a new chiral binuclear copper(II) complex of (S)-2-(diphenylmethanol)-l-(2-pyridylmethyl)pyrrolidine (S-dmpmpH) (prepared by direct electrochemical oxidation of Cu electrode in acetonitrile solutions),[78] metal (M=Mn, Fe, Co, Ni, Cu, Zn, Cd, Pb) complexes with the ligand bis(4-*N*-methylthiosemicarbazone)-2,6-diacetylpyridine, H$_4$DAPTsz-Me,[79] or neutral heteroleptic cobalt complexes [CoL$_2$L′] (L′ = 2,2′-bipyridine or 1,10-phenanthroline) with the anionic form of a series of methyl-substituted *N*-2-pyridylsulfonamide ligands (HL).[80]

In contrast to electrosynthesis, metals can be *electrochemically deposited* from their compound precursors by reduction of metal ions.[81–88] Thus, monodisperse crystalline zero-valent iron, iron–nickel, iron–palladium nanowires were synthesized using template-directed electrodeposition methods. Prior to nanowire fabrication, alumina nanotemplates with controlled pore structure (e.g., pore diameter and porosity) were fabricated by anodizing high-purity aluminum foil in sulfuric acid. After fabrication of alumina nanotemplates, iron, iron–nickel, and iron–palladium nanowires were electrodeposited within the pore structure.[89] The authors of Ref. [90] report the electrosynthesis of polypyrrole films containing gold nanoparticles (PPy/Au) on a glassy carbon electrode. This was done by applying a constant current of 1.43 mA/cm^2 in solutions containing colloidal Au particles and pyrrole monomer, that is, it was electrosynthesized to not a metal but a supporting medium (polypyrrole). A chloroaurate medium with a citrate/tannic acid reducing/protection agent was employed for generating the Au colloids. The same metal (Au) can also be electrochemically formed in nanowire forms with controlled dimensions and crystallinity.[91] By systematically varying the deposition conditions, both polycrystalline and single-crystalline wires with diameters between 20 and 100 nm are successfully synthesized in etched ion-track membranes. Gold sulfite electrolytes lead to a polycrystalline structure at the temperatures and voltages employed. In contrast, gold cyanide solution favors the growth of single crystals at temperatures between 50°C and 65°C under both direct current and reverse pulse current deposition conditions. In addition, gold–silver alloy nanoparticles with various mole ratios were synthesized[92] in aqueous solution by an electrochemical co-reduction of chloroauric acid (HAuCl$_4$) and silver nitrate (AgNO$_3$) in the presence of poly(vinylpyrrolidone) used as a protecting agent.

Additionally, the reported methods of *electrostatical atomization*[93] for nanoparticle/nanopowder production could be attributed to electrochemical methods for metal activation.

22.1.2.1.5 Arc-Discharge Technique

It is examined in Ref. [94]. This method is common for the production of nanomaterials. Thus, a simple, inexpensive and one-step synthesis method of metal-containing carbon nanocapsules using an arc discharge in aqueous solution is reported in Ref. [95]. It is found that Ni, Co, and Fe nanoparticles could be *in situ* encapsulated in carbon shells when the arc is performed respectively in aqueous solutions of NiSO$_4$, CoSO$_4$, and FeSO$_4$. To explain the formation mechanism of metal-containing carbon nanocapsules, a model of discharge in solution is proposed. Iron nanoparticles, coated with graphite nanolayers, are synthesized by annealing mixtures of hematite and carbon under nitrogen atmosphere.[96] Hematite is reduced to Fe$_3$O$_4$, FeO, and finally to Fe completely at 1200°C. High-resolution electron micrographs reveal that the Fe particles are ~200 nm in diameter and are coated with graphite nanolayers of ~30 nm. The method of *electrical explosion of wire* for production of nanopowders is reported in Refs. [97,98].

Intermetallic species can be also prepared by the arc-discharge method, for instance Fe–Sn nanoparticles.[99] The synthesized nanoparticles have a shell/core structure with an SnO_2 shell of 5–10 nm in thickness and a core of polycrystalline intermetallic compounds. It is found that the intermetallic compounds $FeSn_2$ and Fe_3Sn_2 were generated and coexist with the Sn phase as a single nanoparticle.

A combined *arc-evaporation/condensation and plasma polymerization process* is described in Ref. [100]. Thus, copper nanoparticles are synthesized by the low-pressure pulsed arc evaporation of a metal cathode surface, followed by the in-flight deposition of a thin organic layer by capacitively coupled radio-frequency plasma polymerization from a gaseous hydrocarbon monomer. The same metal (copper) can be prepared[101] in mass quantities in a developed pulsed wire discharge apparatus, allowing obtaining Cu nanopowder (with a mean surface diameter of 65 nm) of 2.0 g quantity in N_2 gas at 100 kPa for 90 s. A fundamental work on obtaining ultrafine particles by *plasma transferred arc* is reported in Ref. [102]. A transferred arc is used to produce nanometric particles from the condensation of metallic vapors obtained by controlled evaporation of the anode material, which becomes the solid precursor of the synthesis. The experiments show that particle morphology, size, and shape, for given working parameters, strongly depend on the properties of the material to be vaporized.

22.1.2.1.6 γ-Irradiation- and X-Ray-Induced Preparation of Metal Nanoparticles

Radiation techniques for material (in particular for nanometals) processing are well-established processes and recently reviewed in Ref. [103]. Using *γ-irradiation-induced reduction* in the field of a ^{60}Co γ-ray source, colloidal silver and gold nanoparticles were prepared from their corresponding metal salts in aqueous solution and compared with those by chemical reduction. The radiation-based method provided silver nanoparticles with higher concentration and narrower size distribution than those obtained by chemical reduction method, while there was no significant difference between the two strategies for the preparation of gold nanoparticles. γ-Irradiation of 1.0×10^{-3} M $AgNO_3$ solution resulted in nearly 100 times more highly concentrated silver colloids than those by citrate reduction. Furthermore, the radiation method could lead to more highly concentrated silver colloids by simply increasing the concentration of $AgNO_3$ solution up to 2.0×10^{-2} M.[104,105] Among other metal nanoparticles, obtained by this method, a series of Ni aggregates supported on α-Al_2O_3 at different nickel contents are prepared by ionic exchange of Ni^{2+} from the complex $[Ni(NH_3)_6]^{2+}$ followed by γ-irradiation under inert atmosphere.[106]

Reductant, stabilizer-free colloidal gold solutions were fabricated by a new room-temperature *synchrotron x-ray irradiation method*.[107,108] The characterization included a study of the possible cytotoxicity for the EMT-6 tumor cell line: the negative results indicate that the gold clusters produced with our approach are biocompatible.

22.1.2.1.7 Photosynthesis/UV-Irradiation

It has been reported as a useful tool for obtaining nanometals, for instance palladium nanoparticles,[109] fabricated from K_2PdCl_4 as a precursor. Its aqueous/diethyl ether solution is irradiated with a conventional 45-W fluorescent light for 12 h, slowly forming a thin film at the liquid/liquid interface. This synthetic approach does not require any additional templates and reduction agents and thus offers not only an attractive possibility for the manufacture of thin-film microcircuits and devices but also a very efficient and economic technique for the preparation of nanoparticles. Another example of the photosynthetical obtaining of metal nanoparticles is the production of comparatively high-activity gold-titania catalysts supported on TiO_2.[110] These materials were tested in the carbon monoxide (CO) oxidation reaction and returned markedly higher levels of activity at room temperature, when compared with catalysts prepared by the traditional photodeposition method. Silver nanoparticles are obtained by UV irradiation of the mixture of epoxy–acrylic resin and reactive monomer, as well as $AgNO_3$ in ethylene glycol.[111] The diameter of the formed Ag nanoparticle is 30–50 nm for 1 M and 80–90 nm for 2 and 3 M of $AgNO_3$, in ethylene glycol. Polyacrylonitrile/silver nanoparticle composites are *in situ* synthesized by ultraviolet irradiation of a mixture of AgNO$_3$ and acrylonitrile monomers.[112] Among other metals, there is also information on the photochemical obtaining nickel[113] nanoparticles.

An interesting combined procedure is reported in Ref. [114] Nanocomposite polymers containing bismuth nanoparticles (2 wt.%) have been obtained by photopolymerization of acrylic resins. The bismuth nanoparticles have been synthesized by reduction of BiCl$_3$ with *t*-BuONa activated sodium hydride. The curing process was followed quantitatively by infrared spectroscopy through the decrease upon UV exposure of the IR bands characteristic of the functional groups. The bismuth nanoparticles are found to have no detrimental effect on the photopolymerization kinetics.

22.1.2.1.8   Ultrasonic Activation

It has now become a classic conventional synthetic tool.[115,116] This approach is used, in particular, for the activation of elemental metals in organic/organometallic synthesis. Alkali metals in inert non-aqueous solvents transform to suspensions; in general, reactions of *p*-, *d*-, and *f*-metals with organic or inorganic medium are strongly accelerated by sonication. As an example, indium nanoparticles are easily produced from bulk metal by sonication.[117] Selenium (although it is not, can be considered as a metal) nanorods are formed *via* the reduction of selenious acid induced by ultrasonic irradiation.[118]

Noble metals also have been obtained this way. Thus, gold colloids were sonochemically synthesized[119,120] in different solvents. It showed that different solvents resulted in different morphologies of Au colloids. In the mixture solution of ethanol/water, only spherical Au nanoparticles were obtained. In ethylene glycol, however, nanorod- and platelet-like morphologies were formed together with spherically shaped Au nanoparticles, which resulted in optical absorption different from that of the former. A sonochemical procedure at room temperature for the preparation of palladium nanoparticles loaded within mesoporous silica is described.[121] The formation of Pd nanoparticles (5–6 nm in diameter) was restricted by the coalescence of the sonochemical reduced Pd atoms inside the confined volumes of the porous solid. Aqueous solutions with Au^{3+} and Pt^{4+} ions and additives of surfactants were irradiated with an ultrasound at 200 kHz with an input power of 4.2 W/cm^2, and colloidal nanoparticles were prepared.[122]

22.1.2.1.9   Microwave Treatment

It is used to produce mainly inorganic compounds/composites/materials,[123] and lesser organic/organometallic compounds,[124–126] as an alternative method to classic heating of mixture-precursors. In respect of metals, their transformations in MW field have been generalized in a comprehensive recent book.[127] Metallic nanoparticles in distinct forms can also be produced this way; for instance those of Ag (30–50 nm), Fe, Co, and Ni were produced using a microwave–polyol process in ethyleneglycol at 100°C and 150°C in the presence of polyvinyl pyrrolidone and dodecyl amine.[128] Silver nanorods, in addition to spherical nanoparticles, are prepared by microwave heating of a silver salt (reducing to silver metal) and sodium citrate in the presence of gold seeds.[129,130] The key aspect for the production of nanorods is the addition of an adequate volume of gold seeds. This finding demonstrates that the combination of MW heating synthesis method in the presence of gold seeds is a good approach to prepare templateless and polymerless silver nanorods. Gold nanoparticles in colloidal solutions were shown to be able to assemble on the naked substrates to form a gold monolayer within minutes.[131]

Among other metallic nanoparticles and bimetallic alloys, obtained by MW heating technique, we would like to mention Pt nanoparticles with different mean sizes supported on CNTs (synthesized by MW ethylene glycol solutions of platinum salt with different pH in the presence of CNTs as supports)[132] and copper (by reducing CuSO$_4$ with hydrazine in ethylene glycol).[133] Synthesis of Bi nanotubes by microwave irradiation is reported in Ref. [134]. It is proposed that this method is an easier and less-expensive way to prepare Bi nanotubes. The obtained product is characterized by TEM and

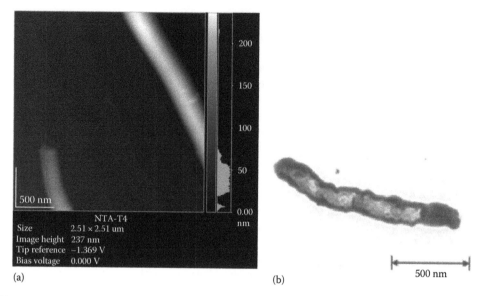

FIGURE 22.4 (a) AFM image of Bi nanotubes obtained by MWCVD for 15 min.; (b) low-magnification TEM images of Bi nanotubes obtained by microwave heating for 15 min.

SEM (Figure 22.4). Single Bi crystalline nanostructure materials are expected to find potential applications in a variety of areas including design of high-efficiency thermoelectric devices.

A variation of microwave heating technique, *microwave plasma synthesis method*, has allowed obtaining nanopowders of nickel, FeNi bimetallic alloy,[135] silver,[136] iron,[137] and a series of Fe–Co (with grain size ranging from micron to nanoscale)[138] and Co–Mo[139] (average metal particle diameters in all cases were less than 10 nm) composites.

22.1.2.1.10   Laser Ablation[140–142]

The main features of ultra-short laser ablation of various materials (metals, semiconductors or insulators) have been studied through the complementary analyses of the plasma plume induced by laser irradiation and of the deposited films. The generation of nanoparticles (in the 10–100 nm range) was observed, and these findings were investigated to obtain information on the relevant parameters governing the formation of these nanoparticles.[143] Nanoparticles of *s*-, *p*-, and *d*-metals are reported to be obtained by laser ablation; some representative examples are as follows. Thus, CaF_2 was exposed to 157 nm excimer laser radiation, and it is shown that several tens of thousands of pulses at fluences near 1 J/cm² can color the material. Absorption spectra of the exposed material confirm the formation of metallic calcium nanoparticles similar to those produced by other forms of energetic radiation. The rate of nanoparticle formation depends on the bulk temperature and displays a local maximum near 50°C.[144] The controllable nanostructuring of thin metal films by nanosecond UV laser pulses is introduced as a novel technique for the production of metal nanoparticles supported on a range of different oxide substrates, including indium tin oxide.[145] This processing is performed at low macroscopic temperatures. Nanopowders of Fe, Al, and intermetallic iron aluminide (particle size of 6–10 nm) have also been synthesized using the laser ablation.[146]

A series of reports are dedicated to noble metals. Thus, results of their femtosecond laser ablation in pure water to synthesize colloidal nanoparticles are reviewed in Ref. [147]. It is shown that such ablation leads to the formation of two different populations of nanoparticles (low and highly size-dispersed ones), while their mean size and relative contribution are strongly affected by the intensity of pumping radiation. Ablation of Ag-, Au-, and Cu-containing solid targets in liquid environments was also studied in nonaqueous solvents (C_2H_5OH, $C_2H_4Cl_2$, etc.).[148] The optical response and morphological changes of gold nanoparticles induced by laser irradiation with single nanosecond laser pulses of different fluences and wavelengths are examined in Ref. [149].

Polymetallic species, obtained by laser ablation, are also examined. Thus, the formation of alloyed Au–Ag and Ag–Cu nanoparticles is reported under laser exposure of a mixture of individual nanoparticles. Catalytic materials containing Rh, bimetallic Rh/Pt, and trimetallic Rh/Pt/Au nanoparticles were synthesized from targets prepared by blending, tableting, and sintering powders of pure metals and by directly collecting the nanoparticles created on support materials.[150] Pulsed excimer laser radiation at 248 nm wavelength was used to ablate $\sim$2 μm feedstock of silver, gold, and permalloy ($Ni_{81}\%:Fe_{19\%}$) under both normal atmospheric conditions and in other gases and pressures.[151]

22.1.2.1.11   Electron-Beam Technique[152]

As well as laser ablation, the Electron-beam (E-beam) method belongs to so-called *rapid thermal processing* (RTP), featuring dynamic control of temperature, which permits high heating and cooling rates that cannot be reached with conventional furnace treatments.[153] High temperatures (up to 1200°C or greater) are reached on a timescale of several seconds or less. Recent experimental approaches to the development of surface nanostructures from nanoparticles are reviewed in Ref. [154].

This is a useful method to produce metallic nanoparticles. Thus, well-protected, isolated bcc-iron nanoparticles embedded in silicon dioxide were prepared by E-beam evaporation and postannealing of multilayers in an ultrahigh vacuum system. The spherical shape and isolation of the particles were confirmed by plan-view and cross-sectional transmission electron microscopy.[155] The formation of nanowires by E-beam writing in films of gold nanoparticles passivated with a specially designed class of ligand molecules (dialkyl sulfides) is presented, together with illustrations of practical nanostructures. Potential applications of this methodology are discussed. By irradiating an E-beam onto silver acetate precursor material, silver nanocrystals with the sizes of 15–40 nm were synthesized.[156] The technique can be useful for mass production of silver nanoparticles and for patterned silver nanoparticle film. Additionally, for silver oxide, the nucleation, growth, and coalescence of its nanoparticles have been investigated dynamically and at high spatial resolution by using the E-beam of a transmission electron microscope to stimulate and to observe the processes.[157] Under the assumption that the particles are hemispherical, the growth rate was found to be proportional to the square root of the electron irradiation time. Among other nanoparticles or supported metals, prepared by E-beam technique, should be noted Pt/TiO_2 films, obtained by E-beam deposition of platinum on TiO_2 films,[158] stable in air nanowires and nanocrystals of cobalt and copper, synthesized from their acetylides CoC_2 and Cu_2C_2.[159]

22.1.2.1.12   Ion-Beam Method[160]

The low-energy metallic ion beams find wide applications in various research fields of materials science. Thus, several metallic ion beams have been developed successfully using the *electron cyclotron resonance* ion source based low-energy ion-beam facility.[161] These metallic ion beams, in particular, of Ni and Fe, were developed by different techniques and utilized for the synthesis of the metal nanoparticles inside various host matrices. Using ion-beam co-sputtering technique, transparent, light brownish, uniform SiO_2 films embedded with spherical Au particles are fabricated on quartz substrates at room temperature and heated in an open furnace at different temperatures from 500°C to 900°C with a 100°C step for 5 min.[162] The synthesis of Co nanoparticles by ion implantation and the effects of postimplantation annealing was carried out.[163] Silica was implanted with 35 keV Co^+ ion beams to doses ranging from 8×10^{15} to 1×10^{17} atoms/cm². The study of nanoparticle size, distribution, and structure *via* TEM measurements revealed the presence of spherical nanoparticles in both as-implanted and annealed samples. Metallic nanoparticles can be deformed by high-energy ion beams, for example metallic cobalt nanoparticles in silica by 200 MeV iodine bombardment.[164]

22.1.2.1.13   Use of Membranes

Polymer-stabilized metal nanoparticles (PSMNPs) can be obtained *in situ* using the ion-exchange membranes as a nanoreactor. Thus, platinum–copper core-shell nanoparticles are reported in Ref. [11]. Membranes were prepared by using sulfonated poly(etherether ketone) of the desired sulfonation degree. The membrane was loaded with metal ions (e.g., Cu^{2+}) or complexes (e.g., $[Pt(NH_3)_4]^{2+}$) followed by metal reduction inside the polymer matrix, resulting in the formation of either monometallic or bimetallic PSMNPs with core-shell structure. The presence of both Cu- and Pt/Cu-PSMNPs inside the membrane substantially improves the electric conductivity of the polymer. α-Fe particles are prepared using a modified metal-membrane incorporation technique based on diffusing metal ions through a dialysis membrane.[165] The diffusion time varies up to 15 min. Single- and poly-crystalline copper wires with diameters down to 30 nm are grown in etched ion-track membranes.[166]

For the first time,[167] polypyrrole-coated electrospun nanofiber mats have been used as separation membranes to electrolessly recover Au from aqueous $[Au(III)Cl_4]^-$ solutions, based on a continuous-flow membrane separation process. With a $[Au(III)Cl_4]^-$ solution passing through the nanofiber membrane, the Au(III) ions were converted into elemental Au. The gold recovered was deposited on the nanofiber membranes in the form of Au particles. It has been found that the polypyrrole-coated electrospun nanofibers are good candidate membrane material for the recovery of Au, and the recovery efficiency is affected by the membrane thickness, the permeate flux rate, and the initial $[Au(III)Cl_4]^-$ concentration.

Metal-containing membranes are applied, in particular, for direct synthesis of hydrogen peroxide on Pd-M (M=Ag, Pt) catalysts prepared by depositing the noble metals by electroless plating deposition or deposition–precipitation methods on α-Al_2O_3 asymmetric ceramic membrane.[168]

22.1.2.2   Chemical Methods

22.1.2.2.1   Decomposition of Metal Complexes

The classic complex $MgC_{14}H_{10} * 3THF$ ("MgA"), obtained by the interaction between $MgBr_2$ with sodium antracenide in THF, is in an equilibrium with free metal and ligand (this complex is called *soluble* magnesium). Active magnesium in powder can be obtained from this compound by heating at 200°C in high vacuum or treating it by ultrasound. Also, finely divided metallic particles can be prepared from their salts and $MgC_{14}H_{10} * 3THF$.[169] In general, the magnesium route to active metals and intermetallics is very interesting and includes the use of such "unusual" forms of magnesium and its compounds as MgA, MgH_2*, MgH_2' and Mg*. Reactivities of magnesium nanopowders are examined in Ref. [170].

In addition to magnesium, many other metal nanoparticles have been prepared this way. Thus, oxidation-resistant iron nanoparticles were produced *via* the decomposition of iron oxalate particles followed by *in situ* passivation with ultrathin alumina films deposited by atomic layer deposition.[171] Decomposition and passivation were carried out in a fluidized bed reactor at low pressure and under mechanical agitation. Heating iron phthalocyanine crystals in vacuum results in the formation of well-defined metal particles embedded in a carbon matrix.[172] Polymeric matrix ferromagnetic nanocomposites, containing Co nanocrystallites, were processed by an innovative fabrication method using frontal polymerization of cobalt acrylamid complex (CoAAm), followed by further thermolysis.[173] The products of thermolysis were in the form of irregular powder particles with a broad range of size distribution, from 10 μm up to 300 μm. The powder particles contain nanocrystallites of Co with a mean of 12–15 nm, depending on the thermolysis temperature. [Bis(salicylidene) cobalt(II)] is used as a precursor to prepare cobalt nanoparticles of average diameter 25–35 nm by thermal decomposition.[174] A process for the synthesis of metallic Ni and Ni–Co alloy nanoparticles at a relatively low temperature is reported,[175] consisting of heating a precursor containing a methyl hydrazine complex of the respective metal ions in nitrogen atmosphere at temperature as low as 400°C. The patent[176] describes obtaining metal nanoparticles by thermal decomposition of a metal acetate $[Fe(OOCCH_3)_2, Ni(OOCCH_3)_2, or Pd(OOCCH_3)_2)]$ or other suitable metal salt, placed in a reaction vessel with a passivating solvent, such as a glycol ether. The contents of the reaction vessel are mixed for a period of time to form a substantially homogenous mixture. The contents of the reaction vessel are then refluxed at a temperature above the melting point of the metal acetate. The desired particle size is achieved by controlling the concentration of metal salt in the passivating solvent and by varying the amount of reflux time.

Copper and noble metal (Ag, Au) nanoparticles were prepared[177] from coinage-metal mesityl (mesityl=$C_6H_2(CH_3)_3$-2,4,6) derivatives ($Cu(C_6H_2(CH_3)_3)_5$, ($Ag(C_6H_2(CH_3)_3)_4$, or ($Au(C_6H_2(CH_3)_3)_5$. Ruthenium nanoparticles are synthesized by pyrolysis of $Ru_3(CO)_{12}$ in sealed ampoules at 190°C for 3 h.[178] Among many other complexes, used for nanometal production by their thermal decomposition, we note aromatic-acetylene containing compound (M=Fe)[179] and metal acetylides MC_2 (M=Sn, Co, Pd, Ni).[180]

22.1.2.2.2   *Pyrolysis[181] ("Co-Solvent-Assisted Spray Pyrolysis Process")*

Related with the previous section, it is used, for example, for generation of metal particles starting from such precursors as [$Cu(NO_3)_2 \cdot 2.5H_2O$], copper acetate $Cu(CH_3COO)_2 \cdot H_2O$], and nickel nitrate [$Ni(NO_3)_2 \cdot 6H_2O$].[182] The process is carried out without the direct addition of hydrogen or other reducing gas over the temperature range of 450°C–1000°C. The addition of ethanol as a co-solvent plays a crucial role in producing phase pure metal powders. Results of a modeling study of ethanol decomposition kinetics suggest that co-solvent decomposition creates a strong reducing atmosphere during spray pyrolysis *via in situ* production of hydrogen and CO. The suggested reaction mechanisms are distinct for each precursor; in the case of, for instance, $Cu[CH_3COO]_2 \cdot H_2O$, the reactions are as follows:

$$Cu[CH_3COO]_2 \cdot H_2O \rightarrow Cu[CH_3COO]_2 + H_2O$$

$$Cu[CH_3COO]_2 \rightarrow Cu + Cu_2O \text{ (major phase)} + \text{gas-phase hydrocarbon products}$$

$$Cu_2O + H_2 \rightarrow 2Cu + H_2O$$

$$Cu_2O + CO \rightarrow 2Cu + CO_2$$

Recent achievements in this area are reported, in particular, in Refs. [183–185]. Among others, we would like to mention obtaining iron nanoparticles from metal phthalocyanine,[173] supported nanoPt/TiO_2 and nanoPd/TiO_2 by *laser pyrolysis*,[186] and carbon-encapsulated Fe nanoparticles (size between 5 and 20 nm) *via* a picric acid-*detonation-induced pyrolysis* of ferrocene, which is characterized by a self-heating and extremely fast process.[187] Nickel particles are prepared[188] by *spray pyrolysis* of a 0.5 M $Ni(NO_3)_2 \cdot 6H_2O$ solution with hydrogen at a residence time of 19 s. All the product particles are composed of nickel oxide and nickel with no trace of nickel nitrate under the condition. The particles are completely reduced to nickel at a furnace set temperature of 410°C. The urea enabled to prepare the nickel particles even without hydrogen though under stricter conditions.

Metallic nanoparticles can be not only produced by pyrolysis, but they also can be used in pyrolysis processes as catalysts. Thus, carbon nanofibers have been formed by polyethylene pyrolysis, using an Ni catalyst and a hydrogen–argon gaseous mixture.[189] The temperature interval of 500°C–700°C was chosen, because small-sized Ni catalyst particles are formed in this case.

22.1.2.2.3   Reactive Alloys, Intermetallics, and Nanometals
             via "Inorganic Grignard Reagents"

This is a research area where some achievements were reported in the 1990s[190,191]; however, currently this is practically a nondeveloping area. The inorganic *Grignard* analogues [$M^1(MgCl)_m$], soluble in THF, allow production of highly reactive nanocrystalline metals, alloys, or intermetallics. Typical examples resulting in the formation of metal alloys are as follows[192]:

$$2NiCl_2 + 3Mg \rightarrow [(NiMgCl)_2MgCl_2]$$

$$n[M^1(MgCl)_m] + mM^2Cl_n \text{ (en THF)} \rightarrow M_n^1M_m^2 + nmMCl_2$$

22.1.2.2.4   Use of Solvents, Acids, and Reductive Agents

Standard chemical methods for elimination of oxide layers are described elsewhere; *depassivation techniques* were reviewed in publications of the last century. A typical classic example is the pretreatment of zinc for further syntheses of organozinc compounds: the metal is washed with diluted acetic acid, water, ethanol, acetone and ether, and then it is dried at 100°C for 2 h. Metal hydrides are also used for zinc pretreatment.[193]

Various reductants, used in classic organic/organometallic chemistry, allow reduction of metal (mainly noble metals) salts to elemental metals, in particular $NaAlH_4$ or $LiAlH_4$. Thus, different phases, in particular Ca-Al alloys, are detected as a result of complex interactions of $MgCl_2$ and $CaCl_2$ with $NaAlH_4$ or $LiAlH_4$.[194] Among other standard reductants, we note sodium borohydride, used for the production of very small gold nanoparticles.[195,196] The influence of the concentration ratio (BH_4^-)/Au and the effect of the pH were also investigated, but no significant trend was observed. The same reductant is used to obtain nanoparticles of metallic nickel from its salts and its derivatives—hydroorganoborates ($NaBEt_3H$, $LiBEt_3H$, $KBPr_3H$, etc.) in THF or DME—are applied for the reduction of MX_n (M=Fe, Cu, Ir, Pt; X=OH, OR, CN, OCN, SCN).[197,198] Bimetallic Ni–Ag (Ni + Ag = 1 wt.%) catalysts supported on crystallized silica are prepared by aqueous chemical reduction with hydrazine at 353 K. The most important feature of the results obtained is the synergistic effect between Ni and Ag, which led to improvement of dispersion and reactivity of nickel in the presence of silver for precipitated catalysts.[199] Other nanoparticle compositions of noble and other metals are reported in Ref. [200] and are produced in a similar way by reacting a salt or complex of a noble metal, such as halide or carboxylate of Au, Ag, Cu, or Pt, with a weak ligand, and a reducing agent (borohydride reagent, hydrazine, or a mixture thereof), in a single liquid phase. Nanocrystals in the size range of 1–20 nm are produced and can be made in a substantially monodisperse form.

Among other reductants, used for obtaining metal nanoparticles, are polyvinyl alcohol-co-vinyl acetate-co-itaconic acid (for synthesis of Fe nanoparticles),[201] polyvinylpyrrolidone (Rh nanoparticles),[202] or short-chain polyethylene glycol (Ag nanoparticles).[203] In the last example, a "one-pot" facile method for environmentally benign production of stable Ag colloids, using short-chain polyethylene glycol as solvent, reducing agent, and stabilizer, is developed. Stable spherical metal particles 15–30 nm in diameter with a well-crystallized structure were obtained at 30°C. Heating to 60°C resulted in increased spherical sizes, and heating over 90°C led to a mixture of shapes: smaller spherical crystallites and bigger triangular and pentagonal nanosized prisms. As reported in Ref. [204], the capacity of the α-synuclein protein to assemble into nanofibers can be used for the synthesis of metallic nanowires. Silver and platinum nanowires with controlled diameters, ranging from 15 to 125 nm, have been synthesized on an α-synuclein protein fiber scaffold.

Synthesis of nickel nanopowders from aqueous solution using a *hydrothermal reduction* method with hydrazine hydrate as a reducing agent and cetyl trimethyl ammonium bromide as a surfactant is investigated in Ref. [205]. The formation of nickel single phase was revealed from XRD patterns. On the other hand, SEM showed that the nickel particles are in nanosized ranges from 55 to 250 nm.

In a similar publication, sea urchin-like nanobelt-based and nanorod-based metallic nickel nano-crystals have been selective synthesized[206] *via* the same hydrothermal reduction technique, in which sodium hydroxide serves as alkaline reagent and hydrazine hydrate as reducing agent.

Using aqueous–organic interface (water–oleic acid) reduction of Cu^{2+} by ascorbic acid, a hydro-phobic copper monolayer and copper particles have been prepared and characterized.[207] The resultant monolayer could be transferred from the interface onto a solid substrate or be dissolved to yield an organosol and copper nanoparticles. Pure submicrometer-sized copper and silver crystallites have been directly synthesized[208] *via solvothermal treatment* of $CuCl_2 \cdot 2H_2O$ or $AgNO_3$ in ethylene-diamine (EDA) at 80°C–180°C for 15–20 h. The authors suggested that the formation of copper and silver crystallites in this solvothermal system is through a typical complexation–reduction process, in which EDA serves not only as a reducing reagent but also as a complexing solvent:

$$CuCl_2 \cdot 2H_2O + x\,EDA \rightarrow [Cu(EDA)_x]^{2+} \rightarrow \text{solvothermal process leading to Cu}$$

$$AgNO_3 + EDA \rightarrow [Ag(EDA)]^+ \rightarrow \text{solvothermal process leading to Ag}$$

22.1.2.3 Biological Synthesis

The development of a reliable and eco-friendly process for the synthesis of metallic nanoparticles is an important step in the filed of application of nanotechnology. One of the options to achieve this objec-tive is to use natural processes such as the use of biological systems.[209] Thus, the fungus *Aspergillus flavus* when challenged with $AgNO_3$ solution accumulates silver nanoparticles (average particle size: 8.92 ± 1.61 nm) on the surface of its cell wall in 72 h. The Fourier transform infrared spectroscopy confirms the presence of protein as the stabilizing agent surrounding the silver nanoparticles. The use of fungus for silver nanoparticles synthesis offers the benefits of eco-friendliness and amenability for large-scale production.[210] The same research group reports another fungus, *Phanerochaete chryso-sporium*, also for obtaining Ag nanoparticles.[211] A bioreduction method is also applied for obtaining Ti/Ni[212] and Zn[213] (clusters up to 4000 atoms) nanoparticles.

Cell isolates of *Bacillus sphaericus* JG-A12 from a uranium mining waste pile in Germany are able to accumulate high amounts of toxic metals such as U, Cu, Pd(II), Pt(II) and Au(III), which are also bound by the highly ordered paracrystalline proteinaceous surface layer (S-layer) that enve-lopes the cells of this strain.[214] These special capabilities of the cells and the S-layer proteins of *B. sphaericus* JG-A12 are highly interesting, in particular, for the production of metal nanoclusters.

REFERENCES

1. Mackerle, J. Nanomaterials, nanomechanics and finite elements, a bibliography (1994–2004). *Modelling and Simulation in Materials Science and Engineering*, 2005, *13* (1), 123–158.
2. Fürstner, A. (Ed.). *Active Metals. Preparation, Characterization, Applications*. VCH, Weinheim, Germany, 1996, 465 pp.
3. Cintas, P. *Activated Metals in Organic Synthesis*. CRC Press, Boca Raton, FL, 1993, 250 pp.
4. Garnovskii, A. D.; Kharisov, B. I. (Eds.). *Direct Synthesis of Coordination and Organometallic Compounds*. Elsevier Science, Amsterdam, the Netherlands, 1999, 244 pp.
5. Garnovskii, A. D.; Kharisov, B. I. (Eds.). *Synthetic Coordination and Organometallic Chemistry*. Marcel Dekker, New York, 2003, 513 pp.
6. Fedlheim, D. L.; Foss, C. A. *Metal Nanoparticles, Synthesis Characterization and Application*. Marcel Dekker, New York, 2002, 360 pp.
7. Klabunde, K. J. *Nanoscale Materials in Chemistry*. VCH, Weinheim, Germany, 2001, 304 pp.
8. Fürstner, A. Chemistry of and with highly reactive metals. *Angewandte Chemie International Edition* (English), 1993, *32* (2), 164–189. Published on-line in 2003.
9. Garnovskii, A. D.; Kharisov, B. I.; Gójon-Zorrilla, G.; Garnovskii, D. A. Direct synthesis of coordina-tion compounds starting from zero-valent metals and organic ligands. *Russian Chemical Reviews*, 1995, *64* (3), 201–221.

10. Muraviev, D. N.; Macanás, J.; Parrondo, J.; Muñoz, M.; Alonso, A.; Alegret, S.; Ortuetac, M.; Mijangos, F. Cation-exchange membrane as nanoreactor: Intermatrix synthesis of platinum-copper core-shell nanoparticles. *Reactive and Functional Polymers*, 2007, *67* (12), 1612–1621.

11. Rao, C. N. R.; Kulkarni, G. U.; Govindaraj, A.; Satishkumar, B. C.; Thomas, P. J. Metal nanoparticles, nanowires, and carbon nanotubes. *Pure and Applied Chemistry*, 2000, *72* (1–2), 21–33.

12. Park, K.; Kim, H. J.; Suh, Y. J. Preparation of tantalum nanopowders through hydrogen reduction of TaCl$_5$ vapor. *Powder Technology*, 2007, *172* (3), 144–148.

13. Rao, C. N. R.; Deepak, F. L.; Gundiah, G.; Govindaraj, A. Inorganic nanowires. *Progress in Solid State Chemistry*, 2003, *31* (1), 5–147.

14. Wee Shong, C.; Chenmin, L. Method of preparing nanowire(s) and product(s) obtained therefrom. U.S. Patent 20070221917, 2007.

15. Owen, J. H. G.; Miki, K.; Bowler, D. R. Self-assembled nanowires on semiconductor surfaces. *Journal of Materials Science*, 2006, *41* (14), 4568–4603.

16. Huang, C.-J.; Chiu, P.-H.; Wang, Y.-H.; Chen, W.-R.; Meen, T.-H.; Yang, C.-F. Preparation and characterization of gold nanodumbbells. *Nanotechnology*, 2006, *17* (21), 5355–5362.

17. Aiken, J. D.; Finke, R. G. A review of modern transition-metal nanoclusters, their synthesis, characterization, and applications in catalysis. *Journal of Molecular Catalysis A: Chemical*, 1999, *145* (1), 1–44.

18. Finney, E. E.; Finke, R. G. Nanocluster nucleation and growth kinetic and mechanistic studies: A review emphasizing transition-metal nanoclusters. *Journal of Colloid Interface Science*, 2008, *317* (2), 351–374.

19. Bansmann, J.; Baker, S. H.; Binns, C.; Blackman, J. A.; Bucher, J.-P.; Dorantes-Dávila, J.; Dupuis, V. et al. Magnetic and structural properties of isolated and assembled clusters. *Surface Science Reports*, 2005, *56* (6), 189–275.

20. He, Y.; Wu, X.; Lu, G.; Shi, G. A facile route to silver nanosheets. *Materials Chemistry and Physics*, 2006, *98* (1), 178–182.

21. Kuchibhatla, S. V. N. T.; Karakoti, A. S.; Bera, D.; Seal, S. One dimensional nanostructured materials. *Progress in Materials Science*, 2007, *52* (5), 699–913.

22. Perez-Juste, J.; Pastoriza-Santos, I.; Liz-Marzan, L. M.; Mulvaney, P. Gold nanorods, synthesis, characterization and applications. *Coordination Chemistry Reviews*, 2005, *249* (17), 1870–1901.

23. Wang, P. I.; Zhao, Y. P.; Wang, G. C.; Lu, T. M. Novel growth mechanism of single crystalline Cu nanorods by electron beam irradiation. *Nanotechnology*, 2004, *15* (1), 218–222.

24. Rodríguez-López, J. L.; Montejano-Carrizales, J. M.; Pal, U.; Sánchez-Ramírez, J. F.; Troiani, H. E.; García, D.; Miki-Yoshida, M.; Yacamán, J. M. Surface reconstruction and decahedral structure of bimetallic nanoparticles. *Physical Review Letters*, 2004, *92* (19), 196102.

25. Kim, W.-B. P.; Suh, J.-S.; Daejeon, C.-Y.; Kil, D.-S.; Lee, J.-C. Method for manufacturing alloy nanopowders. U.S. Patent 20070209477A1, 2007.

26. Chen, Y.; Gong, R.; Zhang, W.; Xu, X.; Fan, Y.; Liu, W. Synthesis of single-crystalline bismuth nanobelts and nanosheets. *Materials Letters*, 2005, *59* (8), 909–911.

27. Zhang, X.; Xie, H.; Fujii, M.; Takahashi, K.; Ikuta, T.; Ago, H.; Abe, H.; Shimizu, T. Experimental study on thermal characteristics of suspended platinum nanofilm sensors. *International Journal of Heat and Mass Transfer*, 2006, *49* (21), 3879–3883.

28. Yamada, Y. M. A.; Uozumi, Y. Development of a convoluted polymeric nanopalladium catalyst, α-alkylation of ketones and ring-opening alkylation of cyclic 1,3-diketones with primary alcohols. *Tetrahedron*, 2007, *63* (35), 8492–8498.

29. Niu, Y.; Crooks, R. M. Dendrimer-encapsulated metal nanoparticles and their applications to catalysis. *Comptes Rendus Chimie*, 2003, *6* (8), 1049–1059.

30. Liu, H.; Song, C.; Zhang, L.; Zhang, J.; Wang, H.; Wilkinson, D. P. A review of anode catalysis in the direct methanol fuel cell. *Journal of Power Sources*, 2006, *155* (2), 95–110.

31. Karen, B.; Brian, H.; Christopher, L.; Thierry Palladium alloy catalysts for fuel cell cathodes. WO07042841A1, 2007.

32. Yurkov, G.; Baranov, D. A.; Dotsenko, I. P.; Gubin, S. P. New magnetic materials based on cobalt and iron-containing nanoparticles. *Composites Part B*, 2006, *37* (6), 413–417.

33. Armelao, L.; Barreca, D.; Bottaro, G.; Gasparotto, A.; Gross, S.; Maragno, C.; Tondello, E. Recent trends on nanocomposites based on Cu, Ag and Au clusters: A closer look. *Coordination Chemistry Reviews*, 2006, *250* (11), 1294–1314.

34. Atanassova, P.; Bhatia, R.; Sun, Y.; Hampden-Smith, M. J.; Brewster, J.; Napolitano, P. Alloy catalyst compositions and processes for manufacturing. U.S. Patent 20070160899A1, 2007.

35. Huang, X. J.; Choi, Y. K. Chemical sensors based on nanostructured materials. *Sensors and Actuators B: Chemical*, 2007, *122* (2), 659–671.

36. Atashbar, M. Z.; Singamaneni, S. Room temperature gas sensor based on metallic nanowires. *Sensors and Actuators B: Chemical*, 2005, *111*, 13–21.

37. Brust, M.; Kiely, C. J. Some recent advances in nanostructure preparation from gold and silver particles: A short topical review. *Colloids and Surfaces A: Physicochemical and Engineering Aspects*, 2002, *202* (2), 175–186.

38. McDowall, L. *Degradation of Toxic Chemicals by Zero-Valent Metal Nanoparticles—A Literature Review*. Human Protection and Performance Division, DSTO Defence Science and Technology Organisation, Melbourne, Victoria, Australia, 2005.

39. Kucka, J.; Hruby, M.; Konak, C.; Kozempel, J.; Lebeda, O. Astatination of nanoparticles containing silver as possible carriers of [211]At. *Applied Radiation and Isotopes*, 2006, *64* (2), 201–206.

40. Klabunde, K. J.; Cardenas-Trivino, G. Chapter 6. Metal atom/vapor approaches to active metal cluster/particles. In *Active Metals. Preparation, Characterization, Applications*. Fürstner A. (Ed.). VCH, Weinheim, Germany, 1996, pp. 237–278.

41. Klabunde, K. J. *Free Atoms, Clusters, and Nanoscale Particles*. Academic Press, San Diego, CA, 1994, 311 pp.

42. Trakhtenberg, L. I.; Gerasimov, G. N.; Aleksandrova, L. N.; Potapov, V. K. Photo and radiation cryochemical synthesis of metal-polymer films, structure, sensor and catalytic properties. *Radiation Physics and Chemistry*, 2002, *65* (4), 479–485.

43. Trakhtenberg, L. I.; Axelrod, E.; Gerasimov, G. N.; Nikolaeva, E. V.; Smirnova, E. I. New nano-composite metal-polymer materials, dielectric behaviour. *Journal of Non-Crystalline Solids*, 2002, *305* (1), 190–196.

44. Pomogailo, A. D. Polymer-immobilized nanoscale and cluster metal particles. *Nanostructured Materials*, 1999, *12* (1), 291–294.

45. Belotti, F.; Fratoddi, I.; La Groia A.; Martra, G.; Mustarelli, P.; Panziera, N.; Pertici, P.; Russo, M. V. Preparation of nanostructured organometallic polymer/palladium hybrids by metal vapour synthesis, structure and morphology. *Nanotechnology*, 2005, *16* (11), 2575–2581.

46. Cloke, F. G. N.; Hitchcock, P. B.; Nixon, J. F.; Vickers, D. M. Metal-vapour synthesis of phospha-organometallic compounds, reaction of nickel atoms with the phospha-alkyne ^{t}BuCP. Structural characterisation of [Ni(η^3-PC$_2$^{t}Bu$_2$)(η^5-P$_3$C$_2$^{t}Bu$_2$)], a derivative of the novel, aromatic phosphirenyl cation. *Comptes Rendus Chimie*, 2004, *7* (8), 931–940.

47. Wann, D. A.; Hinchley, S. L.; Robertson, H. E.; Francis, M. D.; Nixon, J. F.; Rankin, D. W. H. The molecular structure of In(P$_3$C$_2$Bu$_2'$) using gas-phase electron diffraction and ab initio and DFT calculations. *Journal of Organometallic Chemistry*, 2007, *692* (5), 1161–1167.

48. Arnold, P. L.; Petrukhina, M. A.; Bochenkov, V. E.; Shabatina, T. I.; Zagorskii, V. V.; Sergeev, G. B.; Cloke, F. G. N. Arene complexation of Sm, Eu, Tm and Yb atoms, a variable temperature spectroscopic investigation. *Journal of Organometallic Chemistry*, 2003, *688* (1), 49–55.

49. Shabatina, T. I.; Vovk, E. V.; Ozhegova, N. V.; Morosov, Y. N.; Nemukhin, A. V.; Sergeev, G. B. Synthesis and properties of metal-mesogenic nanostructures. *Materials Science and Engineering C*, 1999, *8*, 53–56.

50. Mendoza, O.; Muller-Bunz, H.; Tacke, M. Cocondensation reactions of nitrogen-containing heterocycles with lithium atoms at 77K. *Journal of Organometallic Chemistry*, 2005, *690* (14), 3357–3365.

51. Mendoza, O.; Tacke, M. Cocondensation reactions of heterocyclic aromatic compounds with lithium, calcium and magnesium atoms at 77 K. *Journal of Organometallic Chemistry*, 2006, *691* (6), 1110–1116.

52. Sergeev, G. B.; Shabatina, T. I. Low temperature surface chemistry and nanostructures. *Surface Science*, 2002, *500* (1), 628–655.

53. Champion, Y.; Bigot, J. Characterization of nanocrystalline copper powders prepared by melting in a cryogenic liquid. *Materials Science and Engineering A*, 1996, *217*, 58–63.

54. Wegner, K.; Walker, B.; Tsantilis, S.; Pratsinis, S. E. Design of metal nanoparticle synthesis by vapor flow condensation. *Chemical Engineering Science*, 2002, *57* (10), 1753–1762.

55. Vahlas, C.; Caussat, B.; Serp, P. Angelopoulos, G. N. Principles and applications of CVD powder technology. *Materials Science and Engineering Reports*, 2006, *53* (1), 1–72.

56. Wegner, K.; Piseri, P.; Vahedi, T. H.; Milani, P. Cluster beam deposition, a tool for nanoscale science and technology. *Journal of Physics D: Applied Physics*, 2006, *39* (22), R439–R459.

57. Hahn, H. Gas phase synthesis of nanocrystalline materials. *Nanostructured Materials*, 1997, *9* (1), 3–12.

58. Ebara, A.; Kuramochi, K.; Yamazaki, T.; Hashimoto, I.; Watanabe, K. Structural study of graphite-encapsulated iron nanoparticles via chemical vapor deposition combined with spray method. *Carbon*, 2007, *45* (4), 898–902.

59. Enz, T.; Winterer, M.; Stahl, B.; Bhattacharya, S.; Miehe, G.; Foster, K.; Fasel, C.; Hahn, H. Structure and magnetic properties of iron nanoparticles stabilized in carbon. *Journal of Applied Physics*, 2006, *99*, 044306, 8 pp.

60. Young Lee H.; Gyu Kim S. Kinetic study on the hydrogen reduction of ferrous chloride vapor for preparation of iron powder. *Powder Technology*, 2005, *152* (1), 16–23.

61. Choy, K. L. Chemical vapour deposition of coatings. *Progress in Materials Science*, 2003, *48* (2), 57–170.

62. Lee, J. Y.; Liao, Y.; Nagahata, R.; Horiuchi, S. Effect of metal nanoparticles on thermal stabilization of polymer/metal nanocomposites prepared by a one-step dry process. *Polymer*, 2006, *47* (23), 7970–7979.

63. Chen, X.; Wang, R.; Xu, J.; Yu, D. TEM investigation on the growth mechanism of carbon nanotubes synthesized by hot-filament chemical vapor deposition. *Micron*, 2004, *35* (6), 455–460.

64. Li, X.; Yuan, G.; Brown, A.; Westwood, A.; Brydson, R.; Rand, B. The removal of encapsulated catalyst particles from carbon nanotubes using molten salts. *Carbon*, 2006, *44* (9), 1699–1705.

65. Sarangi, D.; Karimi, A. Bias-enhanced growth of carbon nanotubes directly on metallic wires. *Nanotechnology*, 2003, *14* (2), 109–112.

66. Yoshinaga, M.; Takahashi, H.; Yamamoto, K.; Muramatsu, A.; Morikawa, T. Formation of metallic Ni nanoparticles on titania surfaces by chemical vapor reductive deposition method. *Journal of Colloid and Interface Science*, 2007, *309* (1), 149–154.

67. Granqvist, C. G.; Kish, L. B.; Marlow, W. H. *Gas Phase Nanoparticle Synthesis*. Springer, Berlin, Germany, 2004, 186 pp.

68. Roth, P. Particle synthesis in flames. *Proceedings of the Combustion Institute*, 2007, *31* (2), 1773–1788.

69. Athanassiou, E. K.; Grass, R. N.; Stark, W. J. Large-scale production of carbon-coated copper nanoparticles for sensor applications. *Nanotechnology*, 2006, *17*, 1668–1673.

70. Grass, R. N.; Stark, W. J. *Reducing Flame Synthesis, Production of Metallic Nanoparticles by Cost Efficient Flame Processing*. PARTEC, Nuremberg, Germany, 2007, 3 pp. and references therein.

71. Chiarello, G. L.; Grunwaldt, J. D.; Ferri, D.; Krumeich, F.; Oliva, C.; Forni, L.; Baiker, A. Flame-synthesized LaCoO$_3$-supported Pd$_1$. Structure, thermal stability and reducibility. *Journal of Catalysis*, 2007, *252* (2), 127–136.

72. Moreno, B.; Chinarro, E.; Perez, J. C.; Jurado, J. R. Combustion synthesis and electrochemical characterisation of Pt-Ru-Ni anode electrocatalyst for PEMFC. *Applied Catalysis B: Environmental*, 2007, *76* (3), 368–374.

73. Bakrania, S. D.; Miller, T. A.; Perez, C.; Wooldridge, M. S. Combustion of multiphase reactants for the synthesis of nanocomposite materials. *Combustion and Flame*, 2007, *148* (1), 76–87.

74. Nersisyan, H. H.; Lee, J. H.; Won, C. W. Combustion of TiO$_2$-Mg and TiO$_2$-Mg-C systems in the presence of NaCl to synthesize nanocrystalline Ti and TiC powders. *Materials Research Bulletin*, 2003, *38* (7), 1135–1146.

75. Vander Wal, R. L. Flame synthesis of Ni-catalyzed nanofibers. *Carbon*, 2002, *40* (12), 2101–2107.

76. Athanassiou, E. K.; Grossmann, P.; Grass, R. N.; Stark, W. J. Template free, large scale synthesis of cobalt nanowires using magnetic fields for alignment. *Nanotechnology*, 2007, *18*, 165606, 7 pp.

77. Garnovskii, A. D.; Blanco, L. M.; Kharisov, B. I.; Garnovskii, D. A.; Burlov, A. S. Electrosynthesis of metal complexes, state of the art. *Journal of Coordination Chemistry*, 1999, *48*, 219–263.

78. Yuan, Y.; Yao, J.; Lu, J.; Zhang, Y.; Gu, R. Direct electrochemical synthesis and crystal structure of a copper(II) complex with a chiral (S)-2-(diphenylmethanol)-1-(2-pyridylmethyl)pyrrolidine. *Inorganic Chemistry*, 2005, *8* (11), 1014–1017.

79. Pedrido, R.; Bermejo, M. R.; Romero M. J.; Vázquez, M.; González-Noya, A. M.; Maneiro, M.; Jesús Rodríguez, M.; Isabel Fernández, M. Syntheses and x-ray characterization of metal complexes with the pentadentate thiosemicarbazone ligand bis(4-*N*-methylthiosemicarbazone)-2,6-diacetylpyridine. The first pentacoordinate lead(II) complex with a pentagonal geometry. *Dalton Transactions*, 2005, (3), 572–579.

80. Beloso, I.; Castro, J.; Garcia-Vazquez, J. A.; Perez-Lourido, P.; Romero, J.; Sousa, A. Electrochemical synthesis and structural characterization of heteroleptic cobalt(II) complexes of N-2-pyridyl sulfonamide ligands. *Polyhedron*, 2006, *25* (14), 2673–2682.

81. Lemay, S. G.; Dekker, L. C.; Quinn, B. M. Electrodeposition of noble metal nanoparticles on carbon nanotubes. *Journal of the American Chemical Society*, 2005, *127* (17), 6146–6147.

82. Li, L.; Dai, L. Template-free electrodeposition of multicomponent metal nanoparticles for region-specific growth of interposed carbon nanotube micropatterns. *Nanotechnology*, 2005, *16*, 2111–2117.

83. Hirsch, T.; Zharnikov, M.; Shaporenko, A.; Stahl, J.; Weiss, D.; Wolfbeis, O. S.; Mirsky, V. M. Size-controlled electrochemical synthesis of metal nanoparticles on monomolecular templates. *Angewandte Chemie International Edition*, 2005, *44*, 6775–6778.

84. Jayashree, R. S.; Spendelow, J. S.; Yeom, J.; Rastogi, C.; Shannon, M. A.; Kenis, P. J. A. Characterization and application of electrodeposited Pt, Pt/Pd, and Pd catalyst structures for direct formic acid micro fuel cells. *Electrochimica Acta*, 2005, *50*, 4674–4682.

85. Ogata, Y. H.; Kobayashi, K.; Motoyama, M. Electrochemical metal deposition on silicon. *Current Opinion in Solid State and Materials Science*, 2006, *10* (3), 163–172.

86. Allongue, P.; Maroun, F. Metal electrodeposition on single crystal metal surfaces mechanisms, structure and applications. *Current Opinion in Solid State and Materials Science*, 2006, *10* (3), 173–181.

87. Hoshino, K.; Hitsuoka, Y. One-step template-free electrosynthesis of cobalt nanowires from aqueous $[Co(NH_3)_6]Cl_3$ solution. *Electrochemistry Communications*, 2005, *7*, 821–828.

88. Lu, Y.; Wang, D. Process for the preparation of metal-containing nanostructured film. U.S. Patent 20040118698A1, 2004 and U.S. Patent 7001669, 2006.

89. Yoo, B. Y.; Hernandez, S. C.; Koo, B.; Rheem, Y.; Myung, N. V. Electrochemically fabricated zero-valent iron, iron-nickel, and iron-palladium nanowires for environmental remediation applications. *Water Science and Technology*, 2007, *55* (1–2), 149–156. http://www.iwaponline.com/wst/05501/0149/055010149.pdf

90. Chen, W.; Li, C. M.; Chen, P.; Sun, C. Q. Electrosynthesis and characterization of polypyrrole/Au nano-composite. *Electrochimica Acta*, 2007, *52* (8), 2845–2849.

91. Liu, J.; Duan, J. L.; Toimil-Molares, M. E.; Karim, S.; Cornelius, T. W.; Dobrev, D.; Yao, H. J. et al. Electrochemical fabrication of single-crystalline and polycrystalline Au nanowires, the influence of deposition parameters. *Nanotechnology*, 2006, *17* (8), 1922–1926.

92. Zhou, M.; Chen, S.; Zhao, S.; Ma, H. One-step synthesis of Au–Ag alloy nanoparticles by a convenient electrochemical method. *Physica E, Low-dimensional Systems and Nanostructures*, 2006, *33* (1), 28–34.

93. Lohmann, M.; Schmidt-Ott, A. Production of metallic nanoparticles via electrostatic atomization. *Journal of Aerosol Science*, 1995, *26*, S829–S830.

94. Rao, C. N. R.; Thomas, P. J.; Kulkarni, G. U. *Nanocrystals, Synthesis, Properties and Applications*, Springer Series in Materials Science. Springer, Berlin, Germany, 2007, 182 pp.

95. Xua, B.; Guo, J.; Wang, X.; Liu, X.; Ichinose, H. Synthesis of carbon nanocapsules containing Fe, Ni or Co by arc discharge in aqueous solution. *Carbon*, 2006, *14* (13), 2631–2634.

96. Tokoro, H.; Fujii, S.; Oku, T. Iron nanoparticles coated with graphite nanolayers and carbon nanotubes. *Diamond and Related Materials*, 2004, *13* (4), 1270–1273.

97. Method and apparatus for manufacturing metal nanopowder by electrical explosion of wire. KR2090657A, 2002.

98. Apparatus for production of ultra fine powder from wire comprises a closed loop recirculating gas path, a wire source, two electrodes and an electrical energy source which supplies discharge current between the electrodes. WO0117671A1, 2001.

99. Lei, J. P.; Dong, X. L.; Zhu, X. G.; Lei, M. K.; Huang, H.; Zhang, X. F.; Lu, B.; Park, W. J.; Chung, H. S. Formation and characterization of intermetallic Fe–Sn nanoparticles synthesized by an arc discharge method. *Intermetallics*, 2007, *15* (12), 1589–1594.

100. Qin, C.; Coulombe, S. Organic layer-coated metal nanoparticles prepared by a combined arc evaporation/condensation and plasma polymerization process. *Plasma Sources Science and Technology*, 2007, *16* (2), 240–249.

101. Suematsu, H.; Nishimura, S.; Murai, K.; Hayashi, Y.; Suzuki, T.; Nakayama, T.; Jiang, W.; Yamazaki, A.; Seki, K.; Niihara, K. Pulsed wire discharge apparatus for mass production of copper nanopowders. *Review of Scientific Instruments*, 2007, *78*, 056105, 3 pp.

102. Chazelas, C.; Coudert, J. F.; Jarrige, J.; Fauchais, P. Synthesis of ultra fine particles by plasma transferred arc: Influence of anode material on particle properties. *Journal of the European Ceramic Society*, 2006, *26* (16), 3499–3507.

103. Chmielewski, A. G.; Haji-Saeid M. Radiation technologies, past, present and future. *Radiation Physics and Chemistry*, 2004, *71* (1), 17–21.

104. Li, T.; Parka, H. G.; Choi, S.-H. γ-Irradiation-induced preparation of Ag and Au nanoparticles and their characterizations. *Materials Chemistry and Physics*, 2007, *105* (2–3), 325–330.

105. Liu, F. K.; Hsu, Y. C.; Tsai M. H.; Chu, T. C. Using γ-irradiation to synthesize Ag nanoparticles. *Materials Letters*, 2007, *61* (11), 2402–2405.

106. Keghouche, N.; Chettibi, S.; Latreche, F.; Bettahar, M. M.; Belloni, J.; Marignier, J. L. Radiation-induced synthesis of α-Al_2O_3 supported nickel clusters: Characterization and catalytic properties. *Radiation Physics and Chemistry*, 2005, *74* (3), 185–200.

107. Wang, C.-H.; Hua, T.-E.; Chien, C.-C.; Yu, Y.-L.; Yang, T.-Y.; Liu, C.-J.; Leng, W.-H. et al. Aqueous gold nanosols stabilized by electrostatic protection generated by x-ray irradiation assisted radical reduction. *Materials Chemistry and Physics*, 2007, *106* (2), 323–329.

108. Yang, Y. C.; Wang, C. H.; Hwu, Y. K.; Je, J. H. Synchrotron x-ray synthesis of colloidal gold particles for drug delivery. *Materials Chemistry and Physics*, 2006, *100* (1), 72–76.

109. Lee, K. Y.; Byeon, H.-S.; Yang, J.-K.; Cheong, G.-W.; Han, S. W. Photosynthesis of palladium nanoparticles at the water/oil interface. *Bulletin of the Korean Chemical Society*, 2007, *28* (5), 880–882.

110. Kydd, R.; Chiang, K.; Scott, J.; Amal, R. Low energy photosynthesis of gold-titania catalysts. *Photochemistry and Photobiological Sciences*, 2007, *6*, 829–832.

111. Cheng, W. T.; Chih, Y. W.; Yeh, W. T. In situ fabrication of photocurable conductive adhesives with silver nano-particles in the absence of capping agent. *International Journal of Adhesion and Adhesive*, 2007, *27* (3), 236–243.

112. Zhang, Z.; Zhang, L.; Wang, S.; Chen, W.; Lei, Y. A convenient route to polyacrylonitrile/silver nanoparticle composite by simultaneous polymerization-reduction approach. *Polymer*, 2001, *42* (19), 8315–8318.

113. Smirnova, N. V.; Boitsova, T. B.; Gorbunova, V. V.; Alekseeva L. V.; Pronin, V. P.; Kon'uhov, G. S. Nickel films: Nonselective and selective photochemical deposition and properties. *Thin Solid Films*, 2006, *513* (1), 25–30.

114. Balan, L.; Burget, D. Synthesis of metal/polymer nanocomposite by UV-radiation curing. *European Polymer Journal*, 2006, *42* (12), 3180–3189.

115. Suslick, K. S. (Ed.). *Ultrasound, its Chemical, Physical, and Biological Effects*. VCH, Weinheim, Germany, 1988.

116. Mason, T. J. *Advances in Sonochemistry*, Vol. 1. JAI Press LTD, London, U.K., 1990,.

117. Li, Z.; Tao, X.; Cheng, Y.; Wu, Z.; Zhang, Z.; Dang, H. A simple and rapid method for preparing indium nanoparticles from bulk indium via ultrasound irradiation. *Materials Science and Engineering A*, 2005, *407* (1), 7–10.

118. Wang, X.; Zheng, X.; Lu, J.; Xie, Y. Reduction of selenious acid induced by ultrasonic irradiation—Formation of Se nanorods. *Ultrasonics Sonochemistry*, 2004, *11* (5), 307–310.

119. Li, C.; Cai, W.; Kan, C.; Fu, G.; Zhang, L. Ultrasonic solvent induced morphological change of Au colloids. *Materials Letters*, 2004, *58* (1), 196–199.

120. Reed, J. A.; Cook, A.; Halaas, D. J.; Parazzoli, P.; Robinson, A.; Matula, T. J.; Grieser, F. The effects of microgravity on nanoparticle size distributions generated by the ultrasonic reduction of an aqueous gold-chloride solution. *Ultrasonics Sonochemistry*, 2003, *10* (4), 285–289.

121. Chen, W.; Cai, W.; Lei, Y.; Zhang, L. A sonochemical approach to the confined synthesis of palladium nanoparticles in mesoporous silica. *Materials Letters*, 2001, *50* (2), 53–56.

122. Nakanishi, M.; Takatani, H.; Kobayashi, Y.; Hori, F.; Taniguchi, R.; Iwase, A.; Oshima, R. Characterization of binary gold/platinum nanoparticles prepared by sonochemistry technique. *Applied Surface Science*, 2005, *241* (1), 209–212.

123. Schubert, U.; Hüsing, N. *Synthesis of Inorganic Materials*. Wiley-VCH, Weinheim, Germany, 2001, 413 pp.

124. Lidstrom, P.; Tierney, J.; Wathey, B.; Westman, J. Microwave assisted organic synthesis—A review. *Tetrahedron*, 2001, *57* (45), 9225–9283.

125. Loupy, A. (Ed.). *Microwaves in Organic Synthesis*. Wiley-VCH, Weinheim, Germany, 2003, 523 pp.

126. Bougrin, K.; Loupy, A.; Soufiaoui, M. Microwave-assisted solvent-free heterocyclic synthesis. *Journal of Photochemistry and Photobiology C: Photochemistry Reviews*, 2005, *6* (2), 139–167.

127. Gupta, M.; Wong, E.; Leong, W. *Microwaves and Metals*. Wiley-Interscience, New York, 2007, 256 pp.

128. Komarneni, S.; Katsuki, H.; Li, D.; Bhalla, A. S. Microwave–polyol process for metal nanophases. *Journal of Physics: Condensed Matter*, 2004, *16*, S1305–S1312.

129. Liu, F.-K.; Huang, P.-W.; Chang, Y.-C.; Ko, C.-J.; Ko, F.-H.; Chu, T.-C. Formation of silver nanorods by microwave heating in the presence of gold seeds. *Journal of Crystal Growth*, 2005, *273* (3–4), 439–445.

130. Liu, F. K.; Huang, P. W.; Chu, T. C.; Ko, F. H. Gold seed-assisted synthesis of silver nanomaterials under microwave heating. *Materials Letters*, 2005, *59* (8), 940–944.

131. Huang, H.; Zhang, S.; Qi, L.; Yu, X.; Chen, Y. Microwave-assisted deposition of uniform thin gold film on glass surface. *Surface and Coatings Technology*, 2006, *200* (14), 4389–4396.

132. Li, X.; Chen, W. X.; Zhao, J.; Xing, W.; Xu, Z. D. Microwave polyol synthesis of Pt/CNTs catalysts: Effects of pH on particle size and electrocatalytic activity for methanol electrooxidization. *Carbon*, 2005, *43* (10), 2168–2174.

133. Zhu, H.; Zhang, C.; Yin, Y. Novel synthesis of copper nanoparticles, influence of the synthesis conditions on the particle size. *Nanotechnology*, 2005, *16* (12), 3079–3083.

134. Kharissova, O. V.; Osorio, M.; Garza, M. Synthesis of bismuth by microwave irradiation. *MRS Fall Meeting*, Boston, MA, November 26–30, 2007. Abstract II5.42. p. 773.

135. Chau, J. L. H. Synthesis of Ni and bimetallic FeNi nanopowders by microwave plasma method. *Materials Letters*, 2007, *61* (13), 2753–2756.

136. Chau, J. L. H.; Hsu, M. K.; Hsieh, C. C.; Kao, C. C. Microwave plasma synthesis of silver nanopowders. *Materials Letters*, 2005, *59* (8), 905–908.

137. Kalyanaraman, R.; Yoo, S.; Krupashankara, M. S.; Sudarshan, T. S.; Dowding, R. J. Synthesis and consolidation of iron nanopowders. *Nanostructured Materials*, 1998, *10* (8), 1379–1392.

138. Poddar, P.; Wilson, J. L.; Srikanth, H.; Ravi, B. G.; Wachsmuth, J.; Sudarshan, T. S. Grain size influence on soft ferromagnetic properties in Fe-Co nanoparticles. *Materials Science and Engineering B*, 2004, *106* (1), 95–100.

139. Brenner, J. R.; Harkness, J. B. L.; Knickelbein, M. B.; Krumdick, G. K.; Marshall, C. L. Microwave plasma synthesis of carbon-supported ultrafine metal particles. *Nanostructured Materials*, 1997, *8* (1), 1–17.

140. Yang, G. W. Laser ablation in liquids: Applications in the synthesis of nanocrystals. *Progress in Materials Science*, 2007, *52* (4), 648–698.

141. Gaertner, G. F.; Lydtin, H. Review of ultrafine particle generation by laser ablation from solid targets in gas flows. *Nanostructured Materials*, 1994, *4* (5), 559–568.

142. Phipps, C. (Ed.). *Laser Ablation and its Applications*, Springer Series in Optical Sciences, 1st edn., Springer, Berlin, Germany, 2006, 588 pp.

143. Perrière, J.; Boulmer-Leborgne, C.; Benzerga, R.; Tricot, S. Nanoparticle formation by femtosecond laser ablation. *Journal of Physics D: Applied Physics*, 2007, *40* (22), 7069–7076.

144. Cramer, L. P.; Langford, S. C.; Dickinson, J. T. The formation of metallic nanoparticles in single crystal CaF_2 under 157 nm excimer laser irradiation. *Journal of Applied Physics*, 2006, *99*, 054305, 5 pp.

145. Henley, S. J.; Carey, J. D.; Silva, S. R. P. Metal nanoparticle production by pulsed laser nanostructuring of thin metal films. *Applied Surface Science*, 2007, *253* (19), 8080–8085.

146. Pithawalla, Y. B.; Deevi, S. C.; El-Shall, M. S. Preparation of ultrafine and nanocrystalline FeAl powders. *Materials Science and Engineering A*, 2002, *329*, 92–98.

147. Kabashin, A. V.; Meunier, M. Femtosecond laser ablation in aqueous solutions, a novel method to synthesize non-toxic metal colloids with controllable size. *Journal of Physics: Conference Series*, 2007, *59*, 354–359.

148. Kazakevich, P. V.; Simakin, A. V.; Voronov, V. V.; Shafeev, G. A. Laser induced synthesis of nanoparticles in liquids. *Applied Surface Science*, 2006, *252* (13), 4373–4380.

149. Resta, V.; Siegel, J.; Bonse, J.; Gonzalo, J.; Afonso, C. N.; Piscopiello, E.; Van Tenedeloo G. Sharpening the shape distribution of gold nanoparticles by laser irradiation. *Journal of Applied Physics*, 2006, *100*, 084311.

150. Senkan, S.; Kahn, M.; Duan, S.; Ly, A.; Leidholm, C. High-throughput metal nanoparticle catalysis by pulsed laser ablation. *Catalysis Today*, 2006, *117* (1), 291–296.

151. Becker, M. F.; Brock, J. R.; Cai, H.; Henneke, D. E.; Keto, J. W.; Lee, J.; Nichols, W. T.; Glicksman, H. D. Metal nanoparticles generated by laser ablation. *Nanostructured Materials*, 1998, *10* (5), 853–863.

152. Molokovsky, S. I.; Sushkov, A. D. *Intense Electron and Ion Beams (Particle Acceleration and Detection)*, 1st edn., Springer, Berlin, Germany, 2005, 281 pp.

153. Jin, Z. Q.; Liu, J. P. Rapid thermal processing of magnetic materials. *Journal of Physics D: Applied Physics*, 2006, *39* (14), R227–R244.

154. Mendes, P. M.; Chen, Y.; Palmer, R. E.; Nikitin, K.; Fitzmaurice, D.; Preece, J. A. Nanostructures from nanoparticles. *Journal of Physics: Condensed Matter*, 2003, *15*, S3047–S3063.

155. Wang, F.; Malac, M.; Egerton, R. F.; Meldrum, A.; Zhu, X.; Liu, Z.; Macdonald, N. et al. Multilayer route to iron nanoparticle formation in an insulating matrix. *Journal of Applied Physics*, 2007, *101*, 034314, 7 pp.

156. Li, Y.; Kim, Y. N.; Lee, E. J.; Cai, W. P.; Cho, S. O. Synthesis of silver nanoparticles by electron irradiation of silver acetate. *Nuclear Instruments and Methods in Physics Research Section B: Beam Interactions with Materials and Atoms*, 2006, *251* (2), 425–428.

157. Li, C. M.; Robertson, I. M.; Jenkins, M. L.; Hutchison, J. L.; Doole, R. C. In situ TEM observation of the nucleation and growth of silver oxide nanoparticles. *Micron*, 2005, *36* (1), 9–15.

158. Hou, X. G.; Gu, X. N.; Hu, Y.; Zhang, J. F.; Liu, A. D. Enhanced Pt/TiO$_2$ thin films prepared by electron beam irradiation. *Nuclear Instruments and Methods in Physics Research Section B: Beam Interactions with Materials and Atoms*, 2006, *251* (2), 429–434.

159. Judai, K.; Nishijo, J.; Okabe, C.; Ohishi, O.; Sawa, H.; Nishi, N. Carbon-skinned metallic wires and magnetic nanocrystals prepared from metal acetylides. *Synthetic Metals*, 2005, *155* (2), 352–356.

160. Giannuzzi, L. A.; Stevie, F. A. (Eds.). *Introduction to Focused Ion Beams, Instrumentation, Theory, Techniques and Practice*, 1st edn., Springer, Berlin, Germany, 2004, 358 pp.

161. Kumar, P.; Rodrigues, G.; Lakshmy, P. S.; Kanjilal, D.; Singh, B. P.; Kumar, R. Development of metallic ion beams using ECRIS. *Nuclear Instruments and Methods in Physics Research Section B: Beam Interactions with Materials and Atoms*, 2006, *252* (2), 354–360.

162. Yu, G. Q.; Tay, B. K.; Zhao, Z. W.; Sun, X. W.; Fu, Y. Q. Ion beam co-sputtering deposition of Au/SiO$_2$ nanocomposites. *Physica E, Low-Dimensional Systems and Nanostructures*, 2005, *27* (3), 362–368.

163. Jacobsohn, L. G.; Hawley, M. E.; Cooke, D. W.; Hundley, M. F.; Thompson, J. D.; Schulze, R. K.; Nastasi, M. Synthesis of cobalt nanoparticles by ion implantation and effects of postimplantation annealing. *Journal of Applied Physics*, 2004, *96*, 4444–4450.

164. Klaumunzer, S. Modification of nanostructures by high-energy ion beams. *Nuclear Instruments and Methods in Physics Research Section B: Beam Interactions with Materials and Atoms*, 2006, *244* (1), 1–7.

165. Rodrigues, A. R.; Soares, J. M.; Machado, F. L. A.; de Azevedo, W. M.; de Carvalho, D. D. Synthesis of α-Fe particles using a modified metal-membrane incorporation technique. *Journal of Magnetism and Magnetic Materials*, 2007, *310* (2), Part 3, 2497–2499.

166. Toimil Molares M. E.; Höhberger, E. M.; Schaeflein, Ch.; Blick, R. H.; Neumann, R.; Trautmann, C. Electrical characterization of electrochemically grown single copper nanowires. *Applied Physics Letters*, 2003, *82* (13), 2139–2141.

167. Wang, H.; Ding, J.; Lee, B.; Wang, X.; Lin, T. Polypyrrole-coated electrospun nanofibre membranes for recovery of Au(III) from aqueous solution. *Journal of Membrane Science*, 2007, *303* (1–2), 119–125.

168. Abate, S.; Melada, S.; Centi, G.; Perathoner, S.; Pinna, F.; Strukul, G. Performances of Pd-Me (Me=Ag, Pt) catalysts in the direct synthesis of H$_2$O$_2$ on catalytic membranes. *Catalysis Today*, 2006, *117* (1), 193–198.

169. Bogdanovic, B.; Bönnemann, H. Ger. Offen DE 3 541 633, 1987; U.S. Patent 4 713 110, 1987.

170. Zhang, Y.; Liao, S.; Fan, Y.; Xu, J.; Wang, F. Chemical reactivities of magnesium nanopowders. *Journal of Nanoparticle Research*, 2001, *3* (1), 23–26.

171. Hakim, L. F.; Vaughn, C. L.; Dunsheath, H. J.; Carney, C. S.; Liang, X.; Li, P.; Weimer, A. W. Synthesis of oxidation-resistant metal nanoparticles via atomic layer deposition. *Nanotechnology*, 2007, *18*, 345603, 7 pp.

172. Klinke, C.; Kern, K. Iron nanoparticle formation in a metal–organic matrix, from ripening to gluttony. *Nanotechnology*, 2007, *18* (21), 215601, 4 pp.

173. Sowka, E.; Leonowicz, M.; Andrzejewski, B.; Pomogailo, A. D.; Dzhardimalieva, G. I. Processing and properties of composite magnetic powders containing Co nanoparticles in polymeric matrix. *Journal of Alloys and Compounds*, 2006, *423* (1), 123–127.

174. Salavati-Niasari, M.; Davar, F.; Mazaheri, M.; Shaterian, M. Preparation of cobalt nanoparticles from [bis(salicylidene)cobalt(II)]-oleylamine complex by thermal decomposition. *Journal of Magnetism and Magnetic Materials*, 2008, *320* (3), 575–578.

175. Syukri; Ban, T.; Ohya, Y.; Takahashi, Y. A simple synthesis of metallic Ni and Ni-Co alloy fine powders from a mixed-metal acetate precursor. *Materials Chemistry and Physics*, 2003, *78* (3), 645–649.

176. Harutyunyan, A.; Grigorian, L.; Tokune, T. Method for synthesis of metal nanoparticles. U.S. Patent 6974493, 2005. http://www.freepatentsonline.com/6974493.html

177. Bunge, S. D.; Boyle, T. J. Synthesis metal nanoparticle. U.S. Patent 6929675, 2005. http://www.freepatentsonline.com/6929675.html

178. Duron, S.; Rivera-Noriega, R.; Nkeng, P.; Poillerat, G.; Solorza-Feria, O. Kinetic study of oxygen reduction on nanoparticles of ruthenium synthesized by pyrolysis of Ru$_3$(CO)$_{12}$. *Journal of Electroanalytical Chemistry*, 2004, *566* (2), 281–289.

179. Keller, T. M; Perrin, J.; Qadri, S. B. Metal nanoparticle thermoset and carbon compositions from mixtures of metallocene-aromatic-acetylene compounds. U.S. Patent 6884861, 2005. http://www.freepatentsonline.com/6884861.html

180. Nishijo, J.; Okabe, C.; Oishi, O.; Nishi, N. Synthesis, structures and magnetic properties of carbon-encapsulated nanoparticles via thermal decomposition of metal acetylide. *Carbon*, 2006, *44* (14), 2943–2949.

181. Wampler, T. (Ed.). *Applied Pyrolysis Handbook*, 2nd edn., CRC Press, Boca Raton, FL, 2006, 304 pp.

182. Kim, J. H.; Babushok, V. I.; Germer, T. A.; Mulholland G. W.; Ehrman, S. H. Co-solvent assisted spray pyrolysis for the generation of metal particles. *Journal of Materials Research*, 2003, *18* (7), 1614–1622.

183. Kim, J. H.; Germer, T. M.; Mulholland, G.; Ehrman, S. H. Size-monodisperse metal nanoparticles via hydrogen-free spray pyrolysis. *Advanced Materials*, 2002, *14*, 518–521.

184. Keller, T. M.; Qadri, S. B. Synthesis of metal nanoparticle compositions from metallic and ethynyl compounds. U.S. Patent 7273509, 2007. http://www.freepatentsonline.com/7273509.html

185. Yang, H.; Ito, F.; Hasegawa, D.; Ogawa, T.; Takahashi, M. Facile large-scale synthesis of monodisperse Fe nanoparticles by modest-temperature decomposition of iron carbonyl. *Journal of Applied Physics*, 2007, *101*, 09J112, 3 pp.

186. Maskrot, H.; Leconte, Y.; Herlin-Boime, N.; Reynaud, C.; Guelou, E.; Pinard, L.; Valange, S.; Barrault, J.; Gervaise, M. Synthesis of nanostructured catalysts by laser pyrolysis. *Catalysis Today*, 2006, *116* (1), 6–11.

187. Lu, Y.; Zhu, Z.; Liu, Z. Carbon-encapsulated Fe nanoparticles from detonation-induced pyrolysis of ferrocene. *Carbon*, 2005, *43* (2), 369–374.

188. Kim, K. N.; Kim, S. G. Nickel particles prepared from nickel nitrate with and without urea by spray pyrolysis. *Powder Technology*, 2004, *145* (3), 155–162.

189. Blank, V. D.; Alshevskiy, Yu. L.; Belousov, Yu. A.; Kazennov, N. V.; Perezhogin, I. A.; Kulnitskiy, B. A. TEM studies of carbon nanofibres formed on Ni catalyst by polyethylene pyrolysis. *Nanotechnology*, 2006, *17* (8), 1862–1866.

190. Aleandri, L. E.; Bogdanovic, B.; Bons, P.; Duerr, C.; Gaidies, A.; Hartwig, T.; Huckett, S. C.; Lagarden, M.; Wilczok, U.; Brand, R. A. Inorganic grignard reagents. Preparation and their application for the synthesis of highly active metals, intermetallics, and alloys. *Chemistry of Materials*, 1995, *7*, 1153–1170.

191. Aleandri, L. E.; Bogdanovic, B.; Dürr, C.; Jones, D. J.; Rozière, J.; Wilczok, U. Nanoparticular intermetallics and alloys via inorganic Grignard reagents. *Advanced Materials*, 1996, *8* (7), 600–604.

192. Aleandri, L. E.; Bogdanovic, B. The magnesium route to active metals and intermetalics. In *Active Metals. Preparation, Characterization, Applications*. Fürstner A. (Ed.). VCH, Weinheim, Germany, 1996, pp. 299–338.

193. Frommeyer, G.; Knott, W.; Weier, A.; Weiss, J.; Windbiel, D. Use of zinc treated with metal hydride in organometallic synthesis. U.S. Patent 6521771, 2003. http://www.wikipatents.com/6521771.html

194. Mamatha, M.; Bogdanović, B.; Felderhoff, M.; Pommerin, A.; Schmidt, W.; Schüth, F.; Weidenthaler, C. Mechanochemical preparation and investigation of properties of magnesium, calcium and lithium–magnesium alanates. *Journal of Alloys and Compounds*, 2006, *407* (1–2), 78–86.

195. Murphy, C. J.; Sau, T. K.; Gole, A. M.; Orendorff, C. J.; Gao, J.; Gou, L.; Hunyadi, S. E.; Li, T. Anisotropic metal nanoparticles, synthesis, assembly, and optical applications. *Journal of Physical Chemistry B*, 2005, 109, 13857–13870.

196. Wagner, J.; Tshikhudo, T. R.; Köhler, J. M. Microfluidic generation of metal nanoparticles by borohydride reduction. *Chemical Engineering Journal*, 2008, *135* (1), S104–S109.

197. Bönnemann, H.; Brijoux, W.; Joussen, T. Herstellung feinverteilter Metall- und Legierungspulver. *Angewandte Chemie*, 1990, 102, 324–326; *Angewandte Chemie International Edition* (English), 1990, *29*, 273.

198. Bönnemann, H.; Brijoux, W.; Brinkmann, R.; Dinjus, E.; Joussen, T.; Korall, B. Erzeugung von kolloiden Übergangsmetallen in organischer Phase und ihre Anwendung in der Katalyse. *Angewandte Chemie*, 1991, *103*, 1344–1346; *Angewandte Chemie International Edition* (English), 1991, *30*, 1312.

199. Wojcieszak, R.; Monteverdi, S.; Ghanbaja, J.; Bettahar, M. M. Study of Ni–Ag/SiO$_2$ catalysts prepared by reduction in aqueous hydrazine. *Journal of Colloid Interface Science*, 2008, *317* (1), 166–174.

200. Peng, X.; Lin Song, L.; Jana, N. Monodisperse noble metal nanocrystals. U.S. Patent 7160525, 2007. http://www.freepatentsonline.com/7160525.html.

201. Sun, Y.-P.; Lia, X.-Q.; Zhanga, W.-X.; Wang, H. P. A method for the preparation of stable dispersion of zero-valent iron nanoparticles. *Colloids and Surfaces A: Physicochemical and Engineering Aspects*, 2007, *308* (1–3), 60–66.

202. Ashida, T.; Miura, K.; Nomoto, T.; Yagia, S.; Sumida, H.; Kutluk, G.; Soda, K.; Namatame, H.; Taniguchi, M. Synthesis and characterization of Rh(PVP) nanoparticles studied by XPS and NEXAFS. *Surface Science*, 2007, *601* (18), 3898–3901.

203. Popa, M.; Pradell, T.; Cresp, D.; Calderón-Moreno J. M. Stable silver colloidal dispersions using short chain polyethylene glycol. *Colloids and Surfaces A: Physicochemical and Engineering Aspects*, 2007, *303* (3), 184–190.

204. Padalkar, S.; Hulleman, J.; Deb, P.; Cunzeman, K.; Rochet, J. C.; Stach, E. A.; Stanciu, L. Alpha-synuclein as a template for the synthesis of metallic nanowires. *Nanotechnology*, 2007, *18*, 055609, 9 pp.

205. Abdel-Aal, E. A.; Malekzadeh, S. M.; Rashad, M. M.; El-Midany, A. A.; El-Shall, H. Effect of synthesis conditions on preparation of nickel metal nanopowders via hydrothermal reduction technique. *Powder Technology*, 2007, *171* (1), 63–68.

206. Liu, X.; Liang, X.; Zhang, N.; Qiu, G.; Yi, R. Selective synthesis and characterization of sea urchin-like metallic nickel nanocrystals. *Materials Science and Engineering B*, 2006, *132* (3), 272–277.
207. Yang, J.-G.; Yang, S.-H.; Okamoto, T.; Bessho, T.; Satake, S.; Ichino, R.; Okido, M. Synthesis of copper monolayer and particles at aqueous–organic interface. *Surface Science*, 2006, *600* (24), L318–L320.
208. Zhang, Y. C.; Wang, G. Y.; Hu, X. Y.; Xing, R. Preparation of submicrometer-sized copper and silver crystallites by a facile solvothermal complexation-reduction route. *Journal of Solid State Chemistry*, 2005, *178* (5), 1609–1613.
209. Bhainsa, K. C.; D'Souza, S. F. Extracellular biosynthesis of silver nanoparticles using the fungus *Aspergillus fumigatus*. *Colloids and Surfaces B: Biointerfaces*, 2006, *47* (2), 160–164.
210. Vigneshwaran, N.; Ashtaputre, N. M.; Varadarajan, P. V.; Nachane, R. P.; Paralikar, K. M.; Balasubramanya, R. H. Biological synthesis of silver nanoparticles using the fungus *Aspergillus flavus*. *Materials Letters*, 2007, *61* (6), 1413–1418.
211. Vigneshwaran, N.; Kathe, A. A.; Varadarajan, P. V.; Nachane, R. P.; Balasubramanya R. H. Biomimetics of silver nanoparticles by white rot fungus, *Phaenerochaete chrysosporium*. *Colloids and Surfaces B: Biointerfaces*, 2006, *53* (1), 55–59.
212. Schabes-Retchkiman P. S.; Canizal, G.; Herrera-Becerra, R.; Zorrilla, C.; Liu, H. B.; Ascencio, J. A. Biosynthesis and characterization of Ti/Ni bimetallic nanoparticles. *Optical Materials*, 2006, *29* (1), 95–99.
213. Canizal, G.; Schabes-Retchkiman, P. S.; Pal, U.; Liu, H. B.; Ascencio, J. A. Controlled synthesis of Zn nanoparticles by bioreduction. *Materials Chemistry and Physics*, 2006, *97* (2), 321–329.
214. Pollmann, K.; Raff, J.; Merroun, M.; Fahmy, K.; Selenska-Pobell, S. Metal binding by bacteria from uranium mining waste piles and its technological applications. *Biotechnology Advances*, 2006, *24* (1), 58–68.

23 Activated Micro- and Nanostructured *Rieke* Metals*

Reactions of inorganic and organic substances on elemental metal surfaces are an important area of chemistry from the point of view of catalysis by metals. Throughout the history of chemistry and chemical technology, the reactivity of metals that are naturally "insufficiently active" has been subjected to continuous successful attempts to increase them. The main objective of these research activities is to allow known reactions to be carried out under softer conditions, with higher yields and lesser energetic expenses. One form of achieving this goal is the development of special techniques for elemental metal activation.

The analysis of available literature on the preparation of activated metals reveals that at least eight basic methods can be applied[1–3]:

1. Mechanical reduction in an inert atmosphere or solvents
2. Formation of alloys, for example, Zn-Cu
3. Addition of catalysts or activators to reaction mixture, for example, addition of I_2 or R_3N
4. Chemical purification of metallic surface, for example, with $BrCH_2CH_2Br$
5. Reduction of metal salts with hydrogen or hydrides
6. Ultrasonic treatment
7. Codeposition (or co-condensation) of metal vapor and ligand vapor (or deposition of metal vapor to a cooled surface with a frozen ligand)
8. *Rieke* procedure for the reduction of metal salts

These methods and others (in particular, combustion synthesis, electrochemical deposition, the use of γ- and x-rays, laser ablation, biological techniques, etc.) have been described in a recent review.[4] In this section, we explore reported synthetic procedures for generation of activated metallic powders via *Rieke* techniques. The classic term *Rieke* metal was adopted from original articles in which it has been referenced by their authors.

23.1 HISTORY

Rieke method for the generation of highly reactive metallic powders was discovered, as almost always, accidently while resolving another problem. *Rieke* and *Hundall* studied the formation of a radical anion derived from the reduction of 1,2-dibromobenzocyclobuthene via the interaction of solvated electrons generated by the dissolution of metallic potassium in bis(2-methoxyethyl)ether (diglime) a −78°C forming benzocyclobuthadiene, which could further react with more solvated electrons and be reduced to radical anion (Scheme 23.1). According to the physicochemical studies, the presence of dianion was suggested, which could react with a magnesium salt, forming a di-*Grignard* reagent.

The generation of solvated electrons via dissolution of metallic potassium resulted in the inconveniently slow formation of the di-Grignard, so the use of potassium biphenylide was suggested. Its addition to magnesium salt *before* the addition of 1,2-dibromobenzocyclobutane led to a black solid containing zero-valent magnesium of an elevated reactivity.

---

* This chapter has been contributed by Dr. Luis Angel Garza-Rodríguez (UANL).

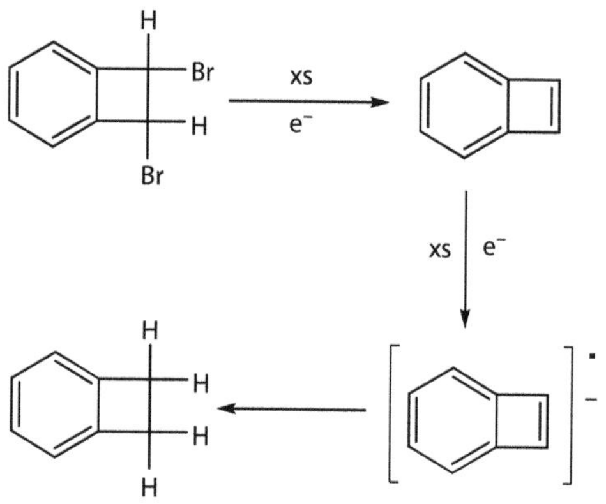

SCHEME 23.1   Intermediate formation of the radical-anion.

Early reports of *Rieke* did not mention reductive organic agents (electron carriers) due to the skepticism of researchers of that time about the use of organic substances as reductors, so all publications of the 1970s made references to the use of alkali metals (Na, K) in solvents, whose boiling point is higher than the melting point of an alkali metal. The said procedure functions excellently for the synthesis of Mg*, Al*, In*, and Tl* but is worse for transition metals. For these last metals, the first attempts to produce them in the activated form required a slight modification: in the synthesis of Ni*, Pd*, Pt*, Co*, Fe*, and Cr*, the adduct of the metal salt with a phosphine ligand (triethylphosphine, triphenylphosphine, etc.) is first prepared and then it is reduced with potassium in tetrahydrofuran (THF). The presence of phosphines is not always desirable: they have a bad smell and generally it is difficult to separate them from reaction products. Therefore, at the beginning of the 1980s, researchers started to use electron carriers in catalytic quantities (5%–10% on the basis of alkali metal), which are easily eliminated from the reaction medium, allow reactions at low temperatures, and facilitate the incorporation of metallic lithium.

It is noted that *Reuben D. Rieke* was not the first researcher who utilized alkali metals for the reduction of metal salts in obtaining elemental-state species. With the isolation of metallic sodium by *Davy*, *Oersted* discovered its first practical use for reduction of aluminum chloride for the production of metallic aluminum. The majority of metals and metalloids have been isolated from their halides by reduction with sodium. Therefore, these reduction processes could be considered as a universal method for their preparation.[5] In 1925, *Kraus* and *Kurtz* showed that metallic sodium in liquid ammonia can reduce halides of those metals, which form alloys with sodium.[6] Scott et al.[7] reported the formation of nickel via the reaction of sodium, solution of naphthalene in 1,2-dimethoxyethane, and nickel chloride (particle size <20 μm). Chu[8] conductometrically studied the reduction of $CoCl_2$ with sodium naphthalenide in THF solutions and obtained relations of the reaction course and the data on magnetic susceptibilities and light absorption. In 1946, Whaley[5] carried out the synthesis of finally divided particles of Fe, Ni, Co, Mn, Cd, Zn, Sn, Ag, and Cu via the reduction of halide dispersions with sodium dispersions in hydrocarbon solvents or ether. Olive[9] in 1967 reported the reduction of VCl_3, $CrCl_3$, $NiBr_2$, and $PtCl_4$ with lithium naphthalenide in THF. An elevated stability of reaction solutions was interpreted by the formation of a complex of the metal in reduced form with naphthalene and hydrogen transfer from this aromatic hydrocarbon to the metal. In 1977, Arnold[10] obtained Zn* and Mg* by *Rieke* method via the reduction of their chlorides with a stoichiometric quantity of sodium naphthalenide.

Despite this, *Reuben D. Rieke* was the *first* researcher (and really the only enthusiastic investigator worldwide) who *systematically developed* synthetic procedures (which at present have his name) for obtaining highly reactive metal powders; moreover, he was the first who showed their applications (nowadays numerous) in organic and organometallic reactions.

23.2 GENERAL CHARACTERISTICS OF *RIEKE* METHODS

23.2.1 PRECAUTIONS

The activation of metals according to *Rieke* procedures generates powders with various grades of reactivity. An ideal methodology for metal activation should satisfy a series of criteria: first, a highly reactive metallic form could be generated, which, among others, can produce organometallic compounds under mild conditions (room temperature [r.t.] or lower); second, the method involves equipment and common reagents available in any laboratory; third, the method can be general for a wide range of transition metals; finally, both active metal and organometallic compound can be generated for a reasonable period of time. Other additional factors can influence the ability to use a wide variety of solvents and the ability to store the formed metal powders without a considerable loss of their reactivity.

Even a researcher is familiar with synthetic practices; it is highly recommended to take necessary measures while working with active metals. During reduction processes, the reaction medium should be exposed to Ar atmosphere to avoid reactions with oxygen and water of air. It was reported that if Mg* and Al* powders are dehydrated before being exposed to air, they possess pyrophoric properties. A potential problem could be self-heating of the reaction mixture during the reduction process; fortunately, this problem has been observed only for the reduction of $ZnCl_2$ and $FeCl_3$ in THF. Additionally, the use of alloys Na/K as reductive agents is not recommended due to possible explosions.

23.2.2 REDUCTION METHODS OF SYNTHESIS

Although the initial discovery of Mg* was carried out by the reduction of $MgCl_2$ with lithium biphenylide, the first investigations related to the formation of activated metals involved the reduction of solutions of metal salts with alkali metals in solvents. Three main methods for generation of activated powders have been applied as follows.

23.2.2.1 Method 1

The reduction is carried out in the presence of an alkali metal in solvents (Scheme 23.2), whose boiling point exceeds the melting point of alkali metal. The metal salts should be partially or totally soluble in the solvent; the reactions are carried out in inert atmosphere, preferably in Ar. Reduction processes are generally exothermal and complete for some hours. In addition to the activated metal, a stoichiometric quantity of alkali metal salt is produced. The common systems, employed by this method, are shown in Table 23.1.

23.2.2.2 Method 2

This technique employs an alkali metal and an electron carrier, such as naphthalene, which is added to the reaction medium in a quantity lower than the stoichiometric (about 5%–10% on the basis of mols of the metal salt to reduce) (Scheme 23.3). These reactions are carried out at r.t. or lower. The presence of a catalytical quantity of electron carrier (naphthalene) continuously regenerates lithium naphthalenide through the reaction shown in Scheme 23.4.

Among alkali metals, lithium is the most preferred for this method, not only due to its safety but also because of the high reactivity of the formed *Rieke* metals in comparison with those obtained using other alkali metals.

$$MX_n + nK \longrightarrow M^* + nKX$$

$$X = Cl, Br, I$$

SCHEME 23.2 General synthesis of activated metal powders in the presence of metallic potassium.

TABLE 23.1
Mostly Employed Solvents for Sodium and Potassium in the Synthesis of Activated *Rieke* Metal Powders (Method 1)

Alkali Metal	Solvent
K	THF
Na	DME
Na	Benzene
K	Benzene
Na	Toluene
K	Toluene

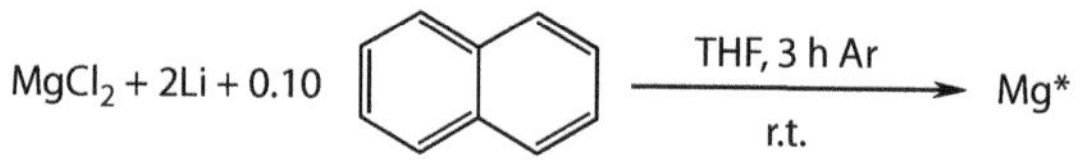

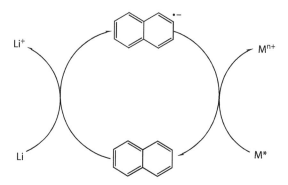

SCHEME 23.3 Reduction of magnesium salt in the presence of a catalytical quantity of naphthalene.

SCHEME 23.4 Regeneration cycle of lithium naphthalenide.

23.2.2.3 Method 3

The third reduction method utilizes a stoichiometric quantity of the electron carrier; the technique requires preliminary preparation of a solution of the anion, for example, lithium naphthalenide. Using this method, the generation of metal powder is fast and controlled by diffusion processes; temperature oscillates in the range from zero to r.t. In some cases, reductions are slow at low temperature due to the poor solubility of metal salts. Reduction time is generally smaller and this way restricts the growth of metal particles, so the obtained powders are more reactive.

If reduction subproducts (alkali metal salts, electron carrier, etc.) affect the course of further reactions of activated metals, it is recommended to allow sedimentation of metal powder, extract floating substance with syringe, and repeat several times with the same freshly distilled solvent or more adequate solvent.

23.2.3 BASIC PRINCIPLES OF THE SYNTHESIS

The operation conditions for obtaining *Rieke* metals are relatively mild and moderate; elevated temperature or pressure is not required.[1,11] The conditions are strict, since inert and dry medium is required; metal salts and solvents should be completely dry and free of oxygen. For the majority of prepared metals, at the end of reaction a finally divided black powder is formed, which sediments slowly (for some hours) and in colored liquid phase. For some activated powders, the reactivity level is identical with commercially available metal powder. However, for other metals the use of the methods mentioned earlier results in superior reactivity.

23.2.4 ADVANTAGES AND DISADVANTAGES OF METHOD OF SYNTHESIS OF *RIEKE* METALS

Among other *advantages* in the preparation of activated metal powders, we note the low cost of utilized equipment and simplicity of its obtaining: flasks with one to three necks, dry box, magnetic stirring, refrigerator, etc., that is, a standard chemical laboratory material. The great advantage of *Rieke* metals is the possibility to carry out with their use a wide spectrum of organic/organometallic reactions unavailable by conventional methods. Moreover, yields using activated metals are almost always higher than those in the case of the application of nonactivated metal powders. The high number of defects and imperfections on the activated metal surface and the absence of protective oxide layer and nano- and microsize of particles provide high reactivity in *Rieke* metals. The main *disadvantages* are related to the necessity to work with completely anhydrous metal salts and solvents freshly distilled in Ar atmosphere and free of water and oxygen. Additionally, the reactivity of obtained activated metal powders decreases with respect to time due to aggregation of particles leading to loss of their reactivity. The use of hydrated salts is possible after their thermal treatment to exclude crystallization and coordination type of water molecules; however, very frequently mixtures of oxides and hydroxides (for example, under thermal decomposition of $MgCl_2 * 6H_2O$) are formed. Therefore, one of the safest techniques to get anhydrous salts (halides) is a direct halogenation of elemental metals in strict dry and inert conditions.

23.3 SYNTHESIS OF HIGHLY REACTIVE *RIEKE* METAL POWDERS

23.3.1 GENERAL CHARACTERISTICS

Rieke metals are powders with particle size between 1 and 2 μm, depending on the metal and preparative method. Scanning electron microscope (SEM) studies indicate their structure in the range from sponge-type to polycrystalline material.[12,13] For some metals, such as Al* and In*, x-ray diffraction studies showed lines, which correspond to metal and alkali metal salt; for Mg* and Co*, only alkali salt lines were found, so in this case the metal is amorphous or possesses particle size corresponding to nanoscale (<0.1 μm). Activated metals could contain considerable quantities of such elements as carbon, hydrogen, oxygen, halogens, and alkali metals (Na, K, or Li). Auger spectroscopy data showed that for many activated metals the oxidation number is zero. The surface area varies depending on the metal and applied technique; as an example, for Ni* this magnitude corresponds to 32.7 m²/g.[1]

23.3.2 ANALYSIS OF REPORTED SYNTHESIS OF *RIEKE* METALS

23.3.2.1 *Rieke* Indium

Rieke In*-activated powder[12–17] is obtained by direct reduction of anhydrous indium chloride ($InCl_3$) in the presence of metallic potassium (molar ratio 1:3) in freshly distilled (in N_2 or Ar atmosphere in the presence of $NaAlH_4$) xylene and benzene (among other solvents). The use of ethereal solvents (THF or 1,2-dimethoxyethane [DME]) is not recommended because of their reductive cleavage reactions in the presence of indium. In microphotographs, In* represents a conglomerate of very small crystals (mixture of In* and KCl; particle size 10–0.2 μm, medium size 4.0 μm); this is also confirmed by x-ray data.

Synthesis of In in the system $InCl_3$-K-Xylene (Scheme 23.5):* Add 200 mL of xylene and 4 g of $NaAlH_4$ into a 500 mL flask, distill to 150 mL, weight 2.21 g (10 mmol) of anhydrous $InCl_3$ and

$$InCl_3 + 3K \xrightarrow[\text{Ar}]{\text{Xylene}} In* + 3KCl$$

SCHEME 23.5 Reduction of $InCl_3$ with metallic potassium in xylene.

$$2Tl* + 3 \quad \text{[iodobenzene]} \xrightarrow[\text{140°C, 24 h}]{\text{Xylene}} \text{[biphenyl]}$$

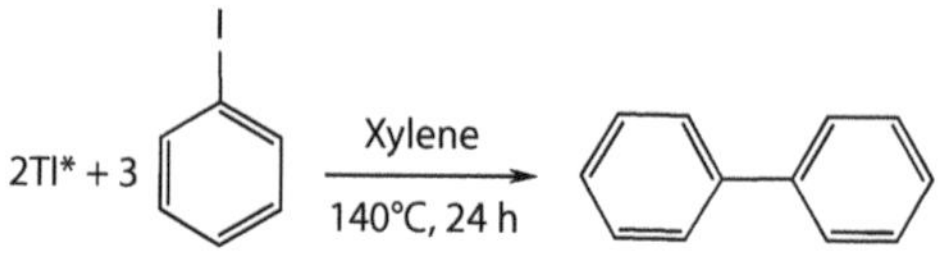

SCHEME 23.6 Synthesis of biphenyl by interaction of Tl* with iodobenzene.

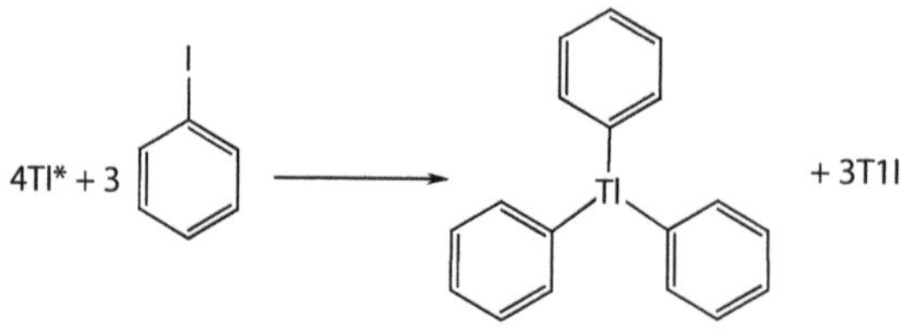

$$4Tl* + 3 \quad \text{[iodobenzene]} \longrightarrow \text{[triphenylthallium]} \quad + 3TlI$$

SCHEME 23.7 Synthesis of triphenylthallium via modification of molar ratio Tl*:iodobenzene.

$$TlCl + K \xrightarrow[\text{Ar, 24 h}]{\text{Xylene}} Tl* + KCl$$

SCHEME 23.8 Synthesis of Tl* via reduction of TlCl with metallic potassium in xylene.

1.17 g (30 mmol) of metallic potassium, and stir at reflux for 4.4 h. Decantate solvent, add 25 mL of xylene, stir, and leave; repeat this step once again.

23.3.2.2 *Rieke* Thallium

Rieke[18] found that the reduction of TlCl with metallic potassium in xylene led to a finally divided black powder (Tl*) with exceptional reactivity. An example of the use of this activated form is its reaction with iodobenzene (molar ratio 1:1.5) forming biphenyl with high yields (Scheme 23.6):

An interesting effect was observed when the reaction was carried out with Tl* excess: for example, a 30% excess in respect of iodobenzene produced mainly triphenylthallium (Scheme 23.7).

Analogous to the synthesis of *Rieke* indium (see earlier), the preparation of Tl* requires the use of nonethereal solvents.

Synthesis of Tl* in the system $TlCl_3$-K-Xylene (Scheme 23.8): Add 300 mL of xylene and 5 g $NaAlH_4$ to a 500 mL flask and distill to 250 mL. Add 1 g (4 mmol) of anhydrous TlCl, 0.16 g (4 mmol) of metallic potassium, and xylene to the reaction flask; stir at reflux for 24 h and wash with xylene.

23.3.2.3 *Rieke* Aluminum

Activated *Rieke* aluminum is prepared by the reduction of its anhydrous halides in the presence of lithium, sodium, and potassium in ethereal solvents and hydrocarbons (THF, xylene, triethylamine); the use of pyridine, DME, and ether leads to incomplete reduction. The melting point of alkali metal defines the solvent to be employed: the solvent should possess a higher boiling point than the melting point of the alkali metal.[1,19] We note that the use of metallic aluminum for reduction of AlX_3 leads to aluminum dihalides AlX_2 (X = Cl, Br).[20] In 1981, Percy Tzu-Jung Li[21] reduced $AlCl_3$ taking such variables as solvent, alkali metal, and electron carrier and found the best conditions (Table 23.2).

According to his results, the combination $AlCl_3$/K/xylene requires 6–15 h of reflux; the use of alloy Na/K allows r.t. reduction. Employment of lithium in the absence or presence of a catalytical quantity of naphthalene generally results an incomplete reduction of the salt. Among other interesting unexpected or obvious observations, the metal (Al*) reactivity can be reduced by the addition

TABLE 23.2
Successful Combinations of Reaction Variables for Obtaining Al*

Alkali Metal	Solvent
K	Xylene
Na/K	Xylene
K + naphthalene	Xylene
NaK + naphthalene	Xylene
Li + naphthalene	THF

$$AlCl_3 + 3Li \xrightarrow[\text{Reflux, Ar}]{\text{THF, xylene}} Al^* + 3LiCl$$

SCHEME 23.9 Synthesis of Al* via reduction of $AlCl_3$ with metallic Li in xylene.

$$AlCl_3 + 3K \xrightarrow[\text{r.t.}]{\text{THF, xylene}} Al^*$$

SCHEME 23.10 Synthesis of Al* via reduction of $AlCl_3$ with metallic K in xylene.

of transition metal salts at the beginning of the reduction process; a 10% excess of $AlCl_3$ in relation with potassium leads to a major reactivity of Al* (synthesis in xylene); $AlCl_3$ is the best precursor among all Al halides forming higher-reactivity Al*; long reduction time and elevated temperature tend to reduce Al* reactivity.

Synthesis of Al* in the system AlCl$_3$-Li-THF-Reflux[21,22] *(Scheme 23.9)*: Add 200 mL of THF to a 500 mL flask, then 2 g Na,* distill to 150 mL. Add 0.53 g (4 mmol) of anhydrous $AlCl_3$,[†] and 0.084 g (12 mmol) of metallic Li[‡] to a 100 mL flask with 15 mL of recently distilled THF, stir and heat until reflux in a dry box for 1–2 h, then leave for some hours. After sedimentation of active powder, decant solvent, add 20 mL of fresh THF, stir and leave for some hours, then decant again.[§]

Synthesis of Al* in the system AlCl$_3$-K-Xylene at r.t.[21] *(Scheme 23.10)*: A potassium piece (1.34 g, 34.3 mmol) and 18 mL of xylene are collocated into a 50 mL flask; the mixture is put into a dry box (Ar), and is heated until the melting of K. 1.52 g (11.4 mmol) of $AlCl_3$ in 15 mL of THF, and collocated into a 25 mL flask. The mixture is stirred until total dissolution, slowly transferred to the flask with melted K for 2.5 h, and stirred at r.t. for 25 h.

23.3.2.4 *Rieke* Calcium

Activated metallic *Rieke* calcium[23,24] is prepared via the reduction of calcium halides dissolved or suspended in THF; the aforementioned salt is transferred slowly to another flask containing a pre-formed dissolution of lithium biphenylide (Scheme 23.11) in THF in Ar atmosphere at r.t. The lithium biphenylide has a double function: (1) as carrier of electrons, of such form that improves the transference of electrons during reduction with the notable decrease in the required time (catalyzing the reaction) and (2) improving the reactivity of the prepared metal.

---

* Metallic Na reacts with water traces and destroys peroxides which could be generated during solvent storage.
† $AlCl_3$ is highly hygroscopic; do not expose it to humidity, and use drying material under weighting.
‡ Metallic Li reacts with atmospheric N_2; minimum exposition to environment is recommended.
§ To avoid formation of Li_3N, Ar atmosphere is recommended instead of nitrogen as reported in Ref. [22].

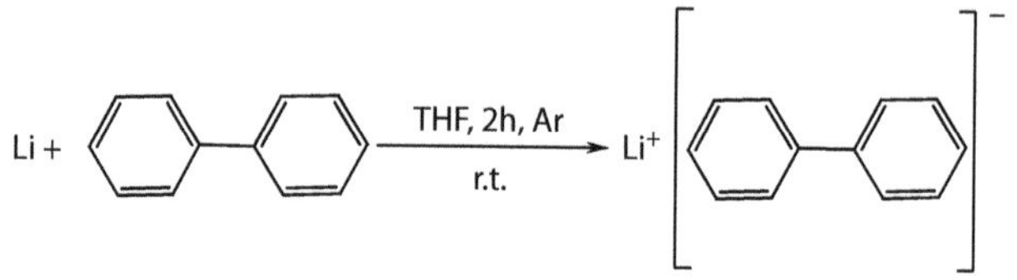

SCHEME 23.11 Synthesis of lithium biphenylide.

In addition to biphenyl, other similar compounds (naphthalene,[25] anthracene,[26] etc.) can be used to produce effective lithium reductants.

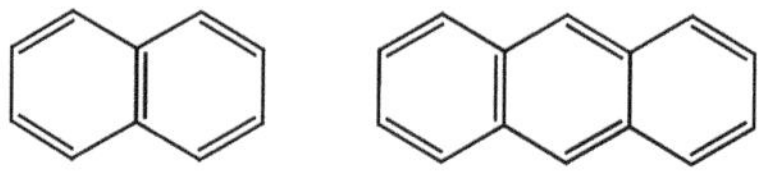

Naphthalene   Anthracene

The synthesis of Ca* presents very interesting variations, when lithium biphenylide is applied (Scheme 23.12): a dark-green *solution* of *activated calcium* is generated, that is, active species of calcium complex with biphenyl is produced, which is very soluble in THF and presents elevated reactivity.[23,27–30]

In the second step of synthesis, biphenyl is replaced by naphthalene,[1,23] which results in the formation of a highly reactive black solid, insoluble in THF and other solvents (such as DMSO and DMF); employing acidic hydrolysis of Ca*, the formation of naphthalene and THF was found. The structure of Ca-naphthalene-THF complex is similar to the soluble species Mg-anthracene-THF (Scheme 23.13).[31–36]

Synthesis of Ca* in the system CaI₂-Li-biphenyl-THF[23,26,27,30]: In a dry box collocate a 500 mL flask with 350 mL of THF and 1 g of Na/K alloy, distill to 250 mL THF in Ar atmosphere, weight 1.51 g (9.8 mmol) of biphenyl and 62.46 mg (9.0 mmol) of metallic Li, and rapidly transfer all to a 100 mL flask (flask A). Collocate the flask A to a dry box, add 20 mL of freshly distilled THF, stir at r.t. until observation of a total dissolution of Li.* In a dry box weigh 1.3224 g (4.5 mmol) of CaI₂ or 0.89959 g (4.5 mmol) of CaBr₂ (4.5 mmol CaBr₂),† collocate it to the flask B, add 20 mL of THF, and stir to

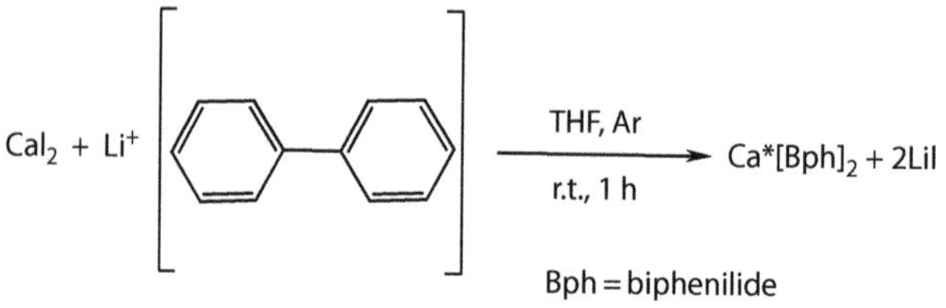

Bph = biphenilide

SCHEME 23.12 Synthesis of a soluble complex of Ca* with biphenyl.

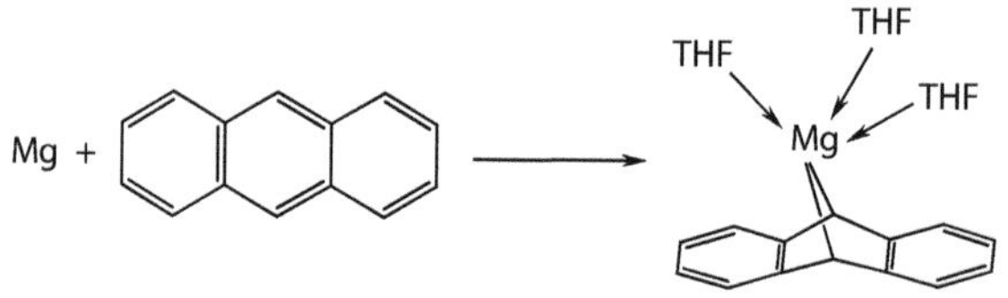

SCHEME 23.13 Synthesis of the adduct of Mg-anthracene (MgA).

---

* Two hours are required for a complete disappearing of lithium; during the reaction of alkali metal with electron carrier, dark green color should be seen, indicating formation of lithium biphenylide.

† Various authors recommend employing an excess of the salt to provide a total absence of reductive agent which could take part in subsequent reactions of the activated metal; a molar ratio 1.5–2.0 of salt per molar equivalent of alkali metal is recommended.

maintain in suspension. Transfer the content from the flask A to B in inert atmosphere and stir at r.t. for 1 h, observing a totally homogeneous dark green solution, consisting of an active metal in the form of a soluble complex in THF.

Synthesis of Ca* from CaI$_2$-Li-naphthalene-THF [23,24]: Collocate in a dry box a 1 L flask with 650 mL of THF, and heat and distill to 500 mL of THF in Ar atmosphere in the presence of Na/K alloy. Transfer 8.46 g of naphthalene (66 mmol C$_{10}$H$_8$) and 0.416 g (60 mmol) of metallic lithium to a 250 mL flask (flask A). Collocate the flask A into a dry box, add 40 mL of freshly distilled THF, and stir until total dissolution of Li (flask A). Transfer 8.816 g (30 mmol) of CaI$_2$ to a 250 mL flask (flask B), add 40 mL of THF, and stir to suspend the metal salt. Transfer the content of the flask A into B in inert atmosphere. Stir at r.t. for 1 h; a black solid is formed, highly reactive and insoluble in THF.

Synthesis of Ca* from CaI$_2$-Li-anthracene-THF [37]: Anthracene is an electron carrier, used for the reduction of calcium salts, generating an activated solution of *Rieke* calcium. Basically, metallic lithium is dissolved in the presence of anthracene in freshly distilled THF (approximately for 3 h); then the mix is added to a suspension of an anhydrous calcium salt (CaI$_2$) in THF for 1 h at r.t.

23.3.2.5 *Rieke* Uranium and Thorium

Starting from 1972, *Rieke* et al have reported procedures for the synthesis of highly reactive metallic powders and their applications in the preparation of organic and organometallic compounds, many of which have been prepared employing ethereal solvents. In the case of actinides, the reduction of uranium anhydrous salts (Scheme 23.14) using Li, Na, and K did not proceed completely due to the passivation of alkali metal surface.[38,39]

In the first reports on *Rieke* uranium, UCl$_4$ was used as a precursor, together with Na/K alloy and DME or THF as a solvent;[38,39] however, solvent decomposition reactions, even at low temperature, were observed. U* exhibits approximately the same reactivity that the pyrophoric activated powder of uranium, obtained via decomposition of uranium hydride, does. The advantage of U* is its simple preparation and it is not pyrophoric. Some applications of the obtained activated uranium are shown in Scheme 23.15.

Later on, a novel synthetic route was offered for preparation of U*, in which its reactivity is not lost due to a reaction with ethereal solvent. This new procedure applies solvents without a carbon-oxygen bond; similarly as for Al* and In* synthesis, nonpolar solvents (benzene, toluene, xylene, decane, etc.) are used. Rieke et al. observed that the most active metals were obtained employing reductive agents, soluble in the reaction solvent. However, a few reductive agents exist, which are soluble in nonpolar solvents. Fujita et al.[40] reported in 1983 the synthesis of a strong reductive agent, soluble in nonpolar solvents; this reductant is a crystalline derivative of lithium naphthalenide dianion [(TMEDA)Li]$_2$[Nap] formed by the reaction of metallic lithium, naphthalene, and *N,N,N',N'*-tetramethyethylendiamine (TMEDA) in toluene in an ultrasonic bath. Khan et al.[41] obtained the crystal of the complex [(TMEDA)Li]$_2$[Nap] by sonication reaction of 1.6 M solution of TMEDA, Li, and naphthalene (2:2:1) in toluene at 42°C. This reductant was used for the preparation of activated *Rieke* uranium; Th* was prepared similarly to U*.

Synthesis of U* in the system UCl$_4$-Na/K-Naphthalene-DME [42] *(Scheme 23.16)*: Distill 200 mL of DME in Ar atmosphere in the presence of a blue solution of sodium cetylbenzophenone. Collocate 0.0269 g (1.17 mmol) of metallic sodium, 0.1177 g (3.010 mmol) of metallic potassium, and 0.3575 g (0.9415 mmol) of anhydrous UCl$_4$, 0.0900 g (0.7022 mmol) of naphthalene* (0.7022 mmol of C$_{10}$H$_8$)

$$UCl_4 + 4M° \xrightarrow[\text{Ar}]{\text{THF or DME}} (1-x)U* + 4 \times M°$$

$$M° = Li, Na, K$$

SCHEME 23.14 Generalized synthesis of U* via reduction of UCl$_4$ with alkali metals in ethereal solvents.

---

* Naphthalene quantity is about 5%–10% molar ratio in respect of alkali metal (Na/K).

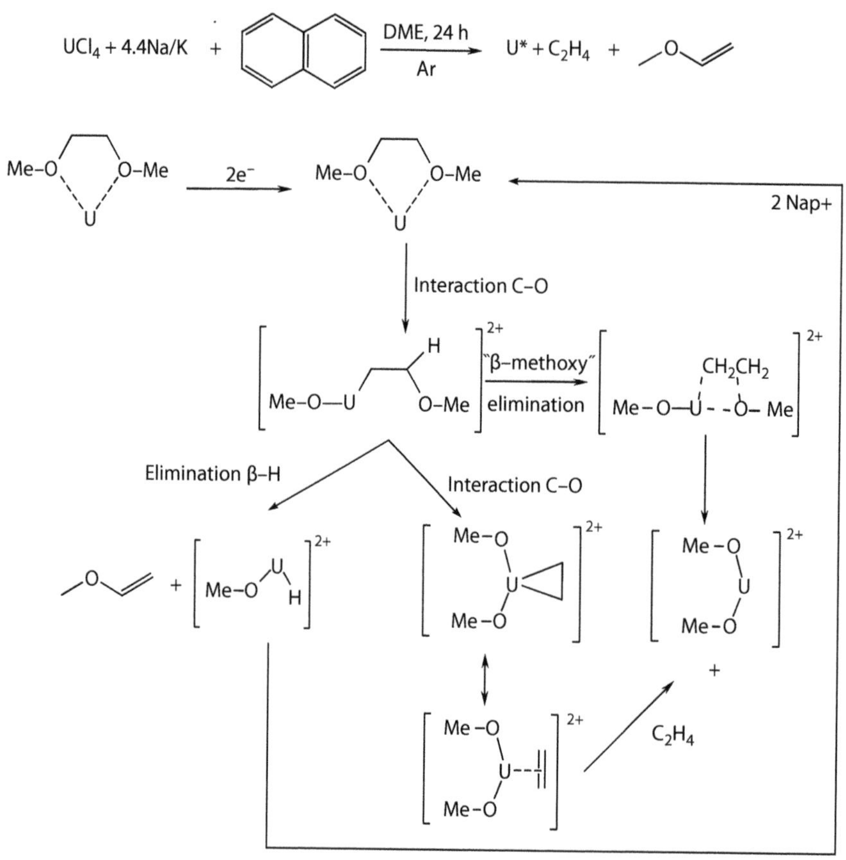

SCHEME 23.15 Applications of U* in organic synthesis.

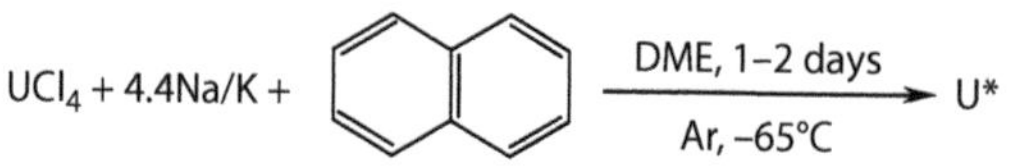

SCHEME 23.16 Synthesis of U* via reduction of UCl_4 with an alloy Na/K in 1,2-dimetoxyethane.

into a 50 mL flask, heat slowly in inert atmosphere until reaching the melting point of potassium, and stir a little to form an alloy.* Collocate the flask into a cooling bath at −60°C and add 5.5 mL of freshly distilled DME. Stir for 1–2 days at −65°C.† Allow the mixture then to reach r.t. A black powder and colorless solvent are formed.

Preparation of the complex [(TMEDA)Li]₂[Nap][42] (Scheme 23.17): Distill in Ar atmosphere 300 mL of TMEDA in the presence of KOH and then in the presence of K. Distill in Ar atmosphere 400 mL of toluene in the presence of sodium cetylbenzophenone. Distill in Ar atmosphere 800 mL of pentane. Collocate in a dry box metallic lithium (5.3144 g, 0.76587 mol) and 49.08 g of naphthalene into a 500 mL flask with three necks, which should be put into an ultrasonic bath.‡ Add 115 mL (0.762 mol) of freshly distilled TMEDA and 200 mL of toluene. Stir 48–92 h until all lithium disappears, cool to −50°C and add 100 mL of freshly distilled pentane, filter and wash

---

* Na/K alloys are highly reactive and do not suffer passivation during reaction. Their melting points vary in function of percent of potassium (50 wt.%K + 15°C, 77.8 wt.%K − 12.6°C).
† Stop stirring once a complete solution of Na/K alloy has been obtained.
‡ Ultrasonic equipment is of low power, which is usually used for washing flasks in laboratory.

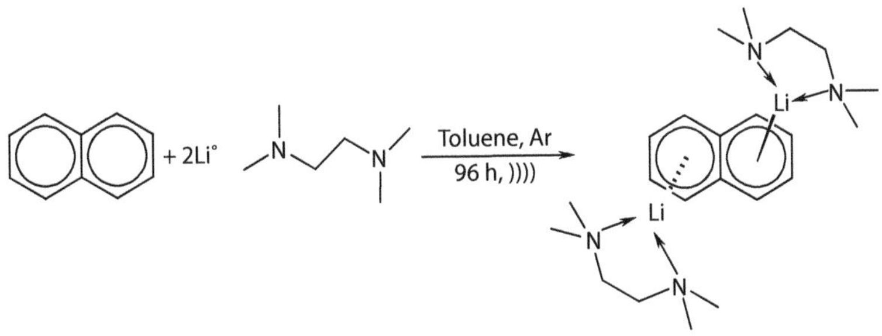

SCHEME 23.17   Synthesis of [(TMEDA)Li]$_2$[Nap].

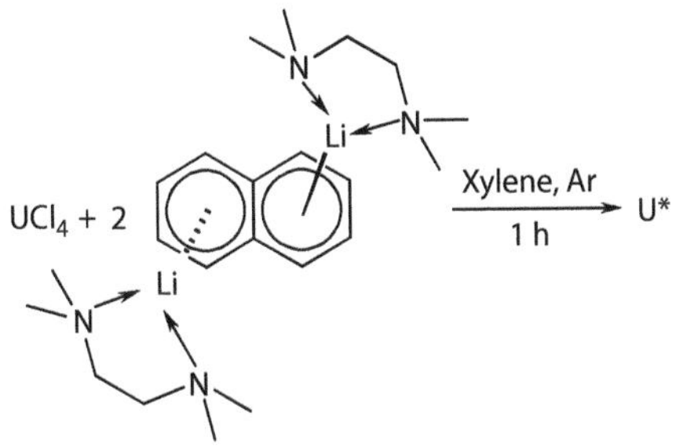

SCHEME 23.18   Synthesis of U* from [(TMEDA)Li]$_2$[Nap].

crystals with two portions of cold pentane (275 mL), and remove a major part of pentane in Ar atmosphere. The yield is 50%.*

Synthesis of U* in the system UCl$_4$-[(TMEDA)Li]$_2$[Nap]-xylene[42] (Scheme 23.18): Distill 100 mL of xylene in Ar atmosphere in the presence of sodium cetylbenzophenone. Collocate 0.502 g (1.322 mmol) of UCl$_4$ and 1.0143 g (2.7086 mmol) of [(TMEDA)Li]$_2$[Nap] into a 50 mL flask, add 20 mL of freshly distilled xylene, and stir at r.t. for 1 h. A dark powder is formed.

23.3.2.6   *Rieke* Copper

Numerous techniques for *Rieke* copper (one of the most important *Rieke* metals) preparation have been reported; the differences between them consist basically in the substitution of insoluble copper salts by the complexes soluble in THF and/or DME. Temperature varies in the range −110 to +70°C, and it plays an essential role in the synthesis.

23.3.3   Use of Phosphine Ligands for Obtaining Cu*

Employment of phosphine ligands generates activated Cu* powders and renders them more reactive and for considerably lesser time, since the complex CuI*PR$_3$ is very soluble in ethereal solvents. The reduction is carried out almost instantly, obtaining metallic copper for 10 min to 2 h. In general, a phosphine ligand with major donor tendency produces Cu* more simply in oxidative addition reactions; moreover, the organocopper complex is more nucleophilic in comparison with classic copper salts. The decrease in reactivity of phosphine ligands is presented as follows: P(NMe$_2$)$_3$ > PEt$_3$ > P(CH$_2$NMe$_2$)$_3$ > P(cyclohexyl)$_3$ > PBu$_3$ > PPh$_3$ > ligand "Diphos" > P(OEt)$_3$.

Ginaj and *Rieke* in diverse articles employed different periods of reduction time varying from 1 min,[43] 20 min,[44] 30 min,[45] to 1 h.[46] Another situation is related to the physical state of the formed

---

* The complex decomposes in nitrogen atmosphere.[42]

activated metal, for example, black-red solution of highly reactive copper or solution/suspension of zero-valent highly reactive copper.

Such phosphines as P(Cy)$_3$ [tri(cyclohexylphosphine)] and HMPT [tris(dimethylamino-phosphine)] also have been employed as complexing agents for CuI forming activated copper solutions of a similar reactivity than those mentioned earlier. HMPT has the advantage of being soluble in water, so it could be purified via a simple extraction with diluted acid; the complex of CuI and P(Cy)$_3$ (similarly with PPh$_3$) is insoluble in THF, so as a disadvantage to work with a heterogeneous reduction system appears.

23.3.4 POLYMER-SUPPORTED CU*

Sometimes, the presence of phosphine as a subproduct in the synthesis of Cu* could provoke interferences in further uses of the metal, since it is very difficult to separate phosphine from the active metal solution. Rieke et al. concluded that the synthesis of activated metal should be modified using the complex of CuI with phosphine bound with a polymeric support. The formed activated metal in this case is "adhered" to the support, and the phosphine could be eliminated by successive washings.

23.3.5 PREPARATION OF THE SUPPORTED COMPLEX

In 1979 Schwartz et al.[47] developed a method for the preparation of complexes of CuI with supported phosphines, specifically triphenylphosphine; Rieke et al. offered a similar method, where CuI solution and supported triphenylphosphine PPh$_3$ are mixed in a molar ratio 1:1 in THF and maintained for 1 week at r.t. According to elemental analysis data, 88%–96% of CuI is incorporated inside the polymeric matrix (Scheme 23.19).

The complexes CuCN*LiX (X = Cl, Br) also represent complexing agents for production of Cu*; in contrast with corresponding phosphine complexes, the main advantage is simplicity to eliminate them. A general reduction method for CuCN*LiX consists of using freshly distilled THF and metallic lithium at −100 to −110°C in inert atmosphere in the presence of lithium naphthalenide.

Synthesis of Cu in the system CuI-K-naphthalene-DME*[39] *(Scheme 23.20)*: Add 0.3522 g of metallic potassium (9.006 mmol), 1.7075 g (8.996 mmol) of CuI, 0.1204 g (0.9393 mmol) of naphthalene, and 10 mL of freshly distilled DME to a 50 mL flask.* Stir vigorously for 8 h in Ar atmosphere until total decoloration of solution. *Rieke* copper is a slowly precipitating black-gray powder.[†]

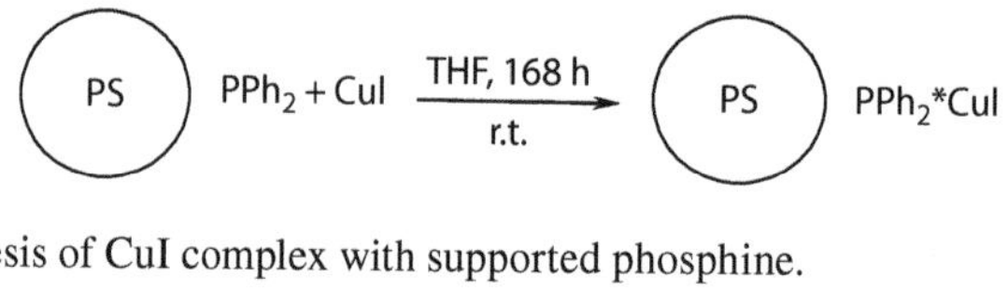

SCHEME 23.19 Synthesis of CuI complex with supported phosphine.

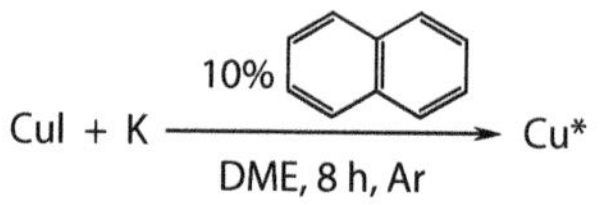

SCHEME 23.20 Synthesis of Cu* via reduction of CuI with metallic K in DME.

---

* DME should be distilled in presence of metallic sodium or sodium/potassium alloy in Ar atmosphere.

[†] Due to the presence of naphthalene in the reaction medium, the potassium forms an anion which reacts with Cu^I; when green color of solution (corresponding to potassium naphthalenide) disappears, it means that all potassium has been consumed and the reduction is complete.

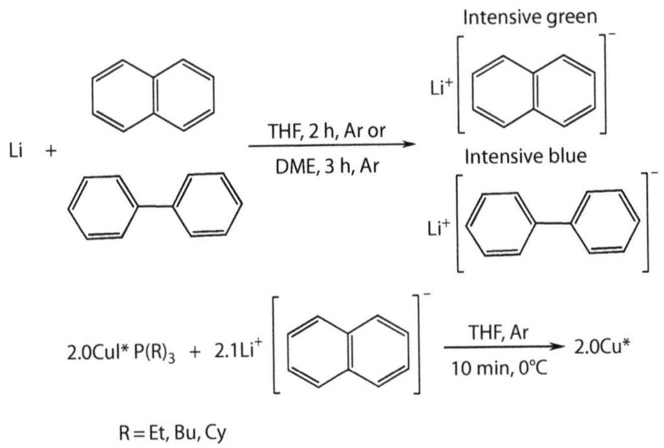

SCHEME 23.21 Synthesis of Cu* from CuI·PR$_3$ and preformed lithium complex.

Synthesis of Cu via reduction of soluble complexes CuI*PR$_3$ employing LiNp^{48} (Scheme 23.21):*

a. *Typical procedure for reduction of CuI*PEt$_3$*

Solution A. In a dry box, add in Ar atmosphere 145.76 mg (21 mmol) of metallic lithium, 2.819 g (22 mmol) of naphthalene or 3.392 g (22 mmol) of biphenyl to a 50 mL flask previously dried at 115°C. Add in Ar atmosphere 20 mL of freshly distilled DME or THF. Lithium naphthalenide or biphenylide (electron carriers) are formed for 2–3 h; the reaction is considered complete when all lithium disappears.*

Solution B. Add 5 mL of solution of CuI*PEt$_3$ (20 mmol) in DME or THF into a 25 mL flask (solution B).^{49} Add the solution B into A and stir for 10 min at 0°C.†

The key to high reactivity of this type of activated metal is its low-temperature synthesis, which allows avoiding the loss of surface area due to synthesization of copper particles at higher temperature (reflux).

b. *Typical procedure for reduction of CuI*PBu$_3$*

The use of PBu$_3$ is preferred due to its lower cost in comparison with PEt$_3$; moreover, Cu* obtained this way gives higher yields in some organic and organometallic reactions.

Solution A. Add 20.82 g (10.3 mmol) of metallic lithium, 1.589 g (12.4 mmol) of naphthalene, and 10 mL of freshly distilled THF to a 25 mL flask in Ar atmosphere and stir until Li completely disappears (approximately for 2 h).

Solution B. Add 5 mL of CuI*PBu$_3$ (9.33 mmol) in THF and excess (14.3 mmol) of tributylphosphine (PBu$_3$) in THF (solution B). Add the solution B into A and stir for 20 min at 0°C. A black-red solution of activated copper is formed.‡

*Synthesis of CuI*PBu$_3$:* Dissolve 0.069 mol of CuI together with 130 g of KI in 100 mL of distilled water (if solution is not incolorus, (i.e., without color, add 1 g of activated carbon, stir, and filter), stir vigorously with freshly distilled PBu$_3$ (12.5 mL, 0.05 mol) for 5 min, maintain stirring until

---

* An indicator for formation of the electron carrier is highly colored solutions (lithium naphthalenide presents dark green color and lithium biphenylide—deep blue).

† Maintain solutions A and B in Ar atmosphere and free of humidity.

‡ Frequently, lithium pieces are covered with metal powder. In this case, Ar flow is increased and lithium pieces are destroyed mechanically on the walls of the flask.

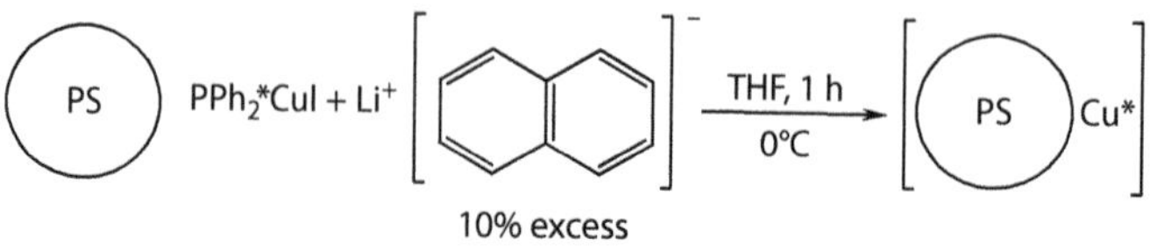

SCHEME 23.22 Synthesis of Cu* supported in polymer.

crystallization, and filter *in vacuo*. Wash with a saturated solution of KI to dissolve traces of CuI, distilled water, and EtOH (95%) and dry in air. Recrystallize from 115 mL etanol/75 mL isopropanol mixture. The yield is 50%.

Synthesis of Cu* using supported CuI*PR$_3$[45] (Scheme 23.22): Add 15 mg (2.17 mmol) of metallic lithium and 0.324 g (2.53 mmol) of naphthalene in 15 mL of freshly distilled THF into a 50 mL flask, stir vigorously in Ar atmosphere for 2 h until metallic lithium disappears. A 10% excess of lithium naphthalenide is used for copper reduction. The complex PS-PPh$_2$-CuI (1.76 mmol of CuI based on 2.69 mmol of CuI/g polymer) is stirred in 30 mL of THF for 1 night before use. Transfer lithium naphthalenide solution into the complex PS-PPh$_2$-CuI and stir at 0°C for 1 h. Wash formed black solid with some portions of THF (10 mL) to eliminate naphthalene and phosphine, leaving Cu* supported in polymer.

Synthesis of Cu* via reduction of complexes CuCN*LiX (Scheme 23.23): Add 37.68 g (5.43 mmol) of metallic lithium, 0.827 g (6.45 mmol) of naphthalene, and 15 mL of freshly distilled THF, stir at 0°C for 2 h (a 10% excess of lithium naphthalenide is used). Cool the mixture to −100°C in N$_2$ bath. Add to another flask 450.8 g (5.03 mmol) of CuCN, 445 mg (10.5 mmol) of LiCl, and 5 mL of freshly distilled THF and stir for 30 min, then cool to 0°C, and transfer the formed solution of CuCN*LiCl to the flask with lithium naphthalenide at −100°C.

Rieke et al. found that the reduction of lithium 2-thienylcyanocuprate (commercially available) produces a highly active metal powder.[50,51]

Synthesis of Cu* via reduction of lithium thienylcyanocuprate (Scheme 23.24): Add 58.29 mg (8.40 mmol) of metallic lithium and 1.18 g (9.2 mmol) of naphthalene to a 100 mL flask, stir in Ar atmosphere until metal disappears (2 h), and cool to −78°C. To another flask (50 mL) in inert

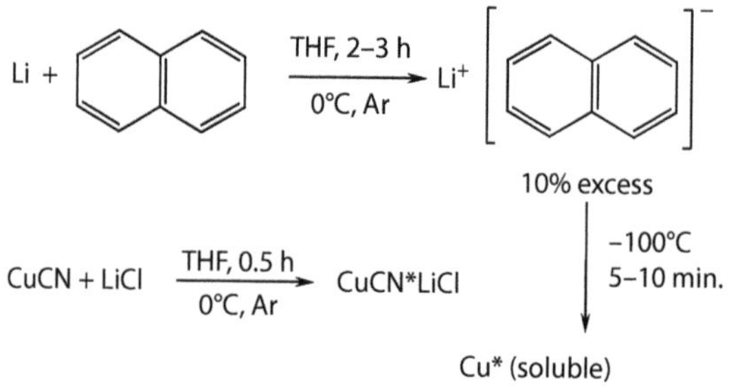

SCHEME 23.23 Synthesis of soluble Cu* via reduction of CuCN*LiCl with preformed lithium naphthalenide.

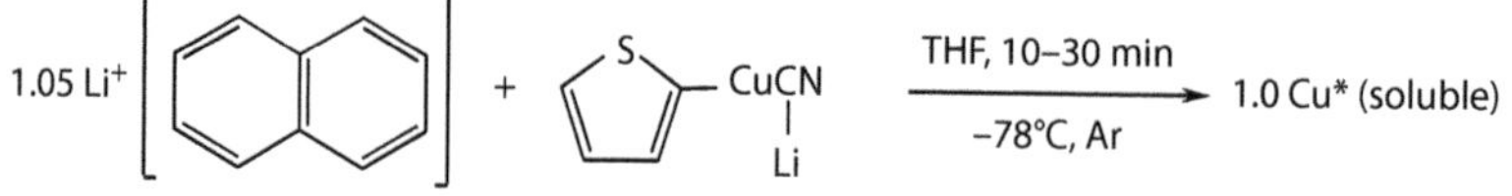

SCHEME 23.24 Synthesis of soluble Cu* from lithium 2-thienylcyanocuprate and preformed lithium naphthalenide.

atmosphere, add 8 mmol of lithium 2-thienylcyanocuprate, cool to −60°C and transfer to the flask with lithium naphthalenide, and stir for 10–30 min at −78°C.

23.3.5.1 *Rieke* Cobalt

Upon analyzing reports on *Rieke* cobalt, three particular methods were found describing the preparation of its activated powders.[52] In the first one ("*Method 1*"), 1,2-dimethoxyethane, metallic lithium, cobalt(II) chloride, and catalytic quantity of naphthalene (6 mol.% on Li basis) were used as precursors. The reduction is carried out at r.t.; a part of Li reacts with solvent, so a slight excess of lithium is added to reaction mixture to ensure a complete reduction. At the end of the reaction, the metal excess can be simply removed from the solvent surface or it can react with the electron carried generating corresponding dianion (lithium naphthalenide), which can be removed by subsequent washings with solvent. The formed activated powder is not pyrophoric. The use of $CoBr_2$ and CoI_2 leads to less-active powders. In 1983 Rieke et al. discovered that substitution of Li by K in the synthesis of Co* generated a metallic powder with considerably decreased reactivity; the replacement of DME by THF improved the solubility of naphthalenide anion and cobalt salt and increased the speed of formation of naphthalenide anion. The procedures for all three methods are presented later.

Synthesis of Co in 1,2-dimethoxyethane - (Li excess) – r.t. (Method 1)*[53] *(Scheme 23.25)*: Add 1.259 g (9.70 mmol) of anhydrous $CoCl_2$, 0.159 g (22.9 mmol Li) of lithium, 0.154 g (1.20 mmol) of naphthalene, and 16 mL of DME to a 50 mL flask and stir for 19 h. The solution obtains blue and further black color. To wash a solid, remove almost all solvent, add 6–8 mL of fresh DME, mix and maintain some hours, then remove the liquid with syringe, and repeat this step two times. A dark gray solid is formed.

Synthesis of Co in 1,2-dimethoxyethane-Li-naphthalene (in excess) at −78°C (Method 2)*[54] *(Scheme 23.26)*: Add 0.3342 g (48.2 mmol) of metallic lithium, 8.0600 g (62.9 mmol) of naphthalene, and 70 mL of freshly distilled DME into a 200 mL flask, stir vigorously until complete dissolution of metal, cool to −78°C, add a suspension of 3.0673 g (23.6 mmol) of anhydrous $CoCl_2$ in 30 mL of DME, and stir for 0.5 h at the same temperature and 3 h at r.t. Cobalt-activated powder, generated by *Method 2*, is a nonferromagnetic solid and pyrophoric (when it is dry) with a crystal size of <30 Å; heating it to 300°C, a highly ferromagnetic material is formed.

Synthesis of Co in the system bis(2-methoxyethyl)ether-K-naphthalene in excess at −45°C (Method 3)*[55]: Add 4.2592 g (108.9 mmol) of metallic potassium, 20.0189 (156.2 mmol) of naphthalene, and 100 mL of freshly distilled diglime (bis(2-methoxyethyl)ether) into a 250 mL flask, stir until complete dissolution of potassium, and cool to −60°C. Add in small portions 6.0669 g (46.7 mmol) of anhydrous $CoCl_2$ to the mixture of potassium naphthalenide (dark green solution).

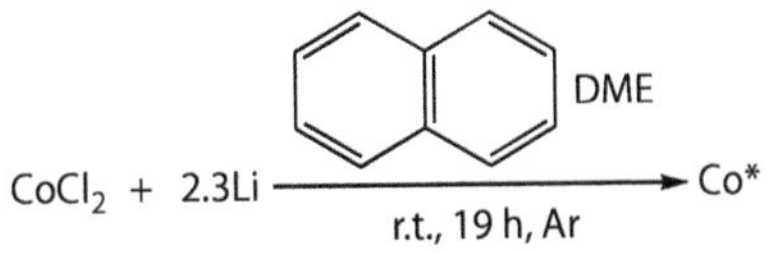

SCHEME 23.25 Synthesis of Co* from $CoCl_2$, metallic Li, and naphthalene at r.t.

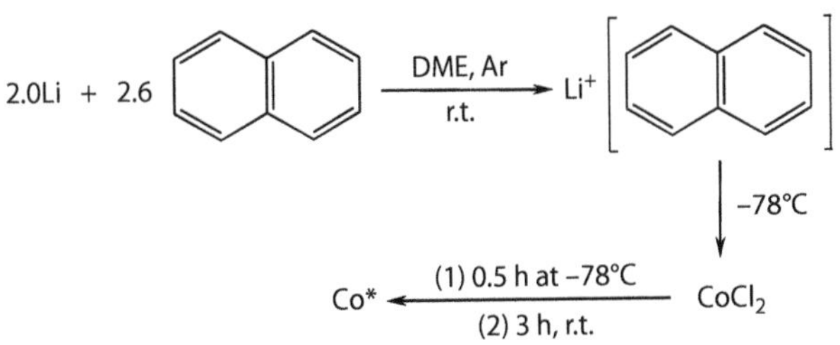

SCHEME 23.26 Synthesis of Co* by reduction of $CoCl_2$ with preformed lithium naphthalenide.

A dark brown solution is formed. Heat to −45°C and stir for 14 h, wash three times with 75 mL of diglime and four times with 90 mL of hexane, and dry *in vacuo*.

23.3.5.2 *Rieke* Iron

Iron, chromium, and manganese have not yet caught the attention of researchers from the point of view of their activated forms. Among the most recent investigations on their synthesis and use, reactive forms of Fe, prepared by the *Rieke* method, were found to combine with dithiols under a CO atm. to give $Fe_2(S_2C_nH_{2n})(CO)_6$ in modest yields under mild conditions.[56]

Synthesis of Fe* in 1,2-dimethoxyethane-Li at r.t.[54]: Add 1.221 g (9.63 mmol) of anhydrous $FeCl_2$, 0.154 g (22.2 mmol) of metallic lithium, 0.187 g (1.46 mmol) of naphthalene, and 18 mL of DME into a 50 mL flask, stir for 24 h, and wash with DME two times.

23.3.5.3 *Rieke* Chromium

Intents to carry out the reduction of chromium salts in ethereal solvents failed due to the high reactivity of Cr^{2+} ion in respect of these solvents. However, the reduction of the adduct $CrCl_3 * 3THF$ in benzene generates a highly active powder (Scheme 23.27); on the other hand, the reduction in the presence of alkali metal salts, in particular KI, increases notably the reactivity of Cr*-activated powder. The process is carried out using potassium for 2 h in reflux.[57]

23.3.5.4 *Rieke* Manganese

One of the most recent applications of *Rieke* metals was direct preparation of various heteroaryl manganese reagents, performed using highly active manganese (Mn*) and heteroaryl halides.[58,59] The resulting organomanganese reagents were coupled with electrophiles such as aryl halides, vinyl halides, and benzoyl chlorides under mild reaction conditions with good yields. Also, the active Mn* was found to readily react with α-haloester in the presence of aldehydes and ketones to yield the corresponding β-hydroxy esters by a modified *Reformatsky* reaction at r.t. in the absence of *Lewis* acid or trapping agents[60] and to produce organomanganese tosylates and mesylates.[61] Mn*-promoted sequential process directed toward the synthesis of (Z)-α-halo-α,β-unsaturated esters or amides is described in Ref. [62]. In both cases, the process takes place with complete Z-stereoselectivity.

A classic method for obtaining Mn* is shown in Scheme 23.28.

Synthesis of Mn* in the system MnI$_2$-Li-Naphthalene-THF[63] *(Scheme 23.28)*: Add 20 mmol of Li, 2 mmol of naphthalene, 10 mmol of MnI_2, and 10 mL of freshly distilled THF into a 20 mL flask and stir for 30 min at r.t. MnI_2 is expensive and difficult to obtain; $MnCl_2$[64] is more common and economic, however, the reaction time is increased to 1–3 h.

23.3.5.5 *Rieke* Platinum and Palladium

These two metals, due to their permanent high importance for organic and organometallic synthesis, have been well studied in respect of their activated forms, in contrast with, for example, iron or chromium. Thus, reduction of palladium and platinum triethylphosphine complexes $[P(C_2H_5)_3]_2MX_2$ by potassium is carried out in ethereal solvents (DME y THF) generating fine highly reactive

$$CrCl_3 * 3THF + 3K \xrightarrow[\text{Reflux, 2h}]{C_6H_6, KI} Cr*$$

SCHEME 23.27   Synthesis of Cr* by reduction of the adduct $CrCl_3 * 3THF$ with potassium.

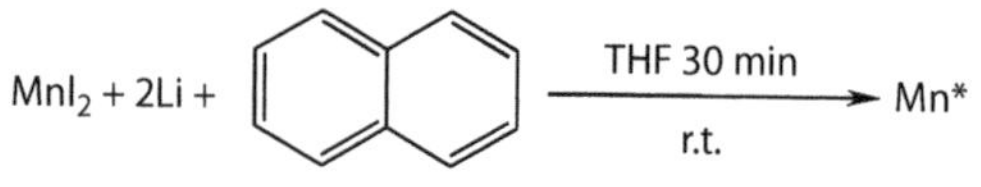

$$MnI_2 + 2Li + \text{(naphthalene)} \xrightarrow[\text{r.t.}]{\text{THF 30 min}} Mn*$$

SCHEME 23.28   Synthesis of Mn* via reduction of MnI_2 with lithium.

TABLE 23.3
Lithium Molar Equivalents
Needed for Pd and Pt
Reduction in Different
Oxidation States

Metal Ion	Lithium
Pd^{2+}	2.6
Pt^{2+}	2.8
Pt^{4+}	5.1

metal powders. The use of metal anhydrous salt or phosphine complex obtains similar results; a phosphine complex is formed for some minutes. Reduction times applying potassium depend on the compound to reduce, for example, solutions of the complex $[P(C_2H_5)_3]_2PtI_2$ require 1 h, meanwhile others take 20 h. During the reduction process, it is recommended to take small aliquots for probes of inflammation with water. For reactions with Li y naphthalene, the lithium pieces float without reaction in the reaction medium.

The activated powders, obtained from complexes $[P(C_2H_5)_3]MX_2$, present major reactivity; in particular in the case of palladium, the activated powder Pd* exists in yellow solvents, undoubtedly containing metal tris- and tetrakis(triethylphosphines) in low oxidation states of metal. It was observed that a part of lithium reacts with the solvent. To ensure a complete reduction of a metal salt, it is recommended to employ an excess of lithium in respect of stoichiometry (Table 23.3).

Synthesis of Pd* in the system trans-[P(C2H5)3]2PdCl2-THF-K in reflux[65] *(Scheme 23.29)*: Add 0.959 g (2.32 mmol) of *trans*-$[P(C_2H_5)_3]_2PdCl_2$, 0.181 g (4.63 mmol) of pure metallic potassium, and 10 mL of freshly distilled THF to a 25 mL flask, and heat in reflux for 22 h. In the case of using $PdBr_2$ (instead of palladium chloride) forming *trans*-$[P(C_2H_5)_3]_2PdBr_2$, the reaction time decreases from 22 to 5 h.

Procedure for preparation of activated Pd* for surface analysis.[54]: After completion of the reduction process, the Pd*-activated powder is allowed to precipitate, the liquid is eliminated with a syringe, and the rest of solvent is removed *in vacuo*. In a dry box (Ar), the metal is transferred to Soxhlet equipment, which is then connected to Ar line and DME is injected. The extraction is carried out for 26 h; the solvent is removed with a syringe, and the metal is dried *in vacuo* for 18 h. The surface analysis is made opening ampoules with metal in dry box (N_2), mounted in Al surface, and transferred into the vacuum camera of the system ESCA.

Synthesis of the complex trans-[P(C_2H_5)_3]_2PdBr_2: Add 0.840 g (3.16 mmol) of $PdBr_2$, 10 mL of freshly distilled DME, and 0.94 mL (6.4 mmol) of $P(C_2H_5)_3$ into a 50 mL flask. Reaction is completed in 10 min.

Synthesis of Pd* in the system trans-[P(C_2H_5)_3]_2PdCl_2-THF-Li-naphthalene at r.t. (Scheme 23.30): Add 1.104 g (2.67 mol) of *trans*-$[P(C_2H_5)_3]_2PdCl_2$, 0.0374 g (0.543 mmol) of metallic lithium, 0.0324 g (0.253 mmol) of naphthalene, and 10 mL of DME to a 50 mL flask and stir for 70 h.

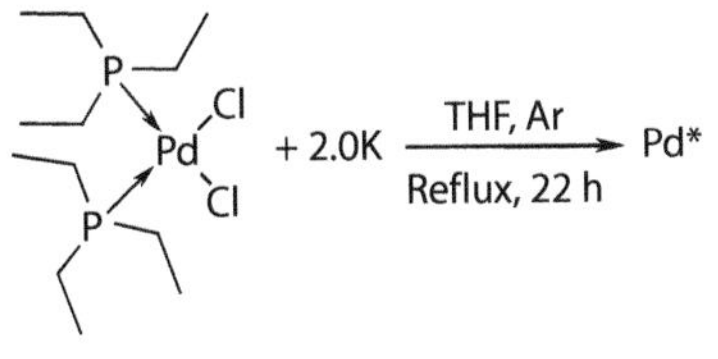

SCHEME 23.29 Synthesis of Pd* by reduction of the complex *trans*-$[P(C_2H_5)_3]_2PdCl_2$ with potassium.

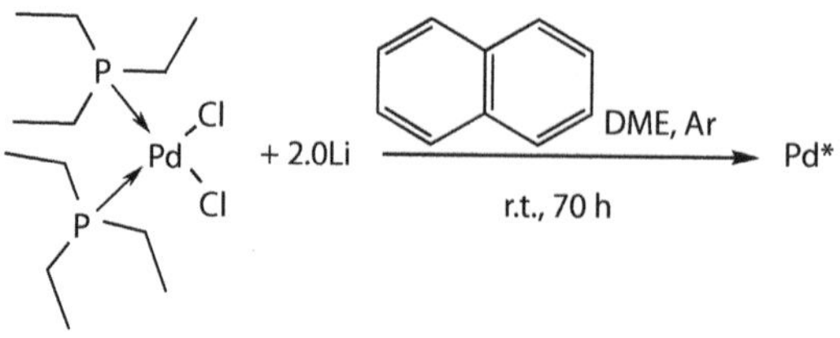

SCHEME 23.30 Synthesis of Pd* by reduction of the complex *trans*-[P(C$_2$H$_5$)$_3$]$_2$PdCl$_2$ with Li and naphthalene.

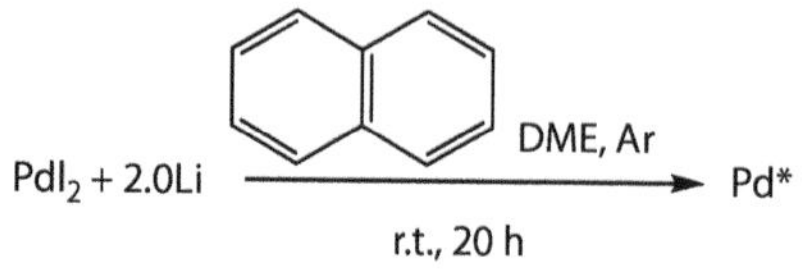

SCHEME 23.31 Synthesis of Pd* from PdI$_2$ using lithium and naphthalene.

Synthesis of Pd in the system PdI$_2$-THF-Li-naphthalene at 40°C (Scheme 23.31):* Add 0.0412 g (5.94 mmol) of metallic lithium, 0.9707 g (2.65 mmol) of PdI$_2$, 0.0293 g (0.229 mmol) of naphthalene, and 10 mL of DME into a 50 mL flask connected to Ar line, and stir at r.t. for 20 h, when the presence of Li without a reaction is observed. Heat mixture at 40°C for some hours until the formation of a black powder in incolored solution.

Synthesis of Pt in the system PtI$_2$-THF-K-P(C$_2$H$_5$)$_3$ at 40°C:* Add 1.406 g (3.13 mmol) of anhydrous PtI$_2$, 8 mL of freshly distilled THF, 0.93 mL (6.3 mmol) of P(C$_2$H$_5$)$_3$, and 0.245 g (6.2 mmol) of metallic potassium into a 50 mL flask, and reflux in Ar atmosphere for 1 h. A black solid in incolored solution is formed. In the case of PtCl$_2$, reflux for 20 h.

23.3.5.6 *Rieke* Zinc

Zinc, as well as Pd and Pt, discussed here, is also a very important metal for organic chemistry, so obtaining and storage[66] of its activated form have been well developed. A series of reported experiments confirmed that Zn* powder, generated through reduction with lithium, presents a major reactivity in comparison with Zn* obtained using Na and K. In the case of using Li and naphthalene, a vigorous stirring is needed to prevent covering lithium surface with the formed activated powder and consequently a decrease of reaction speed. Employment of a preformed solution of lithium naphthalenide reduces the reaction time from 10 to 3 h and eliminates the problem discussed earlier. In this case, the method requires the use of two flasks, one with zinc salt solution and another for preformation of lithium naphthalenide. It is recommended to add drop by drop (1 drop per 3 s) a zinc salt solution to the redactor using a syringe; in this case, a very fine activated powder is formed. When the addition speed increases to 1 drop per 1 s, the Zn*-activated powder presents a sponge structure. In addition to naphthalene, the biphenyl and anthracene can also be used as electron carriers. Using an electron carried (e.g., naphthalene), lithium is a better reductor in comparison with potassium (59% yield against 28% for 10 h, respectively); additionally, K is better than Na (54% against 27% for 18 h, respectively).

The activated Zn*, obtained from ZnCl$_2$, ZnBr$_2$, or ZnI$_2$, presents similar reactivity. No differences for obtained products were observed using THF and DME; however, it is necessary to take into account that: (1) lithium naphthalenide formation in THF (2 h) is faster than in DME (4 h); (2) ZnCl$_2$ is more soluble in THF than in DME; (3) in case of using DME, zinc salt solution should be added more slowly, otherwise powder reactivity could decrease. It is necessary to consider that the quantity of electron carrier plays an important role in reaction speed; generally 10% (based on moles of an alkali metal) is sufficient. Zn*, prepared in THF, is lesser reactive than that obtained in DME.[21] Addition of Al$_2$O$_3$ to the reaction medium reduces Zn* reactivity in DME as in THF. Because of the elevated hygroscopicity of zinc salts, it is possible to use its complex

$ZnCl_2$-N,N,N',N'-tetramethylethylenediamine ($ZnCl_2$-TMEDA) that is commercially available; unfortunately, this complex is low soluble in THF (in contrast with $ZnCl_2$) and there is no evidence of increasing Zn* reactivity.

Among recent applications of activated zinc, we note water-soluble polythiophenes and thin films prepared therefrom having alkylCO_2A, $(CH_2)_n$-aryl$(CH_2)_nCO_2A$, $(CH_2)_n$-heterocycle$(CH_2)_nCO_2A$, or $(CH_2)_n$-cycloalkyl$(CH_2)_nCO_2A$ groups (n = 0–200), and A is an alkali metal cation, attached to the chain[67] and transistors on the basis of polythiophenes.[68] A typical polymer was obtained by the reaction of Et 4-(2,5-dibromothiophene-3-yl)butyrate with Zn* in THF at 0°C, reaction of the organozinc intermediate 4 h at r.t. with Ni(II) 1,2-bis(diphenylphosphino)ethane, and further treatment of the resulting polymer with KOH. The ability of *Rieke* zinc to reduce common organic functional groups was reported in Ref. [69]. It was shown that nitrobenzene, conjugated aldehydes, arylacetylenes, and phenylpropiolates are readily reduced under mild conditions (e.g., CH_3–C_6H_4–$C{\equiv}CH$ is reduced to CH_3–C_6H_4–$CH{=}CH_2$ with Zn* in THF-MeOH-H_2O 7:5:1 mixture) and benzonitrile, alkylacetylenes, ketones, unconjugated aldehydes, and alkenes are not reduced.

Synthesis of Zn* in the system $ZnCl_2$-K-THF[70] (Scheme 23.32): Add 9.54 g (70.0 mmol) of $ZnCl_2$, 40 mL of THF, and 5.47 g (140.0 mmol) of metallic potassium to a 100 mL flask (N_2), heat slightly without stirring before the start of the reaction on the K surface, remove heating, and leave the system (the reaction is highly exothermal) permanently cooled in a water bath. When the reaction is finished, stir slightly to maintain it under control, and then reflux for 25 h with vigorous stirring. Particle size of the formed powder is about 17 μm.[71]

Synthesis of Zn* in the system $ZnCl_2$-Li-naphthalene-DME[72] (Scheme 23.33): Add 0.35 g (50.5 mmol) of metallic lithium, 3.27 g (24.0 mmol) of anhydrous $ZnCl_2$, 0.65 g (5.07 mmol) of naphthalene, and 16 mL of DME into a 50 mL flask and stir at r.t. for about 15 h.

Synthesis of Zn* in the system $ZnCl_2$-Li-naphthalene (preformed)-THF[73] (Scheme 23.34): Connect two 50 mL flasks (A and B) to Ar line. Add to A flask 0.213 g (30.63 mmol) of metallic lithium, 3.983 g (31.15 mmol) of naphthalene, and 15 mL of freshly distilled THF and stir at r.t. until metal dissolution (2 h). Add to flask B 2.09 g (15.37 mmol) of anhydrous $ZnCl_2$, 15 mL of THF and stir

$$ZnCl_2 + 2K \xrightarrow[\text{Reflux } N_2]{\text{THF, 25h}} Zn* + 2KCl$$

SCHEME 23.32   Synthesis of Zn* via reduction of $ZnCl_2$ with potassium in THF.

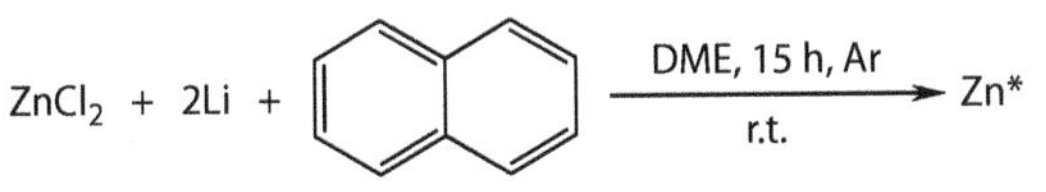

SCHEME 23.33   Synthesis of Zn* via reduction of $ZnCl_2$ with Li and naphthalene in DME.

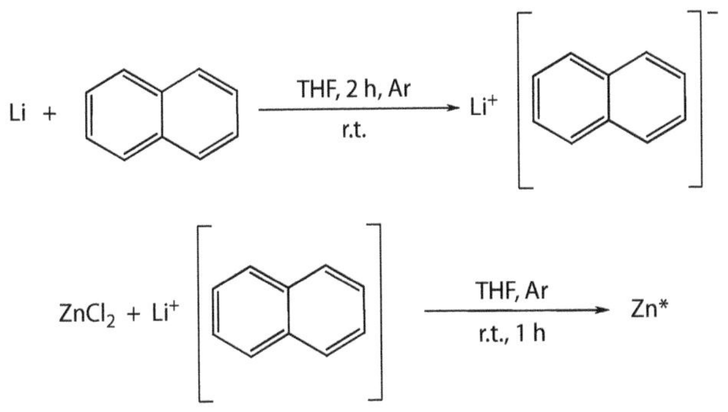

SCHEME 23.34   Synthesis of Zn* via reduction of $ZnCl_2$ with lithium naphthalenide in THF.

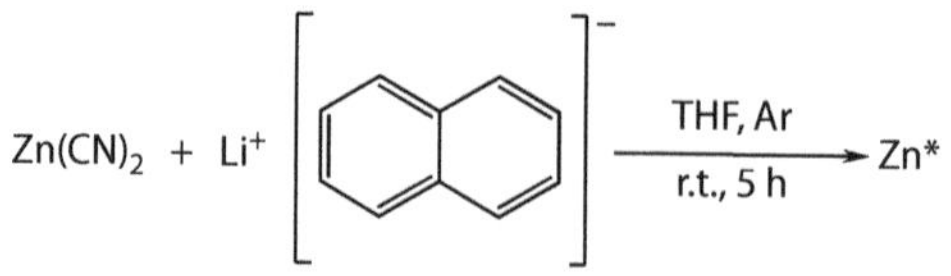

SCHEME 23.35 Synthesis of Zn* via reduction of $Zn(CN)_2$ with lithium naphthalenide in THF.

at r.t. until complete dissolution. With a syringe, transfer B into A very slowly (drop by drop) for 15 min, and wash the formed activated powder of Zn* with various portions of fresh THF to remove naphthalene and LiCl.

Synthesis of Zn* in the system $Zn(CN)_2$-Li-Naphthalene-THF[74] (Scheme 23.35): Add 0.152 g (21.90 mmol) of metallic lithium, 1.35 g (11.49 mmol) of $Zn(CN)_2$, 0.144 g (1.25 mmol) of naphthalene, and 25 mL of freshly distilled THF into a 50 mL flask and stir at r.t. for 5 h until disappearing intensive green color (lithium naphthalenide).

23.3.5.7 *Rieke* Barium

Rieke barium is usually obtained using a lithium-biphenyl reductant in THF according to the following procedure.

Synthesis of Ba* in the system BaI_2-Li-biphenyl-THF[75] (Scheme 23.36): Add 210 mg (30.3 mmol) of metallic lithium and 4.7 g (30.5 mmol) of biphenyl in 80 mL of THF into a 300 mL flask (Ar), stir at r.t. until complete dissolution of metal (for 2–3 h) and the formation of intensive blue color. Add to another flask (100 mL) 6.0 g (15.3 mmol) of BaI_2 (BaI_2 [white] is obtained from $BaI_2 * 2H_2O$ [yellow] by heating it alter pulverization at 150°C for 12–24 h.) under a layer of dry THF, stir at r.t. for 5 min, transfer the mixture with a syringe to the solution of lithium biphenylide, and stir at r.t. for 1 h.

23.3.5.8 *Rieke* Cadmium

There are several techniques for obtaining *Rieke* cadmium, and all of them are based on the use of alkali metal–naphthalene intermediate complex as follows.[76]

Synthesis of Cd* in the system $CdCl_2$-lithium naphthalenide[76] (Scheme 23.37): Add 36.84 mmol of metallic lithium, 38.58 mmol of naphthalene (Ar), and 30 mL of DME or THF in a 200 mL flask and stir for 4 h. Transfer the mixture with syringe to a flask containing anhydrous $CdCl_2$ (16.01 mmol) in 30 mL of THF and stir at r.t. for 30 min. In 6–12 h black powder precipitates and can be washed with fresh solvent.

Synthesis of Cd* in the system $CdCl_2$-lithium naphthalenide-TMEDA[76] (Scheme 23.38): Add 9.993 mmol of metallic lithium, 11.833 mmol of naphthalene, and 11.512 mmol of TMEDA

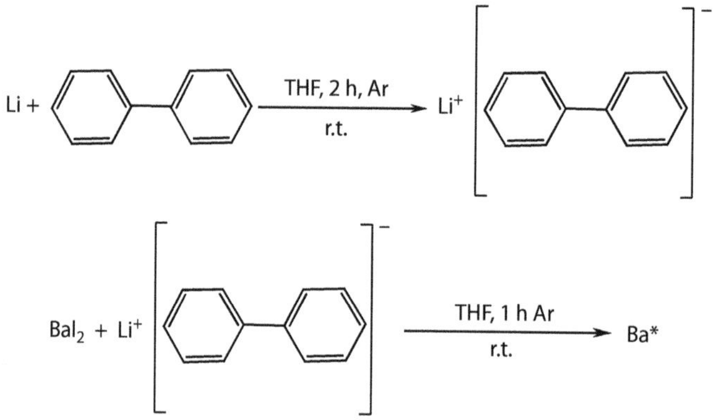

SCHEME 23.36 Synthesis of Ba* via reduction of BaI_2 with lithium naphthalenide in THF.

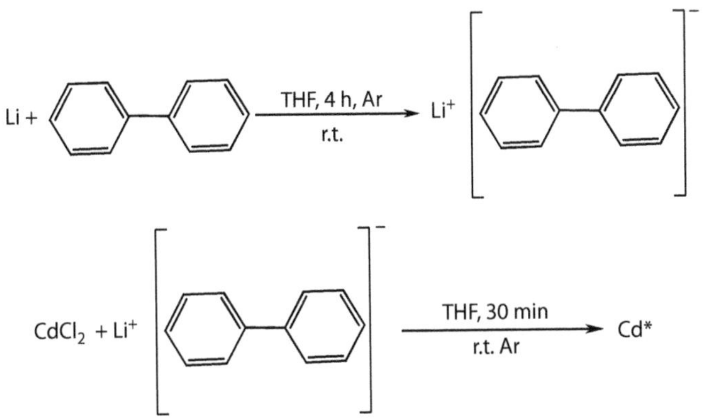

SCHEME 23.37 Synthesis of Cd* via reduction of CdCl$_2$ with lithium naphthalenide in THF.

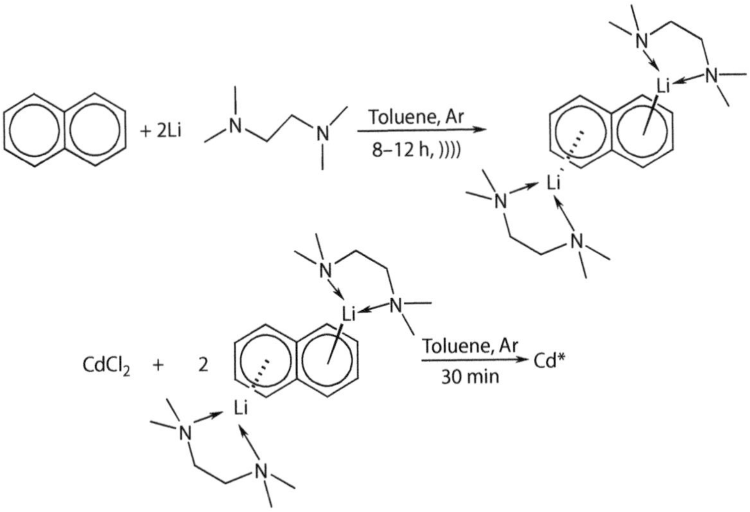

SCHEME 23.38 Synthesis of Cd* via reduction of CdCl$_2$ by the complex system *N,N,N',N'*-tetramethylethyldiamine–lithium naphthalenide in toluene.

(*N,N,N',N'*-tetrametiletildiamina) in 25 mL of toluene into a flask in Ar atmosphere and sonicate for 8–12 h.[40] Transfer the formed purpur solution with a syringe into a flask with 4.982 mol of CdCl$_2$, and stir for 30–60 min. The powder precipitates for 6–12 h. It was found that more reactive powder is formed by the use of hydrocarbon solvents.

Synthesis of Cd from CdCl$_2$-Li-naphthalene*[76]: Add 28.94 mmol of metallic lithium, 13.39 mmol of CdCl$_2$, and a catalytical quantity of naphthalene to a flask containing 20 mL of DME or THF, stir, and maintain at reflux for 3–4 h. The final activated powder contains Cd$_3$Li, whose treatment with 1 equivalent of I$_2$ leads to a reaction of I$_2$ with a major part of the liquid forming an activated Cd* powder.

Synthesis of Cd from CdI$_2$-Na-naphthalene-THF*[77]: In a typical experiment, 6 mmol of CdI$_2$ is collocated in part B of the equipment (Figure 23.1). Part A contains 12 mmol of naphthalene, and the tube D is filled with 20 mmol of Na. The system is evacuated; the naphthalene zone is maintained at −78°C. Sodium is distilled into A, and tube D is sealed at point 1. About 25 mL of freshly distilled THF is collocated into A, removing the vacuum line. The solution of sodium naphthalenide, resulting from the reaction of Na and naphthalene in THF, is transferred to B. The activated metal is

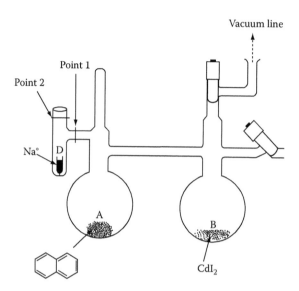

FIGURE 23.1   Scheme of the equipment for synthesis of Cd*. (Reproduced with permission from Stevenson, Ch.D. et al., Electron transfer from *Rieke* cadmium yielding Cd²⁺ coordinated and ion associated anion radical systems, *Inorg. Chem.*, 34(6), 1368–1372. Copyright 1995 American Chemical Society.)

sedimented and the liquid is moved to A. The solid is washed various times with THF and distilled from A to remove NaI residues.

23.3.5.9   *Rieke* Nickel

Rieke et al. found that, in the presence of complexing agents (e.g., triethylphosphine), reduction reactions generated activated nickel powders with particle size lesser than in their absence. Change of R in PR$_3$ leads to distinct reactivities of the formed metal; thus, reduction of NiI$_2$ with triphenylphosphine produced more reactive activated nickel than the use of triethylphosphine.

Synthesis of Ni* in the system NiCl$_2$-K-THF [78] *(Scheme 23.39)*: Add 9.8 g (250 mmol) of metallic potassium, 17 g (130 mmol) of previously dried (at 130°C) NiCl$_2$, and 350 mL de THF into a 500 mL flask, stir and heat to reflux (N$_2$), then reflux for 12 h. After finishing, the metal precipitates for 12 h; then wash 2 × 250 mL with fresh THF.

Synthesis of *Rieke* nickel, as well of other *Rieke* metals, can be carried out with the designed equipment (Figure 23.2).

Synthesis of Ni* in the system NiI$_2$-Li-naphthalene-DME [79] *(Scheme 23.40)*: Add 3.84 g (12.3 mmol) of anhydrous NiI$_2$, 0.196 g (28.2 mmol) of metallic lithium and 0.157 g (1.22 mmol) of naphthalene to a flask, evacuate, and inject Ar (repeat this step two to three times). Add with a syringe 25 mL of freshly distilled DME, and stir at r.t. for 12 h until metallic lithium disappears.

Synthesis of Ni* in the system NiI$_2$-K-THF-phosphine [80] *(Scheme 23.41)*: Addition of triethylphosphine to a THF solution of NiI$_2$ generates a soluble complex diiodobis(triethylphosphine)nickel(II). Refluxing its mixture with potassium in THF for 2 h produces a more reactive metallic powder in comparison with the absence of the complexing agent. PEt$_3$ and PPh$_3$ decrease the reaction time from 12 to 2 h.

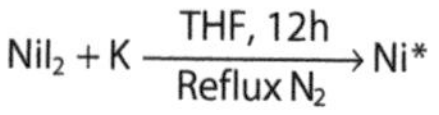

$$NiI_2 + K \xrightarrow[\text{Reflux N}_2]{\text{THF, 12h}} Ni^*$$

SCHEME 23.39   Synthesis of Ni* via reduction of NiI$_2$ with potassium in THF.

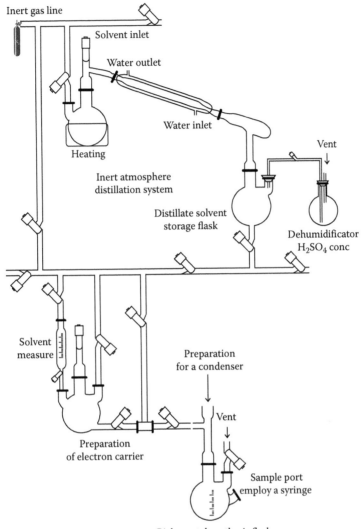

FIGURE 23.2 Equipment used for synthesis of *Rieke* nickel, copper, magnesium, and zinc. (From Rieke, R.D., *J. Org. Chem.*, 46, 4323, 1981.)

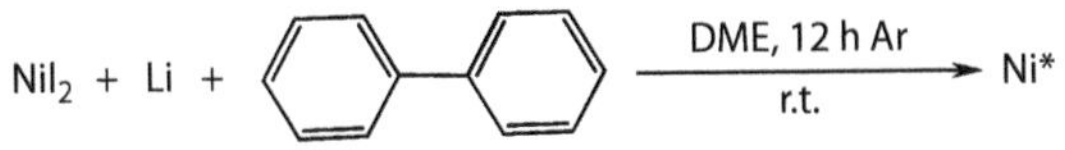

$$NiI_2 + Li + \text{[biphenyl]} \xrightarrow[\text{r.t.}]{\text{DME, 12 h Ar}} Ni^*$$

SCHEME 23.40 Synthesis of Ni* via reduction of NiI_2 with lithium and naphthalene in DME.

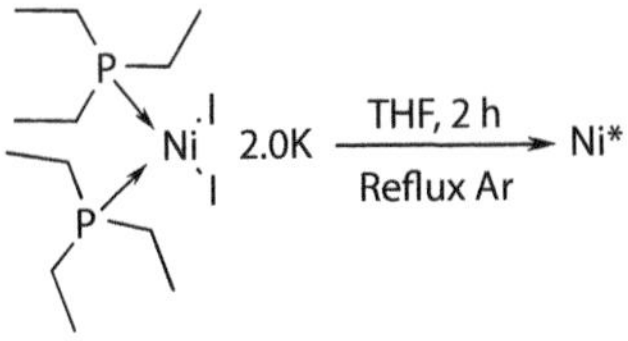

$$\text{[(Et}_3\text{P)}_2\text{NiI}_2] \; 2.0K \xrightarrow[\text{Reflux Ar}]{\text{THF, 2 h}} Ni^*$$

SCHEME 23.41 Synthesis of Ni* via reduction of tte complex NiI_2 and triethylphosphine using potassium in THF.

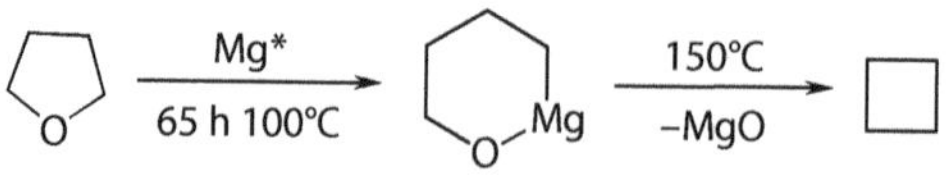

SCHEME 23.42 Degradation products of THF in presence of magnesium.

Among the most recent applications of *Rieke* nickel, it was found that the Ni* generated *in situ* is able to promote the pinacol coupling of various carbonyls efficiently.[81] Based on this information, a catalytically effective, cheaper, and more convenient $NiCl_2(Cat.)/Mg/TMSCl$ system was designed and developed further successfully.

23.3.5.10 *Rieke* Magnesium[82]

As well as experimental techniques mentioned earlier for *Rieke* cadmium, zinc, copper, etc., corresponding methods for magnesium are also well developed due to its high importance in organic and especially organometallic (*Grignard* and related reagents)[83–85] chemistry. The general procedure consists of the reduction of an anhydrous magnesium salt ($MgCl_2$ or $MgBr_2$) in inert atmosphere in an ethereal solvent. The recommended systems for reduction are K-THF and 2-methoxyethylether (diglime)-Na; the reaction time varies from 1 to 2 h for THF-K-MgX_2 to 5–6 h for diglime-Na-MgX_2. It was found that at elevated temperature Mg* inserts into THF, forming (4-oxidobuthyl) magnesium, which in appropriate conditions could generate cyclobutane (Scheme 23.42).

Reduction reactions of $MgCl_2$ applying potassium are highly exothermal; usually a 10% excess of $MgCl_2$ is used to prevent any problem of unreacted K. Generally, preparation of Mg* is carried out from $MgCl_2$, $MgBr_2$ or MgI_2 as precursors in THF, DME, or diglime; this is the basis for the preparation of very active slurries of Mg.[86] The use of MgF_2 and $MgSO_4$ did not produce positive results due to very low solubility of these salts in employed solvents.

Rieke and *Bales* found that addition of inorganic salts before reduction generated Mg* of a major reactivity, specifically addition of NaI and KI at molar ratio between 0.05 and 2.0 of MI in respect of $MgCl_2$. The use of solvent mixtures (THF-triethylamine 1:1) in the system $MgCl_2$/K generated activated Mg* powders having activity comparable with the obtained using lithium.[87] In general, the use of lithium is highly recommended due to the process safety: lithium has an elevated melting point (180°C), so this is impractically to seek a solvent possessing a higher valor. In this case, electron carriers (naphthalene and biphenyl, among others) are used; reductions are carried out at r.t., reaction time varies from 6 to 12 h. Another advantage of lithium employment is generation of Mg* possessing a higher reactivity in comparison with other methods. Insufficient stirring during reduction process leads to the passivation of Li particles with formed Mg*, so in this case reductions are incomplete and very slow. Using Li-electron carrier system, a solution of lithium naphthalenide in THF is first prepared (for 2 h) and then added in equimolar quantity to magnesium salt solution in THF and stir for 30 min at r.t.

Commercial magnesium salts are generally hydrated, their drying leads to magnesium oxyhalides. A simple technique to produce anhydrous salts is based on the decomposition of 1,2-dichloro(or bromo)ethane with elemental magnesium resulting ethylene and anhydrous MgX_2.[13]

Synthesis of Mg in the system $MgBr_2$-K-THF* [82] (*Scheme 23.43*): Add 18.86 g (10.25 mmol) of anhydrous $MgBr_2$, 50 mL of freshly distilled THF, and 0.8 g (20.5 mmol) of metallic potassium to a 100 mL flask and reflux in inert atmosphere for 3 h. As an alternative variant, this reaction can be carried out using Et_2O inside a tube at elevated pressure at 50°C for 10 h.

$MgBr_2$ is prepared according to Scheme 23.44.

$$MgBr_2 + 2K \xrightarrow[\text{Reflux Ar}]{\text{THF, 3h}} Mg^*$$

SCHEME 23.43 Synthesis of Mg* via reduction of $MgBr_2$ using potassium in THF.

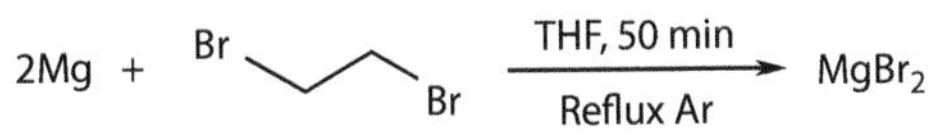

$$2Mg + \underset{Br}{\overset{Br}{\diagup\diagdown}} \xrightarrow[\text{Reflux Ar}]{\text{THF, 50 min}} MgBr_2$$

SCHEME 23.44 Synthesis of anhydrous $MgBr_2$ via decomposition of 1,2-dibromoethane with elemental Mg.

$$MgCl_2 + 2K \xrightarrow[\text{Reflux Ar}]{\text{THF, 3h KI}} Mg*$$

SCHEME 23.45 Synthesis of Mg* via reduction of $MgCl_2$ using potassium in THF.

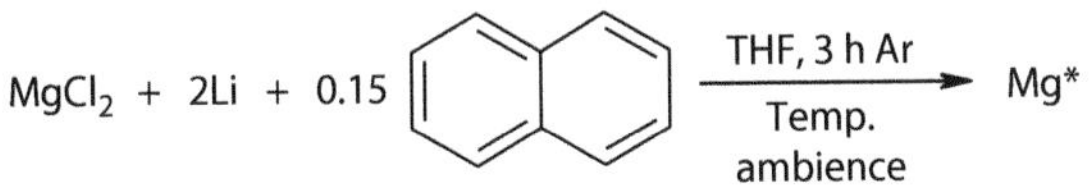

$$MgCl_2 + 2Li + 0.15 \quad \text{[naphthalene]} \quad \xrightarrow[\substack{\text{Temp.} \\ \text{ambience}}]{\text{THF, 3 h Ar}} Mg*$$

SCHEME 23.46 Synthesis of Mg* via reduction of $MgCl_2$ with metallic lithium and naphthalene in THF.

Synthesis of Mg in the system $MgCl_2$-K-KI-THF*[88] *(Scheme 23.45)*: Add 1.5 g (38 mmol) of metallic potassium (washed with hexane), 2.01 g (21.1 mmol) of anhydrous $MgCl_2$, 3.55 g (21.4 mmol) of anhydrous KI (previously dried at 120°C), and 50 mL of freshly distilled THF into a 200 mL flask (Ar*), and stir vigorously at reflux for 3 h.† The reaction is exothermal.

Synthesis of Mg from $MgCl_2$-Li-Naphthalene-THF (Scheme 23.46)*: Add 10 mmol of metallic lithium, 1.5 mmol of naphthalene, 4.88 mmol of anhydrous $MgCl_2$, and 15 ml of freshly distilled THF to a 50 mL flask, stir vigorously at r.t. for 3.5 h until complete dissolution of metal, add 15 mL of THF, and leave to precipitate Mg* for 2 h.

23.3.6 Use of Ultrasound in the Synthesis of *Rieke* Metals

According to some reports, simultaneous use of ultrasound during synthesis of *Rieke* metals allows their formation for lesser time. Thus, Boudjouk et al.[57] in 1986 reported procedures for obtaining some *Rieke* metals. The principal characteristic was decrease in reaction time applying simultaneous ultrasonic treatment (classic ultrasonic bath of 40 kHz) for maximum stirring. Table 23.4 shows necessary conditions for the reduction of magnesium, zinc, chromium, and copper salts in the presence of potassium in reflux and using lithium with ultrasonic assistance. Further use of the obtained *Rieke* metals in organic/organometallic procedures under ultrasonic treatment leads to higher yields and lesser process duration.[71,78]

Analyzing the ultrasonic treatment during *Rieke* synthesis, it is important to mention that the form of lithium (wire or powder) has important effects on the reduction time without ultrasound. Table 23.5 shows variations in reaction time for reduction of diverse metal salts employing lithium in wire or powder forms. Powdered Li in the absence of ultrasound requires major reaction time than wire. In presence of ultrasound, yields are similar to those obtained by classic *Rieke* methods.

23.3.7 Conclusions to This Section

Active investigations in the area of *Rieke* metals synthesis were carried out from the 1970s to the beginning of our century. The most attention has been given to such metals as Pt, Pd, Zn, Cd, Mg, Ni, Al, Ga, Cu, and Co, that is, those intensively used in organic and organometallic synthesis. Definite attention has been put to activated actinides. A series of applications for the obtained activated

---

* Ar atmosphere is recommended, since there are some reports on the interaction of N_2 with Mg*.
† It is important to consider that non-purified pieces of K or Na generate activated powders of Mg* of lower reactivity; so, metallic sodium can be used together with a solvent, whose boiling point is higher than the melting point of alkali metal.

TABLE 23.4

Comparison of Required Time for Reduction of Diverse Metal Salts by *Rieke* Method and Assisted with Ultrasound

M*	MX$_n$	Alkali Metal	Cond.[a]	Reduction Time
Zn*	ZnBr$_2$	K	A	4 h
Zn*	ZnBr$_2$	Li	B	<40 min
Mg*	MgCl$_2$	K	A	1–2 h
Mg*	MgCl$_2$	Li	B	<40 min
Cr*	CrCl$_3$ * 3THF	K	A	2 h
Cr*	CrCl$_3$	Li	B	<40 min
Cu*	CuI$_2$	K	A	8 h
Cu*	CuBr$_2$	Li	B	<40 min

[a] A, reduction in reflux in THF; B, reduction by Li powder in THF with ultrasound at r.t.

TABLE 23.5

Reduction in Presence of Li in the Form of Powder and Wire: A Comparison of Ultrasonic and Conventional Stirring Techniques

MX$_n$	Ultrasound (min)	Conventional Stirring (h)	Cond.
NiCl$_2$	<40	18	Li wire
FeCl$_2$	<40	24	
PdCl$_2$	<40	26	
CoCl$_2$	<40	19	
PbCl$_2$	<40	—	
NiCl$_2$	<40	14	Li powder
NiI$_2$	<40	12	
CuBr$_2$	<40	13	
CuI	<40	12	

metals has been offered. The most frequent reported techniques involve the use of "lithium–electron carrier (biphenyl, naphthalene, anthracene)" precursor-reductant, preformed or appearing in situ in the reaction medium.

From 2004 to 2011, the observed activities in this field were considerably lower (practically in "stand-by" condition) than in the last century and did not involve development of fabrication of activated metals; their applications in some separate investigations have been reported.[4,56,58,62,71,78,85,89] Therefore, at present century of nanotechnology development, when modern chemical, physico-chemical and biological techniques allow preparation of metallic nanoparticles, nanowires, nanorods and many other nanostructures, it seems that "simple" reduction of anhydrous metal salts with alkali metals is already "past millennium" and does not have considerable importance. However, precisely because of the simplicity of this method and its availability for every chemical laboratory, these classic techniques do not lose their significance and, undoubtedly, should be continuously developed according to the well-known rule that "the new is well-forgotten old."

REFERENCES

1. Furstner, A. (Ed.). *Active Metals: Preparation, Characterization, Applications.* VCH, Weinheim, Germany, 1996, pp. 1–59.
2. Klabunde, K. J.; Murdock, T. O. Active metal slurries by metal vapor techniques. Reactions with alkyl and aryl halides. *Journal of Organic Chemistry*, 1979, *44*, 3901–3908.
3. Luche, J. L.; Damiano, J. C. Ultrasounds in organic syntheses. 1. Effect on the formation of lithium organometallic reagents. *Journal of the American Chemical Society*, 1980, *102*, 7926–7927.
4. Kharissova, O. V.; Kharisov, B. I. Synthetic techniques and applications of activated nanostructurized metals: Highlights up to 2008. *Recent Patents on Nanotechnology*, 2008, *2* (2), 103–119.
5. Whaley, T. P. Preparation of metal powders by sodium reduction chap 14, pp. 129–137, Ethyl corp, Baton Rouge, LA.
6. Kraus, C. A.; Kurtz, H. F. The reduction of metals from their salts by means of other metals in liquid ammonia solution. *Journal of the American Chemical Society*, 1925, *47*, 43–60.
7. Scott, N. D.; Frederick, W. J. Process for reducing metal compounds. U.S. Patent 2177412, 1939.
8. Chu, T. L.; Friel, J. V. Reducing action of sodium naphthalide in tetrahydrofuran solution. I. The reduction of cobalt (II) chloride. *Journal of the American Chemical Society*, 1955, *77* (22), 5838–5840.
9. Henrici-Olive, G.; Olive, S. Intramolecular hydrogen transfer in arene complexes of V, Cr and Ni. *Journal of Organometallic Chemistry*, 1967, *9*, 325.
10. Arnold, R. T.; Kulenovic, S. T. Activated metals. A procedure for the preparation of activated magnesium and zinc. *Synthetic Communications*, 1977, *7*, 223–232.
11. Garza-Rodríguez, L. A.; Kharisov, B. I. Metales Activados de Rieke. Parte I. Preparación de metales de Rieke y su empleo en la obtención de reactivos de Grignard. *Ingenierías*, 2005, *8* (26), 66–74.
12. Rieke, R. D. Use of activated metals in organic and organometallic synthesis. *Topics in Current Chemistry*, 1975, *59*, 1–31.
13. Rieke, R. D. Preparation of highly reactive metal powders and their use in organic and organometallic synthesis. *Accounts of Chemical Research*, 1977, *10*, 301–306.
14. Chao, L. C.; Rieke, R. D.; Kenan, W.R. Jr. Activated metals: VI. Preparation and reactions of highly reactive indium metal. *Journal of Organometallic Chemistry*, 1974, *67* (3), C64–C66.
15. Chao, L. C.; Rieke, R. D. Activated metals. VII. Improved method for the synthesis of trialkyl and triaryl indium compounds. *Synthesis and Reactivity in Inorganic and Metal-Organic Chemistry*, 1974, *4* (4), 373–378.
16. Chao, L. C.; Rieke, R. D. Activated metals. X. Direct synthesis of diphenylindium iodide and ditolylindium iodide from activated indium metal. *Synthesis and Reactivity in Inorganic and Metal-Organic Chemistry*, 5 (3), 1975, 165–173.
17. Chao, L. C.; Rieke, R. D. Activated metals. IX. New Reformatsky reagent involving activated indium for the preparation of beta-hydroxy esters. *Journal of Organic Chemistry*, 1975, *40*, 2253–2255.
18. Rieke, R. D. Preparation of organometallic compounds from highly reactive metal powders. *Science*, 1989, *246*, 1260–1264.
19. Chao, L. C.; Rieke, R. D. Activated metals. IV. Preparation of highly reactive aluminum metal and the direct synthesis of phenylaluminum halides. *Synthesis and Reactivity in Inorganic and Metal-Organic Chemistry*, 1974, *4* (2), 101–105.
20. Olah, G. A.; Morteza, S.; Faria, F.; Bruce, M. R.; Clouet, F. L.; Morton, P. R.; Surya Prakash, G. K. et al. Aluminum dichloride and dibromide. Preparation, spectroscopic (including matrix isolation) study, reactions, and role (together with alkyl(aryl)aluminum monohalides) in the preparation of organoaluminum compounds. *Journal of the American Chemical Society*, 1988, *110*, 3231–3238.
21. Li, P. T.-J. Preparation and reactions of Rieke metals activated aluminum and zinc. PhD Dissertation, University of Nebraska, Lincoln, NE, 1981, AAT 8228151, 236 pp., http://digitalcommons.unl.edu/dissertations/AAI8228151/
22. Pyo, S. H.; Han, B. H. Reduction of nitroarenes by activated metals. *Bulletin of the Korean Chemical Society*, 1995, *16*, 181–183.
23. Wu, T. C.; Xiong, H.; Rieke, R. D. Organocalcium chemistry: Preparation and reactions of highly reactive calcium. *Journal of Organic Chemistry*, 1990, *55*, 5045–5051.
24. Cintas, P. *Activated Metals in Organic Synthesis.* CRC Press, Boca Raton, FL, 1993.
25. Rieke, R. D.; Lee, J.-S.; Kye, Y.-S.; Harbison, G. S. Preparation of lithium stannide mixtures in organic solvents. An alternate source of lithium in organolithium chemistry. *Journal of Organometallic Chemistry*, 2004, *689* (21), 3421–3425.
26. Rieke, R. D. Soluble highly reactive form of calcium and reagents thereof. U.S. Patent 5,211, 889, 1993.

27. Rieke, R. D.; Chen, T. A. Facile synthesis of poly (phenyl carbyne): A precursor for diamondlike carbon. *Chemistry of Materials*, 1994, *6*, 576–577.

28. Rieke, R. D.; Wu, T.-C.; Rieke, L. I. Highly reactive calcium for the preparation of organocalcium reagents: 1-Adamantyl calcium halides and their addition to ketones: 1-(1-adamantyl)cyclohexanol. *Organic Syntheses*, 1995, *72*, 147–153.

29. Rieke, R. D. The preparation of highly reactive metals and the development of novel organometallic reagents. *Aldrichimica Acta*, 2000, *33* (2), 52–60.

30. Pyor, L.; Kiessling, A. Synthesis and reactions of pyridinylcalcium bromides. *American Journal of Undergraduate Research*, 2002, *1* (1), 25–28.

31. Freeman. P. K.; Hutchinson, L. L. Magnesium anthracene dianion. *Journal of Organic Chemistry*, 1983, *48*, 879–881.

32. Raston, C. L.; Salem, G. J. Magnesium anthracene: An alternative to magnesium in the high yield synthesis of Grignard reagents. *Journal of the Chemical Society, Chemical Communications*, 1984, 1702–1703.

33. Harvey, S.; Junk, P. C.; Raston, C. L.; Salem, G. Main group conjugated organic anion chemistry. 3. Application of magnesium-anthracene compounds in the synthesis of Grignard reagents. *Journal of Organic Chemistry*, 1988, *53*, 3134–3140.

34. Bogdanovic, B. Magnesium anthracene systems and their application in synthesis and catalysis. *Accounts of Chemical Research*, 1988, *21*, 261–267.

35. Ramsden, H. E. Magnesium and tin derivatives of fusedring hydrocarbons and the preparation thereof. U.S. Patent 3,354,190, 1967.

36. Ramsden, H. E. Organomagnesium olefin addition compounds. U.S. Patent 3388179, 1968, 7 pp.; *Chemical Abstracts*, 1968, *68*, 114744.

37. O'Brien, R. A.; Chen, T.-A.; Rieke, R. D. Chemical modification of halogenated polystyrene resins utilizing highly reactive calcium and the formation of calcium cuprate reagents in the preparation of functionalized polymers. *Journal of Organic Chemistry*, 1992, *57* (9), 2667–2677.

38. Rhyne, I. L. D. Activated nickel, copper, and uranium: II. Molecular orbital investigations of cyclobutadienoid compounds: III. Attempted synthesis of a new electron acceptor. PhD thesis. The University of North Carolina, Chapel Hill, NC, 1980.

39. Rieke, R. D.; Rhyne, L. D. Preparation of highly reactive metal powders. Activated copper and uranium. The Ullmann coupling and preparation of organometallic species. *Journal of Organic Chemistry*, 1979, *44*, 3445–3446.

40. Fujita. T.; Watanaba, S.; Suga, K.; Sugahara, K.; Tsuchimoto, K. *Chemistry and Industry (London)*, 1983, *4*, 167.

41. Khan, B. R. The preparation and reactivity of active uranium. PhD thesis. The University of Nebraska–Lincoln, Lincoln, NE, 1987.

42. Kahn, B. E.; Rieke, R. D. Reaction of active uranium and thorium with aromatic carbonyls and pinacols in hydrocarbon solvents. *Organometallics*, 1988, *7* (2), 463–469.

43. Ginah, F. O.; Donovan, T. A.; Suchan, S. D., Jr.; Pfening, D. R.; Ebert, G. W. Homocoupling of alkyl halides and cyclization of α,ω-dihaloalkanes *via* activated copper. *Journal of Organic Chemistry*, 1990, *55* (2), 584–589.

44. Wehmeyer, R. M.; Rieke, R. D. Novel functionalized organocopper compounds by direct oxidative addition of zerovalent copper to organic halides and some of their reactions with epoxides. *Journal of Organic Chemistry*, 1987, *52*, 5057–5059.

45. O'Brien, R. A.; Rieke, R. D. Direct metalation of *p*-bromopolystyrene using highly reactive copper and the preparation and reaction of highly reactive copper bound to an insoluble polymer. *Journal of Organic Chemistry*, 1990, *55*, 788–790.

46. Wu, T.-C.; Wehmeyer, R. M.; Rieke, R. D. Direct formation of functionalized alkylcopper reagents from alkyl halides using activated copper. Conjugate addition reactions with 2-cyclohexen-1-one. *Journal of Organic Chemistry*, 1987, *52* (22), 5056–5057.

47. Schwartz, R. H.; San Filippo, J., Jr. Comparative study of some selected reactions of homogeneous and polymer-supported lithium diorganocuprates. *Journal of Organic Chemistry*, 1979, *44*, 2705–2712.

48. Ebert, G. W.; Rieke, R. D. Direct formation of organocopper compounds by oxidative addition of zerovalent copper to organic halides. *Journal of Organic Chemistry*, 1984, *49* (26), 5280–5282.

49. Kauffman, G.; Teter, L.; Ichniowski, T. C.; Clifford, A. F. Tetrakis[iodo(tri-*n*-butylphosphine)-copper(I)] and iodo-(2,2-bipyridine)-(tri-*n*-butylphosphine)copper(I). *Inorganic Syntheses*, 1963, *7*, 9–12.

50. Klein, W. R.; Rieke, R. D. Direct formation and reactions of allylic thienyl-based organocopper reagents. *Synthetic Communications*, 1992, *22* (18), 2635–2644.

51. Rieke, R. D.; Wu, T.-C.; Stinn, D. E.; Wehmeyer, R. M. Direct formation and reaction of functionalized thienyl-based organocopper. *Synthetic Communications*, 1989, *19* (11–12), 1833–1840.

52. Rochfort, G. L.; Rieke, R. D. Preparation, characterization, and chemistry of activated cobalt. *Inorganic Chemistry*, 1986, *25*, 348–355.

53. Kavaliunas, A. V.; Rieke, R. D. Preparation of highly reactive metal powders. Direct reaction of nickel, cobalt, and iron metal powders with arene halides. *Journal of the American Chemical Society*, 1980, *102*, 5944–5945.

54. Kavaliunas, A. V.; Taylor, A.; Rieke, R. D. Preparation of highly reactive metal powders. Preparation, characterization, and chemistry of iron, cobalt, nickel, palladium, and platinum microparticles. *Organometallics*, 1983, *2* (3), 377–383.

55. Rochfort, G. L.; Rieke, R. D. Preparation of activated cobalt and its use for the preparation of octacarbon-yldicobalt. *Inorganic Chemistry*, 1984, *23* (6), 787–789.

56. Volkers, P. I.; Boyke, C. A.; Chen, J.; Rauchfuss, T. B.; Whaley, C. M.; Wilson, S. R.; Yao, H. Precursors to [FeFe]-hydrogenase models: Syntheses of $Fe_2(SR)_2(CO)_6$ from CO-free iron sources. *Inorganic Chemistry*, 2008, *47* (15), 7002–7008.

57. Boudjouk, P.; Thompson, D. P.; Ohrbom, W. H.; Han, B.H. Effects of ultrasonic waves on the generation and reactivities of some metal powders. *Organometallics*, 1986, *5* (6), 1257–1260.

58. Rieke, R. D.; Suh, Y. S.; Kim, S.-H. Heteroaryl manganese reagents: Direct preparation and reactivity studies. *Tetrahedron Letters*, 2005, *46* (35), 5961–5964.

59. Suh, Y. S.; Lee, J.-S.; Kim, S.-H.; Rieke, R. D. Direct preparation of benzylic manganese reagents from benzyl halides, sulfonates, and phosphates and their reactions: Applications in organic synthesis. *Journal of Organometallic Chemistry*, 2003, *684* (1–2), 20–36.

60. Suh, Y. S.; Rieke, R. D. Synthesis of β-hydroxy esters using highly active manganese. *Tetrahedron Letters*, 2004, *45* (8), 1807–1809.

61. Kim, S.-H.; Rieke, R. D. A new synthetic protocol for the direct preparation of organomanganese reagents; organomanganese tosylates and mesylates. *Tetrahedron Letters*, 1999, *40* (27), 4931–4934.

62. Concellón, J. M.; Rodríguez-Solla, H.; Díaz, P. Stereoselective synthesis of (Z)-alpha-haloacrylic acid derivatives, and (Z)-haloallylic alcohols from aldehydes and trihaloesters or amides promoted by Rieke manganese. *Organic and Biomolecular Chemistry*, 2008, *6* (16), 2934–2940.

63. Kim, S. H.; Taylor, A.; Rieke, R. D. Benzylic manganese halides, sulfonates, and phosphates: preparation, coupling reactions, and applications in organic synthesis. *Journal of Organic Chemistry*, 2000, *65* (8), 2322–2330.

64. Kim, S. H.; Taylor, A.; Rieke, R. D. New reagent for reductive coupling of carbonyl and imine compounds: Highly reactive manganese-mediated pinacol coupling of aryl aldehydes, aryl ketones, and aldimines. *Journal of Organic Chemistry*, 1998, *63* (15), 5235–5239.

65. Rieke, R. D.; Kavaliunas, A. Preparation of highly reactive metal powders. Preparation and reactions of highly reactive palladium and platinum metal slurries. *Journal of Organic Chemistry*, 1979, *44* (17), 3069–3072.

66. Rieke; R. D. Method of storing active zero valent zinc metal and applications in organic synthesis. U.S. Patent 5964919, 1999, 33 pp.

67. Rieke, R. D. Water-soluble polythiophenes. WO 2005066243, 2005, 22 pp.

68. Kawamura, S.; Yoshida, M.; Hoshino, S.; Kamata, T. Effective dopant analysis for the high performance poly (3-hexylthiophene) field-effect transistors. *MRS Symposium Proceedings (Organic Thin-Film Electronics)*, San Francisco, Vol. 871E, 2005.

69. Kroemer, J.; Kirkpatrick, C.; Maricle. B.; Gawrych, R.; Mosher, M. D.; Kaufman, D. *Rieke* zinc as a reducing agent for common organic functional groups. *Tetrahedron Letters*, 2006, *47* (36), 6339–6341.

70. Rieke, R. D.; Sung, J. U. Activated metals. XI. An improved procedure for the preparation of β-hydroxy esters using activated zinc. *Communication Synthesis*, 1975, (7), 452–453.

71. Kharisov, B. I.; Garza-Rodríguez, L. A.; Leija Gutiérrez, H. M.; Ortiz Méndez, U.; García, R. C. Preparation of non-substituted metal phthalocyanines at low temperature using activated Rieke zinc and magnesium. *Synthesis and Reactivity in Inorganic, Metal-Organic and Nano-Metal Chemistry*, 2005, *35*, 755–760.

72. Rieke, R. D.; Tzu-Jung Li, P.; Burns. T. P.; Uhm, S. T. Preparation of highly reactive metal powders. New procedure for the preparation of highly reactive zinc and magnesium metal powders. *Journal of Organic Chemistry*, 1981, *46*, 4323–4324.

73. Zhu, L.; Wehmeyer, R. M.; Rieke, R. D. The direct formation of functionalized alkyl (aryl) zinc halides by oxidative addition of highly reactive zinc with organic halides and their reactions with acid chlorides, α,β-unsaturated ketones, and allylic, aryl, and vinyl halides. *Journal of Organic Chemistry*, 1991, *56* (4), 1445–1453.

74. Rieke, R. D. Highly reactive zerovalent metals from metal cyanides. U.S. Patent 5,507,973, 1996.
75. Yanagisawa, A.; Yase, K.; Yamamoto, H. Regio- and stereoselective carboxylation of allylic barium reagents: (E)-4,8-dimethyl-3,7-nonadienoic acid. *Organic Syntheses*, 1998, Coll. Vol. *9*, 317; 1997, *74*, 178.
76. Burkhardt, E. R.; Rieke, R. D. The direct preparation of organocadmium compounds from highly reactive cadmium metal powders. *Journal of Organic Chemistry*, 1985, *50*, 416–417.
77. Stevenson, Ch. D.; Reiter, R. C.; Burton, R. D.; Halvorsen, T. D. Electron transfer from Rieke cadmium Yielding Cd^{2+} coordinated and ion associated anion radical systems. *Inorganic Chemistry*, 1995, *34* (6), 1368–1372.
78. Kharisov, B. I.; Ortiz Méndez, U.; Garza-Rodríguez, L. A.; Leija Gutiérrez, H. M.; Medina Medina, A.; Berdonosov, S. S. Use of various activated forms of elemental nickel and copper for the synthesis of phthalocyanine at low temperature. *Journal of Coordination Chemistry*, 2006, *59* (15), 1657–1666.
79. Inaba, S.-I.; Rieke, R. D. Facile preparation of 3-arylpropanenitriles by the reaction of benzylic halides with haloacetonitriles mediated by metallic nickel. *Synthesis*, 1984, 842–844.
80. Rieke, R. D.; Wolf, W. J.; Kujundzic, N.; Kavaliunas, A. V. Highly reactive transition metal powders. Oxidative insertion of nickel, palladium, and platinum metal powders into aryl-halide bonds. *Journal of the American Chemical Society*, 1977, *99* (12), 4159–4160.
81. Shi, L.; Fan, C.-A.; Tu, Y.-Q.; Wang, M.; Zhang, F.-M. Novel and efficient Ni-mediated pinacol coupling of carbonyl compounds. *Tetrahedron*, 2004, *60* (12), 2851–2855.
82. Rieke, R. D.; Hudnall, P. M. Activated metals. I. Preparation of highly reactive magnesium metal. *Journal of the American Chemical Society*, 1972, *94* (20), 7178–7179.
83. Lee, J.-S.; Velarde-Ortiz, R.; Guijarro, A.; Wurst, J. R.; Rieke, R. D. Low-temperature formation of functionalized *Grignard* reagents from direct oxidative addition of active magnesium to aryl bromides. *Journal of Organic Chemistry*, 2000, *65* (17), 5428–5430.
84. Wurst, J. R.; Velarde-Ortiz, R.; Lee, J.; Guijarro, A.; Rieke, R. D. Low temperature generation of functionalized *Grignard* reagents in organic synthesis. *Book of Abstracts, 217th ACS National Meeting*, Anaheim, CA, March 21–25, 1999.
85. Kvicala, J.; Stambasky, J.; Skalicky, M.; Paleta, O. Preparation of fluorohalomethylmagnesium halides using highly active magnesium metal and their reactions. *Journal of Fluorine Chemistry*, 2005, *126* (9–10), 1390–1395.
86. Hazimeh, H.; Kanoufi, F.; Combellas, C.; Mattalia, J.-M.; Marchi-Delapierre, C.; Chanon, M. Radical clocks, solvated electrons, and magnesium. Heterogeneous versus homogeneous electron transfer. Selectivity at interfaces. *Journal of Physical Chemistry C*, 2008, *112* (7), 2545–2557.
87. Sell, M. S. A study of HT regioregular polythiophenes and 3-thienyl organometallic reagents: Using Rieke metals as an important synthetic tool. PhD thesis. University of Nebraska, Lincoln, NE, 1995.
88. Rieke, R. D.; Bales, S.; Hudnall, T. P.; Poindexer, G. S. Highly reactive magnesium for the preparation of Grignard reagents: 1-Norbornanecarboxylic acid. *Organic Syntheses*, 1988, Coll. Vol. *6*, *845*; 1979, *59*, 85.
89. Kharisov, B. I.; Ortiz Mendez, U.; Rivera de la Rosa, J. Low-temperature synthesis of phthalocyanine and its metal complexes. *Russian Journal of Coordination Chemistry*, 2006, *32* (9), 617–631.

24 Bi- and Trimetallic Nanoparticles Based on Gold and Iron*

Finely divided iron has long been known to be pyrophoric, which is a major reason that Fe nanoparticles (Nps) have not been more fully studied to date.[1] This extreme reactivity has traditionally made Fe Nps difficult to study and inconvenient for practical applications. However, iron has a great deal to offer at the nanoscale, including very potent magnetic, catalytic, and medical applications, so methods for reducing its reactivity with simultaneous retention of magnetic and catalytic properties have been developed. Although pure metal iron Nps are unstable in air, by a coating of noble metal on Nps surface, these formed air-stable Nps are protected from the oxidation and retain most of the favorable magnetic properties, which possess the potential applications mentioned here, as well as in high density memory devices by forming self-assembling nanoarrays.[2a]

Gold coating is very promising for magnetic and other particles to be functionalized for targeted drug delivery.[2b] This metal (Au), as well as iron oxides Fe_2O_3 and Fe_3O_4, is biocompatible with human organisms,[3,4] could easily be tracked in it, and can be functionalized with organic and bioorganic molecules, such as proteins or enzymes, and bind bacteria; at the same time, Nps of iron and its oxides possess magnetic properties facilitating their delivery to an organ, tissue, or cancer tumor by using an external magnetic field. Therefore, the unique combination of the nanoscale magnetic core and the functional shell makes Fe-doped Au Nps ideal for biological and biomedical applications due to their conjugation chemistry, optical properties, and surface chemistry.

During the last decade, considerable experimental efforts have been dedicated to biomedical applications of coated magnetic Nps, such as in magnetic resonance imaging contrast enhancement, tissue repair, immunoassay, detoxification of biological fluid, hyperthermia, targeted drug delivery, and cell separation. Results of the investigations in the area of gold-coated iron (iron oxides) core Nps are generalized in a series of monographs,[3–11] as well as in reviews[12,13] and patents[14,15] describing strategies for the synthesis of these Nps, characterization of the core/shell nanostructures, and exploration of potential applications of the core/shell nanomaterials in terms of biological and catalytic interfacial reactivities. In this review, we present the main achievements in the selected area of Fe (Fe_2O_3, Fe_3O_4) core/Au shell Nps from the beginning of this century up to the present.

24.1 METALLIC IRON CORE/AU SHELL AND ALLOY NANOPARTICLES

24.1.1 TYPES OF FE/AU NPS

Water-soluble Fe/Au *alloy* Nps (Fe/Au alloy Nps are solid solutions where iron atoms substitute gold sites in the face of center cubic lattice) have been reported.[16] The diameter of these alloy Nps was 4.9 ± 1.0 and 3.8 ± 1.0 nm for two different precursors of iron, ferrous sulfate heptahydrate, and iron pentacarbonyl[17]; the estimated particle's iron content was 14.8 ± 4.7 mol.%. The formation of alloy nanostructure with a narrow distribution of particle sizes and three compositions $Au_{0.25}Fe_{0.75}$, $Au_{0.5}Fe_{0.5}$, and $Au_{0.75}Fe_{0.25}$ by a polyol process that retains the optical and magnetic properties of the

---

* This chapter has been prepared with the participation of Dr. Miguel José-Yacamán (UTSA).

individual components was discussed in Ref. [18]. Fe/Au Nps of *LI_0 type* (mean composition Fe-22 at.% Au) were fabricated by annealing Fe/Au particles.[19] Au-rich (>32 at.%) and Au-poor (<16 at.%) regions were formed in most of the particles.

The majority of publications have been devoted to Fe core coated with Au shell (so-called *core-shell Nps*).[20–23] For instance, nanosized iron–gold magnetic Nps with an average particle size of 5–25 nm were prepared by a *reverse micelle* method, which is the principal technique in obtaining core-shell Nps, together with *polylol technique* and *homogenous solution reduction*, efficiently used in many reports. The magnetic properties measurements confirmed behavior typical of a superparamagnetic system.[24,25] The XPS studies of the core-shell Fe/Au Nps indicate that besides Fe^0 inside the cores, small amounts of Fe(II,III), located onto the gold surface, were also formed during the sample preparation. A certain number of reports have mentioned pure iron core, for instance Fe core/Au shell (11 nm Fe core, 2.5 nm Au shell) without iron oxides.[26]

24.1.2 METHODS OF SYNTHESIS AND CHARACTERIZATION

General method for obtaining Fe core/Au shell Nps (Figure 24.1) consists of subsequent reduction of iron and Au salts by $NaBH_4$ and other reductants in aqueous or organic media in the presence of surfactants, preventing agglomeration of the formed Nps. In a series of reports, the characterized Nps were considered as containing *pure metallic iron* as a core; however, it should be especially emphasized that the authors of the communications, who applied Mössbauer and x-ray absorption (XAS) spectroscopy, noticed that, together with SEM, TEM, and other techniques, only additional use of two of the aforementioned methods can give complete information on the state of iron in the core. Thus, gold passivated Fe Nps were prepared in a reverse micelle of cetyltrimethylammonium bromide (CTAB),[27–29] and their studies by a series of aforementioned methods, including Mössbauer spectral study (^{57}Fe), showed the presence of iron in the forms of α-Fe (major part), unexpected $Fe_{1-x}B_x$ alloy, and several poorly crystallized ordered Fe_2O_3 components, as well as residual paramagnetic Fe(II) and Fe(III), indicating a more complex structure than had been believed, although the powder x-ray diffraction of Fe/Au Nps-containing samples revealed both the presence of crystalline α-Fe and gold and the absence of any crystalline iron oxides or other crystalline products. In another report,[30] oxidation of Fe in Fe/Au core-shell Nps was analyzed showing that, in contrast to a lot of previous reports assuming the metallic state of iron in Fe/Au Nps, obtained from reverse micelles, the iron component was found to be fully oxidized, according to XAS, as "an ideal tool for characterizing Nps."

In addition to the reverse micelle technique, other methods have also been developed for obtaining Fe/Au, selected examples of which are shown in Table 24.1. Thus, Fe/Au Nps (8 nm) were prepared

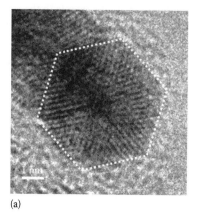
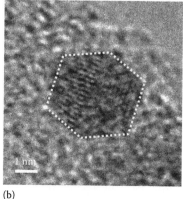

(a)   (b)

FIGURE 24.1 HRTEM image of the Fe-Au alloy Nps. These particles are oriented: (a) almost along the threefold axis and (b) close to the twofold axis, respectively. (Reproduced with permission from Dahal, N. et al., *Chem. Mater.*, 20(20), 6389–6395. Copyright 2008 American Chemical Society.)

TABLE 24.1
Selected Examples of Synthesis of the Nps on the Basis of Gold and Iron

Precursors and Conditions	Product and Its Properties	References
Fe/Au Nps		
Reduction of gold acetate by 1,2-hexadecanethiol and the thermal decomposition of $Fe(CO)_5$ in the presence of stabilizers (oleic acid and oleylamine).	Monodisperse superparamagnetic FeAu Nps possessing the optical properties of Au Nps and the magnetic properties of Fe Nps. The incorporation of Au into Fe Nps leads to a structural change from body-centered cubic to face-centered cubic. The Nps can be self-assembled into parallel strips in the direction of an applied magnetic field.	[149]
Cetyltrimethylammonium bromide, 1-butanol, octane, and water as reverse micelles. Using $NaBH_4$ as reductant, Fe Nps are formed inside the micelles and then $HAuCl_4$ solution addition increases their size and results in formation of gold layer on the Fe Nps.	Fe/Au (1:3 wt. ratio) core-shell Nps (27 nm).	[150]
1. *Preparation from iron sulfate heptahydrate.* (Di-*n*-dodecyl)dimethylammonium bromide, toluene, iron sulfate heptahydrate, $NaBH_4$, $AuCl_3$ solution ($HAuCl_4 \cdot 3H_2O$), 3-mercapto-1-propanesulfonic acid sodium salt. 110°C. 2. *Preparation from Fe(CO)$_5$.* $Fe(CO)_5$, toluene, $AuCl_3$ solution ($HAuCl_4 \cdot 3H_2O$), 3-mercapto-1-propanesulfonic acid sodium salt, $NaBH_4$. 110°C.	Fe/Au Nps (particle size 3.8 [method 1] and 4.9 [method 2, more stable Nps]) with relative low iron content (~15%).	[17]
$FeCl_3$, water, disulfide neoglycoconjugate, MeOH, $HAuCl_4$, $NaBH_4$.	GlycoNps Fe/Au, insoluble in MeOH, but soluble in water.	[151]
Fe(II), $NaBH_4$, water, Au(III).	Fe core/Au shell Nps.	[152]
Fe$_x$O$_y$/Au Nps		
Iron oxide particle (Feridex) was mixed with $HAuCl_4$ solution, polyvinylalcohol, and 2-propanol, and treated with *electron beam*.	Gold particle-attached Feridex particle (primary diameter 6 nm and secondary diameter 250 nm).	[153]
Nps prepared in *reverse micelle* of cetyltrimethylammonium bromide. 1-butanol (cosurfactant), octane (oil phase). Mixture of ferric chloride hexahydrate and ferrous chloride tetrahydrate, $NaBH_4$, $HAuCl_4$, 60°C, 15 min.	Fe_3O_4/Au magnetic Nps.	[154]
Fe_3O_4 Nps, prepared by co-precipitation, were introduced into H_2O droplets as nucleation centers for subsequent formation of metallic shell by $NaBH_4$ reduction of Au^{3+} on the surface of Nps. 3-Mercaptopropionic acid was used to separate the as-prepared Nps from destroyed *reverse micelles* system and stabilize the Nps in organic media (CTAB/octane/butanol/water).	Ultrasmall Fe_3O_4/Au Nps (6.7 nm mean particle size).	[155]
Sonolysis of a solution mixture of $HAuCl_4$ and (3-aminopropyl)triethoxysilane (APTES)-coated Fe_3O_4 Nps with further drop-addition of sodium citrate.	Air-stable Fe_3O_4/Au Nps (30 nm). Nps of Fe_3O_4/Au possess a very high saturation magnetization of about 63 emu/g.	[156]
Citrate reduction of $HAuCl_4$ on the surface of SPION that served as the seeds. Alkanethiol-modified oligonucleotides were bound with the self-made Fe_3O_4 (core)/Au (shell) Nps to form Np hepatitis B virus (HBV) DNA gene probes through Au-S covalent binding.	Well-dispersed (Fe_3O_4, core)/Au (shell) Nps (15 nm).	[157]

(continued)

TABLE 24.1 (continued)

Selected Examples of Synthesis of the Nps on the Basis of Gold and Iron

Precursors and Conditions	Product and Its Properties	References
One-step reaction; aqueous solution of thiol-functionalized neoglycoconjugates (**1** and **2**) of lactose and maltose disaccharides with different linkers and either Au salts or both Au and Fe salts.	Au and Fe/Au_2O_3 glycoNps (GNPs) (average core-size diameters 1.5–2.5 nm) (Figure 24.2), water-dispersible and stable. Iron-free gold GNPs: permanent magnetism; gold-iron oxide GNPs: absence of a permanent magnetism.	[158]
$FeCl_3 \cdot 6H_2O$, ethylene glycol, octylamine, 190°C, stirring (maghemite synthesis). $HAuCl_4$, octylamine, ethylene glycol, 150°C.	Maghemite-gold core-shell Nps.	[159]
Au^{3+}, hydroxylamine, excess of Fe_3O_4 as seeds.	Fe_3O_4/Au Nps (<100 nm diameter).	[160]
Au^{3+}, iron oxides, method of *iterative hydroxylamine seeding*.	Water-soluble γ-Fe_2O_3 or Fe_3O_4 (9 nm diameter)/Au Nps (60 nm diameter). The morphology and optical properties depend on the quantity of deposited Au; magnetic properties are independent.	[161]
Decomposition of $Fe(CO)_5$ on the surface of the Au Nps followed by oxidation in 1-octadecene as solvent.	Dumbbell-like Fe/Au_3O_4 Nps, formed through epitaxial growth of iron oxide on the Au seeds.	[162]
Trimetallic Nps containing Fe and Au		
$Fe(acac)_3$, $Pt(acac)_2$ and $AuCl_3$, oleylamine (reducing agent), paraffin (high-boiling solvent), and surfactants. 200°C, a N_2 atmosphere, 20 min, and then the temperature was raised to reflux conditions (390°C–400°C) for 3 h.	$L1_0$-type FePt and FePtAu ($Fe_{45}Pt_{44}Au_{11}$) Nps.	[163]
Simultaneous decomposition of $Fe(CO)_5$ and reduction of gold acetate and platinum acetylacetonate in high-boiling hexadecylamine at 300°C–360°C (with 1-adamantanecarboxylic acid and HDA as stabilizers).	Partially ordered FePtAu Nps (8 nm).	[164–166]
Three steps: Functionalization of the $CoFe_2O_4$ magnetic Nps with amine and thiol groups, synthesis and attachment of gold Nps (Au-NPs) on amino/thiol-functionalized magnetic Nps, and formation of a gold shell on the magnetic Nps.	Au/$CoFe_2O_4$ Nps	[167]

by sequential high-temperature *decomposition of organometallic compounds* in a coordinating solvent and functionalized with alkanethiolate ligands preventing aggregation, enabling solubility in a series of both hydrophilic and hydrophobic solvents and allowing further derivatization *via* ligand exchange reactions.[31] Fe/Au Nps (Figure 24.3), confirmed to be solid solutions, were *electrodeposited* on an amorphous carbon electrode in a three-compartment glass-made electrochemical cell from aqueous electrolytes at room temperature using 1 mM $FeSO_4$ and $HAuCl_4$ solutions as precursors and $CsClO_4$ as supporting electrolyte in Ar atmosphere.[32] It was proved that these Nps were homogeneously alloyed and the composition of the alloyed Nps was dependent on the electrodepositing potential. In comparison with similarly prepared Au/Ni and Au/Co Nps, the chain length for Fe/Au Nps is lesser: Fe/Au < Ni/Au < Co/Au (the largest length was observed for pure Au Nps); this suggests that the transition metals in the alloyed Nps may prohibit the agglomeration of the alloyed particles. *Radiolytical reduction* of mixed Au^{III}/Fe^{II} ethylene glycol solutions led to two kinds of particles: 2 nm Fe rich Nps as well as large rods (a few tens of nm) and faceted particles rich in gold and containing a small amount of iron.[33] The adjunction of iron to gold enhances remarkably its electro-catalytic

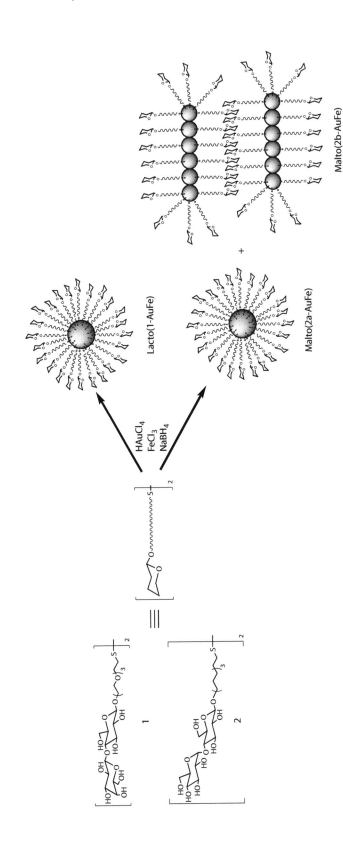

FIGURE 24.2 Preparation of lactose and maltose gold-iron oxide glycoNps. (Reproduced with permission from De la Fuente, J.M. et al., *J. Phys. Chem. B*, 110(26), 13021–13028. Copyright 2006 American Chemical Society.)

10 nm

FIGURE 24.3 Fe/Au decahedral nanoparticle. (Reproduced with permission from Lu, D.-L. et al., *Langmuir*, 18(8), 3226–3232. Copyright 2002 American Chemical Society.)

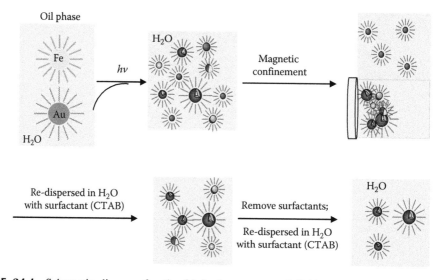

FIGURE 24.4 Schematic diagram for the fabrication process of Fe/Au magnetic core-shell particles. (Reproduced with permission from Zhang, J. et al., *J. Phys. Chem. B*, 110(14), 7122–7128. Copyright 2006 American Chemical Society.)

properties toward oxygen and proton reduction. The authors noted that Au-Fe system is very promising for application in fuel cells. *Pulsed laser* deposition of the Fe-Au alloy (35% Fe) in a mesoporous alumina membrane template allowed the formation of strongly paramagnetic Fe/Au Nps (17% Fe, 46 ± 13 nm), which can be easily functionalized with biological macromolecules.[34]

A schematic diagram for the formation of Fe/Au core-shell Nps by laser treatment is shown in Figure 24.4.[35]

24.1.3 STUDIES OF GROWTH MECHANISM AND OXIDATION

A density-functional theory study of Au-coated Fe nanoclusters of various sizes to understand magnetic properties of this core-shell structure was carried out.[36] It was shown that the magnetic moment of Fe nanocore with Au coating is considerably higher than that in bulk Fe; the coupling between Fe atoms is ferromagnetic and is insensitive to the thickness of the Au coating. Au shell protects the Fe core from oxidation, coalescence, and formation of thromboses in life organisms. In general, growth mechanisms and oxidation resistance for Fe/Au Nps were studied in Refs. [37–39]. It was found that, in particular, the Au shells grow by nucleating on the Fe-core surfaces before coalescing; the magnetic moments of Fe/Au Nps decrease over time due to oxidation. Some Nps with off-centered Fe cores are more susceptible to oxidation.

Among other recent investigations on various types of Fe–Au interactions and growth on the surface of Au, Fe, or other elements, we note the study of formation of Fe and FeO Nps and thin films on the reconstructed Au(111) surface at different growth conditions. Fe grew as 1 monolayer high triangular particles on the Au(111) reconstruction; FeO was grown after room-temperature oxidation by exposing the Fe Nps to molecular oxygen at 323 K, followed by annealing at 500–700 K.[40] Various Fe/Au core-shell and alloy Nps were compared for the formation of iron and gold silicides on Si <111> *via* vapor liquid solid method from the thermal decomposition of tetra-Et silicon. It was shown that Fe/Au Nps are efficient precursors for the formation of iron–gold mixed silicide thin films.[41] Gold/ C_{60}(/C_{70}) nanoclusters, where the pyridine was used as a intermediate to connect and nest the C_{60} (C_{70}) molecule to the gap of Au Nps and iron substrate, were investigated.[42] Inhibition effect of the Au Nps self-assembled films on Fe surface for the substrate in 0.5 M H_2SO_4 solutions was studied.[43]

24.1.4 Current and Proposed Uses

Fe/Au Nps have been investigated for use as catalysts or media for their cleaning,[44] in particular those supported onto HY- and NaY-type zeolites. Thus, interactions of Au with Fe species, introduced into the NaY zeolite by two ways (wet-impregnation and ion exchange), in Fe/Au-imp/ NaY zeolite system, were studied.[45] 1% Au-0.5%Fe/HY sample, possessing high activity in the CO oxidation, was prepared by coexchange in HY zeolite using Fe(II)ethylenediammonium and Au(III)ethylenediamine complex ions.[46] Photoinduced sulfur desorption from the surfaces of Au Nps loaded on a series on metal oxides, in particular Fe_2O_3, was studied.[47] Elemental sulfur S_8 was selectively adsorbed on the Au Nps surfaces of Au/metal oxides in an atomic state. This phenomenon is applicable to the low-temperature cleaning of sulfur-poisoned metal catalysts.

As well as core-shell Fe_xO_y/Au Nps, the Fe/Au Nps have been offered to be applied in biomedicine. Thus, the method for synthesizing antibody-magnetic Nps with gold shell and iron core for recognition and separation of tumor cells was elaborated.[48] Also, their uses have been reported as T1 contrast agents in the unoxidized form,[49] capturers for pathogens,[50] carriers of streptavidin,[51] etc. The 7 nm Fe/Au Nps (1.5% of Fe), obtained[52] by simultaneous reduction of Fe^{3+} and Au^{3+}, can be easily conjugated to thiolated DNA; moreover, they heat in solution to temperatures above 40°C, indicating suitability for hyperthermia.

Fe/Au Nps can serve as precursors for other Nps or functionalized derivatives. Thus, 200–350 nm C-encapsulated magnetic Nps possessing the surface with functional groups like –OH were prepared by hydrothermal heating an aqueous glucose solution of FeAu Nps at 160°C–180°C for 2 h.[53,54] An organometallic approach to the synthesis of CO-protected Fe/Au Nps led to a solution of colloids with hydrodynamic diameters between 4 and 300 nm, from which $[Au_{21}\{Fe(CO)_4\}_{10}]^{5-}$, $[Au_{22}\{Fe(CO)_4\}_{12}]^{6-}$, $[Au_{28}\{Fe(CO)_3\}_4\{Fe(CO)_4\}_{10}]^{8-}$, and $[Au_{34}\{Fe(CO)_3\}_6\{Fe(CO)_4\}_8]^{8-}$ were isolated.[55] Chemical oxidation in a solution of Au Nps, protected by octanethiol and 12-(*N*-pyrrolyl) dodecanethiol, with $FeCl_3$ leads to Au Nps cross-linking with the formation of chain-like structures.[56] The authors proposed that this approach may also be applied for Fe/Au.

24.2 Fe_xO_y/Au NANOPARTICLES

24.2.1 Types of Particles

Among various forms of iron oxides (FeO, iron(II) oxide, wüstite; Fe_3O_4, iron(II,III) oxide, magnetite; Fe_2O_3, iron(III) oxide: its forms are α-Fe_2O_3, hematite, β-Fe_2O_3, γ-Fe_2O_3, maghemite, and ϵ-Fe_2O_3), core-shell Nps with gold have been reported mainly for Fe_3O_4 and γ-Fe_2O_3. The last compound has

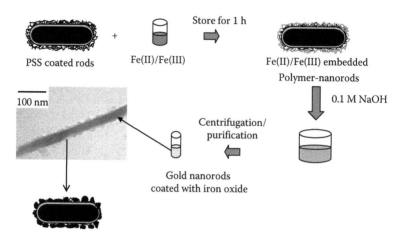

FIGURE 24.5 General protocol used to prepare *in situ* iron oxide–coated gold nanorods. (Reproduced with permission from Gole, A. et al., *Langmuir*, 24(12), 6232–6237. Copyright 2008 American Chemical Society.)

the same structure as magnetite, that is, it is spinel ferrite and is also ferrimagnetic. Maghemite can be considered as an Fe(II)-deficient magnetite with the formula $(Fe_8^{III})_A[Fe_{40/3}^{III}\square_{8/3}]_B O_{32}$, where $\square$ represents a vacancy, A indicates tetrahedral positioning, and B octahedral. The number of reports on Fe_xO_y/Au core-shell Nps in the last decade is considerably higher than that for the case of the Fe/Au Nps, described in the previous section, and the majority of them have medical applications.

In addition to the mostly represented "standard" iron oxide core/gold shell Nps, some interesting unexpected nanostructures have been obtained. Thus, iron oxide–coated gold nanorods were prepared (Figure 24.5) starting from gold nanorods, synthesized by a three-step seed-mediated protocol and coated with a layer of poly(sodium 4-styrenesulfonate).[57] Here, on the contrary, the gold metal acts as a core and the iron oxide as a shell. The negatively charged polymer on the nanorod surface electrostatically attracted a mixture of aqueous Fe(II) and Fe(III) ions. Coprecipitation of these iron salts was used to form uniform coatings of iron oxide Nps on the surface of the gold nanorods. The oxidation state of iron in iron oxide coatings was determined and was consistent with Fe_2O_3 rather than Fe_3O_4. Multifunctional Fe_3O_4/polymer/Au shell Nps[58] with preserved strong magnetization and good NIR absorption display good dispersibility and stability in aqueous solution. Authors supposed that these features would facilitate biomedical applications, combining the benefits of MRI diagnosis, magnetically targeted delivery, and photothermal ablation.

Attention has been paid to *water-soluble* composite Nps. Thus, ligand-protected (with homocysteine) Au-coated Fe oxide Nps (homocys-Au-Fe_3O_4), well-dispersed in water and stable at physiological pH without precipitation, were prepared from Fe_3O_4 Nps by coating with Au layers under hot citrate reduction of $HAuCl_4$ and further place exchange with homocysteine molecules.[59] Water-soluble Fe_xO_y/Au Nps, functionalized with a maltose neoglycoconjugate (neoglycoconjugate 11,11′-dithiobis[undecanyl-β-maltoside]), were fully characterized[60]; it was observed that polymers packed in units about 65 nm in length and 40 nm in width on Au surfaces and the Nps seem to be encapsulated by the organic material. The authors proposed interactions between the sugar residues and the amphiphilic character of the maltose neoglycoconjugate (with a lipophilic undecane spacer) as responsible for the origin of these amazing supramolecular arrangements.

Novel nanocomposite particles consisting of *hollow-type* gold nanoshells with iron oxide Nps inside their interiors were developed.[61] The iron oxide Nps have a magnetic character that enables them to be used as contrast agents for MRI, whereas the hollow-type gold nanostructures that encapsulate the magnetic Nps provide strong absorption (and scattering) of near-IR light. These types of nanocomposite particles were especially designed to be used as multifunctional nanoplatforms for biomedical imaging and targeted cancer therapy.

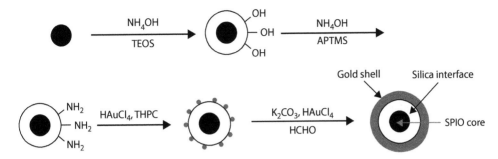

FIGURE 24.6 Synthesis of gold nanoshells with superparamagnetic iron oxide (SPIO) cores, silica inter-face, and gold shell. TEOS, tetraethylorthosilicate; APTMS, 3-aminopropyltrimethoxysilane; THPC, tetrakis(hydroxymethyl)phosphonium chloride. (Reproduced with permission from Ji, X. et al., *J. Phys. Chem. C*, 111(17), 6245–6251. Copyright 2007 American Chemical Society.)

A series of publications have been devoted to *silica-containing or supported* Fe oxide/Au Nps. Thus, hybrid Nps with a superparamagnetic Fe oxide (SPIO)-silica core and a gold nanoshell, suit-able for nanoshell-mediated photothermal therapy, were prepared by a multistep procedure (Figure 24.6).[62] The silica layer served the same role as the silica core in conventional gold nanoshells: it provided a dielectric interface for shifting the plasma resonance to the near-infrared wavelength region. When coated with polyethyleneglycol, these Nps formed a stable solution in water. The effect of gold particle size and Fe/AuO$_x$ interface on the electronic properties and catalytic activity was modeled using samples Fe/AuO$_x$/SiO$_2$/Si(100), FeO$_x$/Au/SiO$_2$/Si(100), etc.[63] It was shown that generally the formation of gold/iron oxide interface increases the catalytic activity in CO oxidation regardless of the sequence of deposition.

Composite Nps with an Au shell, an Fe$_3$O$_4$ inner shell, and a silica core were obtained by the approach utilizing positively charged amino-modified SiO$_2$ particles as templates for the assembly of negatively charged 15 nm superparamagnetic water-soluble Fe$_3$O$_4$ particles.[64] The simple and at the same time brilliant idea of this technique consists of the *electrostatic attraction* of 1–3 nm Au Nps seeds by SiO$_2$-Fe$_3$O$_4$ particles; as a result a continuous gold shell is formed around them upon HAuCl$_4$ reduction. The formed three-layer Nps (Figure 24.7) can be further functionalized with oligonucleotides. In principle, the synthetic procedure can be extended to the preparation of various particle sizes using different-sized SiO$_2$ template particles and other compositions where electro-static assembly can be utilized to generate layered structures analogous to the ones presented here.

FIGURE 24.7 Synthetic scheme for the preparation of the three-layer magnetic Nps. (Reproduced with permission from Stoeva, S.I. et al., *J. Am. Chem. Soc.*, 127(44), 15362–15363. Copyright 2005 American Chemical Society.)

Core-shell microspheres of a polystyrene core (640 nm diameter) covered with a number of Fe_3O_4 layers (12 nm) and SiO_2-coated Au (15 nm) were fabricated using the layer-by-layer method.[65] These spheres can be assembled into chains of <1 mm length by deposition in a magnetic field. Synthesis and characterization of iron oxide/Au Nps in mesoporous silicas was reported in Ref. [66]. It was noted that the dispersion and intimate contact between the gold and iron particles in mesoporous silica are affected not only by the preparation method and pretreatment temperature used, but could be successfully regulated by the pore accessibility of the host matrix.

24.2.2 Synthesis Methods

These methods for core-shell Nps Au/(γ-Fe_2O_3 or Fe_3O_4) are wider[67–72] in comparison with those described here for Fe/Au Nps. Thus, Fe_3O_4/Au Nps (gold shell at the iron oxide nanocrystal cores) with different ratios Fe:Au, high monodispersity, and controllable surface were prepared by reduction of iron(III) acetylacetonate and gold(III) acetate as precursors (Figure 24.8).[73] The *problem of separation* of Fe_3O_4 and Fe_3O_4/Au Nps was discussed. The approach of authors differs from many Np-to-Np based approaches by the ability to control a combination of thermally activated desorption of the capping layer, deposition of Au on the exposed Fe_3O_4 surface, and subsequent re-encapsulation of the Au surface by the capping agent. Fe_xO_y/Au Nps are widely prepared by sequential reduction of Fe^{2+} and Au^{3+} in *reversed micelles*; according to Ref. [74], Au exists in them as metal and magnetic phase was Fe_3O_4.

Composite Nps of gold and iron oxide were obtained by a high-energy *electron beam*, whose irradiation induced the formation of Au Nps, which were firmly immobilized on the surface of the support iron oxide Nps fully coating them. Their size depended on the concentration of gold ions, polymers, and iron oxide Nps before the irradiation.[75] Fe oxide/Au magnetic composite Nps were synthesized with γ-*rays* or *ultrasonics* and further functionalized with thiol-modified oligonucleotides.[76] Due to large adsorbing capacity of the oligonucleotides, adsorbed onto the Au/γ-Fe_2O_3 composite Nps, which was synthesized by γ-ray irradiation, these Nps are found suitable for the separation of biomolecules.[77] Air-stable Nps of Fe_3O_4/Au (30 nm diameter) were prepared

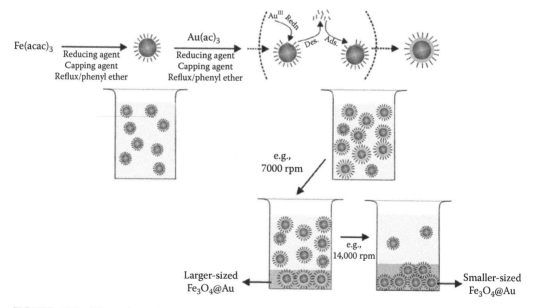

FIGURE 24.8 Illustration of the chemistry and processes involved in the synthesis of the Fe_3O_4 and Fe_3O_4/Au Nps. (Reproduced with permission from Wang, L. et al., *J. Phys. Chem. B*, 109(46), 21593–21601. Copyright 2005 American Chemical Society.)

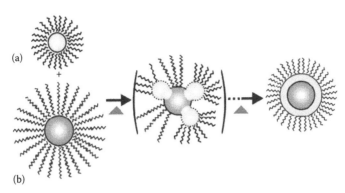

FIGURE 24.9 Idealized illustration of hetero-interparticle coalescence of Au (a) and Fe_2O_3 (b) Nps leading to core/shell Nps. (Reproduced with permission from Park, H.-Y. et al., *Langmuir*, 23(17), 9050–9056. Copyright 2007 American Chemical Society.)

via sonolysis of a solution mixture of $HAuCl_4$ and (3-aminopropyl)triethoxysilane-coated Fe_3O_4 Nps with further drop-addition of sodium citrate.[78] Fe_3O_4/Au composite Nps were obtained by a combination of unique *laser processes* in water (pulsed laser ablation (Nd:YAG laser, 532 nm) in liquid phase leading to pure Au Nps and a *nanosoldering effect* yielding to an Au/magnetite composite).[79,80] Another example of a combinative method consists of the preparation of γ-Fe_2O_3/Au core-shell Nps by a union of high-temperature organic solution synthesis and colloidal microemulsion method.[81] Fe_2O_3 core/Au shell monodisperse Nps with controllable sizes of 5–100 nm and high monodispersity were obtained using a novel strategy (Figure 24.9) based on the *thermally activated heterocoalescence* between Au and Fe_2O_3 Nps; similar results were observed for Fe_3O_4 Nps cores.[82] Each of these core/shell Nps consists of a magnetically active Fe oxide core and a thiolate-active Au shell, which were shown to exhibit the Au surface binding properties for interfacial biological reactivity and Fe oxide core magnetism for magnetic bioseparation. According to the authors' opinion, these magnetic core-shell Nps have shown viability for utilizing both the magnetic core and gold shell properties for interfacial bioassay and magnetic bioseparation.

Selected examples of syntheses of Fe_xO_y/Au Nps are shown in Table 24.1.

24.2.3 Studies of Structure and Properties

Studies of structure and properties of the formed core-shell Nps on the basis of iron oxides and gold confirmed in a series of reports that they are generally *superparamagnetic* at room temperature, that is, the superparamagnetic fraction is retained after coating with gold. They display a localized surface plasmon resonance peak at 605 nm, and their large scattering cross-section allows them to be individually resolved in dark-field optical microscopy as they undergo Brownian motion in aqueous suspension.[83] The influence of synthetic conditions on the product Fe oxide core/Au shell Nps was studied with the aid of high-gradient magnetic filtration.[84]

In some reports, the term *SPION* (superparamagnetic iron oxide Nps) was used.[85] Thus, SPION-type Fe oxide Nps (6 nm diameter) were prepared by controlled chemical coprecipitations and further coated with Au starting from $FeCl_2$, $FeCl_3$, and $HAuCl_4$ as precursors in the system CTAB (cetyltrimethylammonium bromide)/octane/butanol/water.[86] Its Mössbauer spectrum showed a doublet with an isomer shift of $\delta = 0.632$ mm/s, a line width of $\Gamma = 0.52$ mm/s, and a quadrupole splitting of $\Delta E_0 = 0.616$ mm/s indicating that a doublet associated with the SPM state is retained for Au-coated SPION. Another term for Fe-Au-containing Nps, *GoldMag*, was also observed in original publications; thus, the superparamagnetic Fe_3O_4/Au structure with an irregular shape and rough surface (2–3 µm[87] or <100 nm[88] diameter), named by authors as "*GoldMag* particles with assembled structure," was prepared by interaction between micrometer-sized Fe_3O_4 particles and nanosized gold particles (or by reduction of Au(III) with hydroxylamine in the presence of Fe_3O_4), which results in an assembled structure with the Fe_3O_4 core particles around which are attached nano-sized Au particles. It was found that the GoldMag particles can be used as an ideal carrier in immunoassays.

The selected area electron diffraction (SAED) pattern of magnetic Fe oxide core/Au shell Nps was studied in Ref. [89]. It was shown that the SAED pattern for the composite Nps with mean size less than 10 nm is different from either the pattern of pure Fe oxide Nps or that of pure Au Nps. In the authors' opinion, the Au atoms on surfaces of the concerned particles are packed more tightly than in usual Au Nps, and the driving force of this is the coherency strain, which enables the shell material at the heterostructured interface to adapt the lattice parameters of the core. For monodisperse 30 nm iron oxide-core gold-shell Nps, superparamagnetic at r.t., TEM and absorbance spectra showed that both the magnetic cores and the gold shells are monodisperse in size.[90]

Among other reported studies of Fe_xO_y/Au Nps, zero-field-cooled susceptibility (χZFC) and field-cooled susceptibility (χFC) of the Au-coated Fe_3O_4 Nps were measured as a function of temperature.[91] It was shown that they exhibit a superparamagnetic behavior and the susceptibility is characterized by an increase in χFC below a peak temperature with decreasing temperature. Seeding with binary particles showed certain significant advantages in comparison with seeding with pure-metal Nps.[92] Thus, PbSe branched nanorods were grown using Fe/Au_3O_4 Nps as seeds. It was proved that PbSe nucleates on the Au portion of the seed particle and grows anisotropically as one or more straight or branched nanorods. The deposition of well-defined γ-Fe_2O_3 Nps onto conductive substrates like gold under different conditions in order to vary the interactions particle/substrate especially the electrostatic interactions was studied.[93,94]

24.2.4 MAIN APPLICATIONS

Main applications of Fe_xO_y/Au Nps or gold nanocrystals, absorbed on iron oxides, include catalysis and medical uses, among others. *Catalytic* uses, exhibiting significant activity, have been reported for oxidation at ambient or higher temperature of CO and H_2,[95–97] formaldehyde and methanol,[98] and ethylene, among other catalytic processes. These catalysts were prepared mainly by coprecipitation method, for instance Fe/Au_2O_3 (Au:Fe = 1:50; particle size 1.5 ~ 8 nm), whose results of the study for catalytic properties showed that $T_{1/2}$ for CO and H_2 oxidation were 317 and 405 K, respectively.[99] Other synthesis methods for catalysts include impregnation, deposition, and dispersion methods, showing good catalytic properties, for example, for complete ethylene oxidation.[100] It was established that the activity order of these catalysts with preparation techniques was deposition > coprecipitation > impregnation; the main role of the Au Nps was to promote dissociative absorption of oxygen and to enhance the reoxidation of the catalyst.

CO oxidation has been carried out using various forms of Fe_xO_y/Au and related Nps. Thus, catalytic activity on CO oxidation was compared for the samples $Fe/AuO_x/SiO_2/Si(100)$ (fabricated by pulsed laser deposition), Fe/Au_2O_3 prepared by co-precipitation, and Au-Fe/HY (on the zeolite).[101] The mechanism was suggested why oxygen treatment at 470 K for 1 h increases the rate of CO oxidation on the first two samples as compared to subsequent hydrogen treatment at 470 K for 1 h, whereas on the third sample the effect is reversed. In a related publication, the presence of gold was found to promote the development of weakly bonded $(CO)_{ad}$ species over the surface of Fe/Au_2O_3 catalyst during interaction with CO or its mixture with O_2.[102] These species are not formed for Au Nps larger than 11 nm or over Au-free Fe_2O_3. The oxidation of CO on both Fe_2O_3 and Fe/Au_2O_3 occurs *via* similar redox mechanisms, and the presence of Au Nps promotes these processes, which is attributed to their capacity to absorb CO due to their inherent defective structural sites.

Au Nps supported on Fe_2O_3 catalyzed the chemoselective hydrogenation of functionalized nitroarenes with H_2 under mild reaction conditions that avoided the accumulation of hydroxylamines and their potential exothermic decomposition.[103] Hydrogenation of α,β-unsaturated ketone to α,β-unsaturated alcohol on Fe/Au_2O_3 and other gold-supported catalysts was studied.[104] It was shown that the selectivity toward the hydrogenation of the conjugated C=O bond varied with the structural characteristics of the support, ranking in the order FeO(OH) (commercial goethite) > iron oxy-hydroxide (prepared by precipitation) > γ-Fe_2O_3 (maghemite) > α-Fe_2O_3. Other useful information on the catalytic uses of Au/γ-Fe_2O_3 and Au/Fe_3O_4 Nps was reported in Refs. [105,106].

Biological and biomedical/therapeutic applications of Fe_xO_y/Au Nps are wide and include, in particular, analysis in a turbid medium with light, cancer therapy or biomolecular manipulation using light, contrast agents for magnetic resonance imaging, magnetic hyperthermia treatment, bioseparations, biosensors, immunoassay,[107] and drug delivery guide.[108] Nps that consist of a plasmonic layer and an iron oxide moiety could provide a promising platform for the development of multimodal imaging and therapy approaches in future medicine. The gold layer exhibits a surface plasmon resonance that provides optical contrast due to light scattering in the visible region and also presents a convenient surface for conjugating targeting moieties, as well as it makes it possible to track the positions of individual particles, even when they are smaller than the optical diffraction limit, while the iron oxide cores give strong T2 (spin–spin relaxation time) contrast.[109,110]

For *biosensoric* purposes, a composite magnetic particulate composed of Fe_3O_4/Au/glucose oxidase (GOx), exhibiting good superparamagnetism, was synthesized.[111] A novel amperometric biosensor has been developed for the determination of glucose-based Fe_3O_4/Au/GOx particulates absorbed on the refitted glassy carbon electrode through magnet. The sensor exhibited high sensitivity, good selectivity, and stability. Homogeneous films of magnetite and Au Nps were deposited on glass surfaces as initial prototypes for biosensors in solid phase.[112] Au-doped magnetic Fe_3O_4 Nps were synthesized and used for fabrication of an acetylcholinesterase immobilized electrochemical sensor for the detection of organophosphorus pesticides, such as dichlorvos. This biosensor detected as low as 4.0×10^{-13} mol/L dichlorvos.[113]

The SPM properties of Fe_xO_y/Au core-shell Nps logically lead to opportunities to resolve separation problems of biological molecules *via* the functionalization of the Nps with them. Thus, the fundamental theoretical study on gold-coated iron oxide nanostructures (Figure 24.10) on the basis of *magnetic separation* experiments of a series of S-containing *amino acids* (cysteine, methionine, and taurine) was carried out in Ref. [114]. It was established that the interaction of cysteine is stronger than methionine; the respective interaction energies (0.738 and 0.712 eV) are intermediate between van der Waals and covalent bonding. Taurine cannot bind to Au/iron oxide Nps and thus can be separated from two other amino acids used. In another series of publications,[115–119] magnetic

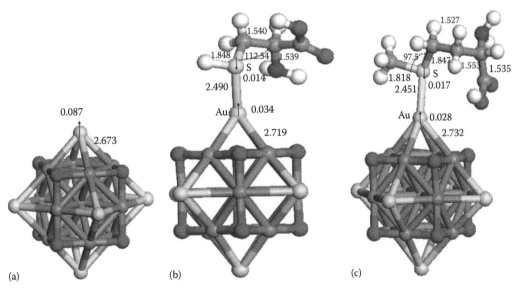

FIGURE 24.10 Geometry of $Au_6Fe_{13}O_8$ (a), cysteine-$Au_6Fe_{13}O_8$ (b), and methionine-$Au_6Fe_{13}O_8$ (c). The numbers close to arrows correspond to the spins on Au and S atoms while the other numbers correspond to bond lengths and bond angles. (Reproduced with permission from Sun, Q. et al., *J. Phys. Chem. C*, 111(11), 4159–4163. Copyright 2007 American Chemical Society.)

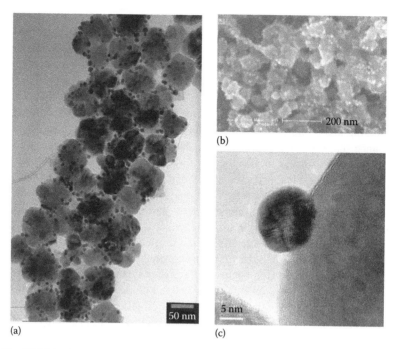

FIGURE 24.11   TEM (a, c) and SEM (b) images of Au-Fe$_3$O$_4$ Nps. (Reproduced with permission from Bao, J. et al., *ACS Nano*, 1(4), 293–298. Copyright 2007 American Chemical Society.)

separations of mixture of the composite Nps with 17 kinds of amino acids, including methionine and cysteine were carried out by similar way. In this case, Au Nps (10 nm) were prepared by sono-chemical reducing Au(III) ions and immobilized on the surface of magnetic γ-Fe$_2$O$_3$ Nps (26 nm). The composite Nps exhibited a high affinity with glutathione (a tripeptide with mercapto group). Additionally, bifunctional Fe/Au$_3$O$_4$ Nps (Figure 24.11), which can be easily modified with other functional molecules to realize various nanobiotechnological separations and detections, were prepared from FeCl$_3$·6H$_2$O and HAuCl$_4$ as precursors. These Nps were modified with nitrilotriacetic acid molecules through Au-S interaction and used to separate *proteins* simply with the assistance of a magnet.[120] The possibility of Fe/Au$_3$O$_4$ nanocomposite particles for the application to the separation and purification of *glutathione biomolecules* by magnetic field was confirmed.[121]

Gold and silica doubly coated γ-Fe$_2$O$_3$ Nps (Figure 24.12) having an efficient photothermal effect and high transverse relaxivities for *MRI applications* (see also Ref. [122]) were extensively characterized structurally and magnetically. It was shown that the Au Nps are dispersed in the outer layer of the silica spheres; average diameters of the Au/SiO$_2$ doubly coated Fe$_2$O$_3$ Nps are about 82 nm; coating of the iron oxide Nps by gold/silica decreases the blocking temperature from 160 to 80 K.[123] Gold-coated iron oxide Np Hepatitis B virus (HBV) DNA probes were prepared by the citrate reduction of HAuCl$_4$ in the presence of iron oxide Nps, which were added as seeds, and their application for HBV *DNA measurement* was studied, confirming that detecting DNA with iron oxide Nps and magnetic separator was feasible and might be an alternative effective method.[124] Magnetically and optically active dumbbell-shaped Au-Fe$_3$O$_4$ Nps are made biocompatible and suitable for *attachment to A431 cells*.[125] Other medical applications are described in Ref. [126].

Among other applications of various Fe$_x$O$_y$-Au systems, the creation of Fe$_2$O$_3$ thin films containing dispersed Au Nps, prepared from Fe(NO$_3$)$_3$·9H$_2$O-HAuCl$_4$·4H$_2$O-MeCOCH$_2$COMe-CH$_3$OC$_2$H$_4$OH system in the presence of polyvinylpyrrolidone, was reported.[127] Also, γ-Fe$_2$O$_3$ maghemite-Au core-shell Nps incorporated into a polyaniline film associated with an electrode led to reversible, magnetoswitchable, controlled charge transport across the polymer and can be used in magnetoswitchable bioelectrocatalysis.[128]

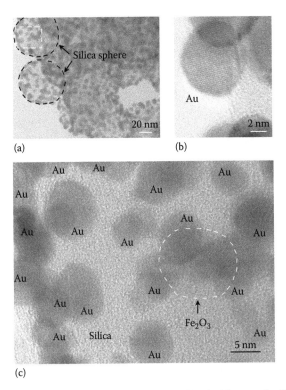

FIGURE 24.12 (a) TEM image in lower magnification for Au/SiO$_2$ doubly coated γ-Fe$_2$O$_3$ Nps; large spherical particles, as marked by the black dashed circles for two of them, are the silica spheres, and the small black dots represent Au particles; (b) enlarged image of several Au particles that clearly show the lattice fringes; (c) enlarged image from the framed area (enclosed by the dashed loop) in panel (a). (Reproduced with permission from Park, K. et al., *J. Phys. Chem. C*, 111(50), 18512–18519. Copyright 2007 American Chemical Society.)

24.3 TRIMETALLIC NANOPARTICLES ON THE FE-AU BASIS

In addition to core-shell (Fe, Fe$_x$O$_y$)/Au Nps and alloys of these metals, trimetallic Nps, alloys, or intermetallic compounds containing Fe (or its oxides) and Au are also well known and have many similar applications like those described here. Attention has been paid to cobalt- and nickel-containing Nps with different compositions. Thus, the strategy for functionalization of CoFe$_2$O$_4$ SPM NPs with a mixture of amino and thiol groups that facilitate the electrostatic attraction and further chemisorption of gold Nps, respectively, is shown in Figure 24.13.[129] Magnetic Nps are functionalized with a mixture of amino and thiol groups that facilitate the electrostatic attraction and further chemisorption of gold Nps, respectively. Using these Nps as seeds, a complete coating shell is achieved by gold salt-iterative reduction leading to monodisperse water-soluble gold-covered magnetic Nps, with an average diameter ranging from 21 to 29 nm. In addition, a combined physical technique, involving an online sputtering/evaporation process with an integrated nanocluster deposition process was used to prepare (Fe$_{60}$Co$_{49}$)core/Au shell superparamagnetic Nps with controllable particle size of 10–20 nm and Au shell thickness of 1–3 nm.[130] Au shell in these Nps is not only functional for potential biocompatibility but also the key to prevent the oxidation of FeCo core. Au$_{80}$Co$_{10}$Fe$_{10}$ granular alloy was characterized in Ref. [131]. Au-coated Nps of Fe$_{20}$Ni$_{80}$ (5 nm) were prepared by a microemulsion process[132] encapsulated in B$_2$O$_3$, and the formed Fe$_{20}$Ni$_{80}$/B$_2$O$_3$ Nps were further encapsulated in 8–20 Au nanospheres. Magnetic data confirmed the existence of a superparamagnetic phase with a blocking temperature of 33 K.

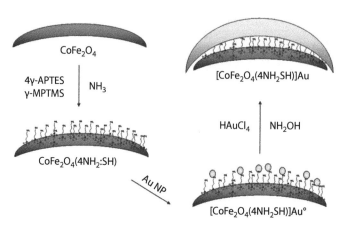

FIGURE 24.13 Reaction strategy showing the successive steps for gold covering process onto 4:1 amino:mercapto functionalized cobalt ferrite Nps. (Reproduced from *J. Colloid Interface Sci.*, 321, Pita, M., Synthesis of cobalt ferrite core/metallic shell nanoparticles for the development of a specific PNA/DNA biosensor, 484–492, Copyright 2008, with permission from Elsevier Science.)

A series of publications have been dedicated to the coating of widely reported FePt Nps, which have useful applications; ratios of Fe, Pt, and Au were different depending on precursors and synthesis conditions. Thus, inorganic heterodimer FePt/Au Nps possessing multifunctional capabilities (catalytic effects of FePt for heteroepitaxial Au growth, high water-solubility, and biocompatibility attained through versatile ligand chemistry of Au-S linkages, Au for chip-based biosensing, and magnetic resonance contrast effects of superparamagnetic FePt) for biological detection were described in Ref. [133]. FePt Nps (6 nm) were synthesized through a transmetalation reaction between Fe^0 and Pt^{2+} precursors and then allowed to react with $AuCl(PPh_3)$ in 1,2-DCB containing 1-hexadecylamine bubbling of 4% H_2/Ar mixture through the solution. Surface modification of hydrophobic FePt/Au Nps by 11-mercaptoundecanoic acid was hydrophilic through ligand exchange and with good biocompatibility for future multifunctional applications.[134] [FePt]$_{95}$Au$_5$ Nps with an average size of about 4 nm were prepared, then spin coated onto silicon substrates and thermally annealed at 250°C–500°C for 30 min.[135] Dynamic coercivity measurements of samples showed that the FePtAu Nps have both higher intrinsic coercivity and higher switching volume at the same annealing temperature than FePt. The effect of ion-beam irradiation (by 300 keV Al ions) on reducing the ordering temperature of FePt and FePt(Au 14%) Nps was studied.[136] It was established that postirradiation annealing at 375°C (much higher than for FePt, 220°C) was required for FePtAu, because Au atoms were trapped at Fe/Pt lattice sites at lower temperatures. FePtCu, FePtAg, and FePtAu Nps were synthesized by simultaneous polyol reduction of platinum complexes and the appropriate metal additive reagent and the decomposition of $Fe(CO)_5$ in the presence of organic surfactants.[137] The effect of the additive metal on the transformation to the L1$_0$ phase was studied showing that during heat treatment Au segregates; the anisotropy of (FePt)$_{76}$Au$_{24}$ annealed at 500°C for 30 min is ~3 × 10^7 erg/cc. The effect of additive Au and other metals on the chemical ordering, magnetic properties, and grain growth of chemically synthesized FePt Nps was elucidated.[138] The Au-coated nanocomposites with magnetic core based on the alloys FePt and Fe$_2$Pt were prepared by reverse micelles technique.[139,140] The Fe$_2$Pt/Au samples showed higher coercivity than FePt/Au (2885 and 500 Oe at 2 K, respectively). Additionally, Fe$_{46}$Pt$_{46}$Au$_8$ Nps were characterized.[141]

Among other trimetallic iron- and gold-containing Nps, we note manganese oxide and iron oxide as carrier and gold Np supported thereon, which were used for CO oxidation into carbon dioxide in a reactive gas containing O_2, CO, H_2, and He.[142] Fe$_3$O$_4$/Au and Fe$_3$O$_4$/Au/Ag Nps were prepared by deposition of Au and Ag on the Fe$_3$O$_4$ Nps surface in aqueous solution at r.t. by reducing HAuCl$_4$ in a chloroform solution of oleylamine (control of Au shell thickness is possible) and adding AgNO$_3$ to the reaction mixture.[143] The resulting Nps are suitable for Np-based diagnostics and therapeutic applications.

Au-Pd Nps on γ-Fe$_2$O$_3$ were prepared in a sonochemical one-step process, and the product is a promising catalyst of high activity.[144] Formally trimetallic Fe/Au/Au onion-like Nps were synthesized using reverse micelles.[145,146] The Nps were formed with 6 nm diameter gold core, 1 nm iron interlayer, and 2 nm gold shell; the shell structure coated on the core is inhomogeneous, but the interlayer iron shell is air-stable, as well as the Fe core in Fe/Au core-shell Nps. High-density Au Nps supported on a [Ru(bpy)$_3$]$^{2+}$-doped silica/Fe$_3$O$_4$ nanocomposite (GNRSF) were prepared by a combined sonication-self-assembled techniques.[147] This hybrid nanomaterial film exhibits a very stable electrochemiluminescence behavior allowing serving as a sensor for tripropylamine. Another recently reported nanocomposite having a complex structure on the basis of hydroxyapatite Ca$_{10}$(PO$_4$)$_6$(OH)$_2$ (HAP) was (HAP)/γ-Fe$_2$O$_3$/Au.[148] This serves as a simple, sensitive, and reusable piezoelectric immunosensor.

Selected examples of the synthesis of trimetallic Nps on the basis of iron and gold are shown in Table 24.1.

REFERENCES

1. Huber, D. L. Synthesis, applications, and applications of iron nanoparticles. *Small*, 2005, *1* (5), 482–501.
2. (a) Zhou, W. L.; Carpenter, E. E.; Lin, J.; Kumbhar, A.; Sims, J.; O'Connor, C. J. Nanostructures of gold coated iron core-shell nanoparticles and the nanobands assembled under magnetic field. *European Physical Journal D: Atomic, Molecular, Optical and Plasma Physics*, 2001, *16* (1–3), 289–292; (b) Oldenburg, S. J.; Averitt, R. D.; Westcott, S. L.; Halas, N. J. Nanoengineering of optical resonances. *Chemical Physics Letters*, 1998, *288*, 243–247.
3. Kotov, N. S. (Ed.). *Nanoparticle Assemblies and Superstructures*. CRC Press, Boca Raton, FL, 2005, 407 pp.
4. de Villiers, M. M.; Aramwit, P.; Kwon, G. S. (Eds.). *Nanotechnology in Drug Delivery*. Springer, Berlin, Germany, 2008, p. 123.
5. Astruc, D. (Ed.). *Nanoparticles and Catalysis*. Wiley-VCH, Weinheim, Germany, 2007, 168 pp.
6. Zhang, X.; Ju, H.; Wang, J. (Eds.). *Electrochemical Sensors, Biosensors and Their Biomedical Applications*. Academic Press, San Diego, CA, 2007, 446 pp.
7. Giersig, M.; Khomutov, G. B. (Eds.). *Nanomaterials for Application in Medicine and Biology*. Springer, Berlin, Germany, 2008, 99 pp.
8. Yao, N.; Wang, Z. L. (Eds.). *Handbook of Microscopy for Nanotechnology*. Springer, Berlin, Germany, 2005, 170 pp.
9. Lalena, J. N.; Cleary, D. A. (Eds.). *Principles of Inorganic Materials Design*. Wiley-Interscience, Hoboken, NJ, 2005, 404 pp.
10. Capek, I. (Ed.). *Nanocomposite Structures and Dispersions. Volume 23 (Studies in Interface Science)*, Elsevier Science, Amsterdam, the Netherlands, 2006, 155 pp.
11. Laguna, A. *Modern Supramolecular Gold Chemistry, Gold-Metal Interactions and Applications*. Wiley-VCH, Weinheim, Germany, 2008, 525 pp.
12. Guczi, L.; Beck, A.; Horvath, A.; Horvath, D. From molecular clusters to metal nanoparticles. *Topics in Catalysis*, 2002, *19* (2), 157–163.
13. Wang, L.; Park, H.-Y.; Lim, S. I.-I.; Schadt, M. J.; Mott, D.; Luo, J.; Wang, X.; Zhong, C. J. Core@shell nanomaterials, gold-coated magnetic oxide nanoparticles. *Journal of Materials Chemistry*, 2008, *18* (23), 2629–2635.
14. Grancharov, S.; O'Brien, S.; Held, G.; Bruce, M. C. Method of preparation of biomagnetic nanoparticles coated with a noble metal layer. U.S. Patent 20060140868, 2006; WO/2004/060580, 2004.
15. Hyeon, T.; Kim, J. Y.; Cho, M.-H.; Kim S. K.; Lee, J. Use of core-shell gold nanoparticle which contains magnetic nanoparticles for MRI T2 contrast agent, cancer diagnostic and therapy. WO 2008048074, 2008.
16. Naitabdi, A.; Cuenya, B. R. Formation, thermal stability, and surface composition of size-selected Au-Fe nanoparticles. *Applied Physics Letters*, 2007, *91* (11), 113110/1–113110/3.
17. Dahal, N.; Chikan, V.; Jasinski, J.; Leppert, V. J. Synthesis of water-soluble iron-gold alloy nanoparticles. *Chemistry of Materials*, 2008, *20* (20), 6389–6395.
18. Liu, H. L.; Wu, J. H.; Min, J. H.; Kim, Y. K. Synthesis of monosized magnetic-optical AuFe alloy nanoparticles. *Journal of Applied Physics*, 2008, *103* (7, Pt. 2), 07D529/1–07D529/3.
19. Sato, K.; Bian, B.; Hirotsu, Y. L10 type ordered phase formation in Fe-Au nanoparticles. *Japanese Journal of Applied Physics Part 2*, 2002, *41* (1A/B), L1–L3.

20. Cho, S.-J.; Kauzlarich, S. M.; Olamit, J.; Liu, K.; Grandjean, F.; Rebbouh, L.; Long, G. J. Characterization and magnetic properties of core-shell structured Fe/Au nanoparticles. *Journal of Applied Physics*, 2004, *95* (11, Pt. 2), 6804–6806.

21. Lin, J.; Zhou, W.; Kumbhar, A.; Wiemann, J.; Fang, J.; Carpenter, E. E.; O'Connor, C. J. Gold-coated iron (Fe/Au) nanoparticles, synthesis, characterization, and magnetic field-induced self-assembly. *Journal of Solid State Chemistry*, 2001, *159* (1), 26–31.

22. Andres, R. P.; Ng, A. T. Fe/Au nanoparticles and methods. WO 2003073444, 2003.

23. Chen, M.; Yamamuro, S.; Farrell, D.; Majetich, S. A. Gold-coated iron nanoparticles for biomedical applications. *Journal of Applied Physics*, 2003, *93* (10, Pt. 2), 7551–7553.

24. Zelenakova, A.; Kovac, J.; Kavecansky, V.; Zelenak, V. Magnetic study of the Fe coated by Au nanoparticles. *Acta Physica Polonica A*, 2008, *113* (1), 533–536.

25. Pana, O.; Teodorescu, C. M.; Chauvet, O.; Payen, C.; Macovei, D.; Turcu, R.; Soran, M. L.; Aldea, N.; Barbu, L. Structure, morphology and magnetic properties of Fe-Au core-shell nanoparticles. *Surface Science*, 2007, *601* (18), 4352–4357.

26. Ban, Z.; Barnakov, Y. A.; Li, F.; Golub, V. O.; O'Connor, C. J. The synthesis of core-shell iron@gold nanoparticles and their characterization. *Journal of Materials Chemistry*, 2005, *15* (43), 4660–4662.

27. Cho, S.; Kauzlarich, S. M.; Susan, M.; Liu, K.; Long, G. J.; Grandjean, F. Fabrication and characterization of gold-coated iron nanoparticles. *Abstracts of Papers, 225th ACS National Meeting*, New Orleáns, LA, March 23–27, 2003, INOR-234.

28. Cho, S.-J.; Shahnin, A. M.; Long, G. J.; Davies, J. E.; Liu, K.; Grandjean, F.; Kauzlarich, S. M. Magnetic and moessbauer study of core/shell Fe/Au nanoparticles. *Chemistry of Materials*, 2006, *18* (4), 960–967.

29. Cho, S.-J.; Shahnin, A. M.; Long, G. J.; Davies, J. E.; Liu, K.; Grandjean, F.; Kauzlarich, S. M. Magnetic and moessbauer study of core/shell Fe/Au nanoparticles. *Condensed Matter*, 2006, *18*, 960–967.

30. Ravel, B.; Carpenter, E. E.; Harris, V. G. Oxidation of iron in iron/gold core/shell nanoparticles. *Journal of Applied Physics*, 2002, *91* (10, Pt. 3), 8195–8197.

31. Fleming, D. A.; Napolitano, M.; Williams, M. E. Chemically functional alkanethiol derivatized magnetic nanoparticles. *MRS Symposium Proceedings (Magnetoelectronics and Magnetic Materials—Novel Phenomena and Advanced Characterization)*, Vol. 746, 2003, pp. 207–212.

32. Lu, D.-L.; Domen, K.; Tanaka, K.-I. Electrodeposited Au-Fe, Au-Ni, and Au-Co alloy nanoparticles from aqueous electrolytes. *Langmuir*, 2002, *18* (8), 3226–3232.

33. Mirdamadi-Esfahani, M.; Mostafavi, M.; Keita, B.; Nadjo, L.; Kooyman, P.; Etcheberry, A.; Imperor, M.; Remita, H. Au-Fe system: Application in electro-catalysis. *Gold Bulletin*, 2008, *41* (2), 98–104.

34. Chang, W.-S.; Park, J.-W.; Rawat, V.; Sands, T.; Lee, G. U. Templated synthesis of gold–iron alloy nanoparticles using pulsed laser deposition. *Nanotechnology*, 2006, *17* (20), 5131–5135.

35. Zhang, J.; Post, M.; Veres, T.; Jakubek, Z. J.; Guan, J.; Wang, D.; Normandin, F.; Deslandes, Y.; Simard, B. Laser-assisted synthesis of superparamagnetic Fe@Au core–shell nanoparticles. *Journal of Physical Chemistry B*, 2006, *110* (14), 7122–7128.

36. Sun, Q.; Kandalam, A. K.; Wang, Q.; Jena, P.; Kawazoe, Y.; Marquez, M. Effect of Au coating on the magnetic and structural properties of Fe nanoclusters for use in biomedical applications: A density-functional theory study. *Physical Review B: Condensed Matter*, 2006, *73* (13), 134409/1–134409/6.

37. Liu, K.; Cho, S.-J.; Kauzlarich, S. M.; Idrobo, J. C.; Joseph, D. E.; Justin, O.; Browning, N. D.; Ahmed, S. M.; John, L. G.; Fernande, G. Fe-core/Au-shell nanoparticles, growth mechanisms, oxidation and aging effects. *MRS Symposium Proceedings (Degradation Processes in Nanostructured Materials)*, Vol. 887, 2006, pp. 121–132.

38. Cho, S.-J.; Hidrobo, J. C.; Olamit, J.; Liu, K.; Browning, N. D.; Kauzlarich, S. M. Growth mechanisms and oxidation resistance of gold-coated iron nanoparticles. *Chemistry of Materials*, 2005, *17* (12), 3181–3186.

39. Cho, S.-J.; Idrobo, J. C.; Olamit, J.; Liu, K.; Browning, N. D.; Kauzlarich, S. M. Growth mechanisms and oxidation-resistance of gold-coated iron nanoparticles. *Condensed Matter*, 2005, *17* (12), 3181–3186.

40. Khan, N. A.; Matranga, C. Nucleation and growth of Fe and FeO nanoparticles and films on Au(111). *Surface Science*, 2008, *602* (4), 932–942.

41. Chikan, V.; Dahal, N. "Gift wrapped" core/shell nanoparticles for the formation of various silicides and radial nanowires. *Abstracts of Papers, 236th ACS National Meeting*, Philadelphia, PA, August 17–21, 2008, INOR-132.

42. Luo, Z.; Yan, F. SERS of gold/C60 (/C70) nano-clusters deposited on iron surface. *Vibrational Spectroscopy*, 2005, *39* (2), 151–156.

43. Ma, H.; Chen, S.; Liu, G.; Xu, J.; Zhou, M. Inhibition effect of self-assembled films formed by gold nanoparticles on iron surface. *Applied Surface Science*, 2006, *252* (12), 4327–4334.

44. Akolekar, D. B.; Bhargava, S. K.; Foran, G.; Takahashi, M. Studies on gold nanoparticles supported on iron, cobalt, manganeso, and cerium oxide catalytic materials. *Journal of Molecular Catalysis A: Chemical*, 2005, *238* (1–2), 78–87.

45. Smolentseva, E.; Pestryakov, A.; Bogdanchikova, N.; Simakov, A.; Avalos, M.; Farias, M.; Diaz, J.; Gurin, V.; Tompos, A. Influence of Fe introduction method on gold state in NaY zeolite. *International Journal of Modern Physics B: Condensed Matter Physics, Statistical Physics, Applied Physics*, 2005, *19* (15–17, Pt. 1), 2496–2501.

46. Horvath, D.; Polisset-Thfoin, M.; Fraissard, J.; Guczi, L. Novel preparation method and characterization of Au-Fe/HY zeolite containing highly stable gold nanoparticles inside zeolite supercages. *Solid State Ionics*, 2001, 141–142, 153–156.

47. Tada, H.; Soejima, T.; Ito, S.; Kobayashi, H. Photoinduced desorption of sulfur from gold nanoparticles loaded on metal surfaces. *Journal of the American Chemical Society*, 2004, *126* (49), 15952–15953.

48. Wang, Z.; Wang, J.; Sun, L. Method for synthesizing antibody-magnetic nanoparticles with gold shell and iron core for recognition and separation of tumor cells. CN 101303342, 2008.

49. Cho, S.-J.; Jarrett, B. R.; Louie, A. Y.; Kauzlarich, S. M. Gold-coated iron nanoparticles, a novel magnetic resonance agent for T1 and T2 weighted imaging. *Nanotechnology*, 2006, *17* (3), 640–644.

50. Jadhav, S. V. Synthesis and characterization of iron core—Gold shell nanoparticles and their application to pathogen detection. *Abstracts of Papers, 39th Central Regional Meeting of the ACS*, Corvington, KY, May 20–23, 2007.

51. Liu, H.; Li, S.; He, N.; Deng, Y. Preparation of streptavidin coated gold magnetic nanoparticles using cysteamine. *Abstracts of Papers, 235th ACS National Meeting*, New Orleans, LA, April 6–10, 2008, ANYL-137.

52. Brown, K. A.; Wijaya, A.; Alper, J. D.; Hamad-Schifferli, K. Synthesis of water-soluble, magnetic Fe/Au nanoparticles. *MRS Symposium Proceedings (Nanoparticles and Nanostructures in Sensors and Catalysis)*, Boston, MA, 2006, Volume Date 2005, Vol. 900 E, Paper # 0900-O07-01.

53. He, N. Y.; Guo, Y. F.; Deng, Y.; Wang, Z. F.; Li, S.; Liu, H. N. Carbon encapsulated magnetic nanoparticles produced by hydrothermal reaction. *Chinese Chemical Letters*, 2007, *18* (4), 487–490.

54. Wang, Z.; Xiao, P.; He, N. Synthesis and characteristics of carbon encapsulated magnetic nanoparticles produced by a hydrothermal reaction. *Carbon*, 2006, *44* (15), 3277–3284.

55. Femoni, C.; Iapalucci, M. C.; Longoni, G.; Tiozzo, C.; Zacchini, S. An organometallic approach to gold nanoparticles, synthesis and x-ray structure of CO-protected $Au_{21}Fe_{10}$, $Au_{22}Fe_{12}$, $Au_{28}Fe_{14}$, and $Au_{34}Fe_{14}$ clusters. *Angewandte Chemie International Edition* (English), 2008, *47* (35), 6666–6669.

56. Wang, T.; Zhang, D.; Xu, W.; Li, S.; Zhu, D. New approach to the assembly of gold nanoparticles, formation of stable gold nanoparticle ensemble with chainlike structures by chemical oxidation in solution. *Langmuir*, 2002, *18* (22), 8655–8659.

57. Gole, A.; Stone, J. W.; Gemmill, W. R.; zur Loye H.-C.; Murphy, C. J. Iron oxide coated gold nanorods, synthesis, characterization, and magnetic manipulation. *Langmuir*, 2008, *24* (12), 6232–6237.

58. Wang, L.; Bai, J.; Li, Y.; Huang, Y. Multifunctional nanoparticles displaying magnetization and near-IR absorption. *Angewandte Chemie International Edition*, 2008, *47* (13), 2439–2442.

59. Lo, C. K.; Xiao, D.; Choi, M. M. F. Homocysteine-protected gold-coated magnetic nanoparticles, synthesis and characterization. *Journal of Materials Chemistry*, 2007, *17* (23), 2418–2427.

60. Fuss, M.; Luna, M.; Alcantara, D.; de la Fuente, J. M.; Penades, S.; Briones, F. Supramolecular self-assembled arrangements of maltose glyconanoparticles. *Langmuir*, 2008, *24* (9), 5124–5128.

61. Lim, Y. T.; Cho, M. Y.; Kim, J. K.; Hwangbo, S.; Chung, B. H. Plasmonic magnetic nanostructure for bimodal imaging and photonic-based therapy of cancer cells. *ChemBioChem*, 2007, *8* (18), 2204–2209.

62. Ji, X.; Shao, R.; Elliot, A. M.; Stafford, R. J.; Esparza-Coss, E.; Liang, G.; Luo, Z. P.; Park, K.; Markert, J. T.; Li, C. Bifunctional gold nanoshells with a superparamagnetic iron-oxide-silica core suitable for both MR imaging and photothermal therapy. *Journal of Physical Chemistry C*, 2007, *111* (17), 6245–6251.

63. Guczi, L.; Paszti, Z.; Frey, K.; Beck, A.; Peto, G.; Daroczy, C. S. Modeling gold/iron oxide interface system. *Topics in Catalysis*, 2006, *39* (3–4), 137–143.

64. Stoeva, S. I.; Huo, F.; Lee, J.-S.; Mirkin, C. A. Three-layer composite magnetic nanoparticle probes for DNA. *Journal of the American Chemical Society*, 2005, *127* (44), 15362–15363.

65. Spasova, M.; Salgueirino-Maceira, V.; Schlachter, A.; Hilgendorff, M.; Giersig, M.; Lez-Marzán, L. M.; Farle, M. Magnetic and optical tunable microspheres with a magnetite/gold nanoparticle shell. *Journal of Materials Chemistry*, 2005, *15* (21), 2095–2098.

66. Tsoncheva, T.; Ivanova, L.; Lotz, A. R.; Smått, J.-H.; Dimitrov, M.; Paneva, D.; Mitov, I.; Linden, M.; Minchev, C.; Fröba, M. Gold and iron nanoparticles in mesoporous silicas: Preparation and characterization. *Catalysis Communications*, 2007, *8* (11), 1573–1577.

67. Chen, M.-Y.; Chang, W.-H.; Lin, C.-I.; Wang, S.-J. J.; Lin, Y.-J. Method for forming superparamagnetic nanoparticles. U.S. Patent 2006141149, 2006.

68. Yao, K.; Lu, Q.; Liu, Z.; Ning, Q.; Xi, D.; Luo, X. Process for preparation for superparamagnetic iron oxide composite nanoparticles. CN 1736881, 2006.

69. Lu, Q. H.; Yao, K. L.; Xi, D.; Liu, Z. L.; Luo, X. P.; Ning, Q. Synthesis and characterization of composite nanoparticles comprised of gold shell and magnetic core/cores. *Journal of Magnetism and Magnetic Materials*, 2006, *301* (1), 44–49.

70. Mandal, M.; Kundu, S.; Ghosh, S. K.; Panigrahi, S.; San, T. K.; Yusuf, S. M.; Pal, T. Magnetite nanoparticles with tunable gold and silver shell. *Colloid Interface Science*, 2005, *286* (1), 187–194.

71. Wang, L.; Luo, J.; Maye, M. M.; Fan, Q.; Rendeng, Q.; Engelhard, M. H.; Wang, C.; Lin, Y.; Zhong, C.-J. Iron oxide-gold core-shell nanoparticles and thin film assembly. *Journal of Materials Chemistry*, 2005, *15* (18), 1821–1832.

72. Wang, L.; Suzuki, M.; Suzuki, I. S.; Luo, J.; Zhong, C.-J. Nanoscale core-shell magnets, síntesis, assembly, and applications. *Abstracts of Papers, 35th Northeast Regional Meeting of the ACS*, Binghamton, NY, October 5–7, 2006.

73. Wang, L.; Luo, J.; Fan, Q.; Suzuki, M.; Suzuki, I. S.; Engelhard, M. H.; Lin, Y. Monodispersed core–shell Fe_3O_4@Au nanoparticles. *Journal of Physical Chemistry B*, 2005, *109* (46), 21593–21601.

74. Kinoshita, T.; Seino. S.; Okitsu, K.; Nakayama, T.; Nakagawa, T.; Yamamoto, T. A. Magnetic evaluation of nanostructure of gold-iron composite particles synthesized by a reverse micelle method. *Journal of Alloys and Compounds*, 2003, *359* (1–2), 46–50.

75. Seino, S.; Kinoshita, T.; Nakagawa, T.; Kojima, T.; Taniguchi, R.; Okuda, S.; Yamamoto, T. A. Radiation induced synthesis of gold/iron-oxide composite nanoparticles using high-energy electron beam. *Journal of Nanoparticle Research*, 2008, *10* (6), 1071–1076.

76. Kinoshita, T.; Seino, S.; Mizukoshi, Y.; Nakagawa, T.; Yamamoto, T. A. Functionalization of magnetic gold/iron-oxide composite nanoparticles with oligonucleotides and magnetic separation of specific target. *Journal of Magnetism and Magnetic Materials*, 2007, *311* (1), 255–258.

77. Kinoshita, T.; Seino, S.; Otome, Y.; Sekino, S.; Seino, Y.; Yamamoto, T. DNA separation using gold/magnetic iron-oxide composite nanoparticles. *MRS Symposium Proceedings (Magnetic Nanoparticles and Nanowires)*, Vol. 877E, 2005, S.30.1.

78. Wu, W.; He, Q.; Chen, H.; Tang, J.; Nie, L. Sonochemical synthesis, structure and magnetic properties of air-stable Fe_3O_4/Au nanoparticles. *Nanotechnology*, 2007, *18* (14), 145609/1–145609/8.

79. Kawaguchi, K.; Jaworski, J.; Ishikawa, Y.; Sasaki, T.; Koshizaki, N. Preparation of gold/iron-oxide composite nanoparticles by a unique laser process in water. *Journal of Magnetism and Magnetic Materials*, 2007, *310* (2, Pt.3), 2369–2371.

80. Kawaguchi, K.; Jaworski, J.; Ishikawa, Y.; Sasaki, T.; Koshizaki, N. Preparation of gold/iron-oxide composite nanoparticles by a laser-soldering method. *IEEE Transactions on Magnetics*, 2006, *42* (10), 3620–3622.

81. Fang, J.; He, J.; Shin, E. Y.; Grimm, D.; O'Connor, C. J.; Jun, M.-J. Colloidal preparation of γ-Fe_2O_3/Au core-shell nanoparticles. *MRS Symposium Proceedings (Materials Inspired by Biology)*, Vol. 774, 2003, pp. 149–154.

82. Park, H.-Y.; Schadt, M. J.; Wang, L.; Lim, I. I.; Njoki, P. N.; Kim, S. H.; Jang, M. Y.; Luo, J.; Zhong, C. J. Fabrication of magnetic core@shell Fe oxide@Au nanoparticles for interfacial bioactivity and bio-separation. *Langmuir*, 2007, *23* (17), 9050–9056.

83. Lim, J. K.; Eggeman, A.; Lanni, F.; Tilton, R. D.; Majetich, S. A. Synthesis and single-particle optical detection of low-polydispersity plasmonic-superparamagnetic nanoparticles. *Advanced Materials*, 2008, *20* (9), 1721–1726.

84. Lu, Q. H.; Yao, K. L.; Xi, D.; Lui, Z. L.; Luo, X. P.; Ning, Q. A magnetic separation study on synthesis of magnetic Fe oxide core/Au shell nanoparticles. *Nanoscience*, 2006, *11* (4), 241–248.

85. Mikhailova, M.; Jo, Y. S.; Kim, D. K.; Bobrysheva, N.; Andersson, Y.; Eriksson, T.; Osmolowsky, M.; Semenov, V.; Muhammed, M. The effect of biocompatible coating layers on magnetic properties of superparamagnetic iron oxide nanoparticles. *Hyperfine Interactions*, 2004, *156/157* (1–4), 257–263.

86. Mikhaylova, M.; Kim, D. K.; Bobrysheva, N.; Osmolowsky, M.; Semenov, V.; Tsakalakos, T.; Muhammed, M. Superparamagnetism of magnetite nanoparticles, dependence on surface modification. *Langmuir*, 2004, *20* (6), 2472–2477.

87. Cui, Y.; Zhang, L.; Su, L.; Zhang, C.; Li, Q.; Cui, T.; Jin, B.; Chen, C. Synthesis of GoldMag particles with assembled structure and their applications in immunoassay. *Science in China, Series B: Chemistry*, 2006, *49* (6), 534–540.
88. Cui, Y.; Wang, Y.; Hui, W.; Zhang, Z.; Xin, X.; Chen, C. The synthesis of GoldMag nanoparticles and their application for antibody immobilization. *Biomedical Microdevices*, 2005, *7* (2), 153–156.
89. Lu, Q.; Yao, K.; Dong, L.; Luo, X.; Ning, Q. A comparative study on the selected area electron diffraction pattern of Fe oxide/Au core-shell structured nanoparticles. *Journal of Materials Science and Technology*, 2007, *23* (2), 189–192.
90. Lim, J. K.; Eggeman, A.; Majetich, S.; Tilton, R. D. Plasmonicmagnetic nanoparticles with low polydispersity. *Abstracts of papers, 234th ACS National Meeting*, Boston, MA, August 19–23, 2007.
91. Suzuki, M.; Fullem, S. I.; Suzuki, I. S.; Wang, L.; Zhong, C.-J. Observation of superspin-glass behavior in Fe_3O_4 nanoparticles. *Condensed Matter*, 2006; 1–10, in press, cond-mat/0608297.
92. Shi, W.; Sahoo, Y.; Zeng, H.; Ding, Y.; Swihart, M. T.; Prasad, P. N. Anisotropic growth of PbSe nanocrystals on Au-Fe_3O_4 hybrid nanoparticles. *Advanced Materials*, 2006, *18* (14), 1889–1894.
93. Lucas, I. T.; Dubois, E.; Chevalet, J.; Durand-Vidal, S. Reactivity of nanocolloidal particles γ-Fe_2O_3 at the charged interfaces. Part 1. The approach of particles to an electrode. *Physical Chemistry Chemical Physics*, 2008, *10* (22), 3263–3273.
94. Lucas, I. T.; Dubois, E.; Chevalet, J.; Durand-Vidal, S.; Joiret, S. Reactivity of nanocolloidal particles gamma-Fe_2O_3 at charged interfaces. Part 2. Electrochemical conversion. Role of the electrode material. *Physical Chemistry Chemical Physics*, 2008, *10* (22), 3274–3286.
95. Nguyen, C. T.; Tran, T. M.; Nguyet, N.; Nguyen, Q. H.; Lai, X. N.; Nguygen, D. T.; Do, T. C.; Tran, Q. C.; Nguyen, Q. T. Study on the preparation and catalytic activity of gold nanoparticles supported on Fe_2O_3. *Tap Chi Hoa Hoc*, 2007, *45* (6), 671–675.
96. Herzing, A. A.; Kiely, C. J.; Carley, A. F.; Landon, P.; Hutchings, G. J. Identification of active gold nanoclusters on iron oxide supports for CO oxidation. *Science*, 2008, *321* (5894), 1331–1335.
97. Moreau, F.; Bond, G. C. CO oxidation activity of gold catalysts supported on various oxides and their improvement by inclusion of an iron component. *Catalysis Today*, 2006, *114* (4), 362–368.
98. Chen, X.; Zhu, H.-Y.; Zhao, J.-C.; Zheng, Z.-F.; Gao, X.-P. Visible-light-driven oxidation of organic contaminants in air with gold nanoparticle catalysts on oxide supports. *Angewandte Chemie International Edition* (English), 2008, *47* (29), 5353–5356.
99. Nguyet, T. T. M.; Trang, N. C.; Huan, N. Q.; Xuan, N.; Hung, L. T.; Date, M. Preparation of gold nanoparticles, Fe/Au_2O_3 by using a co-precipitation method and their catalytic activity. *Journal of the Korean Chemical Society*, 2008, *52* (5), 1345–1349.
100. Ahn, H.-G.; Choi, B.-M.; Lee, D.-J. Complete oxidation of ethylene over supported gold nanoparticle catalysts. *Journal of Nanoscience and Nanotechnology*, 2006, *6* (11), 3599–3603.
101. Guczi, L.; Horvath, D.; Paszti, Z.; Peto, G. Effect of treatment on gold nanoparticles. Relation between morphology, electron structure and catalytic activity in CO oxidation. *Catalysis Today*, 2002, *72* (1–2), 101–105.
102. Gupta, N. M.; Tripathi, A. K. The role of nanosized gold particles in adsorption and oxidation of carbon monooxide over Fe/Au_2O_3 catalyst. *Gold Bulletin*, 2001, *34* (4), 120–128.
103. Corma, A.; Serna, P. Chemoselective hydrogenation of nitro compounds with supported gold catalysts. *Science*, 2006, *313* (5785), 332–334.
104. Milone, C.; Ingoglia, R.; Schipilliti, L.; Crisafulli, C.; Neri, G.; Galvagno, S. Selective hydrogenation of α,β-unsaturated ketone to α,β-unsaturated alcohol on gold-supported iron-oxide catalysts, role of the support. *Journal of Catalysis*, 2005, *236* (1), 80–90.
105. Wang, L.; Han, L.; Luo, J.; Zhong, C.-I. Iron oxide and composite nanoparticles for catalysis and sensors. *Abstracts of Papers, 228th ACS National Meeting*, Philadelphia, PA, August 22–26, 2004.
106. Peto, G.; Geszti, O.; Molnár, G.; Daróczi, Cs. S.; Karacs, A.; Guczi, L.; Beck, A.; Frey, K. Valence band and catalytic activity of Au nanoparticles in Fe_2O_3/SiO_2/Si(100) environment. *Materials Science and Engineering C*, 2003, *C23* (6–8), 733–736.
107. Omoto, I.; Oka, T.; Yamamoto, T.; Nakagawa, T.; Kiyono, S. Gold-iron oxide complex magnetic particles, magnetic particles for immunoassay, and immunoassay. JP 2007205911, 2007.
108. Chung, B. H.; Lim, Y. T.; Kim, J. K. Gold nanocages containing magnetic nanoparticles. WO 2008018707, 2008.
109. Larson, T. A.; Bankson, J.; Aaron, J.; Sokolov, K. Hybrid plasmonic magnetic nanoparticles as molecular specific agents for MRI/optical imaging and photothermal therapy of cancer cells. *Nanotechnology*, 2007, *18* (32), 325101/1–325101/8.

110. Lim, J. K.; Tilton, R. D.; Eggeman, A.; Majetich, S. A. Design and synthesis of plasmonic magnetic nanoparticles. *Journal of Magnetism and Magnetic Materials*, 2007, *311* (1), 78–83.

111. Li, J.-P.; Chen, X.-Z. A novel glucose sensor based on sensitive film composed of Fe_3O_4/Au/GOx magnetic particulates. *Huaxue Xuebao*, 2008, *66* (1), 84–90.

112. Salgueirino-Maceira, M.; Correa-Duante, M. A.; Lopez-Quintela, M. A.; Rivas, J. Surface plasmon resonance in gold/magnetite nanoparticulated layers onto planar substrates. *Sensor Letters*, 2007, *5* (1), 113–117.

113. Min, H.; Qu, Y.-H.; Li, X.-H.; Xie, Z.-H.; Wei, Y.-Y.; Jin, L.-T. Au-doped Fe_3O_4 nanoparticle immobilized acetylcholinesterase sensor for the detection of organophosphorus pesticide. *Huaxue Xuebao*, 2007, *65* (20), 2303–2308.

114. Sun, Q.; Reddy, B. V.; Marquez, M.; Jena, P.; Gonzalez, C.; Wang, Q. Theoretical study on gold-coated iron-oxide nanostructure, magnetism and bioselectivity for amino acids. *Journal of Physical Chemistry, C*, 2007, *111* (11), 4159–4163.

115. Mizukoshi, Y.; Seino, S.; Okitsu, K.; Kinoshita, T.; Nakagawa, T.; Yamamoto, T. Sonochemical preparation of gold/iron oxide composite magnetic nanoparticles and selective magnetic separation of biomolecules. *International Journal of Nanoscience*, 2006, *5* (2–3), 359–363.

116. Mizukoshi, Y.; Seino, S.; Kinoshita, T.; Nakagawa, T.; Yamamoto, T.; Tanabe, S. Selective magnetic separation of sulfur-containing amino acids by sonochemically prepared Au/γ-Fe_2O_3 composite nanoparticles. *Scripta Materialia*, 2005 (volume date 2006), *54* (4), 609–613.

117. Seino, S.; Kinoshita, T.; Otome, Y.; Nakagawa, T.; Okitsu, K.; Mizukoshi, Y.; Nakayama, T.; Sekino, T.; Niihara, K.; Yamamoto, T. A. Gamma-ray synthesis of magnetic nanocarrier composed of gold and magnetic iron oxide. *Journal of Magnetism and Magnetic Materials*, 2005, *293* (1), 144–150.

118. Seino, S.; Kinoshita, T.; Otome, Y.; Maki, T.; Nakagawa, T.; Okitsu, K.; Mizukoshi, Y.; Nakagawa, T.; Okitsu, K,; Yamamoto, T. A. γ-Ray synthesis of composite nanoparticles noble metals and magnetic iron oxides. *Scripta Materialia*, 2004, *51* (6), 467–472.

119. Kinoshita, T.; Seino, S.; Mizukoshi, Y. et al. Magnetic separation of amino acids by gold/iron-oxide composite nanoparticles synthesized by gamma-ray irradiation. *Journal of Magnetism and Magnetic Materials*, 2005, *293* (1), 106–110.

120. Bao, J.; Chen, W.; Liu, T.; Zhut, Y.; Jint, P.; Wang, L.; Liu, J.; Wei, Y.; Li, Y. Bifunctional Au-Fe_3O_4 nanoparticles for protein separation. *ACS Nano*, 2007, *1* (4), 293–298.

121. Kim, S.-H.; Kim, M.-J.; Choa, Y.-H. Fabrication and estimation of Au-coated Fe_3O_4 nanocomposite powders for the separation and purification of biomolecules. *Materials Science and Engineering A, Structural Materials, Properties, Microstructure and Processing*, 2007, A449–A451, 386–388.

122. Saini, G.; Shenoy, D.; Nagesha, D. K.; Kautz, R.; Sridhar, S.; Amiji, M. Superparamagnetic iron oxide-gold core-shell nanoparticles for biomedical applications. In *NSTI Nanotech 2005, NSTI Nanotechnology Conference and Trade Show*, Anaheim, CA, May 8–12, 2005, Vol. 1. Laudon, M.; Romanowicz, B. (Eds.). Nano Science and Technology Institute, Cambridge, MA, pp. 328–331.

123. Park, K.; Liang, G.; Ji, X.; Luo, Z.; Li, C.; Croft, M. C.; Markert, J. T. Structural and magnetic properties of gold and silica doubly coated γ-Fe_2O_3 nanoparticles. *Journal of Physical Chemistry C*, 2007, *111* (50), 18512–18519.

124. Xi, D.; Luo, X. P.; Lu, Q. H.; Yao, K. L.; Liu, Z. L.; Ning, Q. The detection of HBV DNA with gold-coated iron oxide nanoparticle gene probes. *Journal of Nanoparticle Research*, 2008, *10* (3), 393–400.

125. Xu, C.; Xie, J.; Ho, D.; Xie, J.; Wang, C.; Kohler, N.; Walsh, E. G.; Morgan, J. R.; Chin, Y. E.; Sun, S. Au-Fe_3O_4 dumbbell nanoparticles as dual-functional probes. *Angewandte Chemie International Edition* (English), 2008, *47* (1), 173–176.

126. Aaron, J. S.; Oh, J.; Larson, T. A.; Kumar, S.; Milner, T. E.; Sokolov, K. V. Increased optical contrast in imaging of epidermal growth factor receptor using magnetically actuated hybrid gold/iron oxide nanoparticles. *Optics Express*, 2006, *14* (26), 12930–12943.

127. Hida, Y.; Kozuka, H. Photoanodic properties of sol-gel-derived iron oxide thin films with embedded gold nanoparticles, effects of polyvinylpyrrolidone in coating solutions. *Thin Solid Films*, 2005, *476* (2), 264–274.

128. Riskin, M.; Basnar, B.; Huang, Y.; Willner, I. Magnetoswitchable charge transport and bioelectrocatalysis using maghemite-Au core-shell nanoparticle/polyaniline composites. *Advanced Materials*, 2007, *19* (18), 2691–2695.

129. Pita, M.; Abad, J. M.; Vaz-Dominguez, C.; Briones, C.; Mateo-Marti, E.; Martin-Gago, J. A.; Morales, M. P.; Fernandez, V. M. Synthesis of cobalt ferrite core/metallic shell nanoparticles for the development of a specific PNA/DNA biosensor. *Journal of Colloid Interface Science*, 2008, *321*, 484–492.

130. Bai, J.; Wang, J.-P. High-magnetic-moment core-shell-type FeCo-Au/Ag nanoparticles. *Applied Physics Letters*, 2005, *87* (15), 152502/1–152502/3.

131. Kooi, B. J.; Vystavel, T.; De Hosson, J. T. M. Structure and giant magneto-resistive properties of Co and CoFe nanoparticles in a Au matrix. *MRS Symposium Proceedings (Synthesis, Functional Properties and Applications of Nanostructures)*, Vol. 676, 2002, pp. Y8.2.1–Y8.2.6.

132. Curshing, B. L.; Golub, V.; O'Connor, C. J. Synthesis and magnetic properties of Au-coated amorphous $Fe_{20}Ni_{80}$ nanoparticles. *Journal of Physics and Chemistry of Solids*, 2004, *65* (4), 825–829.

133. Choi, J.-S.; Jun, Y.-W.; Yeon, S.-I.; Kim, H. C.; Shin, J.-S.; Cheon, J. Biocompatible heterostructured nanoparticles for multimodal biological detection. *Journal of the American Chemical Society*, 2006, *128* (50), 15982–15983.

134. Wei, D. H.; Hung, D. S.; Ho, C. S.; Wang, J. W.; Yao, Y. D. Fabrication of monodisperse FePt@Au core-shell nanoparticles. *Physica Status Solidi C, Current Topics in Solid State Physics*, 2007, *4* (12), 4421–4424.

135. Kang, S.; Jia, Z.; Zoto, I.; Reed, D.; Nikles, D. E.; Harrell, J. W.; Thompson, G.; Mankey, G.; Krishna-Murthy, V.; Porcor, L. *Applied Physics*, 2006, *99* (8, Pt. 3), 08N704/1–08N704/3.

136. Seetala, N. V.; Harrell, J. W.; Lawson, J.; Nikles, D. E.; Williams, J. R.; Isaacs-Smith T. Ion-irradiation induced chemical ordering of FePt and FePtAu nanoparticles. *Nuclear Instruments and Methods in Physics Research Section B: Beam Interactions with Materials and Atoms*, 2005, *241* (1–4), 583–588.

137. Harrell, J. W.; Nikles, D. E.; Kang, S. S.; Sun, X. C.; Jia, Z. Effect of additive Cu, Ag, and Au on L10 ordering in chemically synthesized FePt nanoparticles. *Nippon Oyo Gakkaishi*, 2004, *28* (7), 847–852.

138. Harrell, J. W.; Nicles, D. E.; Kang, S. S.; Sun, X. C.; Jia, Z.; Lawson, S. J.; Thompson, G. B.; Srivatsava, C.; Seetalea, N. V. Effect of metal additives on L10 ordering of chemically synthesised FePt nanoparticles. *Scripta Materialia*, 2005, *53* (4), 411–416.

139. O'Connor, C.; Sims, J.; Kumbhar, A.; Kolesnichenko, V. L.; Zhou, W. L.; Wiemann, J. A. Magnetic properties of FePt*x*/Au and CoPt*x*/Au core-shell nanoparticles. *Journal of Magnetism and Magnetic Materials*, 2001, *226–230* (Pt. 2), 1915–1917.

140. O'Connor, C. J.; Kolesnichenko, V.; Carpenter, E.; Sangregorio, C.; Zhou, W. L.; Kumbhar, A.; Sims, J.; Agnoli, F. Fabrication and properties of magnetic materials with nanometer dimensions. *Synthetic Metals*, 2001, *122* (3), 547–557.

141. Butera, A.; Kang, S. S.; Nikles, D. E.; Harrell, J. W. Self-assembled arrays of high anisotropy FePt-Au nanoparticles. *Physica B, Condensed Matter*, 2004, *354* (1–4), 108–112.

142. Chen, Y. W.; Lin, M. H.; Hsu, H. C.; Lin, J. H. Nanoscale gold supported manganese oxide/iron oxide catalyst, and preparation method and application thereof. CN 101284239, 2008.

143. Xu, Z.; Hou, Y.; Sun, S. Magnetic core/shell Fe_3O_4/Au and Fe_3O_4/Au/Ag nanoparticles with tunable plasmonic properties. *Journal of the American Chemical Society*, 2007, *129* (28), 8698–8699.

144. Nitani, H.; Yuya, M.; Ono, T.; Nakagawa, T.; Seino, S.; Okitsu, K.; Mizukoshi, Y.; Emura, S.; Yamamoto, T. A. Sonochemically synthesized core-shell structured Au-pd nanoparticles supported on y-Fe_2O_3 particles. *Nanoparticle Research*, 2006, *8* (6), 951–958.

145. Zhou, W. L.; Carpenter, E. E.; Sims, J.; Kumbhar, A.; O'Connor, C. J. *MRS Symposium Proceedings (Nanophase and Nanocomposites Materials III)*, Vol. 581, 2000, pp. 107–112.

146. Srikanth, H.; Carpenter, E. E.; Spinu, L.; Wiggins, J.; Zhou, W. L.; O'Connor, C. J. Dynamic transverse susceptibility in Au-Fe-Au nanoparticles. *Materials Science and Engineering A, Structural Materials: Properties, Microstructure and Processing*, 2001, A304–A306, 901–904.

147. Guo, S.; Li, J.; Wang, E. High-density gold nanoparticles supported on a [Ru(bpy)3]2+-doped silica/Fe_3O_4 nanocomposite, facile preparation, magnetically induced immobilization, and application in ECL detection. *Chemistry—An Asian Journal*, 2008, *3* (8–9), 1544–1548.

148. Zhang, Y.; Wang, H.; Yan, B.; Zhang, Y.; Li, J.; Shen, G.; Yu, R. A reusable piezoelectric immunosensor using antibody-adsorbed magnetic nanocomposite. *Journal of Immunology Methods*, 2008, *332* (1–2), 103–111.

149. Chiang, I.-C.; Chen, D.-H. Synthesis of monodisperse FeAu nanoparticles with tunable magnetic and optical properties. *Advanced Functional Materials*, 2007, *17* (8), 1311–1316.

150. Duanmu, Y.; Zhang, Y.; Ma, M.; Xu, A.; Gu, N. Preparation and characterization of Fe/Au core-shell nanoparticles. *Dongnan Daxue Xuebao, Ziran Kexueban*, 2004, *34* (3), 340–343.

151. Penades, S.; Martin-Lomas, M.; Martines de la Fuente, J.; Rademacher, T. W. Magnetic nanoparticles for diagnosis and therapy. WO 2004108165, 2004.

152. Cho, S.-J.; Lui, K.; Kauzlarich, S. M. *Abstracts of Papers, 39th Western Regional Meeting of the ACS*, Sacramento, CA, October 27–30, 2004.

153. Nakagawa, S.; Okada, N.; Yoshikawa, T. et al. Gold-iron oxide composite fine particles, and MRI contrast agents containing the same. JP 2008266214, 2008.

154. Kouassi, G. K.; Irudayaraj, J. Magnetic and gold-coated magnetic nanoparticles as a DNA sensor. *Analytical Chemistry*, 2006, *78* (10), 3234–3241.

155. Liu, Z.; Peng, L.; Yao, K.; Lu, Q.; Wang, H. Preparation and character of ultrasmall Fe_3O_4/Au nanoparticles. *Gongneng Cailiao*, 2005, *36* (2), 196–199.

156. Wu, W. H.; He, Q.-G.; Chen, H.; Tang, J.-X.; Nie, L.-B. Sonochemical gold coating of Fe_3O_4 nanoparticles and its characterization. *Huaxue Xuebao*, 2007, *65* (13), 1273–1279.

157. Xi, D.; Ning, Q.; Lu, Q.; Yao, K.; Liu, Z.; Luo, X. HBV DNA detection with oligodeoxynucleotide-modified Fe_3O_4 (core)/Au (shell) nanoparticle probes. *Zhonghua Jianyan Yixue Zazhi*, 2006, *29* (4), 339–345.

158. De la Fuente, J. M.; Alcantara, D.; Eaton, P.; Crespo, P.; Rojas, T. C.; Fernandez, A.; Hernando, A.; Penades, S. Gold and gold-iron oxide magnetic glyconanoparticles, synthesis, characterization and magnetic properties. *Journal of Physical Chemistry B*, 2006, *110* (26), 13021–13028.

159. Persoons, A.; Verbiest, T.; Gangopadhyay, P. Targeted delivery of biologically active substances using iron oxide/gold core-shell nanoparticles. GB 2415374, 2005.

160. Cui, Y.; Hui, W.; Su, J.; Wang, Y.; Chen, C. Fe_3O_4/Au composite nanoparticles and their optical properties. *Science in China, Series B: Chemistry*, 2005, *48* (4), 273–278.

161. Lyon, J. L.; Fleming, D. A.; Stone, M. B.; Schiffer, P.; Williams, M. E. Synthesis of Fe oxide core/Au shell nanoparticles by iterative hydroxylamine seeding. *Nano Letters*, 2004, *4* (4), 719–723.

162. Yu, H.; Chen, M.; Rice, P. M.; Wang, S. X.; White, R. L.; Sun, S. Dumbbell-like bifunctional Au-Fe_3O_4 nanoparticles. *Nano Letters*, 2005, *5* (2), 379–382.

163. Zafiropoulou, I.; Devlin, E.; Boukos, N.; Niarchos, D.; Petridis, D.; Tzitzios, V. Direct Chemical Synthesis of L10 FePt Nanostructures. *Chemistry of Materials*, 2007, *19* (8), 1898–1900.

164. Jia, Z.; Kang, S.; Nikles, D. E.; Harrell, J. W. Synthesis of FePtAu nanoparticles in high-boiling-point solvents. *IEEE Transactions on Magnetics*, 2005, *41* (10), 3385–3387.

165. Kang, S.; Jia, Z.; Nikles, D. E.; Harrell, J. W. Synthesis, self-assembly, and magnetic properties of (FePt)1-xAux nanoparticles. *IEEE Transactions on Magnetics*, 2003, *39* (5, Pt. 2), 2753–2757.

166. Jia, Z.; Kang, S.; Shi, S.; Nikles, D. E.; Harrell, J. W. Size effects on L10 ordering and magnetic properties of chemically synthesized FePt and FePtAu nanoparticles. *Journal of Applied Physics*, 2005, *97* (10, Pt. 3), 10J310/1–10J310/3.

167. Pita, M.; Kramer, M.; Zhou, J.; Poghossian, A.; Schoning, M. J.; Fernandez, V. M.; Katz, E. Optoelectronic properties of nanostructured ensembles controlled by biomolecular logic systems. *ACS Nano*, 2008, *2* (10), 2160–2166.

25 Metallic Nanoalloys*

As shown in previous sections, the research field of nanometals in the last decades rapidly developed, due to the fact that physical and chemical properties of the metal nanoparticles differ considerably from those of both bulk metal and isolated atoms.[1,2] Metallic and bimetallic nanoparticles possess unique properties allowing them to be explored for a variety of applications, including nanocatalysis particularly for efficient selective catalysts, sensors, optical markers, and filters, among many other applications. Nanoalloys as a part of the "nanometals" refers to metallic clusters composed of two or more metal elements, and their physical and chemical properties are defined not only by their size and stoichiometry (atomic arrangement) but also by their challenging composition.[3–9] In fact, the field of nanoalloys appeared as far back as in the nineteenth century, much before the field of nanotechnology: the oldest topic in nanoscience is the size-dependent optical properties of Au and Ag colloids or nanoparticles, first studied scientifically by *Michael Faraday* in 1857.[10] Recently, the area of nanoalloys has expanded; however, a predominant number of reports correspond to nickel and noble metals (e.g., AuRe[11] or CoPt[12]) due to their catalytic applications.

Uniform nanoalloys should be formally distinguished from core-shell metallic nanoparticles containing two or more metals and built of a nucleus/kernel (e.g., of metallic iron) and a covering shell (e.g., of Au), although sometimes these two terms are not strictly distinguished in different publications. Computer simulations, carried out[13] to study the mechanism of formation of nanoparticles synthesized in microemulsions, allowed to express the main idea of obtaining nanoalloys or core-shell nanoparticles as follows: a *core-shell structure* is obtained when the reduction rates of both metals are very different, whereas a *nanoalloy* is obtained if both reaction rates are similar.

As for bulk alloys, a wide range of combinations and compositions are also possible for nanoalloys. Bimetallic nanoalloys (A_mB_n) can be generated with, more or less, controlled size ($m + n$) and composition (m/n). The cluster structures and degree of A–B segregation or mixing may depend on the method and conditions of cluster generation (type of cluster source, temperature, pressure, etc.).[5] Several methods based on physical and chemical approaches have been developed for the synthesis of controlled size and shape nanostructures including nanoalloys, such as chemical reduction,[14,15] thermal decomposition[16] (in particular, thermal destruction of bimetallic carbonyl precursors[17]), electrochemical synthesis,[7,18,19] microwave synthesis,[20,21] and molecular beams.[22,23] Recently, the chemistry of nanoalloys has been widely investigated due to their interesting and useful properties and applications in catalysis, photonics, and electronic devices, among many others.[24,25] The applications of nanoparticle alloys are expected to enhance many fields of advanced materials and relevant technology particularly in the areas mentioned earlier and chemical and biological sensors, in optoelectronics, among others.[26,27] The present review is dedicated to recent achievements in the obtaining and main applications of nanoalloys in various areas of chemistry and chemical technology, omitting core-shell metallic nanoparticles.

25.1 THEORETICAL CALCULATIONS

Nanoalloys have been an object of systematic theoretical investigations to reveal their stable structures and compositions and associate these properties with possible applications. Thus, a systematic *ab initio* study of the structures and magnetic properties of $Mo_{4-x}Fe_x$ clusters ($x = 1–3$) was reported.[28] In a related report,[29] *ab initio* calculations of the structure and electronic density of states

---

* This chapter has been contributed by Dr. Blanca Flores Muñoz (UANL).

FIGURE 25.1 (Top row) Chiral structure of $Ag_{107}Cu_{85}$. (Middle row) Achiral penta I_h structure of $Ag_{90}Cu_{56}$. (Bottom row) Achiral penta I_h structure of $Ag_{102}Cu_{75}$. From left to right, for all structures the first and second snapshots are side views (the second showing the structure of the inner Cu core), and the third is a top view. (Reproduced with permission from Bochicchio, D. and Ferrando, R., Size-dependent transition to high-symmetry chiral structures in AgCu, AgCo, AgNi, and AuNi nanoalloys, *Nano Lett.*, 10, 4211–4216. Copyright 2010 American Chemical Society.)

(DOS) of the perfect core-shell $Ag_{27}Cu_7$ nanoalloy attested to its D5h symmetry and confirmed that it has only six nonequivalent (two Cu and four Ag) atoms. In addition, the global structural optimizations for Ni–Al nanoalloy clusters at different compounds were investigated using particle swarm optimization combined with the simulated annealing method.[30] Some stable structures were described for Ni_xAl_x ($x = 1$–8), $Ni_{3x}Al_x$ ($x = 1$–4), and Ni_xAl_{3x} ($x = 1$–4) nanoalloy clusters. In addition, a class of nanomaterials (core-shell nanoalloys with a Cu, Ni, or Co core and a chiral Ag or Au shell of monatomic thickness) possessing the highest degree of chiral symmetry, the chiral icosahedral symmetry, was predicted[31] by a combination of global optimization searches and first-principle calculations. High-symmetry chiral nanoalloys associated strong energetic stability with potential for applications in optics, catalysis, and magnetism. Figure 25.1 shows an example of such compounds.

25.2 MAIN METHODS OF THE SYNTHESIS OF NANOALLOYS

25.2.1 MICROWAVE SYNTHESIS

25.2.1.1 Microwave Heating

The fabrication of nanoalloys by the method of microwave heating (irradiation) (MWI) is a facile and fast method that has been developed to prepare a wide variety of pure metallic and bimetallic alloy nanoparticles with controlled size and shape.[14] The MWI approach provides simple and fast routes to the synthesis of nanomaterials since no high temperature or high pressure is needed. Furthermore, MWI is particularly useful for a controlled large-scale synthesis that minimizes the thermal gradient effects.[15,32] As a result of the difference in the solvent and reactant dielectric constant, selective dielectric heating can provide significant enhancement in reaction rates. The most important advantage of microwave dielectric heating over convective heating is that the reactants can be added at room temperature (or slightly higher temperatures) without the need for

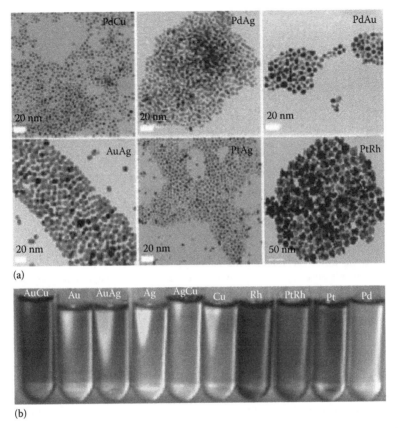

FIGURE 25.2 (a) TEM images of bimetallic nanoalloys prepared by the MWI method. (b) Digital photographs of metallic and bimetallic nanocrystals dispersed in toluene solutions. (Reproduced with permission from Abdelsayed, V. et al., Microwave synthesis of bimetallic nanoalloys and CO oxidation on ceria-supported nanoalloys, *Chem. Mater.*, 21, 2825–2834. Copyright 2009 American Chemical Society.)

high-temperature injection. Additionally, nanoparticle size can be tuned by varying the concentration of the precursor and the MWI duration, and the shape of the formed nanostructures is controlled by varying the concentration and composition of the ligating solvents.

Among recent reports, Abdelsayed described[20] the synthesis and characterization of bimetallic alloys of the Au, Pt, Pd, Rh, Ag, Cu, and Ni; MWI approach was established as a general procedure for the synthesis of the variety of high-quality crystalline bimetallic nanoalloys with controlled size and shape. Figure 25.2a shows TEM images of several examples of bimetallic nanoalloys. Figure 25.2b shows digital photographs of metallic nanocrystals and bimetallic nanoalloys dispersed in toluene solutions. Different colors of nanoalloys as compared to individual metals are clearly visible in all cases. The same method can be used to synthesize bimetallic nanoalloys supported on ceria nanoparticles as nanocatalysts. Thus, it was revealed that the CuPd-, CuRh-, and AuPd-supported nanoalloys exhibited a high activity for CO oxidation.

Parts a and b of Figure 25.3 displays the XRD patterns of the PtAu and PtRh (a) and CuAu and CuPt (b) nanoalloys, respectively. Comparing the XRD pattern of the nanoalloy to the patterns of the individual metals, it is clear that the diffraction peaks of the nanoalloy are located in between the corresponding peaks of the individual metals. This suggests the formation of a solid solution corresponding to the specific nanoalloy. The nanoalloy patterns also show no evidence of any pure metallic peaks, which indicates that binary nucleation has been the major nucleation process involved in the formation of these alloys.

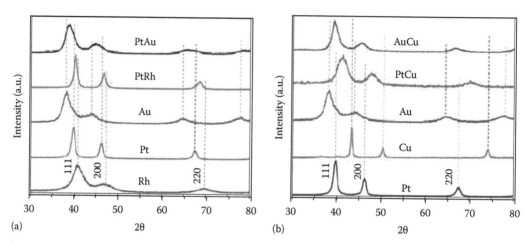

FIGURE 25.3 Comparisons of the XRD patterns of nanoalloys: (a) PtAu and PtRh and (b) AuCu and PtCu, with the patterns of the individual metal nanocrystals prepared by the MWI method. (Reproduced with permission from Abdelsayed, V. et al., Microwave synthesis of bimetallic nanoalloys and CO oxidation on ceria-supported nanoalloys, *Chem. Mater.*, 21, 2825–2834. Copyright 2009 American Chemical Society.)

Lai et al.[33] synthesized Pt–Co binary alloys by MWI of aqueous solutions containing Pt and Co precursors, and a significant enhancement in activity for ORR was observed for compositions such as $Pt_{50}Co_{50}$ as a result of a higher degree of alloying between Pt and Co. In addition, the relationship between the variations in alloying extent and Pt d–band vacancies in Pt–Co/C electrocatalysts was established,[21] which was tunable with Pt and Co composition and had a strong impact on the catalytic activity for the ORR. It was found that the catalytic activity of Pt_1Co_1/C for ORR was higher than that of Pt/C, Pt_1Co_3/C, and Pt_3Co_1/C due to its higher alloying extent, smaller particle size, and proper composition.

25.2.1.2 Microwave Plasma Treatment

In this method, high-frequency microwaves are applied through a resonator to generate plasma. Microwave plasma sources are used increasingly for thin-film deposition.[34] In particular, plasmas produced by resonant absorption at the electron cyclotron frequency, in the so-called ECR (electron-cyclotron resonance) devices, offer a number of desirable characteristics, including high plasma density, low-pressure operation, efficient gas utilization, and uniform plasma with a high degree of ionization and decomposition of gases.[35] The microwave power (2.45 GHz, 1.3 kW maximum) is provided by a homemade source using a conventional magnetron from commercial microwave ovens.[36] A schematic illustration of two different elemental apparatuses for the production of plasma *via* microwave[37,38] is shown in Figure 25.4.

A series of carbon-supported monometallic (Fe, Co) and bimetallic (Co–Mo) materials (average metal particle diameters <10 nm [Ar carrier gas] or <2 nm [H₂/Ar carrier gas]) were synthesized by microwave plasma decomposition of metal carbonyls.[39] The ultrafine metal particles were dispersed on moderate surface area amorphous carbon support matrixes derived from the concomitant microwave decomposition of the toluene solvent. Metallic iron and iron oxide particles were produced by injecting ferrocene into the afterglow region of a low-pressure, low-power, plasma generated using a microwave power source. It was found that two parameters had the largest impact on the particles: injection point and plasma composition. Low yields of small particles (ca. 10 nm) resulted from injection into the afterglow region, whereas much higher yields of large particles (ca. 50 nm) formed if the ferrocene was injected through the coupler. In all cases, significant amounts of graphitic carbon formed around the metal particles.[40]

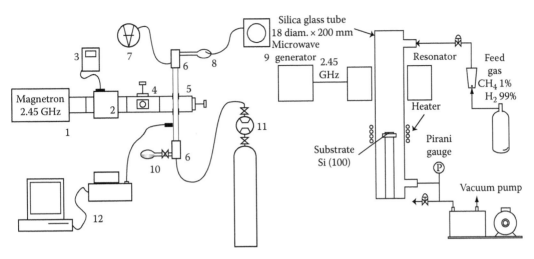

FIGURE 25.4 Schematic illustration of two different apparatuses. (From Saito, A. et al., *J. Mater. Sci. Lett.*, 5, 565–568, 1986; Hoder, B. et al., *Proceedings of Contributed Papers, Part II*, 2005, pp. 300–305. Reproduced with permission from Masaryk University and Springer.)

25.2.1.3 Microwave-Assisted Solvothermal Technique

The microwave-assisted solvothermal technique (MW-ST) approach involves uniform heating (up to 900°C) of polar solvents by absorbing microwave energy and subsequent transfer of the heat selectively to the reactants, which reduces thermal gradients inside the reaction vessel and increases the reaction kinetics. The MW-ST method provides a uniform nucleation environment and offers highly crystalline monodisperse multimetallic nanoparticles. Among nanoalloys, prepared this way, results of the systematic investigation of a series of Pt-encapsulated Pd–Co and alloyed Pt–Pd–Co electrocatalysts were discussed.[41] The particle size and distribution were characterized by TEM and are presented in Figure 25.5a and b. The TEM as shown in the inset of Figure 25.5a demonstrates the crystalline nature of the nanoparticles. Figure 25.5b shows TEM image and particle size distribution of the 75 wt.% Pd + 25 wt.% Pt sample after heat treatment at 900°C. It indicates a considerable increase in the mean particle diameter accompanied by a broadening of the particle size distribution and particle agglomeration after heat treatment at 900°C.

The MW-ST method (15 min, $t < 300°C$) was also applied to synthesize carbon-supported multimetallic nanostructured alloys of Pt, Pd, and Co with high crystallinity and homogeneity for electrocatalytic application in fuel cells.[42] These multimetallic alloys were explored for their electrocatalytic applications as cathode catalysts for oxygen reduction reaction (ORR), showing much higher ORR activity than that of their counterparts synthesized by the conventional borohydride reduction method. Although the ORR activity of $Pt_{70}Pd_{20}Co_{10}$ was comparable to that of commercial Pt, the ORR activity of $Pt_{50}Pd_{30}Co_{20}$ in direct methanol fuel cells (DMFC) was superior to that of commercial Pt at high methanol concentrations due to its high tolerance to methanol that may crossover from the anode to the cathode.

25.2.2 Molecular/Ion/Atom Beams

The development of a great variety molecular beam techniques has allowed obtaining an extend variety of metallic clusters. Actually, diverse types of metal sources exist, allowing the use of molecular beams; due to such properties of metals as volatility or refractory properties, it is possible to use some specific sources. Depending on the source nature and reaction conditions, clusters widely varying in size and distribution can be obtained. Nowadays, laser vaporization, pulsed arc deposition, ion sputtering, and magnetron sputtering are used to produce metallic clusters. Here we discuss the formation of clusters *via* gas aggregation and LA.

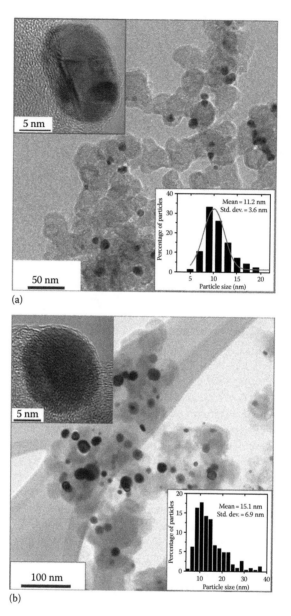

FIGURE 25.5 TEM images of the (a) as-synthesized and (b) 900°C heat-treated 75 wt% $Pd_{80}Co_{20}$ + 25 wt% Pt samples. The insets show the particle size distributions and HR-TEM images. (Reproduced with permission from Sarkar, A. et al., Pt-encapsulated Pd-Co nanoalloy electrocatalysts for oxygen reduction reaction in fuel cells, *Langmuir*, 26, 2894–2903. Copyright 2010 American Chemical Society.)

25.2.2.1 Formation of Clusters via Gas Aggregation

This technique is based on the ability to dynamically control the nucleation and growth of clusters. The result is a narrow and directed beam of clusters, with low size dispersion, that are deposited or incorporated in nanostructures under ultrahigh vacuum conditions. This method is often more suitable, since it offers advantages in terms of the chemical composition, cluster size, and cluster dispersion over a surface. The clusters resulting from substrate-free growth can be structurally totally different from those found in bulk materials, resulting in important modifications in their magnetic properties.

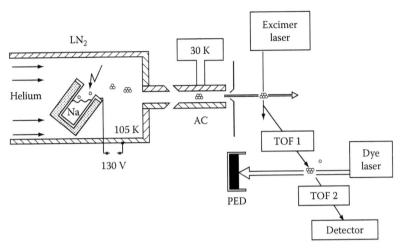

FIGURE 25.6 Gas aggregation source to produce hot and cold sodium clusters. (Reproduced from Ellert, C. et al., *Phys. Rev. Lett.*, 75(9), 1731–1734, 1995. Copyright 1995 by the American Physical Society.)

Among a series of nanoalloys produced in this route, NaK nanoalloy clusters were fabricated by gas aggregation.[22] The gas aggregation source, shown in Figure 25.6, was used to produce hot and cold sodium cluster ions.[23]

The homebuilt cluster source, based on the gas-aggregation principle,[31,43] earlier showed the capability of producing a beam of alkali-metal clusters containing thousands of atoms with concentrations high enough to be probed by SR-based XPS. Figure 25.7 presents typical XPS spectra of

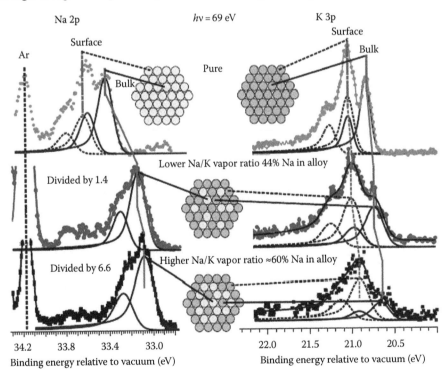

FIGURE 25.7 Na 2p and K 3p XPS spectra for pure single-component Na and K clusters (top), clusters produced from the vapor of lower Na/K ratio (middle), and clusters produced from the vapor of higher Na/K ratio (bottom). (Reproduced from Tchaplyguine, M. et al., *Phys. Rev. B*, 80, 033405/1–033405/4, 2009. Copyright 2009 by the American Physical Society.)

the Na $2p$ and the K $3p$ regions for pure single-component Na and K clusters (top), snapshot spectra for the clusters produced from the vapor of lower Na/K ratio (middle), and for the clusters produced from the vapor of higher Na/K ratio (bottom), together with schematic illustrations of the deduced cluster structures.[22]

25.2.2.1.1 Clusters via Laser-Assisted Synthesis

25.2.2.1.1.1 Laser Vaporization Controlled Condensation Method

Laser vaporization provides several advantages over other heating methods such as the production of a high-density vapor of any metal, the generation of a directional high-speed metal vapor from the solid target, which can be useful for directional deposition of the particles, the control of the evaporation from specific spots on the target, and the simultaneous or sequential evaporation of several different targets.[44]

A sketch of the chamber with the relevant components for the production of nanoparticles by the laser vaporization controlled condensation (LVCC) method is shown in Figure 25.8. The chamber consists of two horizontal, circular, stainless steel plates, separated by a glass ring. A metal target of interest is set on the lower plate, and the chamber is filled with a pure carrier gas such as Ar (99.99% pure). The metal target and the lower plate are maintained at a temperature higher than that of the upper one. The top plate can be cooled to less than 150 K by circulating liquid nitrogen. The large temperature gradient between the bottom and top plates results in a steady convection current, which can be enhanced by using a heavy carrier gas such as Ar under high-pressure conditions (10^3 Torr). The metal vapor is generated by pulsed laser vaporization using the second harmonic (532 nm) of an Nd-YAG laser (15–30 mJ/pulse, 10–8 s/pulse). The role of convection in the experiments is to remove small particles away from the nucleation zone (once condensed out of the vapor phase) before they can grow into larger particles. The rate of convection increases with the temperature gradient in the chamber. Therefore, by controlling the temperature gradient, the total pressure, and the laser power (which determines the number density of the metal atoms released in the vapor phase), it is possible to control the size of the condensing particles. No particles are found anywhere else in the chamber except on the top plate; this supports the assumption that nucleation takes place in the upper half of the chamber and that convection carries the particles to the top plate where deposition occurs.[45,46]

The vapor-phase synthesis of Au–Ag, Au–Pd, and Au–Pt nanoparticle alloys was carried out using LVCC method. The formation of nanoparticle alloys and not simply mixtures of the two metal

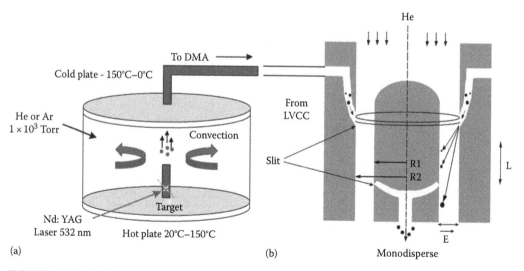

FIGURE 25.8 (a) Experimental setup for the synthesis of nanoparticles using the LVCC method. (b) Experimental setup for the LVCC method coupled with a DMA. (From Glaspell, G. et al., *Pure Appl. Chem.*, 76(9), 1667–1689, 2006. Reproduced from IUPAC.)

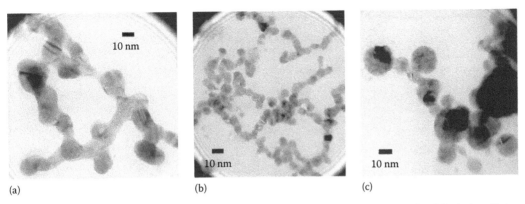

FIGURE 25.9 TEM images of the $Ag_{0.71}Au_{0.29}$ nanoparticles: (a) as-prepared, (b) after 20 min irradiation with 532 nm, and (c) after 20 min irradiation with 1064 nm. (Abdelsayed, V. et al., Laser synthesis of bimetallic nanoalloys in the vapor and liquid phases and the magnetic properties of PdM and PtM nanoparticles (M = Fe, Co and Ni). *Faraday Discuss.*, 138, 163–180, 2008. Reproduced by permission of The Royal Society of Chemistry.)

nanoparticles was confirmed. Irradiation of a mixture of Au/Ag nanoparticles dispersed in water with the 532 nm unfocused laser resulted in efficient alloying while the 1064 nm laser radiation resulted only in evaporation and size reduction of the unalloyed nanoparticles (Figure 25.9). The synthesis of palladium and platinum nanoparticles alloyed with transition metals such as iron and nickel using the LVCC method was also presented. The alloyed nanoparticles (FePd, FePt, NiPd, NiPt, and FeNi) were found to be superparamagnetic.[47]

In 2006, Abdelsayed et al. reported the vapor-phase synthesis of intermetallic aluminide nanoparticles. Specifically, FeAl and NiAl nanoparticles were synthesized *via* LVCC from their bulk powders. The NiAl nanoparticles were found to be paramagnetic at room temperature, with a blocking temperature of approximately 15 K. They focused on the synthesis of intermetallic NiAl and FeAl nanoparticles and the characterization of their magnetic properties. Intermetallic aluminides of nickel and iron alloys have many important applications, especially for high-temperature structural applications due to their high melting points and thermal conductivity.[48]

These materials are further characterized by low density, high strength-to-weight ratio, and corrosion and oxidation resistance, especially at high temperatures.[49,50] However, their poor room-temperature ductility and low high-temperature strength limit their usage.[51] Interestingly, nanoparticles prepared from these materials were proposed to overcome the limitations of bulk materials. For example, iron aluminide and nickel aluminide nanoparticles exhibit room-temperature ductility and superplasticity.[52] Consolidated nanoparticles may show enhanced plasticity, that is, they may exhibit exceptionally large tensile elongation during stretching compared with conventional materials.[52,53] With smaller grain sizes, it is expected that superior mechanical properties can be achieved, which would enhance the room-temperature ductility and the high-temperature strength of the iron and nickel aluminides.

Figure 25.10 displays typical TEM images of the FeAl and NiAl nanoparticles. Most of the primary particles have average diameters in the range of 5–7 nm with a few percent (~2%–4%) larger particles that are approximately 20–30 nm in diameter. Figure 25.11 shows several TEM images of 30 and 14 nm selected FeAl nanoparticles prepared using the LVCC-DMA system.[54,55] Although the nanoparticles are still aggregated (Figure 25.11a), the primary particles appear to exhibit identical sizes. Individual monodisperse particles can be deposited only if the number density of the particles is kept very low to avoid the aggregation of the particles as shown in Figure 25.11b and d.

25.2.2.1.1.2 Laser Ablation Method Laser ablation (LA) is considered as one of the most interesting methods of preparation of metallic nanoparticles in liquids.[56] Following the pioneering work of Patil et al.,[57] this technique was further developed by Henglein, Cotton, and their co-workers.[58,59]

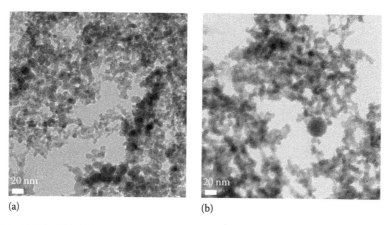

(a)   (b)

FIGURE 25.10 TEM of (a) FeAl and (b) NiAl nanoparticles. (From Glaspell, G. et al., *Pure Appl. Chem.*, 76(9), 1667–1689, 2006. Reproduced from IUPAC.)

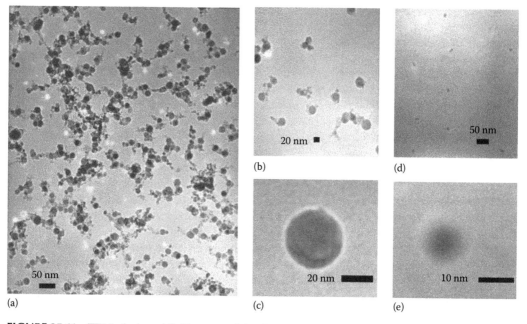

FIGURE 25.11 TEM of selected FeAl nanoparticles, 30 and 14 nm shown as (a through c) and (d, e), respectively. (From Glaspell, G. et al., *Pure Appl. Chem.*, 76(9), 1667–1689, 2006. Reproduced from IUPAC.)

This method can provide a pure solution of small nanoparticles without any contamination of chemical reagent and parasitic ions compared with the chemical ways to produce metallic nanoparticles in solution.

Recently, Mahfouz et al.[60] presented results concerning bimetallic nanoparticles prepared by LA of $Au_{75}Ag_{25}$ and $Ni_{75}Pd_{25}$ targets in distilled water. In conclusion, bimetallic nanoparticles of Au–Ag and Ni–Pd alloy with a homogeneous composition can be produced by LA in water. However, the evolution of the particle size with the ablation time shows different behavior. The size of Au–Ag particles slightly increases with the ablation time, whereas Pd–Ni particle size does not change much. This could result from different interactions between the nanoparticles in solution and the laser at a wavelength of 532 nm.

The sample is attached onto a Teflon plug support and onto a glass vessel containing distilled water, and is closed by a welded glass (or quartz) window (Figure 25.12). The metallic target is

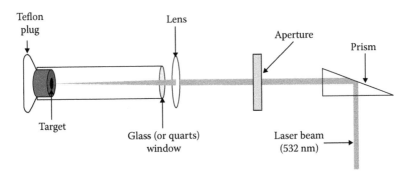

FIGURE 25.12 Experimental setup for laser ablation. (Reproduced with permission from Springer Science+Business Media: *J. Nanopart. Res.*, Elaboration and characterization of bimetallic nanoparticles obtained by laser ablation of Ni75Pd25 and Au75Ag25 targets in water, 12(8), 2010, 3123–3136, Mahfouz, R. et al.)

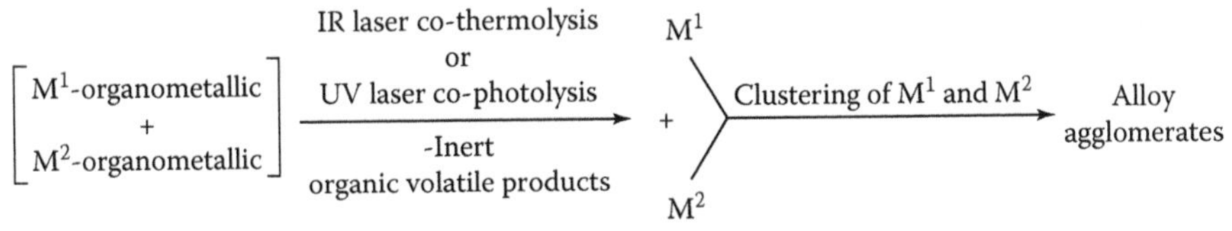

$$\begin{bmatrix} M^1\text{-organometallic} \\ + \\ M^2\text{-organometallic} \end{bmatrix} \xrightarrow[\substack{-\text{Inert} \\ \text{organic volatile products}}]{\substack{\text{IR laser co-thermolysis} \\ \text{or} \\ \text{UV laser co-photolysis}}} + \begin{array}{c} M^1 \\ \diagdown \\ M^2 \end{array} \xrightarrow{\text{Clustering of } M^1 \text{ and } M^2} \begin{array}{c} \text{Alloy} \\ \text{agglomerates} \end{array}$$

FIGURE 25.13 Laser-induced changes in the gas (M^1 and M^2 designate different metal atoms).

completely immersed in liquid. An Nd:YAG laser operating at the second harmonic is focused on the target. The laser energy is fixed. The ablation time could vary depending on the sample.

Alloys were isolated by classic LA methods from mixtures of organometallics as precursors (Figure 25.13). Several nanoalloys have been synthesized this way; thus, Au/Ag colloidal nanoalloys were fabricated with a wide range of compositions by LA of single metal targets in water and a re-irradiation of mixed colloidal suspensions.[61] The properties of the final colloidal alloy and the kinetics of formation can be quantitatively followed by using the extinction spectra fitted with the Mie theory and a simple modified Drude–Lorentz model for optical functions.

In a related report,[62] a femtosecond laser irradiation approach was developed for the production of homogeneous Au–Ag nanoalloys (5–7 nm in size) of various compositions. At gold fractions above 0.4, most of the nanoparticle oxidation was quenched, inhibiting the release of toxic silver ions in solution. The oxidation of the produced nanoparticles was mainly attributed to the generation of free radicals (O, H, OH) and of molecular reactive oxygen species (O_2, H_2, H_2O_2) formed by the decomposition of the water molecules through femtosecond laser-induced optical breakdown. The authors anticipated that Au–Ag nanoalloys would be the best compromise in terms of chemical stability and plasmonic response, as they possess much better resistance to oxidation in comparison to pure silver and a much stronger and narrower plasmon peak in comparison to pure gold. In addition, Sn/Ag/C was also obtained *via* LA.[63]

25.2.3 Chemical Vapor Deposition

As an example of application of the chemical vapor deposition (CVD) technique to fabricate nanoalloys, carbon-coated nanoparticles of Ni–Ru alloys (average Ni–Ru core particle size of 8 ± 3 nm and carbon shell thickness of 2–3 nm) containing 7.5–19 wt.% Ru were synthesized by the high-pressure CVD method using metallocenes of both metals.[64] The magnetic properties of the coated nanoparticles strongly depended on the synthesis parameters so that the superparamagnetic blocking temperature and saturation magnetization can be tailored. Also, the growth of nickel, Ni–Cu

Ni$_{83.6}$Cu$_{16.4}$ Ni$_{52}$Cu$_{48}$

Ni$_{29.57}$Cu$_{70.43}$

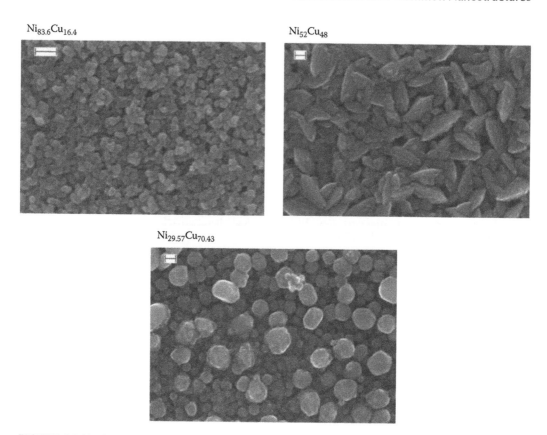

FIGURE 25.14 SEM micrographs of selected Ni-Cu alloy compositions. The scale bar corresponds to 200 min in the first micrograph (Ni$_{83.6}$Cu$_{16.4}$) and to 100 nm in other cases. (Reproduced with permission from Bahlawane, N. et al., Nickel and nickel-based nanoalloy thin films from alcohol-assisted chemical vapor deposition, *Chem. Mater.*, 22(1), 92–100. Copyright 2010 American Chemical Society.)

(Figure 25.14), and Ni–Co alloys was investigated by using the pulsed-spray evaporation CVD process in a hydrogen-free atmosphere.[65] The CVD process (Figure 25.15) enabled the deposition of Ni–Cu alloy thin films with a fully controlled composition, whereas Ni–Ag formed polycrystalline silver in addition to an amorphous phase containing silver and nickel.

25.2.4 Electrochemical Synthesis

The electrochemical synthesis has been widely used as an attractive method for obtaining nanoalloys; the most common technique is *electrochemical deposition* (ECD), offering great advantages due to the possibility to control the process easily. A certain attention has been paid to the electrochemical preparation of Cu–Sn nanoalloys. Thus, the electrochemical synthesis (*via* electrodeposition under spontaneous current oscillations from a single bath containing both sulfate salts and the specific inhibitor substance) and characterization of films of nanometric Cu–Sn intermetallic compounds were reported.[66] It was shown that the lower the frequency/amplitude of the electrodepositing current, the higher the amount of tin. The applied approach led to a nanometric material showing modulation in composition ranging between the two copper-rich high-temperature solid solutions δ-Cu$_{41}$Sn$_{11}$ and ζ-Cu$_{10}$Sn$_{3}$. In addition, the brightness of the studied Cu–Sn coatings was conditioned by the surface morphology.[67] The surface morphology of the Cu–Sn deposits (Figure 25.16), obtained at different potentials, varied with the cathodic potential from fine-grained at $Ec = 0.0$ V (Figure 25.16, image a) toward granular with the constantly increasing grain size (Figure 25.16, images b through e). This regularity broke when Ec became negative to the reversible

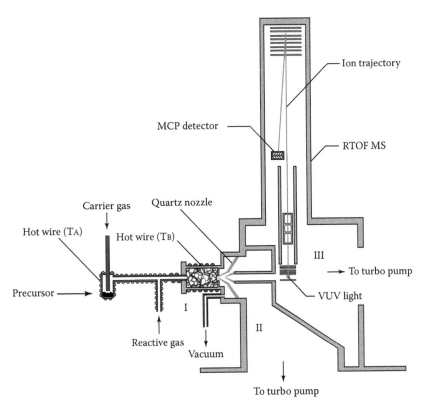

FIGURE 25.15 A schematic diagram of the CVD reactor with *in situ* molecular-beam mass spectrometry. Parts I, II, and III are the CVD reactor, the differential pumped chamber, and the photoionization chamber, respectively. The synchrotron VUV light crosses vertically with the molecular beam. Part I was modified to accommodate a pulsed-spray evaporation unit. (Reproduced with permission from Bahlawane, N. et al., Nickel and nickel-based nanoalloy thin films from alcohol-assisted chemical vapor deposition, *Chem. Mater.*, 22(1), 92–100. Copyright 2010 American Chemical Society.)

potential of the Sn/Sn^{2+} electrode (−0.2 V). The deposit obtained at Ec = −0.23 V (Figure 25.16, image f) was fine-grained, smooth, and bright, similar to that deposited at Ec = 0.0 to −0.1V.

Regarding the electrochemical performance vs. Li, these phases showed better capacity retention by comparison with an Li/Sn cell. In a related research,[68] electrodeposition and heat treatment were attempted to directly obtain a Sn–Cu alloy anode with fine grains of crystals for lithium ion batteries. It was revealed that copper and tin were partially alloyed to form Cu_6Sn_5 and Cu_3Sn after annealing. The Cu-coated electrode presented the first cycle coulomb efficiency reaching 95% and good cycleability.

In addition to Cu–Sn nanoalloys, copper nanoalloys with some transition metals have been fabricated and characterized. Thus, the effect of alkyl polyglucoside (APG) surfactant on the electrodeposition of Co–Ni–Cu alloy nanoparticles was investigated.[69] It was found that characteristics of Co–Ni–Cu alloy nanoparticles, such as size homogeneity, density, dispersion on the electrode substrate, and the chemical composition, depended strongly on the concentration of APG used in the reaction as well as the applied deposition potential. Copper was found to have a high percentage in the nanoalloy deposits. Ordered arrays of cobalt–nickel alloy nanowires (Figure 25.17) were electrodeposited into the pores of anodic alumina templates.[70] Nanowires with different compositional ratio of cobalt and nickel showed a nonlinear dependence of coercivity as a function of cobalt concentration. These magnetic nanoalloys with controlled properties could have various applications, such as high-density magnetic storage or nanoelectrode arrays.

FIGURE 25.16 SEM images of Cu–Sn deposits obtained at *Ec*: 0.0 V (a), −0.1 V (b), −0.14 V (c), −0.17 V (d), −0.185 V (e), and −0.23 V (f). (Reproduced from *Electrochim. Acta*, 52(3), Juskenas, R. et al., XRD studies of the phase composition of the electrodeposited copper-rich Cu-Sn alloys, 928–935, Copyright 2006, with permission from Elsevier.)

Pt–Pd bimetallic nanocomposites with various compositions, structure, and properties have been intensively investigated as the catalysts for electrooxidation of formic acid (FA).[71–74] Thus, Pd-modified Pt nanoparticles,[71] unsupported Pt–Pd bimetallic nanoparticles,[72] and supported Pd–Pt nanoalloys with Pt/Pd molar ratios ranging from 4:1 to 1:2 were synthesized to be applied in FA oxidation.[72,73] However, the reported conclusions are rather controversial regarding the optimal bimetallic composition, and the catalysts either contained relatively high Pt contents or had very limited improvement for electrocatalysis as compared to Pd/C.[71–74] Surprisingly, no prior reports have been found on screening Pd_xPt_{1-x}/C with much lower Pt contents and yet with superior electrocatalysis toward FA oxidation, in light of the existing theories, that is, the "third-body" effect[75] and the *d*-band center theory.[76] Thus, a series of Pd-based Pd_xPt_{1-x} ($x = 0.5$–1) nanoparticles dispersed on carbon black support were initially synthesized[77] and screened for FA electrooxidation, and the unprecedented synergetic activity was pinpointed to the Pt/Pd atomic ratio as low as ca. 1/9. The "third-body" effect for the optimal catalyst was illustrated

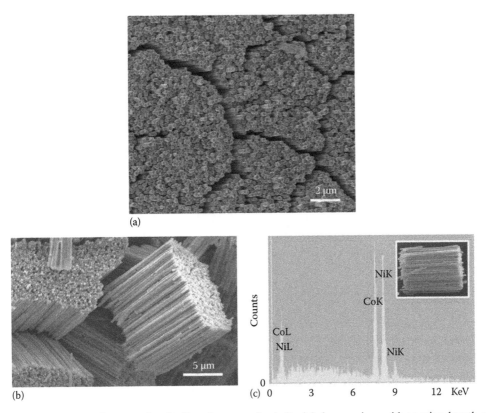

(a)

(b)

(c)

FIGURE 25.17 SEM images of well-aligned arrays of cobalt nickel nanowires with varying length and diameters. (a) A large area showing alloy nanowire growth from the AAO. (b) Nanowire arrays completely etched out of the AAO templates showing typical length achieved for 1 h electrodeposition. (c) Energy dispersive spectra (EDS) showing the presence of cobalt and nickel. CoK, NiK, and NiL lines are clearly seen in the spectra. Inset shows the nanowire bundle from where the EDS were acquired. (Reproduced with permission from Springer Science+Business Media: *J. Mater. Sci.*, Synthesis and characterization of cobalt-nickel alloy nanowires, 44(9), 2009, 2271–2275, Talapatra, S. et al.).

with ball models and the effect of the *d*-band center shift of Pd in Pd_xPt_{1-x} demonstrated by positive correlation of different electrochemical measurements.

25.2.5 THERMAL DECOMPOSITION

Thermal decomposition, also called thermolysis, is defined for nanoalloys as chemical reaction(s) in which substance(s), containing two or more metals, break up into at least two elemental metals when heated. Precursors can be mixtures of metal-containing compounds or polymetallic complexes. This basis is commonly used to obtain certain nanoalloys, in particular from volatile precursors by CVD method (see the following). A row of nanoalloys have been obtained by thermolysis of compounds, low stable at elevated temperatures. Thus, different types of bimetallic magnetic nanoparticles were prepared starting from molecular bimetallic carbonyl cluster anions, ($[Fe-Co_3(CO)_{12}]^-$, $[Fe_3Pt_3(CO)_{15}]^-$, $[FeNi_5(CO)_{13}]$ and $[Fe_4Pt(CO)_{16}]$).[17]

These experiments, based on the thermal decomposition of the anionic bimetallic carbonyl clusters, systematically resulted in the formation of a series of alloy nanoparticles, with a composition reflecting that of the precursor ($FeCo_3$, FePt, $FeNi_4$, and Fe_4Pt). Figure 25.18a, for example, shows a TEM image of $FeCo_3$ NPs, synthesized from $[NEt_4][FeCo_3(CO)_{12}]$, with an average diameter of 6.8 ± 0.8 nm, whereas Figure 25.18b demonstrates a TEM image of $FeNi_4$ NPs, produced under the

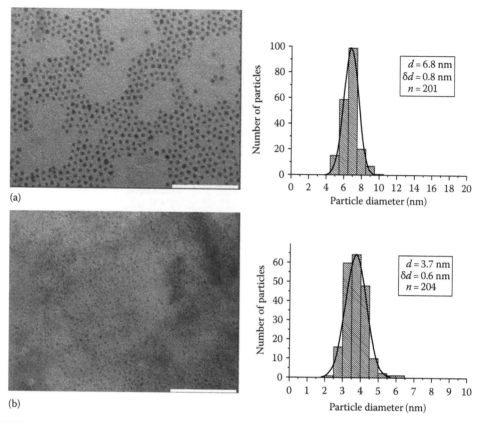

FIGURE 25.18 Typical TEM images and size distributions of (a) FeCo$_3$ and (b) FeNi$_4$ NPs synthesized by the thermal decomposition of [NEt$_4$][FeCo$_3$(CO)$_{12}$] and [NMe$_4$]$_2$[FeNi$_5$(CO)$_{13}$], respectively. Bar, 100 nm. (Reproduced with permission from Robinson, I. et al., Synthesis and characterization of magnetic nanoalloys from bimetallic carbonyl clusters, *Chem. Mater.*, 21, 3021–3026. Copyright 2009 American Chemical Society.)

same conditions (reactant concentrations, reaction temperature, and time), except using [NMe$_4$]$_2$[Fe-Ni$_5$(CO)$_{13}$] as a precursor, with an average diameter of approximately 4.1 ± 0.7 nm.

The principal advantage of this method was to eliminate the problems of phase separation and composition control. Additionally, palladium-based bimetallic and tri-metallic catalysts were explored as alternatives to lower the cost while maintaining the catalytic ability of platinum. Thus, Pd$_{100-x}$W$_x$ (0 ≤ x ≤ 30) nanoalloy electrocatalysts were synthesized[78] by simultaneous thermal decomposition of palladium acetylacetonate and tungsten carbonyl in *o*-xylene in the presence of Vulcan XC-72R carbon, followed by annealing up to 800°C in H$_2$ atmosphere. It was indicated that the formation of single phase face centered cubic (*fcc*) solid solutions for 0 ≤ x ≤ 20 with controlled composition and morphology took place; also, particle size increased with increasing heat treatment temperature. The alloying of Pd with W enhanced the catalytic activity for the oxygen reduction reaction (ORR) as well as the stability (durability) of the electrocatalyst compared with the unalloyed Pd; the composition Pd$_{95}$W$_5$ exhibited the maximum activity for ORR in the Pd–W system. In addition, iron–palladium nanoalloy in the particle size range of 15–30 nm was synthesized by the relatively low-temperature thermal decomposition of coprecipitated [Fe(Bipy)$_3$]Cl$_2$ and [Pd(Bipy)$_3$]Cl$_2$ in an inert ambient of dry argon gas.[79] This superparamagnetic Fe–Pd nanoalloy was found to be in single phase and contained iron sites with up to 11 nearest-neighbor atoms.

25.2.6 CHEMICAL REDUCTION

This method is one of the most important due to its simplicity and the possibility of its application in any chemical laboratory. Thus, a series of amorphous NiCoB catalysts with various modifiers (Mo, La, Fe, and W, existing both in elemental and oxidized states, acted as the spacer to prevent the NiCoB particles from aggregation) were prepared by chemical reduction method using $NaBH_4$ as the reducing agent at 303 K.[80] particle sizes of the modified catalysts were less than that of the unmodified NiCoB catalyst. These catalysts were found to be suitable for p-chloronitrobenzene (p-CNB) hydrogenation, influencing the product distributions (the relative contents of p-chloronitroaniline, aniline, and nitrobenzene) in the p-CNB hydrogenation since the oxidized state of each modifier had its unique Lewis acidity. These effects were responsible for the different behaviors of Mo-, La-, Fe-, and W-modified NiCoB in catalytical processes. It was also established[81] that W-NiCoB was more amorphous in nature and had more defect sites than NiCoB catalyst. The tungsten oxide located among the particles acted as a spacer that could separate the NiCoB particles from their neighbors and inhibit their aggregation. W-NiCoB catalysts had higher activities than NiCoB in the hydrogenation of p-CNB. An HRTEM image of W(0.7)-NiCoB catalyst is shown in Figure 25.19,

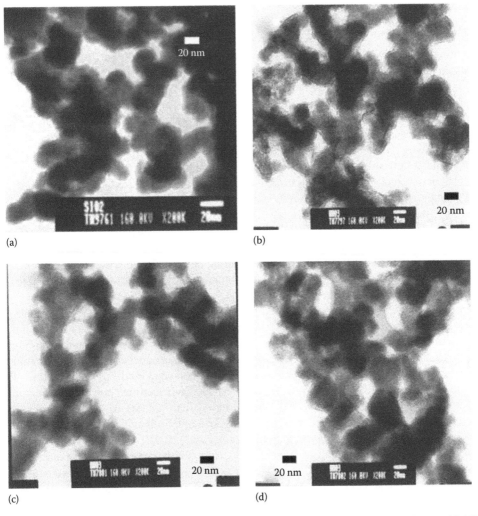

FIGURE 25.19 TEM pictures of the samples: (a) NiB, (b) NiFeB (1:0.1), (c) NiFeB (1:0.2), and (d) NiFeB (1:0.3). (Reproduced with permission from Zhao, B. et al., Hydrogenation of p-chloronitrobenzene on tungsten-modified NiCoB catalyst, *Ind. Eng. Chem. Res.*, 49, 1669–1676. Copyright 2010 American Chemical Society.)

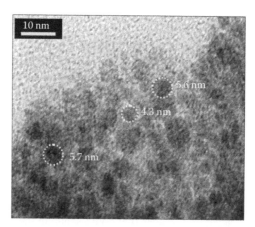

FIGURE 25.20   HR-TEM image of W(0.7)-NiCoB (particles less than 6 nm). (Reproduced with permission from Chen, Y.-W. and Sasirekha, N., Preparation of NiFeB nanoalloy catalysts and their applications in liquid-phase hydrogenation of *p*-chloronitrobenzene, *Ind. Eng. Chem. Res.*, 48, 6248–6255. Copyright 2009 American Chemical Society.)

confirming that the W-NiCoB catalysts were composed of many small particles and that the size of the primary particles was less than 6 nm. This is much smaller than the sizes reported in the literature (~30 nm) for NiCoB.[82] A series of related NiFeB nanoalloy catalysts (Figure 25.20) were prepared by chemical reduction method using metal acetates as the starting materials and NaBH$_4$ as the reducing agent.[83] The amorphous metal materials, combining the features of amorphous and nanometer materials, have more surface atoms and a higher concentration of coordinately highly unsaturated sites. The Ni/Fe molar ratios in the starting materials affected the concentration of boron bounded to the nickel and iron metals, resulting in the difference in surface area, electronic structures of the metals, and the catalytic activities of the catalysts. The NiFeB catalysts were effective during the hydrogenation of *p*-CNB (*p*-chloronitrobenzene). In a related investigation on boron-containing alloys,[84] an Fe$_{50-x}$Cu$_{50-x}$B$_{2x}$ alloy (<10 nm particles) was obtained by an aqueous chemical reduction at pH 7.0 using rapid and slow NaBH$_4$ solution addition methods. The room-temperature Mossbauer spectrum of this Fe–Cu–B nanocrystalline alloy showed a quadrupole doublet with an isomer shift of 0.355(1) mm/s and quadrupole splitting of 0.772(2) mm/s and broadened magnetic sextet with quadrupole splitting of 0.020(3) mm/s and isomer shift of 0.436(5) mm/s, exhibiting a distinctive high average magnetic field of 40.57(5) T.

Aqueously dispersed ferromagnetic NiPd nanoparticle alloys with various Ni/Pd molar ratios were synthesized by a polyol process.[85] The formation of NiPd nanoalloys and not simply mixtures of the two metal nanoparticles was confirmed. The PVP, used as a capping agent to stabilize the nanoparticles and prevent them from oxidation or agglomeration, accounted for ~30% of the total mass fraction of the nanoparticles. Synthesis of the carbon-supported Pd$_4$Co nanoalloy (Figure 25.21, left) samples (denoted as Pd$_4$Co/C) was carried out[86] by a modified polyol reduction process, involving a mixture of ethylene glycol (a polyol) and a small amount of sodium borohydride as reducing agents. The products (with average particle size of 4.6 nm and a narrow size distribution, which increased to 7.3 and 11.0 nm on annealing, respectively, at 350°C and 500°C in 10% H$_2$–90% Ar) were evaluated as electrocatalysts (Figure 25.21, right) for the oxygen reduction reaction (ORR) in fuel cells. It was revealed that, while both Pd/C and Pt/C exhibited a drastic decline in catalytic activity on annealing at 350°C and 500°C due to a significant increase in particle size, the Pd$_4$Co/C sample maintained high catalytic activity even after annealing at 350°C due to an increase in the degree of alloying. The authors noted that Pd$_4$Co/C catalyst offers an added advantage of excellent tolerance to methanol poisoning and lower cost compared with the conventional Pt/C catalyst.

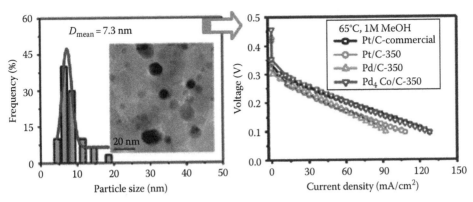

FIGURE 25.21 Pd_4Co nanoalloy. (Left) The products (with average particle size of 4.6 nm and a narrow size distribution, which increased to 7.3 and 11.0 nm on annealing, respectively, at 350°C and 500°C in 10% H_2–90% Ar). (Right) Evaluated as electrocatalysts for the oxygen reduction reaction (ORR) in fuel cells. (Liu, H. and Manthiram. A., Controlled synthesis and characterization of carbon-supported Pd4Co nanoalloy electrocatalysts for oxygen reduction reaction in fuel cells, *Energy Environ. Sci.*, 2, 124–132, 2009. Reproduced by permission of The Royal Society of Chemistry.)

In a related work,[87] carbon-supported Pd–Ni nanoalloy electrocatalysts ($Pd_{80}Ni_{20}$) with controlled particle size were similarly obtained and used as cathode catalysts in DMFC. One more example is nanosized tin telluride compounds, prepared by chemical reduction process using hydrazine dihydrate and hydrothermal methods.[88] The SnTe nanoalloy obtained by a chemical reduction process presented a quasi-spherical morphology (40–50 nm) with aggregation, whereas the powder prepared by a hydrothermal process was nearly nanospheres (30–40 nm of particle size with narrow distribution).

Uniformly distributed carbon-coated $Sn_{92}Co_8$ nanoalloys (Figures 25.22 and 25.23) with a particle size of 50 and 10 nm were prepared using a butyllithium and naphthalide solution at 500°C and 700°C, respectively.[89] The samples annealed at 500°C were covered with an amorphous carbon layer with a thickness of about 40 nm, but its thickness was shrunk to 8 nm when annealed at 700°C. The much improved capacity retention observed of the carbon-coated nanoalloys was found to be associated with the thicker carbon layer on the particle surface, which acts as a more effective buffer layer for volume expansion during lithium alloying/dealloying than that of the bulk alloy.

An effective strategy of composition-dependent assembly was reported to synthesize length-controllable amorphous $(Fe_{1-x}Ni_x)_{0.5}Pt_{0.5}$ nanoalloys (nanoparticles, nanorods, and nanothreads, Figure 25.24) through a phase-transfer process in autoclave from $Fe_2(C_2O_4)_3 \cdot 5H_2O$, $NiCl_2 \cdot 6H_2O$, and $H_2PtCl_6 \cdot 6H_2O$ as precursors.[90] The mechanism of their formation is shown in Figure 25.25. The morphologies of amorphous nanoalloys showed composition-dependent nanothread length variation from 8 μm to 600 nm, which resulted in different phase-transformation behaviors and magnetic properties. FeNiPt nanothreads presented a larger magnetic anisotropy with higher coercivity of 3 kOe than that of its nanoparticles with coercivity of 2.4 kOe.

25.3 MAIN APPLICATIONS OF NANOALLOYS

Catalytic applications of nanoalloys are of a great importance. It is known that classic noble metal catalysts, such as palladium, platinum, and ruthenium, are the most used in syntheses; however, their implementation in industrial applications is certainly limited due to their high cost. In this respect, the synthesis of such nanoalloyed catalysts as M_1M_2, where $M_1 = Pd$, Pt, Au and less expensive $M_2 = Ni$, Cu, and Co, could decrease their price without reduction of high catalytic activity due to the presence of M_1. Thus, the lowest-energy structures of binary $(PtPd)_n$, $(PtNi)_m$,

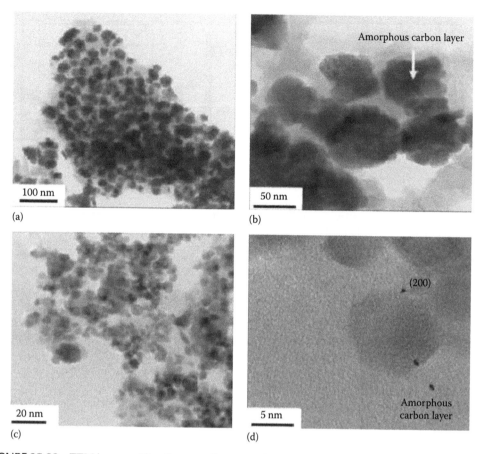

FIGURE 25.22 TEM images of $Sn_{92}Co_8$ nanoalloys annealed at (a, b) 500°C and (c, d) 700°C. (b) and (d) are expanded images of (a) and (c), respectively. (Kim, H. and Cho, J., *J. Electrochem. Soc.*, 154(5), A462–A466, 2007. Reproduced with permission.)

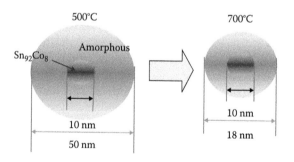

FIGURE 25.23 Schematic diagrams of the $Sn_{92}Co_8$ nanoalloys annealed at 500°C and 700°C for 5 h. (Kim, H. and Cho, J., *J. Electrochem. Soc.*, 154(5), A462–A466, 2007. Reproduced with permission.)

$(PtNi_3)_s$, and $(Pt_3Ni)_s$ nanoclusters (Figure 25.26), with $n = 2$–28, $m = 2$–20, and $s = 4$–6, modeled by the many-body Gupta potential, were obtained by using a genetic-symbiotic algorithm.[91] It was found that the $Pt_{0.75}Ni_{0.25}$ supported nanoparticles presented a higher catalytic activity for the selective oxidation of CO in the presence of hydrogen than the $Pt_{0.5}Ni_{0.5}$ and $Pt_{0.25}Ni_{0.75}$ nanoparticles.

Nickel-containing catalysts are objects of permanent attention. Thus, a nanosized Ni catalyst modified with boron was reported to be a good catalyst for the hydrogenation of nitrobenzene

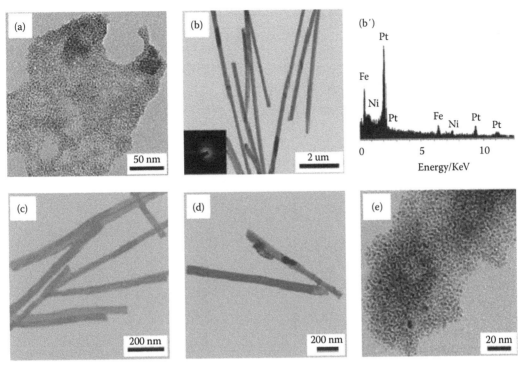

FIGURE 25.24 TEM images of FeNiPt nanoalloys obtained at a stopping temperature of 170°C for 3 h with heating rates of 2°C/min: (a) $Fe_{52}Pt_{48}$ nanoparticles, (b) $Fe_{41}Ni_{12}Pt_{47}$ nanothreads (inset: selected-area electron diffraction pattern), (b′) EDS spectrum of $Fe_{41}Ni_{12}Pt_{47}$ nanothreads, (c) $Fe_{16}Ni_{34}Pt_{45}$ nanothreads, (d) $Fe_6Ni_{44}Pt_{50}$ nanothreads, and (e) $Ni_{54}Pt_{46}$ nanoparticles. (Reproduced with permission from Wen, M. et al., Composition-dependent assembly and magnetic specificity of $(Fe1-xNix)0.5Pt0.5$ amorphous nanothreads through substitution of Ni for Fe in an FePt system, *J. Phys. Chem. C*, 113(46), 19883–19890. Copyright 2009 American Chemical Society.)

and furfural.[92] The catalytic properties were found to be highly dependent upon the preparation method.[93,94] A catalyst for the hydrogenation of *p*-chloronitrobenzene (*p*-CNB) to *p*-chloroaniline (*p*-CAN) was developed with the objective to study the effects of Mo content on the catalytic properties of NiCoB in the hydrogenation of *p*-CNB.[95] Its catalytic properties were tested for liquid-phase hydrogenation of *p*-CNB at 353 K and 1.2 MPa H_2 pressure. The main product was *p*-CAN, and there were two byproducts, nitrobenzene and aniline. The simplified reaction route is displayed in Figure 25.27. It was established that the thermal stability of NiCoB increased upon the addition of Mo. Most of Mo was in the form of hydroxide and acted as a spacer to prevent NiCoB from aggregation/agglomeration. The sample of 0.6 Mo–NiCoB owned the higher concentration of molybdenum, resulting in the higher activity and selectivity for the hydrogenation of *p*-CNB (Table 25.1).[95] In addition, bimetallic nanoalloys of 3*d*-series {Ni–Cu (34 nm), Ni–Co (43 nm), and Ni–Zn (30 nm)} were prepared by hydrazine reduction of respective metal chloride in ethylene glycol at 60°C.[96] The thermolysis of ammonium perchlorate was found to be catalyzed with these nanoalloys, and the burning rate was found to be enhanced considerably.

The first liquid–liquid Ullmann etherification process (Ullmann ether synthesis of 4-phenoxy-pyridine with and without the addition of 18-crown-6, Figure 25.28) mediated not only by oxidatively stable Cu but also by CuZn and CuSn nanoparticle catalysts in conjunction with microwave heating, which also avoided the use of solid and expensive bases, was proposed.[97] Conditions led to improved turnovers and excellent yields in heteroaromatic Ullmann-type coupling reactions.

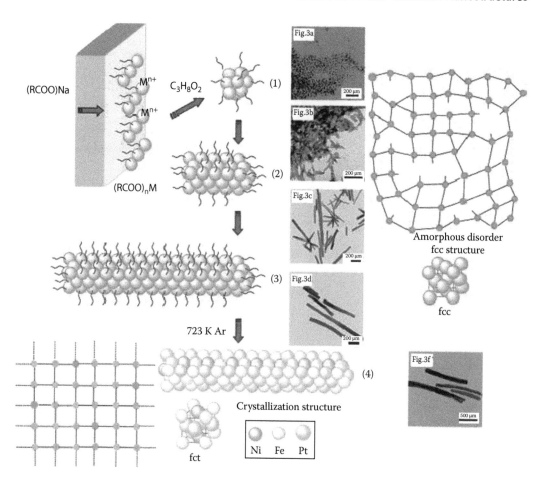

FIGURE 25.25 Schematic illustration of the assembly process of FeNiPt nanothreads. (Reproduced with permission from Wen, M. et al., Composition-dependent assembly and magnetic specificity of (Fe1-xNix)0.5Pt0.5 amorphous nanothreads through substitution of Ni for Fe in an FePt system, *J. Phys. Chem. C*, 113(46), 19883–19890. Copyright 2009 American Chemical Society.)

Low-temperature CO oxidation over a compositional series of SiO_2 fume-supported Pd/Au nanoalloy catalysts (Figure 25.29) was assessed.[98] A bifunctional mechanism was assumed to be responsible for CO oxidation, in which O_{ads} (and/or CO_{ads}) migrates from Pd to neighboring atoms or Au clusters through "spillover," as represented by Equations 25.1 through 25.4, in a mechanism similar to electrooxidation of CO. The authors termed this as the "mixed metal" case, with the key factor being exchange of surface oxygen between Pd and Au. Pd–Au alloy catalysts showed significant activity and stability toward low-temperature CO oxidation even at 300 K for bulk compositions from Pd to Pd_4Au_1, but they showed degraded performance at higher Au contents due to higher activation that is accompanied by greater surface coverage by coverage of Au. In addition, the factors that control the electrocatalytic activity for the ORR in fuel cells were investigated systematically with the low-cost carbon-supported $Pd_{100-x}Co_x$ ($0 \leq x \leq 50$) nanoalloys,[99] prepared by a solution-based reduction procedure and annealed at 350°C and 500°C. It was found, in particular, that below 30 at.% nominal Co content, the catalytic activity of the annealed samples increased with increasing nominal Co content due to both a decreasing crystallite size and increasing degree of alloying.

$$Pd + CO \rightarrow Pd\text{-}CO \tag{25.1}$$

$$Pd + O \rightarrow Pd\text{-}O \tag{25.2}$$

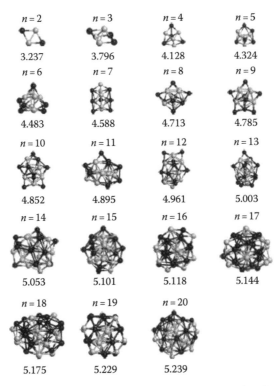

FIGURE 25.26 Geometric structures and binding energies (in eV) for the most stable isomers of $(PtNi)_n$, with $n = 2$–20. The dark spheres represent the Pt atoms. (Reproduced with permission from Springer Science+Business Media: *Eur. Phys. J. D*, Structural and electronic properties of PtPd and PtNi nanoalloys, 52(1–3), 2009, 127–130, Radillo-Diaz, A. et al.)

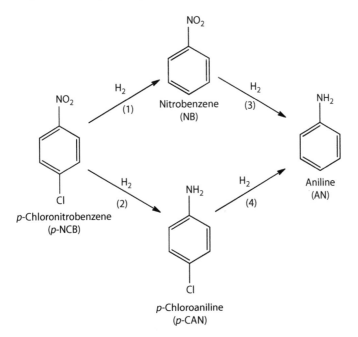

FIGURE 25.27 The simplified reaction scheme of hydrogenation of *p*-CNB. (Reproduced from Lin, M.-H. et al., Hydrogenation of *p*-chloronitrobenzene over Mo-modified NiCoB nanoalloy catalysts: Effect of Mo content, *Ind. Eng. Chem. Res.*, 48, 7037–7043. Copyright 2009 American Chemical Society.)

TABLE 25.1

The Effect of Mo Content on the Hydrogenation of *p*-CNB over Mo–NiCoB Catalyst

Sample	Reaction Time (min) to 100% Conversion	*p*-CAN	AN	NB
NiCoB	60	88.7	11.3	0
0.1 Mo–NiCoB	50	90.3	9.7	0
0.2 Mo–NiCoB	40	86.1	13.9	0
0.3 Mo–NiCoB	40	87.6	12.4	0
0.5 Mo–NiCoB	20	95.6	4.2	0.2
0.6 Mo–NiCoB	10	98.2	1.8	0
1.0 Mo–NiCoB	30	96.7	4.3	0

Source: Reproduced with permission from Lin, M.-H. et al., Hydrogenation of *p*-chloronitrobenzene over Mo-modified NiCoB nanoalloy catalysts: Effect of Mo content, *Ind. Eng. Chem. Res.*, 48, 7037–7043. Copyright 2009 American Chemical Society.

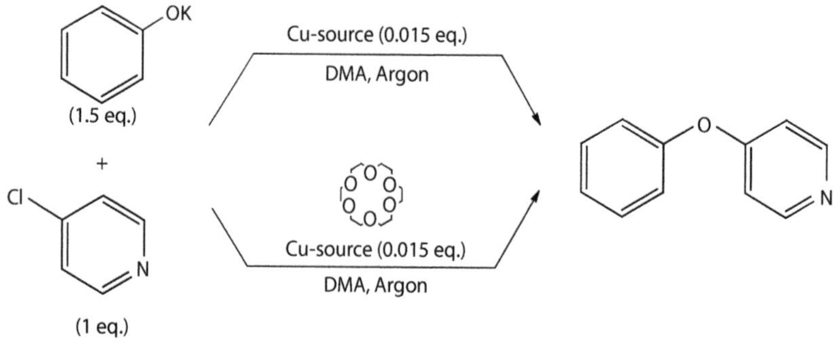

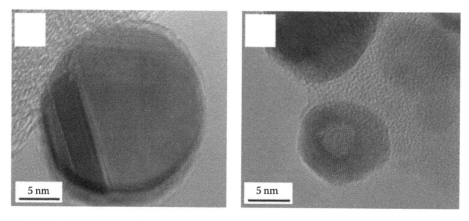

FIGURE 25.28 Ullmann ether synthesis of 4-phenoxypyridine with and without addition of 18-crown-6a. (Reproduced from Engels, V. et al., Cu-based nanoalloys in the base-free Ullmann heterocyle-aryl ether synthesis, *Org. Process Res. Dev.*, 14, 644–649. Copyright 2010 American Chemical Society.)

FIGURE 25.29 (a) Multiply twinned gold crystal from a catalyst of bulk composition Pd:Au = 1:3.4. (b) Hollow sphere Pd-rich crystal from material of bulk composition Pd:Au = 16:1. (Reproduced with permission from Xu, J. et al., Biphasic Pd-Au alloy catalyst for low-temperature CO oxidation, *J. Am. Chem. Soc.*, 132(30), 10398–10406. Copyright 2010 American Chemical Society.)

$$Pd\text{-}O + Au \rightarrow Pd + Au\text{-}O \tag{25.3}$$

$$Au\text{-}O + Pd\text{-}CO \rightarrow Au\text{-}Pd + CO_2 \tag{25.4}$$

Other applications of nanoalloys are not as frequent as in the catalysis. They can be used, in particular, as fuel additives,[100] in batteries and sensors, or for destruction of organic compounds. Thus, carbon-coated $Si_{70}Sn_{30}$ nanoalloys with a particle size <10 nm were prepared from butyl-capped analogues by firing at 900°C under a vacuum and used as a lithium battery anode material.[101] The nanocrystalline single-phase γ-Zn_3Ni alloy modified composite graphite (nano-γ-Zn_3Ni/CG) was employed for the detection of uric acid (UA) by voltammetric techniques,[102] showing an excellent response and specificity for the electrocatalytic oxidation of UA in 0.2 M phosphate buffer (pH 7.0) at 280 mV vs. Ag/AgCl as compared with bare CG electrode at 360 mV. The nano-γ-Zn_3Ni/CG was used for the detection of UA in untreated human urine and serum samples with satisfactory results. In addition, removal of indigo blue in water solutions was evaluated using Fe/Cu nanoparticles and composites of C/Fe–Cu nanoalloy.[103] Synthesis of materials was carried out by the sodium borohydride reduction of $FeSO_4 \cdot 7H_2O$ and $CuSO_4 \cdot 5H_2O$ aqueous mixture.

REFERENCES

1. Muraviev, D. N.; Macanas, J.; Parrondo, J.; Muñoz, A.; Alonso, A.; Alegret, M.; Ortueta, M.; Mijangos, F. Cation-exchange membrane as nanoreactor: Intermatrix synthesis of platinum–copper core–shell nanoparticles. *Reactive and Functional Polymers*, 2007, *67*, 1612–1621.
2. Mackerle, J. Nanomaterial, nanomechanics and finite elements: A bibliography (1994–2004). *Modelling and Simulation in Materials Science and Engineering*, 2005, *3* (1), 123–158.
3. Sinfelt, J. H. *Bimetallic Catalysts: Discoveries, Concepts and Applications*. Wiley, New York, 1983.
4. Jellinek, J.; Krissinel, E. B. Alloy clusters: Structural classes, mining and phase changes. *Theory of Atomic and Molecular Clusters*. Jellinek, J. (Ed.). Springer, Berlin, Germany, 1999, 277–308.
5. Ferrando, R.; Jellinek, J.; Johnston, R. L. Nanoalloys: From theory to applications of alloy clusters and nanoparticles. *Chemical Reviews*, 2008, *108*, 845–904.
6. Wang, R.; Zi, X.; Liu, L.; Dai, H.; He, H. The study and application of core-shell structure bimetallic nanoparticles. *Huaxue Jinzhan*, 2010, *22* (2/3), 358–366.
7. Johnston, R. L; Ferrando, R. Nanoalloys: From theory to application. Preface. *Faraday Discussions*, 2008, *138*, 9–10.
8. Tu, Z.; Hu, H.; Li, N.; Cao, L.; An, M. New evolution of electrodeposition of nanoalloys. *Biaomian Jishu*, 2008, *37* (1), 67–70, 79.
9. Lu, X.; Lu, D.; Chen, S.; Wang, Y. Application of mechanochemistry in preparation of inorganic materials. *Guisuanyan Tongbao*, 2004, *23* (6), 66–70.
10. Wilcoxon, J. Optical absorption properties of dispersed gold and silver alloy nanoparticles. *Journal of Physical Chemistry B*, 2009, *113* (9), 2647–2656.
11. Jiang, Z.; Zhang, J.; Wen, G.; Liang, A.; Liu, Q.; Kang, C.; He, X. Aptamer-modified AuRe nanoalloy probe for trace Hg^{2+} using resonance scattering as detection technique. *Chinese Journal of Chemistry*, 2010, *28* (7), 1159–1164.
12. Demortiere, A.; Losno, R.; Petit, C.; Quisefit, J.-P. Composition study of CoPt bimetallic nanocrystals of 2 nm. *Analytical and Bioanalytical Chemistry*, 2010, *397* (4), 1485–1491.
13. De Dios, M.; Barroso, F.; Tojo, C.; Lopez-Quintela, M. A. Bimetallic nanoparticle formation in microemulsions: Mechanism and Monte Carlo simulations. *Nanoscience and Nanotechnology*, 2008, *8*, 11–16.
14. Bönnemann, H.; Richards, R. M. Nanoscopic metal particles synthetic methods and potential applications. *European Journal of Inorganic Chemistry*, 2001, 2455–2480.
15. Toshima, N.; Yonezawa, T. Bimetallic nanoparticles-novel materials for chemical and physical applications. *New Journal of Chemistry*, 1998, *22*, 1179–1201.
16. Zubris, M.; King, R. B.; Garmestani, H.; Tannenbaum, R. J. FeCo nanoalloy formation by decomposition of their carbonyl precursors. *Journal of Materials Chemistry*, 2005, *15*, 1277–1285.
17. Robinson, I.; Zacchini, S.; Tung, L. D.; Maenosono, S.; Nguyen, T. K. Synthesis and characterization of magnetic nanoalloys from bimetallic carbonyl clusters. *Chemistry of Materials*, 2009, *21*, 3021–3026.

18. Kolb, U.; Quaiser, S. A.; M Winter, M.; Reetz, T. Investigation of tetraalkylammonium bromide stabilized palladium/platinum bimetallic clusters using extended x-ray absorption fine structure spectroscopy. *Chemistry of Materials*, 1996, *8*, 1889–1894.

19. Reetz, M. T.; Helbig, W. Size-selective synthesis of nanostructured transition metal clusters. *Journal of the American Chemical Society*, 1994, *116*, 7401–7402.

20. Abdelsayed, V.; Aljarash, A.; El-Shall, M. S.; Zeid, A.; Othnman, A.; Alghamdi, A. H. Microwave synthesis of bimetallic nanoalloys and CO oxidation on ceria-supported nanoalloys. *Chemistry of Materials*, 2009, *21*, 2825–2834.

21. Gerbec, J. A.; Magana, D.; Washington, A.; Strouse, G. F. Microwave-enhanced reaction rates for nanoparticle synthesis. *Journal of the American Chemical Society*, 2005, *127*, 15791–15800.

22. Tchaplyguine, M.; Legendre, S.; Rosso, A.; Bradeanu, I.; Öhrwall, G.; Canton, S. E.; Anderson, T.; Mårtensson, N.; Svensson, S.; Björneholm, O. Single-component surface in binary self-assembled NaK nanoalloy clusters. *Physical Review Letters B*, 2009, *80*, 033405/1–033405/4.

23. Ellert, C.; Schmidt, M.; Schmitt, C.; Reiners, T.; Haberland, H. Temperature dependence of the optical response of small, open shell sodium clusters. *Physical Review Letters*, 1995, *75*, 1731–1734.

24. Johnston, R. L. *Atomic and Molecular Clusters*. Taylor & Francis, London, U.K., 2002.

25. Alvarez-Puebla, R. A.; Bravo-Vasquez, J. P.; Cheben, P.; Xu, D.-X.; Waldron, P.; Fenniri, H. SERS-active Ag/Au bimetallic nanoalloys on Si/SiO$_x$. *Journal of Colloid and Interface Science*, 2009, *333* (1), 237–241.

26. Schrinner, M.; Ballauff, M.; Talmon, Y.; Kauffmann, Y.; Thun, J.; Möller, M.; Breu, J. Single nanocrystals of platinum prepared by partial dissolution of Au-Pt nanoalloys. *Science*, 2009, *323*, 617–620.

27. Burda, C.; Chen, X.; Narayanan, R.; El-Sayed, M. A. Chemistry and properties of nanocrystals of different shapes. *Chemical Reviews*, 2005, *105*, 1025–1102.

28. Garcia-Fuente, A.; Vega, A.; Aguilera-Granja, F.; Gallego, L. J. Mo$_{4-x}$Fe$_x$ nanoalloy: Structural transition and electronic structure of interest in spintronics. *Physical Review B: Condensed Matter and Materials Physics*, 2009, *79* (18), 184403/1–184403/7.

29. Alcantara-Ortigoza, M.; Rahman, T. S. First principles calculations of the electronic and geometric structure of Ag$_{27}$Cu$_7$ nanoalloy. *Physical Review B: Condensed Matter and Materials Physics*, 2008, *77* (19), 195404/1–195404/14.

30. Zhou, J. C.; Li, W. J.; Zhu, J. B. Theoretical study of Ni-Al nanoalloy clusters using particle swarm optimisation algorithm. *Materials Science and Technology*, 2008, *24* (7), 870–874.

31. Bochicchio, D.; Ferrando, R. Size-dependent transition to high-symmetry chiral structures in AgCu, AgCo, AgNi, and AuNi nanoalloys. *Nano Letters*, 2010, *10*, 4211–4216.

32. Chen, W.; Zhao, J.; Lee, J. Y.; Liu, Z. Microwave heated polyol synthesis of carbon nanotubes supported Pt nanoparticles for methanol electrooxidation. *Materials Chemistry and Physics*, 2005, *91*, 124–129.

33. Lai, F.-J.; Sarma, L. K.; Chou, H.-L.; Liu, D.-G.; Hsieh, C.-A.; Lee, J.-F.; Hwang, B.-J. Architecture of bimetallic Pt$_x$Co$_{1-x}$ electrocatalysts for oxygen reduction reaction as investigated by x-ray absorption spectroscopy. *Journal of Physical Chemistry C*, 2009, *113*, 12674–12681.

34. Chapman, B. *Glow Discharge Processes*. Wiley, New York, 1980.

35. Kawai, Y.; Ueda, Y. Electromagnetic wave propagation in an ECR plasma. *Surface and Coatings Technology*, 2000, *131*, 12–19.

36. Da Mata, J. A. S.; Galvão, R. M. O.; Ruchko, L.; Fantini M. C. A.; Kiyohara, P. K. Description and characterization of a ECR plasma device developed for thin film deposition. *Brazilian Journal of Physics*, 2003, *3*, 123–127

37. Hoder, T.; Kudrle, V.; Frgala, Z.; Janča, J. *Proceedings of Contributed Papers, Pragne. Part II*, 2005, pp. 300–305.

38. Saito, Y.; Matsuda, S.; Nogita, S. Synthesis of diamond by decomposition of methane in microwave plasma. *Journal of Materials Science Letters*, 1986, *5*, 565–568.

39. Brenner, J. R.; Harkness, J. B. L.; Knickelbein, M. B.; Krumdick G. K.; Marshall, C. L. Microwave plasma synthesis of carbon-supported ultrafine metal particles. *Nanostructured Materials*, 1997, *8* (1), 1–17.

40. Hao C.; Phillips, J. Plasma production of metallic nanoparticles. *Journal of Materials Research*, 1992, *7* (8), 2107–2113.

41. Sarkar, A.; Murugan, A. V.; Manthiram, A. Pt-encapsulated Pd-Co nanoalloy electrocatalysts for oxygen reduction reaction in fuel cells. *Langmuir*, 2010, *26*, 2894–2903.

42. Sarkar, A.; Murugan, A. V.; Manthiram, A. Rapid microwave-assisted solvothermal synthesis of methanol tolerant Pt-Pd-Co nanoalloy electrocatalysts. *Fuel Cells*, 2010, *10* (3), 375–383.

43. Sattler, K.; Mühlbach, J. Recknagel, E. Generation of metal clusters containing from 2 to 500 atoms. *Physical Review Letters*, 1980, *45* (10), 821–824.

44. Hadijipanyis, G. C.; Siegel, R. W. *Nanophase Materials: Synthesis, Properties, Applications*. Kluwer Academic, London, U.K., 1994.

45. El-Shall, M. S.; Li, S. In synthesis and characterisation of metal and semiconductor nanoparticles. *Advances in Metal and Semiconductor Clusters*. M. A. Duncan (Ed.). JAI Press, London, U.K., 1998, p. 115.

46. El-Shall, M. S.; Li, S.; Graiver, D.; Pernisz, U. C. Chapter 5: Synthesis of nanostructured materials using a laser Vaporization-Condensation technique. pp. 79–99. In *Nanotechnology: Molecularly Designed Nanostructural Materials, ACS Symposium Series No. 622*. Chow, G.; Gonsalves, K. E. (Eds.). American Chemical Society, Washington, DC, 1996, p. 79.

47. Abdelsayed, V.; Glaspell, G.; Nguyen, M.; Howe, J. M.; El-Shall, M. S. Laser synthesis of bimetallic nanoalloys in the vapor and liquid phases and the magnetic properties of PdM and PtM nanoparticles (M = Fe, Co and Ni). *Faraday Discussions*, 2008, *138*, 163–180.

48. Glaspell, G.; Abdelsayed, V.; Saoud, K. M.; El-Shall, M. S. Vapor-phase synthesis of metallic and intermetallic nanoparticles and nanowires: Magnetic and catalytic properties. *Pure and Applied Chemistry*, 2006, *76* (9), 1667–1689.

49. Stoloff, N. S.; Sikka, V. K. *Physical Metallurgy and Processing of Intermetallic Compounds*. Chapman & Hall, London, U.K., 1996, p. 217.

50. Morsi, K. Review: Reaction synthesis processing of Ni–Al intermetallic materials. *Materials Science and Engineering A*, 2001, *299* (1–2) 1–15.

51. Bose, A.; Moore, B.; German, R. M.; Stoloff, N. S. Elemental powder approaches to Ni_3Al-matrix composites. *Journal of Metals*, 1988, *40* (9), 14–17.

52. Maziasz, P. J.; Goodwin, G. M.; Alexander, D. J.; Viswanathan, S. *International Symposium on Nickel and Iron Aluminides: Processing, Properties, and Applications, Proceedings from Materials Week '96*, Cincinnati, OH, October 7–9, 1996–1997, p. 157.

53. McFadden, S. X.; Valiev, R. Z.; Mukherjee, A. K. Superplasticity in nanocrystalline Ni_3Al. *Materials Science and Engineering A*, 2001, 319–321, 849–853.

54. Abdelsayed, V.; El-Shall, M. S.; Seto, T. Differential mobility analysis of nanoparticles generated by laser vaporization and controlled condensation (LVCC). *Journal of Nanoparticle Research*, 2006, *8* (3.4), 361–369.

55. Abdelsayed, V.; Saoud, K. M.; El-Shall, M. S. Vapor phase synthesis and characterization of bimetallic alloy and supported nanoparticle catalysts. *Journal of Nanoparticle Research*, 2006, *8* (3–4), 519–531.

56. Yang, G. W. Laser ablation in liquids: Applications in the synthesis of nanocrystals. *Progress in Materials Science*, 2007, *52* (4), 648–698.

57. Patil, P. P.; Phase, D. M.; Kulkarni, S. A.; Ghaisas, S. V.; Kulkarni, S. K.; Kanetkar, S. M.; Ogale, S. B.; Bhide, V. G. Pulsed-laser-induced reactive quenching at a liquid–solid interface: aqueous oxidation of iron. *Physical Review Letters*, 1987, *58* (3), 238–241.

58. Fojtik, A.; Henglein, A. Laser vaporization of films and suspended particles in a solvent: Formation of cluster and colloid solutions. *Berichte der Bunsen-Gesellschaft fur Physikalische Chemie*, 1993, *97*, 252–254.

59. Sibbald, M. S.; Chumanov, G.; Cotton, M. Reduction of cytochrome *c* by halide-modified, laser-ablated silver colloids. *Journal of Physical Chemistry*, 1996, *100* (11), 4672–4678.

60. Mahfouz, R.; Cadete Santos Aires, F. J.; Brenier, A.; Ehret, E.; Roumie, M.; Nsouli, B.; Jacquier, B.; Bertolini, J. C. Elaboration and characterization of bimetallic nanoparticles obtained by laser ablation of $Ni_{75}Pd_{25}$ and $Au_{75}Ag_{25}$ targets in water. *Journal of Nanoparticle Research*, 2010, *12* (8), 3123–3136.

61. Compagnini, G.; Messina, E.; Puglisi, O.; Nicolosi, V. Laser synthesis of Au/Ag colloidal nano-alloys: Optical properties, structure and composition. *Applied Surface Science*, 2007, *254*, 1007–1011.

62. Besner, S.; Meunier, M. Femtosecond laser synthesis of AuAg nanoalloys: Photoinduced oxidation and ions release. *Journal of Physical Chemistry C*, 2010, *114* (23), 10403–10409.

63. Murafa, N.; Krenek, T.; Pola, J.; Subrt, J.; Bezdicka, P. Formation of Sn/Ag/C nanoalloy through laser ablation of Ag target and simultaneous decomposition of TMT in the gas phase. *Microscopy and Microanalysis*, 2010, *16* (Suppl. 2), 1254–1255.

64. El-Gendy, A. A.; Khavrus, V. O.; Hampel, S.; Leonhardt, A.; Buchner, B.; Klingeler, R. Morphology, structural control, and magnetic properties of carbon-coated nanoscaled Ni-Ru alloys. *Journal of Physical Chemistry C*, 2010, *114* (24), 10745–10749.

65. Bahlawane, N.; Premkumar, P. A.; Tian, Z.; Hong, X.; Qi, F.; Kohse-Hoinghaus, K. Nickel and nickel-based nanoalloy thin films from alcohol-assisted chemical vapor deposition. *Chemistry of Materials*, 2010, *22* (1), 92–100.

66. Finke, A.; Poizot, P.; Guéry, C.; Tarascon, J. M. Characterization and Li reactivity of electrodeposited copper-tin nanoalloys prepared under spontaneous current oscillations. *Journal of the Electrochemical Society*, 2005, *152* (12), A2364–A2368.

67. Juskenas, R.; Mockus, Z.; Kanapeckaite, S.; Stalnionis, G.; Survila, A. XRD studies of the phase composition of the electrodeposited copper-rich Cu-Sn alloys. *Electrochimica Acta*, 2006, *52* (3), 928–935.

68. Weihua, P.; Xiangming, H.; Jianguo, R.; Chunrong, W.; Changyin, J. Electrodeposition of Sn–Cu alloy anodes for lithium batteries. *Electrochimica Acta*, 2005, *50* (20), 4140–4145.

69. Budi, S.; Daud, A. R.; Radiman, S.; Umar, A. A. Effective electrodeposition of Co-Ni-Cu alloys nanoparticles in the presence of alkyl polyglucoside surfactant. *Applied Surface Science*, 2010, *257* (3), 1027–1033.

70. Talapatra, S.; Tang, X.; Padi, M.; Kim, T.; Vajtai, R.; Sastry, G. V. S.; Shima, M.; Deevi, S. C.; Ajayan, P. M. Synthesis and characterization of cobalt-nickel alloy nanowires. *Journal of Materials Science*, 2009, *44* (9), 2271–2275.

71. Rice, C.; Ha, S. R.; Masel, I.; Wieckowski, A. Catalysts for direct formic acid fuel cells. *Journal of Power Sources*, 2003, *115*, 229–235.

72. Blair, S.; Lycke, D.; Iordache, C. A. Palladium-platinum alloy anode catalysts for direct formic acid fuels. *ECS Transactions*, 2006, *3*, 1325–1332.

73. Li, X. G.; Hsing, I. M. Electrooxidation of formic acid on carbon supported Pt_xPd_{1-x} ($x = 0$–1) nanocatalysts. *Electrochimica Acta*, 2006, *51* (17), 3477–3483.

74. Liu, B.; Li, H. Y.; Die, L.; Zhang, X. H.; Fan, Z.; Chen, J. H. Carbon nanotubes supported PtPd hollow nanospheres for formic acid electrooxidation. *Journal of Power Sources*, 2009, *186*, 62–66.

75. Leiva, E.; Iwasita, T.; Herrero, E.; Feliu, J. M. Effect of adatoms in the electrocatalysis of HCOOH oxidation. A theoretical model. *Langmuir*, 1997, *13*, 6287–6293.

76. Demirci, U. B. Theoretical means for searching bimetallic alloys as anode electrocatalysts for direct liquid-feed fuel cells. *Journal of Power Sources*, 2007, *173*, 11–18.

77. Zhang, H.-X.; Wang, C.; Wang, J.-Y.; Zhai, J.-J.; Cai, W.-B. Carbon-supported Pd–Pt nanoalloy with low Pt content and superior catalysis for formic acid electro-oxidation. *Journal of Physical Chemistry C*, 2010, *114*, 6446–6451.

78. Sarkar, A.; Murugan, A. V.; Manthiram, A. Low cost Pd–W nanoalloy electrocatalysts for oxygen reduction reaction in fuel cells. *Journal of Materials Chemistry*, 2009, *19*, 159–165.

79. Nazir, R.; Mazhar, M.; Akhtar, M. J.; Shah, M. R.; Khan, N. A.; Nadeem, M.; Siddique, M.; Mehmood, M.; Butt, N. M. Superparamagnetic bimetallic iron-palladium nanoalloy: synthesis and characterization. *Nanotechnology*, 2008, *19* (18), 185608/1–185608/6.

80. Zhao, B.; Chou, C. J.; Chen Y.-W. Hydrogenation of p-chloronitrobenzene on Mo, La, Fe, and W-modified NiCoB nanoalloy catalysts. *Journal of Non-Crystalline Solids*, 2010, *356* (18–19), 839–847.

81. Zhao, B.; Chou, C. J.; Chen Y.-W. Hydrogenation of p-chloronitrobenzene on tungsten-modified NiCoB catalyst. *Industrial and Engineering Chemistry Research*, 2010, *49*, 1669–1676.

82. Dai, W.-L.; Qiao, M.-H.; Deng, J.-F. XPS studies on a novel amorphous Ni–Co–W–B alloy powder. *Applied Surface Science*, 1997, *120* (1–2), 119–124.

83. Chen, Y.-W.; Sasirekha, N. Preparation of NiFeB nanoalloy catalysts and their applications in liquid-phase hydrogenation of p-chloronitrobenzene. *Industrial and Engineering Chemistry Research*, 2009, *48*, 6248–6255.

84. Morales-Luckie, R. A.; Sanchez-Mendieta, V.; Arenas-Alatorre, J. A.; Lopez-Castanares, R.; Perez-Mazariego, J. L.; Marquina-Fabrega, V.; Gomez, R. W. One-step aqueous synthesis of stoichiometric Fe-Cu nanoalloy. *Materials Letters*, 2008, *62* (26), 4195–4197.

85. Ai, F.; Yao, A.; Huang, W.; Wang, D.; Zhang, X. Synthesis of PVP-protected NiPd nanoalloys by modified polyol process and their magnetic properties. *Physica E: Low-Dimensional Systems and Nanostructures*, 2010, *42* (5), 1281–1286.

86. Liu, H.; Manthiram. A. Controlled synthesis and characterization of carbon-supported Pd_4Co nanoalloy electrocatalysts for oxygen reduction reaction in fuel cells. *Energy and Environmental Science*, 2009, *2*, 124–132.

87. Zhao, J.; Sarkar, A.; Manthiram, A. Synthesis and characterization of Pd-Ni nanoalloy electrocatalysts for oxygen reduction reaction in fuel cells. *Electrochimica Acta*, 2010, *55*, 1756–1765.

88. Salavati-Niasari, M.; Bazarganipour, M.; Davar, F.; Fazl, A. A. Simple routes to synthesis and characterization of nanosized tin telluride compounds. *Applied Surface Science*, 2010, *257* (3), 781–785.

89. Kim, H.; Cho, J. Synthesis and morphological, electrochemical characterization of $Sn_{92}Co_8$ nanoalloys for anode materials in Li secondary batteries. *Journal of the Electrochemical Society*, 2007, *154* (5), A462–A466.

90. Wen, M.; Zhu, Y.-Z.; Wu, Q.-S.; Zhang, F.; Zhang, T. Composition-dependent assembly and magnetic specificity of $(Fe_{1-x}Ni_x)_{0.5}Pt_{0.5}$ amorphous nanothreads through substitution of Ni for Fe in an FePt system. *Journal of Physical Chemistry C*, 2009, *113* (46), 19883–19890.

91. Radillo-Diaz, A.; Coronado, Y.; Perez, L. A.; Garzon, I. L. Structural and electronic properties of PtPd and PtNi nanoalloys. *European Physical Journal D: Atomic, Molecular, Optical and Plasma Physics*, 2009, *52* (1–3), 127–130.

92. Liu, I. H.; Chang, C. Y.; Liu, S. C.; Chang, I. C.; Shih, S. M. Absorption removal of sulfur dioxide by falling water droplets in the presence of inert solid particles. *Atmospheric Environment*, 1994, *28*, 3409–3415.

93. Shen, J.; Hu, Z.; Zhang, Q.; Zhang, L.; Chen, Y. Investigation of Ni-P-B ultrafine amorphous alloy particles produced by chemical reduction. *Journal of Applied Physics*, 1992, *71*, 5217–5221.

94. Liu, Y. C.; Chen, Y. W. Hydrogenation of *p*-chloronitrobenzene on lanthanum-promoted NiB nanometal catalysts. *Industrial and Engineering Chemistry Research*, 2006, *45*, 2973–2980.

95. Lin, M.-H.; Zhao, B.; Chen, Y.-W. Hydrogenation of *p*-chloronitrobenzene over Mo-modified NiCoB nanoalloy catalysts: Effect of Mo content. *Industrial and Engineering Chemistry Research*, 2009, *48*, 7037–7043.

96. Singh, G.; Kapoor, I. P. S.; Dubey, S. Bimetallic nanoalloys: Preparation, characterization and their catalytic activity. *Journal of Alloys and Compounds*, 2009, *480* (2), 270–274.

97. Engels, V.; Benaskar, F.; Patil, N.; Rebrov, E. V.; Hessel, V.; Hulshof, L. A.; Jefferson, D. A. et al. Cu-based nanoalloys in the base-free Ullmann heterocyle-aryl ether synthesis. *Organic Process Research and Development*, 2010, *14*, 644–649.

98. Xu, J.; White, T.; Li, P.; He, C.; Yu, J.; Yuan, W.; Han, Y.-F. Biphasic Pd-Au alloy catalyst for low-temperature CO oxidation. *Journal of the American Chemical Society*, 2010, *132* (30), 10398–10406.

99. Liu, H.; Li, W.; Manthiram, A. Factors influencing the electrocatalytic activity of $Pd_{100-x}Co_x$ ($0 \leq x \leq 50$) nanoalloys for oxygen reduction reaction in fuel cells. *Applied Catalysis, B: Environmental*, 2009, *90* (1–2), 184–194.

100. Aradi, A. A.; Esche, C. K., Jr.; McIntosh, K.; Jao, T.-C. Nanoalloy fuel additives. European Patent Application EP 1889895, 2008, 23 pp.

101. Kwon, Y. J.; Cho, J. High capacity carbon-coated $Si_{70}Sn_{30}$ nanoalloys for lithium battery anode material. *Chemical Communications*, 2008, *9*, 1109–1111.

102. Tehrani, R. M. A.; Ab Ghani, S. Voltammetric analysis of uric acid by zinc-nickel nanoalloy coated composite graphite. *Sensors and Actuators, B: Chemical*, 2010, *B145* (1), 20–24.

103. Trujillo-Reyes, J.; Sanchez-Mendieta, V.; Colin-Cruz, A.; Morales-Luckie, R. A. Removal of indigo blue in aqueous solution using Fe/Cu nanoparticles and C/Fe-Cu nanoalloy composites. *Water, Air, and Soil Pollution*, 2010, *207* (1–4), 307–317.

26 Nanostructured Forms of Bismuth

Metallic bismuth is an important element with many distinct industrial applications as a component of low-melting alloys, catalysts, and for production of polonium in nuclear reactors and tetrafluorohydrazine among others.[1,2] High-purity metal is used for measuring super-strong magnetic fields. Bismuth in nanostructured forms has been mentioned in some recent monographs,[3-7] reviews,[8-10] patents,[11,12] and a lot of experimental articles. This section is devoted to highlights in obtaining its distinct nanoforms by a series of classic and novel techniques.

26.1 SYNTHESIS OF VARIOUS BISMUTH NANOFORMS

Bismuth nanoparticles, nanopowders, nanowires, nanofilms, and other nanostructured forms have been produced by a host of methods; some representative examples are shown in Table 26.1. The techniques examined for obtaining its nanoforms include chemical reduction of bismuth salts, electrochemical deposition, laser ablation, ultrasonic and microwave treatment, and electron- and ion-beam methods among other methods. As a result of reactions from the same precursors, bismuth can be obtained in various forms depending on reaction conditions, which may influence considerably the form and particle size in the resulting phase. Metallic bulk bismuth, its salts, or complexes usually serve as precursors in these reactions.

In 2000–2008, among other Bi nanoforms, a considerable part of experimental reports is devoted to bismuth *nanoparticles*. Thus, Kim et al.[29] applied *electron-beam* technique in the TEM for obtaining nanometer-sized bismuth particles. It was shown that the size of the crystalline Bi nanoparticles could be controlled by adjusting the irradiation time of the electron beam and the Bi nanoparticles exist in rhombic structure, similar to bulk Bi. Later, Jose-Yacaman et al.[17] reported similar results, preparing bismuth nanoparticles from sodium bismuthate as a precursor. A KrF (248 nm) excimer laser (200 mJ power) was used for preparing 3–10 nm Bi particles in Ar atmosphere by *laser ablation*.[30] Among other physical methods, used for the production of bismuth nanoparticles, we note *cluster beam deposition,* an important tool for nanoscale science and technology.[31]

Vapor-phase techniques are well-established methods for nanoparticle production[32-34] and have been applied to the preparation of bismuth nanoparticles.[35] Bismuth nanoparticles of average diameter of 12–37 nm were made by controlling the quenching gas flow rate, carrier gas flow rate, and process pressure. The modified *flame spray process* (the traditional method has been established as a state-of-the-art method for the production of high-purity mixed metal oxides) allowed the cost-efficient and scalable production of metallic bismuth nanoparticles (70 g/h) of over 98% purity.[36,37] The equipment[37] designed for reducing flame spray synthesis differs from the equipment for conventional flame spray synthesis by the presence of a porous wall around the flame; inert gas is introduced through its holes, preventing further oxidation of the formed metal (Figure 26.1). Bi *nanopowder* (spherical in shape with a size of nearly 50 nm) was synthesized by *levitational gas condensation* method.[38] Any other phase (an oxide or an impurity) is absent in the product (rhombohedral structure of Bi). On the basis of this nanopowder, a more sensitive and conveniently usable electrode sensor for a trace analysis of heavy metals (Cd and Pb) was developed.

γ- or *accelerated electron irradiation* was used for the first time for obtaining bismuth nanoparticles starting from bismuth perchlorate $BiOClO_4$ in an aqueous solution with or without stabilizing

TABLE 26.1
Examples of Production of Distinct Bismuth Nanoforms

Precursor	Conditions/Techniques	Properties of Formed Nanometal and Observations	References
Bismuth powder	Microwave treatment. The samples were MW warmed up in vacuum during 5, 10, and 15 min in a conventional microwave oven of 2.45 GHz and 800 W.	Bismuth *nanotubes* with diameters about 100 nm are formed.	[13]
Bismuth granules	Bi granules are added to paraffin oil, and the system is sealed and stirred above 280°C for 10 h.	Spherical shape of the *nanoparticles* with an average diameter of about 50 nm; the product includes bismuth oxide phase too. Good oil solubility of the nanoparticles.	[14]
Molten bismuth	Molten Bi was injected into the anodic aluminum oxide template using a hydraulic pressure method.	The Bi *nanowires* are dense and continuous with uniform diameter throughout the length. The majority of the individual nanowires were single crystalline, with preferred orientation of growth along the [011] zone axis of the pseudo-cubic structure.	[15,16]
Sodium bismuthate	*In situ* electron-beam irradiation in a TEM.	A rhombohedral structure; diameter of 6 nm. *Nanoparticle* size can be controlled by the irradiation time.	[17]
	Reduction of sodium bismuthate with ethylene glycol in the presence of poly(vinylpyrrolidone) or acetone.	Single-crystalline bismuth *nanowires* and *nanospheres* are formed; bismuth *nanobelts* and Bi/Bi$_2$O$_3$ *nanocables* could also be obtained by changing some reaction parameters.	[18]
BiCl$_3$	Reduction of BiCl$_3$ with *t*-BuONa and sodium hydride at 65°C.	Large quantities of colloidal Bi(0) *nanoparticles* with a diameter in the range of 1.8–3.0 nm are formed. Nanocomposite polymers containing bismuth nanoparticles (2 wt.%) have been obtained by photopolymerization of acrylic resins using the same reduction technique. *In situ* *t*-BuONa stabilization protects the metallic particles against aggregation. TEM analysis has shown that the bismuth nanoparticles are well dispersed in the acrylic resin.	[19,20]
	Reduction of BiCl$_3$ with KBH$_4$ at room temperature.	Average diameter of about 30–80 nm.	[21]
	Reduction with Fe powder, 150°C, 12 h.	Dandelion-shape *microspheres* of ordered bismuth *nanowires*.	[22]
[Bi(NO$_3$)$_3$·5H$_2$O], ethylene diamine	Solvothermal method.	The diameter of the single-crystalline metallic bismuth *nanowires* is about 20–30 nm and lengths range from 0.2 to 2.5 μm.	[23]
Bi$_2$O$_3$, ethylene glycol		The diameter of the bismuth *nanotubes* is about 3–6 nm and length is up to 500 nm.	[24]

TABLE 26.1 (continued)
Examples of Production of Distinct Bismuth Nanoforms

Precursor	Conditions/Techniques	Properties of Formed Nanometal and Observations	References
Bismuth perchlorate	Radiolytic reduction of aqueous solution of the Bi salt with or without polymers (polyacrylic acid [PAA], polyacrylamide carboxyl [PAA-70], polyethyleneimine [PEI] in *i*-propanol).	The polymers are used for stabilization of formed *nanoparticles*.	[25]
[Ni(COD)$_2$] and [Bi$_2$Ph$_4$]	The simultaneous decomposition of bis(cyclooctadiene)nickel (0) [Ni(COD)$_2$] and tetraphenyldibismuthine [Bi$_2$Ph$_4$] in tetrahydrofuran (THF) at reflux temperature.	Nickel–bismuth alloy *nanoparticles* with an average size of 8–10 nm are formed, adopting the hexagonal structure of β-NiBi. A superparamagnetic behavior is observed at temperatures above 45 K and the presence of antiferromagnetic dipolar interactions between the particles. When the decomposition is carried out at a higher temperature (200°C) in the presence of trioctylamine and oleic acid, NiBi *nanowires* were observed in addition to the nanoparticles. A *hydrothermal reduction* method at 150°C leads to pure Bi *nanorods* with diameters of about 50 nm and superconducting NiBi *particles*.	[26,27]
Layered Bi(SC$_{12}$H$_{25}$)$_3$ with a 31.49 Å spacing.	Structure-controlling solventless thermolysis.	Bismuth *nanofilms* with an average thickness of 0.6 nm and monodisperse layered Bi *nanorhombuses* with an average edge length of 21.5 nm and thickness of 0.9 nm are formed.	[28]

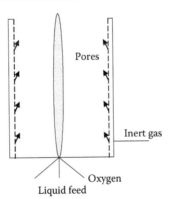

FIGURE 26.1 Equipment for reducing flame spray synthesis. (From Grass, R.N. http://aiche.confex.com/aiche/2006/techprogram/P57873.HTM, 2006. Reproduced from American Institute of Chemical Engineers.)

polymers in quartz ampoules.[25] The maximum irradiation level applied was 200 krad/h for γ-rays and 3000 Mrad/h for accelerated electron irradiation. In general, the formed free radicals and solvated electron take part in reduction of the metal ions; reduction steps include a decrease in metal oxidation number (Reaction 26.1):

$$M^{n+} + (e_{aq}^-, H^{\bullet}) \rightarrow M^{(n-1)+} \tag{26.1}$$

up to the formation of metal atoms (Reaction 26.2):

$$M^+ + (e_{aq}^-, H^\cdot) \rightarrow M \tag{26.2}$$

and their further aggregation into nanoparticles (Reactions 26.3 through 26.5):

$$M + M \rightarrow M_2 \tag{26.3}$$

$$M_2 + M \rightarrow M_3 \text{ etc.} \tag{26.4}$$

$$M_m^{p+} + M_n^{q+} \rightarrow M_{m+n}^{(p+q)+} \tag{26.5}$$

A detailed mechanism for Bi nanoparticle (5 nm) formation is proposed[25] based on the formation at different pHs of distinct metal hydrolyzed forms, such as $Bi_9(OH)_{22}^{5+}$, its reduction, and further agglomeration (Reaction 26.6):

$$Bi_9(OH)_{22}^{5+} + e_{aq}^- \rightarrow nBi^0 \rightarrow Bi_n \tag{26.6}$$

Microwave treatment of bulk bismuth in air in a domestic MW oven (power 800 W and frequency 2.45 GHz) leads to the formation of bismuth nanoparticles (60–70 nm) with Bi_2O_3 impurities,[39] in contrast with a similar treatment in vacuum,[13] when Bi nanotubes forms. The optimal process time is 60 min; the process is highly reproducible and easy. Figure 26.2 shows the 3D image of bismuth nanoparticle agglomerates. Differences between the products, obtained at distinct heating times, are presented in Figure 26.3.

Wet-chemistry methods are widely used to produce Bi nanoparticles and other forms from its various salts. Thus, $BiCl_3$ has been reported as a precursor in a series of publications.[19–22] Its simple *chemical reduction* with, for instance, sodium borohydride ($NaBH_4$) in the presence of poly(vinylpyrrolidone) (PVP) at room temperature in DMF leads to semimetal bismuth nanoparticles (5–500 nm),[40] studied by electron energy loss spectroscopy.[41] The strong interaction observed between the carboxyl oxygen (C=O) of PVP and Bi^{3+} ion and a weak interaction between the carboxyl oxygen (C=O) of PVP and the Bi atom in nanoparticles indicates that PVP serves as an effective capping ligand, which prevents the nanoparticles from aggregation. A similar contribution of polymeric network was studied, where an *inverse microemulsion* method was found to help in obtaining very highly crystalline air-sensitive bismuth particles (nanocrystallites) on the order

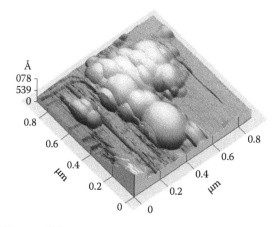

FIGURE 26.2   3D AFM image of bismuth nanoparticle agglomerates.

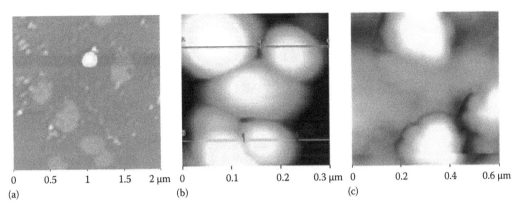

| 0 | 0.5 | 1 | 1.5 | 2 µm | 0 | 0.1 | 0.2 | 0.3 µm | 0 | 0.2 | 0.4 | 0.6 µm |

(a) (b) (c)

FIGURE 26.3 AFM images of the bismuth nanoparticles obtained[39] by MW heating for (a) 60 min, (b) 75 min, and (c) 90 min.

of 20 nm. To prevent air oxidation, an *in situ* polymerization technique using methyl methacrylate (monomer) and 2-hydroxyethyl methacrylate (co-monomer) with cross-link agent,[42] as well as PVP,[43] was employed. It was shown that the polymeric network protected the bismuth particles against oxidation, especially during postsynthesis annealing.[41] Among other classic bismuth salts used, we note $Bi(NO_3)_3 \cdot 5H_2O$.[23,44]

In addition to Bi nanoparticles, its *nanowires* and *nanorods* of bismuth have been given major attention in comparison with other bismuth nanoforms and have been obtained and characterized by a series of modern methods. Thus, a *pulsed electrodeposition technique* has been frequently applied for the production of single- and poly-crystalline Bi nanowire arrays with the diameters from 10 to 250 nm, fabricated within the porous anodic alumina membranes. Different temperature dependencies of lattice parameter (measured using the *in situ* high-temperature XRD method), thermal expansion coefficient, and other important characteristics were found for Bi nanowires having distinct diameters.[45–50] The diameters of Bi nanowires increase with increasing the pulse on–time at a constant pulse off–time.[51] The orientations of nanowires also depend on the electrochemical deposition parameters. The electrodeposition method was also used for the preparation of multi-metal In–Pb–Bi[52] or Sb–Bi[53] alloy nanowires from solutions containing different concentrations of various metal ions. Such synthesis of compositionally encoded nanowire tags is substantially faster and simpler than the preparation of striped nanowires based on sequential plating steps from different metal solutions and leads to high identification accuracy.[52]

Single bismuth nanowires with cross-sectional dimensions of 40×30 and $40 \times 50 \, nm^2$ were fabricated by the low-energy *electron-beam lithography* using the silver/silicon nanowire shadow masks. Their electrical conductivity was measured; its semiconductor-like temperature dependence was found, which is strikingly different from that of the bulk bismuth.[54] The synthesis of ~10 nm diameter Bi nanorods using a *pulsed laser vaporization* method is reported by Reppert et al.[55] HR TEM studies showed a crystalline Bi core oriented along <012> direction and coated with a thin amorphous Bi_2O_3 layer.

An intriguing technique to obtain metal nanowires is reported in Ref. [56]. Bismuth nanowires with diameters ranging from 30 to 200 nm and lengths up to several millimeters were extruded spontaneously at the rate of a few micrometers per second at room temperature from the surfaces of freshly grown composite thin films consisting of Bi and chrome–nitride. The *high compressive stress* in these composite thin films is the driving force responsible for the nanowire formation. In addition, bismuth nanowire arrays have been synthesized within the pores of ordered mesoporous silica templates using another rarely used method for nanowire production—*supercritical fluid inclusion* technique.[57] It was established that the formation of the bismuth nanowire arrays occurred through the initial binding of the bismuth precursor to the inner walls of the mesoporous channels, forming bismuth crystal seeds, which subsequently developed into wire-like

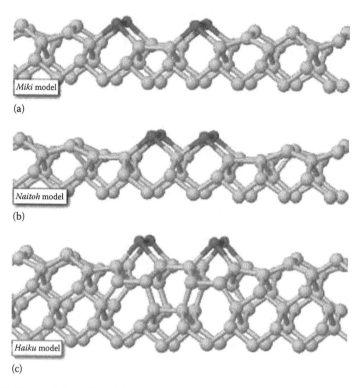

(a)

(b)

(c)

FIGURE 26.4 Ball-and-stick models of the three proposed models for the Bi nanoline: (a) The *Miki* model, (b) the *Naitoh* model, and (c) the *Haiku* model. (Reproduced with permission from Owen, J.H.G. et al., *J. Mater. Sci.*, 41(14), 4568–4603, 2006.)

structures. The highest loading of bismuth nanocrystals inside the mesopores was obtained at temperatures near the critical point of toluene.

Bismuth *nanolines* are examined in detail in a series of recent publications, listed in Ref. [58] among which we note an excellent review of Owen et al.[59] Bismuth nanolines are 1D features, which are very long (>500 nm), perfectly straight, and only 1.5 nm wide. The Bi nanolines form by competitive adsorption and evaporation. If Bi is deposited onto the surface at around 820–870 K, small islands of Bi form. At the end of the deposition time, the islands start to evaporate over the course of 10–15 min. However, not all the Bi evaporates, some of it remains on the surface and forms the nanolines. Proposed models for Bi nanolines are shown in Figure 26.4. It is necessary to mention that the *Haiku* model is the only one that fits all the experimental data and it is far and away the lowest energy of the three in DFT. This model explains that one reason for the Bi dimer stability is that the bond angles are close to 90°. This means that the orbitals can dehybridize—the bonding orbitals are p^3 configuration, while the lone pair is s-type. This also accounts for the extreme inertness of the Bi dimers to atomic H, molecular O_2, O_3, NH_3, and other species.[59] A cross-sectional perspective view of the bismuth nanoline (of the *Haiku* model) is given in Figure 26.5, Bi-nanoline templating—in Figure 26.6, different surfaces due to distinct growth mechanisms of Bi nanolines—in Figures 26.7 and 26.8.

Bismuth nanolines in the Bi:Si(001) nanoline system grow perfectly straight along the <110> directions on the Si(001) surface for hundreds of nanometers, apparently limited only by the terrace size of the underlying substrate. When they encounter a step edge, they will either grow out over lower terraces in a long, narrow promontory, or they will burrow into higher terraces, until a deep inlet is formed. They have a constant width, 1.5 nm or four substrate dimers, and are very stable. As long as the temperature is not so high that the Bi can evaporate, they are stable against prolonged annealing, maintaining the same width. Structurally, the Bi nanoline is unique amongst

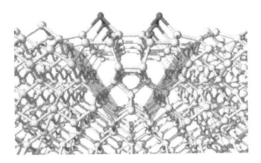

FIGURE 26.5 A cross-sectional perspective view of the Bi nanoline. The nanoline is 1.5 nm wide, occupying the space of four surface dimers, and is built around a pair of Bi dimers, on top of a complex Si substructure. The shaded (111) planes define the interface between the nanoline core and the silicon substrate. (Reproduced with permission from Owen, J.H.G. et al., *J. Mater. Sci.*, 41(14), 4568–4603, 2006.)

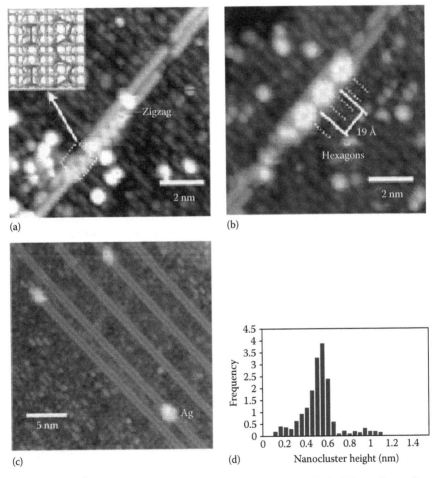

FIGURE 26.6 Examples of Bi-nanoline templating. (a) Indium reacts with the Bi nanoline to form a zigzag feature. (b) A two-layer In island shows hexagonal features, which are 19 Å or 2.5 dimer rows wide, revealing an epitaxial relationship with the Bi nanoline. (c) Isolated 0.5 nm Ag nanoclusters form after deposition of Ag atoms at room temperature onto the Bi nanoline surface. (d) Height analysis reveals a strongly peaked size distribution, with the peak around 0.5 nm, which may correspond to a magic cluster of 13 atoms. (Reproduced with permission from Owen, J.H.G. et al., *J. Mater. Sci.*, 41(14), 4568–4603, 2006.)

(a)   (b)

(c)   (d)

FIGURE 26.7 Different growth procedures for the formation of Bi nanolines, and the resulting surfaces: (a) deposition of 1 ML Bi at a lower temperature, followed by a 20 min anneal at 520°C; (b) deposition of 1 ML Bi at around 550°C, followed by annealing at the same temperature; (c) continuous exposure of 0.1 ML/min Bi flux for 40 min at 570°C; and (d) as (c), but on a 0.5° miscut sample. (Reproduced with permission from Owen, J.H.G. et al., *J. Mater. Sci.*, 41(14), 4568–4603, 2006.)

FIGURE 26.8 Left to right: Bi nanolines growing in both directions, one domain dominates on vicinal surfaces, and defect exclusion zones—patches clear of dark defects—indicate a repulsion between Bi nanoline and missing dimer defects. (Reproduced from Owen, J.H.G., http://homepage.mac.com/jhgowen/research/Bismuth/Biline.html, 2008.)

nanoline systems—it is neither a periodic reconstruction of the surface, nor the result of anisotropically strained heteroepitaxial growth of one bulk structure on another. In fact, it is somewhere in
between. The nanoline structure is built around a pair of Bi ad-dimers, on top of a complex subsurface reconstruction which is responsible for many of the nanoline's remarkable properties, such as
the extreme straightness.[59] Metal deposition onto the Bi nanolines is currently under investigation.
Examples of each type of behavior for different metals, In and Ag,[59] are shown in Figure 26.6. For
the In-Bi chain structure,[60–62] the In is almost planar, that is, sp^2, while the Bi is again almost 90°,
that is, p^3.

Nanotubes and other bismuth nanoforms have also been synthesized and characterized. Thus,
highly oriented and single-crystalline bismuth nanotubes, which may find applications in thermoelectric nanodevices, were produced by *pulsed electrodeposition*.[63] It was shown that there is a
metal–semiconductor transition of Bi nanotube arrays with the decrease of the wall thickness of the
nanotubes, and this transition depends only on the wall thickness and is independent of the diameter
of the nanotubes. Synthesis of Bi nanotubes by *microwave heating* of bismuth powder in vacuum is
reported in Ref. [13]. It is proposed that this method as an easier and less-expensive way to prepare
Bi nanotubes in comparison with traditional techniques. The obtained product is characterized by
TEM and SEM techniques (Figure 26.9). We note that Bi nanoparticles form by MW heating in air
(see earlier section).[39] Low-temperature-controlled *hydrothermal reduction* method for production
of bismuth nanotubes is reported in Ref. [64] and a 3D model for formed bismuth nanotubes was
proposed. Reaction 26.7 takes place in a Teflon-lined stainless steel autoclave at 120°C for 12h as
follows:

$$4Bi(NO_3)_3 + 3N_2H_4 + 12NH_3 \cdot H_2O \rightarrow 4Bi + 3N_2 + 12H_2O + 12NH_4NO_3 \qquad (26.7)$$

Single-crystalline bismuth *nanobelts* have been synthesized successfully by a solution-phase route,
using ethylene glycol as a reductant and solvent.[65] The authors proposed that the final morphology
resulted from the layered structure of rhombohedral bismuth. Monodispersed bismuth was also successfully prepared as triangular *nanoplates via* a simple thermal process.[66] Refluxing provided the

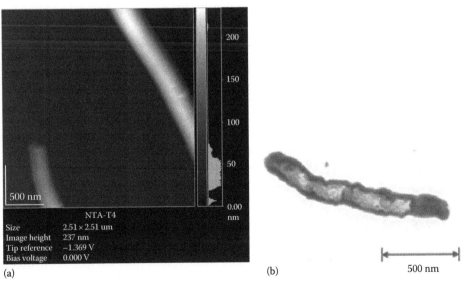

FIGURE 26.9 (a) AFM image of Bi nanotubes obtained by MWCVD for 15min and (b) low-magnification
TEM images of Bi nanotubes obtained by microwave heating for 15min.

driving force to facilitate the Ostwald ripening process growth of the plate-like materials. All the nanoplates were highly oriented single crystals with the (001) planes as the basal planes. Bismuth *nanoclusters* of 40–60 nm diameter, which can serve as precursors/building blocks for Bi nanowires, were studied by HR-TEM.[67] It was established that β-Bi$_2$O$_3$ is the dominant oxide phase covering hexagonal bismuth nanoclusters produced in an inert gas aggregation source (the oxide is 20 ± 5 Å thick on average, at the surface of 320 ± 40 Å diameter clusters).[68] Bismuth *nanodot* structures (100 nm in diameter)[69] have been fabricated using the proximity effects of *electron-beam* writing technique on an oxidized silicon substrate with a 200 nm thick silicon oxide layer.

Bismuth *nanofilms* have also been produced and studied. Thus, the successful growth of Bi(111) epitaxial films by the molecular beam epitaxy method on Si(111) substrates is reported.[70] Structure and stability of the quasicubic Bi{012} film, formed in the initial stage of the bismuth deposition on the Si(111)-7 × 7 surface at room temperature, was examined by scanning tunneling microscopy (STM) investigations and *ab initio* calculations.[71] The authors showed that a paired-layer Bi{012} film grows on top of the initially formed wetting layer, with the Si 7 × 7 lattice preserved underneath. The pairing of the layers in the {012} film leads to the substantial stabilization of the film when it consists of an even number of layers, and only even-number layered Bi{012} islands are observed to be stable. The same authors reported the use of well-ordered Bi(001) films as substrates for the growth of pentacene (Pn).[72] Pn nucleates on Bi(001) into a highly ordered, crystalline layer, with Pn molecules "standing up" on the Bi surface and the (001) plane on the growth front. Moreover, the Pn layer is aligned with the Bi(001) surface having a "point-on-line" commensurate relationship with the substrate. The Pn/Bi(001) film crystallizes in a bulk-like structure directly from the first Pn layer. For these Bi(001) films, a clear evidence of large surface-state conductivity was found by performed *in situ* microscopic-four-point probe conductivity measurements.[73] Additionally, observation of temperature-driven drastic morphology evolution and surface premelting of the Bi(001) nanofilm deposited on the Si(111)-7 × 7 surface by the use of spot-profile-analyzing low-energy electron diffraction is reported.[74]

26.2 MAIN PROPERTIES OF BISMUTH NANOFORMS

Structure and physical properties of formed bismuth nanowires and its other nanoforms have been intensively studied. Thus, for the crystallites of Bi nanorods, it was found[75] to have a higher orientation proportional to the degree of confinement, meanwhile an alteration of the crystal orientation between electrodeposited and recrystallized Bi nanorods was observed. A structural difference between the as-deposited and recrystallized nanorods and that a crystallization mechanism where nucleation determines the overall rate of crystallization occurred during recrystallization were confirmed. Stabilities and electronic properties of bismuth nanotubes were studied by the first-principles molecular dynamics method. The authors[76] concluded that bismuth nanotubes are seen as potential semiconductor nanomaterials for future nanoelectronic applications. The strain energies of Bi(n, n) nanotubes follow the classical $1/D^2$ strain law. For small Bi(n, 0) nanotubes, the strain energies show a nonlinear dependence on $1/D^2$. Bismuth nanotubes are predicted to be semiconducting in both (n, n) and (n, 0) forms.[76] Single Bi crystalline nanostructure materials are expected to find potential applications in a variety of areas including design of high efficiency thermoelectric devices.[13]

The *optical absorption* of bismuth nanowires in the energy (wavenumber) range of 600–4000 cm^{-1} is reported.[77] The enhanced optical absorption in nanowires over bulk bismuth is attributed to a surface term in the matrix element, which results from the spacial gradient of the dielectric function and the large dielectric mismatch between bismuth and the surrounding alumina or air. In another research,[78] Z-scan limiting measurements of ~10 nm wide bismuth nanorods suspended in chloroform, revealed that optical limiting under 532 nm excitation stems from a strong nonlinear scattering subsequent to nonlinear absorption by suspension. Agrawal et al.[79,80] applied the density functional theory in the local density approximation to characterize bismuth nanowires Bi_n and found that the stable structures for the Bi_n wires with $n = 1–18$ form two groups: nonhelical and helical configurations. In addition to the most stable nonhelical 5-Bi pentagonal, 6-Bi hexagonal and 6-Bi triple-zigzag wire configurations, there are three more nonhelical structures, namely the nonhelical 7-Bi hexagonal, 8-Bi heptagonal and 11-Bi pentagonal cross-sectional wire configurations.

The *melting* of Bi nanoparticles was examined by nanocalorimetry technique.[81] The particles were formed by evaporating Bi onto a silicon nitride substrate, which was then heated. The particles self-assemble into truncated spherical particles. The measured melting temperatures for particles below ~7 nm in radius, however, were ~50 K above the value predicted by the homogeneous melting model. A very large melting–solidification hysteresis of Bi nanoparticles embedded in a bulk alkali germanate glass (it contains nanocrystals of elementary Bi of a few tenths of a nanometer in size and could be used in nonlinear optical applications as well as an optical thermo-sensor) is reported in Ref. [82]. Heating and cooling cycles are reproducible and show reversible transitions; upon heating above the Bi melting temperature the glass transmission increases up to 10% with respect to the initial value, which is most likely related to Bi melting. Among other similar recent investigations, melting and solidification of as-deposited and recrystallized Bi crystallites, deposited on highly oriented 002-graphite at 423 K,[83] and thin films formed by Bi nanostructures embedded in amorphous dielectric Al_2O_3[84] or semiconducting Ge,[85] were studied by various methods.

Quantum confinement and *surface-state effects* in bismuth nanowires were studied in Ref. [86]. It was shown that for nanowires of diameters below the critical diameter of around 50 nm, electronic states can be considered to be one dimensional and therefore the thermopower can be very large. Study of the Fermi surface of Bi nanowires of diameters ranging between 200 and 30 nm employing the Shubnikov–de Haas method was made; for 30 nm nanowires the Fermi surface is spherical, that the carriers have high effective mass. *Galvanomagnetic properties* of single-crystal bismuth nanowires, with diameters of 7–200 nm, embedded in an amorphous porous anodic alumina matrix, are examined by Heremans et al.[87] The results confirm the existence of the semimetal-semiconductor phase transition seen in the magnetoresistance. *Magnetoresistivity measurements* of polycrystalline bismuth microwire arrays with a range of different diameters and a bulk sample were performed[88] under applied magnetic fields between 0 and 2 T and temperatures 50–300 K. The measured resistivities were in very good agreement with those of polycrystalline bulk bismuth over the entire measured temperature range, even when the resistances had different values. Longitudinal magnetoresistance of 270 nm diameter bismuth nanowire arrays (embedded in an alumina matrix which are capped with layers of pure Bi that have low contact resistance) shows a broad maximum at intermediate fields.[89] *Electrical transport properties* of Bi nanowires (60–110 nm) are studied for a wide range of temperatures (2.0–300 K) and magnetic fields (0–5.4 T).[90] The temperature dependences of the zero-field resistivity and of the longitudinal magneto-coefficient of the Bi nanowires are sensitive to the wire diameter. For single bismuth nanowires with diameters ranging between 150 nm and 1 μm, electrical resistivity measurements were carried out[91] for each nanowire individually; the obtained results allowed assort the Bi nanowires in three different groups characterized by three distinct mean specific electrical resistivities and conclude that the resistivity is determined mainly by electron scattering at grain boundaries. A lot of additional electrical resistivity/transport and magnetic behavior studies are reported in Refs. [92–101].

26.3 MAIN APPLICATIONS OF BISMUTH NANOFORMS

Bismuth nanowires, nanoparticles, and other nanoforms have several useful applications. Thus, δ-Bi_2O_3 nanowires were successfully fabricated by the oxidation of electroplated Bi nanowires at 350°C,[102] showing a facile way to synthesize functional oxide with desired nanostructure. The authors found that partially oxidized nanowires showed core-shell structure composed of metallic Bi and δ-Bi_2O_3. In a similar research,[103] single-crystalline Bi nanowires were converted to Bi-Bi_2O_3 core-shell nanowires *via* a multistep, slow oxidation method, and then controlling their further conversion to a single-crystalline Bi_2O_3 nanotube array *via* fast oxidation. Such process will allow in future fabricate an easily oxidized metal-metal oxide nanowire and metal oxide nanotube array, which may have applications in nanoscale optics, electronics, and magnetics. In addition to Bi_2O_3 nanowires, bismuth sulfide (Bi_2S_3) nanorods (about 60 nm in diameter and 1–2 μm in length) have been successfully synthesized by a solvothermal process using bismuth particles and $Na_2S_2O_3$ as source materials.[104] It was established that bismuth nanorods with the diameter of below ~10 nm and length below ~35 nm would be extremely useful for thermoelectric devices and applications.[105]

Bismuth nanocrystals were used as crystallization seeds for solution-liquid-solid obtaining crystalline InAs, GaP, GaAs, and InP nanowires at temperatures 300°C–340°C in trioctylphosphine and trioctylphosphine oxide, and trioctylamine.[106] Heating solid bismuth nanoparticles with single-walled carbon nanotubes can produce bismuth-filled carbon nanotubes.[107] Most of the filled nanotubes are 1D nanowires with high aspect ratios. Some of the nanotubes have a second layer formed during the filling process. The use of Bi(001) films as substrates for growth of pentacene (Pn) was already mentioned earlier.[72]

As an example of a biological application is a new method for rapid detection of *Escherichia coli*, which was developed by flow injection analysis (FIA) using bismuth nanofilm modified glassy carbon electrode (BiNFE) in this paper. The BiNFE was fabricated by an electrodeposition of metallic bismuth onto a glassy carbon electrode, which showed a high sensitivity in determination of 4-nitrophenol when used in conjunction with FIA system.[108]

Nanomaterials, in particular metal nanoparticles, can serve as radiation sensors. The so-called nanoparticle-enhanced x-ray therapy (NEXT) uses nanomaterials as radiosensitizers to enhance electromagnetic radiation absorption in specific cells or tissues. The nanomaterial radiosensitizers emit Auger electrons and generate radicals in response to electromagnetic radiation, which can cause localized damage to DNA or other cellular structures such as membranes. Bismuth nanoparticles can be used as such nanomaterials.[12] A microscale Bi electrodeposition process was developed and applied to the 3D fabrication of a highly sensitive x-ray imaging sensor, so called x-ray microcalorimeter, which has an array of sensing elements with a mushroom-shaped x-ray absorber.[109]

Starting from bismuth nanoforms, a series of nanoalloys and eutectic compositions has been successfully prepared. Thus, Sn–Bi nanoalloy was prepared directly from bulk Sn–Bi alloy by a *sonochemical method*[110] or in an aluminum matrix by rapid solidification.[111] The formed nanoparticles are monodispersed and the size distribution is influenced by the ultrasonic power. It was found in both investigations that the Sn–Bi eutectic alloy nanoparticles consisted of the tetragonal phase of tin and the rhombohedral phase of bismuth. Additionally, another phase, primarily a tin-based solid solution, gets stabilized and co-exists with the equilibrium bismuth phase in some of the smaller particles.[111] The prepared alloy powder has excellent antiwear properties through tribological test results.[110] The preparation of nanograin-sized $Bi_{1-x}Co_x$ thin films containing different percentages of bismuth and cobalt by galvanostatic electrodeposition on Cu substrates in organic baths was described.[112] The prepared amorphous $Bi_{1-x}Co_x$ thin films (with a superparamagnetism-like behavior) consisted almost entirely of nanograins with sizes of 20–80 nm. After heat treatment of crystallization at 280°C for 1 h in a protecting atmosphere, Bi(110), Cu(111), $Bi_{10}Co_{16}O_{36}$(532), Co(102) and Co(110) phases were found. The magnetic and transport properties of a nanostructured Mn–Bi eutectic composition (~Mn_5Bi_{95}) produced by melt spinning and low-temperature/short time vacuum annealing were studied in Ref. [113]. A hysteretic magnetostructural transformation from low-temperature phase to high-temperature phase MnBi is confirmed at 520 K. The fact that the transition temperature is lower than that reported for bulk MnBi (633 K) is tentatively attributed to interfacial strain between MnBi and the Bi matrix. Bi–Sb single crystals with a changed Sb composition to 18 at.% have been grown by a modified Czochralski method with Sb feed.[114] Gradient single crystals with a lattice parameter gradient of $\Delta c/\check{c}/\Delta z = 0.972\%/cm$ have been obtained this way.

REFERENCES

1. Norman, N. C. *Chemistry of Arsenic, Antimony and Bismuth*, 1st edn. Springer, Berlin, Germany, 1997, 496 pp.
2. Cotton, F. A.; Wilkinson, G.; Murillo, C. A.; Bochmann, M. *Advanced Inorganic Chemistry*, 6th edn. Wiley-Interscience, New York, 1999, 1376 pp.
3. Sergeev, G. B. *Nanochemistry*, 1st edn. Elsevier Science, Amsterdam, the Netherlands, 2006, 262 pp.
4. Koch, C.; Ovid'ko, I.; Seal, S.; Veprek, S. *Structural Nanocrystalline Materials: Fundamentals and Applications*, 1st edn. Cambridge University Press, Cambridge, U.K., 2007, 364 pp.
5. Fryxell, G. E.; Cao, G. *Environmental Applications of Nanomaterials: Synthesis, Sorbents and Sensors.* Imperial College Press, London, U.K., 2007, 520 pp.
6. Asthana, R.; Kumar, A.; Dahotre, N. B. *Materials Processing and Manufacturing Science*, 1st edn. Butterworth-Heinemann, Oxford, U.K., 2005, 656 pp.
7. Sakka, S. *Handbook of Sol-Gel Science and Technology: Processing Characterization and Applications*, 1st edn. Springer, Berlin, Germany, 2004, 1980 pp.
8. Soderberg, B. C. G. Transition metals in organic synthesis: Highlights for the year 2000. *Coordination Chemistry Reviews*, 2003, *241* (1), 147–247.
9. Cao, G.; Liu, D. Template-based synthesis of nanorod, nanowire, and nanotube arrays. *Advances in Colloid and Interface Science*, 2008, *136* (1), 45–64.
10. Dresselhaus, M. S.; Lin, Y. M.; Rabin, O.; Jorio, A.; Souza Filho, A. G.; Pimenta, M. A.; Saito, R. et al. Nanowires and nanotubes. *Materials Science and Engineering C*, 2003, *23* (1), 129–140.
11. Penner, R. M.; Zach, M. P.; Favier, F. Methods for fabricating metal nanowires. U.S. Patent 7220346, 2007. http://www.freepatentsonline.com/7220346.html
12. Guo, T. Nanoparticle radiosensitizers. Patent WO2006037081, 2006.
13. Kharissova, O. V.; Osorio, M.; Garza, M. Synthesis of bismuth by microwave irradiation. *MRS Fall Meeting*, Boston, MA, November 26–30, 2007, Abstract II5, Vol. 42, p. 773.
14. Zhao, Y.; Zhang, Z.; Dang, H. A simple way to prepare bismuth nanoparticles. *Materials Letters*, 2004, *58* (5), 790–793.
15. Bisrat, Y.; Luo, Z. P.; Davis, D.; Lagoudas, D. Highly ordered uniform single-crystal Bi nanowires: Fabrication and characterization. *Nanotechnology*, 2007, *18* (39), 395601.
16. Zhang, Z.; Gekhtman, D.; Dresselhaus, M. S.; Ying, J. Y. Processing and characterization of single-crystalline ultrafine bismuth nanowires. *Chemistry of Materials*, 1999, *11* (7), 1659–1665.
17. Sepulveda-Guzman, S.; Elizondo-Villarreal, N.; Ferrer, D.; Torres-Castro, A.; Gao, X.; Zhou, J. P.; Jose-Yacaman, M. In situ formation of bismuth nanoparticles through electron-beam irradiation in a transmission electron microscope. *Nanotechnology*, 2007, *18* (33), 335604.
18. Wang, J.; Wang, X.; Peng, Q.; Li, Y. Synthesis and characterization of bismuth single-crystalline nanowires and nanospheres. *Inorganic Chemistry*, 2004, *43* (23), 7552–7556.
19. Balan, L.; Schneider, R.; Billaud, D.; Fort, Y.; Ghanbaja, J. A new synthesis of ultrafine nanometre-sized bismuth particles. *Nanotechnology*, 2004, *15* (8), 940–944.
20. Balan, L.; Burget, D. Synthesis of metal/polymer nanocomposite by UV-radiation curing. *European Polymer Journal*, 2006, *42* (12), 3180–3189.
21. Zhong, G. Q.; Zhou, H. L.; Zhang, J. R.; Jia, Y. Q. A simple method for preparation of Bi and Sb metal nanocrystalline particles. *Materials Letters*, 2005, *59* (18), 2252–2256.
22. Wang, Q.; Jiang, C.; Cao, D.; Chen, Q. Growth of dendritic bismuth microspheres by solution-phase process. *Materials Letters*, 2007, *61* (14), 3037–3040.
23. Gao, Y.; Niu, H.; Zeng, C.; Chen, Q. Preparation and characterization of single-crystalline bismuth nanowires by a low-temperature solvothermal process. *Chemical Physics Letters*, 2003, *367* (1), 141–144.
24. Liu, X.-Y.; Zeng, J.-H.; Zhang, S.-Y.; Zheng, R.-B.; Liu, X.-M.; Qian, Y.-T. Novel bismuth nanotube arrays synthesized by solvothermal method. *Chemical Physics Letters*, 2003, *374* (3), 348–352.
25. Benoit, R. Nanoparticules de bismuth: Synthese, caracterisation et nouvelles proprietes. PhD thesis. Universite D'Orleans, Orleans, France, 2005. http://crmd.cnrs-orleans.fr/theses/Rapports-pdf/th%E8se_Roland_Benoit.pdf
26. Ould-Ely, T.; Thurston, J. H.; Kumar, A.; Respaud, M.; Guo, W.; Weidenthaler, C.; Whitmire, K. H. Wet-chemistry synthesis of nickel-bismuth bimetallic nanoparticles and nanowires. *Chemistry of Materials*, 2005, *17* (18), 4750–4754.
27. Park, S.; Kang, K.; Han, W.; Vogt, T. Synthesis and characterization of Bi nanorods and superconducting NiBi particles. *Journal of Alloys and Compounds*, 2005, *400* (1), 88–91.

28. Chen, J.; Wu, L.-M.; Chen, L. Syntheses and characterizations of bismuth nanofilms and nanorhombuses by the structure-controlling solventless method. *Inorganic Chemistry*, 2007, *46* (2), 586–591.

29. Kim, S. H.; Choi, Y. S.; Kang, K.; Yang, S. I. Controlled growth of bismuth nanoparticles by electron beam irradiation in TEM. *Journal of Alloys and Compounds*, 2007, *427* (1), 330–332.

30. Onari, S.; Miura, M.; Matsuishi, K. Raman spectroscopic studies on bismuth nanoparticles prepared by laser ablation technique. *Applied Surface Science*, 2002, *197*, 615–618.

31. Wegner, K.; Piseri, P.; Vahedi Tafreshi, H.; Milani, P. Cluster beam deposition: A tool for nanoscale science and technology. *Journal of Physics D: Applied Physics*, 2006, *39* (22), R439–R459.

32. Klabunde, K. J. *Free Atoms, Clusters, and Nanoscale Particles*. Academic Press, San Diego, CA, 1994, 311 pp.

33. Klabunde, K. J.; Cardenas-Trivino, G. Chapter 6. Metal atom/vapor approaches to active metal cluster/particles. In *Active Metals. Preparation, Characterization, Applications*. Fürstner, A. (Ed.). VCH, Weinheim, Germany, 1996, pp. 237–278.

34. Swihart, M. T. Vapor-phase synthesis of nanoparticles. *Current Opinion in Colloids and Interface Science*, 2003, *8* (1), 127–133.

35. Wegner, K.; Walker, B.; Tsantilis, S.; Pratsinis, S. E. Design of metal nanoparticle synthesis by vapor flow condensation. *Chemical Engineering Science*, 2002, *57* (10), 1753–1762.

36. Grass, R. N.; Stark, W. J. Flame spray synthesis under a non-oxidizing atmosphere: Preparation of metallic bismuth nanoparticles and nanocrystalline bulk bismuth metal. *Journal of Nanoparticle Research*, 2006, *8* (5), 729–736.

37. Grass, R. N. Production of Metallic Bismuth Nanoparticles by Reducing Flame Spray Synthesis. http://aiche.confex.com/aiche/2006/techprogram/P57873.HTM, 2006.

38. Lee, G. J.; Lee, H. M.; Rhee, C. K. Bismuth nano-powder electrode for trace analysis of heavy metals using anodic stripping voltammetry. *Electrochemistry Communications*, 2007, *9* (10), 2514–2518.

39. Kharissova, O. V.; Rangel Cardenas, J. *The Microwave Heating Technique for Obtaining Bismuth Nanoparticles, in Physics, Chemistry and Application of Nanostructures*. World Scientific, Singapore, 2007, pp. 443–446.

40. Wang, Y. W.; Hong, B. H.; Kim, K. S. Size control of semimetal bismuth nanoparticles and the UV-visible and IR absorption spectra. *Journal of Physical Chemistry B*, 2005, *109* (15), 7067–7072.

41. Wang, Y. W.; Kim, J. S.; Kim, G. H.; Kim, K. S. Quantum size effects in the volume plasmon excitation of bismuth nanoparticles investigated by electron energy loss spectroscopy. *Applied Physics Letters*, 2006, *88*, 143106.

42. Fang, J.; Stokes, K. L.; Wiemann, J.; Zhou, W. Nanocrystalline bismuth synthesized via an in situ polymerization-microemulsion process. *Materials Letters*, 2000, *42* (1), 113–120.

43. Fang, J.; Stokes, K. L.; Wiemann, J. A.; Zhou, W. L.; Dai, J.; Chen, F.; O'Connor, C. J. Microemulsion-processed bismuth nanoparticles. *Materials Science and Engineering B*, 2001, *83* (1), 254–257.

44. Waki, K.; Hattori, Y. Method of producing nanoparticle. EP20030003149 20030218, 2003.

45. Li, L.; Zhang, Y.; Yang.; Y. W.; Huang, X. H.; Li, G. H.; Zhang, L. D. Diameter-depended thermal expansion properties of Bi nanowire arrays. *Applied Physics Letters*, 2005, *87*, 031912.

46. Li, L.; Zhang, Y.; Li, G.; Zhang, L. A route to fabricate single crystalline bismuth nanowire arrays with different diameters. *Chemical Physics Letters*, 2003, *378* (3), 244–249.

47. Cornelius, T. W.; Brötz, J.; Chtanko, N.; Dobrev, D.; Miehe, G.; Neumann, R.; Toimil Molares, M. E. Controlled fabrication of poly- and single-crystalline bismuth nanowires, *Nanotechnology*, 2005, *16* (5), S246–S249.

48. Toimil Molares, M. E.; Chtanko, N.; Cornelius, T. W.; Dobrev, D.; Enculescu, I.; Blick, R. H.; Neumann, R. Fabrication and contacting of single Bi nanowires. *Nanotechnology*, 2004, *15* (4), S201–S207.

49. Tian, Y. T.; Meng, G. M.; Wang, G. Z.; Phillipp, F.; Sun, S. H.; Zhang, L. D. Step-shaped bismuth nanowires with metal–semiconductor junction characteristics. *Nanotechnology*, 2006, *17* (4), 1041–1045.

50. Tian, Y.; Meng, G.; Biswas, S. K.; Ajanyan, P. M.; Sun, S. H.; Zhang, L. D. Y-branched Bi nanowires with metal–semiconductor junction behavior. *Applied Physics Letters*, 2004, *85*, 967.

51. Li, L.; Zhang, Y.; Li, G.; Wang, X.; Zhang, L. Synthetic control of large-area, ordered bismuth nanowire arrays. *Materials Letters*, 2005, *59* (10), 1223–1226.

52. Wang, J.; Liu, G. Templated one-step synthesis of compositionally encoded nanowire tags. *Analytical Chemistry*, 2006, *78* (7), 2461–2464.

53. Zhang, Y.; Li, L.; Li, G. H. Fabrication and anomalous transport properties of an Sb/Bi segment nanowire nanojunction array. *Nanotechnology*, 2005, *16* (10), 2096–2099.

54. Choi, D. S.; Balandin, A. A.; Leung, M. S.; Stupian, G. W.; Presser, N.; Chung, S. W.; Heath, J. R. et al. Transport study of a single bismuth nanowire fabricated by the silver and silicon nanowire shadow masks. *Applied Physics Letters*, 2006, *89*, 141503.

55. Reppert, J.; Rao, R.; Skove, M.; He, J.; Craps, M.; Tritt, T.; Rao, A. M. Laser-assisted synthesis and optical properties of bismuth nanorods. *Chemical Physics Letters*, 2007, *442* (4), 334–338.

56. Cheng, Y.-T.; Weiner, A. M.; Wong, C. A.; Balogh, M. P.; Lukitsch, M. J. Stress-induced growth of bismuth nanowires. *Applied Physics Letters*, 2002, *81*, 3248.

57. Xu, J.; Zhang, W.; Morris, M. A.; Holmes, J. D. The formation of ordered bismuth nanowire arrays within mesoporous silica templates. *Materials Chemistry and Physics*, 2007, *104* (1), 50–55.

58. Owen, J. H. G. Bi Nanolines: A Triumph of Serendipity. http://homepage.mac.com/jhgowen/research/Bismuth/Biline.html, 2008.

59. Owen, J. H. G.; Miki, K.; Bowler, D. R. Self-assembled nanowires on semiconductor surfaces. *Journal of Materials Science*, 2006, *41* (14), 4568–4603.

60. Bowler, D. R.; Bird, C. F.; Owen, J. H. G. 1-D semiconducting atomic chain of In and Bi on Si<001>. *Journal of Physics: Condensed Matter*, 2006, *18*, L241–L249.

61. Miki, K.; Owen, J. H. G.-D. Epitaxial growth of indium on a self-assembled atomic-scale bismuth template. *Nanotechnology*, 2006, *17*, 430–433.

62. Ferhat, M.; Zaoui, A. Structural and electronic properties of III-V bismuth compounds. *Physical Review B*, 2006, *73* (11), 115107.

63. Li, L.; Yang, Y. W.; Huang, X. H.; Li, G. H.; Ang, R.; Zhang, L. D. Fabrication and electronic transport properties of Bi nanotube arrays. *Applied Physics Letters*, 2006, *88*, 103119.

64. Li, Y.; Wang, J.; Deng, Z.; Wu, Y.; Sun, X.; Yu, D.; Yang, P. Bismuth nanotubes: A rational low-temperature synthetic route. *Journal of the American Chemical Society*, 2001, *123* (40), 9904–9905.

65. Chen, Y.; Gong, R.; Zhang, W.; Xu, X.; Fan, Y.; Liu, W. Synthesis of single-crystalline bismuth nanobelts and nanosheets. *Materials Letters*, 2005, *59* (8), 909–911.

66. Fu, R.; Xu, S.; Lu, Y.-N.; Zhu, J.-J. Synthesis and characterization of triangular bismuth nanoplates. *Crystal Growth and Design*, 2005, *5* (4), 1379–1385.

67. Stevens, K. J.; Cheong, K. S.; Knowles, D. M.; Laycock, N. J.; Ayesh, A.; Partridge, J.; Brown, S. A.; Hendy, S. C. Electron microscopy of bismuth building blocks for self-assembled nanowires. *Current Applied Physics*, 2006, *6* (3), 453–456.

68. Stevens, K. J.; Ingham, B.; Toney, M. F.; Brown, S. A.; Partridge, J.; Ayesh, A.; Natali, F. Structure of oxidized bismuth nanoclusters. *Acta Crystallographica Section B: Structural Science*, 2007, *B63*, 569–576.

69. Chiu, P. H. P.; Shih, I. Nonlinear current-voltage characteristics of bismuth nanodot structures. *Applied Physics Letters*, 2006, *88*, 072110.

70. Tanaka, A.; Hatano, M.; Takahashi, K.; Sasaki, H.; Suzuki, S.; Sato, S. Growth and angle-resolved photoemission studies of bismuth epitaxial films. *Surface Science*, 1999, *433*, 647–651.

71. Sadowski, J. T.; Nagao, T.; Yaginuma, S.; Fujikawa, Y.; Sakurai, T.; Oreshkin, A.; Saito, M.; Ohno, T. Stability of the quasicubic phase in the initial stage of the growth of bismuth films on Si(111)-7 × 7. *Journal of Applied Physics*, 2006, *99*, 014904.

72. Sadowski, J. T.; Nagao, T.; Yaginuma, S.; Fujikawa, Y.; Al-Mahboob, A.; Nakajima, K.; Sakurai, T. et al. Thin bismuth film as a template for pentacene growth. *Applied Physics Letters*, 2005, *86*, 073109.

73. Hirahara, T.; Matsuda, I.; Yamazaki, S.; Miyata, N.; Hasegawa, S.; Nagao, T. Large surface-state conductivity in ultrathin Bi films. *Applied Physics Letters*, 2007, *91*, 202106.

74. Yaginuma, S.; Nagao, T.; Sadowski, J. T.; Pucci, A.; Fujikawa, Y.; Sakurai, T. Surface pre-melting and surface flattening of Bi nanofilms on Si(111)-7 × 7. *Surface Science*, 2003, *547* (3), L877–L881.

75. Noh, K. W.; Woo, E.; Shin, K. Alteration of crystal structure of bismuth confined in cylindrical nanopores. *Chemical Physics Letters*, 2007, *444* (1), 130–134.

76. Su, C.; Liu, H.-T.; Li, J.-M. Bismuth nanotubes: Potential semiconducting nanomaterials. *Nanotechnology*, 2002, *13* (6), 746–749.

77. Black, M. R. The optical properties of bismuth nanowires. PhD thesis. Cambridge, MA, 2003, 177 pp.

78. Sivaramakrishnan, S.; Muthukumar, V. S.; Sivasankara Sai, S.; Venkataramaniah, K.; Reppert, J.; Rao, A. M.; Anija, M. et al. Nonlinear optical scattering and absorption in bismuth nanorod suspensions. *Applied Physics Letters*, 2007, *91*, 093104.

79. Agrawal, B. K.; Singh, V.; Srivastava, R.; Agrawal, S. Effect of spin–orbit interaction on the electronic and optical properties of ultrathin bismuth nanowires—A density functional approach. *Nanotechnology*, 2007, *18* (41), 415705.

80. Agrawal, B. K.; Singh, V.; Srivastava, R.; Agrawal, S. Ab initio study of the structural, electronic and optical properties of ultrathin bismuth nanowires. *Nanotechnology*, 2006, *17* (9), 2340–2349.

81. Olson, E. A.; Efremov, M. Y.; Zhang, M.; Zhang, Z.; Allen, L. H. Size-dependent melting of Bi nanoparticles. *Journal of Applied Physics*, 2005, *97*, 034304.

82. Haro-Poniatowski, E.; Jiménez de Castro, M.; Fernández Navarro, J. M.; Morhange, J. F.; Ricolleau, C. Melting and solidification of Bi nanoparticles in a germanate glass. *Nanotechnology*, 2007, *18* (31), 315703.

83. Zayed, M. K.; Elsayed-Ali, H. E. Melting and solidification study of as-deposited and recrystallized Bi thin films. *Journal of Applied Physics*, 2006, *99*, 123516.

84. Haro-Poniatowski, E.; Serna, R.; Suárez-García, A.; Afonso, C. N. Thermally driven optical switching in Bi nanostructures. *Nanotechnology*, 2005, *16* (12), 3142–3145.

85. Serna, R.; de Sande, J. C. G.; Ballesteros, J. M.; Afonso, C. N. Spectroscopic ellipsometry of composite thin films with embedded Bi nanocrystals. *Journal of Applied Physics*, 1998, *84*, 4509.

86. Huber, T. E.; Nikolaeva, A.; Gitsu, D.; Konopko, L.; Graf, M. J. Quantum confinement and surface-state effects in bismuth nanowires. *Physica E: Low-dimensional Systems and Nanostructures*, 2007, *37* (1), 194–199.

87. Heremans, J.; Thrush, C. M.; Dresselhaus, M. S.; Mansfield, J. F. Bismuth nanowire arrays: Synthesis and galvanomagnetic properties. *Physical Review B*, 2000, *61* (4), 2921–2930.

88. Hasegawa, Y.; Nakano, H.; Morita, H.; Kurokouchi, A.; Wada, K.; Komine, T.; Nakamura, H. Aspect ratio dependence of magnetoresistivity in polycrystalline bismuth microwire arrays. *Journal of Applied Physics*, 2007, *101*, 033704.

89. Huber, T. E.; Graf, M. J.; Celestine, K. Longitudinal magneto resistance of 270 nm-diameter Bismuth nanowires. *Physica E*, 2003, *18* (1), 223–224.

90. Zhang, Z.; Sun, X.; Dresselhaus, M. S.; Ying, J. Y.; Heremans, J. Electronic transport properties of single-crystal bismuth nanowire arrays. *Physical Review B*, 2000, *61* (7), 4850–4861.

91. Cornelius, T. W.; Toimil-Molares, M. E.; Neumann, R.; Karim, S. Finite-size effects in the electrical transport properties of single bismuth nanowires. *Journal of Applied Physics*, 2006, *100*, 114307.

92. Hasegawa, Y.; Ishikawa, Y.; Shirai, H.; Morita, H.; Kurokouchi, A.; Wada, K.; Komine, T.; Nakamura, H. Reduction of contact resistance at terminations of bismuth wire arrays. *Review of Scientific Instruments*, 2005, *76*, 113902.

93. Ishikawa, Y.; Hasegawa, Y.; Morita, H.; Kurokouchi, A.; Wada, K.; Komine, T.; Nakamura, H. Resistivity and Seebeck coefficient measurements of a bismuth microwire array. *Physica B: Physics of Condensed Matter*, 2005, *368* (1), 163–167.

94. Hasegawa, Y.; Ishikawa, Y.; Morita, H.; Komine, T.; Shirai, H.; Nakamura, H. Electronic transport properties of a bismuth microwire array in a magnetic field. *Journal of Applied Physics*, 2005, *97*, 083907.

95. Hasegawa, Y.; Ishikawa, Y.; Morita, H.; Komine, T.; Shirai, H.; Nakamura, H. Magneto-Seebeck coefficient of a bismuth microwire array in a magnetic field. *Applied Physics Letters*, 2004, *85*, 917.

96. Hasegawa, Y.; Ishikawa, Y.; Morita, H.; Komine, T.; Shirai, H.; Nakamura, H. SdH oscillations in the contact resistance of bismuth nanowires. *Materials Science and Engineering C*, 2003, *23* (6), 1099–1101.

97. Chiu, P.; Shih, I. A study of the size effect on the temperature-dependent resistivity of bismuth nanowires with rectangular cross-sections. *Nanotechnology*, 2004, *5* (11), 1489–1492.

98. Kai, L.; Chien, C. L.; Searson, P. C.; Kui Y.-Z. Large positive magnetoresistance and finite-size effects in arrays of semimetallic bismuth nanowires. http://medusa.pha.jhu.edu/Research/Biwire.html, 1999.

99. Wang, J.; Cao, G.; Li, Y. Giant positive magnetoresistance in non-magnetic bismuth nanoparticles. *Materials Research Bulletin*, 2003, *38* (11), 1645–1651.

100. Grozav, A. D.; Condrea, E. Positive thermopower of single bismuth nanowires. *Journal of Physics: Condensed Matter*, 2004, *6* (36), 6507–6518.

101. Graf, M. J.; Huber, T. E. Electronic transport in a 3-D network of 1-D Bi and Te-doped Bi quantum wires. *Physica E*, 2003, *18* (1), 260–261.

102. Huang, C. C.; Fung, K. Z. Effect of the surface configuration on the oxidation of bismuth nanowire. *Materials Research Bulletin*, 2006, *41* (9), 1604–1611.

103. Li, L.; Yang, Y.-W.; Li, G.-H.; Zhang, L.-D. Conversion of a Bi nanowire array to an array of Bi-Bi$_2$O$_3$ core-shell nanowires and Bi$_2$O$_3$ nanotubes. *Small* (Weinheim an der Bergstrasse, Germany), 2006, *2* (4), 548–553.

104. Wei, F.; Zhang, J.; Wang, L.; Zhang, Z.-K. Solvothermal growth of single-crystal bismuth sulfide nanorods using bismuth particles as source material. *Crystal Growth and Design*, 2006, *6* (8), 1942–1944.

105. Wang, Y.; Kim, J.-S.; Lee, J. Y.; Kim, G. H.; Kim, K. S. Diameter- and length-dependent volume plasmon excitation of bismuth nanorods investigated by electron energy loss spectroscopy. *Chemistry of Materials*, 2007, *19* (16), 3912–3916.

106. Fanfair, D. D.; Korgel, B. A. Bismuth nanocrystal-seeded III–V semiconductor nanowire synthesis. *Crystal Growth and Design*, 2005, *5* (5), 1971–1976.

107. Kiang, C.-H. Electron irradiation induced dimensional change in bismuth filled carbon nanotubes. *Carbon*, 2000, *38* (11), 1699–1701.

108. Zhang, W.; Tang, H.; Geng, P.; Wang, Q.; Jin, L.; Wu, Z. Amperometric method for rapid detection of *Escherichia coli* by flow injection analysis using a bismuth nano-film modified glassy carbon electrode. *Electrochemistry Communications*, 2007, *9* (4), 833–838.

109. Sato, H.; Homma, T.; Kudo, H.; Izumi, T.; Osaka, T.; Shoji, S. Three-dimensional microfabrication process using Bi electrodeposition for a highly sensitive x-ray imaging sensor. *Journal of Electroanalytical Chemistry*, 2005, *584* (1), 28–33.

110. Chen, H.; Li, Z.; Wu, Z.; Zhang, Z. A novel route to prepare and characterize Sn-Bi nanoparticles. *Journal of Alloys and Compounds*, 2005, *394* (1), 282–285.

111. Bhattacharya, V.; Yamasue, E.; Ishihara, K. N.; Chattopadhyay, K. On the origin and stability of the metastable phase in rapidly solidified Sn-Bi alloy particles embedded in Al matrix. *Acta Materialia*, 2005, *53* (17), 4593–4603.

112. Li, G. R.; Ke, Q. F.; Liu, G. K.; Liu, P.; Tong, Y. X. Electrodeposition of nano-grain sized Bi-Co thin films in organic bath and their magnetism. *Materials Letters*, 2007, *61* (3), 884–888.

113. Kang, K.; Lewis, L. H.; Hu, Y. F.; Li, Q.; Moodenbaugh, A. R.; Choi, Y.-S. Magnetic and transport properties of MnBi/Bi nanocomposites. *Journal of Applied Physics*, 2006, *99*, 08N703.

114. Kozhemyakin, G. N.; Nalivkin, M. A.; Rom, M. A.; Mateychenko, P. V. Growing Bi-Sb gradient single crystals by a modified Czochralski method. *Journal of Crystal Growth*, 2004, *263* (1), 148–155.

Final Remarks (Conclusions and Future Outlook)

As observed in the previous chapters, a host of less-common nanostructures have been synthesized and characterized by a series of modern techniques. Electronic microscopy methods have the principal importance in these studies; it is impossible now to imagine the description of the fabrication of a new rare nanostructure without using SEM or TEM. Currently, nanostructures can be obtained in any appropriate chemical laboratory worldwide even without possessing sophisticated equipment with a cost of hundreds of thousands of dollars or more (for instance, the CVD equipment, applied to produce carbon nanotubes or nanodiamonds (NDs), sputtering equipment, or the use of radiation sources for radiation-assisted fabrication of nanocomposites) due to the availability of various wet-chemical methods. Both classic and less-common nanostructures, described throughout the book, can be obtained frequently together. Applying certain strategies, it is possible (of course, not in all cases) to get purpose-directed nanostructures of desirable dimensionality, shape, length, width, etc. In addition, a considerable number of reported nanostructures have been obtained accidentally. Size and shape of formed nanostructures obviously depend on reaction conditions as in standard chemical reactions: concentration of precursors, temperature, pressure, process duration, and mixing rate.

Among a row of modern sophisticated techniques for nanostructure fabrication, we note a classic, simple, and, at the same time, very useful *ultrasound-assisted method*. Ultrasound is currently extensively used in different areas of nanochemistry and related fields of nanotechnology as a principal or assisting technique. Its main advantages in comparison with conventional mixing are faster processes, higher yields, and the possibility of fabrication of novel compounds, substances or materials, which are impossible or difficult to obtain *via* classic interactions. Main applications of ultrasound in nanochemistry include synthesis of novel nanocatalysts, antibacterial composites, and agents for drug delivery and degradation of dangerous organic contaminants. Several reports on successful united ultrasound/microwave treatments testify about the importance of combined techniques. Main nanoproducts of ultrasonic use are nanomaterials or nanocomposites based on carbon nanoforms, elemental metals, their oxides, salts and complexes, and polymers. *Microwave irradiation*, applied in various areas of chemical synthesis[1] as a heating source, is also a useful tool in nanostructure preparation. At last, various materials and composites, as well as nanostructures, such as nanoparticles, nanotubes, nanowires, and nanofibrils, can be successfully fabricated or modified by different *irradiation techniques*, as with nuclear-nature ionizing radiation (β-particles, γ-rays, etc.) and with those corresponding to visible and near-wave regions (UV and laser irradiation).[2]

Which chemical compounds are most frequently used to create nanostructures? After an analysis of thousands of reports, we note that the range of these chemicals is rather limited by mainly *inorganic substances*: elemental metals, silicon, carbon allotropes, metal oxides, oxo salts, sulfides, selenides, and oxygen-containing salts. Zinc oxide can be considered the most "productive" compound: ZnO can be used for the fabrication of a major part of nanostructures.

Classic *carbon nanostructures* (single-wall and multi-wall carbon nanotubes, NDs, fullerenes, etc.) are obviously outdated and inefficient. In addition to CNTs, their composites are widely fabricated as well, in particular those with polymers, possessing improved properties (hardness and mechanical and chemical stability). In addition, a series of methods are currently applied to solubilize both MWCNTs and SWCNTs in water or organic solvents. Solutions formed can be true or dispersive. In the last decade, much attention has been paid to soluble composites of CNTs with polymers and biomolecules due to the extreme importance of these areas and many potential applications of synthesized composites in medicine and technology.

Graphene, the youngest[3] allotropic form of carbon, can be considered most promising in future applications among other carbon forms. At present, we observe all attributes of an approaching boom due to its unique properties and possible applications.[4–7] At the same time, though this material is of great interest to physicians and technologists, chemists show less interest, probably due to the absence of a well-established industrial method for graphene production. We suppose that, in the near future, the most important contribution of chemists could be the elaboration of a cheap and easy wet-chemical synthetic method for obtaining graphene, as well as chemical modification by insertion of various organic functional groups into its structure in order to manipulate properties of this valuable material. A dramatic increase in experimental publications and patents is expected next year, dedicated to the optimization of synthetic methods for graphene and its applications in various technological areas (composite materials, batteries, gas sensors, field emitters, etc.).

Despite the fact that the diamond in its nanoforms has been researched for several decades, it continues being a field of interest for researchers working in materials chemistry, biology, and medicine. As one of numerous carbon allotropic forms, which are much more biocompatible with human organism in comparison with others, *nanodiamonds* (NDs), functionalized with biomolecules, have been used in a series of useful applications for drug delivery and as fluorescent labels in cells. Additionally, the use of NDs for reinforcing polymers and the creation of ultrahard composites are prevalent in industrial applications. The functionalization capacity of NDs, as well as that of such other carbon forms as graphene, fullerenes,[8] or carbon nanotubes, is very rich and, undoubtedly, is subject to rapid progress in the near future.

Analyzing fullerenes and their derivatives, we pay attention to two groups of intriguing compounds. First, *small fullerenes* $C_{20<n<60}$ and endohedral fullerene complexes $M@C_n$ (M = metal or H, Hal, C, BN), which are represented by a relatively little number of examples in comparison with classic C_{60}, caused by their instability. C_{20}, C_{28}, C_{32}, C_{36}, and C_{50} were found to be the most stable small fullerenes in the range of C_{20}–C_{50}. It has also been established that fullerenes prefer geometries that separate the pentagonal rings as far apart as possible. Another class of fullerene derivatives with multiple applications correspond to the research area of *fulleropyrrolidine*, which continues intensively developing during the first decade of the present century, paying close attention to the synthesis of a series of novel dyads and triads with porphyrins and their metal complexes, ferrocene, S-containing ligands, calixarenes, crown-ethers, polymers, and relatively simple ligand systems. Attention (see Chapter 19) is also given to fulleropyrrolidine poly-adducts, multifullereno-pyrrolidines, and functionalization of carbon nanotubes with pyrrolidines. According to the general tendency in fullerene chemistry, the main part of investigations is carried out on the basis of C_{60}, whereas the progress of C_{70} and higher fullerenes are almost absent in the area of fulleropyrrolidine. A considerable part of synthesized products has got academic or, in some cases, industrial applications.

Describing other exotic nanostructures, we note that *nanomesh/honeycomb* nanoforms are reported by a limited number of examples. These nanostructures are mainly represented by binary compounds (especially boron nitride), various carbon modifications (nanotubes, NDs, graphene, and fullerenes), and several metal O-containing salts. Nanomesh/honeycomb nanostructures are applied or can be applied in diverse fields, including templating and superhydrophobic coating, nanocatalysis, surface functionalization, spintronics, quantum computing and data (or energy) storage media, gas sensors, optoelectronic devices, sensors, fuel cells, field emitter or other thin-film, functional polymer composites, and superconductivity, among many others.

Nanometal area obviously needs to be well discussed due to an extreme importance of elemental metals in standard and reduced-size forms (micro- and nano-) in chemistry and technology. This research field has dramatically developed in the last 20 years. A series of classic and "non-standard" synthetic routes have found their applications in the production of metallic nanoparticles in distinct forms. Among other metals, those with long-life catalytic applications (i.e., Ni, Co, Pt, Pd, and Rh) have been most frequently reported in relation to their synthesis in nano-forms. A considerable attention in current research is given to noble metals (Ag, Au, Pt, Pd, Rh, Ru, etc.), although methods for obtaining other *d*-(W, Fe, Sn) and even *f*-(U, Th) metals are also reported. Among other methods, main attention is given to classic tools (for instance, decomposition of the reduction of metal complexes and salts) as to such relatively new promising techniques as the CVD or laser ablation.

Among other types of activated metals, *Rieke metals* were investigated from the 1970s to the beginning of the present century. The most attention has been paid to such metals as Pt, Pd, Zn, Cd, Mg, Ni, Al, Ga, Cu, and Co, that is, those intensively used in organic and organometallic synthesis. A definite attention has been given to activated actinides. A series of applications for activated metals obtained have been offered. The most frequently reported techniques involve the use of "lithium—electron carrier (biphenyl, naphthalene, and anthracene)," precursor-reductant, preformed, or appearing *in situ* in the reaction medium. At present, the observed activities in this field are considerably lower (practically in "stand-by" condition) than in the last century and do not show development in fabrication of activated metals; their applications have been reported only in some separate investigations.[9] Therefore, at present century of nanotechnology development, when modern chemical, physicochemical, and biological techniques allow preparation of metallic nanoparticles, nanowires, nanorods, and many other nanostructures, it seems that "simple" reduction of anhydrous metal salts with alkali metals is already "past millennium" and does not have a considerable importance. However, precisely to the simplicity of this method and its availability in every chemical laboratory, these classic techniques do not lose their significance and, undoubtedly, should be continuously developed according to the well-known rule that "the new is well-forgotten old."

Core-shell bi- and trimetallic nanoparticles represent a special case of polymetallic nanosystems and are discussed in this book in the example of *gold and iron*. The nanostructures on their basis are generally obtained by various techniques, including reduction in reverse micelles, decomposition of organometallic compounds, electron-beam, laser and γ-irradiation, ultrasound treatment, and electrochemical deposition. (Fe or Fe_xO_y)/Au core-shell Nps are usually studied by a series of various standard methods, in particular microscopy techniques (SEM and TEM), application of low-frequency Raman modes,[10] Mössbauer and x-ray absorption spectroscopies, magnetic field heating study,[11] x-ray diffraction, etc. It is emphasized that real state and metal oxidation number of iron core can be determined by Mössbauer spectroscopy, which was strongly recommended to be a standard technique in the investigations in this field. Main applications of these Nps consist of catalysis (oxidation of CO and H_2, formaldehyde, methanol, and ethylene) and biological and biomedical uses (cancer therapy or biomolecular manipulation using light, contrast agents for magnetic resonance imaging, magnetic hyperthermia treatment, bioseparations, biosensors, immunoassay, and drug delivery guide).

The bi- and trimetallic nanoparticles examined are generally superparamagnetic at room temperature (the superparamagnetic fraction is retained after coating with gold), and the iron core is protected by a gold shell against oxidation, so this is a unique basis for having a series of useful biomedical applications. Nanoparticles consisting of a gold layer (exhibiting a surface plasmon resonance that provides optical contrast due to light scattering in the visible region, presenting a convenient surface for conjugating targeting moieties, and making tracking the positions of individual particles possible) and an SPM iron oxide moiety could provide a promising platform for the development of multimodal imaging and therapy approaches in future medicine. Among other biological and biomedical applications, (Fe or Fe_xO_y)/Au Nps are currently used as biosensors, for separation of proteins and amino acids and DNA measurement. It is expected that in the near future new polymetallic nanoparticles on the basis of gold and iron oxides will be synthesized, resulting in useful catalytic and medical applications.

The area of intermetallics and alloys is also successfully developed. Thus, a series of distinct experimental routes have been recently developed to isolate a large variety of *nanoalloys*. This research field is growing rapidly and it is promising and attractive due to a host of applications in catalysis, as well as fabrication of optics, electronics, and magnetic materials. In many cases, advances in experimental techniques are combined with theoretical studies of structure–property relationships for novel fabricated nanoalloys.

In this book, nanostructures and properties and applications of nanometals are examined in the example of *bismuth nanostructures*. As shown earlier, main efforts of researchers are given to obtaining and characterizing Bi nanowires,[12–15] nanoparticles,[16,17] and nanoalloys[18,19] in comparison with its other nanoforms. One of the most frequently used techniques for their preparation is pulsed electrodeposition, as well as reduction of bismuth salts or complexes in solution. Main applications of bismuth nanoforms include obtaining nanoparticles of Bi compounds, nanoalloys, and semiconductors. The previous discussion allows concluding that metal-nanoparticle research area is a rapidly progressing field having excellent theoretical and practical aspects in the future.

Although coordination and organometallic compounds are generally poorly reported as nanostructures, a variety of metal complexes are widely used in creation of *coordination and organometallic nanomaterials and nanocomposites*, which can be used in many applications: precursors of inorganic nanomaterials, bone cements, dental restorative composites possessing antibacterial properties,[20] catalysis, drug delivery, functional nanomolecular systems,[21] luminescent materials, materials for photoelectrical conversion purposes, electrochemical/electric detectors and sensors, nanoscale electronic devices, and diagnostics of diseases, among many others. A number of reports have been devoted to polymers containing metal complexes. Inorganic nanocomposites, obtained by further treatment of coordination nanomaterials, have also got a host of well-established applications in the areas discussed earlier, among others.

The properties and applications of nanostructures have been observed to depend directly or indirectly on their *shape*. Thus, in the case of animal-like nanoforms, the segmented worm-like nanostructures composed of magnetic iron oxide and coated with a polymer are found to attach to tumors due to their major contact with a cell surface and, therefore, more effective interactions compared with nanospherical particles. Superparamagnetic Fe_3O_4-polydivinylbenzene nanoworms, capable of easy separation by an external magnetic field, have potential applications in drug delivery/targeting, magnetic resonance imaging, and nanoprobes for diagnosis and disease treatment. Nanostructured wormhole-like non-crystalline cresol-formaldehyde material NCF-1 has exhibited photoluminescence property at room temperature, which could be utilized for the fabrication of novel organic optical devices. Hollow, urchin-like ferromagnetic carbon spheres with electromagnetic function and high conductivity at room temperature could be used as a reversible dye adsorbent. Similarly, urchin-like nickel chains exhibited a best absorption property in contrast with other as-synthesized samples and other reported nickel structures, which can be attributed to geometrical effect, high initial permeability, point discharge effect, and multiple absorption; thus, prepared nickel nanomaterials can be applied as promising absorbing materials. As another example, for magnetic materials, the magnetic response is quite sensitive to variability in the size, shape, and spacing of the nanostructured materials. In addition, applications of nanostructures as potential catalysts strongly depend on their shapes. In conclusion, shape of nanostructures, in addition to surface area, is obviously responsible for their properties.

Magnetic properties have been observed to be related to nanostructural type. Thus, a composite hierarchical hollow structure, consisting of discrete WO_2 hollow core spheres with $W_{18}O_{49}$ nanorod shells (hollow urchins), showed unusual magnetic behavior. Additionally, magnetic measurement indicated that urchin-like $MnWO_4$ microspheres showed a weak ferromagnetic ordering at low temperature due to spin-canting and surface spins of microspheres, while much shorter $MnWO_4$ nanorods showed antiferromagnetism at low temperature.

Catalytic applications are classic for nanostructures of several types, as for conventional as less-common nanoforms. An example is heterogeneous molybdenum catalysts, applied for efficient epoxidation of olefins using *t*-butyl hydroperoxide as oxidant, synthesized using sea urchin-like polyaniline (PANI) hollow microspheres constructed with their own oriented nanofiber arrays for support. The catalytic activity of these PANI microsphere–supported catalysts (95% conversion) has been found to be higher than that observed in its corresponding homogeneous catalyst (85% conversion) and the conventional PANI-supported catalyst (65% conversion). Among other catalytic applications, Fe-doped mesoporous TiO_2 microspheres forming disordered wormhole-like mesostructure are suitable for photodegradation of methyl orange. A series of hierarchical squama-like nanostructured/porous titania materials (with wormhole-like mesopores of nanoparticle assembly in each squama) doped with different contents of cerium has exhibited a catalytic activity in the photodegradation of Rhodamine B. Urchin-like Pt nanostructures have shown excellent electrocatalytic activity toward the reduction of dioxygen and oxidation of methanol. At last, urchin-like ZnS has exhibited excellent photocatalytic activity for degradation of *Acid fuchsine*.

In this respect, a careful examination of the so-called less-common nanostructures can open up opportunities for a series of new applications in these and other areas of chemistry and nanotechnology.

REFERENCES

1. Kharissova, O. V.; Kharisov, B. I.; Ortiz Mendez, U. Microwave-assisted synthesis of coordination and organometallic compounds. In *Advances in Induction and Microwave Heating*, Chapter 17. Grundas, S. (Ed.). INTECH, Rijeka, Croatia, ISBN: 978-953-307-522-8, 2011, pp. 345–390.
2. Kharissova, O. V.; Kharisov, B. I.; Ortiz Méndez, U. Radiation-assisted synthesis of composites, materials, compounds, and nanostructures. In *Wiley Encyclopedia of Composites*, in press.
3. Novoselov, K. S.; Geim, A. K.; Morozov, S. V.; Jiang, D.; Zhang, Y.; Dubonos, S. V.; Grigorieva, I. V.; Firsov, A. A. Electric filed effect in atomically thin carbon films. *Science*, 2004, *306*, 666–669.
4. Li, X.; Wang, X.; Zhang, L.; Lee, S.; Dai, H. Chemically derived, ultrasmooth graphene nanoribbon semiconductors. *Science*, 2008, *319* (5867), 1229–1232, doi: 10.1126/science.1150878.
5. Wilson, M. Electrons in atomically thin carbon sheets behave like massless particles. *Physics Today*, 2006, *59* (January), 21–23.
6. Castro Neto, A.; Guinea, F.; Peres, N. M. Drawing conclusions from graphene. *Physics World*, 2006, November, 1–5.
7. Geim, A. K.; MacDonald, A. H. Graphene: Exploring carbon flatland. *Physics Today*, 2007, August, 35–41.
8. Jiménez Gómez, M. A.; Garza Castañón, M.; Kharissova, O. V.; Kharisov, B. I.; Ortiz Méndez, U. Synthesis by Prato reaction and in situ UV-characterization of several fulleropyrrolidine derivatives. *International Journal of Green Nanotechnology*, 2009, *1*, M43–M51.
9. Kharisov, B. I.; Ortiz Mendez, U.; Rivera de la Rosa, J. Low-temperature synthesis of phthalocyanine and its metal complexes. *Russian Journal of Coordination Chemistry*, 2006, *32* (9), 617–631.
10. Cataliotti, R. S.; Compagnini, G.; Crisafulli, C., Minicò, S.; Pignataro, B.; Sassi, P.; Scirè, S. Low-frequency Raman modes and atomic force microscopy for the size determination of catalytic gold clusters supported on iron oxide. *Surface Science*, 2001, *494* (2), 75–82.
11. Wijaya, A.; Brown, K. A.; Alper, J. D.; Hamad-Schifferli, K. *Journal of Magnetism and Magnetic Materials*, 2007, *309* (1), 15–19.
12. Liu, C.-P.; Wang, R.-C.; Kuo, C.-L.; Liang, Y.-H.; Chen, W.-Y. Recent patents on fabrication of nanowires. *Recent Patents on Nanotechnology*, 2007, *1*, 11–20.
13. Li, L.; Yang, Y.; Fang, X.; Kong, M.; Li, G.; Zhang, L. Diameter-dependent electrical transport properties of bismuth nanowire arrays. *Solid State Communications*, 2007, *141*, 492–496.
14. Black, M. R.; Hagelstein, P. L.; Cronin, S. B.; Lin, Y. M.; Dresselhaus, M. S. Optical absorption from an indirect transition in bismuth nanowires. *Physical Review B: Condensed Matter*, 2003, *68*, 235417.
15. Tian, M.; Wang, J.; Kumar, N.; Han, T.; Kobayashi, Y.; Liu, Y.; Mallouk, T.-E.; Chan, M. H. W. Observation of superconductivity in granular Bi nanowires fabricated by electrodeposition. *Nano Letters*, 2006, *6* (12), 2773–2780.
16. Goia, C.; Matijevic, E.; Goia, D. V. Preparation of colloidal bismuth particles in polyols. *Journal of Materials Research*, 2005, *20* (6), 1507–1514.
17. Dellinger, T. M.; Braun, P. V. Lyotropic liquid crystals as nanoreactors for nanoparticle synthesis. *Chemistry of Materials*, 2004, *16*, 2201–2207.
18. Jesser, W. A.; Shneck, R. Z.; Gile, W. W. Solid-liquid equilibria in nanoparticles of Pb-Bi alloys. *Physical Review B: Condensed Matter*, 2004, *69*, 144121.

19. Fang, J.; Stokes, K. L.; He, J.; Zhou, W. L.; O'Connor, C. J. Pattern shape-controlled self-assembly of $Bi_{0.90}Si_{0.10}$ nanocrystallites. *MRS Symposium Proceedings*, Warrendale, PA, 2002, Vol. 691, pp. 365–370 (G10.3.1–G10.3.6).

20. Slenters, T. V.; Hauser-Gerspach, I.; Daniels, A. U.; Fromm, K. M. Silver coordination compounds as light-stable, nano-structured and anti-bacterial coatings for dental implant and restorative materials. *Journal of Materials Chemistry*, 2008, *18* (44), 5359–5362.

21. Otsuki, J. Metal complex based functional nanomolecular systems. *The First Symposium of Nanotechnology Excellence*, Nihon University, Tokyo, Japan, 2009. http://www.chem.cst.nihon-u.ac.jp/~otsuki/otsuki-e.../090918_N.Symp_P30.pdf

Author Index

Subject Index

Milton Keynes UK
Ingram Content Group UK Ltd.
UKHW051858071024
449327UK00025B/2009